T0248614

CRC World Dictionary of

PLANT NAMES

Common Names, Scientific Names, Eponyms, Synonyms, and Etymology

Volume IV R — Z

CRC World Dictionary of
PLANT NAMES

Common Names, Scientific Names, Eponyms, Synonyms, and Etymology

Volume IV R — Z

Umberto Quattrocchi, F.L.S.

CRC Press
Taylor & Francis Group
Boca Raton London New York

CRC Press is an imprint of the
Taylor & Francis Group, an **informa** business

CRC Press
Taylor & Francis Group
6000 Broken Sound Parkway NW, Suite 300
Boca Raton, FL 33487-2742

© 2000 by Taylor & Francis Group, LLC
CRC Press is an imprint of Taylor & Francis Group, an Informa business

First issued in paperback 2019
No claim to original U.S. Government works

ISBN 13: 978-0-367-44750-2 (pbk)
ISBN 13: 978-0-8493-2678-3 (hbk)
ISBN 13: 978-0-8493-2673-8 (set)

This book contains information obtained from authentic and highly regarded sources. Reason-able efforts have been made to publish reliable data and information, but the author and publisher cannot assume responsibility for the validity of all materials or the consequences of their use. The authors and publishers have attempted to trace the copyright holders of all material reproduced in this publication and apologize to copyright holders if permission to publish in this form has not been obtained. If any copyright material has not been acknowledged please write and let us know so we may rectify in any future reprint.

Except as permitted under U.S. Copyright Law, no part of this book may be reprinted, reproduced, transmitted, or utilized in any form by any electronic, mechanical, or other means, now known or hereafter invented, including photocopying, microfilming, and recording, or in any information storage or retrieval system, without written permission from the publishers.

For permission to photocopy or use material electronically from this work, please access www. copyright.com (http://www.copyright.com/) or contact the Copyright Clearance Center, Inc. (CCC), 222 Rosewood Drive, Danvers, MA 01923, 978-750-8400. CCC is a not-for-profit organiza-tion that provides licenses and registration for a variety of users. For organizations that have been granted a photocopy license by the CCC, a separate system of payment has been arranged.

Trademark Notice: Product or corporate names may be trademarks or registered trademarks, and are used only for identification and explanation without intent to infringe.

Visit the Taylor & Francis Web site at
http://www.taylorandfrancis.com

and the CRC Press Web site at
http://www.crcpress.com

R

Rabdochloa P. Beauv. Gramineae

Origins:
From the Greek *rhabdos* "a rod, stick, a magic wand" and *chloe, chloa* "grass"; see also *Leptochloa* P. Beauv.; see Ambroise Palisot de Beauvois, *Essai d'une nouvelle Agrostographie*, ou nouveaux genres des Graminées. 84, 176. Paris (Dec.) 1812.

Rabdosia Hassk. Labiatae

Origins:
Greek *rhabdos* "a rod, stick, a magic wand," referring to the spikes or the branches.

Rabdosiella Codd Labiatae

Origins:
The diminutive of the genus *Rabdosia* Hassk.

Rabenhorstia Reichb. Bruniaceae

Origins:
For the German botanist Gottlob (Gottlieb) Ludwig Rabenhorst, 1806-1881, pharmacist, cryptogamist, botanical collector, 1852-1878 founded and edited *Hedwigia*. See Stafleu and Cowan, *Taxonomic Literature*. 4: 460-473 and 474-528. Utrecht 1983; J.H. Barnhart, *Biographical Notes upon Botanists*. 3: 121. 1965; T.W. Bossert, *Biographical Dictionary of Botanists Represented in the Hunt Institute Portrait Collection*. 321. 1972; Ethelyn Maria Tucker, *Catalogue of the Library of the Arnold Arboretum of Harvard University*. Cambridge, Massachusetts 1917-1933; H.N. Clokie, *Account of the Herbaria of the Department of Botany in the University of Oxford*. 229. Oxford 1964; J. Ewan, ed., *A Short History of Botany in the United States*. New York and London 1969; G. Murray, *History of the Collections Contained in the Natural History Departments of the British Museum*. 1: 175. London 1904; I.C. Hedge and J.M. Lamond, *Index of Collectors in the Edinburgh Herbarium*. Edinburgh 1970.

Rabiea N.E. Br. Aizoaceae

Origins:
After the South African plant collector W.A. Rabie.

Racapa M. Roemer Meliaceae

Origins:
From the genus *Carapa* Aublet, an anagram.

Racemaria Raf. Convallariaceae (Liliaceae)

Origins:
Latin *racemus, i* "the stalk of a cluster, a bunch, cluster"; see C.S. Rafinesque, *Atlantic Journal*, and friend of knowledge. 1: 151. Philadelphia 1832.

Racemobambos Holttum Gramineae

Origins:
Latin *racemus, i* "the stalk of a cluster, a bunch, cluster" plus *Bambos* Retz.

Rachicallis DC. Rubiaceae

Origins:
From the Greek *rhachis* "rachis, axis, midrib of a leaf" and *kalli* "beautiful," *kallos* "beauty"; see also *Rhachicallis* DC.

Raclathris Raf. Boraginaceae

Origins:
See C.S. Rafinesque, *Sylva Telluriana*. 167. 1838.

Racosperma C. Martius Mimosaceae

Origins:
Greek *rhakos* "ragged," *rhakodes* "wrinkled" and *sperma* "seed"; see Carl F.P. von Martius (1794-1868), *Hortus*

regius monacensis. 188, 206 (Index). München, Leipzig 1829.

Raddia Bertoloni Gramineae

Origins:

After the Italian botanist Giuseppe Raddi (Josephus Raddius), 1770-1829, cryptogamist, traveler, explorer, plant collector (Brazil, Madeira, and Egypt), mycologist. He wrote *Synopsis filicum brasiliensium.* Bononiae [Bologna] 1819. See J.H. Barnhart, *Biographical Notes upon Botanists.* 3: 122. 1965; T.W. Bossert, *Biographical Dictionary of Botanists Represented in the Hunt Institute Portrait Collection.* 322. 1972; Ethelyn Maria Tucker, *Catalogue of the Library of the Arnold Arboretum of Harvard University.* Cambridge, Massachusetts 1917-1933; Ida Kaplan Langman, *A Selected Guide to the Literature on the Flowering Plants of Mexico.* University of Pennsylvania Press, Philadelphia 1964; Johannes Proskauer, in *Dictionary of Scientific Biography* 10: 13. 1981; Mariella Azzarello Di Misa, a cura di, *Il Fondo Antico della Biblioteca dell'Orto Botanico di Palermo.* 223. Regione Siciliana, Palermo 1988; R. Zander, F. Encke, G. Buchheim and S. Seybold, *Handwörterbuch der Pflanzennamen.* 14. Aufl. Stuttgart 1993.

Raddia Mazziari Gramineae

Origins:

After the Italian botanist Giuseppe Raddi, 1770-1829.

Raddiella Swallen Gramineae

Origins:

The diminutive of *Raddia.*

Radermachera Zollinger & Moritzi Bignoniaceae

Origins:

Named after the Dutch botanist Jacobus Cornelius Matthaeus Radermacher, 1741-1783, amateur plant collector in Java, official in the Dutch East Indies, founder of the *Bataviaasch Genootschap van Kunsten en Wetenschappen,* author of *Naamlijst der planten.* Batavia 1780-1782. See J.H. Barnhart, *Biographical Notes upon Botanists.* 3: 122. 1965; Jonas C. Dryander, *Catalogus bibliothecae historiconaturalis Josephi Banks.* 3: 181. London 1800; A. Lasègue, *Musée botanique de Benjamin Delessert.* Paris 1845.

Species/Vernacular Names:

R. glandulosa (Bl.) Miq. (*Radermachera stricta* Zoll. & Moritzi ex Zoll.; *Spathodea glandulosa* Bl.; *Stereospermum glandulosum* Miq.)

English: hill fox-glove tree

Malaya: lempoyan

R. pinnata (Blanco) Seem. (*Radermachera acuminata* Merrill; *Radermachera banaibana* Bur.; *Radermachera fenicis* Merrill; *Radermachera mindorensis* Merrill; *Radermachera quadripinna* Seem.; *Stereospermum banaibanai* Rolfe; *Stereospermum pinnatum* Fernandez-Villar; *Stereospermum seemannii* Rolfe)

English: lowland fox-glove tree

R. sinica (Hance) Hemsl. (*Stereospermum sinicum* Hance)

English: Asia bell-tree

Japan: sendan-ki-sasage

Radinocion Ridley Orchidaceae

Origins:

Greek *rhadinos* "thin, slender, delicate" and *kion* "a column," the slender column.

Radinosiphon N.E. Br. Iridaceae

Origins:

Greek *rhadinos* "thin, slender" and *siphon* "tube."

Radiola Hill Linaceae

Origins:

Latin *radiolus* "a small sunbeam, a kind of long olive," Apuleius applied to a plant resembling a fern.

Species/Vernacular Names:

R. linoides Roth

English: allseed

Radlkofera Gilg Sapindaceae

Origins:

For the German botanist Ludwig Adolph Timotheus Radlkofer, 1829-1927, physician, M.D. 1854 München, 1908-1927 Director of the Botanical Museum and State Herbarium at München, professor of botany, author of "Sapindaceae." in *Natürl. Pflanzenfam.* 3(5): 277-366. 1895, "Sapindaceae." in *Pflanzenr.* 4(165) Heft 98: 1-1539. 1931-1934 and *New and Noteworthy Hawaiian Plants* ... Honolulu 1911 (co-author Joseph Francis Charles Rock, 1881-1962). See J.H. Barnhart, *Biographical Notes upon Botanists*. 3: 123. 1965; T.W. Bossert, *Biographical Dictionary of Botanists Represented in the Hunt Institute Portrait Collection.* 322. 1972; E.M. Tucker, *Catalogue of the Library of the Arnold Arboretum of Harvard University.* 1917-1933; Frederico Carlos Hoehne, M. Kuhlmann and Oswaldo Handro, *O jardim botânico de São Paulo.* 1941; Elmer Drew Merrill, in *Contr. U.S. Natl. Herb.* 30(1): 247-248. 1947 and in *Bernice P. Bishop Mus. Bull.* 144: 153-154. 1937; Ida Kaplan Langman, *A Selected Guide to the Literature on the Flowering Plants of Mexico.* 605-606. Philadelphia 1964; R. Zander, F. Encke, G. Buchheim and S. Seybold, *Handwörterbuch der Pflanzennamen.* 14. Aufl. Stuttgart 1993.

Radlkoferella Pierre Sapotaceae

Origins:

For the German botanist Ludwig Adolph Timotheus Radlkofer, 1829-1927.

Radlkoferotoma Kuntze Asteraceae

Origins:

For the German botanist Ludwig Adolph Timotheus Radlkofer, 1829-1927.

Radyera Bullock Malvaceae

Origins:

Named after the South African botanist Robert Allen Dyer, 1900-1987 (d. Pretoria), Director of the Botanical Research Institute at Pretoria (1944-1963). He is best known for his *Ceropegia, Brachystelma and Riocreuxia in Southern Africa.* Rotterdam 1983, *The Vegetation of the Divisions of Albany and Bathurst.* Pretoria 1937, "The cycads of Southern Africa." *Bothalia.* 8: 405-516. 1965 and *The Genera of Southern African Flowering Plants.* Pretoria 1975-1976. See Cythna Letty (1895-1985), *Wild Flowers of the Transvaal.* [Pretoria] 1962; Alain Campbell White (b. 1880), *The Succulent Euphorbieae* (Southern Africa). Pasadena 1941; Arthur Allman Bullock (1906-1980), in *Kew Bulletin for 1956.* 23: 454. (Feb.) 1957; Sima Eliovson, *Discovering Wild Flowers in Southern Africa.* Cape Town 1962; Gordon Douglas Rowley, *A History of Succulent Plants.* Strawberry Press, Mill Valley, California 1997; Mary Gunn and Leslie E. Codd, *Botanical Exploration of Southern Africa.* Cape Town 1981.

Species/Vernacular Names:

R. farragei (F. Muell.) Fryxell & Hashmi (*Hibiscus farragei* F. Muell.)

English: desert rose mallow, desert mallow, bush hibiscus, knobby hibiscus, desert rose

Raffenaldia Godron Brassicaceae

Origins:

After the French botanist Alire Delile (Raffeneau-Delile), 1778-1850 (Montpellier), physician, with Napoléon to Egypt, traveler in North Carolina, 1819-1850 professor of botany at Montpellier. His writings include *Nouveaux Cristaux parmi les grains de pollen* du etc. Montpellier [1836], *Centurie de plantes d'Afrique du voyage à Méroé, recueillies par M. Cailliaud.* Paris 1827, *Notice sur un voyage horticole et botanique en Belgique et en Hollande.* Montpellier 1838, *Description de l'Egypte. Histoire naturelle.* Tom. second. Paris 1812, *Dissertation sur les effets d'un poison de Java, appelé Upas tieuté, et sur la Noix vomique, la Fève de St. Ignace, le Strychnos potatorum, et la Pomme de Vontac,* qui sont du même genre de plantes que l'Upas tieuté. Paris 1809 and *Flore de l'Arabie pétrée.* Plantes recueillies par M. Léon de Laborde, nommées, classées, et décrites par M. Delile. Paris 1830. See John H. Barnhart, *Biographical Notes upon Botanists.* 1: 438. Boston 1965; Léon de Laborde et Linant, *Voyage de l'Arabie pétrée.* Paris 1830 [-1833]; T.W. Bossert, *Biographical Dictionary of Botanists Represented in the Hunt Institute Portrait Collection.* 98. 1972; Friedrich Wilhelm Heinrich Alexander von Humboldt (1769-1859) and Aimé Jacques Alexandre Bonpland (1773-1858), *Plantae aequinoctiales.* 1808; Jean Motte, in *Dictionary of Scientific Biography* 4: 21-22. 1981; Lucia Sala Simion, "1798, campagna scientifica d'Egitto." *Corriere della Sera.* 12 Aprile 1998; Mariella Azzarello Di Misa, a cura di, *Il Fondo Antico della Biblioteca dell'Orto Botanico di Palermo.* 224. Regione Siciliana, Palermo 1988.

Rafflesia R. Br. Rafflesiaceae (Magnoliidae, Aristolochiales)

Origins:

For the British naturalist Sir Thomas Stamford Bingley Raffles, 1781-1826 (d. Middx.), the founder of Singapore, traveler, explorer, plant collector, colonial administrator (Netherlands East Indies, Bencoolen, Sumatra). In 1817 elected a Fellow of the Royal Society, in 1825 became a Fellow of the Linnean Society. His most important work is *History of Java.* London 1817. See Diana M. Simpkins, in *Dictionary of Scientific Biography* 11: 261-262. 1981; J.H. Barnhart, *Biographical Notes upon Botanists.* 3: 123. 1965;

A. Lasègue, *Musée botanique de Benjamin Delessert*. 1845; Lady Sophia Raffles, *Memoir of the Life and Public Services of Sir Thomas Stamford Raffles*. London 1830; Colin Clair, *Sir Stamford Raffles*. Founder of Singapore. Herts. 1936; Emily Hahn, *Raffles of Singapore: A Biography*. New York 1946 (1st American edition.); C.E. Wurtzburg, *Raffles of the Eastern Isles*. London 1954; Ray Desmond, *Dictionary of British & Irish Botanists and Horticulturists*. 570. London 1994; M. Archer, *Natural History Drawings in the India Office Library*. London 1962; C.T. Onions, *The Oxford Dictionary of English Etymology*. Oxford University Press 1966; Georg Christian Wittstein, *Etymologisch-botanisches Handwörterbuch*. 750. 1852.

Rafinesquia Nutt. Asteraceae

Origins:

In honor of the great botanist Constantine (Constantin) Samuel Rafinesque (-Schmaltz), 1783-1840 (b. near Constantinople, d. Philadelphia, Pennsylvania), economist, plant collector, naturalist, traveler, botanical explorer, conchologist, archaeologist, spent many years of his life in Sicily (1804-1815) and in USA, a prolific writer. Among his works are *Caratteri di alcuni nuovi generi e nuove specie di animali e piante della Sicilia*, con varie osservazioni sopra i medesimi. Palermo 1810, *Florula ludoviciana*; or, a flora of the state of Louisiana. New York 1817, *Neogenyton*. [Lexington, Ky.] 1825 and *Specchio delle scienze o giornale enciclopedico della Sicilia*, deposito letterario delle moderne cognizioni, etc. Palermo 1814; see J.H. Barnhart, *Biographical Notes upon Botanists*. 3: 123. 1965; Joseph Ewan, in *Dictionary of Scientific Biography* 11: 262-264. 1981; R. Zander, F. Encke, G. Buchheim and S. Seybold, *Handwörterbuch der Pflanzennamen*. 14. Aufl. 1993; Stafleu and Cowan, *Taxonomic Literature*. 4: 549-563. 1983; Alexander B. Adams, *Eternal Quest. The Story of the Great Naturalists*. New York 1969; William Darlington, *Reliquiae Baldwinianae*. Philadelphia 1843 and *Memorials of John Bartram and Humphry Marshall*. 1849; Edward Lee Greene, *Landmarks of Botanical History*. Edited by Frank N. Egerton. 74-75. Stanford University Press, Stanford, California 1983; T.W. Bossert, *Biographical Dictionary of Botanists Represented in the Hunt Institute Portrait Collection*. 322. 1972; H.N. Clokie, *Account of the Herbaria of the Department of Botany in the University of Oxford*. 229. Oxford 19642; Jeannette Elizabeth Graustein, *Thomas Nuttall, Naturalist. Explorations in America, 1808-1841*. Cambridge, Harvard University Press 1967; William Jay Youmans, ed., *Pioneers of Science in America*. 182-195. New York 1896; E.M. Tucker, *Catalogue of the Library of the Arnold Arboretum of Harvard University*. 1917-1933; Alex Berman, "C.S. Rafinesque (1783-1840): A challenge to the historian of pharmacy." *American Journal of Pharmaceutical*

Education. 16: 409-418. 1952; A. Lasègue, *Musée botanique de Benjamin Delessert*. Paris 1845; Joseph Ewan, *Rocky Mountain Naturalists*. The University of Denver Press 1950; Michael A. Flannery, "The medicine and medicinal plants of C.S. Rafinesque." *Economic Botany*. 52(1): 27-43. 1998; H.B. Haag, "Rafinesque's interests — a century later: medicinal plants." *Science*. 94: 403-406. 1941; E.D. Merrill (1876-1956), *Index rafinesquianus*. The plant names published by C.S. Rafinesque, etc. Jamaica Plain, Massachusetts, USA 1949; Mariella Azzarello Di Misa, a cura di, *Il Fondo Antico della Biblioteca dell'Orto Botanico di Palermo*. 224. Palermo 1988; Garrison and Morton, *Medical Bibliography*. 1849. 1961; Giuseppe M. Mira, *Bibliografia Siciliana*. 2: 260. ["Carlo Rafanesque Schmaltz", sic!] Palermo 1881; Ida Kaplan Langman, *A Selected Guide to the Literature on the Flowering Plants of Mexico*. 606. University of Pennsylvania Press, Philadelphia 1964; S. Lenley et al., *Catalog of the Manuscript and Archival Collections and Index to the Correspondence of John Torrey*. Library of the New York Botanical Garden. 467. 1973.

Species/Vernacular Names:

R. californica Nutt.

English: California chicory

R. neomexicana A. Gray

English: desert chicory

Rafinesquia Raf. Bignoniaceae

Origins:

Named for Constantine (Constantin) Samuel Rafinesque (-Schmaltz), 1783-1840 (Philadelphia), botanist, naturalist, traveler, botanical explorer.

Rafnia Thunberg Fabaceae

Origins:

For the Danish botanist Carl Gottlob Rafn, 1769-1808, school teacher, author of *Udkast til en Plantenphysiologie, grundet paa de nyere Begreber i Physik og Chemie*. Kjøbenhavn 1796 and *Danmarks og Holsteens flora*. København [Copenhagen] 1796-1800, with Johan Daniel Herholdt wrote *An Attempt at an Historical Survey of Life-saving Measures for Drowning Persons*. Århus 1960 and *Experiments with the Metallic Tractors* ... as published by Surgeons Herholdt and Rafn, etc. 1799; see J.H. Barnhart, *Biographical Notes upon Botanists*. 3: 123. 1965; Jens Wilken Hornemann (1770-1841), *Naturh. Tidsskr*. 1: 578. 1837; Sir William Hamilton [Ambassador to the Court of Naples] (1730-1803), *Efterretning om det sidste Udbrud af*

... *Vesuvius* ... oversat ... af C.G.R., etc. ["Account of the last eruption of Mount Vesuvius."] Kopenhagen 1796; C.P. Thunberg, *Nova Genera Plantarum.* 144. 1800 and *Prodromus plantarum Capensium.* 123. Upsaliae 1800; E.M. Tucker, *Catalogue of the Library of the Arnold Arboretum of Harvard University.* 1917-1933; R. Zander, F. Encke, G. Buchheim and S. Seybold, *Handwörterbuch der Pflanzennamen.* 14. Aufl. Stuttgart 1993; Carl Frederik Albert Christensen, *Den danske Botaniks Historie med tilhørende Bibliografi.* Copenhagen 1924-1926.

Ragiopteris C. Presl Dryopteridaceae (Onocleaceae, Aspleniaceae, Woodsiaceae)

Origins:
Possibly from the Greek *rax, ragos, rox* "grape, berry," *ragion* "small berry" and *pteris* "fern."

Raillardella (A. Gray) Benth. Asteraceae

Origins:
Small *Raillardia.*

Species/Vernacular Names:
R. argentea (A. Gray) A. Gray
English: silky raillardella
R. pringlei E. Greene
English: showy raillardella

Raillardiopsis Rydb. Asteraceae

Origins:
Like *Raillardia* Gaudich.

Species/Vernacular Names:
R. muirii (A. Gray) Rydb.
English: Muir's raillardella
R. scabrida (Eastw.) Rydb.
English: scabrid raillardella

Raimannia Rose Onagraceae

Origins:
After the Austrian botanist Rudolf Raimann, 1863-1896, professor of natural history; see J.H. Barnhart, *Biographical Notes upon Botanists.* 3: 124. 1965; E.M. Tucker, *Catalogue of the Library of the Arnold Arboretum of Harvard University.* 1917-1933; R. Zander, F. Encke, G. Buchheim and S.

Seybold, *Handwörterbuch der Pflanzennamen.* 14. Aufl. Stuttgart 1993.

Raimondia Safford Annonaceae

Origins:
For the Italian botanist Antonio Raimondi, 1826-1890, naturalist and geologist, from 1850 in Peru, professor of botany, botanical collector. His writings include *Apuntes sobra la provincia litoral de Loreto.* Lima 1862, *El Departamento de Ancachs y sus riquezas minerales.* Lima 1873, *Elementos de Botanica aplicada á la medicina y á la industria.* Lima 1857, *Minerales del Perú.* Lima 1878, *El Perú. Estudios mineralógicos y geológicos.* Lima 1874-1902 and *El Perú. Itinerarios de viajes.* 1929. See Manuel Rouaud y Paz-Soldan, *Dos ilustres Sabios* [S. Lorente and A. Raimondi] *vindicados* [from the criticisms of E. Desjardins]. Lima 1868; J.H. Barnhart, *Biographical Notes upon Botanists.* 3: 124. 1965; August Weberbauer, *Die Pflanzenwelt der peruanischen Andes in ihren Grundzügen dargestellt.* 13-14, 35. Leipzig 1911; Ettore Janni, [Vita di Antonio Raimondi] *Vida de Antonio Raimondi.* Lima 1942; R. Zander, F. Encke, G. Buchheim and S. Seybold, *Handwörterbuch der Pflanzennamen.* 14. Aufl. 1993; William Edwin Safford (1859-1926), "*Raimondia,* a new genus of Annonaceae from Colombia." *Contr. U.S. Natl. Herb.* 16: 217-219. 1913.

Raimondianthus Harms Fabaceae

Origins:
For the Italian botanist Antonio Raimondi, 1826-1890, naturalist and geologist, from 1850 in Peru.

Rajania L. Dioscoreaceae

Origins:
For the British (b. Black Notley, Essex) naturalist John Ray (Wray), 1627-1705 (d. Black Notley, Essex), traveler, in 1667 elected a Fellow of the Royal Society. His major works are *Catalogus plantarum circa Cantabrigiam nascentium.* Cambridge 1660, *Appendix ad catalogum plantarum.* Cambridge 1663, *Francisci Willughbeii ... ornithologiae.* London 1676 and *Catalogus plantarum Angliae et insularum adjacentium.* London 1670. See Charles Webster, in *Dictionary of Scientific Biography* 11: 313-318. 1981; G.L. Keynes, *John Ray, a Bibliography.* London 1951; John H. Barnhart, *Biographical Notes upon Botanists.* 3: 132. 1965; T.W. Bossert, *Biographical Dictionary of Botanists Represented in the Hunt Institute Portrait Collection.* 326. 1972; A. Lasègue, *Musée botanique de Benjamin Delessert.* Paris 1845; H.N. Clokie, *Account of the Herbaria of the Department of Botany in the University of Oxford.* 230. Oxford

1964; E.M. Tucker, *Catalogue of the Library of the Arnold Arboretum of Harvard University.* 1917-1933; Jonas C. Dryander, *Catalogus bibliothecae historico-naturalis Josephi Banks.* 1800; James Britten, *The Sloane Herbarium,* revised and edited by J.E. Dandy. 1958; Blanche Henrey, *British Botanical and Horticultural Literature before 1800.* 1975; Ray Desmond, *Dictionary of British & Irish Botanists and Horticulturists.* 574-575. 1994; R.T. Gunther, *Early Science in Cambridge.* Oxford 1937; F.D. Drewitt, *The Romance of the Apothecaries' Garden at Chelsea.* London 1924; M. Hadfield et al., *British Gardeners: A Biographical Dictionary.* London 1980; Alexander B. Adams, *Eternal Quest. The Story of the Great Naturalists.* New York 1969; B. Glass et al., eds., *Forerunners of Darwin: 1745-1859.* Baltimore 1959; Elisabeth Leedham-Green, *A Concise History of the University of Cambridge.* Cambridge, University Press 1996; Stafleu and Cowan, *Taxonomic Literature.* 4: 604-610. Utrecht 1983; Emil Bretschneider (1833-1901), *History of European Botanical Discoveries in China.* [Reprint of the original edition 1898.] Leipzig 1981; D.E. Allen, *The Botanists: A History of the Botanical Society of the British Isles through a Hundred and Fifty Years.* Winchester 1986; Mary Gunn and Leslie E. Codd, *Botanical Exploration of Southern Africa.* 31. Cape Town 1981; Mia C. Karsten, *The Old Company's Garden at the Cape and Its Superintendents*: involving an historical account of early Cape botany. Cape Town 1951; G. Murray, *History of the Collections Contained in the Natural History Departments of the British Museum.* 1: 176. London 1904; R. Pulteney, *Historical and Biographical Sketches of the Progress of Botany in England.* London 1790; C.E. Raven, *English Naturalists from Neckam to Ray.* Cambridge 1947; C.E. Raven, *John Ray Naturalist, His Life and Works.* Cambridge 1942; Frans A. Stafleu, *Linnaeus and the Linnaeans.* The spreading of their ideas in systematic botany, 1735-1789. Utrecht 1971; Francis Wall Oliver, ed., *Makers of British Botany.* Cambridge 1913.

Ramatuela Kunth Combretaceae

Origins:

For the French botanist Thomas Albin Joseph d'Audibert de Ramatuelle (Audibert Ramatuelle), 1750-1794, clergyman; see J.H. Barnhart, *Biographical Notes upon Botanists.* 3: 125. 1965; Jonas C. Dryander, *Catalogus bibliothecae historico-naturalis Josephi Banks.* London 1800; R. Zander, F. Encke, G. Buchheim and S. Seybold, *Handwörterbuch der Pflanzennamen.* 14. Aufl. 1993.

Ramatuella Kunth Combretaceae

Origins:

For the French botanist Thomas Albin Joseph d'Audibert de Ramatuelle, 1750-1794.

Ramirezella Rose Fabaceae

Origins:

For the Mexican botanist José Ramírez, 1852-1904. His writings include "El peyote." *An. Inst. Méd. Nac. Méx.* 4: 203-213, 233-249. 1900 and "Los escritos inéditos de Martín Sessé y José Mariano Moçiño." *An. Inst. Méd. Nac. Méx.* 4: 24-32. 1900, with Gabriel V. Alcocer (1852-1916) wrote *Sinonimia vulgar y científica de las plantas mexicanas.* México 1902. See Oscar Sánchez Sánchez, *La Flora del Valle de Mexico.* México, D.F. 1984; Maximino Martínez (1888-1964), *Catálogo de nombres vulgares y científicos de plantas mexicanas.* México 1987; J.H. Barnhart, *Biographical Notes upon Botanists.* 3: 125. 1965; T.W. Bossert, *Biographical Dictionary of Botanists Represented in the Hunt Institute Portrait Collection.* 323. 1972; Ida Kaplan Langman, *A Selected Guide to the Literature on the Flowering Plants of Mexico.* 607-609. University of Pennsylvania Press, Philadelphia 1964; E.M. Tucker, *Catalogue of the Library of the Arnold Arboretum of Harvard University.* 1917-1933; Gordon Douglas Rowley, *A History of Succulent Plants.* Strawberry Press, Mill Valley, California 1997; Irving William Knobloch, compil., "A preliminary verified list of plant collectors in Mexico." *Phytologia Memoirs.* VI. 1983.

Ramischia Opiz ex Garcke Ericaceae

Origins:

For Franz Xavier Ramisch, 1798-1859, botanist, physician, M.D. Prague 1825, professor of medicine; see J.H. Barnhart, *Biographical Notes upon Botanists.* 3: 125. 1965.

Ramonda Richard Gesneriaceae

Origins:

For the French (b. Strasbourg) botanist Louis François Élisabeth Ramond de Carbonnières, 1755-1827 (d. Paris), geologist, traveler, politician, mineralogist, administrator (Prefect of Puy-de-Dôme), plant collector, professor of natural history. His works include *Observations faites dans les Pyrénées.* Paris 1789 and *Voyage au sommet du Mont-Perdu.* Paris 1803; see John H. Barnhart, *Biographical Notes upon Botanists.* 3: 126. Boston 1965; T.W. Bossert, *Biographical Dictionary of Botanists Represented in the Hunt Institute Portrait Collection.* 323. 1972; A. Lasègue, *Musée botanique de Benjamin Delessert.* Paris 1845; M. Colmeiro y Penido, *La Botánica y los Botánicos de la Peninsula Hispano-Lusitana.* Madrid 1858; Giulio Giorello & Agnese Grieco, a cura di, *Goethe scienziato.* Einaudi Editore, Torino 1998; Jonas C. Dryander, *Catalogus bibliothecae historico-naturalis Josephi Banks.* London 1800;

E.M. Tucker, *Catalogue of the Library of the Arnold Arboretum of Harvard University.* 1917-1933; G. Schmid, *Goethe und die Naturwissenschaften.* Halle 1940; Joan M. Eyles, in *Dictionary of Scientific Biography* 11: 272-273. [1755-1827] 1981; F. Boerner & G. Kunkel, *Taschenwörterbuch der botanischen Pflanzennamen.* 4. Aufl. 159 and 458. 1989; William Coxe, *Sketches of the Natural, Civil and Political State of Swisserland* in a Series of Letters to William Melmoth. London 1779; R. Zander, F. Encke, G. Buchheim and S. Seybold, *Handwörterbuch der Pflanzennamen.* 14. Aufl. 767. Stuttgart 1993; Brummitt and Powell, *Authors of Plant Names.* [1753-1827] 1992; Stafleu and Cowan, *Taxonomic Literature.* 4: 572. [b. 1753, d. 1829] Utrecht 1983.

Ramondia Mirbel Schizaeaceae

Origins:

For the French botanist Louis François Élisabeth de Carbonnières Ramond, 1755 (1753?)-1827 (1829?).

Ramonia Schlechter Orchidaceae

Origins:

San Ramón, Costa Rica.

Ramosmania Tirveng. & Verdc. Rubiaceae

Origins:

See D.D. Tirvengadum, "*Ramosmania*, a new monotypic genus of Mascarene Rubiaceae." in *Nord. J. Bot.* 2: 323-327. 1982.

Species/Vernacular Names:

R. heterophylla (Balf.f.) Tirvengadum & Verdc.

Rodrigues Island: café marron

Ramphidia Miq. Orchidaceae

Origins:

Greek *rhamphos* "beak, bill, crooked beak," *rhamphodes* "beak-shaped," indicating the lip.

Ranalisma Stapf Alismataceae

Origins:

Flowers and fruits of this genus resembling certain species of *Ranunculus.*

Randia L. Rubiaceae

Origins:

After the British botanist Isaac Rand, d. 1743, London apothecary, gardener, a member of the Botanical Society, in 1719 was elected a Fellow of the Royal Society, from 1724 to 1743 *Praefectus Horti Chelsiani* and Demonstrator of plants of the Chelsea Physic Garden. His writings include *Index plantarum officinalium ... in Horto chelseiano.* Londini [London] 1730 and *Horti medici Chelseiani ...* Londini 1739. See John Martyn (1699-1768), *Historia plantarum rariorum.* Londini 1728 ("Isaaci Rand sermo de hocce libro habitus coram societate anglica reperitur in *Philos. Transact.* no. 407"); H. Field, *Memoirs of the Botanic Garden of Chelsea* belonging to the Society of Apothecaries of London. London 1878 [R.H. Semple, *Memoirs of the Botanic Garden at Chelsea ...* by the Late Henry Field. London 1878]; F.D. Drewitt, *The Romance of the Apothecaries' Garden at Chelsea.* London 1924; Ida Kaplan Langman, *A Selected Guide to the Literature on the Flowering Plants of Mexico.* 612. Philadelphia 1964; J.H. Barnhart, *Biographical Notes upon Botanists.* 3: 127. 1965; Blanche Henrey, *No Ordinary Gardener — Thomas Knowlton, 1691-1781.* Edited by A.O. Chater. British Museum (Natural History). London 1986; R. Pulteney, *Historical and Biographical Sketches of the Progress of Botany in England.* 2: 102. London 1790; Carl Linnaeus, *Species Plantarum.* 1192. 1753 and *Genera Plantarum.* Ed. 5. 74. 1754; L. Plukenet, *Almagesti botanici mantissa.* Londini 1700; H.N. Clokie, *Account of the Herbaria of the Department of Botany in the University of Oxford.* 229. Oxford 1964; E.M. Tucker, *Catalogue of the Library of the Arnold Arboretum of Harvard University.* 1917-1933; Jonas C. Dryander, *Catalogus bibliothecae historico-naturalis Josephi Banks.* London 1800; Blanche Elizabeth Edith Henrey (1906-1983), *British Botanical and Horticultural Literature before 1800.* Oxford 1975; A. Lasègue, *Musée botanique de Benjamin Delessert.* Paris 1845; James Britten, *The Sloane Herbarium,* revised and edited by J.E. Dandy. 1958; G. Murray, *History of the Collections Contained in the Natural History Departments of the British Museum.* 1: 176. London 1904; D.E. Allen, *The Botanists.* Winchester 1986.

Species/Vernacular Names:

R. spp.

Nigeria: banjeje, bijaje, bijaji, bijagi

Mexico: yaga tzatze

R. armata (Sw.) DC.

Bolivia: crucecito del monte, grano duro, tinajerito, espino blanco, espino, pata de pollo, quichi

R. boliviana Rusby

Bolivia: arrayán, muko-muko

R. calycina Cham.

Bolivia: espino, manzanita

R. fitzalani (F. Muell.) Benth. (after the Australian (Irish-born, Londonderry) botanist Eugene F. Albini Fitzalan, 1830-1911 (d. Brisbane, Queensland, Australia), seedsman, nurseryman, gardener to the Earl of Enniskillen and at Veitch nursery, in 1849 to Victoria, plant collector with John Dallachy (1820-1871), in 1860 a botanical collector on Burdekin River Expedition; see J. Lanjouw and F.A. Stafleu, *Index Herbariorum. Part II (2), Collectors E-H.* Regnum Vegetabile vol. 9. 1957; Ray Desmond, *Dictionary of British & Irish Botanists and Horticulturists.* 248-249. 1994)

English: brown gardenia

R. spinosa (Thunb.) Poir.

English: Malabar randia, spiny randia

Malaya: thorn randa, duri timun tahil, duri timbang tahil

Ranevea L.H. Bailey Palmae

Origins:

An anagram of the name *Ravenea*.

Rangaeris (Schltr.) Summerhayes Orchidaceae

Origins:

An anagram of the generic name *Aerangis*.

Rangia Griseb. Rubiaceae

Origins:

Orthographic variant of *Randia* L.

Ranugia (Schltdl.) Post & Kuntze Cucurbitaceae

Origins:

From *Gurania* (Schltdl.) Cogn.

Ranunculus L. Ranunculaceae

Origins:

Latin *ranunculus, i* "a little frog," the diminutive of *rana, ae,* Greek *batrachion* "a small frog," referring to the habitat (some species grow near the marshes or in damp places) or indicating the shape of the roots; see Carl Linnaeus, *Species Plantarum.* 548. 1753 and *Genera Plantarum.* Ed. 5. 243. 1754; G. Volpi, "Le falsificazioni di Francesco Redi nel Vocabolario della Crusca." in *Atti della R. Accademia della Crusca per la lingua d'Italia.* 33-136. 1915-1916.

Species/Vernacular Names:

R. sp.

Hawaii: makou, 'awa Kanaloa

Tibetan: chu-rug dman-pa, lce-tsha dman-pa

China: yeh chin tsai, lang tu

R. acaulis DC.

English: coast buttercup

R. acris L.

English: meadow buttercup

Peru: chua-chua

China: mao ken, mao chin

R. anemoneus F. Muell.

English: anemone buttercup

R. aquatilis L.

English: water crowsfoot

R. arvensis L.

English: corn buttercup

R. collinus F. Muell.

English: strawberry buttercup, mountain spring buttercup

R. dissectifolius Benth.

English: feather buttercup

R. fascicularis Muhlenb. ex Bigelow

English: early buttercup

R. ficaria L.

English: lesser celandine, pilewort

R. flabellaris Raf.

English: yellow water crowfoot

R. flammula L.

English: lesser spearwort

R. graniticola Melville

English: granite buttercup

R. gunnianus Hook.

English: Gunn's alpine buttercup, Gunn's mountain buttercup

R. hydrocharoides A. Gray

English: frog's-bit buttercup

R. inundatus DC.

English: large river buttercup

R. lappaceus Smith

English: common buttercup, Austral buttercup

R. lingua L.

English: greater spearwort

R. lobbii (Hiern) A. Gray

English: Lobb's aquatic buttercup

R. lyallii Hook.f. (for the British [b. Kincardineshire] naturalist David Lyall, 1817-1895 [d. Cheltenham, Glos.], surgeon, explorer, 1839-1842 on Ross's Antarctic Voyage, 1847 New Zealand, 1852 with Belcher in the Arctic, 1862 Fellow of the Linnean Society; see J.H. Barnhart, *Biographical Notes upon Botanists.* 2: 415. 1965; Thomas Frederick Cheeseman, *Manual of the New Zealand Flora.* xxvii. Wellington 1906; R. Glenn, *The Botanical Explorers of New Zealand.* 175. Wellington 1950; G.A. Doumani, ed., *Antarctic Bibliography.* Washington, Library of Congress 1965-1979; Captain Sir Edward Belcher, *The Last of the Arctic Voyages,* being a narrative of the expedition in HMS *Assistance ...* in search of Sir John Franklin. London 1855; J. Ewan, *Rocky Mountain Naturalists.* The University of Denver Press 1950; Leonard Huxley, *Life and Letters of Sir J.D. Hooker.* London 1918; John T. Walbran, *British Columbia Coast Names, 1592-1906.* To Which are Added a Few Names in Adjacent United States Territory, Their Origin and History. First edition. Ottawa: Government Printing Bureau, 1909; G.P.V. Akrigg & Helen B. Akrigg, *British Columbia Place Names.* Victoria: Sono Nis Press, 1986)

English: Mount Cook lily, mountain lily

R. millanii F. Muell.

English: dwarf buttercup

R. muelleri Benth.

English: felted buttercup

R. multifidus Forssk. (*Ranunculus pinnatus* Poir.; *Ranunculus pubescens* Thunb.)

English: buttercup, wild buttercup

Southern Africa: botterblom, brandblare, geelbotterblom, kankerblare, rhenoster; hlapi (Sotho)

R. muricatus L.

English: sharp buttercup

R. niphophilus B. Briggs

English: Kosciusko buttercup

R. occidentalis Nutt.

English: western buttercup

R. papulentus Melville

English: Nortehrn Rivers buttercup

R. pascuinus (Hook.f.) Melville

English: hairy buttercup

R. pimpinellifolius Hook.

English: bog buttercup

R. plebeius DC.

English: forest buttercup

R. pumilio DC.

English: fan-leaf buttercup, ferny small-flower buttercup

R. repens L.

English: creeping buttercup, butter daisy

Peru: botón de oro, chapo-chapo, chapu-chapu

R. robertsonii Benth.

English: slender buttercup

Ranzania T. Ito Berberidaceae

Origins:

For the Japanese botanist Ranzan Oto, 1729-1810; see F. Boerner & G. Kunkel, *Taschenwörterbuch der botanischen Pflanzennamen.* 4. Aufl. 159. Berlin & Hamburg 1989.

Raoulia Hook.f. ex Raoul Asteraceae

Origins:

Named for the French naval surgeon Édouard Fiacre Louis Raoul, 1815-1852, botanist, from 1840 to 1842 on the *Allier* (with Joseph-Fidèle-Eugène du Bouzet, 1805-1867), from 1842 to 1843 on the *Aube,* from 1840 to 1846 collected and studied New Zealand plants, from 1849 professor of medicine at Brest, France. Wrote *Choix de plantes de la Nouvelle-Zélande* recueillies and décrites par M.E. Raoul. Paris, Leipzig 1846 and *Des rapports des maladies aigües et chroniques du coeur avec les affections dites rhumatismales.* Paris 1839. See John H. Barnhart, *Biographical Notes upon Botanists.* 3: 129. 1965; John Dunmore, *Who's Who in Pacific Navigation.* Honolulu 1991; Thomas Frederick Cheeseman (1846-1923), *Manual of the New Zealand Flora.* Wellington 1906 and 1925; A. Gazel, *French Navigators and the Early History of New Zealand.* Wellington 1946; A. Lasègue, *Musée botanique de Benjamin Delessert.* Paris 1845; Ethelyn Maria Tucker, *Catalogue of the Library of the Arnold Arboretum of Harvard University.* Cambridge, Massachusetts 1917-1933; R. Zander, F. Encke, G. Buchheim and S. Seybold, *Handwörterbuch der Pflanzennamen.* 14. Aufl. 767. Stuttgart 1993.

Raouliopsis S.F. Blake Asteraceae

Origins:
Resembling *Raoulia,* for the French naval surgeon Édouard Fiacre Louis Raoul, 1815-1852.

Rapanea Aublet Myrsinaceae

Origins:
The generic name derives from *rapana,* a common name used in tropical America; see Jean Baptiste Christophore Fusée Aublet (1720-1778), *Histoire des Plantes de la Guiane Françoise.* 1: 121, t. 46. Paris 1775.

Species/Vernacular Names:

R. ferruginea (Ruíz & Pav.) Mez

Brazil: capororoca, capororoca mirim, capororoca assu, pororoca, azeitona do mato

Argentina: lanza blanca, palo San Antonio, canelón blanco

R. howittiana Mez (after Alfred William Howitt, 1830-1908 [d. Victoria], bushman, plant collector, farmer, to Australia 1852, expedition in search of Burke and Wills 1861, Fellow of the Linnean Society 1882; see Ray Desmond, *Dictionary of British & Irish Botanists and Horticulturists.* 360. 1994; N. Hall, *Botanists of the Eucalypts.* Melbourne 1978 and Supplement 1980; John McKinlay (1819-1872), *McKinlay's Journal of Exploration in the Interior of Australia.* (Burke Relief Expedition) Melbourne [1862], "Diary of Mr. J. McKinlay, Leader of the Burke Relief Expedition fitted out by the Government of South Australia." from the *Journal of the Royal Geographical Society of London* for 1863, *Exploration. McKinlay's Diary of His Journey across the Continent of Australia.* Melbourne 1863 and *J. McKinlay's Northern Territory Explorations, 1866.* [J. McKinlay's Journal and Report of Explorations, 1866.] 1867; John Davis, *Tracks of McKinlay and Party across Australia.* By John Davis, one of the expedition. London 1863; Tom Bergin, *In the Steps of Burke & Wills.* Sydney 1981; Jonathan Wantrup, *Australian Rare Books, 1788-1900.* 234-239. Hordern House, Sydney 1987)

English: turnipwood

R. melanophloeos (L.) Mez (*Myrsine melanophloeos* (L.) R. Br.; *Rapanea neurophylla* (Gilg) Mez) (the specific name means "dark bark," from the Greek *melanos* and *phloios*)

English: Cape beech

Southern Africa: boekenhout, witboekenhout; chiKuma, muKwiramakoko, muRgwite, muTomo (Shona); iGcolo, uDzilidzili (Swazi); isiQwane sehlathi (= the protea of the forest) (Xhosa); isiCalabi, umaPhipha, isiQalaba-sehlathi, umHluthi-wentaba, uMaphipha, iKhubalwane (Zulu)

R. umbellata (Mart. ex A. DC.) Mez

Brazil: capororocão, capororoca verdadeira, capororoca branca

Argentina: canelón

R. variabilis (R. Br.) Mez

English: muttonwood

Rapatea Aublet Rapateaceae

Origins:

A vernacular name.

Raphanocarpus Hook.f. Cucurbitaceae

Origins:

Greek *rhaphanos* "cabbage, the radish" and *karpos* "fruit."

Raphanorhyncha Rollins Brassicaceae

Origins:

Greek *rhaphanos, rhaphanis, rhaphanidos* "cabbage, the radish" and *rhynchos* "horn, beak."

Raphanus L. Brassicaceae

Origins:

Latin *raphanus*, Greek *rhaphanos, rhaphanis, rhaphanidos, rhaphane, rhephane* "cabbage, the radish" (Theophrastus, Hippocrates, Aristophanes), Latin *raphanos agria* for a sort of wild radish (Plinius), Greek *rhaphis* and *rhapis*, Latin *rapum, i* "a knob," Akkadian *rabu* "enlarged, swollen"; see Carl Linnaeus, *Species Plantarum.* 669. 1753 and *Genera Plantarum.* Ed. 5. 300. 1754; Giovanni Semerano, *Le origini della cultura europea.* Dizionario della lingua Latina e di voci moderne. 2(2): 542. 1994; Salvatore Battaglia, *Grande dizionario della lingua italiana.* XV: 288. Torino 1994; G. Semerano, *Le origini della cultura europea.* Dizionario della lingua Greca. 2(1): 247. Leo S. Olschki Editore, Firenze 1994.

Species/Vernacular Names:

R. caudatus L.

English: rat-tail radish

R. raphanistrum L.

English: wild radish, wild raddish, field wall flower, jointed charlock, charlock, white charlock, wild mustard, runch

South Africa: knopherik, ramenas, ramnas, wilde mostert, wilde radys

R. sativus L.

English: radish, common radish, wild radish, Japanese white radish, Chinese radish, common cultivated radish

India: muli

China: lai fu zi, lai fu, lo po

Tibetan: la-phug

Japan: daikon

Mexico: guu guiña castilla, coo guiña nagati castilla

The Philippines: labanos, rabanos

Malaya: lobak

Tunisia: fjel

Arabic: figle

Raphia Beauv. Palmae

Origins:

French *raphia*, English *raffia*, based on the Malagasy local plant names, also *rofia*, *raffia*, *ruffia*, *raphia*; see Ambroise Marie François Joseph Palisot de Beauvois, *Flore d'Oware et de Bénin*, en Afrique. 1: 75. Paris 1809; C.T. Onions, *The Oxford Dictionary of English Etymology*. Oxford University Press 1966; Georg Christian Wittstein, *Etymologisch-botanisches Handwörterbuch*. 752. Ansbach 1852; Helmut Genaust, *Etymologisches Wörterbuch der botanischen Pflanzennamen*. 528. Basel 1996; A. Bonavilla, *Dizionario etimologico di tutti i vocaboli usati nelle scienze, arti e mestieri, che traggono origine dal greco*. Milano 1819-1821; A. de Théis, "Spiegazione etimologica de' nomi generici delle piante." tratta dal *Glossario di botanica* di A. de Théis. Vicenza 1815; Ernest Weekley, *An Etymological Dictionary of Modern English*. 2: 1192. New York 1967; Salvatore Battaglia, *Grande dizionario della lingua italiana*. XV: 309. 1994.

Species/Vernacular Names:

R. australis Obermeyer & Strey (*Raphia vinifera* Beauv.) (see Rudolf Georg Strey (1907-1988) and Anna Amelia Obermeyer-Mauve, in *Bothalia* 10(1): 29-37. 1969; M. Gunn and L.E. Codd, *Botanical Exploration of Southern Africa*. 264, 337. Cape Town 1981)

English: Kosi Palm (= found in swamps around the Kosi Bay area, west of Lake Amanzimnyana, northeastern South Africa, close to the Mozambique boundary), raphia palm, giant palm

Southern Africa: Kosipalm, palmboom; umVuma (Zulu)

R. farinifera (Gaertner) N. Hylander (*Raphia kirkii* Becc.; *Raphia lyciosa* Kunth; *Raphia pedunculata* Beauv.; *Raphia ruffia* (Jacq.) Mart.; *Raphia polymita* Kunth; *Sagus farinifera* Gaertn.; *Sagus ruffia* Jacq.)

English: raffia palm, raphia palm

Brazil: palmeira ráfia

Japan: rafia-yashi

China: kuang lang, so mu

R. hookeri G. Mann & H. Wendland

English: wine palm, piassava palm

Nigeria: aiko, angor

Yoruba: aiko

R. hookeri G. Mann & H. Wendl. var. *planifolia* Otedoh

Nigeria: ovie-ogoro (Urhobo)

R. sudanica A. Chev.

Nigeria: tukuruwa (Hausa)

R. taedigera (Mart.) Mart.

Brazil: jupatí, jubati, jurubati, yupati

R. vinifera P. Beauv. var. *nigerica*

Nigeria: olala odumuhol (Ijaw); ovie-ibodje (Urhobo); ala bou (Kalabri)

R. vinifera P. Beauv. var. *vinifera*

Congo: pike, olengue

Nigeria: ako, tukuruwa, angor, emaho-ogoro, ogoro, gongola, gwangola, okut; oha (Edo); ekian (Urhobo); ebeje (Itsekiri); ngwo ude (Igbo)

Yoruba: iko, apako, pako, oguro, igi oguro, eyin agbigbo, eyin arigbo

Raphiacme K. Schumann Asclepiadaceae

Origins:

Greek *rhaphis*, *rhaphidos* "a needle" and *akme* "the top, highest point," see also *Raphionacme* Harvey.

Raphidiocystis Hook.f. Cucurbitaceae

Origins:

Greek *rhaphis*, *rhaphidos* "a needle" and *kystis* "a bladder."

Raphidophyllum Hochst. Scrophulariaceae

Origins:

From the Greek *rhaphis*, *rhaphidos* and *phyllon* "leaf."

Raphiocarpus Chun Gesneriaceae

Origins:

Greek *rhaphis*, *rhaphidos* and *karpos* "a fruit."

Raphiolepis Lindley Rosaceae

Origins:

Orthographic variant of *Rhaphiolepis* Lindley.

Raphionacme Harvey Asclepiadaceae

Origins:

Greek *rhaphis*, *rhaphidos* "a needle" and *akme* "the top, highest point."

Species/Vernacular Names:

R. hirsuta (E. Meyer) R.A. Dyer ex Phill. (*Raphionacme divaricata* Harv.)

English: false gentian

Southern Africa: inTsema (Xhosa)

R. utilis N.E. Br. & Stapf

English: Bitinga rubber

Raphisanthe Lilja Loasaceae

Origins:

Greek *rhaphis* "a needle" and *anthos* "flower."

Raphistemma Wallich Asclepiadaceae

Origins:

Greek *rhaphis* and *stemma* "garland," an allusion to the shape of the lobes of the corolla.

Species/Vernacular Names:

R. pulchellum (Roxburgh) Wallich

English: common raphistemma

China: da hua teng

Rapicactus Buxb. & Oehme Cactaceae

Origins:

From the Greek *rhapis* "a rod" and *Cactus*, see also *Turbinicarpus* (Backeb.) F. Buxb. & Backeb.

Rapinia Loureiro Campanulaceae

Origins:

For the French botanist René Rapin, 1621-1687. His writings include *Carmina*. Parisiis 1723, *Christus patiens*, carmen heroicum. Londini 1713, Rapini *hortorum lib. IV, et disputatio de cultura hortensi*. Paris 1665, *Hortorum* libri IIII. Lugduni-Batav. 1668 and 1672, *Hortorum libri IV, et cultura hortensis*. Parisiis 1780 and *Oeuvres diverses* de P. Rapin, nouv. édition, augmentée du poëme des Jardins. La Haye 1725.

Rapinia Montrouz. Labiatae (Verbenaceae)

Origins:

For the French botanist René Rapin, 1621-1687.

Rapistrum Crantz Brassicaceae

Origins:

Latin *rapistrum, i* "the wild rape" (Junius Moderatus Columella), *rapum, i* and *rapa, ae* "a turnip"; see the Austrian physician and botanist Heinrich Johann Nepomuk von Crantz (1722-1799), author of *Classis cruciformium emendata*. 105. Leipzig 1769.

Species/Vernacular Names:

R. perenne (L.) All.

English: steppe-cabbage

R. rugosum (L.) All. (*Myagrum rugosum* L.)

English: wild mustard, giant mustard, turnip-weed, wild turnip, short-fruited wild turnip

South Africa: wilde mostert

Rapunculus Mill. Campanulaceae

Origins:

Latin *rapum, i* and *rapa, ae* "a turnip."

Rapuntia Chevall. Campanulaceae

Origins:

The diminutive of the Latin *rapum, i* and *rapa, ae* "a turnip."

Rapuntium Mill. Campanulaceae

Origins:

The diminutive of the Latin *rapum, i* and *rapa, ae* "a turnip."

Rapuntium Post & Kuntze Campanulaceae

Origins:

The diminutive of the Latin *rapum, i* and *rapa, ae* "a turnip."

Raputia Aublet Rutaceae

Origins:

From Orapù (Guiana), the place where the plant was discovered; see A. de Théis, "Spiegazione etimologica de' nomi generici delle piante." tratta dal *Glossario di botanica* di A. de Théis. Vicenza 1815; Salvatore Battaglia, *Grande dizionario della lingua italiana*. XV: 493. 1994.

Raputiarana Emmerich Rutaceae

Origins:
Referring to the genus *Raputia* Aubl.

Raram Adans. Gramineae

Origins:
Latin *rarus, rara, rarum* "thin, rare, scanty, scattered, few, sparsely."

Raspailia C. Presl Gramineae

Origins:
For the French scientist François Vincent Raspail, 1794-1878.

Raspalia Brongn. Bruniaceae

Origins:
For the French (b. Carpentras) scientist François Vincent Raspail, 1794-1878 (d. Arcueil, near Paris), botanist, politician, chemist, naturalist, one of the founders of cytochemistry and cellular pathology, determined the agent of scabies (*Sarcoptes scabiei*), self-educated in science. Among his numerous publications are *Manuel annuaire de la Santé pour 1847, ou médecine et pharmacie domestiques*. Paris 1847, *Mémoires sur la famille des Graminées*. Paris 1825 [-1826], *De la Pologne sur les bords de la Vistule et dans l'émigration*. Paris 1839, *L'Ami du Peuple en 1848*. [1848], *Le Choléra en 1865*. Paris 1865 and *Nouveau Système de Physiologie Végétale et de Botanique*. Paris 1837. See Marc Klein, in *Dictionary of Scientific Biography* 11: 300-302. [b. 1794] 1981; John H. Barnhart, *Biographical Notes upon Botanists*. 3: 129. Boston 1965; [François Vincent Raspail], *Procès et defense de F.V. Raspail, poursuivi ... en exercise illégal de la médecine ... sur la dénonciation formelle des sieurs Fouquier ... et Orfila ... agissant comme vice-président et président d'une association anonyme de médecins*. Paris 1846; R. Zander, F. Encke, G. Buchheim and S. Seybold, *Handwörterbuch der Pflanzennamen*. 14. Aufl. Stuttgart 1993; Stafleu and Cowan, *Taxonomic Literature*. 4: 582-583. ["born 29 Jan 1791"] Utrecht 1983; T.W. Bossert, *Biographical Dictionary of Botanists Represented in the Hunt Institute Portrait Collection*. 325. 1972; Ida Kaplan Langman, *A Selected Guide to the Literature on the Flowering Plants of Mexico*. 612-613. University of Pennsylvania Press, Philadelphia 1964.

Rastrophyllum Wild & Pope Asteraceae

Origins:
From the Latin *rastrum* "a toothed hoe" and Greek *phyllon* "leaf."

Rathbunia Britton & Rose Cactaceae

Origins:
Named for Richard Rathbun, 1852-1918, Smithsonian Institute, Washington, D.C. His writings include *The Columbian Institute for the Promotion of Arts and Sciences*: A Washington Society of 1816-1838. Washington 1917, *A Descriptive Account of the Building Recently Erected for the Departments of Natural History of the United States National Museum*. Washington 1913, *The National Gallery of Art*, Department of Fine Arts of the National Museum. Washington 1909 and *The United States National Museum*. Washington 1905.

Ratibida Raf. Asteraceae

Origins:
See Constantine Samuel Rafinesque, *Florula ludoviciana*. 73. New York 1817, *Am. Monthly Mag. Crit. Rev*. 2: 268. 1818 and 4: 189, 195. 1819, *Jour. Phys. Chim. Hist. Nat*. 89: 100. 1819; E.D. Merrill, *Index rafinesquianus*. 239, 240, 241. Jamaica Plain, Massachusetts, USA 1949.

Rattraya J.B. Phipps Gramineae

Origins:
After James Mcfarlane Rattray, 1907-1974, botanist. With the South African botanist Sydney Margaret Stent (1875-1942) wrote "The grasses of Southern Rhodesia." *Proc. Rhod. Sci. Ass*. 32: 1-64. 1933; see Stafleu and Cowan, *Taxonomic Literature*. 4: 586. 1983; Mary Gunn and Leslie E. Codd, *Botanical Exploration of Southern Africa*. 290 [for George Rattray, 1872-1941], 334 [for Sydney Margaret Stent]. Cape Town 1981; John H. Barnhart, *Biographical Notes upon Botanists*. 3: 324. 1965; F.N. Hepper and Fiona Neate, *Plant Collectors in West Africa*. 67. [genus named for the Scottish diatomist John Rattray, 1858-1900] 1971; Ray Desmond, *Dictionary of British & Irish Botanists and Horticulturists*. 573. London 1994; A. White and B.L. Sloane, *The Stapelieae*. Pasadena 1937.

Ratzeburgia Kunth Gramineae

Origins:

Named for the German botanist Julius Theodor Christian Ratzeburg, 1801-1871, entomologist, physician, forester, zoologist, M.D. Berlin 1825, professor of natural sciences. His publications include *Die Forst-Insecten*. Berlin 1837-1844, *Die Waldverderbniss*, etc. Berlin 1866-1868, *Forstnaturwissenschaftliche Reisen durch verschiedene Gegenden Deutschlands ...* Im Anhage, Gebirgsboden-Analysen vom Professor Dr. F. Schulze zu Eldena, etc. Berlin 1842, *Forstwissenschaftliches Schriftsteller-Lexikon*. Berlin 1872. See Stafleu and Cowan, *Taxonomic Literature*. 4: 586-587. Utrecht 1983; G.W. von Viebahn, *Statistik des zollvereinten und nördlichen Deutschlands*. In Verbindung mit ... Ratzeburg ... herausgegeben von G. von Viebahn. 1858; John H. Barnhart, *Biographical Notes upon Botanists*. 3: 130. 1965; Ethelyn Maria Tucker, *Catalogue of the Library of the Arnold Arboretum of Harvard University*. Cambridge, Massachusetts 1917-1933.

Rauhia Traub Amaryllidaceae

Origins:

For the German botanist Werner Rauh, 1913-1997 (Heidelberg), explorer, collector, botanical writer, bryologist and specialist in succulent plants and Bromeliaceae, traveler (Africa and South America, Madagascar), professor and former Director of the Institute for Systematic Botany at the University of Heidelberg (Inst. für Systematische Botanik, Ruprecht-Karls-Universität, Heidelberg, Germany) and the associated Botanical Garden, author of numerous books and articles pertaining to succulent plants. He is commemorated in numerous specific and generic names, contributed to *The Euphorbia Journal*. Among his writings are "The Didiereaceae." *Cact. Succ. J. Amer.* 48: 75. 1976, "Uber einige interessante Sukkulenten aus Kenia." *Sukkulentenkunde*. 7/8: 108-127. 1963, *Kakteen an ihren Standorten*. Berlin & Hamburg 1979 and *Succulent and Xerophytic Plants of Madagascar*. Two volumes. Strawberry Press 1995-1998. See T.W. Bossert, *Biographical Dictionary of Botanists Represented in the Hunt Institute Portrait Collection*. 325. 1972; *The Euphorbia Journal*. vol. 10: 225-226. 1996; Andrew Cowin, "The *Hortus Palatinus*: Heidelberg's eighth wonder of the world." in *Hortus*. 24: 44-54. 1992; Gordon Douglas Rowley, *A History of Succulent Plants*. Strawberry Press, Mill Valley, California 1997; R. Zander, F. Encke, G. Buchheim and S. Seybold, *Handwörterbuch der Pflanzennamen*. 14. Aufl. Stuttgart 1993; Mary Gunn and Leslie E. Codd, *Botanical Exploration of Southern Africa*. 290. Cape Town 1981; Irving William Knobloch, compil., "A preliminary verified list of plant collectors in Mexico." *Phytologia Memoirs*. VI. 1983.

Rauhiella Pabst & Braga Orchidaceae

Origins:

For the German botanist Werner Rauh, 1913-1997.

Rauhocereus Backeberg Cactaceae

Origins:

For the German botanist Werner Rauh, 1913-1997.

Rauia Nees & Martius Rutaceae

Origins:

For the German botanist Ambrosius Rau, 1784-1830, mineralogist, naturalist, botanical collector. Wrote *Enumeratio rosarum circa Wirceburgum*. Norimbergae 1816 and *Ueber den technischen Theil der Salzwerkskunde. Ein Programm*, etc. Würzburg 1809; see John H. Barnhart, *Biographical Notes upon Botanists*. 3: 130. 1965; Ethelyn Maria Tucker, *Catalogue of the Library of the Arnold Arboretum of Harvard University*. Cambridge, Massachusetts 1917-1933.

Rautanenia Buchenau Alismataceae

Origins:

After a Finnish missionary, the Rev. Martti (Martin) Rautanen, 1845-1926 (d. Olukonda, SW Africa), collected 1886-91 SW Africa (Ovamboland); see I.H. Vegter, *Index Herbariorum*. Part II (5), *Collectors N-R*. Regnum Vegetabile vol. 109. 1983; Mary Gunn and Leslie E. Codd, *Botanical Exploration of Southern Africa*. 290-291. Cape Town 1981.

Rauvolfia L. Apocynaceae

Origins:

The generic name honors the German (b. Augsburg, Bavaria) physician and botanist Leonhart (Leonhardus, Leonhard), Rauwolff (Rauwolf, Rawolff, Rauvolfius), 1535-1596 (d. Waitzen/Vac, Hungary, fighting the Turks), traveler, plant collector, studied at the universities of Wittenberg, Montpellier and Valence, M.D. Valence 1562, in Zürich met the naturalist Conrad (Konrad) von Gesner (1516-1565), author of the famous travel account and the first publication to herald coffee in Europe *Aigentliche Beschreibung der Raiss, so er vor diser zeit gegen Auffgang inn die Morgenlaender, fürnemlich Syriam, Iudaeam, Arabiam, Mesopotamiam, Babyloniam, Assyriam, Armeniam* etc. Laugingen [sic] 1582, from 1573 to 1576 visited the Near East, described the riparian flora of the Euphrates. See Johan Frederik (Jan Fredrik, Johannes Fridericus) Gronovius (1686-1762), *Flora orientalis, sive recensio* plantarum, quas botanicorum coryphaeus *Leonhardus Rauwolf* annis 1573-75 in Syria, Arabia, Mesopotamia, Babylonia, Assyria, Armenia et Judaea crescentes observavit et collegit ... Lugduni Batavorum [Leyden] 1755; John H. Barnhart, *Biographical Notes upon Botanists*. 3: 131. 1965; Frans A. Stafleu, *Linnaeus and the Linnaeans*. The spreading of their ideas in systematic botany, 1735-1789. Utrecht 1971; Karl H. Dannenfeldt, in *Dictionary of Scientific Biography* 11: 311-312. 1981; Georg Christian Wittstein, *Etymologisch-botanisches Handwörterbuch*. 753. Ansbach 1852; K.H. Dannenfeldt, *Leonard Rauwolf, Sixteenth Century Physician, Botanist and Traveler*. Cambridge, Massachusetts 1968; E.M. Tucker, *Catalogue of the Library of the Arnold Arboretum of Harvard University*. 1917-1933; A. Lasègue, *Musée botanique de Benjamin Delessert*. Paris 1845; Carl Linnaeus, *Species Plantarum*. 208. 1753 and *Genera Plantarum*. Ed. 5. 98. 1754; Richard J. Durling, comp., *A Catalogue of Sixteenth Century Printed Books in the National Library of Medicine*. 1967.

Species/Vernacular Names:

R. amsoniaefolia A. DC.

The Philippines: sibakong, banogan, maladita

R. caffra Sond. (*Rauvolfia natalensis* Sond.; *Rauvolfia macrophylla* Stapf)

English: quinine tree

Nigeria: wada (Hausa); awa (Yoruba)

Yoruba: alagba

Central Africa: esoma, esombi, sambo

Southern Africa: kinaboom, umHlambamanzi, koorsboom; muKadhlu, muGaururu, iDzurungu, muKashu, muKaururu, muSingwizi (Shona); umKhadluvungu, umHlambamasi (meaning the sour milk cleanser), umJele (Zulu); nchongo (Tsonga); munadzi (Venda); umThundisa, umJelo, umJela, umHlamb'amasi (Xhosa)

R. cubana A. DC.

English: Cuba devil pepper

China: guba luo fu mu

R. inebrians K. Schum.

Southern Africa: Dzurungu, muKaururu, muZungurwi (Shona)

R. samarensis Merr.

The Philippines: kanda sa tahok, bitak-bitak, butbuta, kolanos

R. sandwicensis A. DC. (*Rauvolfia degeneri* Sherff; *Rauvolfia forbesii* Sherff; *Rauvolfia helleri* Sherff; *Rauvolfia mauiensis* Sherff; *Rauvolfia molokaiensis* Sherff; *Rauvolfia remotiflora* Degener & Sherff; *Ochrosia sandwicensis* A. DC.)

Hawaii: hao

R. serpentina (L.) Bentham ex Kurz

English: Java devil pepper

Japan: Indo-zaboku

China: she gen mu

India: sarpagandha, sarpgandha (the roots), chandrika, sarpa-gandha, chota-chand, chandra, naya, dhan-marna, dhan-barua, harkai, patala-gandhi, chivan melpodi, covannamilpori, chivan avelpori, sutranabi, patalagarud, patala-garudada-beru

R. sumatrana Jack

English: Sumatra devil pepper

China: su men da la luo fu mu

Malaya: sumbu badak, tumpul badak, pelir kambing

R. tetraphylla L.

English: four-leaf devil pepper

China: si ye luo fu mu

Central America: chalchupa, alcotán, amatillo, cabamuc, curarina, matacoyote, señorita, viborilla

R. verticillata (Loureiro) Baillon (*Rauvolfia latifrons* Tsiang; *Rauvolfia perakensis* King & Gamble; *Rauvolfia verticillata* var. *hainanensis* Tsiang; *Rauvolfia verticillata* var. *officinalis* Tsiang; *Rauvolfia yunnanensis* Tsiang)

English: broad-leaf devil pepper, common devil pepper, medicinal devil pepper, Perak devil pepper, Hainan devil pepper, Yunnan devil pepper

China: luo fu mu

Malaya: batu pelir kambing, pelir kambing

R. vomitoria Afzel.

English: emetic devil pepper

Central Africa: esoma, esombi, sambo

Congo: ompepe, nduli, ndouli, onduele, ondole

Sierra Leone: kowogae

Nigeria: akata, akanta, atapara, asofeyeje, ira, ira-igbo, bandonge, awgor, bayejokorok, ekam, elongolongo, itongotongo, hinpa, lindondongo, wadda, minjan-minjanga, ayekoje, essembi-sembi; wada (Hausa); asofeyeje (Yoruba); akata (Edo); akanta (Igbo); uto enyin (Efik)

Yoruba: asofeyeje, dodo, awowere, ira, ira igbo, oora, dodo dudu, akanta, apawere, oloragbo

China: cui tu luo fu mu

Rauwenhoffia R. Scheffer Annonaceae

Origins:

After the Dutch botanist Dr. Nicolaas Willem Pieter Rauwenhoff, 1826-1909, plant physiologist, in 1871 successor to F.A.W. Miquel at Utrecht, from 1871 to 1896 professor of botany and Director of the Botanical Garden at Utrecht. His works include *Charles Robert Darwin*. Utrecht 1882 and *La génération sexuée des Gleicheniacées*. Haarlem 1890. See Rudolph Herman Christiaan Carel Scheffer (1844-1880), *Annales du Jardin Botanique de Buitenzorg*. 2: 21. 1885; John H. Barnhart, *Biographical Notes upon Botanists*. 3: 131. 1965; T.W. Bossert, *Biographical Dictionary of Botanists Represented in the Hunt Institute Portrait Collection*. 325. 1972; E.M. Tucker, *Catalogue of the Library of the Arnold Arboretum of Harvard University*. 1917-1933; Stafleu and Cowan, *Taxonomic Literature*. 4: 593-595. Utrecht 1983.

Rauwolfia L. Apocynaceae

Origins:
For Leonhart Rauwolff (Leonhard Rauwolf), 1535-1596.

Ravenala Adans. Strelitziaceae

Origins:
From the native name in Madagascar.

Species/Vernacular Names:
R. madagascariensis Sonn.

English: traveler's tree, traveler's palm

Japan: ogi-bashô

Ravenea Bouché Palmae

Origins:
For Louis Ravené, an official in Berlin, 19th century.

Species/Vernacular Names:
R. hildebrandtii Bouché ex H.A. Wendland

Brazil: falsa guaricanga

Ravenia Vell. Conc. Rutaceae

Origins:
For the French professor J.F. Ravin.

Ravensara Sonn. Lauraceae

Origins:
From the native name in Madagascar.

Ravnia Oersted Rubiaceae

Origins:
For the Norwegian plant collector Peter Ravn, d.1839, surgeon, to the Danish West Indies (1819); see Stafleu and Cowan, *Taxonomic Literature*. 4: 603. Utrecht 1983.

Rawsonia Harvey & Sonder Flacourtiaceae

Origins:
For the British (b. London) pteridologist Sir Rawson William Rawson, 1812-1899 (d. London), traveler, colonial administrator, 1842 Government Secretary Canada, from 1854 to 1864 Colonial Secretary of the Cape of Good Hope, Governor Bahamas (1864), of Jamaica (1865) and Windward Islands (1869-1875). His works include *Synopsis of the Tariffs and Trade of the British Empire*. London 1888, *An Account of the State of Education within the District of Nattore*, etc. London 1838, *Analysis of the Maritime Trade of the United Kingdom, 1889-1891*. London 1892, *The Gospel Narrative*. 1895 and *Report on the Bahamas' Hurricane of October 1866*. With a description of the city of Nassau, N.P. [Nassau 1866?], with Karl (Carl) Wilhelm Ludwig Pappe (1803-1862) wrote *Synopsis filicum Africae australis*. Cape Town 1858. See J.H. Barnhart, *Biographical Notes upon Botanists*. 3: 132. 1965; G. Murray, *History of the Collections Contained in the Natural History Departments of the British Museum*. 1: 176. 1904; Mary Gunn and Leslie E. Codd, *Botanical Exploration of Southern Africa*. 291. Cape Town 1981.

Species/Vernacular Names:
R. lucida Harvey & Sonder

English: forest peach

Southern Africa: bosperske; murambo, tshilala-nari (meaning buffalo bed) (Venda); iThambo, uMahlekasaphule, umHlabamkhwaba, iNanga, umThelelo-omncane, umSathanina, umPindangulube, umKumgwinqa, umDhlunye, umPhambatho (Zulu); umNqayi maphuthi, umPitshi wehlati, iPitshi wehlathi (= the forest peach), umNqandabukuku, umLongo, umNqayi masende, umNqayi weputi (Xhosa); muChekemanu, muNaba (Shona)

Rayania Raf. Dioscoreaceae

Origins:
Orthographic variant of *Rajania* L., dedicated to the British naturalist John Ray (Wray), 1627-1705, traveler, 1667 Fellow of the Royal Society. Among his writings are *Catalogus plantarum circa Cantabrigiam nascentium*. Cambridge 1660 and *Catalogus plantarum Angliae et insularum adjacentium*. London 1670; see Charles Webster, in *Dictionary of Scientific Biography* 11: 313-318. 1981; G.L. Keynes, *John Ray, a Bibliography*. London 1951; John H. Barnhart, *Biographical Notes upon Botanists*. 3: 132. 1965; Blanche Henrey, *British Botanical and Horticultural Literature before 1800*. 1975; Ray Desmond, *Dictionary of British & Irish Botanists and Horticulturists*. 574-575. 1994; Stafleu and Cowan, *Taxonomic Literature*. 4: 604-610. Utrecht 1983; C.E. Raven, *John Ray Naturalist, His Life and Works*.

Cambridge 1942; Frans A. Stafleu, *Linnaeus and the Linnaeans. The spreading of their ideas in systematic botany, 1735-1789.* Utrecht 1971; C.S. Rafinesque, *Autikon botanikon.* Icones plantarum select. nov. vel rariorum, etc. 125. Philadelphia 1840; Mariella Azzarello Di Misa, ed., *Il Fondo Antico della Biblioteca dell'Orto Botanico di Palermo.* 225-226. Regione Siciliana, Palermo 1988; Blanche Henrey, *No Ordinary Gardener — Thomas Knowlton, 1691-1781.* Edited by A.O. Chater. British Museum (Natural History). London 1986.

Raycadenco Dodson Orchidaceae

Origins:

From the Christian names of Raymond McCoullough, Carl Whitner, Dennis D'Alessandro and Cornelia Head.

Raynia Raf. Dioscoreaceae

Origins:

Orthographic variant of *Rajania* L.; see C.S. Rafinesque, *Anal. Nat. Tabl. Univ.* 199. 1815.

Rea Bertero ex Decaisne Asteraceae

Origins:

After the Italian botanist Giovanni Francesco Re, 1773-1833, physician, professor of botany; see John H. Barnhart, *Biographical Notes upon Botanists.* 3: 133. 1965; O. Mattirolo, *Cronistoria dell'Orto Botanico della Regia Università di Torino.* in *Studi sulla vegetazione nel Piemonte* pubblicati a ricordo del II Centenario della fondazione dell'Orto Botanico della R. Università di Torino. Torino 1929.

Reana Brign. Gramineae

Origins:

Possibly after the Italian botanist Giovanni Francesco Re, 1773-1833, physician, professor of botany.

Reaumuria Hasselq. ex L. Tamaricaceae

Origins:

For the French (b. La Rochelle) zoologist René-Antoine Ferchault de Réaumur, 1683-1757 (d. near St.-Julien-du-Terroux), entomologist, naturalist, studied philosophy. Among his major works are *L'art de convertir le fer forgé*

en acier. Paris 1722 and *Mémoires pour servir à l'histoire des insectes.* Paris 1734-1742; see J.B. Gough, in *Dictionary of Scientific Biography* 11: 327-335. 1981; Jean Torlais, *Réaumur, un esprit encyclopédique en dehors de "l'Encyclopédie."* Paris 1936.

Rebis Spach Grossulariaceae

Origins:

From *Ribes* L.

Reboudia Cosson & Durieu Brassicaceae

Origins:

After the French botanist Victor Constant Reboud, 1821-1889, physician, in Africa and Algeria; see John H. Barnhart, *Biographical Notes upon Botanists.* 3: 134. 1965; E.M. Tucker, *Catalogue of the Library of the Arnold Arboretum of Harvard University.* 1917-1933.

Reboulea Kunth Gramineae

Origins:

For the French botanist Eugène de Reboul, 1781-1851; see J.H. Barnhart, *Biographical Notes upon Botanists.* 3: 134. 1965; E.M. Tucker, *Catalogue of the Library of the Arnold Arboretum of Harvard University.* 1917-1933; Helmut Genaust, *Etymologisches Wörterbuch der botanischen Pflanzennamen.* 529. [genus *Reboulia* dedicated the French naturalist to Henri-Paul-Irénée Reboul, 1763-1839] Basel 1996.

Rebutia K. Schumann Cactaceae

Origins:

For the French cactus nurseryman Pierre Rebut, 1830-1898; see Gordon Douglas Rowley, *A History of Succulent Plants.* Strawberry Press, Mill Valley, California 1997; R. Zander, F. Encke, G. Buchheim and S. Seybold, *Handwörterbuch der Pflanzennamen.* 14. Aufl. 768. Stuttgart 1993; F. Boerner & G. Kunkel, *Taschenwörterbuch der botanischen Pflanzennamen.* 4. Aufl. 159 and 458. Berlin & Hamburg 1989.

Recordia Moldenke Verbenaceae

Origins:

For the American botanist Samuel James Record, 1881-1945, dendrologist, wood anatomist, author of *The Mechanical*

Properties of Wood. New York 1914. With Clayton Dissinger Mell (1875-1945) wrote *Timbers of Tropical America*. New Haven 1924. See J.H. Barnhart, *Biographical Notes upon Botanists*. 3: 135. 1965; T.W. Bossert, *Biographical Dictionary of Botanists Represented in the Hunt Institute Portrait Collection*. 326. 1972; Ida Kaplan Langman, *A Selected Guide to the Literature on the Flowering Plants of Mexico*. 614-615. University of Pennsylvania Press, Philadelphia 1964; S. Lenley et al., *Catalog of the Manuscript and Archival Collections and Index to the Correspondence of John Torrey*. Library of the New York Botanical Garden. 340. 1973; E.M. Tucker, *Catalogue of the Library of the Arnold Arboretum of Harvard University*. 1917-1933.

Recordoxylon Ducke Caesalpiniaceae

Origins:

For the American botanist Samuel James Record, 1881-1945, dendrologist, wood anatomist.

Rectanthera O. Degener Commelinaceae

Origins:

From the Latin *rectus* "upright" and *anthera*.

Rectomitra Blume Melastomataceae

Origins:

Latin *rectus* "upright" and *mitra* "a turban, rope, headband."

Redfieldia Vasey Gramineae

Origins:

For the American botanist John Howard Redfield, 1815-1895, zoologist, conchologist, plant collector, palaeontologist, author of *Recollections of John Howard Redfield*. [privately printed] 1900, with the American botanist Edward Lothrop Rand (1859-1924) wrote *Flora of Mount Desert Island*. Cambridge 1894. He was son of the American meteorologist and palaeontologist William C. Redfield (1789-1857); see Harold L. Burstyn, in *Dictionary of Scientific Biography* 11: 340-341. 1981; John H. Barnhart, *Biographical Notes upon Botanists*. 3: 127, 135. 1965; J.W. Harshberger, *The Botanists of Philadelphia and Their Work*. 1899; T.W. Bossert, *Biographical Dictionary of Botanists Represented in the Hunt Institute Portrait Collection*. 327. 1972; S. Lenley et al., *Catalog of the Manuscript and Archival Collections and Index to the Correspondence of John Torrey*. Library

of the New York Botanical Garden. 1973; E.M. Tucker, *Catalogue of the Library of the Arnold Arboretum of Harvard University*. 1917-1933; Joseph Ewan, *Rocky Mountain Naturalists*. The University of Denver Press 1950.

Reedia F. Muell. Cyperaceae

Origins:

After the Australian architect Joseph Reed (probably born 1822), a member of the Royal Society of Victoria; see Ferdinand von Mueller, *Fragmenta Phytographiae Australiae*. 1: 239, t. 10. Melbourne 1864.

Reedrollinsia J.W. Walker Annonaceae

Origins:

For the American botanist Reed Clarke Rollins, b. 1911, a specialist in Cruciferae (alt. Brassicaceae); see J.H. Barnhart, *Biographical Notes upon Botanists*. 3: 174. 1965; T.W. Bossert, *Biographical Dictionary of Botanists Represented in the Hunt Institute Portrait Collection*. 337. 1972; Ida Kaplan Langman, *A Selected Guide to the Literature on the Flowering Plants of Mexico*. 641-642. University of Pennsylvania Press, Philadelphia 1964; S. Lenley et al., *Catalog of the Manuscript and Archival Collections and Index to the Correspondence of John Torrey*. Library of the New York Botanical Garden. 353. 1973; Irving William Knobloch, compil., "A preliminary verified list of plant collectors in Mexico." *Phytologia Memoirs*. VI. 1983.

Reesia Ewart Caryophyllaceae

Origins:

The generic name honors the Australian botanist Bertha Rees, a lecturer on botany at the University of Melbourne, Australia; see Alfred James Ewart (1872-1937) and Bertha Rees, in *Proceedings of the Royal Society of Victoria*. Ser. 2, 26(1): 9, t. II. (Aug.) 1913.

Reevesia Lindley Sterculiaceae

Origins:

For the British (b. Essex) naturalist John Reeves, 1774-1856 (d. Clapham, Surrey), traveler and plant collector, tea specialist, 1812 China, 1817 elected a Fellow of the Royal Society and of the Linnean Society, corresponded with Sir Joseph Banks; see E.H.M. Cox, *Plant-Hunting in China. A history of botanical exploration in China and the Tibetan marches*. London 1945; Robert Morrison, *A Dictionary of the Chinese Language*, in three parts. (Chinese names of stars and constellations collected by J. Reeves. A synopsis of various forms of the Chinese character.) Macao 1815, 1822, 1823; Ray Desmond, *Dictionary of British & Irish Botanists and Horticulturists*. 577. London 1994; Emil Bretschneider, *History of European Botanical Discoveries in China*. Leipzig 1981; Mary Gunn and Leslie E. Codd, *Botanical Exploration of Southern Africa*. 291-292. A.A. Balkema Cape Town 1981; James Britten and George E. Simonds Boulger, *A Biographical Index of Deceased British and Irish Botanists*. 1931.

Regelia Schauer Myrtaceae

Origins:

For the German botanist Eduard August von Regel, 1815-1892, from 1842 to 1855 head gardener at the University Botanical Garden of Zürich, from 1857 to 1867 scientific Director of the Imperial Botanical Gardens at St. Petersburg, and from 1875 to 1892 Director (succeeded Ernst Rudolf von Trautvetter, 1809-1889), a prolific writer, from 1852 to 1884 founder and editor of *Gartenflora*. Among his works are *Florula ajanensis*. Moskwa 1858 and *Cycadearum generum specierumque revisio*. St. Petersburg 1876, with the German botanist Johann Joseph Schmitz (1813-1845) wrote *Flora bonnensis*. [Introduction by Ludolf Christian Treviranus, 1779-1864.] Bonnae [Bonn] 1841. He was father of the Russian physician and botanist Johann Albert von Regel (1845-1908). See L. Wittmack, "Eduard August Regel." *Gartenflora*. 41: 261-269. 1892; Stafleu and Cowan, *Taxonomic Literature*. 4: 638-648. 1983; John H. Barnhart, *Biographical Notes upon Botanists*. 3: 138. 1965; Charles J. Édouard Morren, *Correspondance botanique*. Liège 1874 and 1884; J.C. Schauer, *Dissertatio Phytographica de Regelia, Beaufortia et Calothamno, generibus plantarum Myrtacearum*. Vratislaviae 1843; R. Zander, F. Encke, G. Buchheim and S. Seybold, *Handwörterbuch der Pflanzennamen*. 14. Aufl. 768. Stuttgart 1993; Emil Bretschneider, *History of European Botanical Discoveries in China*. Leipzig 1981; T.W. Bossert, *Biographical Dictionary of Botanists Represented in the Hunt Institute Portrait Collection*. 327. 1972; Ida Kaplan Langman, *A Selected Guide to the Literature on the Flowering Plants of Mexico*. 617-618. University of Pennsylvania Press, Philadelphia 1964; S. Lenley et al., *Catalog of the Manuscript and Archival Collections and Index to the Correspondence of John Torrey*. Library of the New York Botanical Garden. 467. 1973; E.M. Tucker, *Catalogue of the Library of the Arnold Arboretum of Harvard University*. 1917-1933; Ernest Nelmes and William Cuthbertson, *Curtis's Botanical Magazine Dedications, 1827-1927*. 230-232. [1931]; Gordon Douglas Rowley, *A History of Succulent Plants*. Strawberry Press, Mill Valley, California 1997; Georg Christian Wittstein, *Etymologisch-botanisches Handwörterbuch*. 754. Ansbach 1852.

Species/Vernacular Names:

R. velutina (Turcz.) C. Gardner

English: velvet regelia, Barren's regelia

Regnellia Barb. Rodr. Orchidaceae

Origins:

After the Swedish botanist Anders Fredrik (André Frederick) Regnell, 1807-1884, plant collector (with Gustaf Anders Lindberg (1832-1900) and Salomon Eberhard Henschen, 1847-1930), physician, lichenologist, 1837 M.D. Uppsala, 1840 in Brazil, botanical explorer; see John H. Barnhart, *Biographical Notes upon Botanists*. 3: 138. 1965; T.W. Bossert, *Biographical Dictionary of Botanists Represented in the Hunt Institute Portrait Collection*. 327. 1972; A. Lasègue, *Musée botanique de Benjamin Delessert*. Paris 1845; Gustaf Oskar Andersson Malme (*né* Andersson) (1864-1937), *Ex herbario Regnelliano*. Stockholm 1898-1901 and "Die Compositen der zweiten Regnellschen Reise. III. Puente del Inca und Las Cuevas (Mendoza)." *Ark. Bot.* 24A(8): 58-66. 1932.

Regnellidium Lindman Marsileaceae

Origins:

After the Swedish botanist Anders Fredrik (André Frederick) Regnell, 1807-1884.

Rehdera Moldenke Verbenaceae

Origins:

After the German botanist Alfred Rehder, 1863-1949, gardener, dendrologist and professor of dendrology. His writings include *Bibliography of Cultivated Trees and Shrubs Hardy in the Cooler Temperatures of the Northern Hemisphere*. Jamaica Plain, Massachusetts 1949 and *The Bradley Bibliography*. 1911-1918, 1919-1940 founder and editor of the *Journal of the Arnold Arboretum*. See John H. Barnhart, *Biographical Notes upon Botanists*. 3: 138. 1965; T.W. Bossert, *Biographical Dictionary of Botanists Represented in the Hunt Institute Portrait Collection*. 328. 1972; R. Zander, F. Encke, G. Buchheim and S. Seybold, *Handwörterbuch der Pflanzennamen*. 14. Aufl. Stuttgart 1993; Joseph Ewan, *Rocky Mountain Naturalists*. The University of Denver Press 1950; H.R. Fletcher, *Story of the Royal Horticultural Society, 1804-1968*. Oxford 1969; E.M. Tucker, *Catalogue of the Library of the Arnold Arboretum of Harvard University*. 1917-1933; Ida Kaplan Langman, *A Selected Guide to the Literature on the Flowering Plants of Mexico*. 618. University of Pennsylvania Press, Philadelphia 1964; J. Ewan, ed., *A Short History of Botany in the United States*. 136. 1969; S. Lenley et al., *Catalog of the Manuscript and Archival Collections and Index to the*

Correspondence of John Torrey. Library of the New York Botanical Garden. 341. 1973.

Rehderodendron Hu Styracaceae

Origins:
After the German botanist Alfred Rehder, 1863-1949.

Species/Vernacular Names:
R. gongshanense Y.C. Tang
English: Gongshan Xian rehdertree
China: gong shan mu gua hong
R. kwangtungense Chun
English: Kwangtung rehdertree
China: guang dong mu gua hong
R. kweichowense Hu
English: Kweichow rehdertree, hairyfruit rehdertree
China: gui zhou mu gua hong
R. macrocarpum Hu
English: large-fruit rehdertree
China: mu gua hong

Rehderophoenix Burret Palmae

Origins:
After the German botanist Alfred Rehder, 1863-1949.

Rehia Fijten Gramineae

Origins:
After the British botanist Richard Eric Holttum, 1895-1990, traveler and botanical explorer (Malay Peninsula); see J.H. Barnhart, *Biographical Notes upon Botanists.* 2: 197. 1965; T.W. Bossert, *Biographical Dictionary of Botanists Represented in the Hunt Institute Portrait Collection.* 180. 1972; Ida Kaplan Langman, *A Selected Guide to the Literature on the Flowering Plants of Mexico.* Philadelphia 1964; R. Zander, F. Encke, G. Buchheim and S. Seybold, *Handwörterbuch der Pflanzennamen.* 14. Aufl. Stuttgart 1993; Ray Desmond, *Dictionary of British & Irish Botanists and Horticulturists.* 351. London 1994.

Rehmannia Libosch. ex Fischer & C.A. Meyer Gesneriaceae (Scrophulariaceae)

Origins:
After the Russian physician Joseph Rehmann, 1779 (or 1753, in Pritzel)-1831; see H. Genaust, *Etymologisches Wörterbuch der botanischen Pflanzennamen.* 530. 1996;

Georg Christian Wittstein, *Etymologisch-botanisches Handwörterbuch.* 754. 1852.

Species/Vernacular Names:
R. elata N.E. Br.
English: Chinese foxglove
R. glutinosa Libosch. ex Fischer & C. Meyer
English: glutinous rehmannia
China: di huang, ti huang, gan di huang, shu di huang
Vietnam: sinh dia, dia hoang

Reichardia Roth Asteraceae

Origins:
After the German botanist Johann Jacob (Jakob) Reichard, 1743-1782, physician, from 1773 to 1782 supervisor of the botanical garden and library of the Senckenberg Foundation. His works include *Flora moeno-francofurtana.* Francofurti ad Moenum [Frankfurt am Main] 1772-1778 and *Enumeratio stirpium horti botanici senckenbergiani,* qui Francofurti ad Moenum est. Francofurti ad Moenum 1782; see John H. Barnhart, *Biographical Notes upon Botanists.* 3: 139. 1965; Albrecht Wilhelm Roth (1757-1834), in *Botanische Abhandlungen und Beobachtungen.* 35. 1787; G. Schmid, *Goethe und die Naturwissenschaften.* Halle 1940; Jonas C. Dryander, *Catalogus bibliothecae historico-naturalis Josephi Banks.* London 1800; A. Lasègue, *Musée botanique de Benjamin Delessert.* Paris 1845; E.M. Tucker, *Catalogue of the Library of the Arnold Arboretum of Harvard University.* 1917-1933; R. Zander, F. Encke, G. Buchheim and S. Seybold, *Handwörterbuch der Pflanzennamen.* 14. Aufl. Stuttgart 1993; Mariella Azzarello Di Misa, ed., *Il Fondo Antico della Biblioteca dell'Orto Botanico di Palermo.* 227. Regione Siciliana, Palermo 1988.

Species/Vernacular Names:
R. tingitana (L.) Roth (*Scorzonera tingitana* L.; *Picridium tingitanum* (L.) Desf.)
English: false sow-thistle, reichardia
Arabic: huwwa

Reichea Kausel Myrtaceae

Origins:
For the German botanist Karl (Carl, Carlos Federico) Friedrich Reiche, 1860-1929, naturalist, traveler, 1889 to Chile, botanical explorer, plant collector, professor of botany. His works include "Memoria jeneral sobre la espedición esploradora del Rio Palena. Diciembre 1893-Marzo 1894. Informe del señor doctor Karl Reiche, naturalista de la espedición." *Anales Univ. Chile.* 90: 715-747. 1895, "Die

Vegetations-Verhältnisse am Unterlaufe des Rio Maule (Chile)." *Bot. Jahrb. Syst.* 21: 1-52. 1895, "Die botanischen Ergebnisse meiner Reise in die Cordilleren von Nahuelbuta und von Chillan." *Bot. Jahrb. Syst.* 22: 1-16. 1895, *Flora de Chile.* Santiago de Chile 1896-1911 and "Rudolf Amandus Philippi." *Ber. Deutsch. Bot. Ges.* 22: 68-83. 1905. See Clodomiro Marticorena, *Bibliografía Botánica Taxonómica de Chile.* Missouri Botanical Garden 1992; Irving William Knobloch, compil., "A preliminary verified list of plant collectors in Mexico." *Phytologia Memoirs.* VI. 1983; J.H. Barnhart, *Biographical Notes upon Botanists.* 3: 139. 1965; E.M. Tucker, *Catalogue of the Library of the Arnold Arboretum of Harvard University.* 1917-1933; T.W. Bossert, *Biographical Dictionary of Botanists Represented in the Hunt Institute Portrait Collection.* 328. 1972; Ida Kaplan Langman, *A Selected Guide to the Literature on the Flowering Plants of Mexico.* 618-619. University of Pennsylvania Press, Philadelphia 1964; H. Ross, "Karl Reiche." *Ber. Deutsch. Bot. Ges.* 47(2): 103-110. 1930.

Reicheella Pax Caryophyllaceae (Hectorellaceae)

Origins:

For the German botanist Karl (Carl, Carlos Federico) Friedrich Reiche, 1860-1929.

Reicheia Kausel Myrtaceae

Origins:

For the German botanist Karl (Carl, Carlos Federico) Friedrich Reiche, 1860-1929.

Reichembachanthus Barb. Rodr. Orchidaceae

Origins:

See *Reichenbachanthus.*

Reichenbachanthus Barb. Rodr. Orchidaceae

Origins:

Named for the German botanist Heinrich Gustav Reichenbach, 1824-1889, orchidologist, professor of natural history, Director of the Botanical Garden at Hamburg, founder of the Reichenbach Herbarium (he bequeathed it to the herbarium in Vienna), foreign member of the Royal Horticultural Society and of the Linnean Society. His writings include *De pollinis Orchidearum.* Lipsiae 1852 and *Beiträge zu einer Orchideenkunde Central-Amerika's.*

Hamburg 1866; see J.H. Barnhart, *Biographical Notes upon Botanists.* Boston 1965; Stafleu and Cowan, *Taxonomic Literature.* 4: 689-693. 1983; T.W. Bossert, *Biographical Dictionary of Botanists Represented in the Hunt Institute Portrait Collection.* 328. 1972; Frederico Carlos Hoehne, M. Kuhlmann and Oswaldo Handro, *O jardim botânico de São Paulo.* 171-172. 1941; R. Zander, F. Encke, G. Buchheim and S. Seybold, *Handwörterbuch der Pflanzennamen.* 14. Aufl. 767. 1993; S. Lenley et al., *Catalog of the Manuscript and Archival Collections and Index to the Correspondence of John Torrey.* Library of the New York Botanical Garden. 1973; Elmer Drew Merrill, *Contr. U.S. Natl. Herb.* 30(1): 251-252. 1947; Ernest Nelmes and William Cuthbertson, *Curtis's Botanical Magazine Dedications, 1827-1927.* [1931]; Graham Yearsley, "Heinrich Gustav Reichenbach. The autocratic taxonomist." in *The Orchid Review.* 105(1214): 78-80. [b. 1823] 1997; Ida Kaplan Langman, *A Selected Guide to the Literature on the Flowering Plants of Mexico.* 619. Philadelphia 1964; Merle A. Reinikka, *A History of the Orchid.* Timber Press 1996.

Reichenbachia Sprengel Nyctaginaceae

Origins:

After the German botanist Heinrich Gottlieb Ludwig Reichenbach, 1793-1879, naturalist, physician, M.D. Leipzig 1817, ornithologist, Director of the Botanical Garden of Dresden, professor of natural history, prolific author. Among his numerous works are *Iconographia botanica exotica.* Lipsiae [1824-] 1827-1830 and *Flora exotica.* Leipzig 1834-1836. He was father of Heinrich Gustav Reichenbach (1824-1889), brother of Anton Benedict Reichenbach (1807-1860). See John H. Barnhart, *Biographical Notes upon Botanists.* Boston 1965; Stafleu and Cowan, *Taxonomic Literature.* 4: 666-689. Utrecht 1983; Ida Kaplan Langman, *A Selected Guide to the Literature on the Flowering Plants of Mexico.* 619. Philadelphia 1964; A. Lasègue, *Musée botanique de Benjamin Delessert.* Paris 1845; T.W. Bossert, *Biographical Dictionary of Botanists Represented in the Hunt Institute Portrait Collection.* 328. 1972; H.N. Clokie, *Account of the Herbaria of the Department of Botany in the University of Oxford.* 231. Oxford 1964; E.M. Tucker, *Catalogue of the Library of the Arnold Arboretum of Harvard University.* 1917-1933; R. Zander, F. Encke, G. Buchheim and S. Seybold, *Handwörterbuch der Pflanzennamen.* 14. Aufl. 767. 1993; Günther Schmid, *Chamisso als Naturforscher.* Eine Bibliographie. Leipzig 1942; G. Schmid, *Goethe und die Naturwissenschaften.* Halle 1940; Mariella Azzarello Di Misa, ed., *Il Fondo Antico della Biblioteca dell'Orto Botanico di Palermo.* 228. Palermo 1988.

Reidia Wight Euphorbiaceae

Origins:

For Liet.-Colonel Francis Alexander Reid, d. 1862, Madras Horticultural Society.

Reifferscheidia Presl Dilleniaceae

Origins:

For the German botanist Joseph Franz Maria Anton Hubert Ignatz Fürst zu Salm-Reifferscheid-Dyck, 1773-1861 (d. Düsseldorf), horticulturist and artist. He is remembered for *Monographia generum Aloes et Mesembryanthemi.* Bonnae 1836-1863, *Index plantarum succulentarum* in horto dyckensi cultarum. Anno 1829. Aachen 1829 and *Observationes botanicae* in Horto Dyckensi notatae. Anno 1820. Coloniae 1820; see J.H. Barnhart, *Biographical Notes upon Botanists.* 3: 203. 1965; T.W. Bossert, *Biographical Dictionary of Botanists Represented in the Hunt Institute Portrait Collection.* 346. 1972; H.N. Clokie, *Account of the Herbaria of the Department of Botany in the University of Oxford.* 236. Oxford 1964; E.M. Tucker, *Catalogue of the Library of the Arnold Arboretum of Harvard University.* 1917-1933; Ernest Nelmes (1895-1959) and William Cuthbertson (c. 1859-1934), *Curtis's Botanical Magazine Dedications, 1827-1927.* 66-68. [1931]; Gilbert Westacott Reynolds (1895-1967), *The Aloes of South Africa.* 94-95. Balkema, Rotterdam 1982; A. White and B.L. Sloane, *The Stapelieae.* Pasadena 1937; Georg Christian Wittstein, *Etymologisch-botanisches Handwörterbuch.* 307, 781. Ansbach 1852; R. Zander, F. Encke, G. Buchheim and S. Seybold, *Handwörterbuch der Pflanzennamen.* 14. Aufl. 773. Stuttgart 1993; Gordon Douglas Rowley, *A History of Succulent Plants.* Strawberry Press, Mill Valley, California 1997.

Reineckea Karst. Palmae

Origins:

For the German botanist Johann Heinrich Julius Reinecke, 1799-1871, horticulturist.

Reineckea Kunth Convallariaceae (Liliaceae)

Origins:

For the German botanist Johann Heinrich Julius Reinecke, 1799-1871, horticulturist.

Species/Vernacular Names:

R. carnea (Andrews) Kunth

China: chi hsiang tsao (= plant of felicity)

Reinssekia Endlicher Rhamnaceae

Origins:

After the Austrian botanist Siegfried Reissek (Reisseck), 1819-1871; see John H. Barnhart, *Biographical Notes upon Botanists.* 3: 143. 1965; E.M. Tucker, *Catalogue of the Library of the Arnold Arboretum of Harvard University.* 1917-1933; R. Zander, F. Encke, G. Buchheim & S. Seybold, *Handwörterbuch der Pflanzennamen.* 14 Aufl. Stuttgart 1993.

Reinwardtia Dumortier Linaceae

Origins:

To remember the Dutch scientist Caspar Georg Carl Reinwardt, 1773-1854 (d. Leiden, Holland), botanist, plant collector at the Cape (Jan.-Feb. 1816), traveler throughout Indonesia, 1817 founder and first Director of Bogor (Buitenzorg) Botanic Gardens, Java, from 1823 to 1845 professor of natural history at the University of Leyden. Among his works are *Über den Charakter der Vegetation auf den Inseln des Indischen Archipels.* Berlin 1828, *Enumeratio plantarum in horto Lugduno-Batavo coluntur.* [Leyden] 1831 and *Reis naar het oostelijk gedeelte van den Indischen Archipel in het jaar 1821* ... met bijlagen vermeerderd door W. de Vriese. Amsterdam 1858; see J.H. Barnhart, *Biographical Notes upon Botanists.* 3: 142. 1965; P.M.W. Dakkus, *An Alphabetical List of Plants Cultivated in the Botanic Gardens, Buitenzorg.* Buitenzorg, Archipel Drukkerij's Lands Plantentuin (Botanic Gardens) Buitenzorg-Java, Dutch-East Indies 1930; [C.A. Backer] *On Some Results of the Botanical Investigation of Java (1911-1913).* Buitenzorg 1913; *Flora von Buitenzorg.* Leyden 1898-1914; Karl Ludwig von Blume (1796-1862), *Rumphia.* Lugduni Batavorum 1836; Barthélemy Charles Joseph Dumortier (1797-1878), *Commentationes Botanicae.* 19. Tournay 1822; P.A. Tiele, *Mémoire Bibliographique sur les Journaux des Navigateurs Néerlandais* ... la plupart en la possession de Frederick Muller à Amsterdam. Amsterdam 1867; Frederick Muller, *Four Catalogues.* [*Les Indes Orientales: Catalogue de Livres sur les Possessions Néerlandaises aux Indes*; etc.] Amsterdam 1882; T. W. Bossert, *Biographical Dictionary of Botanists Represented in the Hunt Institute Portrait Collection.* 329. Boston, Massachusetts 1972; A. Lasègue, *Musée botanique de Benjamin Delessert.* Paris 1845; E.M. Tucker, *Catalogue of the Library of the Arnold Arboretum of Harvard University.* 1917-1933; G.C. Wittstein, *Etymologisch-botanisches Handwörterbuch.* 755. 1852; R. Zander, F. Encke, G. Buchheim & S. Seybold, *Handwörterbuch der Pflanzennamen.* 14 Aufl. 768. 1993; N. Hall, *Botanists of the Eucalypts.* Melbourne 1978 and Supplement 1980; Mary Gunn and

Leslie E. Codd, *Botanical Exploration of Southern Africa.* 294. Cape Town 1981.

Species/Vernacular Names:
R. indica Dumort. (*Reinwardtia tetragyna* Planch.; *Reinwardtia trigyna* (Roxb.) Planch.)

English: yellow flax

Japan: ki-bana-ama

Nepal: piunli

Reinwardtia Korthals Theaceae

Origins:
For the Dutch scientist Caspar Georg Carl Reinwardt, 1773-1854.

Reinwardtiodendron Koorders Meliaceae

Origins:
For the Dutch scientist Caspar Georg Carl Reinwardt, 1773-1854.

Relbunium (Endl.) Hook.f. Rubiaceae

Origins:
A Mapuche name for a Chilean species, *relbu* or *relvun*.

Species/Vernacular Names:
R. sp.

Peru: jaca huanuseck, uquia, unquia

Relchela Steud. Gramineae

Origins:
An anagram of *Lechlera* Steud.

Relhania L'Hérit. Asteraceae

Origins:
For the clergyman Rev. Richard Relhan, (b. Dublin) 1754-1823, botanist, plant collector, bryologist, lichenologist, 1787 elected a Fellow of the Royal Society, in 1789 became a Fellow of the Linnean Society (one of the founders in 1788), in 1791 Rector at Hemingby, Lincoln (Cambridge), in 1798 Associate of Linnean Society, contributed to James Sowerby (1757-1822) and J.E. Smith's *English Botany*, author of *Flora Cantabrigiensis*, exhibens plantas agro

Cantabrigiensi indigenas, secundum systema sexuale digestas. Cantabrigiae [Cambridge] 1785. See John H. Barnhart, *Biographical Notes upon Botanists.* 3: 143. 1965; G.C. Gorham, *Memoirs of John Martyn ... and Thomas Martyn.* London 1830; F. Boerner & G. Kunkel, *Taschenwörterbuch der botanischen Pflanzennamen.* 4. Aufl. 160. Berlin & Hamburg 1989; Jonas C. Dryander, *Catalogus bibliothecae historico-naturalis Josephi Banks.* London 1800; Blanche Henrey, *British Botanical and Horticultural Literature before 1800.* 1975; E.M. Tucker, *Catalogue of the Library of the Arnold Arboretum of Harvard University.* 1917-1933.

Species/Vernacular Names:
R. genistifolia (L.) L'Hérit.

English: pepperbush, sticky shrublet

South Africa: ghombossie, peperbos

Remirea Aublet Cyperaceae

Origins:
From the name of the plant in Guiana; see Jean Baptiste Christophore Fusée Aublet, *Histoire des Plantes de la Guiane Françoise.* 1: 44, t. 16. Paris 1775.

Remusatia Schott Araceae

Origins:
After the French botanist Jean Pierre Abel Rémusat, 1788-1832, physician, orientalist, author of *Élemens de la Grammaire Chinoise.* Paris 1822 and *Mélanges Asiatiques.* Paris 1825, 1826.

Remya W.F. Hillebrand ex Bentham Asteraceae

Origins:
Named in honor of the French botanist Ezechiel Jules Rémy, 1826-1893, traveler, naturalist, ethnologist, from 1851 to 1855 plant collector in Hawaii (all of the main islands). Among his writings are *Ka Mooolelo [Mo'olelo] Hawaii. Histoire de L'Archipel Havaiien (Iles Sandwich).* Paris 1862, *Analecta boliviana.* Parisiis 1846, *Monografía de las Compuestas de Chile.* Paris 1849, *Pèlerinage d'un curieux au monastère bouddique de Pemmiantsi.* Châlons-sur-Marne 1880, *Récits d'un Vieux Sauvage*; pour servir à l'histoire ancienne de Hawaii. Notes d'un voyageur. Châlons-sur-Marne 1859 and *Voyage au pays des Mormons.* Paris 1860. See J.H. Barnhart, *Biographical Notes upon Botanists.* 3: 143. 1965; Claude Gay (1800-1873), *Historia fisica y politica de Chile.* Paris & Santiago [1843-] 1844-1871;

Julius Lucius Brenchley, *Jottings During the Cruise of H.M.S. Curaçoa among the South Sea Islands in 1865.* London 1873; I.H. Vegter, *Index Herbariorum.* Part II (5), *Collectors N-R.* Regnum Vegetabile vol. 109. 1983; T. W. Bossert, *Biographical Dictionary of Botanists Represented in the Hunt Institute Portrait Collection.* 329. Boston, Massachusetts 1972; E.M. Tucker, *Catalogue of the Library of the Arnold Arboretum of Harvard University.* 1917-1933.

Renanthera Lour. Orchidaceae

Origins:
Latin *renes* "kidneys" and *anthera* "anther," referring to the shape of the anthers, alluding to the kidney-shaped pollinia; see B.A. Lewis and P.J. Cribb, *Orchids of the Solomon Islands and Bougainville.* Royal Botanic Gardens, Kew 1991.

Species/Vernacular Names:
R. endsfeldtii F. Muell. & Kraenzlin
Bougainville Island: papino poku

Renantherella Ridley Orchidaceae

Origins:
The diminutive of the genus *Renanthera* Lour.

Renata Ruschi Orchidaceae

Origins:
For Miss Renata Aurelia Ruschi.

Rendlia Chiovenda Gramineae

Origins:
After the British (b. London) botanist Alfred Barton Rendle, 1865-1938 (d. Surrey), traveler, plant collector, in 1888 a Fellow of the Linnean Society, Keeper of the Botany Department of the British Museum (Natural History), 1923-1927 President of the Linnean Society, 1909 Fellow of the Royal Society, editor of the *Journal of Botany.* See Alfred Barton Rendle et al., "Catalogue of the Plants collected by Mr. and Mrs. P.A. Talbot in the Oban District of South Nigeria." *British Museum Trustees, Natural History.* [the English botanist Spencer Le Marchant Moore, 1850-1931, was co-author.] London 1913; John H. Barnhart, *Biographical Notes upon Botanists.* 3: 144. 1965; T. W. Bossert, *Biographical Dictionary of Botanists Represented in the Hunt Institute Portrait Collection.* 329. Boston, Massachusetts 1972; S. Lenley et al., *Catalog of the Manuscript and Archival Collections and Index to the Correspondence of*

John Torrey. Library of the New York Botanical Garden. 342. 1973; E.M. Tucker, *Catalogue of the Library of the Arnold Arboretum of Harvard University.* 1917-1933; Elmer Drew Merrill, *Contr. U.S. Natl. Herb.* 30(1): 253-254. 1947 and *Bernice P. Bishop Mus. Bull.* 144: 157. 1937; Fawcett & Rendle, *Flora of Jamaica.* 1910-1936; William Philip Hiern (1839-1925), *Catalogue of the African Plants Collected by F. Welwitsch in 1853-61.* London 1896-1901; R. Zander, F. Encke, G. Buchheim & S. Seybold, *Handwörterbuch der Pflanzennamen.* 14 Aufl. Stuttgart 1993; M. Hadfield et al., *British Gardeners: A Biographical Dictionary.* London 1980; Gilbert Westacott Reynolds, *The Aloes of South Africa.* Rotterdam 1982.

Renealmia L.f. Zingiberaceae

Origins:
For the French physician Paul Reneaulme (Paulus Renealmus), 1560-1624, botanist, author of *Specimen Historiae Plantarum.* Parisiis 1611, *P. Renealmi ... ad medicorum quorundam libellum responsio.* [Paris? 1615?] and *Ex curationibus observationes quibus videre ets* [sic] *morbus tuto ... possa debellari: si praecipue Galenicis praeceptis chymica veniant subsidio.* Parisiis 1606, editor of J.A. de Thou, the Elder, *I.A. Th[ou] Crambe, Viola, Lilium,* etc. 1611. See John H. Barnhart, *Biographical Notes upon Botanists.* 3: 144. 1965; Jonas C. Dryander, *Catalogus bibliothecae historico-naturalis Josephi Banks.* London 1800; A. Lasègue, *Musée botanique de Benjamin Delessert.* Paris 1845; Ethelyn Maria Tucker, *Catalogue of the Library of the Arnold Arboretum of Harvard University.* Cambridge, Massachusetts 1917-1933.

Species/Vernacular Names:
R. sp.
Peru: ajeng

Renealmia R. Br. Iridaceae

Origins:
For the French physician Paul Reneaulme (Paulus Renealmus), 1560-1624, botanist; see Robert Brown (1773-1858), *Prodromus florae Novae Hollandiae et Insulae van-Diemen.* 591. [Addenda] London 1810.

Rennellia Korth. Rubiaceae

Origins:
For the English (b. Devon) geographer James Rennell, 1742-1830 (d. London), a former naval officer, 1764-1777 Surveyor General to the East India Company in Bengal, traveler and explorer, 1781 Fellow of the Royal Society.

His works include *An Account of the Ganges and Burram-pooter Rivers* ... Read at the Royal Society January 25, 1780. London 1781, *A Chart of the Bank of Lagullus*. [A detailed chart of the Agulhas Bank and Cape showing the coast line from Table Bay to Algoa Bay, giving soundings and prevalent currents.] London 1778, *War with France, the only Security of Britain* ... By an Old Englishman. 1794, *The Geographical System of Herodotus* examined and explained by a comparison with those of other ancient authors. London 1800, *Memoir of a Map of Hindoostan, or, the Mogul Empire*. London 1788 and *The Marches of the British Armies in the Peninsula of India during the Campaigns of 1790 and 1791* ... explained by reference to a map. London 1792. See Sir Clements Robert Markham (1830-1916), *Major James Rennell and the Rise of Modern English Geography*. London 1895; Mungo Park (1771-1806), *Abstracts of Mungo Park's Travels ... in the Years 1795-1797*, with geographical illustrations ... by Major J. Rennell. 1790; Charles Athanase Walckenaer, *Notice historique sur la vie et les ouvrages de M. le Major Rennell*. 1842; Joan M. Eyles, in *Dictionary of Scientific Biography* (Editor in Chief Charles Coulston Gillispie.) 11: 376. 1981.

Rennera Merxm. Asteraceae

Origins:

After the German botanist Otto Renner, 1883-1960, professor of botany, Director of the Botanical Garden of Jena, traveler, 1933-1944 editor of *Flora*. See John H. Barnhart, *Biographical Notes upon Botanists*. 3: 144. 1965; T. W. Bossert, *Biographical Dictionary of Botanists Represented in the Hunt Institute Portrait Collection*. 329. Boston, Massachusetts 1972; S. Lenley et al., *Catalog of the Manuscript and Archival Collections and Index to the Correspondence of John Torrey*. Library of the New York Botanical Garden. 342. 1973.

Rephesis Raf. Moraceae

Origins:

See Constantine Samuel Rafinesque (1783-1840), *Sylva Telluriana*. 59. 1838.

Requienia DC. Fabaceae

Origins:

For the French botanist Esprit Requien, 1788-1851 (Corsica), malacologist, traveler and botanical explorer, from 1809 Director of the Botanical Garden of Avignon, correspondent of Jean Baptiste Mougeot (1776-1858) and Filippo

Parlatore, author of *Catalogue des végétaux ligneux* qui croissent naturellement en Corse ou qui y sont généralement cultivés. Avignon 1868; see J.H. Barnhart, *Biographical Notes upon Botanists*. 3: 145. 1965; H.N. Clokie, *Account of the Herbaria of the Department of Botany in the University of Oxford*. 231. Oxford 1964; A. Lasègue, *Musée botanique de Benjamin Delessert*. Paris 1845; T. W. Bossert, *Biographical Dictionary of Botanists Represented in the Hunt Institute Portrait Collection*. 329. Boston, Massachusetts 1972; R. Zander, F. Encke, G. Buchheim & S. Seybold, *Handwörterbuch der Pflanzennamen*. 14 Aufl. Stuttgart 1993.

Reseda L. Resedaceae

Origins:

Plinius: "Circa Ariminum nota est herba quam resedam vocant. Discutit collectiones inflammationesque omnes. Qui curant ea addunt haec verba: Reseda, morbis reseda, ... Haec ter dicunt totiensque despuunt." Latin *reseda, ae* for *Reseda alba* L., *resedo, are* (*sedo, avi, atum, are*) "to heal, calm," possibly referring to its use for healing wounds or as a balm; see Carl Linnaeus, *Species Plantarum*. 448. 1753 and *Genera Plantarum*. Ed. 5. 207. 1754.

Species/Vernacular Names:

R. alba L.

English: wild mignonette, white upright, white mignonette

Arabic: baasous khrouf

R. lutea L.

English: dyer's rocket, wild rocket, dyer's weed, weld, wild mignonette, cut-leaf mignonette, cut-leaved mignonette, yellow mignonette

South Africa: katstertbossie

R. luteola L.

English: dyer's weld, dyer's rocket, dyer's weed, yellow weed

Arabic: bliha, asfar

R. odorata L.

English: mignonette, sweet mignonette, common mignonette, garden mignonette, sweet reseda, bastard rocket

Resinaria Comm. ex Lam. Combretaceae

Origins:

Latin *resina* "resin."

Resnova J.J.M. v.d. Merwe Hyacinthaceae (Liliaceae)

Origins:
Latin *res* "thing, event" and *novus* "recent, young, fresh."

Restella Pobed. Thymelaeaceae

Origins:
An anagram of *Stellera*.

Restiaria Lour. Rubiaceae

Origins:
Latin *restiarius, ii* "a rope-maker."

Restio Rottb. Restionaceae

Origins:
Latin *restio, restionis* (*restis, is* "a rope, cord") "a rope-maker, rope-seller," early settlers in South Africa used restios for making cords; see Christen Friis Rottboell (1727-1797), *Descriptiones plantarum rariorum*. 9. Havniae [Copenhagen] 1772.

Species/Vernacular Names:
R. australis R. Br.
English: mountain cord-rush, Austral cord-rush

R. complanatus R. Br.
English: flat cord-rush

R. dimorphus R. Br.
English: two-formed cord-rush

R. fimbriatus R. Br.
English: fringed cord-rush

R. gracilis R. Br.
English: slender cord-rush

R. tetraphylla Labill.
English: tassel cord-rush, feather plant

Restrepia Kunth Orchidaceae

Origins:
Named in honor of a Spanish geographer, José E. (or Juan Manuel?) Restrepo, naturalist in South America, in Colombia; see F. Boerner & G. Kunkel, *Taschenwörterbuch der botanischen Pflanzennamen*. 4. Aufl. 160. Berlin & Hamburg 1989; Georg Christian Wittstein, *Etymologisch-botanisches Handwörterbuch*. 756. Ansbach 1852; Helmut Genaust, *Etymologisches Wörterbuch der botanischen Pflanzennamen*. 532. Basel 1996.

Restrepiella Garay & Dunsterville Orchidaceae

Origins:
The diminutive of the orchid genus *Restrepia* Kunth.

Restrepiopsis Luer Orchidaceae

Origins:
Resembling the genus *Restrepia* Kunth.

Resupinaria Raf. Fabaceae

Origins:
Latin *resupinus* "bent back"; see Constantine Samuel Rafinesque, *Sylva Telluriana*. 115. 1838.

Retama Raf. Fabaceae

Origins:
From the Arabic *retem* or *retam, ratama* for the broom bush; see Constantine Samuel Rafinesque, *Sylva Telluriana*. 22. 1838.

Species/Vernacular Names:
R. raetam (Forsskål) Webb (*Genista raetam* Forsskål; *Genista retama* Nicholson; *Genista rhodorhizoides* Webb & Berth.)
English: white weeping broom, white Spanish broom
Arabic: r'tem, retem, retem behan

Retiniphyllum Bonpl. Rubiaceae

Origins:
Greek *rhetine* "resin" and *phyllon* "leaf," referring to the appearance of the leaves.

Retinispora Siebold & Zucc. Cupressaceae

Origins:
Greek *rhetine* and *spora, sporos* "seed, spore."

Retinodendron Korth. Dipterocarpaceae

Origins:

From the Greek *rhetine* "resin" and *dendron* "a tree."

Retispatha J. Dransf. Palmae

Origins:

Latin *rete, retis* "a net" and *spatha, ae* "spathe of a palm-tree," referring to the netlike bracts subtending the flowering branches.

Retrophyllum C.N. Page Podocarpaceae

Origins:

Latin *retro* "behind" and Greek *phyllon* "leaf."

Retzia Thunberg Stilbaceae (Retziaceae, Loganiaceae)

Origins:

Named for the Swedish scientist Anders Jahan (Johan) Retzius, 1742-1821, botanist, lichenologist, bryologist, naturalist, entomologist, professor of natural history at the University of Lund. He did work in chemistry, botany, zoology, mineralogy and paleontology. His writings include *Lectiones publicae de vermibus intestinalibus, imprimis humanis.* Holmiae 1786. He was the grandfather of the Swedish anatomist Magnus Gustaf Retzius (1842-1919) and the father of Anders Adolf Retzius (1796-1860); see J.H. Barnhart, *Biographical Notes upon Botanists.* 3: 146. 1965; T. W. Bossert, *Biographical Dictionary of Botanists Represented in the Hunt Institute Portrait Collection.* 329. Boston, Massachusetts 1972; Jonas C. Dryander, *Catalogus bibliothecae historico-naturalis Josephi Banks.* London 1800; A. Lasègue, *Musée botanique de Benjamin Delessert.* Paris 1845; Ethelyn Maria Tucker, *Catalogue of the Library of the Arnold Arboretum of Harvard University.* Cambridge, Massachusetts 1917-1933; Vladislav Kruta, in *Dictionary of Scientific Biography* 11: 379-381. 1981; Gerhard Rudolph, in *Dictionary of Scientific Biography* 11: 381-383. 1981; Mariella Azzarello Di Misa, ed., *Il Fondo Antico della Biblioteca dell'Orto Botanico di Palermo.* 229. Regione Siciliana, Palermo 1988; R. Zander, F. Encke, G. Buchheim & S. Seybold, *Handwörterbuch der Pflanzennamen.* 14 Aufl. Stuttgart 1993; J.D. Milner, *Catalogue of Portraits of Botanists Exhibited in the Museums of the Royal Botanic Gardens.* Royal Botanic Gardens, Kew, London 1906; Stafleu and Cowan, *Taxonomic Literature.* 4: 735-738. Utrecht 1983.

Reutera Boissier Umbelliferae

Origins:

After the Swiss botanist George François Reuter, 1805-1872, botanical collector; see J.H. Barnhart, *Biographical Notes upon Botanists.* 3: 147. 1965; T. W. Bossert, *Biographical Dictionary of Botanists Represented in the Hunt Institute Portrait Collection.* 330. Boston, Massachusetts 1972; A. Lasègue, *Musée botanique de Benjamin Delessert.* Paris 1845; Ethelyn Maria Tucker, *Catalogue of the Library of the Arnold Arboretum of Harvard University.* Cambridge, Massachusetts 1917-1933; H.N. Clokie, *Account of the Herbaria of the Department of Botany in the University of Oxford.* 231. Oxford 1964; M. Colmeiro y Penido, *La Botánica y los Botánicos de la Peninsula Hispano-Lusitana.* Madrid 1858; R. Zander, F. Encke, G. Buchheim & S. Seybold, *Handwörterbuch der Pflanzennamen.* 14 Aufl. 1993; Stafleu and Cowan, *Taxonomic Literature.* 4: 742-744. 1983.

Reverchonia A. Gray Euphorbiaceae

Origins:

After the French botanist Julien Reverchon, 1834-1905, plant collector (Dallas, Texas), traveler, brother of the French plant collector Elisée Reverchon (1835-1914); see J.H. Barnhart, *Biographical Notes upon Botanists.* 3: 147. Boston 1965; T. W. Bossert, *Biographical Dictionary of Botanists Represented in the Hunt Institute Portrait Collection.* 330. Boston, Massachusetts 1972; I.C. Hedge and J.M. Lamond, *Index of Collectors in the Edinburgh Herbarium.* Edinburgh 1970.

Reyesia Gay Solanaceae

Origins:

For Ant. Garcia Reyes, 19th century Chilean botanist.

Reymondia Karsten & Kuntze Orchidaceae

Origins:

After the German (b. Berlin) physiologist Emil Heinrich Dubois-Reymond (Du Bois-Reymond, du Bois-Reymond), 1818-1896 (Berlin), a pioneer of electrophysiology, investigator of animal electricity and electrical phenomena, studied at the University of Berlin, 1845 founded the Physikalische Gesellschaft with Brücke, Heinrich Wilhelm Dove (1803-1879) and others, with Reichert editor of *Archiv für Anatomie, Physiologie und wissenschaftliche Medizin.* His writings include "Vorläufiger Abriss einer Untersuchung über den sogenannten Froschstrom und über die elektrischen Fische." in *Annalen der Physik und Chemie.*

58: 1-30. 1843 and *Abhandlungen zur allgemeinen Muskel-und Nervenphysik*. Leipzig 1875-1877, friend of Ernst Brücke, Hermann Helmholtz, Karl Reichert, and Karl Ludwig, brother of Paul David Gustav Du Bois-Reymond (1831-1889); see Heinrich Boruttau, *Emil du Bois-Reymond*. Vienna 1922; K.E. Rothschuh, in *Dictionary of Scientific Biography* 4: 200-205. 1981; Lubos Novy, in *Dictionary of Scientific Biography* 4: 205-206. 1981.

Reynoldsia A. Gray Araliaceae

Origins:

Named in honor of Jeremiah N. Reynolds, plant collector in Chile in the early 19th century, John Downes' (1784-1854) secretary during the Pacific cruise of the *Potomac* which lasted from 1831 to 1834, promoter of the South Sea Exploring Expedition, author of *Voyage of the United States Frigate "Potomac" during the Circumnavigation of the Globe*. New York 1835.

Species/Vernacular Names:

R. sandwicensis A. Gray (*Reynoldsia degeneri* Sherff; *Reynoldsia hillebrandii* Sherff; *Reynoldsia oblonga* Sherff; *Reynoldsia hosakana* Sherff; *Reynoldsia mauiensis* Sherff; *Reynoldsia mauiensis* var. *macrocarpa* Degener & Sherff; *Reynoldsia huehuensis* Sherff; *Reynoldsia huehuensis* var. *brevipes* Sherff; *Reynoldsia huehuensis* var. *intermedia* Sherff; *Reynoldsia venusta* Sherff; *Reynoldsia venusta* var. *lanaiensis* Sherff; *Reynoldsia sandwicensis* var. *intercedens* Sherff; *Reynoldsia sandwicensis* var. *molokaiensis* Sherff)

Hawaii: 'ohe, 'ohe kukuluae'o, 'ohe makai, 'ohe'ohe, 'oheokai

Reynoutria Houtt. Polygonaceae

Origins:

After a not well-known Reynoutre or van Reynoutre, Dutch or French botanist and naturalist; see Maarten (Martin) Houttuyn (1720-1798), *Natuurlijke historie*. 2(8): 639, t. 51. Amsterdam (Dec.) 1777.

Species/Vernacular Names:

R. japonica Houtt.

English: Japanese knotweed

R. sachalinensis (F. Schmidt) Nakai (*Polygonum sachalinense* F. Schmidt)

English: giant knotweed, sacaline

Rhabdadenia Müll. Arg. Apocynaceae

Origins:

Greek *rhabdos* "a rod, stick, a magic wand" and *aden* "gland."

Rhabdia Mart. Boraginaceae

Origins:

Greek *rhabdos* "a rod, stick, a magic wand," referring to the branches; see Salvatore Battaglia, *Grande dizionario della lingua italiana*. XV: 189. 1994.

Rhabdocalyx Lindl. Boraginaceae

Origins:

From the Greek *rhabdos* and *kalyx* "calyx."

Rhabdocaulon (Benth.) Epling Labiatae

Origins:

From the Greek *rhabdos* and *kaulos* "a stem, a branch or stalk."

Rhabdochloa Kunth Gramineae

Origins:

Orthographic variant of *Rabdochloa* P. Beauv.

Rhabdodendron Gilg & Pilger Rhabdodendraceae

Origins:

From the Greek *rhabdos* "a rod, stick" and *dendron* "tree."

Rhabdophyllum Tieghem Ochnaceae

Origins:

Greek *rhabdos* "a rod, stick" and *phyllon* "leaf"; see also *Gomphia* Schreb.

Rhabdosciadium Boiss. Umbelliferae

Origins:

Greek *rhabdos* "a rod, stick" and *skias, skiados* "a canopy, umbel," *skiadion, skiadeion* "umbel, parasol."

Rhabdostigma Hook.f. Rubiaceae

Origins:

From the Greek *rhabdos* "a rod, stick" and *stigma*.

Rhabdothamnopsis Hemsley Gesneriaceae

Origins:
Resembling *Rhabdothamnus* A. Cunn.

Rhabdothamnus A. Cunn. Gesneriaceae

Origins:
From the Greek *rhabdos* and *thamnos* "shrub."

Rhabdotosperma Hartl Scrophulariaceae

Origins:
Greek *rhabdotos* "made with rods, ribbed" and *sperma* "seed."

Rhachicallis DC. Rubiaceae

Origins:
Greek *rhachis* "rachis, axis" and *kallos* "beauty."

Rhachidosorus Ching Dryopteridaceae (Aspleniaceae, Woodsiaceae)

Origins:
Greek *rhachis* "rachis, axis" and *soros* "a vessel, a spore case, a heap."

Rhachidospermum Vasey Gramineae

Origins:
From the Greek *rhachis* "rachis, axis" and *sperma* "a seed."

Rhacodiscus Lindau Acanthaceae

Origins:
Greek *rhakos* "ragged, wrinkles" and *diskos* "a disc."

Rhacoma P. Browne ex L. Celastraceae

Origins:
Latin *rhacoma* or *rhecoma* "a root, rhubarb" (Plinius), Greek *rhakoma* "rags," *rhakos* "tattered garment."

Rhadamanthus Salisb. Hyacinthaceae

Origins:
A son of Jupiter and Europa; see G. Semerano, *Le origini della cultura europea. Dizionari Etimologici. Dizionario della lingua Greca.* 2(1): 246. Firenze 1994.

Rhadinopus S. Moore Rubiaceae

Origins:
Greek *rhadinos* "slender, slim, delicate" and *pous, podos* "a foot."

Rhadinothamnus Paul G. Wilson Rutaceae

Origins:
Greek *rhadinos* "slender, slim, delicate, tender" and *thamnos* "bush," referring to the habit and nature of these plants; see Paul G. Wilson, in *Nuytsia.* 1: 197. (May) 1971.

Rhaesteria Summerhayes Orchidaceae

Origins:
Greek *rhaister* "hammer, smasher," referring to the shape of the rostellum.

Rhagadiolus Scop. Asteraceae

Origins:
Greek *rhagas, rhagados* "fissure," referring to the nature of the ovary or of the inner phyllaries.

Rhagodia R. Br. Chenopodiaceae

Origins:
Resembling berries, from the Greek *rhax, rhagos* "a berry, grape"; see Robert Brown (1773-1858), *Prodromus florae Novae Hollandiae et Insulae van-Diemen.* 408. London 1810; Paul G. Wilson, "A taxonomic revision of the tribe Chenopodieae (Chenopodiaceae) in Australia." *Nuytsia.* 4(2): 135-262. 1983.

Species/Vernacular Names:
R. baccata (Labill.) Moq.
Australia: seaberry saltbush
R. candolleana Moq.
Australia: seaberry saltbush

R. crassifolia R. Br.

Australia: fleshy saltbush, thick-leaf saltbush

R. parabolica R. Br.

Australia: fragrant saltbush

R. spinescens R. Br. (*Rhagodia spinescens* R. Br. var. *deltophylla* F. Muell.; *Rhagodia deltophylla* (R. Br.) A.J. Scott)

Australia: spiny saltbush, thorny saltbush, hedge saltbush, creeping saltbush

Rhammatophyllum O.E. Schulz Brassicaceae

Origins:

Greek *rhamma, rhammatos* "thread, seam, suture" and *phyllon* "leaf."

Rhamnella Miq. Rhamnaceae

Origins:

The diminutive of the genus *Rhamnus* L.

Species/Vernacular Names:

R. franguloides (Maxim.) Weberb. (*Microrhamnus franguloides* Maxim.; *Rhamnella inaequilatera* Ohwi; *Rhamnella franguloides* var. *inaequilatera* (Ohwi) Hatusima)

Japan: neko-no-chichi

Rhamnidium Reissek Rhamnaceae

Origins:

Referring to the genus *Rhamnus*.

Rhamnoluma Baillon Sapotaceae

Origins:

From the Greek *rhamnos*, name of various prickly shrubs, plus *luma*.

Rhamnoneuron Gilg Thymelaeaceae

Origins:

Greek *rhamnos*, name of various prickly shrubs and *neuron* "nerve."

Rhamnus L. Rhamnaceae

Origins:

Greek *rhamnos* "a kind of prickly plant or spiny shrub," Latin *rhamnos* "buckthorn, Christ's-thorn" (Plinius), perhaps from the root of the Greek *rhabdos* "a rod, stick, a magic wand," Akkadian *rapasu* "to beat," *rappu*, Sumerian *rab* "stick, branch"; see Carl Linnaeus, *Species Plantarum*. 193. 1753 and *Genera Plantarum*. Ed. 5. 89. 1754; Manlio Cortelazzo & Paolo Zolli, *Dizionario etimologico della lingua italiana*. 4: 1029. Zanichelli, Bologna 1985; E. Vuolo, in *Cultura neolatina*. XVI: 170-171. Modena 1956; E. Monaci, *Crestomazia italiana dei primi secoli*. Nuova edizione riveduta e aumentata per cura di F. Arese. [13th century, *Mare amoroso*: "Degio pur tacendo consumare, / sicome l'albero ch'à nome ranno, / che face uscire de le sue spine foco / e arde se medesimo in questo modo."] Roma-Napoli-Città di Castello 1955; Giovanni Semerano, *Le origini della cultura europea*. Dizionari Etimologici. Basi semitiche delle lingue indeuropee. Dizionario della lingua Greca. 2(1): 246-247. Leo S. Olschki Editore, Firenze 1994; Salvatore Battaglia, *Grande dizionario della lingua italiana*. XV: 410. Torino 1994.

Species/Vernacular Names:

R. spp.

Mexico: capulín cimarrón, capulincillo

R. alaternus L.

English: blow-fly bush, Italian buckthorn, buckthorn, barren privet

Italian: alaterno

Arabic: oud el khir, qased, 'oud el-kheir, zafrin

R. alnifolia L'Hér.

English: alder buckthorn

R. californicus Eschsch.

English: coffeeberry, California coffeeberry

R. carolinianus Walter

English: Carolina buckthorn

R. cathartica L.

English: common buckthorn, purging buckthorn, European buckthorn, Rhine berries

Italian: ramno catartico, ramno purgativo

R. crenatus Sieb. & Zucc.

China: li la gen

R. crocea Nutt.

English: spiny redberry, redberry

R. erythroxyloides Hoffsgg.

English: Pallas buckthorn

R. frangula L.

English: alder buckthorn

Brazil: frangula, amieiro-preto

Italian: frangola

R. ilicifolia Kellogg

English: holly-leaf redberry

R. japonica Maxim.

China: shu li

R. pirifolia E. Greene

English: Island redberry (Guadalupe Island)

R. prinoides L'Hérit. (*Celtis rhamnifolia* Presl, nom. illegit.) (Greek *prinos* "the scarlet oak," Latin *prinus* "ilex, the holm-oak, great scarlet oak")

English: dogwood

Southern Africa: blinkblaar, mofifi (= darkness), dark blinkblaar, hondepishout, hondepis (it is an ancient pioneer name which alludes to the unpleasant taste of the fruits); umNyenye, uNyenye, umGilindi, umHlinye (Zulu); liNyenye (Swazi); umNyenye, umGlindi (Xhosa); mofifi (South Sotho); muBariro, muBerere, Mandara, Sukuchuma, Tsonga, Vunambezu

R. purshiana DC.

Spanish: cascara sagrada

R. rubra E. Greene

English: Sierra coffeeberry

R. saxatilis Jacq.

English: Avignon berry, yellow berry, Persian berry

Italian: ramno dei tintori, tintorio

China: lu chai

R. sectipetala Martius

Brazil: tapixaba, tapichiba, canjica, canjuqueira

R. tomentella Benth.

English: hoary coffeeberry

Rhamphicarpa Benth. Scrophulariaceae

Origins:
Greek *rhamphos* "a beak" and *karpos* "fruit," referring to the nature of the fruits; see W.J. Hooker, *Companion to the Botanical Magazine*. 1: 368. (Jul.) 1836.

Rhamphidia Lindley Orchidaceae

Origins:
Greek *rhamphos* "a beak, crooked beak," referring to the bent and beak-like lip; see J. Lindley, in *Journal of Proceedings of the Linnean Society*. Botany. 1: 181. (Mar.) 1857.

Rhamphocarya Kuang Juglandaceae

Origins:
From the Greek *rhamphos* and *karyon* "nut," see also *Carya* Nutt.

Rhamphogyne S. Moore Asteraceae

Origins:
Greek *rhamphos* "a beak" and *gyne* "woman, female."

Rhampholepis Stapf Gramineae

Origins:
Greek *rhamphos* "a beak" and *lepis* "a scale," see also *Sacciolepis* Nash.

Rhamphorhynchus Garay Orchidaceae

Origins:
Greek *rhamphos* "a beak, crooked beak" and *rhynchos* "snout."

Rhanteriopsis Rauschert Asteraceae

Origins:
Resembling the genus *Rhanterium* Desf.

Rhanterium Desf. Asteraceae

Origins:
Greek *rhanterion* "fit for sprinkling," *rhanter* "one who sprinkles," *rhaino* "to sprinkle," referring to the seeds.

Rhaphidanthe Hiern ex Gürke Ebenaceae

Origins:
From the Greek *rhaphis, rhaphidos* "a needle" and *anthos* "flower."

Rhaphidophora Hassk. Araceae

Origins:
Greek *rhaphis, rhaphidos* "a needle, pin" and *phoros* "carrying," referring to the points on the fruits; see Justus Carl Hasskarl (1811-1894), *Flora oder allgemeine Botanische Zeitung*. 25(2) Beibl. 1. (Jul.) 1842; D.H. Nicolson, "Derivation of aroid generic names." *Aroideana*. 10: 15-25. 1988.

Species/Vernacular Names:
R. celatocaulis (N.E. Br.) Knoll
English: shingle plant

Rhaphidophyton Iljin Chenopodiaceae

Origins:

Greek *rhaphis, rhaphidos* "a needle, pin" and *phyton* "plant."

Rhaphidorhynchus Finet Orchidaceae

Origins:

Greek *rhaphis, rhaphidos* and *rhynchos* "horn, beak," referring to the shape of the rostellum.

Rhaphidospora Nees Acanthaceae

Origins:

Greek *rhaphis, rhaphidos* and *spora, sporos* "seed, spore"; see Nathaniel Wallich (1786-1854), *Plantae Asiaticae rariores*. 3: 115. London 1832.

Rhaphidura Bremek. Rubiaceae

Origins:

Greek *rhaphis, rhaphidos* "a needle, pin" and *oura* "a tail."

Rhaphiodon Schauer Labiatae

Origins:

From the Greek *rhaphis, rhaphidos* and *odous, odontos* "a tooth."

Rhaphiolepis Lindley Rosaceae

Origins:

Greek *rhaphis* "a needle" and *lepis* "a scale," referring to the narrow bracteoles on the inflorescence; see John Bellenden Ker Gawler (1764-1842), in *Edwards's Botanical Register*. 6, t. 468. London (Jul.) 1820.

Species/Vernacular Names:
R. indica (L.) Lindley (*Crataegus indica* L.)
English: Indian hawthorn

R. umbellata (Thunb.) Makino (*Rhaphiolepis ovata* Briot; *Rhaphiolepis japonica* Sieb. & Zucc.)
Japan: sharin-bay, tikachi

Rhaphiophallus Schott Araceae

Origins:

Greek *rhaphis* "a needle" and *phallos* "a penis, wooden club," see also *Amorphophallus* Blume ex Decne.

Rhaphiostylis Planchon ex Benth. Icacinaceae

Origins:

Greek *rhaphis* and *stylos* "pillar, column, style."

Species/Vernacular Names:
R. beninensis (Hook.f.) Planch. ex Benth.
Yoruba: ajasile, itapara, igbehin

Rhaphis Lour. Gramineae

Origins:

From the Greek *rhaphis* "a needle."

Rhaphispermum Benth. Scrophulariaceae

Origins:

Greek *rhaphis* "a needle" and *sperma* "seed."

Rhaphithamnus Miers Verbenaceae

Origins:

From the Greek *rhaphis* and *thamnos* "a shrub," with reference to the spiny character of some of the species.

Rhapidophyllum H.A. Wendland & Drude Palmae

Origins:

From the genus *Rhapis* L.f. ex Aiton and *phyllon* "leaf," referring to the elongate needle-like spines on persistent leaf-sheaths.

Species/Vernacular Names:
R. hystrix (Pursh) H.A. Wendl. & Drude (*Chamaerops hystrix* Pursh)

English: blue palmetto, needle palmetto, needle palm

Rhapis L.f. ex Aiton Palmae

Origins:

Greek *rhapis* "a rod," referring to the leaf segments or to the awns of the corolla.

Species/Vernacular Names:
R. excelsa (Thunb.) Henry (*Rhapis flabelliformis* L'Hérit. ex Aiton)

English: ground rattan, bamboo palm, lady palm, large lady palm, miniature fan palm, Buddha's bamboo, dwarf ground rattan

Brazil: palmeira ráfia, palmeira rápis

Japan: Kannon-chiku (= Buddha's bamboo, from Kannon-zan, Ryukyu Islands) (chiku = bamboo)

Okinawa: kwan-nun-chiku, urada-chingu

R. humilis Blume

English: reed rhapis, slender lady palm, lady palm

Brazil: palmeira ráfia, palmeira rápis

Japan: shuro-chiku, shyuro-chiku

Rhaponticum Ludw. Asteraceae

Origins:

Latin *radix Pontica*, Greek *rheon*, *rha* "roots and rhizomes (from Iran)," the *rha* of Pontus, Greek *Rha* (said to be from the ancient name of the river Volga/Wolga), Latin *Rha* "the Volga, on whose banks grew the radix pontica, *Rha ponticum*, rhubarb, *Rheum rhaponticum* L., which thence received its name"; see H. Genaust, *Etymologisches Wörterbuch der botanischen Pflanzennamen*. 532, 533-534. 1996; Ernest Weekley, *An Etymological Dictionary of Modern English*. 2: 1234. [rhubarb] 1967; C.T. Onions, *The Oxford Dictionary of English Etymology*. Oxford University Press 1966; R. Zander, F. Encke, G. Buchheim & S. Seybold, *Handwörterbuch der Pflanzennamen*. 14 Aufl. 475. 1993; Manlio Cortelazzo & Paolo Zolli, *Dizionario etimologico della lingua italiana*. 4: 1017. Bologna 1985; Salvatore Battaglia, *Grande dizionario della lingua italiana*. XV: 176, 966. 1994.

Rhaptonema Miers Menispermaceae

Origins:

Greek *rhapto* "to sew, to devise" and *nema* "thread, filament."

Rhaptopetalum Oliver Scytopetalaceae

Origins:

Greek *rhapto* "to sew, to devise," *rhaptos* "sewn together" and *petalum* or *petalon* "petal," the corolla is valvate.

Rhaptostylum Humb. & Bonpland Olacaceae

Origins:

From the Greek *rhaptos* "sewn together" and *stylos* "pillar, column."

Rhazya Decne. Apocynaceae

Origins:

Possibly from the Greek *rhazein, rhazo* "snarl as a dog."

Species/Vernacular Names:

R. stricta Decne.

Arabic: harmal

Rheedia L. Guttiferae

Origins:

For the Dutch botanist Hendrik (Henricum, Henricus, Henric) Adriaan (Adrien) van Rheede (Reede) tot Draakestein (Draakensteen, Drakestein, Drakenstein) (Rheedius a Drackenstein), 1637-1691, colonial administrator, with the Dutch East India Company, 1669-1676 Malabar, from 1684 India. Among his works is the famous *Hortus indicus malabaricus*. Amstelodami 1678[-1703]; see John H. Barnhart, *Biographical Notes upon Botanists*. 3: 149. 1965; R. Pulteney, *Historical and Biographical Sketches of the Progress of Botany in England*. London 1790; T. W. Bossert, *Biographical Dictionary of Botanists Represented in the Hunt Institute Portrait Collection*. 330. Boston, Massachusetts 1972; Jonas C. Dryander, *Catalogus bibliothecae historico-naturalis Josephi Banks*. London 1800; A. Lasègue, *Musée botanique de Benjamin Delessert*. Paris 1845; Ethelyn Maria Tucker, *Catalogue of the Library of the Arnold Arboretum of Harvard University*. Cambridge, Massachusetts 1917-1933; Isaac Henry Burkill, *Chapters on the History of Botany in India*. Delhi 1965; Stafleu and Cowan, *Taxonomic Literature*. 4: 750-753. Utrecht 1983; Johannes Heniger, *Hendrik Adriaan van Reede tot Drakenstein (1636-1691) and Hortus Malabaricus. A contribution to the history of Dutch colonial botany*. Rotterdam 1986; Justus Carl Hasskarl (1811-1894), *Horti malabarici Rheedeani clavis locupletissima*. Dresden 1867; Lewis Weston Dillwyn (1778-1855), *A Review of the References to the Hortus malabaricus of Henry van Rheede van Draakenstein*. Swansea 1839; Mariella Azzarello Di Misa, a cura di, *Il Fondo Antico della Biblioteca dell'Orto Botanico di Palermo*. 230. Regione Siciliana, Palermo 1988; John Landwehr, *VOC: A Bibliography of Publications Relating to the Dutch East India Company, 1602-1800*. Ed. Peter van der Krogt. HES Publisher, Utrecht 1991; G.C. Wittstein, *Etymologisch-botanisches Handwörterbuch*. 759. Ansbach 1852; Frans A. Stafleu, *Linnaeus and the Linnaeans. The spreading of their ideas in systematic botany, 1735-1789*. Utrecht 1971.

Species/Vernacular Names:

R. gardneriana Planchon & Triana (*Lamprophyllum gardnerianum* Miers)

Brazil: bacupari (= a vernacular name for many species of Brazilian fruits), bacoparé, bacopari, bacopari miúdo, mangostão amarelo, scropari, bacuri miúdo

Rheithrophyllum Hassk. Gesneriaceae

Origins:
Greek *rheithron* "a stream, river, that which flows" and *phyllon* "a leaf," see also *Aeschynanthus* Jack.

Rhektophyllum N.E. Br. Araceae

Origins:
Greek *rhektos* "penetrable, rent" and *phyllon* "a leaf," referring to the leaves, cut and perforated; see D.H. Nicolson, "Derivation of aroid generic names." *Aroideana.* 10: 15-25. 1988.

Rheome Goldblatt Iridaceae

Origins:
Partial anagram of *Homeria.*

Rheopteris Alston Vittariaceae (Adiantaceae)

Origins:
Greek *rheo* "to flow, run, stream" and *pteris* "a fern."

Rhetinodendron Meissner Asteraceae

Origins:
Greek *rhetine* "resin" and *dendron* "a tree."

Rhetinolepis Cosson Asteraceae

Origins:
Greek *rhetine* "resin" and *lepis* "a scale."

Rhetinosperma Radlk. Meliaceae

Origins:
Greek *rhetine* and *sperma* "a seed"; see H.G.A. Engler & K.A.E. Prantl (1849-1893), *Die Natürlichen Pflanzenfamilien.* Nachtr. III to 3(5): 204. (Jul.) 1907.

Rheum L. Polygonaceae

Origins:
Greek *rheon, rha* "roots and rhizomes (from Iran)," Dioscorides used *rha* for the rhubarb, the *rha* of Pontus, Greek *Rha* (said to be from the ancient name of the river Volga/Wolga), Latin *Rha* "the Volga, on whose banks grew the radix pontica, *Rha ponticum*, rhubarb, *Rheum rhaponticum* L., which thence received its name"; see H. Genaust, *Etymologisches Wörterbuch der botanischen Pflanzennamen.* 532, 533-534. 1996; Ernest Weekley, *An Etymological Dictionary of Modern English.* 2: 1234. [rhubarb] 1967; C.T. Onions, *The Oxford Dictionary of English Etymology.* Oxford University Press 1966; R. Zander, F. Encke, G. Buchheim & S. Seybold, *Handwörterbuch der Pflanzennamen.* 14 Aufl. 475. 1993; M. Cortelazzo & P. Zolli, *Dizionario etimologico della lingua italiana.* 4: 1017. Bologna 1985; S. Battaglia, *Grande dizionario della lingua italiana.* XV: 176, 966. 1994; Georg Christian Wittstein, *Etymologisch-botanisches Handwörterbuch.* 757, 759. Ansbach 1852.

Species/Vernacular Names:
R. australe D. Don
English: Himalyan rhubarb, Indian rhubarb
Nepal: padamchal, amalbed
R. officinale Baillon
English: rhubarb, Chinese rhubarb
China: da huang, ta huang, huang liang (= yellow efficacy), chiang chun (= captain general)
Mexico: nocuana lanini castilla
R. palmatum L.
English: Chinese rhubarb, Turkish rhubarb, rhubarb
Tibetan: lcum-rtsa
Ladakhi: lchumtsa

Rhexia L. Melastomataceae

Origins:
An ancient Greek name, *rhegnymi* "to break, break asunder, rend," *rhexis* "breaking, bursting, cleft"; Plinius applied Latin *rhexia, ae* to a plant, also called *onochilis*; see H. Genaust, *Etymologisches Wörterbuch der botanischen Pflanzennamen.* 534-535. 1996; Georg Christian Wittstein, *Etymologisch-botanisches Handwörterbuch.* 759. 1852.

Rhigiocarya Miers Menispermaceae

Origins:
From the Greek *rhigos* "cold" and *karyon* "nut."

Species/Vernacular Names:
R. racemifera Miers
Yoruba: atiba, ebo dudu

Rhigiophyllum Hochst. Campanulaceae

Origins:
Greek *rhigos* and *phyllon* "leaf."

Rhigospira Miers Apocynaceae

Origins:
Greek *rhigoo* "to be rigid, to be cold," *rhigos* "cold, frost" and *speira* "a spiral."

Rhigozum Burchell Bignoniaceae

Origins:
Greek *rhigos* "cold, frost" and *ozos* "branch, knot," the branches are rigid.

Species/Vernacular Names:
R. brevispinosum Kuntze (*Rhigozum linifolium* S. Moore; *Rhigozum spinosum* Burch. ex Sprague)
Southern Africa: mohurukwana (Ngwaketse dialect, Botswana)
R. obovatum Burch.
English: wild pomegranate
South Africa: wildegranaat, geelgranaat, geelberggranaat, berggranaat, granaatbos, ystervarkbos, driedoring
R. trichotomum Burch.
English: three-thorn rhigozum
R. zambesiacum Bak.
English: scrambled-egg bush, Transvaal wild pomegranate
Southern Africa: mopane pomegranate, mopaniegranaat; iQaqamba (Zulu)

Rhinacanthus Nees Acanthaceae

Origins:
Greek *rhis, rhinos* "snout, nose" and *akantha, akanthos* "thorn," alluding to the shape of the corolla or to the nature of the thorns.

Rhinactinidia Novopokr. Asteraceae

Origins:
Greek *rhine* "a file, rasp" and *aktin* "ray," some suggest from the Greek *rhis, rhinos* "snout, nose."

Rhinanthus L. Scrophulariaceae

Origins:
Greek *rhine* "a file, rasp" and *anthos* "flower," referring to the appearance of the corolla.

Species/Vernacular Names:
R. minor L.
English: yellow rattle, hay rattle

Rhinephyllum N.E. Br. Aizoaceae

Origins:
Greek *rhine* "a file, rasp" and *phyllon* "a leaf."

Rhinerrhiza Rupp Orchidaceae

Origins:
Greek *rhine* and *rhiza* "a root," the roots resembling a rasp; see Herman Montague Rucker Rupp (1872-1956), in *The Victorian Naturalist*. 67: 206. (Feb.) 1951.

Species/Vernacular Names:
R. divitiflora (Benth.) Rupp
English: raspy root
R. moorei (H.G. Reichb.) M. Clements, B. Wallace & D. Jones (*Sarcochilus moorei* (H.G. Reichb.) Schlechter)
English: Moore's sarcochilus

Rhiniachne Steud. Gramineae

Origins:
Greek *rhine* "a file, rasp" or *rhis, rhinos* "snout, nose" and *achne* "chaff, glume"; see also *Thelepogon* Roth ex Roemer & Schult.

Rhinopetalum Fischer ex D. Don Liliaceae

Origins:
From the Greek *rhis, rhinos* "snout, nose" and *pteryx* "wing"; see also *Fritillaria* L.

Rhinopterys Niedenzu Malpighiaceae

Origins:
Greek *rhis, rhinos* "snout, nose" and *pteryx* "wing"; see also *Acridocarpus* Guill. & Perr.

Rhipidantha Bremek. Rubiaceae

Origins:

Greek *rhipis, rhipidos* "a bellows, a lady's fan, a fan" and *anthos* "flower."

Rhipidia Markgr. Apocynaceae

Origins:

Greek *rhipis, rhipidos* "a bellows, a lady's fan, a fan"; see also *Condylocarpon* Desf.

Rhipidocladum McClure Gramineae

Origins:

Greek *rhipis, rhipidos* and *klados* "branch."

Rhipidoglossum Schltr. Orchidaceae

Origins:

Greek *rhipis, rhipidos* "a bellows, a fan" and *glossa* "a tongue," the flabellate lip.

Rhipidopteris Fée Lomariopsidaceae

Origins:

From the Greek *rhipis, rhipidos* and *pteris* "fern," referring to the fronds; see also *Peltapteris* Link.

Rhipogonum Sprengel Smilacaceae (Ripogonaceae, Rhipogonaceae, Liliaceae)

Origins:

Orthographic variant of *Ripogonum* Forst. & Forst.f.

Rhipsalidopsis Britton & Rose Cactaceae

Origins:

Resembling *Rhipsalis* Gaertner.

Rhipsalis Gaertner Cactaceae

Origins:

Greek *rhips, rhipos* "wicker-work, willow branch, mat, young twig," referring to the flexible branches; Akkadian *rapa'u* "to heal," Hebrew *rafa* "to bind: a wound to heal."

Species/Vernacular Names:

R. baccifera (J.S. Miller) Stearn

La Réunion Island: la perle, cactus-gui

Rhizanota Lour. ex Gomes Tiliaceae

Origins:

Greek *rhiza* "root" and *notos* "back"; see Giovanni Semerano, *Le origini della cultura europea*. Dizionari Etimologici. Basi semitiche delle lingue indeuropee. Dizionario della lingua Greca. 2(1): 249. Leo S. Olschki Editore, Firenze 1994.

Rhizanthella R. Rogers Orchidaceae

Origins:

The diminutive of the saprophytic genus *Rhizanthes* Dumort. (Rafflesiaceae), referring to the inflorescences and to the subterranean rhizomes.

Species/Vernacular Names:

R. gardneri R. Rogers (after Charles Austin Gardner, 1896-1970, from 1928 to 1960 Government Botanist and Curator of Western Australia Herbarium, with H.W. Bennetts wrote *The Toxic Plants of Western Australia*. 1956)

English: Western Australia underground orchid

Rhizanthemum Tieghem Loranthaceae

Origins:

Greek *rhiza* "root" and *anthemon* "a flower."

Rhizanthes Dumort. Rafflesiaceae

Origins:

From the Greek *rhiza* "root" and *anthos* "a flower."

Rhizobotrya Tausch Brassicaceae

Origins:

Greek *rhiza* "root" and *botrys* "cluster, a bunch of grapes, a cluster of grapes."

Rhizocephalum Wedd. Campanulaceae

Origins:

Greek *rhiza* and *kephale* "head."

Rhizocephalus Boiss. Gramineae

Origins:

From the Greek *rhiza* "root" and *kephale* "head."

Rhizoglossum C. Presl Ophioglossaceae

Origins:

Greek *rhiza* and *glossa* "tongue," see also *Ophioglossum* L.

Rhizomatopteris Khokhr. Dryopteridaceae (Aspleniaceae, Woodsiaceae)

Origins:

Greek *rhizoma, rhizomatos* "the mass of roots, stem" and *pteris* "fern," see also *Cystopteris* Bernh.

Rhizomonanthes Danser Loranthaceae

Origins:

Greek *rhizoma* "the mass of roots, stem," *monos* "single" and *anthos* "flower"; see B. Migliorini, *Parole d'autore (onomaturgia)*. Firenze 1975.

Rhizophora L. Rhizophoraceae

Origins:

Greek *rhiza* "a root" and *phoros* "bearing, carrying," *rhizophorus* "root bearing," referring to the aerial roots from stem and branches; see Carl Linnaeus, *Species Plantarum*. 443. 1753 and *Genera Plantarum*. Ed. 5. 202. 1754; William Griffith (1810-1845), *Notulae ad plantas asiaticas*. 665. 1854; Ding Hou, "A review of the genus *Rhizophora* with special reference to the Pacific species." in *Blumea*. 10(2): 625-634. 1960.

Species/Vernacular Names:

R. apiculata Blume

English: tall-stilted mangrove

India: naya kandal

The Philippines: bakawan, bakaw

R. mangle L.

English: American mangrove, mangrove, red mangrove

R. mucronata Lam.

English: red mangrove, four-petaled mangrove, long-fruited red mangrove, true mangrove

Southern Africa: rooiwortelboom, beebasboom; umNgom-bamkhonto, umHlume, umHluma (Zulu); umHluma (Xhosa)

Japan: Yaeyama-hirugi, ôba-hirugi, pushiki, funiki

India: kamo, bhora, uppuponna, adaviponna, peykkandal, kandal, kandaale, paniccha kandal, pikandal, venkandal, rai, rohi, bairada, jumuda, pyu, sorapinnai

Malaya: bakau belukap, bakau jangkar, bakau kurap, belukap, lenggayong

R. racemosa G.F.W. Meyer

French: palétuvier

Yoruba: egba, igba dudu

Nigeria: agala, egba, igba-dudu, ngala, odonowe, odo nowe, tanda, litanda, urher-nwere; egba (Yoruba); odo nowe (Edo); odo (Itsekiri); urheruwerim (Urhobo); agala (Ijaw); ngala (Igbo); nunung (Efik); nunung (Ibibio)

Central Africa: ntan, ntana, tanda

Cameroon: tanda

Gabon: ntan

R. stylosa Griffith

English: spotted-leaved red mangrove

Rhizophyllum Newman Polypodiaceae

Origins:

Greek *rhizophyllos* "with leaves from the root, with rooting leaves."

Rhizosperma Meyen Azollaceae

Origins:

Greek *rhiza* "a root" and *sperma* "seed."

Rhodalsine J. Gay Caryophyllaceae

Origins:

Greek *rhodon* "rose" and *alsine*, applied to a plant perhaps a chickweed, or the genus *Alsine* Gaertner, Caryophyllaceae.

Rhodamnia Jack Myrtaceae

Origins:

Greek *rhodon* "rose" and *amnion* "the amnion, a bowl, the membrane around the fetus," referring to the red unripe fruits, or from Greek *rhodamnos, rhadamnos* "a young branch, sprout, shoot," from the size of the plant; see William Jack (1795-1822), in *Malayan Miscellanies*. 2(7): 48. Bencoolen 1822.

Species/Vernacular Names:
R. argentea Benth.
English: white myrtle, brown malletwood
R. maideniana C. White
English: smooth-leaved scrub turpentine
R. rubescens (Benth.) Miq. (*Rhodamnia trinervia* Blume)
English: silver back, scrub turpentine, scrub stringybark, brown mallet-wood
Malaya: poyan, mempoyan, empoyan, tempoyai, mengkoyan
R. whiteana Guymer & Jessup
English: white malletwood

Rhodanthe Lindley Asteraceae

Origins:
With rose-colored flowers, from the Greek *rhodon* "rose" and *anthos* "flower," Akkadian *wurdinu* "rose"; see John Lindley (1799-1865), *Edwards's Botanical Register*. 20, t. 1703. (Sep.) 1834.

Rhodanthemum (Vogt) Bremer & Humphries Asteraceae

Origins:
Greek *rhodon* "rose" and *anthemon* "flower."

Rhodax Spach Cistaceae

Origins:
Greek *rhodon* "rose," rock-roses.

Rhodiola L. Crassulaceae

Origins:
The diminutive from the Greek *rhodon* "rose," referring to the scented roots.

Species/Vernacular Names:
R. sp.
Tibetan: tshan byi-mjug

Rhodocactus (A. Berger) F. Knuth Cactaceae

Origins:
Greek *rhodon* "rose" plus *cactus*.

Rhodocalyx Müll. Arg. Apocynaceae

Origins:
Greek *rhodon* "rose, red" and *kalyx, kalykos* "calyx," the calyx is colored.

Rhodochiton Zucc. ex Otto & Dietr. Scrophulariaceae

Origins:
Greek *rhodon* "red" and *chiton* "a tunic, covering," referring to the calyx.

Rhodocodon Baker Hyacinthaceae (Liliaceae)

Origins:
Greek *rhodon* "red, rose" and *kodon* "a bell."

Species/Vernacular Names:
R. madagascariensis Baker
Madagascar: famonototozy, kilobaloba, sahondra, tapabatana

Rhodocolea Baillon Bignoniaceae

Origins:
Greek *rhodon* "red, rose" and *koleos* "a sheath."

Rhodocoma Nees Restionaceae

Origins:
From the Greek *rhodon* and *kome* "hair of the head."

Rhododendron L. Ericaceae

Origins:
Greek *rhodon* "a rose, rose garden" and *dendron* "a tree," possibly referring to the bunches of flowers, Latin *rhododendros* and *rhododendron* and *rhododaphne* for the rosebay, oleander (Plinius); see Carl Linnaeus, *Species Plantarum*. 392. 1753 and *Genera Plantarum*. Ed. 5. 185. 1754;

A.J.C. Grierson & D.G. Long, *Flora of Bhutan.* 2(1): 357-387. Edinburgh 1991; V. Bertoldi, " Per la storia del lessico botanico popolare." in *Archivum romanicum.* XI: 14-30. 1927; P.E. Guerriero, "La rosa delle alpi." in *Studi letterari e linguistici dedicati a Pio Rajna.* Firenze 1911; M. Cortelazzo & P. Zolli, *Dizionario etimologico della lingua italiana.* 4: 1101. Zanichelli, Bologna 1985; Helmut Genaust, *Etymologisches Wörterbuch der botanischen Pflanzennamen.* 536. 1996; E. Weekley, *An Etymological Dictionary of Modern English.* 2: 1234. 1967.

Species/Vernacular Names:

R. anthopogon D. Don

India: nichni, rattankat, nera, kai zaban, morua, talisa, talisri, tazaktsum, talis-faz, dhoop, palu

R. arboreum Sm.

English: tree-rhododendron, rose-tree

Nepal: lali gurans, gurass, gurans

India: burans, bhorans, guras, lalguras, brus, baras, ghonas, taggu, ardawal, mandal, chiu, aru, broa, chacheon, kamri, chhan, tin-saw, dieng-tin-thuin, etok, dotial

R. campanulatum D. Don

Nepal: nilo chimal

India: cherailu, cheraidhu, cheriala, ghentaboras, sarnagar, shinwala, shargar, simrung, gaggar yurmi, gaggar vurmi, nichnai, chimul, chimura, simris, teotosa, nilo chimal

R. cinnabarinum Hook.f.

Nepal: sanu chimal

India: kema, kechung, balu, sanu chimal

R. dalhousiae Hook.f.

Nepal: lahare chimal

R. dauricum L.

English: daurian rhododendron

China: man shan hong, yeh tu chuan

R. edgeworthii Hook.f.

Nepal: lahare chimal

R. falconeri Hook.f.

Nepal: korlinga

India: kegu, kalma, korlinga

R. ferrugineum L.

Europe: alpenrose, alpine rose

R. grande Wight

Nepal: patle korlinga

R. griffithianum Wight

Nepal: seto chimal

R. hodgsonii Hook.f.

Nepal: korlinga, korling

R. indicum (L.) Sweet

Japan: satsuki

China: tu chuan, ying shan hung, hung chih chu, hung tu chuan

R. kesangiae Long & Rushforth

Bhutan: tala

R. lepidotum G. Don

Nepal: balu sun pate, bhale sunpate, bahle, sunpatie, saluma

R. lindleyi Moore

Nepal: lahare chimal

R. lochiae F. Muell. (or *Lochae*) (after Lady Loch, wife of Sir Henry Brogham Loch, former Governor of Victoria, Australia; see [Great Britain — South Africa], *Correspondence Relating to the Liquor Traffic in Certain Native Territories in South Africa.* Presented to both Houses of Parliament by Command of Her Majesty, July 1890. [Correspondence between Sir Henry B. Loch, Governor; Sir George Baden-Powell; and others concerning Basutoland and British Bechuanaland.] London 1890)

English: Australian rhododendron

R. macrophyllum D. Don

English: California rose-bay

R. molle (Bl.) G. Don

English: Chinese azalea, yellow azalea

China: nao yang hua

R. occidentale (Torrey & A. Gray) A. Gray

English: western azalea

R. simsii Planch. (*Rhododendron eriocarpum* (Hay.) Nakai; *Rhododendron indicum* (L.) Sweet var. *eriocarpum* Hayata)

English: Sims's azalea, red azalea

China: du juan hua

Japan: maruba-satsuki, kikazô

R. stenopetalum (Hogg) Mabb.

English: spider azalea

R. tomentosum Harmaja

English: Labrador tea

R. triflorum Hook.f.

Nepal: pahenle chimal

R. viscosum (L.) Torrey

English: swamp honeysuckle

Rhododon Epling Labiatae

Origins:

Greek *rhodon* "a rose, red" and *odous, odontos* "a tooth."

Rhodogeron Griseb. Asteraceae

Origins:

Greek *rhodon* and *geron* "an old man," possibly referring to the pappus.

Rhodognaphalon (Ulbr.) Roberty Bombacaceae

Origins:

Greek *rhodon* "red" and *gnaphalon*, *knaphallon* "flock of wool, pillow."

Species/Vernacular Names:

R. brevicuspe (Sprague) Roberty (*Bombax brevicuspe* Sprague)

Nigeria: awori (Yoruba); ogiukpogha (Edo); akpudele (Igbo); nyamenyok (Boki)

Congo: n'demo

Gabon: koma, alone

Ivory Coast: kondroti

Cameroon: ovong, ovonga, enonu, tenonu

Rhodognaphalopsis A. Robyns Bombacaceae

Origins:

Resembling *Rhodognaphalon* (Ulbr.) Roberty, see also *Bombax* L.

Rhodohypoxis Nel Hypoxidaceae (Liliaceae)

Origins:

Greek *rhodo*, *rhodon* and the genus *Hypoxis*, from the color of the flowers.

Species/Vernacular Names:

R. milloides (Bak.) Hilliard & Burtt (*Rhodohypoxis palustris* Killick)

South Africa: rooisterretjie

Rhodolaena Thouars Sarcolaenaceae

Origins:

From the Greek *rhodon* "red" and *chlaena*, *laina* "cloak, blanket."

Rhodoleia Champ. ex Hook. Hamamelidaceae (Rhodoleiaceae)

Origins:

Greek *rhodo*, *rhodon* and *leios* "smooth."

Rhodomyrtus (DC.) Reichenbach Myrtaceae

Origins:

Myrtle-like flowers, from the Greek *rhodon* and *myrtos* "myrtle, branch of myrtle," Akkadian *murdudu* and Sumerian *mur-du-du* "a plant"; see Heinrich G.Ludwig Reichenbach (1793-1879), *Der Deutsche Botaniker ... Herbarienbuch*. 1: 177. (Jul.) 1841.

Species/Vernacular Names:

R. macrocarpa Benth.

English: finger cherry, Cooktown loquat

R. psidioides (G. Don.) Benth. (*Nelitris psidioides* G. Don) (resembling *Psidium*)

English: native guava

R. tomentosa (Ait.) Hassk. (*Myrtus tomentosa* Ait.; *Rhodomyrtus parviflora* Alston)

English: downy myrtle, hill gooseberry, hill guava, rose myrtle, downy rosemyrtle

China: shan ren zi

Japan: tennin-ka, satagi-ima

Malaya: kemunting

Vietnam: pieu nim, sim, hong sim

Rhodophiala C. Presl Amaryllidaceae (Liliaceae)

Origins:

From the Greek *rhodon* "red, rose" and *phiale* "a vial."

Rhodopis Urban Fabaceae

Origins:

Greek *rhodopos* "rosy, rosy-faced," poet. fem. *rhodopis*, *rhodopidos*.

Rhodosciadium S. Watson Umbelliferae

Origins:

For the American botanist Joseph Nelson Rose, 1862-1928, botanical explorer, traveler, plant collector (southwestern

USA, Mexico, and South America), author of "Notes on useful plants of Mexico." *Contr. U.S. Natl. Herb.* 5(4): 209-259. 1899, with Nathaniel Lord Britton (1859-1934) wrote *The Cactaceae*. Washington 1919-1923; see John H. Barnhart, *Biographical Notes upon Botanists.* 3: 177. 1965; T.W. Bossert, *Biographical Dictionary of Botanists Represented in the Hunt Institute Portrait Collection.* 338. 1972; Ethelyn Maria Tucker, *Catalogue of the Library of the Arnold Arboretum of Harvard University.* Cambridge, Massachusetts 1917-1933; Ida Kaplan Langman, *A Selected Guide to the Literature on the Flowering Plants of Mexico.* 645-650. Philadelphia 1964; S. Lenley et al., *Catalog of the Manuscript and Archival Collections and Index to the Correspondence of John Torrey.* Library of the New York Botanical Garden. 354. 1973; Lyman David Benson (1909-1993), *The Cacti of the United States and Canada.* Stanford, California 1982; Larry W. Mitich, *Cactus and Succulent Journal.* Vol. 53(6): 299-303. 1981; Ira L. Wiggins, *Flora of Baja California.* 43. Stanford, California 1980; Gordon Douglas Rowley, *A History of Succulent Plants.* Strawberry Press, Mill Valley, California 1997; Irving William Knobloch, compil., "A preliminary verified list of plant collectors in Mexico." *Phytologia Memoirs.* VI. 1983; J. Ewan & Nesta Dunn Ewan, "Biographical dictionary of Rocky Mountain naturalists." *Reg. Veget.* 107: 1-253. 1981; R. Zander, F. Encke, G. Buchheim & S. Seybold, *Handwörterbuch der Pflanzennamen.* 14 Aufl. Stuttgart 1993.

Rhodosepala Baker Melastomataceae

Origins:
From the Greek *rhodon* "red, rose" and *sepalon* "sepal."

Rhodospatha Poeppig Araceae

Origins:
Greek *rhodon* and *spatha* "spathe," referring to the color of the spathe in some species; see D.H. Nicolson, "Derivation of aroid generic names." *Aroideana.* 10: 15-25. 1988.

Rhodosphaera Engler Anacardiaceae

Origins:
Greek *rhodon* and *sphaira* "a globe, ball," referring to the globular and reddish fruits; see H.G.A. Engler, in *Botanische Jahrbücher.* 1: 423, t. XII, fig. 107. (Jan.) 1881.

Species/Vernacular Names:
R. rhodanthema (F. Muell.) Engl. (*Rhus rhodanthema* F. Muell.; *Rhus elegans* W. Hill)
English: deep yellow wood, yellow cedar, tulip satinwood

Rhodostachys Philippi Bromeliaceae

Origins:
Greek *rhodon* "red, rose" and *stachys* "a spike."

Rhodostegiella Li Asclepiadaceae

Origins:
Greek *rhodon* "red, rose" and *stege, stegos* "roof, shelter."

Rhodostemonodaphne Rohwer & Kubitzki Lauraceae

Origins:
Greek *rhodon* "red, rose," *stemon* "pillar, stamen," *monos* "single, one" plus *Daphne*; Latin *rhododaphne* applied by Plinius to the rose-bay, oleander.

Rhodostoma Scheidw. Rubiaceae

Origins:
From the Greek *rhodon* and *stoma* "mouth."

Rhodothamnus Reichb. Ericaceae

Origins:
Greek *rhodon* "red, rose" and *thamnos* "a shrub," referring to the flowers.

Rhodotypos Siebold & Zucc. Rosaceae

Origins:
Greek *rhodon* "a rose" and *typos* "a type"; Akkadian *wurdinu* "rose."

Species/Vernacular Names:
R. scandens (Thunb.) Makino (*Rhodotypos kerrioides* Sieb. & Zucc.; *Rhodotypos tetrapetala* (Sieb.) Makino; *Kerria tetrapetala* Sieb.)
Japan: shiro-yama-buki (= white *Kerria japonica*)

Rhodoxylon Raf. Convolvulaceae

Origins:
Greek *rhodon* "rose, red" and *xylon* "wood"; see Constantine Samuel Rafinesque (1783-1840), *Flora Telluriana.* 4: 79. 1836 [1838].

Rhoeo Hance Commelinaceae

Origins:

Derivation of the name unknown, perhaps from the Latin *rhoeas, adis* or *rhoea, ae* used by Plinius for the wild poppy, Greek *rhoias* applied by Theophrastus (*HP.* 9.12.4) and Dioscorides to the corn poppy, a species of *Papaver*, or from Greek *rhoe, rhoa* "river, stream, flowing of sap."

Rhoiacarpos A. DC. Santalaceae

Origins:

Greek *rhoia, rhoa, rhoie* "pomegranate, pomegranate-tree" and *karpos* "a fruit," referring to the dark red fruits.

Species/Vernacular Names:

R. capensis (Harv.) A. DC.

South Africa: wildegranaat

Rhoicissus Planchon Vitaceae

Origins:

Presumably from the Latin *rhoicus, a, um* "belonging to the *Rhus*, to the sumac" and *cissos, i* "ivy," or from the Greek *rhoia* "pomegranate" and *kissos* "ivy."

Species/Vernacular Names:

R. digitata (L.f.) Gilg & Brandt (*Cissus thunbergii* Eckl. & Zeyh. nom. illegit.; *Rhoicissus cirrhiflora* (L.f.) Gilg & Brandt p.p.)

English: baboon grape, wild grape

Southern Africa: bobbejaandruif, bobbejaanbolle, Boesmansdruif; isiNwazi, umThwazi, umNangwazi, umPhambane (Zulu); chiDzamahosi, chiNgamahosi (Shona)

R. revoilii Planch. (*Rhoicissus cymbifolius* C.A. Sm.; *Rhoicissus schlechteri* Gilg & Brandt)

English: bitter forest grape

Southern Africa: bitterbosdruif; isiNwasi (Zulu); isAqoni, isaQoni (Xhosa)

R. rhomboidea (E. Meyer ex Harv.) Planchon (*Cissus rhomboidea* E. Mey. ex Harv.)

English: glossy forest grape

Southern Africa: blinkblaarbosdruif; isiNwazi (Zulu); umThwazi (Xhosa)

R. tomentosa (Lam.) Wild & Drum. (*Cissus capensis* Willd.; *Rhoicissus capensis* (Willd.) Planch.; *Vitis capensis* (Willd.) Thunb. non Burm.f.)

English: common forest grape, monkey-rope, simple-leaved grape, wild grape, wild vine

Southern Africa: gewone bosdruif, bosdruif, bosdruiwe, wildedruif, wildedruiwe, bobbejaantou; isiNwazi, isiwazana (Zulu); isaQoni (Xhosa); kundzu (Tsonga)

R. tridentata (L.f.) Wild & Drum. subsp. *cuneifolia* (Eckl. & Zeyh.) N.R. Urton (*Cissus cuneifolia* Eckl. & Zeyh.; *Rhoicissus cuneifolia* (Eckl. & Zeyh.) Planch.; *Rhoicissus erythrodes* (Fresen.) Planch.)

English: Bushman's grape, wild grape, bitter grape, common forest grape

Southern Africa: Boesmansdruif, droog-my-keel grape, bobbejaantou, wildedruiwe, wildedruif; isiNwazi, umThwazi (Zulu); isaQoni (Xhosa); murumbula-mbudzana (= pricks the kid) (Venda)

R. tridentata (L.f.) Wild & Drum. subsp. *tridentata* (*Rhoicissus cirrhiflora* sensu Gilg & Brandt p.p. excl. syn. *Rhus cirrhiflora* L.f.)

Southern Africa: bitterdruif, bobbejaantou, Boesmansdruif, droog-my-keel, wildedruiwe, wildedruif; lumbu (Tsonga); morarana (Sotho)

Rhoiptelea Diels & Hand.-Mazz. Rhoipteleaceae (Hamamelidae, Juglandales)

Origins:

Latin *rhoicus, a, um* "belonging to the *Rhus*, to the sumac" and *ptelea* "an elm tree," aromatic tree with aromatic resinous glands.

Rhombochlamys Lindau Acanthaceae

Origins:

Greek *rhombos* "rhombus, lozenge" and *chlamys, chlamydos* "cloak."

Rhomboda Lindley Orchidaceae

Origins:

Greek *rhombos* "rhombus, lozenge," referring to the base of the lip.

Rhombolythrum Airy Shaw Gramineae

Origins:

See *Rhombolytrum* Link.

Rhombolytrum Link Gramineae

Origins:

Greek *rhombos* and *elytron* "sheath, scale, husk."

Rhombonema Schltr. Asclepiadaceae

Origins:
Greek *rhombos* "rhombus" and *nema* "thread, filament."

Rhombophyllum (Schwantes) Schwantes Aizoaceae

Origins:
Greek *rhombos* and *phyllon* "leaf."

Rhombospora Korth. Rubiaceae

Origins:
Greek *rhombos* "rhombus, lozenge" and *spora* "a seed."

Rhoogeton Leeuwenb. Gesneriaceae

Origins:
Greek *rhoos* "stream, current" and *geiton* "a neighbor."

Rhopalephora Hassk. Commelinaceae

Origins:
Greek *rhopalon* "a club" and *phoros* "bearing, carrying."

Rhopaloblaste Scheffer Palmae

Origins:
Greek *rhopalon* and *blastos* "bud, sprout, germ, ovary, sucker," referring to the embryo with a club-shaped appendage.

Rhopalobrachium Schltr. & K. Krause Rubiaceae

Origins:
From the Greek *rhopalon* and *brachion* "the arm, the forearm."

Rhopalocarpus Bojer Ochnaceae (Diegodendraceae, Sphaerosepalaceae)

Origins:
Greek *rhopalon* "a club" and *karpos* "fruit."

Rhopalocarpus Teijsm. & Binn. ex Miq. Annonaceae

Origins:
Greek *rhopalon* "a club" and *karpos* "fruit."

Rhopalocnemis Jungh. Balanophoraceae

Origins:
Greek *rhopalon* "a club" and *kneme* "limb, leg."

Rhopalocyclus Schwantes Aizoaceae

Origins:
Greek *rhopalon* "a club" and *kyklos* "a circle, ring."

Rhopalopilia Pierre Opiliaceae

Origins:
Greek *rhopalon* plus the genus *Opilia* Roxb.

Rhopalopodium Ulbr. Ranunculaceae

Origins:
From the Greek *rhopalon* "a club" and *podion* "a little foot, stalk."

Rhopalosciadium Rech.f. Umbelliferae

Origins:
From the Greek *rhopalon* and *skiadion* "umbel, parasol."

Rhopalostigma Phil. Solanaceae

Origins:
From the Greek *rhopalon* "a club" and *stigma* "stigma."

Rhopalostylis H.A. Wendland & Drude Palmae

Origins:
Greek *rhopalon* and *stylos* "pillar, column, style," an allusion to the spadix, the female organ in the male flower is club-shaped; see H.E. Connor and E. Edgar, "Name changes in the indigenous New Zealand flora, 1960-1986 and Nom-

ina Nova IV, 1983-1986." *New Zealand Journal of Botany.* Vol. 25: 115-170. 1987; W.R. Sykes, in *New Zealand DSIR Bull.* 219: 184-186. 1977.

Species/Vernacular Names:

R. baueri (Hook.f.) H.A. Wendland & Drude var. *cheesemanii* (Cheeseman) Sykes

English: Kermadec I. nikau palm

R. sapida H. Wendland & Drude

English: nikau palm

New Zealand: nikau (Maori name)

Rhopalostylis Klotzsch ex Baillon Euphorbiaceae

Origins:

Greek *rhopalon* "a club" and *stylos* "pillar."

Rhopalota N.E. Br. Crassulaceae

Origins:

Greek *rhopalon* "a club," referring to stem and branches.

Rhopium Schreb. Euphorbiaceae

Origins:

Greek *rhopion, rhopeion* "bush, twig, bough," see also *Phyllanthus* L.

Rhus L. Anacardiaceae (Part I)

Origins:

Latin *rhus, rhois* and *roris* for a bushy shrub, sumac, Greek *rhous* (probably from *rhodo, rhodos* "red"), the ancient name used by the Greek philosopher Theophrastus (-os) for one species, *Rhus coriaria* L.; see Carl Linnaeus, *Species Plantarum.* 265. 1753 and *Genera Plantarum.* Ed. 5. 129. 1754; Shri S.P. Ambasta, ed., *The Useful Plants of India.* Council of Scientific & Industrial Research, New Delhi 1986. Theophrastus, born at Eresus in Lesbos c. 370 BC., after Plato's death (347) joined Aristotle at Assos, and succeeded him as head of the Lyceum in 322.

Species/Vernacular Names:

R. sp.

English: sumac, sumach

South America: birringo

R. acocksii Moffett

South Africa: taaibos

R. angustifolia L.

South Africa: willow taaibos

R. aromatica Ait.

English: fragrant sumac, lemon sumac, polecat-bush, squaw-bush, sweet-scented sumac

R. chinensis Miller (*Rhus semialata* Murray)

English: Chinese gall, nutgall, Chinese sumac, nutgall tree, Chinese nutgall tree

Bhutan: datick

Nepal: bhakimlo, bakimilo

China: wu bei zi, wu pei tzu, fu yang, yan fu zi, yen fu tzu, fu mu

India: tatri, arkhar, chechar, dudla, tetri, thissa, hulug, rashtu, wansh, dasmila, dharmil, tibri, arkhoi, naga-tenga, dieng-soh-sma, soh-ma, khitma, gimbao, thanghaerkung, takhrit, bhaimlo, bhagm ili

R. chirindensis Bak.f. (*Rhus chirindensis* Bak.f. forma *legatii* (Schönl.) R. & A. Fernandes; *Rhus legatii* Schönl.) (the typical form was first collected near the Chirinda forest in southeast Rhodesia near the border)

English: red currant, bush jarrah

Southern Africa: bostaaibos, bloedhout, westelike essenhout, bosganna; monaatlou (North Sotho); pegasudza (Shona); inHlokoshiyane-enkulu, umHlabamvubu, isi-Banda, iKhathabane, umDwendwelencuba (Zulu); muvhadela-phanga (= wood for knife handles) (Venda); umHlakothi, umHlakothi omkhulu, inTlokolotshane enkulu (Xhosa)

R. copallina L.

English: dwarf sumac, mountain sumac, shining sumac, flame-tree sumac, wing-rib sumac, winged sumac

R. coriaria L.

English: Sicilian sumac, tanner's sumac, elm leaved sumac, sumac, sumach, tanning sumach

Portuguese: sumagre

India: samaka, timtima, tatrak, sumok, sumak

R. crenata Thunb.

English: dune crow-berry

South Africa: duinekraaibessie, kraaibessie

R. dentata Thunb. (*Rhus dentata* Thunb. var. *dentata* forma *sparsepilosa* R. Fernandes; *Rhus dentata* Thunb. var. *parvifolia* (Harv. ex Sond.) Schönl.; *Rhus dentata* Thunb. var. *parvifolia* Eckl. & Zeyh. forma *parvifolia*; *Rhus dentata* Thunb. var. *parvifolia* Eckl. & Zeyh. forma *villosissima* R. Fernandes; *Rhus dentata* Thunb. var. *puberula* Sond. forma *glabra* (Schönl.) R. Fernandes; *Rhus dentata* Thunb. var. *puberula* Sond. forma *pilosissima* (Engl.) R. Fernandes; *Rhus dentata* Thunb. var. *puberula* Sond. forma *puberula*)

English: nana berry

Southern Africa: nanabessie; inHlokoshiyane, umHlalamvubu (Zulu); iNtlakotshana (Xhosa); iAmbashira, muBikasadza, muFungamfura, Ngaranga (Shona); lebelebele (South Sotho)

R. diversiloba Torr. & Gray (*Rhus toxicodendron* subsp. *diversiloba* (Torrey & Gray) Engler; *Toxicodendron diversilobum* (Torr. & Gray) Greene)

English: Pacific poison-oak, wester poison-oak

Mexico: hiedra, hiegra (Sonora and Chihuahua)

R. engleri Britten (the species honors the German botanist and leader of the Berlin school of plant taxonomy Heinrich Gustav Adolf Engler, 1844-1930; see J.H. Barnhart, *Biographical Notes upon Botanists*. 1: 510. Boston 1965; Simon Mayo, Josef Bogner and Peter Boyce, "The Acolytes of the Araceae." *Curtis's Botanical Magazine*. Volume 12. 3: 153-168. August 1995; Stafleu and Cowan, *Taxonomic Literature*. 1: 757-797. Utrecht 1976; T.W. Bossert, *Biographical Dictionary of Botanists Represented in the Hunt Institute Portrait Collection*. 117. 1972; Ida Kaplan Langman, *A Selected Guide to the Literature on the Flowering Plants of Mexico*. Philadelphia 1964; S. Lenley et al., *Catalog of the Manuscript and Archival Collections and Index to the Correspondence of John Torrey*. Library of the New York Botanical Garden. 154-155. 1973)

South Africa: karaa

R. erosa Thunb.

Southern Africa: besembos (= broom bush), soettaaibos, rosyntjiebos; tshilabele (South Sotho)

R. fastigiata Eckl. & Zeyh.

English: broom currant

Southern Africa: besemtaaibos; iNtlakotshane yedobo, iTlokolotshane yedobo, iTlokolotshane encinane, iNtlakotshane encinane (Xhosa)

R. glabra L.

English: smooth sumac, scarlet sumac, vinegar tree

R. glauca Thunb.

English: blue Kuni-bush

South Africa: bloukoeniebos, grey-green taaibos, blinkblaar, korentebessie, rosyntjiebos, suurbessie, taaibos

R. gueinzii Sond. (*Rhus crispa* (Harv. ex Engl.) Schönl.; *Rhus simii* Schönl.; *Rhus simii* Schönl. var. *lydenburgensis* Schönl.; *Rhus spinescens* Diels) (for the German apothecary and collector Wilhelm Gueinzius; see Gustav Kunze, 1793-1851, "Filicum in Promontorio Bonae Spei et ad Portum Natalensum a Gueinzio nuperius collectarum." *Linnaea*. 18: 113-124. 1844, and also Philipp Bruch, 1781-1847, (with G.W. Bischoff and J.B.W. Lindenberg) "*Musci* et *Hepatici* Kraussiani." *Flora*. 29: 132-136. 1846)

English: thorny karree

Southern Africa: doringkaree, thorny taaibos, taaibosdoring; iNhlokoshiyane, umPhondo (Zulu); inHlangutshane (Swazi); motshotlho (Western Transvaal, northern Cape, Botswana); nsasane (Kalanga, northern Botswana)

R. hookeri Sahni & Bahadur

Nepal: khag bhalayo, bhalaye

India: momphulai, sehr-kung, kagphulai

R. integrifolia (Nutt.) Benth. & Hook.f. ex S. Watson

English: lemonade berry, lemonade sumac, sourberry

R. krebsiana Presl ex Engl. (After the German apothecary and botanical collector Georg Ludwig Engelhard (Englehard) Krebs, 1792-1844, naturalist, from 1817 to 1844 at the Cape; see J.H. Barnhart, *Biographical Notes upon Botanists*. 2: 320. 1965; H.N. Clokie, *Account of the Herbaria of the Department of Botany in the University of Oxford*. 195. Oxford 1964; Ignatz Urban, *Geschichte des Königlichen Botanischen Museums zu Berlin-Dahlem (1815-1913). Nebst Aufzählung seiner Sammlungen*. Dresden 1916; Günther Schmid, *Chamisso als Naturforscher*. Eine Bibliographie. Leipzig 1942; Stafleu and Cowan, *Taxonomic Literature*. 2: 669-670. 1979)

English: false sour currant

South Africa: bastersuurtaaibos, Kreb's taaibos

R. laevigata L. (*Rhus incana* Mill.; *Rhus mucronata* Thunb.; *Rhus viminalis* sensu Vahl non Ait.)

English: dune taaibos, western karee, dune currant, red currant

Southern Africa: duinetaaibos, korentebos, witkareeboom, wilderosyntjieboom, bosganna, bostaaibos, kieriehout, taaibos; garas (Nama: southern southwest Africa); umHlakothi (Xhosa); ugcane (Ndebele); muKungu (Shona)

R. lancea L.f. (*Rhus viminalis* Ait.)

English: bastard willow, karoo tree, karee, karree, common karee, willow rhus

Southern Africa: rooikaree, rivierkaree, kareeboom, gewone karee, hoenderspoorkaree, rooikareehout, kareehout, makkaree, soetkaree, waterkaree, grootkareehout, taaibos, krieboom; iQunguwe (Xhosa); inHlokoshiyane (Zulu); inHlangutshane (Swazi); mosilabele, mosilabelo (South Sotho); motlhotlho, mohlwehlwe, mokalabata (North Sotho); mosotlhoana (Sotho); mosilabele (Tswana: Western Transvaal, northern Cape, Botswana); moshabela (Malete dialect, Botswana); moshilabele (Tlokwa dialect, Botswana); oruso (Herero); garas (Nama: southern southwest Africa); mutepe (Shona); ucane (Ndebele)

R. leptodictya Diels (*Rhus amerina* Meike; *Rhus gueinzii* sensu Schonl. non Sond.; *Rhus rhombocarpa* R. & A. Fernandes) (the specific name from the Greek *leptos* "delicate, thin, slender" and *diktyon* "net," probably referring to the venation of the leaves)

English: mountain karee, false karee, rock rhus

Southern Africa: bergkaree, basterkaree, plat kareebessie-boom, taaibos; mohlwehlwe (North Sotho); mushakaladza (Venda)

R. longispina Eckl. & Zeyh.

English: long-spined rhus

South Africa: karoo thorny taaibos, buffalsdoring, wol-wedoring

R. lucida L. (*Rhus lucida* L. var. *outeniquensis* (Szyszyl.) Schönl.; *Rhus lucida* L. var. *tipica* Schönl.; *Rhus outeniquensis* Szyszyl.; *Rhus schlechteri* Diels; *Rhus africana* Mill.; *Rhus cavanillesii* DC.; *Searsia lucida* (L.) F.A. Barkley; *Toxicodendron lucidum* (L.) Kuntze)

English: glossy currant, wild currant, shiny rhus, shiny leaved rhus

Southern Africa: korentbos, besembos, besemkaree, kraaibessie, rosyntjiebos, taaibos, glossy taaibos, blinktaaibos, slaptaaibos; iNhlokoshiyane (Zulu); inTlakotshan 'ebomvu (Xhosa)

Rhus L. Anacardiaceae (Part II)

Origins:

Latin *rhus*, *rhois* and *roris* for a bushy shrub, sumac, Greek *rhous* (probably from *rhodo*, *rhodos* "red"), the ancient name used by the Greek philosopher Theophrastus (-os) for one species, *Rhus coriaria* L.; see Carl Linnaeus, *Species Plantarum*. 265. 1753 and *Genera Plantarum*. Ed. 5. 129. 1754; Shri S.P. Ambasta, ed., *The Useful Plants of India*. Council of Scientific & Industrial Research, New Delhi 1986. Theophrastus, born at Eresus in Lesbos c. 370 BC., after Plato's death (347) joined Aristotle at Assos, and succeeded him as head of the Lyceum in 322.

Species/Vernacular Names:

R. marlothii Engl. (the species after the South African (b. Germany) botanist Hermann W. Rudolph Marloth, 1855-1931, author of "Das Südöstliche Kalahari-Gebiet. Ein Beitrag zur Pflanzen-Geographie Südafrikas." *Bot. Jb.* 8: 247-260, 1887, and "Notes on the vegetation of Southern Rhodesia." *S. Afr. J. Sci.* 2: 300-307, 1904, and "The vegetation of the southern Namib." *S. Afr. J. Sci.* 6: 80-87, 1910; see John H. Barnhart, *Biographical Notes upon Botanists*. 2: 449. 1965; R. Zander, F. Encke, G. Buchheim and S. Seybold, *Handwörterbuch der Pflanzennamen*. 14. Aufl. Stuttgart 1993; Hans Herre, *The Genera of the Mesembryanthemaceae*. 49-50. Cape Town 1971; I.C. Hedge and J.M. Lamond, *Index of Collectors in the Edinburgh Herbarium*. Edinburgh 1970; A. White and B.L. Sloane, *The Stapelieae*. Pasadena 1937; A. Engler et al., "Plantae Marlothianae." in *Bot. Jahrb.* 10: 1-50, 242-285. 1889; Mary Gunn and Leslie E. Codd, *Botanical Exploration of Southern Africa*. Cape Town 1981; Gordon Douglas

Rowley, *A History of Succulent Plants*. Strawberry Press, Mill Valley, California 1997)

English: bitter karree

South Africa: bitter karee, suurkaree

R. metopium L. (*Metopium metopium* Small)

Cuba: guao de costa

R. microphylla Engl. (*Toxicodendron microphyllum* O. Kuntze; *Rhoeidium microphyllum* Greene; *Schmaltzia microphylla* Small)

English: desert sumac, scrub sumac

Mexico: lima de la sierra; agrillo (Durango and Coahuila); correosa (Coahuila, Durango and San Luis Potosí); salado (Coahuila); limilla de la sierra (Rio Bavispe, northeast Sonora)

R. mollis Kunth (*Toxicodendron molle* (Kunth) O. Ktze; *Rhus standleyi* Barkley; *Schmaltzia mollis* (Kunth) Barkley)

Mexico: tnu-ndé; zumaqui (Oaxaca, Hidalgo); yucucaya, cimarron, sumaco (Oaxaca); xoxoco (Tlaxcala; see Diego Muñoz Camargo, Manuscript, "Fragmentos de la Historia de Tlaxcala." Mexico City 1852, the Tlaxcalans were Cortés's chief allies in conquering the Aztec empire); vinagrillo (Michoacán)

R. montana Diels (*Rhus gerrardii* (Harv. ex Engl.) Schönl. var. *montana* (Diels) Schönl.)

English: Drakensberg karree

Southern Africa: Drakensbergkaree, mountain taaibos; inHlokoshiyane (Zulu)

R. natalensis Bernh. ex Krauss

English: Natal karree

Southern Africa: Natalkaree, Nataltaaibos; iNhlokoshiyane, inHlokoshiyane (Zulu); umGwele (Xhosa); bikasaza, muKungu, muPuma, muTsonha, muZazati (Shona)

Tanzania: msigiyo, bujorori, busigyo

Yoruba: jin, orijin

R. nebulosa Schonl.

South Africa: sand taaibos

R. ovata S. Watson

English: sugar bush, sugar bush sumac, sugar sumac

R. pachyrrhachis Hemsl. (*Rhus sempervirens* var. *pachyrrhachis* (Hemsl.) Engler; *Schmaltzia pachyrrhachis* (Hemsl.) Barkley)

Mexico: lantrisco (Guadalcazar, San Luis Potosí); copal lantrisco (San Luis Potosí)

R. paniculata Hook.f.

Bhutan: khir khobtang, khai roptang shing, prekoptang shing

R. pentheri Zahlbr. (for the Austrian botanical collector and naturalist Arnold Penther, 1865-1931, he published (with the Austrian botanist Emmerich Zederbauer, 1877-1950) *Ergebnisse einer naturwissenschaftlichen Reise zum Erds-chias-Dag (Kleinasien)*. Vienna 1905-1906; see also A. Zahlbruckner, *Plantae Pentherianae. Aufzählung der von Dr. A. Penther ... in SüdAfrika gesammelten Pflanzen*. Wien [Annalen der K.K. Naturhistorischen Hofmuseums] 1900 [-1905])

English: common crow-berry

Southern Africa: gewone kraaibessie, thornveld taaibos; iNhlokoshiyane, inHlokoshiyane (Zulu); mutasiri (Venda)

R. pyroides Burch. var. *pyroides* (*Rhus baurii* Schönl.; *Rhus intermedia* Schönl.; *Rhus vulgaris* Meikle) (Greek *pyr* "fire" and *eidos*, *oides* "resemblance")

English: common wild currant, fire thorn

Southern Africa: gewone taaibos, brandtaaibos, rooi karee-boom, taaibos; iNhlokoshiyane, inHlokoshiyane (Zulu); koditshane (South Sotho); mogodiri (Hebron dialect, central Transvaal); mogodiri, mogwediri (Ngwaketse dialect, Botswana); modupaphiri (Tawana dialect, Ngamiland); mogodiri, mogweriri (North Sotho)

R. radicans L. (*Rhus toxicodendron* var. *radicans* (L.) Torr.; *Rhus toxicodendron* f. *radicans* (L.) Engler; *Toxicodendron radicans* (L.) O. Ktze; *Rhus toxicodendron* f. *radicans* (L.) McNair; *Rhus toxicodendron* subsp. *radicans* (L.) R.T. Clausen)

English: poison ivy

Mexico: dominguilla, guau, hiedra, hiedra mala, hiedra venenosa, hincha huevos, mala mujer, zumaque, sumaque; mexie, meye (Otomì l., Hidalgo); fuego (Pueblo Nuevo, Solistahuacan, Chiapas); betz-tzaj (Huasteca l., southeast San Luis Potosí; bemberecua, huembereua (Michoacán); chechen (Yucatan); guadalagua (Jalisco); lachi-golilla, lachi-cobilla, yaga-beche-topa, yaga-peche-topa (Zapoteca l., Oaxaca; see Laura Nader, *Harmony Ideology: Justice and Control in a Zapotec Mountain Village*. Stanford 1990)

R. rehmanniana Engl. (*Rhus macowanii* Schonl.; *Rhus macowanii* Schonl. forma *rehmanniana* (Engl.) Schönl.; *Rhus vulgaris* auct. mult. non Meike) (named for the Aus-trian-Polish plant collector Anton(i) Rehmann (Rehman), 1840-1917, botanist, geographer, author of *Einige Notizen über die Vegetation der nördlichen Gestade des Schwarzen Meeres*. Brünn 1871 [1872]; see J.H. Barnhart, *Biographi-cal Notes upon Botanists*. 3: 139. 1965; Stafleu and Cowan, *Taxonomic Literature*. 4: 655-656. 1983; G. Murray, *His-tory of the Collections Contained in the Natural History Departments of the British Museum*. 1: 176. 1904; T.W. Bossert, *Biographical Dictionary of Botanists Represented in the Hunt Institute Portrait Collection*. 328. 1972; H.N. Clokie, *Account of the Herbaria of the Department of Bot-any in the University of Oxford*. 230. 1964; R. Zander, F. Encke, G. Buchheim and S. Seybold, *Handwörterbuch der Pflanzennamen*. 14. Aufl. 1993; Mary Gunn and Leslie E.

Codd, *Botanical Exploration of Southern Africa*. 292-294. 1981)

English: blunt-leaved currant

Southern Africa: stompblaartaaibos, MacOwan's taaibos, Rehmann's taaibos; iNhlokoshiyane, inHlokoshiyane (Zulu); inTlokolotshane ephakathi (Xhosa); inHlanguts-hane (Swazi)

R. sandwicensis A. Gray (*Rhus chinensis* Mill. var. *sand-wicensis* (A. Gray) Degener & Greenwell; *Rhus semialata* J.A. Murray var. *sandwicensis* Engl.)

India: tatri

Hawaii: neleau, neneleau

R. schiedeana Schlechtend. (*Toxicodendron schiedeanum* (Schlechtend.) O. Ktze; *Schmaltzia schiedeana* (Schlecht-end.) Barkley)

Mexico: pajulul, palo de pajulul (Las Margaritas, Chiapas); agrin, palo agrin (Berriozabal, Chiapas)

R. striata Ruíz & Pav. (*Toxicodendron striata* O. Ktze)

Mexico: mala mujer; yagalache, hincha huevos (Zapoteca l., Oaxaca); cuyulté (Tzotzil, Simojovel, Chiapas); amté (Tojolobal l., Montebello, Chiapas); yayalché (El Cafetal, Oaxaca); palo de viruela (Simojovel, Chiapas)

R. succedanea L. (*Rhus succedanea* var. *japonica* Engl.)

English: wax tree, Japanese wax tree, Japanese lacquer tree, wild varnish tree, galls

South Africa: wasboom, sumac, scarlet rhus

East Asia: haze

China: lin bei zi, huang lu

Nepal: rani bhaloyo, rani bhalayo, rani walai, raniwhalayo

India: sumak, kakra-singi, kakada-shingi, kakarashingi, kak-kata-shingi, kakadsingi, takada-singi, kakur-singhi, karkata-shringi, karkatashringi, kakrasringi, kakeera-sryngi, kakain, karkkaada-gasurgi, choklu, hala, holashi, lakhar, rikhul, habatul-khizra, shah, ding-keon, bol-micheng, bol-khat-thi, serhnyok, raniwhalayo

Japan: haze-no-ki, haze, Ryûkyû-fushi-no-ki, hajigi

R. tenuinervis Engl. (*Rhus commiphoroides* Engl. & Gilg.)

Southern Africa: hyaena taaibos; modupaphiri (= scented by hyaena), morupapiri (Western Transvaal, northern Cape, Botswana)

R. tepetate Standl. & Barkley (*Schmaltzia tepetate* (Standl. & Barkley) Barkley)

Mexico: tepetate (Sierra de Charuco, Sonora)

R. terebinthifolia Schlecht. & Cham. (*Toxicodendron tere-binthifolium* (Schlecht. & Cham.) O. Ktze; *Schmaltzia tere-binthifolia* (Schlecht. & Cham.) Barkley)

Mexico: catzuundu, sal de venado; hierba del temazcal, temazcal (Oaxaca); jaguay, paguay (Sinaloa); zumaqui cimarron (Chiapas); li-mu-le-ma-fou-nol (Chontal l., Oaxaca); yagabiche, yagabichi, yaga-exee (Zapoteca l., Oaxaca); temazcalchihual (Veracruz)

R. tomentosa L.

English: real wild currant, wild currant

Southern Africa: korentebos, korentebessie, hairy taaibos; umHlakothi (Zulu)

R. transvaalensis Engl. (*Rhus eburnea* Schonl.)

English: Transvaal currant, Transvaal rhus

Southern Africa: Transvaaltaaibos; iNhlokoshiyane (Zulu); mutzhaku-tzhaku, muthaku-thaku, mutasiri, tshitasiri (Venda)

R. trilobata Nutt. (*Schmaltzia trilobata* Small; *Toxicodendron triphyllum* var. *trilobatum* O. Kuntze; *Toxicodendron trilobatum* O. Kuntze)

English: skunkbush, skunkbrush

Mexico: lambrisco; agrito (Guadalcazar, San Luis Potosí); landrisco (Tamaulipas)

R. tumulicola S. Moore (*Rhus culminum* R. & A. Fernandes; *Rhus dura* Schönl.; *Rhus synstylica* R. & A. Fernandes var. *synstylica*)

English: hard-leaved currant

South Africa: hardetaaibos, hard taaibos

R. typhina L. (*Rhus hirta* Sudw., not Engl.; *Rhus hirta* var. *typhina* Farwell; *Toxicodendron typhinum* O. Ktze)

English: sumach, dyer's sumach, staghorn sumach, Virginian sumac, velvet sumac

Canada: sumac amaranthe, sumac de Virginie, vinaigrier

R. undulata Jacq. (*Rhus undulata* Jacq. var. *celastroides* (Sond.) Schönl. p.p.; *Rhus undulata* Jacq. var. *genuina* Schönl. forma *contracta* Schönl. p.p.; *Rhus undulata* Jacq. var. *genuina* Schönl. forma *excisa* (Thunb.) Schönl. p.p.; *Rhus undulata* Jacq. var. *genuina* Schönl. forma *undulata* Schönl.; *Rhus undulata* Jacq. var. *undulata* Schönl. forma *contracta* (Schönl.) R. Fernandes p.p.)

South Africa: kunibos, blinkblaar taaibos

R. verniciflua Stokes

English: Japanese varnish tree, lacquer tree, Japanese lacquer tree, Chinese lacquer tree

Japan: urushi

China: qan qi, kan chi, chi

R. vernix L.

English: poison elder

R. virens Lindh. (*Schmaltzia virens* Small)

English: evergreen sumac, tobacco sumac, lentisco

Mexico: lentisco; lambrisco, lantrisco (Tamaulipas); capulin (Durango)

R. zeyheri Sond. (*Rhus zeyheri* Sond. var. *parvifolia* Burtt Davy) (after the German botanist and collector Carl Ludwig

Philip(p) Zeyher, 1799-1858. He published, with Christian Friedrich Ecklon (1795-1868), *Enumeratio plantarum Africae Australis* 1835-1837)

South Africa: bloublaar taaibos

Rhyacophila Hochst. Lythraceae

Origins:
Greek *rhyax, rhyakos* "rushing stream, mountain torrent, torrent" and *philos* "lover, loving."

Rhynchadenia A. Richard Orchidaceae

Origins:
Greek *rhynchos* "beak" and *aden* "gland," indicating the shape of the elongate rostellum, see also *Macradenia* R. Br.

Rhynchandra Reichb.f. Orchidaceae

Origins:
Greek *rhynchos* "beak" and *aner, andros* "male, man, stamen," the pointed rostellum, see also *Corymborkis* Thouars.

Rhynchanthera Blume Orchidaceae

Origins:
From the Greek *rhynchos* and *aner, andros* "male, man, stamen," the shape of the anther, see also *Corymborkis* Thouars.

Rhynchanthera DC. Melastomataceae

Origins:
Beaked anthers, from the Greek *rhynchos* "horn, beak" and *anthera* "anther."

Rhynchanthera F. Muell. ex Benth. Asclepiadaceae

Origins:
Orthographic variant of *Rhyncharrhena* F. Muell.; see Arthur D. Chapman, ed., *Australian Plant Name Index*. 2513. 1991

Rhynchanthus Hook.f. Zingiberaceae

Origins:
From the Greek *rhynchos* and *anthos* "flower."

Rhyncharrhena F. Muell. Asclepiadaceae

Origins:

Greek *rhynchos* and *arrhen* "male," referring to the inner appendages; F. von Mueller, *Fragmenta Phytographiae Australiae*. 1: 128. (Apr.) 1859.

Species/Vernacular Names:
R. linearis (Decne.) K.L. Wilson (*Pentatropis linearis* Decne.; *Pentatropis atropurpurea* (F. Muell.) Benth.; *Pentatropis kempeana* F. Muell.; *Rhyncharrhena atropurpurea* F. Muell.; *Daemia kempeana* (F. Muell.) F. Muell.)

English: climbing purple-star

Rhynchelythrum Nees Gramineae

Origins:
See *Rhynchelytrum* Nees.

Rhynchelytrum Nees Gramineae

Origins:

Greek *rhynchos* "horn, beak, snout" and *elytron* "sheath, cover, scale, husk," referring to the upper glume; see John Lindley, *A Natural System of Botany*. Second edition. 446. London (Oct.) 1836.

Species/Vernacular Names:
R. nerviglume (Franch.) Chiov.

English: bristle-leaved red top, red top grass

Southern Africa: blinkgras, ferweelgras, steekblaarblinkgras; letsoiri-le-lenyenyane (Sotho)

R. repens (Willd.) C.E. Hubb. (*Saccharum repens* Willd.; *Tricholaena repens* (Willd.) Hitchc.)

English: Natal grass, red Natal grass, Natal red top grass, Natal red top, red top grass, ruby grass, fairy grass, red top

Yoruba: eeran eye, owu, sokodoya

South Africa: bergrooigras, blinkgras, ferweelgras, haargras, hangegras, kopersaadgras, Natal blinkgras, Natalse rooipluim, rooihaargras, rooiwolsaadgras, wolgras

Rhynchocalyx Oliver Rhynchocalycaceae (Crypteroniaceae, Rosidae, Myrtales)

Origins:
Greek *rhynchos* "horn, beak, snout" and *kalyx, kalykos* "calyx."

Species/Vernacular Names:
R. lawsonioides Oliv. (the specific name means that the tree resembles a member of the genus *Lawsonia*)

English: Natal privet, rhynchocalyx

South Africa: Natalliguster

Rhynchocarpa Backer ex Heyne Caesalpiniaceae

Origins:
Greek *rhynchos* "horn, beak, snout" and *karpos* "fruit."

Rhynchocarpa Becc. Palmae

Origins:

Greek *rhynchos* "horn, beak, snout" and *karpos* "fruit," referring to the irregularly shouldered and pebbled fruit, to the sculptured endocarp.

Rhynchocarpa Schrad. ex Endl. Cucurbitaceae

Origins:

Greek *rhynchos* and *karpos* "fruit," see also *Kedrostis* Medik.

Rhynchocladium T. Koyama Cyperaceae

Origins:

Greek *rhynchos* and *kladion* "a branchlet, a small branch," *klados* "a branch."

Rhynchocorys Griseb. Scrophulariaceae

Origins:
Greek *rhynchos* and *korys, korythos* "helmet."

Rhynchodia Benth. Apocynaceae

Origins:
Referring to the Greek *rhynchos* "beak, snout."

Rhynchodium C. Presl Fabaceae

Origins:

Like a *rhynchos* "beak, snout."

Rhynchoglossum Blume Gesneriaceae

Origins:

From the Greek *rhynchos* "beak, snout" and *glossa* "tongue."

Rhynchogyna Seidenf. & Garay Orchidaceae

Origins:

From the Greek *rhynchos* "horn, beak, snout" and *gyne* "female, woman, female organs."

Rhyncholacis Tul. Podostemaceae

Origins:

Greek *rhynchos* and genus *Lacis* Lindl.

Rhyncholaelia Schltr. Orchidaceae

Origins:

Greek *rhynchos* "horn, snout" plus the closely allied orchid genus *Laelia*, referring to the rostrate fruit.

Rhynchopera Klotzsch Orchidaceae

Origins:

From the Greek *rhynchos* "horn, snout" and *pera* "a pouch," perhaps referring to the column, the lip is basally concave.

Rhynchophora Arènes Malpighiaceae

Origins:

Greek *rhynchos* and *phoros* "bearing, carrying."

Rhynchophreatia Schltr. Orchidaceae

Origins:

Greek *rhynchos* and the genus *Phreatia* Lindl., referring to the long rostellum.

Rhynchopsidium DC. & A. DC. Asteraceae

Origins:

Greek *rhynchos* "horn, beak, snout."

Rhynchopyle Engl. Araceae

Origins:

Greek *rhynchos* and *pyle* "door, gate," see also *Piptospatha* N.E. Br., see D.H. Nicolson, "Derivation of aroid generic names." *Aroideana.* 10: 15-25. 1988.

Rhynchoryza Baillon Gramineae

Origins:

From the Greek *rhynchos* "horn, beak, snout" and *oryza* "rice," referring to the beaked lemma.

Rhynchosia Lour. Fabaceae

Origins:

Greek *rhynchos* "horn, beak, snout," referring to the style or the flowers with beaked keels; see J. de Loureiro, *Flora cochinchinensis.* 460. [Lisboa] (Sept.) 1790.

Species/Vernacular Names:

R. sp.

Yoruba: ajadi

R. minima (L.) DC. (*Dolichos minimus* L.; *Dolicholus minimus* (L.) Medikus; *Rhynchosia candollei* Decne.; *Rhynchosia rhombifolia* (Willd.) DC. var. *timoriensis* DC.)

English: rhynchosia, rhyncho, ryncho

Japan: hime-no-azuki

R. minima (L.) DC. var. *australis* (Benth.) C. Moore & E. Betche

English: native rock trefoil

R. villosa (Meisn.) Druce (*Sigmodostyles villosa* Meisn.; *Rhynchosia sigmoides* Benth. ex Harv.; *Eriosema villosum* (Meisn.) C.A. Sm.)

English: hairy rhynchosia

R. volubilis Lour.

Japan: tankiri-mame

China: lu huo, lu tou, lao tou, yeh lu tou

Rhynchosida Fryxell Malvaceae

Origins:

From the Greek *rhynchos* plus the genus *Sida* L.

Rhynchosinapis Hayek Brassicaceae

Origins:

From the Greek *rhynchos* "horn, snout" plus the genus *Sinapis* L.

Rhynchospermum Reinw. Asteraceae

Origins:

From the Greek *rhynchos* "horn, beak, snout" and *sperma* "seed."

Rhynchospora Vahl Cyperaceae

Origins:

Greek *rhynchos* and *spora, sporos* "seed, spore," referring to the achenes; see Martin H. Vahl (1749-1804), *M. Vahlii ... Enumeratio Plantarum.* 2: 229. Hauniae (& Lipsiae) 1806.

Species/Vernacular Names:
R. sp.
Hawaii: mau'u, kaluhaluha, kuolohia
R. alba (L.) Vahl
English: white beaked rush
R. californica Gale
English: California beaked rush
R. chinensis Nees
Japan: inu-no-hana-hige
R. corymbosa (L.) Britton (*Scirpus corymbosus* L.)
Japan: Yaeyama-abura-suge, oni-no-hige
Yoruba: ajitadi, ewe adi
R. nervosa (Vahl) Boeck
English: star rush, star grass, beak rush
R. rubra (Lour.) Makino (*Schoenus ruber* Lour.)
Japan: iga-kusa
R. rugosa (Vahl) Gale (*Schoenus rugosa* Vahl)
Hawaii: pu'uko'a

Rhynchostele Reichb.f. Orchidaceae

Origins:

Greek *rhynchos* "snout" and *stele* "a pillar, trunk, central part of stem," referring to the pseudobulbous stems or to the elongated rostellum, see also *Leochilus* Knowles & Westc.

Rhynchostemon Steetz Sterculiaceae

Origins:

Greek *rhynchos* and *stemon* "stamen"; see Johann G.C. Lehmann, *Plantae Preissianae.* 2: 333. 1848.

Rhynchostigma Benth. Asclepiadaceae

Origins:

From the Greek *rhynchos* "horn, beak, snout" and *stigma* "a stigma."

Rhynchostylis Blume Orchidaceae

Origins:

Greek *rhynchos* and *stylos* "a pillar, column," referring to the shape of the column or the flowers.

Species/Vernacular Names:
R. coelestis Reichb.f.
Thailand: ueang kao kae
R. gigantea (Lindley) Ridl.
Thailand: chang dang, chang kra, chang puek
R. retusa Blume
English: foxtail orchid
Thailand: iyaret

Rhynchotechum Blume Gesneriaceae

Origins:

Greek *rhynchos* "horn, a beak" and *theke* "a box, case, capsule," referring to the fruits.

Rhynchotheca Ruíz & Pavón Geraniaceae (Ledocarpaceae, Rhynchothecaceae)

Origins:

From the Greek *rhynchos* and *theke* "a box, case, capsule."

Rhynchotoechum Blume Gesneriaceae

Origins:

From the Greek *rhynchos* "horn, a beak" and *theke* "a box, case, capsule," referring to the fruits, see also *Rhynchotechum.*

Rhynchotropis Harms Fabaceae

Origins:

From the Greek *rhynchos* "horn, a beak" and *tropis, tropidos* "a keel."

Rhysolepis S.F. Blake Asteraceae

Origins:

From the Greek *rhysos* "wrinkled" and *lepis* "a scale."

Rhysopterus J.M. Coulter & Rose Umbelliferae

Origins:

From the Greek *rhysos* "wrinkled" and *pteron* "a wing."

Rhysospermum C.F. Gaertner Oleaceae

Origins:

From the Greek *rhysos, rhyssos* "wrinkled, shrivelled" and *sperma* "a seed"; see Carl (Karl) Friedrich von Gaertner (1772-1850), *Supplementum carpologicae.* 3, t. 224, fig. 2. Leipzig 1807.

Rhysotoechia Radlk. Sapindaceae

Origins:

Greek *rhysos, rhyssos* and *toichos* "a wall," possibly referring to the wrinkled fruit walls; see L.A.T. Radlkofer, ["Ueber die Sapindaceen Holländisch-Indiens."] in *Actes du Congrès International de Botanistes.* Amsterdam (for 1877). 131. 1879.

Rhyssocarpus Endl. Rubiaceae

Origins:

From the Greek *rhyssos* and *karpos* "fruit."

Rhyssolobium E. Meyer Asclepiadaceae

Origins:

From the Greek *rhyssos* and *lobion, lobos* "lobe, pod, small pod, fruit," referring to the follicles.

Rhyssopteris Blume ex A. Juss. Malpighiaceae

Origins:

Greek *rhysos, rhyssos* "wrinkled, shrivelled" and *pteryx* "wing," referring to the fruit; see Arthur D. Chapman, ed., *Australian Plant Name Index.* 2518. Canberra 1991.

Species/Vernacular Names:

R. timoriensis (DC.) Adr. Juss. (*Banisteria timorense* DC.)

Japan: Sasaki-kazura (possibly named to honor Syun'iti Sasaki, 1888-1960, botanist and plant collector in Taiwan, author of "A catalogue of the Government Herbarium." *Rep. Res. Inst. Formosa.* 9: 1-592. 1930)

Rhyssopterys Blume ex A. Juss. Malpighiaceae

Origins:

Greek *rhyssos* "wrinkled" and *pteryx* "wing," referring to the fruit carpels or samaras; see Jules Paul Benjamin Delessert (1773-1847), *Icones selectae Plantarum.* 3: 21. Paris 1838; Arthur D. Chapman, ed., *Australian Plant Name Index.* 2518. 1991.

Rhyssostelma Decne. Asclepiadaceae

Origins:

From the Greek *rhyssos* and *stelma, stelmatos* (*stello* "to bring together, to bind, to set") "a girdle, belt."

Rhytachne Desv. Gramineae

Origins:

From the Greek *rhytis* "a wrinkle" and *achne* "chaff, glume."

Rhyticalymma Bremek. Acanthaceae

Origins:

From the Greek *rhytis* "a wrinkle" and *kalymma* "a covering."

Rhyticarpus Sonder Umbelliferae

Origins:

From the Greek *rhytis* and *karpos* "a fruit"; see Constantine Samuel Rafinesque (1783-1840), *The Good Book.* 56. Philadelphia 1840; E.D. Merrill, *Index rafinesquianus.* 178. 1949.

Rhyticaryum Beccari Icacinaceae

Origins:

Orthographic variant of *Ryticaryum* Beccari, Greek *rhytis* "a wrinkle" and *karyon* "a nut."

Rhyticocos Beccari Palmae

Origins:
From the Greek *rhytis* plus *Cocos*, referring to the albumen of the seed.

Rhytidachne K. Schumann Gramineae

Origins:
From the Greek *rhytis, rhytidos* "a wrinkle" and *achne* "chaff, glume," see also *Rhytachne* Desv.

Rhytidandra A. Gray Alangiaceae

Origins:
Greek *rhytis, rhytidos* "a wrinkle" and *aner, andros* "male"; see A. Gray, in *United States Exploring Expedition ... 1838-1842*. Under the command of C. Wilkes. Botany. *Phanerogamia*. [Phanerogamia collected on the islands and shores of the Pacific Ocean.] 1: 302, fig. 28. Philadelphia 1854.

Rhytidanthe Benth. Asteraceae

Origins:
Greek *rhytis, rhytidos* "a wrinkle" and *anthos* "flower"; see George Bentham et al., *Enumeratio Plantarum quas in Novae Hollandiae ...* collegit C. de Hügel. 63. Vindobonae 1837.

Rhytidanthera (Planchon) Tieghem Ochnaceae

Origins:
From the Greek *rhytis, rhytidos* "a wrinkle" and *anthera* "anther."

Rhytidocaulon Bally Asclepiadaceae

Origins:
From the Greek *rhytis, rhytidos* "a wrinkle" and *kaulos* "a stem, a branch, stalk."

Rhytidomene Rydberg Fabaceae

Origins:
From the Greek *rhytis, rhytidos* and *mene* "moon, the crescent moon," referring to the fruit; see C.S. Rafinesque, *Atl.*

Jour. 1: 145. 1832; Elmer D. Merrill, *Index rafinesquianus.* 147. 1949.

Rhytidophyllum Martius Gesneriaceae

Origins:
From the Greek *rhytis, rhytidos* and *phyllon* "leaf," referring to the rugose leaves.

Rhytidosporum F. Muell. Pittosporaceae

Origins:
Greek *rhytis, rhytidos* and *spora* "seed"; see Sir Ferdinand Jacob Heinrich von Mueller (1825-1896), *The Plants Indigenous to the Colony of Victoria.* 1: 75. Melbourne 1860-1865.

Species/Vernacular Names:
R. procumbens (Hook.) F. Muell.
English: white marianth

Rhytidostylis Reichb. Cucurbitaceae

Origins:
From the Greek *rhytis, rhytidos* "a wrinkle" and *stylos* "style, column," see also *Rytidostylis* Hook. & Arn.

Rhytidotus Hook.f. Rubiaceae

Origins:
From the Greek *rhytis, rhytidos* "a wrinkle" and *ous, otos* "an ear."

Rhytiglossa Nees ex Lindl. Acanthaceae

Origins:
From the Greek *rhytis* and *glossa* "tongue," see also *Isoglossa* Oerst.

Rhytis Lour. Euphorbiaceae

Origins:
From the Greek *rhytis, rhytidos* "a wrinkle."

Rhytispermum Link Boraginaceae

Origins:
From the Greek *rhytis, rhytidos* "a wrinkle" and *sperma* "a seed."

Ribes L. Grossulariaceae (Saxifragaceae)

Origins:

Possibly from the Arabic name *ribas* "acid-tasting, sorrel, rhubarb"; see Carl Linnaeus, *Species Plantarum*. 200. 1753 and *Genera Plantarum*. Ed. 5. 94. 1754; Yuhanna ibn Sarabiyun [Joannes Serapion], *Liber aggregatus in medicinis simplicibus*. Venetijs 1479; Serapiom, *El libro agregà de Serapiom*. A cura di G. Ineichen. Venezia-Roma 1962-1966; G.B. Pellegrini, *Gli arabismi nelle lingue neolatine con speciale riguardo all'Italia*. Brescia 1972; E. Weekley, *An Etymological Dictionary of Modern English*. 2: 1235. 1967; Salvatore Battaglia, *Grande dizionario della lingua italiana*. XVI. 22. Torino 1995.

Species/Vernacular Names:

R. amarum McClatchie

English: bitter gooseberry

R. americanum Miller

English: American blackcurrant

R. aureum Pursh

English: golden currant, buffalo currant

R. binominatum A.A. Heller

English: trailing gooseberry

R. bracteosum Douglas ex Hook.

English: stink currant

R. californicum Hook. & Arn.

English: hillside gooseberry

R. canthariforme Wiggins

English: Moreno currant (near Moreno Dam, San Diego)

R. cereum Douglas

English: wax currant

R. curvatum Small

English: granite gooseberry

R. cynosbati L.

English: American gooseberry, prickly gooseberry

R. divaricatum Douglas

English: Worcesterberry

R. glandulosum Grauer ex Weber

English: skunk currant

R. grossularioides Maxim.

English: catberry

R. indecorum Eastw.

English: white flowering currant

R. inerme Rydb.

English: white-stemmed gooseberry

R. lacustre (Pers.) Poiret

English: swamp currant

R. laxiflorum Pursh

English: trailing black currant

R. lobbii A. Gray

English: gummy gooseberry

R. marshallii E. Greene

English: Marshall's gooseberry

R. menziesii Pursh

English: canyon gooseberry

R. montigenum McClatchie

English: mountain gooseberry

R. nevadense Kellogg

English: mountain pink currant

R. nigrum L.

English: blackcurrant, currants

India: nabar, papear, muradh, nabar beli, skaktekas

Italian: ribes nero

R. odoratum H.L. Wendl.

English: buffalo currant

R. quercetorum E. Greene

English: oak gooseberry

R. roezlii Regel

English: Sierra gooseberry

R. rubrum L.

English: redcurrant, garnetberry, red currants

India: dak, phulanch, kinkolia

R. sanguineum Pursh

English: red flowering currant, flowering currant, American currant

R. speciosum Pursh

English: fuchsia-flowered gooseberry, California fuchsia

R. tularense (Cov.) Fedde

English: sequoia gooseberry

R. uva-crispa L. (*Ribes grossularia* L.)

English: gooseberry, feaberry

India: amlanch, baikunti

R. viburnifolium A. Gray

English: evergreen currant

R. viscosissimum Pursh

English: sticky currant

Ribesiodes O. Kuntze Myrsinaceae

Origins:

The genus *Ribes* L. and *-odes* "resembling, of the nature of"; see Otto Kuntze (1843-1907), *Revisio generum Plantarum Vascularium*. 2: 403. Leipzig 1891.

Ribesium Medik. Grossulariaceae

Origins:

Referring to the genus *Ribes* L.

Ricardia Adans. Rubiaceae

Origins:

After the English botanist Richard Richardson, 1663-1741, physician, see also *Richardia* L.

Richardia Kunth Araceae

Origins:

After the French botanist Louis Claude Marie Richard, 1754-1821, explorer and traveler, horticulturist, naturalist and zoologist, botanical and zoological collector, a pupil of Bernard de Jussieu (1699-1777), from 1781 to 1785 in French Guyana and the Antilles (sent by Louis XVI on the recommendation of the Academy of Sciences), in 1785 in Brazil, from 1795 to 1821 professor of botany in the school of medicine of Paris, edited the fourth edition of Jean Baptiste François Bulliard (1752-1793), *Dictionnaire élémentaire de botanique*. Paris an vii [1798]. His works include *De Orchideis europaeis annotationes*. Parisiis 1817 and *De Musaceis*. Vratislaviae et Bonnae. 1831; see Antoine Laurent de Jussieu, Institut Royal de France. *Funérailles de M. Richard*. [Discourse pronounced by A.L.J.] [Paris 1821]; J.H. Barnhart, *Biographical Notes upon Botanists*. 3: 151. 1965; R. Zander, F. Encke, G. Buchheim & S. Seybold, *Handwörterbuch der Pflanzennamen*. 14 Aufl. Stuttgart 1993; Mariella Azzarello Di Misa, ed., *Il Fondo Antico della Biblioteca dell'Orto Botanico di Palermo*. 232. Palermo 1988; Stafleu and Cowan, *Taxonomic Literature*. 4: 764-767. 1983; C.S. Kunth, in *Mémoires du Muséum d'Histoire Naturelle*. 4: 433, t. 20. 1815; J.D. Milner, *Catalogue of Portraits of Botanists Exhibited in the Museums of the Royal Botanic Gardens*. Royal Botanic Gardens, Kew, London 1906; Samuel J. Hough and Penelope R. Hough, *The Beinecke Lesser Antilles Collection at Hamilton College: A Catalogue of Books, Manuscripts, Prints, Maps, and Drawings, 1521-1860*. Gainesville [1994]; T.W. Bossert, *Biographical Dictionary of Botanists Represented in the Hunt Institute Portrait Collection*. 331. 1972; E.M. Tucker, *Catalogue of the Library of the Arnold Arboretum of Harvard University*. 1917-1933; Jonas C. Dryander, *Catalogus bibliothecae historico-naturalis Josephi Banks*. London 1800; A. Lasègue, *Musée botanique de Benjamin Delessert*. 1845; Ida Kaplan Langman, *A Selected Guide to the Literature on the Flowering Plants of Mexico*. 625. 1964.

Richardia L. Rubiaceae

Origins:

After the English botanist Richard Richardson, 1663-1741 (d. North Bierly, Yorkshire), physician, M.D. Leyden 1690, collected lichens, antiquary and book collector, studied under Paul Hermann and Boerhaave, 1712 Fellow of the Royal Society, a friend of Sir John Franklin, plant hunter. See Dawson Turner (1775-1858), *Extracts from the Literary and Scientific Correspondence of R. Richardson, of Bierly, Yorkshire: Illustrative of the State and Progress of Botany* [Edited by D. Turner. — Extracted from the memoir of the Richardson family, by Mrs. D. Richardson] Yarmouth 1835; R. Pulteney, *Historical and Biographical Sketches of the Progress of Botany in England*. 2: 185-188. London 1790; Carl Linnaeus, *Species Plantarum*. 330. 1753 and *Genera Plantarum*. Ed. 5. 153. 1754; H.N. Clokie, *Account of the Herbaria of the Department of Botany in the University of Oxford*. 232. Oxford 1964; John H. Barnhart, *Biographical Notes upon Botanists*. 3: 152. 1965; Edm. and D.S. Berkeley, *John Clayton, Pioneer of American Botany*. Chapel Hill 1963; T. W. Bossert, *Biographical Dictionary of Botanists Represented in the Hunt Institute Portrait Collection*. 331. 1972; James Britten, *The Sloane Herbarium*, revised and edited by J.E. Dandy. 1958; Joseph Ewan, in *Dictionary of Scientific Biography* 1: 431-432. 1981; Ray Desmond, *Dictionary of British & Irish Botanists and Horticulturists*. 582. London 1994; Blanche Henrey, *No Ordinary Gardener — Thomas Knowlton, 1691-1781*. Edited by A.O. Chater. British Museum (Natural History), London 1986.

Species/Vernacular Names:

R. brasiliensis Gomes (*Richardsonia brasiliensis* (Gomes) Hayne)

English: Mexican clover, Mexican richardia, tropical richardia, Brazilian pusley

South Africa: Meksikaanse klawer, Meksikaanse richardia, tropiese richardia

R. humistrata (Cham. & Schltdl.) Steud. (*Richardsonia humistrata* Cham. & Schltdl.)

English: Peelton-richardia, Peelton weed

South Africa: Peelton-richardia

R. scabra L. (*Richardsonia scabra* (L.) St.-Hil. p.p.)

English: Mexican clover, Florida pusley

Richardsiella Elffers & Kennedy-O'Byrne Gramineae

Origins:

For the botanist Mary Alice Eleanor Richards (*née* Stokes), 1885-1977, collected African plants for Kew; see Ray Desmond, *Dictionary of British & Irish Botanists and*

Horticulturists. 581. London 1994; Stafleu and Cowan, *Taxonomic Literature*. 4: 769. [genus named for Mrs. H.M. Richards, collector of African plants.] Utrecht 1983.

Richardsonia Kunth Rubiaceae

Origins:

After the English botanist Richard Richardson, 1663-1741, physician, M.D. Leyden 1690, antiquary and book collector, in 1712 Fellow of the Royal Society, a friend of Sir John Franklin. See Dawson Turner (1775-1858), *Extracts from the Literary and Scientific Correspondence of R. Richardson, of Bierly, Yorkshire: Illustrative of the State and Progress of Botany* [Edited by D. Turner. — Extracted from the memoir of the Richardson family, by Mrs. D. Richardson] Yarmouth 1835; John H. Barnhart, *Biographical Notes upon Botanists*. 3: 152. 1965; C.S. Kunth, in *Mémoires du Muséum d'Histoire Naturelle*. 4: 430. Paris 1815.

Richea Labill. Epacridaceae

Origins:

After the French physician Claude Antoine Gaspard Riche, 1762-1797.

Richea R. Br. Epacridaceae

Origins:

Named after the French physician Claude Antoine Gaspard Riche, 1762-1797, naturalist, participant in the voyage of *Espérance* (commanded by Jean-Michel Huon de Kermadec, 1748-1793) with d'Entrecasteaux in search of the lost expedition of La Pérouse. Wrote *De Chemia Vegetabilium*. Avenione 1786-1787. See [Société Philomathique de Paris], *Rapports généraux des travaux de la société ...* par les citoyens Riche et Silvestre, etc. Paris [1793, etc.]; Robert Brown, *Prodromus florae Novae Hollandiae et Insulae van-Diemen*. 555. London 1810; Georges Léopold Chrétien Frédéric Dagobert Cuvier (1769-1832), *Recueil des éloges historiques* lus dans les séances publiques de l'Institut royal de France. Strasbourg & Paris 1819-1827; Elisabeth-Paul-Édouard de Rossel (1765-1829), ed., *Voyage de Dentrecasteaux Envoyé à la "Recherche" de la Pérouse*. Paris 1808; John Dunmore, *Who's Who in Pacific Navigation*. Honolulu 1991; Jonathan Wantrup, *Australian Rare Books, 1788-1900*. Hordern House, Sydney 1987.

Species/Vernacular Names:

R. acerosa (Lindley) F. Muell.

English: slender richea

R. angustifolia B.L. Burtt

English: narrow-leaf richea

R. continentis B.L. Burtt

English: candle heath, mainland richea

R. milliganii (Hook.f.) F. Muell.

English: Milligan's richea

R. procera (F. Muell.) F. Muell.

English: lowland richea

R. sprengelioides (R. Br.) F. Muell.

English: yellow richea

Richeria Vahl Euphorbiaceae

Origins:

For the French (b. Châlons-sur-Marne) botanist Pierre Richer de Belleval, *circa* 1564-1632 (d. Montpellier), physician, M.D. Avignon 1587, horticulturist, 1593 one of the founders of the Jardin des Plantes de Montpellier, wrote *Opuscules ... auxquels on a joint un traité d'Olivier de Serres sur la manière de travailler l'écorce du mûrier blanc*. Nouvelle édition ... par Broussonet. Paris 1785; see John H. Barnhart, *Biographical Notes upon Botanists*. 3: 153. 1965; Antoine Lasègue (1793-1873), *Musée botanique de M. Benjamin Delessert*. 518. Paris, Leipzig 1845; Jacques Anselme d'Orthes, *Éloge historique* de P. Richer de Belleval, Instituteur du Jardin-Royal de Botanique de Montpellier sous Henri iv. Mémoire qui a remporté le prix de la Société Royale de Sciences en 1788. Montpellier 1788; Louis Dulieu, in *Dictionary of Scientific Biography* 1: 592. 1981.

Richeriella Pax & K. Hoffmann Euphorbiaceae

Origins:

The diminutive of *Richeria* Vahl.

Richteria Karelin & Kirilov Asteraceae

Origins:

For Alexander Richter, see Stafleu and Cowan, *Taxonomic Literature*. 4: 776. 1983.

Ricinella Müll. Arg. Euphorbiaceae

Origins:

The diminutive of the genus *Ricinus*.

Ricinocarpos Desf. Euphorbiaceae

Origins:

The genus *Ricinus* plus the Greek *karpos* "a fruit," referring to the fruits; see René Louiche Desfontaines (1750-1833), in *Mémoires du Muséum d'Histoire Naturelle*. 3: 459, t. 22. 1817.

Species/Vernacular Names:

R. pinifolia Desf.

English: wedding bush

R. velutinus F. Muell.

English: velvety wedding bush

Ricinocarpus Kuntze Euphorbiaceae

Origins:

The genus *Ricinus* and Greek *karpos* "a fruit."

Ricinodendron Müll. Arg. Euphorbiaceae

Origins:

Ricinus plus the Greek *dendron* "a tree," referring to the leaves.

Species/Vernacular Names:

R. sp.

Nigeria: okwein

N. Rhodesia: mulela, mungongo

R. heudelotii (Baillon) Heckel subsp. *africanum* (Müll. Arg.) J. Léonard (*Ricinodendron africanum* Müll. Arg.)

Cameroon: njangsang, nyangsang, njansang, bosisang, esang, essang, essesang, essessang, wonjangasanga, wonjasanga, djansan, andjejang, avemp, eko, n'gaha, gobo, n'zoo aille, osok, timboa mounganga, njouli

Central Africa: musodo

Congo: sanga sanga

Gabon: essegang, essang, essasang, essessang, essesang, engessan, ensesang, enzesang, esoang, osongosongo, nsasanga, gesanga, gesantgala, issanguila, mosongo, mugela, mundzangala, mungembe, ngele, ozaneguilia, ozangilya

Ghana: anwama, onwama, owama, onwana, ngwama, nwuama, wama, wamba, oua oua, assomah, epuwi, ewai, ewan, kpedi, okao koodo, okuduru, sosali

Ivory Coast: eho, agqui, haipi, hapi, hobo, hoho, mbob, nbor, oho, poposi, popossi, sosau, tsain

Liberia: gbolei, koor

Yoruba: erinmado, erinmodan, oro, omodan, potopoto, putuputu, ologbo igbo, ajagbo

Nigeria: bonjasanga, bonjangsang, esang, es-es-ang, ishangi, bissi, bisse, okhui, okue, okwe, okwen, okhuen, okhuen nebo, okwenseba, erinmodan, okingbo, okengbo, okponum, okpunum, jan-jang, omodon, ovovo, poposi, putu-putu, potopoto, putuputu, wawankurmi, ogbodo, wonjangasanga, eke, ekwo, erimado, erinmado, isisang, njansang, nsa-sana, gayio kimi, awmawdan, ookwe; wawanputu kurmi (Hausa); erinmado (Yoruba); okhuen (Edo); eke (Urhobo); okengbo (Ijaw); okwe (Igbo); okue (Itsekiri)

Uganda: kishongo, musodo

W. Africa: assomah, bon jasanga, kingele

Zaire: moboto, peke, kingela, kitililundu, mongongome, mulela, sanga-sanga, musonga

Ricinoides Moench Euphorbiaceae

Origins:

Resembling the genus *Ricinus* L.

Ricinus L. Euphorbiaceae

Origins:

Latin *ricinus* for a tick, a louse, a kind of vermin that infests sheep and dogs, etc. (see Marcus Porcius Cato and Marcus Terentius Varro), referring to the seeds; C. Plinius Secundus used the name for this genus, for a plant called also *cici* and *croton*, and for the germ of the mulberry; Latin *cicinus* "an aperient oil expressed from the fruit of the cici, castor-oil," *cici* "an Egyptian tree, *palma Christi* or castor-oil tree, also called *croton*"; Greek *kiki* "castor-oil, the castor-oil tree," *kikinos* "made from the kiki-tree," *kikinon elaion*; see Carl Linnaeus, *Species Plantarum*. 1007. 1753 and *Genera Plantarum*. Ed. 5. 437. 1754; George Bennett (1804-1893), *Gatherings of a Naturalist in Australasia*. 332. London 1860; G.P. Bergantini, *Voci italiane d'autori approvati dalla Crusca*, nel Vocabolario d'essa non registrate ... Venezia 1745; Salvatore Battaglia, *Grande dizionario della lingua italiana*. XVI: 98. Torino 1995.

Species/Vernacular Names:

R. communis L.

English: castor oil, castor bean, castor bean tree, castor oil plant, castor oil tree, castor oil bush, palma Christi, wonder tree, red castor oil plant

Arabic: arash, kharwa, kherwa', shemouga

Peru: girgilla, higerilla negra, higuereta, higuerilla, higuerillo, mamona, ricina, ricino

Brazil: carrapateiro, carapateiro, carrapateira, balureira, figo-do-inferno, mamoeira, mamoneira, mamona, rícino, figueira do inferno, nhambuguaçu

Mexico: higuerilla, higuera del diablo, palma Cristi, ricino; tsajtüima'ant (Mixe language, San Lucas Camotlan, Oaxaca); thiquela (Huasteca 1., southeast San Luis Potosí); alpai-ue (Chontal 1., Oaxaca); yuntu-nduchi-dzaha, nduchidxaha (Mixteca 1., Oaxaca); chashilandacui (Zoque 1., Tapalapa, Chiapas); cashtilenque (Totonaca 1., El Tajin, Veracruz); k'ooch, xkoch, x-k'ooch (Maya 1., Yucatan); degha (Otomi 1., Hidalgo); guechi-beyo, quechi-peyo-Castilla, yaga-bilape, yaga-queze-aho, yaga-higo, yaga-hiigo, yaga-hijco (Zapoteca 1., Oaxaca); tzapolotl (Azteca 1., southeast San Luis Potosí); quebe'enogua (Sonora); xoxapajtzi (Mexican dialect, Tetelcingo)

The Philippines: hawa, tañgan-tañgan, tangan-tangan, tantangan, sina, lansina, lingang-sina, taca-taca, tawa-tawa, taua-taua, casla, katana, gatlaoua

Japan: tô-goma, chanda-kashi

China: bi ma zi, pei ma

Vietnam: ma puong si, thau dau, ty ma

Tibetan: dan-khra

India: digherandi, bindi, risa jaradaru, ranga bindi, ranga-nedjara, era, karbis, erandah, erand, arid, avanakku, velutta avanakku, amanakku, avanacu, avanacoe, citavanacu, raktairandah, cuvanna avanakku, lal erand, pandi avanacu

Nepal: ander, dhatura

Malaya: jarak, jarak besar, minyak jarak

Hawaii: pa'aila, ka'apeha, kamakou, koli, la'au 'aila

Madagascar: kimanga, kinamena, kinana, tanantanamanga, tanatanamanga, tanatanamena, tseroka, ricin, palma Christi

Nigeria: zurma

Sierra Leone: kasta weh, ngele bondoi

Southern Africa: kasterolieboom, bloubottelboom, bosluisboom, olieboom, oliepitboom, wonderboom; nhlampfura (Tsonga); mufuta, muFude (Shona); muplure (Venda); mbariki (Swahili); mohlafotha (Sotho); mokhura (North Sotho); mokhura (Pedi); umHlakuva (Zulu); umfude (Ndebele); umhlakuva (Xhosa)

Yoruba: lara, lara pupa, ilara, ilarun, lapalapa adete, ara pupa

Riddelia Raf. Sterculiaceae

Origins:

After the American botanist John Leonard Riddell, 1807-1865; see C.S. Rafinesque, *New Fl. N. Am.* 4: 15. 1836 [1838]; E.D. Merrill, *Index rafinesquianus*. The plant names published by C.S. Rafinesque, etc. 166, 167. 1949.

Riddellia Nutt. Asteraceae

Origins:

After the American botanist John Leonard Riddell, 1807-1865, physician; see John H. Barnhart, *Biographical Notes upon Botanists*. 3: 155. 1965; H.N. Clokie, *Account of the Herbaria of the Department of Botany in the University of Oxford*. 232. Oxford 1964; E.M. Tucker, *Catalogue of the Library of the Arnold Arboretum of Harvard University*. 1917-1933; Howard Atwood Kelly and Walter Lincoln Burrage, *Dictionary of American Medical Biography*. New York 1928; S. Lenley et al., *Catalog of the Manuscript and Archival Collections and Index to the Correspondence of John Torrey*. Library of the New York Botanical Garden. 1973; J. Ewan, ed., *A Short History of Botany in the United States*. New York and London 1969; R. Zander, F. Encke, G. Buchheim & S. Seybold, *Handwörterbuch der Pflanzennamen*. 14 Aufl. Stuttgart 1993.

Riddellia Raf. Sterculiaceae

Origins:

For the American botanist John Leonard Riddell, 1807-1865, physician; see Constantine S. Rafinesque, *Sylva Telluriana*. 113. 1838, *Autikon botanikon*. Icones plantarum select. nov. vel rariorum, etc. 5. Philadelphia 1840 and *The Good Book*. 46. 1840.

Ridleya (Hook.f.) Pfitzer Orchidaceae

Origins:

For the British botanist Henry Nicholas Ridley, 1855-1956, British Museum (Botany Department 1880-1888), from 1888 to 1912 Director of the Botanic Gardens of Singapore. His writings include *The Flora of the Malay Peninsula*. London 1922-1925; see John H. Barnhart, *Biographical Notes upon Botanists*. 3: 155. 1965.

Ridleyella Schlechter Orchidaceae

Origins:

For the British botanist Henry Nicholas Ridley, 1855-1956 (Kew, Surrey), plant collector, botanical explorer, traveler, Botany Department British Museum 1880-1888, in 1888 was appointed Singapore's first Scientific Director of the Botanic Gardens. He was largely responsible for establishing the rubber industry in the Malay Peninsula; he fathered the region's rubber industry by convincing Malaya's planters of the superior worth of *Hevea brasiliensis*. (Eleven seedlings arrived at the Singapore Botanical Garden where H.N. Ridley developed a method to propagate them rapidly and tap them for their lucrative white sap; today the method of collecting the latex from the trees is basically the same as when devised by H.N. Ridley over a century ago). His writings include *The Flora of the Malay Peninsula*. London 1922-1925, *The Dispersal of Plants Throughout the World*. Ashford, Kent 1930 and "A monograph of the genus *Liparis*." *J. Linn. Soc., Bot*. 22: 244-297. 1886; see John H. Barnhart, *Biographical Notes upon Botanists*. 3: 155. 1965; T. W. Bossert, *Biographical Dictionary of Botanists Represented in the Hunt Institute Portrait Collection*. 332. 1972; E.M. Tucker, *Catalogue of the Library of the Arnold Arboretum of Harvard University*. 1917-1933; S. Lenley et al., *Catalog of the Manuscript and Archival Collections and Index to the Correspondence of John Torrey*. Library of the New York Botanical Garden. 1973; Ida Kaplan Langman, *A Selected Guide to the Literature on the Flowering Plants of Mexico*. Philadelphia 1964; E.D. Merrill, *An Enumeration of Philippine Flowering Plants*. 4: 221-222. Manila 1925-1926; I.C. Hedge and J.M. Lamond, *Index of Collectors in the Edinburgh Herbarium*. Edinburgh 1970; G. Murray, *History of the Collections Contained in the Natural History Departments of the British Museum*. 177. London 1904; Ernest Nelmes and William Cuthbertson, *Curtis's Botanical Magazine Dedications, 1827-1927*. 314-316. [1931]; Merle A. Reinikka, *A History of the Orchid*. Timber Press 1996; R. Zander, F. Encke, G. Buchheim & S. Seybold, *Handwörterbuch der Pflanzennamen*. 14 Aufl. Stuttgart 1993.

Ridolfia Moris Umbelliferae

Origins:

After the Italian scientist Cosimo Ridolfi, 1794-1865 (Florence), statesman, botanist, friend of Gino Capponi (1792-1876), Raffaello Lambruschini (1788-1873) and Giovan Pietro Vieusseux (1779-1863), Director of the Museo di Fisica e Storia Naturale di Firenze. His writings include *Lezioni orali di agraria* ... Pubblicate ... per cura dell'Accademia Empolese di Scienze economiche. Firenze 1857, 1858 and *Saggio di Agrologia*. Firenze 1865; see Roberto Ridolfi (1899-1991), *Cosimo Ridolfi e gli istituti del suo tempo*. Firenze 1900 and *L'opera agraria di Cosimo Ridolfi*. Firenze 1903; F. Bettini, *Meleto, Cosimo Ridolfi e la scuola del lavoro*. Brescia 1941; R. Ciampini, *Due campagnoli dell'Ottocento: Lambruschini e Ridolfi*. Firenze 1947. According to Pritzel the genus was named for the Italian Marquis Cosimo Ridolfi, 1769-1844 (Pisa), author of *Catalogo delle piante coltivate a Bibbiani*, etc. Firenze 1843; J.H. Moris, *Enum. Seminum Regii Horti Bot. Taurin*. 1841: 43. Nov-Dec 1841.

Riedelia Oliver Zingiberaceae

Origins:

For a Mr. Riedel, see Stafleu and Cowan, *Taxonomic Literature*. 4: 789. Utrecht 1983.

Riedeliella Harms Fabaceae

Origins:

For the German traveler Ludwig Riedel, 1790-1861, plant collector in Brazil (with the German surgeon Georg Heinrich von Langsdorff, 1774-1852); see Carlos Augusto Taunay, *Manual do Agricultor Brazileiro*. Rio de Janeiro 1839; John H. Barnhart, *Biographical Notes upon Botanists*. 3: 156. Boston 1965; H.N. Clokie, *Account of the Herbaria of the Department of Botany in the University of Oxford*. 233. Oxford 1964; Antoine Lasègue, *Musée botanique de M. Benjamin Delessert*. Paris 1845; Günther Schmid, *Chamisso als Naturforscher. Eine Bibliographie*. Leipzig 1942; Stafleu and Cowan, *Taxonomic Literature*. 4: 789-790. 1983.

Riedlea Vent. Sterculiaceae

Origins:

After the French gardener Anselme Riedle, 1775-1801, at Paris Botanic Garden, from 1801 to 1804 chief gardener on Baudin's Expedition; see Étienne Pierre Ventenat (1757-1808), in *Mémoires de la classe des sciences mathématiques et physiques de l'Institut National de France*. 1: 2. Paris 1807.

Rigidella Lindley Iridaceae

Origins:

Latin *rigidus, a, um* "stiff, inflexible."

Rigiolepis Hook.f. Ericaceae

Origins:

From the Latin *rigeo* "to be stiff" and *lepis* "a scale," see also *Vaccinium* L.

Rigiopappus A. Gray Asteraceae

Origins:

Latin *rigeo* and *pappus*, stiff pappus.

Rigiostachys Planchon Simaroubaceae

Origins:

Latin *rigeo* "to be stiff" and Greek *stachys* "a spike."

Rikliella J. Raynal Cyperaceae

Origins:

Named for the Swiss botanist Martin Albert Rikli, 1868-1951, plant geographer, traveler and botanical collector, studied at Berlin under A. Engler and Simon Schwendener (1829-1919), from 1906 to 1930 Curator of the Botanical Museum of the E.T.H. Zürich. Among his numerous writings are *Die Arve in der Schweiz.* Zürich 1907 and *Das Pflanzenkleid der Mittelmeerländer.* Bern [1942-1948]. With Carl Schröter (1855-1939) wrote *Botanische Excursionen* im Bedretto, Formazza- und Bosco-Tal ... Zürich 1904 and *Vom Mittelmeer zum Nordrand der algerischen Sahara.* Zürich 1912; see John H. Barnhart, *Biographical Notes upon Botanists.* 3: 157. 1965; Jean Raynal (1933-1979), in *Adansonia.* Ser. 2, 13: 154. 1973; T.W. Bossert, *Biographical Dictionary of Botanists Represented in the Hunt Institute Portrait Collection.* 332. 1972; E.M. Tucker, *Catalogue of the Library of the Arnold Arboretum of Harvard University.* 1917-1933; R. Zander, F. Encke, G. Buchheim & S. Seybold, *Handwörterbuch der Pflanzennamen.* 14 Aufl. Stuttgart 1993.

Rimacola Rupp Orchidaceae

Origins:

Latin *rima, rimae* "a crack, cleft, fissure," referring to the habitat and to the habit of growing in rock fissures; see Herman Montague Rucker Rupp (1872-1956), in *The Victorian Naturalist.* 58: 188. (Apr.) 1942.

Species/Vernacular Names:

R. elliptica (R. Br.) Rupp (*Lyperanthus ellipticus* R. Br.)

English: green rock-orchid

Rimaria L. Bolus Aizoaceae

Origins:

Latin *rima, rimae* "a crack, cleft, fissure."

Rimaria N.E. Br. Aizoaceae

Origins:

Latin *rima, rimae* "a crack, cleft, fissure," possibly referring to the habitat.

Rindera Pallas Boraginaceae

Origins:

For A. Rinder, Russian physician and plant collector; see G.C. Wittstein, *Etymologisch-botanisches Handwörterbuch.* 767. Ansbach 1852.

Species/Vernacular Names:

R. tetraspis Pallas

China: chi guo cao

Ringentiarum Nakai Araceae

Origins:

Latin *ringor, rictus* "to open wide the mouth," *ringens* "gaping," possibly referring to the flowers; see D.H. Nicolson, "Derivation of aroid generic names." *Aroideana.* 10: 15-25. 1988.

Rinorea Aublet Violaceae

Origins:

Meaning quite obscure, or from a vernacular name in French Guinea, or from the Greek *rhis, rhinos* "nose" and *oros* "hill, mountain," or referring to the anthers; see Jean Baptiste Christophore Fusée Aublet (1720-1778), *Histoire des Plantes de la Guiane Françoise.* 1: 235, t. 93. Paris 1775.

Species/Vernacular Names:

R. sp.

Peru: yurac varilla, canilla de vieja, palo vena vena

Yoruba: nakenake, abafin

R. angustifolia (Thouars) Baill. (*Alsodeia angustifolia* Thouars; *Alsodeia natalensis* (Engl.) Bak.f.; *Rinorea ardisiiflora* (Oliv.) Kuntze; *Rinorea natalensis* Engl.)

English: white violet-bush, sweet rinorea

Southern Africa: witviooltjiebos; iThwakela, iThwakele, imPicamaguma (Zulu); umZungulu, umGudlamfene (meaning baboon's rubbing post), ukuTyakwemfene (meaning baboon's breakfast) (Xhosa); muputa-mazhana (meaning honey-grub wrapper) (Venda)

R. dentata (P. Beauv.) Kuntze

Nigeria: oloborobo (Yoruba); iyokheze (Edo)

Yoruba: oloborobo, abosulolo

R. domatiosa Van Wyk (Latin *doma, atis* "a roof, house," *domatium, domatia, domatiis*; Greek *doma*)

South Africa: bebaarde witviooltjiebos

English: bearded white violet-bush

R. ilicifolia (Welw. ex Oliv.) Kuntze (*Alsodeia ilicifolia* Welw. ex Oliv.; *Alsodeia spinosa* Tul.; *Rinorea spinosa* (Tul.) Baillon; *Rinorea angolensis* Exell)

English: yellow violet-bush, holly-leaved rinorea

Southern Africa: geelviooltjiebos; unKonkomade, umKokomane (Zulu)

Nigeria: abeko (Yoruba)

R. welwitschii (Oliv.) Kuntze (*Rinorea elliotii* Engl.)

Nigeria: iparoko, orinkase (Yoruba); iyokheze (Edo); aweigbesa (Igbo)

Yoruba: iparoko

Rinoreocarpus Ducke Violaceae

Origins:

The genus *Rinorea* Aublet and Greek *karpos* "fruit."

Rinzia Schauer Myrtaceae

Origins:

Honoring the early 19th-century German botanist Sebastian Rinz and his son Jacob (in 1856 *Mitherausgeber für Deutschland* of *Gartenflora*, a co-editor), horticulturists of Frankfurt, introduced many exotic plants in Germany; see J.C. Schauer, in *Linnaea*. 17: 239. ["Sebastianus et Jacobus Rinz, pater et filius, hortulani Francofurtenses, viri de plantarum exoticarum cultu in Germania eximie meriti."] (post May) 1843; Malcolm E. Trudgen, "Reinstatement and revision of *Rinzia* Schauer (Myrtaceae, Leptospermeae, Baeckeinae)." *Nuytsia*. 5(3): 415-439. 1986.

Riocreuxia Decaisne Asclepiadaceae

Origins:

For the French artist Alfred Riocreux, 1820-1912, botanical illustrator. See Sacheverell Sitwell & Wilfrid Blunt, *Great Flower Books 1700-1900*. A bibliographical record of two centuries of finely-illustrated flower books. [Collaborators: Patrick M. Synge, W.T. Stearn, Sabine Wilson and Handasyde Buchanan.] New York 1990; John H. Barnhart, *Biographical Notes upon Botanists*. 3: 159. 1965; Antoine Lasègue, *Musée botanique de M. Benjamin Delessert*. 554. Paris 1845; S. Lenley et al., *Catalog of the Manuscript and Archival Collections and Index to the Correspondence of John Torrey*. Library of the New York Botanical Garden. 467. 1973; E.M. Tucker, *Catalogue of the Library of the Arnold Arboretum of Harvard University*. 1917-1933.

Ripidium Bernh. Schizaeaceae

Origins:

Rhipidion, the diminutive the Greek *rhipis, rhipidos* "a fan," see also *Schizaea* Sm.

Ripidium Trin. Gramineae

Origins:

Rhipidion, the diminutive the Greek *rhipis, rhipidos* "a fan," see also *Saccharum* L.

Ripogonum Forst. & Forst.f. Smilacaceae (Ripogonaceae, Rhipogonaceae, Liliaceae)

Origins:

Greek *rhips, rhipos* "wicker-work, willow branch, mat, young twig" and *gony* "a joint, a knee," referring to the many jointed stalks, or from *gone* "offspring, generation, descent"; see Johann Reinhold Forster (1729-1798) and his son Johann Georg Adam (1754-1794), *Characteres generum plantarum*. 49, t. 25. Londini [London] (Aug.) 1776; H.E. Connor and E. Edgar, "Name changes in the indigenous New Zealand flora, 1960-1986 and Nomina Nova IV, 1983-1986." *New Zealand Journal of Botany*. Vol. 25: 115-170. 1987.

Species/Vernacular Names:

R. album R. Br.

English: white supplejack

R. brevifolium Conran & Cliff.

English: small-leaved supplejack

R. discolor F. Muell.

English: prickly supplejack

R. elseyanum F. Muell. (after the Australian [b. England] surgeon Joseph Ravenscroft Elsey, 1834-1857 (d. Springfield, St. Kitts), explorer, naturalist and plant collector, collected plants for August Heinrich Rudolf Grisebach (1814-1879); 1856-1857 with Augustus Charles Gregory (1819-1905), Expedition to North Western and Northern Australia; see Jonathan Wantrup, *Australian Rare Books, 1788-1900*. Hordern House, Sydney 1987)

English: hairy supplejack, Elsey's supplejack

R. fawcettianum Benth.

English: small supplejack

R. scandens Forst. & Forst.f.

English: supplejack

New Zealand: kareao, pirita (Maori names)

Risleya King & Pantling Orchidaceae

Origins:

Dedicated to the Hon. Herbert Hope Risley, of the Bengal Civil Service, ethnologist, editor of *The Imperial Gazetteer of India*. His works include *The People of India*. Calcutta, London 1908, *The Tribes and Castes of Bengal. Anthropometric Data*. Calcutta 1891 and *The Tribes and Castes of Bengal. Ethnographic Glossary*. Calcutta 1891.

Ristantia P.G. Wilson & Waterhouse Myrtaceae

Origins:

See *Tristania* R. Br., an anagram, in *Australian Journal of Botany*. 30: 442. 1982.

Ritaia King & Pantling Orchidaceae

Origins:

Dedicated to Mr. Rita, of the Khasi Commission, collector of orchids of Khasi hills (India).

Ritchiea R. Br. ex G. Don f. Capparidaceae (Capparaceae)

Origins:

Named after the British explorer and traveler Joseph Ritchie, d. 1821, surgeon and plant collector in Africa, who attempted to determine the course of the Niger River; see Robert Huish, *The Travels of Richard and John Lander ... for the Discovery of the Course ... of the Niger, ...* with a prefatory analysis of the previous travels of Park, Denham, Clapperton, Adams, G.F. Lyon, J. Ritchie, etc., into the hitherto unexplored countries of Africa. London 1836; Denham, Clapperton and Oudney, *Narrative of Travels and Discoveries in Northern and Central Africa. 1822-1824.* [Botany by R. Brown.] London 1826; Christopher Lloyd, *The Search for the Niger.* n.d.; Hugh Clapperton and Richard Lander, *Journal of a Second Expedition into the Interior of Africa from the Bight of Benin to Soccatoo ...* to which is added, the *Journal of Richard Lander from Kano to the Sea-coast.* Philadelphia 1829 (1st American edition; first published in London in the same year); Richard and John Lander, *Journal of an Expedition to Explore the Course and Termination of the Niger.* NY: Harpers Family Library 1836; William Henry Giles Kingston, *Travels of Mungo Park, Denham, and Clapperton.* London [1886]; George Francis Lyon (1795-1832), *A Narrative of Travels in Northern Africa, in the Years 1818, 1819, and 20,* accompanied by geographical notices of Soudan, and of the course of the Niger. London 1821.

Species/Vernacular Names:

R. duchesnei (De Wild.) F. White (*Maerua duchesnei* (De Wild.) F. White)

Nigeria: isuru (Yoruba); sabere (Edo)

Ritterocereus Backeb. Cactaceae

Origins:

For the German botanist Friedrich Ritter, 1898-1989 (Tenerife, Canaries), traveler, plant collector, cactus hunter, botanical explorer (Mexico and South America); see Gordon Douglas Rowley, *A History of Succulent Plants*. Strawberry Press, Mill Valley, California 1997.

Rivasgodaya Esteve Fabaceae

Origins:

For the Spanish botanist Salvador Rivas Goday, 1905-1981; see R. Zander, F. Encke, G. Buchheim & S. Seybold, *Handwörterbuch der Pflanzennamen*. 14 Aufl. Stuttgart 1993.

Rivea Choisy Convolvulaceae

Origins:

For the Swiss physician Auguste de la Rive or from the Latin *rivus* "a small stream of water"; see M.P. Nayar, *Meaning of Indian Flowering Plant Names*. 300. Dehra Dun 1985; R. Gordon Wasson, "Notes on the present status of Ololiuhqui and the other hallucinogens of Mexico." from *Botanical Museum Leaflets, Harvard University*. Vol. 20(6): 161-212. Nov. 22, 1963; Blas Pablo Reko, *Mitobotánica Zapoteca*. [Appended by an analysis of "Lienzo de Santiago Guevea"] Tacubaya 1945.

Species/Vernacular Names:

R. corymbosa Hall.f.

Mexico: ololiuhqui (= round thing), coaxihuitl (= snake plant), hiedra, bejicco, semilla de la Virgen

Rivina L. Phytolaccaceae

Origins:

Named for the German (b. Leipzig) botanist Augustus Quirinus Rivinus (Bachmann), 1652-1723 (d. Leipzig), physician, M.D. Helmstedt 1676 and Leipzig 1677, professor of botany and physiology, professor of therapy, author of *Introductio generalis in rem herbariam*. Lipsiae 1690. He was the son of the German physician Andreas Bachmann (1600-1656). See *Bibliotheca riviniana* ... Praemissa est vita Rivini descripta per M. Georg Samuel Hermann. Lipsiae [1727]; Joseph Ruland, *Ueber das botanische System des Rivinus*. Würzburg 1832; Ernst Huth (1845-1897), *Clavis riviniana*. Frankfurt a.O. 1891; John H. Barnhart, *Biographical Notes upon Botanists*. 3: 161. 1965; Carl Linnaeus, *Species Plantarum*. 121. 1753 and *Genera Plantarum*. Ed. 5. 57. 1754; T.W. Bossert, *Biographical Dictionary of Botanists Represented in the Hunt Institute Portrait Collection*. 333. 1972; Edward Lee Greene, *Landmarks of Botanical History*. Edited by Frank N. Egerton. Stanford, California 1983; Blanche Henrey, *No Ordinary Gardener — Thomas Knowlton, 1691-1781*. Edited by A.O. Chater. British Museum (Natural History). London 1986; Alexander B. Adams, *Eternal Quest. The Story of the Great Naturalists*. New York 1969; E.M. Tucker, *Catalogue of the Library of the Arnold Arboretum of Harvard University*. 1917-1933; Jonas C. Dryander, *Catalogus bibliothecae historico-naturalis Josephi Banks*. London 1800; A. Lasègue, *Musée botanique de Benjamin Delessert*. Paris 1845; Blanche Henrey, *British Botanical and Horticultural Literature before 1800*. 1975; Huldrych M. Koelbing, in *Dictionary of Scientific Biography* 1: 368-370. 1981; Frans A. Stafleu, *Linnaeus and the Linnaeans*. The spreading of their ideas in systematic botany, 1735-1789. Utrecht 1971.

Species/Vernacular Names:

R. humilis L. (*Rivina laevis* L.)

English: baby pepper, bloodberry, rouge-plant, coralberry

Rivinoides Afzel ex Prain Euphorbiaceae

Origins:
Resembling *Rivina* (a genus with red fruits), see also *Erythrococca* Benth., the fruits are bright orange-red when ripe.

Robbia A. DC. Apocynaceae

Origins:
Possibly from the Latin *ruber* "red."

Robertia Scop. Sapotaceae

Origins:
After the French Nicolas (Nicolass, Nicholas) Robert, 1610-1684 (or 1685), botanical artist. See Lys de Bray, *The Art of Botanical Illustration*. The classic illustrators and their achievements from 1550 to 1900. 90, 91, 96, 101. Bromley, Kent 1989; Wilfrid Blunt and W.T. Stearn,

The Art of Botanical Illustration. London 1950; J.H. Barnhart, *Biographical Notes upon Botanists*. 3: 162. 1965; E.M. Tucker, *Catalogue of the Library of the Arnold Arboretum of Harvard University*. 1917-1933; Jonas C. Dryander, *Catalogus bibliothecae historico-naturalis Josephi Banks*. London 1800; A. Lasègue, *Musée botanique de Benjamin Delessert*. Paris 1845; Sacheverell Sitwell & Wilfrid Blunt, *Great Flower Books 1700-1900*. A bibliographical record of two centuries of finely-illustrated flower books. [Collaborators: Patrick M. Synge, W.T. Stearn, Sabine Wilson and Handasyde Buchanan.] 10-11, 15, 16, 27, 29-30, 91. New York 1990; W. Blunt, *Tulipomania*. Penguin Books 1950.

Robinia L. Fabaceae

Origins:
After the French botanist Jean Robin, 1550-1629, royal gardener, herbalist to Henry IV of France, from 1590 at the Jardin des Plantes, Paris, author of *Catalogus Stirpium tam indigenarum quam exoticarum quae Lutetiae coluntur*. Paris 1601 and *Histoire des Plantes trouvées en l'Isle Virgine* ... Paris 1620. See Pierre Vallet (c. 1575-c. 1657), *Le Jardin du Roy tres chrestien Henry IV*. [Paris] 1608; Carl Linnaeus, *Species Plantarum*. 722. 1753 and *Genera Plantarum*. Ed. 5. 322. 1754; John H. Barnhart, *Biographical Notes upon Botanists*. 3: 164. 1965; Kenneth Lemmon, *The Golden Age of Plant Hunters*. London 1968; Frances L. Jewett, *Plant Hunters*. Boston 1958; Charles Lyte, *The Plant Hunters*. London 1983; Michael S. Tyler-Whittle, *The Plant Hunters*. Philadelphia 1970; T.W. Bossert, *Biographical Dictionary of Botanists Represented in the Hunt Institute Portrait Collection*. 334. 1972; E.M. Tucker, *Catalogue of the Library of the Arnold Arboretum of Harvard University*. 1917-1933; Salvatore Battaglia, *Grande dizionario della lingua italiana*. XVII: 3. [d. 1628] Torino 1995.

Species/Vernacular Names:

R. hispida L.

English: bristly locust, moss locust, rose acacia

Japan: hana-akashia

R. kelseyi Kelsey ex Hutch.

English: Alleghany moss

R. neomexicana A. Gray

English: desert locust, New Mexico locust

R. pseudoacacia L. (*Robinia pseudoacacia* L. f. *erecta* Rehder; *Robinia pseudoacacia* L. f. *microphylla* (Lodd. ex Loudon) Rehder; *Robinia pseudoacacia* L. f. *pyramidalis* (Pepin) Rehder; *Robinia pseudoacacia* L. f. *rehderi* C. Schneider; *Robinia pseudoacacia* L. f. *rozynskiana* (Spaeth) Rehder; *Robinia pseudoacacia* L. f. *semperflorens* (Carrière) Voss; *Robinia pseudoacacia* L. var. *microphylla* Lodd. ex Loudon; *Robinia pseudoacacia* L. var. *pyramidalis* Pepin;

Robinia pseudoacacia L. var. *rozynskiana* Spaeth; *Robinia pseudoacacia* L. var. *semperflorens* Carrière)

English: black locust, false acacia, locust tree, yellow locust, white acacia

Italian: pseudoacacia, falsa acacia

South Africa: vals-akasia

Japan: hari-enju

China: ci huai hua

R. viscosa L.

English: clammy locust, rose acacia

Robinsonella J.N. Rose & Baker f. Malvaceae

Origins:

For the American botanist Benjamin Lincoln Robinson, 1864-1935, plant collector, with Sereno Watson, professor of systematic botany, a specialist in *Eupatorium*, editor of *Rhodora*. His writings include *Flora of the Galapagos Islands*. 1902 and "A recension of the *Eupatoriums* of Peru." *Proc. Amer. Acad. Arts*. 55(1): 42-88. 1919. See John H. Barnhart, *Biographical Notes upon Botanists*. 3: 164. 1965; R. Zander, F. Encke, G. Buchheim & S. Seybold, *Handwörterbuch der Pflanzennamen*. 14 Aufl. 1993; T.W. Bossert, *Biographical Dictionary of Botanists Represented in the Hunt Institute Portrait Collection*. 334. 1972; J. Ewan, ed., *A Short History of Botany in the United States*. 1969; Ethelyn Maria Tucker, *Catalogue of the Library of the Arnold Arboretum of Harvard University*. Cambridge, Massachusetts 1917-1933; Ida Kaplan Langman, *A Selected Guide to the Literature on the Flowering Plants of Mexico*. 633-635. Philadelphia 1964; S. Lenley et al., *Catalog of the Manuscript and Archival Collections and Index to the Correspondence of John Torrey*. Library of the New York Botanical Garden. 349. 1973.

Robinsonia DC. Asteraceae

Origins:

Dedicated to Robinson Crusoe, see Daniel Defoe (1660-1731), *The Life and Adventures of Robinson Crusoe*; Blanche Henrey, *No Ordinary Gardener — Thomas Knowlton, 1691-1781*. Edited by A.O. Chater. 43-44, 67. British Museum (Natural History). London 1986

Robinsoniodendron Merrill Urticaceae

Origins:

For the Canadian botanist Charles Budd Robinson, Jr., 1871-1913 (murdered), economic botanist, plant collector,

bryologist. Among his writings are *The Characeae of North America*. New York 1906 and "Roxburgh's Hortus Bengalensis." *Philippine Journ. Sci. Bot*. 7: 411-419. 1912. See John H. Barnhart, *Biographical Notes upon Botanists*. 3: 164. 1965; T.W. Bossert, *Biographical Dictionary of Botanists Represented in the Hunt Institute Portrait Collection*. 334. 1972; J. Ewan, ed., *A Short History of Botany in the United States*. New York and London 1969; Ethelyn Maria Tucker, *Catalogue of the Library of the Arnold Arboretum of Harvard University*. Cambridge, Massachusetts 1917-1933.

Robiquetia Gaudich. Orchidaceae

Origins:

For the French (b. Rennes) chemist Pierre-Jean Robiquet, 1780-1840 (d. Paris), scientist, in 1806 with Louis Nicolas Vauquelin (1763-1829) first isolated an amino acid, asparagine, from asparagus, from 1811 taught at the École Polytechnique and the École of Pharmacie in Paris, in 1821 discovered caffeine, 1830 Pierre-Jean Robiquet and Boutron (also Antoine François Boutron-Charlard) discovered the hydrolytic splitting of amygdalin by an extract of defatted bitter almonds; the agent was named "emulsin" by Justus von Liebig (1803-1873) and Friedrich Wöhler (1800-1882) in 1837; in 1832 codeine or methylmorphine was first isolated from opium by Robiquet. See L.C.D. de Freycinet, *Voyage autour du Monde entrepris par ordre du Roi ... sur les corvettes de S.M. "L'Uranie" et "La Physicienne"*. Paris 1826[-1830]; for the botanical results of Freycinet's voyages see Charles Gaudichaud-Beaupré, 426, t. 34. Paris 1829; Alex Berman, in *Dictionary of Scientific Biography* 11: 494-495. 1981.

Robynsia Hutchinson Rubiaceae

Origins:

After the Belgian botanist Frans Hubert Édouard Arthur Walter Robyns, 1901-1986, traveler and plant collector (Ivory Coast), from 1954 to 1964 President of the International Association for Plant Taxonomy. Among his writings are *Flore agrostologique du Congo belge et du Ruanda-Urundi ...* Bruxelles 1929-1934, *Tentamen monographiae Vangueriae generumque affinium*. [in *Bull. Jard. Bot. Brux*. 11: 1- 359. 1928] Bruxelles 1928 and "Essai de révision des espèces africaines du genre *Annona* L." *Bull. Soc. Bot. Belg*. 67: 7-50. 1934, co-author of *Flore générale de Belgique*. Bruxelles 1950, etc.; see John H. Barnhart, *Biographical Notes upon Botanists*. 3: 166. 1965; T.W. Bossert, *Biographical Dictionary of Botanists Represented in the Hunt Institute Portrait Collection*. 334. 1972; S. Lenley et al., *Catalog of the Manuscript and Archival Collections and*

Index to the Correspondence of John Torrey. Library of the New York Botanical Garden. 351. 1973; F.N. Hepper and Fiona Neate, *Plant Collectors in West Africa.* 69. 1971.

Robynsiella Suess. Amaranthaceae

Origins:

After the Belgian botanist Frans Hubert Édouard Arthur Walter Robyns, 1901-1986; see John H. Barnhart, *Biographical Notes upon Botanists.* 3: 166. 1965.

Robynsiochloa Jacq.-Fél. Gramineae

Origins:

After the Belgian botanist Frans Hubert Édouard Arthur Walter Robyns, 1901-1986, author of *Flore agrostologique du Congo belge et du Ruanda-Urundi* ... Bruxelles 1929-1934 and *Les espèces congolaises du genre Panicum L.* Bruxelles 1932; see John H. Barnhart, *Biographical Notes upon Botanists.* 3: 166. 1965.

Robynsiophyton R. Wilczek Fabaceae

Origins:

After the Belgian botanist Frans Hubert Édouard Arthur Walter Robyns, 1901-1986; see J.H. Barnhart, *Biographical Notes upon Botanists.* 3: 166. 1965.

Rochea DC. Crassulaceae

Origins:

For the Swiss botanist Daniel de la Roche (Delaroche, de Laroche), 1743-1813, physician, M.D. Leiden 1766 and his son François (1780-1813), botanist and physician. Their writings include *Specimen botanicum inaugurale sistens descriptiones plantarum aliquot novarum.* Lugd. Batavorum 1766, *Analyse des fonctions du système nerveux pour servir d'introduction à un examen des maux des nerfs.* Genève 1778 and *Eryngiorum nec non generis novi Alepideae historia.* Paris 1808, (Daniel) with Brewer wrote *Essai sur le galvanisme.* [1802]; see J.H. Barnhart, *Biographical Notes upon Botanists.* 1: 437. 1965; Ida Kaplan Langman, *A Selected Guide to the Literature on the Flowering Plants of Mexico.* Philadelphia 1964; Stafleu and Cowan, *Taxonomic Literature.* 1: 613 ["François Delaroche, 1780-1813, Daniel Delaroche, 1743-1812."] 1976 and 4: 821-822. ["F. de la Roche, 1782-1814."] 1985.

Rochelia Reichenbach Boraginaceae

Origins:

After the Austrian botanist Anton Rochel, 1770-1847, surgeon, from 1788 to 1798 in the Austrian army, 1820-1840 Curator of the Pest Botanical Garden, traveler. Among his works are *Pflanzen-Umrisse.* Wien 1820, *Plantae Banatus rariores.* Pestini [Pest, Budapest] 1828 and *Botanische Reise in das Banat im Jahre 1835.* Pesth, Leipzig 1838; see J.H. Barnhart, *Biographical Notes upon Botanists.* 3: 166. 1965; Heinrich Gottlieb Ludwig Reichenbach (1793-1879), in *Flora oder allgemeine Botanische Zeitung.* 7: 243. (Apr.) 1824; Antoine Lasègue, *Musée botanique de M. Benjamin Delessert.* Paris 1845; Ethelyn Maria Tucker, *Catalogue of the Library of the Arnold Arboretum of Harvard University.* Cambridge, Massachusetts 1917-1933; R. Zander, F. Encke, G. Buchheim & S. Seybold, *Handwörterbuch der Pflanzennamen.* 14 Aufl. 770. Stuttgart 1993; Mariella Azzarello Di Misa, ed., *Il Fondo Antico della Biblioteca dell'Orto Botanico di Palermo.* 233. Palermo 1988.

Species/Vernacular Names:

R. bungei Trautvetter

English: Bunge rochelia

China: luan guo he shi

Rochetia Delile Meliaceae

Origins:

For the French chemist C.E.X. (or C.E.K., C.L.X) Rochet d'Héricourt, traveler and explorer, geographer, wrote *Voyage sur la côte orientale de la Mer Rouge, dans le pays d'Adel et le Royaume de Choa.* Paris 1841 and *Second voyage sur les duex rives de la Mer Rouge dans le pays des Adels et le royaume de Choa.* Paris 1846; see J.H. Barnhart, *Biographical Notes upon Botanists.* 3: 167. 1965; Stafleu and Cowan, *Taxonomic Literature.* 4: 824. 1983.

Rockia Heimerl Nyctaginaceae

Origins:

For the legendary American (b. Vienna, he emigrated to America in 1905) explorer Joseph Francis Charles (Joseph Franz Karl) Rock, 1884-1962, photographer, botanist, ethnologist, author, geographer, traveler (1907-1920 Hawaii, between 1922 and 1949 he spent the majority of his time in China and Tibet). Among his numerous publications are "The poisonous plants of Hawaii." *Hawaiian Forester Agric.* 17: 59-62 and 17: 97-101. 1920, "Hunting the Chaulmoogra Tree." *National Geographic.* 243-276. Mar. 1922, *The Indigenous Trees of the Hawaiian Islands.* Honolulu 1913 and "Trees recommended for planting." *Hawaiian Pl.*

Rec. 18: 414–421. 1918; see J.H. Barnhart, *Biographical Notes upon Botanists.* 3: 167. 1965; T.W. Bossert, *Biographical Dictionary of Botanists Represented in the Hunt Institute Portrait Collection.* 335. 1972; S. Lenley et al., *Catalog of the Manuscript and Archival Collections and Index to the Correspondence of John Torrey.* Library of the New York Botanical Garden. 351. 1973; E.M. Tucker, *Catalogue of the Library of the Arnold Arboretum of Harvard University.* 1917-1933; J. Ewan, ed., *A Short History of Botany in the United States.* 1969; H.R. Fletcher, *Story of the Royal Horticultural Society, 1804-1968.* Oxford 1969; Elmer Drew Merrill, in *Contr. U.S. Natl. Herb.* 30(1): 255-257. 1947 and *Bernice P. Bishop Mus. Bull.* 144: 158-159. 1937; I.C. Hedge and J.M. Lamond, *Index of Collectors in the Edinburgh Herbarium.* Edinburgh 1970; R. Zander, F. Encke, G. Buchheim & S. Seybold, *Handwörterbuch der Pflanzennamen.* 14 Aufl. 1993; Alexandra David-Néel (1868-1969), *Journal de voyage,* vol. 2. Plon, Paris 1976; Barbara M. Foster & Michael Foster, *Forbidden Journey — The Life of Alexandra David-Néel.* San Francisco 1987; http://rbge-sun1.rbge.org.uk/inverleith-house/rock.htm.

Rockingamia Airy Shaw Euphorbiaceae

Origins:
Named for the English statesman Charles Watson Wentworth, 2nd Marquis of Rockingham, 1730-1782, in 1750 was created Earl of Malton in the Irish peerange, twice held the office Prime Minister to George III. See George H. Guttridge, *The Early Career of Lord Rockingham, 1730-1765.* [University of California Publications in History. vol. 44.] 1952; George Thomas Keppel, Earl of Albemarle, *Memoirs of the Marquis of Rockingham and His Contemporaries.* London 1852; Ross John Swartz Hoffman, *The Marquis. A Study of Lord Rockingham, 1730-1782.* New York 1973; Francis O'Gorman, *The Rise of Party in England. The Rockingham Whigs 1760-1782.* London 1975; Paul Langford, *The First Rockingham Administration, 1765-1766.* London 1973; Airy Shaw, in *Kew Bulletin.* 20: 29. (Jun.) 1966.

Rodatia Raf. Acanthaceae

Origins:
For the Italian botanist Luigi (Aloysio, Aloysius) Rodati, 1762-1832, from 1792 to 1802 Prefect of the Bologna Botanical Garden; see Constantine Samuel Rafinesque, *Autikon botanikon.* Icones plantarum select. nov. vel rariorum, etc. 31. Philadelphia 1840; E.D. Merrill, *Index rafinesquianus.* 224. 1949; Mariella Azzarello Di Misa, ed., *Il Fondo Antico della Biblioteca dell'Orto Botanico di Palermo.* 233. Palermo 1988.

Rodentiophila F. Ritter ex Backeb. Cactaceae

Origins:
Latin *rodo, -is, -si, -sum, -rodere* "to gnaw" and Greek *philos* "lover, loving," see also *Eriosyce* Phil.

Rodetia Moq. Amaranthaceae

Origins:
For the French botanist Henri Jean Antoine Rodet, 1810-1875, wrote *Botanique agricole et médicale.* Paris 1872; see J.H. Barnhart, *Biographical Notes upon Botanists.* 3: 167. 1965.

Rodgersia A. Gray Saxifragaceae

Origins:
For Comm. John Rodgers, 1812-1882 (Washington), American explorer, traveler, admiral, 1840-1842 commanded the *Boxer,* 1852 he was appointed to the North Pacific Exploring Expedition (the leader was Capt. Cadwalader Ringgold (1802-1867), one of the botanical collectors was the American botanist Charles (Carlos) Wright, 1811-1885), 1861-1865 Civil War, in 1865 sailed to the Pacific with the *Vanderbilt,* in 1871 sailed to Korea in the *Colorado,* President of the Naval Institute. See F. Boerner & G. Kunkel, *Taschenwörterbuch der botanischen Pflanzennamen.* 4. Aufl. 162. Berlin & Hamburg 1989; John Dunmore, *Who's Who in Pacific Navigation.* 40, 207-208. Honolulu 1991; W.P. Cummings, S.E. Hillier, D.B. Quinn and G. Williams, *The Exploration of North America 1630-1776.* London 1974; R.G. Ward, *American Activities in the Central Pacific 1790-1870.* Ridgewood 1966-1970.

Species/Vernacular Names:
R. aesculifolia Batal.
China: mu he

Rodriguezia Ruíz & Pavón Orchidaceae

Origins:
Named for a Don Manuel Antonio Rodríguez de Vera (1780-1846) or José Demetrio Rodríguez (1780-1847), or Don Manuel Rodríguez, a Spanish botanist and apothecary; see Richard Evans Schultes and Arthur Stanley Pease, *Generic Names of Orchids. Their Origin and Meaning.* 272. 1963; Hubert Mayr, *Orchid Names and Their Meanings.* Vaduz 1998; Georg Christian Wittstein, *Etymologisch-botanisches Handwörterbuch.* 769. Ansbach 1852; F. Boerner & G. Kunkel, *Taschenwörterbuch der botanischen Pflanzennamen.* 4. Aufl. 162. 1989; Helmut Genaust, *Etymologisches*

Wörterbuch der botanischen Pflanzennamen. 541. 1996; Stafleu and Cowan, *Taxonomic Literature*. 4: 830. Utrecht 1983.

Rodrigueziella Kuntze Orchidaceae

Origins:

After the Brazilian botanist João Barbosa Rodrigues, 1842-1909, traveler, plant collector. Among his numerous publications are *Genera et Species Orchidearum novarum*. Sebastianopolis 1877-1881[-1882], *Plantas novas cultivadas no Jardim botânico do Rio de Janeiro*. Rio de Janeiro 1891-1898 and *Sertum palmarum brasiliensium*. Bruxelles 1903, from 1883 to 1889 Director of the Manaos Botanical Garden, from 1889 to 1909 Director of the Botanical Garden of Rio de Janeiro; see J.H. Barnhart, *Biographical Notes upon Botanists*. 1: 120. 1965; T.W. Bossert, *Biographical Dictionary of Botanists Represented in the Hunt Institute Portrait Collection*. 24. 1972; E.M. Tucker, *Catalogue of the Library of the Arnold Arboretum of Harvard University*. 1917-1933; Ida Kaplan Langman, *A Selected Guide to the Literature on the Flowering Plants of Mexico*. Philadelphia 1964; R. Zander, F. Encke, G. Buchheim & S. Seybold, *Handwörterbuch der Pflanzennamen*. 14 Aufl. Stuttgart 1993.

Rodrigueziopsis Schltr. Orchidaceae

Origins:

Resembling *Rodriguezia* Ruíz & Pav.

Roea Hügel ex Benth. Fabaceae

Origins:

Named after the English (b. Berks.) naval officer John Septimus Roe, 1797-1878 (d. Perth, Western Australia), explorer and naturalist, botanical collector in Australia, Persian Gulf and Sri Lanka, from 1813 to 1827 in the Royal Navy, 1828 Fellow of the Linnean Society, from 1828/1829 to 1870 Surveyor General, W. Australia, author of *Report of a Journey of Discover into the Interior of Western Australia Between 8th September 1848 and 3rd February, 1849*. [in W.J. Hooker, *J. Bot. Kew Gard. Misc.* 1854-1855]. See Frederick Royston Mercer, *Amazing Career. The Story of Western Australia's First Surveyor-General* (John Septimus Roe). Perth [1962]; John H. Barnhart, *Biographical Notes upon Botanists*. 3: 169. 1965; Ida Lee (afterwards Marriott), *Early Explorers in Australia*. London 1925; I.H. Vegter, *Index Herbariorum*. Part II (5), *Collectors N-R*. Regnum Vegetabile vol. 109. 1983; George Bentham et al., *Enumeratio Plantarum quas in Novae Hollandiae ...* collegit C. de

Hügel. 34. Vindobonae (Apr.) 1837; H.N. Clokie, *Account of the Herbaria of the Department of Botany in the University of Oxford*. 233. Oxford 1964; Antoine Lasègue (1793-1873), *Musée botanique de M. Benjamin Delessert*. Paris 1845; N. Hall, *Botanists of the Eucalypts*. Melbourne 1978 and Supplement 1980.

Roella L. Campanulaceae

Origins:

After Wilhelm [Gulielmus] Roell, professor of anatomy in Amsterdam in the 18th century, horticulturist, author of *Disputatio ... de ventriculi fabrica et actione musculari*. Lugduni Batavorum [Leyden 1725]; see Carl Linnaeus, *Species Plantarum*. 170. 1753 and *Genera Plantarum*. Ed. 5. 77. 1754.

Species/Vernacular Names:

R. glomerata A. DC.

Southern Africa: iBhosisi

Roemeria Medikus Papaveraceae

Origins:

For the Swiss botanist Johann Jakob Roemer (Römer), 1763-1819, physician and naturalist, studied at the Universities of Zürich and Göttingen, from 1797 to 1819 Director of the Botanical Garden of the Naturforschende Gesellschaft Zürich, co-editor with Paul Usteri (1768-1831) of *Magazin für die Botanik*. Zürich 1787-1791. Among his many works are *Scriptores de plantis hispanicis, lusitanicis, brasiliensibus*. Norimbergae [Nürnberg] 1796 and *Catalogus horti botanici societatis physicae turicensis*. Zürich 1802; see Paul Usteri (1768-1831), in *Annalen der Botanick*. 1(3): 15. (Jul.) 1792; J.H. Barnhart, *Biographical Notes upon Botanists*. 3: 169. 1965; Georg Christian Wittstein, *Etymologisch-botanisches Handwörterbuch*. 769. 1852; T.W. Bossert, *Biographical Dictionary of Botanists Represented in the Hunt Institute Portrait Collection*. 335. 1972; R. Zander, F. Encke, G. Buchheim & S. Seybold, *Handwörterbuch der Pflanzennamen*. 14 Aufl. 770f. 1993; E. Bretschneider, *History of European Botanical Discoveries in China*. Leipzig 1981; H.N. Clokie, *Account of the Herbaria of the Department of Botany in the University of Oxford*. 233-234. 1964; E.M. Tucker, *Catalogue of the Library of the Arnold Arboretum of Harvard University*. 1917-1933; Ida Kaplan Langman, *A Selected Guide to the Literature on the Flowering Plants of Mexico*. 638. 1964; Jonas C. Dryander, *Catalogus bibliothecae historico-naturalis Josephi Banks*. London 1800; A. Lasègue, *Musée botanique de Benjamin Delessert*. 1845; G. Murray, *History of the Collections Contained in the Natural History Departments of the British Museum.* 1: 177. 1904; Rodolfo E.G. Pichi Sermolli, "The publication of Roth's genera *Athyrium* and *Polystichum.*" *Webbia.* 8(2): 437-442. 1952.

Species/Vernacular Names:

R. hybrida (L.) DC. (*Chelidonium hybridum* L.; *Roemeria hybrida* (L.) DC. var. *velutino-eriocarpa* Fedde)

English: violet horned-poppy

Roemeria Roem. & Schultes Gramineae

Origins:

For the Swiss botanist Johann Jakob Roemer, 1763-1819, physician and naturalist.

Roemeria Thunberg Sapotaceae

Origins:

For the Swiss botanist Johann Jakob Roemer, 1763-1819, physician and naturalist.

Roentgenia Urban Bignoniaceae

Origins:

After the German physicist Wilhelm Konrad (Conrad) von Röntgen (Roentgen), 1845-1923 (Munich), professor of physics at Strasbourg (1876-1879), Giessen (1879-1888), Würzburg (1888-1890) and Munich (1900-1920), 1895 discovered the electromagnetic rays (X-rays), in 1901 received the Nobel Prize for Physics. See Otto Glasser, *Wilhelm Conrad Röntgen and the History of the Roentgen Rays.* London 1933; G. L'E. Turner, in *Dictionary of Scientific Biography* 11: 529-531. 1981.

Roepera A. Juss. Zygophyllaceae

Origins:

After the German botanist Johannes August Christian Roeper (Röper), 1801-1885, physician, M.D. Göttingen 1823, traveler, studied with A.P. de Candolle, professor of botany at Basel and Rostock, University librarian at Rostock, correspondent of the German botanist Diederich Franz Leonhard von Schlechtendal (1794-1866), contributor to C.F.P. von Martius *Flora Brasiliensis* (Euphorbiaceae). His publications include *Zur Flora Mecklenburgs.* Rostock 1843 [-1844], *Enumeratio Euphorbiarum quae in Germania et Pannonia gignuntur.* Gottingae 1824, *Vorgefasste botanische Meinungen.* Rostock 1860 and *De floribus et affinitatibus Balsaminearum.* Basiliae [Basel] 1830; see

John H. Barnhart, *Biographical Notes upon Botanists.* 3: 170. 1965; G. Schmid, *Goethe und die Naturwissenschaften.* Halle 1940; Charles J. Édouard Morren, *Correspondance botanique.* Liège 1874 and 1884; A.H.L. de Jussieu, in *Mémoires du Muséum d'Histoire Naturelle.* 12: 454, t. 15, no. 3. [Mémoires sur les Rutacées.] Paris 1825; T.W. Bossert, *Biographical Dictionary of Botanists Represented in the Hunt Institute Portrait Collection.* 335. 1972; Antoine Lasègue, *Musée botanique de M. Benjamin Delessert.* Paris 1845; E.M. Tucker, *Catalogue of the Library of the Arnold Arboretum of Harvard University.* 1917-1933; Mariella Azzarello Di Misa, ed., *Il Fondo Antico della Biblioteca dell'Orto Botanico di Palermo.* 234. Palermo 1988.

Roeperia Sprengel Euphorbiaceae

Origins:

For the German botanist Johannes August Christian Roeper (Röper), 1801-1885, physician; see John H. Barnhart, *Biographical Notes upon Botanists.* 3: 170. 1965; Kurt Polycarp Joachim Sprengel (1766-1833), *Systema Vegetabilium.* 3: 13, 147. (Jan.-Mar.) 1826.

Roeperocharis Reichb.f. Orchidaceae

Origins:

After the German botanist Johannes August Christian Roeper (Röper), 1801-1885, physician, M.D. Göttingen 1823, traveler; see John H. Barnhart, *Biographical Notes upon Botanists.* 3: 170. 1965.

Roettlera Vahl Gesneriaceae

Origins:

For the Alsatian-born Danish missionary Johan (John) Peter Rottler, 1749-1836 (d. Madras, India), traveler and botanist, orientalist, collector, from 1776 to 1806 at the Danish Mission at Tranquebar (Madras, Coromandel coast, India), in 1788 in the Ganges region and in 1795 at Ceylon, wrote *A Dictionary of the Tamil and English Languages.* Madras 1834-1841, in 1828 translated *The Book of Common Prayer* in Tamil. See [Bible] [Tamil], *The Old Testament* [revised by C.G.E. Rhenius, with J.P. Rottler and others. 1827; John H. Barnhart, *Biographical Notes upon Botanists.* 3: 183. 1965; Mary Gunn and Leslie E. Codd, *Botanical Exploration of Southern Africa.* 301. A.A. Balkema Cape Town 1981; Martin H. Vahl (1749-1804), *M. Vahlii ... Enumeratio Plantarum.* 1: 87. Hauniae 1804; Bishop James M. Thoburn, *The Christian Conquest of India.* [Forward Mission Study Courses, edited under the auspices of the Young

People's Missionary Movement.] Eaton & Mains, NY n.d.; Antoine Lasègue, *Musée botanique de M. Benjamin Delessert.* Paris 1845; Isaac Henry Burkill, *Chapters on the History of Botany in India.* Delhi 1965; R. Zander, F. Encke, G. Buchheim & S. Seybold, *Handwörterbuch der Pflanzennamen.* 14 Aufl. Stuttgart 1993; Carl Frederik Albert Christensen, *Den danske Botaniks Historie med tilhørende Bibliografi.* Copenhagen 1924-1926.

Roezlia Regel Melastomataceae

Origins:

Named for the Bohemian botanist Benedict (Benedikt, Benito) Roezl, 1824-1885, gardener, traveler, 1854-1875 plant and orchid collector in South America, Mexico (Michoacán, Oaxaca, Veracruz and Yucatán) and Cuba, importer of cacti, orchids and lilies. See John H. Barnhart, *Biographical Notes upon Botanists.* 3: 171. Boston 1965; E.M. Tucker, *Catalogue of the Library of the Arnold Arboretum of Harvard University.* 1917-1933; Ida Kaplan Langman, *A Selected Guide to the Literature on the Flowering Plants of Mexico.* 638-639. University of Pennsylvania Press, Philadelphia 1964; Merle A. Reinikka, *A History of the Orchid.* Timber Press 1996; Graham Yearsley, "Benedict Roezl. The indefatigable orchid hunter." in *The Orchid Review.* 104(1212): 357-359. [b. 1823] 1996; Gordon Douglas Rowley, *A History of Succulent Plants.* Strawberry Press, Mill Valley, California 1997; R. Zander, F. Encke, G. Buchheim & S. Seybold, *Handwörterbuch der Pflanzennamen.* 14 Aufl. Stuttgart 1993; Irving William Knobloch, compil., "A preliminary verified list of plant collectors in Mexico." *Phytologia Memoirs.* VI. 1983.

Roezliella Schltr. Orchidaceae

Origins:

For the Bohemian botanist Benedict (Benedikt, Benito) Roezl (Rözl), 1824-1885, gardener, traveler, 1854-1875 plant and orchid collector in South America, Mexico (Michoacán, Oaxaca, Veracruz and Yucatán) and Cuba; see John H. Barnhart, *Biographical Notes upon Botanists.* 3: 171. Boston 1965.

Rogeonella A. Chev. Sapotaceae

Origins:

After a Mr. Rogeon, plant collector in Mali and Niger; see Auguste Jean Baptiste Chevalier, *Flore vivante de l'Afrique Occidentale Française.* 1: xxvii-xxx. Paris 1938; F.N. Hepper and F. Neate, *Plant Collectors in West Africa.* 69. 1971.

Rogeria J. Gay ex Delile Pedaliaceae

Origins:

After a Mr. Roger, French plant collector (Senegal); see Auguste Jean Baptiste Chevalier, *Flore vivante de l'Afrique Occidentale Française.* 1: xxvii-xxx. Paris 1938; F.N. Hepper and F. Neate, *Plant Collectors in West Africa.* 69. 1971.

Roggeveldia Goldblatt Iridaceae

Origins:

Found in Roggeveld, Cape Province, South Africa.

Rohdea Roth Convallariaceae (Liliaceae)

Origins:

For the German botanist Michael Rohde, 1782-1812, botanist, M.D. Göttingen 1804, author of *Monographiae Cinchonae generis tentamen.* Gottingae 1804; see John H. Barnhart, *Biographical Notes upon Botanists.* 3: 172. Boston 1965; Ethelyn Maria Tucker, *Catalogue of the Library of the Arnold Arboretum of Harvard University.* 1917-1933; F. Boerner & G. Kunkel, *Taschenwörterbuch der botanischen Pflanzennamen.* 4. Aufl. 162. 1989; Georg Christian Wittstein, *Etymologisch-botanisches Handwörterbuch.* 770. 1852.

Species/Vernacular Names:

R. japonica (Thunb.) Roth

English: lily of China

China: wan nian qing gen

Japan: omoto

Roigia Britton Euphorbiaceae

Origins:

For the Cuban botanist Juan Tomás Roig y Mesa, b. 1878, *una gloria de la Ciencia cubana,* author of *Diccionario Botánico de Nombres Vulgares Cubanos.* La Habana 1988; see John H. Barnhart, *Biographical Notes upon Botanists.* 3: 173. Boston 1965; Ida Kaplan Langman, *A Selected Guide to the Literature on the Flowering Plants of Mexico.* 639. University of Pennsylvania Press, Philadelphia 1964; S. Lenley et al., *Catalog of the Manuscript and Archival Collections and Index to the Correspondence of John Torrey.* Library of the New York Botanical Garden. 352. 1973.

Rojasia Malme Asclepiadaceae

Origins:

For the Paraguayan botanist Teodor Rojas, 1877-1954, plant collector with E. Hassler; see Stafleu and Cowan, *Taxonomic Literature.* 4: 868. Utrecht 1983.

Rojasianthe Standley & Steyerm. Asteraceae

Origins:
For Ulises Rojas, professor of botany in Guatemala; see Stafleu and Cowan, *Taxonomic Literature*. 4: 868. Utrecht 1983.

Rojasiophyton Hassler Bignoniaceae

Origins:
For the Paraguayan botanist Teodor (Teodoro) Rojas, 1877-1954.

Rojasiophytum Hassler Bignoniaceae

Origins:
After the Paraguayan botanist Teodor (Teodoro) Rojas, 1877-1954, plant collector with Emilio (Émile) Hassler (1861-1937); see Stafleu and Cowan, *Taxonomic Literature*. 4: 868. Utrecht 1983.

Rolandra Rottb. Asteraceae

Origins:
For the Swedish botanist Daniel Rolander, 1725-1793, traveler and plant collector; see John H. Barnhart, *Biographical Notes upon Botanists*. 3: 173. Boston 1965; Jonas C. Dryander, *Catalogus bibliothecae historico-naturalis Josephi Banks*. London 1800; A. Lasègue, *Musée botanique de Benjamin Delessert*. Paris 1845; Carl Frederik Albert Christensen, *Den danske Botaniks Historie med tilhørende Bibliografi*. Copenhagen 1924-1926.

Rolfea Zahlbr. Orchidaceae

Origins:
Dedicated to the English (born at Ruddington, near Nottingham) botanist Robert Allen Rolfe, 1855-1921 (Kew, Surrey), authority on orchids, plant collector, 1879 to Kew as a gardener, 1880 appointed the first Curator of the Orchid Herbarium by Sir Joseph Hooker (then Director of Kew Gardens), 1885 he was elected an Associate of the Linnean Society, 1893-1920 founder and editor of *The Orchid Review*, wrote "Revision of the Genus *Phalaenopsis*." in the *Gardeners' Chronicle*. 1886, monographed Orchidaceae for volume 7 of the *Flora of Tropical Africa*. 1898, edited the English edition of *Lindenia*, with Charles Chamberlain Hurst (1870-1949) wrote *The Orchid Stud-Book*. Kew 1909. See John H. Barnhart, *Biographical Notes upon Botanists*. 3: 173. 1965; R. Zander, F. Encke, G. Buchheim & S. Seybold, *Handwörterbuch der Pflanzennamen*. 14 Aufl. 1993; Frederico Carlos Hoehne, M. Kuhlmann and

Oswaldo Handro, *O jardim botânico de São Paulo*. 175. 1941; Elmer Drew Merrill, *Contr. U.S. Natl. Herb.* 30(1): 258. 1947 and in *Bernice P. Bishop Mus. Bull.* 144: 160. 1937; [Anon.], "Robert Allen Rolfe." *Gardeners' Chronicle*. 69: 74, 204. 1921; Alec M. Pridgeon, "Robert Allen Rolfe (1855-1921)." *The Kew Magazine*. 10(1): 46-51. February 1993; Ida Kaplan Langman, *A Selected Guide to the Literature on the Flowering Plants of Mexico*. 641. Philadelphia 1964; Emil Bretschneider, *History of European Botanical Discoveries in China*. Leipzig 1981; Merle A. Reinikka, *A History of the Orchid*. Timber Press 1996.

Rolfeella Schltr. Orchidaceae

Origins:
For the English botanist Robert Allen Rolfe, 1855-1921, orchidologist, plant collector, 1893-1920 founder and editor of *The Orchid Review*; see John H. Barnhart, *Biographical Notes upon Botanists*. 3: 173. Boston 1965.

Rollandia Gaudich. Campanulaceae

Origins:
Named after a M. Rolland, one of the members of the Freycinet expedition of 1817-1820. See Ludolf Karl Adelbert von Chamisso and D.F.L. von Schlechtendal, "De plantis in expeditione speculatoria Romanzoffiana observatis." *Linnaea*. 8: 201-223. 1833; L.C.D. de Freycinet, *Voyage autour du Monde entrepris par ordre du Roi ... sur les corvettes de S.M. "L'Uranie" et "La Physicienne"*. Paris 1826[-1830]; Franz Julius Ferdinand Meyen (1804-1840), *Observationes botanicas* in itinere circum terram institutas. Opus posthumum, sociorum Academiae curis suppletum. Vratislaviae et Bonnae [Breslau and Bonn] 1843; A.N. Vaillant, *Voyage autour du Monde exécuté pendant les Années 1836 et 1837 sur la corvette "La Bonite"*. Paris 1845-1852.

Rollinia A. St.-Hil. Annonaceae

Origins:
Named for the French historian Charles Rollin, 1661-1741, professor of eloquence in the Royal College. Among his writings are Ad illustrissimum virum Franciscum Michaelem Le Tellier, marchionem de Loubois ... cum ejus filius Camillus de Louvois ... *de Theocrito publice responderet*. Paris 1689, *De la Manière d'enseigner et d'étudier les Belles-Lettres*, par raport à l'esprit et au coeur. Paris 1726-1728, *Histoire Romaine*. Paris 1738-1741 and *Histoire ancienne* des Égyptiens, des Carthaginois, des Assyriens, des Babyloniens, des Mèdes, et des Perses, des Macédoniens, des Grecs. Paris 1730-1738; see H. Ferté,

Rollin. Sa vie, ses oeuvres. 1902; F.M. de Marsy, Histoire moderne de Chinois, des Japonnois ... pour servir de suite à l'Histoire ancienne de M. Rollin. 1755; Luíz Caldas Tibiriçá, Dicionário Guarani-Português. 34. Traço Editora, Liberdade 1989.

Species/Vernacular Names:

R. sp.

Peru: anona, anonilla, biribá, envira, manotigre

Bolivia: chirimoya

R. exalbida (Vell.) Mart.

Brazil: araticum (= fruto mole), araticum do mato, araticum alvadio, araticum de Santa Catarina, cortiça, cortiça de comer, cortiça de ouriço, fruta de conde pequena, quaresma, imbira

R. herzogii R.E. Fries

Bolivia: chirimoya de gato

R. mucosa (Jacq.) Baillon

Peru: anón cimarrón, cachiman morveux

Tropical America: biribá

Rolliniopsis Saff. Annonaceae

Origins:
Resembling Rollinia.

Rollinsia Al-Shehbaz Brassicaceae

Origins:
For the American botanist Reed Clarke Rollins, b. 1911, specialist in Cruciferae (Brassicaceae); see J.H. Barnhart, Biographical Notes upon Botanists. 3: 174. 1965; T.W. Bossert, Biographical Dictionary of Botanists Represented in the Hunt Institute Portrait Collection. 337. 1972; Irving William Knobloch, compil., "A preliminary verified list of plant collectors in Mexico." Phytologia Memoirs. VI. 1983; Ida Kaplan Langman, A Selected Guide to the Literature on the Flowering Plants of Mexico. 641-642. University of Pennsylvania Press, Philadelphia 1964; S. Lenley et al., Catalog of the Manuscript and Archival Collections and Index to the Correspondence of John Torrey. Library of the New York Botanical Garden. 353. 1973.

Romanschulzia O.E. Schulz Brassicaceae

Origins:
After to the German botanist Roman Schulz, 1873-1926, brother of Otto Eugen Schulz (1874-1936); see John H. Barnhart, Biographical Notes upon Botanists. 3: 246. 1965;

T.W. Bossert, Biographical Dictionary of Botanists Represented in the Hunt Institute Portrait Collection. 356. 1972.

Romanzoffia Cham. Hydrophyllaceae

Origins:
After Count Nikolaj Petrovic Rumjancëv (Romanzoff, Romanzov), 1754-1826, the grand Chancellor, promoter of a Russian expedition seeking the Northwest Passage through the Bering Strait. See Christian Gottfried Ehrenberg (1795-1876), Fungos a clarissimo Adalberto de Chamisso ... sub auspiciis Romanzoffianis in itinere circa terrarum globum collectos enumeravit novosque descripsit et pinxit Dr. C.G. Ehrenberg. [Bonn 1820]; Ludolf Karl Adelbert von Chamisso and D.F.L. von Schlechtendal, "De plantis in expeditione speculatoria Romanzoffiana observatis rationem dicunt." Linnaea. i-x. 1826-1836; L.K.A. von Chamisso, Reise um die Welt mit der Romanzoffischen Entdeckungsexpedition in den Jahren 1815-1818 auf der Brigg Rurik, Cpt. Otto von Kotzebue. Leipzig 1836; John Dunmore, Who's Who in Pacific Navigation. 144-145. Honolulu 1991.

Romneya Harvey Papaveraceae

Origins:
In honor of Thomas Coulter's friend Robinson Romney (1792-1822), or dedicated to the Irish astronomer Thomas Romney Robinson (1792-1833); see Philip McMillan Browse, "Romneya." in The Plantsman. 11(2): 121-124. September 1989; F. Boerner & G. Kunkel, Taschenwörterbuch der botanischen Pflanzennamen. 4. Aufl. 163. 1989; F. Boerner, Taschenwörterbuch der botanischen Pflanzennamen. 2. Aufl. 168. Berlin & Hamburg 1966; Helmut Genaust, Etymologisches Wörterbuch der botanischen Pflanzennamen. 541-542. 1996.

Species/Vernacular Names:

R. coulteri Harvey

California: Matilija poppy, Matillija poppy, Coulter's Matilija poppy

Romulea Maratti Iridaceae

Origins:
After Romulus, legendary founder and first King of Rome, son of Mars and Rhea Sylvia, and twin brother of Remus; see Giovanni Francesco Maratti (1723-1777), Plantarum Romulae et Saturniae in agro Romano existentium specificas notas describit inventor. 13. Romae 1772.

Species/Vernacular Names:

R. flava (Lam.) De Vos var. flava

South Africa: frutang, froetang, knikkertjies

R. flava (Lam.) De Vos var. *minor* (Bég.) De Vos

English: onion grass

R. minutiflora Klatt

English: lesser Guildford grass, small-flowered onion-grass

South Africa: froetang, frutang, knikkertjie

R. rosea (L.) Ecklon var. *australis* (Ewart) De Vos (*Romulea cruciata* (Jacq.) Bég. var. *australis* Ewart; *Romulea longifolia* (Salisb.) Baker; *Trichonema longifolium* Salisb.)

English: Guildford grass, pink romulea, onion weed, onion grass

South Africa: frutang, froetang, perdefroetang, knikkertjies, spruitjie

R. rosea (L.) Ecklon var. *rosea*

South Africa: frutang, froetang, knikkertjies

R. sabulosa Schltr. ex Beg.

South Africa: satin flower, silk flower, satynblom, syblom

Rondeletia L. Rubiaceae

Origins:

Named for the French (b. Montpellier) physician Guillaume Rondelet (Gulielmus Rondeletius), 1507-1566 (d. Réalmont, Tarn, France), botanist, zoologist, ichthyologist, 1537 graduated M.D. Montpellier, professor of medicine. Rondelet's major work is *Libri de piscibus marinis in quibus verae piscium effigies expressae sunt: Universae aquatilium historiae pars altera*. Lyons 1554-1555. He was a friend of the French physician and writer François Rabelais (*circa* 1494-1553). See J.H. Barnhart, *Biographical Notes upon Botanists*. 3: 175. 1965; T. W. Bossert, *Biographical Dictionary of Botanists Represented in the Hunt Institute Portrait Collection*. 337. 1972; E.M. Tucker, *Catalogue of the Library of the Arnold Arboretum of Harvard University*. 1917-1933; Jules Émile Planchon (1823-1888), *Rondelet et ses disciples*. Montpellier 1866; A.G. Keller, in *Dictionary of Scientific Biography* 11: 527-528. 1981; Garrison and Morton, *Medical Bibliography*. 282. 1961; Richard J. Durling, compil., *A Catalogue of Sixteenth Century Printed Books in the National Library of Medicine*. 1967; Pierre Huard & Marie-José Imbault-Huart, in *Dictionary of Scientific Biography* 11: 253-254. 1981.

Roodia N.E. Br. Aizoaceae

Origins:

After the South African plant collector Petrusa Benjamina Rood, 1861-1946 (*née* Van Rhyn), collected seeds, succulents and bulbous plants; see Mary Gunn and Leslie E. Codd, *Botanical Exploration of Southern Africa*. 299. Cape Town 1981.

Rooksbya Backeb. Cactaceae

Origins:

After Ellen Rooksby, 1873-1952, from 1929 to 1952 founded and ran *Desert/Desert Plant Life*; see Gordon Douglas Rowley, *A History of Succulent Plants*. California 1997.

Ropalopetalum Griff. Annonaceae

Origins:

Greek *rhopalon* "a club" and *petalon* "leaf, petal," see also *Artabotrys* R. Br.

Rophostemon Endl. Orchidaceae

Origins:

Greek *rhopheo, rhophao* "drain dry, gulp down, empty, live on slops" and *stemon* "a stamen."

Roptrostemon Blume Orchidaceae

Origins:

Greek *rhoptron* "the wood in a trap, a tambourine, a knocker" and *stemon* "a stamen, thread, pillar," possibly referring to the clavate and elongate column.

Roraimanthus Gleason Ochnaceae

Origins:

Roraima, at the junction of Brazil, Venezuela and British Guiana; see Ernst Heinrich Georg Ule (1854-1915), *Die Vegetation des Roraima*. Leipzig und Berlin 1914.

Rorida J.F. Gmelin Capparidaceae (Capparaceae)

Origins:

Latin *roridus, a, um* "dewy, wet with dew," *ros, roris* "dew, moisture."

Roridula Burm.f. ex L. Byblidaceae (Roridulaceae)

Origins:

Latin *roridus, a, um* "dewy, wet with dew," referring to the long-stalked oily or mucilaginous glands.

Roridula Forssk. Capparidaceae (Capparaceae)

Origins:

From the Latin *roridus, a, um* "dewy, wet with dew."

Roripella (Maire) Greuter & Burdet Brassicaceae

Origins:

The diminutive of the generic name *Rorippa* Scop.

Rorippa Scop. Brassicaceae

Origins:

Possibly from a Saxon vernacular name, *rorippen*. See Euricius Cordus (1486-1535), *Botanologicon*. Coloniae 1534; William Curtis (1746-1799), *Flora Londinensis*. London [1775-] 1777-1798; Giovanni Antonio Scopoli (1723-1788), *Flora Carniolica*. 520. Viennae 1760; F. Boerner & G. Kunkel, *Taschenwörterbuch der botanischen Pflanzennamen*. 4. Aufl. 163. 1989.

Species/Vernacular Names:

R. amphibia (L.) Bess.

English: great yellow cress

R. austriaca (Crantz) Besser

English: Austrian fieldcress

R. columbiae (Robinson) Howell

English: Columbia yellow cress

R. fluviatilis (E. Mey. ex Sond.) Thell. var. *fluviatilis* (*Nasturtium fluviatile* E. Mey. ex Sond.; *Rorippa fluviatilis* (E. Mey. ex Sond.) Thell.)

South Africa: waterkervel

R. microphylla (Boenn.) Hylander

English: one-rowed watercress

R. nasturtium-aquaticum (L.) Hayek (*Nasturtium officinale* R. Br.; *Sisymbrium nasturtium-aquaticum* L.)

English: watercress, common watercress

Peru: berro, chijchi, occoruro

Southern Africa: brongras, bronkhorstslaai, sterkkos, stercors; kerese (Sotho)

Japan: Oranda-garashi (= Holland mustard)

R. palustris (L.) Besser (*Nasturtium palustre* (L.) DC.; *Rorippa islandica* (Oeder) Borbás)

English: marshcress, yellow cress, marsh watercress, yellow marshcress, marsh yellow cress, yellow swampcress

R. sinuata (Torrey & A. Gray) Hitchc.

English: spreading yellow cress

R. subumbellata Rollins

English: Tahoe watercress

R. sylvestris (L.) Bess.

English: creeping yellow cress, yellow fieldcress

R. teres (Michx.) Stuckey

English: terete yellow cress

Rosa L. Rosaceae

Origins:

The ancient Latin name *rosa, ae*, Akkadian *russu* "red"; see Carl Linnaeus, *Species Plantarum*. 491. 1753 and *Genera Plantarum*. Ed. 5. 217. 1754; G. Semerano, *Le origini della cultura europea*. Dizionario della lingua Latina e di voci moderne. 2(2): 547. 1994.

Species/Vernacular Names:

R. sp.

Indonesia: mawar

Bali: bungan mawa

R. acicularis Lindl.

English: bristly rose, prickly rose

R. x *alba* L.

English: common English dog rose, white cottage rose, Indian white rose

India: bhringeshtha, gulab, shwet gulab, swet gulab, mullusevantige, seboti, sevati, gulseoti, gulchini

R. arkansana Porter

English: prairie rose, Arkansas rose, dwarf prairie rose, prairie wild rose

R. arvensis Huds.

English: field rose

R. banksiae Ait.

English: banksian rose, banksia rose

China: mu hsiang

R. blanda Aiton

English: smooth rose

R. bracteata H. Wendl.

English: Macartney's rose

Japan: yaeyama-no-ibara

R. brunonii Lindl.

English: Himalayan musk rose

India: sewati, kuji, kunja, karer, kwiala, tarni, phulwari, chal

R. californica Cham. & Schltdl.

English: California rose, California wild rose

R. canina L.

English: dog rose, common brier, dog brier

Arabic: nisri, nisrin, nesri, ward es-sini

R. carolina L.

English: Carolina rose, pasture rose

R. x centifolia L.

English: Provence rose, cabbage rose, Holland rose, hundred-leaved rose

India: devataruni, gulab, irosa, satapatri, troja, pannir

Japan: seiyô-ibara, bara

Mexico: guie becohua, quije pecohua castilla

R. chinensis Jacq.

English: China rose, Bengal rose, monthly rose, Chinese tea rose

India: kanta gulab

Japan: kôshin-ibara

China: xiao jin ying, chiang wei

R. cymosa Tratt.

China: xiao jin ying

R. x damascena Mill.

English: Damask rose, summer Damask rose, Persian rose

Arabic: ward, ward djouri, zirr el-ward

India: panniruppu, shatapatri, satapatri, gul, gulab, fasli gulab, gulabshavante, gulabihuvu, gulisurkh, roja-puvu, golappu, rojappu, bussorah, gulab-ke-phul, golap-phul, gulabnu-phul

R. eglanteria L. (*Rosa rubiginosa* L.)

English: eglantine, sweetbriar, sweetbrier, wild rose

R. foetida Herrm.

English: Austrian briar, Austrian yellow rose

R. foliolosa Nutt. ex Torr. & A. Gray

English: leafy rose, white prairie rose

R. gallica L.

English: French rose, red rose

India: gulap

Arabic: ouard

R. gymnocarpa Nutt.

English: wood rose

R. hemispherica Herrm.

English: sulphur rose

R. laevigata Michaux

English: Cherokee rose

China: jin ying zi, chin ying tsu

Vietnam: kim anh, thich le tu

R. luciae Franch. & Rochebr. & Crépin

Japan: ô-fuji-bara

R. majalis Herrm.

English: cinnamon rose, may rose

R. minutifolia Engelm.

English: small-leaved rose

R. moschata Herrm.

English: musk rose, musk-scented rose

India: kubjaka, kujai, kuja

R. multiflora Thunb.

English: Japanese rose, baby rose, multiflora rose, tea rose, seven sisters rose

China: qiang wei hua, chiang wei, chiang m (= wall rose)

R. nutkana C. Presl

English: Nootka rose, Nutka rose

R. odorata (Andr.) Sweet

English: tea rose

R. palustris Marshall

English: swamp rose

R. pimpinellifolia L.

English: Burnet rose, Scotch rose

R. pinetorum A.A. Heller

English: pine rose

R. pisocarpa A. Gray

English: cluster rose

R. roxburghii Tratt.

English: chestnut rose, Chinquapin rose

Tibetan: se-ba'i me-tog

R. rubiginosa L.

English: sweet briar

R. rugosa Thunb.

English: Japanese rose, Turkestan rose, rugosa rose

China: mei gui hua, mei kuei hua

R. sempervirens L.

English: evergreen rose

China: yueh chi hua

R. setigera Michx.

English: prairie rose, sunshine rose, climbing rose, climbing prairie rose

R. spinosissima L.

English: Burnet's rose, Scotch rose

R. virginiana Mill.

English: Virginia rose

R. wichuraiana Crépin

English: memorial rose

Japan: teri-ha-no-ibara

R. woodsii Lindl.

English: western wild rose, western rose, Wood's rose

Rosanowia Regel gesneriaceae

Origins:

For the Russian botanist Sergei Matveevich (Matveevic) Rosanoff (Rozanov), 1840-1870 (died at sea); see John H. Barnhart (1871-1949), *Biographical Notes upon Botanists.* 3: 177. Boston 1965; Stafleu and Cowan, *Taxonomic Literature.* 4: 881. 1983.

Rosanthus Small Malpighiaceae

Origins:

For the American botanist Joseph Nelson Rose, 1862-1928, botanical explorer, traveler, plant collector (southwestern USA, Mexico and South America). Among his works are "Notes on useful plants of Mexico." *Contr. U.S. Natl. Herb.* 5(4): 209-259. 1899 and "List of plants collected by the *U.S.S. Albatross* in 1887-1891 along the western coast of America." *Contr. U.S. Natl. Herb.* 1(5): 135-142. 1892, with Nathaniel Lord Britton (1859-1934) wrote *The Cactaceae.* Washington 1919-1923; see John H. Barnhart, *Biographical Notes upon Botanists.* 3: 177. 1965; T.W. Bossert, *Biographical Dictionary of Botanists Represented in the Hunt Institute Portrait Collection.* 338. 1972; E.M. Tucker, *Catalogue of the Library of the Arnold Arboretum of Harvard University.* 1917-1933; Ida Kaplan Langman, *A Selected Guide to the Literature on the Flowering Plants of Mexico.* 645-650. Philadelphia 1964; S. Lenley et al., *Catalog of the Manuscript and Archival Collections and Index to the Correspondence of John Torrey.* Library of the New York Botanical Garden. 354. 1973; Lyman David Benson (1909-1993), *The Cacti of the United States and Canada.* Stanford, California 1982; Larry W. Mitich, *Cactus and Succulent Journal.* Vol. 53, no. 6: 299-303. 1981; Gordon Douglas Rowley, *A History of Succulent Plants.* Strawberry Press, Mill Valley, California 1997; R. Zander, F. Encke, G. Buchheim & S. Seybold, *Handwörterbuch der Pflanzennamen.* 14 Aufl. Stuttgart 1993; J. Ewan & Nesta Dunn Ewan, "Biographical

dictionary of Rocky mountain naturalists." *Reg. Veget.* 107: 1-253. 1981; Irving William Knobloch, compil., "A preliminary verified list of plant collectors in Mexico." *Phytologia Memoirs.* VI. 1983; Ira L. Wiggins, *Flora of Baja California.* 43. California 1980.

Roscheria H. Wendland ex Balf.f. Palmae

Origins:

To commemorate Dr. Albrecht Roscher, 1836-1860 (murdered, L. Nyassa), a German traveler and explorer, geologist, collected algae in Zanzibar, collected for the German botanist Julius Rudolph Theodor Vogel (1812-1841), with Burton, Speke and Grant in exploring East Africa; see *Principes.* Volume 21(3): 140. 1977; Carl Claus (Karl Klaus) von der Decken (1833-1865), *Baron C.C. von der Decken's Reisen in Ost Afrika in 1859-61.* Leipzig & Heidelberg 1869-1879; F.A. Maximilian Kuhn, *Filices africanae ... Accedunt filices Deckenianae et Petersianae.* Lipsiae [Leipzig] 1868; F.N. Hepper and Fiona Neate, *Plant Collectors in West Africa.* 69. Utrecht 1971; John H. Barnhart, *Biographical Notes upon Botanists.* Boston 1965; William Allen (1793-1864) and Thomas Richard Heywood Thomson, *A Narrative of the Expedition ... to the River Niger, in 1841.* London 1848; J.F. Schön & S. Crowther, *Journals of the Expedition up the Niger in 1841.* London 1848; I.H. Vegter, *Index Herbariorum.* Part II (5), *Collectors N-R.* Regnum Vegetabile vol. 109. 1983.

Roscoea Smith Zingiberaceae

Origins:

Named for William Roscoe, 1753-1831 (Liverpool), botanist, historian, in 1802 founded the Liverpool Botanical Garden, 1804 a Fellow of the Linnean Society, correspondent of J.E. Smith. See Henry Roscoe, *The Life of William Roscoe,* by his son. Boston 1833; John H. Barnhart, *Biographical Notes upon Botanists.* 3: 177. Boston 1965; T. W. Bossert, *Biographical Dictionary of Botanists Represented in the Hunt Institute Portrait Collection.* 338. Boston, Massachusetts 1972; Warren R. Dawson, *The Banks Letters, A Calendar of the Manuscript Correspondence of Sir Joseph Banks.* London 1958; Antoine Lasègue (1793-1873), *Musée botanique de M. Benjamin Delessert.* 539. Paris, Leipzig 1845; S. Lenley et al., *Catalog of the Manuscript and Archival Collections and Index to the Correspondence of John Torrey.* Library of the New York Botanical Garden. 354. 1973; M. Hadfield et al., *British Gardeners: A Biographical Dictionary.* London 1980; *Memoir and Correspondence of ... Sir J.E. Smith ...* Edited by Lady Pleasance Smith. London 1832; Jeannette Elizabeth Graustein, *Thomas Nuttall, Naturalist. Explorations in America, 1808-1841.* Cambridge,

Harvard University Press 1967; R. Zander, F. Encke, G. Buchheim & S. Seybold, *Handwörterbuch der Pflanzennamen*. 14 Aufl. 771. Stuttgart 1993; Ida Kaplan Langman, *A Selected Guide to the Literature on the Flowering Plants of Mexico*. Philadelphia 1964; J.D. Milner, *Catalogue of Portraits of Botanists Exhibited in the Museums of the Royal Botanic Gardens*. Royal Botanic Gardens, Kew, London 1906; Jill Cowley & Richard Wilford, "*Roscoea tumjensis.*" in *Curtis's Botanical Magazine*. 15(4): 220-225. November 1998; Jill Cowley & Richard Wilford, "*Roscoea capitata.*" in *Curtis's Botanical Magazine*. 15(4): 226-230. November 1998.

Rosea Klotzsch Rubiaceae

Origins:

After the German apothecary Valentin Rose; see Berlin. *Berlinisches Jahrbuch für die Pharmacie*, etc. [Edited by V.R., etc.] 1818 etc.; A. Pabst, in *Dictionary of Scientific Biography* 11: 539-540. 1981; Stuart Pierson, in *Dictionary of Scientific Biography* 11: 540-542. 1981.

Roseanthus Cogniaux Cucurbitaceae

Origins:

For the American botanist Joseph Nelson Rose, 1862-1928.

Roseia Fric Cactaceae

Origins:

Probably for the American botanist Joseph Nelson Rose, 1862-1928.

Rosenbergiodendron Fagerl. Rubiaceae

Origins:

Dedicated to the Swedish botanist Gustaf Otto Rosenberg, 1872-1948, embryologist, professor of botany, author of *Apogamie und Parthenogenesis bei Pflanzen*. Berlin 1930; see John H. Barnhart, *Biographical Notes upon Botanists*. 3: 178. 1965; T.W. Bossert, *Biographical Dictionary of Botanists Represented in the Hunt Institute Portrait Collection*. 338. 1972.

Rosenia Thunberg Asteraceae

Origins:

Named for the Swedish physician and botanist Eberhard Rosén (Rosenblad) (1714-1796) and his brother Nils Rosén (later N. Rosén von Rosenstein) (1706-1773); see John H. Barnhart, *Biographical Notes upon Botanists*. 3: 178, 179. [see Rosenblad and Rosenstein] 1965; Ethelyn Maria Tucker, *Catalogue of the Library of the Arnold Arboretum of Harvard University*. Cambridge, Massachusetts 1917-1933; Jonas C. Dryander, *Catalogus bibliothecae historico-naturalis Josephi Banks*. London 1800.

Rosenstockia Copeland Hymenophyllaceae

Origins:

After the German botanist Eduard Rosenstock, 1856-1938, pteridologist, fern collector, author of "Beiträge zur Pteridophytenflora Südbrasiliens ... II." *Hedwigia*. 46: 57-167. 1906; see John H. Barnhart, *Biographical Notes upon Botanists*. 3: 179. 1965; H.N. Clokie, *Account of the Herbaria of the Department of Botany in the University of Oxford*. 234. Oxford 1964; Elmer Drew Merrill, in *Bernice P. Bishop Mus. Bull*. 144: 160. 1937; I.C. Hedge and J.M. Lamond, *Index of Collectors in the Edinburgh Herbarium*. Edinburgh 1970.

Roseocactus A. Berger Cactaceae

Origins:

For the American botanist Joseph Nelson Rose, 1862-1928, botanical explorer, traveler, plant collector (southwestern USA, Mexico and South America), author of "Notes on useful plants of Mexico." *Contr. U.S. Natl. Herb*. 5(4): 209-259. 1899, with Nathaniel Lord Britton (1859-1934) wrote *The Cactaceae*. Washington 1919-1923. See John H. Barnhart, *Biographical Notes upon Botanists*. 3: 177. 1965; T.W. Bossert, *Biographical Dictionary of Botanists Represented in the Hunt Institute Portrait Collection*. 338. 1972; Ethelyn Maria Tucker, *Catalogue of the Library of the Arnold Arboretum of Harvard University*. Cambridge, Massachusetts 1917-1933; Ida Kaplan Langman, *A Selected Guide to the Literature on the Flowering Plants of Mexico*. 645-650. Philadelphia 1964; S. Lenley et al., *Catalog of the Manuscript and Archival Collections and Index to the Correspondence of John Torrey*. Library of the New York Botanical Garden. 354. 1973; Lyman David Benson (1909-1993), *The Cacti of the United States and Canada*. Stanford, California 1982; Larry W. Mitich, *Cactus and Succulent Journal*. Vol. 53(6): 299-303. 1981; Ira L. Wiggins, *Flora of Baja California*. 43. Stanford, California 1980; Gordon Douglas Rowley, *A History of Succulent Plants*. 1997.

Roseocereus Backeb. Cactaceae

Origins:

For the American botanist Joseph Nelson Rose, 1862-1928.

Roseodendron Miranda Bignoniaceae

Origins:

For the American botanist Joseph Nelson Rose, 1862-1928.

Roshevitzia Tzvelev Gramineae

Origins:

For the Russian botanist Roman (Romain) Julievich Roshevitz, 1882-1949, agrostologist; see J.H. Barnhart, *Biographical Notes upon Botanists*. 3: 180. 1965; T.W. Bossert, *Biographical Dictionary of Botanists Represented in the Hunt Institute Portrait Collection*. 338. 1972.

Rosifax C.C. Townsend Amaranthaceae

Origins:

Latin *roseus, a, um* "rose-colored, rosy" and *fax, facis* "a torch."

Rosmarinus L. Labiatae

Origins:

Latin name for the plant, from *ros, roris* "dew" and *marinus* "maritime," *ros marinus, ros maris, marinus ros*; see Carl Linnaeus, *Species Plantarum*. 23. 1753 and *Genera Plantarum*. Ed. 5. 14. 1754; H. Genaust, *Etymologisches Wörterbuch der botanischen Pflanzennamen*. 543. 1996; Pietro Bubani, *Flora Virgiliana*. 97-98. [Ristampa dell'edizione di Bologna 1870] Bologna 1978; Manlio Cortelazzo & Paolo Zolli, *Dizionario etimologico della lingua italiana*. 4: 1106. 1985.

Species/Vernacular Names:

R. officinalis L.

English: rosemary, common rosemary

Arabic: klil, iklil, iklil el-gabal

Italian: rosmarino

India: rusmari

China: mi die xiang, mi tieh hsiang

Mexico: guixi cicanaca yala-rillaa, quixi cicanaca yala-tillaa

Rossioglossum (Schltr.) Garay & G.C. Kennedy Orchidaceae

Origins:

In honor of the English orchid collector John Ross, in Mexico between 1830 and 1840, referring to the genus *Odontoglossum*; see R. Zander, F. Encke, G. Buchheim & S. Seybold, *Handwörterbuch der Pflanzennamen*. 14 Aufl. 771. [the genus is dedicated to the Italian botanist M.L. Rossi, 1850-1932] 1993; Richard Evans Schultes and Arthur Stanley Pease, *Generic Names of Orchids. Their Origin and Meaning*. 1963; Hubert Mayr, *Orchid Names and Their Meanings*. 1998; Ray Desmond, *Dictionary of British & Irish Botanists and Horticulturists*. 595. 1994.

Rostellaria Gaertner Sapotaceae

Origins:

Latin *rostellum, i* "a little beak, a small snout."

Rostellaria Nees Acanthaceae

Origins:

From the Latin *rostellum, i* "a little beak, a small snout."

Rostellularia Reichb. Acanthaceae

Origins:

Latin *rostellum*, a little beak, a small snout, *rostrum, i* "the beak of a bird," *rodo, is, si, sum, ere* "to gnaw, eat away, corrode," referring to a small basal appendage on the lower anther cell; see Heinrich Gottlieb L. Reichenbach (1793-1879), *Handbuch des natürlichen Pflanzensystems nach allen seinen Classen, Ordnungen und Familien, etc*. 190. Dresden & Leipzig 1837.

Species/Vernacular Names:

R. adscendens (R. Br.) R.M. Barker (*Justicia adscendens* R. Br.)

English: pink tongues, red trumpet

R. obtusa Nees (*Rostellularia peploides* (Nees) Nees)

English: white trumpet

Rostkovia Desvaux Juncaceae

Origins:

For the German botanist Friedrich Wilhelm Gottlieb (Theophilus) Rostkovius, 1770-1848, physician, wrote *De Junco*. Halae [1801]; see John H. Barnhart, *Biographical Notes upon Botanists*. 3: 181. 1965; Ethelyn Maria Tucker, *Catalogue of the Library of the Arnold Arboretum of Harvard University*. Cambridge, Massachusetts 1917-1933.

Rostraria Trin. Gramineae

Origins:

Latin *rostrum, i* "the beak of a bird."

Species/Vernacular Names:
R. cristata (L.) Tzvelev
English: cat-tail grass

Rostrinucula Kudô Labiatae

Origins:
Latin *rostrum, i* and *nucula, ae* "a small nut."

Species/Vernacular Names:
R. dependens (Rehder) Kudô
English: hooked rostrinucula
China: goi zi mu
R. sinensis (Hemsley) C.Y. Wu
English: Chinese rostrinucula
China: chang ye goi zi mu

Rosularia (DC.) Stapf Crassulaceae

Origins:
Latin *rosula, ae* "a little rose," referring to the leaf-rosettes.

Rotala L. Lythraceae

Origins:
Latin *rotalis* "wheeled, wheel-like, having a wheel, having wheels," *rota* "a wheel," referring to the whorled leaves; see C. Linnaeus, *Mantissa Plantarum.* 2: 143, 175. 1771.

Species/Vernacular Names:
R. indica (Willd.) Koehne
English: tooth-cup

Rotang Adans. Palmae

Origins:
From the common names *rattan, rotan*; see C.T. Onions, *The Oxford Dictionary of English Etymology.* Oxford University Press 1966; Helmut Genaust, *Etymologisches Wörterbuch der botanischen Pflanzennamen.* 544. 1996.

Rotanga Boehm. Palmae

Origins:
From the common names *rattan, rotan, rotan(g)*, see also *Calamus* L.

Rotantha Baker Lythraceae

Origins:
Latin *rota, ae* "a wheel" and Greek *anthos* "flower."

Rotantha Small Campanulaceae

Origins:
From the Latin *rota, ae* and Greek *anthos* "flower," see also *Campanula* L.

Rotbolla Zumaglini Gramineae

Origins:
Orthographic variant, see *Rottboellia* L.f.

Rotbollia T. Nuttall Gramineae

Origins:
Orthographic variant, see *Rottboellia* L.f.

Rotheca Raf. Labiatae

Origins:
See Constantine Samuel Rafinesque, *Flora Telluriana.* 4: 69. 1836 [1838]; E.D. Merrill, *Index rafinesquianus.* The plant names published by C.S. Rafinesque, etc. 204. Jamaica Plain, Massachusetts, USA 1949.

Rothia Borkh. Gramineae

Origins:
For the German botanist Albrecht Wilhelm Roth, 1757-1834, physician.

Rothia Pers. Fabaceae

Origins:
For the German botanist Albrecht Wilhelm Roth, 1757-1834, physician, studied at Halle and Erlangen, M.D. Erlangen 1778. His works include Dissertatio ... *de diaeta puerperarum bene instituenda.* Erlangae [1778], *Catalecta botanica.* Lipsiae 1797-1806, *Tentamen Florae Germanicae.* Lipsiae 1788, etc., *Botanische Abhandlungen und Beobachtungen.* Nürnberg 1787 and *Novae plantarum species praesertim Indiae orientalis.* Ex collectione doct. Benj. Heynii. Halberstadii 1821, father of the German botanist C.W. Roth (1810-1881); see John H. Barnhart, *Biographical Notes upon Botanists.* 3: 182. 1965; R. Zander, F. Encke, G. Buchheim & S. Seybold, *Handwörterbuch der Pflanzennamen.* 14

Aufl. Stuttgart 1993; Mariella Azzarello Di Misa, ed., *Il Fondo Antico della Biblioteca dell'Orto Botanico di Palermo*. 235-236. Regione Siciliana, Palermo 1988; Christiaan Hendrik Persoon (1761-1836), *Synopsis plantarum*. 2: 638 & corrigenda. Paris et Tubingae 1807; T.W. Bossert, *Biographical Dictionary of Botanists Represented in the Hunt Institute Portrait Collection*. 339. 1972; Ethelyn Maria Tucker, *Catalogue of the Library of the Arnold Arboretum of Harvard University*. Cambridge, Massachusetts 1917-1933; J.D. Milner, *Catalogue of Portraits of Botanists Exhibited in the Museums of the Royal Botanic Gardens*. Royal Botanic Gardens, Kew, London 1906; Jonas C. Dryander, *Catalogus bibliothecae historico-naturalis Josephi Banks*. London 1800; A. Lasègue, *Musée botanique de Benjamin Delessert*. Paris 1845; Isaac Henry Burkill, *Chapters on the History of Botany in India*. Delhi 1965; G. Schmid, *Goethe und die Naturwissenschaften*. Halle 1940; Giulio Giorello & Agnese Grieco, a cura di, *Goethe scienziato*. Einaudi Editore, Torino 1998.

Rothmaleria Font Quer Asteraceae

Origins:

For the German botanist Werner Hugo Paul Rothmaler, 1908-1962, traveler, professor of botany, professor of systematic botany and agricultural biology. His writings include "Sobre algunas *Alchemilla* de Sudamérica." *Revista Sudamer. Bot*. 2(1): 11-14. 1935 and "*Alchemillae* novae." *Bull. Misc. Inform*. 1938: 269-274. 1938. See J.H. Barnhart, *Biographical Notes upon Botanists*. 3: 183. [d. 1962] 1965; R. Zander, F. Encke, G. Buchheim & S. Seybold, *Handwörterbuch der Pflanzennamen*. 14 Aufl. 1993; T.W. Bossert, *Biographical Dictionary of Botanists Represented in the Hunt Institute Portrait Collection*. 340. 1972; Stafleu and Cowan, *Taxonomic Literature*. 4: 924-925. [d. 1963] Utrecht 1983; Ida Kaplan Langman, *A Selected Guide to the Literature on the Flowering Plants of Mexico*. 651. 1964.

Rothmannia Thunberg Rubiaceae

Origins:

After the Swedish physician and botanist Dr. Georg (Göran) Rothman, 1739-1778, M.D. Uppsala 1763, a friend of Thunberg and a student of Linnaeus, traveler and plant collector through North Africa, author of *De Raphania dissertatio medica ...* Praes. C. v. Linné, etc. Upsaliae [May 1763]; see Johan Gottschalk Wallerius (Wallerio) (1709-1785), *De origine oleorum in vegetabilibus*. Upsaliae (Mar.) 1761; John H. Barnhart, *Biographical Notes upon Botanists*. 3: 183. 1965; C.P. Thunberg, in *Kongl. Vetenskaps Academiens Handlingar*. 37: 65. Stockholm 1776; [Librairie Paul Jammes — Paris], Cabinets de Curiosités. Collections. Collectionneurs. [Item no. 381.] 1998.

Species/Vernacular Names:

R. capensis Thunberg (*Gardenia rothmannia* L.f.; *Gardenia capensis* (Thunb.) Druce)

English: Cape gardenia, candlewood

Southern Africa: Kaapse katjiepiering, aapsekos, kershout (kers means both "cherry" and "candle"), bobbejaanappel, bergkatjiepiering; umPhazane-mkhulu, umPhazana omkhulu (Zulu); umGupa, isiThebe, umZukuza, iBolo (Xhosa); modula-tshwene (North Sotho); mukubudu, muratha-mapfene (Venda)

R. fischeri (K. Schumann) Bullock subsp. *fischeri* (*Randia fischeri* K. Schum.) (for the German physician Dr. Gustav Adolf Fischer, 1848-1886, botanical collector in Masailand, East Africa)

English: woodland rothmannia, Rhodesian gardenia

Southern Africa: Zimbabwese katjiepiering, Rhodesiese katjiepiering; umKotshi-wenyama, isiPhindemuva (Zulu); muratha-mapfene (= baboon bridge) (Venda)

R. globosa (Hochst.) Keay (*Gardenia globosa* Hochst.)

English: september bells

Southern Africa: septemberklokkies, kafferklapper; umPhazane, isiCelankobe, isiGcatha-inkobe, isiCathangobe, imBhikihla, isiQhoba (Zulu); umGubhe (Xhosa)

R. hispida (K. Schum.) Fagerlind

Nigeria: asogbodun (Yoruba); asun (Edo); odi okong (Igbo); obong (Efik); eton (Boki)

R. lujae (De Wild.) Keay

Nigeria: asun nokwa (Edo); mbong (Efik)

Cameroon: endon

Gabon: etoum-essale

R. manganjae (Hiern) Keay

English: scented bells

R. urcelliformis (Schweinf. ex Hiern) Bullock ex Robyns

English: forest rothmannia

Southern Africa: tamba we bungu (Shona)

Nigeria: uli obwe (Igbo)

R. whitfieldii (Lindley) Dandy

Nigeria: buje-nla (Yoruba); uli oba (Igbo)

Yoruba: buje nla

Rothrockia A. Gray Asclepiadaceae

Origins:

For the American botanist Joseph Trimble Rothrock, 1839-1922, surgeon, explorer, plant collector, professor of botany, forester, one-time student of Asa Gray, with Robert Kennicott and Major Frank Pope exploring expedition to British Columbia and Alaska; see J.H. Barnhart, *Biographical Notes upon Botanists*. 3: 183. 1965; T.W. Bossert, *Biographical Dictionary of Botanists Represented in the Hunt Institute Portrait Collection*. 340. 1972; Ethelyn Maria Tucker, *Catalogue of the Library of the Arnold Arboretum of Harvard University*. Cambridge, Massachusetts 1917-1933; Ida Kaplan Langman, *A Selected Guide to the Literature on the Flowering Plants of Mexico*. 651-652. 1964; S. Lenley et al., *Catalog of the Manuscript and Archival Collections and Index to the Correspondence of John Torrey*. Library of the New York Botanical Garden. 355. 1973; J.W. Harshberger, *The Botanists of Philadelphia and Their Work*. 305-313. 1899; Howard Atwood Kelly and Walter Lincoln Burrage, *Dictionary of American Medical Biography*. New York 1928; Joseph Ewan, *Rocky Mountain Naturalists*. The University of Denver Press 1950; John T. Walbran, *British Columbia Coast Names, 1592-1906. To Which are Added a Few Names in Adjacent United States Territory, Their Origin and History*. First edition. Ottawa: Government Printing Bureau, 1909; Irving William Knobloch, compil., "A preliminary verified list of plant collectors in Mexico." *Phytologia Memoirs*. VI. 1983; G.P.V. Akrigg & Helen B. Akrigg, *British Columbia Place Names*. Sono Nis Press, Victoria 1986.

Rottboelia Dumort. Gramineae

Origins:

Orthographic variant, see *Rottboellia* L.f.

Rottboelia Scop. Olacaceae

Origins:

After the Danish botanist Christen Friis Rottboell (Rottbøll), 1727-1797; see J.H. Barnhart, *Biographical Notes upon Botanists*. 3: 183. 1965.

Rottboella L.f. Gramineae

Origins:

Orthographic variant, see *Rottboellia* L.f.

Rottboellia L.f. Gramineae

Origins:

After the Danish botanist Christen Friis Rottboell (Rottbøll), 1727-1797, physician, M.D. Copenhagen 1755, pupil of Linnaeus, traveler, 1770-1797 Director of the Copenhagen Botanical Garden, 1776-1797 professor of medicine. Among his writings are *Descriptiones plantarum rariorum*. Havniae [Copenhagen] 1772, *Plantas Horti Universitatis rariores programmate ... describit C.F. Rottböll*, etc. Hafniae 1773 and *Descriptiones rariorum plantarum*, nec non materiae medicae atque oeconomicae *e terra surinamensi ...* Havniae [1776]; see Johann G. König (1728-1785), Dissertationem inauguralem *de remediorum indigenorum ad morbos cuivis regioni endemicos expugnandos efficacia*, praeside C.F. Rottböll. Hafniae 1773; Mariella Azzarello Di Misa, a cura di, *Il Fondo Antico della Biblioteca dell'Orto Botanico di Palermo*. 236-237. Regione Siciliana, Palermo 1988; Carl Frederik Albert Christensen, *Den danske Botaniks Historie med tilhørende Bibliografi*. Copenhagen 1924-1926; J.H. Barnhart, *Biographical Notes upon Botanists*. 3: 183. 1965; C. Linnaeus (filius), *Supplementum Plantarum*. 114. 1782; J.W. Hornemann, *Naturh. Tidsskr*. 1: 566-567. 1837; T.W. Bossert, *Biographical Dictionary of Botanists Represented in the Hunt Institute Portrait Collection*. 340. 1972; R. Zander, F. Encke, G. Buchheim & S. Seybold, *Handwörterbuch der Pflanzennamen*. 14 Aufl. Stuttgart 1993; Ethelyn Maria Tucker, *Catalogue of the Library of the Arnold Arboretum of Harvard University*. Cambridge, Massachusetts 1917-1933; Jonas C. Dryander, *Catalogus bibliothecae historico-naturalis Josephi Banks*. London 1800; A. Lasègue, *Musée botanique de Benjamin Delessert*. Paris 1845.

Species/Vernacular Names:

R. cochinchinensis (Lour.) W. Clayton (*Rottboellia exaltata* L.f.)

English: Guinea-fowl grass, itch grass, kokoma grass, shamva grass, Raoul grass, Kelly grass

East Africa: ewokiwok, mbaya, mwamba nyama, nyamrungru

Yoruba: holo

South Africa: tarentaalgras

Japan: tsuno-ai-ashi

Okinawa: yamatsu-gusa

Rottbolla Lam. Gramineae

Origins:

Orthographic variant, see *Rottboellia* L.f.

Rottbollia Juss. Gramineae

Origins:

Orthographic variant, see *Rottboellia* L.f.

Rottlera Roemer & Schultes Gesneriaceae

Origins:

Orthographic variant of *Roettlera* Vahl.

Rottlera Roxb. Euphorbiaceae

Origins:

For the Alsatian-born Danish missionary Johan Peter Rottler, 1749-1836, traveler and botanist, from 1776 to 1806 at Tranquebar (Madras, Coromandel coast, India), in 1788 in the Ganges region and in 1795 at Ceylon, wrote *A Dictionary of the Tamil and English Languages*. Madras 1834-1841, in 1828 translated *The Book of Common Prayer* in Tamil. See [Bible] [Tamil], *The Old Testament* [revised by C.G.E. Rhenius, with J.P. Rottler and others. 1827; John H. Barnhart, *Biographical Notes upon Botanists*. 3: 183. 1965; Mary Gunn and Leslie E. Codd, *Botanical Exploration of Southern Africa*. 301. Cape Town 1981; Martin H. Vahl (1749-1804), *M. Vahlii ... Enumeratio Plantarum*. 1: 87. Hauniae 1804; Bishop James M. Thoburn, *The Christian Conquest of India*. [Forward Mission Study Courses, edited under the auspices of the Young People's Missionary Movement.] Eaton & Mains, NY n.d.; Antoine Lasègue, *Musée botanique de M. Benjamin Delessert*. Paris 1845; Isaac Henry Burkill, *Chapters on the History of Botany in India*. Delhi 1965; R. Zander, F. Encke, G. Buchheim & S. Seybold, *Handwörterbuch der Pflanzennamen*. 14 Aufl. Stuttgart 1993; Carl Frederik Albert Christensen, *Den danske Botaniks Historie med tilhørende Bibliografi*. Copenhagen 1924-1926; William Roxburgh (1751-1815), *Plants of the Coast of Coromandel*. 2: 36, t. 168. London 1802.

Rottlera Willd. Euphorbiaceae

Origins:

After the Danish (b. Alsatia) missionary Johan Peter Rottler, 1749-1836, traveler and botanist.

Rotula Lour. Boraginaceae (Ehretiaceae)

Origins:

Latin *rotula, ae* "a little wheel."

Species/Vernacular Names:

R. aquatica Loureiro

English: water rotula

China: lun guan mu

Roucheria Planchon Linaceae

Origins:

For the French poet Jean Antoine Roucher, 1745-1794. His works include *Hymne funèbre chanté au Champ de la Fédération*, le 3 juin 1792 ... pour honorer la mémoire de J.G. Simoneau, Maire d'Estampes. Paris [1792], *Maximilien Jules Léopold Duc de Brunswick-Lunebourg*, poëme. Paris 1786 and *Les Mois*, poème en douze chants. Paris 1779; see Antoine Guillois, *Pendant le Terreur*. Le poète Roucher, 1745-1794. Paris 1890.

Rouliniella Vail Asclepiadaceae

Origins:

The diminutive of *Roulinia* Decne.

Roupala Aublet Proteaceae

Origins:

From the native name in Guiana; see Jean B.C.F. Aublet (1720-1778), *Histoire des Plantes de la Guiane Françoise*. 83, t. 32. Paris 1775.

Species/Vernacular Names:

R. sp.

Peru: ají, cedro bordado, cedrorana, louro faia

Roupellia Wallich & Hooker Apocynaceae

Origins:

Dedicated to the Roupell family; see Stafleu and Cowan, *Taxonomic Literature*. 4: 938. Utrecht 1983.

Roupellina (Baillon) Pichon Apocynaceae

Origins:

For the British (b. Shropshire) botanical artist Arabella Elizabeth Roupell (*née* Piggott), 1817-1914 (d. Berkshire), painter of flowers, traveler, author of *Specimens of the Flora of South Africa by a Lady*. [London 1850] (the botanical text is by William Henry Harvey, 1811-1866); see Allan Bird, *The Lady of the Cape Flowers*. Paintings by Arabella Roupell. Johannesburg. The South African Natural History Publ. [1964]; Mary Gunn and Leslie E. Codd, *Botanical Exploration of Southern Africa*. 301-302. Cape Town 1981; Allan Bird, *More Cape Flowers by a Lady*. Paintings by Arabella Roupell. Johannesburg. The South African Natural

History Publ. 1964; John H. Barnhart, *Biographical Notes upon Botanists*. 3: 184. 1965.

Rourea Aublet Connaraceae

Origins:

From the native name in Guiana; see Jean B.C.F. Aublet (1720-1778), *Histoire des Plantes de la Guiane Françoise*. 1: 467, t. 187. Paris 1775.

Roureopsis Planch. Connaraceae

Origins:

Resembling *Rourea* Aubl.

Roussaea A. DC. Grossulariaceae (Brexiaceae, Escalloniaceae)

Origins:

Orthographic variant of *Roussea* Smith.

Roussea Smith Grossulariaceae (Brexiaceae, Escalloniaceae)

Origins:

For the Swiss political philosopher Jean Jacques Rousseau, 1712-1778, writer, plant collector; see T.W. Bossert, *Biographical Dictionary of Botanists Represented in the Hunt Institute Portrait Collection*. 340. 1972; Ethelyn Maria Tucker, *Catalogue of the Library of the Arnold Arboretum of Harvard University*. Cambridge, Massachusetts 1917-1933; D.E. Allen, *The Naturalist in Britain*. 1970; Jonas C. Dryander, *Catalogus bibliothecae historico-naturalis Josephi Banks*. London 1800; A. Lasègue, *Musée botanique de Benjamin Delessert*. Paris 1845; E. Earnest, *John and William Bartram, Botanists and Explorers 1699-1777, 1739-1823*. Philadelphia 1940; Bentley Glass et al., eds., *Forerunners of Darwin: 1745-1859*. Edited by ... O. Temkin, William Strauss, Jr. First edition. John Hopkins Press, Baltimore 1959; Frans A. Stafleu, *Linnaeus and the Linnaeans*. The spreading of their ideas in systematic botany, 1735-1789. Utrecht 1971; Stafleu and Cowan, *Taxonomic Literature*. 4: 938-941. Utrecht 1983.

Rousseaua Post & Kuntze Grossulariaceae (Brexiaceae, Escalloniaceae)

Origins:

Orthographic variant of *Roussea* Smith.

Rousseauvia Bojer Grossulariaceae (Brexiaceae, Escalloniaceae)

Origins:

Orthographic variant of *Roussea* Smith.

Rousselia Gaudich. Urticaceae

Origins:

For the French botanist Alexandre Victor Roussel, 1795-1874, pharmacist, 1835-1838 military pharmacist in Algeria; see T.W. Bossert, *Biographical Dictionary of Botanists Represented in the Hunt Institute Portrait Collection*. 340. 1972.

Roussoa Roemer & Schultes Grossulariaceae (Brexiaceae, Escalloniaceae)

Origins:

Orthographic variant of *Roussea* Smith.

Rouxia Husnot Gramineae

Origins:

After the French botanist Nisius Roux, 1854-1923; see John H. Barnhart, *Biographical Notes upon Botanists*. 3: 185. 1965.

Rouya Coincy Umbelliferae

Origins:

After the French botanist Georges C. Ch. Rouy, 1851-1924; see John H. Barnhart, *Biographical Notes upon Botanists*. 3: 185. 1965; T.W. Bossert, *Biographical Dictionary of Botanists Represented in the Hunt Institute Portrait Collection*. 340. 1972; E.M. Tucker, *Catalogue of the Library of the Arnold Arboretum of Harvard University*. 1917-1933; R. Zander, F. Encke, G. Buchheim and S. Seybold, *Handwörterbuch der Pflanzennamen*. 14. Aufl. Stuttgart 1993.

Roxburghia W. Jones ex Roxburgh Stemonaceae

Origins:

For the Scottish (b. Ayrshire) botanist William Roxburgh, 1751-1815 (d. Edinburgh), physician, M.D. Edinburgh, traveler and plant collector, 1776-1780 with the East India Company in the Madras Medical Service, from 1781

Superintendent of the Samalkot Botanic Garden, 1793-1813 Superintendent of the Calcutta Botanic Garden and Chief Botanist of East India Company, from 1798 to 1799 at Cape, 1799 Fellow of the Linnean Society, 1814 St. Helena. Among his writings are *Account of the Chermes Lacca ... From the Philosophical Transactions.* [London 1791], *Flora Indica;* or descriptions of Indian plants, etc. [Edited by William Carey, 1761-1834] Serampore 1832, *Plants of the Coast of Coromandel.* London 1795-1820 and *Hortus bengalensis.* Serampore 1814. See Joseph François Charpentier-Cossigny de Palma, *Essai sur la fabrique de l'indigo.* Isle de France 1779, and *Memoir Containing an Abridged Treatise, on the Cultivation and Manufacture of Indigo.* [*Process of Making Indigo on the Coast of Ingeram,* by William Roxburgh.] Calcutta 1789; D.G. Crawford, *A History of the Indian Medical Service, 1600-1913.* London 1914; M. Hadfield et al., *British Gardeners: A Biographical Dictionary.* London 1980; Stafleu and Cowan, *Taxonomic Literature.* 4: 954-958. 1983; Leonard Huxley, *Life and Letters of Sir Joseph Dalton Hooker.* London 1918; G. Murray, *History of the Collections Contained in the Natural History Departments of the British Museum.* 1: 46, 178. 1904; Ray Desmond, *Dictionary of British & Irish Botanists and Horticulturists.* 597-598. 1994; R. Zander, F. Encke, G. Buchheim and S. Seybold, *Handwörterbuch der Pflanzennamen.* 14. Aufl. 1993; N. Hall, *Botanists of the Eucalypts.* 1978 and Supplement 1980; John H. Barnhart, *Biographical Notes upon Botanists.* 3: 187. 1965; K. Biswas, ed., *The Original Correspondence of Sir Joseph Banks Relating to the Foundation of the Royal Botanic Garden, Calcutta* and *The Summary of the 150th Anniversary Volume of the Royal Botanic Garden, Calcutta.* Calcutta 1950; Alice Margaret Coats, *The Quest for Plants. A History of the Horticultural Explorers.* London 1969; Mary Gunn and L.E. Codd, *Botanical Exploration of Southern Africa.* 303. A.A. Balkema Cape Town 1981; Mariella Azzarello Di Misa, ed., *Il Fondo Antico della Biblioteca dell'Orto Botanico di Palermo.* 237. Regione Siciliana, Palermo 1988; K. Lemmon, *Golden Age of Plant Hunters.* London. 1968; A. White and B.L. Sloane, *The Stapelieae.* Pasadena 1937; Blanche Henrey, *British Botanical and Horticultural Literature before 1800.* 1975; M. Archer, *Natural History Drawings in the India Office Library.* London 1962; R. Desmond, *The European Discovery of the Indian Flora.* Oxford 1992; T.W. Bossert, *Biographical Dictionary of Botanists Represented in the Hunt Institute Portrait Collection.* 341. 1972; Ethelyn Maria Tucker, *Catalogue of the Library of the Arnold Arboretum of Harvard University.* Cambridge, Massachusetts 1917-1933; Jonas C. Dryander, *Catalogus bibliothecae historico-naturalis Josephi Banks.* London 1800; A. Lasègue, *Musée botanique de Benjamin Delessert.* Paris 1845; Emil Bretschneider (1833-1901), *History of European Botanical Discoveries in China.* Leipzig 1981.

Roycea C.A. Gardner Chenopodiaceae

Origins:

After Robert Dunlop Royce, born 1914, plant collector, from 1960 to 1974 Curator of the Perth Herbarium, Western Australia; see C.A. Gardner, in *Journal of the Royal Society of Western Australia.* 32: 77, t. II. (Jul.) 1948.

Roydsia Roxb. Capparidaceae (Capparaceae)

Origins:

After Sir John Royds, 1750-1817 (India), botanist, Judge Supreme Court of Bengal, a patron of science; see William Roxburgh, *Plants of the Coast of Coromandel.* 3: 87. London 1819.

Royena L. Ebenaceae

Origins:

For the Dutch botanist Adriaan (Adrianus) van Royen, 1704-1779, physician, M.D. Leyden 1728, professor of botany and medicine, 1730-1754 Director of Botanic Garden at Leyden, friend of Linnaeus, correspondent of A. V. Haller. Among his writings are Dissertatio ... *de anatome et oeconomia plantarum.* Lugduni Batavorum [Leyden] [1728], *Oratio qua ... medicinae cultoribus commendatur doctrina botanica.* Lugduni Batavorum 1729, Adriani van Royen *poemata.* (Carmen elegiacum in lapsum prae fracto femoris osse, etc.) Lugduni Batavorum 1778, *Florae Leydensis Prodromus,* exhibens plantas quae in horto academico Lugduno-Batavo aluntur. Lugduni Batavorum 1740 and *Ericetum africanum.* 40 tab. absque descriptione. (J. v. d. Spyk fecit. J.P. Catell del.). Opus ineditum. Lugdunum Batavorum circa 1760 (illustrations of 40 unnamed species of *Erica,* without text or title page, in the library of the Royal Botanic Gardens, Kew), uncle of the Dutch botanist and physician David van Royen (1727-1799). See Martin Wilhelm Schwencke (1707-1785), *Novae plantae Schwenckia* dictae a celeberrimo Linnaeo in *Gen. plant.* ed. VI. p. 567 ex celeb. Davidiis van Rooijen Charact. mss. 1761 communicata brevis descriptio et delineatio cum notis characteristicis. Hagae Comitum [The Hague], typ. van Karnebeek. 1766; H. Veendorp and L.G.M. Baas Becking, *Hortus Academicus Lugduno-Batavus 1587-1937.* The development of the gardens of Leyden University. Harlemi [Haarlem] 1938; D.O. Wijnands, *The Botany of the Commelins.* Rotterdam 1983; J.H. Barnhart, *Biographical Notes upon Botanists.* 3: 187. 1965; H.N. Clokie, *Account of the Herbaria of the Department of Botany in the University of Oxford.* 1964; Ethelyn Maria Tucker, *Catalogue of the Library of the Arnold Arboretum of Harvard University.* 1917-1933; G. Murray, *History of the Collections Contained in the Natural*

History Departments of the British Museum. 1: 178. 1904; Jonas C. Dryander, *Catalogus bibliothecae historico-naturalis Josephi Banks.* London 1800; A. Lasègue, *Musée botanique de Benjamin Delessert.* 1845; R. Zander, F. Encke, G. Buchheim and S. Seybold, *Handwörterbuch der Pflanzennamen.* 14. Aufl. 1993; Frans A. Stafleu, *Linnaeus and the Linnaeans.* 1971; Blanche Henrey, *No Ordinary Gardener — Thomas Knowlton, 1691-1781.* Edited by A.O. Chater. British Museum (Natural History). London 1986.

Royenia Spreng. Ebenaceae

Origins:

Orthographic variation of *Royena* L.

Roylea Wallich ex Bentham Labiatae

Origins:

After the British (b. India) botanist John Forbes Royle, 1800-1858 (d. Middx.), traveler, physician, M.D. München 1833, plant collector, 1833 Fellow of the Linnean Society, 1837 Fellow of the Royal Society, in Bengal (surgeon, East India Company), professor of materia medica, Curator Saharanpur Botanical Garden. His works include *The Cultivation of Cotton in India.* [London 1840], *Review of the Measures Which Have Been Adopted in India for the Improved Culture in Cotton.* London 1857, *An Essay on the Antiquity of Hindoo Medicine.* London 1837, *Medical Education.* London 1845 and *A Manual of Materia Medica and Therapeutics.* London 1847; see M. Archer, *Natural History Drawings in the India Office Library.* London 1962; R. Desmond, *The European Discovery of the Indian Flora.* Oxford 1992; J.H. Barnhart, *Biographical Notes upon Botanists.* 3: 187. 1965; E.M. Tucker, *Catalogue of the Library of the Arnold Arboretum of Harvard University.* 1917-1933; Mea Allan, *The Hookers of Kew.* London 1967; Isaac Henry Burkill, *Chapters on the History of Botany in India.* Delhi 1965; A. Lasègue, *Musée botanique de Benjamin Delessert.* Paris 1845; Ida Kaplan Langman, *A Selected Guide to the Literature on the Flowering Plants of Mexico.* 654. Philadelphia 1964; D.G. Crawford, *A History of the Indian Medical Service, 1600-1913.* London 1914; H.R. Fletcher, *Story of the Royal Horticultural Society, 1804-1968.* Oxford 1969; Mariella Azzarello Di Misa, a cura di, *Il Fondo Antico della Biblioteca dell'Orto Botanico di Palermo.* 237. Palermo 1988; Ray Desmond, *Dictionary of British & Irish Botanists and Horticulturists.* 598. 1994; R. Zander, F. Encke, G. Buchheim and S. Seybold, *Handwörterbuch der Pflanzennamen.* 14. Aufl. Stuttgart 1993; Leonard Huxley, *Life and Letters of Sir Joseph Dalton Hooker.* 1918.

Roystonea O.F. Cook Palmae

Origins:

For General Roy Stone, 1836-1905, American Army Engineer in Puerto Rico; see Claire C. Coons, "General Roy Stone: Portrait of a Gentleman." *Principes.* 18(3): 99-104. 1974; S. Zona, "*Roystonea* (Arecaceae: Arecoideae)." *Flora Neotropica.* Volume 71. The New York Botanical Garden. 1996; F. Boerner & G. Kunkel, *Taschenwörterbuch der botanischen Pflanzennamen.* 4. Aufl. 163. Berlin & Hamburg 1989.

Species/Vernacular Names:

R. borinquena Cook

English: Puerto Rican royal

Brazil: palmeira imperial de Porto Rico, palmeira coca cola

R. elata (Bartr.) F. Harper

English: Florida royal palm

R. hispaniolana Bail.

English: Hispaniolan royal palm

R. oleracea (Jacq.) Cook

English: feathery cabbage palm

Brazil: palmeira real

R. princeps (Becc.) Burret

English: morass royal

Brazil: palmeira imperial do brejo

R. regia (Kunth) Cook

English: Cuban royal, Cuban royal palm, royal palm

Brazil: palmeira real, palmeira imperial de Cuba

Spanish: palma real

Japan: daiwo-yashi

Ruagea Karsten Meliaceae

Origins:

An anagram of the generic name *Guarea* L.

Species/Vernacular Names:

R. insignis (C. DC.) Pennington (*Ruagea surutensis* Harms)

Bolivia: cedro blanco

Rubeola Hill Rubiaceae

Origins:

Latin *rubeo, ere* "to be red, blushing," see also *Sherardia* L.

Rubeola Mill. Rubiaceae

Origins:

From the Latin *rubeo, ere* "to be red, blushing," see also *Crucianella* L.

Rubia L. Rubiaceae

Origins:

Latin *herbam rubiam,* Latin *ruber, rubra, rubrum* "red," a reddish dye from the roots is used in dyeing; Plinius and Vitruvius used *rubia, ae* for madder; see Carl Linnaeus, *Species Plantarum.* 109. 1753 and *Genera Plantarum.* Ed. 5. 47. 1754; Salvatore Battaglia, *Grande dizionario della lingua italiana.* XVII: 1-2. UTET, Torino 1995.

Species/Vernacular Names:

R. cordifolia L.

English: Indian madder, dyer's madder

Italian: robbia d'India

India: madder, manjito, manjista, manjistha, manjeshta, manjushtha, manjestateega, manditta, manjit, manjitti, manjetti, majith, majathi, mitu, taamaravalli, tamravalli, chitravalli, chiranji, shevelli, kala-meshika, siomalate, siragatti, poon, poont, barheipani, dandu, kukarphali, tiuru, itari, sheni, runang, ryhoi, soh-misem, enhu, chenhu, moyum, vhyem, soth, mandastic

Tibetan: btsod

China: qian cao gen, chien tsao

R. peregrina L.

English: Levant madder, wild madder

Arabic: fuwwa

R. tinctorum L.

English: madder, European madder

Arabic: fowwa

Italian: robbia, robia, rubbia, rubea, rubia, ruggia

India: alizari, bacho, manyunth

Rubiteucris Kudô Labiatae

Origins:

Perhaps from the Latin *ruber, rubra, rubrum* "red" and the genus *Teucrium.*

Species/Vernacular Names:

R. palmata (Bentham ex Hook.f.) Kudô

English: palm germander

China: zhang ye shi can

Rubrivena M. Král Polygonaceae

Origins:

Latin *ruber, rubra, rubrum* "red" and *vena, ae* "a vein," see also *Persicaria* (L.) Mill.

Rubus L. Rosaceae

Origins:

Latin *rubus* for the blackberry-bush, bramble bush, a blackberry (see in Plinius, Vergilius, Horatius, Ovidius, etc.), adj. *rubeus,* Latin *ruber, rubra, rubrum* "red"; see Carl Linnaeus, *Species Plantarum.* 492. 1753 and *Genera Plantarum.* Ed. 5. 218. 1754; Giovanni Semerano, *Le origini della cultura europea.* Dizionario della lingua Latina e di voci moderne. 2(2): 548. Firenze 1994.

Species/Vernacular Names:

R. sp.

Peru: huaricchapi

Mexico: beyo zaa guixi, peyo zaa quixi, looba beyo zaa, loba bio zaa, zoba chiba guechi, yaga beyo zaa guixi, nitica peyo zaa quixi

R. affinis Wight & Arn.

English: blackberry

R. allegheniensis L.H. Bailey

English: Alleghany blackberry, sow-teat blackberry, Allegheny blackberry

R. almus (L.H. Bail.) L.H. Bail.

English: Mayes dewberry blackberry

R. apetalus Poir. (*Rubus adolfi-friedericii* Engl.; *Rubus adolfi-friedericii* Engl. var. *rubristylus* C.E. Gust.; *Rubus ecklonii* Focke; *Rubus exsuccus* Steud.; *Rubus interjungens* C.E. Gust.; *Rubus pinnatiformis* C.E. Gust.)

English: raspberry

R. arcticus L.

English: crimson bramble, Arctic bramble, Arctic blackberry, crimson-berry

R. argutus Link (*Rubus koehnei* H. Lév.; *Rubus mauicola* Focke; *Rubus penetrans* L.H. Bailey)

English: prickly Florida blackberry, southern blackberry, high-bush blackberry

Hawaii: 'ohelo 'ele'ele

R. australis Forst.f.

English: New Zealand lawyer

R. bellobatus L.H. Bail.

Australia: Kittatinny blackberry

R. buergeri Miq.

China: han mei

R. caesius L.

English: dewberry, European dewberry

R. canadensis L.

English: American dewberry, smooth blackberry, smooth bramble

R. chamaemorus L.

English: cloudberry, salmonberry, yellow berry, bake apple, baked apple, baked apple berry, malka

R. chloocladus W.C.R. Watson

English: blackberry

R. cissoides A. Cunn.

English: bush lawyer

New Zealand: tataramoa

R. coreanus Miq.

English: Korean raspberry

China: fu pen zi, dao sheng gen, cha tien piao

R. cuneifolius Pursh (*Rubus parviflorus* Walt. non L.)

English: sand blackberry, American bramble, Gozard's curse, sand bramble

South Africa: Amerikaanse braambos, sandbraam

R. deliciosus Torr.

English: Rocky Mountain flowering raspberry, Rocky Mountain raspberry

R. discolor Weihe & Nees

English: blackberry, Himalayan blackberry

R. ellipticus Smith (*Rubus ellipticus* var. *obcordatus* Focke)

English: yellow Himalayan raspberry, yellow raspberry, oval-leaf bramble

Nepal: ainselu

R. flagellaris Willd.

English: dewberry, American dewberry, northern dewberry, running blackberry, trailing bramble

R. fruticosus L.

English: blackberry, European blackberry

R. gunnianus Hook.

English: alpine bramble

R. hawaiensis A. Gray

Hawaii: 'akala, 'akalakala, kala

R. hillii F. Muell.

English: Molucca bramble, broad-leaf bramble

R. hispidus L.

English: swamp dewberry, running blackberry, swamp blackberry, bristly dewberry

R. ichangensis Hemsl. & Kuntze

China: tung piao tzu

R. idaeus L.

English: raspberry, wild raspberry, red raspberry, European raspberry, European red raspberry

French: framboise

R. idaeus L. var. *strigosus* (Michx.) Maxim.

English: American raspberry

R. laciniatus Willd.

English: cut-leaved bramble, parsley-leaved bramble, cut-leaf bramble, cut-leaved blackberry, cut-leaf blackberry, evergreen blackberry, Italian blackberry, parsley-leaf blackberry

R. leightonii Lees & Leighton

English: blackberry

R. leucodermis Torrey & A. Gray

English: blackcap blackberry

R. loganobaccus L.H. Bailey

English: loganberry, boysenberry

R. longepedicellatus (C.E. Gust.) C.H. Stirton (*Rubus rigidus* J.E. Sm. var. *longepedicellatus* C.E. Gust.)

English: raspberry

R. macraei A. Gray

Hawaii: 'akala, 'akalakala, kala

R. microphyllus L.f. (*Rubus incisus* Thunb.)

China: hsuan kou tzu, shan mei, mu mei

R. mirus L.H. Bail.

English: marvel dewberry

R. moluccanus L.

English: Molucca raspberry, blackcherry, bramble

India: katsol, katson, sufokji, katsoi, bipen kanta

R. moorei F. Muell.

English: silky bramble, bush lawyer

R. muelleri L.H. Bail.

English: Queensland raspberry

R. nivalis Douglas

English: snow dwarf bramble

R. niveus Thunb. (*Rubus albescens* Roxb.; *Rubus lasiocarpus* J.E. Sm.)

English: hill raspberry, Java bramble, Mysore raspberry, black raspberry, raspberry, Mahabaleshwar raspberry

Tibetan: kandakari

India: gauriphal, kala hinsalu, kalianchhi, gowriphal, gunacha, kandiari, kalo aselu

R. occidentalis L.

English: black raspberry, black cap, thimbleberry

R. odoratus L.

English: flowering raspberry, purple-flowering raspberry, thimbleberry

R. parviflorus Nutt. (*Rubus triphyllus* Thunb.)

English: salmonberry, thimbleberry, native raspberry, small-leaf bramble, western thimbleberry, Chinese raspberry

Japan: nawa-shiro-ichigo, munjuru-ichubi

China: hao tian biao

R. parvifolius L.

China: nou tien piao

R. phoenicolasius Maxim.

English: wineberry, wine raspberry

R. pinnatus Willd. (*Rubus kinginsis* Engl.; *Rubus pappei* Eckl. & Zeyh.; *Rubus pinnatus* Willd. subsp. *afrotropicus* Engl.; *Rubus pinnatus* Willd. var. *afrotropicus* (Engl.) C.E. Gust.; *Rubus pinnatus* Willd. var. *defensus* (Engl.) C.E. Gust.; *Rubus pinnatus* Willd. var. *subglandulosus* (C.E. Gust.) R.A. Grah.)

English: bramble, Cape bramble, South African blackberry, South African bramble, South African raspberry

Southern Africa: braambos, braamwortels, Kaapse braam; iqunube, iGqunube (Xhosa); mfongosi (Zulu); molopilopi (South Sotho); motopilope (Pedi)

R. pubescens Raf.

English: dwarf red raspberry, dwarf raspberry

R. pyramidalis Kaltenb.

English: blackberry

R. radula Weihe & Boenn.

English: blackberry

R. rigidus J.E. Sm. (*Rubus atrocoeruleus* C.E. Gust.; *Rubus chrysocarpus* Mundt; *Rubus discolor* E. Mey.; *Rubus inedulis* Rolfe; *Rubus mundtii* Cham. & Schltdl.; *Rubus rigidus* J.E. Sm. var. *buchananii* Focke; *Rubus rigidus* J.E. Sm. var. *chrysocarpus* (Mundt) Focke; *Rubus rigidus* J.E. Sm. var. *incisus* C.E. Gust.; *Rubus rigidus* J.E. Sm. var. *mundtii* (Cham. & Schltdl.) Focke; *Rubus rigidus* J.E. Sm. var. *rigidus*)

English: bramble

Southern Africa: braambos; ijingijolo (Zulu); monokotswai-wabanna (Sotho); gato, ruWhato (Shona)

R. roribaccus (L.H. Bail.) Rydb.

English: North American dewberry

R. rosifolius Smith

English: Mauritius raspberry, thimbleberry, rose-leaf bramble, forest bramble, wild raspberry, rose-leaved raspberry

Hawaii: 'akala, 'akalakala, ola'a

China: dao chu san, tu mi

R. spectabilis Pursh

English: salmonberry

R. strigosus Michx. (Latin *strigosus* "covered with strigae, with stiff bristles")

English: American red raspberry

R. thibetanus Franch.

English: ghost bramble

R. trifidus Thunb.

China: hao ying piao

R. trivialis Michx.

English: southern dewberry, coastal plain dewberry

R. ulmifolius Schott

English: bramble, blackberry

Arabic: alliq, tout el alliq, 'allaiq

R. ursinus Cham. & Schltdl.

English: loganberry, California blackberry, Pacific blackberry, Pacific dewberry

R. vestitus Weihe & Nees

English: blackberry

Rudbeckia L. Asteraceae

Origins:

After the Swedish botanist Olaus (Olof) Olai Rudbeck (Rudbeckius), 1660-1740, physician, M.D. Utrecht 1690, professor of anatomy and botany, traveler, teacher of Linnaeus, author of *De fundamentali plantarum notitia.* Trajecti ad Rhenum [Utrecht] 1790. He was father of the Swedish natural scientist Johan Olof Rudbeck (1711-1790); according to Stearn the name also commemorates the Swedish physician and botanist Olaus (Olof) Johannis Rudbeck (1630-1702, b. Västerås, d. Uppsala), father of O.O. Rudbeck; see J.H. Barnhart, *Biographical Notes upon Botanists.* 3: 188. 1965; Alexander B. Adams, *Eternal Quest. The Story of the Great Naturalists.* New York 1969; Sten Lindroth, in *Dictionary of Scientific Biography* 11: 586-588. 1981; Stafleu and Cowan, *Taxonomic Literature.* 4: 968-970. Utrecht 1983; T.W. Bossert, *Biographical Dictionary of Botanists Represented in the Hunt Institute Portrait Collection.* 341. 1972; E.M. Tucker, *Catalogue of the Library of the Arnold Arboretum of Harvard University.* 1917-1933; Jonas C. Dryander, *Catalogus bibliothecae historico-naturalis Josephi Banks.* London 1800; A. Lasègue, *Musée botanique de Benjamin Delessert.* Paris 1845; James Britten, *The Sloane Herbarium,* revised and edited by J.E. Dandy. 1958; Blanche Henrey, *British Botanical and Horticultural Literature before 1800.* 1975; Garrison and Morton, *Medical Bibliography.* 1097, 1098. New York 1961; Frans A. Stafleu, *Linnaeus and the Linnaeans. The spreading of their ideas in systematic botany, 1735-1789.* Utrecht 1971.

Species/Vernacular Names:

R. californica A. Gray

English: California cone-flower

R. occidentalis Nutt. var. *occidentalis*

English: western cone-flower

Rudgea Salisbury Rubiaceae

Origins:

For the English (b. Evesham, Worcs.) botanist Edward Rudge, 1763-1846 (d. Evesham), antiquary, 1802 Fellow of the Linnean Society, 1805 Fellow of the Royal Society, author of *Plantarum Guianae rariorum icones et descriptiones*. Londini 1805[-1806] and (presumably) of *Memoir* [of Anna Rudge] London [1836?], the botanist Samuel Rudge (1727-1817) was his uncle; see J.H. Barnhart, *Biographical Notes upon Botanists*. 3: 188. Boston 1965; Ethelyn Maria Tucker, *Catalogue of the Library of the Arnold Arboretum of Harvard University*. Cambridge, Massachusetts 1917-1933; G. Murray, *History of the Collections Contained in the Natural History Departments of the British Museum*. 1: 178. London 1904; Ray Desmond, *Dictionary of British & Irish Botanists and Horticulturists*. 598-599. 1994; R. Zander, F. Encke, G. Buchheim and S. Seybold, *Handwörterbuch der Pflanzennamen*. 14. Aufl. Stuttgart 1993.

Species/Vernacular Names:

R. spp.

Peru: cotocoto, mulatinho, rayo caspi, pichico caspi

R. retifolia Standley

Peru: amanga, capinurí, pichico, pichico runtu, raya caspi, sanango de bajo

Rudolfiella Hoehne Orchidaceae

Origins:

After the German (b. Berlin) botanist Friedrich Richard Rudolf Schlechter, 1872-1925 (d. Berlin), traveler, plant collector and orchidologist, student of Engler, assistant to Harry Bolus (1834-1911), botanical explorer (Togo Rep., S. Nigeria and W. Cam.), 1899-1900 leader of the German West Africa Rubber Expedition, 1921-1925 Curator at Berlin-Dahlem, prolific writer. His works include *Die Orchideen; ihre Beschreibung, Kultur und Züchtung*. Berlin 1914-1915, *West Afrikanische Kautschuk-Expedition*. Berlin 1910 and *Die Guttapercha-und Kautschuk-Expedition*. Berlin 1911. He was brother of Max Schlechter (1874-1960, d. Cape Town), who collected with him in South Africa (1896-1897); see John Peter Jessop (b. 1939), "Itinerary of Rudolf Schlechter's collecting trips in Southern Africa." *J.*

S. Afr. Bot. 30(3): 129-146. 1964; Stafleu and Cowan, *Taxonomic Literature*. 5: 205-213. 1985; R. Zander, F. Encke, G. Buchheim and S. Seybold, *Handwörterbuch der Pflanzennamen*. 14. Aufl. 775. 1993; J.H. Barnhart, *Biographical Notes upon Botanists*. 3: 228. 1965; Ludwig E. Theodor Loesener, *Notizbl. Bot. Gart. Berl.* 9: 912-948. 1926; F.N. Hepper and Fiona Neate, *Plant Collectors in West Africa*. 72. Utrecht 1971; Merle A. Reinikka, *A History of the Orchid*. Timber Press 1996; N. Hall, *Botanists of the Eucalypts*. Melbourne 1978 and Supplement 1980; Ida Kaplan Langman, *A Selected Guide to the Literature on the Flowering Plants of Mexico*. Philadelphia 1964; Anthonius Josephus Maria Leeuwenberg, "Isotypes of which holotypes were destroyed in Berlin." *Webbia*. 19: 861-863. 1965; Réné Letouzey (1918-1989), "Les botanistes au Cameroun." in *Flore du Cameroun*. 7: 1-110. Paris 1968; T.W. Bossert, *Biographical Dictionary of Botanists Represented in the Hunt Institute Portrait Collection*. 353. 1972; Ethelyn Maria Tucker, *Catalogue of the Library of the Arnold Arboretum of Harvard University*. Cambridge, Massachusetts 1917-1933; Mary Gunn and Leslie Edward W. Codd, *Botanical Exploration of Southern Africa*. Cape Town 1981; Frederico Carlos Hoehne, M. Kuhlmann and Oswaldo Handro, *O jardim botânico de São Paulo*. 1941; Elmer Drew Merrill, in *Bernice P. Bishop Mus. Bull*. 144: 164. 1937 and in *Contr. U.S. Natl. Herb*. 30(1): 267. 1947; A. White and B.L. Sloane, *The Stapelieae*. Pasadena 1937; Gordon Douglas Rowley, *A History of Succulent Plants*. Strawberry Press, Mill Valley, California 1997; Richard Evans Schultes and Arthur Stanley Pease, *Generic Names of Orchids. Their Origin and Meaning*. 1963.

Ruellia L. Acanthaceae

Origins:

For the French (b. Soissons) botanist Jean Ruel (Joannes Ruellius, Jean de la Ruelle, du Ruel), 1474-1537 (Paris), physician, herbalist to François I of France, translator of Dioscorides into Latin, obtained a canonry at Notre-Dame Cathedral, author of *De natura stirpium libri tres*. Parisiis 1536. See Leodegarius a Quercu [pseud., i.e. Léger Duchesne] [-1588], *In Ruellium De Stirpibus Epitome*. Parisiis [1539]; Scribonius Largus, *De compositionibus medicamentorum liber unus, antehac nusquam excusus: Joanne Ruellio doctore medico castigatore*. Parisiis 1529; Ernst H.F. Meyer, *Geschichte der Botanik*. IV: 249-253 and II: 26-39. Königsberg 1854-1857; Carl Linnaeus, *Species Plantarum*. 634. 1753 and *Genera Plantarum*. Ed. 5. 283. 1754; Richard J. Durling, comp., *A Catalogue of Sixteenth Century Printed Books in the National Library of Medicine*. 1967; Paul Jovet & J.C. Mallet, in *Dictionary of Scientific Biography* 11: 594-595. 1981; Georg Christian Wittstein, *Etymologisch-botanisches Handwörterbuch*. 775. 1852.

Species/Vernacular Names:

R. geminiflora Kunth

Peru: ipecacuanha de flor rosa, ipeca de flor roja

R. macrantha Mart. ex Nees

English: Christmas pride

R. makoyana hort. Makoy ex Closon

English: monkey plant, trailing velvet plant

R. tuberosa L.

English: meadow-weed, menow weed

Peru: ipecacuanha de flor rosa, ipeca de flor roja

India: chetapatakaayala mokka, tapas kaaya, ote sirka ba

Ruelliola Baillon Acanthaceae

Origins:

The diminutive of *Ruellia*.

Ruelliopsis C.B. Clarke Acanthaceae

Origins:

Resembling the genus *Ruellia* L.

Rufodorsia Wiehler Gesneriaceae

Origins:

Latin *rufus* "red, reddish" and *dorsum, dorsi* "the back, the ridge."

Rugelia Shuttleworth ex Chapman Asteraceae

Origins:

After the American (b. Germany) botanist Ferdinand Ignatius Xavier Rugel, 1806-1879, surgeon, botanical explorer, pharmacist, plant collector (in southwestern USA, Europe and Cuba); see J.H. Barnhart, *Biographical Notes upon Botanists.* 3: 190. 1965; T.W. Bossert, *Biographical Dictionary of Botanists Represented in the Hunt Institute Portrait Collection.* 235. 1972; G. Murray, *History of the Collections Contained in the Natural History Departments of the British Museum.* 1: 178. London 1904; Stafleu and Cowan, *Taxonomic Literature.* 4: 979-980. [d. 1878] 1983.

Ruilopezia Cuatrec. Asteraceae

Origins:

For the Spanish botanist and traveler Hipólito Ruíz López, 1754-1815 (or 1816, see *Dictionary of Scientific Biography* 11: 605-606. 1981).

Ruizia Cav. Sterculiaceae

Origins:

For the eminent Spanish (b. Burgos Province) botanist and traveler Hipólito Ruíz López, 1754-1815 (d. Madrid), plant collector, explorer, studied under the Spanish botanist Casimiro Gómez de Ortega (1740-1818), 1777-1778 in Peru and Chile (a French-Spanish expedition, with Joseph Dombey). His writings include *Quinología.* Madrid 1792, *Memoria de las virtudes y usos de la raiz de la planta llamada Yallhoy en el Perú.* Madrid 1805 and *Respuesta para desengaño del público á la impugnacion que ha divulgado prematuramente el Presbitero Don Josef Antonio Cavanilles, contra el Pródromo de la Flora del Perú.* Madrid 1796. He is best remembered for [in collaboration with José Antonio Pavón y Jiménez] *Flora peruvianae, et chilensis prodromus.* Madrid 1794 and *Flora peruviana, et chilensis.* Madrid 1798-1802. See Thomas F. Glick, in *Dictionary of Scientific Biography* 11: 605-606. [d. 1816] 1981; R. Zander, F. Encke, G. Buchheim and S. Seybold, *Handwörterbuch der Pflanzennamen.* 14. Aufl. Stuttgart 1993; Stafleu and Cowan, *Taxonomic Literature.* 4: 981-986. [d. 1815] 1983; Arthur R. Steele, *Flowers for the King.* The expedition of Ruíz and Pavón and the Flora of Peru. Durham, N.C. 1964; R.E.G. Pichi Sermolli, "Le collezioni cedute da J. Pavón a F.B. Webb e conservate nell'Herbarium Webbianum." in *Nuovo Giorn. Bot. Ital.*, ser. 2. 56(4): 699-701. 1950 [1949]; E. Alvarez López, "Algunos aspectos de la obra de Ruíz y Pavón." *Anales Inst. Bot. Cavanilles.* 12(1): 5-111. 1954 and "Comentario sobre *Laurus* de Ruíz y Pavón, con notas de Dombey acerca de algunas de sus especies." *Anales Inst. Bot. Cavanilles.* 13: 71-78. 1955; J.H. Barnhart, *Biographical Notes upon Botanists.* 3: 191. 1965; T.W. Bossert, *Biographical Dictionary of Botanists Represented in the Hunt Institute Portrait Collection.* 342. 1972; H.N. Clokie, *Account of the Herbaria of the Department of Botany in the University of Oxford.* 235. Oxford 1964; Ethelyn Maria Tucker, *Catalogue of the Library of the Arnold Arboretum of Harvard University.* Cambridge, Massachusetts 1917-1933; M. Colmeiro y Penido, *La Botánica y los Botánicos de la Peninsula Hispano-Lusitana.* Madrid 1858; G. Murray, *History of the Collections Contained in the Natural History Departments of the British Museum.* 1: 178. London 1904; Jonas C. Dryander, *Catalogus bibliothecae historico-naturalis Josephi Banks.* London 1800; August Weberbauer, *Die Pflanzenwelt der peruanischen Andes in ihren Grundzügen dargestellt.* 2-4. Leipzig 1911; Blanche Henrey, *British Botanical and Horticultural Literature before 1800.* 1975; Ida Kaplan Langman, *A Selected Guide to the Literature on the Flowering Plants of Mexico.* Philadelphia 1964; A. Lasègue, *Musée botanique de Benjamin Delessert.* Paris 1845; Frans A. Stafleu, *Linnaeus and the Linnaeans.* The spreading of their ideas in systematic botany, 1735-1789. 290. [dedicated to C. Gómez de Ortega] Utrecht 1971; Mariella Azzarello Di

Misa, ed., *Il Fondo Antico della Biblioteca dell'Orto Botanico di Palermo*. 238. Palermo 1988.

Ruizodendron R.E. Fries Annonaceae

Origins:

For the Spanish botanist and traveler Hipólito Ruíz López, 1754-1815 (or 1816, in *Dictionary of Scientific Biography* 11: 605-606. 1981).

Ruizterania Marcano-Berti Vochysiaceae

Origins:

For Luis Ruiz-Terán, apothecary and professor of dendrology.

Rulingia R. Br. Sterculiaceae

Origins:

For the German botanist Johann Philipp Rueling, b. 1741, physician, studied medicine at Göttingen. His works include *Commentatio botanica de ordinibus naturalibus plantarum*. Goettingae [1766], *Physikalisch-medicinisch-ökonomische Beschreibung der zum Fürstenthum Göttingen ... Stadt Northeim, und ihrer umliegenden Gegend*. Göttingen 1779 and *Ordines naturales plantarum*. Goettingae 1774; see J. Sims, in *Curtis's Botanical Magazine*. 48, t. 2191. 1820; John H. Barnhart, *Biographical Notes upon Botanists*. 3: 190. 1965; Paul Usteri (1768-1831), *Delectus opusculorum botanicorum*. Argentorati [Strasbourg] 1790.

Species/Vernacular Names:

R. dasyphylla (Andrews) Sweet

Australia: kerrawang

R. hermanniifolia (Gay) Endl.

Australia: wrinkled kerrawang

R. procumbens Maiden & E. Betche

Australia: mat kerrawang, spreading kerrawang

R. prostrata Maiden & E. Betche

Australia: prostrate kerrawang

R. salvifolia (Steetz) Benth.

Australia: velvety kerrawang

Rumex L. Polygonaceae

Origins:

From the ancient Latin name *rumex, icis* for sorrel (Plautus et al.) or for a kind of missile ("genus teli" Sext. Pompeius Festus, grammarian); Akkadian *ramaku* "to pour, to pour out"; Hebrew *romak* "javelin, spear"; see Carl Linnaeus, *Species Plantarum*. 333. 1753 and *Genera Plantarum*. Ed. 5. 156. 1754; G. Semerano, *Le origini della cultura europea*. Dizionario della lingua Latina e di voci moderne. 2(2): 549. 1994.

Species/Vernacular Names:

R. sp.

English: dock

Peru: huacha, mula chuchu ckora

Tibetan: lug-sho, sho-mang

Maori names: paewhenua, runa

R. abyssinicus Jacq.

English: spinach rhubarb

R. acetosa L.

English: garden sorrel, sour dock, green sorrel

Italian: acetosa maggiore, romice acetosa

Tibetan: rgya-sho

China: suan mo, suan mu, shan yang ti, shan ta huang

India: khatta palak, chukapalam

R. acetosella L. (*Acetosella angiocarpa* (Murb.) A. Löve; *Acetosella vulgaris* Fourr.; *Rumex angiocarpus* Murb.)

English: sheep sorrel, sorrel, field sorrel

India: chutrika, chuka-palam, chukapalam, chuk

R. acetosella L. subsp. *angiocarpus* (Murb.) Murb.

English: field sorrel, dock, sheep sorrel

South Africa: boksuring, steenboksuring

R. albescens Hillebr.

Hawaii: hu'ahu'ako

R. alpinus L.

English: monk's rhubarb, mountain rhubarb

Italian: romice alpina

R. bidens R. Br.

English: mud dock

R. brownii Campd.

English: slender dock, swamp dock

R. conglomeratus Murray (*Rumex acutus* Sm.; *Rumex glomeratus* Schreb.; *Rumex nemolapathum* Ehrh.; *Rumex paludosus* Withering)

English: dock, cluster dock, clustered dock

Italian: romice dei fossi, romice selvatica

South Africa: maksuring, suring, tongblaar

India: jangli-palak, jal-palam, bun-palung, banpalang, bijband, khattikan, hulaobul

R. crispus L.

English: curled sorrel, curled dock, curly dock, dock, narrow dock, narrow-leaved dock, sour dock, yellow dock

Spanish: acitosa, lengua de vaca

Italian: romice crespa, lapazio

Peru: moztaza

Japan: nagaba-gishigishi

Tibetan: rgya-sho

China: niu er da huang

India: amla-vedasa, shula-vedhi-chukra, bijband, endranee, suk-gu-kire, chuka-bija, shukku, chukkah, kuraku, hummaz, pdlivanchi, kala khen-boun, turshah

Southern Africa: beesslaai, krulblaarrumex, krulsuring, krultongblaar, tongblaar, suring, weeblaar, wildespinasie; ubuklunga (Xhosa); ubuklunga (Zulu)

R. crystallinus Lange

English: shiny dock, glistening dock

R. dumosus Meissner

English: wiry dock, wing dock

R. giganteus Ait.

Hawaii: pawale, uhauhako

R. hastatus D. Don

Nepal: amile jhar

R. hydrolapathum Huds.

English: great water dock

R. hymenosepalus Torrey

English: tanner's dock, wild rhubarb, canaigre

R. lanceolatus Thunb. (*Rumex ecklonianus* Meisn.; *Rumex ecklonii* Meisn.; *Rumex linearis* Campd.; *Rumex meyeri* Meisn.; *Rumex meyerianus* Meisn.)

English: common dock, smaller dock, smooth dock

Southern Africa: gladdetongblaar, tongblaar; dolonyana (Xhosa); dolonyana (Zulu); kxamane (Sotho)

R. maritimus L.

English: golden dock

India: jangli-palak, jungli palak, jal-palam, jub-palum, bunpalung, banpalang, bijband, khattikan, hulaobul

R. nepalensis Spreng.

Nepal: halhale

R. obtusifolius L.

English: bitter dock, broadleaf dock

Italian: romice dei prati

Peru: paico, chuchu ckora, ppaico, urcko chuchuckora

R. occidentalis S. Watson

English: western dock

R. patientia L.

English: patience dock, spinach dock, patience herb, monk's rhubarb

Italian: romice domestica, erba pazienza, lapazio

Peru: huacho, lengua de vaca

R. pulcher L.

English: fiddle dock

Italian: romice selvatica

R. sagittatus Thunb. (*Acetosa sagittatus* (Thunb.) L. Johnson & B. Briggs)

English: red sorrel, rambling dock

Southern Africa: rooisuring; tshitamba-tshedzi (Venda)

R. salicifolius J.A. Weinm.

English: willow dock

R. scutatus L.

English: French sorrel, garden sorrel

India: changeri, amrula, ambavati

R. skottsbergii Degener & I. Degener

Hawaii: pawale

R. tenax K.H. Rech.

English: shiny dock

R. venosus Pursh

English: wild begonia, wild hydrangea, sour greens, veiny dock, winged dock

R. violascens Rech.f.

English: Mexican dock

Rumicastrum Ulbr. Portulacaceae

Origins:

An incomplete resemblance with the genus *Rumex* L.

Rumicicarpus Chiov. Tiliaceae

Origins:

Latin *rumex, rumicis* and Greek *karpos* "fruit."

Rumohra Raddi Dryopteridaceae (Davalliaceae)

Origins:

In honor of the German art expert and writer Karl Friedrich Ludwig Felix von Rumohr, 1785-1843, a collector of antiquities, a patron of botany, studied painting under Fiorillo at Göttingen, visited Italy. Among his writings are *Ein Band Novellen.* München 1833-1835, *Drei Reisen nach Italien. Erinnerungen.* Leipzig 1832, *Italienische Forschungen.* Berlin und Stettin 1827-1831, *Kynalopekomachia. Der Hunde Fuchsenstreit.* Lübeck 1835 and *Ursprung der Besitzlosigkeit des Colonen im neueren Toscana. Aus den*

Urkunden. Hamburg 1830. See the Italian botanist Giuseppe Raddi (1770-1829) and his *Synopsis filicum brasiliensium*. Bononiae [Bologna] 1819; Heinrich Wilhelm Schulz, *K.F. von Rumohr, sein Leben und seine Schriften*. Leipzig 1844; Georg Christian Wittstein, *Etymologisch-botanisches Handwörterbuch*. 775. Ansbach 1852.

Species/Vernacular Names:
R. adiantiformis (G. Forster) Ching
Chile: pereq, calahuala, yerba del lagarto

Runcina Allam. Gramineae

Origins:
Runcina was a rural goddess presiding over weeding, Latin *runcino, are* "to plane off," *runco, are* "to weed out, root up, to pluck, to mow," Latin *runcina, ae* and Greek *rhykane* for a plane.

Rungia Nees Acanthaceae

Origins:
After Rungia, a well-known Indian botanical artist of Robert Wight; see Nathaniel Wallich, *Plantae Asiaticae rariores*. 3: 77, 109. London (Aug.) 1832; Robert Wight, *Icones plantarum Indiae orientalis*, or figures of Indian plants. Madras [1838-] 1840-1853; M.P. Nayar, *Meaning of Indian Flowering Plant Names*. 304. Dehra Dun 1985.

Species/Vernacular Names:
R. parviflora (Retz.) Nees
Nepal: bisaune jhar

Rupestrina Prov. Gramineae

Origins:
Latin *rupes, is* "a rock."

Rupicapnos Pomel Fumariaceae

Origins:
Growing on cliffs, Latin *rupes, is* "a rock" and *kapnos* "smoke," Dioscorides used *kapnos* for the fumitory, *Fumaria officinalis*.

Rupicola J.H. Maiden & E. Betche Epacridaceae

Origins:
Latin *rupes, is* "a rock," referring to the rocky habitat; see Joseph Henry Maiden (1859-1925) and Ernst Betche

(1851-1913), in *Proceedings of the Linnean Society of New South Wales*. 23: 774, t. xxviii. 1899.

Rupifraga Raf. Saxifragaceae

Origins:
Latin *rupes, is* "a rock" and *frango* "to break"; see C.S. Rafinesque, *Flora Telluriana*. 2: 67, 68. 1836 [1837], 4: 121. 1836 [1838].

Rupiphila Pimenov & Lavrova Umbelliferae

Origins:
Latin *rupes, is* "a rock" and Greek *philos* "lover, loving."

Ruppia L. Ruppiaceae (Potamogetonaceae)

Origins:
For the German botanist Heinrich Bernard Ruppius (Rupp), 1688-1719, author of *Flora jenensis*. Francofurti & Lipsiae 1718; see J.H. Barnhart, *Biographical Notes upon Botanists*. 3: 192. 1965; Carl Linnaeus, *Species Plantarum*. 127. 1753 and *Genera Plantarum*. Ed. 5. 61. 1754; A. White and B.L. Sloane, *The Stapelieae*. Pasadena 1937; Jacques Julien Houtton de Labillardière (1755-1834), *Novae Hollandiae plantarum specimen*. Parisiis 1804-1806 [1807]; Ethelyn Maria Tucker, *Catalogue of the Library of the Arnold Arboretum of Harvard University*. Cambridge, Massachusetts 1917-1933; Jonas C. Dryander, *Catalogus bibliothecae historico-naturalis Josephi Banks*. London 1800; Mariella Azzarello Di Misa, a cura di, *Il Fondo Antico della Biblioteca dell'Orto Botanico di Palermo*. 240. Regione Siciliana, Palermo 1988; G.C. Wittstein, *Etymologisch-botanisches Handwörterbuch*. 776. 1852; Emil Bretschneider, *History of European Botanical Discoveries in China*. Leipzig 1981; R. Zander, F. Encke, G. Buchheim and S. Seybold, *Handwörterbuch der Pflanzennamen*. 14. Aufl. Stuttgart 1993; Frans A. Stafleu, *Linnaeus and the Linnaeans*. The spreading of their ideas in systematic botany, 1735-1789. Utrecht 1971.

Species/Vernacular Names:
R. cirrhosa (Petagna) Grande (*Ruppia spiralis* L. ex Dumort.)
English: ditch grass

R. maritima L. (*Ruppia rostellata* Koch ex Reichb.; *Ruppia maritima* var. *rostellata* Agardh)
English: widgeon grass, sea tassel, ditch grass
Japan: kawa-tsuru-mo

Ruprechtia C.A. Meyer Polygonaceae

Origins:

For the Austrian botanist Franz Josef Ruprecht, 1814-1870, physician; see J.H. Barnhart, *Biographical Notes upon Botanists*. 3: 192. 1965; R. Zander, F. Encke, G. Buchheim and S. Seybold, *Handwörterbuch der Pflanzennamen*. 14. Aufl. 1993; Ethelyn Maria Tucker, *Catalogue of the Library of the Arnold Arboretum of Harvard University*. Cambridge, Massachusetts 1917-1933; Emil Bretschneider, *History of European Botanical Discoveries in China*. Leipzig 1981; Stafleu and Cowan, *Taxonomic Literature*. 4: 993-997. 1983; T.W. Bossert, *Biographical Dictionary of Botanists Represented in the Hunt Institute Portrait Collection*. 342. 1972; Mariella Azzarello Di Misa, ed., *Il Fondo Antico della Biblioteca dell'Orto Botanico di Palermo*. 240. Palermo 1988.

Rusbya Britton Ericaceae

Origins:

For the American physician Henry Hurd Rusby, 1855-1940, botanist, botanical explorer and plant collector in South and Central America (Bolivia, Peru, Amazon region, Venezuela, Colombia, Mexico, Brazil), M.D. New York 1884, professor of botany and materia medica at New York, President of the Torrey Botanical Club, one of the founders of the New York Botanical Garden. His works include *Descriptions of Three Hundred New Species of South American Plants*. New York 1920, "New species from Bolivia, collected by R.S. Williams, I & II." *Bull. New York Bot. Gard.* 6(22): 487-517. 1910 and 8(28): 89-135. 1912, "Report of work on the Mulford Biological Exploration of 1921-1922." *Mem. New York Bot. Gard.* 29: 101-112. 1922, "Description of new genera and species of plants collected on the Mulford Biological Exploration of the Amazon Valley, 1921-1922." *Mem. New York Bot. Gard.* 7(3): 205-387. 1927 and "New species of plants of the Ladew Expedition to Bolivia." *Phytologia*. 1: 49-80. 1934. See John H. Barnhart, *Biographical Notes upon Botanists*. 3: 192. 1965; T.W. Bossert, *Biographical Dictionary of Botanists Represented in the Hunt Institute Portrait Collection*. 342. 1972; Ethelyn Maria Tucker, *Catalogue of the Library of the Arnold Arboretum of Harvard University*. Cambridge, Massachusetts 1917-1933; Joseph Ewan, *Rocky Mountain Naturalists*. The University of Denver Press 1950; S. Lenley et al., *Catalog of the Manuscript and Archival Collections and Index to the Correspondence of John Torrey*. Library of the New York Botanical Garden. 356. 1973; J. Ewan, ed., *A Short History of Botany in the United States*. New York and London 1969; Ida Kaplan Langman, *A Selected Guide to the Literature on the Flowering Plants of Mexico*. 657-658. Philadelphia 1964; R. Zander, F. Encke, G. Buchheim and S. Seybold,

Handwörterbuch der Pflanzennamen. 14. Aufl. Stuttgart 1993; Irving William Knobloch, compil., "A preliminary verified list of plant collectors in Mexico." *Phytologia Memoirs*. VI. 1983.

Rusbyanthus Gilg Gentianaceae

Origins:

For the American physician and botanist Henry Hurd Rusby, 1855-1940.

Rusbyella Rolfe ex Rusby Orchidaceae

Origins:

For the American physician Henry Hurd Rusby, 1855-1940, botanist, medical botanist.

Ruschia Schwantes Aizoaceae

Origins:

After the South African botanist Ernst Julius Rusch, 1867-1957, plant collector, farmer and succulent plant nurseryman at Lichtenstein, near Windhoek, in South Africa, father of Ernst Franz Theodor Rusch (1897-1964); see Martin Heinrich Gustav (Georg) Schwantes (1891-1960), in *Zeitschrift für Sukkulentenkunde*. 2: 186. Berlin (Apr.) 1926; Gordon Douglas Rowley, *A History of Succulent Plants*. Strawberry Press, Mill Valley, California 1997; Mary Gunn and Leslie Edward W. Codd, *Botanical Exploration of Southern Africa*. 304-305. Cape Town 1981; F. Boerner & G. Kunkel, *Taschenwörterbuch der botanischen Pflanzennamen*. 4. Aufl. 163. Berlin & Hamburg 1989.

Species/Vernacular Names:

R. canonotata (L. Bolus) Schwantes

English: cattle vygie

South Africa: beesvygie

Ruschianthemum Friedrich Aizoaceae

Origins:

For Ernst Franz Theodor Rusch (1897-1964), son of the South African botanist Ernst Julius Rusch (1867-1957); see Gordon Douglas Rowley, *A History of Succulent Plants*. 1997; F. Boerner & G. Kunkel, *Taschenwörterbuch der botanischen Pflanzennamen*. 4. Aufl. 163. 1989.

Ruschianthus L. Bolus Aizoaceae

Origins:

For Ernst Franz Theodor Rusch (1897-1964), son of the South African botanist Ernst Julius Rusch (1867-1957); see Gordon Douglas Rowley, *A History of Succulent Plants.* 380. Mill Valley, California 1997.

Ruscus L. Asparagaceae (Ruscaceae, Liliaceae)

Origins:

Latin *ruscum, rustum, rustum ex rubus* "butcher's broom"; see G. Semerano, *Le origini della cultura europea.* Dizionario della lingua Latina e di voci moderne. 2(2): 549-550. Firenze 1994.

Species/Vernacular Names:

R. aculeatus L.

English: butcher's broom, knee holly, box-holly

Arabic: khizana

Ruspolia Lindau Acanthaceae

Origins:

For the Italian (b. Tiganesti, Romania) explorer Eugenio Ruspoli, 1866-1893 (Burgi, Somalia, killed by an elephant), ethnologist, botanical and zoological collector, naturalist, son of Emanuele Ruspoli (1837-1899).

Russea J.F. Gmelin Grossulariaceae (Brexiaceae, Escalloniaceae)

Origins:

Orthographic variant of *Roussea* Smith.

Russelia Jacq. Scrophulariaceae

Origins:

For the Scottish (b. Edinburgh) naturalist and physician Alexander Russell, c. 1715-1768 (d. London), botanist, M.D. Glasgow, from *circa* 1740 to 1753 physician to the English factory at Aleppo (Syria), 1756 Fellow of the Royal Society, physician to St. Thomas's Hospital (London), author of *The Natural History of Aleppo*, and parts adjacent. London 1756, sent seeds to Peter Collinson (1694-1768). He was half-brother of the Scottish (b. Edinburgh) physician and naturalist Patrick Russell (1727-1805, d. London); see E. Smith, *The Life of Sir Joseph Banks.* London 1911; John

H. Barnhart, *Biographical Notes upon Botanists.* 3: 193. 1965; John Coakley Lettsom (1744-1815), *Memoirs of John Fothergill, M.D.* London 1786; William Munk, *The Roll of the Royal College of Physicians of London.* London 1878; G. Murray, *History of the Collections Contained in the Natural History Departments of the British Museum.* London 1904; Jonas C. Dryander, *Catalogus bibliothecae historico-naturalis Josephi Banks.* London 1800; A. Lasègue, *Musée botanique de Benjamin Delessert.* Paris 1845; Ethelyn Maria Tucker, *Catalogue of the Library of the Arnold Arboretum of Harvard University.* Cambridge, Massachusetts 1917-1933; William Roxburgh (1751-1815), *Plants of the Coast of Coromandel.* London 1795-1820; R. Desmond, *The European Discovery of the Indian Flora.* Oxford 1992; Ray Desmond, *Dictionary of British & Irish Botanists and Horticulturists.* 599, 600. London 1994; Stafleu and Cowan, *Taxonomic Literature.* 4: 1002-1003, 1004-1005. Utrecht 1983.

Species/Vernacular Names:

R. equisetiformis Schltdl. & Cham. (*Russelia juncea* Zucc.)

English: coral plant, fountain plant, firecracker plant, russelia

Japan: hana-chôji

Portuguese: lágrimas de amor

Russowia C. Winkler Asteraceae

Origins:

After the Estonian botanist Edmund August Friedrich Russow, 1841-1897, professor of botany, Director of the Botanical Garden at Dorpat; see J.H. Barnhart, *Biographical Notes upon Botanists.* 3: 194. 1965; T.W. Bossert, *Biographical Dictionary of Botanists Represented in the Hunt Institute Portrait Collection.* 343. 1972; Ida Kaplan Langman, *A Selected Guide to the Literature on the Flowering Plants of Mexico.* 658. Philadelphia 1964.

Ruta L. Rutaceae

Origins:

Latin *ruta, ae* "rue, bitterness, a bitter herb, unpleasantness" (Plinius), Greek *rhyte, peganon*; Akkadian *ratu, retu* "to make firm, fortify," Hebrew *ratam* "to bind fast"; Akkadian *pahu, pehu* "to close"; see Giovanni Semerano, *Le origini della cultura europea.* Dizionario della lingua Latina e di voci moderne. 2(2): 550. Firenze 1994.

Species/Vernacular Names:

R. chalepensis L.

Arabic: fidjel, fidjla

R. graveolens L.

English: rue, common rue, garden rue, countryman's treacle, herb of grace, herb of repentance

Southern Africa: binnewortel, wynruit, wynryk; tloro (Sotho)

Japan: henruda

China: chou cao, yun hsiang tsao

Arabic: fijel

R. montana (L.) L.

Arabic: fidjla el-djebeli, fidjel

Rutaea Roemer Meliaceae

Origins:

For the Italian botanist and physician Antonio Turra (1730-1796), mineralogist, author of *Istoria del arbore della China*. Livorno 1764 and *Florae italicae prodromus*. Vicetiae [Vicenza] 1780; see C. Linnaeus, *Mantissa Plantarum*. 2: 150, 237. 1771; J.H. Barnhart, *Biographical Notes upon Botanists*. 3: 410. 1965.

Rutamuraria Ortega Aspleniaceae

Origins:

Wall-rue, from the genus *Ruta* L. and Latin *murus* "a wall."

Rutaneblina Steyerm. & Luteyn Rutaceae

Origins:

Pico da Neblina, Brazil (the country's highest elevation, at 3,014 m), in the Amazon Plains, in Guiana Highlands in the north on the border with Venezuela.

Ruthea Bolle Umbelliferae

Origins:

For the German botanist Johannes (Johann) Friedrich Ruthe, 1788-1859, naturalist, zoologist, entomologist, with the German zoologist Arend Friedrich August Wiegmann (1802-1841) wrote *Handbuch der Zoologie*. Zweite Auflage. Berlin 1843; see J.H. Barnhart, *Biographical Notes upon Botanists*. 3: 195. 1965; Ethelyn Maria Tucker, *Catalogue of the Library of the Arnold Arboretum of Harvard University*. Cambridge, Massachusetts 1917-1933; Günther Schmid, *Chamisso als Naturforscher. Eine Bibliographie*. Leipzig 1942.

Rutheopsis Hansen & Kunkel Umbelliferae

Origins:

For the German botanist Johannes Friedrich Ruthe, 1788-1859.

Ruthiella Steenis Campanulaceae

Origins:

Named for Miss Ruth van Crevel, botanical illustrator; see Stafleu and Cowan, *Taxonomic Literature*. 4: 1009. Utrecht 1983.

Rutidea DC. Rubiaceae

Origins:

Greek *rhytis, rhytidos* "a wrinkle."

Rutidochlamys Sonder Asteraceae

Origins:

Greek *rhytis, rhytidos* "a wrinkle" and *chlamys, chlamydos* "cloak"; see O.W. Sonder, in *Linnaea*. 25: 497. 1853.

Rutidosis DC. Asteraceae

Origins:

Greek *rhytis, rhytidos* "a wrinkle," in reference to the involucral bracts; see A.P. de Candolle, *Prodromus*. 6: 158. 1838.

Species/Vernacular Names:

R. helichrysoides DC.

English: grey wrinklewort, yellow top

R. leiolepis F. Muell.

English: tufted wrinklewort

R. leptorrhynchoides F. Muell.

English: button wrinklewort

R. leucantha F. Muell.

English: white-flowered wrinklewort

R. multiflora (Nees) Robinson (*Styloncereus multiflorus* Nees; *Pumilo argyrolepis* Schltdl.; *Rutidosis pumilo* Benth.)

English: small wrinklewort

R. murchisonii F. Muell.

English: Murchison's wrinklewort

Rutosma A. Gray Rutaceae

Origins:

The genus *Ruta* and Greek *osme* "smell, odor, perfume," see also *Thamnosma* Torr. & Frém.

Ruttya Harvey Acanthaceae

Origins:

For the British (b. Wilts.) physician John Rutty, 1697-1775 (d. Dublin), naturalist, entomologist, lichenologist, M.D. Leyden 1723. His writings include Dissertatio ... *de diarrhoea*. Lugduni Batavorum 1723, *An Account of Some New Experiments and Observations on Joanna Stephen's Medicine for the Stone*. London 1742, *A Chronological History of the Weather and Seasons*, and of ... diseases in Dublin. London 1770, *The Liberty of the Spirit and of the Flesh Distinguished*. Dublin 1756, *A Spiritual Diary and Soliloquies*. London 1776, *An Essay Towards a Natural History of the County of Dublin*. Dublin 1772 and *A Methodical Synopsis of Mineral Waters*. London 1757; see John H. Barnhart, *Biographical Notes upon Botanists*. 3: 195. 1965; Jonas C. Dryander, *Catalogus bibliothecae historico-naturalis Josephi Banks*. London 1800; J.D.H. Widdess, *A History of the Royal College of Physicians of Ireland 1654-1963*. London 1963; William Fryer Harvey, *John Rutty of Dublin, Quaker Physician*. [The Lister Lecture ... 1933. Reprinted from "The Friends' Quarterly Examiner".] [London 1934]; Garrison and Morton, *Medical Bibliography*. 1772, 5309. [b. 1698] New York 1961.

Ruyschia Jacquin Marcgraviaceae

Origins:

For the celebrated Dutch (b. The Hague) physician and anatomist Frederik (Fredrik, Frédéric) Ruysch (Ruijsch, Ruisch) (Fridericus Ruischius), 1638-1731 (d. Amsterdam), M.D. Leiden 1664, plant collector, professor of botany at Amsterdam. His writings include *Thesaurus anatomicus*. i-x. Amstelaedami 1701-1716, *Observationum anatomico-chirurgicarum centuria*. Amsterdam 1691, *Thesaurus animalium* primus. Amstelaedami 1710 and *Dilucidatio valvularum in vasis lymphaticis et lacteis*. Hagae-Comitiae 1665; see G.A. Lindeboom, in *Dictionary of Scientific Biography* 12: 39-42. 1981; Garrison and Morton, *Medical Bibliography*. 389, 1099. New York 1961; Richard Eimas, comp. and ed., *Heirs of Hippocrates*. Iowa City 1990; Johan Commelin, *Horti Medici Amstelodamensis* ... Opus posthumum Latinitate donatum ... a F. Ruyschio et F. Kiggelario. 1697; John H. Barnhart, *Biographical Notes upon Botanists*. 3: 504. 1965; T.W. Bossert, *Biographical Dictionary of Botanists Represented in the Hunt Institute Portrait Collection*.

343. 1972; H.N. Clokie, *Account of the Herbaria of the Department of Botany in the University of Oxford*. 236. Oxford 1964; James Britten, *The Sloane Herbarium*, revised and edited by J.E. Dandy. 1958; G. Murray, *History of the Collections Contained in the Natural History Departments of the British Museum*. 1: 179. London 1904; Blanche Henrey, *British Botanical and Horticultural Literature before 1800*. 1975; Frans A. Stafleu, *Linnaeus and the Linnaeans*. The spreading of their ideas in systematic botany, 1735-1789. Utrecht 1971.

Ryania Vahl Flacourtiaceae

Origins:

After J. Ryan, a plant collector.

Rylstonea R.T. Baker Myrtaceae

Origins:

From the township of Rylstone, on the Cudgegong River, Mudgee, New South Wales, Australia; see Richard Thomas Baker (1854-1941), in *Proceedings of the Linnean Society of New South Wales*. 23: 768, t. xxvii. 1899.

Ryncholeucaena Britton & Rose Mimosaceae

Origins:

Greek *rhynchos* "horn, beak, snout" plus *Leucaena* Benth.

Ryparosa Blume Flacourtiaceae

Origins:

Greek *rhyparos* "dirty," possibly referring to the dirty hairs; see Karl Ludwig von Blume (1796-1862), *Bijdragen tot de flora van Nederlandsch Indië*. 600. (Jan.) 1826.

Ryssopterys Blume ex A. Juss. Malpighiaceae

Origins:

Greek *rhysos*, *rhyssos* "wrinkled, shrivelled" and *pteryx* "wing," referring to the fruit.

Species/Vernacular Names:

R. timoriensis (DC.) Adr. Juss. (*Banisteria timorense* DC.)

Japan: Sasaki-kazura (possibly named to honor Syun'iti Sasaki, 1888-1960, botanist and plant collector in Taiwan, author of "A catalogue of the Government Herbarium." *Rep. Res. Inst. Formosa*. 9: 1-592. 1930)

Ryticaryum Becc. Icacinaceae

Origins:
Greek *rhytis, rhytidos* "a wrinkle" and *karyon* "walnut, nut," referring to the fruit; see Odoardo Beccari (1843-1920), *Malesia.* Firenze-Roma 1877-1890.

Rytidea Spreng. Rubiaceae

Origins:
Greek *rhytis, rhytidos* "a wrinkle," see *Rutidea* DC.

Rytidocarpus Cosson Brassicaceae

Origins:
Greek *rhytis, rhytidos* "a wrinkle" and *karpos* "fruit."

Rytidosperma Steudel Gramineae

Origins:
Greek *rhytis, rhytidos* and *sperma* "a seed"; see Ernst Gottlieb von Steudel (1783-1856), *Synopsis plantarum glumacearum.* 1: 425. 1854.

Rytidostylis Hooker & Arn. Cucurbitaceae

Origins:
Greek *rhytis, rhytidos* and *stylos* "a style."

Rytidotus Hook.f. Rubiaceae

Origins:
Greek *rhytis, rhytidos* "a wrinkle" and *ous, otos* "an ear."

Rytigynia Blume Rubiaceae

Origins:
Greek *rhytis, rhytidos* "a wrinkle" and *gyne* "a woman, female," referring to the ovary; see J.R. Tennant, in *Kew Bulletin.* 19(2): 279-280. 1965.

Rytilix Raf. ex Hitchc. Gramineae

Origins:
See C.S. Rafinesque, *Seringe Bull. Bot.* 1: 219. 1830; E.D. Merrill, *Index rafinesquianus.* The plant names published by C.S. Rafinesque, etc. 76. Jamaica Plain, Massachusetts 1949.

S

Saba (Pichon) Pichon Apocynaceae

Origins:
Saba is a vernacular name in Mali for *Saba senegalensis.*

Sabal Adanson Palmae

Origins:
From a South American vernacular name for these palms.

Species/Vernacular Names:
S. bermudana L.H. Bailey
English: Bermuda palmetto
Brazil: sabal das Bermudas
S. blackburniana (Cook) Glazebr. ex Schult. & Schult.f.
non Hemsl. (*Sabal domingensis* Becc.; *Sabal haitensis*
Becc.; *Sabal neglecta* Becc.; *Sabal umbraculifera* (Jacq.)
Mart.)
English: Hispaniolan palmetto
S. causiarum (Cook) Beccari
English: Puerto Rican hat palm
Brazil: sabal de Porto Rico
S. etonia Swingle ex Nash (*Sabal megacarpa* Small)
English: scrub palmetto
S. jamaicensis Becc.
English: Jamaican palmetto, bull thatch
S. mauritiiformis (Karsten) Grisebach & H.A. Wendland
Brazil: palmeira leque
S. mexicana Martius (*Sabal guatemalensis* Becc.; *Sabal*
texana (Cook) Becc.)
English: Texas palmetto, Oaxaca palmetto
Brazil: sabal do México
S. minor (Jacq.) Persoon (*Sabal deeringiana* Small; *Sabal*
glabra Sarg.; *Sabal minima* Nutt.; *Sabal pumila* Elliott)
English: dwarf palmetto, scrub palmetto, bush palmetto
Brazil: sabal acaule, sabal anão
Japan: miki-nashi-sabaru-yashi
S. palmetto (Walter) Lodd. ex Schult. & Schult.f. (*Sabal*
jamesiana Small; *Sabal schwarzii* Becc.)
English: blue palmetto, cabbage palmetto, cabbage tree,
common palmetto, serpent palm
Brazil: sabal da Flórida, palmeto

S. parviflora Becc. (*Sabal florida* Becc.)
English: Cuban palmetto
S. uresana Trel.
English: Sonoran palmetto

Sabatia Adanson Gentianaceae

Origins:
Presumably for the Italian botanist Liberato Sabbati, b.
1714, gardener; see J.H. Barnhart, *Biographical Notes upon*
Botanists. 3: 197. 1965; E.M. Tucker, *Catalogue of the*
Library of the Arnold Arboretum of Harvard University.
1917-1933; Stafleu and Cowan, *Taxonomic Literature.* 4:
1021. 1983; G. Murray, *History of the Collections Con-*
tained in the Natural History Departments of the British
Museum. 1: 179. London 1904; F. Boerner & G. Kunkel,
Taschenwörterbuch der botanischen Pflanzennamen. 4.
Aufl. 164. 1989; Jonas C. Dryander, *Catalogus bibliothecae*
historico-naturalis Josephi Banks. London 1800; G.C.
Wittstein, *Etymologisch-botanisches Handwörterbuch.*
778. 1852; Helmut Genaust, *Etymologisches Wörterbuch*
der botanischen Pflanzennamen. 549. 1996; Mariella Azzarello Di Misa, ed., *Il Fondo Antico della Biblioteca*
dell'Orto Botanico di Palermo. 240. Regione Siciliana, Palermo 1988.

Sabia Colebr. Sabiaceae

Origins:
A Hindi name.

Saccanthus Herz. Scrophulariaceae

Origins:
From the Greek *sakkos* "sack" and *anthos* "a flower."

Saccardophytum Spegazzini Solanaceae

Origins:
For the Italian botanist Pier Andrea Saccardo, 1845-1920,
botanical collector, mycologist, professor of natural sciences, professor of botany, 1879-1915 Director of the
Botanical Garden of the University of Padova. His writings

include *Di un'operetta sulla flora della Corsica di autore pseudonimo e plagiario*. Venezia 1908, *Spegazzinia*. Patavii 1879, "Fungilli novi Europaei et Asiatici." *Grevillea*. 21: 65-69. 1893 and *Fungi italici*. Patavii 1877-1886, with Roberto de Visiani (1800-1878) wrote *Catalogo delle piante vascolari del Veneto*. Venezia 1869. He was the father of the Italian botanist Domenico Saccardo (1872-1952) and the uncle of the Italian botanist Francesco Saccardo (1869-1896); see J.H. Barnhart, *Biographical Notes upon Botanists*. 3: 197. 1965; E.M. Tucker, *Catalogue of the Library of the Arnold Arboretum of Harvard University*. 1917-1933; T.W. Bossert, *Biographical Dictionary of Botanists Represented in the Hunt Institute Portrait Collection*. 344. 1972; Ida Kaplan Langman, *A Selected Guide to the Literature on the Flowering Plants of Mexico*. 661. Philadelphia 1964; S. Lenley et al., *Catalog of the Manuscript and Archival Collections and Index to the Correspondence of John Torrey*. Library of the New York Botanical Garden. 357. 1973; Stafleu and Cowan, *Taxonomic Literature*. 4: 1024-1040. Utrecht 1983.

Saccellium Bonpl. Boraginaceae (Ehretiaceae)

Origins:

From the Latin *saccellus, i* "a little bag."

Saccharifera Stokes Gramineae

Origins:

From *Saccharum* L. and *fero, fers, tuli, latum, ferre* "to bear, carry," see also *Saccharum* L.

Saccharum L. Gramineae

Origins:

Greek *sakcharon* "sugar"; Indian *sarkara* "the juice prepared from sugar cane," *uch, uchari* "sugar"; Malay *singkara*; Sanskrit *hascha* "pleasure," *ha, hu* "good," *ikshu, ikshuka, ikshuraka* "sugarcane," *ikshurasa* "juice of sugar cane." See Carl Linnaeus, *Species Plantarum*. 54. 1753 and *Genera Plantarum*. Ed. 5. 28. 1754.

Species/Vernacular Names:

S. alopecuroideum (L.) Nutt.

English: silver plume grass

S. arundinaceum Retz.

India: sarkanda, adava

Japan: yoshi-susuki

Okinawa: to-guchi-chi

Malaya: tebrau

S. bengalense Retz. (*Saccharum munja* Roxb.; *Saccharum ciliare* Anderss.; *Saccharum sara* Roxb.)

English: pin reed grass

India: sarkara, gundra, ramsar, sar, tilanaka, kanra, garba ganda, karkana, palawar, sara, bellu-ponik

S. brevibarbe (Michx.) Pers.

English: brown plume grass

S. contortum (Baldwin ex Elliott) Nutt.

English: bent-awn plume grass

S. giganteum (Walter) Pers.

English: sugar-cane plume grass

S. officinarum L. (*Saccharum barberi* Jesw.; *Saccharum sinense* Roxb.)

English: sugar cane, noble cane, North Indian sugarcane

Italian: canna da zucchero

Spanish: caña, caña de azúcar, caña japonesa

Mexico: caen, coba-gui-naxi, coba-qui-naxi, gubaguinashi, maca, pepa, pacab, uhuatl, caña de azúcar, nite, nito, nito naxi, nito guia baa xtilla, goba gui naxi

Peru: impuco, itica, mishi quiro, ñaamura, paat, pagat, pagad, pochoasiri, pochwaksuru, sabi, senoorr, senorr, taa vata, xai

Brazil: cana de açúcar

South Africa: suikerriet

Tanzania: moba

Yoruba: ireke

India: ikshu, rasalah, khanda, kandya, ganna, kamand, sarkara, pundia, pundya, paunda, cheruku, karimbu, karumbu, poovan karumbu, kabbu, patta patti kabbu, uukh, kajali, ak, shakir surkh, uus, sherdi, kabirya, malbari, kala kalbari, kavangiri, wansi

Tibetan: bu-ram, ka-ra

China: gan zhe, kan che

South Laos: (people Nya Hön) grao

Japan: kara-satô-kibi, shin-satô-kibi (shin = true)

Okinawa: satô-kibi, uji

The Philippines: tubo, atbo, agbo, caña dulce, tubu, una, unas, unat

Malaya: tebu

Hawaii: ko

S. ravennae (L.) Murray

English: Ravenna's grass, Ravennagrass

S. spontaneum L.

English: wild sugar cane, wild cane, thatch grass

India: kasa, chhote-kase

Japan: nan-goku-wase-obana (meaning early-flowering *Miscanthus*)

The Philippines: talahib, lidda, bogang, tigbau, salin, sikai, tibayo

Saccia Naudin Convolvulaceae

Origins:
Greek *sakkos* "sack."

Saccidium Lindley Orchidaceae

Origins:
Greek *sakkos* "sack," *sakkidion* "a small sack," the saccate lateral sepals.

Saccifolium Maguire & Pires Gentianaceae (Saccifoliaceae)

Origins:
Latin *saccus, i* "sack, bag" and *folium, ii* "a leaf," the leaves are saccate-vaginate.

Sacciolepis Nash Gramineae

Origins:
Greek *sakkos* "sack" and *lepis* "a scale"; see Nathaniel Lord Britton (1859-1934), *Manual of the Flora of the Northern States and Canada*. New York 1901.

Species/Vernacular Names:
S. indica (L.) Chase

English: Glenwood grass, Indian cupscale grass

The Philippines: buntot-pusa, sabsabung, salaimaya, sangu-mai, sikuan

Malaya: rumput bidis, bontoh darat

Saccocalyx Cosson & Durieu Labiatae

Origins:
From the Greek *sakkos* "sack" and *kalyx* "a calyx."

Saccochilus Blume Orchidaceae

Origins:
From the Greek *sakkos* "sack" and *cheilos* "a lip," the shape of the lip.

Saccoglossum Schltr. Orchidaceae

Origins:
From the Greek *sakkos* and *glossa* "tongue," the shape of the lip.

Saccoglottis Walp. Humiriaceae

Origins:
See *Sacoglottis* Mart.

Saccolabiopsis J.J. Smith Orchidaceae

Origins:
Resembling the genus *Saccolabium* Blume; see J.J. Smith, in *Bulletin du Jardin Botanique de Buitenzorg*. Ser. 2, 26: 93. (Feb.) 1918; Alick William Dockrill, *Australasian Sarcanthinae*. 1967, and *Australian Indigenous Orchids*. Sydney 1969.

Saccolabium Blume Orchidaceae

Origins:
Latin *saccus* "bag" and *labium* "lip," referring to the shape of the labellum; see Karl Ludwig von Blume, *Bijdragen tot de flora van Nederlandsch Indië*. 292. (Sep.-Dec.) 1825; Alick William Dockrill, *Australasian Sarcanthinae*. 1967, and *Australian Indigenous Orchids*. Sydney 1969.

Saccolaria Kuhlm. Lentibulariaceae

Origins:
From the Latin *saccus* "bag," *sacculus* "a little bag," see also *Utricularia* L.

Saccolena Gleason Melastomataceae

Origins:
Greek *sakkos* "sack" and *chlaena, laina* "cloak, blanket."

Saccoloma Kaulf. Dennstaedtiaceae

Origins:
From the Greek *sakkos* and *loma* "border, margin, fringe, edge."

Saccopetalum Bennett Annonaceae

Origins:

Greek *sakkos* and *petalon* "petal," Latin *saccus* and *petalum* "a petal," referring to the shape of the petals, bagged or saccate; see John Joseph Bennett (1801-1876) and Robert Brown, *Plantae Javanicae rariores*. 165, t. 35. London 1840.

Saccoplectus Oerst. Gesneriaceae

Origins:

From the Greek *sakkos* "sack" and *plektos* "twisted, plaited."

Saccularia Kellogg Scrophulariaceae

Origins:

Latin *saccus* "bag," *sacculus* "a little bag."

Sacculina Bosser Lentibulariaceae

Origins:

From the Latin *saccus* "bag," *sacculus* "a little bag," see also *Utricularia* L.

Sacharodendron Raf. Aceraceae

Origins:

Based on *Acer* L. and *Acer saccharinum* L., from the Greek *sakcharon* "sugar" and *dendron* "tree"; see Constantine Samuel Rafinesque, *New Fl. N. Am.* 1: 47. 1836.

Sachsia Grisebach Asteraceae

Origins:

After the German (b. Breslau, now Wroclaw, Poland) botanist Ferdinand Gustav Julius von Sachs, 1832-1897 (d. Würzburg), plant physiologist; see Martin Bopp, in *Dictionary of Scientific Biography* 12: 58-60. 1981; T.W. Bossert, *Biographical Dictionary of Botanists Represented in the Hunt Institute Portrait Collection*. 344. 1972; S. Lenley et al., *Catalog of the Manuscript and Archival Collections and Index to the Correspondence of John Torrey*. Library of the New York Botanical Garden. 357. 1973; E.M. Tucker, *Catalogue of the Library of the Arnold Arboretum of Harvard University*. 1917-1933; H.R. Fletcher and W.H. Brown, *Royal Botanic Garden Edinburgh, 1670-1970*. Edinburgh 1970; B. Glass et al., eds., *Forerunners of Darwin: 1745-1859*. Baltimore 1959; J.D. Milner, *Catalogue of Portraits of Botanists Exhibited in the Museums of the Royal Botanic Gardens*. Royal Botanic Gardens, Kew, London 1906; G. Schmid, *Goethe und die Naturwissenschaften*. Halle 1940; Frans A. Stafleu and Richard S. Cowan, *Taxonomic Literature*. 4: 1040-1044. Utrecht 1983.

Sacodon Raf. Orchidaceae

Origins:

Greek *sakkos, sakos* "bag, pouch" and *odous, odontos* "a tooth," referring to the mouth of the lip; see C.S. Rafinesque, *Flora Telluriana*. 4: 45. 1836 [1838].

Sacoglottis C. Martius Humiriaceae

Origins:

Greek *sakkos, sakos* and *glotta* "tongue," referring to the anthers.

Species/Vernacular Names:

S. gabonensis (Baillon) Urban

English: Liberian cherry

Yoruba: atala

Nigeria: atala, edet, eruk, nchi, nche, ndat, osuga, ozouga, okia, okpi-uta, tala, ugu, amwan, amouan, eyete, iheti, ewete, edat; atala (Yoruba); ugu (Edo); itala (Itsekiri); nche (Igbo); edat (Boki); ndat (Efik); edat (Ibibio)

Congo: niuka

Gabon: lombi, ossuga, ozouga, esoua, essona, essoua, issoua, mosoukouga

Cameroon: zole, biawa, bebvo, bedwa, bidou, bidu, idou, ilouye, bodoua, edoue, eloue, ozongo, ozouga, ozougo

Ivory Coast: akohia, efeuna, amuan, akouapo

Liberia: cherry, dauh, mahogany

Sacoila Raf. Orchidaceae

Origins:

Greek *sakkos, sakos* and *koilos* "hollow," referring to the hollow spur; see C.S. Rafinesque, *Flora Telluriana*. 2: 86. 1836 [1837].

Sacosperma G. Taylor Rubiaceae

Origins:

From the Greek *sakkos, sakos* "bag, pouch" and *sperma* "seed."

Sadiria Mez Myrsinaceae

Origins:

The anagram of the generic name *Ardisia* Sw.

Sadleria Kaulfuss Blechnaceae (Blechnoideae)

Origins:

For the Hungarian botanist Joseph Sadler, 1791-1849, physician, M.D. Budapest 1820, professor of botany, naturalist. He is remembered for *De filicibus veris hungariae.* Budae [1830] and *Flora Comitatus Pestiensis.* [Octavo.] Pestini 1825-1826; see John H. Barnhart, *Biographical Notes upon Botanists.* 3: 198. 1965; T.W. Bossert, *Biographical Dictionary of Botanists Represented in the Hunt Institute Portrait Collection.* 344. 1972.

Saffordia Maxon Pteridaceae (Adiantaceae)

Origins:

Named after the American botanist William Edwin Safford, 1859-1926, naturalist, conchologist, 1899-1900 Guam, 1902-1926 U.S. Department of Agriculture. His works include *The Useful Plants of the Island of Guam.* Washington 1905, *Cactaceae of Northeastern and Central Mexico.* Washington 1909, "An Aztec narcotic." *J. Hered.* 6: 291-311. 1915, "Cultivated plants of Polynesia and their vernacular names, an index to the origin and migration of the Polynesians." *Proc. 1st Pan-Pacific Sci. Conf., Honolulu.* 535-536. 1921, "Synopsis of the genus *Datura.*" *J. Wash. Acad. Sci.* 11: 173-189. 1921 and "*Daturas* of the Old World and New. An account of their narcotic properties and their use in oracular and initiatory ceremonies." *Annual Rep. Board Regents Smithsonian Inst.* 1920: 537-567. 1922. See Henry Elwood Baum, *The Breadfruit* ... Together with a biographical sketch of the author by W.E. Safford. Washington 1904; John H. Barnhart, *Biographical Notes upon Botanists.* 3: 199. 1965; T.W. Bossert, *Biographical Dictionary of Botanists Represented in the Hunt Institute Portrait Collection.* 345. 1972; S. Lenley et al., *Catalog of the Manuscript and Archival Collections and Index to the Correspondence of John Torrey.* Library of the New York Botanical Garden. 357. 1973; Elmer Drew Merrill, in *Bernice P. Bishop Mus. Bull.* 144: 161. 1937; Ida Kaplan Langman, *A Selected Guide to the Literature on the Flowering Plants of Mexico.* University of Pennsylvania Press, Philadelphia 1964; Irving William Knobloch, compil., "A preliminary verified list of plant collectors in Mexico." *Phytologia Memoirs.* VI. 1983; R. Zander, F. Encke, G. Buchheim and S. Seybold, *Handwörterbuch der Pflanzennamen.* 14. Aufl. Stuttgart 1993.

Saffordiella Merrill Myrtaceae

Origins:

For the American botanist William Edwin Safford, 1859-1926.

Sagenia C. Presl Dryopteridaceae (Aspleniaceae)

Origins:

Greek *sagene* "a net"; see Karl (Carl) B. Presl, *Tentamen Pteridographiae, seu genera Filicacearum.* 86, t. II. Prague 1836.

Sagenopteris Trevis. Dryopteridaceae (Aspleniaceae)

Origins:

Greek *sagene* "a net" and *pteris* "fern."

Sageraea Dalzell Annonaceae

Origins:

Sagare or *sajeri*, Kannada and Marathi names, India.

Sageretia Brongn. Rhamnaceae

Origins:

For the French agriculturist Augustin Sageret, 1763-1851, landowner, botanist, author of *Mémoire sur les Cucurbitacées.* Paris 1826; see J.H. Barnhart, *Biographical Notes upon Botanists.* 3: 199. 1965; Adolphe Théodore (de) Brongniart (1801-1876), in *Annales des Sciences Naturelles.* 10: 359, t. 13, fig. 2. (Apr.) 1827; Georg Christian Wittstein, *Etymologisch-botanisches Handwörterbuch.* 779f. 1852; E.M. Tucker, *Catalogue of the Library of the Arnold Arboretum of Harvard University.* 1917-1933; see also Herbert F. Roberts, *Plant Hybridization before Mendel.* Princeton 1929; F. Boerner & G. Kunkel, *Taschenwörterbuch der botanischen Pflanzennamen.* 4. Aufl. 164. Berlin & Hamburg 1989.

Species/Vernacular Names:

S. thea (Osbeck) M. Johnston (*Sageretia theezans* (L.) Brongn.; *Rhamnus theezans* L.)

India: ankol, thum, dargola, kutku, khadgu, burlcha

Japan: kuro-ige

China: que mei teng

Sagina L. Caryophyllaceae

Origins:

Latin *sagina, ae* "a feeding, fattening, food, a fatted animal," Akkadian *sahu* "swine"; see Carl Linnaeus, *Species Plantarum*. 128. 1753 and *Genera Plantarum*. Ed. 5. 62. 1754.

Species/Vernacular Names:

S. apetala L.

English: common pearlwort, annual pearlwort

S. japonica (Swartz) Ohwi (*Spergula japonica* Swartz)

English: pearlwort, Japanese pearlwort

Japan: tsume-kusa

China: qi gu cao, chi ku tsao

S. maritima G. Don

English: sea pearlwort

S. procumbens L.

English: spreading pearlwort, Arctic pearlwort

Sagittaria L. Alismataceae

Origins:

Latin *sagittarius, a, um* "belonging to an arrow," *sagitta, ae* "an arrow, shaft," referring to the form of the leaves; see Carl Linnaeus, *Species Plantarum*. 993. 1753 and *Genera Plantarum*. Ed. 5. 429. 1754.

Species/Vernacular Names:

S. graminea Michaux var. *platyphylla* Engelm.

English: sagittaria

S. latifolia Willd. (*Sagittaria simplex* Pursh; *Sagittaria gracilis* Pursh; *Sagittaria hastata* Pursh)

English: wapato, duck potato

S. montevidensis Chamisso & Schltdl.

English: arrowhead

S. pygmaea Miq. (*Blyxa coreana* Nakai)

Japan: uri-kawa

S. sagittifolia L. (*Sagittaria trifolia* L.)

English: arrowhead, old world arrowhead

Japan: omo-daka

China: ci gu, chieh ku, pai ti li, tzu ku

S. sanfordii E. Greene

English: Sanford's arrowhead

Sagittipetalum Merr. Rhizophoraceae

Origins:

Latin *sagitta, ae* "an arrow, shaft" and *petalum* or *petalon* "petal."

Sagotanthus Tieghem Olacaceae

Origins:

For the French botanist Paul Antoine Sagot, 1821-1888, physician.

Sagotia Baillon Euphorbiaceae

Origins:

For the French botanist Paul Antoine Sagot, 1821-1888, physician, M.D. Paris 1848, traveler, plant collector (French Antilles and French Guyana), professor of natural sciences, surgeon with the French navy. His writings include "Catalogue des plantes ... de la Guyane française." in *Ann. Sci. nat., Bot.* 1880-1885, *Du malaise ... dans les maladies fébriles.* in Collection des thèses soutenues à la Faculté de Médecine de Paris. an 1848, tom. 12. Paris 1839-1878 and *De l'état sauvage et des résultats de la culture et de la domestication.* Nantes 1865, with Victor Pérez wrote *Le Tagasaste, cytisus proliferus varietas, fourrage important.* Paris 1892. See Jules Nicolas Crevaux, P. Sagot & L. Adam, *Grammaires et vocabulaires roucouyenne, arrouague, piapoco et d'autres langues de la région des Guyanes.* [in Bibliothèque linguistique américaine. Paris 1871-1903.] Paris 1882; J.H. Barnhart, *Biographical Notes upon Botanists.* 3: 199. 1965; T.W. Bossert, *Biographical Dictionary of Botanists Represented in the Hunt Institute Portrait Collection.* 345. 1972; S. Lenley et al., *Catalog of the Manuscript and Archival Collections and Index to the Correspondence of John Torrey.* Library of the New York Botanical Garden. 468. 1973; E.M. Tucker, *Catalogue of the Library of the Arnold Arboretum of Harvard University.* 1917-1933; Frans A. Stafleu and Richard S. Cowan, *Taxonomic Literature.* 4: 1053-1055. 1983.

Sagraea DC. Melastomataceae

Origins:

After the Spanish botanist Ramón de la Sagra, 1798-1871, economist, traveler, professor of botany, agriculturist, plant collector (Cuba), bryologist. Among his works are *Tablas necrologicas del Colera-Morbus en la Ciudad de la Habana y sus arrabales.* Habana 1833, *Le Mal et le Remède: aphorismes sociaux.* Paris 1859, *Description et culture de l'Ortie de la Chine.* Paris [1870], *Introduction à la Philosophie religieuse.* Paris 1869, *Historia económico-política y estadística de la isla de Cuba.* Habana 1831, *Aforismos sociales con aplicacion a España.* Madrid 1854-1855, *Memoria sobre el bejuco del Guace.* Habana 1833 and *Historia física, política y natural de la isla de Cuba.* Paris 1842-1861. See Aurelio Mitjans, *Estudio sobre el movimiento científico y literario de Cuba.* Habana 1890; M.

Núñez de Arena, *D. Ramón de la Sagra, reformador social.* 1924; T.W. Bossert, *Biographical Dictionary of Botanists Represented in the Hunt Institute Portrait Collection.* 345. 1972; Francisco Calcagno, *Diccionario biográfico cubano.* New York 1878-[1886]; A. Lasègue, *Musée botanique de Benjamin Delessert.* Paris 1845; M. Colmeiro y Penido, *La Botánica y los Botánicos de la Peninsula Hispano-Lusitana.* Madrid 1858; Ida Kaplan Langman, *A Selected Guide to the Literature on the Flowering Plants of Mexico.* 663. University of Pennsylvania Press, Philadelphia 1964; E.M. Tucker, *Catalogue of the Library of the Arnold Arboretum of Harvard University.* 1917-1933; S. Lenley et al., *Catalog of the Manuscript and Archival Collections and Index to the Correspondence of John Torrey.* Library of the New York Botanical Garden. 357. 1973; Frans A. Stafleu and Richard S. Cowan, *Taxonomic Literature.* 4: 1055-1057. 1983.

Saguaster Kuntze Palmae

Origins:

See Otto Kuntze (1843-1907), *Revisio generum plantarum.* 2: 734. Leipzig etc. 1891, see also *Drymophloeus* Zipp.

Saguerus Steck Palmae

Origins:

See Abraham Steck, Dissertatio inauguralis medica *de Sagu.* Argentorati [Strasbourg] [1757], see also *Arenga* Labill.

Sagus Gaertner Palmae

Origins:

The Indonesian word *sagu*; see Joseph Gaertner (1732-1791), *De fructibus et seminibus plantarum.* 1: 27. 1788, see also *Raphia* P. Beauv.

Sagus Steck Palmae

Origins:

See Abraham Steck, Dissertatio inauguralis medica *de Sagu.* Argentorati [Strasbourg] [1757], see also *Metroxylon* Rottb.

Saintpaulia H.A. Wendland Gesneriaceae

Origins:

For the German Walter Freiherr von Saint Paul Illaire, 1860-1910, traveler, plant collector in East Africa, author of *Suaheli Handbuch.* [in Berlin–Friedrich–Wilhelms Universität — Seminar für orientalische Sprachen. Lehrbuucher, etc. Bd. 2.] Berlin 1890 and *Swahili Sprachführer.* Dar-es-Salaam 1896. See F. Boerner & G. Kunkel, *Taschenwörterbuch der botanischen Pflanzennamen.* 4. Aufl. 164. Berlin & Hamburg 1989; O. Schmeil & A. Seybold, *Lehrbuch der Botanik.* Heidelberg 1958; Helmut Genaust, *Etymologisches Wörterbuch der botanischen Pflanzennamen.* 551. [genus dedicated to Walter Freiherr von Saint-Paul Hilaire, 1860-1910] Basel 1996; Antonia Eastwood, Benny Bytebier, Hilary Tye, Alan Tye, Ann Robertson and Mike Maunder, "The conservation status of *Saintpaulia*." in *Curtis's Botanical Magazine.* 15(1): 49-62. [" Baron Adalbert Emil Walter Radcliffe von Saint Paul-Illaire, 1860-1940"] February 1998; William T. Stearn, *Stearn's Dictionary of Plant Names for Gardeners.* 264. ["Baron Walter von Saint Paul Illaire, 1860-1910"] Cassell 1993.

Saintpauliopsis Staner Acanthaceae

Origins:
Resembling *Saintpaulia.*

Sakersia Hook.f. Melastomataceae

Origins:
For Saker, British missionary, in F. Po with Gustav Mann; see F.N. Hepper and Fiona Neate, *Plant Collectors in West Africa.* 53, 71. Utrecht 1971.

Salacca Reinwardt Palmae

Origins:
A vernacular name, *salak.*

Species/Vernacular Names:
S. clemensiana Becc. (*Zalacca clemensiana* Beccari)
The Philippines: lakaubi

Salacia L. Celastraceae (Hippocrateaceae)

Origins:
After Salacia, the Roman goddess of the spring water; when Neptunus, the Roman god of freshwater, by 399 BC, was identified with the Greek Poseidon, the god of the sea, she was identified and equated with the latter's wife Amphitrite; Greek *salos* "any unsteady motion," Latin *salum, i* "the sea, the high sea, the stream"; see C. Linnaeus, *Mantissa Plantarum.* 2: 159, 293. 1771.

Species/Vernacular Names:

S. sp.

Yoruba: okundun, asobiori, igbadu, igba adaloju

S. gerrardii Harv.

South Africa: climbing salacia

S. leptoclada Tul. (*Salacia wardii* Verdoorn; *Salacia baumannii* Loes.) (Greek *leptos* "slender" and *klados* "branch")

English: lemon rope, ward's salacia

Southern Africa: lemoentjietou; iSundwane, amaPunzana, uHlangahomo (Zulu)

Salacicratea Loes. Celastraceae (Hippocrateaceae)

Origins:

From the genera *Salacia* L. and *Hippocratea* L.; see Ludwig E. Theodor Loesener (1865-1941), in Carl Ernst Arthur Wichmann (b. 1851), *Nova Guinea*. [Botanique. Vol. 8: 281.] Leiden 1910.

Salaciopsis Baker f. Celastraceae

Origins:

Resembling the genus *Salacia* L.

Salacistis Reichb.f. Orchidaceae

Origins:

Possibly from Mount Salak, Java.

Salaxis Salisb. Ericaceae

Origins:

Origin unknown, Latin *salax, salacis* "lustful," Greek *salax, akos* "miner's sieve, miner's riddle."

Salazaria Torrey Labiatae

Origins:

For José Salazar Ylarregui (y Larregui, y Larrequi), Mexican astronomer, US Mexican Boundary Survey. Among his works are *Datos de los trabajos astronomicos y topographicos ... practicados ... por la Comision de Limites Mexicana en la linea que divide esta republica de la de los Estados Unidos*. Mexico 1850.

Saldanha Vellozo Rubiaceae

Origins:

After the Brazilian politician Martin Lopez Saldanha; see Frans A. Stafleu and Richard S. Cowan, *Taxonomic Literature*. 5: 1. Utrecht 1985.

Saldanhaea Bureau Bignoniaceae

Origins:

After the Brazilian botanist José de Saldanha da Gama, 1893-1905, professor of botany, plant collector; see John H. Barnhart, *Biographical Notes upon Botanists*. 3: 202. 1965; T.W. Bossert, *Biographical Dictionary of Botanists Represented in the Hunt Institute Portrait Collection*. 346. 1972; E.M. Tucker, *Catalogue of the Library of the Arnold Arboretum of Harvard University*. 1917-1933; R. Zander, F. Encke, G. Buchheim and S. Seybold, *Handwörterbuch der Pflanzennamen*. 14. Aufl. Stuttgart 1993.

Salicaria Adans. Lythraceae

Origins:

Latin *salix, icis* "willow, a willow-tree, sallow."

Salicornia L. Chenopodiaceae (Salicorniaceae)

Origins:

Latin *salicornia*, French *salicorne*, from *salicor*, the name of a plant, from the Arabic *sala al-qarab*; see Carl Linnaeus, *Species Plantarum*. 3. 1753 and *Genera Plantarum*. Ed. 5. 4. 1754; Salvatore Battaglia, *Grande dizionario della lingua italiana*. XVII: 402. Torino 1995; Helmut Genaust, *Etymologisches Wörterbuch der botanischen Pflanzennamen*. 551-552. 1996; Georg Christian Wittstein, *Etymologisch-botanisches Handwörterbuch*. 780. Ansbach 1852.

Species/Vernacular Names:

S. arabica L.

Arabic: hmadha

French: salicorne

S. europaea L.

English: glasswort, marsh samphire, chicken claws

Salisia Lindley Myrtaceae

Origins:

Possibly from Adalbert Ulysses von Salis-Marschlins, 1795-1886, wrote about the flora of Corsica; see John Lindley (1799-1865), *Edwards's Botanical Register*. Appendix to Vols. 1-23: *A Sketch of the Vegetation of the Swan River Colony*. London (November) 1839.

Salix L. Salicaceae

Origins:

Latin *salix*, *salicis* "willow, sallow" (Marcus Terentius Varro, 116-27 BC., et al.); Akkadian *salihu* "sprinkler of water," *salahu* "to moisten"; Celtic *sal* "near" and *lis* "water"; Irish *saile* "saliva." See Carl Linnaeus, *Species Plantarum*. 1015. 1753 and *Genera Plantarum*. Ed. 5. 447. 1754; G. Semerano, *Dizionario della lingua Latina e di voci moderne*. 2(2): 553. 1994.

Species/Vernacular Names:

S. sp.

English: weeping willow, willow

Mexico: sauce, ahuexotl, quetzalhuexotl (= sauce precioso), yaga gueza, yaga queza, yaga queza lachi

Tibetan: lcang-ma, sgro-ba

Nigeria: ba-ruwana, rimni

S. alba L.

English: cricket-bat willow, Huntingdon willow, swallow-tailed willow, white willow

Arabic: khilaf, safsaf abiad

S. arctica Pallas

English: Arctic willow

S. babylonica L. (*Salix lasiogyne* Seem.)

English: Babylon weeping willow, weeping willow, willow tea, Chinese weeping willow

Nepal: bainsh

China: liu zhi, liu, hsiao yang, yang liu

Japan: shidare-yanagi

Okinawa: tai-yanagi

Southern Africa: huilwilgerboom, makwilgerboom, treurboom, treurwilger, wilgerboom, wilgerhout; moluoane (Sotho)

S. bebbiana Sarg.

English: cricket-bat willow

S. cinerea L.

English: pussy willow, common sallow

S. exigua Nutt.

English: narrow-leaved willow

S. fragilis L.

English: crack willow, brittle willow

S. humilis Marshall

English: prairie willow

S. koriyanagi Kimura

Japan: kôri-yanagi

S. integra Thunb.

Japan: inu-kôri-yanagi

S. laevigata Bebb

English: red willow

S. ligulifolia (C. Ball) C. Ball

English: strap-leafed willow

S. lucida Muhlenb.

English: shining willow

S. lutea Nutt.

English: yellow willow

S. matsudana Koidz.

English: Hankow willow

Japan: unryû-yanagi

S. melanopsis Nutt.

English: dusky willow

S. mucronata Thunb. subsp. *capensis* (Thunb.) Immelman (*Salix capensis* Thunb.; *Salix capensis* Thunb. var. *gariepina* (Burch.) Anders.; *Salix mucronata* Thunb. var. *caffra* Burtt Davy; *Salix mucronata* Thunb. var. *integra* Burtt Davy)

English: Vaal River willow, Cape willow, Cape river, river willow, wild willow

Southern Africa: Vaal willow, Vaalwilger, wildewilgeboom, wilgeboom, rivierwilg; umNgcunube, umSwi, umThenstema, umGcunube, umBhenya (Xhosa); umGcunube, umNyezane (Zulu); moduane (South Sotho); mogokaro, modubu-noka (North Sotho)

S. mucronata Thunb. subsp. *hirsuta* (Thunb.) Immelman (*Salix hirsuta* Thunb.)

English: wild willow

South Africa: wilgeboom

S. mucronata Thunb. subsp. *mucronata* (*Salix capensis* Thunb. var. *mucronata* (Thunb.) Anders.; *Salix safsaf* Forssk. ex Trautv.; *Salix subserrata* Willd.)

English: wild willow

South Africa: wild willow, wilgeboom; muPuma, muSamangwena, muSambangwena, muTepe (Shona)

S. mucronata Thunb. subsp. *woodii* (Seemen) Immelman (*Salix woodii* Seemen)

English: wild willow, wood's willow, Natal willow

Southern Africa: Natalwilger, wilgeboom, hangtakkies; umNyezane, umZekana (Zulu); umNyetane (Swazi); modunae (South Sotho); mogokaro (North Sotho); munengeledzi, munangalendzi (Venda)

S. polaris Wahlenb.

English: polar willow

S. purpurea L.

English: purple willow, basket willow, purple osier

China: shui yang gen, shui yang, pu liu, ching yang

S. pyrifolia Anderss.

English: balsam willow

S. repens L.

English: creeping willow

S. reptans Rupr.

English: Arctic creeping willow

S. rotundifolia Trautv.

English: round leaved willow

S. silesiaca Willd.

English: Silesian willow

S. tetrasperma Roxb.

English: fourseed willow, Indian willow

Malaya: dalu dalu, dedalu, medalu, mendalu

S. triandra L.

English: almond-leaved willow

S. viminalis L.

English: common osier, basket willow

Salmalia Schott & Endl. Bombacaceae

Origins:
Salmali, a Sanskrit name for *Salmalia malabarica* (DC.) Schott & Endl.

Salmea DC. Asteraceae

Origins:
For the German botanist Joseph Franz Maria Anton Hubert Ignatz Fürst zu Salm-Reifferscheid-Dyck, 1773-1861, horticulturist and artist. He is remembered for *Monographia generum Aloes et Mesembryanthemi*. Bonnae 1836-1863 and *Observationes botanicae* in Horto Dyckensi notatae. Anno 1820. Coloniae 1820; see J.H. Barnhart, *Biographical Notes upon Botanists*. 3: 203. 1965; T.W. Bossert, *Biographical Dictionary of Botanists Represented in the Hunt Institute Portrait Collection*. 346. 1972; H.N. Clokie, *Account of the Herbaria of the Department of Botany in the University of Oxford*. 236. Oxford 1964; Ernest Nelmes (1895-1959) and William Cuthbertson (c. 1859-1934), *Curtis's Botanical Magazine Dedications, 1827-1927*. 66-68. [1931]; Gilbert Westacott Reynolds (1895-1967), *The Aloes of South Africa*. 94-95. Balkema, Rotterdam 1982; A. White

and B.L. Sloane, *The Stapelieae*. Pasadena 1937; Georg Christian Wittstein, *Etymologisch-botanisches Handwörterbuch*. 307, 781. Ansbach 1852; E.M. Tucker, *Catalogue of the Library of the Arnold Arboretum of Harvard University*. 1917-1933; R. Zander, F. Encke, G. Buchheim and S. Seybold, *Handwörterbuch der Pflanzennamen*. 14. Aufl. 773. Stuttgart 1993; Gordon Douglas Rowley, *A History of Succulent Plants*. Strawberry Press, Mill Valley, California 1997; Ida Kaplan Langman, *A Selected Guide to the Literature on the Flowering Plants of Mexico*. 1964; Stafleu and Cowan, *Taxonomic Literature*. 5: 7-9. Utrecht 1985.

Salmeopsis Benth. Asteraceae

Origins:
Resembling *Salmea* DC.

Salmiopuntia Fric Cactaceae

Origins:
For the German botanist J.F.M.A.H Ignatz Fürst zu Salm-Reifferscheid-Dyck, 1773-1861.

Salmonia Scopoli Vochysiaceae

Origins:
For the English physician William Salmon, 1644-1712 (or 1713), professor of medicine, a prolific writer, empiric in Smithfield (London). His works include *Collectanea Medica*, the Country Physician. London 1703, *Doron Medicum*. London 1683, *A Discourse against Transubstantiation*. London 1690, *Ars Anatomica*, or the Anatomy of the Humane Bodies. London 1714, *Botanologia*. London 1710-1711, *A Dissertation upon Water-Baptism*. 1700, *Iatrica*; seu Praxis medendi. London 1694, *Medicina Practica*; or the Practical Physician. London 1707, *Phylaxa Medicina*. London 1688, *Seplasium*. The compleat English Physician. London 1693 and *Synopsis medicinae* or compendium of Astrological, Galenical and Chymical physick. London 1671; see T. Ballard, *Bibliotheca Salmoneana*, pars prima. London [1713]; R. Pulteney, *Historical and Biographical Sketches of the Progress of Botany in England*. 1: 185-188. London 1790; Blanche Henrey, *British Botanical and Horticultural Literature before 1800*. 1975.

Salomonia Lour. Polygalaceae

Origins:
Greek *salos* "any unsteady motion" and *monos* "single, one"; some suggest from Salomon, King of Israel from c.

973 BC. to c. 937 BC.; see João (Joannes) de Loureiro, *Flora cochinchinensis*: 14. [Lisboa] 1790.

Salpianthus Bonpl. Nyctaginaceae

Origins:
Greek *salpinx* "a trumpet, tube" and *anthos* "flower," referring to the calyx.

Salpichlaena J. Sm. Blechnaceae (Blechnoideae)

Origins:
Greek *salpinx* and *chlaena* "a cloak, blanket."

Salpichroa Miers Solanaceae

Origins:
From the Greek *salpinx* and *chroa* "color," referring to the flowers; see W.J. Hooker, in *Hooker's London Journal of Botany*. 4: 321. 1845.

Species/Vernacular Names:
S. origanifolia (Lam.) Baillon (*Physalis origanifolia* Lam.)
English: Pampas lily of the valley, cock's eggs

Salpiglossis Ruíz & Pavón Solanaceae

Origins:
Greek *salpinx* "a trumpet, tube" and *glossa* "tongue," presumably referring to the style.

Salpinctes Woodson Apocynaceae

Origins:
Greek *salpinx* "a trumpet."

Salpinctium T.J. Edwards Acanthaceae

Origins:
Greek *salpinx* "a trumpet," Latin *salpicta, ae* and *salpista* "trumpeter," referring to the flowers.

Species/Vernacular Names:
S. natalense (C.B. Cl.) T.J. Edwards (*Asystasia natalensis* C.B. Cl.)
English: Natal asystasia

Salpinga Martius ex DC. Melastomataceae

Origins:
Greek *salpinx, salpingos* "a trumpet," indicating the calyx.

Salpingantha Lem. Acanthaceae

Origins:
Greek *salpinx, salpingos* "a trumpet" and *anthos* "flower," see also *Salpixantha* Hook.

Salpingia (Torrey & A. Gray) Raim. Onagraceae

Origins:
From the Greek *salpinx, salpingos* "a trumpet."

Salpingolobivia Y. Ito Cactaceae

Origins:
From the Greek *salpinx, salpingos* "a trumpet" plus *Lobivia*.

Salpingostylis Small Iridaceae

Origins:
From the Greek *salpinx, salpingos* and *stylos* "style."

Salpinxantha Urb. Acanthaceae

Origins:
Greek *salpinx, salpingos* "a trumpet" and *xanthos* "yellow" and *anthos* "flower," see also *Salpixantha* Hook.

Salpistele Dressler Orchidaceae

Origins:
Greek *salpinx, salpingos* and *stele* "trunk, central part of stem."

Salpixantha Hook. Acanthaceae

Origins:
Greek *salpinx, salpingos* and *xanthos* "yellow" or *anthos* "flower."

Salsola L. Chenopodiaceae

Origins:

Latin *sal, is* "salt," *salsus, a, um* "salted, salty," possibly referring to *Halogeton sativus* (L.) C. Meyer and to the habitat of these plants, sea coasts and other saline habitats; see Carl Linnaeus, *Species Plantarum*. 222. 1753 and *Genera Plantarum*. Ed. 5. 104. 1754.

Species/Vernacular Names:

S. sp.

Namibia: ganna, brakbos (Afrikaans)

German: brackbusch, salzbusch

S. aphylla (R. Br.) Sprengel (*Salsola aphylla* L.f. var. *canescens* Fenzl ex Drège; *Salsola caffra* Sparrman; *Salsola caroxylon* Moq.; *Caroxylon brevifolium* St.-Lag.; *Caroxylon brevifolium* St.-Lag.; *Kochia aphylla* R. Br.)

English: lye bush, lye ganna

South Africa: asbossie, beesganna, brakganna, channa, gannabos, ganna-asbos, gewone ganna, gona, grootganna, kannabos, kortbeenganna, langbeenganna, seepbossie, seepbosganna, seepganna, soutbossie, soutganna

S. australis R. Br.

English: Russian thistle, salt-bush

S. kali L. (*Salsola australis* R. Br.; *Salsola pestifer* A. Nels.; *Salsola australis* R. Br. var. *strobilifera* (Benth.) Domin; *Salsola kali* L. var. *strobilifera* (Benth.); *Salsola kali* L. var. *leptophylla* Benth.; *Salsola kali* L. var. *tenuifolia* Tausch; *Salsola kali* L. subsp. *austroafricana* Aellen)

English: tumbleweed, rush-and-tumbleweed, Russian tumbleweed, glasswort, prickly glasswort, prickly saltwort, saltwort, rolypoly, soft rolypoly, Russian thistle, buckbush

India: sajjibuti

South Africa: kakiebos, kakiedissel, rolbossie, Russiese rolbossie, silwerbossie, tolbossie, waaibossie

S. paulsenii Litv.

English: barb-wire Russian thistle

S. tragus L.

English: Russian thistle, tumbleweed

S. tuberculatiformis Botsch. (*Salsola tuberculata* (Moq.) Fenzl var. *tomentosa* Aell.)

English: cauliflower saltwort

South Africa: blomkoolbossie, blomkoolganna, bosganna, brakbos, koolganna, koolpol

S. vermiculata L.

English: Mediterranean saltwort

Saltera Bullock Penaeaceae

Origins:

For the English (b. Glos.) botanist Terence Macleane Salter, 1883-1969 (d. Cape Town), traveler, 1930-1960 Bolus Herbarium, plant collector in South Africa (Cape), wrote *The Genus Oxalis in South Africa: A Taxonomic Revision*. Cape Town 1944, with Robert Stephen Adamson (1855-1965) edited *Flora of the Cape Peninsula*. Cape Town, Johannesburg 1950; see Mary Maytham Kidd, *Wild Flowers of the Cape Peninsula*. Cape Town 1950; John H. Barnhart, *Biographical Notes upon Botanists*. 3: 204. 1965; T.W. Bossert, *Biographical Dictionary of Botanists Represented in the Hunt Institute Portrait Collection*. 346. 1972; Mary Gunn and Leslie Edward W. Codd, *Botanical Exploration of Southern Africa*. 306-307. Cape Town 1981.

Saltia R. Br. ex Moq. Amaranthaceae

Origins:

After the English (b. Staffs.) traveler Henry Salt, 1780-1827 (d. Alexandria), 1802-1805 India and Africa with Lord Valentia (George Annesley, 1769-1844, 2nd Earl of Mountnorris, Viscount Valentia), 1805 and 1809-1810 in Abyssinia, in 1812 a Fellow of the Royal Society of London and of the Linnean Society, 1815-1827 British Consul-General in Egypt, sent algae to D. Turner. His writings include *A Voyage to Abyssinia and Travels into ... that Country ... in ... 1809 and 1810*. (Appendix. Zoology, with descriptions of the Birds by J. Latham, and additional remarks on them by Lord Stanley, afterwards 13th Earl of Derby. Botany, list of new and rare plants by R. Brown) London 1814, *Egypt, A Descriptive Poem with Notes by a Traveller*. 1824 and *Essays on Dr. Young's and M. Champollion's Phonetic System of Hieroglyphics*. London 1825. See Joseph Vallot, "Études sur la flore du Sénégal." in *Bull. Soc. Bot. de France*. 29: 189. Paris 1882; A. Lasègue, *Musée botanique de Benjamin Delessert*. Paris 1845; Alexander Murray (1798-1838), *Account of the Life and Writings of James Bruce*. [or rather, by H. Salt, with considerable additions and emendations by A. Murray.] Edinburgh 1808; Sir Alan Gardiner, *Egyptian Grammar. Being an introduction to the study of hieroglyphs*. Third edition, revised. Griffith Institute, Ashmolean Museum, Oxford 1982; J.H. Barnhart, *Biographical Notes upon Botanists*. 3: 204. 1965; Ethelyn Maria Tucker, *Catalogue of the Library of the Arnold Arboretum of Harvard University*. Cambridge, Massachusetts 1917-1933; G. Murray, *History of the Collections Contained in the Natural History Departments of the British Museum*. 1: 179. London 1904; Ray Desmond, *Dictionary of British & Irish Botanists and Horticulturists*. 604. 1994.

Salvadora L. Salvadoraceae

Origins:

The generic name after the Spanish apothecary and botanist Juan Salvador y Bosca, 1598-1681, plant collector; see *Noticia historica de la familla de Salvador de la ciudad de Barcelona* por Don Pedro Andres Pourret. Barcelona 1796.

Species/Vernacular Names:

S. australis Schweick. (*Salvadora angustifolia* Turrill var. *australis* (Schweick.) Verdoorn)

English: Transvaal mustard tree, narrow-leaved salvadora of the south

Southern Africa: Transvaalse mosterdboom, leubos; umPheme, iChithamuzi (Zulu)

S. persica L. (*Salvadora paniculata* Zucc. ex Steud.; *Salvadora crassinervia* Hochst. ex T. Anders.)

English: mustard tree, tooth-brush tree, saltbush

Arabic: arak, siwak, miswak

India: darakht-i-miswak, kharjal, kharijal, kharijar, jhal, jhak, rhakhan, khabhar, kakham, khakhin, kickni, vivay, miraj, mirajoli, pelu, chhota-pilu, pilu, piludi, pilva, pilvu, thorapilu, toboto, motijalya, ghunia, goni-mara, varagogu, kalawa, karkol, perungoli, ughaiputtai, kotungo

Southern Africa: omungambu (Herero)

Namibia: kerriebos (Afrikaans); ozongambu (fruit), otjingambu (Herero); khoris (Nama/Damara); okatunguya, omumkavu (Ndonga); enghadu (Kwanyama); omungavo, omungavu (Mbalantu); omunkavu (Kwambi, Nkolonkadhi, Ngandjera); omgavo, omungavu (Eunda); omungavu (Kwaluudhi)

Nigeria: asawaki (Hausa); kighir (Kanuri)

German: löwenbusch

Salvadoropsis H. Perrier Celastraceae

Origins:
Resembling *Salvadora* L.

Salvertia A. St.-Hil. Vochysiaceae

Origins:
For Dutour de Salvert, brother-in-law of Auguste de Saint-Hilaire.

Salvia L. Labiatae

Origins:

Latin *salvia, ae* "the herb sage" (Plinius), Latin *salvus, a, um* "whole, well-preserved, safe"; Akkadian *salwu, salmu* "healthy"; ancient Indian *sarvah*, ancient Persian *haruva-*; see Carl Linnaeus, *Species Plantarum*. 23. 1753 and *Genera Plantarum*. Ed. 5. 15. 1754; R. Gordon Wasson, "Notes on the Present Status of Ololiuhqui and the Other Hallucinogens of Mexico." from *Botanical Museum Leaflets, Harvard University*. Vol. 20(6): 161-212. Nov. 1963; Blas Pablo Reko, *Mitobotánica Zapoteca*. [Appended by an analysis of "Lienzo de Santiago Guevea"] Tacubaya 1945.

Species/Vernacular Names:

S. spp.

Mexico: salvia, chía, chian, mirto, sangre de toro; tepechía (Valle de Bravo); hierba del barretero (Coahuila)

S. aethiopis L.

English: African sage, woolly sage, woolly salvia, Mediterranean sage

S. apiana Jepson

English: bee sage, California white sage, white sage

S. azurea Michaux ex Lam.

English: blue sage

S. bowleyana Dunn

English: Bowley sage

China: nan dan shen

S. brandegei Munz

English: Brandegee's sage

S. cacaliifolia Benth. (*Salvia atriplicifolia* Fern.; *Salvia hempsteadiana* Blake)

English: cacalia-leaved salvia

S. carduacea Bentham

English: thistle sage

S. cavaleriei H. Léveillé

English: Kweichow sage

China: gui zhou shu wei cao

S. cavaleriei H. Léveillé var. *simplicifolia* E. Peter

English: unileaf sage

China: xue pen cao

S. chinensis Bentham

English: Chinese sage

China: hua shu wei cao, shi jian chuan

S. clevelandii (A. Gray) E. Greene

English: Jim sage

S. coccinea Jussieu ex Murray (*Salvia pseudococcinea* Jacq.; *Salvia rosea* Vahl; *Salvia glaucescens* Pohl; *Salvia ciliata* Benth.; *Salvia galeottii* Mart.)

English: red salvia, scarlet salvia, South American sage, Texas sage, Texas salvia

Mexico: mirto, cardenales, salvia cardenal, cardinal salvia, salvia roja; chak-tsits, tsab-tsits, tsab-xiu, ts'unum-pak (Yucatán); macancachauat (El Tajín, Ver.)

Madagascar: afolava, sangasanga, sangasanganandevolahy, sauge coccinée

South Africa: maksalie, rooisalie, vra-vir-pa

Japan: beni-bana-sarubiya

Hawaii: lililehua

S. deserta Schangin

English: desert sage

China: xin jiang shu wei cao

S. digitaloides Diels

English: foxglove-like sage

China: mao di huang shu wei cao

S. divinorum Epling & Játiva

Mexico: pipiltzintzintli

S. dorrii (Kellogg) Abrams

English: purple sage, desert sage, grey ball sage

S. elegans Vahl (*Salvia incarnata* Cav.; *Salvia microcalyx* Scheele; *Salvia longiflora* Sessé et Moçiño; *Salvia camertoni* Regel)

English: pineapple sage

S. eremostachya Jepson

English: desert sage

S. farinacea Benth.

English: mealy sage

S. fulgens Cav.

English: fiery sage

S. funerea M.E. Jones

English: Death Valley sage

S. glutinosa L.

English: Jupiter's distaff

S. greatae Brandegee

English: Orocopia sage (Orocopia, Chocolate Mountains)

S. greggii A. Gray

English: autumn sage, Gregg's salvia

S. hispanica L.

English: wild sage

Mexico: chía, chía comercial, chía del campo, chía del monte, romerillo, salvia; chaa, gueeza chaa, queza chaa, guia belaga, quije pelaga (Oaxaca)

S. japonica Thunberg

English: Japanese sage

China: shu wei cao, shu wei tsao, wu tsao, shui ching

S. leucantha Cav.

English: white-blooming sage

Mexico: pluma de Santa Teresa, salvia real; hierba de la playa (Veracruz); Santa María (Oaxaca)

S. leucophylla E. Greene

English: chaparral sage, San Luis purple sage, purple sage

S. lyrata L.

English: cancer weed

S. mellifera E. Greene

English: California black sage, black sage

S. mexicana L.

Mexico: tacote (Sinaloa); tapachichi (Morelos)

S. microphylla Kunth (*Salvia grahami* Benth.; *Salvia gasteracantha* Briq.)

English: myrtle sage, cardinal sage

Mexico: mirto, cardenales, salvia cardenal, cardinal salvia, salvia roja, pabellón mexicano, salvia del monte; bandera mexicana (Oaxaca); toronjil (Guerrero)

S. miltiorrhiza Bunge

English: red-rooted sage

China: dan shen, tan shen

S. munzii Epling

English: Munz's sage

S. officinalis L.

English: common sage, sage, garden sage

Arabic: salima, salma, selmia, mufassiha, mofassa

India: salvia sefakuss, salbia-sefakuss

Brazil: sálvia, salva

S. patens Cav.

Mexico: almaraduj, pájaros azules; almoraduz (San Luis Potosí); flor del gallito, gallitos (Hidalgo); quiquiriquí (Guanajuato)

S. plebeia R. Br.

English: Austral sage, common sage

India: kaka-buradi, kokaburadi, bhu-tulasi, bhui-tulsi, sathi, shati, samundar sok, sumandarsaka, kammar kas, jingiba, kinro, bijabuda, nirvisham, kasturi manjal, kachoralu, kachora, kichili-baddalu, pulam-kizhanma, jadvar, thamwen

China: li zhi cao, ching chieh

Nepal: kalo pati

S. plectranthoides Griffith

English: long-coronate sage

China: chang guan shu wei cao

S. pratensis L.

English: meadow clary

S. prionitis Hance

English: hispid sage

China: hong gen cao

S. przewalskii Maxim.

English: Przewalski sage

China: gang xi shu wei cao

S. reflexa Hornem.

English: mintweed, sage

South Dakota: maká ceyáka (= earth mint) (see Dilwyn J. Rogers, *Lakota Names and Traditional Uses of Native Plants by Sicangu (Brule) People in the Rosebud Area, South Dakota*. A study based on Father Eugene Buechel's collection of plants of Rosebud around 1920. 50. Buechel Memorial Lakota Museum, St. Francis, SD 1980)

Mexico: hierba del pajarito (Guadalcázar); mimititán (Puebla)

S. runcinata L.f. (*Salvia monticola* Benth. var. *monticola*; *Salvia runcinata* L.f. var. *grandiflora* Skan; *Salvia runcinata* L.f. var. *nana* Skan; *Salvia sisymbrifolia* Skan)

Southern Africa: hardesalie, wilde salie, setlepetlepe; mangoakoane (Sotho); mosisili (South Sotho)

S. rutilans Carr.

English: pineapple-scented sage

S. scapiformis Hance

English: scape-like sage

China: di geng shu wei cao

S. sclarea L.

English: clary-sage, clary, clear-eye

Arabic: kaff ed-dubb

S. sesseis Benth.

English: Mexican sage

S. sonomensis E. Greene

English: creeping sage

S. spathacea E. Greene

English: pitcher sage

S. splendens Sellow ex Roemer & Schultes

English: scarlet sage, scarlet salvia

Japan: hi-goromo-sô

China: yi chuan hong

Portuguese: camarões

S. stenophylla Burch. ex Benth. (*Salvia chlorophylla* Briq.; *Salvia pallida* Dinter ex Engl.; *Salvia stenophylla* Burch. ex Benth. var. *subintegra* Skan)

English: wild sage

Southern Africa: fynblaarsalie, salie, wildesalie; mangoakoane (Sotho); mosisili (South Sotho)

S. substolonifera E. Peter

English: creeping sage

China: fo guang cao

S. umbratica Hance

English: shady sage

China: yin sheng shu wei cao

S. verbenaca L. (*Salvia clandestina* L. var. *angustifolia* Benth.; *Salvia clandestina* L. var. *clandestina*)

English: salvia, vervain salvia, wild clary, vervain, wild sage

South Africa: salie, wildesalie

S. viridis L.

English: bluebeard

S. yunnanensis C.H. Wright

English: Yunnan sage

China: yun nan shu wei cao

Salviastrum Scheele Labiatae

Origins:
Indicating inferiority or incomplete resemblance to the genus *Salvia* L.

Salvinia Séguier Salviniaceae

Origins:
After the Italian Antonio Maria Salvini, 1633-1729, professor of Greek in Florence, a friend of the Italian botanist Pier'Antonio Micheli (1679-1737). Among his writings are *Discorsi accademici sopra alcuni dubbj proposti nell'Accademia degli Apatisti*. Firenze 1695, *Prose toscane*. Firenze 1715-35 and *Prose sacre*. Firenze 1716; see Jean François Séguier (1703-1784), *Plantae veronenses*. 3: 52. Veronae 1754.

Species/Vernacular Names:
S. minima Baker

English: water spangles, floating fern

S. molesta D.S. Mitchell (*Salvinia auriculata* auct. non Aubl.)

English: African pyle, Kariba weed (see Elizabeth Colson, *The Social Organization of the Gwembe Tonga. Human Problems of Kariba*. Manchester 1960; Barrie Reynolds, *The Material Culture of the Peoples of the Gwembe Valley*. Kariba Studies Vol. III. Manchester, University Press for the National Museums of Zambia 1968), water fern

South Africa: watervaring

S. natans (L.) All.

China: tzu ping

Salzmannia DC. Rubiaceae

Origins:

For the German botanist Philipp Salzmann, 1781-1851, physician, entomologist, plant collector and traveler (Brazil and North Africa), wrote *Enumeratio plantarum rariorum, in Gallia australi sponte nascentium*. Monspelii 1818. See John Hendley Barnhart (1871-1949), *Biographical Notes upon Botanists*. Boston 1965; H.N. Clokie, *Account of the Herbaria of the Department of Botany in the University of Oxford*. 237. Oxford 1964; A. Lasègue, *Musée botanique de Benjamin Delessert*. Paris 1845; E.M. Tucker, *Catalogue of the Library of the Arnold Arboretum of Harvard University*. Cambridge, Massachusetts 1917-1933.

Samadera Gaertner Simaroubaceae

Origins:

From *simandara*, a Sinhalese name for *Samadera indica* Gaertner; see Joseph Gaertner (1732-1791), *De fructibus et seminibus plantarum*. 2: 352. 1791.

Samanea (DC.) Merr. Mimosaceae

Origins:

From the native Spanish or South American name *saman* or *zaman*.

Samara L. Myrsinaceae

Origins:

From the Latin *samara* or *samera, ae* "the seed of the elm."

Samaropyxis Miq. Euphorbiaceae

Origins:

Latin *samara* or *samera, ae* "the seed of the elm" and Greek *pyxis* "a small box."

Samarpses Raf. Oleaceae

Origins:

See C.S. Rafinesque, *New Fl. N. Am.* 3: 93. 1836 [1838], *Alsographia americana*. 39. Philadelphia 1838, *Sylva Telluriana*. 10. 1838 and *The Good Book*. 46. 1840; E.D. Merrill, *Index rafinesquianus*. 190. Jamaica Plain, Massachusetts, USA 1949.

Sambucus L. Caprifoliaceae (Sambucaceae, Adoxaceae)

Origins:

Etymology uncertain, Latin *sambucus, i* and *sabucus*, used by Plinius for an elder tree; *sambucum, i* "the fruit of the elder, elder-berries"; see J. Hubschmid in *Revue de linguistique romane*. 27: 380-381. Paris 1963; Carl Linnaeus, *Species Plantarum*. 269. 1753 and *Genera Plantarum*. Ed. 5. 130. 1754; P. Aebischer, in *Vox Romanica*. XII: 82-94. 1951-1952; G. Bonfante, in *Bollettino del centro di studi filologici e linguistici siciliani*. I: 58-60. Palermo 1953; G. Rohlfs, in *Bollettino del centro di studi filologici e linguistici siciliani*. IX: 97. Palermo 1965; Manlio Cortelazzo & Paolo Zolli, *Dizionario etimologico della lingua italiana*. 5: 1125. Zanichelli, Bologna 1988; H. Genaust, *Etymologisches Wörterbuch der botanischen Pflanzennamen*. 553-554. Basel 1996.

Species/Vernacular Names:

S. australasica (Lindley) Fritsch

English: yellow elderberry, native elderberry

S. caerulea Raf. (*Sambucus glauca* Nutt. ex Torrey & A. Gray)

English: blue elder, blue elderberry

S. callicarpa Greene

English: Pacific red coast elder

S. canadensis L.

English: American elder, sweet elder, garden elder

Central America: sauco, bajman, sacatsun, tzoloj, tzolokquen

Bolivia: sauco

S. chinensis Lindl.

China: so tiao, chieh ku tsao

S. ebulus L.

English: dwarf elder, Dane's elder, danewort, wallwort, ground elder

French: petit sureau

Arabic: khaman saghir, rourawa, khelwan

S. gaudichaudiana DC.

English: white elder, white elderberry

S. mexicana C. Presl ex A. DC. (*Sambucus mexicana* var. *bipinnata* (Cham. & Schltdl.) Schwer.)

English: Mexican elder

Central America: sauco, bajman, sacatsun, tzoloj, tzolokquen

Mexico: saúco, anshiquel, azumiatl, bixhumi, cundumbo, cundemba, condemba, coyapa, shauc, yaga zulache, yaga bixhumi

S. nigra L.

English: common elder, black elder, bourtree, elderberry, European elder, common European elder, elder tree, arn tree

French: sureau noir, sureau

Italian: sambuco

Arabic: okkez sidi moussa, bilasan, khelwan, senbouqa, kaman kabir

Portuguese: sabugueiro negro

S. peruviana Kunth

Peru: arrayan, ccola, kjola, layán, rayán, ramrash, sauce, sauco, yalan

Bolivia: sauco

S. pubens Michx. (*Sambucus racemosa* var. *pubens* (Michx.) Koehne)

English: American red elder, red-berried elder, stinking elder

S. racemosa L. (*Sambucus miquelii* (Nakai) Kom. & Klob.-Alis.)

English: European red elder, red-berried elder, red elderberry

China: chieh ku mu, hsu ku mu, mu so tiao

S. sieboldiana (Miq.) Graebn.

China: chieh ku mu

Sameraria Desv. Brassicaceae

Origins:

Latin *samara* or *samera, ae* "the seed of the elm"; see Richard W. Spjut, *A Systematic Treatment of Fruit Types*. Memoirs of the New York Botanical Garden. Vol. 70: 106-108. New York 1994.

Samolus L. Primulaceae

Origins:

From *samolus, i* a Latin name for a plant, presumably a species of *Pulsatilla* or *Samolus valerandii* L. (Plinius) or *Anemone pulsatilla* L. or brook-weed; see Carl Linnaeus, *Species Plantarum*. 171. 1753 and *Genera Plantarum*. Ed. 5. 78. 1754; R. Zander, F. Encke, G. Buchheim and S. Seybold, *Handwörterbuch der Pflanzennamen*. 14. Aufl. 497. Stuttgart 1993.

Species/Vernacular Names:

S. eremaeus S.W.L. Jacobs

English: brookweed

S. repens (Forster & Forster f.) Pers. (*Sheffieldia repens* Forster & Forster f.)

English: creeping brookweed

S. valerandii L.

English: common brookweed, brookweed, water pimpernel

China: shui hui ca

Arabic: oudhina, oudhana

Sampaiella J.C. Gomes Bignoniaceae

Origins:

For the Brazilian botanist Alberto José de Sampaio, 1881-1946. His works include *A flora de Matto Grosso*. 1916 and *Nomes vulgares de plantas da Amazonia*. Rio de Janeiro 1934; see John H. Barnhart, *Biographical Notes upon Botanists*. 3: 205. 1965; E.M. Tucker, *Catalogue of the Library of the Arnold Arboretum of Harvard University*. Cambridge, Massachusetts 1917-1933.

Samuela Trel. Agavaceae

Origins:

For the American botanist Samuel Farlow Trelease, 1892-1958, plant physiologist; see J. Ewan, ed., *A Short History of Botany in the United States*. 68. New York and London 1969.

Samuelssonia Urban & Ekman Acanthaceae

Origins:

For the Swedish botanist Gunnar Samuelsson, 1885-1944; see John H. Barnhart, *Biographical Notes upon Botanists*. 3: 206. 1965; S. Lenley et al., *Catalog of the Manuscript and Archival Collections and Index to the Correspondence of John Torrey*. Library of the New York Botanical Garden. 359. 1973; T.W. Bossert, *Biographical Dictionary of Botanists Represented in the Hunt Institute Portrait Collection*. 347. 1972; Stafleu and Cowan, *Taxonomic Literature*. 5: 23-25. Utrecht 1985; R. Zander, F. Encke, G. Buchheim and S. Seybold, *Handwörterbuch der Pflanzennamen*. 14. Aufl. Stuttgart 1993.

Sanango Bunting & Ducke Buddlejaceae (Loganiaceae)

Origins:

A Quechua vernacular name for different species and genera (*Abuta, Anisomeris, Brunfelsia, Himatanthus, Malpighia, Petrea, Rauvolfia, Tabernaemontana*, etc.), *sanangu*.

Species/Vernacular Names:

S. racemosum (Ruíz & Pav.) Barringer (*Gomara racemosa* Ruíz & Pav.; *Gomaranthus racemosus* (Ruíz & Pav.) Rauschert; *Sanango durum* Bunting & Ducke)

Peru: sanango, sanangu

Sanchezia Ruíz & Pavón Acanthaceae

Origins:

For José Sánchez, a Spanish professor of botany.

Sanctambrosia Skottsb. Caryophyllaceae

Origins:

San Ambrosio Isla, Chile.

Sandbergia Greene Brassicaceae

Origins:

After the American botanist John Herman Sandberg, 1848-1917, born in Sweden, went to America in 1868, plant collector; see John H. Barnhart, *Biographical Notes upon Botanists*. 3: 206. 1965; H.N. Clokie, *Account of the Herbaria of the Department of Botany in the University of Oxford*. 237. Oxford 1964; John Michael Holzinger (1853-1929), *Report on a Collection of Plants made by J.H. Sandberg* and assistants in Northern Idaho in 1892. [in *Contr. U.S. Natl. Herb*. 3(4): 205-287. 1895] 1895; Joseph Ewan, *Rocky Mountain Naturalists*. [Sandberg's birthyear 1849] The University of Denver Press 1950; Joseph William Blankinship (1862-1938), "A century of botanical exploration in Montana, 1805-1905: collectors, herbaria and bibliography." in *Montana Agric. Coll. Sci. Studies Bot*. 1: 1-31. 1904.

Sandemania Gleason Melastomataceae

Origins:

For the English (b. London) botanist Christopher Albert Walter Sandeman, 1882-1951, traveler and plant collector in tropical South America. Among his writings are *No Music in Particular*. Letters from a friend. Edited and annotated by C. Sandeman. 1943, *A Forgotten River. A Book of Peruvian Travel and Botanical Notes*. London 1039, *Thyme and Bergamot*. London 1947 and *A Wanderer in Inca-land*. London 1948; see Gertrude Angela Mary Konstam [calling herself Gertrude Kingston], *Private and Confidential*. [A series of letters between Gertrude Kingston and Christopher

Sandeman, covering the period June 1932-June 1933.] Edited by Brian Grayson. London 1935.

Sanderella O. Kuntze Orchidaceae

Origins:

For the German-born British horticulturist Henry Frederick (Heinrich Friedrich) Conrad Sander, 1847-1920 (Belgium), in 1886 a Fellow of the Linnean Society, seedsman, orchidologist and orchid grower and hybridizer. Among his works are *Reichenbachia*. London [1886] 1887-1890 and *Orchid hybrids*. St. Albans [1906]. See John H. Barnhart, *Biographical Notes upon Botanists*. 3: 207. 1965; S. Lenley et al., *Catalog of the Manuscript and Archival Collections and Index to the Correspondence of John Torrey*. Library of the New York Botanical Garden. 440. 1973; A. Swinson, *The Orchid King*. London 1970; M. Hadfield et al., *British Gardeners: A Biographical Dictionary*. London 1980; R. Zander, F. Encke, G. Buchheim and S. Seybold, *Handwörterbuch der Pflanzennamen*. 14. Aufl. Stuttgart 1993; Merle A. Reinikka, *A History of the Orchid*. Timber Press 1996; Stafleu and Cowan, *Taxonomic Literature*. 5: 30-31. 1985.

Sandersonia Hook. Colchicaceae (Liliaceae)

Origins:

After the Scottish (born near Glasgow) horticulturist John Sanderson, 1820/21-1881 (d. Durban, South Africa), 1850 to Durban, botanical collector in South Africa, first discoverer and original collector of the plant, journalist, member of the Agricultural and Horticultural Society of Natal, sent plants and collections to William Henry Harvey (1811-1866); see James Chapman (1831-1872), *Travels in the Interior of South Africa, 1849-1863*. [botanical appendix by J. Sanderson, *Rough Notes on the Botany of Natal*, Edward Armitage (1822-1906), *Observations on the Botany of Natal* and John Croumbie Brown, *List of Natal Trees*] London 1868; Mary Gunn and Leslie Edward W. Codd, *Botanical Exploration of Southern Africa*. 307-308. Cape Town 1981; John H. Barnhart, *Biographical Notes upon Botanists*. 3: 207. 1965; A. White and B.L. Sloane, *The Stapelieae*. Pasadena 1937; F. Boerner & G. Kunkel, *Taschenwörterbuch der botanischen Pflanzennamen*. 4. Aufl. 164. Berlin & Hamburg 1989.

Species/Vernacular Names:

S. aurantiaca Hook.

English: Chinese lantern lily, Christmas bells

South Africa: Christmas bells, geelklokkie

Sandoricum Cav. Meliaceae

Origins:

Sandori or *santoor* is a Moluccan plant name for *Sandoricum koetjape* (Burm.f.) Merr.; see Antonio José Cavanilles (1745-1804), *Monadelphiae classis dissertationes decem.* 7: 359, tt. 202, 203. Matriti 1786-1787.

Species/Vernacular Names:

S. koetjape (Burm.f.) Merr.

Malaya: kechapi, sentol, setoi, setieh, seteh, setia, red sentol, yellow sentol, santol, sentul

The Philippines: santol, santor, katul

Sandwithia Lanjouw Euphorbiaceae

Origins:

After the English (Notts.) botanist Noel Yvri Sandwith, 1901-1965 (d. Kew, Surrey), from 1924 Kew Gardens, plant collector and traveler, 1936 Fellow of the Linnean Society, son of the botanical collector Mrs. Cecil Ivry Sandwith (1871-1961); see John H. Barnhart, *Biographical Notes upon Botanists.* 3: 208. 1965; T.W. Bossert, *Biographical Dictionary of Botanists Represented in the Hunt Institute Portrait Collection.* 347. 1972; H.N. Clokie, *Account of the Herbaria of the Department of Botany in the University of Oxford.* 237. Oxford 1964; Ida Kaplan Langman, *A Selected Guide to the Literature on the Flowering Plants of Mexico.* 671. 1964.

Sandwithiodoxa Aubrév. & Pellegr. Sapotaceae

Origins:

After the English botanist Noel Yvri Sandwith, 1901-1965.

Sanfordia Drumm. ex Harvey Rutaceae

Origins:

For Major Sandford, plant collector in W. Australia, botanist and naturalist; see Ray Desmond, *Dictionary of British & Irish Botanists and Horticulturists.* 607. London 1994; *Hooker's Journal of Botany & Kew Garden Miscellany.* 7: 53. London (Feb.) 1855.

Sanguinaria Bubani Gramineae

Origins:

Latin *sanguinarius, a, um* "blood-, belonging to blood," *sanguis, inis* "blood."

Sanguinaria L. Papaveraceae

Origins:

Latin *sanguinarius, a, um* "blood-, belonging to blood," *sanguis, inis* "blood," referring to the color of the sap; *sanguinaria* or *sanguinalis herba*, the name of an herb that stanches blood (see Greek *polygonon*).

Species/Vernacular Names:

S. canadensis L.

English: bloodroot, red puccoon

Sanguinella Gleichen Gramineae

Origins:

From the Latin *sanguis, inis* "blood."

Sanguisorba L. Rosaceae

Origins:

Latin *sanguis, sanguinis* "blood" and *sorbeo, bui, (ptum), ere* "to absorb, soak up," the rootstock has astringent and styptic qualities and properties; see Carl Linnaeus, *Species Plantarum.* 116. 1753 and *Genera Plantarum.* Ed. 5. 53. 1754.

Species/Vernacular Names:

S. canadensis L.

English: Canadian burnet, American burnet

S. minor Scop. (*Poterium sanguisorba* L.)

English: burnet, garden burnet, salad burnet

China: ti yu

S. minor Scop. subsp. *muricata* (Spach) Briq. (*Poterium polygamum* Waldst. & Kitaib.; *Poterium sanguisorba* sensu J. Black)

Australia: sheeps burnet, salad burnet

S. occidentalis Pursh

English: western burnet

S. officinalis L.

English: great burnet, burnet bloodwort, garden burnet, burnet

China: di yu, ti yu

S. stipulata Raf.

English: Sitka burnet (Baranof Island, in SE Alaska)

S. tenuifolia Fisch. ex Link

Japan: shiro-ware-moko

Sanhilaria Baillon Bignoniaceae

Origins:

Dedicated to the French botanist Auguste (Augustin) François César Prouvençal de Saint-Hilaire, 1779-1853 (Loiret, France), plant collector, explorer, entomologist, naturalist, traveler (Brazil and Uruguay). His writings include *Voyage dans le district des diamans et sur le littoral du Brésil*. Paris 1833 and *Flora Brasiliae Meridionalis*. [Written with A. de Jussieu and J. Cambessèdes] Parisiis 1824[-1833]; see Francisco Guerra, in *Dictionary of Scientific Biography* (Editor in Chief Charles Coulston Gillispie.) 12: 72. 1981; John H. Barnhart, *Biographical Notes upon Botanists*. 3: 200. 1965; T.W. Bossert, *Biographical Dictionary of Botanists Represented in the Hunt Institute Portrait Collection*. 345. 1972; A. Lasègue, *Musée botanique de Benjamin Delessert*. Paris 1845; S. Lenley et al., *Catalog of the Manuscript and Archival Collections and Index to the Correspondence of John Torrey*. Library of the New York Botanical Garden. 358. Boston, Massachusetts 1973; E.M. Tucker, *Catalogue of the Library of the Arnold Arboretum of Harvard University*. Cambridge, Massachusetts 1917-1933; Mariella Azzarello Di Misa, ed., *Il Fondo Antico della Biblioteca dell'Orto Botanico di Palermo*. 241. Regione Siciliana, Palermo 1988; Stafleu and Cowan, *Taxonomic Literature*. 4: 1064-1071. 1983; R. Zander, F. Encke, G. Buchheim and S. Seybold, *Handwörterbuch der Pflanzennamen*. 14. Aufl. Stuttgart 1993; Frederico Carlos Hoehne, M. Kuhlmann and Oswaldo Handro, *O jardim botânico de São Paulo*. 1941.

Sanicula L. Umbelliferae

Origins:

Latin *sano, avi, atum, are* (*sanus*) "to heal, cure," alluding to its healing properties; see Helmut Genaust, *Etymologisches Wörterbuch der botanischen Pflanzennamen*. 555-556. Basel 1996.

Species/Vernacular Names:

S. arctopoides Hook. & Arn.

English: footsteps of spring, yellow mats

S. bipinnata Hook. & Arn.

English: poison sanicle

S. bipinnatifida Hook.

English: purple sanicle, shoes buttons

S. europaea L.

English: sanicle, wood sanicle

S. hoffmannii (Munz) C. Bell

English: Hoffmann's sanicle

S. maritima S. Watson

English: adobe sanicle

S. peckiana J.F. Macbr.

English: Peck's sanicle

S. saxatilis E. Greene

English: rock sanicle

S. tracyi Shan & Constance

English: Tracy's sanicle

Saniculophyllum C.Y. Wu & Ku Saxifragaceae (Saniculophyllaceae)

Origins:

Sanicula and Greek *phyllon* "leaf."

Sanidophyllum Small Guttiferae

Origins:

Presumably from the Greek *sanis, sanidos* "board, seat, plank" and *phyllon* "leaf."

Saniella Hilliard & B.L. Burtt Amaryllidaceae (Hypoxydaceae, Liliaceae)

Origins:

Sani Pass, Drakensberg.

Sansevieria Thunb. Dracaenaceae (Agavaceae)

Origins:

After the Italian Raimondo di Sangro, Prince of Sansevero, 1710-1771, author of *Breve nota di quel che si vede in casa del Principe di Sansevero ... nella città di Napoli*. [Naples] 1766 and *Dissertation sur una Lampe antique trouvée à Munich en 1753 ... par Mr. le Prince de St. Sevère, pour servir de suite à la première partie de ses lettres à Mr. l'Abbé Nollet ... sur une découverte qu'il a faite dans la Chimie*, etc. Naples 1756; see Carl Peter Thunberg (1743-1828), *Prodromus plantarum Capensium*. 1794.

Species/Vernacular Names:

S. sp.

Peru: rabo de camaleão

Yoruba: oja ikooko, ida orisa

S. aethiopica Thunb. (*Sansevieria scabrifolia* Dinter)

English: mother-in-law's tongue

S. cylindrica Bojer

English: sansevieria

Japan: yari-ran, tsutsu-chitose-ran

S. grandis Hook.f.

English: Somali hemp

S. hyacinthoides (L.) Druce

English: African bowstring hemp

S. liberica Gérôme & Labroy

Yoruba: oja koriko, oja ikooko, pasan koriko, agbo-molowoibi

Congo: ilanga la ngovi

S. roxburghiana Schult.

Japan: chitose-ran

S. trifasciata Prain

English: mother-in-law's tongue, snake plant, bowstring hemp, African bowstring hemp, African hemp

Congo: ilanga

China: hu wei lan

S. zeylanica (L.) Willd.

English: bowstring hemp, Ceylon bowstring hemp, Indian bowstring hemp

India: murva, muruva, maurvi, marul, murba, murga, murahri, morwa, murvel, marul-kalang, gorachakra, ghona-saphan, nagfan, niyanda, sagal, mottamanji, manjinanaru, goddumanji, heggurutike, ishaura-koda-udr, katukapel

The Philippines: buntot-tigre, sinawa, rabo de leon, lengua de leon, aspe-aspe

Peru: cola de lobo, espadas, lengua de suegra

Santalina Baillon Rubiaceae

Origins:
Greek *santalinos, sagalinos* referring to the sandalwood tree.

Santalodes Kuntze Connaraceae

Origins:
Resembling the genus *Santalum* L.

Santaloidella Schellenb. Connaraceae

Origins:
The diminutive of the genus *Santaloides* Schellenb.

Santaloides Schellenb. Connaraceae

Origins:
Resembling the genus *Santalum* L.; see Gustav August Ludwig David Schellenberg (1882-1963), *Beiträge zur vergle-* *ichenden Anatomie und zur Systematik der Connaraceen.* Wiesbaden 1910.

Santalum L. Santalaceae

Origins:
Greek *santalon* "sandalwood tree," Arabic and Persian *shandal*, Sanskrit *chandana* "fragrant"; see Carl Linnaeus, *Species Plantarum.* 349. 1753 and *Genera Plantarum.* Ed. 5. 165. 1754; G.B. Pellegrini, *Gli arabismi nelle lingue neolatine con speciale riguardo all'Italia.* 120. Brescia 1972.

Species/Vernacular Names:
S. sp.

Hawaii: 'iliahi, 'a'ahi, 'aoa, la'au, 'ala, wahie 'ala

S. acuminatum (R. Br.) A. DC. (*Fusanus acuminatus* R. Br.; *Eucarya acuminata* (R. Br.) Sprague & Summerh.)

Australia: quandong, native peach, sweet quandong, katunga, burn-burn

S. album L.

English: true sandalwood, white sandalwood tree, sandalwood, sandal tree

Tibetan: tsan-dan dkarpo, suru tsan-dan

China: tan xiang, tan hsiang, chen tan, chan tan

India: ananditam, srigandha, swet chandan, chandan, chandana, miniak chandana, sadachandan, chandanam, candanam, shandanak-kattai, chandana-kattai, chandena-maram, sandanamaram, pitchandan, peetchandan, safed chandan, sufaid-chandan, gandashrah, bhadra shree, sufeed sandal, sukhada, sukhad, sukhet, gandhapu-chekka, srigandapu-manu, shrigandhada-mara, srikhanda, taliaparnam, gandha-chakoda

Malaya: chendana

S. ellipticum Gaud. (*Santalum cuneatum* (Hillebr.) Rock; *Santalum littorale* (Hillebr.) Rock)

English: coast sandalwood

Hawaii: 'iliahialo'e

S. lanceolatum (R. Br.) (*Fusanus acuminatus* R. Br. var. *angustifolius* (A. DC.) Benth.; *Santalum angustifolium* A. DC.)

Australia: plumbush, native plum bush, cherry bush, northern sandalwood, bolan (N. Queensland Aborigines)

S. murrayanum (Mitchell) C. Gardn. (*Eucarya murrayana* Mitchell)

Australia: bitter quandong

S. obtusifolium R. Br.

English: sandalwood

S. ovatum R. Br.

English: sandalwood

S. spicatum (R. Br.) A. DC. (*Fusanus spicatus* R. Br.; *Santalum cygnorum* Miq.)

English: sandalwood, western sandalwood

Santapaua Balakr. & Subramanyam Acanthaceae

Origins:

For the Spanish botanist Hermenegild Santapau, 1903-1970, clergyman, 1928 to India, Director of the Botanical Survey of India, professor of botany, author of *The Flora of Khandala on the Western Ghats of India*. Delhi 1953, with A.N. Henry wrote *A Dictionary of the Flowering Plants in India*. New Delhi 1975.

Santia Savi Gramineae

Origins:

For the Italian botanist Giorgio Santi, 1746-1822, professor of botany.

Santia Wight & Arnott Rubiaceae

Origins:

For the Italian botanist Giorgio Santi, 1746-1822, professor of botany and chemistry, 1782-1814 Director of the Botanical Garden of Pisa (succeeded Angelo Attilio Tilli, Director from 1740 to 1782). His writings include *Analisi chimica delle acque dei bagni Pisani, e dell'acqua acidula di Asciano*. Pisa 1789 and *Viaggio al Montamiata*. Pisa 1795. See L. Amadei, "Note sull'*Herbarium Horti Pisani*: l'origine delle collezioni." *Museol. Scient.* 4(1-2): 119-129. 1987; Fabio Garbari, L. Tomasi Tongiorgi and A. Tosi, *Giardino dei Semplici: L'Orto Botanico di Pisa dal XVI al XX secolo.* Pisa 1991; F.A. Stafleu, "Die Geschichte der Herbarien." *Bot. Jahrb. Syst.* 108(2/3): 155-166. 1987; John H. Barnhart, *Biographical Notes upon Botanists*. 3: 210. 1965; Ethelyn Maria Tucker, *Catalogue of the Library of the Arnold Arboretum of Harvard University*. Cambridge, Massachusetts 1917-1933; R. Zander, F. Encke, G. Buchheim and S. Seybold, *Handwörterbuch der Pflanzennamen*. 14. Aufl. Stuttgart 1993.

Santiria Blume Burseraceae

Origins:

After Bapa Santir of Java; see Alfred Russel Wallace (1823-1913), *The Malay Archipelago*. Singapore 1986;

E.R. Scidmore, *Java. The Garden of East*. Singapore 1986; Herman Johannes Lam (1892-1977), "The Burseraceae of the Malay Archipelago and Peninsula, with annotations concerning extra-Malayan species especially of *Dacryodes*, *Santiria* and *Canarium*." *Bull. Jard. Bot. Buitenzorg*, sér. III. 12(3-4): 281-561, pl. 1-14. 1932.

Species/Vernacular Names:

S. trimera (Oliv.) Aubréville

Nigeria: gologolo (Edo)

Yoruba: sawawa, kangara

Congo: toab

Gabon: ebo

Cameroon: ebap, libaba, poba

Ivory Coast: adjouaba à racines aériennes

Santiriopsis Engl. Burseraceae

Origins:

Resembling *Santiria* Blume.

Santisukia Brummitt Bignoniaceae

Origins:

Thawatchai Santisuk, born 1944; see *Natural History Bulletin of the Siam Society*. 33: 82. 1986.

Santolina L. Asteraceae

Origins:

Latin *santolina*, from *santonina*, Latin *santonicus* "belonging to the Santoni or Santones, Santonian," Greek *santonikon* for worm-seed, *Artemisia maritima*; see Salvatore Battaglia, *Grande dizionario della lingua italiana*. XVII: 541. UTET, Torino 1995; H. Genaust, *Etymologisches Wörterbuch der botanischen Pflanzennamen*. 556. Basel 1996.

Sanvitalia Lam. Asteraceae

Origins:

Probably for the Italian naturalist Federico Sanvitali, 1704-1761, author of *Elementi di architettura civile ...* Opera postuma. Brescia 1765 and "Dissertazione sopra il passaggio degli uccelli." in Giovanni Battista Chiaramonti, *Dissertazioni ... recitate ... nell'adunanza letteraria del Conte* G.M. Mazzuchelli. tom. I. Brescia 1765.

Species/Vernacular Names:

S. abertii A. Gray (for James William Abert, 1820-1897, see Joseph Ewan, *Rocky Mountain Naturalists*. The University of Denver Press 1950)

English: Abert's sanvitalia

Saphesia N.E. Br. Aizoaceae

Origins:

Greek *saphes* "clear, distinct, different," probably because this genus is unlike the others.

Sapindopsis F.C. How & C.N. Ho Sapindaceae

Origins:

Resembling *Sapindus* L.

Sapindus L. Sapindaceae

Origins:

Latin *sapo, saponis* (Akkadian *sapu* "to bathe," *sapu* "dyer") "soap" and *indicus* "Indian, of India"; see Carl Linnaeus, *Species Plantarum*. 367. 1753 and *Genera Plantarum*. Ed. 5. 171. 1754.

Species/Vernacular Names:

S. drumondii Hooker & Arnott (*Sapindus acuminatus* S. Watson & Coult.; *Sapindus saponaria* Torr. non L.)

English: wild China tree

S. mukorossi Gaertner (*Sapindus abruptus* Lour.; *Sapindus detergens* Roxb.)

English: Chinese soapberry, soapberry tree, soap nut, soap nut tree, soap nut tree of North India, bodhi seeds

Nepal: ritha, rithha

China: wu huan zi, wu huan tzu, mu huan tzu

India: arishta, urista, phenila, ita, aritha, ritha, reetha, bararitha, rintya-rooku, rathoh, dodan, kanmar, thali, kunduduchettu, kungitikaya, puvandi, ponnan-kottai, punnangkottai, poongan-kottay, poongankottai, chavakayimaram, kookatakayi, noorekayi, kudale-kaye, urvanjik-kaya, finduk-i-hindi, keeltha, hlingsi

Japan: mukuro-ju

Malaya: buah lerak

S. oahuensis Hillebr. ex Radlk. (*Sapindus lonomea* St. John)

Hawaii: lonomea, aulu, kaulu

S. rarak DC.

India: am selenga

Nepal: ritha

Bhutan: nakapani

Malaya: lerak

S. saponaria L. (*Sapindus indicus* Poir.; *Sapindus saponaria* Lour. non L.; *Sapindus thurstonii* Rock)

English: soapberry, false dogwood

Tropical America: jaboncillo

Mexico: amole, sihom, amolli, yaga biaa, yaga piaa, bibi, pibi, pipi

Hawaii: a'e, manele

S. trifoliatus L.

English: soap nut tree of South India

India: arishta, phenila, reetha, bara-ritha, rithe, ardal, pitha, ringin, aritha, kunkudu-chettu, kunkullu, kunkudu-kayalu, homie, puvamkottai, antawala, kunkatekaye, kugatemara, urvanjikaya, pasakotta, muktamaya, muktimonjro, rettia

Sapiopsis Müll. Arg. Euphorbiaceae

Origins:

Resembling *Sapium*.

Sapium Jacq. Euphorbiaceae

Origins:

Origins obscure, probably from the Latin *sappinus, sapinus, sappium* "a kind of fir-tree, pine-tree, the lower smooth part of the fir-tree" (Plinius); some suggest from the Latin *sapio, ii* "savory, witty, tasty," or from Celtic *sap* "fat," in allusion to the greasy exudation from the wounded trunk, or from Latin *sapo, saponis* "soap," etc.; see Patrick Browne (1720-1790), *The Civil and Natural History of Jamaica in Three Parts*. 338. London 1756; Salvatore Battaglia, *Grande dizionario della lingua italiana*. XVII: 556. Torino 1995; G.C. Wittstein, *Etymologisch-botanisches Handwörterbuch*. 784. Ansbach 1852; in *Enum. Syst. Pl.* 9, 31. Aug-Sep. 1760.

Species/Vernacular Names:

S. sp.

English: tallow tree, Mexican jumping bean, arrow-wood

Central America: arrow-wood, palo de la flecha, hierba de la flecha, palo lechon

Peru: burra lechera, caucho blanco, caucho claspi, chichis, jihuigene, sheringa mashan

Salvador: chilamate

Venezuela: lechero

Argentina: lecheron

Costa Rica: yos

Mexico: matahiza

Indochina: co pen, soi

Nepal: ban peepal

S. ellipticum (Krauss) Pax (*Sapium mannianum* (Müll. Arg.) Benth.; *Sclerocroton ellipticus* Hochst.)

English: jumping-seed tree, jumping seeds

Cameroon: ebusok, osiemvot, dambu

Central Africa: m'bombasse

Ivory Coast: tomi

Gabon: asep, asep-eli

N. Rhodesia: mutundu

Southern Africa: springsaadboom; umDlampunzi, umHlepha, umHlepa (Zulu); umWongolo, umHongolo (Xhosa); muNyeredzi, mugarahwiriti (Shona)

Uganda: muruku, musanvuma, musasa, mushasha, muzanvuma

S. glandulatum (Vell.) Pax (*Omphalea glandulata* Vell.; *Sapium petiolare* (Müll. Arg.) Huber; *Sapium longipes* (Müll. Arg.) Huber)

Brazil: mata olho, leiteira, pau de leite, péla cavalo, árvore de borracha, currupiteira

S. glandulosum (L.) Morong (*Hippomane biglandulosum* L.; *Sapium aereum* Klotsch ex Müll. Arg.; *Sapium biglandulosum* var. *hamatum* Poeppig ex Müll. Arg.; *Sapium biglandulosum* var. *pavonianum* Poeppig ex Müll. Arg.; *Sapium hamatum* (Poeppig ex Müll. Arg.) Pax & Hoffmann; *Sapium hippomane* G. Meyer; *Sapium ixiamasense* Jablonski; *Sapium pavonianum* (Müll. Arg.) Huber; *Sapium poeppigii* Hemsley; *Sapium taburu* Ule)

English: Colombian rubber tree

Brazil: leiteiro, pao de leite

Ecuador: palo de leche

Colombia: pinique

Peru: árbol de leche, burra leiteira, caucho, caucho blanco, caucho mashan, gutapercha, murupita, santoncate, seringarana, sheringa rana, shiringa masha, tapurú

S. haematospermum Müll. Arg.

Bolivia: leche leche

S. indicum Willd.

English: mock willow

India: hurua, hurna, batul, batan, pencolum, karmmatti, venkshiri

Malaya: gurah, guring, gayan, buah saminyak

S. insigne (Royle) Benth. ex Hook.

Nepal: khirra

S. integerrimum (Hochst.) J. Léonard (*Sapium reticulatum* Pax; *Sclerocroton integerrimus* Hochst.)

Southern Africa: duikerbessie, duikerberry; umHlalampunzi, umQathampunzi, umDlampunzi, umHlepha, umHlepa (Zulu)

S. japonicum (Siebold & Zucc.) Pax & Hoffmann (*Sapium japonicum* var. *ryukyuensis* Masam.; *Stillingia japonica* Sieb. & Zucc.; *Triadica japonica* (Sieb. & Zucc.) Nakai; *Shirakia japonica* (Sieb. & Zucc.) Hurusawa)

Japan: shira-ki

S. marmierii Müll. Arg.

Bolivia: leche leche

S. sebiferum (L.) Roxb. (*Croton sebiferum* L.; *Seborium sebiferum* (L.) Hurusawa; *Triadica sebiferum* (L.) Small)

English: Chinese tallow tree, vegetable tallow tree, tallow tree, Chinese vegetable tallow

Japan: Nankin-haze, tokaji

India: vilaiti shisham, vilayati-shisham, pahari-shisham, tarcharbi, toyapippali, pippalyang, momchina, meenadabattimara, ronojita

China: wu jiu mu gen pi

Sapium P. Browne Euphorbiaceae

Origins:

See *Sapium* Jacquin; see P. Browne, *Civ. Nat. Hist. Jamaica.* 338. 10 Mar. 1756.

Saponaria L. Caryophyllaceae

Origins:

From the Medieval Latin *sapo, saponis* "soap," referring to the mucilaginous juice, Akkadian *sapu* "to bathe, dyer," Anglo-Saxon *sape*; Old Norse *sapa*; see Carl Linnaeus, *Species Plantarum.* 408. 1753 and *Genera Plantarum.* Ed. 5. 191. 1754; Salvatore Battaglia, *Grande dizionario della lingua italiana.* XVII: 556-557. 1995.

Species/Vernacular Names:

S. officinalis L.

English: bladder soapwort, bouncing bet, China cockle, cockle, cow basil, cow cockle, cow foot, cow herb, coa soapwort, glond, soapwort, spring cockle, fuller's herb

South Africa: akkerkoeikruid

Sapota Miller Sapotaceae

Origins:

From native South American name; see Philip Miller (1691-1771), *The Gardeners Dictionary.* Abr. ed. 4. London (28 Jan.) 1754.

Sapranthus Seemann Annonaceae

Origins:

Greek *sapros* "putrid, rotten" and *anthos* "flower."

Sapria Griffith Rafflesiaceae

Origins:

Latin and Greek *sapros* "putrid, rotten," referring to the nature of the plant.

Saprosma Blume Rubiaceae

Origins:

Greek *sapros* and *osme* "smell, odor, perfume."

Sapucaya Knuth Lecythidaceae

Origins:

A South American vernacular name for some species of *Lecythis* Loefl., *Lecythis zabucajo*.

Saraca L. Caesalpiniaceae

Origins:

From the Indian word *asoka*, Sanskrit *sara* "colored, spotted."

Species/Vernacular Names:

S. asoca (Roxb.) Wilde (*Jonesia asoca* Roxb.)

English: asoka tree, sorrowless tree, Indian sorrowless tree, Indian asoka

India: asokah, asokam, asoka, asokada, acokam, asok, asjogam, asogam, asogha, ashoka, asupala, ashopalava, kenkalimara, kankelli, vichitrah, gandapushpa, anganapriya, thawgabo

S. declinata (Jack) Miq. (*Jonesia declinata* Jack; *Saraca cauliflora* Baker)

English: red saraca, Sumatra asoka

Malaya: gapis kunyit, talan kunyit

S. indica L.

China: wu yu hua

S. thaipingensis Cantley ex Prain

English: yellow saraca

Saracha Ruíz & Pav. Solanaceae

Origins:

Named for a monk, Isidoro Saracha, 1733-1803, botanist.

Sararanga Hemsley Pandanaceae

Origins:

Probably from the Sanskrit *sara* "colored, spotted" and *raga* "the red color, a color."

Sarawakodendron Ding Hou Celastraceae

Origins:

From Sarawak and Greek *dendron* "tree."

Sarcandra Gardner Chloranthaceae

Origins:

From the Greek *sarx*, *sarkos* "flesh" and *aner*, *andros* "male," referring to the stamens.

Species/Vernacular Names:
S. glabra (Thunb.) Nakai (*Chloranthus glaber* (Thunb.) Makino)

Japan: sen-ryô, senryo (= 1,000 ryo, an old gold coin)

China: jiu jie cha

Sarcanthidion Baillon Icacinaceae

Origins:

Greek *sarx*, *sarkos* "flesh" and *anthos* "flower.

Sarcanthopsis Garay Orchidaceae

Origins:

Resembling the genus *Sarcanthus* Lindl.

Sarcanthus Lindley Orchidaceae

Origins:

Greek *sarx*, *sarkos* "flesh" and *anthos* "flower," referring to the fleshy flowers; see John Lindley, *The Botanical Register.* subt. 817. (Aug.) 1824; Alick William Dockrill, *Australasian*

Sarcanthinae. 1967, and *Australian Indigenous Orchids.* Sydney 1969; Herman Montague Rucker Rupp (1872-1956), in *The Victorian Naturalist.* 57: 218. (Apr.) 1941.

Species/Vernacular Names:
S. micranthus Ames
Japan: Urai-muyo-ran

Sarcaulus Radlk. Sapotaceae

Origins:
Greek *sarx, sarkos* and *kaulos* "a stem, a branch or stalk."

Sarcobatus Nees Chenopodiaceae

Origins:
Greek *sarx, sarkos* "flesh" and *batia, batos* "a bush, thicket, a kind of cup," referring to the bark.

Species/Vernacular Names:
S. vermiculatus (Hook.) Torrey
English: greasewood

Sarcobodium Beer Orchidaceae

Origins:
Origin and meaning not clear, possibly from the Greek *sarx, sarkos* "flesh" and *bodion, boidion* "a little ox," *bous* "an ox."

Sarcobotrya R. Vig. Fabaceae

Origins:
From the Greek *sarx, sarkos* and *botrys* "cluster, a bunch of grapes."

Sarcocapnos DC. Fumariaceae (Papaveraceae)

Origins:
From the Greek *sarx, sarkos* "flesh" and *kapnos* "smoke."

Sarcocaulon (DC.) Sweet Geraniaceae

Origins:
Greek *sarx, sarkos* and *kaulos* "a stem, a branch or stalk," fleshy shrublets.

Species/Vernacular Names:
S. camdeboense Moffett (*Sarcocaulon patersonii* sensu Harv., sensu R. Knuth, non (DC.) G. Don; *Sarcocaulon*

vanderietiae sensu R.A. Dyer, non L. Bolus.) (Camdebo is a Khoi name meaning *green elevations*; it is applied to the mountains and river near Aberdeen, South Africa)

South Africa: candle bush, kersbos

S. inerme Rehm

Southern Africa: bushman candle, boesmankers

S. marlothii Engl. (*Sarcocaulon mossamedense* sensu auctt. pro parte, R. Knuth, Rehm, Merxm. & A. Schreib., non (Welw. ex Oliv.) Hiern) (for the South African botanist Hermann Wilhelm Rudolph Marloth, 1855-1931, botanical explorer, plant collector, author of "The vegetation of the southern Namib." *S. Afr. J. Sci.* 6: 80-87. 1910, *The Flora of South Africa.* Cape Town and London 1913-1932, *Dictionary of the Common Names of Plants* with a list of foreign plants cultivated in the open. Cape Town 1917 and "On the means of distribution of seeds in the South African flora." *Trans. S. Afr. Phil. Soc.* 8: lxxiv-lxxx. 1896; see John Hutchinson, *A Botanist in Southern Africa.* 645-646. 1946; A. Engler et al., "Plantae Marlothianae." in *Bot. Jahrb.* 10: 1-50, 242-285. 1889; J.H. Barnhart, *Biographical Notes upon Botanists.* 2: 449. 1965; R. Zander, F. Encke, G. Buchheim and S. Seybold, *Handwörterbuch der Pflanzennamen.* 14. Aufl. 1993; T.W. Bossert, *Biographical Dictionary of Botanists Represented in the Hunt Institute Portrait Collection.* 254. 1972; H.N. Clokie, *Account of the Herbaria of the Department of Botany in the University of Oxford.* 206. 1964; E.M. Tucker, *Catalogue of the Library of the Arnold Arboretum of Harvard University.* 1917-1933; Hans Herre, *The Genera of the Mesembryanthemaceae.* 49-50. Cape Town 1971; I.C. Hedge and J.M. Lamond, *Index of Collectors in the Edinburgh Herbarium.* 1970; A. White and B.L. Sloane, *The Stapelieae.* Pasadena 1937; Mary Gunn and Leslie E. Codd, *Botanical Exploration of Southern Africa.* 1981; Gordon Douglas Rowley, *A History of Succulent Plants.* 1997)

English: bushman's candle

Southern Africa: kersbos; boesmankers (Namibia, former Southwest Africa)

German: buschmannskerze

S. mossamedense (Welw. ex Oliv.) Hiern (*Monsonia mossamedensis* Welw. ex Oliv.) (the species occurs in southern Angola, just north and east of the port of Mossâmedes)

English: candle bush

S. multifidum E. Mey. ex R. Knuth (*Monsonia multifida* E. Mey.) (the leaves are multifid or segmented, Latin *multifidus, a, um* "many-cleft, many-parted")

Southern Africa: bushman candle, boesmankers

S. patersonii (DC.) G. Don (*Sarcocaulon rigidum* Schinz)
English: candle bush

S. peniculinum Moffett (Latin *peniculus, i* (dim. *penis*) "a little tail, a brush," when not in flower the sausage-like branches are very similar to a shoe brush)

English: candle bush

S. salmoniflorum Moffett (*Sarcocaulon l'heritieri* DC. var. *brevimucronatum* Schinz; *Sarcocaulon patersonii* Eckl. & Zeyh.; *Sarcocaulon patersonii* (DC.) G. Don subsp. *badium* Rehm; *Sarcocaulon patersonii* (DC.) G. Don subsp. *curvatum* Rehm)

English: candle bush, bushman candle

Southern Africa: boesmankers, kersbossie, heldoring, maagdoring; qorab (Southern Namibia)

S. vanderietiae L. Bol.

English: candle bush

Sarcocephalus Afzel. ex Sabine Rubiaceae

Origins:

Greek *sarx, sarkos* and *kephale* "head," referring to the edible fruits; see Joseph Sabine (1770-1837), in *Transactions of the Horticultural Society*. 5: 442. 1824.

Species/Vernacular Names:
S. sp.

Nigeria: egbere, ekeng, lalawe, akondok, andinding, badi, bade

S. latifolius (Sm.) E.A. Bruce

English: Africa peach, Guinea peach

Nigeria: opepe-ira, obiache-eze, andinding, badi, bade

Yoruba: egbesi, agbesi abisi, agbesi ogun

Sarcochilus R. Br. Orchidaceae

Origins:

Greek *sarx, sarkos* and *cheilos* "a lip," referring to the labellum, to the fleshy lip; see Robert Brown (1773-1858), *Prodromus florae Novae Hollandiae et Insulae van-Diemen*. 332. London 1810.

Species/Vernacular Names:
S. australis (Lindley) H.G. Reichb.

English: butterfly orchid, Gunn's orchid

S. ceciliae F. Muell.

English: fairy bells, Cecilia's sarcochilus

S. falcatus R. Br.

English: orange blossom orchid

S. fitzgeraldii F. Muell.

English: ravine orchid

S. hartmannii F. Muell.

English: waxy sarcochilus, Hartmann's sarcochilus

S. hillii (F. Muell.) F. Muell.

English: Hill's sarcochilus

S. olivaceus Lindley

English: Olive sarcochilus

Sarcochlamys Gaudich. Urticaceae

Origins:

From the Greek *sarx, sarkos* "flesh" and *chlamys, chlamydos* "cloak," referring to the female flowers.

Sarcoclinium Wight Euphorbiaceae

Origins:

From the Greek *sarx, sarkos* "flesh" and *kline* "a bed."

Sarcococca Lindley Buxaceae

Origins:

From the Greek *sarx, sarkos* and *kokkos* "a berry," referring to the fleshy kernel.

Sarcocolla Kunth Penaeaceae

Origins:

Latin *sarcocolla* for a Persian gum (Plinius), Greek *sarx, sarkos* "flesh" and *kolla* "glue," referring to the fleshy stem below the inflorescence; sarcocolla or gum sarcocolla, common names applied to *Astracantha gummifera* (Lab.) Podlech, Fabaceae.

Sarcocornia A.J. Scott Chenopodiaceae

Origins:

Greek *sarx, sarkos* "flesh" and Latin *cornu, us* "horn"; see Andrew John Scott (1950-), in *Botanical Journal of the Linnean Society*. 75: 366. (Apr.) 1978.

Species/Vernacular Names:
S. blackiana (Ulbr.) A.J. Scott

English: Black's glasswort

S. natalensis (Ung.-Sternb.) A.J. Scott (*Arthrocnemum affine* Moss; *Arthrocnemum africanum* Moss; *Arthrocnemum natalense* (Bunge ex Ung.-Sternb.) Moss; *Salicornia natelensis* Bunge ex Ung.-Sternb.)

South Africa: seekoraal

S. quinqueflora (Ung.-Sternb.) A.J. Scott (*Salicornia quinqueflora* Ung.-Sternb.)

Australia: beaded glasswort

Sarcodes Torrey Ericaceae (Monotropaceae)

Origins:
Flesh-like, from the Greek *sarx, sarkos* and *eidos* "resemblance," referring to the inflorescences.

Species/Vernacular Names:
S. sanguinea Torrey
English: snow plant

Sarcodraba Gilg & Muschler Brassicaceae

Origins:
From the Greek *sarx, sarkos* and the genus *Draba* L.

Sarcodrimys (Baill.) Baum.-Bod. Winteraceae

Origins:
From the Greek *sarx, sarkos* "flesh" and the genus *Drimys* Forst. & Forst.f.

Sarcoglossum Beer Orchidaceae

Origins:
From the Greek *sarx, sarkos* meaning "flesh" and *glossa* meaning "tongue."

Sarcoglottis C. Presl Orchidaceae

Origins:
Greek *sarx, sarkos* and *glossa, glotta* "tongue," indicating the fleshy lip of the flowers.

Sarcoglyphis Garay Orchidaceae

Origins:
From the Greek *sarx, sarkos* "flesh" and *glypho* "to carve, engrave."

Sarcogonum G. Don ex Sweet Polygonaceae

Origins:
Greek *sarx, sarkos* "flesh" and *gony* "a knee, bend, joint," *gonia* "an angle."

Sarcolaena Thouars Sarcolaenaceae

Origins:
Greek *sarx, sarkos* and *chlaena* "a cloak, blanket," stellate indumentum and mucilaginous cells.

Sarcolobus R. Br. Asclepiadaceae

Origins:
From the Greek *sarx, sarkos* "flesh" and *lobos* "a pod"; see R. Brown, "On the Asclepiadeae." *Memoirs of the Wernerian Natural History Society.* 1: 34. Edinburgh 1811.

Sarcolophium Troupin Menispermaceae

Origins:
From the Greek *sarx, sarkos* "flesh" and *lophos* "a crest."

Sarcomelicope Engler Rutaceae

Origins:
From the Greek *sarx, sarkos* "flesh" and the genus *Melicope* Forst. & Forst.f.; see H.G.A. Engler & K.A.E. Prantl (1849-1893), *Die Natürlichen Pflanzenfamilien.* 3(4): 122. (Mar.) 1896.

Sarcomphalus P. Browne Rhamnaceae

Origins:
From the Greek *sarx, sarkos* "flesh" and *omphalos* "umbilicus."

Sarcopetalum F. Muell. Menispermaceae

Origins:
From the Greek *sarx, sarkos* and *petalum* or *petalon* "petal," the petals are fleshy and thick; see Ferdinand Jacob Heinrich von Mueller (1825-1896), *The Plants Indigenous to the Colony of Victoria.* 1: 26. Melbourne 1862.

Species/Vernacular Names:

S. harveyanum F. Muell. (after the Irish botanist William Henry Harvey, 1811-1866)

English: pearl vine, big-leaf vine

Sarcophagophilus Dinter Asclepiadaceae

Origins:

Greek *sarkophagos* "eating flesh, carnivorous, cannibal, coffin" and *philos* "loving."

Sarcopharyngia (Stapf) Boiteau Apocynaceae

Origins:

From the Greek *sarx, sarkos* "flesh" and *pharynx, pharyngos* "pharynx."

Sarcophrynium K. Schumann Marantaceae

Origins:

From the Greek *sarx, sarkos* "flesh" and *phrynos* "a frog."

Species/Vernacular Names:

S. brachystachys (Benth.) K. Schum.

Yoruba: gbodogi

Sarcophyllus Thunb. Fabaceae

Origins:

From the Greek *sarx, sarkos* "flesh" and *phyllon* "leaf."

Sarcophysa Miers Solanaceae

Origins:

From the Greek *sarx, sarkos* "flesh" and *physa* "bladder."

Sarcophyte Sparrman Balanophoraceae (Sarcophytaceae)

Origins:

From the Greek *sarx, sarkos* and *phyton* "a plant"; see Anders Sparrmann (1748-1820), in *Kongl. Vetenskaps Academiens Handlingar.* 27: 300, t. 7. Stockholm 1776.

Sarcophyton Garay Orchidaceae

Origins:

Greek *sarx, sarkos* "flesh" and *phyton* "a plant."

Sarcopilea Urban Urticaceae

Origins:

From the Greek *sarx, sarkos* "flesh" and the genus *Pilea* Lindley.

Sarcopodium Lindley Orchidaceae

Origins:

From the Greek *sarx, sarkos* "flesh" and *podion* "a little foot, stalk," referring to the base of the column.

Sarcopoterium Spach Rosaceae

Origins:

From the Greek *sarkos* "flesh" and the genus *Poterium* L., Greek *poterion* "a drinking cup, a drinking vessel, a goblet, cup.

Sarcopteryx Radlk. Sapindaceae

Origins:

Greek *sarx, sarkos* and *pteryx* "wing," referring to the wings of the fruits or to the leaflet stalks; see L.A.T. Radlkofer, ["Ueber die Sapindaceen Holländisch-Indiens." 57. (Jan.-Feb.) 1879] in *Actes du Congrès International de Botanistes.* Amsterdam (for 1877). 127. 1879.

Species/Vernacular Names:

S. stipata (F. Muell.) Radlk. (*Cupania stipata* F. Muell.)

English: corduroy, corduroy tree

Sarcopygme Setch. & Christoph. Rubiaceae

Origins:

From the Greek *sarkos* "flesh" and *pygmaios* "dwarfish."

Sarcopyramis Wallich Melastomataceae

Origins:

From the Greek *sarx, sarkos* "flesh" and *pyramis* "a pyramid," referring to the shape of the fruits.

Sarcorhachis Trel. Piperaceae

Origins:

From the Greek *sarx, sarkos* "flesh" and *rhachis* "rachis, axis, midrib of a leaf."

Sarcorhynchus Schltr. Orchidaceae

Origins:
From the Greek *sarx, sarkos* "flesh" and *rhynchos* "horn, beak, snout," the fleshy rostellum.

Sarcorrhiza Bullock Asclepiadaceae (Periplocaceae)

Origins:
From the Greek *sarx, sarkos* "flesh" and *rhiza* "root."

Sarcosiphon Blume Burmanniaceae

Origins:
Greek *sarx, sarkos* and *siphon* "a tube"; see Karl Ludwig von Blume, *Museum Botanicum Lugduno-Batavum.* 1: 65. Lugduni-Batavorum (Apr.) 1850.

Sarcosperma Hook.f. Sapotaceae (Sarcospermataceae)

Origins:
From the Greek *sarx, sarkos* "flesh" and *sperma* "a seed," referring to the nature of the seeds.

Species/Vernacular Names:
S. arboreum Buch.-Ham. ex C.B. Clarke
English: tree-like fleshseed tree
China: da rou shi shu

Sarcostemma R. Br. Asclepiadaceae

Origins:
Greek *sarx, sarkos* and *stemma, stemmatos* "a garland, crown," the corona is fleshy, the shrubs are succulent; see Robert Brown, "On the Asclepiadeae." *Memoirs of the Wernerian Natural History Society.* 1: 50. Edinburgh 1811 and *Prodromus florae Novae Hollandiae et Insulae van-Diemen.* 463. London 1810.

Species/Vernacular Names:
S. acidum (Roxburgh) Voigt
English: acid flesh coral
China: rou shan hu
S. australe R. Br.
Australia: caustic bush, caustic vine, Tableland caustic vine, sarcostemma
S. hirtellum (A. Gray) R. Holm

English: trailing townula
S. secamone (L.) Bennett
Egypt: libbein
S. viminale (L.) R. Br.
English: caustic bush, caustic creeper, caustic vine
Rodrigues Island: liane calé
Southern Africa: melktou, melkbos, spantoumelkbos, spantou, wolfsmelk; ingotsha (Zulu); ma belabela (Ndebele); umbelebele, umBelenele (Xhosa); ntlalamela (Sotho); morarwane (Tswana); mutungu (Venda); nneta (Tsonga); nyokudomba (Shona)

Sarcostigma Wight & Arnott Icacinaceae

Origins:
From the Greek *sarx, sarkos* "flesh" and *stigma* "stigma," referring to the nature of the stigma.

Sarcostoma Blume Orchidaceae

Origins:
From the Greek *sarx, sarkos* and *stoma* "mouth," the midlobe of the lip.

Sarcostyles C. Presl ex DC. Hydrangeaceae

Origins:
Greek *sarx, sarkos* "flesh" and *stylos* "style."

Sarcotheca Blume Oxalidaceae

Origins:
Greek *sarx, sarkos* "flesh" and *theke* "a box, case," referring to the edible and acid fruits.

Species/Vernacular Names:
S. sp.
Malaya: piang, belimbing burong

Sarcotoechia Radlk. Sapindaceae

Origins:
Greek *sarx, sarkos* and *toichos* "a wall," possibly referring to the fruits; see Ludwig Adolph Timotheus Radlkofer (1829-1927), in *Sitzungsberichte der mathematisch-physikalischen Classe der k.b. Akademie der Wissenschaften zu München.* 9: 501, 659. (Jul.) 1879.

Sarcoyucca (W. Trelease) L. Lindinger Agavaceae

Origins:

From the Greek *sarx*, *sarkos* "flesh" and the genus *Yucca* L.

Sarcozona J.M. Black Aizoaceae

Origins:

From the Greek *sarx*, *sarkos* "flesh" and *zone* "a belt, armor, girdle," referring to the involucre; see John McConnell Black (1855-1951), in *Transactions and Proceedings of the Royal Society of South Australia*. 58: 176, t. XI, fig. 2. (Dec.) 1934.

Species/Vernacular Names:

S. praecox (F. Muell.) S.T. Blake (*Mesembryanthemum praecox* F. Muell.; *Carpobrotus pulleinei* J. Black; *Sarcozona pulleinei* (J. Black) J. Black)

Australia: sarcozona, pigface, ridged noonflower

Sarcozygium Bunge Zygophyllaceae

Origins:

From the Greek *sarx*, *sarkos* and *zygon*, *zygos* "yoke."

Sarga Ewart Gramineae

Origins:

Meaning not explained, possibly from the Greek *sargane* "a basket, a plait, a band" or derived from *sorghum*, or from Latin *sargus* "a kind of sea-fish" see A.J. Ewart et al., *Proceedings of the Royal Society of Victoria*. 1909.

Sargentia H.A. Wendland & Drude ex Salomon Palmae

Origins:

Dedicated to the American botanist Charles Sprague Sargent, 1841-1927.

Sargentia S. Watson Rutaceae

Origins:

Dedicated to the eminent American botanist Charles Sprague Sargent, 1841-1927, dendrologist, traveler and plant collector, from 1872 (first) Director of the Arnold Arboretum of Harvard University; see Stafleu and Cowan, *Taxonomic Literature*. 5: 51-56. Utrecht 1985; R. Zander, F. Encke, G. Buchheim and S. Seybold, *Handwörterbuch der Pflanzennamen*. 14. Aufl. 1993; John H. Barnhart, *Bio-*

graphical Notes upon Botanists. 3: 211. 1965; T.W. Bossert, *Biographical Dictionary of Botanists Represented in the Hunt Institute Portrait Collection*. 348. 1972; S. Lenley et al., *Catalog of the Manuscript and Archival Collections and Index to the Correspondence of John Torrey*. Library of the New York Botanical Garden. 359. 1973; J.D. Milner, *Catalogue of Portraits of Botanists Exhibited in the Museums of the Royal Botanic Gardens*. Royal Botanic Gardens, Kew, London 1906; Ida Kaplan Langman, *A Selected Guide to the Literature on the Flowering Plants of Mexico*. 673-674. University of Pennsylvania Press, Philadelphia 1964; Ethelyn Maria Tucker, *Catalogue of the Library of the Arnold Arboretum of Harvard University*. Cambridge, Massachusetts 1917-1933; J. Ewan, ed., *A Short History of Botany in the United States*. New York and London 1969; Ernest Nelmes and William Cuthbertson, *Curtis's Botanical Magazine Dedications, 1827-1927*. [1931]; Irving William Knobloch, compil., "A preliminary verified list of plant collectors in Mexico." *Phytologia Memoirs*. VI. 1983; Emil Bretschneider, *History of European Botanical Discoveries in China*. [Reprint of the original edition 1898.] Leipzig 1981; Joseph Ewan, *Rocky Mountain Naturalists*. The University of Denver Press 1950; Leonard Huxley, *Life and Letters of Sir Joseph Dalton Hooker*. London 1918.

Sargentodoxa Rehder & E.H. Wilson Lardizabalaceae (Sargentodoxaceae)

Origins:

For the American botanist Charles Sprague Sargent, 1841-1927.

Species/Vernacular Names:

S. cuneata Rehd. & Wils.

China: da xue teng

Saribus Blume Palmae

Origins:

From an Indonesian vernacular name, *sariboe*.

Sarissus Gaertner Rubiaceae

Origins:

Greek *sarisa*, *sarissa*, used for the sarissa, Macedonian pike, a long pike used in a Macedonian phalanx.

Saritaea Dugand Bignoniaceae

Origins:

Named in honor of the author's wife.

Species/Vernacular Names:
S. magnifica (W. Bull) Dugand (*Bignonia magnifica* hort.; *Arrabidaea magnifica* Sprague ex Steenis)

Japan: murasaki-bigunonia

Sarmentaria Naudin Melastomataceae

Origins:
Latin *sarmentum, i* "twigs, brushwood, light branches," *sarpo* "to cut off, trim, prune," see also *Adelobotrys* DC., Greek *adelos* "obscure" and *botrys* "a cluster."

Sarmienta Ruíz & Pav. Gesneriaceae

Origins:
Spanish *sarmiento* "a shoot or branch of a vine," a liane.

Sarojusticia Bremek. Acanthaceae

Origins:
Greek *saron* "a broom" and the genus *Justicia* L., referring to the appearance of the plants; see Cornelis Eliza Bertus Bremekamp (1888-1984), in *Acta Botanica Neerlandica*. 11: 199. (Aug.) 1962.

Sarothamnus Wimmer Fabaceae

Origins:
Greek *saron* "a broom" and *thamnos* "shrub," referring to the broom-like shrub; see Christian Friedrich Heinrich Wimmer (1803-1868), *Flora von Schlesien*. 278. Berlin 1832.

Sarotheca Nees Acanthaceae

Origins:
From the Greek *saron* and *theke* "a box, case."

Sarothra L. Guttiferae

Origins:
From the Greek *sarotron, sarothron* "a broom, besom."

Sarothrochilus Schlechter Orchidaceae

Origins:
From the Greek *sarotron, sarothron* "a broom" and *cheilos* "lip," see also *Staurochilus* Ridley ex Pfitzer.

Sarothrostachys Klotzsch Euphorbiaceae

Origins:
From the Greek *sarotron, sarothron* and *stachys* "a spike."

Sarracenia L. Sarraceniaceae

Origins:
Named by Tournefort from the French physician Michel Sarrazin (Sarracenus), 1659-1734 (or 1735/1736), naturalist, in Québec, plant collector, who sent him the plant; see Frère Marie-Victorin, *Un Manuscrit botanique prélinnéen*. L'"Histoire des Plantes de Canada," etc. Montréal 1936; B. Boivin, "La flore du Canada en 1708, étude d'un manuscrit de Michel Sarrazin et Sébastien Vaillant." *Provancheria*. no. 9. 1978; Arthur Vallée, *Michel Sarrazin, 1659-1735*. (Un biologiste canadien) Sa vie, ses travaux et son temps. Québec. 1927; C.T. Onions, *The Oxford Dictionary of English Etymology*. Oxford University Press 1966; Georg Christian Wittstein, *Etymologisch-botanisches Handwörterbuch*. 787. 1852. According to Pritzel the genus was dedicated to the French physician Jean Antoine Sarrasin (Lat. Janus Antonius Saracenus or Sarracenus), 1547-1598, author of *de Peste, commentarius*. Lugduni 1572, editor of G. Fabricii Hildani ... *selectae observationes chirurgicae quinque et viginti*, etc. [Author Fabricius von Hilden, 1560-1634; translated from the French by J. Rheterius.] 1598; see Richard J. Durling, comp., *A Catalogue of Sixteenth Century Printed Books in the National Library of Medicine*. 4067-4068. 1967.

Species/Vernacular Names:
S. purpurea L.

English: huntsman's cup

Italian: sarracenia porporina

Sartoria Boissier & Heldreich Fabaceae

Origins:
After the German botanist Joseph Sartori, 1809-1880, physician, botanical explorer (in Greece); see John H. Barnhart, *Biographical Notes upon Botanists*. 3: 212. 1965.

Sartorina R. King & H. Robinson Asteraceae

Origins:
For the German botanist Carl (Karl, Carlos) Christian Wilhelm Sartorius, 1796-1872, plant collector in Mexico, father of Florentin Sartorius; see Ida K. Langman, "Dos figuras casi olvidadas en la historia de la botánica Méxicana." *Rev. Soc. Méx. Hist. Nat*. 10: 329-336. 1949; T.W. Bossert, *Bio-*

graphical *Dictionary of Botanists Represented in the Hunt Institute Portrait Collection.* 348. 1972; Irving William Knobloch, compil., "A preliminary verified list of plant collectors in Mexico." *Phytologia Memoirs.* VI. 1983.

Sartwellia A. Gray Asteraceae

Origins:
After the American botanist Henry Parker Sartwell, 1792-1867, physician; see John H. Barnhart, *Biographical Notes upon Botanists.* 3: 212. 1965; H.N. Clokie, *Account of the Herbaria of the Department of Botany in the University of Oxford.* 237. Oxford 1964; S. Lenley et al., *Catalog of the Manuscript and Archival Collections and Index to the Correspondence of John Torrey.* Library of the New York Botanical Garden. 468. 1973; E.M. Tucker, *Catalogue of the Library of the Arnold Arboretum of Harvard University.* 1917-1933; Howard Atwood Kelly and Walter Lincoln Burrage, *Dictionary of American Medical Biography.* New York 1928.

Sarx St. John Cucurbitaceae

Origins:
Greek *sarx, sarkos* "flesh."

Sasa Makino & Shibata Gramineae

Origins:
The Japanese name.

Species/Vernacular Names:
S. japonica (Siebold & Zucc.) Makino (*Arundinaria japonica* Siebold & Zucc.; *Pseudosasa japonica* (Siebold & Zucc.) Makino)

English: arrow bamboo

Japan: ya-dake

Sasaella Makino Gramineae

Origins:
The diminutive of *Sasa* Makino & Shibata.

Sasamorpha Nakai Gramineae

Origins:
From the genus *Sasa* and Greek *morphe* "a form, shape."

Sasanqua Raf. Theaceae

Origins:
For *Sasanqua* Nees = *Camellia* L., from a Japanese name, *sasankwa*; see Constantine Samuel Rafinesque, *Sylva Telluriana.* 140. 1838.

Sassafras Nees Lauraceae

Origins:
Spanish *sasafràs*, perhaps from the Latin *saxifragus, a, um* "stone-breaking, stone-crushing," or derived from an Indian/American name; see E. Weekley, *An Etymological Dictionary of Modern English.* 2: 1279. New York 1967; S. Battaglia, *Grande dizionario della lingua italiana.* XVII: 584. Torino 1995; Helmut Genaust, *Etymologisches Wörterbuch der botanischen Pflanzennamen.* 560-561. Basel 1996.

Species/Vernacular Names:
S. albidum (Nutt.) Nees
English: sassafras
S. tzumu (Hemsley) Hemsley
English: Chinese sassafras

Sassafridium Meissner Lauraceae

Origins:
Resembling *Sassafras*.

Satakentia H.E. Moore Palmae

Origins:
After Toshihiko Satake, a Japanese manufacturer of rice-milling machinery of Saijo-machi (near Hiroshima, Japan), an expert on palms.

Species/Vernacular Names:
S. liukiuensis (Hatusima) H.E. Moore (*Gulubia liukiuensis* Hatusima; *Exorrhiza savoryana* (Rehd. & Wils.) Burret)

Japan: no-yashi (= field palm), Yaeyama-yashi

Okinawa: binro

Brazil: palmeira satake

Satanocrater Schweinf. Acanthaceae

Origins:
For Satanas and Satan "the devil" and *crater* "a vessel"; Latin *herba satanaria* was a plant, also called *peucedanos*, sulphurwort, hog's fennel.

Satorkis Thouars Orchidaceae

Origins:

The genus *Satyrium* and *orchis* "orchid," a relationship to *Satyrium*.

Satureja L. Labiatae

Origins:

Latin *satureia, ae*, the common name for the pot-herb *cunila*, savory (L. Junius Moderatus Columella, Plinius et al.).

Species/Vernacular Names:

S. chandleri (Brandegee) Druce

English: San Miguel savory

S. douglasii (Benth.) Briq.

Spanish: yerba buena

S. hortensis L.

English: summer savory

S. montana L.

English: winter savory

Satyria Klotzsch Ericaceae

Origins:

Greek *satyros* "satyr," Latin *satyriasis* "priapism, lascivious madness."

Species/Vernacular Names:

S. spp.

Mexico: yaga yooba dau

Satyridium Lindley Orchidaceae

Origins:

The diminutive of the generic name *Satyrium* Swartz, *satyridion* is the diminutive of the Greek *satyros* "satyr."

Satyrium Swartz Orchidaceae

Origins:

From the Greek *satyrion* (*satyros* "satyr"), an ancient name used by Dioscorides and Plinius for an orchid, the man orchid, *Aceras anthropophorum*, probably because of the supposed aphrodisiacal properties of some species; the satyrs were creatures of the woods and mountains, half man, half beast, lascivious, closely connected with Dionysus.

Species/Vernacular Names:

S. carneum (Dryand.) Sims

South Africa: rooikoppie, rooitrewwa

S. coriifolium Swartz

South Africa: ewwa-trewwa

S. nepalense D. Don

Nepal: gamdol

Saugetia A. Hitchcock & Chase Gramineae

Origins:

Named for the French-born Cuban botanist Joseph Sylvestre (José Silvestre) Sauget y Barbis (aka Frère Léon, Hermano León), 1871-1955, Havana 1905, author of *Las exploraciones botanicas de Cuba*. Habana 1918 and "Nouveaux Anastraphia de la flore cubaine." *Contr. Inst. Bot. Univ. Montreal*. 49: 77-86. 1944, with Hermano Alain (Dr. E.E. Lioger) wrote *Flora de Cuba*. Habana 1946-1963; see John H. Barnhart, *Biographical Notes upon Botanists*. 3: 210. 1965; T.W. Bossert, *Biographical Dictionary of Botanists Represented in the Hunt Institute Portrait Collection*. 234. 1972; S. Lenley et al., *Catalog of the Manuscript and Archival Collections and Index to the Correspondence of John Torrey*. Library of the New York Botanical Garden. 262. 1973.

Saundersia Reichb.f. Orchidaceae

Origins:

For the English (b. Bucks.) botanist William Wilson Saunders, 1809-1879 (d. Sussex), entomologist, plant collector, horticulturist, 1830 with the East India Company, 1833 Fellow of the Linnean Society, President of the Entomological Society, 1853 Fellow of the Royal Society, 1863 succeeded J. Lindley as Secretary Horticultural Society of London. His works include *An Address Delivered at the Anniversary Meeting of the Entomological Society* 24th January 1843. London 1843, *Insecta Saudersiana: or the Characters of Undescribed Insects in the Collection of W.W. Saunders*. London 1850-1869, *Refugium botanicum*. Edited by W. Wilson Saunders, ... the descriptions by H.G. Reichenbach and John Gilbert Baker (1834-1920), figures and dissections by Walter Hood Fitch (1817-1892). London [1868-]1869-1873 and *Mycological Illustrations*. Edited by W. Wilson Saunders and Worthington George Smith (1835-1917), assisted by Alfred William Bennett (1833-1902); see John H. Barnhart, *Biographical Notes upon Botanists*. 3: 215. 1965; R. Zander, F. Encke, G. Buchheim and S. Seybold, *Handwörterbuch der Pflanzennamen*. 14. Aufl. Stuttgart 1993; Ray Desmond, *Dictionary of British & Irish Botanists and Horticulturists*. 610. 1994; E.M. Tucker, *Cat-*

alogue of the Library of the Arnold Arboretum of Harvard University. 1917-1933; A. White and B.L. Sloane, *The Stapelieae.* Pasadena 1937; M. Hadfield et al., *British Gardeners: A Biographical Dictionary.* London 1980; Ernest Nelmes and William Cuthbertson, *Curtis's Botanical Magazine Dedications, 1827-1927.* 154-156. [1931]; H.N. Clokie, *Account of the Herbaria of the Department of Botany in the University of Oxford.* 238. Oxford 1964; H.R. Fletcher, *Story of the Royal Horticultural Society, 1804-1968.* Oxford 1969; Ida Kaplan Langman, *A Selected Guide to the Literature on the Flowering Plants of Mexico.* University of Pennsylvania Press, Philadelphia 1964.

Saurauia Willd. Actinidiaceae (Saurauiaceae)

Origins:

In honor of the Austrian J. von Saurau or Count Friedrich von Saurau, 1760-*circa* 1832/1840, a patron of arts and science and natural sciences, friend of Willdenow; see Carl Ludwig von Willdenow (1765-1812), in *Der Gesellschaft naturforschender Freunde zu Berlin, neue Schriften.* 3: 407, t. 4. Berlin 1801.

Species/Vernacular Names:

S. spp.

Mexico: taa beecho, taa peecho

S. napaulensis (Royle) Benth. ex Hook.

Nepal: goban

S. tristyla DC. (*Saurauia oldhami* Hemsley; *Saurauia kawagoeana* Hatus.)

Japan: Takasago-shira-tama

Malaya: jelatang gajah

Sauroglossum Lindley Orchidaceae

Origins:

From the Greek *sauros* "lizard" and *glossa* "a tongue."

Sauromatum Schott Araceae

Origins:

Greek *sauros* "lizard," referring to the spotted cataphylls and spathe; see Henry Noltie, "*Sauromatum venosum*: the Monarch of the East, or Voodoo Lily." in *The New Plantsman.* 4(2): 120-127. June 1997; Helmut Genaust, *Etymologisches Wörterbuch der botanischen Pflanzennamen.* 562. Basel 1996; R. Zander, F. Encke, G. Buchheim and S. Seybold, *Handwörterbuch der Pflanzennamen.* 14. Aufl. 501. 1993; D.H. Nicolson, "Derivation of aroid generic names." *Aroideana.* 10: 15-25. 1988.

Species/Vernacular Names:

S. venosum (Aiton) Kunth

English: purple-flowered arum, spotted sauromatium, monarch of the East, voodoo lily, red calla

Sauropus Blume Euphorbiaceae

Origins:

Greek *sauros* and *pous* "a foot"; see Karl Ludwig von Blume, *Bijdragen tot de flora van Nederlandsch Indië.* 595. (Jan.) 1826.

Species/Vernacular Names:

S. rigens (F. Muell.) Airy Shaw (*Synostemon rigens* F. Muell.; *Phyllanthus rigens* (F. Muell.) Müll. Arg.; *Glochidion rigens* (F. Muell.) H. Eichler)

English: stiff spurge

Saururus L. Saururaceae

Origins:

From the Greek *sauros* "lizard" and *oura* "a tail."

Species/Vernacular Names:

S. cernuus L.

English: swamp lily

S. chinensis (Lour.) Baill. (*Spathium chinense* Lour.; *Saururus loureirii* Decne.)

English: Chinese lizard's tail, Chinese lizardtail

Japan: han-geshô

Okinawa: fiiragusa

China: san bai cao, san pai tsao

Saussurea DC. Asteraceae

Origins:

For the Swiss scientist Horace Bénédict de Saussure, 1740-1799 (Geneva), philosopher and botanist, mountaineer, experimental petrologist and geologist, meteorologist, naturalist, traveler, a Fellow of the Royal Society, 1762-1786 professor at Academy of Geneva (succeeded by his pupil, the Swiss scientist Marc-Auguste Pictet, 1752-1825), joint founder and editor of *Journal de Genève*), a member of the Council of Two Hundred, in 1798 was chosen a member of the National Assembly. Among his writings are *Voyages dans les Alpes.* Neuchâtel 1779-1796, *Lettres de H.-B. de Saussure à sa femme.* Chambéry 1937, *Essais sur l'Hygrométrie.* Neuchâtel 1783, *Observations sur l'écorce des feuilles et pétales.* Genève 1762 and numerous papers in the *Journal de Physique* and the *Journal de Genève*; the generic name also honors the Swiss botanist and chemist

Nicolas Théodore de Saussure (1767-1845, d. Geneva), son of Horace Bénédict de Saussure, plant physiologist, in 1815 he was a founding member of the Société Helvétique des Sciences Naturelles, author of *Recherches chimiques sur la végétation*. Paris [1804]; see Douglas W. Freshfield, *Life of Horace Bénédict de Saussure*. London 1920; A.P. de Candolle, in *Annales du Muséum National d'Histoire Naturelle*. 16: 156, 196, 198. Paris 1810; T.W. Bossert, *Biographical Dictionary of Botanists Represented in the Hunt Institute Portrait Collection*. 350. 1972; Georges Léopold Chrétien Frédéric Dagobert Cuvier (1769-1832), *Recueil des éloges historiques* lus dans les séances publiques de l'Institut royal de France. Strasbourg & Paris 1819-1827; G. Schmid, *Goethe und die Naturwissenschaften*. Halle 1940; John H. Barnhart, *Biographical Notes upon Botanists*. 3: 215. 1965; Ethelyn Maria Tucker, *Catalogue of the Library of the Arnold Arboretum of Harvard University*. Cambridge, Massachusetts 1917-1933; Albert V. Carozzi, in *Dictionary of Scientific Biography* 12: 119-123. 1981; P.E. Pilet, in *Dictionary of Scientific Biography* 12: 123-124. 1981; Georg Christian Wittstein, *Etymologisch-botanisches Handwörterbuch*. 788. Ansbach 1852; Stafleu and Cowan, *Taxonomic Literature*. 5: 70-73. 1985.

Species/Vernacular Names:
S. americana D. Eaton
English: American sawwort
S. costus (Falc.) Lipsch.
English: costus root, costus
China: mu xiang

Sauvagesia L. Ochnaceae (Sauvagesiaceae) ·

Origins:
For the French naturalist François Boissier de la Croix de Sauvages (Franciscus Sauvagesius, Boissier de Sauvages de la Croix), 1706-1767, physician, botanist, M.D. Montpellier 1726, professor of medicine and botany at the University of Montpellier, correspondent of Linnaeus. His works include *Pathologia methodica, seu de cognoscendis morbis*. Amstelodami 1752, Tractatus methodici practici in duos tomos divisi ... Tomus primus de morbis puerorum. (Tomus secundus de morbis cutaneis.) Neapoli 1778, *De venenatis Galliae animalibus, & venenorum in ipsis fideli observatione compertorum indole atque antidotis*. Monspelii 1764, *Nosologia methodica, sistens morborum classes, juxtà Sydenhami mentem & botanicorum ordinem*. Amstelodami 1768 and *Methodus foliorum, seu plantae florae monspeliensis, juxta foliorum ordinem*. 'Gravenhage [La Haye] 1751. Brother of the French clergyman and botanist Pierre Augustin Boissier de Sauvages de la Croix (1710-1795); see Jules Grasset, *Le Médecin de l'amour au temps de Marivaux. Étude sur Boissier de Sauvages*, etc. Montpellier

1896; Joannes Carolus Boecking, *De motu cordis ab aucta vasorum resistentia aucto. Dissertatio opposita argumentis celeberrimi De Sauvages*, etc. Praes. J.P. Eberhard. Halae ad Salam [Halle an der Saale] [1757]; J.H. Barnhart, *Biographical Notes upon Botanists*. 3: 216. 1965; Thomas Sydenham (1624-1689), *Methodus curandi febres*. [8vo, second edition, the book is dedicated to the Irish scientist Robert Boyle, 1627-1691.] London 1668; T.W. Bossert, *Biographical Dictionary of Botanists Represented in the Hunt Institute Portrait Collection*. 350. 1972; Garrison and Morton, *Medical Bibliography*. 2203. New York 1961; Mariella Azzarello Di Misa, ed., *Il Fondo Antico della Biblioteca dell'Orto Botanico di Palermo*. 243. Regione Siciliana, Palermo 1988; Frans A. Stafleu, *Linnaeus and the Linnaeans*. The spreading of their ideas in systematic botany, 1735-1789. Utrecht 1971.

Species/Vernacular Names:
S. erecta L.
English: Creole tea
Peru: goma huayo, hierba de San Martín

Sauvallea W. Wright Commelinaceae

Origins:
After the American botanist Francisco (Franciscus) Adolfo Sauvalle, 1807-1879, malacologist, from 1824 in Cuba. See John H. Barnhart, *Biographical Notes upon Botanists*. 3: 216. 1965; T.W. Bossert, *Biographical Dictionary of Botanists Represented in the Hunt Institute Portrait Collection*. 350. 1972; Ida Kaplan Langman, *A Selected Guide to the Literature on the Flowering Plants of Mexico*. 675. Philadelphia 1964; E.M. Tucker, *Catalogue of the Library of the Arnold Arboretum of Harvard University*. Cambridge, Massachusetts 1917-1933.

Sauvallella Rydb. Fabaceae

Origins:
After the American botanist Francisco Adolfo Sauvalle, 1807-1879, malacologist.

Savannosiphon Goldblatt & Marais Iridaceae

Origins:
Savanna or savannah, a treeless plain, an area of tropical or subtropical grassland, from Spanish *zavana, sabana*.

Savia Willdenow Euphorbiaceae

Origins:

For the Italian botanist Gaetano Savi, 1769-1844, physician, 1814-1843 Director of the Third (founded 1591 by Lorenzo Mazzanga and Giuseppe Casabona) Botanical Garden of Pisa (succeeded Giorgio Santi). Among his writings are *Memoria sopra una pianta cucurbitacea.* Milano 1818 [Reprinted from *Biblioteca Italiana.* 9: 158-165. 1818] and *Flora pisana.* Pisa 1798, father of the Italian zoologist and geologist Paolo Savi (1798-1871) and of the Italian botanist Pietro Savi (1811-1871). See John H. Barnhart, *Biographical Notes upon Botanists.* 3: 216. 1965; T.W. Bossert, *Biographical Dictionary of Botanists Represented in the Hunt Institute Portrait Collection.* 350. 1972; P.E. Tomei and Carlo Del Prete, "The Botanical Garden of the University of Pisa." *Herbarist.* 49: 47-71. Concord, Massachusetts 1983; E.M. Tucker, *Catalogue of the Library of the Arnold Arboretum of Harvard University.* 1917-1933; Jonas C. Dryander, *Catalogus bibliothecae historico-naturalis Josephi Banks.* London 1800; A. Lasègue, *Musée botanique de Benjamin Delessert.* Paris 1845; L. Amadei, "Note sull'*Herbarium Horti Pisani*: l'origine delle collezioni." *Museol. Scient.* 4(1-2): 119-129. 1987; Fabio Garbari, L. Tomasi Tongiorgi and A. Tosi, *Giardino dei Semplici: L'Orto Botanico di Pisa dal XVI al XX secolo.* Pisa 1991; F.A. Stafleu, "Die Geschichte der Herbarien." *Bot. Jahrb. Syst.* 108(2/3): 155-166. 1987; Mariella Azzarello Di Misa, a cura di, *Il Fondo Antico della Biblioteca dell'Orto Botanico di Palermo.* 244-246. Palermo 1988; R. Zander, F. Encke, G. Buchheim and S. Seybold, *Handwörterbuch der Pflanzennamen.* 14. Aufl. Stuttgart 1993.

Savignya DC. Brassicaceae

Origins:

After the French (b. Brie) botanist Marie Jules César Lelorgne de Savigny, 1777-1851 (Versailles), zoologist, entomologist, with Napoléon-le-Grand in Egypt, biologist. His works include *Histoire naturelle et mythologique de l'Ibis.* Paris 1805, *Mémoires sur les animaux sans vertèbres.* Paris 1816 and *Description de l'Égypte*, ou recueil des observations ... faites en Égypte pendant l'expédition de l'armée française. [Some of the memoirs and plates on Natural History are by M.J.C. Lelorgne de Savigny.] 1809-[1826]. See Paul Pallary, *Explication des planches de J.C. Savigny.* Paris 1926; P. Pallary, *Marie Jules César Savigny. Sa vie et son oeuvre.* Paris 1931; John H. Barnhart, *Biographical Notes upon Botanists.* Boston 1965; Mary P. Winsor, in *Dictionary of Scientific Biography* 12: 130-131. 1981; R. Zander, F. Encke, G. Buchheim and S. Seybold, *Handwörterbuch der Pflanzennamen.* 14. Aufl. 1993.

Saxegothaea Lindley Podocarpaceae

Origins:

From Saxe-Gothaea, referring to Albert Prinz v. Sachsen-Coburg-Gotha, Prince Albert; see Audrey Le Lièvre, "Saxe-Coburg and Gotha: an Unexpected Glance at a Ducal House." *Hortus.* 28: 12-23. 1993.

Species/Vernacular Names:

S. conspicua Lindley

English: Prince Albert's yew

Chile: mañío hembra

Saxicolella Engl. Podostemaceae

Origins:

Latin *saxum, i* "a rock, stone" and *-cola* "inhabitant, resident."

Saxifraga L. Saxifragaceae

Origins:

Latin *saxifragus, a, um* (*saxum frango*) "stone-breaking, stone-crushing"; see Carl Linnaeus, *Species Plantarum.* 398. 1753 and *Genera Plantarum.* Ed. 5. 189. 1754.

Species/Vernacular Names:

S. aizoides L.

English: yellow mountain saxifrage, yellow saxifrage

S. aspera L.

English: rough saxifrage, stiff-haired saxifrage

S. bryoides L.

English: mossy saxifrage

S. burseriana L.

English: Burser's saxifrage

S. caesia L.

English: blue green saxifrage

S. callosa Sm.

English: limestone saxifrage

S. cotyledon L.

English: great alpine rockfoil, greater evergreen saxifrage

S. cuneifolia L.

English: shield-leaved saxifrage

S. cuscutiformis Lodd

English: dodder-like saxifrage

S. exarata Vill.

English: furrowed saxifrage

S. granulata L.

English: meadow saxifrage, bulbous saxifrage, fair maids of France

S. hirsuta L.

English: kidney saxifrage

S. howellii E. Greene

English: Howell's saxifrage

S. hypnoides L.

English: mossy saxifrage

S. longifolia Lapeyr.

English: Pyrenean saxifrage

S. maderensis D. Don

English: Madeira saxifrage, Madeira breakstone

S. oppositifolia L.

English: purple saxifrage

S. paniculata Mill.

English: lifelong saxifrage

S. rotundifolia L.

English: round-leaved saxifrage

S. stellaris L.

English: starry saxifrage

S. stolonifera Meerb. (*Saxifraga sarmentosa* L.f.)

English: mother of thousands, strawberry geranium, wandering Jew

Japan: yuki-no-shita, minjai-gusa

China: hu er cao, hu erh tsao, shih ho yeh

Saxifragella Engl. Saxifragaceae

Origins:
The diminutive of the genus *Saxifraga* L.

Saxifragites Gagnep. Hamamelidaceae

Origins:
Latin *saxifragus, a, um* "stone-breaking."

Saxifragodes D.M. Moore Saxifragaceae

Origins:
Resembling *Saxifraga* L.

Saxifragopsis Small Saxifragaceae

Origins:
Resembling *Saxifraga* L.

Saxiglossum Ching Polypodiaceae

Origins:
Latin *saxum, i* "a rock, stone" and *glossa* "tongue."

Sayeria Kraenzlin Orchidaceae

Origins:
After W.A. Sayer, orchid collector in New Guinea.

Scabiosa L. Dipsacaceae

Origins:
Latin *scabiosus, a, um* (*scabies, em* "scurf," *scabo, scabi* "to scratch") "mangy, itchy, rough," referring to its medicinal uses; see Carl Linnaeus, *Species Plantarum*. 98. 1753 and *Genera Plantarum*. Ed. 5. 43. 1754.

Species/Vernacular Names:
S. albanensis R.A. Dyer

Southern Africa: isiLawu esikhulu, isiLawu esimhlophe (Xhosa)

S. atropurpurea L. (*Scabiosa maritima* L.; *Scabiosa atropurpurea* L. subsp. *maritima* (L.) Arcang.; *Scabiosa atropurpurea* L. var. *maritima* (L.) Bég.)

English: mournful widow, sweet scabious, pincushion flower, Egyptian rose, scabious, purple pincushion, mourning-bride

Japan: seiyô-matsu-mushi-sô

S. buekiana Eckl. & Zeyh.

English: scabious

Southern Africa: isiLawu esikhulu, isiLawu esimhlophe (Xhosa)

S. columbaria L. (*Scabiosa anthemifolia* Eckl. & Zeyh.; *Scabiosa austro-africana* Heine)

English: riceflower

Southern Africa: isiLawu esikhulu, isiLawu esimhlophe (Xhosa)

S. incisa Mill.

Southern Africa: wild scabious, koringblommetjie; isiLawu esikhulu, isiLawu esimhlophe (Xhosa)

S. japonica Miq.

Japan: matsu-mushi-sô

S. prolifera L.

English: carmel daisy

S. tysonii L. Bol.

English: wild scabious

Southern Africa: isiLawu esikhulu, isiLawu esimhlophe (Xhosa)

Scabiosella Tieghem Dipsacaceae

Origins:
The diminutive of *Scabiosa* L.

Scabiosiopsis Rech.f. Dipsacaceae

Origins:
Resembling *Scabiosa* L.

Scadoxus Raf. Amaryllidaceae (Liliaceae)

Origins:
Greek *skiadion* "umbel, parasol" and *doxa* "glory"; the genus was named by Constantine Samuel Rafinesque (1783-1840): "umb. glor." (glorious umbel?), see C.S. Rafinesque, *Flora Telluriana*. 4: 19. 1836 [1838].

Species/Vernacular Names:
S. sp.

Yoruba: oka

S. membranaceus (Bak.) Friis & Nordal (*Haemanthus puniceus* L. var. *membranaceus* (Bak.) Bak.)

English: april fool, paintbrush

South Africa: seerogblom

S. multiflorus (Martyn) Raf.

English: blood flower

S. multiflorus (Martyn) Raf. subsp. *katharinae* (Bak.) Friis & Nordal (*Haemanthus katharinae* Bak.)

English: Catherine wheel

S. puniceus (L.) Friis & Nordal (*Haemanthus magnificus* Herb.; *Haemanthus natalensis* Pappe ex Hook.)

English: blood flower, blood lily, paintbrush, royal paintbrush, sore eye flower, snake lily, snake flower, april fool_

South Africa: bloedlelie, seerogblom, skeerkwas, poeierkwas

Scaevola L. Goodeniaceae

Origins:
After the Roman hero C. Mucius Scaevola, whose surname means "left-handed," Latin *scaevus, a, um* "left," referring to the left-handed twisting and to the incomplete appearance of the corolla; see C. Linnaeus, *Mantissa Plantarum*. 2: 145. 1771; Arthur D. Chapman, ed., *Australian Plant Name Index*. 2571-2579. Canberra 1991.

Species/Vernacular Names:
S. aemula R. Br. (*Lobelia aemula* (R. Br.) Kuntze; *Scaevola sinuata* R. Br.)

English: fairy fanflower, common fanflower

S. albida (Smith) Druce (*Goodenia albida* Smith; *Goodenia laevigata* Curtis; *Scaevola microcarpa* Cav.; *Scaevola pallida* R. Br.)

English: small-fruited fanflower, pale fanflower, smaller fanflower

S. calendulacea (Andrews) Druce (*Goodenia calendulacea* Andrews; *Lobelia calendulacea* (Kenn.) Kuntze; *Merkusia suaveolens* (R. Br.) Vriese; *Scaevola suaveolens* R. Br.)

English: dune fanflower, scented fanflower, sweet fanflower

S. calliptera Benth. (*Scaevola macropoda* DC.; *Scaevola benthamea* Vriese)

English: royal robe

S. chamissoniana Gaud. (*Lobelia chamissoniana* (Gaud.) Kuntze; *Lobelia cylindrocarpa* (Hillebr.) Kuntze; *Scaevola cylindrocarpa* Hillebr.; *Temminckia chamissoniana* (Gaud.) de Vriese)

Hawaii: naupaka kuahiwi

S. coriacea Nutt. (*Lobelia coriacea* (Nutt.) Kuntze)

Hawaii: dwarf naupaka

S. crassifolia Labill. (*Merkusia crassifolia* (Labill.) Vriese; *Lobelia crassifolia* (Labill.) Kuntze)

English: thick-leaved fanflower

S. enanthophylla F. Muell. (*Lobelia enanthophylla* (F. Muell.) Kuntze; *Scaevola scandens* Bailey) (opposite-leaved, from the Greek *enantion* "opposite, against" and *phyllon* "leaf")

English: climbing fanflower

S. gaudichaudiana Cham. (*Scaevola pubescens* Nutt.; *Scaevola ligustrifolia* Nutt.; *Scaevola skottsbergii* St. John; *Temminckia ciliata* (G. Don) de Vriese)

Hawaii: naupaka kuahiwi

S. gaudichaudii Hook. & Arnott (*Lobelia gaudichaudii* (Hook. & Arnott) Kuntze; *Scaevola menziesiana* Cham.; *Scaevola swezeyana* Rock; *Temminckia gaudichaudii* (Hook. & Arnott) de Vriese; *Temminckia menziesiana* (Cham.) de Vriese)

Hawaii: naupaka kuahiwi

S. glabra Hook. & Arnott (*Lobelia glabra* (Hook. & Arnott) Kuntze; *Scaevola kauaiensis* (Degener) St. John; *Scaevola wahiawaensis* St. John; *Camphusia kauaiensis* (Degener) Degener & I. Degener)

Hawaii: 'ohe naupaka

S. hookeri (Vriese) Hook.f. (*Merkusia hookeri* Vriese; *Lobelia hookeri* (Vriese) Kuntze)

English: creeping fanflower

S. kilaueae Degener (*Scaevola kilaueae* var. *powersii* Degener & I. Degener)

Hawaii: naupaka kuahiwi, huahekili uka, papa'ahekili

S. laciniata F.M. Bailey

English: fanflower

S. linearis R. Br. (*Merkusia linearis* (R. Br.) Vriese; *Lobelia linearis* (R. Br.) Kuntze)

English: rough fanflower

S. mollis Hook. & Arnott (*Lobelia mollis* (Hook. & Arnott) Kuntze; *Scaevola mollis* f. *albiflora* Degener & Greenwell; *Camphusia glabra* (Hook. & Arnott) de Vriese; *Temminckia mollis* (Hook. & Arnott) de Vriese)

Hawaii: naupaka kuahiwi

S. nitida R. Br. (*Merkusia nitida* (R. Br.) Vriese; *Lobelia nitida* (R. Br.) Kuntze)

English: shrubby fanflower

S. plumieri (L.) Vahl (*Scaevola lobelia* Murr.; *Scaevola thunbergii* Eckl. & Zeyh.)

Southern Africa: umQhaphu (Xhosa)

S. procera Hillebr. (*Lobelia procera* (Hillebr.) Kuntze; *Scaevola procera* f. *dolichocarpa* Skottsb.)

Hawaii: naupaka kuahiwi

S. ramosissima (Smith) K. Krause (*Goodenia ramosissima* Smith; *Scaevola hispida* Cav.; *Merkusia hispida* (Cav.) Vriese)

English: hairy fanflower

S. sericea Vahl (*Scaevola taccada* (Gaertn.) Roxb., nom. illeg.; *Scaevola taccada* Roxb.; *Lobelia taccada* Gaertn.; *Scaevola fauriei* H. Lév.; *Scaevola frutescens* K. Krause)

English: beach berry, sea-lettuce tree

Japan: teriha-kusa-tobera, sukî

Malaya: ambong ambong, merambong, pelambong

Hawaii: naupaka kahakai, aupaka, huahekili, naupaka kai

S. spinescens R. Br. (*Crossotoma spinescens* (R. Br.) Vriese; *Scaevola oleoides* DC.)

Australia: spiny fanflower, prickly fanflower, currant bush, poontoo, maroon bush

Scambopus O.E. Schulz Brassicaceae

Origins:

Greek *skambos* "curved, crooked, bent (of the legs)" and *pous* "foot," referring to the curved pedicels; see Otto Eugen Schulz (1874-1936), *Das Pflanzenreich*. Heft 86. 259. (July) 1924.

Scandia J.W. Dawson Umbelliferae

Origins:

See H.E. Connor and E. Edgar, "Name changes in the indigenous New Zealand flora, 1960-1986 and Nomina Nova IV, 1983-1986." *New Zealand Journal of Botany*. Vol. 25: 115-170. 1987; J.W. Dawson, *New Zeal. J. Bot.* 5: 407. 1967.

Species/Vernacular Names:

S. rosifolia (Hook.) J.W. Dawson (*Angelica rosifolia* Hook.)

English: rose-leaved anise

Maori names: koheriki, kohepiro

Scandicium Thell. Umbelliferae

Origins:

Greek *skandix*, *skandikos* applied by Aristophanes and Theophrastus (*HP*. 7.7.1 and 7.8.1) to chervil, Latin *scandix, icis* for the herb chervil, *Scandix caerefolium* L. (Plinius); see H.E. Connor and E. Edgar, "Name changes in the indigenous New Zealand flora, 1960-1986 and Nomina Nova IV, 1983-1986." *New Zealand Journal of Botany*. Vol. 25: 115-170. 1987.

Scandix L. Umbelliferae

Origins:

From the Greek *skandix*, *skandikos* applied by Aristophanes and Theophrastus (*HP*. 7.7.1 and 7.8.1) to chervil, Latin *scandix, icis* for the herb chervil, *Scandix caerefolium* L.; see Carl Linnaeus, *Species Plantarum*. 256. 1753 and *Genera Plantarum*. Ed. 5. 124. 1754.

Species/Vernacular Names:

S. pecten-veneris L.

English: shepherd's needle, Venus' needle, Venus's comb, Lady's comb

Scaphiophora Schltr. Burmanniaceae

Origins:

Greek *skaphion* "a small boat" and *phoros* "bearing, carrying."

Scaphispatha Brongn. ex Schott Araceae

Origins:

Greek *skaphion*, *skaphis* "a small boat, small bowl," *skaphe* "a light boat, tub, bowl" and *spathe* "spathe"; see D.H. Nicolson, "Derivation of aroid generic names." *Aroideana*. 10: 15-25. 1988.

Scaphium Schott & Endl. Sterculiaceae

Origins:

Latin *scaphium* (*scapium*), *ii* "a hollow vessel," Greek *skaphion* "a small boat," *skaphe* "a light boat," referring to the shape of the fruits.

Species/Vernacular Names:
S. sp.

Malaya: kembang sa-mangkok, kembang semangkok

Scaphocalyx Ridley Flacourtiaceae

Origins:

Greek *skaphe* "a light boat, boat," *skaphos* "ship, hollow" and *kalyx* "calyx."

Scaphochlamys Baker Zingiberaceae

Origins:

Greek *skaphe* "a light boat, boat," *skaphos* "ship" and *chlamys, chlamydos* "cloak," referring to the bracts.

Scaphopetalum Mast. Sterculiaceae

Origins:

Latin *scaphium* (*scapium*), *ii* "a hollow vessel" and *petalum* or *petalon* "petal," referring to a boat-shaped petal; Greek *skaphion* "a small boat," *skaphe* "a light boat."

Scaphosepalum Pfitzer Orchidaceae

Origins:

Greek *skaphe* and Latin *sepalum*, referring to the concave form of the connate lateral sepals.

Scaphospermum Korovin Umbelliferae

Origins:

From the Greek *skaphe* "a light boat, boat," *skaphos* "ship" and *sperma* "seed."

Scaphyglottis Poeppig & Endl. Orchidaceae

Origins:

Greek *skaphis* "bowl, spade, shovel," the diminutive of *skaphe*, and *glotta* "tongue" referring to the lip, to the form of the concave labellum.

Scapicephalus Ovcz. & Chukavina Boraginaceae

Origins:

Latin *scapus, i* "a shaft, stem, stalk, trunk" and Greek *kephale* "head," Greek *skepto* "stay one thing against another," *skeptron* "staff, stick, sceptre."

Scaredederis Thouars Orchidaceae

Origins:

Origin and meaning obscure, possibly from the Greek *skaris, askaris* "intestinal worm."

Scariola F.W. Schmidt Asteraceae

Origins:

Latin *escariolam*, from *escarius* "pertaining to food, eating," *escaria, escariorum* "fit for eating," *esca* "food"; see Helmut Genaust, *Etymologisches Wörterbuch der botanischen Pflanzennamen.* 565. 1996; Manlio Cortelazzo & Paolo Zolli, *Dizionario etimologico della lingua italiana.* 5: 1145. Bologna 1988; Salvatore Battaglia, *Grande dizionario della lingua italiana.* XVII: 863. Torino 1995; G. Manuzzi, ed., *Libro della cura delle malattie.* Testo del buon secolo della lingua allegato nel Vocabolario della Crusca, ora per la prima volta posto in luce dal cav. Abate G. Manuzzi. Firenze 1863.

Sceletium N.E. Br. Aizoaceae

Origins:

Greek *skeletos* "dried up, dry, lean, mummy, skeleton," Latin *sceletus* "a skeleton," referring to the poisonous dried fermented leaves.

Scelochiloides Dodson & M.W. Chase Orchidaceae

Origins:

Resembling the genus *Scelochilus* Klotzsch.

Scelochilus Klotzsch Orchidaceae

Origins:

Indicating the appendages of the lip, from the Greek *skelos* "leg" and *cheilos* "lip," referring to the divided lip or to the horned structures on the margin of the lip.

Scepa Lindley Euphorbiaceae

Origins:
Presumably from the Greek *skepas* "covering, shelter."

Scepasma Blume Euphorbiaceae

Origins:
From the Greek *skepasma* "covering, covering membrane."

Scepseothamnus Cham. Rubiaceae

Origins:
Greek *skepsis, skepseos* "viewing" and *thamnos* "shrub."

Sceptridium Lyon Ophioglossaceae

Origins:
Greek *skepto* "stay one thing against another," *skeptron* "staff, stick, sceptre," Latin *sceptrum, i* "a royal staff, a sceptre."

Sceptrocnide Maxim. Urticaceae

Origins:
Greek *skeptron* "staff, stick, sceptre" and *knide* "nettle."

Schachtia Karsten Rubiaceae

Origins:
For the German botanist Hermann Schacht, 1814-1864, plant anatomist, professor of botany, Director of the Botanical Garden of Bonn, traveler, botanical collector; see John H. Barnhart, *Biographical Notes upon Botanists*. 3: 219. 1965; E.M. Tucker, *Catalogue of the Library of the Arnold Arboretum of Harvard University*. 1917-1933.

Schaenomorphus Thorel ex Gagnepain Orchidaceae

Origins:
Greek *schoinos* "rush, reed, cord" and *morphe* "a form, shape," the rush-like habit.

Schaffnera Bentham Gramineae

Origins:
After the German pharmacist Wilhelm Schaffner (J. Guillermo Schaffner), 1830-1882, plant collector, from 1856 in Mexico.

Schaffnerella Nash Gramineae

Origins:
For the German pharmacist Wilhelm Schaffner (J. Guillermo S.), 1830-1882.

Schaffneria Fée ex T. Moore Aspleniaceae

Origins:
For the collector of the first specimen, the German pharmacist Wilhelm Schaffner (J. Guillermo, J. Wilhelm), 1830-1882, from 1856 collected in Mexico. See Manuel Villada, "El Sr. Dr. Guillermo Schaffner." *Naturaleza*. 4: 32. [Sociedad Mexicana de Historia Natural.] Mexico City 1878; Nicolas Jean Baptiste Gaston Guibourt (1790-1861), "Observations sur les productions du Mexique." *Jour. Pharm. et Chim*. 4: 95-108. 1866; John H. Barnhart, *Biographical Notes upon Botanists*. 3: 224. 1965; Berthold Carl Seemann (1825-1871), "*Hanburia*, eine neue Cucurbitaceen gattung von Mexico." *Bonplandia*. 6: 293. 1858, 7: 2-3. 1859, 10: 189-190. 1862; T.W. Bossert, *Biographical Dictionary of Botanists Represented in the Hunt Institute Portrait Collection*. 351. 1972; Irving William Knobloch, compil., "A preliminary verified list of plant collectors in Mexico." *Phytologia Memoirs*. VI. 1983; Ida Kaplan Langman, *A Selected Guide to the Literature on the Flowering Plants of Mexico*. Philadelphia 1964; Stafleu and Cowan, *Taxonomic Literature*. 5: 113. 1985.

Schaueria Hasskarl Labiatae

Origins:
After the German botanist Johannes Conrad (Johann Konrad) Schauer, 1813-1848; see J.H. Barnhart, *Biographical Notes upon Botanists*. 3: 222. 1965; Ida Kaplan Langman, *A Selected Guide to the Literature on the Flowering Plants of Mexico*. 677. Philadelphia 1964; E.M. Tucker, *Catalogue of the Library of the Arnold Arboretum of Harvard University*. Cambridge, Massachusetts 1917-1933; A. Lasègue, *Musée botanique de Benjamin Delessert*. Paris 1845; R. Zander, F. Encke, G. Buchheim and S. Seybold, *Handwörterbuch der Pflanzennamen*. 14. Aufl. 774. Stuttgart 1993.

Schedonnardus Steudel Gramineae

Origins:
Greek *schedon* "near, nearby" and *nardos* "spikenard."

Species/Vernacular Names:
S. paniculatus (Nutt.) Trel.

English: tumble grass

Schedonorus P. Beauv. Gramineae

Origins:

Greek *schedon* and *oros* "mountain," possibly referring to the habitat; see Ambroise Marie François Joseph Palisot de Beauvois (1752-1820), *Essai d'une nouvelle Agrostographie*. 99, 177. Paris (Dec.) 1812.

Scheelea Karsten Palmae

Origins:

Dedicated to the Swedish (b. Stralsund, Swedish Pomerania) apothecary Carl (Karl) Wilhelm Scheele (Carolus Gulielmus Scheelius), 1742-1786 (d. Köping, Sweden), one of the most eminent chemists of the eighteenth century, 1772 discovered oxygen, chlorine (1774), glycerine and hydrogen sulfide, acquainted with Johan Gottlieb Gahn, in 1775 he was elected to the Royal Academy of Sciences; see Uno Boklund, in *Dictionary of Scientific Biography* 12: 143-150. 1981; R.A. DeFilipps, "Carl Wilhelm Scheele, of the Palm Genus *Scheelea*." *Principes*. 31(3): 107-109. 1987; Helmut Genaust, *Etymologisches Wörterbuch der botanischen Pflanzennamen*. 566. [genus named for the German botanist Georg Heinrich Adolf Scheele, 1808-1864] Basel 1996; R. Zander, F. Encke, G. Buchheim and S. Seybold, *Handwörterbuch der Pflanzennamen*. 14. Aufl. 774. Stuttgart 1993; F. Boerner & G. Kunkel, *Taschenwörterbuch der botanischen Pflanzennamen*. 4. Aufl. 166. 1989.

Scheeria Seemann Gesneriaceae

Origins:

For the German botanist Friedrich (Frederick) Scheer (pseud. Diogenes), 1792-1868 (Kent), merchant, a supporter of Kew, wrote *Kew and Its Gardens*. London 1840, *The Cape of Good Hope versus Egypt*. London 1839 and *A Letter to Thomas Baring, Esq., on the Effects of the Californian and Australian Gold Discoveries*. London 1852, translated Julius R.T. Vogel, *Journal of the Voyage to the Niger*. 1849; see J.H. Barnhart, *Biographical Notes upon Botanists*. 3: 223. 1965; T.W. Bossert, *Biographical Dictionary of Botanists Represented in the Hunt Institute Portrait Collection*. 351. 1972; Gordon Douglas Rowley, *A History of Succulent Plants*. Strawberry Press, California 1997; Ray Desmond, *Dictionary of British & Irish Botanists and Horticulturists*. 612. London 1994; Erwin G. Gudde, *California Place Names. The Origin and Etymology of Current Geographical Names*. University of California Press, Berkeley 1974 [1949]; R. Zander, F. Encke, G. Buchheim

and S. Seybold, *Handwörterbuch der Pflanzennamen*. 14. Aufl. Stuttgart 1993.

Schefferella Pierre Sapotaceae

Origins:

After the Dutch botanist Rudolph Herman Christiaan Carel (Karel) Scheffer, 1844-1880, from 1868 Botanic Gardens Buitenzorg (Bogor, Indonesia), founded and edited *Annales du Jardin Botanique de Buitenzorg* [superseded by *Annals of the Botanic Gardens, Buitenzorg*]; see J.H. Barnhart, *Biographical Notes upon Botanists*. 3: 223. 1965; T.W. Bossert, *Biographical Dictionary of Botanists Represented in the Hunt Institute Portrait Collection*. 351. 1972; E.M. Tucker, *Catalogue of the Library of the Arnold Arboretum of Harvard University*. 1917-1933; Elmer Drew Merrill, in *Contr. U.S. Natl. Herb*. 30(1): 265. 1947; G. Murray, *History of the Collections Contained in the Natural History Departments of the British Museum*. 1: 180. London 1904.

Schefferomitra Diels Annonaceae

Origins:

After the Dutch botanist Rudolph Herman Christiaan Carel Scheffer, 1844-1880.

Schefflera Forster & Forster f. Araliaceae

Origins:

The genus was named to honor the botanist Jakob (Jacob) Christoph Scheffler (Jacobus Christophorus S.), physician in Danzig, Poland, possibly author of Disputatio botanico-medica inauguralis *De Asaro*. D. Altdorfi 1721 and *De mortibus ex baccis Solani*. 1733; see Albrecht von Haller, *Disputationes ad morborum historiam et curationem facientes*, quas collegit, edidit, et recensuit A.H. Lausanne 1757-1760; Christianus Jacobi, *Historiam hydropis saccati*. Altorfii Noricorum [1724]; Johann Reinhold Forster (1729-1798) and his son Johann Georg Adam (1754-1794), *Characteres generum plantarum*. 45, t. 23. Londini [London] (Nov.-Dec.) 1775; Georg Christian Wittstein, *Etymologisch-botanisches Handwörterbuch*. 791. 1852; F. Boerner & G. Kunkel, *Taschenwörterbuch der botanischen Pflanzennamen*. 4. Aufl. 166. 1989; H. Genaust, *Etymologisches Wörterbuch der botanischen Pflanzennamen*. 566. [the generic name is dedicated to a Jacob Christian Scheffler] Basel 1996.

Species/Vernacular Names:

S. actinophylla (Endl.) Harms (*Brassaia actinophylla* Endl.)

English: Australian ivy palm, octopus tree, Queensland umbrella tree, umbrella tree

S. arboricola (Hayata) Hayata (*Heptapleurum arboricolum* Hayata)

English: scandent schefflera

China: qi ye lian

Japan: yadori-fuka-no-ki

S. bodinieri (Lévl.) Rehd.

English: Bodinier schefflera

S. cephalotes (Clarke) Harms

English: Indian schefflera

S. delavayi (Franch.) Harms

English: Delavay schefflera

China: da pao tong

S. digitata Forst. & Forst.f.

English: seven finger

Maori name: patete

S. elata (Clarke) Harms

English: high schefflera

S. elegantissima (Veitch ex Mast.) Lowry & Frodin

English: false aralia

French: petit boux caledonien

S. elliptica (Bl.) Harms

Asia: galamai-amo

S. glomerulata Li

English: glomerulose schefflera

S. heptaphylla (L.) Frodin (*Schefflera octophylla* (Lour.) Harms; *Aralia octophylla* Lour.; *Agalma lutchuense* Nakai)

English: ivy tree

Japan: Fukanoki, tûjingii, asaguru, asanguru

East Asia: chiang mao

S. hoi (Dunn) Vig.

English: Ho schefflera

S. hypoleuca (Kurz) Harms

English: whiteback schefflera

S. khasiana (Clarke) Vig.

English: Khas schefflera

S. kwangsiensis Merr. ex Li

English: Kwangsi schefflera

S. littorea (Seem.) Frodin

English: Rumphius's ivy-palm

S. morototonii (Aublet) Maguire, Steyerm. & Frodin

English: mountain trumpet

South Africa: Lucifershout

Tropical America: mandioqueira, jereton

S. octophylla (Lour.) Harms

English: common schefflera, schefflera, ivy tree

China: ya jiao mu pi

Vietnam: la lang, chan nim

S. odorata (Blanco) Merr. & Rolfe

The Philippines: arasagot, galamay-amo, lima-lima, kalakang, karangkang, tarangkang, kayangkang, kokontimbalun, palan, panayang, tagima, tagilima, tuglima, tughik

S. umbellifera (Sond.) Baill. (*Cussonia umbellifera* Sond.; *Neocussonia umbellifera* (Sond.) Hutch.)

English: forest cabbage tree, bastard cabbage tree, cabbage wood, cabbage tree

Southern Africa: sambreelboom, forest kiepersol; umSenge, umKisiso (Xhosa); umSenge, umGezisa, umBumbu, umBegele, umSengembuzi (Zulu); umSengambuti (Swazi); mosetshe (North Sotho); muChaka, muChenje, muShenjetsoka (Shona)

Scheffleropsis Ridley Araliaceae

Origins:
Resembling *Schefflera*.

Schelhammera R. Br. Convallariaceae (Liliaceae)

Origins:
After the German physician Günther Christoph Schelhammer (Guntherus Christophorus Schelhamerus or Schelhammerus), 1649-1716, professor of medicine at Helmstedt, Jena and Kiel. Among his most valuable writings are *Anatomes Xiphiae piscis*. Hamburgi 1707, *Catalogus Plantarum*, maximam partem rariorum, quas per hoc biennium in hortulo domestico aluit, et, paucis exceptis, etiam his vernis aestivisque mensibus poterit exhibere G.C. Schelhammer. Helmestadii 1683, *De genuina febres curandi methodo dissertatio*. Jenae 1692, G.C. Schelhammeri ... *Programma Rei Herbariae* professioni in horto medico auspicandae praemissum quo medicae artis ... studiosos ad demonstrationes plantarum ... invitat. Jenae [1690]; see Christianus Stephanus Scheffelius, *Virorum clarissimorum ad G.C. Schelhammerum Epistolae selectiores ... vitam Schelhammeri cum indice scriptorum eius* ... praemisit C.S. Scheffelius. 1727; Robert Brown, *Prodromus florae Novae Hollandiae et Insulae van-Diemen*. 273. London 1810.

Species/Vernacular Names:
S. undulata R. Br. (*Parduyna undulata* (R. Br.) Dandy)

English: lilac lily

Schellenbergia C.E. Parkinson Connaraceae

Origins:

After the German botanist Gustav August Ludwig David Schellenberg, 1882-1963, a specialist in Connaraceae; see Ida Kaplan Langman, *A Selected Guide to the Literature on the Flowering Plants of Mexico*. 677. Philadelphia 1964; E.M. Tucker, *Catalogue of the Library of the Arnold Arboretum of Harvard University*. Cambridge, Massachusetts 1917-1933; Elmer Drew Merrill, *Contr. U.S. Natl. Herb.* 30(1): 265. 1947 and *Bernice P. Bishop Mus. Bull.* 144: 163. 1937.

Schellolepis J. Smith Polypodiaceae

Origins:

Greek *skello* "to dry, parch" and *lepis* "scale"; see John Smith (1798-1888), *Ferns: British and Foreign*. London 1866.

Schenckia K. Schumann Rubiaceae

Origins:

After the German botanist Johann Heinrich Rudolf Schenck, 1860-1927, traveler, professor of botany, Director of the Botanical Garden of Darmstadt, plant collector in Brazil and Mexico; see J.H. Barnhart, *Biographical Notes upon Botanists*. 3: 224. 1965; T.W. Bossert, *Biographical Dictionary of Botanists Represented in the Hunt Institute Portrait Collection*. 352. 1972; E.M. Tucker, *Catalogue of the Library of the Arnold Arboretum of Harvard University*. 1917-1933; Stafleu and Cowan, *Taxonomic Literature*. 5: 130-133. 1985; Ida Kaplan Langman, *A Selected Guide to the Literature on the Flowering Plants of Mexico*. Philadelphia 1964; Irving William Knobloch, compil., "A preliminary verified list of plant collectors in Mexico." *Phytologia Memoirs*. VI. 1983.

Schenodorus P. Beauv. Gramineae

Origins:

Orthographic variant of *Schedonorus* P. Beauv.

Scherya R.M. King & H. Robinson Asteraceae

Origins:

For Robert Walter Schery, 1917-1987; see Irving William Knobloch, compil., "A preliminary verified list of plant collectors in Mexico." *Phytologia Memoirs*. VI. 1983; Ida Kaplan Langman, *A Selected Guide to the Literature on the Flowering Plants of Mexico*. 678. Philadelphia 1964.

Scheuchzeria L. Scheuchzeriaceae (Juncaginaceae)

Origins:

In honor of the Swiss botanist Johannes Gaspar Scheuchzer, 1684-1738, agrostologist, author of *Agrostographiae helveticae prodromus*. [Zürich] 1708, father of the Swiss botanist Johannes Scheuchzer (1738-1815) and brother of the Swiss naturalist and physician Johann Jakob (Jean-Jacques) Scheuchzer (1672-1733); see J.H. Barnhart, *Biographical Notes upon Botanists*. 3: 227. 1965; T.W. Bossert, *Biographical Dictionary of Botanists Represented in the Hunt Institute Portrait Collection*. 352. 1972; Mariella Azzarello Di Misa, ed., *Il Fondo Antico della Biblioteca dell'Orto Botanico di Palermo*. 247-248. Palermo 1988; H.N. Clokie, *Account of the Herbaria of the Department of Botany in the University of Oxford*. 238, 279. Oxford 1964; E.M. Tucker, *Catalogue of the Library of the Arnold Arboretum of Harvard University*. 1917-1933; Jonas C. Dryander, *Catalogus bibliothecae historico-naturalis Josephi Banks*. London 1800; A. Lasègue, *Musée botanique de Benjamin Delessert*. 347. Paris 1845; James Britten, *The Sloane Herbarium*, revised and edited by J.E. Dandy. 1958; P.E. Pilet, in *Dictionary of Scientific Biography* 12: 159. 1981; Johann Jakob Scheuchzer, *Herbarium diluvianum* collectum à Johanne Jacobo Scheuchzero. Leyde 1723.

Schidiomyrtus Schauer Myrtaceae

Origins:

Greek *schizo, schizein* "to split, divide" and the genus *Myrtus*, Latin *schidia, ae* "a chip, splinter of wood"; see J.C. Schauer, in *Linnaea*. 17: 237. 1843.

Schiedea A. Rich. Rubiaceae

Origins:

Named for the German botanist Christian Julius Wilhelm Schiede, 1798-1836.

Schiedea Bartl. Rubiaceae

Origins:

For the German botanist Christian Julius Wilhelm Schiede, 1798-1836.

Schiedea Chamisso & Schlechtendal Caryophyllaceae

Origins:

For the German botanist Christian Julius Wilhelm Schiede, 1798-1836, physician, traveler, gardener, traveling companion of Ferdinand Deppe (1794-1861), from 1828 to 1836 plant collector and botanical explorer in Mexico. His writings include *De plantis hybridis sponte natis*. Cassellis Cattorum [Kassel] 1825, with A. de Chamisso wrote "Plantarum mexicanarum a cl. viris Schiede et Deppe collectarum recensio brevis." *Linnaea*. 1830-1831; see John H. Barnhart, *Biographical Notes upon Botanists*. 3: 224. 1965; Ludolf Karl Adelbert von Chamisso and D.F.L. von Schlechtendal, "De plantis in expeditione speculatoria Romanzoffiana observatis." *Linnaea*. 1: 1-73. 1826; Stephan F. Ladislaus Endlicher, *Atakta botanika*. Nova genera et species plantarum descripta et iconibus illustrata. Vienna 1833; I.H. Vegter, *Index Herbariorum*. Part II (6), *Collectors S*. Regnum Vegetabile vol. 114. 1986; H.N. Clokie, *Account of the Herbaria of the Department of Botany in the University of Oxford*. 239. Oxford 1964; A. Lasègue, *Musée botanique de Benjamin Delessert*. Paris 1845; Günther Schmid, *Chamisso als Naturforscher*. Eine Bibliographie. Leipzig 1942; Ida Kaplan Langman, *A Selected Guide to the Literature on the Flowering Plants of Mexico*. 678. 1964; Gordon Douglas Rowley, *A History of Succulent Plants*. Strawberry Press, Mill Valley, California 1997; R. Zander, F. Encke, G. Buchheim and S. Seybold, *Handwörterbuch der Pflanzennamen*. 14. Aufl. Stuttgart 1993; Irving William Knobloch, compil., "A preliminary verified list of plant collectors in Mexico." *Phytologia Memoirs*. VI. 1983.

Schiedeella Schlechter Orchidaceae

Origins:

Named for the German botanist Christian Julius Wilhelm Schiede, 1798-1836.

Schiedophytum H. Wolff Umbelliferae

Origins:

For the German botanist Christian Julius Wilhelm Schiede, 1798-1836, physician, traveler, gardener, from 1828 to 1836 plant collector and botanical explorer in Mexico.

Schima Reinw. ex Blume Theaceae

Origins:

Greek *skiasma* "a shade, shadow, shelter," with reference to the dense crown of the tree.

Species/Vernacular Names:

S. wallichii (DC.) Korth. (*Schima superba* Gardn. & Champ.; *Schima argentea* Pritz.; *Schima kankaoensis* Hayata; *Schima khasiana* Dyer; *Schima noronhae* Reinw. ex Blume) (after the Danish physician and botanical collector Nathaniel Wallich (originally Nathan Wolff or Wulff), 1786-1854, in 1807 went out to India as surgeon, in 1813 with the Hon. East India Company, in 1833 visited Assam, from 1814 to 1841 Superintendent of the Calcutta Botanic Garden, 1820-1822 plant collector in Nepal; see K. Biswas, ed., *The Original Correspondence of Sir Joseph Banks Relating to the Foundation of the Royal Botanic Garden, Calcutta* and *The Summary of the 150th Anniversary Volume of the Royal Botanic Garden, Calcutta*. Calcutta 1950; Isaac Henry Burkill, *Chapters on the History of Botany in India*. Delhi 1965; John H. Barnhart, *Biographical Notes upon Botanists*. 3: 454. 1965; Andrew Thomas Gage, *A History of the Linnean Society of London*. London 1938; Mary Gunn and Leslie E. Codd, *Botanical Exploration of Southern Africa*. Cape Town 1981; E.M. Tucker, *Catalogue of the Library of the Arnold Arboretum of Harvard University*. Cambridge, Massachusetts 1917-1933; Antoine Lasègue, *Musée botanique de M. Benjamin Delessert*. 1845; I.H. Vegter, *Index Herbariorum*. Part II (7), *Collectors T-Z*. Regnum Vegetabile vol. 117. 1988; I.H. Vegter, *Index Herbariorum*. Part II (5), *Collectors N-R*. Regnum Vegetabile vol. 109. 1983; Emil Bretschneider, *History of European Botanical Discoveries in China*. [Reprint of the original edition 1898.] Leipzig 1981; R. Desmond, *The European Discovery of the Indian Flora*. Oxford 1992; T. W. Bossert, *Biographical Dictionary of Botanists Represented in the Hunt Institute Portrait Collection*. 425. Boston, Massachusetts 1972; Ray Desmond, *Dictionary of British & Irish Botanists and Horticulturists*. London 1994)

English: schima, Darjeeling gugertree, needle wood, chilauni

Malaya: cha antan, medang, changkoh, kelat gelugor, medang bekawi

India: chilauni, kanak, makusal, makrisal, cheloni, makriya, makria chilauni, makriasal, nogabhe, gugera, boldak, diengngan, dingan, bonak, jam, khiang, ingkhia-chin, sambrangkung

China: chilauni

Nepal: chilaune, aule-chilaune, goe-chassi, sule-chilauni

Japan: hime-tsuba-ki, iju, njunuki

Schimpera Hochstetter & Steudel ex Endlicher Brassicaceae

Origins:

For the German botanist Georg Heinrich Wilhelm Schimper, 1804-1878, plant collector, brother of the German botanist and zoologist Karl Friedrich Schimper (1803-1867, d. Schwetzingen, near Heidelberg, Germany), 1831 collected in S. France and Algeria, 1834-1836 Egypt and Arabia, from 1837 in Abyssinia, 1840-1855 Governor of the district of Antitscho (Abyssinia). His writings include W. Schimper's *Reise nach Algier in ... 1831 und 1832 ...* Herausgegeben von [C.F. Hochstetter and E. Steudel], etc. Stuttgart 1834, he was the cousin of the Alsatian bryologist and palaeontologist Wilhelm Philipp (Guillaume Philippe) Schimper (1808-1880); see J.H. Barnhart, *Biographical Notes upon Botanists*. 3: 226. 1965; H.N. Clokie, *Account of the Herbaria of the Department of Botany in the University of Oxford*. 239. 1964; A. Lasègue, *Musée botanique de Benjamin Delessert*. 1845; J.B. Gillett, "W.G. Schimper's botanical collecting localities in Ethiopia." *Kew Bulletin*. 27(1): 115-128. 1972; J.B. Gillett, *Kew Bulletin*. 12: 460. 1958; Stafleu and Cowan, *Taxonomic Literature*. 5: 163-165. 1985; A.P.M. Sanders, in *Dictionary of Scientific Biography* 12: 165-167. 1981; Heinz Tobien, in *Dictionary of Scientific Biography* 12: 167-168. 1981; P.W. Richards, in *Dictionary of Scientific Biography* 12: 168-169. 1981; Mariella Azzarello Di Misa, ed., *Il Fondo Antico della Biblioteca dell'Orto Botanico di Palermo*. 249. Regione Siciliana, Palermo 1988.

Schimperella H. Wolff Umbelliferae

Origins:

For the German botanist Georg Heinrich Wilhelm Schimper, 1804-1878, plant collector; see J.H. Barnhart, *Biographical Notes upon Botanists*. 3: 226. 1965.

Schimperina Tieghem Loranthaceae

Origins:

For the German botanist Georg Heinrich Wilhelm Schimper, 1804-1878, plant collector; see John H. Barnhart, *Biographical Notes upon Botanists*. 3: 226. 1965.

Schinopsis Engl. Anacardiaceae

Origins:

Resembling the genus *Schinus* L.

Schinus L. Anacardiaceae

Origins:

From *schinos*, the Greek name for the mastic tree, *Pistacia lentiscus* L., Latin *schinos* or *schinus, i,* as some species yield mastic-like juices or resin; see Carl Linnaeus, *Species Plantarum*. 388. 1753 and *Genera Plantarum*. Ed. 5. 184. 1754.

Species/Vernacular Names:

S. sp.

Brazil: aroeira do campo, aroeira blanca

Argentina: chichita mononi, litre, molle de curtir

S. areira L. (*Schinus molle* L. var. *areira* (L.) DC.)

English: pepper tree, pepperina

S. latifolius (Gillies) Engl. (*Schinus dependens* var. *alpha* Hook.; *Schinus dependens* var. *crenatus* Arech.; *Schinus dependens* var. *latifolius* Marchand)

Argentina: molle morado

S. longifolius (Lindl.) Speg. (*Schinus dependens* DC.; *Schinus dependens* DC. var. *subintegra*)

Argentina: molle, molle guazu, trementina

S. molle L. (*Schinus huigan* Molina)

English: peppermint tree, pepper tree, Peruvian mastic tree, Peruvian pepper tree, pepper tree of Peru, California pepper tree, Australian pepper, mastic tree, peppercorn tree, pepperina, American mastic

French: poivrier d'Amérique, poivrier du Pérou, faux poivrier

Italian: falso pepe

Southern Africa: peperboom; umpelempele (Zulu)

Argentina: aguaribay, balsamo, curanguay, gualeguay, molle de Bolivia, molle de Castilla, molle del Peru, pimienta del diablo, terebinto

Peru: árbol de la vida, cullash, falsa pimienta, maera, molle, mulli, orcco mulli (see José María Arguedas, *Musica, danze e riti degli indios del Perú*. Giulio Einaudi Editore, Torino 1991)

Chile: pimentero

Costa Rica: pimento de California

Mexico: árbol de la pimienta, árbol del Perú, pirul, pirwi, tsactumi, tzactumi, tzantuni; pimienta de America; pirú (Valle de Mexico); xasa, xaza (Otomí l.); peloncuahuitl (Azteca l.); yaga-cica, yaga-lache (Zapoteca l., Oaxaca)

Brazil: aroeira (= araroeira, from arara plus -eira), aroeira molle, aroeira periquita, aroeira folha de salso, aroeira mole, corneita, anacauíta, fruto de sabiá, pimenta del diablo, pimentero, árbol de la pimenta, molle, terebinto, bálsamo, aguaribay, gualeguay

Uruguay: aruera

Ecuador: molle

Colombia: muelle, pimiento

India: hucchu menasina mara

S. pearcei Engler

Peru: chinamulli, china molle, orcco mulli, atojlloque, molle, mulli, orcco mulle

S. terebinthifolius Raddi

English: pepper tree, broad-leaf pepper tree, Brazilian pepper, Brazilian pepper tree, Brazil pepper tree, Japanese pepper tree, South American pepper, Christmas berry tree, Christmas berry, Florida holly

South Africa: peperboom

Rodrigues Island: poivrier sauvage

New Caledonia: poivrier

Brazil: aroeira, aroeira vermelha, aroeira precoce, aroeira da praia, aroeira mansa, aroeira rasteira, aroeira do brejo, aroeira negra, aroeira branca, aroeira do campo, aroeira do sertão, aroeira do paraná, fruto de sabiá, fruto de raposa, coração de bugre, bálsamo, cambuy, aguaraiba

Hawaii: wilelaiki, naniohilo

Schinzafra Kuntze Bruniaceae

Origins:
After the Swiss botanist Hans Schinz, 1858-1941, traveler, professor of botany.

Schinziella Gilg Gentianaceae

Origins:
After the Swiss botanist Hans Schinz, 1858-1941, traveler, professor of botany.

Schinziophyton Hutch. ex Radcliffe-Sm. Euphorbiaceae

Origins:
For the Swiss (b. Zurich) botanist Hans Schinz, 1858-1941 (d. Zurich), traveler, professor of botany, botanical collector and explorer. Among his writings are *Plantae meny-harthianae ein Beitrag zur Kenntniss der Flora des Unteren Sambesi ...* Wien, 1905 [Collector: Ladislav Menyharth, 1849-1897], "Durch Südwestafrika." *Verh. Ges. Erdk. Berl.* 14: 322-324. 1887, *Mein Lebenslauf.* Zürich 1940, *Observations sur une collection de plantes du Transvaal.* Genève 1891; see Albert Thellung (1881-1928), "Verzeichnis der Veröffentlichungen von Prof. Dr. Hans Schinz." *Beibl. Viertelj.-Schr. naturf. Ges. Zürich* 15 (Jahrg. 73): 773-783. [the bibliography of works and papers by Schinz] 1928; Théophile Alexis Durand (1855-1912), *Conspectus florae Africae*, ou, Enumération des plantes d'Afrique, par T.D. et Hans Schinz. Bruxelles 1895-1898; John H. Barnhart, *Biographical Notes upon Botanists.* 3: 227. 1965; Mary Gunn and Leslie E. Codd, *Botanical Exploration of Southern Africa.* 311-313. Cape Town 1981; A. White and B.L.

Sloane, *The Stapelieae.* Pasadena 1937; T.W. Bossert, *Biographical Dictionary of Botanists Represented in the Hunt Institute Portrait Collection.* 353. 1972; E.M. Tucker, *Catalogue of the Library of the Arnold Arboretum of Harvard University.* 1917-1933; Elmer Drew Merrill, *Bernice P. Bishop Mus. Bull.* 144: 163. 1937; F.N. Hepper and Fiona Neate, *Plant Collectors in West Africa.* 72. Utrecht 1971; R. Zander, F. Encke, G. Buchheim and S. Seybold, *Handwörterbuch der Pflanzennamen.* 14. Aufl. Stuttgart 1993; Stafleu and Cowan, *Taxonomic Literature.* 5: 175-181. 1985.

Species/Vernacular Names:

S. rautanenii (Schinz) Radcliffe-Sm. (*Ricinodendron rautanenii* Schinz) (after a Finnish missionary, the Rev. Rautanen)

English: mongongo tree, manketti nut tree, featherweight tree

Zaire: mukusu

Southern Africa: wilde akkerneut; mokongwa (Mangwato dialect, Botswana); mugongo (W. Caprivi); mugongo, ugongo (Deiriku); ugongo (Sambui); omungeta (Herero); mugongo (Ovambo)

Rhodesia: m'gongo

Zambia: mkusu, mungongo

Schippia Burret Palmae

Origins:
After the Australian botanist William August Schipp, 1891-1967, botanical explorer, plant collector (N. Australia, New Guinea, Java and British Honduras), traveler, wrote *Flora of British Honduras.* 1934; see John H. Barnhart, *Biographical Notes upon Botanists.* 3: 227. 1965.

Species/Vernacular Names:
S. concolor Burret
Brazil: palmeira leque

Schirostachyum Vriese Gramineae

Origins:
See *Schizostachyum* Nees.

Schisandra Michx. Schisandraceae

Origins:
Greek *schisis* "partage, cleavage, curdling" and *aner, andros* "man, stamen, male, seed," referring to the anthers; see Salvatore Battaglia, *Grande dizionario della lingua italiana.* XVII: 1035. Torino 1995.

Species/Vernacular Names:

S. chinensis (Turcz.) Baillon

English: Chinese magnolia vine, schisandra

China: wu wei zi, wu wei tzu

Schischkinia Iljin Asteraceae

Origins:

For the Russian botanist Boris Konstantinovich Schischkin (Shishkin), 1886-1963, traveler, botanical explorer, professor of botany. See John H. Barnhart, *Biographical Notes upon Botanists.* 3: 228. 1965; T.W. Bossert, *Biographical Dictionary of Botanists Represented in the Hunt Institute Portrait Collection.* 353. 1972; R. Zander, F. Encke, G. Buchheim and S. Seybold, *Handwörterbuch der Pflanzennamen.* 14. Aufl. Stuttgart 1993.

Schischkiniella Steenis Caryophyllaceae

Origins:

For the Russian botanist Boris Konstantinovich Schischkin (Shishkin), 1886-1963.

Schismatoclada Baker Rubiaceae

Origins:

Greek *schisma, schismatos* "division, cleft, schism" and *klados* "branch."

Schismatoglottis Zoll. & Moritzi Araceae

Origins:

From the Greek *schisma, schismatos* "division, cleft, schism" and *glotta* "tongue"; see D.H. Nicolson, "Derivation of aroid generic names." *Aroideana.* 10: 15-25. 1988.

Schismatopera Klotzsch Euphorbiaceae

Origins:

From the Greek *schisma, schismatos* "division, cleft" and *pera* "a pouch."

Schismocarpus S.F. Blake Loasaceae

Origins:

Greek *schisma, schismatos* "division, cleft" and *karpos* "a fruit."

Schismoceras C. Presl Orchidaceae

Origins:

From the Greek *schisma, schismatos* and *keras* "a horn," the shape of the sepals.

Schismus P. Beauv. Gramineae

Origins:

Greek *schismos* "cleaving," the lemma is split; see A.M.F.J. Palisot de Beauvois, *Essai d'une nouvelle Agrostographie.* 73, 177, t. 15, fig. 4. Paris (Dec.) 1812.

Species/Vernacular Names:

S. arabicus Nees

English: Mediterranean grass

S. barbatus (L.) Thell. (*Festuca barbata* L.; *Festuca calycina* Loefl.; *Schismus calycinus* Loefl.) C. Koch; *Schismus marginatus* P. Beauv.)

Australia: Arabian grass, kelch grass, mulcha grass

Schistachne Figari & De Not. Gramineae

Origins:

From the Greek *schistos* "cut, divided" and *achne* "chaff, glume."

Schistanthe Kunze Scrophulariaceae

Origins:

Greek *schistos* "cut, divided" and *anthos* "flower."

Schistocarpaea F. Muell. Rhamnaceae

Origins:

From the Greek *schistos* "cut, divided," *schizein* "to divide" and *karpos* "fruit," referring to the divided fruit; see F. von Mueller, in *The Victorian Naturalist.* 7:182. (Mar.) 1891.

Schistocarpha Less. Asteraceae

Origins:

From the Greek *schistos* and *karphos* "chip of straw, chip of wood."

Schistocaryum Franchet Boraginaceae

Origins:

Greek *schistos* "cut, divided" and *karyon* "nut."

Schistogyne Hooker & Arnott Asclepiadaceae

Origins:

Greek *schistos* and *gyne* "female, woman, female organs."

Schistolobos W.T. Wang Gesneriaceae

Origins:

Greek *schistos* "cut, divided" and *lobos* "a pod."

Schistonema Schltr. Asclepiadaceae

Origins:

Greek *schistos* "cut, divided" and *nema* "thread, filament."

Schistophragma Benth. ex Endl. Scrophulariaceae

Origins:

Greek *schistos* and *phragma* "a partition, compartment, wall, fence."

Schistophyllidium (Juz. ex Fedorov) Ikonn. Rosaceae

Origins:

Greek *schistos* "cut, divided" and *phyllon* "a leaf," Latin *schistos* "split, cleft, curdled."

Schistostemon (Urban) Cuatrec. Humiriaceae

Origins:

From the Greek *schistos* "cut, divided" and *stemon* "a stamen."

Schistostephium Less. Asteraceae

Origins:

Greek *schistos* and *stephos* "a crown," referring to the ray florets.

Schistostigma Lauterb. Euphorbiaceae

Origins:

From the Greek *schistos* "divided" and *stigma* "a stigma," see also Cleistanthus Hook.f. ex Planchon.

Schistotylus Dockrill Orchidaceae

Origins:

Greek *schistos* and *tylos* "lump, knob," referring to the trilobate labellum; see Alick William Dockrill, *Australasian Sarcanthinae*. 30, t. 43. (Sept.) 1967, and *Australian Indigenous Orchids*. Sydney 1969.

Schivereckia Andrz. ex DC. Brassicaceae

Origins:

After A. Schiwereck, d. 1806, see Georg Christian Wittstein, *Etymologisch-botanisches Handwörterbuch*. 792. Ansbach 1852; or for S.B. Schiwereck, 1782-1815, see F. Boerner & G. Kunkel, *Taschenwörterbuch der botanischen Pflanzennamen*. 4. Aufl. 166. Berlin & Hamburg 1989; see H. Genaust, *Etymologisches Wörterbuch der botanischen Pflanzennamen*. 566. Basel 1996.

Schizachne Hackel Gramineae

Origins:

Greek *schizo*, *schizein* "to split, divide" and *achne* "chaff, glume."

Schizachyrium Nees Gramineae

Origins:

Greek *schizo*, *schizein* and *achyron* "chaff, husk," alluding to the corona glume, referring to the toothed lemma; see Nees von Esenbeck, *Agrostologia Brasiliensis*. 331. 1829 in C.F.P. von Martius *Flora Brasiliensis*.

Species/Vernacular Names:

S. sp.

English: little bluestem, beardgrass

S. scoparium (Michaux) Nash (*Schizachyrium littorale* (Nash) C. Bickn.)

English: bluestem, broom beardgrass, prairie grass, wire grass, bunchgrass, little bluestem

Schizaea J.E. Smith Schizaeaceae

Origins:

Greek *schizo* "to split, divide," referring to the lobes; see J.E. Smith, in *Mémoires de l'Académie Royale de Sciences de Turin.* 5: 419, t. 9, fig. 9. Turin 1793.

Species/Vernacular Names:
S. bifida Willd.
English: forked comb-fern
S. dichotoma (L.) Smith
English: branched comb-fern
S. digitata (L.) Swartz (*Acrostichum digitatum* L.)
Japan: fusa-shida (= fringe fern)

Schizangium Bartl. ex DC. Rubiaceae

Origins:

From the Greek *schizo* "to split, divide" and *angeion, aggeion* "a vessel, capsule."

Schizanthus Ruíz & Pavón Solanaceae

Origins:

Greek *schizo, schizein* "to split, divide" and *anthos* "flower," referring to the deeply cut flowers; see Salvatore Battaglia, *Grande dizionario della lingua italiana.* XVII: 1035-1036. Torino 1995.

Schizeilema (Hook.f.) Domin Umbelliferae

Origins:

Greek *schizo, schizein* and *lemma* "rind, sheath" or *eilema* "a veil, covering, involucre," referring to the calyx-teeth; see Karel Domin, in *Botanische Jahrbücher.* 40: 573. (May) 1908.

Species/Vernacular Names:
S. fragosea (F. Muell.) Domin
English: alpine pennywort

Schizenterospermum Homolle ex Arènes Rubiaceae

Origins:

Greek *schizo, schizein* "to split, divide" and the genus *Enterospermum* Hiern, Greek *enteron* "intestine" and *sperma* "seed"; in *Notul. Syst.* (Paris) 16: 9. Oct 1960.

Schizobasis Baker Hyacinthaceae (Liliaceae)

Origins:

From the Greek *schizo* and *basis* "base, pedestal."

Schizoboea (Fritsch) B.L. Burtt Gesneriaceae

Origins:

From the Greek *schizo, schizein* and the genus *Boea* Comm. ex Lam.

Schizocalomyrtus Kausel Myrtaceae

Origins:

From the Greek *schizo* "to split, divide," *kalos* "beautiful" plus *myrtus*.

Schizocalyx Hochst. Salvadoraceae

Origins:

Greek *schizo, schizein* and *kalyx* "calyx."

Schizocalyx O. Berg Myrtaceae

Origins:

Greek *schizo, schizein* and *kalyx* "calyx."

Schizocalyx Wedd. Rubiaceae

Origins:

From the Greek *schizo, schizein* "to split, divide" and *kalyx* "calyx."

Schizocapsa Hance Taccaceae

Origins:

Greek *schizo* "to split, divide" and Latin *capsa, ae* "box, case."

Schizocardia A.C. Sm. & Standl. Cyrillaceae

Origins:

From the Greek *schizo* "to split, divide" and *kardia* "heart," probably referring to the anthers.

Schizocarphus J.J.M. v.d. Merwe Hyacinthaceae (Liliaceae)

Origins:
Greek *schizo* and *karphos* "chip of straw, chip of wood."

Schizocarpum Schrader Cucurbitaceae

Origins:
Greek *schizo* "to split, divide" and *karpos* "fruit."

Schizocarya Spach Onagraceae

Origins:
From the Greek *schizein* "to split, divide" and *karyon* "nut."

Schizocasia Schott Araceae

Origins:
Greek *schizo* plus the genus *Colocasia* Schott.

Schizocentron Meissner Melastomataceae

Origins:
From the Greek *schizein* and *kentron* "a spur, prickle," see also *Heterocentron* Hook. & Arn.

Schizochilus Sonder Orchidaceae

Origins:
From the Greek *schizo* "to split, divide" and *cheilos* "lip," referring to the lobed lip.

Schizochiton Spreng. Meliaceae

Origins:
Greek *schizo, schizein* "to split, divide" and *chiton* "a tunic, covering," see also *Chisocheton* Blume.

Schizococcus Eastw. Ericaceae

Origins:
From the Greek *schizo* and *kokkos* "a berry."

Schizocodon Siebold & Zucc. Diapensiaceae

Origins:
Greek *schizo* "to split, divide" and *kodon* "a bell."

Schizocolea Bremek. Rubiaceae

Origins:
Greek *schizo* and *koleos* "a sheath."

Schizocorona F. Muell. Asclepiadaceae

Origins:
Greek *schizo* "to split, divide" and Latin *corona, ae* "a crown"; see W.J. Hooker, in *Hooker's Journal of Botany & Kew Garden Miscellany.* 5: 107. London (Apr.) 1853.

Schizodium Lindley Orchidaceae

Origins:
Greek *schizein* and *eidos* "resemblance," referring to the column or to the shape of the petals.

Schizoglossum E. Meyer Asclepiadaceae

Origins:
From the Greek *schizo* "to split, divide" and *glossa* "tongue," referring to the corolla.

Schizogramma Link Pteridaceae (Adiantaceae)

Origins:
Greek *schizo* and *gramma* "a letter, line, character."

Schizogyne Cass. Asteraceae

Origins:
Greek *schizo* "to split, divide" and *gyne* "a woman, female."

Schizogyne Ehrenb. ex Pax Euphorbiaceae

Origins:
Greek *schizo* "to split, divide" and *gyne* "a woman, female, female organs, pistil."

Schizolaena Thouars Sarcolaenaceae

Origins:
Greek *schizo* "to split, divide" and *chlaena* "a cloak, blanket."

Schizolegnia Alston Dennstaedtiaceae (Lindsaeoideae, Lindsaeaceae)

Origins:
Greek *schizo, schizein* and *legnon* "border, colored edging"; see A.H.G. Alston, in *Boletim da Sociedade Broteriana*. Ser. 2, 30: 23. Coimbra, Portugal 1956.

Schizolepton Fée Pteridaceae (Adiantaceae, Taenitidaceae)

Origins:
Greek *schizo* "to split, to divide" and *leptos* "slender, small."

Schizolobium Vogel Caesalpiniaceae

Origins:
From the Greek *schizo* "to split, to divide" and *lobion* "a little pod."

Schizoloma Gaudich. Dennstaedtiaceae (Lindsaeoideae)

Origins:
Greek *schizo* and *loma* "border, margin, fringe, edge"; see Charles Gaudichaud-Beaupré (1789-1854), in *Annales des Sciences Naturelles. Botanique*. 3: 507. 1824.

Schizomeria D. Don Cunoniaceae

Origins:
From the Greek *schizein* "to split, divide" and *meris* "a portion, part," referring to the petals; see David Don (1799-1841), in *Edinburgh New Philosophical Journal*. 9: 94. (Apr.-Jun.) 1830.

Species/Vernacular Names:
S. ovata D. Don
English: white cherry, crab-apple

Schizomeryta R. Vig. Araliaceae

Origins:
Greek *schizo* "to split, divide" plus *Meryta* Forst. & Forst.f.

Schizomussaenda H.L. Li Rubiaceae

Origins:
Greek *schizein* "to split, divide" plus *Mussaenda* L.

Schizonepeta (Benth.) Briq. Labiatae

Origins:
Greek *schizo* "to split, divide" plus the genus *Nepeta* L.

Species/Vernacular Names:
S. tenuifolia (L.) Briq.
English: fine-leaved schizonepeta, Japanese catnip
China: jingjie, jing jie

Schizonotus A. Gray Asclepiadaceae

Origins:
From the Greek *schizo* and *notos* "back."

Schizopedium Salisb. Orchidaceae

Origins:
From the Greek *schizo* "to split, divide" and *pedilon* "a slipper, sandal," referring to *Cypripedium* L.

Schizopepon Maxim. Cucurbitaceae

Origins:
Greek *schizein* "to split" and Latin *pepo, peponis* "a species of large melon," *pepon* "ripe," related genus *Pepo* Mill., flowers bisexual.

Schizopetalon Sims Brassicaceae

Origins:
From the Greek *schizo* plus *petalon* "leaf, petal."

Schizophragma Siebold & Zucc. Hydrangeaceae

Origins:

From the Greek *schizo* "to split, divide" and *phragma* "wall, fence," referring to the habit of the plant.

Species/Vernacular Names:

S. integrifolium Oliv.

China: zuan di feng

Schizopleura (Lindley) Endlicher Myrtaceae

Origins:

Greek *schizo* "to split, divide" and *pleura, pleuro, pleuron* "side, rib"; see Stephan Friedrich Ladislaus Endlicher, *Genera Plantarum*. 2: 1228. 1840; Arthur D. Chapman, ed., *Australian Plant Name Index*. 2585. Canberra 1991.

Schizopogon Reichb. ex Spreng. Gramineae

Origins:

From the Greek *schizo* "to split, divide" and *pogon* "beard," see also *Schizachyrium* Nees.

Schizopremna Baillon Labiatae (Verbenaceae)

Origins:

From the Greek *schizo* and *premnon* "trunk, stem, root."

Schizopsis Bureau Bignoniaceae

Origins:

Greek *schizo* and *opsis* "aspect, resemblance, appearance," see also *Tynnanthus* Miers, Greek *tynnos* "so small" and *anthos* "flower."

Schizoscyphus K. Schumann ex Taubert Caesalpiniaceae

Origins:

Greek *schizo* "to split, divide" and *skyphos* "cup, jug."

Schizosepala G.M. Barroso Scrophulariaceae

Origins:

From the Greek *schizo* and Latin *sepalum* "sepal."

Schizosiphon K. Schumann Caesalpiniaceae

Origins:

Greek *schizo* "to split, divide" and *siphon* "a tube."

Schizospatha Furtado Palmae

Origins:

Greek *schizo, schizein* and *spathe* "spathe," Latin *spatha, ae* "spattle, a spathe."

Schizospermum Boivin ex Baill. Rubiaceae

Origins:

Greek *schizo* and *sperma* "seed," see also *Cremaspora* Benth., Greek *kremao* "to hang, to let hang down" and *sporos, spora* "seed, offspring."

Schizostachyum Nees Gramineae

Origins:

From the Greek *schizo, schizein* "to split, divide" and *stachys* "spike, ear of corn."

Species/Vernacular Names:

S. brachycladum Kurz

The Philippines: buho, kauayang

S. curranii Gamble

The Philippines: kauayan

S. dielsianum (Pilger) Merr.

The Philippines: bikal baboy

S. diffusum (Blanco) Merrill

The Philippines: baliaro, bolikao, bikil, mangnau, bongbong, loob, indi, inri, hindi, babui, butor, lilit, bikal-babi, bikal-babui, gimak, tagisi, usiu

S. fenixii Gamble

The Philippines: bolo, paua, puser

S. glaucifolium (Rupr.) Munro (*Bambusa glaucifolia* Rupr.)

Hawaii: 'ohe

S. lima (Blanco) Merrill

The Philippines: anos, bagakai, bitu, bolo, buho, lakap, naonap, sumibiling

S. lumampao (Blanco) Merrill

The Philippines: lumanpao, bokaui, babakan, boho, bolo, bulo, caña de hoho, daso, napnap, oras, balu, vulu

S. polymorphum (Munro) Majumdar

The Philippines: bayto

S. toppingii Gamble

The Philippines: usiu

Schizostege W.F. Hillebr. Pteridaceae (Adiantaceae)

Origins:
From the Greek *schizo* "to split, divide" and *stege*, *stegos* "roof, shelter."

Schizostegeopsis Copel. Pteridaceae (Adiantaceae)

Origins:
Resembling *Schizostege*.

Schizostephanus Hochst. ex K. Schumann Asclepiadaceae

Origins:
From the Greek *schizo* "divide" and *stephanos* "a crown."

Schizostigma Arn. Cucurbitaceae

Origins:
Greek *schizo* and *stigma* "a stigma."

Schizostigma Arn. ex Meissner Rubiaceae

Origins:
From the Greek *schizo* "divide" and *stigma* "a stigma."

Schizostylis Backh. & Harvey Iridaceae

Origins:
Greek *schizo*, *schizein* "to split, to divide" and *stylis* "a column, style"; the style is divided into three parts.

Species/Vernacular Names:
S. coccinea Backh. & Harvey (*Schizostylis pauciflora* Klatt)
South Africa: scarlet river lily, kaffir lily

Schizotorenia Yamaz. Scrophulariaceae

Origins:
Greek *schizo*, *schizein* "to split" plus the genus *Torenia*.

Schizotrichia Bentham Asteraceae

Origins:
From the Greek *schizo* "to split" plus *thrix*, *trichos* "hair."

Schizozygia Baillon Apocynaceae

Origins:
From the Greek *schizo*, *schizein* "to split, to divide" and *zygon*, *zygos* "yoke."

Schkuhria A.W. Roth Asteraceae

Origins:
After the German botanist Christian Schkuhr, 1741-1811. His writings include *Botanisches Handbuch*. Wittenberg [1787-] 1791-1803 and *Enchiridion botanicum*. Lipsiae [Leipzig] 1805; see Gustav Kunze (1793-1851), *Die Farrnkräuter in kolorirten Abbildungen naturgetreu erläutert und beschrieben ... Schkuhr's Farrnkräuter*, Supplement. Leipzig 1840-47; John H. Barnhart, *Biographical Notes upon Botanists*. 3: 228. 1965; Albrecht Wilhelm Roth (1757-1834), *Catalecta botanica*. 1: 116. Lipsiae 1797; T.W. Bossert, *Biographical Dictionary of Botanists Represented in the Hunt Institute Portrait Collection*. 353. 1972; R. Zander, F. Encke, G. Buchheim and S. Seybold, *Handwörterbuch der Pflanzennamen*. 14. Aufl. Stuttgart 1993; Mariella Azzarello Di Misa, a cura di, *Il Fondo Antico della Biblioteca dell'Orto Botanico di Palermo*. 249. Palermo 1988; Jonas C. Dryander, *Catalogus bibliothecae historico-naturalis Josephi Banks*. London 1800; Ida Kaplan Langman, *A Selected Guide to the Literature on the Flowering Plants of Mexico*. 679. Philadelphia 1964; E.M. Tucker, *Catalogue of the Library of the Arnold Arboretum of Harvard University*. Cambridge, Massachusetts 1917-1933; A. Lasègue, *Musée botanique de Benjamin Delessert*. Paris 1845.

Species/Vernacular Names:
S. pinnata (Lam.) Thell. (*Schkuhria bonariensis* Hook. & Arn.; *Schkuhria isopappa* Bentham)

English: dwarf marigold, dwarf Mexican marigold, khaki bush, yellow tumbleweed

East Africa: onyalobiro

Southern Africa: bitterbossie, bokrambossie, dwerg-goudsblom, hardebossie, kakiebossie, klei-gousblom, kleinkakiebos, kousbossie, rolbossie, rolkakiebossie, waaibossie; letapiso (Sotho)

Schlagintweitiella Ulbr. Ranunculaceae

Origins:

For the German travelers, explorers and plant collectors Adolf von Schlagintweit (1829-1857) and his brother Hermann Alfred Rudolph von Schlagintweit-Sakünlünski (1826-1882); see John H. Barnhart, *Biographical Notes upon Botanists*. 3: 228. 1965; T.W. Bossert, *Biographical Dictionary of Botanists Represented in the Hunt Institute Portrait Collection*. 353. 1972; Emil Bretschneider (1833-1901), *History of European Botanical Discoveries in China*. [Reprint of the original edition 1898.] Leipzig 1981; E.M. Tucker, *Catalogue of the Library of the Arnold Arboretum of Harvard University*. Cambridge, Massachusetts 1917-1933; Stafleu and Cowan, *Taxonomic Literature*. 5: 187-189. 1985.

Schlechtendalia Lessing Asteraceae

Origins:

After the German botanist Diederich Franz Leonhard von Schlechtendal, 1794-1866, plant collector, traveler, 1819 M.D. Berlin, close friend and traveling companion of A. von Chamisso, 1833-1866 professor of botany and Director of the Botanical Garden of Halle University, from 1826 to 1866 editor of *Linnaea*, from 1843 to 1866 co-editor with Hugo von Mohl (1805-1872) of the *Botanische Zeitung*. Among his many works are *Flora berolinensis*. Berolini [Berlin] 1823-1824 and *Florae insulae Sti. Thomae Indiae occidentalis*. 1828-1831; see John H. Barnhart, *Biographical Notes upon Botanists*. 3: 228. 1965; I.H. Vegter, *Index Herbariorum*. Part II (6), *Collectors S*. Regnum Vegetabile vol. 114. 1986; Stafleu and Cowan, *Taxonomic Literature*. 5: 190-204. 1985; T.W. Bossert, *Biographical Dictionary of Botanists Represented in the Hunt Institute Portrait Collection*. 353. 1972; Mariella Azzarello Di Misa, ed., *Il Fondo Antico della Biblioteca dell'Orto Botanico di Palermo*. 249. Regione Siciliana, Palermo 1988; S. Lenley et al., *Catalog of the Manuscript and Archival Collections and Index to the Correspondence of John Torrey*. Library of the New York Botanical Garden. 468. 1973; Ida Kaplan Langman, *A Selected Guide to the Literature on the Flowering Plants of Mexico*. 679-681. Philadelphia 1964; E.M. Tucker, *Catalogue of the Library of the Arnold Arboretum of Harvard University*. Cambridge, Massachusetts 1917-1933; G. Schmid, *Goethe und die Naturwissenschaften*. Halle 1940; Günther Schmid, *Chamisso als Naturforscher. Eine Bibliographie*. Leipzig 1942; Giulio Giorello & Agnese Grieco, a cura di, *Goethe scienziato*. Einaudi Editore, Torino 1998; R. Zander, F. Encke, G. Buchheim and S. Seybold, *Handwörterbuch der Pflanzennamen*. 14. Aufl. 1993; N. Hall, *Botanists of the Eucalypts*. Melbourne 1978 and Supplement 1980; Irving William Knobloch, compil., "A preliminary

verified list of plant collectors in Mexico." *Phytologia Memoirs*. VI. 1983.

Schlechteranthus Schwantes Aizoaceae

Origins:

After the German (b. Berlin) botanist Friedrich Richard Rudolf Schlechter, 1872-1925 (d. Berlin), traveler, plant collector in Africa and orchidologist, student of Engler, assistant to Harry Bolus (1834-1911), 1899-1900 leader of the German West Africa Rubber Expedition, 1921-1925 Curator at Berlin-Dahlem, a specialist in Asclepiadaceae and succulents. His writings include *Die Orchideen; ihre Beschreibung, Kultur und Züchtung*. Berlin 1914-1915, *Beiträge zur Kenntniss südafrikanischer Asclepiadaceen*. *Bot. Jahrb*. 1894-1895 and *Die Guttapercha-und Kautschuk-Expedition*. Berlin 1911, brother of Max Schlechter (1874-1960, d. Cape Town), who collected with him in South Africa (1896-1897); see Ludwig E. Theodor Loesener, *Notizbl. Bot. Gart. Berl*. 9: 912-948. 1926; Anthonius Josephus Maria Leeuwenberg, "Isotypes of which holotypes were destroyed in Berlin." *Webbia*. 19: 861-863. 1965; Réné Letouzey (1918-1989), "Les botanistes au Cameroun." in *Flore du Cameroun*. 7: 1-110. Paris 1968; J.H. Barnhart, *Biographical Notes upon Botanists*. 3: 228. 1965; T.W. Bossert, *Biographical Dictionary of Botanists Represented in the Hunt Institute Portrait Collection*. 353. 1972; Ida Kaplan Langman, *A Selected Guide to the Literature on the Flowering Plants of Mexico*. 681. Philadelphia 1964; Ethelyn Maria Tucker, *Catalogue of the Library of the Arnold Arboretum of Harvard University*. Cambridge, Massachusetts 1917-1933; Mary Gunn and Leslie E. Codd, *Botanical Exploration of Southern Africa*. 313-315. A.A. Balkema Cape Town 1981; Elmer Drew Merrill, *Contr. U.S. Natl. Herb*. 30(1): 267. 1947 and *Bernice P. Bishop Mus. Bull*. 144: 164. 1937; F.N. Hepper and Fiona Neate, *Plant Collectors in West Africa*. 72. 1971; Gordon Douglas Rowley, *A History of Succulent Plants*. Strawberry Press, Mill Valley, California 1997; John Peter Jessop (b. 1939), "Itinerary of Rudolf Schlechter's collecting trips in Southern Africa." *J. S. Afr. Bot*. 30(3): 129-146. 1964; Stafleu and Cowan, *Taxonomic Literature*. 5: 205-213. 1985; R. Zander, F. Encke, G. Buchheim and S. Seybold, *Handwörterbuch der Pflanzennamen*. 14. Aufl. 775. 1993; A. White and B.L. Sloane, *The Stapelieae*. Pasadena 1937; Merle A. Reinikka, *A History of the Orchid*. Timber Press 1996; N. Hall, *Botanists of the Eucalypts*. Melbourne 1978 and Supplement 1980.

Schlechterella Hoehne Orchidaceae

Origins:

After the German botanist Friedrich Richard Rudolf Schlechter, 1872-1925, see also *Rudolfiella* Hoehne.

Schlechterella K. Schumann Asclepiadaceae

Origins:

After the German botanist Friedrich Richard Rudolf Schlechter, 1872-1925, traveler, plant collector and orchidologist, in 1899-1900 leader of the German West Africa Rubber Expedition, from 1921 to 1925 Curator at Berlin-Dahlem Botanical Garden. His writings include *Beiträge zur Kenntniss südafrikanischer Asclepiadaceen. Bot. Jahrb.* 1894-1895, "Contributions to South African Asclepiadology." *J. Bot.* 1894-1895 and "Revision of extra-tropical South-African Asclepiadaceae." *J. Bot.* 1898.

Schlechteria Bolus Brassicaceae

Origins:

After the German botanist Friedrich Richard Rudolf Schlechter, 1872-1925, traveler, plant collector and orchidologist, assistant to Harry Bolus (1834-1911), in 1899-1900 leader of the German West Africa Rubber Expedition. His writings include *Beiträge zur Kenntniss südafrikanischer Asclepiadaceen. Bot. Jahrb.* 1894-1895, "Contributions to South African Asclepiadology." *J. Bot.* 1894-1895 and "Revision of extra-tropical South-African Asclepiadaceae." *J. Bot.* 1898; see J.H. Barnhart, *Biographical Notes upon Botanists.* 3: 228. 1965.

Schlechterianthus Quisumb. Asclepiadaceae

Origins:

Dedicated to the German botanist Friedrich Richard Rudolf Schlechter, 1872-1925, traveler, plant collector and orchidologist, student of Engler, assistant to Harry Bolus (1834-1911), in 1899-1900 leader of the German West Africa Rubber Expedition, from 1921 to 1925 Curator at Berlin-Dahlem. His writings include *Die Orchideen; ihre Beschreibung, Kultur und Züchtung.* Berlin 1914-1915, *West Afrikanische Kautschuk-Expedition.* Berlin 1910 and *Die Guttapercha-und Kautschuk-Expedition.* Berlin 1911. See J.H. Barnhart, *Biographical Notes upon Botanists.* 3: 228. 1965; Ludwig E. Theodor Loesener, *Notizbl. Bot. Gart. Berl.* 9: 912-948. 1926; Anthonius Josephus Maria Leeuwenberg, "Isotypes of which holotypes were destroyed in Berlin." *Webbia.* 19: 861-863. 1965; Réné Letouzey (1918-1989), "Les botanistes au Cameroun." in *Flore du Cameroun.* 7: 1-110. Paris 1968.

Schlechterina Harms Passifloraceae

Origins:

Dedicated to the German botanist Friedrich Richard Rudolf Schlechter, 1872-1925, traveler, plant collector and

orchidologist, student of Engler, assistant to Harry Bolus (1834-1911), in 1899-1900 leader of the German West Africa Rubber Expedition, from 1921 to 1925 Curator at Berlin-Dahlem. His writings include *Die Orchideen; ihre Beschreibung, Kultur und Züchtung.* Berlin 1914-1915, *West Afrikanische Kautschuk-Expedition.* Berlin 1910 and *Die Guttapercha-und Kautschuk-Expedition.* Berlin 1911; see J.H. Barnhart, *Biographical Notes upon Botanists.* 3: 228. 1965.

Schlechterosciadium H. Wolff Umbelliferae

Origins:

Dedicated to the German botanist Friedrich Richard Rudolf Schlechter, 1872-1925, traveler, plant collector and orchidologist, student of Engler, assistant to Harry Bolus (1834-1911); see J.H. Barnhart, *Biographical Notes upon Botanists.* 3: 228. 1965.

Schleichera Willdenow Sapindaceae

Origins:

In honor of the German botanist Johann Christoph Schleicher, 1768-1834, author of *Catalogus plantarum in Helvetia cis- et transalpina sponte nascentium.* Bex [1800]; see Carl (Karl) L. von Willdenow (1765-1812), *Species Plantarum.* Ed. 4, (4)2: 1096. (Apr.) 1806; John H. Barnhart, *Biographical Notes upon Botanists.* 3: 229. 1965; Jonas C. Dryander, *Catalogus bibliothecae historico-naturalis Josephi Banks.* London 1800; E.M. Tucker, *Catalogue of the Library of the Arnold Arboretum of Harvard University.* 1917-1933; R. Zander, F. Encke, G. Buchheim and S. Seybold, *Handwörterbuch der Pflanzennamen.* 14. Aufl. Stuttgart 1993; H.N. Clokie, *Account of the Herbaria of the Department of Botany in the University of Oxford.* 239. 1964.

Species/Vernacular Names:

S. oleosa (Lour.) Oken (*Schleichera trijuga* Willd.)

English: Ceylon oak, honey tree, lac tree

Schleinitzia Warburg ex Nevling & Niezgoda Mimosaceae

Origins:

For Georg von Schleinitz, the first administrator of German New Guinea; see *Adansonia.* ser. 2. 18: 356. 28 Dec. 1978.

Schliebenia Mildbr. Acanthaceae

Origins:

After the German (b. Saxony) botanist Hans-Joachim Eberhardt Schlieben, 1902-1975 (d. Essen, Germany), plant collector (in Tanzania), photographer and writer; see John H. Barnhart, *Biographical Notes upon Botanists*. 3: 229. 1965; Mary Gunn and Leslie E. Codd, *Botanical Exploration of Southern Africa*. 316. Cape Town 1981; Anthonius Josephus Maria Leeuwenberg, "Isotypes of which holotypes were destroyed in Berlin." in *Webbia*. 19: 861-863. 1965.

Schlimmia Planchon & Linden Orchidaceae

Origins:

Dedicated to the Belgian Louis Joseph Schlim, plant collector in South and Central America, 1841-1844 (with Jean Jules Linden and Nicolas Funk, 1816-1896) and 1845 (with N. Funk) collected in Venezuela. He was half-brother of J.J. Linden (1817-1898); see A. Lasègue, *Musée botanique de Benjamin Delessert*. 414. Paris 1845; Henri François Pittier, *Manual de las Plantas Usuales de Venezuela y su Suplemento*. Caracas 1971; H.N. Clokie, *Account of the Herbaria of the Department of Botany in the University of Oxford*. 240. Oxford 1964.

Schlumbergera Lem. Cactaceae

Origins:

Named for the horticulturist Frédéric Schlumberger, 1823-1893; see Helmut Genaust, *Etymologisches Wörterbuch der botanischen Pflanzennamen*. 566. [for the Belgian gardener Frédéric Schlumberger, 1804-1865] Basel 1996; Gordon Douglas Rowley, *A History of Succulent Plants*. Strawberry Press, California 1997.

Species/Vernacular Names:

S. x buckleyi (T. Moore) Tjaden (*Schlumbergera bridgesii* (Lem.) Loefgr.)

English: Christmas cactus

S. russelliana (Hook.) Britton & Rose (*Cereus russelliana* Gardn.)

Japan: shako-saboten

S. truncata (Haw.) Moran (*Epiphyllum truncatum* Haw.; *Zygocactus truncatus* (Haw.) K. Schum.; *Zygocactus delicatus* N.E. Br.)

English: crab cactus, claw cactus, yoke cactus, Thanksgiving cactus, linkleaf, Christmas cactus

Japan: kani-saboten

Schmalhausenia C. Winkler Asteraceae

Origins:

After the Russian botanist Johannes Theodor Schmalhausen, 1849-1894; see J.H. Barnhart, *Biographical Notes upon Botanists*. 3: 230. 1965; T.W. Bossert, *Biographical Dictionary of Botanists Represented in the Hunt Institute Portrait Collection*. 353. 1972; E.M. Tucker, *Catalogue of the Library of the Arnold Arboretum of Harvard University*. Cambridge, Massachusetts 1917-1933; R. Zander, F. Encke, G. Buchheim and S. Seybold, *Handwörterbuch der Pflanzennamen*. 14. Aufl. Stuttgart 1993; Stafleu and Cowan, *Taxonomic Literature*. 5: 230-231. 1985.

Schmaltzia Desvaux ex Small Anacardiaceae

Origins:

In honor of the great botanist Constantine (Constantin) Samuel Rafinesque (-Schmaltz), 1783-1840 (Philadelphia), economist, naturalist, traveler, botanical explorer, conchologist, archaeologist, spent many years of his life in Sicily (1804-1815) and in USA, a prolific writer. Among his works are *Caratteri di alcuni nuovi generi e nuove specie di animali e piante della Sicilia*, con varie osservazioni sopra i medesimi. Palermo 1810 and *Specchio delle scienze o giornale enciclopedico della Sicilia*, deposito letterario delle moderne cognizioni, etc. Palermo 1814; see J.H. Barnhart, *Biographical Notes upon Botanists*. 3: 123. 1965; T.W. Bossert, *Biographical Dictionary of Botanists Represented in the Hunt Institute Portrait Collection*. 322. 1972; Garrison and Morton, *Medical Bibliography*. 1849. 1961; Joseph Ewan, in *Dictionary of Scientific Biography* 11: 262-264. 1981; Edward Lee Greene, *Landmarks of Botanical History*. Edited by Frank N. Egerton. 74-75. Stanford University Press, Stanford, California 1983; H.N. Clokie, *Account of the Herbaria of the Department of Botany in the University of Oxford*. 229. Oxford 1964; E.M. Tucker, *Catalogue of the Library of the Arnold Arboretum of Harvard University*. 1917-1933; E.D. Merrill (1876-1956), *Index rafinesquianus*. The plant names published by C.S. Rafinesque, etc. Jamaica Plain, Massachusetts, USA 1949; Mariella Azzarello Di Misa, a cura di, *Il Fondo Antico della Biblioteca dell'Orto Botanico di Palermo*. 224. Palermo 1988.

Schmidelia L. Sapindaceae

Origins:

After the German (b. Bayreuth) physician Casimir (Kasimir) Christoph (Casimirus Christophorus) Schmidel (Schmiedel), 1718-1792 (d. Ansbach), naturalist, received M.D. at Jena 1742, from 1743 to 1763 professor of medicine and pharmacology at the University of Bayreuth/Erlangen, editor of Conrad (Konrad) von Gesner (1516-1565), *Opera botanica*. Norimbergae 1751-1771. Among his many essays are *Icones plantarum* ... curante et edente Georgio Wolfgang Knorrio ... [Nürnberg] 1747 [-1751], Dissertatio inauguralis botanica de *Buxbaumia*. Erlangae [Erlangen] [1758] and *Descriptio itineris per Helvetiam Galliam et Germaniae* partem ann. [1773 et 1774] instituti ... Erlangae 1794. See A. Geus, in *Dictionary of Scientific Biography* 12: 185-187. 1981; C. Linnaeus, *Systema Naturae*. 2: 274. 1767 and *Mantissa Plantarum*. 1: 10, 67. 1767; John H. Barnhart, *Biographical Notes upon Botanists*. 3: 231. 1965; Mariella Azzarello Di Misa, a cura di, *Il Fondo Antico della Biblioteca dell'Orto Botanico di Palermo*. 250. Palermo 1988; Jonas C. Dryander, *Catalogus bibliothecae historico-naturalis Josephi Banks*. London 1800; A. Lasègue, *Musée botanique de Benjamin Delessert*. 1845; E.M. Tucker, *Catalogue of the Library of the Arnold Arboretum of Harvard University*. Cambridge, Massachusetts 1917-1933; R. Zander, F. Encke, G. Buchheim and S. Seybold, *Handwörterbuch der Pflanzennamen*. 14. Aufl. Stuttgart 1993.

Schmidtia Steudel ex J.A. Schmidt Gramineae

Origins:

In honor of the German botanist Johann Anton Schmidt, 1823-1905, professor of botany, traveler, plant collector, 1851 Capo Verde Islands, author of *Beiträge zur Flora der Cap Verdischen Inseln*. Heidelberg 1852 and *Flora von Heidelberg*. Heidelberg 1857, and contributor to C.F.P. von Martius, *Flora Brasiliensis* (Labiatae, Scrophulariaceae, Phytolaccaceae, Nyctaginaceae, Plumbaginaceae, Plantaginaceae); see David Porter (1780-1843), *Journal of a Cruise Made to the Pacific Ocean ... in the United States Frigate Essex, in the Years 1812, 1813, and 1814*. Containing descriptions of the Cape Verd [sic] Islands, etc. Philadelphia 1815; John H. Barnhart, *Biographical Notes upon Botanists*. 3: 232. 1965; R. Zander, F. Encke, G. Buchheim and S. Seybold, *Handwörterbuch der Pflanzennamen*. 14. Aufl. Stuttgart 1993; T.W. Bossert, *Biographical Dictionary of Botanists Represented in the Hunt Institute Portrait Collection*. 353. 1972; Ethelyn Maria Tucker, *Catalogue of the Library of the Arnold Arboretum of Harvard University*. Cambridge, Massachusetts 1917-1933.

Species/Vernacular Names:

S. pappophoroides Steud. (*Schmidtia bulbosa* Stapf)

English: sand quick grass

South Africa: blougras, suurgras, kopgras, lidjiesgras, vaalgras, vaalkrulgras, krulgras, blouplatsaadjie, Kalahari sandkweek, sandkweek, kalkweek

Schmidtia Tratt. Gramineae

Origins:

Named in honor of the German botanist Franz Wilibald Schmidt, 1764-1796, professor of botany, physician; see John H. Barnhart, *Biographical Notes upon Botanists*. 3: 231. 1965; Jonas C. Dryander, *Catalogus bibliothecae historico-naturalis Josephi Banks*. London 1800; E.M. Tucker, *Catalogue of the Library of the Arnold Arboretum of Harvard University*. Cambridge, Massachusetts 1917-1933; R. Zander, F. Encke, G. Buchheim and S. Seybold, *Handwörterbuch der Pflanzennamen*. 14. Aufl. Stuttgart 1993.

Schmidtottia Urban Rubiaceae

Origins:

After the German botanist Otto Christian Schmidt, 1900-1951, professor of pharmacognosy (Universities of Berlin and Münster), author of *Die marine Vegetation der Azoren*. Stuttgart 1931; see John H. Barnhart, *Biographical Notes upon Botanists*. 1965; Elmer Drew Merrill, *Contr. U.S. Natl. Herb*. 30(1): 268. 1947; Stafleu and Cowan, *Taxonomic Literature*. 5: 259-260. 1985.

Schmiedtia Raf. Gramineae

Origins:

For the German botanist Franz Wilibald Schmidt, 1764-1796, see C.S. Rafinesque, *Autikon botanikon*. Icones plantarum select. nov. vel rariorum, etc. 187. Philadelphia 1840; E.D. Merrill, *Index rafinesquianus*. 76. 1949.

Schnittspahnia Reichb. Annonaceae

Origins:

For the German botanist Georg Friedrich Schnittspahn, 1810-1865; see John H. Barnhart, *Biographical Notes upon Botanists*. 3: 236. 1965; Ida Kaplan Langman, *A Selected Guide to the Literature on the Flowering Plants of Mexico*. Philadelphia 1964; E.M. Tucker, *Catalogue of the Library of the Arnold Arboretum of Harvard University*. Cambridge, Massachusetts 1917-1933; R. Zander, F. Encke, G. Buchheim and S. Seybold, *Handwörterbuch der Pflanzennamen*. 14. Aufl. 1993.

Schnizleinia Steudel Gramineae

Origins:

See Stafleu and Cowan, *Taxonomic Literature*. 5: 279-284. 1985.

Schoberia C. Meyer Chenopodiaceae

Origins:

After Gottlieb (Gottlob) Schober, c. 1675-1739, physician and hydrotherapist in Northern Germany, St. Petersburg and Moscow, (from 1717) on a commission from Peter the Great (1672-1725) travelled along the Volga, the Caspian Sea and Iran, wrote *Disputatio ... de cholera.* Trajecti ad Rhenum [Utrecht] 1696; see Carl (Karl) Friedrich von Ledebour (1785-1851), *Icones Plantarum.* 1: 11. (May-Dec.) 1829.

Schoenanthus Adans. Gramineae

Origins:

Greek *schoinos* "a rush, reed, cord" and *anthos* "flower," *schoinanthe* "flower of *schoinos*."

Schoenefeldia Kunth Gramineae

Origins:

For the German botanist Wladimir de Schoenefeld (Schoenfeld), 1816-1875, a student of Adr. de Jussieu, one of the founders of the Société Botanique de France, wrote "Rapport sur une excursion faite en août 1860 ... au Bourg-d'Oisans." *Bull. Soc. Bot. France.* 7(8): 804-815. Paris 1863; see J.H. Barnhart, *Biographical Notes upon Botanists.* 3: 236. 1965; E.M. Tucker, *Catalogue of the Library of the Arnold Arboretum of Harvard University.* Cambridge, Massachusetts 1917-1933; Stafleu and Cowan, *Taxonomic Literature.* 5: 286-287. 1985.

Schoenia Steetz Asteraceae

Origins:

For the German physician Johann(es) Matthias Albrecht Schoen (pseud. J. Krohn), ophthalmologist, botanical illustrator and a friend of Joachim Steetz (1804-1862), wrote *Beiträge zur praktischen Augenheilkunde.* Hamburg 1861, and *Neendörp Plattdüütsche Rymels.* Hamburg 1856; see Johann G.C. Lehmann (1792-1860), *Plantae Preissianae ... Plantarum quas in Australasia occidentali et meridionali-occidentali annis 1838-41 collegit L. Preiss.* 1: 480. Hamburgi (Aug.) 1845.

Schoenlandia Cornu Cyanastraceae (Tecophilaeaceae, Liliaceae)

Origins:

For the German botanist Selmar Schönland (Schonland, Schoenland), 1860-1940 (d. Grahamstown, C.P.), professor of botany, traveler, plant collector, in South Africa, pupil of Engler, interested in the Crassulaceae, 1890 Fellow of the Linnean Society; see J.H. Barnhart, *Biographical Notes upon Botanists.* 3: 237. 1965; T.W. Bossert, *Biographical Dictionary of Botanists Represented in the Hunt Institute Portrait Collection.* 354. 1972; Ethelyn Maria Tucker, *Catalogue of the Library of the Arnold Arboretum of Harvard University.* Cambridge, Massachusetts 1917-1933; Mary Gunn and Leslie E. Codd, *Botanical Exploration of Southern Africa.* 318. Cape Town 1981; A. White and B.L. Sloane, *The Stapelieae.* Pasadena 1937; Botanical Survey of South Africa, *A Guide to Botanical Survey Work.* [S. Schonland, *Hints on the Photographing of Plants.*] Pretoria 1922; Gordon Douglas Rowley, *A History of Succulent Plants.* Strawberry Press, Mill Valley, California 1997; R. Zander, F. Encke, G. Buchheim and S. Seybold, *Handwörterbuch der Pflanzennamen.* 14. Aufl. Stuttgart 1993.

Schoenleinia Klotzsch Rubiaceae

Origins:

For the German (b. Bamberg) botanist Johann Lucas Schönlein, 1793-1864 (d. Bamberg), physician, professor of medicine, father of the German botanist, traveler and plant collector in West Africa Philipp Schönlein (1834-1856); see Gloria Robinson, in *Dictionary of Scientific Biography* 12: 202-203. 1981; J.H. Barnhart, *Biographical Notes upon Botanists.* 3: 237. 1965; T.W. Bossert, *Biographical Dictionary of Botanists Represented in the Hunt Institute Portrait Collection.* 355. 1972; Garrison and Morton, *Medical Bibliography.* 3058, 4029. New York 1961; F.N. Hepper and Fiona Neate, *Plant Collectors in West Africa.* 73. 1971; Joseph Vallot, "Études sur la flore du Sénégal." in *Bull. Soc. Bot. de France.* 29: 168-238. Paris 1882.

Schoenobiblus Mart. Thymelaeaceae

Origins:

Greek *schoinos* "rush, reed, cord" and *biblos, byblos* "bark, rind," *biblion, byblion* "strip of *byblos*."

Schoenocaulon A. Gray Melanthiaceae (Liliaceae)

Origins:

Greek *schoinos* and *kaulos* "stem, stalk."

Species/Vernacular Names:

S. officinale A. Gray

Mexico: sabadilla, cevadilla

Schoenocephalium Seub. Rapateaceae

Origins:

Greek *schoinos* and *kephale* "head."

Schoenocrambe E. Greene Brassicaceae

Origins:

Greek *schoinos* "rush, reed, cord" plus the genus *Crambe* L., *krambe* "cabbage."

Schoenodendron Engl. Cyperaceae

Origins:

From the genus *Schoenus* L. plus the Greek *dendron* "tree."

Schoenodorus Roemer & Schultes Gramineae

Origins:

From the Greek *schoinos* "rush, reed, cord" plus *doron* "a gift."

Schoenodum Labill. Restionaceae

Origins:

Resembling the genus *Schoenus* L.; see Jacques J. Houtton de Labillardière (1755-1834), *Novae Hollandiae plantarum specimen*. 2: 79, t. 29. Parisiis (Aug.) 1806.

Schoenoides Seberg Cyperaceae

Origins:

Resembling *Schoenus* L.; see Ole Seberg (1952-), in *Willdenowia*. 16: 181. (Aug.) 1986.

Schoenolaena Bunge Umbelliferae

Origins:

From the Greek *schoinos* and *chlaena* "a cloak, blanket," referring to the involucral bracts; see Johann G.C. Lehmann, *Plantae Preissianae*. 1: 289. Hamburgi (Feb.) 1845.

Schoenolirion Torrey Hyacinthaceae (Liliaceae)

Origins:

From the Greek *schoinos* and *leirion* "a lily."

Schoenomorphus auctt. Orchidaceae

Origins:

Greek *schoinos* "rush, reed, cord" and *morphe* "a form, shape."

Schoenoplectus (Reichb.) Palla Cyperaceae

Origins:

Greek *schoinos* and *plektos* "twisted, plaited"; Latin *schoenus, i* "a rush"; Akkadian *qannu*, Hebrew *qane*, Arabian *qana* "reed, cane, sweet cane, calamus, measuring rod"; see Eduard Palla (1864-1922), in *Verhandlungen der Kaiserlich-Königlichen Zoologisch-Botanischen Gesellschaft in Wien*. Vienna (Mar.) 1888.

Species/Vernacular Names:

S. sp.

Hawaii: mau'u, kaluhaluha

S. californicus (C.A. Meyer) Palla (*Scirpus californicus* (C.A. Meyer) Steud.)

Hawaii: kaluha, 'aka'akai

S. juncoides (Roxb.) Palla (*Scirpus juncoides* Roxb.)

Hawaii: kaluha

S. lacustris (L.) Palla (*Scirpus lacustris* L.)

English: great bulrush, clubrush, bulrush, lake scirpus

Hawaii: kaluha, 'aka'akai, 'aka'akai naku, naku, nanaku, neki

Maori name: wawa

S. litoralis (Schrader) Palla (*Scirpus litoralis* Schrad.)

English: sedge

Arabic: khabb

South Africa: biesie, papgras, steekbiesie, steekrietjie

S. pungens (Vahl) Palla (*Scirpus pungens* Vahl)

English: American clubrush, sharp-leaved rush

Schoenorchis Blume Orchidaceae

Origins:

Greek *schoinos* and *orchis* "orchid, testicle," referring to the rush-like leaves; see K.L. von Blume, *Bijdragen tot de flora van Nederlandsch Indië*. 361, t. 3, fig. 23. (Sep.-Dec.) 1825.

Schoenoxiphium Nees Cyperaceae

Origins:

Greek *schoinos* "rush, reed, cord" and *xiphos* "a sword"; Latin *xiphias, ae* "sword-shaped," *xiphion, ii* "sword-flag, gladiole."

Species/Vernacular Names:

S. lehmannii (Nees) Steud. (*Carex uhligii* K. Schum. ex C.B. Cl.; *Kobresia lehmannii* (Nees) Koyama var. *lehmannii*; *Kobresia lehmannii* (Nees) Koyama var. *schimperiana* (Boeck.) Koyama)

South Africa: watergras

Schoenus L. Cyperaceae

Origins:

Greek *schoinos* "rush, reed, cord," Latin *schoenus, i* for a rush; see Carl Linnaeus, *Species Plantarum*. 42. 1753 and *Genera Plantarum*. Ed. 5. 26. 1754.

Species/Vernacular Names:

S. apogon Roemer & Schultes (*Chaetospora imberbis* R. Br.; *Schoenus brownii* Hook. f.) (Greek *a* "without" and *pogon* "beard")

English: common bog-rush, fluke bog-rush

Japan: hige-kusa, no-gusa (= field grass)

S. breviculmis Benth.

English: matted bog-rush

S. brevifolius R. Br.

English: zigzag bog-rush, short-leaf bog-rush

S. calyptratus Kük.

English: alpine bog-rush

S. falcatus R. Br.

English: sickle bog-rush

S. fluitans R. Br.

English: floating bog-rush

S. imberbis R. Br.

English: beardless bog-rush

S. kennyi (Bailey) S.T. Blake

English: Kenny's bog-rush

S. latelaminatus Kük.

English: Medusa bog-rush, Gimlet bog-rush

S. maschalinus Roemer & Schultes

English: leafy bog-rush, dwarf bog-rush

S. melanostachys R. Br.

English: black bog-rush

S. nanus (Lehm.) Benth.

English: tiny bog-rush

S. nigricans L.

English: black bog-rush, black sedge

S. nitens (R. Br.) Roemer & Schultes

English: shiny bog-rush

S. ornithopodioides (Kük.) S.T. Blake

English: bird's foot bog-rush

S. pygmaeus S.T. Blake

English: pigmy bog-rush

S. racemosus J. Black

English: branched bog-rush

S. sculptus (Nees) Boeckeler

English: Gimlet foot bog-rush

S. sparteus R. Br.

English: broom bog-rush

S. subaphyllus Kük.

English: desert bog-rush, leafless bog-rush

Schoepfia Schreber Olacaceae (Schoepfiaceae)

Origins:

Probably named for the German botanist Johann David Schoepf (Schöpf, Schoepff), 1752-1800, naturalist and physician, M.D. Erlangen 1776, from 1777 to 1783 army physician (as a field surgeon to the Hessian mercenaries) in the English Army during the American War of Liberation, traveler, in 1795 and 1797 President of the medical councils of Ansbach and Bayreuth. His writings include *Materia medica americana potissimum regni vegetabilis*. [8vo, first edition.] Erlangae [Erlangen] 1787, *Beyträge zur mineralogischen Kenntniss des östlichen Theils von Nordamerica und seiner Gebürge*. Erlangen 1787, *Reise durch einige der mittlern und südlichen Vereinigten nordamerikanischen Staaten nach Ost-Florida und den Bahama-Inseln unternommen in den Jahren 1783 und 1784 ...* Erlangen 1788 and Joannis Davidis Schoepff *Historia Testudinum iconibus illustrata*. Erlangae 1792 [-1802]; see John H. Barnhart, *Biographical Notes upon Botanists*. 3: 237. 1965; Garrison and Morton, *Medical Bibliography*. 1837. New York 1961; Jonas C. Dryander, *Catalogus bibliothecae historico-naturalis Josephi Banks*. London 1800; Ida Kaplan Langman, *A Selected Guide to the Literature on the Flowering Plants of Mexico*. 683. Philadelphia 1964; Ethelyn Maria Tucker, *Catalogue of the Library of the Arnold Arboretum of Harvard University*. Cambridge, Massachusetts 1917-1933; A. Lasègue, *Musée botanique de Benjamin Delessert*. 216-219. 1845.

Species/Vernacular Names:

S. jasminodora Siebold & Zucc.

Japan: boroboro-no-ki

Schoepfiopsis Miers Olacaceae

Origins:

Resembling *Schoepfia*, named for the German botanist Johann David Schoepf, 1752-1800.

Scholleropsis H. Perrier Pontederiaceae

Origins:

For the German botanist Friedrich Adam Scholler, 1718-1795; see John H. Barnhart, *Biographical Notes upon Botanists*. 3: 237. 1965; Jonas C. Dryander, *Catalogus bibliothecae historico-naturalis Josephi Banks*. London 1800; E.M. Tucker, *Catalogue of the Library of the Arnold Arboretum of Harvard University*. Cambridge, Massachusetts 1917-1933; R. Zander, F. Encke, G. Buchheim and S. Seybold, *Handwörterbuch der Pflanzennamen*. 14. Aufl. Stuttgart 1993.

Scholtzia Schauer Myrtaceae

Origins:

After the German botanist Johann Eduard Heinrich Scholtz, 1812-1859, zoologist, physician, M.D. Breslau 1836, wrote *Enumeratio filicum in Silesia sponte crescentium*. Vratislaviae [1836] and *Flora der Umgegend von Breslau*. Breslau 1843; see John H. Barnhart, *Biographical Notes upon Botanists*. 3: 237. 1965; J.C. Schauer, in *Linnaea*. 17: 241. 1843.

Schomburgkia Lindley Orchidaceae

Origins:

For the German (b. Freiburg) botanist Sir Robert Hermann Schomburgk, 1804-1865 (d. Berlin), traveler, explorer, in British Guiana (with his brother Richard), plant collector in Siam and Venezuela, 1844 knighted, 1859 Fellow of the Royal Society, British Consul in Santo Domingo and Bangkok, sent plants to Reichenbach. His writings include *A Description of British Guiana*. London 1840, *Reisen in Guayana u. am Orinoko*. Leipzig 1845, *The Natural History of the Fishes of Guiana*. 1841-1843 and *The History of Barbados*. London 1848, brother of the German botanist and explorer Moritz Richard Schomburgk (1811-1891); see A. Lasègue, *Musée botanique de Benjamin Delessert*. 216-219. 1845; R. Zander, F. Encke, G. Buchheim and S. Seybold, *Handwörterbuch der Pflanzennamen*. 14. Aufl. 776. 1993; J.H. Barnhart, *Biographical Notes upon Botanists*. 3: 238. 1965; T.W. Bossert, *Biographical Dictionary of Botanists Represented in the Hunt Institute Portrait Collection*. 354. 1972; S. Lenley et al., *Catalog of the Manuscript and Archival Collections and Index to the Correspondence of John Torrey*. Library of the New York Botanical Garden. 361. 1973; Ida Kaplan Langman, *A Selected Guide to the Literature on the Flowering Plants of Mexico*. 1964; E.M. Tucker, *Catalogue of the Library of the Arnold Arboretum of Harvard University*. Cambridge, Massachusetts 1917-1933; J.W. Harshberger, *The Botanists of Philadelphia and Their Work*. 190. 1899; Ray Desmond, *Dictionary of British & Irish Botanists and Horticulturists*. 612. 1994; Georg Christian Wittstein, *Etymologisch-botanisches Handwörterbuch*. 796. 1852; G. Murray, *History of the Collections Contained in the Natural History Departments of the British Museum*. 1: 181. 1904; Henri Pittier, *Manual de las Plantas Usuales de Venezuela* y su Suplemento. Caracas 1978; J.D. Milner, *Catalogue of Portraits of Botanists Exhibited in the Museums of the Royal Botanic Gardens*. Royal Botanic Gardens, Kew, London 1906; D.J. Carr and S.G.M. Carr, eds., *People and Plants in Australia*. 1981; Stafleu and Cowan, *Taxonomic Literature*. 5: 295-301. 1985.

Schonlandia L. Bolus Aizoaceae

Origins:

For the German botanist Selmar Schönland (Schonland), 1860-1940.

Schotia Jacq. Caesalpiniaceae

Origins:

After the Dutch traveler and naturalist Richard van der Schot, c. 1730-1819, born in Holland of French parents, head gardener at the palace of Schönbrunn near Vienna, pupil of Jussieu, friend and traveling companion (between 1754-1756) of the Dutch-born Austrian botanist Nicolaus (Nicolaas or Nikolaus) Joseph von Jacquin (1727-1817).

Species/Vernacular Names:

S. afra (L.) Thunb. var. *afra* (*Guaiacum afrum* L.; *Schotia speciosa* Jacq.; *Schotia speciosa* Jacq. var. *tamarindifolia* (Afzel. ex Sims) Harv.; *Schotia parvifolia* Jacq.; *Schotia tamarindifolia* Afzel. ex Sims; *Theodora speciosa* Medik.)

English: African schotia

Southern Africa: karooboerboon, karroo boer-bean; umQongci (Xhosa)

S. afra (L.) Thunb. var. *angustifolia* (E. Meyer) Harvey (*Schotia angustifolia* E. Meyer)

English: Hottentots bean

Southern Africa: karoo boerboon; umGxam (Xhosa)

S. brachypetala Sonder (*Schotia brachypetala* Sond. var. *pubescens* Burtt Davy; *Schotia rogersii* Burtt Davy; *Schotia semireducta* Merxm.)

English: fuchsia-tree, tree-fuchsia, African walnut

Southern Africa: huilboerboon, weeping boer-bean, weeping boerboon; muSamba, muTondochuru, muTondozwi (Shona); umGxamu, uVovovo, iHluze (Zulu); umGxam, iShimnumyane (Xhosa); uVovovo (Swazi); n'wavilombe (eastern Transvaal); molope (North Sotho: north and north-east Transvaal); mununzwa (Venda)

S. capitata Bolle (*Schotia transvaalensis* Rolfe; *Theodora capitata* (Bolle) Taub.)

Southern Africa: Transvaal boerboon, kleinboerboon, dwarf boerboon, dwarf boer-bean; uVovovwana, isiNcasha, umGxamu (Zulu)

S. latifolia Jacq. (*Omphalobium schotia* Jacq.f. ex DC.; *Schotia cuneifolia* Gand.; *Schotia diversifolia* Walp.; *Theodora stipulata* Ait.f.) Eckl. & Zeyh.)

Southern Africa: bosboerboon, bush boer-bean, forest boerboon; umGxam, umXamo (Xhosa); umGxamu (Zulu)

S. romii De Wild.

Zaire: banke, lutepa

Schousboea Schumacher & Thonn. Euphorbiaceae

Origins:

For the Danish botanist Peder Kofod Anker Schousboe, 1766-1832, traveler, plant collector, Consul at Tangier, collected algae; see John H. Barnhart, *Biographical Notes upon Botanists*. 3: 239. 1965; A. Lasègue, *Musée botanique de Benjamin Delessert*. 1845; E.M. Tucker, *Catalogue of the Library of the Arnold Arboretum of Harvard University.* Cambridge, Massachusetts 1917-1933; R. Zander, F. Encke, G. Buchheim and S. Seybold, *Handwörterbuch der Pflanzennamen*. 14. Aufl. Stuttgart 1993; Carl Frederik Albert Christensen, *Den danske Botaniks Historie med tilhørende Bibliografi*. Copenhagen 1924-1926.

Schousboea Willdenow Combretaceae

Origins:

For the Danish botanist Peder Kofod Anker Schousboe, 1766-1832.

Schoutenia Korth. Tiliaceae

Origins:

Dedicated to the Dutch navigator Willem Corneliszoon Schouten, 1567?-1625 (Antongil Bay, Madagascar), explorer, in 1615-1616 captained the ship *Eendracht* (Unity, or Concord), discovered the Cape Horn route (Dutch: Kaap Hoorn) and islands in the Tuamotu archipelago, discovered the Drake Passage (through the Le Maire Strait between Tierra Del Fuego and Staten Island and into the Pacific), around the southern tip of South America, connecting the Atlantic Ocean with the Pacific, in October 1616 reached Batavia and Java, wrote *Journal ou Description du Merveilleux Voyage de Guillaume Schouten*. Amsterdam 1618. See W.A. Engelbrecht and P.J. Van Herweden, *De Ontdekkingsreis van Jacob Le Maire en Willem Cornelisz Schouten in de jaren 1615-1617*. The Hague 1945; Alexander Dalrymple (1737-1808), *An Account of the Discoveries made in the South Pacifick Ocean previous to 1764*. London 1767 and *An Historical Collection of the Several Voyages and Discoveries in the South Pacific Ocean*. London 1770-1771; John Callander, *Terra Australis Cognita*; or, Voyages to the Terra Australis, or Southern Hemisphere, during the sixteenth, seventeenth, and eighteenth centuries. [A translation, with additions, of the *Histoire des navigations aux terres australes* of Charles de Brosses, Count de Tournay. Paris 1756, the most important general histories of early voyages to the Pacific, from 1502 through 1747] Edinburgh 1766-1768; R. Major, *Early Voyages to Terra Australis*. London 1859; Pieter Willem Korthals (1807-1892), in *Nederlandsch kruidkundig Archief*. 1: 312. Amsterdam 1848; John M. Cooper, *Analytical and Critical Bibliography of the Tribes of Tierra del Fuego and Adjacent Territory*. Bureau of American Ethnology (Smithsonian) Bulletins [relating to the native people of North and South America]. Bul. #63. 1917.

Schouwia DC. Brassicaceae

Origins:

For the Danish (b. Copenhagen) botanist Joakim (Joachim) Frederik Schouw, 1789-1852 (d. Copenhagen), traveler, professor of botany; see A.P.M. Sanders, in *Dictionary of Scientific Biography* 12: 214-215. 1981; J.H. Barnhart, *Biographical Notes upon Botanists*. 3: 239. 1965; T.W. Bossert, *Biographical Dictionary of Botanists Represented in the Hunt Institute Portrait Collection*. 355. 1972; E.M. Tucker, *Catalogue of the Library of the Arnold Arboretum of Harvard University*. Cambridge, Massachusetts 1917-1933; Ida Kaplan Langman, *A Selected Guide to the Literature on the Flowering Plants of Mexico*. 684. University of Pennsylvania Press, Philadelphia 1964; A. Lasègue, *Musée botanique de Benjamin Delessert*. 1845; Carl Frederik Albert Christensen, *Den danske Botaniks Historie med tilhørende Bibliografi*. Copenhagen 1924-1926.

Schradera Vahl Rubiaceae

Origins:

For the German botanist Heinrich Adolph Schrader, 1767-1836, physician, M.D. Göttingen 1795, professor of botany; see John H. Barnhart, *Biographical Notes upon Botanists*. 3: 239. 1965; T.W. Bossert, *Biographical Dictionary of Botanists Represented in the Hunt Institute Portrait Collection*.

355. 1972; Jonas C. Dryander, *Catalogus bibliothecae historico-naturalis Josephi Banks*. 1800; A. Lasègue, *Musée botanique de Benjamin Delessert*. 1845; E.M. Tucker, *Catalogue of the Library of the Arnold Arboretum of Harvard University*. Cambridge, Massachusetts 1917-1933; Ida Kaplan Langman, *A Selected Guide to the Literature on the Flowering Plants of Mexico*. 684. University of Pennsylvania Press, Philadelphia 1964; R. Zander, F. Encke, G. Buchheim and S. Seybold, *Handwörterbuch der Pflanzennamen*. 14. Aufl. 1993; G. Schmid, *Goethe und die Naturwissenschaften*. Halle 1940.

Schradera Willd. Euphorbiaceae

Origins:

For the German botanist Heinrich Adolph Schrader, 1767-1836, physician, M.D. Göttingen 1795, professor of botany.

Schraderia Medik. Labiatae

Origins:

Possibly dedicated to the German botanist Heinrich Adolph Schrader, 1767-1836.

Schrammia Britton & Rose Caesalpiniaceae

Origins:

Possibly dedicated to the American biologist J.R. Schramm.

Schranckiastrum Hassler Mimosaceae

Origins:

For the German botanist Franz von Paula von Schrank, 1747-1835.

Schrankia Willdenow Mimosaceae

Origins:

For the German botanist Franz von Paula von Schrank, 1747-1835, entomologist, professor of botany and agronomy, 1809-1832 Director Botanical Garden München; see John H. Barnhart, *Biographical Notes upon Botanists*. 3: 240. 1965; T.W. Bossert, *Biographical Dictionary of Botanists Represented in the Hunt Institute Portrait Collection*. 355. 1972; Jonas C. Dryander, *Catalogus bibliothecae historico-naturalis Josephi Banks*. 1800; A. Lasègue, *Musée botanique de Benjamin Delessert*. 1845; E.M. Tucker, *Catalogue of the Library of the Arnold Arboretum of Harvard University*. Cambridge, Massachusetts 1917-1933; Ida Kaplan Langman, *A Selected Guide to the Literature on the Flowering Plants of Mexico*. 684. Philadelphia 1964; Stafleu and Cowan, *Taxonomic Literature*. 5: 323-328. 1985; R. Zander, F. Encke, G. Buchheim and S. Seybold, *Handwörterbuch der Pflanzennamen*. 14. Aufl. 777. Stuttgart 1993; G. Schmid, *Goethe und die Naturwissenschaften*. Halle 1940.

Schrebera Roxburgh Oleaceae

Origins:

The genus was named after the German botanist and zoologist Johann Christian Daniel von Schreber, 1739-1810, correspondent of Linnaeus, author of *De Phasco observationes*, quibus hoc genus muscorum vindicatur atque illustratur ... Cum tabulis aeri incisis. Lipsiae 1770 and *Theses medicae*, quas ... praeside ... Carolo Linnaeo ... publico submittet examini Jo. Christ. Dan. Schreber ... ad diem xiv junii anni mdcclx ... Upsaliae 1760; see Michel Adanson (1727-1806), *A Voyage to Senegal, the Isle of Gorée, and the River Gambia*. London. 1759 (First edition in English), or *Nachricht von seiner Reise nach Senegal*; übersetzt von Schreber. Leipzig 1773; Ernst Wilhelm Martius (1756-1849), *Erinnerungen aus meinem 90jährigen Leben*. Leipzig 1847; John H. Barnhart, *Biographical Notes upon Botanists*. 3: 240. 1965; T.W. Bossert, *Biographical Dictionary of Botanists Represented in the Hunt Institute Portrait Collection*. 355. 1972; Jonas C. Dryander, *Catalogus bibliothecae historico-naturalis Josephi Banks*. 1800; A. Lasègue, *Musée botanique de Benjamin Delessert*. 1845; E.M. Tucker, *Catalogue of the Library of the Arnold Arboretum of Harvard University*. Cambridge, Massachusetts 1917-1933; R. Zander, F. Encke, G. Buchheim and S. Seybold, *Handwörterbuch der Pflanzennamen*. 14. Aufl. 777. Stuttgart 1993; William Darlington (1782-1863), *Reliquiae Baldwinianae*. Philadelphia 1843 and *Memorials of John Bartram and Humphry Marshall*. 1849; Frans Antonie Stafleu (1921-1997), *Linnaeus and the Linnaeans. The spreading of their ideas in systematic botany, 1735-1789*. Utrecht 1971; Mariella Azzarello Di Misa, ed., *Il Fondo Antico della Biblioteca dell'Orto Botanico di Palermo*. 251. Palermo 1988.

Species/Vernacular Names:

S. alata (Hochst.) Welw. (*Schrebera argyrotricha* Gilg; *Schrebera saundersiae* Harv.; *Schrebera latialata* Gilg; *Schrebera gilgiana* Lingelsh.; *Schrebera mazoensis* S. Moore; *Natusia alata* Hochst.)

English: wild jasmine, bushveld tree-jasmine, wing-leaved wooden-pear, forest tree-jasmine

Southern Africa: wildejasmyn, sandjasmyn; muTisi (Shona); umGwenyahlungulu, umGwenyahlungula, isiHlulamanye, umSishane wehlanze, Loziphungwane, umTshwatshwala

(Zulu); sEhlulamanye (Swazi); mokauke (Tawana dialect, Ngamiland); mokauke (Subya); mokauke (Kololo, Barotseland)

S. arborea A. Chev.

Nigeria: opele (Yoruba)

Yoruba: opele

Cameroon: kambe, ngolombe, oban

Ivory Coast: oualio

Congo: were

S. golungensis Welw.

Cameroon: oban

Schreibersia Pohl Rubiaceae

Origins:

For the Austrian (b. Pressburg, Hungary, now Bratislava) naturalist and zoologist Karl (Carl) Franz Anton von Schreibers, 1775-1852 (d. Vienna), physician, scientist; see J.H. Barnhart, *Biographical Notes upon Botanists*. 3: 240. 1965; T.W. Bossert, *Biographical Dictionary of Botanists Represented in the Hunt Institute Portrait Collection*. 355. 1972; E.M. Tucker, *Catalogue of the Library of the Arnold Arboretum of Harvard University*. Cambridge, Massachusetts 1917-1933; Nils Spjeldnaes, in *Dictionary of Scientific Biography* 12: 215-216. 1981; G. Schmid, *Goethe und die Naturwissenschaften*. Halle 1940.

Schrenkia Fischer & C. Meyer Umbelliferae

Origins:

For the German botanist Alexander Gustav von Schrenk (Schrenck), 1816-1876, traveler; see John H. Barnhart, *Biographical Notes upon Botanists*. 3: 241. 1965; H.N. Clokie, *Account of the Herbaria of the Department of Botany in the University of Oxford*. 240. Oxford 1964; E.M. Tucker, *Catalogue of the Library of the Arnold Arboretum of Harvard University*. Cambridge, Massachusetts 1917-1933; A. Lasègue, *Musée botanique de Benjamin Delessert*. 1845; R. Zander, F. Encke, G. Buchheim and S. Seybold, *Handwörterbuch der Pflanzennamen*. 14. Aufl. Stuttgart 1993; Mariella Azzarello Di Misa, ed., *Il Fondo Antico della Biblioteca dell'Orto Botanico di Palermo*. 252. Regione Siciliana, Palermo 1988.

Schroeterella Briquet Zygophyllaceae

Origins:

For the Swiss botanist Carl (Karl) Joseph Schröter, 1855-1939, traveler, professor of systematic botany; see John H.

Barnhart, *Biographical Notes upon Botanists*. 3: 520. 1965; T.W. Bossert, *Biographical Dictionary of Botanists Represented in the Hunt Institute Portrait Collection*. 355. 1972; E.M. Tucker, *Catalogue of the Library of the Arnold Arboretum of Harvard University*. Cambridge, Massachusetts 1917-1933; Mary Gunn and Leslie E. Codd, *Botanical Exploration of Southern Africa*. 318. Cape Town 1981; Stafleu and Cowan, *Taxonomic Literature*. 5: 339-348. 1985.

Schubertia Martius Asclepiadaceae

Origins:

For the German physician Gotthilf Heinrich von Schubert, 1780-1860, scientist; see John H. Barnhart, *Biographical Notes upon Botanists*. 3: 242. 1965; T.W. Bossert, *Biographical Dictionary of Botanists Represented in the Hunt Institute Portrait Collection*. 356. 1972; G. Schmid, *Goethe und die Naturwissenschaften*. Halle 1940.

Schuenkia Raf. Solanaceae

Origins:

After the Dutch physician Martin Wilhelm (Martinus Wilhelmus) Schwencke, 1707-1785; see C.S. Rafinesque, in *Princ. Somiol*. 30. 1814.

Schufia Spach Onagraceae

Origins:

The anagram of the generic name *Fuchsia* L., after the German botanist Leonhart Fuchs, 1501-1566; see J.H. Barnhart, *Biographical Notes upon Botanists*. 2: 15. 1965.

Schultesia Martius Gentianaceae

Origins:

For the Austrian botanist Josef August Schultes, 1773-1831, M.D. Wien 1796, naturalist, plant collector. Among his writings are *Reise auf den Glockner*. Wien 1804, *Ratio medendi in schola clinica medica*. Lipsiae 1828, *Reisen durch Oberösterreich in den Jahren 1794-1808*. Tübingen 1809 and *Observationes botanicae*. Oeniponti [Innsbruck] 1809, with Johann Jakob Roemer (1763-1819) wrote *Systema vegetabilium*. 1807-1830; see John H. Barnhart, *Biographical Notes upon Botanists*. 3: 244. 1965; H.N. Clokie, *Account of the Herbaria of the Department of Botany in the University of Oxford*. 240. Oxford 1964; E.M. Tucker, *Catalogue of the Library of the Arnold Arboretum of Harvard*

University. Cambridge, Massachusetts 1917-1933; R. Zander, F. Encke, G. Buchheim and S. Seybold, *Handwörterbuch der Pflanzennamen*. 14. Aufl. Stuttgart 1993; Günther Schmid, *Chamisso als Naturforscher*. Eine Bibliographie. Leipzig 1942; Emil Bretschneider (1833-1901), *History of European Botanical Discoveries in China*. Leipzig 1981; G. Schmid, *Goethe und die Naturwissenschaften*. Halle 1940.

Schultesia Roth Campanulaceae

Origins:

For the Austrian botanist Josef August Schultes, 1773-1831.

Schultesia Sprengel Gramineae

Origins:

For the Austrian botanist Josef August Schultes, 1773-1831.

Schultesianthus Hunz. Solanaceae

Origins:

For the American botanist Richard Evans Schultes, b. 1915, botanical explorer, ethnobotanist, plant collector, Jeffrey Professor of Biology and Director of the Botanical Museum of Harvard University, member of the Linnean Society of London, author of *Where the Gods Reign. Plants and People of the Colombian Amazon*. London 1988, with Robert F. Raffauf wrote *Vine of the Soul. Medicine Men, their Plants and Rituals in the Colombian Amazonia*. Synergetic Press, Oracle, Arizona 1992 and *The Healing Forest. Medicinal and Toxic Plants of the Northwest Amazonia*. Dioscorides Press, Portland, Oregon 1992.

Schultesiophytum Harling Cyclanthaceae

Origins:

For the American botanist Richard Evans Schultes, b. 1915.

Schulzia Sprengel Umbelliferae

Origins:

After the German physicians and botanists Karl Friedrich Schultz (1765-1837) and Johann Heinrich Schulze (1687-1744); see John H. Barnhart, *Biographical Notes upon Botanists*. 3: 244. 1965; A. Lasègue, *Musée botanique de Benjamin Delessert*. 1845; Jonas C. Dryander, *Catalogus bibliothecae historico-naturalis Josephi Banks*. London 1800; E.M. Tucker, *Catalogue of the Library of the Arnold Arboretum of Harvard University*. Cambridge, Massachusetts 1917-1933; R. Zander, F. Encke, G. Buchheim and S.

Seybold, *Handwörterbuch der Pflanzennamen*. 14. Aufl. Stuttgart 1993; T.W. Bossert, *Biographical Dictionary of Botanists Represented in the Hunt Institute Portrait Collection*. 356. 1972.

Schumacheria Vahl Dilleniaceae

Origins:

For the Danish physician Heinrich Christian Friederich (Cristen Frederic) Schumacher, 1757-1830, botanist, professor of surgery and professor of anatomy; see John H. Barnhart, *Biographical Notes upon Botanists*. 3: 246. 1965; Jonas C. Dryander, *Catalogus bibliothecae historico-naturalis Josephi Banks*. London 1800; A. Lasègue, *Musée botanique de Benjamin Delessert*. Paris 1845; E.M. Tucker, *Catalogue of the Library of the Arnold Arboretum of Harvard University*. Cambridge, Massachusetts 1917-1933; R. Zander, F. Encke, G. Buchheim and S. Seybold, *Handwörterbuch der Pflanzennamen*. 14. Aufl. Stuttgart 1993; Carl Frederik Albert Christensen, *Den danske Botaniks Historie med tilhørende Bibliografi*. Copenhagen 1924-1926; Mariella Azzarello Di Misa, ed., *Il Fondo Antico della Biblioteca dell'Orto Botanico di Palermo*. 253. Regione Siciliana, Palermo 1988.

Schumannia Kuntze Umbelliferae

Origins:

For the German botanist Karl Moritz Schumann, 1851-1904, taxonomist, botanical collector, from 1891 onwards edited the *German Cactus Society's Journal*; see John H. Barnhart, *Biographical Notes upon Botanists*. 3: 247. Boston 1965; T.W. Bossert, *Biographical Dictionary of Botanists Represented in the Hunt Institute Portrait Collection*. 356. 1972; Ida Kaplan Langman, *A Selected Guide to the Literature on the Flowering Plants of Mexico*. 687-688. University of Pennsylvania Press, Philadelphia 1964; E.M. Tucker, *Catalogue of the Library of the Arnold Arboretum of Harvard University*. Cambridge, Massachusetts 1917-1933; A. White and B.L. Sloane, *The Stapelieae*. Pasadena 1937; Stafleu and Cowan, *Taxonomic Literature*. 5: 400-408. 1985; Gordon Douglas Rowley, *A History of Succulent Plants*. Strawberry Press, Mill Valley, California 1997; R. Zander, F. Encke, G. Buchheim and S. Seybold, *Handwörterbuch der Pflanzennamen*. 14. Aufl. Stuttgart 1993.

Schumannianthus Gagnepain Marantaceae

Origins:

For the German botanist Karl Moritz Schumann, 1851-1904.

Schumanniophyton Harms Rubiaceae

Origins:

For the German botanist Karl Moritz Schumann, 1851-1904.

Schunda-Pana Adans. Palmae

Origins:

A vernacular name.

Schuurmansia Blume Ochnaceae

Origins:

For the Dutch botanist Jacobus Hermanus (Herman) Schuurmans Stekhoven, 1792-1855; see E.M. Tucker, *Catalogue of the Library of the Arnold Arboretum of Harvard University*. Cambridge, Massachusetts 1917-1933.

Schuurmansiella Hallier Ochnaceae

Origins:

For the Dutch botanist Jacobus Hermanus (Herman) Schuurmans Stekhoven, 1792-1855.

Schwabea Endlicher & Fenzl Acanthaceae

Origins:

For the German (b. Dessau) botanist Samuel Heinrich Schwabe, 1789-1875 (d. Dessau), pharmacist, astronomer, he was elected a member of the Royal Society of London in 1868; see Sister Maureen Farrell, F.C.J., in *Dictionary of Scientific Biography* 12: 239-240. 1981; John H. Barnhart, *Biographical Notes upon Botanists*. 3: 248. 1965; E.M. Tucker, *Catalogue of the Library of the Arnold Arboretum of Harvard University*. Cambridge, Massachusetts 1917-1933.

Schwackaea Cogniaux Melastomataceae

Origins:

For the German botanist Carl (Karl) August Wilhelm Schwacke, 1848-1904, naturalist, traveler, scientist, professor of botany, from 1872 in Brazil; see John H. Barnhart, *Biographical Notes upon Botanists*. 3: 248. 1965; E.M. Tucker, *Catalogue of the Library of the Arnold Arboretum of Harvard University*. Cambridge, Massachusetts 1917-1933; T.W. Bossert, *Biographical Dictionary of Botanists*

Represented in the Hunt Institute Portrait Collection. 357. 1972.

Schwaegerichenia Sprengel ex Steudel Haemodoraceae

Origins:

Orthographic variant of *Schwaegrichenia* Sprengel; see Arthur D. Chapman, ed., *Australian Plant Name Index*. 2599. Canberra 1991.

Schwaegrichenia Sprengel Haemodoraceae

Origins:

After the German botanist Christian Friedrich Schwägrichen, 1775-1853, physician, M.D. Leipzig 1799, professor of natural history and botany at the University of Leipzig and from 1806 to 1873 Director of the Botanical Garden. Among his writings are *Topographiae botanicae et etymologicae lipsiensis specimen primum*. Lipsiae [1799] and "Bemerkungen über einige Stellen in der Flora brasiliensis von Endlicher und Martius." *Linnaea*. 14(5): 517-528. (Feb.) 1841, editor of Johann Hedwig (1730/1733-1799), *Species muscorum frondosorum*. Opus posthumum editum a F.S. 1801; see J.H. Barnhart, *Biographical Notes upon Botanists*. 3: 248. Boston 1965; William Darlington, *Reliquiae Baldwinianae*. Philadelphia 1843; Arthur D. Chapman, ed., *Australian Plant Name Index*. 2599. 1991; G. Schmid, *Goethe und die Naturwissenschaften*. Halle 1940; E.M. Tucker, *Catalogue of the Library of the Arnold Arboretum of Harvard University*. Cambridge, Massachusetts 1917-1933; Kurt Polycarp Joachim Sprengel, *Plantarum Minus Cognitarum Pugillus*. 2: 58. 1815; A. Lasègue, *Musée botanique de Benjamin Delessert*. 1845.

Schwalbea L. Scrophulariaceae

Origins:

After Christian Georg (Christianus Georgius) Schwalbe, author of *De China officinarum*. Lugduni Batavorum [1715].

Schwannia Endlicher Malpighiaceae

Origins:

After the German (b. Neuss) physician Theodor Ambrose Hubert Schwann, 1810-1882 (d. Cologne, Germany), a founder of modern histology, microscopist, professor of anatomy and professor of physiology, discovered the enzyme pepsin; see Marcel Florkin, in *Dictionary of Scientific Biography* (Editor in Chief Charles Coulston Gillispie.) 12: 240-245. New York 1981; J.H. Barnhart, *Biographical Notes upon Botanists*. 3: 248. Boston 1965; Garrison and Morton, *Medical Bibliography*. 113, 481, 674, 990-991 and 993. 1961; Stafleu and Cowan, *Taxonomic Literature*. 5: 421-422. 1985.

Schwantesia Dinter Aizoaceae

Origins:

For the German botanist Gustav Schwantes, 1891-1960, archeologist, professor of pre-history (Kiel); see Adolar Gottlieb Julius Hans Herre (1895-1979), *The Genera of the Mesembryanthemaceae*, ... illustrations by Harry Bolus, Beatrice Carter, Mary Page and Maisie Walgate. Cape Town 1971; Gordon Douglas Rowley, *A History of Succulent Plants*. Strawberry Press, Mill Valley, California 1997.

Schwantesia L. Bolus Aizoaceae

Origins:

For the German botanist Gustav Schwantes, 1891-1960.

Schwartzkopffia Kraenzlin Orchidaceae

Origins:

Dedicated to the brothers Ernest and Philip Schwarzkoppf (Schwartzkoppf), friends of the German orchidologist Friedrich Fritz Wilhelm Ludwig Kraenzlin (Kränzlin) (1847-1934); see Richard Evans Schultes and Arthur Stanley Pease, *Generic Names of Orchids. Their Origin and Meaning*. 284. Academic Press, New York and London 1963; Hubert Mayr, *Orchid Names and Their Meanings*. 178. Vaduz 1998; Engler, *Bot. Jahrb.* 28: 177. 1901.

Schweiggeria Sprengel Violaceae

Origins:

For the German botanist August Friedrich Schweigger, 1783-1821 (murdered, Agrigento, Sicily), zoologist, physician, professor of botany, traveler; see John H. Barnhart, *Biographical Notes upon Botanists*. 3: 249. 1965; E.M. Tucker, *Catalogue of the Library of the Arnold Arboretum of Harvard University*. Cambridge, Massachusetts 1917-1933; Ida Kaplan Langman, *A Selected Guide to the Literature on the Flowering Plants of Mexico*. 689. 1964; G. Schmid, *Goethe und die Naturwissenschaften*. Halle 1940; Stafleu and Cowan, *Taxonomic Literature*. 5: 428-430.

1985; Kenneth L. Caneva, in *Dictionary of Scientific Biography* 12: 253-255. [A Friedrich Schweigger was professor of theology at the Protestant University of Erlangen, archdeacon of a parish in that city and father of the German scientist Johann Salomo Christoph Schweigger (b. Erlangen, d. Halle), 1779-1857] 1981; Giulio Giorello & Agnese Grieco, a cura di, *Goethe scienziato*. Einaudi Editore, Torino 1998; R. Zander, F. Encke, G. Buchheim and S. Seybold, *Handwörterbuch der Pflanzennamen*. 14. Aufl. Stuttgart 1993.

Schweinfurthia A. Braun Scrophulariaceae

Origins:

For the German botanist Georg August Schweinfurth, 1836-1925, explorer, traveler, botanical collector in east and north Africa (tropical Africa and Arabia), ethnologist. Among many other works, he was the author of *Illustration de la flore d'Egypt*. Le Caire 1887, *Sur la flore des anciens jardins arabes d'Egypt*. Le Caire 1888, *The Heart of Africa*. London 1873, *Beitrag zur Flora Aethiopiens*. Berlin 1867 and *Novae species aethiopicae*. [Wien 1868]; see John H. Barnhart, *Biographical Notes upon Botanists*. 3: 249. 1965; R. Zander, F. Encke, G. Buchheim and S. Seybold, *Handwörterbuch der Pflanzennamen*. 14. Aufl. 1993; E.M. Tucker, *Catalogue of the Library of the Arnold Arboretum of Harvard University*. Cambridge, Massachusetts 1917-1933; T.W. Bossert, *Biographical Dictionary of Botanists Represented in the Hunt Institute Portrait Collection*. 357. 1972; Stafleu and Cowan, *Taxonomic Literature*. 5: 430-437. 1985; G.E. Wickens, "Dr. G. Schweinfurth's journeys in the Sudan." *Kew Bulletin*. 27(1): 129-146. 1972; Anthonius Josephus Maria Leeuwenberg, "Isotypes of which holotypes were destroyed in Berlin." in *Webbia*. 19: 861-863. 1965; G.E. Wickens, "J.D.C. Pfund, a botanist in the Sudan with the Egyptian Military Expeditions, 1875-1876." *Kew Bulletin*. 24(1): 191-216. 1970; Gordon Douglas Rowley, *A History of Succulent Plants*. Strawberry Press, Mill Valley, California 1997; Joseph Vallot, "Études sur la flore du Sénégal." in *Bull. Soc. Bot. de France*. 29: 191-192. Paris 1882; A. White and B.L. Sloane, *The Stapelieae*. Pasadena 1937.

Schweinitzia Elliott ex Nuttall Ericaceae

Origins:

For the American botanist Lewis David von Schweinitz, 1780-1835, mycologist, plant collector; see J.H. Barnhart, *Biographical Notes upon Botanists*. 3: 250. 1965; T.W. Bossert, *Biographical Dictionary of Botanists Represented in the Hunt Institute Portrait Collection*. 357. 1972; E.M. Tucker, *Catalogue of the Library of the Arnold Arboretum* of *Harvard University*. Cambridge, Massachusetts 1917-1933; H.N. Clokie, *Account of the Herbaria of the Department of Botany in the University of Oxford*. 241. Oxford 1964; S. Lenley et al., *Catalog of the Manuscript and Archival Collections and Index to the Correspondence of John Torrey*. Library of the New York Botanical Garden. 1973; Stafleu and Cowan, *Taxonomic Literature*. 5: 437-442. 1985; J. Ewan, ed., *A Short History of Botany in the United States*. New York and London 1969; J.W. Harshberger, *The Botanists of Philadelphia and Their Work*. 127-132. 1899; R. Zander, F. Encke, G. Buchheim and S. Seybold, *Handwörterbuch der Pflanzennamen*. 14. Aufl. Stuttgart 1993; William Darlington, *Reliquiae Baldwinianae*. Philadelphia 1843; Jeannette Elizabeth Graustein, *Thomas Nuttall, Naturalist. Explorations in America, 1808-1841*. Cambridge, Harvard University Press 1967; William Jay Youmans, ed., *Pioneers of Science in America*. New York 1896.

Schwenckia Vahl Solanaceae

Origins:

After the Dutch physician Martin Wilhelm (Martinus Wilhelmus) Schwencke, 1707-1785, botanist. His works include *Dissertatio ... de operatione inguinali*. Lugduni Batavorum 1731, *Novae plantae Schwenckia dictae a celeberrimo Linnaeo in Gen. plant.* ed. VI. p. 567 ex celeb. Davidiis van Rooijen Charact. mss. 1761 communicata brevis descriptio et delineatio cum notis characteristicis. Hagae Comitum [The Hague] [typ. van Karnebeek] 1766 and *Officinalium plantarum catalogus*. Hagae-Comitum 1752; see John H. Barnhart, *Biographical Notes upon Botanists*. 3: 250. 1965; Jonas C. Dryander, *Catalogus bibliothecae historico-naturalis Josephi Banks*. London 1800; H. Heine, in *Kew Bulletin*. 16(3): 465-469. 1963.

Schwenckiopsis Dammer Solanaceae

Origins:

After the Dutch physician Martin Wilhelm (Martinus Wilhelmus) Schwencke, 1707-1785.

Schwendenera Schumann Rubiaceae

Origins:

After the Swiss (b. St. Gallen) botanist Simon Schwendener, 1829-1919 (d. Berlin), morphologist, physiologist, professor of botany, father of the theory of the symbiotic nature of the lichens; see John H. Barnhart, *Biographical Notes upon Botanists*. 3: 250. 1965; T.W. Bossert, *Biographical Dictionary of Botanists Represented in the Hunt Institute Portrait Collection*. 357. 1972; Ethelyn Maria Tucker,

Catalogue of the Library of the Arnold Arboretum of Harvard University. Cambridge, Massachusetts 1917-1933; A.P.M. Sanders, in *Dictionary of Scientific Biography* 12: 255-256. 1981.

Schwenkfelda Schreb. Rubiaceae

Origins:

For Kaspar (Caspar) Schwenckfelt, 1563-1609, author of *Stirpium & fossilium Silesiae catalogus*. Lipsiae 1600.

Schwenkfeldia Willd. Rubiaceae

Origins:

For Kaspar Schwenckfelt, 1563-1609; see *Pritzel*, 292.

Schwenkia L. Solanaceae

Origins:

After the Dutch physician Martin Wilhelm (Martinus Wilhelmus) Schwencke, 1707-1785; see H. Heine, in *Kew Bulletin*. 16(3): 465-469. 1963.

Species/Vernacular Names:

S. americana L.

Yoruba: igbale odan, ale odan, odan adija, sasara

Congo: lantsuni

Schwenkiopsis Dammer Solanaceae

Origins:

After the Dutch physician Martin Wilhelm (Martinus Wilhelmus) Schwencke, 1707-1785.

Sciacassia Britton Caesalpiniaceae

Origins:

Greek *skias, skiados* "umbel, canopy" plus *Cassia*, see also *Senna* Mill.

Sciadiodaphne Reichb. Lauraceae

Origins:

From the Greek *skias, skiados* "umbel, canopy" plus *Daphne*, see also *Umbellularia* (Nees) Nutt.

Sciadocalyx Regel Gesneriaceae

Origins:
Greek *skias, skiados* "umbel, canopy" and *kalyx* "calyx."

Sciadocephala Mattf. Asteraceae

Origins:
From the Greek *skias, skiados* and *kephale* "head."

Sciadodendron Griseb. Araliaceae

Origins:
Greek *skias, skiados* "canopy, umbel" and *dendron* "a tree."

Sciadonardus Steud. Gramineae

Origins:
Greek *skias, skiados* "canopy, umbel of plants" and *nardos* "spikenard."

Sciadopanax Seem. Araliaceae

Origins:
From the Greek *skias, skiados* plus *Panax*, see also *Polyscias* Forst. & Forst.f.

Sciadophyllum P. Browne Araliaceae

Origins:
From the Greek *skias, skiados* "canopy, umbel" and *phyllon* "leaf."

Sciadopitys Siebold & Zucc. Sciadopityaceae (Taxodiaceae)

Origins:
Parasol pine, from the Greek *skias, skiados* "canopy, umbel" and *pitys* "pine, a fir tree."

Species/Vernacular Names:
S. verticillata (Thunb.) Siebold & Zucc.

English: parasol pine, Japanese umbrella pine

Japan: koya-maki

Sciadotenia Miers Menispermaceae

Origins:
From the Greek *skias, skiados* and *tainia* "fillet."

Sciaphila Blume Triuridaceae

Origins:
Presumably from the Greek *skia* "shade, shadow" and *philos* "loving," referring to the habitat; see Karl Ludwig von Blume, *Bijdragen tot de flora van Nederlandsch Indië.* 514. (Dec. 1825-Mar. 1826) 1825-1826.

Species/Vernacular Names:
S. takakumensis Ohwi (named for Mount Takakuma, Kimotsuki-gun, Kagoshima Pref. Japan)
Japan: Takakuma-sô

Sciaphyllum Bremek. Acanthaceae

Origins:
Presumably from the Greek *skia* "shade, shadow" and *phyllon* "leaf."

Sciaplea Rauschert Caesalpiniaceae

Origins:
Greek *skia* "shade, shadow" and *pleios* "many," *pleos* "full."

Scilla L. Hyacinthaceae (Liliaceae)

Origins:
Latin *scilla* and *squilla* "a sea-onion, sea-leek, squill" (Plinius), Greek *skilla* "scilla," Latin *scillinos* and Greek *skillinos* for sea-onions or squills, Greek *oxos skillitikos, oxos skillinos*, Akkadian *sikillu* for a plant, Sanskrit *kshira* "milk, sap," *kshiri* "water" or *kshi* "to destroy," the bulbs of several *Scilla* species are used medicinally by indigenous people and are potentially poisonous; see Carl Linnaeus, *Species Plantarum.* 308. 1753 and *Genera Plantarum.* Ed. 5. 146. 1754; G. Semerano, *Le origini della cultura europea.* Dizionari Etimologici. Dizionario della lingua Greca. 2(1): 267. Firenze 1994; E. Weekley, *An Etymological Dictionary of Modern English.* 2: 1292, 1401. New York 1967; H. Genaust, *Etymologisches Wörterbuch der botanischen Pflanzennamen.* 568. Basel 1996.

Species/Vernacular Names:
S. autumnalis L.

English: autumn squill

S. natalensis Planchon (*Scilla dracomontana* Hilliard & Burtt; *Scilla kraussii* Bak.)

English: blue squill, blue hyacinth, wild squill

Southern Africa: blouberglelie, blouslangkop; matunga (Sotho)

S. nervosa (Burch.) Jessop (*Schizocarphus gerrardii* (Bak.) v.d. Merwe; *Schizocarphus nervosus* (Bak.) v.d. Merwe; *Schizocarphus rigidifolius* (Kunth) v.d. Merwe; *Schizocarphus gerrardii* Bak.; *Schizocarphus hispidula* Bak.; *Schizocarphus rigidifolia* Bak.; *Schizocarphus gerrardii* Bak. var. *acerosa* v.d. Merwe)

English: sand lily, scilla, wild squill

Southern Africa: sand lelie; ditsonya (Tswana)

S. peruviana L.

English: Cuban lily

Italian: scilla del Perù

S. plumbea Lindley

South Africa: blouberglelie

S. scilloides (Lindley) Druce (*Barnardia scilloides* Lindl.; *Scilla sinensis* (Lour.) Merrill)

English: squill, Chinese squill

Japan: tsurubo, surubo

China: mian zao er

South Africa: blouberglelie

S. verna L.

English: spring squill

Scindapsus Schott Araceae

Origins:
From *skindapsos*, ancient Greek name for a four-stringed musical instrument, applied also to an ivy-like plant; see Salvatore Battaglia, *Grande dizionario della lingua italiana*. XVIII: 46. UTET, Torino 1997.

Species/Vernacular Names:
S. pictus Hassk.

Malaya: sireh chechawai

Sciothamnus Endl. Umbelliferae

Origins:
Possibly from the Greek *skias, skiados* "a canopy, umbel, shadow" and *thamnos* "shrub."

Scirpidiella Rauschert Cyperaceae

Origins:
Referring to the genus *Scirpus* L.

Scirpobambus Kuntze Gramineae

Origins:
Latin *scirpus, i* "a rush, bulrush" plus *bambos, bambusa.*

Scirpodendron Zipp. ex Kurz Cyperaceae

Origins:
Scirpus L. with Greek *dendron* "tree"; see Wilhelm Sulpiz Kurz (1834-1878), in *Journal of the Asiatic Society of Bengal*. Part 2. Natural History. 38: 84. 1869.

Species/Vernacular Names:
S. ghaeri (Gaertner) Merr.

Malaya: selinsing

Scirpoides Séguier Cyperaceae

Origins:
Resembling the genus *Scirpus* L.; see Jean François Séguier (1703-1784), *Plantae veronenses*. 3: 73. Veronae 1754.

Scirpus L. Cyperaceae

Origins:
Latin *scirpus* or sometimes *sirpus* for a rush, bulrush (Plinius), Hebrew *habar* "to string together, to compose," *haber* "joined, united"; see Carl Linnaeus, *Species Plantarum*. 47. 1753 and *Genera Plantarum*. Ed. 5. 26. 1754; Giovanni Semerano, *Le origini della cultura europea*. Dizionario della lingua Latina e di voci moderne. 2(2): 557. 1994.

Species/Vernacular Names:
S. spp.

Mexico: yaga tzego, ya yoguero

S. americanus Pers.

English: sword grass, chairmaker's rush

S. ciliaris L. (*Fuirena ciliaris* (L.) Roxb.)

Japan: kuro-tama-gaya-tsuri

S. clementis M.E. Jones

English: Yosemite bulrush

S. cyperinus (L.) Kunth

English: wool grass

China: kuai tsao

S. fluviatilis (Torrey) A. Gray

English: river bulrush

China: san leng

S. fuirena T. Koyama (*Fuirena umbellata* Rott.)

Japan: hiro-ha-no-kuro-tama-gaya-tsuri (= broad-leaved *Fuirena*)

S. heterochaetus A. Chase

English: slender bulrush

S. paludosus Nelson

English: bayonet grass

S. pumilus Vahl

English: dwarf bulrush

S. pungens Vahl

English: common threesquare

S. subterminalis Torrey

English: water bulrush

S. sylvaticus L.

English: wood club rush

S. triqueter L.

Japan: san-kaku-i (= triangular rush)

Sclerachne R. Br. Gramineae

Origins:
Greek *skleros* "hard, dry" and *achne* "chaff, glume"; see John Joseph Bennett and Robert Brown, *Plantae Javanicae rariores*. 15, t. 4. London (Jul.) 1838.

Sclerachne Trin. Gramineae

Origins:
Greek *skleros* and *achne* "chaff, glume."

Sclerandrium Stapf & C.E. Hubb. Gramineae

Origins:
Greek *skleros* "hard, dry" and *aner, andros* "male, man," referring to the lower glume of the male spikelets; see O. Stapf and Charles Edward Hubbard, *Hooker's Icones Plantarum*. Ser. 5. 3, t. 3262. (Sep.) 1938.

Scleranthera Pichon Apocynaceae

Origins:
Greek *skleros* "hard, harsh, stiff" and *anthera* "anther."

Scleranthopsis Rech.f. Caryophyllaceae

Origins:
Resembling *Scleranthus* L.

Scleranthus L. Caryophyllaceae

Origins:
Greek *skleros* and *anthos* "flower," referring to the fruiting calyx; see Carl Linnaeus, *Species Plantarum*. 406. 1753 and *Genera Plantarum*. Ed. 5. 190. 1754.

Species/Vernacular Names:
S. annuus L.

English: annual scleranthus, knawel, annual knawel

South Africa: knawel

S. biflorus (Forst. & Forst.f.) Hook.f.

English: twin-flower knawel, alpine moss

New Zealand: mossy scabweed

S. diander R. Br.

English: tufted knawel

S. minusculus F. Muell.

English: cushion knawel

S. pungens R. Br.

English: prickly knawel

S. singuliflorus (F. Muell.) Mattf.

English: mossy knawel

Scleria Bergius Cyperaceae

Origins:
Greek *skleros* "hard, dry," referring to the fruits; see Peter Jonas Bergius (1730-1790), in *Kongl. Vetenskaps Academiens Handlingar*. 26: 142, t. 4, 5. Stockholm 1765.

Species/Vernacular Names:
S. sp.

Hawaii: mau'u, kaluhaluha

S. barteri C.B. Cl.

French: liane rasoir

Sierra Leone: njaewae

Congo: kingeri

S. biflora Roxb. (*Scleria ferruginea* Ohwi)

Japan: hosoba-shinju-gaya (= narrow-leaved *Scleria*)

S. terrestris (L.) Fassett (*Zizania terrestris* L.; *Scleria doederleiniana* Böckeler)

English: nut rush

Japan: ô-shinju-gaya (= large *Scleria*)

Sclerobassia Ulbr. Chenopodiaceae

Origins:

Greek *skleros* plus the genus *Bassia* All.; see H.G.A. Engler & K.A.E. Prantl, *Die Natürlichen Pflanzenfamilien*. Ed. 2. 16c: 539. 1934.

Scleroblitum Ulbr. Chenopodiaceae

Origins:

Greek *skleros* with *Blitum*; see H.G.A. Engler & K.A.E. Prantl, *Die Natürlichen Pflanzenfamilien*. Ed. 2. 16c: 495. (Jan.-Mar.) 1934; Paul G. Wilson, "A taxonomic revision of the tribe Chenopodieae (Chenopodiaceae) in Australia." *Nuytsia*. 4(2): 135-262. 1983.

Species/Vernacular Names:

S. atriplicinum (F. Muell.) Ulbr. (*Blitum atriplicinum* F. Muell.; *Chenopodium atriplicinum* (F. Muell.) F. Muell.)

Australia: purple goosefoot, starry goosefoot, purple-leaved goosefoot, lamb's tongue

Sclerocactus Britton & Rose Cactaceae

Origins:

From the Greek *skleros* "hard, dry" and *cactus*.

Species/Vernacular Names:

S. polyancistrus (Engelm. & J. Bigelow) Britton & Rose

English: Mojave fish-hook cactus

Sclerocalyx Nees Acanthaceae

Origins:

Greek *skleros* and *kalyx* "calyx."

Sclerocarpus Jacq. Asteraceae

Origins:

From the Greek *skleros* and *karpos* "a fruit," referring to the covering of the seeds.

Sclerocarya Hochst. Anacardiaceae

Origins:

From the Greek *skleros* "hard, dry" and *karyon* "walnut, nut," referring to the nature of the seeds.

Species/Vernacular Names:

S. birrea (A. Rich.) Hochst. (*Spondias birrea* A. Rich.)

English: cider tree, marula, maroola plum

Mali: nkunan, ngana, nkudanga

Nigeria: danya (Hausa); heri (Fula)

Tanzania: ntoregani

Madagascar: sakoa

Eastern Africa: mungongo (Tabora)

S. birrea (A. Rich.) Hochst. subsp. *caffra* (Sond.) Kokwaro (*Sclerocarya birrea* sensu Van der Veken non (A. Rich.) Hochst.; *Sclerocarya caffra* Sond.; *Sclerocarya schweinfurthiana* Schinz) (the specific name based on the word *birr*, the common name for the tree in Senegambia)

English: cider tree, marula, maroola plum

Southern Africa: meroola, maroela; mufula (Venda); mfula (Kalanga, northern Botswana); muFuna, muFura, muGanu, iKanyi, muKwakwa, muPfura, muShomo, muSomo, muTsomo (Shona); umGanu (Zulu); umganu (Swazi); nkanyi (Tsonga); morula (Western Transvaal, northern Cape, Botswana); morula (North Sotho: north and northeast Transvaal); morwa (Yei, Ngamiland); uge, muge (Deiriku: Okavango Native Territory); omuongo (Herero); omuongo (northern southwest Africa)

Sclerocaryopsis Brand Boraginaceae

Origins:

Greek *skleros* "hard, dry" and *karyon* "walnut, nut."

Sclerocephalus Boiss. Caryophyllaceae (Illecebraceae)

Origins:

From the Greek *skleros* and *kephale* "head."

Sclerochiton Harvey Acanthaceae

Origins:

Greek *skleros* "hard, dry" and *chiton* "a tunic, covering," referring to the hard and woody capsules.

Sclerochlamys F. Muell. Chenopodiaceae

Origins:

Greek *skleros* and *chlamys, chlamydos* "cloak," referring to the perianth; see F. von Mueller, in *Transactions of the Philosophical Institute of Victoria*. 2: 76. (Sep.) 1857.

Species/Vernacular Names:

S. brachyptera F. Muell. (*Echinopsilon brachypterus* (F. Muell.) F. Muell.; *Kochia brachyptera* (F. Muell.) Benth.; *Bassia brachyptera* (F. Muell.) R. Anderson)

Australia: short-winged copperburr, hairy bassia, shortwing saltbush

Sclerochloa P. Beauv. Gramineae

Origins:

Greek *skleros* and *chloe, chloa* "grass," referring to the nature of grass, to the thick glumes; see Ambroise Marie François Joseph Palisot de Beauvois (1752-1820), *Essai d'une nouvelle Agrostographie*. 97, 177, t. 19, fig. 4. Paris (Dec.) 1812.

Species/Vernacular Names:

S. dura (L.) P. Beauv. (*Cynosurus durus* L.)

English: hard meadow-grass, hard grass

Sclerochorton Boiss. Umbelliferae

Origins:

From the Greek *skleros* "hard, dry" and *chortos* "green herbage, grass."

Sclerocladus Raf. Sapotaceae

Origins:

Greek *skleros* "hard, dry" and *klados* "branch"; see C.S. Rafinesque, *Sylva Telluriana*. 35. 1838.

Sclerococcus Bartl. Rubiaceae

Origins:

From the Greek *skleros* "hard, dry" and *kokkos* "a berry."

Sclerocroton Hochst. Euphorbiaceae

Origins:

From the Greek *skleros* "hard, dry" plus *Croton*.

Sclerocyathium Prokh. Euphorbiaceae

Origins:

Greek *skleros* and *kyathos* "a cup, a ladle."

Sclerodactylon Stapf Gramineae

Origins:

From the Greek *skleros* "hard, dry" and *daktylos* "a finger."

Sclerodeyeuxia Pilger Gramineae

Origins:

From the Greek *skleros* "hard, dry" plus the genus *Deyeuxia*, dedicated to the French professor of pharmacology Nicolas Deyeux, 1745-1837. Among his publications are *Analyse des nouvelles eaux minérales de Passy*, communiquée à l'École de médecine de Paris. Paris 1808 and *Considérations chimiques et médicales sur le sang des ictériques*, présentées ... à l'École de médecine de Paris. Paris 1804; see Ambroise Marie François Joseph Palisot de Beauvois, *Essai d'une nouvelle Agrostographie*, ou nouveaux genres des Graminées. 43, pl. IX. Paris 1812.

Scleroglossum Alderw. Grammitidaceae

Origins:

Greek *skleros* "hard, dry" and *glossa* "tongue"; see Cornelis Rugier Willem Karel van Alderwerelt van Rosenburgh (1863-1936), in *Bulletin du Jardin Botanique de Buitenzorg*. Ser. 2, 7: 37. Buitenzorg, Dutch E. Indies [= Bogor, Indonesia] 1912.

Sclerolaena A. Camus Gramineae

Origins:

Greek *skleros* and *chlaena* "a cloak, blanket," see also *Cyphochlaena* Hack.

Sclerolaena R. Br. Chenopodiaceae

Origins:

Greek *skleros* "hard, dry" and *chlaena* "a cloak, blanket"; see Robert Brown (1773-1858), *Prodromus florae Novae Hollandiae et Insulae van-Diemen*. 410. London 1810.

Species/Vernacular Names:

S. anisacanthoides (F. Muell.) Domin

English: yellow burr

S. articulata (J. Black) A.J. Scott (*Bassia articulata* J. Black)

English: jointed poverty-bush

S. bicornis Lindley (*Bassia bicornis* (Lindley) F. Muell.)

English: goathead burr

S. birchii (F. Muell.) Domin

English: galvanized burr

S. calcarata (Ising) A.J. Scott

English: redburr

S. cuneata Paul G. Wilson

English: tangled copperburr

S. diacantha (Nees) Benth. (*Bassia diacantha* (Nees) F. Muell.)

English: grey copperburr

S. divaricata (R. Br.) Smith

English: tangled copperburr

S. lanicuspis (F. Muell) Benth. (*Bassia lanicuspis* (F. Muell) F. Muell.; *Anisacantha lanicuspis* F. Muell.)

English: woolly copperburr

S. muricata (Moq.) Domin

English: black rolypoly

S. tetracuspis (C. White) A.J. Scott

English: Brigalow copperburr

S. tricuspis (F. Muell.) Ulbr.

English: giant redburr

Scleroleima Hook.f. Asteraceae

Origins:
Greek *skleros* "hard, dry" and *leimon* "a meadow, a moist place, a flowery surface"; see J.D. Hooker, in *Hooker's London Journal of Botany.* 5: 444, t. 14. 1846.

Sclerolepis Cass. Asteraceae

Origins:
From the Greek *skleros* and *lepis* "a scale."

Sclerolinon C.M. Rogers Linaceae

Origins:
Greek *skleros* "hard, dry" and *linon* "flax," rough surface of the nutlets.

Sclerolobium Vogel Caesalpiniaceae

Origins:
From the Greek *skleros* "hard, dry" and *lobion* "a little pod."

Scleromelum K. Schumann & Lauterb. Santalaceae

Origins:
Greek *skleros* "hard, dry, stiff" and *melon* "an apple," see also *Scleropyrum* Arn.

Scleromitrion Wight & Arnott Rubiaceae

Origins:
Greek *skleros* "hard, dry, stiff" and *mitrion*, the diminutive of *mitra*, *mitre* "a turban, headband."

Scleronema Benth. Bombacaceae

Origins:
From the Greek *skleros* "hard, dry" and *nema* "filament, thread."

Sclerophylax Miers Solanaceae

Origins:
Greek *skleros* "hard, dry" and *phylax*, *phylakos* "a guardian, protector, vigilant," halophytes.

Sclerophyllum Griff. Gramineae

Origins:
From the Greek *skleros* and *phyllon* "leaf."

Scleropoa Griseb. Gramineae

Origins:
From the Greek *skleros* "hard, dry" and *poa* "grass, pasture grass."

Scleropogon Philippi Gramineae

Origins:
Greek *skleros* "hard, dry" and *pogon* "a beard," referring to the awns.

Species/Vernacular Names:
S. brevifolius Philippi

English: burro grass

Scleropterys Scheidweiler Orchidaceae

Origins:

Greek *skleros* and *pteryx, pteron* "wing," referring to the stigma.

Scleropyrum Arn. Santalaceae

Origins:

Greek *skleros* "hard, dry" and *pyros* "grain, wheat," referring to the drupaceous fruits.

Sclerorhachis (Rech.f.) Rech.f. Asteraceae

Origins:

From the Greek *skleros* "hard, dry" and *rhachis* "rachis, axis, midrib of a leaf."

Sclerosciadium W.D.J. Koch ex DC. Umbelliferae

Origins:

From the Greek *skleros* "hard, dry" and *skiadion, skiadeion* "umbel, parasol," see also *Capnophyllum* Gaertner.

Sclerosperma G. Mann & H.A. Wendland Palmae

Origins:

Greek *skleros* "hard, dry" and *sperma* "seed," the very hard endosperm in the seed.

Sclerostachya (Hackel) A. Camus Gramineae

Origins:

From the Greek *skleros* and *stachys* "spike, an ear of grain, ear of corn."

Sclerostegia Paul G. Wilson Chenopodiaceae

Origins:

Greek *skleros* "hard, dry" and *stege, stegos* "roof, shelter," referring to the pericarp; see Paul G. Wilson, "A revision of the Australian species of Salicornieae (Chenopodiaceae)." in *Nuytsia.* 3(1): 3-154. 1980.

Species/Vernacular Names:

S. arbuscula (R. Br.) Paul G. Wilson (*Salicornia arbuscula* R. Br.; *Arthrocnemum arbuscula* (R. Br.) Moq.)

English: shrubby glasswort

S. tenuis (Benth.) Paul G. Wilson (*Salicornia tenuis* Benth.; *Pachycornia tenuis* (Benth.) J. Black)

English: slender glasswort

Sclerostephane Chiov. Asteraceae

Origins:

Greek *skleros* "hard, dry" and *stephanos* "a crown."

Sclerothamnus R. Br. Fabaceae

Origins:

From the Greek *skleros* "hard, dry" and *thamnos* "shrub"; see W.T. Aiton, *Hortus Kewensis.* Ed. 2. 3: 16. 1811.

Sclerotheca A. DC. Campanulaceae

Origins:

From the Greek *skleros* "hard, dry" and *theke* "a box, case."

Sclerothrix C. Presl Loasaceae

Origins:

From the Greek *skleros* "hard, dry" and *thrix, trichos* "hair," referring to the nature of the plants.

Sclerotiaria Korovin Umbelliferae

Origins:

From the Greek *skleros* and *tiara* "a Persian diadem, a small crown."

Scleroxylon Bertol. Sapotaceae

Origins:

From the Greek *skleros* "hard" and *xylon* "wood."

Sclerozus Raf. Sapotaceae

Origins:

Greek *skleros* and *ozos* "branch, knot"; see C.S. Rafinesque, *Autikon botanikon.* Icones plantarum select. nov. vel rariorum, etc. 73. Philadelphia 1840 and *The Good Book.* 48.

1840; E.D. Merrill, *Index rafinesquianus*. 188. 1949; see also *Scclerocladus* Raf. and *Sideroxylon* L.

Scobinaria Seibert Bignoniaceae

Origins:

Latin *scobina* "a rasp."

Scoliaxon Payson Brassicaceae

Origins:

Greek *skolios* "curved, bent, crooked" and *axon* "axis."

Scoliochilus Reichb.f. Orchidaceae

Origins:

Greek *skolios* and *cheilos* "a lip," see also *Appendicula* Blume, from the Latin *appendicula, ae* "little appendix, a small appendage," alluding to the callus on the lip.

Scoliopus Torr. Trilliaceae (Liliaceae)

Origins:

Greek *skolios* "curved, bent" and *pous* "foot," *skoliopos* "looking askew, oblique," referring to the pedicels.

Scoliosorus T. Moore Vittariaceae (Adiantaceae)

Origins:

From the Greek *skolios* and *soros* "a spore case, a vessel for holding anything, a heap."

Scoliotheca Baillon Gesneriaceae

Origins:

From the Greek *skolios* "curved, bent" and *theke* "a box, case."

Scolochloa Link Gramineae

Origins:

Greek *skolos* "thorn, prickle," *skolops* "anything pointed" and *chloe, chloa* "grass."

Species/Vernacular Names:

S. festucacea (Willd.) Link

English: sprangletop

Scolochloa Mert. & W.D.J. Koch Gramineae

Origins:

From the Greek *skolos* "thorn, prickle," *skolops* "anything pointed" and *chloe, chloa* "grass."

Scolopendrium Adans. Aspleniaceae

Origins:

Latin *scolopendrion* for a plant also called *callitrichos, adiantum, capillus Veneris* and *scolibrochon*, Greek *skolopendrion* (see Dioscorides, *asplenos, asplenon*), *skolopendra* "millipede, millepede, scolopendra," referring to the sori and fronds, to a fancied likeness to the scolopendra; see Salvatore Battaglia, *Grande dizionario della lingua italiana*. XVIII: 122. Torino 1997; H. Genaust, *Etymologisches Wörterbuch der botanischen Pflanzennamen*. 570. 1996.

Scolophyllum Yamaz. Scrophulariaceae

Origins:

From the Greek *skolops* "anything pointed" and *phyllon* "leaf."

Scolopia Schreber Flacourtiaceae

Origins:

Greek *skolops, skolopos* "a thorn, a pointed stick, anything pointed," *skolos* "thorn, prickle," referring to the youngest shoots and the presence of thorns; see Johann Christian Daniel von Schreber (1739-1810), *Genera Plantarum*. 335. (Apr.) 1789.

Species/Vernacular Names:

S. ecklonii (Nees) Harv. (*Scolopia zeyheri* (Nees) Harv.; *Phoberos zeyheri* (Nees) Arn.; *Phoberos ecklonii* (Nees) Arn. ex Harv.; *Eriudaphus zeyheri* Nees; *Eriudaphus ecklonii* Nees) (after the German botanist and plant collector Carl Ludwig Philipp Zeyher, 1799-1858; see C.F. Ecklon, "Nachricht über die von Ecklon und Zeyher unternommenen Reisen und deren Ausbeute in botanischer Hinsicht." *Linnaea*. 8: 390-400. 1833, and Karl Boriwog Presl (1794-1852), *Botanische Bemerkungen*. Prague 1844)

English: thorn pear, Wolf's thorn

Southern Africa: doringpeer, hoenderspoor, wolwedoring, bergsaffran; iDungamuzi-lehlati, uMahlambahlale (meaning washes and does not go off) (Zulu); umQaqoba, iQumza elinameva (Xhosa); umTapana (Swazi); moretlhetle (Ngwaketse dialect, Botswana); mohlono (North Sotho); mutunu (Venda)

S. flanaganii (H. Bol.) Sim (for the plant collector and farmer Henry George Flanagan, 1861-1919)

English: small-leaved thorn pear, Flanagan's thorn pear, bastard mountain saffron

South Africa: fynblaardoringpeer

S. heterophylla (Lam.) Sleumer

Rodrigues Island: goyave marron

La Réunion Island: bois de tisane rouge, bois à fièvre, prune marron, bois de prune, bois d'oiseau, bois de Balai, bois de Quivi

Mauritius: bois de bouchon

S. mundii (Eckl. & Zeyh.) Warb. (*Eriudaphus mundii* Eckl. & Zeyh.; *Phoberos mundii* (Eckl. & Zeyh.) Harv.) (the tree was named after the German pharmacist and collector J.L. Leopold Mund, 1791-1831)

English: red pear, mountain saffron

Southern Africa: rooipeer, bergsaffraan, klipdoring; uLoyiphela, inGqumuza, iHlambahlale, iHlambahlala, iDungamuzi-lehlathi (Zulu); umNqanqa, isiNqandizembe, iQumza (Xhosa); umThabane (Swazi); mohlono (North Sotho)

S. oreophila (Sleum.) Killick (*Scolopia flanaganii* (H. Bol.) Sim var. *oreophila* Sleumer)

English: Wakkerstroom thorn pear

South Africa: Wakkerstroomse doringpeer

S. stolzii Gilg ex Sleum.

English: water thorn pear

South Africa: waterdoringpeer

Scolosanthus Vahl Rubiaceae

Origins:

Greek *skolos* "thorn, prickle, pointed stake" and *anthos* "flower."

Scolymus L. Asteraceae

Origins:

Greek *skolymos* "cardoon, an edible kind of thistle"; *scolymos, i* is the Latin name used by Plinius for *Scolymus hispanicus* L., the Spanish oyster-plant, for an edible kind of thistle, cardoon, *Cynara cardunculus* L.; see Carl Linnaeus, *Species Plantarum*. 813. 1753 and *Genera Plantarum*. Ed. 5. 355. 1754.

Species/Vernacular Names:
S. hispanicus L.
English: golden thistle, Spanish oyster-plant, Spanish oyster

S. maculatus L.
English: spotted thistle, spotted golden thistle
Portuguese: cardo, tigarro

Scoparebutia Fric & Kreuz. ex Buining Cactaceae

Origins:

Latin *scopa, ae* "twigs, shoots, a broom, thin branches" plus *Rebutia.*

Scoparia L. Scrophulariaceae

Origins:

Broomlike, from the Latin *scopa, ae* "twigs, shoots, a broom"; see Carl Linnaeus, *Species Plantarum*. 116. 1753 and *Genera Plantarum*. Ed. 5. 52. 1754.

Species/Vernacular Names:
S. dulcis L.
French: balai doux, herbe à balai
Japan: seitaka-kanabiki-sô
China: ye gan cao
Vietnam: da cam thao, cam thao dat, cam thao nam
Malaya: bunga baik salam, cha padang, dulis, teh makau
Sierra Leone: pondo livali
Yoruba: omisinmisin gogoro, mesaenmesen gogoro, olomu yinrin, bimobimo, mayinmayin
Congo: ginge, oye

Scopelogena L. Bolus Aizoaceae

Origins:

Greek *skopeo* "to contemplate, behold," *skopos* "a watcher, one that watches," *skopelos* "a high rock, peak" and *genos* "people, nation," *gennao* "to generate," referring to the habitat.

Scopolia Jacq. Solanaceae

Origins:

For the Austrian (b. Tirol, Trentino-Alto Adige, Italy) botanist Giovanni Antonio (Joannes Antonius, Johann Anton)

Scopoli, 1723-1788, physician, chemist, professor of chemistry and botany, plant collector, naturalist; see J.H. Barnhart, *Biographical Notes upon Botanists*. 3: 250. 1965; T.W. Bossert, *Biographical Dictionary of Botanists Represented in the Hunt Institute Portrait Collection*. 358. 1972; Jonas C. Dryander, *Catalogus bibliothecae historico-naturalis Josephi Banks*. 1800; A. Lasègue, *Musée botanique de Benjamin Delessert*. 1845; E.M. Tucker, *Catalogue of the Library of the Arnold Arboretum of Harvard University*. Cambridge, Massachusetts 1917-1933; Mariella Azzarello Di Misa, a cura di, *Il Fondo Antico della Biblioteca dell'Orto Botanico di Palermo*. 253-254. Regione Siciliana, Palermo 1988; R. Zander, F. Encke, G. Buchheim and S. Seybold, *Handwörterbuch der Pflanzennamen*. 14. Aufl. 779. Stuttgart 1993; F.A. Stafleu, *Linnaeus and the Linnaeans*. The spreading of their ideas in systematic botany, 1735-1789. Utrecht 1971.

Scopularia Lindley Orchidaceae

Origins:

Latin *scopulae, arum* "a little broom," *scopula* "a broom-twig," referring to the apex of the lip, see also *Holothrix* Rich. ex Lindley.

Scopulophila M.E. Jones Caryophyllaceae (Illecebraceae)

Origins:

Latin *scopulus, i* "a rock, cliff," Greek *skopelos* "peak, promontory" and *philos* "loving," referring to the habitat.

Species/Vernacular Names:

S. rixfordii (Brandegee) Munz & I.M. Johnston

English: Rixford rockwort

Scorodocarpus Becc. Olacaceae

Origins:

Greek *skorodon, skordon* "garlic" and *karpos* "fruit," onion-scented.

Species/Vernacular Names:

S. borneensis (Baillon) Becc.

Malaya: kulim, bawang hutan, ungsunah

Scorodophloeus Harms Caesalpiniaceae

Origins:

Greek *skorodon, skordon* "garlic" and *phloios* "bark of trees."

Species/Vernacular Names:

S. sp.

Zaire: mvenzi, oghidi, tshipanda, ufiri, wenzi, yenghe

S. zenkeri Harms

Zaire: bofiri, bofili, mufili, ufili

Congo: divida, muniengili

Gabon: bongaso, nsigna, esung-ngang, essoun, liviza

Cameroon: bobombi, olom, olan, bobimbi, essoun, fondi, yomi, ngo, mingegne

Scorodosma Bunge Umbelliferae

Origins:

From the Greek *skorodon, skordon* "garlic" and *osme* "smell, odor, perfume."

Scorpia Ewart & A. Petrie Tiliaceae

Origins:

Derivation obscure, possibly from the Greek *skorpios* "a scorpion," referring to the pods or to the flower-cymes; see Alfred James Ewart (1872-1937) & A.H.K. Petrie (fl. 1926), in *Proceedings of the Royal Society of Victoria*. Ser. 2, 38: 169, textfig. 2. (Jul.) 1926; F.A. Sharr, *Western Australian Plant Names and Their Meanings*. 63. University of Western Australia Press 1996.

Scorpiothyrsus H.L. Li Melastomataceae

Origins:

From the Greek *skorpios* "a scorpion" and *thyrsos* "a panicle."

Scorpiurus L. Fabaceae

Origins:

Greek *skorpios* "a scorpion" and *oura* "a tail," alluding to the shape of the twisted pods; see Carl Linnaeus, *Species Plantarum*. 744. 1753 and *Genera Plantarum*. Ed. 5. 332. 1754.

Species/Vernacular Names:

S. muricatus L.

English: scorpiontail

Arabic: zanab el'aqrab

Scortechinia Hook.f. Euphorbiaceae

Origins:

After the Italian botanist Rev. Benedetto (Bertold) Scortechini, 1845-1886 (Calcutta), botanical explorer, clergyman, traveler, in Australia and Malay Peninsula, collected ferns, 1881 Fellow of the Linnean Society, collaborated with Bailey and von Mueller; see J.H. Barnhart, *Biographical Notes upon Botanists*. 3: 251. 1965; T.W. Bossert, *Biographical Dictionary of Botanists Represented in the Hunt Institute Portrait Collection*. 358. 1972; D.J. Carr and S.G.M. Carr, eds., *People and Plants in Australia*. 1981; Henry Nicholas Ridley (1855-1956), *The Flora of the Malay Peninsula*. London 1922-1925; I.H. Vegter, *Index Herbariorum*. Part II (6), *Collectors S*. Regnum Vegetabile vol. 114. 1986; M.N. Chaudhri, I.H. Vegter and C.M. De Wal, *Index Herbariorum*, Part II (3), *Collectors I-L*. Regnum Vegetabile vol. 86. 1972; Ray Desmond, *Dictionary of British & Irish Botanists and Horticulturists*. 613. London 1994.

Scorzonera L. Asteraceae

Origins:

Italian *scorzone, scherzone, scorsone, scurzone, saettone* "a snake, an adder," Spanish *escorzonera* or *escuerçonera*, Catalan *escurçonera* (*escurçó* "viper"), Latin *excurtio, onis*, *curtio, onis*, spoken Latin *excurtionem, curtionem* (*curtus* "short"), referring to its use as antidote to snake-bite; see Carl Linnaeus, *Species Plantarum*. 790. 1753 and *Genera Plantarum*. Ed. 5. 346. 1754; B. Migliorini, *Storia della lingua italiana*. Firenze 1960; Salvatore Battaglia, *Grande dizionario della lingua italiana*. XVIII: 265-266. Torino 1997.

Species/Vernacular Names:

S. hispanica L.

English: Spanish salsify, viper's grass

Italian: scorzonera viperina

Scotanthus Naudin Cucurbitaceae

Origins:

Greek *skotos* "darkness, gloom" and *anthos* "flower."

Scottellia Oliver Flacourtiaceae

Origins:

For the English (b. Calcutta, India) botanist George Francis Scott-Elliot, 1862-1934 (d. Dumfries, Scotland), 1890 Fellow of the Linnean Society, botanist on the Sierra Leone Boundary Commission, plant collector in Sierra Leone (1891-1892) and in East Africa (with the British East Africa Expedition, 1893-1894), President of the Antiquarian Society. Among his writings are *Report on the District Traversed by the Anglo-French Boundary Commission*. Sierra Leone. Botany. 1893, *The Flora of Dumfriesshire*. Dumfries 1896 and *A Naturalist in Mid-Africa*: being an account of a journey to the Mountains of the Moon and Tanganyika. London 1896; see Samuel P. Oliver, *The Life of Philibert Commerson*. Edited by G.F. Scott-Elliot. London 1909; John H. Barnhart, *Biographical Notes upon Botanists*. 1: 504. 1965; Benjamin Daydon Jackson (1846-1927), "A list of the contributors to the herbarium of the Royal Botanic Gardens, Kew, brought down to 31st December 1899." *Bull. Misc. Inf. Kew*. 1901 and "A list of the collectors whose plants are in the herbarium of the Royal Botanic Gardens, Kew, to 31st December 1899." in *Kew Bulletin*. 1-80. 1901; Mary Gunn and Leslie E. Codd, *Botanical Exploration of Southern Africa*. 320. Cape Town 1981; [Sotheby's — Marquess of Bute], *Catalogue of a Portion of the Valuable Library from Dumfries House, Ayrshire*. The Property of the Most Hon. The Marquess of Bute. The First Portion: the important collection of mathematical and scientific books. Sale of 3-4 July 1961. The Second Portion: Americana, early printed books, travel, early Italian literature, bibliography, books on the arts and architecture. Sale of 16-18 Oct. 1961. London; F.N. Hepper and Fiona Neate, *Plant Collectors in West Africa*. 73-74. 1971; G. Murray, *History of the Collections Contained in the Natural History Departments of the British Museum*. London 1904; E.M. Tucker, *Catalogue of the Library of the Arnold Arboretum of Harvard University*. Cambridge, Massachusetts 1917-1933; Auguste Jean Baptiste Chevalier (1873-1956), *Flore vivante de l'Afrique Occidentale Française*. 1: xxvii-xxx. Paris 1938.

Species/Vernacular Names:

S. coriacea A. Chev. ex Hutch. & Dalz.

Central Africa: ngobosolbo, kelembicho

Ivory Coast: akossika

Nigeria: odoko, otoko, emwenfuohai, emuefuohai, eranfohai, odoliyan; odoko (Yoruba); emuefuohai (Edo); eranfohi (Ishan); akporo (Igbo); uguokpa (Engenni)

Yoruba: odoko, eluro-orunge, elure, otoko

Cameroon: ngobisolbo, nkomonome

Gabon: bilogh-bi-nkele

S. mimfiensis Gilg

Nigeria: emuefuohai eze (Edo); ewono (Ijaw)

Scottia Grönblad Chlorophyceae-Desmidiaceae

Origins:

After the English engineer Arthur Moreland Scott, 1888-1963, phycologist; see J.H. Barnhart, *Biographical Notes upon Botanists.* 3: 251. 1965; Rolf Leo Grönblad (1895-1962), *Bot. Not.* 1954: 167. 22 June 1954; T.W. Bossert, *Biographical Dictionary of Botanists Represented in the Hunt Institute Portrait Collection.* 358. 1972; Stafleu and Cowan, *Taxonomic Literature.* 5: 457. 1985.

Scottia R. Brown ex W. T. Aiton Fabaceae

Origins:

In honor of the Irish botanist Robert Scott, 1757-1808 (d. Dublin), bryologist, physician and plant collector, from 1785 to 1808 professor of botany, Trinity College, Dublin, friend of Dawson Turner (1775-1858), contributed to James Sowerby (1757-1822) and J.E. Smith's *English Botany.* See T.P.C. Kirkpatrick, *History of the Medical Teaching in Trinity College, Dublin.* Dublin 1912; W.T. Aiton, *Hortus Kewensis.* Ed. 2. 4: 268. (Dec.) 1812; Lady Pleasance Smith, ed., *Memoir and Correspondence of ... Sir J.E. Smith.* London 1832.

Scottia Thunb. Caesalpiniaceae

Origins:

After the Dutch traveler and naturalist Richard van der Schot, c. 1730-1819, born in Holland of French parents, head gardener at the palace of Schönbrunn near Vienna, pupil of Jussieu, friend and traveling companion (between 1754-1756) of the Dutch-born Austrian botanist Nicolaus (Nicolaas or Nikolaus) Joseph von Jacquin (1727-1817); see also *Schotia* Jacq.

Scribneria Hackel Gramineae

Origins:

For the American botanist Frank Lamson-Scribner (Franklin Pierce Lamson), 1851-1938, agrostologist. His writings include *Useful and Ornamental Grasses.* [see United States of America–Division of Agrostology. *Bulletin*, no. 3.] 1896, "Notes on the grasses in the Bernhardi Herbarium, collected by Thaddeus Haenke, and described by J.S. Presl." *Annual Rep. Missouri Bot. Gard.* 10: 35-59. 1899 and "New or little known grasses." *U.S.D.A. Circ.* 30: 1-8. 1901; see John H. Barnhart, *Biographical Notes upon Botanists.* 2: 339. Boston 1965; J. Ewan, ed., *A Short History of Botany in the United States.* New York and London 1969; S. Lenley et al., *Catalog of the Manuscript and Archival Collections and Index to the Correspondence of John Torrey.* Library of the New York Botanical Garden. 363-364. 1973; J.W. Harshberger, *The Botanists of Philadelphia and Their Work.* 358-360. 1899; Thomas Henry Kearney (1874-1956), in *Leafl. Western Bot.* 8(12): 276. 1958; Ida Kaplan Langman, *A*

Selected Guide to the Literature on the Flowering Plants of Mexico. University of Pennsylvania Press, Philadelphia 1964; Joseph Ewan, *Rocky Mountain Naturalists.* The University of Denver Press 1950; Joseph William Blankinship (1862-1938), "A century of botanical exploration in Montana, 1805-1905: collectors, herbaria and bibliography." in *Montana Agric. Coll. Sci. Studies Bot.* 1: 1-31. 1904.

Scrobicaria Cass. Asteraceae

Origins:

Latin *scrobis* or *scrobs* "a ditch, trench."

Scrobicularia Mansf. Melastomataceae

Origins:

Latin *scrobiculus* "a little ditch," *scrobis* or *scrobs* "a ditch, trench."

Scrofella Maxim. Scrophulariaceae

Origins:

Latin *scrofulae, arum* "scrofula," Latin *scrofa, ae* "a breeding-sow," the diminutive.

Scrophularia L. Scrophulariaceae

Origins:

Latin *scrofulae, arum* "scrofula, swelling of the glands of the neck," *scrofa, ae* "swine-breeding," *herba scrophularia,* Greek *choirades* "scrofula" and *choiros, choira* "a pig," referring to the medicinal properties; see Carl Linnaeus, *Species Plantarum.* 619. 1753 and *Genera Plantarum.* Ed. 5. 271. 1754; H. Genaust, *Etymologisches Wörterbuch der botanischen Pflanzennamen.* 572. 1996; G.C. Wittstein, *Etymologisch-botanisches Handwörterbuch.* 804. 1852; Salvatore Battaglia, *Grande dizionario della lingua italiana.* XVIII: 31-313. Torino 1997.

Species/Vernacular Names:

S. atrata Pennell

English: black-flowered figwort

S. auriculata L.

English: water figwort

S. californica Chamisso & Schltdl.

English: California figwort

S. lanceolata Pursh

English: figwort

S. marilandica L.

English: pilewort, carpenter's-square

S. ningpoensis Hemsl.

English: Ningpo figwort, black figwort

China: xuan shen

Vietnam: hac sam

S. nodosa L.

English: figwort

S. villosa Pennell

English: Santa Catalina figwort

Scrotochloa Judz. Gramineae

Origins:

Possibly from the Latin *scrotum, i* "scrotum" and Greek *chloe, chloa* "grass," see also *Leptaspis* R. Br., Greek *leptos* "slender" and *aspis* "a shield," the fruit is enclosed in the enlarged glume.

Scubulon Raf. Solanaceae

Origins:

See Constantine Samuel Rafinesque (1783-1840), *Autikon botanikon. Icones plantarum select. nov. vel rariorum*, etc. 109. Philadelphia 1840; E.D. Merrill, *Index rafinesquianus.* 212. 1949.

Scubulus Raf. Solanaceae

Origins:
See *Scubulon* Raf.

Scurrula L. Loranthaceae

Origins:

Latin *scurrula, ae* "a little buffoon," *scurra, ae* "a guardsman, a dandy," *scurrantis speciem praebere* "of a buffoonish parasite" (Quintus Horatius Flaccus).

Scutachne Hitchc. & Chase Gramineae

Origins:

Latin *scutum, i* "an oblong shield, a shield" and Greek *achne* "chaff, glume."

Scutellaria L. Labiatae

Origins:

Latin *scutella, ae* "a small shield or dish, saucer," the upper calyx lip has a crest or a pouch; see Carl Linnaeus, *Species Plantarum*. 598. 1753 and *Genera Plantarum*. Ed. 5. 260. 1754.

Species/Vernacular Names:

S. amoena C.H. Wright

English: Yunnan skullcap

China: dian huang qin

S. anhweiensis C.Y. Wu

English: Anhui skullcap

China: an hui huang qin

S. baicalensis Georgi (*Scutellaria macrantha* Fisch.)

English: Baikal skullcap, Chinese skullcap, Baical skullcap

China: huang qin, huang chin

S. barbata D. Don

English: barbed skullcap

China: ban zhi lian

S. caryopteroides Hand.-Mazzetti

English: bluebeard-like skullcap

China: you zhang huang qin

S. chihshuiensis C.Y. Wu & H.W. Li

English: Chihshui skullcap

China: chi shui huang qin

S. chimenensis C.Y. Wu

English: Qimen skullcap

China: qi men huang qin

S. chungtienensis C.Y. Wu

English: Zhongdian skullcap

China: zhong dian huang qin

S. delavayi H. Léveillé

English: Delavay skullcap

China: fang zhi huang qin

S. discolor Wallich ex Bentham

English: discolored skullcap

China: yi se huang qin

S. formosana N.E. Brown

English: blueflower skullcap

China: lan hua huang qin

S. forrestii Diels

English: Forrest skullcap

China: hui yan huang qin

S. franchetiana Lévl.

English: Franchet skullcap

China: yan huo xiang

S. galericulata L.

English: marsh skullcap

S. hainanensis C.Y. Wu

English: Hainan skullcap

China: hai nan huang qin

S. honanensis C.Y. Wu & H.W. Li

English: Henan skullcap

China: he nan huang qin

S. humilis R. Br.

English: skullcap, dwarf skullcap

S. hunanensis C.Y. Wu

English: Hunan skullcap

China: hu nan huang qin

S. hypericifolia H. Léveillé

English: St. John's wort-leaf skullcap

China: lian qiao ye huang qin

S. indica L.

English: Indian skullcap

Japan: tatsu-nami-sô

China: han xin cao

S. lateriflora L.

English: blue skullcap

S. likiangensis Diels

English: Likiang skullcap

China: li jiang huang qin

S. mairei H. Léveillé

English: Maire skullcap

China: mao jing huang qin

S. mollis R. Br.

English: soft skullcap

S. moniliorrhiza Komarov

English: moniliform skullcap

China: nian zhu gen jing huang qin

S. orthocalyx Handel-Mazzetti

English: erect-calyx skullcap

China: zhi e huang qin

S. pekinensis Maxim.

English: Peking skullcap

China: jing huang qin

S. quadrilobulata Sun ex C.H. Hu

English: fourlobed skullcap

China: si lie hua huang qin

S. racemosa Pers.

English: skullcap

S. scordifolia Fischer ex Schrank

English: twinflower skullcap

China: bing tou huang qin

S. sessilifolia Hemsley

English: sessile skullcap

China: shi wu gong cao

S. shansiensis C.Y. Wu & H.W. Li

English: Shanxi skullcap

China: shan xi huang qin

S. taiwanensis C.Y. Wu

English: Taiwan skullcap

China: tai wan huang qin

S. tuberifera Wu & Chen

English: tuberous skullcap

S. tuminensis Nakai

English: Tumen skullcap

China: tu men huang qin

S. viscidula Bunge

English: viscid-hair skullcap

China: zhan mao huang qin

S. wenshanensis C.Y. Wu & H.W. Li

English: Wenshan skullcap

China: wen shan huang qin

S. yangbiense H.W. Li

English: Yangbi skullcap

China: qian ma ye huang qin

S. yunnanensis H. Léveillé

English: Yunnan skullcap

China: hong jin huang qin

Scutia (DC.) Brongn. Rhamnaceae

Origins:
Latin *scutum, i* "a shield," referring to the form of the disk.

Species/Vernacular Names:
S. myrtina (Burm.f.) Kurz (*Scutia commersonii* Brongn.; *Rhamnus myrtina* Burm.f.) (the specific name means "myrtle-like")

English: cat-thorn

Rodrigues Island: bambara

Southern Africa: katdoring, katdoringnaels, katnael, katnaal, droog-my-keel (= dry-my-throat), droog-my-keel bossie, drogies, spingle, spindle, bobbejaantou, bos-wag-'n-bietjie, rank-wag-'n-bietjie, wag-'n-bietjie; uSondela, isiBinda, isiPinga, uSondeza, uSondelanganga (meaning approach and let me kiss), umsondeza (Zulu); isiPhingo, isipinga, umSondezo (that which is brought near), umQapuna, umQaphula, umQokwane (Xhosa)

Scuticaria Lindl. Orchidaceae

Origins:

Latin *scutica, ae* "a lash, whip," Greek *skytikos* "skilled in shoemaking," *skytinos* "made of leather," referring to the shape of the pendant leaves.

Scutinanthe Thwaites Burseraceae

Origins:

Greek *skytinos* "leathern, made of leather" and *anthos* "flower."

Scutula Lour. Melastomataceae

Origins:

Latin *scutula, ae* "a little dish, a lozenge-shaped figure," referring to the calyx.

Scybalium Schott & Endl. Balanophoraceae

Origins:

Greek *skybalon* "dung, excrement." Some suggest an origin from *skyphos* "a cup, goblet, a jug" and *balios* "spotted"; see Helmut Genaust, *Etymologisches Wörterbuch der botanischen Pflanzennamen.* 572. Basel 1996.

Scyopteris C. Presl Polypodiaceae

Origins:

Greek *skia* "shadow, shade" and *pteris* "a fern."

Scyphanthus R. Sweet Loasaceae

Origins:

From the Greek *skyphos* "a cup, goblet, a jug" and *anthos* "flower."

Scyphellandra Thwaites Violaceae

Origins:

Greek *skyphos* and *aner, andros* "man, stamen," Latin *scyphulus, i* "a small cup."

Scyphiphora Gaertn.f. Rubiaceae

Origins:

Greek *skyphos* and *phero, phoreo, phorein* "to bear," *phoros* "bearing, carrying," referring to the flowers or to the ridged fruits; see Carl (Karl) Friedrich von Gaertner (1772-1850), *Supplementum carpologicae* seu continuati operis Josephi Gaertner de fructibus et seminibus plantarum. 91. Mai 1806.

Species/Vernacular Names:

S. hydrophylacea C.F. Gaertner

English: common scyphiphora, yamstick mangrove

Malaya: chengam, chingam, sebasah, singam

Scyphocephalium Warb. Myristicaceae

Origins:

From the Greek *skyphos* and *kephale* "head," referring to the flowers.

Species/Vernacular Names:

S. ochocoa Warb.

Cameroon: eboutzok, goklom, soko, n'jion, nometen, noumeten

Gabon: issombo, isombe, isombo, tsisombo, tsogo, usoro, usozo, izombe, nsoko, soko, n'soro, n'soultou, ochoco, ochoko, otchoco, osoko, ossoko, otsoko, otsuku, sorro, sogo, sogoe, sogoue-sousou, sokoue, sokwe, sorho, sougo

Nigeria: ejinjin

Zaire: tshidungadunga, tshifula

Scyphochlamys Balf.f. Rubiaceae

Origins:

Greek *skyphos* and *chlamys, chlamydos* "cloak, mantle"; see Wendy Strahm, *Plant Red Data Book for Rodrigues.* Published by Koeltz Scientific Books, Königstein for IUCN, The World Conservation Union 1989 and "The vegetation of the Mascarene Islands," "Botanical history of the Mascarene Islands," "Conservation of the flora of the Mascarene Islands," *Curtis's Botanical Magazine.* Volume 13. 4: 214-237. November 1996.

Species/Vernacular Names:

S. revoluta Balf.f.

Rodrigues Island: bois mangue

Scyphocoronis A. Gray Asteraceae

Origins:

Greek *skyphos* and Latin *corona, ae* "a crown," referring to a cup surrounding the achene; see A. Gray, *Hooker's Icones Plantarum*. New Ser. 5. 9, t. 854. (Apr.-Dec.) 1851.

Scyphofilix Thouars Dennstaedtiaceae

Origins:

Latin *scyphus* "a cup, goblet" and *filix* "fern."

Scyphogyne Decne. Ericaceae

Origins:

Greek *skyphos* "a cup, goblet, beaker" and *gyne* "a woman, female," referring to the shape of the stigma.

Scypholepia J. Sm. Dennstaedtiaceae

Origins:

From the Greek *skyphos* and *lepis* "a scale," referring to the indusia.

Scyphonychium Radlk. Sapindaceae

Origins:

From the Greek *skyphos* and *onyx, onychos* "a claw, nail."

Scyphopappus Nordenstam Asteraceae

Origins:

Greek *skyphos* "a cup, beaker" and *pappos* "fluff."

Scyphopetalum Hiern Sapindaceae

Origins:

From the Greek *skyphos* "a cup, beaker" and *petalon* "petal, leaf."

Scyphopteris Raf. Dennstaedtiaceae

Origins:

From the Greek *skyphos* "a cup" and *pteris* "a fern"; see C.S. Rafinesque, *Principes fondamentaux de Somiologie*. 26. Palerme 1814 and *Flora Telluriana*. 1: 84. 1836 [1837]; E.D. Merrill, *Index rafinesquianus*. 72. 1949.

Scyphostachys Thwaites Rubiaceae

Origins:

Greek *skyphos* "a cup" and *stachys* "spike, ear of corn."

Scyphostegia Stapf Scyphostegiaceae (Dilleniidae, Violales)

Origins:

Greek *skyphos* "cup, can" and *stege* "covering, shelter," possibly referring to the tubular bracts or to the large nectary glands.

Scyphostelma Baillon Asclepiadaceae

Origins:

From the Greek *skyphos* and *stelma, stelmatos* "a girdle, belt."

Scyphostigma M. Roemer Meliaceae

Origins:

From the Greek *skyphos* "a cup, goblet" and *stigma* "a stigma."

Scyphostrychnos S. Moore Strychnaceae (Loganiaceae)

Origins:

Greek *skyphos* with the genus *Strychnos* L., Greek *strychnon, strychnos* "acrid, bitter."

Scyphosyce Baillon Moraceae

Origins:

Greek *skyphos* "a cup, goblet, beaker" and *sykon* "fig."

Scyphularia Fée Davalliaceae

Origins:
From the Latin *scyphulus, i* "a small cup."

Scytanthus T. Anderson ex Bentham Acanthaceae

Origins:
From the Greek *skytos* "leather, a skin, hide" and *anthos* "flower."

Scytopetalum Pierre ex Engl. Scytopetalaceae (Dilleniidae, Theales)

Origins:
From the Greek *skytos* "leather, whip, hide" and *petalon* "petal," referring to the linear lobes.

Species/Vernacular Names:
S. pierreanum (De Wild.) Tieghem
Zaire: libenge, bonkombe, sanga

Scytophyllum Ecklon & Zeyher Celastraceae

Origins:
From the Greek *skytos* "leather, a skin, hide" and *phyllon* "leaf."

Scytopteris Presl Polypodiaceae

Origins:
Greek *skytos* "leather, a skin, hide" and *pteris* "fern"; see Karl B. Presl, *Epimeliae botanicae*. 133. Pragae 1851 [reprinted from *Abhandlungen der Königlichen Böhmischen Gesellschaft der Wissenschaften*. Ser. 5, 6: 493. (Oct.) 1851].

Seaforthia R. Br. Palmae

Origins:
After Francis Mackenzie Humberston, Lord Seaforth and Mackenzie, 1754-1815, plant collector, Lieut.-General, from 1800 to 1806 Governor of Barbados, 1794 Fellow of the Royal Society, 1796 Fellow of the Linnean Society, sent algae to Dawson Turner and plants to Aylmer Bourke Lambert (1761-1842); see *Catalogue of the Library which belonged to the Right Honourable the late Lord Seaforth* ... (Arranged by C. Elliot.) It is wished to sell this collection ... before ... 31st January 1816; till which day, offers ...

will be received by Messrs. Mackenzie & Monypenny, W.S. Edinburgh, or Mr. Cochrane, 63 Judd Street, Brunswick Square, London. Edinburgh 1815; R. Desmond, *Dictionary of British & Irish Botanists and Horticulturists*. 364. 1994; *A Catalogue of a Portion of the Library of the late Right Hon. Francis Lord Seaforth*: which will be sold by auction, by Mr. Cochrane, on the 3rd & 4th of May 1816. London 1816; H.S. Miller, "The herbarium of Aylmer Bourke Lambert notes on its acquisition, dispersal, and present whereabouts." *Taxon*. 19: 489-553. 1970; Robert Brown, *Prodromus florae Novae Hollandiae et Insulae van-Diemen*. 267. London 1810; Henry C. Andrews (1794-1830), *The Botanist's Repository*. London 1807-1808.

Sebaea Solander ex R. Br. Gentianaceae

Origins:
For the Dutch naturalist Albertus Seba, 1665-1736, apothecary, botanist, traveler, author of *Locupletissimi rerum naturalium thesauri accurata descriptio, et iconibus artificiosissimis expressio, per universam physices historiam*, etc. Amstelaedami [Amsterdam] 1734-1765; see Robert Brown, *Prodromus florae Novae Hollandiae et Insulae van-Diemen*. 451. London 1810; *Sale Catalogue of the Second Natural History Collection of A. Seba*, sold in lots after his death. [Amsterdam 1737]; F.A. Sharr, *Western Australian Plant Names and Their Meanings*. 63. University of Western Australia Press, Nedlands, Western Australia 1996; James A. Baines, *Australian Plant Genera. An Etymological Dictionary of Australian Plant Genera*. 337. Chipping Norton, N.S.W. 1981; Helmut Genaust, *Etymologisches Wörterbuch der botanischen Pflanzennamen*. 573. [from the Latin *sebum* "tallow, grease"] Basel 1996.

Species/Vernacular Names:
S. albidiflora F. Muell.
English: white sebaea, white centaury
S. hymenosepala Gilg
Southern Africa: iBulawa (Xhosa)
S. microphylla (Edgeworth) Knoblauch
English: littleleaf sebaea
China: xiao huang guan
S. ovata (Labill.) R. Br. (*Exacum ovatum* Labill.)
English: yellow sebaea, yellow centaury

Sebastiania Sprengel Euphorbiaceae

Origins:
For the Italian botanist Francesco Antonio Sebastiani, 1782-1821, physician, professor of botany, from 1815 to 1820 Director of the Rome Botanical Garden (first Hortus, behind

the fountain of Acqua Paola, Gianicolo). Among his publications are *Esposizione del sistema di Linneo*. Roma 1819, *Romanarum plantarum* fasciculus primus. Romae 1813 and *Romanarum plantarum* fasciculus alter. Romae 1815, with Ernesto Mauri (1791-1836) wrote *Florae romanae prodromus*. Romae 1818; see John H. Barnhart, *Biographical Notes upon Botanists*. 3: 254. 1965; Kurt P.J. Sprengel, *Neue Entdeckungen im ganzen Umfang der Pflanzenkunde*. 2: 118. Leipzig 1821; M. Catalano and E. Pellegrini, *L'Orto Botanico di Roma*. Roma 1975; Mariella Azzarello Di Misa, a cura di, *Il Fondo Antico della Biblioteca dell'Orto Botanico di Palermo*. 254-255. Palermo 1988; Ethelyn Maria Tucker, *Catalogue of the Library of the Arnold Arboretum of Harvard University*. Cambridge, Massachusetts 1917-1933; T.W. Bossert, *Biographical Dictionary of Botanists Represented in the Hunt Institute Portrait Collection*. 359. 1972; A. Lasègue, *Musée botanique de Benjamin Delessert*. 1845; R. Zander, F. Encke, G. Buchheim and S. Seybold, *Handwörterbuch der Pflanzennamen*. 14. Aufl. Stuttgart 1993.

Species/Vernacular Names:
S. bilocularis S. Watson
Spanish: yerba de la feche

Sebastiano-schaueria Nees Acanthaceae

Origins:
After Sebastian Schauer, fl. 1847, published with C.G. Nees von Esenbeck; see Stafleu and Cowan, *Taxonomic Literature*. 5: 119. 1985.

Sebertia Pierre ex Engler Sapotaceae

Origins:
For the French Hippolyte Sebert, 1839-1930, traveler, with I. Pancher wrote *Notice sur les bois de la Nouvelle Calédonie*. Paris [1874]; see Ethelyn Maria Tucker, *Catalogue of the Library of the Arnold Arboretum of Harvard University*. Cambridge, Massachusetts 1917-1933; Elmer Drew Merrill, *Bernice P. Bishop Mus. Bull*. 144: 166. 1937.

Seborium Raf. Euphorbiaceae

Origins:
Latin *sebo, are* "to make of tallow," *sebosus, a, um* "tallowy, greasy," possibly referring to *Stillingia sebifera* Michx. and *Sapium sebifera* (L.) Roxb.; see C.S. Rafinesque, *Sylva Telluriana*. 63. 1838; E.D. Merrill, *Index rafinesquianus*. 156. 1949.

Secale L. Gramineae

Origins:
Latin *secale* (*sicale*), *secalis* "a kind of grain, rye, black spelt," *seges, segetis* "a cornfield, a field, ground," Akkadian *se'u, sehu* "grain, corn"; see Carl Linnaeus, *Species Plantarum*. 84. 1753 and *Genera Plantarum*. Ed. 5. 36. 1754; S. Battaglia, *Grande dizionario della lingua italiana*. XVIII: 449-450. 1997; H. Genaust, *Etymologisches Wörterbuch der botanischen Pflanzennamen*. 573. 1996; M. Cortelazzo & P. Zolli, *Dizionario etimologico della lingua italiana*. 5: 1174-1175. 1988; N. Tommaseo & B. Bellini, *Dizionario della lingua italiana*. Torino 1865-1879; Giovanni Semerano, *Le origini della cultura europea*. Dizionario della lingua Latina e di voci moderne. 2(2): 559, 560. Firenze 1994.

Species/Vernacular Names:
S. cereale L.
English: rye, cereal rye
Peru: centeno
Italian: ségale, sécale, sécola, ségala, ségalla, ségla, ségola
China: mai jiao

Secalidium Schur Gramineae

Origins:
Referring to *Secale* L., see also *Dasypyrum* (Cosson & Durieu) T. Durand.

Secamone R. Br. Asclepiadaceae

Origins:
Derived from the Arabic name *squamona* used for *Secamone aegyptiaca*; see Robert Brown, *Prodromus florae Novae Hollandiae et Insulae van-Diemen*. 464. London 1810.

Species/Vernacular Names:
S. afzelii (Schultes) K. Schumann
Yoruba: arilu, ailu, itakunailu
S. alpinii Schultes
Southern Africa: muKangaza
S. bonii Costantin
English: Bon secamone
China: ban pi ji yu teng
S. elliptica R. Br. (*Secamone lanceolata* Blume)
English: secamone, common secamone
China: ji yu teng

S. ferruginea Pierre ex Costantin

English: rusty-hair secamone

China: xiu mao ji yǔ teng

S. minutiflora (Woodson) Tsiang (*Secamone szechuanensis* Tsiang & P.T. Li)

English: Chinese secamone, Szechwan secamone

China: cui tu ji yu teng

S. sinica Handel-Mazzetti

English: Chinese secamone

China: dia shan tao

Secamonopsis Jum. Asclepiadaceae

Origins:
Resembling *Secamone* R. Br.

Sechiopsis Naudin Cucurbitaceae

Origins:
Resembling *Sechium* P. Browne.

Sechium P. Browne Cucurbitaceae

Origins:
From the Greek *sekos* "weight," food for pigs; some suggest from the Greek *sikyos* "wild cucumber, gourd," or from a West Indian name, possibly from vernacular *chacho*; see Patrick Browne (1720-1790), *The Civil and Natural History of Jamaica in Three Parts*. 355. London 1756; S. Battaglia, *Grande dizionario della lingua italiana*. UTET, Torino 1997; Georg Christian Wittstein, *Etymologisch-botanisches Handwörterbuch*. 806. Ansbach 1852.

Species/Vernacular Names:
S. edule (Jacq.) Sw. (*Sycios edulis* Jacq.)

English: chow chow, vegetable pear, Christophine, chayote, Madeira marrow

Central America: chayote, choyote, chocho, pipinella, chocho, choco, choko, güisquil, chimá, pataste, perulero

Mexico: apa, ape, yapa, yape, guu yau, yu yau

Peru: gayota

Japan: hayato-uri, chayote

The Philippines: tsayote

Italian: sechio, zucca centenaria

Secondatia A. DC. Apocynaceae

Origins:
For the French botanist Jean Baptiste de Secondat, 1716-1796. Among his works are *Mémoire sur l'électricité*. 1746 and *Mémoires sur l'histoire naturelle du Chêne*. Paris 1785; see Jonas C. Dryander, *Catalogus bibliothecae historico-naturalis Josephi Banks*. 1800; E.M. Tucker, *Catalogue of the Library of the Arnold Arboretum of Harvard University*. Cambridge, Massachusetts 1917-1933.

Secula Small Fabaceae

Origins:
From the Latin *secula, ae* "a sickle."

Securidaca L. Polygalaceae

Origins:
Latin *securis, is* "hatchet, an axe," *securicula* "a little axe," referring to the shape of dorsal-winged samara; Plinius used *securidaca* or *securiclata* for a weed growing among lentils, the hatchet-vetch.

Species/Vernacular Names:
S. longipedunculata Fres. (*Lophostylis pallida* Klotzsch; *Securidaca spinosa* Sim; *Securidaca longipedunculata* var. *parvifolia* Oliv.)

English: violet tree, fiber tree, tree violet

French: arbre à serpents, arbre aux hachettes

N. Rhodesia: mutata

Yoruba: ipeta

Nigeria: ipeta, sainya, sanya, uwar-magunguna, malaria; sainyia, sanya, womagunguna, uwar magunguna (= mother of medicines) (Hausa); alali (Fula); epeta (Yoruba)

East Africa: mkola, mteyu

W. Africa: alale, joro, juri

Southern Africa: krinkhout; tsatsu (Thonga); mmaba (western Transvaal, northern Cape, Botswana); mmaba (North Sotho); muFubfavana, muFugwe, muFufu, muPupu, muTangeni (Shona); mpesu (Venda); mofufu (Subya: Botswana, eastern Caprivi)

Ghana: foji, fosi

Zaire: moo-aye

Congo: djomba

Gabon: pele

Sudan: youro

Mali: jori, joto

Arabia: sagat

Securigera DC. Fabaceae

Origins:
Latin *securiger, gera, gerum* "axe-bearing."

Securinega Comm. ex Juss. Euphorbiaceae

Origins:
Latin *securis, -is* "a hatchet, an axe" and *nego, -avi, -atum, -are* "to refuse, to negate," the wood is very strong and hard and not easy to cut; see A.L. de Jussieu, *Genera Plantarum.* 388. 1789.

Species/Vernacular Names:
S. sp.
Yoruba: iran owo
S. durissima J. Gmelin
English: Otaheite myrtle
Rodrigues Island: bois rouge, bois dur
S. suffruticosa (Pallas) Rehd. (*Flueggea suffruticosa* (Pallas) Baill.; *Securinega ramiflora* (Ait.) Müll. Arg.; *Securinega flueggeoides* Müll. Arg.; *Pharnaceum suffruticosum* Pallas)
China: yi ye qiu
Japan: hito-tsuba-hagi
S. virosa (Willd.) Baill. (*Flueggea microcarpa* Blume; *Flueggea virosa* Roxb. ex Willd.; *Phyllanthus virosus* Roxb. ex Willd.) (Latin *virosus, -a, -um* "stinking, poisonus, smelly," *virus, i* "slime, poison")
English: white-berry bush, white waterberry
W. Africa: jene, jeme
Yoruba: iranje, iranje eluju, iranje ogan, iranje
Southern Africa: witbessiebos; umYaweyawe, isiBangamhlota sehlati (Zulu); nhlangaume (Thonga); mutangauma (Venda); muDyangombe, muPompoma, muSosoti, muZozoti (Shona)
Japan: Taiwan-hito-tsuba-hagi
Malaya: beti ayer, beti-beti, membeti, bebeti, memeti

Sedastrum Rose Crassulaceae

Origins:
Resembling *Sedum* L.

Seddera Hochst. Convolvulaceae

Origins:
After a mountain near Mecca, Arabia; see Christian Ferdinand Hochstetter (1787-1860), *Flora oder allgemeine Botanische Zeitung.* 27(1): 7. 1844.

Sedderopsis Roberty Convolvulaceae

Origins:
Resembling *Seddera* Hochst.

Sedella Britton & Rose Crassulaceae

Origins:
Referring to *Sedum* L.; see Reid Moran, "The genus *Sedella* Britton & Rose (Crassulaceae)." in *Haseltonia.* 5: 53-60. 1997.

Sedella Fourr. Crassulaceae

Origins:
Referring to *Sedum* L.

Sedirea Garay & H.R. Sweet Orchidaceae

Origins:
An anagram of the genus *Aerides* Lour.

Species/Vernacular Names:
S. japonica (Linden & Reichb.f.) Garay & Sweet (*Aerides japonica* Linden & Reichb.f.)
Japan: Nago-ran
Okinawa: Nagu-ran

Sedopsis (Engler) Exell & Mendonça Portulacaceae

Origins:
Like *Sedum*; see Arthur Wallis Exell (1901-) and Francisco de Ascenção Mendonça (1899-1982), *Conspectus florae angolensis.* 1(1): 116. Lisboa (Jan.) 1937.

Sedum L. Crassulaceae

Origins:
Latin *sedum* for houseleek, called also *aizoon* or *digitellum* (Plinius, L. Junius Moderatus Columella); some suggest from the Latin *sedo, avi, atum, are* "to calm, soothe" or from *sedeo, sedes, sedi, sessum, sedere* "to sit"; see Carl

Linnaeus, *Species Plantarum*. 430. 1753 and *Genera Plantarum*. Ed. 5. 197. 1754; Helmut Genaust, *Etymologisches Wörterbuch der botanischen Pflanzennamen*. 574. Basel 1996; Georg Christian Wittstein, *Etymologisch-botanisches Handwörterbuch*. 806. Ansbach 1852.

Species/Vernacular Names:

S. spp.

Mexico: ceche, ciiche, cijchi, guia ciiche, guia cicha, quije cijche, quije cijcha, quije zijche

S. acre L.

English: stone crop, wall pepper

S. albomarginatum R.T. Clausen

California: Feather River stone crop (Plumas County)

S. brissemoretii Hamet

English: Brissemoret's stonecrop

Portuguese: erva arroz

S. divergens S. Watson

English: cascade stone crop

S. eastwoodiae (Britton) A. Berger

California: Red Mountain stone crop (Red Mountain, Mendocino County)

S. farinosum Lowe

English: mealy stonecrop

Portuguese: erva arroz

S. formosanum N.E. Br. (*Sedum mariae* Hamet)

Japan: hama-mannen-gusa, Takasago-mannen-gusa

S. lineare Thunb.

Japan: o-no-mannen-gusa

China: fo jia cao, fo chia tsao

S. niveum Davidson

English: Davidson's stone crop

S. nudum Ait.

English: smooth stonecrop

Portuguese: erva arroz

S. oblanceolatum R.T. Clausen

English: Applegate stone crop

S. paradisum (Denton) Denton

English: Canyon Creek stone crop

S. praealtum A. DC.

English: giant stonecrop

Portuguese: ensaião gigante

S. telephium L.

English: live-forever

Italian: erba di San Giovanni

Seemannantha Alefeld Fabaceae

Origins:

Dedicated to the German botanist and explorer Berthold Carl Seemann, 1825-1871, naturalist and botanical collector.

Seemannaralia R. Viguier Araliaceae

Origins:

For the German botanist and explorer Berthold Carl Seemann, 1825-1871 (d. Nicaragua, Central America), naturalist, botanical collector, from 1847 to 1851 naturalist to the HMS *Herald*, 1852 Fellow of the Linnean Society, editor of *Bonplandia* 1853-1862, from 1863 to 1869 editor of *The Journal of Botany*. Among his works are *The Botany of the Voyage of H.M.S. Herald*, under the command of Captain Henry Kellett, R.N., C.B., during the years 1845-1851. Published under the authority of the Lords Commissionars of the Admiralty by Berthold Seemann ... naturalist of the expedition. London 1852-1857, *Viti: An Account of a Government Mission to the Vitian or Fijian Islands in the years 1860-61*. Cambridge 1862, *Die Volksnamen der amerikanischen Pflanzen*. Hannover 1851, "Abstract of a journal kept during the voyage of H.M.S. Herald." *Hook., J. Bot. Kew Gdn. Misc.* 4: 18-26, 82-92, 212-217, 238-242. 1852, *Popular History of the Palms and Their Allies*. London 1856 and *Flora vitiensis: A Description of the Plants of the Viti or Fiji Islands*. London 1865-1873, travelled and collected in Nicaragua and Panama, in the Fiji Islands and Venezuela, he was brother of the German botanist and geologist Wilhelm Eduard Gottfried Seeman (d. 1868); see Burtt & Dickinson, *Notes R. Bot. Gdn. Edinb.* 33: 449. 1975; Stafleu and Cowan, *Taxonomic Literature*. 5: 474-481. 1985; Nicolas Leon, *Biblioteca botanico-Mexicana*. Mexico 1895; Ray Desmond, *Dictionary of British & Irish Botanists and Horticulturists*. 617. 1994; J.H. Barnhart, *Biographical Notes upon Botanists*. 3: 255. 1965; R. Zander, F. Encke, G. Buchheim and S. Seybold, *Handwörterbuch der Pflanzennamen*. 14. Aufl. 779. 1993; G. Murray, *History of the Collections Contained in the Natural History Departments of the British Museum*. 1: 181. 1904; Mary Gunn and Leslie E. Codd, *Botanical Exploration of Southern Africa*. 321. 1981; T.W. Bossert, *Biographical Dictionary of Botanists Represented in the Hunt Institute Portrait Collection*. 359. 1972; H.N. Clokie, *Account of the Herbaria of the Department of Botany in the University of Oxford*. 241. 1964; E.M. Tucker, *Catalogue of the Library of the Arnold Arboretum of Harvard University*. 1917-1933; Emil Bretschneider, *History of European Botanical Discoveries in China*. Leipzig 1981; Elmer Drew Merrill, *Contr. U.S. Natl. Herb.* 30(1): 270-273. 1947 and *Bernice P. Bishop Mus. Bull.* 144: 167-168. 1937; Ida Kaplan Langman, *A Selected Guide to the Literature on the Flowering*

Plants of Mexico. Philadelphia 1964; Irving William Knobloch, compil., "A preliminary verified list of plant collectors in Mexico." *Phytologia Memoirs.* VI. 1983.

Species/Vernacular Names:

S. gerrardii (Seemann) Harms

English: wild maple

Southern Africa: wildeahorn, basterkiepersol; uMaweni (Zulu); uMaweni (Xhosa)

Seemannia Regel Gesneriaceae

Origins:

After the German botanist Berthold Carl Seemann, 1825-1871, explorer, naturalist, botanical collector and traveler. His works include *The Botany of the Voyage of H.M.S. Herald,* under the command of Captain Henry Kellett, R.N., C.B., during the years 1845-1851. Published under the authority of the Lords Commissionars of the Admiralty by Berthold Seemann ... naturalist of the expedition. London 1852-1857, *Viti: An Account of a Government Mission to the Vitian or Fijian Islands in the years 1860-61.* Cambridge 1862 and *Flora vitiensis: A Description of the Plants of the Viti or Fiji Islands.* London 1865-1873.

Seetzenia R. Br. ex Decne. Zygophyllaceae

Origins:

For the German traveler Ulrich Jasper Seetzen, 1767-1811 (Yemen), naturalist, botanical collector. His writings include *A Brief Account of the Countries Adjoining the Lake of Tiberias, the Jordan, and the Dead Sea.* Bath 1810, *Systematum generaliorum de morbis plantarum brevis dijudicatio.* Gottingae [1789] and *Verzeichniss der für die orientalische Sammlung in Gotha zu Damask, Jerusalem* etc. Leipzig 1810; see E.M. Tucker, *Catalogue of the Library of the Arnold Arboretum of Harvard University.* Cambridge, Massachusetts 1917-1933; J.H. Barnhart, *Biographical Notes upon Botanists.* 3: 256. 1965; Jonas C. Dryander, *Catalogus bibliothecae historico-naturalis Josephi Banks.* 1800.

Seguiera Reichb. ex Oliv. Combretaceae

Origins:

For the French botanist Jean François Séguier, 1703-1784, historian and botanical bibliographer.

Seguieria Loefling Phytolaccaceae

Origins:

For the French botanist Jean François Séguier (Joannes Franciscus Seguierius), 1703-1784, historian, traveler, archeologist, botanical bibliographer, astronomer, studied with Antoine de Jussieu. His works include *Bibliotheca botanica.* Hagae-Comitum ['s Gravenhage — The Hague] 1740, *Dissertation sur l'ancienne inscription de la Maison-Carrée de Nismes.* Paris 1759, *Osservazioni della cometa di quest'anno 1744 ...* fatte in Verona da G. Guglienzi e da Gianfresco Seguier. 1744 and *Plantae veronenses.* Veronae 1745-1754; see J.H. Barnhart, *Biographical Notes upon Botanists.* 3: 256. 1965; Mariella Azzarello Di Misa, ed., *Il Fondo Antico della Biblioteca dell'Orto Botanico di Palermo.* 255. Palermo 1988; Stafleu and Cowan, *Taxonomic Literature.* 5: 484-488. 1985; R. Zander, F. Encke, G. Buchheim and S. Seybold, *Handwörterbuch der Pflanzennamen.* 14. Aufl. Stuttgart 1993; T.W. Bossert, *Biographical Dictionary of Botanists Represented in the Hunt Institute Portrait Collection.* 359. 1972; Jonas C. Dryander, *Catalogus bibliothecae historico-naturalis Josephi Banks.* 1800; A. Lasègue, *Musée botanique de Benjamin Delessert.* 1845; Blanche Elizabeth Edith Henrey (1906-1983), *British Botanical and Horticultural Literature before 1800.* Oxford 1975; F.A. Stafleu, *Linnaeus and the Linnaeans. The spreading of their ideas in systematic botany, 1735-1789.* Utrecht 1971.

Sehima Forssk. Gramineae

Origins:

From the Arabic *saehim* or *sehim,* vernacular names for *Sehima ischaemoides* Forssk. collected in Yemen, Al Hadiyah ["Yemen in montibus ad Hadîe"], 1763; see P. Forsskål (1732-1763), *Flora aegyptiaco-arabica.* 178. Copenhagen 1775.

Species/Vernacular Names:

S. ischaemoides Forssk.

Arabic: saehim, sehîm

S. nervosum (Rottl.) Stapf

English: rat's-tail grass

Seidelia Baillon Euphorbiaceae

Origins:

For the German botanist Christoph Friedrich Seidel, flourished 1869; see E.M. Tucker, *Catalogue of the Library of the Arnold Arboretum of Harvard University.* Cambridge,

Massachusetts 1917-1933; Stafleu and Cowan, *Taxonomic Literature*. 5: 490. 1985.

Seidenfadenia Garay Orchidaceae

Origins:

Named after the Danish botanist Gunnar Seidenfaden, b. 1908, explorer, orchid taxonomist, politician and diplomat, traveler and plant collector in Baffin Island, Greenland and Thailand. His writings include "Report on the Fourth Thai-Danish Botanical Expedition. (Jan. 16 — Mar. 9, 1964)." *Nat. Hist. Bull. Siam Soc.* 20(4): 227. 1964, with Tem Smitinand (b. 1920) wrote *The Orchids of Thailand.* See John H. Barnhart, *Biographical Notes upon Botanists.* 3: 257. 1965; I.H. Vegter, *Index Herbariorum.* Part II (6), *Collectors S.* Regnum Vegetabile vol. 114. 1986.

Seidlitzia Bunge ex Boissier Chenopodiaceae

Origins:

After the Latvian botanist Nikolai Karlovic (Karl Samuel) von Seidlitz, 1831-1907, traveler, statistician, wrote *Botanische Ergebnisse einer Reise durch das östliche Transkaukasien.* Dorpat [Tartu] 1857; see John H. Barnhart, *Biographical Notes upon Botanists.* 3: 257. 1965; T.W. Bossert, *Biographical Dictionary of Botanists Represented in the Hunt Institute Portrait Collection.* 359. 1972; E.M. Tucker, *Catalogue of the Library of the Arnold Arboretum of Harvard University.* Cambridge, Massachusetts 1917-1933; R.K. Brummitt and C.E. Powell, eds., *Authors of Plant Names.* [dates of birth and death: 1831-1930.] Royal Botanic Gardens, Kew 1992.

Selaginella P. Beauv. Selaginellaceae

Origins:

The diminutive of the Latin *selago, inis*, a plant resembling the savin-tree (Plinius), genus *Selago* L.; see A.M.F.J. Palisot de Beauvois (1752-1820), *Prodrome des cinquième et sixième familles de l'Aethéogamie.* Les mousses. Les lycopodes. 101. Paris 1805.

Species/Vernacular Names:

S. biformis A. Br.

Japan: tsuru-kata-hiba (= vine *Selaginella involvens* (Sw.) Spring)

S. denticulata (L.) Link

Spain: pinchuíta, selaginela

Portuguese: selaginela

S. gracillima (Kunze) Alston (*Lycopodium gracillimum* Kunze; *Selaginella preissiana* Spring)

English: tiny selaginella

S. involvens (Sw.) Spring

Japan: kata-hi-ba (= unilateral *Selaginella*)

China: yan zhou juan bai, chuan po

S. kraussiana (Kunze) A. Braun (*Lycopodium kraussianum* Kunze)

English: staghorn moss, trailing spike moss, garden selaginella, spreading clubmoss, mat spike moss

S. lepidophylla (Hook. & Grev.) Spring

English: rose of Jericho, resurrection plant

S. myosurus (Sw.) Alston

Sierra Leone: ndimoimoi

Congo: larbuoni, larbouoni, malangala

S. pallescens (Presl) Spring

English: moss fern, sweat plant

S. rupestris (L.) Spring

English: rock spike moss, dwarf lycopod

S. uliginosa (Labill.) Spring

English: swamp selaginella

S. uncinata (Desv. ex Poir.) Spring

English: peacock moss, rainbow fern, blue spike moss

S. willdenovii (Desv. ex Poir.) Bak.

English: peacock fern

Selaginoides Séguier Selaginellaceae

Origins:

Referring to the Latin *selago, inis*, a plant resembling the savin-tree (Plinius), see also *Selaginella* P. Beauv.

Selago Hill Lycopodiaceae

Origins:

Selago, ancient Latin name for *Lycopodium selago* L. (Plinius), see also *Huperzia* Bernh.

Selago L. Scrophulariaceae (Globulariaceae)

Origins:

Selago, inis ancient Latin name for *Lycopodium selago* L., a moss-like plant, a plant resembling the savin-tree (Plinius); see Carl Linnaeus, *Species Plantarum.* 629. 1753 and *Genera Plantarum.* Ed. 5. 486. 1754; Salvatore Battaglia,

Grande dizionario della lingua italiana. XVIII: 523. Torino 1997.

Species/Vernacular Names:
S. corymbosa L.
English: poverty bush
South Africa: bitterblombos, blombos
S. spuria L.
English: blue haze

Selago P. Browne Selaginellaceae

Origins:
Selago, inis ancient Latin name for *Lycopodium selago* L. (Plinius), a moss-like plant, see also *Selaginella* P. Beauv.

Selago Schur Lycopodiaceae

Origins:
Selago, inis ancient Latin name for *Lycopodium selago* L. (Plinius), see also *Huperzia* Bernh.

Selatium G. Don Gentianaceae

Origins:
Greek *selas* "brightness, flame," *selasma, selasmatos* "shining," Latin *selas* "a kind of fiery meteor."

Selbya M. Roemer Meliaceae

Origins:
For the British (b. Northumberland) naturalist Prideaux John Selby, 1788-1867 (d. Northumberland), ornithologist and entomologist. His writings include *A Catalogue of the Generic and Sub-Generic Types of the Class Aves, Birds,* etc. Newcastle-upon-Tyne 1840 and *Illustrations of British Ornithology.* Edinburgh 1852-1853; see John H. Barnhart, *Biographical Notes upon Botanists.* 3: 258. 1965; T.W. Bossert, *Biographical Dictionary of Botanists Represented in the Hunt Institute Portrait Collection.* 359. 1972; S. Lenley et al., *Catalog of the Manuscript and Archival Collections and Index to the Correspondence of John Torrey.* Library of the New York Botanical Garden. 365. 1973; E.M. Tucker, *Catalogue of the Library of the Arnold Arboretum of Harvard University.* Cambridge, Massachusetts 1917-1933; Sir William Jardine (1800-1874) and Prideaux John Selby, *Illustrations of Ornithology,* ... With the co-operation of ... Thomas Hardwicke. Edinburgh [1826-1835]; George Montagu, *A Dictionary of British Birds.* 1866.

Selenia Nutt. Brassicaceae

Origins:
Greek *selene* "the moon," referring to the appearance of the flowers or to the shape of the seeds.

Selenicereus (A. Berger) Britton & Rose Cactaceae

Origins:
Night-blooming cereus, from the Greek *selene* "the moon" plus *Cereus*, referring to the nocturnal flowers.

Species/Vernacular Names:
S. grandiflorus (L.) Britton & Rose
English: night flowering cactus, night-blooming cereus
S. megalanthus (K. Schum. ex Vaupel) Moran (*Cereus megalanthus* K. Schum. ex Vaupel; *Mediocactus megalanthus* (K. Schum. ex Vaupel) Britton & Rose)
Latin America: yellow pitaya

Selenipedium Reichb.f. Orchidaceae

Origins:
Greek *selene* "the moon" and *pedilon* "a slipper," referring to the saccate or corniculate lip.

Selenodesmium (Prantl) Copel. Hymenophyllaceae

Origins:
Greek *selene* and *desmis, desmos* "a bond, band, bundle"; see Edwin Bingham Copeland (1873-1964), in *Philippine Journal of Science.* 67: 80. Manila (Oct.) 1938.

Selenothamnus Melville Malvaceae

Origins:
From the Greek *selene* "the moon" and *thamnos* "a shrub"; see Ronald Melville (1903-1985), in *Kew Bulletin.* 20: 514. London (Jan.) 1967.

Selera Ulbrich Malvaceae

Origins:
For the German archeologist and anthropologist Eduard Georg Seler, 1849-1922, traveler (Central America), bota-

nist and plant collector. His writings include *Notice sur les Langues Zapotèque et Mixtèque*. Congrès International des Américanistes. Compte-Rendu. 8th sess. Paris 1890, *Codex Vaticanus no. 3773* ... An old Mexican Pictorial manuscript. 1902-1903, "The Bat God of the Maya Race." *Bureau of American Ethnology*. Bull. 28. 1904, "The Mexican Picture Writings of Alexander von Humboldt." *Bureau of American Ethnology*. Bull. 28: 127-229. 1904, *Codex Borgia*. Berlin 1904-1909 and *Einige Kapitel aus dem Geschichtswerke des Fray Bernardino de Sahagún* aus dem Aztekischen etc. Stuttgart 1927. See Jacques Soustelle, *The Four Suns*. 134. New York 1971; H.R. Harvey and Hanns J. Prem, editors, *Explorations in Ethnohistory*. Indians of Central Mexico in the Sixteenth Century. University of New Mexico Press, Albuquerque 1984; J.H. Barnhart, *Biographical Notes upon Botanists*. 3: 258. 1965; Ida Kaplan Langman, *A Selected Guide to the Literature on the Flowering Plants of Mexico*. 1964; Ludwig Eduard Theodor Loesener (1865-1941), *Plantae selerianae* die von Dr. Eduard Seler [1849-1922] und Frau Caecilie Seler [*née* Sachs, b. 1855] in Mexico gesammelten Pflanzen unter Mitwirkung von Fachmännern veröffentlicht ... [1894-1923]; *Festschrift Eduard Seler*, edited by Walter Lehmann. Stuttgart 1922; Irving William Knobloch, compil., "A preliminary verified list of plant collectors in Mexico." *Phytologia Memoirs*. VI. 1983.

Selinocarpus A. Gray Nyctaginaceae

Origins:

Greek *selinon* "parsley, celery" or *selene* "moon" or Latin *selinas* "a kind of cabbage" and Greek *karpos* "fruit."

Selinopsis Cosson & Durieu ex Batt. & Trab. Umbelliferae

Origins:

From the Greek *selinon* "parsley, celery" and *opsis* "appearance, like."

Selinum L. Umbelliferae

Origins:

Greek *selinon, selinnon* "parsley, celery," Theophrastus, *HP.* 1.2.2 and 9.11.10, Latin *selinon* for parsley, *apium*.

Species/Vernacular Names:

S. sp.

China: mi wu

Selkirkia Hemsley Boraginaceae

Origins:

Dedicated to Alexander Selkirk (1676-1721), the sailing master on whom Daniel Defoe modeled the character of Robinson Crusoe; when Woodes Rogers (1678-1732) (who sailed from Bristol on August 1708) reached Juan Fernández on February 1709, Selkirk was still living there. See Ruben Borba de Moraes, *Bibliographia brasiliana*. University of California 1983; W.H. Bonner, *Captain William Dampier: Buccaneer-Author*. Palo Alto 1934; B. Little, *Crusoe's Captain: Woodes Rogers*. London 1960; J. Ross Brown, *Crusoe's Island*. New York 1864; R.L. Woodward, *Robinson Crusoe's Island: A History of the Juan Fernandez Islands*. Chapel Hill (N. Carolina) 1969; Benjamin Vicuña Mackenna, *Juan Fernández; historia verdadera de la isla de Robinson Crusoe*. Santiago 1883; Woodes Rogers, *A Cruising Voyage Round the World*. ... An account of Alexander Selkirk's living alone four years and four months in an island. The second edition, corrected. London 1718; P.N. Furbank & W.R. Owens, *A Critical Bibliography of Daniel Defoe*. London 1998.

Selliera Cav. Goodeniaceae

Origins:

After the French engraver François Noël Sellier, 1737-1800, between 1780 and 1800 engraved botanical illustrations for the publications of the Spanish clergyman and botanist Antonio José Cavanilles (1745-1804) and of the French botanist René Louiche Desfontaines (1750-1833); see A.J. Cavanilles, in *Anales de Historia Natural Madrid*. 1(1): 41, t. 5, fig. 2. 1799; Georg Christian Wittstein, *Etymologisch-botanisches Handwörterbuch*. 807. Ansbach 1852; F. Boerner & G. Kunkel, *Taschenwörterbuch der botanischen Pflanzennamen*. 4. Aufl. 168. 1989.

Species/Vernacular Names:

S. radicans Cav. (*Goodenia radicans* (Cav.) Pers.; *Goodenia repens* Labill.; *Selliera herpystica* Schltdl.; *Selliera repens* (Labill.) Vriese)

English: selliera, swamp weed, tongue plant

Selloa Kunth Asteraceae

Origins:

After the German gardener Friedrich Sellow (Sello), 1789-1831 (d. by drowning), botanical explorer, naturalist, plant collector in Brazil and Uruguay, in Brazil with the German botanist Maximilian Alexander Philipp zu Wied-Neuwied (1782-1867); see M.A.P. zu Wied-Neuwied, *Reise nach Brasilien in den Jahren 1815 bis 1817*. Frankfurt a.M. 1820-1821; R. Zander, F. Encke, G. Buchheim and S. Seybold,

Handwörterbuch der Pflanzennamen. 14. Aufl. 779. Stuttgart 1993; J.H. Barnhart, *Biographical Notes upon Botanists.* 3: 259. 1965; A. Lasègue, *Musée botanique de Benjamin Delessert.* Paris 1845; Günther Schmid, *Chamisso als Naturforscher.* Eine Bibliographie. Leipzig 1942; Gordon Douglas Rowley, *A History of Succulent Plants.* Strawberry Press, Mill Valley, California 1997.

Sellocharis Taubert Fabaceae

Origins:

After the German gardener Friedrich Sellow (Sello), 1789-1831, botanical explorer.

Selmation T. Durand Asclepiadaceae

Origins:

Probably from the Greek *selma, selmatos* "deck, ship, seat, any timberwork," see also *Metastelma* R. Br.

Selwynia F. Muell. Menispermaceae

Origins:

For the Australian (b. Somerset, England) geologist Alfred Richard Cecil Selwyn, 1824-1902 (d. Vancouver, British Columbia), 1852-1869 Director Geological Survey of Victoria, 1869-1894 Director of the Geological Survey of Canada. His works include *Geology of the Colony of Victoria.* 1861, *A Descriptive Catalogue of the Rock Specimens and Minerals in the National Museum* collected by the Geological Survey of Victoria. 1868 and *Notes and Observations on the Gold Fields of Quebec and Nova Scotia.* Halifax, N.S. [1871], with the field geologist George Henry Frederick Ulrich wrote *Notes on the Physical Geography, Geology and Mineralogy of Victoria.* [Intercolonial Exhibition, Official Record, Essays, no. 3.] Melbourne 1867. See George H.F. Ulrich, *Observations on the Mode of Occurrence and the Treatment of Auriferous, Lead, and Silver Ores, at Schemnitz, Upper Hungary.* Melbourne 1868; Ferdinand von Mueller, *Fragmenta Phytographiae Australiae.* 4: 153. Melbourne 1864; C. Gordon Winder, in *Dictionary of Scientific Biography* 12: 292-294. 1981; John T. Walbran, *British Columbia Coast Names, 1592-1906. To Which are Added a Few Names in Adjacent United States Territory, Their Origin and History.* First edition. Ottawa: Government Printing Bureau, 1909; G.P.V. Akrigg & Helen B. Akrigg, *British Columbia Place Names.* Sono Nis Press, Victoria 1986.

Semecarpus L.f. Anacardiaceae

Origins:

Greek *sema, semeion* "a sign, mark, token" and *karpos* "fruit," the juice of the nuts used as an ink to mark clothes or linen; see C. Linnaeus (filius), *Supplementum Plantarum.* 25, 182. 1782; G.C. Wittstein, *Etymologisch-botanisches Handwörterbuch.* 807f. Ansbach 1852; F. Boerner & G. Kunkel, *Taschenwörterbuch der botanischen Pflanzennamen.* 4. Aufl. 361. 1989.

Species/Vernacular Names:

S. anacardium L.f.

English: marking-nut tree, marking-nut tree of India, Indian marking-nut tree

India: bhilawa, bhelwa, kiro, sosa daru

Nepal: bhalayo

Tibetan: go-byi-la

S. australiensis Engl.

English: marking-nut tree, tar tree, marking-nut

S. cuneiformis Blanco

The Philippines: ligas, libas, agas, anagas, hanagas, langas, longas, ingas, kaming, kamiing, kamiding, kamiling, kamiring, pakan, ameleng

Semeiandra Hook. & Arn. Onagraceae

Origins:

Greek *semeion* "a sign, standard" and *aner, andros* "male, stamen," presumably referring to the expanded petaloid stamen or to the explosive pollen-shedding.

Semeiocardium Zoll. Balsaminaceae

Origins:

Greek *semeion* and *kardia* "heart," possibly referring to the explosive five-valved capsule, see also *Impatiens* L.

Semeiostachys Drobov Gramineae

Origins:

Greek *semeion* and *stachys* "spike, an ear of grain, ear of corn."

Semele Kunth Asparagaceae (Ruscaceae, Liliaceae)

Origins:

After Semele, a daughter of Cadmus (Kadmos) and mother of Bacchus (Dionysos) by Jupiter.

Species/Vernacular Names:

S. androgyna (L.) Kunth

English: climbing butcher's broom

Portuguese: alegra campo

Semetor Raf. Fabaceae

Origins:

See C.S. Rafinesque, *Sylva Telluriana*. 69. 1838; E.D. Merrill, *Index rafinesquianus*. 148. 1949.

Semetum Raf. Brassicaceae

Origins:

See Constantine Samuel Rafinesque, *Autikon botanikon. Icones plantarum select. nov. vel rariorum*, etc. 17. Philadelphia 1840; E.D. Merrill, *Index rafinesquianus*. 130. 1949.

Semialarium N. Hallé Celastraceae

Origins:

Perhaps from the Latin *semi* "half" and *alarius, a, um* "that is upon the wing, of the wing," see also *Hemiangium* A.C. Sm.

Semiaquilegia Makino Ranunculaceae

Origins:

Latin *semi* "half" plus the genus *Aquilegia*, like *Aquilegia*.

Species/Vernacular Names:

S. adoxoides (DC.) Makino

China: tian kui

Semiarundinaria Makino ex Nakai Gramineae

Origins:

Latin *semi* plus the genus *Arundinaria*.

Semibegoniella C.E.C. Fischer Begoniaceae

Origins:

Latin *semi* "half" plus the genus *Begoniella* Oliv.

Semiliquidambar H.T. Chang Hamamelidaceae

Origins:

From the Latin *semi* "half" plus the genus *Liquidambar* L.

Semilta Raf. Euphorbiaceae

Origins:

See C.S. Rafinesque, *Sylva Telluriana*. 63. 1838; E.D. Merrill, *Index rafinesquianus*. 156. 1949.

Semiphajus Gagnep. Orchidaceae

Origins:

Latin *semi* "half" plus the genus of orchids *Phaius* Lour., referring to the number of pollinia.

Semnanthe N.E. Br. Aizoaceae

Origins:

Greek *semnos* "august, stately" and *anthos* "flower," referring to the pink flowers.

Semnostachya Bremek. Acanthaceae

Origins:

Greek *semnos* and *stachys* "spike, an ear of grain, ear of corn."

Semnothyrsus Bremek. Acanthaceae

Origins:

Greek *semnos* "august, stately" and *thyrsos* "a panicle."

Semonvillea Gay Molluginaceae

Origins:

After Hyppolite Boissel, 1794-1863, Baron de Monville, French amateur botanist, plant collector; see R. Zander, F. Encke, G. Buchheim and S. Seybold, *Handwörterbuch der Pflanzennamen*. 14. Aufl. 753. Stuttgart 1993.

Sempervivella Stapf Crassulaceae

Origins:

The genus *Sempervivum* L., the diminutive.

Sempervivum L. Crassulaceae

Origins:

Latin *semper-vivus, a, um* "ever-living," *semperviva, sempervivum,* the plant also called *aizoon,* evergreen, houseleek.

Senckenbergia Post & Kuntze Nyctaginaceae

Origins:

For the German physician Johann Christian Senckeberg, 1707-1772, naturalist, wrote *Dissertatio inauguralis de Lilii Convallium.* Gottingae [1737]; see J.H. Barnhart, *Biographical Notes upon Botanists.* 3: 259. 1965; Jonas C. Dryander, *Catalogus bibliothecae historico-naturalis Josephi Banks.* London 1800; G. Schmid, *Goethe und die Naturwissenschaften.* Halle 1940; Johann Jacob (Jakob) Reichard (1743-1782), *Enumeratio stirpium horti botanici senckenbergiani,* qui Francofurti ad Moenum est. Francofurti ad Moenum 1782.

Senebiera DC. Brassicaceae

Origins:

After the Swiss (b. Geneva) botanist Jean Senebier, 1742-1809 (Geneva), bibliographer and linguist, clergyman, physiologist, from 1773 to 1809 librarian of the city (Republic) of Genève. Among his works are *Histoire littéraire de Genève.* Genève 1786, *Physiologie végétale.* Genève [1800], *Expériences sur l'action de la lumière solaire dans la végétation.* Genève 1788, *Catalogue raisonné des Manuscrits conservés dans la Bibliothèque de la Ville et République de Genève.* Genève 1779, *Mémoire historique sur la vie et les écrits de H.B. Desaussure.* Genève [1801], *L'art d'observer.* Genève 1775 and *Essai sur l'art d'observer et de faire des expériences.* Genève, an x. [1802], with the Italian naturalist and scientist Lazzaro Spallanzani (1729-1799) wrote *Rapports de l'air avec les êtres organisés.* Genève 1807 and *Expériences pour servir à l'histoire de la génération des animaux et des plantes.* Pavia 1787; see A.P. de Candolle, in *Mémoires de la Société d'Histoire Naturelle de Paris.* 140, 142. Paris 1799; T.W. Bossert, *Biographical Dictionary of Botanists Represented in the Hunt Institute Portrait Collection.* 360. 1972; Jonas C. Dryander, *Catalogus bibliothecae historico-naturalis Josephi Banks.* London 1800; John H. Barnhart, *Biographical Notes upon Botanists.* 3: 259. 1965; J.P. Maunoir, *Éloge historique de M. Jean Senebier.* Paris, Genève 1810; P.E. Pilet, in *Dictionary of Scientific Biography* 12: 308-309. 1981; Claude E. Dolman, in *Dictionary of Scientific Biography* 12: 553-567. 1981; Mariella Azzarello Di Misa, a

cura di, *Il Fondo Antico della Biblioteca dell'Orto Botanico di Palermo.* 255. Regione Siciliana, Palermo 1988.

Senecio L. Asteraceae

Origins:

Latin *senecio, senecionis,* ancient name for a plant, called also erigeron, groundsel (Plinius), from Latin *senex, senis* "old, aged," possibly referring to the white pappus; Akkadian *santu, sattu* "year," *sanu* "to change"; Arabic *sanat*; Hebrew *sana* "year," *sin'an* "repetition," *sana* "to do second time, to be different"; see Carl Linnaeus, *Species Plantarum.* 866. 1753 and *Genera Plantarum.* Ed. 5. 373. 1754; S. Battaglia, *Grande dizionario della lingua italiana.* XVIII: 620. 1997; G. Semerano, *Le origini della cultura europea.* Dizionario della lingua Latina e di voci moderne. 2(2): 561-562. Firenze 1994.

Species/Vernacular Names:

S. spp.

South Africa: Dan's cabbage, grass-staggers weed, Molteno disease plant, staggers bush, ragwort, dunsiektebossie, kraakstewel, krakerbossie

Peru: checpas, gallo sisa, huira-huayo, pumpush, ushpahuayta

S. abyssinicus Sch.Bip.

Yoruba: amunimuye

S. ambavilla (Bory) Persoon

La Réunion Island: ambaville, ambaville bleu, Jean Baville

S. ampullaceus Hook.

English: Texas groundsel

S. amygdalifolius F. Muell.

English: broad-leaf fireweed

S. angulatus L.f. (*Senecio macropodus* DC.)

English: Cape ivy

Southern Africa: inDindilili (Xhosa)

S. apiifolius (DC.) Bentham & Hook.f. ex O. Hoffm. (*Senecio peculiaris* Dinter)

English: ragwort

S. articulatus (L.f.) Sch.Bip.

English: candle plant, hot-dog cactus, sausage crassula

S. behrianus Sonder

English: Behr's groundsel

S. bernardinus E. Greene

English: San Bernardino ragwort

S. biafrae Oliv. & Hiern

Yoruba: bologi, gbologi, rorowo, woorowo, ewoio eyonu

S. bipinnatisectus Belcher

English: Commonwealth weed

S. biserratus Belcher

English: jagged fireweed

S. brunonis (Hook.f.) J.H. Willis

English: Tasmanian tree groundsel

S. bupleuroides (DC.) (*Senecio bupleuroides* DC. var. *denticulatus* DC.; *Senecio bupleuroides* DC. var. *falcatus* Sch.Bip.)

English: ragwort, yellow starwort

Southern Africa: lehlongoana (Sotho); lereka (South Sotho); iDwarane (Xhosa)

S. burchellii DC. (*Senecio lichtensteinensis* Dinter)

English: Molteno disease plant, Molteno disease senecio, guanobush, ragwort

Southern Africa: geelgifbossie, Burchell senecio, gifbossie, kovanna, Moltenobossie, sprinkaanbossie; khotolia (Sotho)

S. clevelandii E. Greene

English: Cleveland's ragwort

S. consanguineus DC.

English: ragwort, starvation bush, starvation senecio

South Africa: hongerbossenecio, sprinkaanbossie

S. cunninghamii DC.

English: branching groundsel

S. deltoideus Less. (*Senecio durbanensis* Gandoger; *Senecio fimbrillifer* B.L. Robinson; *Senecio mikaniae* DC.; *Senecio mikaniaeformis* DC.; *Cacalia scandens* Thunb.; *Mikania auriculata* Willd.)

English: Canary creeper, Cape ivy

S. elegans L. (*Senecio elegans* Willd.; *Senecio pseudoelegans* Less.)

English: wild cineraria, purple groundsel, purple ragwort

Southern Africa: wilde cineraria, jacobaea; iZuba (Xhosa)

S. eurycephalus A. Gray var. *lewisrosei* (J. Howell) T. Barkley

English: cut-leaved ragwort

S. fremontii Torr. & A. Gray

English: dwarf mountain butterweed

S. ganderi T. Barkley & Beauch.

English: Gander's ragwort

S. georgianus DC.

English: grey groundsel

S. glastifolius L.f.

English: large senecio

South Africa: waterdissel

S. glomeratus Poiret

English: annual fireweed

S. glossanthus (Sonder) Belcher

English: slender groundsel

S. glutinosus Thunb. (*Senecio brachyglossus* Turcz.; *Senecio inamoenus* DC.)

Southern Africa: dunsiektesenecio, hongerbossie; khotolia (Sotho)

S. gregorii F. Muell.

English: fleshy groundsel, annual yellow-top

S. gunnii (Hook.f.) Belcher

English: mountain fireweed

S. hawortii (Sw.) Steud.

English: cocoon plant

S. hispidulus A. Rich.

English: hill fireweed, rough fireweed

S. hybridus Regel

English: florists' cineraria

S. hydrophiloides Rydb.

English: sweet marsh ragwort

S. ilicifolius L. (*Senecio quercifolius* Thunb.)

English: ragwort

South Africa: ghwanobos, gifbossie, guanobos, kowannabossie, sprinkaanbossie, sprinkaansenecio

S. inaequidens DC. (*Senecio burchellii* DC. p.p.)

English: Canaryweed

Southern Africa: geelopslag; khotolia (Sotho)

S. inornatus DC. (*Senecio caulopterus* DC.; *Senecio diversidentatus* Muschl.; *Senecio fraudulentus* Phill. & C.A. Sm.; *Senecio macroalatus* M.D. Henderson; *Senecio ommanei* DC.)

Southern Africa: lehlomane-le-leholo (Sotho)

S. isatideus DC.

English: Dan's cabbage, poisonous ragwort

Southern Africa: inkanga, blouvleisenecio; lebato (Sotho)

S. jacobaea L.

English: ragwort, tansy ragwort

S. juniperinus L.f. var. *juniperinus*

English: ragwort

S. kleiniiformis Suess.

English: Texas groundsel

S. latifolius DC. (*Senecio sceleratus* Schweick.)

English: Dan's cabbage, groundsel, Molteno disease plant, pictou disease, ragwort, Rhodesian ragwort, senecio, stagger's bush, Winton disease

Southern Africa: gifbossiesenecio, kraakstewel, krakerbossie; chigurungu (Shona); idwara (Xhosa); mabolo (Ndebele)

S. lautus Willd.

English: variable groundsel

S. layneae E. Greene

English: Layne's ragwort

S. leptocarpus DC.

English: branched alpine groundsel

S. macounii E. Greene

English: Siskiyou Mountains ragwort

S. macrocephalus DC. (*Senecio gyrophyllus* Klatt)

English: ragwort

Southern Africa: iHlaba lenkomo (Xhosa)

S. macroglossus DC.

English: Natal ivy, wax vine

S. madagascariensis Poiret

English: fireweed

S. magnificus F. Muell.

English: tall yellow-top

S. mikanioides Walp.

English: German-ivy, climbing groundsel

Portuguese: tasneirinha

S. minimus Poiret

English: shrubby fireweed

S. odoratus Hornem.

English: scented groundsel

S. orarius J. Black

English: small-flowered groundsel

S. papillosus F. Muell.

English: downy groundsel

S. pattersonensis Hoover

English: mono ragwort

S. pectinatus DC.

English: alpine groundsel

S. platylepis DC.

English: flat-scale groundsel

S. primulifolius F. Muell.

English: heart-leaf groundsel

S. pterophorus DC. (*Senecio pterophorus* DC. var. *verus* and var. *apterus* Harvey)

English: African daisy, South African daisy, winged groundsel

South Africa: perdegifbos, perdegifbossenecio

S. purpureus L. (*Senecio mucronulatus* Sch.Bip.)

English: ragwort

S. quinquelobus (Thunb.) DC. (*Cacalia quinqueloba* Thunb.)

Southern Africa: uChantikhulu (Xhosa)

India: morta

S. retrorsus DC. (*Senecio graminicolus* C.A. Sm.; *Senecio latifolius* DC. var. *barbellatus* (DC.) Harv.; *Senecio latifolius* DC. var. *retrorsus* (DC.) Harv.; *Senecio latifolius* DC. var. *subedentulus* DC.)

English: bushweed, Dan's cabbage, grass stagger's weed, Molteno disease plant, poisonous ragwort, stagger's bush, stagger's senecio

South Africa: dunsiektebossie, kraakstewel, kraakstewelsenecio, krakerbossie

S. runcinifolius J.H. Willis

English: tall groundsel

S. scandens Buch.-Ham. ex D. Don

China: quan li guang, chien li chi, chien li kuang

S. serpens G. Rowley

English: blue chalksticks

S. tamoides DC.

English: Canary creeper, climbing cineraria

South Africa: Kanarieklimop

S. vagus F. Muell.

English: saw groundsel

S. velleoides DC.

English: forest groundsel

S. vulgaris L.

English: groundsel, common groundsel

Arabic: murrar, morrar, muraya, eshbet salema

Italian: senecio comune, senacione, senecio, senezio, senezione, calderugia, erba calderina, cardellina, sollecciolla, verzellina

Senecioides Post & Kuntze Asteraceae

Origins:
Resembling *Senecio* L.

Senegalia Raf. Mimosaceae

Origins:
Referring to Senegal and *Acacia senegal* (L.) Willd.; see C.S. Rafinesque, *Sylva Telluriana*. 119. 1838; E.D. Merrill, *Index rafinesquianus*. 148. 1949.

Senna Miller Caesalpiniaceae

Origins:

Arabic *sana* or *sanna*; see Philip Miller (1691-1771), *The Gardeners Dictionary*. Abr. ed. 4. London (28 Jan.) 1754; Timothy J. Killeen, Emilia García E. and Stephan G. Beck, eds., *Guía de Arboles de Bolivia*. 413-418. Herbario Nacional de Bolivia and Missouri Botanical Garden 1993; G.B. Pellegrini, *Gli arabismi nelle lingue neolatine con speciale riguardo all'Italia*. 119. Brescia 1972; H. Genaust, *Etymologisches Wörterbuch der botanischen Pflanzennamen*. 577. 1996; M. Cortelazzo & P. Zolli, *Dizionario etimologico della lingua italiana*. 5: 1180. Bologna 1988; Petrus de Crescentiis, *Il libro dell'agricultura*. per Nicholaum [Laurentii] [In-folio] Florentie 1478.

Species/Vernacular Names:

S. alata (L.) Roxb. (*Cassia alata* L.; *Herpetica alata* (L.) Raf.)

English: ringworm bush, ringworm shrub, ringworm senna, candle bush, seven golden candlesticks, Christmas candle

Japan: hane-senna

Malaya: gelenggang, daun kurap

India: dadmari, dadamardana, dadmurdan, dadrughna, daddupana, dad-ka-patta, vendukolli, anjali, attora, sheemigida, agase gida, simayavisa, sheemaavisi, sheemai-agatti, sima avisl, seemagati, elakajam, mettatamara, vilayati-agati, maizali-gi

Bolivia: kota-kota, yunka mutuillu, cara de caballo

Yoruba: asunwon

S. alexandrina Miller (*Cassia bicapsularis* L.; *Cassia emarginata* L.; *Cassia transversali-seminata* De Wild.; *Cassia angustifolia* M. Vahl; *Cassia lanceolata* Forsskål; *Cassia senna* L.)

English: Alexandrian senna, Indian senna, Tinnevelly senna, true senna, Tinnevelly senna plant

Arabic: senna mekki, senna haram, senna hindi

China: fan xie ye

India: hindisana, hindi-sana, sana-e-hindi, sonamukhi, sona-mukhi, sanna-makki, senamakki, sonamakki, naelaponna, nilavirai, nila virai, nela-tangedu, nila vaka, nilavaka, nila vakai, nilavakai, mulcacha

S. armata (S. Watson) H. Irwin & Barneby

English: spiny senna

S. artemisioides (DC.) Randell

English: silver senna, feathery cassia, wormwood senna

S. auriculata (L.) Roxb. (*Cassia auriculata* L.)

English: senna, tanner's cassia, mature tea tree

India: avaram, tarwar, tangedu, rana-vara, avara, avarike, avarai, avirae, aveeram, ponnaviram, awal, awala, tarwad, taravada, taravada-gida, jimute, chakusina-gida

S. aymara Irwin & Barneby

Bolivia: moto-moto, takarkaya, mutuy mutuy

S. barclayana (Sweet) Randell (*Cassia barclayana* Sweet)

English: smooth senna

S. bicapsularis (L.) Roxb. (*Cassia bicapsularis* L.; *Cassia emarginata* L.; *Cassia transversali-seminata* De Wild.)

English: stiverbush

Belize: barba jolote

Mexico: vainillo, mora hedionda

Guatemala: vainillo

S. burkartiana (Villa) Irwin & Barneby

Bolivia: garbancillo

S. corymbosa (Lam.) H. Irwin & Barneby (*Cassia bonariensis* Colla; *Cassia corymbosa* Lam.)

English: autumn cassia, buttercup tree, golden senna, scrambled eggs, flowering senna

S. covesii (A. Gray) H. Irwin & Barneby

English: Coues' senna

S. didymobotrya (Fresen.) H. Irwin & Barneby (*Cassia didymobotrya* Fresen.; *Cassia nairobensis* Aggeler & Musser)

English: peanut cassia, peanut-butter cassia, wild senna, popcorn bush

Japan: futa-ho-senna

Southern Africa: munwahuku (Shona); utupa (Swahili)

S. x floribunda (Cav.) H. Irwin & Barneby (*Cassia floribunda* Cav.)

English: arsenic bush, hill cassia

Indochina: bo cap nuroc

S. fruticosa (Mill.) H. Irwin & Barneby (*Cassia fruticosa* Mill.)

English: drooping cassia

S. gaudichaudii (Hook. & Arn.) H. Irwin & Barneby (*Cassia gaudichaudii* Hook. & Arn.)

Hawaii: kolomona, heuhiuhi, kalamona, uhiuhi

S. hirsuta (L.) Irwin & Barneby

Yoruba: ejo ogun, rere pupa

S. italica Miller (*Cassia italica* (Miller) Sprengel; *Cassia obovata* Colladon, nom. illeg.)

English: senna, dog senna, Eland's senna, wild senna, Italian senna, Spanish senna, country senna

Arabic: senna, senna ghanami

South Africa: Elandsertjie, grondboontjie, kalwerbossie, swartstormbos

India: bhuitarwar, bhui-tarwad, surati sonamukhi, suna-mukhi, chottataroda, seruvanni, bhutalapota, kattunilavirai, nilavagai, nilavarai, neltangedu

S. macrophylla (Kunth) Irwin & Barneby var. *macrophylla*

Bolivia: xenáne

S. macrophylla (Kunth) Irwin & Barneby var. *gigantifolia* (Britton & Killip) Irwin & Barneby

Bolivia: samomaquexti, cashixopa jahuéhua

S. marilandica (L.) Link

English: American senna, wild senna

S. multiglandulosa (Jacq.) H. Irwin & Barneby (*Cassia multiglandulosa* Jacq.; *Cassia tomentosa* L.f.)

English: woolly senna, wild senna

South Africa: peulbos, wildesenna

Bolivia: iso mutu-mutu, mutu-mutu

S. multijuga (Rich.) H. Irwin & Barneby (*Cassia multijuga* Rich.)

English: leafy senna, cassia

Bolivia: manicillo, ramo, flor de mayo

S. obtusifolia (L.) H. Irwin & Barneby (*Cassia obtusifolia* L.)

English: sicklepod

India: chakunda

Yoruba: ako rere, opa iku, asimawu

S. occidentalis (L.) Link (*Cassia occidentalis* L.; *Ditremexa occidentalis* (L.) Britton & Rose)

English: coffeeweed, negro coffee, coffee senna, stinking weed, wild coffee, styptic weed

French: faux kinkeliba

India: doddatagache, hodu-taikilo, kasamarda, kasondi, kasinda, kasunda, kalakasunda, kalkashunda, kasuvayee, hikal, kasinda, nattutakarai, paeravirai, ponnavirai, ponnaveeram, natrum-takara, natram-takara, nattam-takarai

China: wang jiang nan, wang chiang nan

Japan: habu-so

The Philippines: balatong-aso, kabal-kabalan, tambalisa, andadasi, duda, sumting

Hawaii: 'auko'i, 'au'auko'i, mikipalaoa, pi'hohono

Ghana: amanseidua

Congo: nkrantali

Mali: sumakala, nbalafin

Nigeria: rere

Yoruba: abo rere, rere, adaweresewere, oganlara

Southern Africa: isinyembane (Zulu); lilyanyoka (Swati); mnuka uvundo (Swahili)

Central America: fijolillo, bricho, comida de murciélago, cimarrón, escapacle, furrusca, habilla, hediondío, moquillo

Nicaragua: pigue pajaro

Mexico: vainillo

Brazil: fedegoso, erva federonta, pajamarioba, menuerioba, paieriaba, folha-de-pagé, tavarucu, mamangá, lava-pratos, manjericona, mangeroba

S. petersiana (Bolle) Lock (*Cassia delagoensis* Harv.; *Cassia petersiana* Bolle) (after the German physician and traveler Wilhelm Carl Hartwig Peters, 1815-1883, naturalist, zoologist, co-author of *Naturwissenschaftliche Reise nach Mossambique.* Berlin [1861-] 1862-1864; see J.H. Barnhart, *Biographical Notes upon Botanists.* 3: 73. 1965; Ida Kaplan Langman, *A Selected Guide to the Literature on the Flowering Plants of Mexico.* University of Pennsylvania Press, Philadelphia 1964; A. Lasègue, *Musée botanique de Benjamin Delessert.* Paris 1845; E.M. Tucker, *Catalogue of the Library of the Arnold Arboretum of Harvard University.* Cambridge, Massachusetts 1917-1933; Stafleu and Cowan, *Taxonomic Literature.* 4: 194-195. 1983)

English: monkey pod, eared cassia, dwarf cassia

Southern Africa: apiespeul; umNembenembe, uHwabile (Zulu); liJoyi, li-joyi (Swazi); muKunda, muRemberembe, muRemperempe, muSadzinbuzi (Shona); mukunda (Shona); nnembenembe (Eastern Transvaal); nembenembe (Tonga); munembenembe (Venda)

S. pleurocarpa (F. Muell.) Randell var. *pleurocarpa* (*Cassia pleurocarpa* F. Muell. var. *pleurocarpa*)

English: fire bush

S. podocarpa (Guill. & Perr.) Lock

Yoruba: ajanrere, asunwon dudu, asunwon oyinbo, asunwon funfun, asinarunwale

S. reticulata (Willd.) H. Irwin & Barneby (*Cassia reticulata* Willd.)

Honduras: baraja, barajo

Guatemala: baraja, barajo

Panama: wild senna

Nicaragua: serincontil

Bolivia: mamuri

S. ruiziana (G. Don) Irwin & Barneby var. *micrandra* Irwin & Barneby

Bolivia: chirimoya del monte

S. septemtrionalis (Viv.) H. Irwin & Barneby (*Cassia septemtrionalis* Viv.; *Adipera laevigata* (Willd.) Britton & Rose; *Cassia laevigata* Willd.)

Hawaii: kolomona, kalamona

Honduras: baraja, barajo

S. siamea (Lam.) H. Irwin & Barneby (*Cassia siamea* Lam.)

English: Djoowar, Kassod tree, Siamese senna

Japan: Tagaya-san-no-ki

Indochina: ang kanh, ang kel, bois perdrix, khi lech, muong, muong nui, ong can

India: beati, mezali, ponnavarai, karungkonnai, seematangedu, sima tangedu, kassod, manje-konne

Malaya: jahar, mezali, sibusuk, johar, johor

Sri Lanka: wa

East Africa: iron wood

S. silvestris (Vell. Conc.) Irwin & Barneby var. *silvestris*

Bolivia: samoma queti, ramo del monte, cuqui colorado

S. singueana (Del.) Lock (*Cassia goratensis* Fresen.; *Cassia singueana* Del.; *Cassia sinqueana* sensu auct. non Del., orth. err.; *Cassia zanzibarensis* Vatke) (the specific name possibly refers to the name of the place, in Ethiopia, where the French traveler and naturalist Frédéric Cailliaud (1787-1869) found the plant; see Alire Delile (Raffeneau-Delile) (1778-1850), *Centurie de plantes d'Afrique du voyage à Méroé, recueillies par M. Cailliaud.* Paris 1826)

English: wild cassia

Nigeria: rumfu (Hausa); rumfuhi (Fula); tugulele (Kanuri); shadarat al bashimi (Shuwa Arabic)

South Africa: wild cassia

S. sophera (L.) Roxb. (*Cassia sophera* L.)

English: large-leaved senna, senna sophera, senna esculenta, senna purpurea

India: kasamarda, jangli-takla, sarphoka, ran-tankala, paiditangaedu, peria-takarai, ponnantakara, pounantakara, kalkasunda, kalkashunda, kondakashinda, kasunda, bas-ki-kasunda, kasodi, sularai

China: jiang mang, chueh ming, wang chiang nan

Japan: o-ba-no-senna

Yoruba: ojua

S. spectabilis (DC.) H. Irwin & Barneby var. *spectabilis* (*Cassia spectabilis* DC.; *Cassia carnaval* Speg.)

Honduras: candelillo

Trinidad: casse

Bolivia: limoncillo, hediondillo, pacaisillo, aceitón ordinario

S. splendida (Vogel) H. Irwin & Barneby

English: golden wonder

S. surattensis (Burm.f.) H. Irwin & Barneby (*Cassia suffruticosa* König ex Roth; *Cassia surattensis* Burm.f.; *Cassia surattensis* Burm.f. var. *suffruticosa* (König ex Roth) Chatterjee) (Surat, near Bombay)

English: sickle senna, foetid cassia, glaucous cassia

Hawaii: kolomona, kalamona

Bali: kembang koning, kuning, kemoning

S. timoriensis (DC.) H. Irwin & Barneby (*Cassia timoriensis* DC.)

English: limestone cassia

Sri Lanka: anemene

Indochina: ki-lep-pa, muong, muong do, muong rut, muong trang, muong xoan

Malaya: batai

S. tora (L.) Roxb. (*Cassia tora* L.)

English: sickle senna, foetid cassia, cassia, fetid senna, ringworm plant, sickle pod, coffee weed, tavara

W. Africa: sifla, swire

India: dadamardana, dadamari, kharjugna, taga, ayudham, prabhoonata, kulikul, sanji, chakramarda, chakunda, chakundia, chakaund, panevar, pambar, panwar, kovaria, tarota, kovaraya, tankala, takala, takla, tarota, tantepuchettu, tagarisha-chettu, tagirisa, tantemu, tantiyamu, ushittagarai, thagarai-verai, tagarai, ushit-tagarai, tagache, taragashee, takara, daddupan, tora

Sri Lanka: tora

Japan: hoso-mi-ebisu-gusa, chamami

China: jue ming zi

Tibetan: thalka rdorje

Vietnam: thao quyet minh, muong ngu, dau ma, lac troi, muong lac, nha coc be, diem tap, t'rang

S. versicolor (Meyer ex J. Vogel) Irwin & Barneby

Bolivia: takarquea, alcaparrilla, mutu-mutu

S. weddelliana Irwin & Barneby

Bolivia: orko moto-moto, maranhuay

Sennia Chiov. Caesalpiniaceae

Origins:

After the Italin botanist Lorenzo Senni, 1879-1954, botanical explorer, traveler, forester, Console della milizia forestale; see A. White and B.L. Sloane, *The Stapelieae.* Pasadena 1937.

Sennia Pascher Chenopodiaceae

Origins:

After the Swiss botanist Gustav Alfred Senn, 1875-1945, traveler, plant collector, professor of botany and Director of the Basel Botanical Garden. Among his works are *Über einige Coloniebildende einzellige Algen.* Basel 1899 and *The Upper Limit of Forests in Java.* Manchester 1927; see John H. Barnhart, *Biographical Notes upon Botanists.* 3: 260. 1965; T.W. Bossert, *Biographical Dictionary of Botanists Represented in the Hunt Institute Portrait Collection.* 360. 1972.

Senniella Aellen Chenopodiaceae

Origins:

Diminutive of the genus *Sennia* Pascher in Chenopodiaceae, after the Swiss botanist Gustav Alfred Senn, 1875-1945, traveler, plant collector, professor of botany and Director of the Basel Botanical Garden, author of *Über einige Coloniebildende einzellige Algen.* Basel 1899 and of *The Upper Limit of Forests in Java.* Manchester 1927; see Paul Aellen (1896-1973), in *Botanische Jahrbücher.* 68: 416. 1938; John H. Barnhart, *Biographical Notes upon Botanists.* 3: 260. 1965.

Sentis F. Mueller Myoporaceae

Origins:

Latin *sentis, is* "brier, bramble, a thorn-bush, the dog-rose, wild-brier," *sentix, senticis* "the plant *cynosbatos* or *sentis canis,* dog-rose, wild-brier"; see Ferdinand von Mueller, *Fragmenta Phytographiae Australiae.* 4: 47. Melbourne 1863.

Sepalosaccus Schltr. Orchidaceae

Origins:

Latin *sepalum* "sepal" and *saccus, i* "a bag, sack," referring to the lateral sepals, see also *Maxillaria* Ruíz & Pav.

Sepalosiphon Schltr. Orchidaceae

Origins:

Latin *sepalum* and *siphon, onis* "a siphon, a little pipe," referring to the elongated lateral sepals, see *Glossorhyncha* Ridley.

Separotheca Waterf. Commelinaceae

Origins:

Latin *separo, avi, atum* "to part, divide" and *theca, ae* "an envelope, cover, case, sheath."

Sepikea Schlechter Gesneriaceae

Origins:

Possibly named after Sepik River, Papua, New Guinea.

Septimetula Tieghem Loranthaceae

Origins:

Latin *septum* "an enclosure, fence" and *metula, ae* "a small pyramid, obelisk," presumably referring to the anthers, see also *Phragmanthera* Tieghem.

Septogarcinia Kosterm. Guttiferae

Origins:

Latin *septum* "an enclosure, fence" plus the genus *Garcinia* L.

Septotheca Ulbr. Bombacaceae

Origins:

Latin *septum* "an enclosure, fence" and *theca, ae* "cover, case, sheath."

Septulina Tieghem Loranthaceae

Origins:

From the Latin *septum* "an enclosure, fence," *septulum* "a small septum."

Sequoia Endl. Taxodiaceae

Origins:

Dedicated to George Gist (George Guess or Gess) (also known as Soquoiah, Sequoyah, Sequoiar, Sequoiah, Sequoia, Sequoya, Se-Quo-Yah), 1760/1770-1843, the creator of the Cherokee alphabet and writing system; *sikwayi* is the Cherokee name for the opossum; see Grant Foreman, *Sequoyah.* University of Oklahoma Press, Norman 1938; George Everett Foster, *Se-Quo-Yah, the American Cadmus and Modern Moses.* A complete biography. Office of the Indian Rights Association, Philadelphia 1885; Frances S. Nichols, compil., *Index to Schoolcraft's "Indian Tribes of the United States."* Smithsonian Institution. Bureau of American Ethnology. Bulletin 152. Washington 1954; George Everett Foster, *Literature of the Cherokees,* also bibliography and the story of their genesis. Ithaca, N.Y. 1889; Erwin G. Gudde, *California Place Names.* The Origin and Etymology of Current Geographical Names. University of California Press, Berkeley 1974 [1949]; H. Genaust, *Etymologisches Wörterbuch der botanischen Pflanzennamen.* 578. 1996; E. Weekley, *An Etymological Dictionary of Modern English.* 2: 1317. 1967.

Species/Vernacular Names:

S. sempervirens (D. Don) Endl. (*Taxodium sempervirens* D. Don)

English: redwood, Californian redwood

Italian: sequoia

Sequoiadendron Buchholz Taxodiaceae

Origins:

The genus *Sequoia* and Greek *dendron* "tree."

Species/Vernacular Names:

S. giganteum (Lindley) Buchholz (*Wellingtonia gigantea* Lindley)

English: Sierra redwood, giant sequoia, big tree, mammoth tree, wellingtonia

Seraphyta Fischer & Meyer Orchidaceae

Origins:

Greek *seira* "rope" and *phyton* "plant," referring to the rhizomes.

Serapias L. Orchidaceae

Origins:

Greek *serapias* or *sarapias*, Latin *serapias, serapiadis* and also *serapion*, used by Plinius for a plant, also called *orchis*, an orchid.

Serapiastrum Kuntze Orchidaceae

Origins:

Resembling *Serapias* L.

Serenoa Hook.f. Palmae

Origins:

For the American (b. Connecticut) botanist Sereno Watson, 1826-1892 (d. Cambridge, Massachusetts), attended Yale College, traveler, plant collector, assisted William Whitman Bailey in botanical collections, in March 1868 official botanist of the Clarence King's Geological Exploration of the fortieth parallel, Assistant with Asa Gray at the Gray Herbarium and later Curator, in 1889 elected to the National Academy of Sciences. His most significant works are *Botany of California*. Cambridge, Massachusetts 1876, 1880 and *Bibliographical Index to North American Botany* ... Part I, *Polypetalae*. Washington, D.C. 1878. See *United States Geological Expolration* [sic] of the fortieth parallel. Botany. by Sereno Watson. Washington 1871 and *List of Plants Collected in Nevada and Utah 1867-1869*. Sereno Watson, collector. [Washington, D.C. 1871]; R. Zander, F. Encke, G. Buchheim and S. Seybold, *Handwörterbuch der Pflanzennamen*. 14. Aufl. 517, 797. Stuttgart 1993; Elizabeth Noble Shor, in *Dictionary of Scientific Biography* 14: 192-193. 1981; John H. Barnhart, *Biographical Notes upon Botanists*. 3: 465. 1965; William H. Brewer, *Biographical Memoirs*. National Academy of Sciences. 5: 267-290. 1903; T.W. Bossert, *Biographical Dictionary of Botanists Represented in the Hunt Institute Portrait Collection*. 427. 1972; E.M. Tucker, *Catalogue of the Library of the Arnold Arboretum of Harvard University*. Cambridge, Massachusetts 1917-1933; Ida Kaplan Langman, *A Selected Guide to the Literature on the Flowering Plants of Mexico*. University of Pennsylvania Press, Philadelphia 1964; S. Lenley et al., *Catalog of the Manuscript and Archival Collections and Index to the Correspondence of John Torrey*. Library of the New York Botanical Garden. 1973; J. Ewan, ed., *A Short History of Botany in the United States*. 92, 110. 1969.

Species/Vernacular Names:

S. repens (Bartram) Small

English: saw palmetto

Brazil: serenoa

Seretoberlinia P.A. Duvign. Caesalpiniaceae

Origins:

After the Swedish botanist Andreas (Anders) Henricus (Henricsson) Berlin, 1746-1773, pupil of Linnaeus, 1771-1773 plant collector in Sierra Leone with Henry Smeathman, Adam Afzelius and John Matthews, author of *Dissertatio academica Usus Muscorum, breviter delineatura, quam ... submittit ... A.H. Berlin*. Upsaliae 1766; see Carl Frederik Albert Christensen, *Den danske Botaniks Historie med tilhørende Bibliografi*. Copenhagen 1924-1926; Harry Charles Luke, *A Bibliography of Sierra Leone*. Oxford 1948; R.W.J. Keay, "Botanical collectors in West Africa prior to 1860." in *Comptes Rendus A.E.T.F.A.T.* 55-68. Lisbon 1962; Charles Dennis Adams (b. 1920), "Activities of Danish botanists in Guinea, 1783-1850." *Transactions of the Historical Society of Ghana*. 3: 30-46. Accra 1957; John Matthews, *A Voyage to the River Sierra-Leone*. London 1788; I.H. Vegter, *Index Herbariorum*. Part II (4), *Collectors M*. Regnum Vegetabile vol. 93. 1976; F.N. Hepper and Fiona Neate, *Plant Collectors in West Africa*. 10. Utrecht 1971; John H. Barnhart, *Biographical Notes upon Botanists*. Boston 1965.

Seriana Willd. Sapindaceae

Origins:
Orthographic variant of *Serjania* Miller.

Serianthes Benth. Mimosaceae

Origins:
See George Bentham, in *Hooker's London Journal of Botany*. 3: 225. 1844.

Sericandra Raf. Mimosaceae

Origins:
Greek *serikos* "silky" and *aner, andros* "male," with silky stamens; see Constantine Samuel Rafinesque (1783-1840), *Sylva Telluriana*. 119. Philadelphia 1838.

Sericanthe Robbrecht Rubiaceae

Origins:
With silky flowers, from the Greek *ser* "silkworm," *serikos* "silky" and *anthos* "flower"; Latin *sericus, a, um* "silky" but also "from Sera" a region of eastern Asia famous for silk-making factories; Seres is the name given by ancient writers to the Chinese, according to the Roman historian Ammianus Marcellinus (late 4th century A.D.) the Great Wall of China was *Aggeres Serium*.

Species/Vernacular Names:
S. andongensis (Hiern) Robbrecht var. *andongensis* (*Neorosea andongensis* (Hiern) Halle; *Tricalysia andongensis* Hiern; *Tricalysia legatii* Hutch.; *Tricalysia pachystigma* K. Schum.) (Andonga, in Angola)
Southern Africa: tshipengo (= insanity), tshituku (Venda)
English: shipengo tree

Sericeocassia Britton Caesalpiniaceae

Origins:
Greek *serikos* "silky" plus *Cassia*, see also *Senna* Mill.

Sericocactus Y. Ito Cactaceae

Origins:
From the Greek *serikos* "silky" plus *Cactus*, see also *Parodia* Speg.

Sericocalyx Bremek. Acanthaceae

Origins:
Greek *serikos* and *kalyx, kalykos* "calyx," referring to the silky calyx.

Sericocarpus Nees Asteraceae

Origins:
From the Greek *serikos* "silky" and *karpos* "fruit."

Sericocoma Fenzl Amaranthaceae

Origins:
Greek *serikos* "silky" and *kome* "hair of the head," referring to the woolly ovary.

Sericocomopsis Schinz Amaranthaceae

Origins:
Resembling the genus *Sericocoma* Fenzl.

Sericodes A. Gray Zygophyllaceae

Origins:
From the Greek *serikos* and *-odes* "having the form."

Sericographis Nees Acanthaceae

Origins:
Greek *serikos* "silky" and *graphis* "brush, a style for writing, a needle."

Sericolea Schltr. Elaeocarpaceae

Origins:
From the Greek *serikos* "silky" and *elaia* "olive."

Sericorema (Hook.f.) Lopr. Amaranthaceae

Origins:
From the Greek *serikos* and *korema* "broom," referring to the bracteoles of the sterile flowers.

Sericospora Nees Acanthaceae

Origins:
Greek *serikos* "silky" and *spora, sporos* "seed, spore."

Sericostachys Gilg & Lopr. Amaranthaceae

Origins:
From the Greek *serikos* "silky" and *stachys* "a spike."

Sericostoma Stocks Boraginaceae

Origins:
From the Greek *serikos* "silky" and *stoma* "mouth," referring to the silky or hairy mouth of the corolla.

Sericotheca Raf. Rosaceae

Origins:
From the Greek *serikos* "silky" and *theke* "a box, case"; see C.S. Rafinesque, *Sylva Telluriana*. 152. 1838.

Sericrostis Raf. Gramineae

Origins:
Greek *serikos* "silky" and *Agrostis*; see C.S. Rafinesque, *Neogenyton, or Indication of Sixty-Six New Genera of Plants of North America*. 4. 1825.

Sericura Hassk. Gramineae

Origins:
From the Greek *serikos* "silky" and *oura* "a tail," see also *Pennisetum* Rich.

Serigrostis Steud. Gramineae

Origins:
From the Greek *serikos* "silky" plus *agrostis, agrostidos* "grass, weed, couch grass."

Seringea F. Muell. Sterculiaceae

Origins:
Orthographic variant of *Seringia* Gay; see Sir Ferdinand Jacob Heinrich von Mueller (1825-1896), *Fragmenta Phytographiae Australiae*. 10: 96. 1877.

Seringia J. Gay Sterculiaceae

Origins:
For the French botanist Nicolas Charles Seringe, 1776-1858, plant collector, 1796-1801 Army-surgeon, 1830-1858 professor of botany and Director of the Botanical Garden at Lyon. His works include *Rapport de la commission nommée dans le sein de la Société d'Horticulture pratique du Rhône pour s'occuper de la maladie des pommes de terre.* Lyon 1845, *Saules de la Suisse.* [Bern 1805-1814], *Mélanges botaniques.* Berne 1818-1831, *Bulletin botanique.* Genève 1830 [-1832], *Description, culture et taille des mûriers, leurs espèces et leurs variétés.* Paris, Lyon [printed] 1855, *Flore du pharmacien, du droguiste et de l'herboriste.* Lyon 1851 and *Notice sur le Maclure orangé.* Lyon 1837, contributed to A.P. de Candolle, *Prodromus* (Cucurbitaceae, Leguminosae, Onagrariae, Ranunculaceae, Rosaceae and Saxifragaceae). He was father of the French entomologist Jean Claude Seringe (1810-1833); see J.H. Barnhart, *Biographical Notes upon Botanists.* 3: 260. 1965; L. Boullieux, *Biographie de N.-C. Seringe.* Lyon 1859; Mariella Azzarello Di Misa, a cura di, *Il Fondo Antico della Biblioteca dell'Orto Botanico di Palermo.* 256. Palermo 1988; Georges Roffavier (1775-1866), *Supplement à la flore Lyonnaise.* Lyon 1835; R. Zander, F. Encke, G. Buchheim and S. Seybold, *Handwörterbuch der Pflanzennamen.* 14. Aufl. Stuttgart 1993; Jacques Étienne Gay (1786-1864), in *Mémoires du Muséum d'Histoire Naturelle.* [= Monographie des cinq genres de plantes que comprend la tribu des Lasiopétalées dans la famille des Büttnériacées.] 7: 442, t. 16, 17. Paris 1821; H.N. Clokie, *Account of the Herbaria of the Department of Botany in the University of Oxford.* 241. Oxford 1964; E.M. Tucker, *Catalogue of the Library of the Arnold Arboretum of Harvard University.* Cambridge, Massachusetts 1917-1933; Ida Kaplan Langman, *A Selected Guide to the Literature on the Flowering Plants of Mexico.* Philadelphia 1964.

Seriola L. Asteraceae

Origins:
Latin *seriola, ae* "a small jar," diminutive of *seria, ae* "a cylindrical earthen vessel, a large jar."

Seriphidium (Bess.) Poljak. Asteraceae

Origins:
Latin *Seriphius, a, um* "Seriphian, absinthium."

Species/Vernacular Names:
S. maritimum (L.) Soják
English: sea wormwood

Serissa Comm. ex Juss. Rubiaceae

Origins:

The East Indian name for *Serissa foetida* (L.f.) Lam.

Species/Vernacular Names:

S. foetida (L.f.) Lam. (*Serissa japonica* Thunb.; *Lycium japonicum* Thunb.; *Lycium foetidum* L.f.)

Japan: haku-chôge

China: bai ma gu, chu chieh tsao, liu yueh ling, liu yueh shuang

Serjania Miller Sapindaceae

Origins:

In honor of the Rev. Father Philippe Sergeant of Caux, France, a monk, botanist, or after the French friar and botanist Paul Serjeant; see Philip Miller, *The Gardeners Dictionary*. Abr. ed. 4. London (28 Jan.) 1754; H. Genaust, *Etymologisches Wörterbuch der botanischen Pflanzennamen*. 579. Basel 1996; G.C. Wittstein, *Etymologisch-botanisches Handwörterbuch*. 809. Ansbach 1852.

Serpyllopsis Bosch Hymenophyllaceae

Origins:

Resembling *serpyllum* or *serpillum* "thyme, wild-thyme."

Serrafalcus Parl. Gramineae

Origins:

Latin *serra, ae* "saw" and *falx, falcis* "a sickle, scythe," referring to the awns of *Serrafalcus alopecuros* (Poiret) C. Gardner; see Filippo Parlatore (1816-1877), *Rariorum plantarum et haud cognitarum in Sicilia sponte provenientium*. 2: 14. Panormi [Palermo] 1838-1840.

Serrastylis Rolfe Orchidaceae

Origins:

Latin *serra, ae* and Greek *stylos, stylis* "a style, column, a pillar," referring to the clinandrium, see also *Macradenia* R. Br.

Serratula L. Asteraceae

Origins:

Latin *serratula, ae*, the Italian name for betony (Plinius), the diminutive of the Latin *serratus* "saw-shaped," referring to the leaves.

Species/Vernacular Names:

S. tinctoria L.

English: sawwort

Italian: serratola, serretta, seratula, serratula

Serruria Burm. ex Salisb. Proteaceae

Origins:

The generic name honors Joseph (Josephus) Serrurier, 1668-1742, professor of medicine and botany in Utrecht. Among his works are *Oratio pro Philosophia*. Trajecti ad Rhenum 1706, Disputatio ... *de febribus in genere*. Lugduni Batavorum [Leyden] 1690, Disputatio ... *de gravitate aeris*. Lugduni Batavorum 1690 and *Oratio Funebris in Obitum H. Relandi*. (Carmina virorum eruditorum, qui H. Relandi ... memoriam celebraverunt.) Trajecti ad Rhenum [Utrecht] 1718.

Species/Vernacular Names:

S. florida (Thunb.) Salisb. ex Knight

English: blushing bride

Sersalisia R. Br. Sapotaceae

Origins:

After Jo. Bapt. Sersalis or Sersale, a Neapolitan monk or priest, zoologist, early 17th century; see Robert Brown (1773-1858), *Prodromus florae Novae Hollandiae et Insulae van-Diemen*. 529. London 1810.

Sertifera Lindley & Reichb.f. Orchidaceae

Origins:

Latin *sertus, a, um* "wreath of flowers, garland," *sero, tum* "to bind together, plait" and *fero* "to bear, carry," perhaps referring to the habit of the plant.

Sesamoides Ortega Resedaceae

Origins:

Resembling *Sesamum* L., see Ortega, *Tabulae Bot*. 24. 1773.

Sesamothamnus Welw. Pedaliaceae

Origins:

The name means "the shrubby *Sesamum*," from the Greek *thamnos* "shrub" and the genus *Sesamum* L.

Species/Vernacular Names:

S. benguellensis Welw. (Benguela, Angola; see H. Capello and R. Ivens, *From Benguella to the Territory of Yacca. Description of a journey into Central and West Africa …* by H.C. and R.I., Officers of the Royal Portuguese Navy. Expedition organised in the years 1877-1880. London 1882; James Johnston, *Reality versus Romance in South Central Africa.* An account of a journey across the continent from Benguella on the West, … to the Mouth of Zambesi on the East Coast. London 1893)

South Africa: Benguela sesamothamnus

S. guerichii (Engl.) E.A. Bruce (*Sigmatosiphon guerichii* Engl.) (to honor Georg Julius Ernst Gürich (Guerich), 1859-1938, plant collector in southwest Africa from 1885-1888, Guinea, Tumbo Island, Konakry, he published *Deutsch Südwest-Afrika. Reise bilder und Skizzen aus den Jahren 1888 und 1889, mit einer Original-Routenkarte.* Hamburg 1891; see A. Engler, "Plantae Gürichianae; ein Beitrag zur Kenntnis der Flora von Deutschsüdwestafrika." *Bot. Jb.* 19: 128-154. 1895; F.N. Hepper and Fiona Neate, *Plant Collectors in West Africa.* 34. 1971)

South Africa: spekboom

English: South West African sesamothamnus

S. lugardii N.E. Br. ex Stapf (after the botanical collector and Major of the British Army Edward James Lugard, 1865-1944 (d. Dorking), collected in Bechuanaland, Basutoland and Ngamiland; he was the brother of General Sir Frederick Dealtry Lugard, 1st Baron Lugard (1858-1945), from 1912 to 1919 Governor of Nigeria; he was husband of Charlotte Eleanor Lugard (*née* Howard), 1859-1939, plant collector; see E. Lugard and N.E. Brown, "The flora of Ngamiland." *Kew Bull.* 1909: 81-146; A. White and B.L. Sloane, *The Stapelieae.* Pasadena 1937; J. H. Barnhart, *Biographical Notes upon Botanists.* 1965; Mary Gunn and Leslie E. Codd, *Botanical Exploration of Southern Africa.* 235-236. Cape Town 1981; F.N. Hepper and Fiona Neate, *Plant Collectors in West Africa.* 51. 1971; Gordon Douglas Rowley, *A History of Succulent Plants.* Strawberry Press, Mill Valley, California 1997; Ray Desmond, *Dictionary of British & Irish Botanists and Horticulturists.* London 1994; G. Hedgecock, in *Asklepios.* 64: 27-28. 1995)

English: Transvaal sesame-bush, Transvaal sesamothamnus

Southern Africa: Transvaalse sesambos, kanniedood; moboana, siboana (Western Transvaal, northern Cape, Botswana)

Sesamum L. Pedaliaceae

Origins:

Latin *sesamum, sisamum* "sesame, sesamum, an oily plant, *sesima* or *sesama* "another name for *cici*, the palma Christi, castor oil plant" (Plinius), Greek *sesamon, sasamon, saamon, sesame*; Hebrew *shemen* "oil"; Akkadian *samassammu, samsammu*; see Carl Linnaeus, *Species Plantarum.* 634. 1753 and *Genera Plantarum.* Ed. 5. 282. 1754; Salvatore Battaglia, *Grande dizionario della lingua italiana.* XVIII: 788-789. UTET, Torino 1997; Giovanni Semerano, *Le origini della cultura europea.* Dizionari Etimologici. Basi semitiche delle lingue indeuropee. Dizionario della lingua Greca. 2(1): 260. Leo S. Olschki Editore, Firenze 1994.

Species/Vernacular Names:

S. alatum Thonn.

Southern Africa: motlhomaganyane (Tswana)

S. indicum L. (*Sesamum orientale* L.)

English: Benue oil, gingelly-oil plant, gingelly, gingil, gingelly seed, gingilli, sem-semoil, sesamum, sesam, sesame, sesame oil, teel oil, tilseed, thunderbolt flower, wild foxglove

Arabic: simsim, jelilan, jiljlen, hall

Italian: sesamo, sasamo, sisamo, sisimo

Yoruba: yanmoti, eku igi

Southern Africa: oliebossie; mfuta (Swahili); molekelela (Sotho)

Shaba (formerly Katanga): molende, mulinda

Tanzania: simsim

Congo: bonanguila

Morocco: janjlane, azzanjlane

Tunisia: jiljlèn

India: til-ka-tel, Krishna-tel, Krishna-til, kala-til, chadu-til, rakta-til, sanki-til, barik-til, tallatil, ashadi-tal, mitha-tel, tila-taila, mitho-tel, kala-katwa, purbia, snehapahla, tilmin, tilaha, tila, tir, thirr, tisi, til, tili, tal, tel, teel, mitho-tel, yellu, ellu, yelluchedi, yellu-cheddie, khasa, rashi, nuvvulu, nuvvu, pollanuvvulu, uruellu, guvvulu, sumsum, kunjad, bhunguru, chokhota tela, manchinune, nallenne, ellenne

Malaya: bijan, lenga, gingili

Japan: goma, uguma

Tibetan: til

China: hei zhi ma, hei yu ma, pai yu ma, hu ma, ching jang

Iran: kunjad

Cuba: ajonjoli

S. schinzianum Aschers. (*Sesamum antirrhinoides* Welw. ex Aschers.)

English: sesam

S. triphyllum Welw. ex Aschers. var. *triphyllum* (*Sesamum gibbosum* Bremek. & Oberm.)

English: wild sesame

South Africa: wildesesam

Sesban Adans. Fabaceae

Origins:

Arabic *sisaban, seshban, saisaban* or *sesaban*, Persian *sisaban*, see also *Sesbania* Scop. and *Sesbania sesban* (L.) Merrill.

Sesbania Scop. Fabaceae

Origins:

Arabic *sisaban, seshban, saisaban* or *sesaban*, Persian *sisaban*; see Giovanni Antonio Scopoli, *Introductio ad Historiam Naturalem*. 308. Pragae (Jan.-Apr.) 1777.

Species/Vernacular Names:

S. bispinosa (Jacq.) W. Wight. (*Aeschynomene aculeata* Schreber; *Aeschynomene bispinosa* Jacq.; *Sesban aculeatus* Poiret, nom. illeg.; *Sesbania aculeata* Pers., nom. illeg.)

English: sesbania, spiny sesbania

South Africa: stekel-sesbania

S. cannabina (Retz.) Pers. (*Sesbania sericea* (Willd.) Link var. *glabra*), "Dhaincha," "Dhunchi"

English: sesbania

Spanish: canicha

Japan: densei, tsuno-kusa-nemu

S. cannabina (Retz.) Pers. var. *cannabina* (*Aeschynomene cannabina* Retz.; *Sesbania aculeata* sensu J. Black, non (Schreber) Poiret)

English: yellow pea-bush

S. exaltata (Raf.) A.W. Hill

English: Colorado River hemp, peatree, coffee weed

S. grandiflora (L.) Pers. (*Aeschynomene grandiflora* (L.) L.; *Agati grandiflora* (L.) Desv.; *Robinia grandiflora* L.; *Sesban grandiflorus* (L.) Poiret)

English: vegetable hummingbird, scarlet wistaria tree, sesban

Indochina: ang kea dey, danh ca

Malaya: getih, turi, kachang turi, kelur

The Philippines: kature, katurai, katuray, katuday, katudai, katodai, gauai-gauai, kambang-turi, diana

India: agathi, agati

Tibetan: socha, posocha

S. microphylla Phill. & Hutch. (*Sesbania mossambicensis* sensu Phill. & Hutch.)

Southern Africa: muJekejeke (Shona)

S. punicea (Cav.) Benth. (*Daubentonia punicea* (Cav.) DC.; *Daubentonia tripetii* Poit.; *Piscidia punicea* Cav.; *Sesbania tripetii* (Poit.) hort. ex Hubb.)

English: Brazilian glory pea, coffee weed, glory pea, rattlepod, red sesbania, sesbania, tango

South Africa: Brasiliaanse glorie-ertjie, glorie-ertjie, rooi-sesbania

S. sesban (L.) Merrill

English: sesban, Egyptian rattlepod

Japan: kidachi-densei, Indo-densei

S. sesban (L.) Merr. subsp. *sesban* var. *sesban* (*Sesbania aegyptiaca* (Poiret) Pers.; *Sesbania aegypticus* Poir.; *Aeschynomene sesban* L.)

English: river bean

Southern Africa: rivierboontjie, frother; umSokhosokho, umKhumukhweqwe, umQambuqweqwe (Zulu); mpupunwa (= frother) (Venda)

S. tomentosa Hook. & Arnott (*Agati tomentosa* (Hook. & Arnott) Nutt. ex A. Gray)

Hawaii: 'ohai

S. virgata (Cav.) Pers. (*Aeschynomene virgata* Cav.; *Coursetia virgata* (Cav.) DC.; *Sesban virgatus* (Cav.) Poiret; *Sesbania marginata* Benth.)

Argentina: cafe cimarron, cambay, porotillo, tembetari

Seseli L. Umbelliferae

Origins:

Greek *seselis, seseli*, ancient name for an umbelliferous plant, Latin *seselis, is* for a plant, meadow saxifrage, hartwort, seseli (Plinius), or *sil, sili*, Akkadian *sillu*, Hebrew *sel* "shade: of a tree"; see Carl Linnaeus, *Species Plantarum*. 634. 1753 and *Genera Plantarum*. Ed. 5. 282. 1754.

Species/Vernacular Names:

S. libanotis (L.) K. Koch

China: hsieh hao

Seselopsis Schischk. Umbelliferae

Origins:
Resembling *Seseli* L.

Seshagiria Ansari & Hemadri Asclepiadaceae

Origins:

The name commemorates the Indian botanist Rolla Seshagiri Rao, b. 1921, professor of botany, joint Director of the Botanical Survey of India.

Sesleria Scopoli Gramineae

Origins:
For the Italian (Venetian) naturalist Leonardo (Lionardo) Sesler, d. 1785, physician, owner of a private botanical garden, author of *Lettera intorno ad un nuovo genere di Piante terrestri*, in Vitaliano Donati (1713-1763), *Della storia naturale marina dell'Adriatico*. [Edited by G. Rubbi-Carli.] Venezia 1750; see Giovanni Antonio Scopoli, *Flora Carniolica*. 189. Viennae 1760; R. Zander, F. Encke, G. Buchheim and S. Seybold, *Handwörterbuch der Pflanzennamen*. 14. Aufl. 780. 1993; F. Boerner & G. Kunkel, *Taschenwörterbuch der botanischen Pflanzennamen*. 4. Aufl. 169. 1989; John H. Barnhart, *Biographical Notes upon Botanists*. 3: 261. Boston 1965; T.W. Bossert, *Biographical Dictionary of Botanists Represented in the Hunt Institute Portrait Collection*. 361. 1972; Jonas C. Dryander, *Catalogus bibliothecae historico-naturalis Josephi Banks*. London 1800.

Species/Vernacular Names:
S. caerulea (L.) Ard.
English: blue moorgrass

Sesleriella Deyl Gramineae

Origins:
The diminutive of *Sesleria* Scop.

Sessea Ruíz & Pavón Solanaceae

Origins:
After the Spanish (b. Aragón) physician Martín de Sessé y Lacasta, 1751-1808 (Madrid), botanist, traveler, plant collector, 1785 in Mexico, with the Royal botanical expedition to New Spain (*Expedición botanica de Nueva España*), Director of the Royal Botanical Garden in Mexico, with José Mariano Moçiño (1757-1820) wrote *Plantae Nouae Hispaniae*. Mexici 1887 [-1890]; see John H. Barnhart, *Biographical Notes upon Botanists*. 3: 261. 1965; H.N. Clokie, *Account of the Herbaria of the Department of Botany in the University of Oxford*. 242. Oxford 1964; M. Colmeiro y Penido, *La Botánica y los Botánicos de la Peninsula Hispano-Lusitana*. Madrid 1858; A. Lasègue, *Musée botanique de Benjamin Delessert*. Paris 1845; E.M. Tucker, *Catalogue of the Library of the Arnold Arboretum of Harvard University*. Cambridge, Massachusetts 1917-1933; Ignatz Urban, ed., *Symbolae Antillanae*. Berlin 1902; Rogers Mc Vaugh, in *Dictionary of Scientific Biography* 12: 326-328. 1981; Arthur Robert Steele, *Flowers for the King. The expedition of Ruiz and Pavon and the Flora of Peru*. Durham, N.C. 1964; Gordon Douglas Rowley, *A History of Succulent Plants*. Strawberry Press, Mill Valley, California 1997; Ida Kaplan Langman, *A Selected Guide to the Literature on the Flowering Plants of Mexico*. University of Pennsylvania Press 1964; Irving William Knobloch, compil., "A preliminary verified list of plant collectors in Mexico." *Phytologia Memoirs*. VI. 1983.

Species/Vernacular Names:
S. sp.
Peru: laplacata

Sesseopsis Hassler Solanaceae

Origins:
Resembling *Sessea*.

Sessilanthera Molseed & Cruden Iridaceae

Origins:
Latin *sessilis* "with a broad foot, low, dwarf, stalkless" and Greek *anthera* "anther."

Sessilibulbum Brieger Orchidaceae

Origins:
Latin *sessilis* "stalkless" and *bulbus, i* or *bulbos* "a bulb, bulbous root."

Sessilistigma Goldblatt Iridaceae

Origins:
From the Latin *sessilis* "stalkless" and *stigma* "a stigma."

Sessleria Sprengel Gramineae

Origins:
Orthographic variant of *Sesleria* Scop.

Sestochilos Breda Orchidaceae

Origins:
Greek *setho, sethein* "to sift, bolt" and *cheilos* "lip," referring to the shape of the long lip, see R.E. Schultes and A.S. Pease, *Generic Names of Orchids. Their Origin and Meaning*. 285, 287. Academic Press, New York and London 1963; some suggest from the Latin *sextus* "the sixth" and Greek *cheilos*.

Sesuvium L. Aizoaceae

Origins:

Derived from Latin *sedum*, or perhaps from the Latin *Sesuvii, orum* (*Essui, orum*), a Gallic tribe, west of the Seine; see C. Linnaeus, *Systema Naturae*. Ed. 10. 2: 1052, 1058, 1371. 1759.

Species/Vernacular Names:

S. ayresii Marais

Rodrigues Island: pourpier marin

S. portulacastrum (L.) L. (*Portulaca portulacastrum* L.; *Trianthema portulacastrum* L. var. *hillebrandii* Degener & I. Degener)

English: purslane sesuvium, sea purslane

The Philippines: dampalit, gélang-laut, taraumpalit, bilang-bilang, karam-palit

Japan: miru-suberi-hiyu, hama-mizu-na

Malaya: gelang laut, sepit, sesepit

Hawaii: 'akulikuli

S. verrucosum Raf.

English: western sea-purslane

Setaria P. Beauv. Gramineae

Origins:

Latin *saeta* (*seta*), *ae* "a bristle, hair," the spikelets are bristly; see A.M.F.J. Palisot de Beauvois, *Essai d'une nouvelle Agrostographie*. 51, 178. Paris (Dec.) 1812.

Species/Vernacular Names:

S. australiensis (Scribner & Merr.) Vick.

English: scrub pigeon grass

S. glauca (L.) Beauv.

English: yellow bristle grass, glaucous bristle grass

Arabic: sha'r el-far

S. gracilis Kunth

English: yellow foxtail, perennial foxtail, slender pigeon grass

Hawaii: mau'u Kaleponi

S. incrassata (Hochst.) Hack. (*Setaria eylesii* Stapf & C.E. Hubb.; *Setaria eylesii* Stapf; *Setaria gerrardii* Stapf; *Setaria holstii* Herr.; *Setaria pabularis* Stapf; *Setaria palustris* Stapf; *Setaria perberis* de Wit; *Setaria phragmitoides* Stapf; *Setaria porphyrantha* Stapf; *Setaria rudifolia* Stapf; *Setaria woodii* Hack. subsp. *bechuanica* de Wit; *Setaria woodii* Hack. var. *fonssalutis* de Wit; *Setaria woodii* Hack. var. *woodii*) (Latin *incrassatus* "made stout, fattened")

English: wood bristle grass

South Africa: turfsetaria

S. italica (L.) Beauv. (*Panicum italicum* L.)

English: foxtail, foxtail millet, common millet, Italian millet, Japanese millet, German millet, Hungarian millet, spiked millet, Bengal grass, Hungarian grass

Southern Africa: boermanna, geelboermanna, geelgras, katstertgras, manna, mannakoring, rooimanna, vingermanna, witmanna; joang-ba-lipere (Sotho)

China: su mi, liang, ku tzu

The Philippines: daua, bikakau, borona, rautnokara, sabug, sammang, turai

Malaya: rumput iskoi

S. lindenbergiana (Nees) Stapf (*Setaria phillipsii* de Wet)

English: mountain bristle grass, tussock grass

South Africa: bergsetaria, bosbuffelsgras, koppiesbuffelsgras, randjiesbuffelsgras, watergras

S. megaphylla (Steud.) Dur. & Schinz (*Setaria chevalieri* Stapf ex Stapf & C.E. Hubb.; *Setaria insignis* de Wit)

English: broad-leaved setaria, buffalo grass, bush buffalo grass, fine sword grass, forest buffalo grass, macopo grass, ribbon bristle grass, ribbon grass

Southern Africa: sclitz grass, solitzgrass; bosbuffelsgras, breë-blaar borselgras, breëblaarsetaria, riffelblaarsetaria; mufhafha (Venda)

S. nigrirostris (Nees) Durand & Schinz

English: large seed setaria

South Africa: grootsaad setaria, mannagras

S. pallide-fusca (Schumach.) C.E. Hubb.

English: annual Timothy, cat's tail, garden bristle grass, garden setaria, horse grass, red bristle grass, water setaria

Southern Africa: katstertgras, perdesoetgras, rooiborselgras, rooiborselsaadgras, rooisaadgras, tuinsetaria; lehola-la-lipere (Sotho)

S. palmifolia (Koenig) Stapf (*Panicum palmifolium* Koenig)

English: palm grass

Japan: sasa-kibi

S. plicata (Lam.) Cooke (*Panicum plicatum* Lam.)

Japan: ko-sasa-kibi

Malaya: akar pimpan

S. pumila (Poiret) Roemer & Schultes (*Panicum pumila* Poiret)

English: pale pigeon grass, yellow foxtail, pigeon grass, cat-tail millet

East Africa: aloom, aluum

S. sphacelata (Schumach.) Moss

English: South African pigeon grass, setaria

Yoruba: orisoso

S. sphacelata (Schumach.) Moss var. *sphacelata* (*Setaria decipiens* de Wit; *Setaria flabellata* Stapf subsp. *flabellata*; *Setaria neglecta* de Wit; *Setaria perennis* Hack.; *Setaria sphacelata* (Schumach.) Moss subsp. *aquamontana* de Wit; *Setaria sphacelata* (Schumach.) Moss var. *stolonifera* de Wit; *Setaria stenantha* Stapf)

English: common bristle grass, golden millet, golden Timothy, landgrass, oldlandsgrass, Rhodesian Timothy grass, South African golden millet grass, South African pigeon grass, setaria

Southern Africa: gewone setaria, kanariegras, katstertmannagras, mannagras, oulandgras; ikununu (Zulu)

S. sphacelata (Schumach.) Moss var. *torta* (Stapf) Clayton (*Setaria flabellata* Stapf subsp. *natalensis* de Wit; *Setaria homblei* De Willd.; *Setaria torta* Stapf)

English: small creeping foxtail

South Africa: buffelgras, kleinkruipmannagras, kruipgras, mannagras

S. splendida Stapf

English: splendid grass

S. verticillata (L.) Beauv. (*Panicum verticillatum* L.)

English: rough bristle grass, bristly foxtail, burr bristle grass, cat's tail, lovegrass, sticky bristle grass, whorled pigeon grass, sticky grass

Mexico: belaga ciiti, pelaga cijti, guixi ciiti

Arabic: qamh el-far

East Africa: anaga, ekibebia, etanuka, kiamata, malamata, maramata, rirariria

Yoruba: eemo eye

South Africa: katstertgras, klawergras, kleefgras, klitsgras, klitsborselgras, klitskatstert, klitssetaria, siergras, steekgras

Hawaii: mau'u pilipili

S. viridis (L.) Beauv. (*Panicum viride* L.)

English: green bristle grass, green pigeon grass, rough bristle grass, green foxtail

Arabic: deil el-far

China: gou wei cao, su, hsien su

Setariopsis Scribner ex Millsp. Gramineae

Origins:
Resembling *Setaria*.

Setchellanthus Brandegee Capparidaceae (Capparaceae)

Origins:
For the American (b. Norwich, Connecticut) botanist William Albert Setchell, 1864-1943 (d. Berkeley, California), botanical historian, algologist (marine algae of the northern Pacific and crustaceous algae), professor of botany, traveler and botanical collector, phycologist, bryologist. His writings include "South American sea grasses." *Revista Sudamer. Bot.* 1(4): 107-110. 1934 and "The genus *Ruppia*." *Proc. Calif. Acad. Sci.* ser. 4, 25(18): 469-477. 1946; see John H. Barnhart, *Biographical Notes upon Botanists.* 3: 262. 1965; J. Ewan, ed., *A Short History of Botany in the United States.* New York and London 1969; T.W. Bossert, *Biographical Dictionary of Botanists Represented in the Hunt Institute Portrait Collection.* 361. 1972; S. Lenley et al., *Catalog of the Manuscript and Archival Collections and Index to the Correspondence of John Torrey.* Library of the New York Botanical Garden. 366. 1973; Elmer Drew Merrill, in *Contr. U.S. Natl. Herb.* 30(1): 273-274. 1947 and in *Bernice P. Bishop Mus. Bull.* 144: 168-169. 1937; E.M. Tucker, *Catalogue of the Library of the Arnold Arboretum of Harvard University.* Cambridge, Massachusetts 1917-1933; Stafleu and Cowan, *Taxonomic Literature.* 5: 528-535. Utrecht 1985; Joseph Ewan, in *Dictionary of Scientific Biography* 12: 328-329. 1981.

Setiacis S.L. Chen & Y.X. Jin Gramineae

Origins:
Latin *saeta* (*seta*), *ae* "a bristle, hair" and Greek *ake, akis, akidos* "tip, thorn, a sharp point."

Seticereus Backeb. Cactaceae

Origins:
Latin *saetosus, a, um* "bristly" plus *Cereus.*

Seticleistocactus Backeb. Cactaceae

Origins:
Latin *saetosus, a, um* "bristly" plus *Cleistocactus* Lem.

Setiechinopsis Backeb. Cactaceae

Origins:
From the Latin *saetosus, a, um* "bristly" plus *Echinopsis* Zucc.

Setilobus Baillon Bignoniaceae

Origins:
Latin *saetosus, a, um* "bristly" and Greek *lobos* "a pod."

Setirebutia Fric & Kreuz. Cactaceae

Origins:

From the Latin *saetosus, a, um* "bristly" plus *Rebutia* K. Schum.

Setosa Ewart Gramineae

Origins:

Full of bristles, bristly, from the Latin *s(a)etosus, a, um* (*saeta*); see Alfred James Ewart (1872-1937) and Olive Blanche Davies (fl. 1917), *The Flora of the Northern Territory*. 33. Melbourne 1917.

Sevada Moq. Chenopodiaceae

Origins:

Suiwed mullah, suivada, suwayd, suweid, suwed, suda, sauda, sawad "salty," Arabic names for *Suaeda baccata* or *Suaeda aegyptiaca*.

Seychellaria Hemsley Triuridaceae

Origins:
Seychelles Islands.

Seymeria Pursh Scrophulariaceae

Origins:

For the English botanist Henry Seymer, 1745-1800, gardener, friend of R. Pulteney and D. Solander; see Ray Desmond, *Dictionary of British & Irish Botanists and Horticulturists*. 618. London 1994.

Seymeriopsis Tzvelev Scrophulariaceae

Origins:
Resembling *Seymeria* Pursh.

Shafera Greenman Asteraceae

Origins:

After the American botanist John Adolf (Adolph) Shafer, 1863-1918, plant collector (Cuba, West Indies and South America, with N.L. Britton), pharmacist, traveler; see John H. Barnhart, *Biographical Notes upon Botanists*. 3: 264. 1965; T.W. Bossert, *Biographical Dictionary of Botanists Represented in the Hunt Institute Portrait Collection*. 361. 1972; S. Lenley et al., *Catalog of the Manuscript and Archival Collections and Index to the Correspondence of John Torrey*. Library of the New York Botanical Garden. 366. 1973; E.M. Tucker, *Catalogue of the Library of the Arnold Arboretum of Harvard University*. Cambridge, Massachusetts 1917-1933; Gordon Douglas Rowley, *A History of Succulent Plants*. Strawberry Press, Mill Valley, California 1997; R. Zander, F. Encke, G. Buchheim and S. Seybold, *Handwörterbuch der Pflanzennamen*. 14. Aufl. Stuttgart 1993.

Shaferocharis Urban Rubiaceae

Origins:
After the American botanist John Adolf Shafer, 1863-1918.

Shaferodendron Gilly Sapotaceae

Origins:
After the American botanist John Adolf Shafer, 1863-1918.

Shantzia Lewton Malvaceae

Origins:

For the American botanist Homer LeRoy (Leroy) Shantz, 1876-1958, traveler (Africa) and plant collector, wrote *The Use of Fire as a Tool in the Management of the Brush Ranges of California*. 1947; see J.H. Barnhart, *Biographical Notes upon Botanists*. 3: 265. Boston 1965; Mary Gunn and Leslie E. Codd, *Botanical Exploration of Southern Africa*. Cape Town 1981; J. Ewan, ed., *A Short History of Botany in the United States*. New York and London 1969; Joseph Ewan, *Rocky Mountain Naturalists*. The University of Denver Press 1950; T.W. Bossert, *Biographical Dictionary of Botanists Represented in the Hunt Institute Portrait Collection*. 362. 1972; E.M. Tucker, *Catalogue of the Library of the Arnold Arboretum of Harvard University*. Cambridge, Massachusetts 1917-1933; Ida Kaplan Langman, *A Selected Guide to the Literature on the Flowering Plants of Mexico*. 696. University of Pennsylvania Press, Philadelphia 1964; Irving William Knobloch, compil., "A preliminary verified list of plant collectors in Mexico." *Phytologia Memoirs*. VI. 1983.

Shawia Forst. & Forst.f. Asteraceae

Origins:

For the British Rev. Thomas Shaw, 1694-1751 (Oxford), natural historian and antiquary, traveler in the Levant, plant collector, from 1720 to 1733 appointed Chaplain to an

English factory at Algiers, in 1734 a Fellow of the Royal Society and Vicar of Godshill on the Isle of Wight, in 1740 Principal of St. Edmund Hall (Oxford), in 1741 *Regius Professor* of Greek at Oxford University, from 1742 to 1751 Vicar of Bramley (Hampshire), author of *A Supplement to a Book Entitled, Travels, or Observations*, etc., wherein some objections lately made against it [by R. Pococke] are ... answered, with ... additional remarks and observations. Oxford 1746 and *A Further Vindication of the Book of Travels and the Supplement to it*, in a letter to R. Clayton, Bishop of Clogher [in answer to objections by him]. [Oxford 1747], J.J. Dillenius (1684-1747) assisted in his *Specimen phytographiae africanae.* [Appendix to *Travels, or Observations Relating to Several Parts of Barbary and the Levant.*] Oxford 1738. See J.R. Forster and J.G.A. Forster, *Characteres generum plantarum.* 95, t. 48. (Aug.) 1776; Blanche Henrey, *British Botanical and Horticultural Literature before 1800.* 1975; R. Pulteney, *Historical and Biographical Sketches of the Progress of Botany in England.* 2: 173-174. London 1790; Dawson Turner, *Extracts from the Literary and Scientific Correspondence of R. Richardson, of Bierly, Yorkshire: Illustrative of the State and Progress of Botany.* Yarmouth 1835.

Sheareria S. Moore Asteraceae

Origins:

For George Shearer, physician, plant collector and traveler (China); see Emil Bretschneider (1833-1901), *History of European Botanical Discoveries in China.* [Reprint of the original edition 1898.] Leipzig 1981.

Sheffieldia Forst. & Forst.f. Primulaceae

Origins:

After the British (b. Warwickshire) Rev. William Sheffield, c. 1732-1795 (d. Oxford), from 1772 to 1795 Keeper of the Ashmolean Museum at Oxford, friend of Joseph Banks and of the Rev. Gilbert White (1720-1793); see J.R. Forster and J.G.A. Forster, *Characteres generum plantarum.* 17, t. 9. (Aug.) 1776; Rashleigh Holt White, *The Life and Letters of Gilbert White of Selborne.* Written and edited by his great-grandnephew R. Holt White. London 1901; E.A. Martin, *A Bibliography of Gilbert White ... of Selborne.* 1934; George Claridge Druce (1850-1932), *Flora of Berkshire.* Oxford 1897 and *Flora of Oxfordshire.* Ed. 2. Oxford 1927; Robert Plot (1640-1696), *The Natural History of Oxfordshire.* [Folio, first edition, the classic Baconian natural history.] Oxford and London 1677.

Shepherdia Nuttall Elaeagnaceae

Origins:

For John Shepherd, 1764-1836 (Liverpool), Curator of the Liverpool Botanic Garden, friend of J.E. Smith and T. Nuttall, wrote *A Catalogue of Plants in the Botanic Garden, at Liverpool.* Liverpool 1808; see John H. Barnhart, *Biographical Notes upon Botanists.* 3: 268. 1965; H.N. Clokie, *Account of the Herbaria of the Department of Botany in the University of Oxford.* 242. Oxford 1964; F. Boerner & G. Kunkel, *Taschenwörterbuch der botanischen Pflanzennamen.* 4. Aufl. 169. 1989.

Species/Vernacular Names:

S. argentea (Pursh) Nutt.

English: buffalo berry, beef-suet-tree

Sherardia L. Rubiaceae

Origins:

After the British (b. Bushby, Leicestershire) botanist William Sherard (Sherwood, Sheerwood), 1659-1728 (d. London), antiquarian, plant collector in Greece and Asia Minor, founder of the Sherardian Chair of Botany at Oxford, from 1686 to 1688 studied botany in Paris under Joseph Pitton de Tournefort (1656-1708), from 1688 to 1689 a pupil of H. Boerhaave in Leyden, from 1703 to 1716 Consul at Smyrna, 1720 Fellow of the Royal Society. See Paul Hermann, *Paradisus batavus.* Opus posthumum edidit William Sherard. Lugduni-Batavorum [Leyden] 1698; Carolyn D. Torosian, in *Dictionary of Scientific Biography* 12: 394-395. 1981; Ray Desmond, *Dictionary of British & Irish Botanists and Horticulturists.* 623. 1994; B.D. Jackson, "A sketch of the life of William Sherard." in *Journal of Botany.* 12: 129-138. 1874; John H. Barnhart, *Biographical Notes upon Botanists.* 3: 269. 1965; F. Boerner & G. Kunkel, *Taschenwörterbuch der botanischen Pflanzennamen.* 4. Aufl. 169. 1989; James Britten, *The Sloane Herbarium,* revised and edited by J.E. Dandy. 1958; R. Pulteney, *Historical and Biographical Sketches of the Progress of Botany in England.* London 1790; G.C. Gorham, *Memoirs of John Martyn ... and Thomas Martyn.* London 1830; John Ray, *Historia plantarum.* London 1686-1704; Carl Linnaeus, *Species Plantarum.* 102. 1753 and *Genera Plantarum.* Ed. 5. 45. 1754; William Munk, *The Roll of the Royal College of Physicians of London.* London 1878; George Claridge Druce, *Flora of Berkshire.* Oxford 1897 and *Flora of Oxfordshire.* Ed. 2. Oxford 1927; T.W. Bossert, *Biographical Dictionary of Botanists Represented in the Hunt Institute Portrait Collection.* 364. 1972; H.N. Clokie, *Account of the Herbaria of the Department of Botany in the University of Oxford.* 1964; Blanche Henrey, *British Botanical and Horticultural Literature before 1800.* 1975; Jonas C. Dryander, *Catalogus bibliothecae historico-naturalis*

Josephi Banks. London 1800; A. Lasègue, *Musée botanique de Benjamin Delessert*. Paris 1845; E.M. Tucker, *Catalogue of the Library of the Arnold Arboretum of Harvard University*. Cambridge, Massachusetts 1917-1933; H.R. Fletcher, *Story of the Royal Horticultural Society, 1804-1968*. Oxford 1969; R.H. Semple, *Memoirs of the Botanic Garden at Chelsea ... by the Late Henry Field*. London 1878; B. Henrey, *No Ordinary Gardener — Thomas Knowlton, 1691-1781*. Edited by A.O. Chater. British Museum (Natural History). London 1986.

Species/Vernacular Names:
S. arvensis L.

English: field madder

Shibataea Makino ex Nakai Gramineae

Origins:
For the Japanese botanist Keita Shibata (Shibata Keito), 1877-1949, biochemist, from 1910 to 1911 studied with the German plant physiologist and bryologist Wilhelm Friedrich Philipp Pfeffer (1845-1920) at Leipzig, founder and first editor of *Acta phytochimica*, from 1912 to 1938 professor of botany at Tokyo University, in 1917 worked in Buitenzorg, from 1938 to 1949 Director of the Iwata Institute for Plant Biochemistry, author of *Cytologische Studien über die endotrophen Mykorrhizen*. Leipzig 1902, with the Japanese lichenologist Yasuhiko Asahina (1881-1975) wrote *Chemistry of Lichen Substances*. 1954; see John H. Barnhart, *Biographical Notes upon Botanists*. 3: 270. 1965; T.W. Bossert, *Biographical Dictionary of Botanists Represented in the Hunt Institute Portrait Collection*. 364. 1972; E.M. Tucker, *Catalogue of the Library of the Arnold Arboretum of Harvard University*. Cambridge, Massachusetts 1917-1933; Stafleu and Cowan, *Taxonomic Literature*. 5: 563. Utrecht 1985; R. Zander, F. Encke, G. Buchheim and S. Seybold, *Handwörterbuch der Pflanzennamen*. 14. Aufl. 780. Stuttgart 1993.

Species/Vernacular Names:
S. kumasasa (Steudel) Nakai

Japan: okame-zasa

Shibateranthis Nakai Gramineae

Origins:
The genera *Shibataea* and *Eranthis* Salisb.

Shinnersia R. King & H. Robinson Asteraceae

Origins:
After the American (b. Canada) botanist Lloyd Herbert Shinners, 1918-1971, author of "New varietal names for

New World *Ludwigia* (Onagraceae)." *Sida*. 1(6): 385-386. 1964; see John H. Barnhart, *Biographical Notes upon Botanists*. 3: 271. 1965; T.W. Bossert, *Biographical Dictionary of Botanists Represented in the Hunt Institute Portrait Collection*. 365. 1972; Ida Kaplan Langman, *A Selected Guide to the Literature on the Flowering Plants of Mexico*. University of Pennsylvania Press, Philadelphia 1964; S. Lenley et al., *Catalog of the Manuscript and Archival Collections and Index to the Correspondence of John Torrey*. Library of the New York Botanical Garden. 369. 1973; R. Zander, F. Encke, G. Buchheim and S. Seybold, *Handwörterbuch der Pflanzennamen*. 14. Aufl. Stuttgart 1993.

Shinnersoseris Tomb Asteraceae

Origins:
After the American botanist Lloyd Herbert Shinners, 1918-1971.

Shorea Roxb. ex Gaertner f. Dipterocarpaceae

Origins:
To commemorate Sir John Shore Teignmouth, 1751-1834, first Baron, in 1769 was sent to Bengal as a cadet in the East India Company's civil service, in 1792 was made a baronet, in 1793 appointed successor to Lord Cornwallis and held the office of Governor-General of India till the close of 1797. In 1797 on his return to England he was created an Irish peer by the title of Lord Teignmouth; he was the first President of the British and Foreign Bible Society. His works include *Memoirs of the Life, Writings and Correspondence of Sir William Jones*. London 1804 and *Considerations on the Practicability, Policy and Obligation of Communicating to the Natives of India the Knowledge of Christianity ... By a late resident in Bengal*. 1808. See *The Private Record of an Indian Governor-Generalship*. The correspondence of Sir John Shore, Governor-General, with Henry Dundas, President of the Board of Control, 1793-1798. Edited with an introduction and notes by Holden Furber. Cambridge, Massachusetts 1933; Shore Charles John, *Memoir of the Life and Correspondence of John, Lord Teignmouth*. By his son, Lord Teignmouth. London 1843.

Species/Vernacular Names:
S. affinis (Thw.) Ashton (*Doona affinis* Thw.)

Sri Lanka: beraliya dun

S. congestiflora (Thw.) Ashton (*Doona congestiflora* Thw.)

Sri Lanka: dun, tiniya

S. cordifolia (Thw.) Ashton (*Doona nervosa* Thw.; *Doona cordifolia* Thw.)

Sri Lanka: dun, kotikan beraliya

S. disticha (Thw.) Ashton (*Doona disticha* (Thw.) Pierre; *Doona oblonga* Thw.; *Vateria disticha* Thw.; *Vateria disticha* (Thw.) A. DC.; *Sunaptea disticha* (Thw.) Trimen; *Stemonoporus distichus* (Thw.) Heim)

Sri Lanka: dun

S. dyeri Thw. (*Hopea discolor* sensu Worthington)

Sri Lanka: yakahalu dun

S. gardneri (Thw.) Ashton (*Doona gardneri* Thw.)

Sri Lanka: koongili maram, rata dun

S. lissophylla Thw. (*Isoptera lissophylla* (Thw.) Livera)

Sri Lanka: malmora

S. megistophylla Ashton (*Doona macrophylla* Thw.)

Sri Lanka: honda-beraliya, kana-beraliya, maha-beraliya

S. oblongifolia Thw. (*Doona oblonga* sensu Worthington)

Sri Lanka: dun

S. ovalifolia (Thw.) Ashton (*Doona ovalifolia* Thw.)

Sri Lanka: tiniya dun

S. pallescens Ashton (*Shorea dealbata* Alston)

Sri Lanka: ratu dun

S. robusta Roxb. ex Gaertner f.

English: the sal tree

India: sala, asvakarna, sal, salwa, sakhu, saku, sakher, sarye, sarei, rinjal, gugal, koroh

Nepal: sal, sakhuwa, sal dammar (for the white resin)

Tibetan: spos-dkar

China: so lo mu

S. stipularis Thw.

Sri Lanka: nawadun, nawada, hulan-idda

S. trapezifolia (Thw.) Ashton (*Doona trapezifolia* Thw.)

Sri Lanka: yakahalu dun

S. worthingtoni Ashton (*Doona venulosa* Thw.)

Sri Lanka: dun

S. zeylanica (Thw.) Ashton (*Doona zeylanica* Thw.)

Sri Lanka: koongili, dun

Shortia Raf. Brassicaceae

Origins:

For the American botanist Dr. Charles Wilkins Short, 1794-1863; see C.S. Rafinesque, *Autikon botanikon*. Icones plantarum select. nov. vel rariorum, etc. 16, 17. Philadelphia 1840; E.D. Merrill, *Index rafinesquianus*. 130. 1949.

Shortia Torrey & A. Gray Diapensiaceae

Origins:

The generic name commemorates the American (Kentucky) botanist Charles Wilkins Short, 1794-1863, physician, 1815 M.D. University of Pennsylvania, corresponded with Asa Gray, professor of *materia medica* and medical botany Transylvania University (Lexington, Kentucky), with John Esten Cooke editor of *Transylvania Journal of Medicine and the Associate Sciences*. Lexington 1828-1831, with Robert Peter (1805-1894) and Henry A. Griswold (fl. 1835) wrote *Catalogue of the Native ... Plants and Ferns of Kentucky*. Lexington 1833. Among his works are *Florula lexingtoniensis*. [1828-1829] and *A Sketch of the Progress of Botany in Western America*. Lexington 1836; see John H. Barnhart, *Biographical Notes upon Botanists*. 3: 273. Boston 1965; Jeannette Elizabeth Graustein, *Thomas Nuttall, Naturalist. Explorations in America, 1808-1841*. Cambridge, Harvard University Press 1967; Ronald Lewis Stuckey (b. 1938), ed., *Scientific Publications of Charles Wilkins Short*. New York 1978; J. Ewan, ed., *A Short History of Botany in the United States*. New York and London 1969; T.W. Bossert, *Biographical Dictionary of Botanists Represented in the Hunt Institute Portrait Collection*. 365. 1972; A. Lasègue, *Musée botanique de Benjamin Delessert*. Paris 1845; S. Lenley et al., *Catalog of the Manuscript and Archival Collections and Index to the Correspondence of John Torrey*. Library of the New York Botanical Garden. 1973; E.M. Tucker, *Catalogue of the Library of the Arnold Arboretum of Harvard University*. Cambridge, Massachusetts 1917-1933; Howard Atwood Kelly and Walter Lincoln Burrage, *Dictionary of American Medical Biography*. New York 1928; F. Boerner & G. Kunkel, *Taschenwörterbuch der botanischen Pflanzennamen*. 4. Aufl. 169. 1989.

Species/Vernacular Names:

S. galacifolia Torr. & A. Gray (referring to the resemblance of the leaves to those of *Galax*)

English: Oconee bells (Oconee County, South Carolina)

S. rotundifolia (Maxim.) Makino (*Schizocodon rotundifolius* Maxim.) (*rotundifolia* = round-leaved)

Japan: Shima-iwa-kagami (= island crag mirror)

S. soldanelloides (Siebold & Zucc.) Makino (*Schizocodon soldanelloides* Sieb. & Zucc.) (= like *Soldanella*, Primulaceae)

English: fringed galax, fringed bell, fringe-bell

Japan: Iwa-kagami (= crag mirror)

S. uniflora (Maxim.) Maxim. (*Schizocodon uniflorus* Maxim.) (*uniflora* = one-flowered)

English: Nippon bells

Japan: Iwa-uchiwa (= crag fan)

Shoshonia Evert & Constance Umbelliferae

Origins:
Shoshone, a tribe of North American Indians of northern Shoshonean stock of the Uto-Aztecan family, formerly occupying western Wyoming; Shoshone River, in northwest Wyoming; Shoshone Cavern, on the Shoshone River, southwest of Cody, Wyoming.

Shuteria Wight & Arn. Fabaceae

Origins:
In honor of the British botanist James Shuter, d. *circa* 1827 (or 1834), physician, naturalist and plant collector, a friend of the British surgeon and botanist Robert Wight (1796-1872), in 1819 a Fellow of the Linnean Society, in 1822 Government Naturalist at Madras, India; see Robert Wight and George Arnott Walker Arnott (1799-1868), *Prodromus florae Peninsulae Indiae Orientalis.* 207. London 1834; *Curtis's Botanical Magazine.* t. 3302. 1834.

Siagonanthus Poeppig & Endl. Orchidaceae

Origins:
Greek *siagon, siegon* "jaw, jaw-bone" and *anthos* "flower," the resemblance of the flower to an open jaw; see also *Maxillaria* Ruíz & Pav.

Siamosia Larsen & T.M. Pedersen Amaranthaceae

Origins:
From Siam (now Thailand).

Sibara Greene Brassicaceae

Origins:
An anagram of *Arabis.*

Species/Vernacular Names:
S. filifolia (E. Greene) E. Greene
English: Santa Cruz Island rock cress

Sibbaldia L. Rosaceae

Origins:
For the Scottish (b. Edinburgh) physician Sir Robert Sibbald, 1641-1722 (d. Edinburgh), botanist, naturalist, M.D. Leyden 1661, knighted 1682, professor of medicine. Among his writings are *Scotia Illustrata.* Edinburgi 1683 and 1684, *Portus, Coloniae, et Castella Romana, ad Bodotriam et ad Taum.* Edinburgh 1711, *Phalainologia nova*, sive observationes de rarioribus quibusdam Balaenis in Scotiae littus nuper ejectis, etc. Edinburgi 1692 and *Memoria Balfouriana.* Edinburgi 1699, with Sir Andrew Balfour (1630-1694) founded the Edinburgh Botanic Garden; see John Claudius Loudon, *An Encyclopaedia of Gardening.* London 1822; William Munk, *The Roll of the Royal College of Physicians of London.* 1: 439-441. London 1878; Sir William Jardine, *The Birds of Great Britain and Ireland.* (Memoir of Sir R. Sibbald, W. Smellie, J. Walker, A. Wilson). Edinburgh 1839-1843; H.R. Fletcher and W.H. Brown, *Royal Botanic Garden Edinburgh, 1670-1970.* Edinburgh 1970; James W. Cursiter, *List of Books and Pamphlets Relating to Orkney and Shetland.* 61. Wm. Peace & Son, Kirkwall 1894.

Sibbaldianthe Juz. Rosaceae

Origins:
The genus *Sibbaldia* and Greek *anthos* "flower."

Sibbaldiopsis Rydb. Rosaceae

Origins:
Resembling *Sibbaldia.*

Sibiraea Maxim. Rosaceae

Origins:
From Siberia and the genus *Spiraea.*

Sibthorpia L. Scrophulariaceae

Origins:
For the British (b. Canwick, Lincoln) botanist Humphrey Waldo Sibthorp, 1713-1797 (d. Instow, Devon), physician, professor of botany, father of the British botanist John Sibthorp (b. Oxford 1758-d. Bath, Somerset 1796); see J.H. Barnhart, *Biographical Notes upon Botanists.* 3: 275. 1965; Stafleu and Cowan, *Taxonomic Literature.* 5: 577-580. Utrecht 1985; I.H. Vegter, *Index Herbariorum.* Part II (6), *Collectors S.* Regnum Vegetabile vol. 114. 1986; T.W. Bossert, *Biographical Dictionary of Botanists Represented in the Hunt Institute Portrait Collection.* 366. 1972; H.N. Clokie, *Account of the Herbaria of the Department of Botany in the University of Oxford.* 1964; Blanche Henrey, *British Botanical and Horticultural Literature before 1800.* 1975; William Darlington, *Memorials of John Bartram and*

Humphry Marshall. 428-430. 1849; Humphrey Sibthorp, *To the Gentlemen Delegates of Accounts, and of the Committee for the Physick Garden*. [An Address respecting the Botanic Library and other buildings connected with the Botanic Gardens at Oxford] [Oxford 1778?]; Mariella Azzarello Di Misa, a cura di, *Il Fondo Antico della Biblioteca dell'Orto Botanico di Palermo*. 257. Palermo 1988; H.R. Fletcher, *Story of the Royal Horticultural Society, 1804-1968*. Oxford 1969; Ray Desmond, *Dictionary of British & Irish Botanists and Horticulturists*. 626. 1994; R. Zander, F. Encke, G. Buchheim and S. Seybold, *Handwörterbuch der Pflanzennamen*. 14. Aufl. Stuttgart 1993; Georg Christian Wittstein, *Etymologisch-botanisches Handwörterbuch*. 812. 1852; F. Boerner & G. Kunkel, *Taschenwörterbuch der botanischen Pflanzennamen*. 4. Aufl. 169. 1989.

Species/Vernacular Names:

S. europaea L.

English: Cornish moneywort, pennywort

S. peregrina L.

English: Madeira moneywort

Portuguese: erva redonda

Sicana Naudin Cucurbitaceae

Origins:

Latin *Sicanus, a, um* "Sicanian, Sicilian," or from *secana*, a Peruvian name for this cultivated vine, see F. Boerner & G. Kunkel, *Taschenwörterbuch der botanischen Pflanzennamen*. 4. Aufl. 169. 1989; R. Zander, F. Encke, G. Buchheim and S. Seybold, *Handwörterbuch der Pflanzennamen*. 14. Aufl. 518. Stuttgart 1993; H. Genaust, *Etymologisches Wörterbuch der botanischen Pflanzennamen*. 583. 1996.

Species/Vernacular Names:

S. odorifera (Vell. Conc.) Naudin (*Cucurbita euodicarpa* Hasskarl; *Cucurbita odorifera* Vell. Conc.) (see National Research Council, *Lost Crops of the Incas: Little-Known Plants of the Andes with Promise for Worldwide Cultivation*. National Academy Press, Washington, D.C. 1989)

Tropical America: casabanana, cassabanana

Peru: carua, curaba, curua, curuba, mamuk, olero, padea, secana, secona, upe

Siccobaccatus P.J. Braun & E.Est. Pereira Cactaceae

Origins:

Latin *siccus* "dry" and *baca* or *bacca, ae* "a berry, a small round berry."

Sichuania M. Gilbert & P.T. Li Asclepiadaceae

Origins:

Sichuan (Szechwan) province, southwestern and central western China, capital Chengtu.

Species/Vernacular Names:

S. alterniloba M.G. Gilbert & P.T. Li

China: si chuan teng

Sickmannia Nees Cyperaceae

Origins:

For the German botanist Johann Rudolph Sickmann, 1779-1849; see John H. Barnhart, *Biographical Notes upon Botanists*. 3: 275. 1965; E.M. Tucker, *Catalogue of the Library of the Arnold Arboretum of Harvard University*. Cambridge, Massachusetts 1917-1933.

Sicydium Schltdl. Cucurbitaceae

Origins:

Resembling *sikyos* "wild cucumber, gourd."

Sicyocarya (A. Gray) St. John Cucurbitaceae

Origins:

Greek *sikyos* "wild cucumber, gourd" and *karyon* "nut," see also *Sicyos* L.

Sicyocaulis Wiggins Cucurbitaceae

Origins:

From the Greek *sikyos* "wild cucumber, gourd" and *kaulos* "stem, stalk," see also *Sicyos* L.

Sicyoides Mill. Cucurbitaceae

Origins:

Resembling the genus *Sicyos* L., see also *Sicyos* L.

Sicyos L. Cucurbitaceae

Origins:

Greek *sikyos* "wild cucumber, gourd," *sikyos agrios* "the wild cucumber," Latin *sicyos agrios* "wild cucumber"; see Carl Linnaeus, *Species Plantarum*. 1013. 1753 and *Genera*

Plantarum. Ed. 5. 443. 1754; Giovanni Semerano, *Le origini della cultura europea.* Dizionari Etimologici. Basi semitiche delle lingue indeuropee. Dizionario della lingua Greca. 2(1): 262. Firenze 1994.

Species/Vernacular Names:
S. angulata L. (*Sicyos australis* Endl.)
English: star cucumber, bur cucumber
S. cucumerinus A. Gray
Hawaii: panunukuahiwi
S. maximowiczii Cogn. (*Cladocarpa maximowiczii* (Cogn.) St. John; *Cladocarpa atollensis* (St. John) St. John; *Cladocarpa caumii* (St. John) St. John; *Cladocarpa lamoureuxii* (St. John) St. John; *Cladocarpa lehuaensis* St. John; *Cladocarpa niihauensis* (St. John) St. John; *Sicyos atollensis* St. John; *Sicyos caumii* St. John; *Sicyos lamoureuxii* St. John; *Sicyos niihauensis* St. John)
Hawaii: puaokama
S. pachycarpus Hook. & Arn. (*Sicyos laysanensis* St. John; *Sicyos microcarpus* H. Mann; *Sicyos niger* St. John; *Sicyos remyanus* Cogn.)
Hawaii: kupala

Sicyosperma A. Gray Cucurbitaceae

Origins:
Greek *sikyos* "wild cucumber, gourd" and *sperma* "seed."

Sicyus Clements Cucurbitaceae

Origins:
From the Greek *sikyos* "wild cucumber, gourd," see also *Sicyos* L.

Sida L. Malvaceae

Origins:
Side is the ancient Greek name for the waterlily, *Nymphaea alba* L. (Theophrastus) and for pomegranate too; Akkadian *sedum* "red"; see Carl Linnaeus, *Species Plantarum.* 683. 1753 and *Genera Plantarum.* Ed. 5. 306. 1754; H. Genaust, *Etymologisches Wörterbuch der botanischen Pflanzennamen.* 583. 1996.

Species/Vernacular Names:
S. sp.
Mexico: nocuana lao beyo, nocuana lao peyo, goba ya laza, coba ya laza
Yoruba: jokodagba

S. acuta Burm.f. (*Sida carpinifolia* sensu Mast.)
English: broom plant
East Africa: ekerundu, ongodi, owich
Congo: kissafi
South Africa: Pretoriabossie, taaiman
Yoruba: agidimagbayin, ose potu, ose putu, sanrin
Japan: hosoba-kin-goji-ka
China: huang hua ren
The Philippines: walis-walisan, eskobang-haba, eskubilis, eskuba, eskubilla, silhigo, saliki, higot balato, mamalis, pamalis, taking-baka, takim-baka, attaiba-baka, uaualisan, basbasot, herbara, surusighid, sinaguri-langau

S. alba L.
English: spring sida, spiny sida, prickly sida
Arabic: meloukhiet iblis
Southern Africa: lente sida, stekeltaaiman; sindanibita (Tsonga)

S. argentea Bailey
English: silver sida

S. cordifolia L.
English: flannel weed, heartleaf sida, white burr, country mallow
India: bala
China: huang hua zi
Malaya: kelulut puteh
Southern Africa: hartblaartaaiman, koekbossie, verdompsterk; inama (Ndebele)
Yoruba: ekuru oko
N. Rhodesia: kabuhu, kavuvu
Congo: kissafi, libamu

S. corrugata Lindley
English: dwarf sida

S. corymbosa R.E. Fr.
Yoruba: ose potu, ose putu, sanrin

S. dregei Burtt Davy (*Sida lancifolia* Burtt Davy; *Sida longipes* E. Meyer ex Harvey)
English: spider leg, Sutherland's curse

S. fallax Walp. (*Anoda ovata* Meyen; *Sida diellii* A. Gray; *Sida ledyardii* St. John; *Sida meyeniana* Walp.; *Sida nelsonii* St. John; *Sida sandwicensis* H. Lév.; *Sida sertum* Nutt. ex A. Gray)
Hawaii: 'ilima

S. hermaphrodita (L.) Rusby
English: Virginia mallow

S. intricata F. Muell.
English: tangled sida

S. linifolia Juss. ex Cav.

Yoruba: obonibi, iso, obole, obolokolepon

S. petrophila F. Muell.

English: rock sida

S. platycalyx Benth.

English: lifesaver burr

S. rhombifolia L.

English: arrowleaf sida, spiny sida, Pretoria sida, Queensland hemp, sida hemp, broomweed, Paddy's lucerne, Canary Island tea plant

Southern Africa: Pretoria bossie, Pretoria-sida, smalblaar-taaiman, taaiman; ivivane (Swati)

South Africa: ntswembana, quaquaza, lethlakanye

Yoruba: ifin, ewe ifin

Congo: kili

Madagascar: kasindahorina, kinsindohora, kitsindaloro, sandroky, sindahorona, tsimaitofangady, tsimatipangady, tsindahoro, tsindahory

China: huang hua mu

Japan: kin-goji-ka, chankan-i, shimi-kata-masa

Malaya: bunga padang, jerun, seliguri padang, sendaguri, seguri, sidaguri

S. spinosa L. (*Sida angustifolia* Lam., non Mill. nec Medik.; *Sida spinosa* var. *angustifolia* (Lam.) Griseb.)

English: spiny sida, prickly mallow, prickly sida

Japan: Amerika-kin-goji-ka

S. stipulata Cav.

Sierra Leone: helui

Yoruba: agidimagbayin

Congo: lumvumvu

S. subspicata Benth.

English: Queensland hemp

S. urens L.

Yoruba: esisi gogoro, esisi afin, esinsi ile, keke ekeji

S. veronicifolia Lam.

Yoruba: esisi ile

S. virgata Hook.

English: twiggy sida

Sidalcea A. Gray Malvaceae

Origins:

From the genera *Sida* L. and *Alcea* L., two names for mallow.

Species/Vernacular Names:

S. covillei E. Greene

English: Owens Valley checkerbloom

S. keckii Wiggins

English: Keck's checkerbloom

S. malviflora (DC.) Benth.

English: checker mallow

S. pedata A. Gray

English: bird-footed checkerbloom

S. robusta E.M. Roush

English: Butte County checkerbloom

S. stipularis J. Howell & True

English: Scadden Flat checkerbloom (Scadden Flat, Grass Valley, Nevada County)

Sidasodes Fryx. & Fuertes Malvaceae

Origins:

Referring to the genus *Sida* L.

Sidastrum Baker f. Malvaceae

Origins:

From the genus *Sida* L. and the suffix *-aster* meaning false or an incomplete resemblance.

Siderasis Raf. Commelinaceae

Origins:

See H. Genaust, *Etymologisches Wörterbuch der botanischen Pflanzennamen*. 583. 1996; C.S. Rafinesque, *Flora Telluriana*. 3: 67. 1836 [1837]; E.D. Merrill, *Index rafinesquianus*. 84. 1949; F. Boerner & G. Kunkel, *Taschenwörterbuch der botanischen Pflanzennamen*. 4. Aufl. 169. 1989; F. Boerner, *Taschenwörterbuch der botanischen Pflanzennamen*. 2. Aufl. 174. Berlin & Hamburg 1966.

Sideria Ewart & A.H.K. Petrie Malvaceae

Origins:

Possibly from *siderion*, ancient Greek name for the plant ironwort, vervain, or from Latin *sidus, sideris* "a group of stars, a constellation" in reference to the flowers, or from genus *Sida* L.; see Alfred James Ewart (1872-1937) & A.H.K. Petrie (fl. 1926), in *Proceedings of the Royal Society of Victoria*. Ser. 2, 38: 174, textfig. 3. (Jul.) 1926.

Sideritis L. Labiatae

Origins:
From the Greek and Latin *sideritis* "ironwort, vervain."

Species/Vernacular Names:
S. balansae Boissier
English: Balansa sideritis
China: zi hua du mao cao
S. candicans Ait.
English: Madeira ironwort
Portuguese: erva branca
S. montana L.
English: montane sideritis
China: du mao cao

Siderobombyx Bremek. Rubiaceae

Origins:
Greek *sideros* "iron" and *bombyx* "silk."

Siderocarpos Small Mimosaceae

Origins:
From the Greek *sideros* "iron" and *karpos* "fruit."

Siderocarpus Pierre Sapotaceae

Origins:
Greek *sideros* and *karpos* "fruit."

Siderodendrum Schreb. Rubiaceae

Origins:
From the Greek *sideros* "iron" and *dendron* "tree."

Sideropogon Pichon Bignoniaceae

Origins:
From the Greek *sideros* and *pogon* "beard."

Sideroxyloides Jacq. Rubiaceae

Origins:
Greek *sideros* "iron" and *xylon* "wood."

Sideroxylon L. Sapotaceae

Origins:
Greek *sideros* "iron" and *xylon* "wood," referring to the hard timber of some species; see Carl Linnaeus, *Species Plantarum*. 192. 1753 and *Genera Plantarum*. Ed. 5. 89. 1754.

Species/Vernacular Names:
S. borbonicum DC. var. *borbonicum*
La Réunion Island: bois de fer bâtard, natte bâtard, natte coudine, natte cochon, natte blanc
S. borbonicum DC. var. *capuronii* Aubr.
La Réunion Island: bois de fer bâtard, natte bâtard, bois de fer blanc
S. foetidissimum Jacq.
English: Barbados mastic
S. galeatum (A.W. Hill) Baehni
Rodrigues Island: bois de pomme, bois de fer
S. inerme L. subsp. *inerme*
English: white milkwood, sea oak
Southern Africa: witmelkhout, melkhout, jakkalsbessie, melkbessie; umNweba (Swazi); aMasethole-amhlope, uMakhwelafinqane, umHlahle amaSethole, uMakhwelafingqane, umBhobe (Zulu); aMasethole, umQwashu (Xhosa)
S. majus (Gaertn.f.) Baehni
La Réunion Island: bois de fer blanc
S. obtusifolium (Roemer & Schultes) Pennington subsp. *obtusifolium* (*Bumelia obtusifolia* Roemer & Schultes; *Bumelia obtusifolia* subsp. *excelsa* (A. DC.) Cronq.; *Bumelia sartorum* Martius)
Bolivia: chirimolle, orko molle

Sidopsis Rydb. Malvaceae

Origins:
Resembling *Sida* L., see also *Malvastrum* A. Gray.

Siebera H.G.L. Reichenbach Umbelliferae

Origins:
For the Bohemian naturalist Franz Wilhelm Sieber, 1789-1844 (d. Prague), natural history collector and traveler, botanist, plant collector in South Africa, Mauritius and Australia, in 1823 in New South Wales. Among his publications are *Herbarium florae aegyptiacae*. Wien 1820, *Reise nach der Insel Kreta*. Leipzig und Sorau 1823 and *Reise von Cairo nach Jerusalem*. Prag, Leipzig 1823, sent Andreas

Döllinger, Franz (Francis) Kohaut (d. 1822) and Joseph Schmidt to collect for him in Senegal; see John H. Barnhart, *Biographical Notes upon Botanists*. 3: 275. 1965; R. Zander, F. Encke, G. Buchheim and S. Seybold, *Handwörterbuch der Pflanzennamen*. 14. Aufl. Stuttgart 1993; F.C. Dietrich, in *Jahrb. Konigl. Bot. Gart. Mus. Berlin*. 1: 278-306. 1881; D.J. Carr and S.G.M. Carr, eds., *People and Plants in Australia*. 1981; Mariella Azzarello Di Misa, a cura di, *Il Fondo Antico della Biblioteca dell'Orto Botanico di Palermo*. 257. Palermo 1988; Mary Gunn and Leslie E. Codd, *Botanical Exploration of Southern Africa*. 322-323. A.A. Balkema Cape Town 1981; Joseph Vallot, "Études sur la flore du Sénégal." in *Bull. Soc. Bot. de France*. 29: 191-192. Paris 1882; Heinrich Gottlieb Ludwig Reichenbach (1793-1879), *Conspectus Regni Vegetabilis*. 145. Lipsiae 1828; F.N. Hepper and Fiona Neate, *Plant Collectors in West Africa*. 74. 1971; H.N. Clokie, *Account of the Herbaria of the Department of Botany in the University of Oxford*. 244. Oxford 1964; A. Lasègue, *Musée botanique de Benjamin Delessert*. Paris 1845; E.M. Tucker, *Catalogue of the Library of the Arnold Arboretum of Harvard University*. Cambridge, Massachusetts 1917-1933; N. Hall, *Botanists of the Eucalypts*. Melbourne 1978 and Supplement 1980; G. Schmid, *Goethe und die Naturwissenschaften*. Halle 1940.

Siebera J. Gay Asteraceae

Origins:

For the Bohemian naturalist Franz Wilhelm Sieber, 1789-1844.

Sieberia Sprengel Orchidaceae

Origins:

For the Bohemian naturalist Franz Wilhelm Sieber, 1789-1844; see John H. Barnhart, *Biographical Notes upon Botanists*. 3: 275. 1965.

Siegesbeckia Steudel Asteraceae

Origins:

Orthographic variant of *Sigesbeckia* L.

Siegfriedia C.A. Gardner Rhamnaceae

Origins:

Probably after the German mythical hero Siegfried, probably referring to the hidden flower-heads; see C.A. Gardner,

in *Journal of the Royal Society of Western Australia*. 19: 85. (Jul.) 1933; F.A. Sharr, *Western Australian Plant Names and Their Meanings*. A glossary. 64. University of Western Australia Press 1996; James A. Baines, *Australian Plant Genera. An Etymological Dictionary of Australian Plant Genera*. 342. Chipping Norton, N.S.W. 1981.

Sieglingia Bernh. Gramineae

Origins:

For a botanical collector and botanist named Siegling; see Helmut Genaust, *Etymologisches Wörterbuch der botanischen Pflanzennamen*. 584. Basel 1996.

Siemssenia Steetz Asteraceae

Origins:

Possibly after C.H. Siemssen, fl. 1795; see Johann G.C. Lehmann (1792-1860), *Plantae Preissianae*. 1: 467. Hamburgi 1845.

Sievekingia Reichb.f. Orchidaceae

Origins:

For a German burgomaster of Hamburg, Dr. C. Sieveking.

Sieversia Willd. Rosaceae

Origins:

Dedicated to the German pharmacist Johann August Carl Sievers, d. 1795, botanist, explorer, 1790-1795 traveled in southern Siberia and Mongolia, author of *J. Sievers ... Briefe aus Sibirien an seine Lehrer ... Herrn Brande, ... Herrn Ehrhart, und ... Herrn Westrumb*. St. Petersburg 1796; see Stafleu and Cowan, *Taxonomic Literature*. 5: 596. [genus named for Johann Erasmus Sievers] Utrecht 1985; Jonas C. Dryander, *Catalogus bibliothecae historico-naturalis Josephi Banks*. 3: 274. London 1800; A. Lasègue, *Musée botanique de Benjamin Delessert*. Paris 1845; Emil Bretschneider, *History of European Botanical Discoveries in China*. Leipzig 1981.

Sigesbeckia L. Asteraceae

Origins:

After the German botanist Johann Georg Siegesbeck, 1686-1755, physician, M.D. Wittenberg 1716, Director of the Botanical Garden of St. Petersburg, a critic and opponent of Linnaeus, author of *Chronologiae verioris specimen*, quo aequinoctium vernum die xxi. Martii anni ... 1735 etc. Helmstadii [1735?], *Primitiae florae petropolitanae*. Rigae [Riga] [1736], *Propempticum medico-botanicum de Majanthemo, Lilium convallium officinis vulgo nuncupato*. Petropoli [1736], *Programma medico-botanicum de Tetragono Hippocratis*. Petropoli 1737 and *Botanosophiae verioris brevis sciagraphia* in usum discentium adornata. Petropoli [St. Petersburg] 1737; see Johann G. Gleditsch (1714-1786), *Consideratio epicriseos Siegesbeckianae in Linnaei systema plantarum sexuale et methodum botanicam huic superstructam*. Berolini [Berlin] 1740; J.H. Barnhart, *Biographical Notes upon Botanists*. 3: 276. 1965; Carl Linnaeus, *Species Plantarum*. 900. 1753 and *Genera Plantarum*. Ed. 5. 383. 1754; Jonas C. Dryander, *Catalogus bibliothecae historico-naturalis Josephi Banks*. London 1800.

Species/Vernacular Names:

S. orientalis L.

English: common St. Paul's wort, Indian weed, Saint Paul's herb, St. Paul's wort, small yellow crown-beard

South Africa: gewone St. Paul's kruid, kleefgras, Pauluskruid

Madagascar: satrikoazamaratra, herbe grasse, colle-colle, guérit-vite, herbe divine

Japan: tsukushi-me-namomi

China: xi xian cao, xi xian, hsi lien, kou kao, chu kao mu, hu kao, nien hu tsai

Vietnam: cuc dinh, nu ao ria, co di

Nepal: thulo piryu

Sigmatanthus Huber ex Emmerich Rutaceae

Origins:
Greek *sigma* "S-shaped" and *anthos* "flower."

Sigmatochilus Rolfe Orchidaceae

Origins:
Greek *sigma* "S-shaped" and *cheilos* "a lip, margin," referring to the shape of the lip.

Sigmatogyne Pfitzer Orchidaceae

Origins:
From the Greek *sigma* and *gyne* "a woman, female, female organs, pistil," referring to the gymnostemium.

Sigmatosiphon Engl. Pedaliaceae

Origins:
Greek *sigma* and *siphon* "a tube," referring to the corolla tube.

Sigmatostalix Reichb.f. Orchidaceae

Origins:
Greek *sigma* and *stalix, stalikos* "a stake," alluding to the sigmoid shape of the slender column.

Silaum Miller Umbelliferae

Origins:
From the Latin *silaus, i* "a kind of parsley, small-age," Plinius applied to a species of *Apium*.

Species/Vernacular Names:
S. silaus (L.) Schinz & Thell.

English: pepper saxifrage

Silaus Bernh. Umbelliferae

Origins:
Latin *silaus, i* "a kind of parsley, small-age."

Silene L. Caryophyllaceae

Origins:
From Silenus (Silenos), the Greek woodland deity, the tutor and companion of Bacchus; see Carl Linnaeus, *Species Plantarum*. 416. 1753 and *Genera Plantarum*. Ed. 5. 193. 1754; Salvatore Battaglia, *Grande dizionario della lingua italiana*. XIX: 1. 1998.

Species/Vernacular Names:
S. acaulis (L.) Jacq.

English: moss campion

S. armeria L.

English: sweet William catch-fly, none-so-pretty

Japan: mushi-tori-nadeshiko

S. burchellii Otth var. *burchellii* (*Silene dinteri* Engl.)

English: gunpowder plant

Southern Africa: iYeza lehashe (Xhosa)

S. conoidea L.

English: conoid catchfly

Egypt: za'far

S. dioica (L.) Clairv. (*Lychnis dioica* L.)

English: red campion

S. gallica L. (*Silene anglica* L.)

English: French catchfly, French silene, gunpowder weed, windmill pink

South Africa: eierbossie, Frans silene, jobakraaltjies, jobskrale, kruitbossie

S. invisa C. Hitchc. & Maguire

English: short-petaled campion

S. marmorensis Kruckeb.

English: Marble Mountain campion

S. nocturna L.

English: night-flowering catchfly

S. pratensis (Rafn) Godron & Gren.

English: white campion

S. primuliflora Eckl. & Zeyh.

English: gunpowder weed, bladder campion, gunpowder plant

S. rubella L.

Egypt: abu el-nagaf

S. suksdorfii Robinson

English: cascade alpine campion

S. undulata Ait. (*Silene capensis* Otth)

English: gunpowder plant

S. vulgaris (Moench) Garcke subsp. *macrocarpa* (Marsden) Jones & Turrill

English: bladder campion, maiden's tears

S. vulgaris (Moench) Garcke subsp. *vulgaris* (*Cucubalus behen* L.; *Cucubalus inflatus* Salisb.; *Silene cucubalus* Wibel; *Silene inflata* (Salisb.) Smith; *Behen vulgaris* Moench)

English: bladder campion, blue root

Siler Miller Umbelliferae

Origins:
Latin *siler, sileris* applied by Plinius and Vergilius to a kind of brook-willow.

Silicularia Compton Brassicaceae

Origins:
Latin *silicula, ae* "a little husk or pod, a silicle," *siliqua, ae* "a pod of leguminous plants."

Siliquamomum Baillon Zingiberaceae

Origins:
Latin *siliqua, ae* "a pod, husk" and the genus *Amomum* Roxb.

Siloxerus Labill. Asteraceae

Origins:
Derivation obscure, probably from the Greek *stylos* "a column, pillar, style" and *ogkeros, onkeros* "swollen," referred to the base of the calyx; see Jacques Julien Houtton de Labillardière, *Novae Hollandiae plantarum specimen.* 2: 57. Parisiis (Jun.) 1806.

Silphiosperma Steetz Asteraceae

Origins:
Greek *silphion* for an umbelliferous plant and *sperma* "a seed"; see Johann G.C. Lehmann (1792-1860), *Plantae Preissianae.* 1: 433. Hamburgi (Aug.) 1845.

Silphium L. Asteraceae

Origins:
The Greek name *silphion* for an umbelliferous plant, laserwort; Latin *silphium* for a plant called *laserpitium, laserpicium.*

Silvorchis J.J. Smith Orchidaceae

Origins:
Latin *silva* "forest" and Greek *orchis* "orchid," the habitat.

Silybum Adans. Asteraceae

Origins:
Greek *silybos, silybon* used for a spiny or thistle-like plant; Latin *sillybus, sittybus* "a kind of thistle" (Plinius, M. Tullius Cicero); see Michel Adanson (1727-1806), *Familles des plantes.* 2: 116, 605. Paris 1763 [-1764].

Species/Vernacular Names:
S. marianum (L.) Gaertner (*Carduus marianus* L.)

English: blessed milk-thistle, blessed thistle, holy thistle, Mary's thistle, St. Mary's thistle, lady's thistle, Our Lady's milk thistle, milk thistle, variegated thistle

Arabic: shouk el-gamal, shouk sinnari, shouk el-gemal

Spanish: cardo de la alameda

Central America: cardo de Maria, cardo lechero, lechal

South Africa: boerkwadissel, disseldoring, gevlekte sily-bum, Mariendissel, melkdissel

Simaba Aublet Simaroubaceae

Origins:
Simaba, a native name in Guiana for one species, *Simaba multiflora* A. Juss.; see Jean Baptiste Christophore Fusée Aublet (1720-1778), *Histoire des Plantes de la Guiane Française*. 1: 400. Paris 1775.

Simarouba Aublet Simaroubaceae

Origins:
Simaruba, a native vernacular name in Guiana for *Simarouba amara* Aublet or *Quassia amara* L.

Species/Vernacular Names:
S. versicolor St.-Hil.
Brazil: simaruba, calunga, paraíba, pau paraíba

Simira Aubl. Rubiaceae

Origins:
A native name for *Simira tinctoria* Aublet, called *simira* by the Galibis, French Guiana.

Simmondsia Nutt. Simmondsiaceae (Buxaceae)

Origins:
For the English naturalist Thomas William Simmonds, d. 1804 (Trinidad), physician, with Francis Mackenzie Humberston (1754-1815), Lord Seaforth and Mackenzie (from 1800 to 1806 Governor of Barbados); see Ignatz Urban, ed., *Symbolae Antillanae*. 3: 127. Berlin 1902.

Species/Vernacular Names:
S. chinensis (Link) C. Schneider
English: goat nut, pig nut
Spanish: jojoba

Simocheilus Klotzsch Ericaceae

Origins:
Greek *simos* "flat-nosed, bent-upwards, concave" and *cheilos* "lip."

Simsia Persoon Asteraceae

Origins:
For the British botanist John Sims, 1749-1831, medical man, M.D. Edinburgh 1774, from 1800/1801 to 1826 editor of the *Curtis's Botanical Magazine*, with Karl Dietrich Eberhard Koenig (1774-1851) editor from 1804 to 1806 of *Annals of Botany*, in 1788 a Fellow of the Linnean Society and in 1814 of the Royal Society, author of *Testamen medicum inaugurale de usu aquae frigidae interno*. Edinburgi 1774. See Laurence Dopson, *The Bicentenary of John Sims, M.D., F.R.S.*, etc. London [1950]; N. Hall, *Botanists of the Eucalypts*. Melbourne 1978 and Supplement 1980; A. White and B.L. Sloane, *The Stapelieae*. Pasadena 1937; William Cooke [Honorary Canon of Chester], *An Address to British Females on the Moral Management of Pregnancy and Labour* ... Suggested by the death of Her Royal Highness Princess Charlotte Augusta of Wales. With a vindication of Her Royal Highness's physicians, Sir Richard Croft, Dr. Baillie, and Dr. Sims. London 1817; William Munk, *The Roll of the Royal College of Physicians of London*. 2: 322. London 1878; J.H. Barnhart, *Biographical Notes upon Botanists*. 3: 281. 1965; E.M. Tucker, *Catalogue of the Library of the Arnold Arboretum of Harvard University*. Cambridge, Massachusetts 1917-1933; T.W. Bossert, *Biographical Dictionary of Botanists Represented in the Hunt Institute Portrait Collection*. 368. 1972; A. Lasègue, *Musée botanique de Benjamin Delessert*. 529. Paris 1845; H.R. Fletcher, *Story of the Royal Horticultural Society, 1804-1968*. Oxford 1969; Ray Desmond, *Dictionary of British & Irish Botanists and Horticulturists*. 629. London 1994; Emil Bretschneider, *History of European Botanical Discoveries in China*. Leipzig 1981; R. Zander, F. Encke, G. Buchheim and S. Seybold, *Handwörterbuch der Pflanzennamen*. 14. Aufl. Stuttgart 1993; J.D. Milner, *Catalogue of Portraits of Botanists Exhibited in the Museums of the Royal Botanic Gardens*. Royal Botanic Gardens, Kew, London 1906.

Simsia R. Br. Proteaceae

Origins:
In honor of the British botanist John Sims, 1749-1831; see Robert Brown, in *Transactions of the Linnean Society of London. Botany*. 10: 152. 1810.

Sinacalia H. Robinson & Brettell Asteraceae

Origins:
From China and the genus *Cacalia* L.

Sinadoxa C.Y. Wu, Z.L. Wu & R.F. Huang Adoxaceae

Origins:

From China and the genus *Adoxa* L.

Sinapidendron Lowe Brassicaceae

Origins:

From the genus *Sinapis* L. and Greek *dendron* "tree."

Sinapis L. Brassicaceae

Origins:

Latin *sinapis*, *sinape* or *sinapi* for mustard ("sinapis scelera," T. Maccius Plautus), Greek *sinapi* "mustard"; see Carl Linnaeus, *Species Plantarum*. 668. 1753 and *Genera Plantarum*. Ed. 5. 299. 1754; Giovanni Semerano, *Le origini della cultura europea*. Dizionario della lingua Latina e di voci moderne. 2(2): 566. Firenze 1994; S. Battaglia, *Grande dizionario della lingua italiana*. XVIII: 616-617. Torino 1997; N. Tommaseo & B. Bellini, *Dizionario della lingua italiana*. Torino 1865-1879; G. Volpi, "Le falsificazioni di Francesco Redi nel Vocabolario della Crusca." in *Atti della R. Accademia della Crusca per la lingua d'Italia*. 33-136. 1915-1916; [Crusca], *Vocabolario degli Accademici della Crusca*. Firenze 1729-1738.

Species/Vernacular Names:

S. alba L.

English: white mustard, salad mustard, black mustard, yellow mustard

Italian: senape bianca

China: bai jie zi, bai jie, chieh, hu chieh

Tibetan: yungs-dkar

Arabic: khardal abiad, kabar abiad, khardal

S. arvensis L. (*Brassica sinapis* Vis.; *Brassica sinapistrum* Boiss.)

French: moutarde des champs

English: charlock

Italian: senape selvatica

Tunisia: khardel

Sinarundinaria Nakai Gramineae

Origins:

From China and the genus *Arundinaria*.

Sinclairia Hooker & Arn. Asteraceae

Origins:

For Dr. Sinclair, with Captain Frederick William Beechey (1796-1856) on his voyage; see Sir William J. Hooker and G.A.W. Arnott, *The Botany of Capt. Beechey's Voyage*; comprising an account of the Plants collected by Messrs. Lay and Collie ... during the voyage to the Pacific and Bering's Strait, performed in H.M.S. *Blossom* ... 1825-1828. London [1830-] 1841.

Sindechites Oliv. Apocynaceae

Origins:

Referring to *Echites*.

Species/Vernacular Names:

S. chinensis (Merrill) Markgraf & Tsiang

English: Chinese sindechites

China: ni teng

S. henryi Oliver

English: Henry sindechites

China: mao yao teng

Sindora Miq. Caesalpiniaceae

Origins:

Sindor is a Malay name.

Species/Vernacular Names:

S. sp.

Malaya: sepetir, ketil, meketil, sepeteh, petir

Sindoropsis J. Léonard Caesalpiniaceae

Origins:

Resembling the genus *Sindora*.

Sindroa Jumelle Palmae

Origins:

A vernacular name, Masoala Peninsula, Marambo, Madagascar.

Sineoperculum Jaarsveld Aizoaceae

Origins:

Latin *sine* "without" and *operculum* "lid, cover."

Sinningia Nees Gesneriaceae

Origins:

For the German horticulturist Wilhelm Sinning, 1792-1874, head gardener of the Botanical Garden of the University of Bonn; see J.H. Barnhart, *Biographical Notes upon Botanists.* 3: 282. 1965; E.M. Tucker, *Catalogue of the Library of the Arnold Arboretum of Harvard University.* Cambridge, Massachusetts 1917-1933.

Species/Vernacular Names:

S. speciosa (Lodd.) Hiern

English: gloxinia, Brazilian gloxinia, florist's gloxinia, violet-slipper gloxinia

Sinoadina Ridsdale Rubiaceae

Origins:

Greek *sinai* "the Chinese" and the genus *Adina* Salisb.

Sinoarundinaria Ohwi Gramineae

Origins:

From the Greek *sinai* "the Chinese" and the genus *Arundinaria* Michx.

Sinobacopa D.Y. Hong Scrophulariaceae

Origins:

Greek *sinai* and the genus *Bacopa* Aubl.

Sinobambusa Makino ex Nakai Gramineae

Origins:

From the Greek *sinai* "the Chinese" and the genus *Bambusa* Schreb.

Species/Vernacular Names:

S. kunishii (Hay.) Nakai (*Arundinaria kunishii* Hayata)

Japan: Taiwan-yadake

S. tootsik (Sieb. ex Mak.) Makino (*Arundinaria tootsik* Sieb. ex Makino) (the species named from *too* or *tô* "China, Chinese" and *tsik* or *chiku* meaning bamboo)

English: China bamboo

Japan: tô-chiku

Okinawa: kanso-daki

Sinoboea Chun Gesneriaceae

Origins:

From the Greek *sinai* and the genus *Boea* Comm. ex Lam.

Sinocalamus McClure Gramineae

Origins:

Greek *sinai* "the Chinese" and *kalamos* "reed."

Sinocalycanthus (W.C. Cheng & S.Y. Chang) W.C. Cheng & S.Y. Chang Calycanthaceae

Origins:

Greek *sinai* "the Chinese" plus *Calycanthus* L.

Sinocarum H. Wolff Umbelliferae

Origins:

Greek *sinai* "the Chinese" and *Carum* L.

Sinochasea Keng Gramineae

Origins:

Greek *sinai* and *Chasea* Nieuwl., dedicated to the American botanist Mary Agnes Chase (*née* Merrill), 1869-1963, agrostologist, plant collector, traveler.

Sinocrassula A. Berger Crassulaceae

Origins:

From the Greek *sinai* and the genus *Crassula* L., from the Latin *crassus, a, um* "thick."

Sinodielsia H. Wolff Umbelliferae

Origins:

For the German botanist Friedrich Ludwig Emil Diels, 1874-1945, from 1900-1901 plant collector in Western Australia, Director of Berlin-Dahlem Botanic Garden and Museum, friend and companion of Ernst Georg Pritzel (1875-1946). His works include *Beiträge zur Kenntniss der Vegetation und Flora von Ecuador.* Stuttgart 1937, *Ersatzstoffe aus dem Planzenreich.* [Edited by Diels.] Stuttgart 1918, *Jugendformen und Blütenreife im Pflanzenreich.* Berlin 1906 and *Die Pflanzenwelt von West-Australien südlich des Wendekreises.* Leipzig 1906; see J.H. Barnhart, *Bio-*

graphical *Notes upon Botanists*. 1: 454. 1965; Gottfried Wilhelm Johannes Mildbraed (1879-1954), in *Bot. Jb.* 74(2): 173-198. 1948 (Obituary and bibliography); Friedrich Ludwig Emil Diels (1874-1945) and Ernst Georg Pritzel (1875-1946), in *Botanische Jahrbücher*. 35: 88, fig. 6 A-L. (Apr.) 1904; T.W. Bossert, *Biographical Dictionary of Botanists Represented in the Hunt Institute Portrait Collection*. 102. 1972; Ida Kaplan Langman, *A Selected Guide to the Literature on the Flowering Plants of Mexico*. 241. University of Pennsylvania Press, Philadelphia 1964.

Sinodolichos Verdc. Fabaceae

Origins:
The genus *Dolichos* L. and China.

Sinofranchetia (Diels) Hemsley Lardizabalaceae

Origins:
Dedicated to the French botanist Adrien René Franchet, 1834-1900, traveler, wrote *Plantae davidianae ex sinarum imperio*. Paris 1884-1888 and *Plantae delavayanae*. [Collector the Abbé Pierre Jean Marie Delavay, 1834-1895.] Paris 1889[-1890]; see J.H. Barnhart, *Biographical Notes upon Botanists*. 2: 2. 1965; S. Lenley et al., *Catalog of the Manuscript and Archival Collections and Index to the Correspondence of John Torrey*. Library of the New York Botanical Garden. 173. 1973; Emil Bretschneider, *History of European Botanical Discoveries in China*. Leipzig 1981; Ida Kaplan Langman, *A Selected Guide to the Literature on the Flowering Plants of Mexico*. Philadelphia 1964; F. Boerner & G. Kunkel, *Taschenwörterbuch der botanischen Pflanzennamen*. 4. Aufl. 170. Berlin & Hamburg 1989; R. Zander, F. Encke, G. Buchheim and S. Seybold, *Handwörterbuch der Pflanzennamen*. 14. Aufl. 712. Stuttgart 1993.

Sinoga S.T. Blake Myrtaceae

Origins:
An anagram of the generic name *Agonis* (DC.) Sweet; see S.T. Blake, in *Proceedings of the Royal Society of Queensland*. 69: 80. (Sep.) 1958.

Sinojackia Hu Styracaceae

Origins:
For the American botanist John George Jack, 1861-1949, dendrologist, professor of dendrology, 1891-1935 Arnold Arboretum; see J.H. Barnhart, *Biographical Notes upon*

Botanists. 2: 237. 1965; Ethelyn Maria Tucker, *Catalogue of the Library of the Arnold Arboretum of Harvard University*. Cambridge, Massachusetts 1917-1933; Stafleu and Cowan, *Taxonomic Literature*. 2: 395. Utrecht 1979; R. Zander, F. Encke, G. Buchheim and S. Seybold, *Handwörterbuch der Pflanzennamen*. 14. Aufl. Stuttgart 1993.

Species/Vernacular Names:
S. rehderiana Hu
English: Rehder sinojackia
China: xia guo cheng chui shu
S. xylocarpa Hu
English: xylocarpous sinojackia
China: cheng chui shu

Sinojohnstonia H.H. Hu Boraginaceae

Origins:
For the American botanist Ivan Murray Johnston, 1898-1960, in Chile 1925-1926, professor of botany at Harvard 1938-1960, specialist in Boraginaceae. Among his works are "Expedition of the California Academy of Sciences to the Gulf of California in 1921. The botany (the vascular plants)." *Proc. Calif. Acad. Sci.* Ser. 4, 12: 951-1218. 1924, "The botanical activities of Thomas Bridges." *Contr. Gray Herb*. 81: 98-106. 1928, "Papers on the flora of northern Chile. 1. The coastal flora of the departments of Chañaral and Taltal; 2. The flora of the Nitrate Coast; 3. Undescribed species from the cordilleras of Atacama." *Contr. Gray Herb*. 85: 1-172. 1929 and "A study of Nolanaceae." *Proc. Amer. Acad. Arts*. 71(1): 1-87. [Philadelphia 1936]; see J.H. Barnhart, *Biographical Notes upon Botanists*. 2: 259. 1965; J. Ewan, ed., *A Short History of Botany in the United States*. 43, 122. New York and London 1969; T.W. Bossert, *Biographical Dictionary of Botanists Represented in the Hunt Institute Portrait Collection*. 199. 1972; Ida Kaplan Langman, *A Selected Guide to the Literature on the Flowering Plants of Mexico*. 397-398. University of Pennsylvania Press, Philadelphia 1964; Elmer Drew Merrill, *Contr. U.S. Natl. Herb*. 30(1): 167. 1947; Irving William Knobloch, compil., "A preliminary verified list of plant collectors in Mexico." *Phytologia Memoirs*. VI. 1983.

Species/Vernacular Names:
S. chekiangensis (Migo) W.T. Wang
English: Chekiang sinojohnstonia
China: zhe gan che qian zi cao
S. plantaginea H.H. Hu
English: common sinojohnstonia
China: che qian zi cao

Sinoleontopodium Y.L. Chen Asteraceae

Origins:

From China and generic name *Leontopodium* (Pers.) R. Br.

Sinolimprichtia H. Wolff Umbelliferae

Origins:

For the German botanist Hans Wolfgang Limpricht, b. 1877, traveler, botanical explorer in China and Tibet, wrote *Botanische Reisen in den Hochgebirgen Chinas und Ost-Tibets.* Dahlem bei Berlin 1922. He was son of the German bryologist Karl Gustav Limpricht (1834-1902); see E.D. Merrill & Egbert H. Walker, *A Bibliography of Eastern Asiatic Botany.* 273. The Arnold Arboretum of Harvard University, Jamaica Plain, Massachusetts1938; J.H. Barnhart, *Biographical Notes upon Botanists.* 2: 382. 1965; T.W. Bossert, *Biographical Dictionary of Botanists Represented in the Hunt Institute Portrait Collection.* 238. 1972; S. Lenley et al., *Catalog of the Manuscript and Archival Collections and Index to the Correspondence of John Torrey.* Library of the New York Botanical Garden. 267 and 268. 1973.

Sinomenium Diels Menispermaceae

Origins:

From the Greek *sinai* "the Chinese" and *mene* "the moon."

Species/Vernacular Names:

S. acutum (Thunb.) Rehd. & Wils. (*Menispermum acutum* Thunb.)

English: orient vine

Japan: ô-tsuzura-fuji

China: qing feng teng

Sinomerrillia Hu Convolvulaceae

Origins:

For the American (b. Maine) systematist Elmer Drew Merrill, 1876-1956 (d. Massachusetts), botanist, plant collector, 1927-1929 Director of the California Botanic Garden, 1930-1935 Director of the New York Botanical Garden (succeeding Nathaniel Lord Britton) and professor of botany at Columbia University, 1931 founder of *Brittonia*, 1935-1948 professor of botany Harvard University, from 1937 Director of the Arnold Arboretum. Among his many and valuable works are *The North American Species of Spartina.* Washington 1902, *New or Noteworthy Philippine Plants.* Manila 1903, *A Bibliographic Enumeration of Bornean Plants.* [Singapore] 1921 and *Plant Life of the Pacific World.* New York 1945; see J.H. Barnhart, *Biographical Notes upon Botanists.* 2: 479. 1965; T.W. Bossert, *Biographical Dictionary of Botanists Represented in the Hunt Institute Portrait Collection.* 264. 1972; S. Lenley et al., *Catalog of the Manuscript and Archival Collections and Index to the Correspondence of John Torrey.* Library of the New York Botanical Garden. 286-289. 1973; Ethelyn Maria Tucker, *Catalogue of the Library of the Arnold Arboretum of Harvard University.* Cambridge, Massachusetts 1917-1933; Ida Kaplan Langman, *A Selected Guide to the Literature on the Flowering Plants of Mexico.* 501. Philadelphia 1964; Joseph Ewan, in *Dictionary of Scientific Biography* 15: 421-422. 1981; I.H. Vegter, *Index Herbariorum.* Part II (4), *Collectors M.* Regnum Vegetabile vol. 93. 1976.

Sinopanax H.L. Li Araliaceae

Origins:

China and *Panax.*

Sinopimelodendron Tsiang Euphorbiaceae

Origins:

Chinese *Pimelodendron* Hassk.

Sinoplagiospermum Rauschert Rosaceae

Origins:

Chinese *Plagiospermum* Oliver.

Sinopodophyllum Ying Berberidaceae

Origins:

From the Greek *sinai* "the Chinese" and the genus *Podophyllum* L.

Sinopteris C. Chr. & Ching Pteridaceae (Adiantaceae)

Origins:

From the Greek *sinai* and *pteris* "fern."

Sinoradlkofera F.G. Meyer Sapindaceae

Origins:

For the German botanist Ludwig Adolph Timotheus Radlkofer, 1829-1927, physician, M.D. 1854, professor of botany, from 1908-1927 Director of the Botanical Museum and State Herbarium at München, author of "Sapindaceae." in *Natürl. Pflanzenfam.* 3(5): 277-366. 1895, "Sapindaceae." in *Pflanzenr.* 4(165) Heft 98: 1-1539. 1931-1934 and *New and Noteworthy Hawaiian Plants* ... Honolulu 1911 (co-author Joseph Francis Charles Rock, 1881-1962); see J.H. Barnhart, *Biographical Notes upon Botanists.* 3: 123. 1965; T.W. Bossert, *Biographical Dictionary of Botanists Represented in the Hunt Institute Portrait Collection.* 322. 1972; E.M. Tucker, *Catalogue of the Library of the Arnold Arboretum of Harvard University.* 1917-1933; Frederico Carlos Hoehne, M. Kuhlmann and Oswaldo Handro, *O jardim botânico de São Paulo.* 165. 1941; Elmer Drew Merrill, in *Contr. U.S. Natl. Herb.* 30(1): 247-248. 1947 and "Bibliography of Polynesian botany." *Bernice P. Bishop Mus. Bull.* 144: 153-154. 1937; Ida Kaplan Langman, *A Selected Guide to the Literature on the Flowering Plants of Mexico.* 605-606. University of Pennsylvania Press, Philadelphia 1964; Stafleu and Cowan, *Taxonomic Literature.* 4: 541-546. Utrecht 1983.

Sinorchis S.C. Chen Orchidaceae

Origins:
From the Greek *sinai* "the Chinese" and *orchis.*

Sinosassafras H.W. Li Lauraceae

Origins:
Greek *sinai* plus *Sassafras* Nees & Eberm.

Sinosenecio Nordenstam Asteraceae

Origins:
The Chinese *Senecio* L.

Sinosideroxylon (Engler) Aubréville Sapotaceae

Origins:
The Chinese *Sideroxylon* L.

Species/Vernacular Names:
S. wightianum (Hooker & Arnott) Aubréville

English: Wight sinosideroxylon

China: ge ye tie lan

S. yunnanense (C.Y. Wu) H. Chuang

English: Yunnan sinosideroxylon

China: dan tie lan

Sinowilsonia Hemsley Hamamelidaceae

Origins:
Dedicated to the American (b. England, Chipping Camden, Gloucestershire) botanist Ernest Henry Wilson, 1876-1930 (d. near Worcester, Massachusetts, in a motor accident), traveler and plant collector, Kew gardener, to China in 1907-1908 and 1910 on behalf of the Arnold Arboretum; see J.H. Barnhart, *Biographical Notes upon Botanists.* 3: 503. 1965; T.W. Bossert, *Biographical Dictionary of Botanists Represented in the Hunt Institute Portrait Collection.* 438. 1972; Mary Gunn and Leslie E. Codd, *Botanical Exploration of Southern Africa.* 378. Cape Town 1981; E.M. Tucker, *Catalogue of the Library of the Arnold Arboretum of Harvard University.* Cambridge, Massachusetts 1917-1933; Ida Kaplan Langman, *A Selected Guide to the Literature on the Flowering Plants of Mexico.* University of Pennsylvania Press, Philadelphia 1964; S. Lenley et al., *Catalog of the Manuscript and Archival Collections and Index to the Correspondence of John Torrey.* Library of the New York Botanical Garden. 437. 1973; R. Zander, F. Encke, G. Buchheim and S. Seybold, *Handwörterbuch der Pflanzennamen.* 14. Aufl. Stuttgart 1993; Charles Lyte, *The Plant Hunters.* London 1983; H.R. Fletcher, *Story of the Royal Horticultural Society, 1804-1968.* Oxford 1969; Ernest Nelmes and William Cuthbertson, *Curtis's Botanical Magazine Dedications, 1827-1927.* 367-368. [1931]; J. Ewan, ed., *A Short History of Botany in the United States.* 16. New York and London 1969; E.H.M. Cox, *Plant-Hunting in China.* A history of botanical exploration in China and the Tibetan marches. London 1945; Alice Margaret Coats, *The Quest for Plants. A History of the Horticultural Explorers.* London 1969; M. Hadfield et al., *British Gardeners: A Biographical Dictionary.* London 1980; Charles Sprague Sargent, ed., *Plantae Wilsonianae.* ["the greatest of all plant hunters"] Dioscorides Press, Portland, Oregon 1988; Roy W. Briggs, *A Life of Ernest Wilson.* HMSO, London 1993; Creina Glegg, "*Chinese* Wilson, Plant Hunter." *Hortus.* 39: 31-47. 1996; Ray Desmond, *Dictionary of British & Irish Botanists and Horticulturists.* 746. London 1994.

Sinthroblastes Bremek. Acanthaceae

Origins:
More or less an anagram of the generic name *Strobilanthes* Blume.

Sipaneopsis Steyerm. Rubiaceae

Origins:
Resembling *Sipanea* Aubl.

Siparuna Aublet Monimiaceae (Siparunaceae)

Origins:

A vernacular name for a species of *Perebea* Aublet (Moraceae), *Perebea chimicua* J.F. Macbride (meaning *Perebea xanthochyma* Karsten); see J.R. Perkins, "Monographie der Gattung *Siparuna*." *Bot. Jahrb. Syst.* 29: 660-705. 1901.

Species/Vernacular Names:

S. sp.

Peru: añasquero, xëxë nuitunën ñunshincë ro

S. gilgiana Perkins

Peru: pampa orégano, pampa orégano mashan

S. guianensis Aublet

Peru: curuinsi sacha, curuinshi sacha, isula micuna, isula micunan, isula caspi

Bolivia: shishohuitsa

Brazil: limão bravo, limoeiro do mato, cidreira brava

S. obovata (Gardner) A. DC.

Brazil: catingueira, negra mina, caatinga de mesa, fruto cheiroso

S. plana J.F. Macbride

Peru: isula micuna, isula micunan

S. pyricarpa (Ruíz & Pav.) Perkins (*Citrosma eriocalyx* Tulasne; *Siparuna eriocalyx* (Tulasne) A. DC.)

Peru: limoncillo

S. radiata (Poeppig & Endlicher) A. DC. (*Citrosma radiata* Poeppig & Endl.; *Siparuna magnifica* Perkins)

Peru: isula micuna, isula micunan

S. thecaphora (Poeppig & Endlicher) A. DC. (*Citrosma thecaphora* Poeppig & Endl.)

Peru: 'axa, curuinsi sacha, curuinshi sacha, isula caspi, limón del monte, macúsari, macusaro, sacha lima, sacha limón

Siphanthera Pohl ex DC. Melastomataceae

Origins:

From the Greek *siphon* "a tube" and *anthera* "anther."

Siphantheropsis Brade Melastomataceae

Origins:

Resembling *Siphanthera* Pohl.

Siphaulax Raf. Solanaceae

Origins:

From the Greek *siphon* "a tube" and *aulax, aulakos* "a furrow"; see C.S. Rafinesque, *Flora Telluriana*. 3: 74. 1836 [1837].

Siphoboea Baillon Labiatae (Verbenaceae)

Origins:

Greek *siphon* "a tube" and the genus *Boea* Comm. ex Lam.

Siphocampylus Pohl Campanulaceae

Origins:

From the Greek *siphon* "a tube" and *kampylos* "curved," referring to the corolla.

Siphocodon Turcz. Campanulaceae

Origins:

From the Greek *siphon* and *kodon* "a bell," alluding to the shape of the corolla tube.

Siphocolea Baill. Bignoniaceae

Origins:

Greek *siphon* "a tube" and *koleos* "sheath."

Siphocranion Kudô Labiatae

Origins:

Greek *siphon* and *kranion, kraneion* "head, skull."

Species/Vernacular Names:

S. macranthum (Hook.f.) C.Y. Wu

English: big-flower siphocranion

China: tong guan hua

S. nudipes (Hemsley) Kudô

English: naked-stalk siphocranion

China: guang bing tong guan hua

Siphokentia Burret Palmae

Origins:

Greek *siphon* "a tube, pipe" plus *Kentia* Blume, referring to the sepals and petals of the female flowers; dedicated to the Dutch gardener William Kent, 1779-1827, gardener at Bogor (Buitenzorg) Botanic Gardens (Java), a companion of Reinwardt and Blume on travels through Indian Archipelago. See Caspar Georg Carl Reinwardt (1773-1854), *Reis naar het oostelijk gedeelte van den Indischen Archipel in het jaar 1821 ...* met bijlagen vermeerderd door W. de

Vriese. Amsterdam 1858; Karl Ludwig von Blume (1796-1862), *Rumphia*. Lugduni Batavorum 1836.

Siphomeris Bojer Rubiaceae

Origins:
From the Greek *siphon* "a tube" and *meris* "a part, portion."

Siphonandra Klotzsch Ericaceae

Origins:
From the Greek *siphon* and *andros* "a man, male."

Siphonandra Turcz. Rubiaceae

Origins:
Greek *siphon* "a tube" and *andros* "a man, male."

Siphonandrium K. Schumann Rubiaceae

Origins:
Greek *siphon* "a tube" and *andros* "a man, male."

Siphonanthus L. Labiatae (Verbenaceae)

Origins:
Greek *siphon* "a tube" and *anthos* "flower"; see Carl Linnaeus, *Species Plantarum*. 109. 1753 and *Genera Plantarum*. Ed. 5. 47. 1754.

Siphonanthus Schreb. ex Baillon Euphorbiaceae

Origins:
Greek *siphon* "a tube" and *anthos* "flower."

Siphonella (A. Gray) Heller Polemoniaceae

Origins:
Greek *siphon* "a tube," the diminutive.

Siphonella Small Valerianaceae

Origins:
From the Greek *siphon* "a tube," the diminutive.

Siphonema Raf. Solanaceae

Origins:
From the Greek *siphon* "a tube" and *nema* "filament, thread"; see C.S. Rafinesque, *Flora Telluriana*. 3: 75. 1836 [1837].

Siphoneranthemum (Oersted) Kuntze Acanthaceae

Origins:
From the Greek *siphon* and the genus *Eranthemum* L.; see O. Kuntze, *Revisio generum Plantarum Vascularium*. 1: 494. 1891.

Siphoneugena O. Berg Myrtaceae

Origins:
Greek *siphon* "a tube" and the genus *Eugenia* L.

Siphonia Benth. Rubiaceae

Origins:
Greek *siphon* "a tube."

Siphonia Rich. Euphorbiaceae

Origins:
Greek *siphon* "a tube."

Siphonidium J.B. Armstr. Scrophulariaceae

Origins:
From the Greek *siphon* "a tube," resembling a.

Siphonochilus J.M. Wood & Franks Zingiberaceae

Origins:
From the Greek *siphon* "a tube" and *cheilos* "lip."

Species/Vernacular Names:
S. aethiopicus (Schweinf.) B.L. Burtt
Yoruba: oburo nla, oburo lela

Siphonodiscus F. Muell. Meliaceae

Origins:
Greek *siphon* "a tube" and *diskos* "a disc"; see Ferdinand von Mueller, *Fragmenta Phytographiae Australiae*. 9: 134, t. 10. (Sep.) 1875.

Siphonodon Griffith Celastraceae (Siphonodonthaceae)

Origins:
Greek *siphon* and *odous, odontos* "a tooth," referring to the structure of the flower; see William Griffith (1810-1845), in *Calcutta Journal of Natural History*. 4: 246. Calcutta 1844.

Species/Vernacular Names:
S. australis Benth.
English: weeping ivorywood, ivorywood
S. pendulus Bailey
English: ivorywood

Siphonoglossa Oersted Acanthaceae

Origins:
Greek *siphon* "a tube" and *glossa* "a tongue," referring to the bilabiate corolla.

Siphonosmanthus Stapf Oleaceae

Origins:
From the Greek *siphon* "a tube" and the genus *Osmanthus* Lour.

Siphonostegia Benth. Scrophulariaceae

Origins:
From the Greek *siphon* "a tube" and *stege* "roof, cover."

Species/Vernacular Names:
S. chinensis Benth.
Japan: hiki-yomogi
China: ling yin chen, an lu

Siphonostelma Schltr. Asclepiadaceae

Origins:
Greek *siphon* plus *Brachystelma* R. Br., *stelma, stelmatos* "a girdle, belt," *stello, stellein* "to bring together, to bind, to set."

Siphonostema Griseb. Ericaceae

Origins:
Greek *siphon* "a tube" and *stema* "stamen."

Siphonostylis W. Schulze Iridaceae

Origins:
From the Greek *siphon* "a tube" and *stylos* "style," see also *Iris* L.

Siphonychia Torrey & A. Gray Caryophyllaceae (Illecebraceae)

Origins:
From the Greek *siphon* and *onyx, onychos* "a claw, nail."

Sirhookera O. Kuntze Orchidaceae

Origins:
For the British (b. Suffolk) botanist Sir Joseph Dalton Hooker, 1817-1911 (d. Berks.), traveler, explorer, plant geographer, 1842 Fellow of the Linnean Society, 1847 Fellow of the Royal Society, 1839-1843 to Antarctic on Captain Sir James Clark Ross's expedition, 1865 Director Kew, 1873-1878 President of the Royal Society, a very prolific writer. Among his writings are *Himalayan Journals*. London 1854, *A Century of Indian Orchids*. Calcutta, London 1895 and (assisted by various botanists) *The Flora of British India*. London [1872-] 1875-1897, he was the second child of Sir William Jackson Hooker (1785-1865); see John H. Barnhart, *Biographical Notes upon Botanists*. 2: 200. 1965; Mea Allan, in *Dictionary of Scientific Biography* 6: 492-495. 1981; R. Desmond, in *Dictionary of Scientific Biography* 6: 488-492. 1981; Ray Desmond, *Dictionary of British & Irish Botanists and Horticulturists*. 352-353. London 1994; Stafleu and Cowan, *Taxonomic Literature*. 2: 267-283. 1979; Leonard Huxley, *Life and Letters of Sir Joseph Dalton Hooker*. 1918; Peter Cormack Sutherland (1822-1900), *Journal of a Voyage in Baffin's Bay and Barrow Straits in the Years 1850-51 ... in Search of the Missing Crews of H.M. Ships Erebus and Terror*. London 1852; Mea Allan, *The Hookers of Kew*. 1967; G.A. Doumani, ed., *Antarctic Bibliography*. [12 vols., quarto.] Washington, Library of Congress 1965-1979; T.W. Bossert, *Biographical Dictionary of Botanists Represented in the Hunt Institute Portrait Collection*. 181. 1972; E.M. Tucker, *Catalogue of the Library of the Arnold Arboretum of Harvard University*. Cambridge, Massachusetts 1917-1933; Blanche Henrey, *British Botanical and Horticultural Literature before 1800*. Oxford 1975; Isaac Henry Burkill, *Chapters on the History*

of Botany in India. Delhi 1965; F.D. Drewitt, *The Romance of the Apothecaries' Garden at Chelsea.* London 1924; H.R. Fletcher and W.H. Brown, *Royal Botanic Garden Edinburgh, 1670-1970.* Edinburgh 1970; S. Lenley et al., *Catalog of the Manuscript and Archival Collections and Index to the Correspondence of John Torrey.* Library of the New York Botanical Garden. 1973; Joseph Ewan, *Rocky Mountain Naturalists.* The University of Denver Press 1950; R. Zander, F. Encke, G. Buchheim and S. Seybold, *Handwörterbuch der Pflanzennamen.* 14. Aufl. Stuttgart 1993; J. Ewan, ed., *A Short History of Botany in the United States.* New York and London 1969; R. Glenn, *The Botanical Explorers of New Zealand.* Wellington 1950; J.D. Milner, *Catalogue of Portraits of Botanists Exhibited in the Museums of the Royal Botanic Gardens.* Royal Botanic Gardens, Kew, London 1906; Jeannette Elizabeth Graustein, *Thomas Nuttall, Naturalist. Explorations in America, 1808-1841.* Cambridge, Harvard University Press 1967; N. Hall, *Botanists of the Eucalypts.* Melbourne 1978 and Supplement 1980; Frederico Carlos Hoehne, M. Kuhlmann and Oswaldo Handro, *O jardim botânico de São Paulo.* 1941; E.D. Merrill, *Contr. U.S. Natl. Herb.* 30(1): 153-156. 1947; Ernest Nelmes and William Cuthbertson, *Curtis's Botanical Magazine Dedications, 1827-1927.* [1931].

Sirmuellera Kuntze Proteaceae

Origins:

After the German-born Australian botanist Sir Ferdinand Jacob Heinrich von Mueller, 1825-1896, pharmacist and physician, in 1847 to South Australia, botanical explorer and plant collector, in 1859 a Fellow of the Linnean Society and in 1861 of the Royal Society, 1857-1896 Government Botanist of Victoria, 1857-1873 Director of the Melbourne Botanical Garden, contributed to George Bentham (1800-1884), *Flora Australiensis.* London 1863-1878. Among his numerous writings are *Descriptive Notes on Papuan Plants.* Melbourne 1875[-1890], *Key to the System of Victorian Plants.* Melbourne 1885-1888, *Fragmenta phytographiae Australiae.* [Co-authors: Otto Wilhelm Sonder, 1812-1881; A.C.H. Braun, 1805-1877; G.E.L. Hampe, 1795-1880; C.M. Gottsche, 1808-1892; A. von Krempelhuber, 1813-1882; M. C. Cooke, 1825-1914] Melbourne 1858-1882 and *Plants of North-Western Australia.* Perth 1881; see J.H.L. Cumpston, *Augustus Gregory and the Inland Sea.* Canberra 1972; John H. Barnhart, *Biographical Notes upon Botanists.* 2: 524. 1965; M. Willis, *By Their Fruits. A Life of Ferdinand von Mueller, Botanist and Explorer.* Sydney, London 1949; I.H. Vegter, *Index Herbariorum.* Part II (7), *Collectors T-Z.* Regnum Vegetabile vol. 117. 1988; A.C. Gregory & F.T. Gregory, *Journals of Australian Explorations.* Brisbane 1884; N. Hall, *Botanists of the Eucalypts.* Melbourne 1978 and Supplement 1980; C.T. White, in *Bul-*

letin of Miscellaneous Information, Kew. 234, t. 7. 1930; I.C. Hedge and J.M. Lamond, *Index of Collectors in the Edinburgh Herbarium.* Edinburgh 1970; T.W. Bossert, *Biographical Dictionary of Botanists Represented in the Hunt Institute Portrait Collection.* 276. 1972; H.N. Clokie, *Account of the Herbaria of the Department of Botany in the University of Oxford.* 215. Oxford 1964; Ida Kaplan Langman, *A Selected Guide to the Literature on the Flowering Plants of Mexico.* 528-529. Philadelphia 1964; S. Lenley et al., *Catalog of the Manuscript and Archival Collections and Index to the Correspondence of John Torrey.* Library of the New York Botanical Garden. 1973; Ethelyn Maria Tucker, *Catalogue of the Library of the Arnold Arboretum of Harvard University.* Cambridge, Massachusetts 1917-1933; O. Kuntze, *Revisio generum Plantarum Vascularium.* 2: 581. (Nov.) 1891; Stafleu and Cowan, *Taxonomic Literature.* 3: 615-625. Utrecht 1981; R. Zander, F. Encke, G. Buchheim and S. Seybold, *Handwörterbuch der Pflanzennamen.* 14. Aufl. Stuttgart 1993; E.D. Merrill, in *Bernice P. Bishop Mus. Bull.* 144: 139-140. 1937; J.D. Milner, *Catalogue of Portraits of Botanists Exhibited in the Museums of the Royal Botanic Gardens.* Royal Botanic Gardens, Kew, London 1906; G. Murray, *History of the Collections Contained in the Natural History Departments of the British Museum.* 1: 170. London 1904; Ernest Nelmes and William Cuthbertson, *Curtis's Botanical Magazine Dedications, 1827-1927.* [1931]; Merle A. Reinikka, *A History of the Orchid.* Timber Press 1996; Jonathan Wantrup, *Australian Rare Books, 1788-1900.* Hordern House, Sydney 1987.

Sison L. Umbelliferae

Origins:

Greek *sinon, sison* for stone parsley, *Sison Amomum* (Dioscorides), *sinon agrios* for *peukedanon*; Latin *sison agrion* for a plant, also called *peucedanos* and *pinastellus.*

Sisymbrella Spach Brassicaceae

Origins:

The diminutive of the genus *Sisymbrium* L.

Sisymbrianthus Chevall. Brassicaceae

Origins:

The genus *Sisymbrium* and *anthos* "flower."

Sisymbriopsis Botsch. & Tzvelev Brassicaceae

Origins:

Resembling the genus *Sisymbrium* L.

Sisymbrium L. Brassicaceae

Origins:

From *sisymbrion*, the Greek name for some sweet-smelling plant, a fragrant herb, a watercress, wild thyme, *Mentha* spp.; Hebrew *sis* "flower, festoon, ornament," *sisa* "flower," *bor* "pit, well"; Akkadian *burum* "pond, well"; Latin *sisymbrium, ii* "a herb sacred to Venus" (Plinius, Ovidius); see Carl Linnaeus, *Species Plantarum*. 657. 1753 and *Genera Plantarum*. Ed. 5. 330. 1754.

Species/Vernacular Names:

S. sp.

South America: mostaza prieta

S. altissimum L.

English: tumble, Jim Hill mustard

S. burchellii DC. var. *burchellii* (*Sisymbrium burchellii* DC.; *Sisymbrium gariepinum* Burch. ex DC.)

South Africa: wildemosterd

S. capense Thunb.

English: Cape mustard, Cape wild mustard, wild mustard

Southern Africa: strandmosterd, strandwildemosterd, wildemosterd; tlhako-ea-khomo (South Sotho)

S. erysimoides Desf.

English: smooth mustard

S. irio L.

English: London rocket

Egypt: figl el-gamal

S. officinale (L.) Scop. (*Erysimum officinale* L.)

English: hedge mustard, hedge wild mustard

Arabic: harif, horf, sammarah

South Africa: heiningwildemosterd

S. orientale L. (*Sisymbrium strigosum* Thunb. p.p.; *Sisymbrium columnae* Jacq.)

English: Indian hedge mustard, wild mustard

S. thellungii O.E. Schulz (*Brassica pachypoda* Thell.)

English: common wild mustard, hedge mustard, wild mustard, African turnip weed

Southern Africa: gewone wildemosterd, wildemosterd, isiqwashube; liababa (Sotho)

Sisyranthus E. Meyer ex Sonder Asclepiadaceae

Origins:

Greek *sisyra* "a shaggy goatskin, a thick garment" and *anthos* "a flower."

Sisyrinchium L. Iridaceae

Origins:

Greek *sisyrinchion*, a classical name for a species of iris or a bulbous plant, *sys* "pig" and *rynchos* "a snout, muzzle," referring to the shape of the roots or because the pigs are fond of the roots; see Carl Linnaeus, *Species Plantarum*. 954. 1753 and *Genera Plantarum*. Ed. 5. 409. 1754.

Species/Vernacular Names:

S. acre H. Mann

Hawaii: mau'u la'ili, mau'u ho'ula'ili

S. angustifolium Miller

English: blue-eyed gras

S. atlanticum E. Bickn.

English: blue-eyed grass

Japan: niwa-zeki-sho

S. bellum S. Watson

English: Californian blue-eyed grass, blue-eyed grass

S. bermudiana L.

English: blue-eyed grass

S. californicum (Ker Gawler) Dryander

English: golden-eyed grass

S. iridifolium Kunth

English: blue pigroot, striped rush-leaf

Sisyrolepis Radlk. Sapindaceae

Origins:

From the Greek *sisyra* "a shaggy goatskin, a thick garment" and *lepis* "a scale."

Sitanion Raf. Gramineae

Origins:

Greek *sitos* "grain," *setaneios, setanios, satanios* "of this year, spring-wheat," *sitanias pyros* "a branching cereal" (Theophrastus, *HP.* 8.2.3), Latin *sitanius, a, um* "of this year, of summer wheat"; see C.S. Rafinesque, *Jour. Phys. Chim. Hist. Nat.* 89: 103. 1819; E.D. Merrill, *Index rafinesquianus*. 76. 1949.

Sitella L.H. Bailey Sterculiaceae

Origins:

Latin *sitella* "a kind of urn used in drawing lots," *situla* "a bucket, an urn."

Sitilias Raf. Asteraceae

Origins:

See C.S. Rafinesque, *New Fl. N. Am.* 4: 85. 1836 [1838]; E.D. Merrill, *Index rafinesquianus.* 241. 1949.

Sitodium Parkinson Moraceae

Origins:

Resembling the Greek *sitos* "grain."

Sitolobium Desv. Dennstaedtiaceae

Origins:

Greek *sitos* "grain" and *lobion* "a little pod," referring to the fruits.

Sitolobium J. Smith Dicksoniaceae (Cyatheaceae)

Origins:

Orthographic variant of the genus *Sitobolium* Desvaux.

Sitopsis (Jaub. & Spach) Á. Löve Gramineae

Origins:

From the Greek *sitos* "grain" and *opsis* "like, resemblance."

Sitospelos Adans. Gramineae

Origins:

Greek *sitos* "grain" and *pelos* "black."

Sium L. Umbelliferae

Origins:

From the ancient Greek name *sion* for a marsh herb; see Carl Linnaeus, *Species Plantarum.* 954. 1753 and *Genera Plantarum.* Ed. 5. 409. 1754.

Species/Vernacular Names:

S. sisarum L.

English: crummock, skirret

Siumis Raf. Umbelliferae

Origins:

For *Sium* L.; see C.S. Rafinesque, *New Fl. N. Am.* 4: 31. 1836 [1838].

Sixalix Raf. Dipsacaceae

Origins:

See C.S. Rafinesque, *Flora Telluriana.* 4: 95. 1836 [1838]; E.D. Merrill, *Index rafinesquianus.* 230. 1949.

Skapanthus C.Y. Wu & H.W. Li Labiatae

Origins:

Latin *scapus* "shaft, stem, stalk, trunk," Greek *skepto* "stay one thing against another" and *anthos* "flower."

Species/Vernacular Names:

S. oreophilus (Diels) C.Y. Wu & H.W. Li

English: common skapanthus

China: ting hua

Skeptrostachys Garay Orchidaceae

Origins:

Greek *skeptron* "a sceptre, royal staff" and *stachys* "a spike."

Skiatophytum L. Bolus Aizoaceae

Origins:

From the Greek *skia* "shade, shadow" and *phyton* "plant."

Skimmia Thunb. Rutaceae

Origins:

From the Japanese *shikimi* or *ashiki-mi*, bad or poisonous or harmful fruit; see Helmut Genaust, *Etymologisches Wörterbuch der botanischen Pflanzennamen.* 591. Basel 1996; F. Boerner & G. Kunkel, *Taschenwörterbuch der botanischen Pflanzennamen.* 4. Aufl. 170. Berlin & Hamburg 1989; Georg Christian Wittstein (1810-1887), *Etymologisch-botanisches Handwörterbuch.* 817. Ansbach 1852.

Species/Vernacular Names:

S. japonica Thunb.

Japan: miyama-shikimi

China: yin yu

S. laureola (DC.) Decne.

India: ner, barru, nehar, gurl pata

Skinneria Choisy Convolvulaceae

Origins:

For Stephen Skinner, 1622-1667, physician, author of *Etymologicon Linguae Anglicanae, seu explicatio vocum Anglicarum etymologica ex propriis fontibus ... Accedit Etymologicon Botanicum ...* Londini 1671.

Skiophila Hanst. Gesneriaceae

Origins:

Greek *skia, skie* "shade, shadow" and *philos* "loving."

Skirrhophorus DC. Asteraceae

Origins:

Greek *skirrhos* "scirrhus, induration, swelling" and *phoreo* "to bear"; see A.P. de Candolle, *Prodromus.* 6: 150. 1838.

Skoliopterys Cuatrec. Malpighiaceae

Origins:

From the Greek *skolios* "curved, bent, crooked, not straight" and *pteron* "a wing."

Skoliostigma Lauterb. Anacardiaceae

Origins:

From the Greek *skolios* and *stigma* "a stigma."

Skottsbergianthus Boelcke Brassicaceae

Origins:

After the Swedish botanist Carl Johan Fredrik Skottsberg, 1880-1963, traveler and botanical explorer (South America, Chile, Patagonia), professor of botany, collected cacti. His works include "The Swedish Magellanian expedition, 1907-1909: Preliminary reports." *Geogr. Jour.* London 1908, *The Wilds of Patagonia: A Narrative of the Swedish Expedition to Patagonia, Tierra del Fuego and the Falkland Islands in 1907-1909.* London 1911 and "Observations on the natives of the Patagonian Channel region." *Amer. Anthr.* 4: 578-616. 1913; see Stafleu and Cowan, *Taxonomic Literature.* 5: 627-639. Utrecht 1985; R. Zander, F. Encke, G. Buchheim and S. Seybold, *Handwörterbuch der Pflanzennamen.* 14. Aufl. Stuttgart 1993; John M. Cooper, *Analytical and Critical Bibliography of the Tribes of Tierra del Fuego and Adjacent Territory.* [Smithsonian Institution — Bureau of American Ethnology — Bulletin 63.] Washington 1917; John H. Barnhart, *Biographical Notes upon Botanists.* 3: 285. 1965; T.W. Bossert, *Biographical Dictionary of Botanists Represented in the Hunt Institute Portrait Collection.* 370. 1972; E.M. Tucker, *Catalogue of the Library of the Arnold Arboretum of Harvard University.* Cambridge, Massachusetts 1917-1933; Clodomiro Marticorena, *Bibliografía Botánica Taxonómica de Chile.* 518-521. Missouri Botanical Garden 1992; Gordon Douglas Rowley, *A History of Succulent Plants.* Strawberry Press, Mill Valley, California 1997; J. Ewan, ed., *A Short History of Botany in the United States.* New York and London 1969; E.D. Merrill, in *Bernice P. Bishop Mus. Bull.* 144: 171-173. 1937 and *Contr. U.S. Natl. Herb.* 30(1): 278-282. 1947; Margaret Patricia Henwood Laver, *An Annotated Bibliography of the Falkland Islands and the Falkland Island Dependencies* (as delimited on 3rd March, 1962). Cape Town 1977.

Skottsbergiella Boelcke Brassicaceae

Origins:

After the Swedish botanist Carl Johan Fredrik Skottsberg, 1880-1963, traveler and botanical explorer, professor of botany.

Skottsbergiella Epling Labiatae

Origins:

After the Swedish botanist Carl Johan Fredrik Skottsberg, 1880-1963, traveler and botanical explorer, professor of botany.

Skottsbergiliana St. John Cucurbitaceae

Origins:

After the Swedish botanist Carl Johan Fredrik Skottsberg, 1880-1963, traveler and botanical explorer, professor of botany.

Skutchia Pax & K. Hoffmann ex C. Morton Moraceae

Origins:

After the American botanist Alexander Frank Skutch, traveler and plant collector (Central America).

Skytanthus Meyen Apocynaceae

Origins:

Greek *skytos* "leather, a skin, hide" and *anthos* "flower."

Slackia Griff. Gesneriaceae

Origins:

For Henry Slack, d. before 1845; see Ray Desmond, *Dictionary of British & Irish Botanists and Horticulturists*. 631. London 1994.

Slackia Griff. Palmae

Origins:

For Henry Slack, d. before 1845; see Ray Desmond, *Dictionary of British & Irish Botanists and Horticulturists*. 631. 1994.

Sleumerodendron Virot Proteaceae

Origins:

After the Dutch botanist (of German birth) Hermann Otto Sleumer, traveler, plant collector, pharmacist, taxonomist. His writings include "Taxonomy of the genus *Pernettya* Gaud. (Ericaceae)" *Bot. Jahrb. Syst.* 105(4): 449-480. 1985 and "Revision der Gattung *Azara* R. et P. (Flacourtiaceae)" *Bot. Jahrb. Syst.* 98(2): 151-175. 1977; see T.W. Bossert, *Biographical Dictionary of Botanists Represented in the Hunt Institute Portrait Collection*. 371. 1972; Ida Kaplan Langman, *A Selected Guide to the Literature on the Flowering Plants of Mexico*. 702. University of Pennsylvania Press, Philadelphia 1964; S. Lenley et al., *Catalog of the Manuscript and Archival Collections and Index to the Correspondence of John Torrey*. Library of the New York Botanical Garden. 370. 1973; E.D. Merrill, *Contr. U.S. Natl. Herb.* 30(1): 282-283. 1947; R. Zander, F. Encke, G. Buchheim and S. Seybold, *Handwörterbuch der Pflanzennamen*. 14. Aufl. Stuttgart 1993.

Sloanea L. Elaeocarpaceae

Origins:

Named for Sir Hans Sloane (b. Killyleagh, Northern Ireland) 1660-1753 (d. Chelsea, London), physician and naturalist, a founder of the British Museum, traveler and plant collector in the West Indies, a pupil of Tournefort and Duverney, in 1685 a Fellow of the Royal Society and from 1727 to 1741 President (in 1727, succeeded Sir Isaac Newton), in 1687 a Fellow of the Royal College of Physicians and in 1719 President, 1687 accompanied Christopher Monck (at that time appointed Governor of Jamaica), 1712 physician to Queen Anne, gave Chelsea Physic Garden to the Apothecaries' Company, editor of the *Philosophical Transactions* of the Royal Society. His works include *Catalogus plantarum quae in insula Jamaica sponte proveniunt*. Londini 1696, *An Account of a Most Efficacious Medicine for Soreness, Weakness, and Several Other Distempers of the Eyes*. London 1745 and *A Voyage to the Islands Madera, Barbados, Nieves, S. Christophers and Jamaica*. London 1707-1725, a friend of Samuel Pepys, John Locke, E. Halley and Newton; see G.R. DeBeer, *Sir Hans Sloane and the British Museum*. London 1953; G.R. DeBeer, in *Dictionary of Scientific Biography* 12: 456-459. 1981; Frans Antonie Stafleu, *Linnaeus and the Linnaeans*. The spreading of their ideas in systematic botany, 1735-1789. 1971; William King, *The Present State of Physick in the Island of Cajamai* [i.e. Jamaica]. To the Members of the Royal Society. [London 1710?]; James Britten, *The Sloane Herbarium*, revised and edited by J.E. Dandy. 1958; John H. Barnhart, *Biographical Notes upon Botanists*. 3: 287. 1965; D.E. Allen, *The Botanists*. Winchester 1986; Edmund Berkeley and D.S. Berkeley, *Dr. Alexander Garden of Charles Town*. University of North Carolina Press [1969], and *John Clayton, Pioneer of American Botany*. Chapel Hill 1963; H.J. Braunholtz, *Sir Hans Sloane and Ethnography*. London 1970; E. Earnest, *John and William Bartram, Botanists and Explorers 1699-1777, 1739-1823*. Philadelphia 1940; Mary Gunn and Leslie E. Codd, *Botanical Exploration of Southern Africa*. Cape Town 1981; Mia C. Karsten, *The Old Company's Garden at the Cape and Its Superintendents*: involving an historical account of early Cape botany. Cape Town 1951; W. Eric St. John Brooks, *Sir Hans Sloane. The Great Collector and His Circle*. London 1954; F.D. Drewitt, *The Romance of the Apothecaries' Garden at Chelsea*. London 1924; H. Field, *Memoirs of the Botanic Garden of Chelsea* belonging to the Society of Apothecaries of London. London 1878; R. Pulteney, *Historical and Biographical Sketches of the Progress of Botany in England*. 2: 65-96. London 1790; William Munk, *The Roll of the Royal College of Physicians of London*. 1: 460-467. London 1878; Carl Linnaeus, *Species Plantarum*. 512. 1753 and *Genera Plantarum*. Ed. 5. 228. 1754; W. Darlington, *Memorials of John Bartram and Humphry Marshall*. Philadelphia 1849; E.G. Wheelwright, *The Physick Garden*, medicinal plants and their history. Boston 1935; T.W. Bossert, *Biographical Dictionary of Botanists Represented in the Hunt Institute Portrait Collection*. 371. 1972; H.N. Clokie, *Account of the Herbaria of the Department of Botany in the University of Oxford*. 245. Oxford 1964; Blanche Henrey, *British Botanical and Horticultural Literature before 1800*. Oxford 1975; E.M. Tucker, *Catalogue of the Library of the Arnold Arboretum of Harvard University*. Cambridge, Massachusetts 1917-1933; Mariella Azzarello Di Misa, a cura di, *Il Fondo Antico della Biblioteca dell'Orto Botanico di Palermo*. 258. Regione Siciliana, Palermo 1988; Blanche Henrey, *No Ordinary Gardener — Thomas Knowlton, 1691-1781*. Edited by A.O. Chater. British Museum (Natural History). London 1986.

Species/Vernacular Names:

S. spp.

Peru: cepanchila, huangana casha, huangana caspi

S. australis (Benth.) F. Muell. (*Echinocarpus australis* Benth.)

English: maiden's blush, blush carabeen (or carrabeen)

S. durissima Spruce ex Bentham

Peru: sacha acioto

S. guianensis (Aublet) Bentham (*Ablania guianensis* Aublet)

Peru: cutana cuspi

S. langii F. Muell.

English: white carabeen

S. laurifolia (Bentham) Bentham (*Dasynema laurifolium* Bentham)

Peru: cepanchina cutana

S. laxiflora Spruce ex Bentham (*Sloanea acutiflora* Uittien; *Sloanea polyantha* Ducke)

Peru: cepanchina

S. macbrydei F. Muell.

English: northern yellow carabeen

S. terniflora (Moçiño & Sessé ex DC.) Standley (*Lecostemon terniflorum* Moçiño & Sessé ex DC.; *Sloanea quadrivalvis* Seemann)

Peru: anallo caspi

S. woollsii F. Muell.

English: yellow carabeen, grey carabeen

Sloetiopsis Engl. Moraceae

Origins:

Resembling the related genus *Sloetia* Teijsm. & Binn. ex Kurz.

Smallanthus Mack. Asteraceae

Origins:

For the American taxonomic botanist John Kunkel Small, 1869-1938, Herbarium of Columbia College, Museum of the New York Botanical Garden, traveler and botanical explorer, author of the monumental *Flora of the Southeast-*

ern United States. 1903. His articles appeared regularly in the *Journal of the New York Botanical Gerden*. See John H. Barnhart, *Biographical Notes upon Botanists*. 3: 287. 1965; T.W. Bossert, *Biographical Dictionary of Botanists Represented in the Hunt Institute Portrait Collection*. 245. 1972; E.M. Tucker, *Catalogue of the Library of the Arnold Arboretum of Harvard University*. Cambridge, Massachusetts 1917-1933; Ida Kaplan Langman, *A Selected Guide to the Literature on the Flowering Plants of Mexico*. Philadelphia 1964; S. Lenley et al., *Catalog of the Manuscript and Archival Collections and Index to the Correspondence of John Torrey*. Library of the New York Botanical Garden. 1973; Stafleu and Cowan, *Taxonomic Literature*. 5: 650-657. Utrecht 1985; R. Zander, F. Encke, G. Buchheim and S. Seybold, *Handwörterbuch der Pflanzennamen*. 14. Aufl. Stuttgart 1993; J. Ewan, ed., *A Short History of Botany in the United States*. New York and London 1969.

Species/Vernacular Names:

S. parviceps (S.F. Blake) H. Robinson (*Polymnia parviceps* S.F. Blake)

Peru: taraca

S. sonchifolius (Poeppig & Engler) H. Robinson (*Polymnia edulis* Weddell; *Polymnia sonchifolia* Poeppig) (see National Research Council, *Lost Crops of the Incas: Little-Known Plants of the Andes with Promise for Worldwide Cultivation*. National Academy Press, Washington, D.C. 1989)

English: yacon, yacon strawberry, jiquima

Quechua names: yacón, llakuma

Aymara name: aricoma, aricona

Spanish: yacón, jacón, llacón, llamón, arboloco, puhe, jícama, jíquima, jíkima, jiquimílla

Peru: aricoma, llacjon, llacon, puhe, yacon

Smallia Nieuwl. Orchidaceae

Origins:

For the American botanist John Kunkel Small, 1869-1938.

Smeathmannia Solander ex R. Br. Passifloraceae

Origins:

For the British naturalist Henry Smeathman, d. 1786, botanist, plant collector in Sierra Leone (1771-1772), Madagascar and the West Indies (after 1775), associated with Adam Afzelius (1750-1837) in scheme for settling freed slaves in Freetown, Sierra Leone, author of *Elocution and Polite Literature*. [An announcement of lessons in elocution.] [London

1785?], *Mémoire pour servir à l'histoire de quelques Insectes, connus sous les noms de Termès, ou Fourmis blanches* ... Ouvrage rédigé en François par M. C. Rigaud. Paris 1786, and *Plan of a Settlement to Be Made near Sierra Leona, on the Grain Coast of Africa.* London 1786, friend of John Coakley Lettsom (1744-1815); see William Wilberforce, [Speech on the motion for the abolition of the slave trade, 12 May 1789.] — *Correct Copies of the Twelve Propositions Submitted ... by Mr. Wilberforce, to the Consideration of the Committee.* [England. Parliament. House of Commons. Proceedings II.] London 1789; H.M. Wheeler, *The Slave's Champion, or a Sketch of the Life, Deeds and Historical Days of William Wilberforce.* London 1860; Michael Hennell, *William Wilberforce, 1759-1833. The Liberator of the Slave.* London [1950]; Charles Buxton, ed., *Memoirs of Sir Thomas Fowell Buxton, Baronet.* [Sir T. Buxton was the successor to Wilberforce as leader of the Anti-Slavery party in the House of Commons.] London 1848; Peter Hogg, *Slavery. The Afro-American Experience.* London 1979; C. Bolt, *The Anti-Slavery Movement.* Oxford 1969; Johnson U.I. Asiegbu, *Slavery and the Politics of Liberation 1787-1861.* London 1969; James Pope-Hennessy, *Sins of the Fathers. A Study of the Atlantic Slave Traders 1441-1807.* London 1967; James Ramsay, *An Essay on the Treatment and Conversion of African Slaves in the British Sugar Colonies.* London 1784; Joseph Vallot (1854-1925), "Études sur la flore du Sénégal." in *Bull. Soc. Bot. de France.* 29: 168-238. Paris 1882; *Negro History: 1553-1903.* Philadelphia 1969; John H. Barnhart, *Biographical Notes upon Botanists.* 3: 288. 1965; I.H. Vegter, *Index Herbariorum.* Part II (6), *Collectors S.* Regnum Vegetabile vol. 114. 1986; A. Lasègue, *Musée botanique de Benjamin Delessert.* Paris 1845; J.C. Lettsom, *Memoirs of John Fothergill, M.D.* London 1784; F.N. Hepper and Fiona Neate, *Plant Collectors in West Africa.* 75. 1971; R.W.J. Keay, "Botanical collectors in West Africa prior to 1860." in *Comptes Rendus A.E.T.F.A.T.* 55-68. Lisbon 1962.

Species/Vernacular Names:

S. laevigata Soland. ex R. Br.

Yoruba: moyida

S. pubescens Soland. ex R. Br.

Nigeria: moyida (Yoruba)

Yoruba: moyida

Sierra Leone: ndovote

Smelophyllum Radlk. Sapindaceae

Origins:

Greek *smegma, smema, smama, smele* "soap, unguent" and *phyllon* "leaf"; some suggest from *Smelowskia* C. A. Meyer (Cruciferae alt. Brassicaceae, East Asia, Western North America) ("with leaves like *Smelowskia*"); see Helmut Genaust, *Etymologisches Wörterbuch der botanischen Pflanzennamen.* 591. Basel 1996.

Species/Vernacular Names:

S. capense (Sond.) Radlk. (*Sapindus capensis* Sond.)

South Africa: buig-my-nie (bend-me-not)

Smelowskia C.A. Meyer Brassicaceae

Origins:

After the Russian pharmacist Timotheus Smielowsky (Smelowsky), 1769-1815, botanist at St. Petersburg, author of *Hortus petropolitanus.* Petropoli 1806; see John H. Barnhart, *Biographical Notes upon Botanists.* 3: 288. 1965.

Species/Vernacular Names:

S. ovalis M.E. Jones var. *congesta* Rollins

English: Lassen Peak smelowskia

Smicrostigma N.E. Br. Aizoaceae

Origins:

From the Greek *smikros* "small" and *stigma* "stigma."

Smidetia Raf. Gramineae

Origins:

For the German botanist Franz Wilibald Schmidt, 1764-1796; see C.S. Rafinesque, *Autikon botanikon.* Icones plantarum select. nov. vel rariorum, etc. 187. Philadelphia 1840; E.D. Merrill, *Index rafinesquianus.* 76. 1949.

Smilacina Desf. Convallariaceae (Liliaceae)

Origins:

From the Greek *smilakinos* "of the *smilax, xylon*"; see R. Zander, F. Encke, G. Buchheim and S. Seybold, *Handwörterbuch der Pflanzennamen.* 14. Aufl. 521. Stuttgart 1993.

Smilax L. Smilacaceae

Origins:

Greek *smilax, smilakos* (*milax, milakos*) "a bind-weed"; Sanskrit *mah* "to grow," *mula* "root"; Akkadian *melu* "height, elevation, ascent"; the Latin name *smilax, acis* was used by Plinius for the bindweed, *Smilax aspera* L., but also for the yew-tree and for a species of oak. See Carl Linnaeus, *Species Plantarum.* 1028. 1753 and *Genera Plantarum.* Ed. 5. 455. 1754; Andreas Vesalius (1514-1564), *Radicis Chynae usus.* [16mo, pocket edition.] Lyons 1547; Richard J. Durling, comp., *A Catalogue of Sixteenth Century Printed Books in the National Library of Medicine.* 4589. 1967; Giovanni Semerano, *Le origini della cultura europea.* Dizionario della lingua Greca. 2(1): 271. Leo S. Olschki Editore, Firenze 1994; Alexander B. Adams, *Eternal Quest. The Story of the Great Naturalists.* New York 1969.

Species/Vernacular Names:

S. sp.

English: catbrier, greenbrier

Peru: alambre casha, cuiro sacha, jipota, zarzaparrilla, carnosidad casha, clavo caspi, garrapata casha

Mexico: bejuco de vida, ut, chiapahuac-xíhuitl

Malaya: alak batu, chanar babi

S. anceps Willd. (*Smilax kraussiana* Meisn.) (the specific name means "two-sided, two-headed")

English: wild sarsaparilla

Southern Africa: iyali (Zulu); mugarangurwe, muGwanangurwe, muGwenanguruve (Shona); mukokole (Venda); muwaruwaru (Tsonga)

Yoruba: kan san, ikan san, eekanna magbo, worewore, iwokuwo

S. aspera L.

English: rough bindweed, prickly ivy

Arabic: zeqresh, 'ulliq, 'ushba

Brazil: salsaparrilha, jupicanga, salsaparrilha da índia, salsaparrilha da espanha, salsaparrilha indígena, salsaparrilha da europa

S. australis R. Br.

English: Austral smilax, Austral sarsaparilla, lawyer vine, bush lawyer, sarsaparilla

S. bona-nox L.

English: China brier, China briar

S. brasiliensis Spreng.

Brazil: salsaparrilha do rio, salsaparrilha, japecanga, ingapencanga, juapecanga, japicanga, raiz de quina vermelha, raiz de quina branca, jupicanga

S. china DC.

English: China root, bamboo briar root

Brazil: salsaparrilha

China: ba qia, pa chia, chin kang ken, tieh ling chio, wang kua tsao

India: dwipautra, wacha, madhusnuhi, chobchini, parangichekkai, parinkipatte, shuk-china, paringay, pirangichekka, gali-chekka, China-alla

Malaya: gadong china, restong, ubat raja

S. eucalyptifolia Kunth (*Smilax subinermis* C. Presl)

Peru: purtrampui, putrampui

S. febrifuga Kunth

Peru: puntrampui, palo santo, purampui, santo palo

Brazil: salsaparrilha do Pará

S. glauca Walter

English: wild sarsaparilla

S. glyciphylla Smith

English: thornless smilax, sarsaparilla, sweet sarsaparilla

S. irrorata C. Martius ex Grisebach

Brazil: salsaparrilha, japecanga

S. jamesii G.A. Wallace

English: English peak greenbriar

S. japicanga Griseb.

Brazil: salsaparrilha, salsaparrilha do rio, japecanga, inhapecanga, juapecanga, salsa do campo, japicanga, raiz de quina branca, raiz de quina vermelha

S. kunthii Killip & Morton (*Smilax floribunda* Kunth, nom. illeg.)

Peru: palo chino, chico, palo chico

S. longifolia Richard (*Smilax papyracea* Duhamel)

Peru: sausa, chija sacha, zarza, zarza parilla

Brazil: salsaparrilha, jepecanda

S. melastomifolia Sm. (*Smilax hawaiensis* Seem.)

Hawaii: hoi kuahiwi, aka'awa, pi'oi, uhi, ulehihi

S. obliquata Duhamel

Brazil: salsaparrilha

S. poeppigii Kunth

Peru: chicjasacja, chicja sacha

S. regelii Killip & Morton

Central America: zarzaparrilla, bejuco de la vida, cocolmeca, cuculmeca, diente de chucho, palo de la vida

Brazil: salsaparrilha do pará, salsaparrilha das boticas, japecanga

S. rotundifolia L.

English: bull brier, bull briar

Smirnowia Bunge Fabaceae

Origins:

For the Russian botanist Michael Nikolajewitsch Smirnov (Smirnow, Smirnoff), 1849-1889; see John H. Barnhart, *Biographical Notes upon Botanists.* 3: 289. 1965; T.W. Bossert, *Biographical Dictionary of Botanists Represented in the Hunt Institute Portrait Collection.* 372. 1972.

Smithia Aiton Fabaceae

Origins:

For the British (b. Norwich) botanist Sir James Edward Smith, 1759-1828 (d. Norwich), physician, taxonomist, translated several of Linnaeus' works, 1785 Fellow of the Royal Society, 1786 took the degree of Doctor of Medicine at the University of Leyden, (1788-1828) a founder (with Rev. Samuel Goodenough, 1743-1827, and the entomologist Thomas Marsham, d. 1819) and first President of the Linnean Society of London (re-elected President each year until his death), lectured at the Royal Institution and at Cambridge, knighted 1814, an extraordinary and prolific writer. Among his valuable writings are *Plantarum icones hactenus ineditae*. Londini 1789-1791, *Sketch of a Tour on the Continent*. London 1793, *Tracts Relating to Natural History*. [Octavo, first edition, a collection of 12 essays.] London 1798, *English Botany*. 1790-1814, *Flora britannica*. Londini 1800-1804, *Exotic Botany*. London 1804-1805 [1808] and *The English Flora*. London 1824-1828; see A.M. Geldart, "Sir James Edward Smith and some of his friends." *Trans. Norfolk and Norwich Naturalists' Soc.* 9: 647. 1914; Stafleu and Cowan, *Taxonomic Literature*. 5: 678-691. 1985; Diana M. Simpkins, in *Dictionary of Scientific Biography* (Editor in Chief Charles Coulston Gillispie.) 12: 471-472. New York 1981; F.A. Stafleu, *Linnaeus and the Linnaeans*. The spreading of their ideas in systematic botany, 1735-1789. 1971; R. Zander, F. Encke, G. Buchheim and S. Seybold, *Handwörterbuch der Pflanzennamen*. 14. Aufl. 1993; Blanche Henrey, *British Botanical and Horticultural Literature before 1800*. Oxford 1975; W. Aiton, in *Hortus Kewensis*. 3: 496. 1789; John H. Barnhart, *Biographical Notes upon Botanists*. 3: 293. 1965; Lady Pleasance Smith, ed., *Memoir and Correspondence of ... Sir J.E. Smith*. London 1832; A.T. Gage, *A History of the Linnean Society of London*. London 1938.

Species/Vernacular Names:

S. sensitiva Aiton (*Smithia javanica* Benth.)

Japan: nemuri-hagi

Smithiantha Kuntze Gesneriaceae

Origins:

For the British (b. Bombay, India) botanical artist Matilda Smith, 1854-1926 (d. Kew, Surrey), from 1877 associated with Kew; see J.H. Barnhart, *Biographical Notes upon Botanists*. 3: 536. 1965; Ray Desmond, *Dictionary of British & Irish Botanists and Horticulturists*. 639. 1994; Alain White (1880-1951) and Boyd Lincoln Sloane (1886-1955), *The Stapelieae*. Pasadena 1937; Thomas Frederick Cheeseman (1846-1923) and William Botting Hemsley (1843-1924), *Illustrations of the New Zealand Flora*. [The plates drawn by Miss Matilda Smith, 1854-1926] Wellington 1914; Sir H.H. Johnston, *Liberia*. [24 botanical drawings by Miss M. Smith.] 1906; F. Boerner & G. Kunkel, *Taschenwörterbuch der botanischen Pflanzennamen*. 4. Aufl. 170. 1989.

Smithiella Dunn Urticaceae

Origins:

For the British botanical artist Matilda Smith, 1854-1926, from 1877 associated with the Kew Herbarium.

Smithiodendron Hu Moraceae

Origins:

For the British botanist Sir William Wright Smith, 1875-1956, Regius Keeper of the Royal Botanic Gardens of Edinburgh.

Smithorchis Tang & Wang Orchidaceae

Origins:

For the British (b. Dumfriesshire) botanist Sir William Wright Smith, 1875-1956 (d. Edinburgh), plant collector, specialist in Chinese flora, 1907-1911 herbarium Botanic Garden Calcutta, Regius Keeper and Director of the Royal Botanic Gardens of Edinburgh, 1918 Fellow of the Linnean Society, 1932 knighted, 1945 Fellow of the Royal Society, Director of the Botanical Survey of India. His works include *The Alpine and Sub-alpine Vegetation of South-east Sikkim*. Calcutta 1913 and "Species novae plantarum in herbario Horti Reg. Calcutt. cognitarum." *Rec. Bot. Survey India*. 6(4): 99-104. 1914, with George Forrest (1873-1932) and Harold Roy Fletcher (1907-1978) wrote *The Genus Primula*. Vaduz 1977, with the Kew gardener and plant collector (Tibet, Nepal and Sikkim) George H. Cave (1870-1965) wrote *The Vegetation of the Zemu and Llonakh Valleys of Sikkim*. Calcutta 1911 and *Note on the East Himalayan Species of Alangium*. Calcutta 1914; see J.H. Barnhart, *Biographical Notes upon Botanists*. 3: 296. Boston 1965; H.R. Fletcher and W.H. Brown, *Royal Botanic Garden Edinburgh, 1670-1970*. Edinburgh 1970; S. Lenley et al., *Catalog of the Manuscript and Archival Collections and Index to the Correspondence of John Torrey*. Library of the New York Botanical Garden. 378. 1973; H.R. Fletcher, *Story of the Royal Horticultural Society, 1804-1968*. Oxford 1969; T.W. Bossert, *Biographical Dictionary of Botanists Represented in the Hunt Institute Portrait Collection*. 374. 1972; E.M. Tucker, *Catalogue of the Library of the Arnold Arboretum of Harvard University*. Cambridge, Massachusetts 1917-1933; M. Hadfield et al., *British Gardeners: A Biographical*

Dictionary. London 1980; Isaac Henry Burkill, *Chapters on the History of Botany in India*. 141. Delhi 1965; J.M. Cowan, *The Journeys and Plant Introduction of George Forrest*. R.H.S. 1952; R. Zander, F. Encke, G. Buchheim and S. Seybold, *Handwörterbuch der Pflanzennamen*. 14. Aufl. Stuttgart 1993.

Smithsonia Saldanha Orchidaceae

Origins:

For the American botanist James Smithson, see Hubert Mayr, *Orchid Names and Their Meanings*. Vaduz 1998.

Smitinandia Holttum Orchidaceae

Origins:

Named after the botanist Tem Smitinand, 1920-1995, orchid taxonomist, plant collector with M. Raymond and also with Hermann Otto Sleumer (1906-1993) and B. Hansen, Director of the Royal Thailand Department of Forestry, wrote *Thai Plant Names*. Royal Forest Department. Bangkok 1980, co-author with Gunnar Seidenfaden of *The Orchids of Thailand*. 1959-1965; see I.H. Vegter, *Index Herbariorum*. Part II (6), *Collectors S*. Regnum Vegetabile vol. 114. 1986.

Smodingium E. Meyer ex Sonder Anacardiaceae

Origins:

Greek *smodix*, *smodingos*, *smodiggos* "wheal caused by a blow," referring to swelling (the plant has irritant properties), or also to indurated mark or hardened margins of the hard fruits.

Species/Vernacular Names:

S. argutum E. Meyer ex Sonder (the specific epithet refers to the sharply toothed or notched leaves, from the Latin *argutus*, *a*, *um*)

English: rainbow leaf, African poison oak, pain bush, African poison ivy

Southern Africa: pynbos, tovane, tovana; inTovane (Zulu); inTovane, uTovani (Xhosa); Umtomvane (Xhosa and Pondo); tshilabele, kgolo (South Sotho)

Smyrniopsis Boiss. Umbelliferae

Origins:

Resembling *Smyrnium* L.

Smyrnium L. Umbelliferae

Origins:

Smyrnion, from the Greek name *smyrna* "myrrh," referring to the strong smell; Latin *smyrnion* or *zmyrnium*, *ii* applied by Plinius to a kind of herb like myrrh, the common alexanders; see Carl Linnaeus, *Species Plantarum*. 262. 1753 and *Genera Plantarum*. Ed. 5. 127. 1754.

Species/Vernacular Names:

S. olusatrum L.

English: alexanders

Smythea Seemann ex A. Gray Rhamnaceae

Origins:

For a Col. Smythe; see Stafleu and Cowan, *Taxonomic Literature*. 5: 708. Utrecht 1985.

Snowdenia C.E. Hubbard Gramineae

Origins:

After the British (b. Silverdale, Staffs.) botanist Joseph Davenport Snowden, 1886-1973, Kew gardener, economic botanist, traveler and plant collector (Uganda), wrote *The Cultivated Races of Sorghum*. London 1936; see T.W. Bossert, *Biographical Dictionary of Botanists Represented in the Hunt Institute Portrait Collection*. 375. 1972; Ida Kaplan Langman, *A Selected Guide to the Literature on the Flowering Plants of Mexico*. University of Pennsylvania Press, Philadelphia 1964; Ray Desmond, *Dictionary of British & Irish Botanists and Horticulturists*. 641-642. London 1994; R. Zander, F. Encke, G. Buchheim and S. Seybold, *Handwörterbuch der Pflanzennamen*. 14. Aufl. Stuttgart 1993.

Sobennikoffia Schlechter Orchidaceae

Origins:

After Mrs. Rudolf Schlechter, whose maiden name was Sobennikoff.

Sobolewskia M. Bieb. Brassicaceae

Origins:

For the Russian botanist Gregor Fedorovitch (Grigoriy Fedorowich) Sobolewsky (Sobolewski), 1741-1807, physician, professor of botany; see J.H. Barnhart, *Biographical Notes upon Botanists*. 3: 299. 1965; E.M. Tucker, *Catalogue of the Library of the Arnold Arboretum of Harvard University*. Cambridge, Massachusetts 1917-1933.

Sobralia Ruíz & Pavón Orchidaceae

Origins:
Ruíz & Pavón named the genus after the Spanish physician Dr. Don Francisco Martínez Sobral, d. 1799, a promoter of botany. He was Physician at the Royal Court of the Spanish King, Director of the Botanical Gardens of Madrid.

Species/Vernacular Names:
S. dichotoma Ruíz & Pav.
Spanish: flor de paraíso
Peru: inquil, monte azucena, tahua-tahua
S. macrantha Lindl.
English: bamboo orchid, cane orchid

Socotora Balf.f Asclepiadaceae (Periplocaceae)

Origins:
From Socotra Island.

Socotranthus Kuntze Asclepiadaceae (Periplocaceae)

Origins:
From Socotra Island, Indian Ocean.

Socratea Karsten Palmae

Origins:
Named for the Greek philosopher Socrates (Sokrates).

Species/Vernacular Names:
S. exorrhiza (Mart.) H. Wendl.
English: paxiuba palm
Brazil: paxiubá, castiçal, pona, baxiúba, coqueiro-acunã, pachiuba, pachiubeira, sacha pona
Venezuela: palma de cacho

Sodiroa E.F. André Bromeliaceae

Origins:
For the Italian botanist Luigi (Aloysius, Aloisio, Luis) Sodiro, 1836-1909, clergyman (Jesuit), 1870 Ecuador, professor of botany.

Sodiroella Schlechter Orchidaceae

Origins:
After the Italian botanist Luigi Sodiro, 1836-1909, clergyman, 1870 to Ecuador, professor of botany, botanical explorer, author of *Cryptogamae Vasculares Quitenses*. Quiti 1893; see J.H. Barnhart, *Biographical Notes upon Botanists*. 3: 299. 1965; Stafleu and Cowan, *Taxonomic Literature*. 5: 713-717. 1985; T.W. Bossert, *Biographical Dictionary of Botanists Represented in the Hunt Institute Portrait Collection*. 375. 1972; E.M. Tucker, *Catalogue of the Library of the Arnold Arboretum of Harvard University*. Cambridge, Massachusetts 1917-1933; S. Lenley et al., *Catalog of the Manuscript and Archival Collections and Index to the Correspondence of John Torrey*. Library of the New York Botanical Garden. 379. 1973; H.N. Clokie, *Account of the Herbaria of the Department of Botany in the University of Oxford*. 246. Oxford 1964; R. Zander, F. Encke, G. Buchheim and S. Seybold, *Handwörterbuch der Pflanzennamen*. 14. Aufl. Stuttgart 1993.

Soehrensia Backeb. Cactaceae

Origins:
For Johannes Soehrens, d. 1934, an authority on Chilean cacti, Director of the Santiago Botanical Garden in Chile; see Gordon Douglas Rowley, *A History of Succulent Plants*. Strawberry Press, Mill Valley, California 1997.

Solanastrum Fabr. Solanaceae

Origins:
Resembling *Solanum* L.

Solandra Sw. Solanaceae

Origins:
For the Swedish (b. Piteå, Norrland) botanist Daniel Carl(sson) Solander, 1733-1782 (d. London), naturalist, traveler and plant collector, circumnavigator, friend and companion to Joseph Banks, 1750-1759 a pupil of Linnaeus at the University of Uppsala, 1764 Fellow of the Royal Society, 1768-1771 accompanied Joseph Banks on Captain Cook's first voyage of exploration in the ship *Endeavour*, from 1771 Bank's secretary and librarian, from 1773 Keeper of the Natural History Department of the British Museum (London), author of *Caroli Linnaei Elementa botanica*. 1756, with John Ellis wrote *The Natural History of Many Curious and Uncommon Zoophytes*, collected from various parts of the globe by the late John Ellis, etc. London 1786; see Sir Joseph Banks (1743-1820) and Daniel Solander, *Illustrations of Australian Plants Collected in 1770 during Captain Cook's Voyage Round the World in HMS Endeavour* ... with determinations by James Britten. Trustees of the British Museum. London 1900-1905; John Fothergill (1712-1780), *Some Accounts of the Cortex Winteranus, or*

Magellanicus. London 1779; Gustavus Brander (1720-1787), *Fossilia Hantoniensia* collecta, et in Musaeo Britannico deposita. London 1766; [Bishop Uno von Troil], *Bref rorende en Resa til Island in aaren 1772.* [A collection of letters describing a voyage to Iceland headed by Joseph Banks and Daniel Solander in 1772] Uppsala 1777; Stafleu and Cowan, *Taxonomic Literature.* 5: 721-723. 1985; J.H. Barnhart, *Biographical Notes upon Botanists.* 3: 300. 1965; Roy A. Rauschenberg, in *Dictionary of Scientific Biography* 12: 515-517. 1981; R. Zander, F. Encke, G. Buchheim and S. Seybold, *Handwörterbuch der Pflanzennamen.* 14. Aufl. 783. 1993; M. Hadfield et al., *British Gardeners: A Biographical Dictionary.* London 1980; F.A. Stafleu, *Linnaeus and the Linnaeans.* The spreading of their ideas in systematic botany, 1735-1789. 1971; Blanche Henrey, *British Botanical and Horticultural Literature before 1800.* Oxford 1975; Dennis John Carr and S.G.M. Carr, eds., *People and Plants in Australia.* 1981; Mary Gunn and Leslie E. Codd, *Botanical Exploration of Southern Africa.* 328. [b. 1736] Cape Town 1981; T.W. Bossert, *Biographical Dictionary of Botanists Represented in the Hunt Institute Portrait Collection.* 376. 1972; A. Lasègue, *Musée botanique de Benjamin Delessert.* 1845; R. Glenn, *The Botanical Explorers of New Zealand.* Wellington 1950; Ethelyn Maria Tucker, *Catalogue of the Library of the Arnold Arboretum of Harvard University.* Cambridge, Massachusetts 1917-1933; Jonas C. Dryander, *Catalogus bibliothecae historico-naturalis Josephi Banks.* London 1800; E. Bretschneider, *History of European Botanical Discoveries in China.* Leipzig 1981; Edmund Berkeley and Dorothy Smith Berkeley, *Dr. Alexander Garden of Charles Town.* University of North Carolina Press [1969]; James Britten and George E. Simonds Boulger, *A Biographical Index of Deceased British and Irish Botanists.* London 1931; T.F. Cheeseman (1846-1923), *Manual of the New Zealand Flora.* [collections Banks and Solander] Wellington 1906; W. Darlington, *Memorials of John Bartram and Humphry Marshall.* Philadelphia 1849; N. Hall, *Botanists of the Eucalypts.* Melbourne 1978 and Supplement 1980; Harold B. Carter, *Sir Joseph Banks (1743-1820). A Guide to Biographical and Bibliographical Sources.* Winchester 1987; Mia C. Karsten, *The Old Company's Garden at the Cape and Its Superintendents:* involving an historical account of early Cape botany. Cape Town 1951; John Dunmore, *Who's Who in Pacific Navigation.* 9-10, 231. University of Hawaii Press, Honolulu 1991.

Species/Vernacular Names:

S. sp.

Mexico: lipa-ca-tu-hue

S. grandiflora Sw. (*Solandra nitida* Zuccagni; *Swartzia grandiflora* (Swartz) J.F. Gmel.)

English: silver cup, cup of gold

Spanish: copa de oro, bolsa de Judas

Mexico: cútacua, ndari, tecomaxóchitl, tecomaxóchil, tetona, hueytlaca, hueypatli

Peru: mendieta

S. guttata D. Don (*Swartzia guttata* (D. Don ex Lindl.) Standl.)

English: goldcup, cup of gold, trumpet plant

Spanish: copa de oro, floripundio del monte, floripondio del norte

Mexico: cútacua, perilla, tecomaxóchitl

Solanecio (Sch. Bip.) Walp. Asteraceae

Origins:
Referring to the genera *Solanum* and *Senecio*.

Solanocharis Bitter Solanaceae

Origins:
The genus *Solanum* and *charis* "delight, grace, beauty."

Solanopsis Bitter Solanaceae

Origins:
Resembling *Solanum*.

Solanopteris Copel. Polypodiaceae

Origins:
The genus *Solanum* and *pteris* "fern," some suggest from Latin *solanus* "the east wind."

Species/Vernacular Names:
S. bifrons (Hook.) Copel.
English: potato fern

Solanum L. Solanaceae (Part I)

Origins:
Latin *solanum*, a plant name used probably for *Solanum nigrum* (Aulus Cornelius Celsus, *De medicina.* Florentiae 1478), *solanum* applied by Plinius to a plant, also called *strychnos*, nightshade, from the Latin *sol, solis* "sun," the plants of the sun, see also *solago, inis* "a plant called also heliotropium"; many authors suggest from the Latin *solor, aris, atus sum, ari* "to soothe," in allusion to the soothing properties of the plant, *solamen, inis* "a comfort, consolation," Akkadian *sululu* "happy, glad," *elelu* "jubilation, happiness,"

sulu "to move objects to a higher location," *elu, alu* "to come up, to rise." See Carl Linnaeus, *Species Plantarum*. 184. 1753 and *Genera Plantarum*. Ed. 5. 85. 1754; G. Semerano, *Le origini della cultura europea*. Dizionario della lingua Latina e di voci moderne. 2(2): 570. 1994; M. Lessona & C.A. Valle, *Dizionario universale di scienze, lettere ed arti*. Milano 1875; Manlio Cortelazzo & Paolo Zolli, *Dizionario etimologico della lingua italiana*. 5: 1222. [from the Latin *sol, solis* "sun"] 1988; Salvatore Battaglia, *Grande dizionario della lingua italiana*. XIX: 297. [from the Latin *sol, solis* "sun"] 1997; James A. Baines, *Australian Plant Genera*. An Etymological Dictionary of Australian Plant Genera. 344-345. 1981; H. Genaust, *Etymologisches Wörterbuch der botanischen Pflanzennamen*. 593. Basel 1996; National Research Council, *Lost Crops of the Incas: Little-Known Plants of the Andes with Promise for Worldwide Cultivation*. National Academy Press, Washington, D.C. 1989; William W. Megenney, *A Bahian Heritage*. University of North Carolina at Chapel Hill 1978; Regina Harrison, *Signs, Songs, and Memory in the Andes*. Translating Quechua Language and Culture. 172-195. University of Texas Press, Austin 1989.

Species/Vernacular Names:

S. sp.

Mexico: papa cimarrona, papa coyote, he-beech, huitzcolotli, pichecua, pitzecua, rerósuacua

Peru: apalo, aya mullaca, cocona flotadora, coconilla colorada, cubiu, cuyaca, gombruisha hormis, erva moura, hormis, huyrumsha, intuto huasca, juá jurubeba, komlu, macusari, mashu ricrou, mashu ricron, morera, mullaca, sacha gaito, tituna sacha, veneno huasca

Brazil: jurubeba-grande

Yoruba: ikan pupa, igbagba

S. acaule Bitter

Peru: atoj-papa, atokcpa papan, añaspa-papan, aya-papa

S. acroglossum Juz.

Peru: papita de zorro

S. aculeastrum Dun.

English: goat apple, bitter apple, apple of Sodom, devil's apple, goat bitter apple, poison-apple

Southern Africa: bokappel, bitterappel, bokbitterappel, gifappel, doodenappel; murulwa (Venda); dungwiza, muTura (Shona); shulwa (Venda); umThuma (for all kinds of *Solanum*) (Xhosa)

S. Rhodesia: muTura, dungwiza

S. aculeatissimum Jacquin

English: apple of Sodom, devil's apple, Sodom apple, soda-apple nightshade

Brazil: arrebenta-cavalo

Southern Africa: dungwiza (Shona); thola (Sotho); umthuma (Xhosa); umthuma (Zulu)

China: ka xi qie

Nepal: kantageri

S. albidum Dunal

Peru: anticristo, enves blanca, espina blanca, huaritar, yúrac huacta, yura huasa, envés blanco, yurachuacta, yurahuasa

Bolivia: lavaplato plateado

S. amblophyllum Hook.f.

Peru: shuruco, shambo cimarrón, shushucu

S. americanum Miller (*Solanum fauriei* H. Lév.)

English: glossy nightshade, American black nightshade

Hawaii: popolo, 'olohua, polopolo, popolohua

China: shao hua long kui

Yoruba: egunmo, egunmo agunmo gara

S. anomalum Thonn.

Yoruba: ikan-yanrin, ikan-yinrin, igba-yinrin elegun, igba-nyere

S. angustialatum Bitter

Peru: viuda shambu

S. aviculare Forst.f.

English: kangaroo apple, bullibulli

New Zealand: poroporo, kohoho (Maori names)

S. barbisetum Nees

English: spine-bract nightshade

China: ci bao qie

S. bicolor Roemer & Schultes

Peru: toé mullaca

S. brownii Dunal

English: violet nightshade

S. x *burbankii* Bitter

English: sunberry, wonderberry

S. campanulatum R. Br.

English: prickle-cup

S. capense L.

English: nightshade

Southern Africa: isithumana (Zulu); monyaku (South Sotho); umthumana (Xhosa)

S. capsicastrum Link ex Schauer

English: false Jerusalem cherry, winter cherry

S. capsiciforme (Domin) Baylis

English: capsicum kangaroo apple

S. capsicoides All. (*Solanum ciliatum* Lam.; *Solanum bodinieri* H. Léveillé & Vaniot)

English: cockroach berry, devil's apple

Hawaii: kikania lei, akaaka, akaka

China: niu qie zi

S. caricaefolium Rusby

Bolivia: lavaplato

S. carolinense L.

English: horse nettle, ball nightshade, Carolina horse nettle, Carolina nettle

S. chenopodinum F. Muell.

English: goosefoot potato-bush, goosefoot nightshade

S. chenopodioides Lam. (*Solanum gracile* Dun.)

English: black berry, black nightshade, white tip nightshade

Southern Africa: msobosobo, sobosobo (Xhosa); msobosobo, sobosobo (Zulu)

S. cinereum R. Br.

Australia: Narrawa burr (from Narrawa, Crookwell, New South Wales)

S. citrullifolium A. Braun

English: melon-leaf nightshade

S. coactiliferum J. Black

English: western nightshade

S. coccineum Jacq.

Southern Africa: kleingrysbitterappeltjie, slangappeltjie; morolwana-o-mornye (Tswana)

S. dasyphyllum Schum. & Thonn.

Yoruba: ikanelepon agbo, bamoni, mafowobamomi, bobo, bobo awodi, mafi owo ba omo mi

S. dimidiatum Raf.

English: Torrey's nightshade, robust horse nettle

S. dulcamara L.

English: bittersweet, climbing nightshade, deadly nightshade, poisonous nightshade, woody nightshade, bittersweet nightshade

Arabic: hulwa murra

Italian: dulcamara

Brazil: dulcamara, doce-amergo, dalcamaro, uva-de-cão, erva-moura-de-trepa, vinha-da-índia

Peru: asna panga

Mexico: dulcamara, flor de gloria, gloria, jazmín morado

India: kakmachi, rubabarik, bhalu-mash, anabessalib

China: ku qie, ku chieh, shu yang chuan

S. duplosinuatum Klotzsch

Sierra Leone: kwao gbolo

S. elaeagnifolium Cav. (*Solanum flavidum* Torr.)

English: tomato weed, bitter apple, silver-leaf bitter apple, silver-leaved nettle, silver-leaf nettle, silver-leaved nightshade, silver-leaf nightshade, silver nightshade, white horse nettle, bull nettle, silver horse nettle

Mexico: trompillo, tomatillo, guajtomate, tomatillo del campo, tomatito, buena mujer, pera, tomatito de buena mujer

South Africa: bitterappel, bloubos, satansbos, silwerblaar-bitterappel

Japan: gumi-ba-nasu

S. elegans Poiret

English: spiny kangaroo apple

S. ellipticum R. Br.

English: velvet potato bush

S. erianthum D. Don

English: mullein nightshade, potato tree, tobacco tree

Mexico: berenjena, sacamanteca

Japan: yanbaru-nasubi, tabakugii

China: jia yan ye shu

Vietnam: ngoi, ca hoi, co sa lang

The Philippines: malatabako, ungali, pangau, kasungog, kayok, liuangkag, malatalong, noog-noog, saca manteca, salvadora, hierba de San Pedro

Yoruba: ewuro ijebu, ijebu kogbin, openiniwuni

S. esuriale Lindley

Australia: quena, dwarf nightshade, wild tomato, tomato bush, potato bush, potato weed

S. excisirhombeum Bitter (*Solanum atriplicifolium* Nees)

Peru: hierba de paucar, japichina, kacharichingachari, muyuccaya, ckapichinya, muyacasacha

S. furcatum Dunal (*Solanum arequipense* Bitter)

Peru: ccjaya-ccjaya, manca tojiechi, gapichinia

S. geniculatum E. Meyer

Southern Africa: umThuma (Xhosa)

S. giganteum Jacq. (*Solanum niveum* Thunb.)

English: healing-leaf tree, red bitter-apple, giant bitter-apple, goat bitter-apple, red bittery berry, African holly

Southern Africa: geneesblaarboom (= healing-leaf tree), geneesblaarbos, groot bitterappel, rooibitterbessie; iCuba lasendle (Xhosa)

S. gomphodes Dunal

Bolivia: capuca

S. goodspeedii Roe

Bolivia: yobrini, tabaquillo

S. grandiflorum Ruíz & Pav.

Brazil: fruta-do-lobo

Peru: balsa, mite, bazo, hígado

S. griffithii (Prain) C.Y. Wu & S.C. Huang

English: Griffith nightshade

China: mo e qie

S. hermannii Dun. (*Solanum sodomeum* L. var. *hermannii* (Dun.) Dun.; *Solanum sodomeum* non L.)

English: bitter-apple, Sodom's apple, apple of Sodom

South Africa: bitterappel, gifappel, doodenappel, dortelappel, Sodomsappel

S. hispidum Persoon

English: devil's fig

Peru: campu cassa, huachulla, huirca cassa, huirca san, huircassa, koyrumsha

Mexico: sosa, berenjena

S. hystrix R. Br.

English: Afghan thistle, prickly potato weed, porcupine solanum

S. immite Dunal (*Solanum mathewsii* Bitter)

Peru: papa

S. incanum L. (*Solanum sanctum* L.)

English: bitter-apple, grey bitter-apple, thorn apple, Sodom apple

Arabic: mazg, mazi

East Africa: entengotengo, mutongu, mtula, mtulwa, mtunguja mwitu, ochok, omoratora

Yoruba: igba aja, igba igun, igbo, igba, igba oyinbo, ikan oyinbo, ikan, ikan nla

South Africa: thorn apple, bitterappel, gifappel, grysbitterappel; dinjinsa, Dungwiza, muDulukwa (Shona); morola (Pedi); thola (South Sotho); thola (Tswana); umdulukwa (Ndebele)

S. Rhodesia: uimDulukwa

S. incompletum Dunal

Hawaii: popolo ku mai, popolo

S. insidiosum Martius

Brazil: jurubeba, jurumbeba, juribeba, juripeba, jubeba, jumbeba, juveva, juvena, juúna, joatica

S. insolaesolis Bitter (*Solanum violaceistriatum* Bitter; *Solanum medianiviolaceum* Bitter)

Peru: papa

S. integrifolium Poir.

English: Chinese scarlet egg-plant, fruited egg-plant, scarlet egg-plant, tomato-fruited egg-plant

S. jamaicense Miller

Peru: coronilla

S. jamesii Torr.

English: wild potato

S. japonense Nakai

English: Japanese nightshade

China: ye hai qie

S. jasminoides Paxton

English: potato vine, potato climber, jasmine nightshade

Peru: jazmín

Mexico: gloria, jazmín, jazmín morado

S. kitagawae Schönbeck-Temesy

English: Kitagawa nightshade

China: guang bai ying

S. kwebense N.E. Br. (*Solanum chondropetalum* Damm.; *Solanum luederitzii* Schinz; *Solanum tenuiramosum* Damm.; *Solanum upingtoniae* Schinz)

South Africa: rooibessie, bitterappel

S. laciniatum Aiton

English: kangaroo apple, large kangaroo apple, large-flowered kangaroo apple, cut-leaved nightshade

China: ao zhou qie

S. lacunarium F. Muell.

English: lagoon nightshade

S. lanceolatum Cav.

English: lance-leaf nightshade

S. lasiophyllum Dunal ex Poir.

English: flannel bush

S. leucopogon Huber

Peru: cuburuchi, intuto quiro, cubunuchi

S. linearifolium Symon

English: mountain kangaroo apple

S. linnaeanum Hepper & P. Jaeger (*Solanum sodomeum* L.)

English: apple of Sodom, yellow popola, Dead Sea apple, Sodom apple

French: pomme de Sodome

Arabic: lim nsara

Hawaii: popolo kikania, yellow fruited popolo, thorny popolo

S. luteo-album Persoon

Peru: ama de casa, pajarito, tunya-tunya, amacasa, yujangmis

S. luzoniense Merrill

English: Luzon nightshade

China: lü song qie

S. lyratum Thunberg

English: bittersweet, climbing nightshade

China: bai ying, bai mao teng

Solanum L. Solanaceae (Part II)

Origins:

Latin *solanum*, a plant name used probably for *Solanum nigrum* (Aulus Cornelius Celsus, *De medicina*. Florentiae 1478), *solanum* applied by Plinius to a plant, also called *strychnos*, nightshade, from the Latin *sol, solis* "sun," the plants of the sun, see also *solago, inis* "a plant also called heliotropium"; many authors suggest from the Latin *solor, aris, atus sum, ari* "to soothe," in allusion to the soothing properties of the plant, *solamen, inis* "a comfort, consolation," Akkadian *sululu* "happy, glad," *elelu* "jubilation, happiness," *sulu* "to move objects to a higher location," *elu, alu* "to come up, to rise." See Carl Linnaeus, *Species Plantarum*. 184. 1753 and *Genera Plantarum*. Ed. 5. 85. 1754; G. Semerano, *Le origini della cultura europea*. Dizionario della lingua Latina e di voci moderne. 2(2): 570. 1994; M. Lessona & C.A. Valle, *Dizionario universale di scienze, lettere ed arti*. Milano 1875; Manlio Cortelazzo & Paolo Zolli, *Dizionario etimologico della lingua italiana*. 5: 1222. [from the Latin *sol, solis* "sun"] 1988; Salvatore Battaglia, *Grande dizionario della lingua italiana*. XIX: 297. [from the Latin *sol, solis* "sun"] 1997; James A. Baines, *Australian Plant Genera*. An Etymological Dictionary of Australian Plant Genera. 344-345. 1981; H. Genaust, *Etymologisches Wörterbuch der botanischen Pflanzennamen*. 593. Basel 1996; National Research Council, *Lost Crops of the Incas: Little-Known Plants of the Andes with Promise for Worldwide Cultivation*. National Academy Press, Washington, D.C. 1989; William W. Megenney, *A Bahian Heritage*. University of North Carolina at Chapel Hill 1978; Regina Harrison, *Signs, Songs, and Memory in the Andes*. Translating Quechua Language and Culture. 172-195. University of Texas Press, Austin 1989.

Species/Vernacular Names:

S. macaonense Dunal

English: Macao nightshade

China: shan qie

S. macrocarpon L.

Yoruba: osun, osun bogo etido, igba osun, efo osun

Congo: ngabi

S. mammosum L. (*Solanum globiferum* Dun.)

English: lady nipples, nipple fruit, macaw bush, papillate nightshade

Mexico: berenjena, berenjenita peluda, berenjenita peludita, cuchito, chichiguita, coión de gato, chichigüita, cojón de gato

Peru: coconilla dulce, tinta uma, tintoma, tintonilla, tintuma, veneno

Brazil: peito-de-moça, jurubeba

Japan: tsuno-nasu

China: ru qie

S. marginatum L.f. (*Solanum abyssinicum* Jacq. ex Vitm.)

English: white-margined nightshade

Bolivia: cardo santo

Mexico: sosa

S. mauritianum Scop. (*Solanum auriculatum* Aiton; *Solanum carterianum* Rock)

English: bug weed, bug berry, bug tree, wild tobacco tree, wild tobacco bush, Mauritius nightshade

Madagascar: ambiatibe, seva, sevabe, voaseva, tambakoke, tsiariboamena, voampoabe, tabac marron

Southern Africa: grootbitterappel, luisboom, luisbos; iBongabonga, isiGwayana (Zulu)

Hawaii: pua nana honua

Portuguese: tabaqueira

French: bois de tabac

S. melanocerasum All.

English: garden huckleberry, huckleberry

S. melongena L.

English: aubergine, egg-plant, Jew's apple, mad apple, garden egg-plant

Mexico: berenjena

Peru: berenjena, berengena, melongena

Japan: nasu, nâshibi

China: qie, qie zi, chieh, lo su, kun lun kua (= Kunlun melon)

The Philippines: talong, tarong, tolung, berengena

Malaya: terong, brinjal

South Laos: (people Nya Hön) plää trob, plää trob cäh, plää trob lo'ruay, plää trob dang, plää trob dway, plää trob glää iyan, plää trob kuen, plää trob lôb

India: vartaku, bartaku, peetaphala, begun, baigun, bhanta, bhantaki, baigan, baigana, wangan, vengan, bengan, badangan, vayinge, vayingana, wangi, vankayi, vankaya, vanga, katterikayi, kaththiri, valutina, valuthina, mulukutakali, badanekayi

Italian: melanzana

S. montanum L.

Peru: monte papa, papa de montaña, papas de lomas, sacha papa, tomado de Kichua

S. muricatum Aiton (see Jaime Prohens et al., "The Pepino (*Solanum muricatum*, Solanaceae): A "new" crop with a history." in *Economic Botany*. 50(4): 355-380. 1996)

English: melon pear, melon shrub, pepino, pear melon, Peruvian pepino, tree melon, sweet cucumber, sweet pepino, mellowfruit

Peru: cachum, cachuma, cachun, jem, kachan, kachuma, mata serrano, pepino, pepino dulce

Mexico: pepino de maceta

Latin America: pepino, pepino dulce, huevo de gato, manguena, mataserrano, pera melón, melón pera, pepino amarillo, pepino blanco, pepino de agua, pepino de chupar, pepino de fruta, pepino de las Indias, pepino de la tierra, pepino mango, pepino morado, pepino redondo

China: xiang gua quie

S. myrianthum Rusby

Bolivia: lavaplato

S. nelsonii Dunal (*Solanum laysanense* Bitter)

Hawaii: popolo, 'akia

S. nemorense Dunal

Peru: rocotito del monte, shina xocon, tuxa raho

S. nigrum L.

English: woody nightshade, black berry nightshade, nightshade, black nightshade, common nightshade, deadly nightshade, garden nightshade, duscle, hound's berry, inkberry, petty morel, stubbleberry, black berry, poison berry

Arabic: 'enab el-deib, eneb edhib, 'enab ed-dib, 'inab ed-dib, tmatem kleb, tmitma

Italian: solano nero, morella, erba morella, solatro, erba puzza

Peru: ccajaya-ccajaya, ccaya-ccaya, kaya-kaya, cajaya-ccjaya, ñuchcu, mata gallina, pilli yuyu, hierba mora, yerba mora

Brazil: erva-moura, suê, carachichu, aguaraquiá, maria-preta, pimenta-da-galinha, erva-de-bicho, guariguinha

East Africa: egwangira, managu, mnavu, ol'momoit, osuga, rinagu

Tanzania: nanjoni, mnafu

South Africa: mupya, thotho

Southern Africa: galbessie, inkbossie, nachtschade, nagskaal, nagskade, nagskaalbossie, nagtegaalbossie, nasgalbossie, nastergal, sobosobo, wildenastergal; ixabaxaba (Ndebele); musaka (Shona); seshoa-bohloko (Sotho); umsobo (Swati); umsobosobo (Xhosa); umsobosobo (Zulu)

S. Rhodesia: muSaka, iXabaxaba

Mexico: chichiquelite, hierba mora, veneno de cuervo, bahab-kan, bahalkan, ich-kan, pak'al-kan, bi-tache, bitaxe, la-bithoxi, pettoxe, pittoxe, pitoxe, pitoxi, pittoxi, laa pithoxe, vishate, chichiquilitl, chuchilitas, ha-mung, ixcapul, mambia, maniloche, mutztututi, pahalkan, tojonchichi, tonchichi, toniche, tojonechichi, tucupachexacua, tzopilotlacuatl, ichamal, itztonchichitzi

The Philippines: kunti, onti, kuti, amti, anti, muti, kamaka-matisan, gama-gamatisan, lubi-lubi, bolagtab, hulabhub, lagkakum, kamatis-manok, malasili, natang-ni-aso, nateng

Malaya: terong meranti, terong para chichit, terong perat

India: kakmachi, kakamachi, kachmach, kakmunchi, kamanchi, kamanchi-chettu, kanchi-pundu, kachi, makoi, mako, gurkamai, tulidun, anabusa-thaliba, kamuni, ghati, piludu, manattakkali, munna-takali-pullum, milagu-takkali, tudavalam, kambei

Nepal: paire golbhera

Japan: inu-hôzuki, uwâguwâ-kâtô

China: long kui, lung kuei, tien chieh tzu, tien pao tsao

S. nitidum Ruíz & Pav.

Peru: cahuincho, campu cassa, catruincho, huisca cassa, illaura, illauru, nununya, ñuñunga, ñuñua, ñuñuma, ñuñumea, ñuñuncca, ñuñunccai, ñuñunquia, ñce, tacachilla

S. nodiflorum Jacq.

English: black berry, black nightshade

Peru: ají, ayac mullaca, ccjaya-ccjaya, yerba mora

Southern Africa: inkbessie, nastergal; munyanya porini (Swahili)

S. nutans Ruíz & Pav.

Peru: campucassa, chuculate, huirsca cassa, huiscacassa

S. opacum A. Braun & Bouché

English: green-berry nightshade

S. pallidum Ruisby

Peru: achihuay

S. panduriforme E. Mey.

English: apple of Sodom, bitter apple, poison apple, yellow bitter apple

Southern Africa: bitterappel, geelappel, geelbitterappel, gifappel; morolane (Pedi); morolwana (Tswana); musaka, muTsungutsungu, muTuturwa (Shona); mtunguruja (Swahili); setlwane (South Sotho)

S. Rhodesia: nTuturwa

S. paniculatum L.

Brazil: jarumbeba, jurubeda, jurumbeba, juribeba, juripeba, jubeba, jumbeba, juveva, juvena, juúna, joatica, jupeta, urambeba, jurema, jaúna

S. pensile Sendtner

Bolivia: cashixopa

S. petrophilum F. Muell.

English: rock nightshade

S. pittosporifolium Hemsley

English: pittosporum-leaf bittersweet

China: hai tong ye bai ying

S. prinophyllum Dunal

English: forest nightshade

S. procumbens Loureiro

English: Hainan nightshade

China: hai nan qie

S. pseudocapsicum L.

English: false capsicum, Jerusalem cherry, Madeira winter cherry, Natal cherry, winter cherry

China: shan hu yin

Peru: tomate chino

Mexico: coral, collar de la reina, manzanita de amor

South Africa: bosgifappel, gifappel, gifbessie, wilderissie

S. pungetium R. Br.

English: eastern nightshade, jagged nightshade

S. quaesitum C. Morton

Bolivia: manzano

S. quitoense Lamarck

South America (Andes): naranjilla, lulo

Peru: cocona, lulo, naranjilla, naranjita de Quito, popo

S. radicans L.f.

Peru: cushay, cusmayllu, cuspallo, kusmaillo, kusmaillu

S. retroflexum Dun.

South Africa: nasgal, nastergal

S. rigescens Jacq. (*Solanum rubetorum* Dun.)

South Africa: wildelemoentjie

S. riparium Pers.

Bolivia: lavaplato

S. rostratum Dunal (*Solanum heterandrum* Pursh)

English: buffalo bitter apple, buffalo bur, hedgehog bush, sandbur, pincushion nightshade, buffalo berry, Kansas thistle

Mexico: duraznillo, hierba del sapo, huiztecuatl, acayocahuitle, ayohuistle, ayohuiztle, hierba del gato, mala mujer, rabo de iguana, churuni, ixpa-halkan, quiebra plato, vaquerillo

South Africa: bitterappel, buffelbitterappel, doringappel, ystervarkbos

S. sandwicense Hooker & Arnott (*Solanum hillebrandii* St. John; *Solanum kavaiense* (A. Gray) Hillebr.; *Solanum woahense* Dunal)

Hawaii: popolo 'aiakeakua, popolo

S. saponaceum Dunal

Peru: ama mullaca, casia muru, tululuque, jollmis, ulmish

S. sarrachoides Sendtn.

English: hairy nightshade

S. schlechtendalianum Walp.

Bolivia: hediondilla

S. seaforthianum Andrews

English: potato creeper, St. Vincent lilac, glycine, Italian jasmine, Brazilian nightshade

Mexico: flor de gloria, gloria, guía de jazmincillo, lágrimas de la Virgen, piocha, lazo de amor

Japan: fusanari-tsuru-nasubi

China: nan qing qi

S. septemlobum Bunge

English: seven-lobed nightshade

China: qing qi

S. sessile Ruíz & Pav.

Bolivia: hediondilla

S. sessiliflorum Dunal (*Solanum topiro* Dun.)

Peru: cocona, coconilla, lulo topiro, cona, dyoxooro, kochari, lulo, topiro

S. simile F. Muell.

Australia: kangaroo apple, oondooroo (an Aborigines name, Cloncurry)

S. spirale Roxb.

English: coiled-flower nightshade

China: xuan hua qie

S. stelligerum Smith

English: devil's needles, star nightshade

S. sisymbriifolium Lam. (*Solanum balbisii* Dun.)

English: dense-thorned bitter apple, wild tomato, viscid nightshade, sticky nightshade

Brazil: juá, juá-ti, juciri

Peru: misqui corrota, ocote mullaca

South Africa: digdoringbitterappel, doringtamatie, tamatiedissel, wilde tamatie

China: suan jie qie

S. sturtianum F. Muell.

Australia: Sturt's nightshade, Thargomindah nightshade

S. subinerme Jacquin (*Solanum heterophyllum* Lam.)

Peru: ayac mullaca, ayoc

S. symonii H. Eichler. (*Solanum simile* F. Muell. var. *fasciculatum* (F. Muell.) Domin)

Australia: kangaroo apple, South Australian kangaroo apple

S. torvum Swartz

English: turkey-berry, spiny nightshade, water nightshade

China: shui qie

Malaya: terong mangas, terong pipit, terong rajawali, terong rembang

India: titbaigun, hathibhekuri, sundai, kondavuste, kattuchunta

Mexico: sosa, berenjena, amasclanchi, conoca, che'el-ik, tompaap, friega platos, prendedora, tapdacui, doc'a, berenjenita cimarrona

Yoruba: ikan wewe, ikan igun, igba yinrin elegun

S. triflorum Nutt.

English: three-flowered nightshade, cut-leaf nettle, cut-leaf nightshade, three-flowered nettle, wild tomato

S. tuberosum L. (*Solanum goniocalyx* Juz. & Buk.)

English: Irish potato, potato, white potato, Zulu-potato

French: pomme de terre

Mexico: papa, jroca, nyami-tecuinti, rerogüe, rerohue, rir-oui, ri'rohui, ta'upu'u, xojat-hapec, papa correlona, pöpa

Peru: acacha, accsu, acshu, acso, akso, aksho, apalo, apharu, catzari, catzari tseri, cchoke, cchoque, cchoqque, chaucha, curao kara, kesia, maaona, mojaqui, mosaqui, mosaki, papa, patata, moy papa, papa de gentil, pua, quinqui, tseri

Brazil: amilo de batata

Arabic: batata

Southern Africa: aartappel, ertappel; litapole (Sotho); mazabane (Zulu)

India: alu, golalu, belathi-aloo, batata, papeta, urla-kalangu, uru-laikkizhangu, urlagadda

Japan: Jagatara-imo, jaga-imo

China: yang yu, yang shu, tu yu, tu luan, huang tu

Malaya: ubi benggala, ubi gendang, ubi kentang

Hawaii: 'uala kahiki

Maori name: taewa

S. umbelliferum Eschsch.

English: blue witch

S. uporo Dunal (*Solanum anthropophagorum* Seem.)

English: cannibal's tomato

Pacific Islands: boro dina, poroporo

S. urticans Dunal

Bolivia: manzanillo

S. wallacei (A. Gray) Parish

English: Catalina nightshade, Wallace's nightshade

S. wendlandii Hook.f.

English: potato vine, giant potato creeper, paradise flower, Wendland's nightshade

S. wrightii Bentham

English: Brazilian potato tree, potato tree, Wright nightshade

China: da hua qie

Yoruba: ewuro ijebu, ikan igun, igba yinrin elegun

S. xanti A. Gray (*Solanum cupuliferum* Greene)

English: purple nightshade

Solaria Philippi Alliaceae (Liliaceae)

Origins:

From the Latin *solaris, e* "solar, belonging to the sun."

Soldanella L. Primulaceae

Origins:

Perhaps from the Latin *solidus, soldus* "a gold coin, firm, compact," perhaps referring to the shape of the leaves; see Manlio Cortelazzo & Paolo Zolli, *Dizionario etimologico della lingua italiana.* 5: 1222. 1988; S. Battaglia, *Grande dizionario della lingua italiana.* XIX: 304. 1997; Helmut Genaust, *Etymologisches Wörterbuch der botanischen Pflanzennamen.* 593-594. Basel 1996.

Solea Sprengel Violaceae

Origins:

Dedicated to the British (b. Thetford, Cambridgeshire) botanist William Sole, 1741-1802 (d. Bath, Somerset), apothecary, author of *Mentha britannicae.* Bath 1798; see Heinrich Adolph Schrader (1767-1836), *Journal für die Botanik.* 1800(2): 192. (Oct.-Dec.) 1801; Dawson Turner and Lewis Weston Dillwyn (1778-1855), *The Botanist's Guide Through England and Wales.* 747. London 1805; John H. Barnhart, *Biographical Notes upon Botanists.* 3: 301. 1965; Blanche Henrey, *British Botanical and Horticultural Literature before 1800.* Oxford 1975; H.N. Clokie, *Account of the Herbaria of the Department of Botany in the University of Oxford.* 246. 1964; R. Zander, F. Encke, G. Buchheim and S. Seybold, *Handwörterbuch der Pflanzennamen.* 14. Aufl. 1993. Some suggest from the Latin *solea, ae* "a sandal, a sole," *solum* "ground, earth, soil, land."

Soleirolia Gaudich. Urticaceae

Origins:

Origin not very clear, possibly after the military man Joseph François Soleirol, 1781-1863, plant collector in Corsica; see I.H. Vegter, *Index Herbariorum.* Part II (6), *Collectors S.* Regnum Vegetabile vol. 114. 1986; John Dunmore, *Who's Who in Pacific Navigation.* Honolulu 1991; F. Boerner & G. Kunkel, *Taschenwörterbuch der botanischen Pflanzennamen.* 4. Aufl. 171. Berlin & Hamburg 1989; L.C.D. de Freycinet, *Voyage autour du Monde entrepris par ordre du Roi ... sur les corvettes de S.M. "L'Uranie" et "La Physicienne".* [see Charles Gaudichaud-Beaupré for the Botany of the Voyage.] 504. Paris 1826[-1830]; Esprit Requien, in

Annales des Sciences Naturelles. ["M. Soleirol, capitaine de génie militaire, qui a parcouru toute la Corse et fait un herbier considérable des plantes de cette île a trouvé cette jolie plante à Cervione, dans un lien ombragé ..."] 384. 1825.

Species/Vernacular Names:

S. soleirolii (Req.) Dandy (*Helxine soleirolii* Req.; *Parietaria soleirolii* (Req.) Sprengel)

English: mind-your-own-business, peace-in-the-home, touch-me-not, mother-of-thousands, baby's tears, Pollyanna vine, angel's tears, Irish moss, Japanese moss, Corsican curse, Corsican carpet plant, helxine

Solena Lour. Cucurbitaceae

Origins:
Greek *solen* "a tube, pipe," referring to the nature of the flowers.

Solena Willd. Rubiaceae

Origins:
Greek *solen* "a tube, pipe," referring to the flowers.

Solenachne Steudel Gramineae

Origins:
Greek *solen* "a tube, pipe" and *achne* "chaff, glume."

Solenandra (Reisseck) Kuntze Rhamnaceae

Origins:
Greek *solen* "a tube, pipe" and *aner, andros* "male, man, stamen"; see Arthur D. Chapman, ed., *Australian Plant Name Index.* 2688. Canberra 1991.

Solenandra Hook.f. Rubiaceae

Origins:
From the Greek *solen* and *aner, andros* "male, man, stamen," referring to the stamens, see also *Exostema* (Pers.) Bonpl.

Solenangis Schltr. Orchidaceae

Origins:
From the Greek *solen* and *angeion, aggeion* "a vessel, cup," perhaps referring to the narrowly conical spur.

Solenantha G. Don Rhamnaceae

Origins:
Greek *solen* "a tube, pipe" and *anthos* "flower," referring to the form of the flowers; see George Don (1798-1856), *A General History of the Dichlamydeous Plants.* 2: 22, 39. London 1832; Arthur D. Chapman, ed., *Australian Plant Name Index.* 2709. Canberra 1991.

Solenanthus Ledebour Boraginaceae

Origins:
Greek *solen* and *anthos* "flower," referring to the form of the corolla.

Species/Vernacular Names:
S. circinnatus Ledebour
China: chang riu liu li cao

Solenidiopsis Senghas Orchidaceae

Origins:
Resembling *Solenidium* Lindl.

Solenidium Lindley Orchidaceae

Origins:
The diminutive of the Greek *solen* "tube, box," referring to the form of the lip.

Soleniscia DC. Epacridaceae

Origins:
Greek *soleniskos* "a small tube, pipe," diminutive of *solen*; see A.P. de Candolle, *Prodromus.* 7(2): 737. 1839.

Solenixora Baillon Rubiaceae

Origins:
From the Greek *solen* "tube" and the genus *Ixora* L.

Solenocalyx Tieghem Loranthaceae

Origins:
From the Greek *solen* "tube" and *kalyx* "calyx."

Solenocarpus Wight & Arn. Anacardiaceae

Origins:
From the Greek *solen* "a tube, pipe" and *karpos* "fruit," referring to the nature of the fruits.

Solenocentrum Schltr. Orchidaceae

Origins:
Greek *solen* "a tube, pipe, canal" and *ken..on* "a spur, prickle," the spur on the lip.

Solenochasma Fenzl Acanthaceae

Origins:
From the Greek *solen* and *chasme* "gaping, yawning."

Solenogyne Cass. Asteraceae

Origins:
Greek *solen* and *gyne* "a woman, female," referring to the female tubular florets; see Alexandre Henri Gabriel Comte de Cassini, *Dictionnaire des Sciences Naturelles*. 56: 174. Paris 1828.

Solenomelus Miers Iridaceae

Origins:
From the Greek *solen* "a tube, pipe" and *melos* "limb, part, member," referring to the tubular perianth.

Solenophora Benth. Gesneriaceae

Origins:
From the Greek *solen* "a tube, pipe" and *phoros* "bearing," referring to the tubular corolla.

Solenophyllum Baillon Gramineae

Origins:
From the Greek *solen* "a tube, pipe" and *phyllon* "leaf."

Solenopsis C. Presl Campanulaceae

Origins:
From the Greek *solen* "a tube, pipe" and *opsis* "appearance."

Solenoruellia Baillon Acanthaceae

Origins:
From the Greek *solen* "a tube, pipe" plus the genus *Ruellia* L.

Solenospermum Zoll. Celastraceae

Origins:
Greek *solen* "a tube, pipe" and *sperma* "seed," referring to the shape of the seeds.

Solenostemma Hayne Asclepiadaceae

Origins:
Greek *solen* and *stemma, stemmatos* "garland, crown."

Solenostemon Thonning Labiatae

Origins:
Greek *solen* "a tube, pipe" and *stemon* "stamen," referring to the connate filaments; see Heinrich Christian Friedrich Schumacher (1757-1830), *Beskrivelse af Guineiske Planter som ere fundne af Danske Botanikere isaer af Etatsraad Thonning*. [Copenhagen 1828-1829].

Solenostigma Endlicher Ulmaceae

Origins:
Greek *solen* and *stigma* "stigma"; see Stephan L. Endlicher, *Prodromus florae Norfolkicae*. 41. Vindobonae [Wien] 1833.

Solfia Rech. Palmae

Origins:
See Natalie W. Uhl & John Dransfield, *Genera Palmarum*. 389. Allen Press, Lawrence, Kansas 1987; B.E.V. Parnham, *Plants of Samoa*. Wellington 1972.

Solidago L. Asteraceae

Origins:

Latin *solido, avi, atum, are* "to make firm, to make whole," *solidus* "whole, firm," decoctions were applied to cure wounds and ulcers; see Carl Linnaeus, *Species Plantarum*. 878. 1753 and *Genera Plantarum*. Ed. 5. 374. 1754.

Species/Vernacular Names:

S. altissima L. var. *altissima*

English: late goldenrod

S. californica Nutt.

English: California goldenrod

S. canadensis L.

English: goldenrod, Canadian goldenrod, Canada goldenrod

Japan: Kanada-aki-no-kirin-sô

S. canadensis L. ssp. *elongata* (Nutt.) Keck

English: Canada goldenrod

S. confinis Nutt.

English: southern goldenrod

S. gigantea Aiton

English: smooth goldenrod

S. guiradonis A. Gray

English: Guirado's goldenrod

S. multiradiata Aiton

English: northern goldenrod

S. spathulata DC. ssp. *spathulata*

English: coast goldenrod

S. spectabilis (D. Eaton) A. Gray

English: showy goldenrod

S. virgaurea L.

China: yi zhi huang hua, liu chi nu tsao

Solisia Britton & Rose Cactaceae

Origins:

For the Mexican botanist Octavio Solis.

Soliva Ruíz & Pavón Asteraceae

Origins:

After the Spanish botanist and medical man Salvador Soliva, physician to the Spanish Court, author of *Dissertacion sobre el Sen de España*. Pruebase como específicamente no es distinto del Alexandrino ú Oriental, y explícanse sus virtudes en la medicina ... A que se añade la lámina de la planta. Madrid 1774; see Hipólito Ruíz López (1754-1815) & José Antonio Pavón (1754-1844), *Flora peruvianae, et chilensis prodromus*. 113, t. 24. Madrid 1794.

Species/Vernacular Names:

S. anthemifolia (A.L. Juss.) Loudon (*Gymnostyles anthemifolia* A.L. Juss.)

English: dwarf jo-jo, button burweed

S. pterosperma (A.L. Juss.) Less. (*Gymnostyles pterosperma* A.L. Juss.)

English: jo-jo, bindyi

S. sessilis Ruíz & Pavón

English: jo-jo, bindyi, lawn burweed

S. stolonifera (Brot.) Loudon (*Gymnostyles stolonifera* Brot.)

English: jo-jo, carpet burweed

Sollya Lindley Pittosporaceae

Origins:

For the English (b. London) botanist Richard Horsman Solly, 1778-1858 (London), plant physiologist and anatomist, in 1807 a Fellow of the Royal Society and in 1826 of the Linnean Society, a friend of John Lindley; see John Lindley (1799-1865), *Edwards's Botanical Register*. 17, t. 1466. (Jan.) 1832; F. Boerner & G. Kunkel, *Taschenwörterbuch der botanischen Pflanzennamen*. 4. Aufl. 171. Berlin & Hamburg 1989; Georg Christian Wittstein, *Etymologisch-botanisches Handwörterbuch*. 821. Ansbach 1852.

Species/Vernacular Names:

S. heterophylla Lindley (*Billardiera fusiformis* Labill.; *Labillardiera fusiformis* (Labill.) Schultes; *Pronaya lanceolata* Turcz.; *Sollya fusiformis* (Labill.) Briq.; *Sollya fusiformis* Payer)

English: sollya, Australian bluebell, bluebell creeper

Solms-Laubachia Muschler Brassicaceae

Origins:

For the German botanist Hermann Maximilian Carl Ludwig Friedrich zu Solms-Laubach, 1842-1915, bryologist, professor of botany, plant collector, 1889-1908 co-editor of *Botanische Zeitung*, 1909-1915 co-founder and editor of *Zeitschrift für Botanik*; see John H. Barnhart, *Biographical Notes upon Botanists*. 3: 301. 1965; T.W. Bossert, *Biographical Dictionary of Botanists Represented in the Hunt Institute Portrait Collection*. 376. 1972; E.M. Tucker, *Catalogue of the Library of the Arnold Arboretum of Harvard University*. Cambridge, Massachusetts 1917-1933; S. Lenley et al., *Catalog of the Manuscript and Archival Collections and Index to the Correspondence of John Torrey*. Library of the New York Botanical Garden. 469. 1973; Stafleu and Cowan, *Taxonomic Literature*. 5: 729-735. Utrecht 1985; R. Zander, F. Encke, G. Buchheim and S. Seybold, *Handwörterbuch der Pflanzennamen*. 14. Aufl. Stuttgart 1993.

Solmsia Baillon Thymelaeaceae

Origins:

For the German botanist Hermann Maximilian Carl Ludwig Friedrich zu Solms-Laubach, 1842-1915.

Sommera Schltdl. Rubiaceae

Origins:

For C.N. Sommer, an entomologist; see Stafleu and Cowan, *Taxonomic Literature*. 5: 736. Utrecht 1985.

Sommerfeltia Lessing Asteraceae

Origins:

For the Norwegian botanist Søren (Sören, Severin) Christian (Severinus Christianus) Sommerfelt, 1794-1838, clergyman, author of *Supplementum Florae lapponicae quam edidit Dr. Georgius Wahlenberg*. Oslo 1826; see John H. Barnhart, *Biographical Notes upon Botanists*. 3: 302. 1965; T.W. Bossert, *Biographical Dictionary of Botanists Represented in the Hunt Institute Portrait Collection*. 376. 1972; A. Lasègue, *Musée botanique de Benjamin Delessert*. Paris 1845; E.M. Tucker, *Catalogue of the Library of the Arnold Arboretum of Harvard University*. Cambridge, Massachusetts 1917-1933.

Sommieria Beccari Palmae

Origins:

For the Italian botanist Carlo Pietro Stefano (Stephen) Sommier, 1842-1922, plant collector, traveler. His writings include *Le isole Pelagie*, Lampedusa, Linosa, Lampione e la loro flora, con un elenco completo delle piante di Pantelleria. Firenze 1908 and (with Alfredo Caruana Gatto, 1868-1926) *Flora Melitensis nova*. Firenze 1915; see John H. Barnhart, *Biographical Notes upon Botanists*. 3: 303. 1965; T.W. Bossert, *Biographical Dictionary of Botanists Represented in the Hunt Institute Portrait Collection*. 376. 1972; E.M. Tucker, *Catalogue of the Library of the Arnold Arboretum of Harvard University*. 1917-1933; R. Zander, F. Encke, G. Buchheim and S. Seybold, *Handwörterbuch der Pflanzennamen*. 14. Aufl. Stuttgart 1993.

Somphoxylon Eichler Menispermaceae

Origins:

From the Greek *somphos* "spongy, porous" and *xylon* "wood."

Sonchus L. Asteraceae

Origins:

Greek *sonchos, sogkos, sogchos, sonkos*, the sowthistle (Theophrastus); Latin *sonchus, i* used by Plinius for *Sonchus oleraceus* L., the herb sowthistle; see Carl Linnaeus, *Species Plantarum*. 793. 1753 and *Genera Plantarum*. Ed. 5. 347. 1754.

Species/Vernacular Names:

S. arvensis L.

English: perennial sowthistle

India: sahadevi-bari, sahadevibari, kalabhangra, bonpalang, banpalang, nalla-tapata, nallatapata, bhangra, birbarang

Tibetan: rgya-khur

China: niu she tou

S. asper (L.) Hill subsp. *asper* (*Sonchus asper* (L.) Hill s. str.)

English: common sowthistle, prickly sowthistle, rough sowthistle, annual sowthistle, spiny annual sowthistle, spiny sowthistle, milk thistle, sowthistle, thistle, wild thistle

Maori names: taweke, tawheke

Peru: casha ckaña, ccjana, ckaña, citucasha

South Africa: doringsydissel, gewone sydissel, sydissel

S. dregeanus DC. (*Sonchus ecklonianus* DC.)

English: sowthistle

Southern Africa: leharasoana (Sotho)

S. oleraceus L.

English: sowthistle, common sowthistle, annual sowthistle, milk thistle, thistle, wild thistle

French: laiteron commun

Peru: achcaña, canacho, ccashaccnaña, cerraja, eskana, kanapako, llampu-ccjana

Arabic: godeid, galawein, tifef, difef

East Africa: apuruku, ekijwamate, ekinnyahwa, ekinyamate, kakovu, lapuku, mahiu, orunyamate, riroria, shunga pwapwa

Southern Africa: gewone sydissel, melkdissel, pypdissel, seidissel, suigdissel, sydissel, tuindissel, wilde latuw, mahondzo, khondo, lehondo; lesabe (South Sotho); bonosa-lekhoaba (Sotho); ihahabe (Zulu); ihlaba (Xhosa); ingabe (Swati); lesese (Pedi); shashe (Venda)

India: mhatara, titaliya, dodak, ratrinta

Japan: no-geshi, mââfâ

China: ku cai, tu, ku tsai

Hawaii: pualele

Maori names: rauriki, pororua, puwha, tawheke

S. pinnatus Ait.

English: sow thistle

Sondera Lehmann Droseraceae

Origins:

After the German botanist Otto Wilhelm Sonder, 1812-1881 (d. Hamburg), pharmacist, botanical explorer and collector, joint author with William H. Harvey (1811-1866) of the first three volumes of *Flora capensis*. Dublin, Cape Town 1860-1865, co-editor of Johann G.C. Lehmann, *Plantae Preissianae*. Hamburgi 1844-1848. Among his many works are *Flora hamburgensis*. Hamburg 1851 [1850] and *Die Algen des tropischen Australiens*. [Hamburg 1871], collected algae; see John H. Barnhart, *Biographical Notes upon Botanists*. 3: 303. 1965; Dennis John Carr and S.G.M. Carr, eds., *People and Plants in Australia*. 1981; Mary Gunn and Leslie E. Codd, *Botanical Exploration of Southern Africa*. 328-329. A.A. Balkema Cape Town 1981; Michel Gandoger (1850-1926), "L'herbier africain de Sonder." *Bull. Soc. bot. France*. 60: 414-422, 445-462. 1913; Johann G.C. Lehmann, *Flora oder allgemeine Botanische Zeitung*. 27: 82. (Feb.) 1844; T.W. Bossert, *Biographical Dictionary of Botanists Represented in the Hunt Institute Portrait Collection*. 376. 1972; Ethelyn Maria Tucker, *Catalogue of the Library of the Arnold Arboretum of Harvard University*. Cambridge, Massachusetts 1917-1933; Gordon Douglas Rowley, *A History of Succulent Plants*. Strawberry Press, Mill Valley, California 1997; R. Zander, F. Encke, G. Buchheim and S. Seybold, *Handwörterbuch der Pflanzennamen*. 14. Aufl. 783. Stuttgart 1993; Leonard Huxley, *Life and Letters of Sir Joseph Dalton Hooker*. London 1918.

Sonderina Wolff Umbelliferae

Origins:

After the German botanist Otto Wilhelm Sonder, 1812-1881.

Sonderothamnus R. Dahlgren Penaeaceae

Origins:

After the German botanist Otto Wilhelm Sonder, 1812-1881.

Sondottia P.S. Short Asteraceae

Origins:

After the German botanist Otto Wilhelm Sonder, 1812-1881.

Sonerila Roxb. Melastomataceae

Origins:

From *soneri-la*, a Malayalam name used by van Rheede in *Hortus Indicus Malabaricus*. 9: t. 65. 1689, *suwarna* "beautifully colored" and *ila* "leaf," referring to the leaves of *Sonerila wallichii* Benn., or from *sona* "red" and *aralah* "bent, curved," referring to the leaves.

Sonneratia L.f. Sonneratiaceae (Lythraceae)

Origins:

After the French (b. Lyons) botanist Pierre Sonnerat, 1748-1814 (d. Paris), explorer, draughtsman, naturalist and traveler, natural historian, colonial administrator. He worked in Mauritius, Madagascar, Pondicherry, Malabar and Malacca, and is author of *Voyage à la Nouvelle Guinée*. Paris 1776 and *Voyage aux Indes Orientales et à la Chine*, fait par ordre du Roi, depuis 1774 jusqu'en 1781. Paris 1782; see Eugène Jacob de Cordemoy (1835-1911), *Flore de l'Ile de la Réunion*. Paris 1895 and *La médecine extra médicale à l'Ile de la Réunion*. St. Denis, Réunion 1864; Madeleine Ly-Tio-Fane, *The Career of Pierre Sonnerat*. London 1973; R. Zander, F. Encke, G. Buchheim and S. Seybold, *Handwörterbuch der Pflanzennamen*. 14. Aufl. 748. 1993; Madeleine Ly-Tio-Fane, in *Dictionary of Scientific Biography* 12: 535-538. New York 1981; John H. Barnhart, *Biographical Notes upon Botanists*. 3: 303. 1965; F. Boerner & G. Kunkel, *Taschenwörterbuch der botanischen Pflanzennamen*. 4. Aufl. 171. 1989; Joseph François Charpentier-Cossigny de Palma, *Lettre à Monsieur Sonnerat*. [Correcting some mis-statements in a work by the latter, *Voyage aux Indes Orientales*.] Ile de France 1784; Madeleine Ly-Tio-Fane, "Pierre Poivre et l'Expansion française dans l'Indo-Pacifique." *Bulletin de l'École Française d'Extrême Orient*. 53: 453-511. 1967; C. Linnaeus (filius), *Supplementum Plantarum*. 38, 252. (Apr.) 1782; Emil Bretschneider, *History of European Botanical Discoveries in China*. Leipzig 1981; Mary Gunn and Leslie E. Codd, *Botanical Exploration of Southern Africa*. 329-330. [b. 1745] Cape Town 1981; Carl Peter Thunberg (1743-1828), *Voyage en Afrique et en Asie*, principalement au Japon, pendant les années 1770-1779. Paris 1794; Jonas C. Dryander, *Catalogus bibliothecae historico-naturalis Josephi Banks*. London 1800; A. Lasègue, *Musée botanique de Benjamin Delessert*. Paris 1845; Ethelyn Maria Tucker, *Catalogue of the Library of the Arnold Arboretum of Harvard University.* Cambridge, Massachusetts 1917-1933.

Species/Vernacular Names:

S. alba Smith (*Sonneratia iriomotensis* Masam.)

English: white-flowered Pornupan mangrove

Japan: maya-pushiki

Malaya: perepat, gedabu, pepat

S. caseolaris (L.) Engl.

English: red-flowered Pornupan mangrove

Malaya: beremban

Sooia Pócs Acanthaceae

Origins:

After the Hungarian botanist Károly Rezsö Sóo, 1903-1980, professor of botany; see Stafleu and Cowan, *Taxonomic Literature.* 5: 747-748. 1985; John H. Barnhart, *Biographical Notes upon Botanists.* 3: 303. 1965; T.W. Bossert, *Biographical Dictionary of Botanists Represented in the Hunt Institute Portrait Collection.* 376. 1972; R. Zander, F. Encke, G. Buchheim and S. Seybold, *Handwörterbuch der Pflanzennamen.* 14. Aufl. Stuttgart 1993.

Sophora L. Fabaceae

Origins:

From the Arabian names *sophera* or *sufayra*; Linnaeus used first to describe *Sophora alopecuroides* L. (*Goebelia alopecuroides* (L.) Bunge ex Boiss.; *Vexibia alopecuroides* (L.) Yakovlev); see Carl Linnaeus, *Species Plantarum.* 373. 1753, and *Genera Plantarum.* Ed. 5. 175. 1754; F. Boerner & G. Kunkel, *Taschenwörterbuch der botanischen Pflanzennamen.* 4. Aufl. 171. 1989; R. Zander, F. Encke, G. Buchheim and S. Seybold, *Handwörterbuch der Pflanzennamen.* 14. Aufl. 524. 1993; S. Battaglia, *Grande dizionario della lingua italiana.* XIX: 274. UTET, Torino 1997; H. Genaust, *Etymologisches Wörterbuch der botanischen Pflanzennamen.* 596. 1996; G.C. Wittstein, *Etymologisch-botanisches Handwörterbuch.* 822. Ansbach 1852.

Species/Vernacular Names:

S. affinis Torrey & A. Gray

English: Texas sophora

S. alopecuroides L.

English: foxtail-like sophora

S. chrysophylla (Salisb.) Seemann (*Edwardsia chrysophylla* Salisb.)

Hawaii: mamani, mamane

S. flavescens Ait.

English: lightyellow sophora

China: ku shen

S. glauca Lesch.

English: greyblue sophora

S. inhambanensis Klotzsch (*Sophora nitens* Harv.) (the specific name indicates where the plant was collected, near Inhambane, between Lourenço Marques and Beira, southeastern Mozambique)

English: coast bean bush

Southern Africa: kusboontjiebos; iPhelenjane (Zulu)

S. japonica L. (*Sophora japonica* L. f. *pendula* Zabel; *Styphnolobium japonicum* (L.) Schott)

English: pagoda tree, sofora, Japanese pagoda tree, Chinese scholar tree, tree of success in life

Italian: robinia del Giappone

Argentina: acacia del Japon, sofora llorona, sofora pendula

Japan: emju, en-shu, enju (ju = tree)

China: huai hua, huai

Indochina: hoe

Vietnam: hoe, hoe me, hoe hoa

S. microphylla Aiton (*Sophora tetraptera* J.F. Miller var. *microphylla* (Aiton) Hook.f.)

New Zealand: kowhai

S. secundiflora (Ortega) Lagasca ex DC. (*Broussonetia secundiflora* Ortega)

English: mescalbean, mescalbean sophora, Texas-mountain-laurel

Latin America: frijolito

S. subprostrata Chun & Chen

English: pigeon pea

China: guang dou gen, shan dou gen

S. tetraptera Miller (*Edwardsia grandiflora* Salisb.)

English: the four wing sophora

New Zealand: kowhai (= the Maori's word for the yellow color of the flowers)

Peru: mayú, pelú, pilo

Chile: pelu, pilo

S. tomentosa L.

English: silverbush

Malaya: ki-koetjing, pelotok

Japan: iso-fuji, ikaki

S. toromiro Skottsb. (see Barbara Mckinder and Martin Staniforth, "*Sophora toromiro.*" in *Curtis's Botanical Magazine.* 14(4): 221-226. Nov. 1997; Mike Maunder, "Conservation of the extinct Toromiro tree. *Sophora toromiro.*" in

Curtis's Botanical Magazine. 14(4): 226-231. November 1997)

English: toromiro tree

Sophronanthe Benth. Scrophulariaceae

Origins:
From the Greek *sophron* "modest" and *anthos* "flower."

Sophronia Lindley Orchidaceae

Origins:
Greek *sophron* "modest, chaste," referring to the size of the plant.

Sophronitella Schltr. Orchidaceae

Origins:
The diminutive of the genus *Sophronia* Lindl.

Sophronitis Lindley Orchidaceae

Origins:
The diminutive of *Sophronia* Lindl., Greek *sophron* "modest," small plants but with large flowers; see Carl L. Withner, *The Cattleyas and their Relatives. Vol. III. Schomburgkia, Sophronitis, and Other South American Genera.* Timber Press, Portland, Oregon 1993.

Sopropis Britton & Rose Mimosaceae

Origins:
An anagram of the generic name *Prosopis* L.

Sopubia Buch-Ham. ex D. Don Scrophulariaceae

Origins:
Probably from a native Indian or Nepalese name, or an anagram of *Bopusia* Presl (Scrophulariaceae); see David Don (1799-1841), *Prodromus Florae Nepalensis*, sive enumeratio Vegetabilium quae ... Ann. 1802-1803 detexit atque legit F. Hamilton (*olim* Buchanan). 88. London 1825.

Soranthus Ledeb. Umbelliferae

Origins:
Greek *soros* "a heap, mound" and *anthos* "flower."

Sorbaria (DC.) A. Braun Rosaceae

Origins:
Referring to *Sorbus* L.

Sorbus L. Rosaceae

Origins:
Latin *sorbus, sorbi* "true sorb, service tree," *Sorbus domestica* L.; *sorbum* for the fruit of the *sorbus*, a sorb-apple, sorb, service-berry; see Carl Linnaeus, *Species Plantarum.* 477. 1753 and *Genera Plantarum.* Ed. 5. 213. 1754; S. Battaglia, *Grande dizionario della lingua italiana.* XIX: 478. 1997.

Species/Vernacular Names:
S. americana Marshall
English: roundwood, American mountain ash, dogberry, missey-moosey
S. aria (L.) Crantz
English: whitebeam
Italian: sorbo selvatico, sorbo montano, farinaccio
S. aucuparia L.
English: rowan, mountain ash, quickbeam
Italian: sorbo degli uccellatori
S. decora (Sarg.) Schneid.
English: showy mountain ash
S. domestica L.
English: service tree
Italian: sorbo domestico
S. hedlundii Schneider
English: whitebeam
S. hybrida L.
Norwegian: asal
S. intermedia (Ehrh.) Pers.
English: Swedish whitebeam
S. meinichii (Lindeberg) Hedlund (after Hans Thomas Meinich, 1817-1878, a Member of the Norwegian Parliament, plant collector and botanist, 1860-1869 Chief administrative officer of Hordaland county; see Per H. Salvesen, "*Sorbus meinichii*." in *The Plantsman.* 13(4): 193-198. March 1992)

Norwegian: fagerrogn (= the fair rowan), halvasal (= half bastard whitebeam), sur-asal (= sour bastard whitebeam)

S. sitchensis M. Roem.

English: Pacific mountain ash

S. thibetica (Cardot) Handel-Mazzetti

English: whitebeam

S. torminalis (L.) Crantz

English: service tree

Italian: ciavardello

S. vestita (G. Don) Loddiges

English: whitebeam

Sorghastrum Nash Gramineae

Origins:

From the genus *Sorghum* Moench and *astrum*, a Latin substantival suffix indicating inferiority or incomplete resemblance; see Nathaniel Lord Britton (1859-1934), *Manual of the Flora of the Northern States and Canada*. 71. New York 1901.

Sorghum Moench Gramineae

Origins:

The Italian *sorgo, soricum, surgus, suricum* (12th century), *surico* (10th century), from the spoken Latin *suricum granum* "grain from Syria," from Suria, a variation of Syria; see Conrad Moench (1744-1805), *Methodus plantas horti botanici et agri Marburgensis a staminum situ describendi*. 207. Marburgi Cattorum [Marburg] 1794-1802; S. Battaglia, *Grande dizionario della lingua italiana*. XIX: 492-493. 1997.

Species/Vernacular Names:

S. sp.

English: grain sorghum, millet

South Africa: Amaquaskoorn, graansorghum, ternataanse koring, rooikafferkoring, witkafferkoring

Yoruba: baba, oka baba, oka isi, oka, bomo, boromo, sosoki

S. bicolor (L.) Moench (*Holcus bicolor* L.; *Sorghum vulgare* Pers.)

English: millet, sorghum, cultivated sorghums, grain sorghum, forage sorghum, sweet sorghum, broom millet, great millet, kafir corn, Guinea corn, Rhodesian Sudan grass, Sudan grass, milo

China: gao liang, shu shu

Japan: nami-morokoshi

Okinawa: tonuchin

Tanzania: buhembe

Yoruba: oka pupa, baba, oka baba

S. bicolor (L.) Moench subsp. *arundinaceum* (Desv.) de Wet & Harlan (*Sorghum verticilliflorum* (Steud.) Stapf)

English: common wild sorghum, evergreen millet, Johnson grass, wild grain sorghum

Southern Africa: baster Jonhsongras, Jonhsongras, gewone wildesorghum, grootswartsaadgras, wilde graansorghum; iquangaboto (Ndebele); molémogéle (Tswana)

S. bicolor (L.) Moench subsp. *drummondii* (Steud.) de Wet (*Sorghum sudanense* (Piper) Stapf)

English: broom corn, red kaffir corn, shallu, sugar millet, sugar reed, sugar sorghum, sweet cane, sweet reed, sweet sorghum, wild grain sorghum, chicken corn, Sudan grass, white kaffir corn

Southern Africa: anaguaskoorn, besemkoring, kafferkoring, witkafferkoring, rooikafferkoring, soetriet, suikerriet, wilde graansorghum, kiepiemannakoring, opblaasgras, Soedangras; imfe (Zulu); nabele (Ndebele); mapfunde (Shona)

S. halepense (L.) Pers. (*Sorghum almum* Parodi; *Holcus halepense* L.; *Andropogon halepensis* (L.) Brot.)

English: Columbus grass, Aleppo grass, means grass, Johnson grass, Egyptian millet, evergreen millet, grass sorghum

Peru: grama china

Southern Africa: Columbusgras, Jonhsongras; iquangaboto (Ndebele)

Japan: Seiban-morokoshi

S. leiocladum (Hackel) C.E. Hubb.

English: wild sorghum

S. nitidum (Vahl) Pers.

The Philippines: bagokbok

Sorgum Adans. Gramineae

Origins:

From the Italian *sorgo, soricum, surgus, suricum* (12th century).

Soridium Miers Triuridaceae

Origins:

Greek *soridion*, the diminutive of *soros* "a vessel, a coffin."

Sorindeia Thouars Anacardiaceae

Origins:

From the Madagascan vernacular name for *Sorindeia madagascarensis* DC., or from the Greek *soros* "a heap" and

India; see H. Genaust, *Etymologisches Wörterbuch der botanischen Pflanzennamen*. 597. Basel 1996.

Species/Vernacular Names:
S. sp.
Uganda: muziru
S. gilletii De Wild.
Zaire: okote, dikonda, dongi
S. grandifolia Engl.
Nigeria: ehegogo (Edo)

Sorindeiopsis Engl. Anacardiaceae

Origins:
Resembling *Sorindeia*.

Sorocea A. St.-Hil. Moraceae

Origins:
Soroco is a vernacular name for *Sorocea bonplandii* (Baillon) Burger, Lanj. & Boer.

Species/Vernacular Names:
S. bonplandii (Baillon) Burger, Lanj. & Boer
Brazil: cincho, soroco, carapicica de folha miúda, larangeira do mato, araçari, canxim, resple

Sorocephalus R. Br. Proteaceae

Origins:
Greek *soros* "a heap" and *kephale* "head," referring to the clustered flowers.

Sorolepidium Christ Dryopteridaceae (Aspleniaceae)

Origins:
Greek *soros* "a heap, vessel" and *lepis* "scale," *lepidion* "a little scale."

Soromanes Fée Dryopteridaceae (Aspleniaceae)

Origins:
Greek *soros* "a vessel for holding anything, a heap, a spore case" and *manes* "a cup, a kind of cup."

Soroseris Stebb. Asteraceae

Origins:
Greek *soros* "a heap, mound" and *seris, seridos* "chicory, lettuce," referring to the dwarf habit of the plant.

Species/Vernacular Names:
S. sp.
Tibetan: srol-gong dkar-po/sngon-po

Sorostachys Steudel Cyperaceae

Origins:
From the Greek *soros* and *stachys* "spike."

Souliea Franchet Ranunculaceae

Origins:
For the French botanical collector Jean André Soulié, 1858-1900, missionary, naturalist, zoological collector, traveler; see J.H. Barnhart, *Biographical Notes upon Botanists*. 3: 305. Boston 1965; E. Bretschneider, *History of European Botanical Discoveries in China*. Leipzig 1981.

Sowerbaea Smith Anthericaceae (Liliaceae)

Origins:
After the British (b. London) botanical artist James Sowerby, 1757-1822 (d. Lambeth, London), engraver of plates of flowering plants and fungi, studied at the Royal Academy of Arts, 1793 Fellow of the Linnean Society, collaborator of William Curtis. Among the many works he illustrated were Sir James Edward Smith's 36 volumes on *English Botany* (1790-1814), he was father of the British artists Charles Edward Sowerby (1795-1842), James de Carle Sowerby (1787-1871) and George Brettingham Sowerby (1788-1854); see James Sowerby & George Brettingham Sowerby, [*Of the Genera of Recent and Fossil Shells* ... in monthly numbers, etc. London 1820-1825 and 1834] [Forty-two parts in two volumes. With 265 plates engraved by James Sowerby, after the death of J.S. his second son George continued the work.]; J. Collins, *A Note on the History of the Sowerby Family Archive*. London 1973; A. Wheeler, ed., "Papers on the Sowerby family." *J. Soc. Bibl. nat. Hist.* 6(6): 377-559. 1974; R.J. Cleevely, "A provisional bibliography of natural history works of the Sowerby family." *J. Soc. Bibl. Nat. Hist.* 6(6): 484-492. 1974; R.R. Sowerby, *Sowerby of China*. Kendal 19565; Joan M. Eyles, in *Dictionary of Scientific Biography* 12: 552-553. 1981; R. Zander, F. Encke, G. Buchheim and S. Seybold, *Handwörterbuch der Pflanzennamen*. 14. Aufl. 1993; Stafleu and Cowan, *Taxonomic Literature*. 5: 759-762. 1985; Ray Des-

mond, *Dictionary of British & Irish Botanists and Horticulturists.* 644. 1994; Blanche Henrey, *British Botanical and Horticultural Literature before 1800.* Oxford 1975; John H. Barnhart, *Biographical Notes upon Botanists.* 3: 305. 1965; Sir James Edward Smith (1759-1828), in *Transactions of the Linnean Society of London. Botany.* 4: 218. (May) 1798; J.D. Milner, *Catalogue of Portraits of Botanists Exhibited in the Museums of the Royal Botanic Gardens.* Royal Botanic Gardens, Kew, London 1906; M. Hadfield et al., *British Gardeners: A Biographical Dictionary.* London 1980; Jonas C. Dryander, *Catalogus bibliothecae historico-naturalis Josephi Banks.* London 1800; Ethelyn Maria Tucker, *Catalogue of the Library of the Arnold Arboretum of Harvard University.* Cambridge, Massachusetts 1917-1933; H.R. Fletcher, *Story of the Royal Horticultural Society, 1804-1968.* Oxford 1969; G. Murray, *History of the Collections Contained in the Natural History Departments of the British Museum.* 1: 184. London 1904.

Species/Vernacular Names:

S. juncea Andrews

English: rush lily, vanilla plant, chocolate flower, chocolate lily

Sowerbea Dum.-Cours. Anthericaceae (Liliaceae)

Origins:

Orthographic variant of *Sowerbaea* Smith.

Sowewbia Andrews Anthericaceae (Liliaceae)

Origins:

Orthographic variant of *Sowerbaea* Smith.

Soyauxia Oliver Medusandraceae (Flacourtiaceae)

Origins:

For the German botanist Hermann Soyaux (b. 1852), gardener, traveler and plant collector, between 1873 and 1876 in Western tropical Africa with Loango-Expedition, between 1879 and 1885 in Gabon, 1888 Brazil, author of *Aus West-Afrika 1873-1876.* Leipzig 1879-1880; see Richard P. Wilhelm Guessfeldt, born 1840, *Die Loango-Expedition* ausgesandt von der Deutschen Gesellschaft zur Erfoschung Aequatorial-Africas 1873-76 etc. Abth. I, von P. Guessfeldt, II, von J. Falkenstein, III, von E. Pechuël-Loesche. Leipzig 1879-82; Joseph Vallot (1854-1925), "Études sur la flore du Sénégal." in *Bull. Soc. Bot. de France.* 29: 168-238. Paris 1882; Anthonius Josephus Maria

Leeuwenberg, "Isotypes of which holotypes were destroyed in Berlin." in *Webbia.* 19: 861-863. 1965; John Gossweiler, "Elementos para a Historia da Exploração Botanica de Angola." *Bol. Soc. Broter.* 13: 303-304. 1939; John H. Barnhart, *Biographical Notes upon Botanists.* 3: 306. 1965; Ethelyn Maria Tucker, *Catalogue of the Library of the Arnold Arboretum of Harvard University.* Cambridge, Massachusetts 1917-1933.

Soymida A. Juss. Meliaceae

Origins:

Somida is a Telugu name for *Soymida febrifuga* A. Juss., from *swami* "god."

Species/Vernacular Names:

S. febrifuga (Roxb.) A. Juss.

English: Indian redwood, bastard cedar

Spachea A. Juss. Malpighiaceae

Origins:

For the French botanist Édouard Spach, 1801-1879, naturalist, author of *Histoire naturelle des Végétaux. Phanérogames.* Paris 1834-1838; see John H. Barnhart, *Biographical Notes upon Botanists.* 3: 306. 1965; A. Lasègue, *Musée botanique de Benjamin Delessert.* 319. Paris 1845; Ethelyn Maria Tucker, *Catalogue of the Library of the Arnold Arboretum of Harvard University.* Cambridge, Massachusetts 1917-1933; Ida Kaplan Langman, *A Selected Guide to the Literature on the Flowering Plants of Mexico.* 713. Philadelphia 1964; R. Zander, F. Encke, G. Buchheim and S. Seybold, *Handwörterbuch der Pflanzennamen.* 14. Aufl. 1993; Mariella Azzarello Di Misa, ed., *Il Fondo Antico della Biblioteca dell'Orto Botanico di Palermo.* 259. Palermo 1988.

Spadostyles Benth. Fabaceae

Origins:

Greek *spadon* "a tear, eunuch, a rupture," *spao* "to draw out, to tear away" and *stylos* "a column, style"; see G. Bentham, *Commentationes de Leguminosarum generibus.* 16. Vindobonae (Jun.) 1837 [in *Ann. Wiener Mus. Naturgesch.* 2: 80. Feb. 1838].

Spallanzania DC. Rubiaceae

Origins:

After the Italian (b. Scandiano) scientist Lazzaro Spallanzani, 1729-1799 (Pavia), doctor of philosophy, physiologist, traveler, professor of natural history. His writings include *Nouvelles recherches sur les découvertes microscopiques, et la génération des corps organisés.* Londres et Paris 1769 and *Dissertazioni di fisica animale e vegetabile.* Modena 1780, with Jean Senebier (1742-1809) wrote *Rapports de l'air avec les êtres organisés.* Genève 1807 and *Expériences pour servir à l'histoire de la génération des animaux et des plantes.* Pavia 1787; see Claude E. Dolman, in *Dictionary of Scientific Biography* 12: 553-567. 1981; Giuseppe Montalenti, *Lazzaro Spallanzani.* Milano 1928; P.E. Pilet, in *Dictionary of Scientific Biography* 12: 308-309. 1981; Jean Louis Marie Alibert (1768-1837), *Eloges historiques composés pour la Société Médicale de Paris.* [Octavo, first book edition; referring to Lazzaro Spallanzani, Luigi Galvani (1737-1798) and Pierre Roussel, 1742-1802.] Paris 1806; John H. Barnhart, *Biographical Notes upon Botanists.* 3: 307. 1965; T.W. Bossert, *Biographical Dictionary of Botanists Represented in the Hunt Institute Portrait Collection.* 377. 1972; Jonas C. Dryander, *Catalogus bibliothecae historico-naturalis Josephi Banks.* London 1800; E.M. Tucker, *Catalogue of the Library of the Arnold Arboretum of Harvard University.* Cambridge, Massachusetts 1917-1933; B. Glass et al., eds., *Forerunners of Darwin: 1745-1859.* Baltimore 1959; Dino Prandi, *Bibliografia delle opere di Lazzaro Spallanzani.* Firenze 1951.

Spananthe Jacq. Umbelliferae

Origins:

Greek *spanios* "rare, scarce" and *anthos* "a flower."

Spaniopappus B.L. Robinson Asteraceae

Origins:

Greek *spanios* "rare, scarce" and *pappos* "fluff, downy appendage."

Spanoghea Blume Sapindaceae

Origins:

For the Dutch plant collector Johan Baptist Spanoghe, 1798-1838, from 1816 East Indies (Java and Timor); see J.H. Barnhart, *Biographical Notes upon Botanists.* 3: 307. 1965; A. Lasègue, *Musée botanique de Benjamin Delessert.* Paris 1845; Karl Ludwig von Blume (1796-1862), in *Rumphia.* 3: 172. 1849; R. Zander, F. Encke, G. Buchheim and S. Seybold, *Handwörterbuch der Pflanzennamen.* 14. Aufl. Stuttgart 1993.

Sparattanthelium Martius Hernandiaceae (Gyrocarpaceae)

Origins:

Greek *sparasso, sparassein, sparatto* "to tear, rend" and *anthelion, anthyllion*, a diminutive of *anthos* "flower, blossom."

Sparattosperma Martius ex Meissner Bignoniaceae

Origins:

From the Greek *sparasso, sparatto* "to tear, rend" and *sperma* "seed."

Sparattosyce Bureau Moraceae

Origins:

From the Greek *sparasso, sparatto* and *sykon* "fig."

Sparaxis Ker Gawler Iridaceae

Origins:

Greek *sparassein, sparaxo* "to tear," the bracts or the spathes are lacerated; see John Sims (1749-1831), in *Curtis's Botanical Magazine.* 15, t. 548. (Jan.) 1802.

Species/Vernacular Names:

S. bulbifera (L.) Ker Gawl. (*Ixia bulbifera* L.)

English: Harlequin flower

South Africa: ferweelblom, botterblom

S. grandiflora (Delaroche) Ker Gawl. subsp. *acutiloba* Goldbl.

South Africa: botterblom

S. grandiflora (Delaroche) Ker Gawl. subsp. *fimbriata* (Lam.) Goldbl.

South Africa: witkalossie, botterblom, fluweelblom

S. grandiflora (Delaroche) Ker Gawl. subsp. *grandiflora*

South Africa: perskalkoentjie

S. tricolor (Schneevogt) Ker Gawl. (*Ixia tricolor* Schneev.)

English: tricolor Harlequin flower, Jack Spratts

South Africa: Harlequin flower, fluweeltjie, ferweelblom

Sparganium L. Sparganiaceae (Typhaceae)

Origins:

Greek and Latin *sparganion* for the plant bur-weed, bur-reed (Plinius and Dioscorides), Greek *sparganon* "a diaper, swaddling band, band, a ribbon," the leaves are ribbon-like; see Carl Linnaeus, *Species Plantarum*. 971. 1753 and *Genera Plantarum*. Ed. 5. 418. 1754.

Species/Vernacular Names:

S. angustifolium Michaux

English: narrow-leaved bur-reed

S. erectum L.

English: branching bur-reed

S. natans L.

English: small bur-reed

S. stenophyllum Maxim.

Japan: hime-mi-kuri

S. subglobosum Morong (*Sparganium antipodum* Graebner)

English: floating bur-reed

Sparganophorus Boehmer Asteraceae

Origins:

From the Greek *sparganon* "a fillet, a ribbon" and *phoreo* "to bear."

Sparmannia L.f. Tiliaceae

Origins:

See *Sparrmannia* L.f.

Sparrmannia L.f. Tiliaceae

Origins:

After the Swedish botanist Anders (Andrew) Sparrmann, 1748-1820 (d. Stockholm), naturalist, pupil of Linnaeus, traveler, physician, 1772-1775 doctor on Captain Cook's second expedition (in the ship *Resolution*), with C.P. Thunberg in South Africa, 1765-1767 traveled with Carl Gustav Ekeberg (1716-1784) in East Indies and China, author of *A Voyage to the Cape of Good Hope ... from the Year 1772 to 1776*. London 1785-1786 (first English translation); see C. Linnaeus, *Iter in Chinam*. (Anders Sparrman). Upsaliae 1768; John Hutchinson (1884-1972), *A Botanist in Southern Africa*. 611-613. London 1946; E. Bretschneider, *History of European Botanical Discoveries in China*. Leipzig 1981; J.H. Barnhart, *Biographical Notes upon Botanists*. 3: 307. 1965; G.C. Wittstein, *Etymologisch-botanisches Handwörterbuch*. 824. 1852; F. Boerner & G. Kunkel, *Taschenwörterbuch der botanischen Pflanzennamen*. 4. Aufl. 171. 1989; Mary Gunn and L.E.W. Codd, *Botanical Exploration of Southern Africa*. 330-331. 1981; H. Genaust, *Etymolo-*

gisches Wörterbuch der botanischen Pflanzennamen. 598. [1747-1787] 1996; Gilbert Westacott Reynolds (1895-1967), *The Aloes of South Africa*. 736, 44. Balkema, Rotterdam 1982; Peter [Pyotr] Simon Pallas (1741-1811), *A Naturalist in Russia. Letters from Peter Simon Pallas to Thomas Pennant*. Edited by Carol Urness. Minneapolis [1967]; Peter Simon Pallas, *Flora Rossica*. St. Petersburg 1784-1788; Jonas C. Dryander, *Catalogus bibliothecae historico-naturalis Josephi Banks*. London 1800; A. Lasègue, *Musée botanique de Benjamin Delessert*. 1845; T.W. Bossert, *Biographical Dictionary of Botanists Represented in the Hunt Institute Portrait Collection*. 377. 1972; E.M. Tucker, *Catalogue of the Library of the Arnold Arboretum of Harvard University*. Cambridge, Massachusetts 1917-1933.

Species/Vernacular Names:

S. africana L.f.

English: sock-rose, African hemp, house lime

South Africa: stokroos

Sparteum P. Beauv. Gramineae

Origins:

Latin *sparteus, a, um* "of broom, made of broom."

Spartidium Pomel Fabaceae

Origins:

Resembling *sparton* "a rope, bond," Latin *spartum, i*, for a grass used for cordage, nets and mats.

Spartina Schreber Gramineae

Origins:

Greek *sparton* "a rope, bond"; Latin *spartum, sparton*, a grass used for cordage, nets and mats; Akkadian *sabaru*, Hebrew *safar* "to bind"; see Johann Christian Daniel von Schreber (1739-1810), *Genera Plantarum*. 43. (Apr.) 1789.

Species/Vernacular Names:

S. alterniflora Lois.

English: salt-water cord grass

S. densiflora Brongn.

English: dense-flowered cord grass

S. foliosa Trin.

English: California cord grass

S. gracilis Trin.

English: alkali cord grass

S. maritima (Curtis) Fernald (*Spartina capensis* Nees)

English: Cape cord grass

South Africa: Kaapse slykgras

S. patens (Aiton) Muhlenb.

English: salt-meadow cord grass

S. pectinata Link

English: prairie cord grass, freshwater cord grass, slough grass

Spartium L. Fabaceae

Origins:
Greek *spartion* (*sparton*) "a small cord," a kind of grass or plant used for weaving and cordage, Latin *spartum* or *sparton* for a plant originally growing in Spain, of which ropes and mats were made; see Carl Linnaeus, *Species Plantarum*. 708. 1753 and *Genera Plantarum*. Ed. 5. 317. 1754.

Species/Vernacular Names:
S. junceum L.

English: Spanish broom, weaver's broom

Arabic: kessaba, tertakh, ratam

Spartochloa C.E. Hubb. Gramineae

Origins:
Greek *sparton* "a cord" and *chloe, chloa* "grass"; see C.E. Hubbard, in *Kew Bulletin*. 7: 308. (Oct.) 1952.

Spartocytisus Webb & Berthel. Fabaceae

Origins:
From the Greek *sparton* "a cord" plus *Cytisus*.

Spartothamnella Briquet Labiatae (Dicrastylidaceae, Verbenaceae)

Origins:
Diminutive of the genus *Spartothamnus* Walpers; see H.G.A. Engler & K.A.E. Prantl, *Die Natürlichen Pflanzenfamilien*. 4(3a): 24. (Feb.) 1895.

Species/Vernacular Names:
S. juncea (Walp.) Briq. (*Spartothamnus juncea* Walp.)

English: bead bush, goodnight scrub

S. teucriiflora (F. Muell.) Mold. (*Spartothamnus teucriiflorus* F. Muell.)

English: bead bush, red-berried stick-plant

Spartothamnus Walpers Labiatae (Dicrastylidaceae, Verbenaceae)

Origins:
Greek *sparton* "a cord" and *thamnos* "a bush"; see Wilhelm Gerhard Walpers, in *Repertorium botanices systematicae*. 6: 694. (May) 1847.

Spartum P. Beauv. Gramineae

Origins:
From the Latin *spartum* or *sparton*, Greek *sparton*, Spanish broom, a plant originally growing in Spain, of which ropes and mats were made (Plinius).

Spatalla Salisb. Proteaceae

Origins:
From the Greek *spatalos* "delicate," referring to the flowers.

Spatallopsis E. Phillips Proteaceae

Origins:
Resembling *Spatalla* Salisb.

Spathacanthus Baillon Acanthaceae

Origins:
Greek *spathe* "spathe" and the genus *Acanthus* L., Greek *akantha* "thorn."

Spathandra Guillemin & Perrottet Melastomataceae

Origins:
Greek *spathe* "spathe" and *aner, andros* "male," referring to the stamens.

Species/Vernacular Names:
S. blakeoides (G. Don) Jacq.-Fél. (*Memecylon blakeoides* G. Don)

Nigeria: egbeda (Edo); asumabrakori (Ijaw); anya enyi (Igbo)

Spathantheum Schott Araceae

Origins:

From the Greek *spathe* and *anthos* "flower," referring to the spadix adnate to the spathe, inflorescence epiphyllous; see D.H. Nicolson, "Derivation of aroid generic names." *Aroideana*. 10: 15-25. 1988.

Spathanthus Desv. Rapateaceae

Origins:

From the Greek *spathe* "spathe" and *anthos* "flower."

Spathelia L. Rutaceae

Origins:

Greek *spathe* "spathe," hapaxanthic pachycaul plant with terminal inflorescences

Species/Vernacular Names:

S. sorbifolia (L.) Fawcett & Rendle

English: mountain pride

Spathia Ewart Gramineae

Origins:

Greek *spathe* "a spathe, a blade"; see Alfred James Ewart and Olive Blanche Davies, *The Flora of the Northern Territory*. 26. Melbourne 1917.

Spathicalyx J.C. Gomes Bignoniaceae

Origins:

From the Greek *spathe* "a spathe, a blade" and *kalyx, kalykos* "calyx."

Spathicarpa Hook. Araceae

Origins:

Greek *spathe* "a spathe, a blade" and *karpos* "fruit," the fruits are on the adnate spadix; see D.H. Nicolson, "Derivation of aroid generic names." *Aroideana*. 10: 15-25. 1988.

Species/Vernacular Names:

S. sagittifolia Schott

English: caterpillar plant

Spathichlamys R. Parker Rubiaceae

Origins:

From the Greek *spathe* and *chlamys, chlamydos* "cloak, mantle."

Spathidolepis Schltr. Asclepiadaceae

Origins:

From the Greek *spathe* "a spathe, a blade" and *lepis* "scale."

Spathiger Small Orchidaceae

Origins:

Greek *spathe* "a spathe, a blade," referring to the bracts or leaves.

Spathionema Taubert Fabaceae

Origins:

From the Greek *spathe* "a spathe, a blade" and *nema* "thread, filament."

Spathiostemon Blume Euphorbiaceae

Origins:

From the Greek *spathe* and *stemon* "stamen."

Spathipappus Tzvelev Asteraceae

Origins:

Greek *spathe* and *pappos* "fluff, downy appendage."

Spathiphyllum Schott Araceae

Origins:

Greek *spathe* and *phyllon* "a leaf," referring to the leaflike spathe; see D.H. Nicolson, "Derivation of aroid generic names." *Aroideana*. 10: 15-25. 1988.

Spathodea P. Beauv. Bignoniaceae

Origins:

From the Greek *spathe* "spathe" and *-odes* "resembling, of the nature of," referring to the calyx; see Ambroise Palisot de Beauvois (1752-1820), *Flore d'Oware et de Benin en Afrique*. 1: 46. Paris 1805.

Species/Vernacular Names:

S. campanulata P. Beauv. (*Spathodea nilotica* Seem.)

English: scarlet bell, fountain tree, tulip tree, African tulip tree, flame of the forest, Nile tulip tree, flame tree, Nandi flame, Uganda flame, Nile flame

French: tulipier d'Afrique, tulipier du Gabon

Central Africa: abandiri

Congo: agon, ogon, vulu, voulou

Cameroon: etutu, mbako, mbeleme, essoussouk, mvouley, bwele ba mbongo, bobori, todouk, toduk, evovonne, essussuk

Gabon:evong evong

Ghana: kokonsu, osisiriw, adatsigo, kokoanidua, vatsho, aninsu, akuoko, bie bie, odumanki, osisiri, kokoanisua, adadase, abeni, kokoninsu, akuaka, no-tsho

Ivory Coast: boro, kokonayur, paoutou, se

Uganda: kifabakazi, munyara, ekifurafura, kikussu, omunyara, ekinyara, opal, etukubai, kichubi, kijubu, chemungwa, mungobe, lapengwata, omudungudungu

Nigeria: owewe, bwele-ba-mboji, efonfon, mba-ako, okonko, fuga, etoto, echib, oruru, orudu, totook, imi-ewo, okwokwi; okwekwe, akpoti, obo-omi, mba; dimberi-chamili (Nupe); oruru (Yoruba); owewe (Edo); imi ewu (Igbo); esenim (Efik); kenshie (Boki)

Yoruba: oruru, owewe

Kenya: kibobakasi

Sierra Leone: ngele-gwe, dundunturi, dununtinidi, dumentili, ta gboyo

East Africa: mutsurio

Guinea: tunda

Japan: kaen-boku

Malaya: panchut panchut, panchit

Spathodeopsis Dop Bignoniaceae

Origins:

Resembling *Spathodea* P. Beauv.

Spathoglottis Blume Orchidaceae

Origins:

Greek *spathe* and *glotta* "tongue," referring to the mid-lobe of the lip; see Karl Ludwig von Blume, *Bijdragen tot de flora van Nederlandsch Indië*. 400. (Sep.-Dec.) 1825; B. Lewis and Phillip Cribb, *Orchids of Vanuatu*. Royal Botanic Gardens, Kew 1989; B.A. Lewis and P.J. Cribb, *Orchids of the Solomon Islands and Bougainville*. Royal Botanic Gardens, Kew 1991.

Species/Vernacular Names:

S. plicata Blume (*Bletia angustifolia* Gaud.; *Spathoglottis vieillardii* Reichb.f.)

English: Malayan ground orchid, Philippine ground orchid

Japan: kôtô-shi-ran (named for Kôtôshô Island, Taiwan)

Vanuatu: evére

The Solomon Islands: tongoruha, molamola, laulau ngwane

Spatholirion Ridley Commelinaceae

Origins:

Greek *spathe* "spathe" and *leirion* "a lily."

Spatholobus Hassk. Fabaceae

Origins:

From the Greek *spathe* and *lobos* "a pod," referring to the nature of the pod; see Justus Carl Hasskarl (1811-1894), *Flora*. 25(2) (Beibl.): 52. 28 Sep. 1842.

Spathoscaphe Oerst. Palmae

Origins:

From the Greek *spathe* "spathe" and *scaphis* "a vessel," *skaphe* "a light boat, boat."

Spathulata (Borisova) Á. Löve & D. Löve Crassulaceae

Origins:

From the Latin *spatula, spathula, ae*, the diminutive of *spatha, ae*.

Spathulopetalum Chiov. Asclepiadaceae

Origins:

From the Latin *spatula, spathula, ae* and *petalum* "petal."

Specklinia Lindley Orchidaceae

Origins:

For Veit (Vitus) Rudolf (Rodolph) Speckle, artist, wood engraver, sculptor; see Leonhart Fuchs (1501-1566), *De historia stirpium* commentarii insignes. [V.R. Speckle, with Heinrich Füllmaurer and Albrecht Meyer (the draftsmen for Fuchs, *pictores operis*), prepared the woodcuts for the fine illustrations.] Basel 1542; Edward Lee Greene, *Landmarks of Botanical History.* Edited by Frank N. Egerton. 274-275, 288-290, 487. Stanford University Press, Stanford, California 1983.

Specularia A. DC. Campanulaceae

Origins:

Latin *specularis, e* "like a mirror," *specularia, orum* "a window."

Spegazzinia Backeb. Cactaceae

Origins:

For the Argentine (b. Italy) botanist Carlo (Carlos) Luigi Spegazzini, 1858-1926, naturalist, traveler, professor of botany, botanical explorer, plant collector, mycologist. His writings include "Costumbres de los habitantes de la Tierra de Fuego." *Anales Soc. cient. argent.* Buenos Aires 1882, *Fungi guaranitici pugillus i-ii.* Buenos Aires 1883-1888, "Costumbres de los Patagones." *Anales Soc. cient. argent.* 17: 221-240. Buenos Aires 1884, "Apuntes filológicos sobre las lenguas de la Tierra de Fuego." *Anales Soc. cient. argent.* 18: 131-144. Buenos Aires 1884 and *Fungi guaranitici nonnulli novi v. critici.* Buenos Aires 1891. See John M. Cooper, *Analytical and Critical Bibliography of the Tribes of Tierra del Fuego and Adjacent Territory.* [Smithsonian Institution — Bureau of American Ethnology — Bulletin 63.] Washington 1917; Bartolomé Mitre, *Catálogo razonado de la sección lenguas americanas.* Buenos Aires 1909-1911; Roberto Dabbene, "Viaje á la Tierra del Fuego y á la isla de los Estados." in *Bol. Inst. geogr. argent.* 21: 3-78. Buenos Aires; Giacomo Bove, "Viaggio alla Patagonia ed alla Terra del Fuoco." in *Nuova antologia di scienze, lettere ed arti.* LXVI, seconda ser. XXXVI: 733-801. Roma 1882; Roberto Dabbene, "Los indígenas de la Tierra del Fuego." in *Bol. Inst. geogr. argent.* 25(5-6): 163-226 and 25(7-8): 247-300. Buenos Aires 1911; J.H. Barnhart, *Biographical Notes upon Botanists.* 3: 308. 1965; T.W. Bossert, *Biographical Dictionary of Botanists Represented in the Hunt Institute Portrait Collection.* 378. 1972; E.M. Tucker, *Catalogue of the Library of the Arnold Arboretum of Harvard University.* Cambridge, Massachusetts 1917-1933; Stafleu and Cowan, *Taxonomic Literature.* 5: 776-785. 1985; Gordon Douglas Rowley, *A History of Succulent Plants.* Strawberry Press, Mill Valley, California 1997; R. Zander, F. Encke, G. Buchheim and S. Seybold, *Handwörterbuch der Pflanzennamen.* 14. Aufl. Stuttgart 1993.

Speirantha Baker Convallariaceae (Liliaceae)

Origins:

From the Greek *speira* "a spiral" and *anthos* "a flower," referring to the inflorescence.

Speirema Hook.f. & Thomson Campanulaceae

Origins:

Greek *speirema, speirama* "coil, convolution, twisted thread."

Spelta Wolf Gramineae

Origins:

Latin *spelta, ae* "grains of spelt."

Spenceria Trimen Rosaceae

Origins:

After the English botanist Spencer Le Marchant Moore, 1850-1931 (d. near London), plant collector, explorer, 1872-1879 Kew Herbarium, 1875 Fellow of the Linnean Society, 1880 appointed assistant at British Museum, 1891-1892 to Brazil (Matto Grosso Expedition), 1894-1896 to Western Australia. His writings include "Six new South African plants." *J. Bot., Lond.* 43: 169-173. 1905, "New or noteworthy South African plants." *J. Bot., Lond.* 40: 380-385. 1902, "The genus *Pleiotaxis* Steetz." *J. Bot., Lond.* 63: 43-50. 1925 and "The genus *Lopholaena* DC." *J. Bot., Lond.* 67: 274-276. 1929, co-author of volume seven of *Flora of Jamaica* by William Fawcett and Alfred B. Rendle, editor of John Byrne Leicester Warren (1835-1895), *The Flora of Cheshire.* 1899, 1877-1879 assistant editor of the *Journal of Botany.* See J.H. Barnhart, *Biographical Notes upon Botanists.* 2: 510. 1965; Ray Desmond, *Dictionary of*

British & Irish Botanists and Horticulturists. 498. 1994; Alfred Barton Rendle (1865-1938) et al., "Catalogue of the plants collected by Mr. and Mrs. P.A. Talbot in the Oban District of South Nigeria." *British Museum Trustees, Natural History*. [Spencer Le Marchant Moore was co-author.] London 1913; E.M. Tucker, *Catalogue of the Library of the Arnold Arboretum of Harvard University*. Cambridge, Massachusetts 1917-1933; Emil Bretschneider, *History of European Botanical Discoveries in China*. Leipzig 1981; I. Urban, *Geschichte des Königlichen Botanischen Museums zu Berlin-Dahlem (1815-1913)*. Dresden 1916; R. Zander, F. Encke, G. Buchheim and S. Seybold, *Handwörterbuch der Pflanzennamen*. 14. Aufl. 1993; Ludwig Emil Diels, *Die Pflanzenwelt von West-Australien südlich des Wendekreises*. Leipzig 1906; E.D. Merrill, in *Bernice P. Bishop Mus. Bull.* 144: 138. 1937 and *Contr. U.S. Natl. Herb.* 30(1): 216-220. 1947; G. Murray, *History of the Collections Contained in the Natural History Departments of the British Museum*. 1: 169. London 1904; Ida Kaplan Langman, *A Selected Guide to the Literature on the Flowering Plants of Mexico*. Philadelphia 1964; F. Boerner & G. Kunkel, *Taschenwörterbuch der botanischen Pflanzennamen*. 4. Aufl. 172. 1989; F. Boerner, *Taschenwörterbuch der botanischen Pflanzennamen*. 2. Aufl. 177. Berlin & Hamburg 1966.

Spennera Martius ex DC. Melastomataceae

Origins:

For the German botanist Fridolin Carl (Karl) Leopold Spenner, 1798-1841, physician, professor of medical botany; see John H. Barnhart, *Biographical Notes upon Botanists*. 3: 309. 1965; R. Zander, F. Encke, G. Buchheim and S. Seybold, *Handwörterbuch der Pflanzennamen*. 14. Aufl. Stuttgart 1993; Ethelyn Maria Tucker, *Catalogue of the Library of the Arnold Arboretum of Harvard University*. Cambridge, Massachusetts 1917-1933; Mariella Azzarello Di Misa, ed., *Il Fondo Antico della Biblioteca dell'Orto Botanico di Palermo*. 260. Regione Siciliana, Palermo 1988.

Spergula L. Caryophyllaceae

Origins:

Latin *spargo, is, sparsi, sparsum, ere* "to scatter," referring to the seed dispersal; some suggest from *spergel, spörgel*, the German name for *Sagina spergula*; see Carl Linnaeus, *Species Plantarum*. 440. 1753 and *Genera Plantarum*. Ed. 5. 199. 1754; Salvatore Battaglia, *Grande dizionario della lingua italiana*. XIX: 824. Torino 1997.

Species/Vernacular Names:
S. arvensis L.

English: spurrey, corn spurry, spurry, corn spurrey, starwort, stickwort

Italian: renaiola

Southern Africa: Holladsche gras, perdegras, sporrie; bolepo-ba-seokha-sa-merung (Sotho)

East Africa: chemutwet ab koik

Spergularia (Pers.) J. Presl & C. Presl Caryophyllaceae

Origins:

Referring to *Spergula* L.; see Jan Swatopluk Presl (1791-1849) and K. Presl, *Flora Chechica*. 94. Pragae 1819.

Species/Vernacular Names:
S. bocconii (Scheele) Aschers. & Graebn. (*Spergularia atheniensis* Aschers. & Schweinf.)

English: sand spurry, sea spurrey, sand spurrey

S. diandra (Guss.) Boiss. (*Arenaria diandra* Guss.)

English: lesser sand spurry, small sand spurrey

S. marina (L.) Griseb. (*Arenaria rubra* L. var. *marina* L.; *Spergularia dillenii* Lebel; *Spergularia marina* L. var. *simonii* Degener & I. Degener; *Spergularia salina* (Pers.) J. Presl & C. Presl)

English: saltmarsh sand spurry, sand spurrey

Hawaii: mimi'ilio

Arabic: abu gholam

S. rubra (L.) Presl & C. Presl

English: sand spurrey, red sand-wort

Arabic: sherifa

Spermachiton Llanos Gramineae

Origins:
From the Greek *sperma* "a seed" and *chiton* "a tunic, covering."

Spermacoce L. Rubiaceae

Origins:

Greek *sperma* "a seed" and *akoke* "a point," referring to the fruits; see Carl Linnaeus, *Species Plantarum*. 102. 1753 and *Genera Plantarum*. Ed. 5. 44. 1754.

Species/Vernacular Names:
S. assurgens Ruíz & Pavón

English: buttonweed

Spermacoceodes Kuntze Rubiaceae

Origins:

The genus *Spermacoce* L. and *-odes* "resembling, of the nature of."

Spermacon Raf. Rubiaceae

Origins:

For *Spermacoce* L.; see C.S. Rafinesque, *Act. Soc. Linn. Bordeaux.* 6: 269. 1834.

Spermadictyon Roxb. Rubiaceae

Origins:

From the Greek *sperma* and *diktyon* "a net," referring to the nature of the seeds.

Spermatochiton Pilg. Gramineae

Origins:

Greek *sperma*, *spermatos* "a seed" and *chiton* "a tunic, covering."

Spermaxyrum Labill. Olacaceae

Origins:

Greek *sperma* "a seed" and *xyron* "a rasor"; see Jacques Julien Houtton de Labillardière (1755-1834), *Novae Hollandiae plantarum specimen.* 2: 84, t. 233. Parisiis 1804-1806 [1807].

Spermolepis Brongn. & Gris Myrtaceae

Origins:

Greek *sperma* "seed" and *lepis* "scale"; see also *Arillastrum* Panch. ex Baillon.

Spermolepis Raf. Umbelliferae

Origins:

Greek *sperma* "seed" and *lepis* "scale," referring to the fruit; see C.S. Rafinesque, *Neogenyton, or Indication of Sixty-Six New Genera of Plants of North America.* 2. 1825; E.D. Merrill, *Index rafinesquianus.* 183. 1949.

Sphacanthus Benoist Acanthaceae

Origins:

Greek *sphakos* "sage" and *akantha* "thorn."

Sphacele Benth. Labiatae

Origins:

From the Greek *sphakos* "sage," referring to the foliage.

Sphacophyllum Benth. Asteraceae

Origins:

Greek *sphakos* "sage" and *phyllon* "leaf," Latin *sphacos*, *sphagnos* for a kind of fragrant moss and for elelisphacos, a kind of sage (Plinius); see also *Anisopappus* Hook. & Arn.

Sphaenolobium Pimenov Umbelliferae

Origins:

From the Greek *sphen* "wedge" and *lobos* "pod, lobe."

Sphaeradenia Harling Cyclanthaceae

Origins:

Greek *sphaira* "a globe" and *aden* "gland."

Sphaeralcea A. St.-Hil. Malvaceae

Origins:

Globe mallow, from the Greek *sphaira* and *alkea* "mallow," in reference to the globular fruit; Latin *alcea, ae* for *Malva alcea* L. (Plinius); see Auguste François César Prouvençal de Saint-Hilaire (1779-1853), *Flora Brasiliae Meridionalis.* 163. Parisiis 1825.

Species/Vernacular Names:

S. ambigua A. Gray

English: desert mallow, apricot mallow

S. angustifolia (Cav.) G. Don f. ssp. *cuspidata* (A. Gray) Kearney

English: scarlet globe mallow

Spanish: yerba de la negrita

S. bonariensis (Cav.) Griseb.

English: false mallow

S. coccinea (Pursh) Rydb.

English: prairie mallow, red false mallow

S. fendleri A. Gray

English: Fendler globe mallow

S. obtusiloba (Hook.) G. Don

English: blunt-leaved Chilean mallow

S. orcuttii Rose

English: Carrizo mallow

S. philippiana Krapov.

English: trailing mallow

S. rusbyi A. Gray var. *eremicola* (Jepson) Kearney

English: Rusby's desert mallow, Panamint mallow

Sphaeranthus L. Asteraceae

Origins:

Greek *sphaira* and *anthos* "flower," in reference to the globular clusters of capitula; see Carl Linnaeus, *Species Plantarum*. 927. 1753 and *Genera Plantarum*. Ed. 5. 399. 1754.

Species/Vernacular Names:

S. africanus L.

India: sveta-hapusa, velutta-atakkamaniyan, gorakh mundi

S. indicus L.

English: Indian globe thistle

India: hapusa, atakkamaniyan, gorakh mundi, adaca-manjen, atakkamaniyen

Sphaerantia Peter G. Wilson & B. Hyland Myrtaceae

Origins:

From the Greek *sphaira* "a globe" and *anthos* "flower"; see Peter G. Wilson & Bernard Patrick Matthew Hyland (b. 1937), in *Telopea*. 3: 260. (May) 1988.

Sphaerella Bubani Gramineae

Origins:

Greek *sphaira* "a globe, ball, sphere," the diminutive.

Sphaereupatorium (O. Hoffmann) Robinson Asteraceae

Origins:

Greek *sphaira* "a globe" and the genus *Eupatorium* L.

Sphaerium Kuntze Gramineae

Origins:

Greek *sphaira* "a globe," *sphairion* "a little ball, a pill."

Sphaerobambos S. Dransf. Gramineae

Origins:

From the Greek *sphaira* "a globe" plus *bambos, bambusa*.

Sphaerocardamum Nees & Schauer Brassicaceae

Origins:

Greek *sphaira* "a globe, ball" and *kardamon* "nasturtium, a kind of cress."

Sphaerocaryum Nees ex Hook.f. Gramineae

Origins:

From the Greek *sphaira* "a globe" and *karyon* "nut," referring to the globular grains.

Sphaerocionium C. Presl Hymenophyllaceae

Origins:

Greek *sphaira* and *kion* "a column, pillar"; see Karl B. Presl, in *Abhandlungen der Königlichen Böhmischen Gesellschaft der Wissenschaften*. Ser. 5, 3: 125. 1845.

Sphaeroclinium (DC.) Sch. Bip. Asteraceae

Origins:

From the Greek *sphaira* "a globe" and *klinion* the diminutive of *kline* "a bed, couch, receptacle."

Sphaerocodon Benth. Asclepiadaceae

Origins:

From the Greek *sphaira* and *kodon* "a bell," referring to the shape of the corolla.

Sphaerocoma T. Anderson Caryophyllaceae (Illecebraceae)

Origins:

From the Greek *sphaira* and *kome* "hair of the head, lock of hair."

Sphaerocoryne (Boerl.) Ridley Annonaceae

Origins:
Greek *sphaira* "a globe" and *koryne* "a club."

Sphaerocyperus Lye Cyperaceae

Origins:
From the Greek *sphaira* "a globe" and the genus *Cyperus* L.

Sphaerodendron Seem. Araliaceae

Origins:
From the Greek *sphaira* and *dendron* "tree."

Sphaerodiscus Nakai Celastraceae

Origins:
Greek *sphaira* and *diskos* "a disc."

Sphaerogyne Naudin Melastomataceae

Origins:
From the Greek *sphaira* "a globe" and *gyne* "woman, female."

Sphaerolobium J.E. Smith Fabaceae

Origins:
Greek *sphaira* and *lobos* "a pod," referring to the shape of the pods; see Karl Koenig (1774-1851) and John Sims, *Annals of Botany*. 1: 509. London (Jan.) 1805.

Species/Vernacular Names:
S. alatum Benth.
English: winged globe-pea
S. daviesioides Turcz.
English: leafless globe-pea
S. vimineum Smith
English: prickly globe-pea

Sphaeroma DC. Malvaceae

Origins:
From the Greek *sphaira* "a globe."

Sphaeromariscus E.G. Camus Cyperaceae

Origins:
Latin *sphaera, ae* "a ball" and *mariscos* or *mariscus, i* "a rush," see also *Cyperus* L.

Sphaeromeria Nutt. Asteraceae

Origins:
From the Greek *sphaira* "a sphere" and *meris* "a part, portion."

Sphaeromorphaea DC. Asteraceae

Origins:
Greek *sphaira* and *morphe* "a form, shape," referring to the inflorescences; see A.P. de Candolle, *Prodromus*. 6: 140. (Jan.) 1838.

Sphaerophora Blume Rubiaceae

Origins:
Greek *sphaira* "a globe" and *phoros* "bearing"; see *Mus. Bot*. 1: 179. Oct 1850? (non (A. H. Hassall) J. Lindley 1846).

Sphaerophora C.H. Schultz-Bip. Asteraceae

Origins:
Greek *sphaira* and *phoros* "bearing."

Sphaerophysa DC. Fabaceae

Origins:
From the Greek *sphaira* "a globe" and *physa* "bladder," referring to the fruits; see A. P. de Candolle, *Prodromus*. 2: 270. Nov. (med.) 1825.

Species/Vernacular Names:
S. salsula (Pall.) DC.
English: Austrian peaweed

Sphaeropteris Bernh. Cyatheaceae

Origins:
From the Greek *sphaira* "a globe" and *pteris* "fern"; see Heinrich Adolph Schrader, *Journal für die Botanik*. 1800(2): 122, t. 1(1). (Oct.-Dec.) 1801.

Sphaeropteris R. Br. ex Wall. Dryopteridaceae

Origins:

Greek *sphaira* "a globe" and *pteris* "fern."

Sphaeropus Boeckeler Cyperaceae

Origins:

From the Greek *sphaira* and *pous* "foot"; see Johann Otto Boeckeler (1803-1899), *Flora oder allgemeine Botanische Zeitung.* 56: 89. (Feb.) 1873.

Sphaerosacme Wallich ex M. J. Roemer Meliaceae

Origins:

Greek *sphaira* "a globe" and *akme* "the highest point"; see *Fam. Nat. Syn. Monogr.* 1: 81, 98. 14 Sep-15 Oct 1846.

Sphaerosciadium Pimenov & Kljuykov Umbelliferae

Origins:

From the Greek *sphaira* and *skiadion, skiadeion* "umbel, parasol."

Sphaerosepalum Baker Sphaerosepalaceae (Ochnaceae, Diegodendraceae)

Origins:

Greek *sphaira* and Latin *sepalum* "sepal."

Sphaerosicyos Hook.f. Cucurbitaceae

Origins:

From the Greek *sphaira* "a globe" and *sikyos* "wild cucumber, gourd."

Sphaerostephanos J. Sm. Thelypteridaceae

Origins:

Greek *sphaira* "a globe" and *stephein* "to crown," *stephanos* "a crown," referring to the shape of the indusium; see W.J. Hooker, *Genera Filicum.* t, 24. 1839.

Sphaerostichum C. Presl Polypodiaceae

Origins:

Greek *sphaira* "a globe" and *stichos* "a row, line."

Sphaerostigma (Ser.) Fischer & C.A. Meyer Onagraceae

Origins:

From the Greek *sphaira* "a globe" and *stigma* "a stigma."

Sphaerostylis Baillon Euphorbiaceae

Origins:

From the Greek *sphaira* and *stylos* "a style, column."

Sphaerothalamus Hook.f. Annonaceae

Origins:

Greek *sphaira* and *thalamos* "the base of the flower."

Sphaerothylax Bisch. ex Krauss Podostemaceae

Origins:

From the Greek *sphaira* "a globe" and *thylakos* "bag, sack, pouch, scrotum."

Sphaerotylos Chen Urticaceae

Origins:

Greek *sphaira* "a globe" and *tylos* "lump, knob."

Sphagneticola O. Hoffm. Asteraceae

Origins:

Latin *sphagnos, i* "a kind of fragrant moss" (Plinius).

Sphalanthus Jack Combretaceae

Origins:

Probably from the Greek *sphallos, sphalos* "a round block of wood" and *anthos* "flower."

Sphallerocarpus Besser ex DC. Umbelliferae

Origins:

Greek *sphaleros* "slippery" and *karpos* "fruit."

Sphalmanthus N.E. Br. Aizoaceae

Origins:

Greek *sphalma*, *sphalmatos* "an error, stumble, false step" and *anthos* "flower."

Sphalmium B.G. Briggs, B. Hyland & L.A.S. Johnson Proteaceae

Origins:

Greek *sphalma* "an error," probably referring to the classification of *Orites racemosa* C. White, base name for *Sphalmium racemosum* (C. White) B. Briggs, B. Hyland & L. Johnson; see Barbara Gillian Briggs (1934-), Bernard Patrick Matthew Hyland and Lawrence Alexander Sidney Johnson (1925-), in *Australian Journal of Botany*. 23: 165. (Feb.) 1975; James A. Baines, *Australian Plant Genera. An Etymological Dictionary of Australian Plant Genera*. 349. Chipping Norton, N.S.W. 1981.

Sphedamnocarpus Planchon ex Bentham Malpighiaceae

Origins:

Presumably from the Greek *sphendamnos* "Olympian maple" and *karpos* "seed," referring to the fruit shape.

Species/Vernacular Names:

S. pruriens (Juss.) Szyszyl.

Southern Africa: Ngadzapene, Nyakanyoka (Shona)

Sphenandra Benth. Scrophulariaceae

Origins:

Greek *sphen* "wedge" and *aner, andros* "male, anther," referring to the shape of the anthers.

Sphenantha Schrad. Cucurbitaceae

Origins:

From the Greek *sphen* "wedge" and *anthos* "flower," see also *Cucurbita* L.

Sphenista Raf. Chrysobalanaceae (Rosaceae)

Origins:

See Constantine Samuel Rafinesque, *Sylva Telluriana*. 90. 1838; E.D. Merrill, *Index rafinesquianus*. 140. 1949.

Sphenocarpus Korovin Umbelliferae

Origins:

Greek *sphen* "wedge" and *karpos* "fruit."

Sphenocentrum Pierre Menispermaceae

Origins:

From the Greek *sphen* "wedge" and *kentron* "a spur, prickle."

Species/Vernacular Names:

S. jollyanum Pierre

Nigeria: ijo, ajo, obanabe

Yoruba: akerejupon pupa, obalabi, obanabe, ogbalagbe

Sphenoclea Gaertner Sphenocleaceae (Campanulaceae)

Origins:

From the Greek *sphen* and *kleio* "to shut, enclose," referring to the dehiscence of the fruit or to the shape of the capsule; see Joseph Gaertner, *De fructibus et seminibus plantarum*. 1: 113. 1788.

Sphenodesme Jack Verbenaceae (Symphoremataceae)

Origins:

From the Greek *sphen* "wedge" and *desme* "a bundle," referring to the flowers.

Species/Vernacular Names:

S. involucrata (Presl) Robinson

English: involucrate sphenodesme

China: zhua xie chi teng

S. mollis Craib

English: hairy sphenodesme

China: mao xie chi teng

Sphenogyne R. Br. Asteraceae

Origins:
From the Greek *sphen* and *gyne* "a woman, female," alluding to the flower.

Sphenomeris Maxon Dennstaedtiaceae (Lindsaeoideae)

Origins:
Greek *sphen* "wedge" and *meris* "a part, portion," alluding to the segments of the fronds; see William Ralph Maxon (1877-1948), in *Journal of the Washington Academy of Sciences*. 3: 144. 1913.

Species/Vernacular Names:
S. biflora (Kaulf.) Tagawa (*Davallia biflora* Kaulf.)
Japan: hama-hora-shinobu (= beach *Sphenomeris*)

Sphenopholis Scribner Gramineae

Origins:
Greek *sphen* and *pholis, pholidos* "scale, horny scale," referring to the shape of the upper glume.

Species/Vernacular Names:
S. obtusata (Michaux) Scribner
English: prairie wedgegrass

Sphenopus Trinius Gramineae

Origins:
From the Greek *sphen* "wedge" and *pous* "foot," referring to the thickened pedicels; see Carl Bernhard von Trinius (1778-1844), *Fundamenta Agrostographiae*. 1820.

Sphenosciadium A. Gray Umbelliferae

Origins:
From the Greek *sphen* "wedge" and *skiadion, skiadeion* "umbel, parasol."

Species/Vernacular Names:
S. peckiana J.F. Macbr.
English: swamp white heads, ranger's buttons

Sphenostemon Baill. Aquifoliaceae (Sphenostemonaceae)

Origins:
Greek *sphen* "a wedge" and *stemon* "stamen," referring to the shape of the stamens; see H.E. Baillon, in *Bulletin mensuel de la societé linnéenne de Paris*. 7: 53. Paris 1875.

Species/Vernacular Names:
S. lobosporus (F. Muell.) L.S. Smith (*Phlebocalymna lobospora* F. Muell.)
English: feather beech

Sphenostigma Baker Iridaceae

Origins:
From the Greek *sphen* "a wedge" and *stigma* "stigma."

Sphenostylis E. Meyer Fabaceae

Origins:
From the Greek *sphen* and *stylos* "a column, style," referring to the wedge-shaped style; see E.H.F. Meyer, *Comment. Pl. Africae Austr*. 148. 14 Feb-5 Jun 1836.

Sphenotoma (R. Br.) R. Sweet Epacridaceae

Origins:
Greek *sphen* and *tome, tomos, temno* "division, section, to slice," referring to the corona; see Robert Sweet (1783-1835), *Flora Australasica*. t. 44. (Apr.) 1828.

Spheroidea Dulac Marsileaceae

Origins:
Greek *sphaira* "a globe, ball, sphere," resembling a ball.

Spherosuke Raf. Moraceae

Origins:
Greek *sphaira* and *sykon* "fig"; see Constantine Samuel Rafinesque, in *Sylva Telluriana*. 57. 1838; E.D. Merrill, *Index rafinesquianus*. 112. 1949.

Sphinctacanthus Benth. Acanthaceae

Origins:
From the Greek *sphinktos* "tightly bound," *sphingo, sphiggo, sphiggein, sphingein* "to bind" and *Acanthus.*

Sphinctanthus Benth. Rubiaceae

Origins:
Greek *sphinktos* "tightly bound" and *anthos* "flower."

Sphincterostoma Stschegl. Epacridaceae

Origins:
Greek *sphinkter* "that which binds tight, a lace" and *stoma* "mouth"; see Serge S. Stscheglejew, in *Bulletin de la Société Impériale des Naturalistes de Moscou.* 32(1): 22. 1859.

Sphinctospermum Rose Fabaceae

Origins:
Greek *sphinktos* "tightly bound" and *sperma* "seed."

Sphingiphila A.H. Gentry Bignoniaceae

Origins:
Probably from the Latin *sphingion, ii* "a kind of ape, bracelet, necklace" or Greek *sphingo* "bind tight" and *philos* "loving."

Sphragidia Thwaites Euphorbiaceae

Origins:
Greek *sphragis, sphragidos* "signet, a seal," *sphragites* "the impression of a seal."

Sphyranthera Hook.f. Euphorbiaceae

Origins:
Greek *sphyra* "a hammer" and *anthera* "an anther," referring to the shape of the anthers.

Sphyrarhynchus Mansfield Orchidaceae

Origins:
Greek *sphyra* "a hammer" and *rhynchos* "horn, beak, snout," referring to the shape of the rostellum.

Sphyrastylis Schltr. Orchidaceae

Origins:
From the Greek *sphyra* and *stylos* "style, column," referring to the swelling of the apex of the column.

Sphyrospermum Poeppig & Endl. Ericaceae

Origins:
From the Greek *sphyra* "a hammer" and *sperma* "seed."

Spicanta C. Presl Blechnaceae

Origins:
Etymology not clear, perhaps from an ancient German name ("Spicant aliqui Germani vocant, forte quod radices Indicam Spicam aliquomodo referant" Gesner); see Karl B. Presl, *Epimeliae botanicae.* 114. (Oct.) 1851; Giovanni Semerano, *Le origini della cultura europea.* Dizionario della lingua Latina e di voci moderne. 2(2): 572-573. Firenze 1994; Helmut Genaust, *Etymologisches Wörterbuch der botanischen Pflanzennamen.* 602-603. Basel 1996.

Spicantopsis Nakai Blechnaceae

Origins:
Resembling *Spicanta* C. Presl.

Spicillaria A. Rich. Rubiaceae

Origins:
Latin *spica* or *speca* "a point, spike, tuft"; Latin *specillum* (from *specio* "to look") is a surgical instrument.

Spiciviscum Engelm. Viscaceae

Origins:
Latin *spica, ae* "a point, spike, tuft" plus *Viscum* L., see also *Phoradendron* Nutt.

Spiculaea Lindley Orchidaceae

Origins:

Latin *spiculum* "a little sharp point, a sting," *spiculus, a, um* "pointed," *spicula, ae* "a plant, called also *chamaepitys*, ground-pine," referring to the petals and sepals or to the claw on the labellum; see John Lindley (1799-1865), *Edwards's Botanical Register*. Appendix to Vols. 1-23: *A Sketch of the Vegetation of the Swan River Colony*. lvi. London (Jan.) 1840.

Spielmannia Medikus Scrophulariaceae

Origins:

After the German botanist (b. Alsatia) Jakob (Jacob) Reinhold Spielmann, 1722-1783, traveler, pharmacist, physician. His works include *Pharmacopoea generalis*, edita a Jacobo Reinholdo Spielmann. Argentorati 1783 and *Anleitung zur Kenntniss der Arzneymittel*. Strassburg 1775; see John H. Barnhart, *Biographical Notes upon Botanists*. 3: 310. 1965; G. Schmid, *Goethe und die Naturwissenschaften*. Halle 1940; T.W. Bossert, *Biographical Dictionary of Botanists Represented in the Hunt Institute Portrait Collection*. 378. 1972; Ethelyn Maria Tucker, *Catalogue of the Library of the Arnold Arboretum of Harvard University*. Cambridge, Massachusetts 1917-1933; Jonas C. Dryander, *Catalogus bibliothecae historico-naturalis Josephi Banks*. London 1800.

Spigelia L. Strychnaceae (Loganiaceae)

Origins:

In honor of the Dutch (b. Brussels, Belgium) physician Adriaan (Adrian, Adrianus, Adriano) van der Spiegel (Spiegelius, Spigelius, Spieghel, Spigeli, Spigel), 1578-1625 (d. Padova), botanist, professor of anatomy and surgery at Padua, 1623 he was elected knight of St. Mark. Among his writings are *Isagoges in rem herbariam libri duo*. Patavii 1606, *De semitertiana libri quatuor*. Francofurti 1624, *De formato foetu liber singularis … Epistolae duae anatomicae. Tractatus de Arthritide*. Opera posthuma, studio L. Cremae … edita. Patavii [1626] and *De lumbrico lato liber*. Patavii 1618; see G.A. Lindeboom, in *Dictionary of Scientific Biography* 12: 577-578. 1981; R. Bentley and H. Trimen, "*Spigelia marilandica*." *Medicinal Plants*. 3: 180. 1880; Garrison and Morton, *Medical Bibliography*. 5229. 1961.

Species/Vernacular Names:

S. anthelmia L.

Yoruba: ewe aran

S. marilandica L.

English: pinkroot, Indian pink, wormgrass, worm tea

Spiladocorys Ridl. Asclepiadaceae

Origins:

Greek *spilas, spilados* "a spot" and *korys, korythos* "helmet."

Spilanthes Jacq. Asteraceae

Origins:

Greek *spilos* "a spot, stain, fleck" and *anthos* "flower," in some species the flowers are black or brown spotted; see Nicolaus Joseph von Jacquin (1727-1817), *Enumeratio systematica plantarum*, quas in insulis Caribaeis vicinaque Americes continente detexit novas, aut jam cognitas emendavit. 8, 28. Lugduni Batavorum 1760.

Spilocarpus Lem. Boraginaceae

Origins:

Greek *spilos* "a spot, stain, fleck" and *karpos* "fruit."

Spiloxene Salisb. Hypoxidaceae

Origins:

Greek *spilos* "a spot, stain, fleck" and *oxys* "sharp-pointed, sharp," in some species the corm is usually covered by thick scales; some suggest from *xenos* "a host, foreigner, alien."

Species/Vernacular Names:

S. aquatica (L.f.) Fourc.

South Africa: vleiblommetjie, witsterretjie, vleiuintjie, sterretjie, waterblom, watersterretjie

S. canaliculata Garside

South Africa: geelsterretjie, channelled spiloxene

S. capensis (L.) Garside (*Spiloxene stellata* (Thunb.) Salisb.)

South Africa: peacock flower, golden star, sterretjie, stars, poublommetjie, witsterretjie, geelsterretjie

S. flaccida (Nel) Garside

South Africa: yellow star, sterretjie

S. gracilipes (Schltr.) Garside

South Africa: slender spiloxene

S. ovata (L.f.) Garside

South Africa: geelsterretjie

S. scullyi (Bak.) Garside (for the Irish (b. Dublin) author William Charles Scully, 1855-1943 (d. Umbogintwini, South Africa, Natal), plant collector in eastern Cape and Namaqualand, magistrate, friend of E.E. Galpin; see Mary

Gunn and Leslie Edward W. Codd, *Botanical Exploration of Southern Africa*. 320. Cape Town 1981)

South Africa: sterretjie

S. serrata (Thunb.) Garside var. *serrata*

South Africa: golden star, sterretjie, gouesterretjie

Spinacia L. Chenopodiaceae

Origins:

Arabic *'isfenah, 'isfanak, isfinaj*; Persian *aspanah, ispanak, isfanaj*; Latin *spina, ae* "a prickle, thorn," Medieval Latin *spinacium*; see C.T. Onions, *The Oxford Dictionary of English Etymology*. Oxford University Press 1966; Helmut Genaust, *Etymologisches Wörterbuch der botanischen Pflanzennamen*. 603-604. 1996; Ernest Weekley, *An Etymological Dictionary of Modern English*. 2: 1390. New York 1967.

Species/Vernacular Names:

S. oleracea L.

English: spinach

Italian: spinacio

Japan: hôren-sô

Okinawa: furinno

China: bo cai, po leng tsai, potsai

Spinicalycium Fric Cactaceae

Origins:

Latin *spina, ae* "a thorn, spine" and Greek *kalyx* "calyx," see also *Acanthocalycium* Backeb.

Spinifex L. Gramineae

Origins:

Spiny, from the Latin *spina, ae* "a thorn, spine," referring to the leaves or to the inflorescences; see C. Linnaeus, *Mantissa Plantarum*. 2: 163, 300. 1771.

Species/Vernacular Names:

S. hirsutus Labill. (*Spinifex sericeus* R. Br.)

English: rolling spinifex, spiny rolling-grass, roly-poly, hairy spinifex, coastal spinifex, silver grass

S. littoreus (Burm.f.) Merr. (*Stipa littorea* Burm.f.)

Japan: tsuki-ige, hari-hama-mugi

Okinawa: uma-hara-sa

S. longifolius R. Br.

English: long-leaved spinifex, sea-coast grass, sandstay, porcupine grass

Spiniluma Baillon ex Aubrév. Sapotaceae

Origins:

Latin *spina, ae* "a thorn, spine" plus *luma*.

Spiracantha Kunth Asteraceae

Origins:

From the Greek *speira* "a spiral" and *akantha* "thorn."

Spiradiclis Blume Rubiaceae

Origins:

Greek *speira* "a spiral" and *diklis* "double-folding, two-valved," referring to the fruit.

Spiraea L. Rosaceae

Origins:

Greek *speira* "a spiral," *speiraia* "the herb meadowsweet," Latin *spiraea, ae* "the herb meadowsweet" (Plinius), the plant was once used for garlands and wreaths; see Carl Linnaeus, *Species Plantarum*. 489. 1753 and *Genera Plantarum*. Ed. 5. 216. 1754; G.C. Wittstein, *Etymologisch-botanisches Handwörterbuch*. 832. 1852.

Species/Vernacular Names:

S. alba Duroi

English: meadowsweet

S. x *arguta* Zab.

English: garland spiraea, bridal wreath

S. cantoniensis Lour.

English: Reeves spiraea, Reeve's spirea, Reeves spirea, Cape may, maybush, bridal wreath

Japan: ko-demai, kodemari

S. douglasii Hook.

English: hack-brush, hard-hack·

S. japonica L.f.

English: Japanese spiraea

Japan: shimotsuke

S. latifolia (Ait.) Borkh.

English: meadowsweet

S. x *multiflora* Zab.

English: snow garland spiraea

S. nipponica Maxim.

Japan: iwa-shimotsuke

S. prunifolia Siebold & Zucc.

English: bridal wreath spiraea, bridal wreath

Japan: shijimi-bana

China: xiao ye hua

S. salicifolia L.

English: bridewort

Japan: hozaki-shimotsuke

S. thunbergii Sieb. ex Blume

Japan: yuki-yanagi

S. tomentosa L.

English: hard-hack, steeplebush

Spiraeanthemum A. Gray Cunoniaceae

Origins:

Greek *speira* "a spiral" and *anthos, anthemon* "flower"; see A. Gray, "Characters of some new genera of plants mostly from Polynesia." 128. (May) 1854, in *Proceedings of the American Academy of Arts & Sciences*. 3: 128. 1857.

Spiraeanthus (Fischer & C.A. Meyer) Maxim. Rosaceae

Origins:

The genus *Spiraea* L. and *anthos* "flower."

Spiraeopsis Miq. Cunoniaceae

Origins:

Resembling a *speira* "a spiral."

Spiranthera A. St.-Hil. Rutaceae

Origins:

From the Greek *speira* "a spiral" and *anthera* "anther," referring to the form of the anthers.

Spiranthera Bojer Convolvulaceae

Origins:

Greek *speira* and *anthera* "anther."

Spiranthes L.C. Rich. Orchidaceae

Origins:

Coiled flowers, Greek *speira* "a spiral, coil" and *anthos* "flower," referring to the spiral or twisted inflorescence; see Louis Claude Marie Richard (1754-1821), *De Orchideis europaeis annotationes*. 20, 28, 36. Parisiis 1817.

Species/Vernacular Names:

S. sinensis (Pers.) Ames (*Neottia sinensis* Pers.; *Spiranthes australis* (R. Br.) Lindley; *Spiranthes neocaledonica* Schltr.)

English: ladies tresses, Austral ladies tresses

Japan: shiro-mojizuri, neji-bana

Okinawa: mudiku-bana

China: pan long shen

S. spiralis (L.) Chevallier

English: ladies tresses, autumn lady's tresses

Spirella Costantin Asclepiadaceae

Origins:

For the French physician Camille Joseph Spire, botanist and plant collector, in Indochina; see John H. Barnhart, *Biographical Notes upon Botanists*. 3: 310. 1965; Ethelyn Maria Tucker, *Catalogue of the Library of the Arnold Arboretum of Harvard University*. Cambridge, Massachusetts 1917-1933; Camille Spire, *Les Laotiens. Coutumes, hygiène, pratiques médicales*. Paris 1907.

Spiroceratium H. Wolff Umbelliferae

Origins:

From the Greek *speira* "a spiral, coil" and *keration* "little horn, pod, siliqua."

Spirochloe Lunell Gramineae

Origins:

From the Greek *speira* and *chloe, chloa* "grass."

Spirodela Schleiden Lemnaceae

Origins:

Greek *speira* and *delos* "evident, visible, obvious, conspicuous," referring to the roots; see Matthias Jacob Schleiden (1804-1881), in *Linnaea*. 13: 391. 1839.

Species/Vernacular Names:

S. polyrhiza (L.) Schleid. (*Lemna polyrhiza* L.) (Greek *polys* "much" and *rhiza* "root")

English: great duckweed, greater duckweed, duckweed, large duckweed, water flaxseed, duck's meat, giant duckweed

Japan: uki-kusa

China: fu ping

S. punctata (G. Meyer) Thompson (*Lemna punctata* G. Meyer; *Lemna oligorrhiza* Kurz; *Spirodela oligorrhiza* (Kurz) Hegelm.)

English: duckweed, thin duckweed

Spirogardnera Stauffer Santalaceae

Origins:

Greek *speira* "spiral" (referring to the spiral arrangement of the inflorescence) and Charles Austin Gardner, 1896-1970, in 1909 emigrated to Western Australia, in 1921 botanist to Kimberley Exploration Expedition, from 1928 to 1960 Government Botanist and Curator of the Western Australia Herbarium, plant collector, author of *Wildflowers of Western Australia*. Perth 1975 and *Enumeratio Plantarum Australiae Occidentalis*. 1930, with H.W. Bennetts wrote *The Toxic Plants of Western Australia*. 1956; see the Swiss botanist and botanical explorer Hans Ulrich Stauffer (1929-1965), in *Mitteilungen aus dem Botanischen Museum der Universität Zürich*. 307, fig. 1. 1968; Mary Gunn and Leslie Edward W. Codd, *Botanical Exploration of Southern Africa*. 333. Cape Town 1981.

Spirolobium Baillon Apocynaceae

Origins:

From the Greek *speira* "a spiral, coil" and *lobion* "a small lobe, little pod."

Spironema Lindl. Commelinaceae

Origins:

Greek *speira* and *nema* "thread," referring to the stamen filaments.

Spiropetalum Gilg Connaraceae

Origins:

From the Greek *speira* "spiral" and *petalon* "leaf, petal."

Spiropodium F. Muell. Asteraceae

Origins:

Greek *speira* and *podion* "a little foot, stalk" (*pous, podos* "a foot"); see Ferdinand von Mueller, *Fragmenta Phytographiae Australiae*. 1: 33. 1858.

Spirorhynchus Karelin & Kir. Brassicaceae

Origins:

From the Greek *speira* "spiral" and *rhynchos* "horn, beak, snout."

Spiroseris Rech.f. Asteraceae

Origins:

From the Greek *speira* "spiral" and *seris, seridos* "chicory, lettuce."

Spirospermum Thouars Menispermaceae

Origins:

From the Greek *speira* "spiral" and *sperma* "seed."

Spirostachys S. Watson Chenopodiaceae

Origins:

Greek *speira* and *stachys* "a spike."

Spirostachys Sonder Euphorbiaceae

Origins:

From the Greek *speira* "spiral" and *stachys* "a spike," referring to the flower spikes.

Species/Vernacular Names:

S. africana Sond.

English: jumping-bean tree, African spirostachys, African mahogany tree, African sandalwood, Cape sandalwood, headache tree

Southern Africa: tamboti, tambotie, umtombotie, sandaleen wood, helengomaash; muTivoti, muTomboti (Shona); umThombothi, umtomboti, iJuqu, uBanda (Zulu); umtomboti, umThombothi (Xhosa); umThombotsi, umThombothi (Swazi); ndzopfori (Thonga); morukuru (Western Transvaal, northern Cape, Botswana); morekhure (Kgatla dialect, Botswana); morekuri, morekure (North Sotho); muonze

(Venda); orupapa (Herero); tshopfori (Shangaan); ubande (Ndebele)

Spirostachys Ung.-Sternb. Chenopodiaceae

Origins:

Greek *speira* "spiral" and *stachys* "a spike," see also *Heterostachys* Ung.-Sternb.

Spirostegia Ivanina Scrophulariaceae

Origins:

From the Greek *speira* and *stege* "roof, covering, shelter."

Spirostigma Nees Acanthaceae

Origins:

Greek *speira* "spiral" and *stigma* "stigma."

Spirostylis C. Presl ex Schult. & Schult.f. Loranthaceae

Origins:

From the Greek *speira* "spiral" and *stylos* "a style."

Spirotecoma Baillon ex Dalla Torre & Harms Bignoniaceae

Origins:

Greek *speira* and the genus *Tecoma* Juss.

Spirotheca Ulbr. Bombacaceae

Origins:

From the Greek *speira* "spiral" and *theke* "a box, case."

Spirotheros Raf. Gramineae

Origins:

Possibly from the Greek *speiro, sperro* "sow seed" and *thereios, theros* "summer, in summer, summer-time"; see E.D. Merrill, *Index rafinesquianus*. 76. 1949.

Spirotropis Tul. Fabaceae

Origins:

Greek *speira* "spiral" and *tropis, tropidos* "a keel."

Spixia Leandro Euphorbiaceae

Origins:

For the German (b. Höchstadt an der Aisch) naturalist Johann Baptist von Spix, 1781-1826 (d. Munich), zoologist, traveler and explorer (with Martius in Brazil), zoological collector; see John H. Barnhart, *Biographical Notes upon Botanists*. 3: 311. 1965; Ethelyn Maria Tucker, *Catalogue of the Library of the Arnold Arboretum of Harvard University*. Cambridge, Massachusetts 1917-1933; M. Colmeiro y Penido, *La Botánica y los Botánicos de la Peninsula Hispano-Lusitana*. Madrid 1858; A.P.M. Sanders, in *Dictionary of Scientific Biography* 12: 578-579. 1981; Giulio Giorello & Agnese Grieco, a cura di, *Goethe scienziato*. Einaudi Editore, Torino 1998; João Barbosa Rodrigues (1842-1909), *A flora brasiliensis de Martius*. Rio de Janeiro 1907; Christian Friedrich Schwägrichen (1775-1853), "Bemerkungen über einige Stellen in der Flora brasiliensis von Endlicher und Martius." *Linnaea*. 14(5): 517-528. (Feb.) 1841; G. Schmid, *Goethe und die Naturwissenschaften*. Halle 1940.

Spodiopogon Trin. Gramineae

Origins:

Greek *spodos, spodia* "ashes, ash grey" and *pogon* "a beard," referring to the hairs; see Carl Bernhard von Trinius, *Fundamenta Agrostographiae*. 192. 1820.

Spondianthus Engl. Euphorbiaceae

Origins:

Greek *anthos* "a flower" and the genus *Spondias*, see Ronald William John Keay, *Trees of Nigeria*. 172. Oxford Science Publications, Oxford 1989.

Species/Vernacular Names:

S. sp.

Nigeria: adenja, oboekute

Mexico: biache, piache, yaga biache, yaga piache, yaga yechi, yechi, yetze loa, ya reje

S. preussii Engl.

Nigeria: isena, opolata; obo ekute (= rat poison) (Yoruba); orho (Edo); opolata (Ijaw); ibok-eku (= rat poison) (Efik)

Yoruba: isena, owe, obo ekute

Cameroon: ebai, ategue, mfoum

Ivory Coast: agboboba, djilika, senan

Zaire: tshipanda

Spondias L. Anacardiaceae

Origins:

From *spondias* or *spodias* (*spodos*, *spodia* "ashes"), the Greek classical name for the wild plum tree (Theophrastus, *HP.* 3.6.4) or a tree of the plum kind, in reference to the fruit; see Carl Linnaeus, *Species Plantarum.* 371. 1753 and *Genera Plantarum.* Ed. 5. 174. 1754.

Species/Vernacular Names:

S. cytherea Sonn. (*Spondias dulcis* Parkinson)

English: Otaheite apple, great hog-plum, Aphrodite plum, Jew's plum, golden apple

Rodrigues Island: fruit de Cythère

Malaya: kedongdong, kedondong jawa

China: jen mien tzu

S. mombin L. (*Spondias lutea* L.) (from a French West Indian name)

English: yellow mombin, thorny hog-plum, mombin, Spanish plum, caja fruit

Tropical America: jobo

Bolivia: orocorocillo, cedrillo

Peru: ciruela, ciruelo, ciruela agria, ciruela de la China, hubas, itahuba, hubus, shungu, taperiba, ubo, ubos, ubos colorado, ushum, ushun

Mali: minkon, minkwo

Nigeria: aginiran, iyeye, yeye, anjegonkwi, ijikera, efor, ishinkere, iginihen, kasamango, monganga, tsadar-lama-rudu, tsadar masar, dadum, kumaga, haperriyi, enokeba, ogikan, agak, okiken, ugomugo, akikan, malaria, nganga, odji, ogege, okikan, opon, tomlibeli, ugriya; tsadar masar (Hausa); chabbulimakka (Fula); jinjerechi (Nupe); kakka (Tiv); ekikan (Yoruba); okhighan (Edo); okhikhen (Urhobo); akikan (Itsekiri); aginiran (Ijaw); isikala, ishikere (Igbo); nsukakara (Efik); kechibo (Boki); ekpi (Ekoi)

Yoruba: ekika, okika, iyeye, olosan, ilewo olosan

Tibetan: snying zho-sha

S. pinnata (L.f.) Kurz (*Spondias dulcis* var. *acida* (Bl.) Engl.; *Spondias mangifera* Willd.)

English: common hog-plum

Malaya: kedondong, memberah, emberah, emerah

India: amra

Nepal: amaroo, amaro

S. purpurea L. (*Spondias dulcis* Blanco; *Spondias mombin* Burck)

English: Spanish plum, red mombin

Tropical America: jocote

Guatemala: jobo, jocote

Central America: jocote, abal, canum, chiabal, ciruelo, jondura, k'iis, poon, rum, sismoyo, xúgut

Peru: ajuela, ciruela, ciruela agria

The Philippines: sineguelas, sirihuelas, sarguelas, saraguelas, sereguelas, siriguelas

Spondogona Raf. Sapotaceae

Origins:

See C.S. Rafinesque, *Sylva Telluriana.* 35. 1838; E.D. Merrill, *Index rafinesquianus.* 188. 1949.

Spongiocarpella Yakovlev & N. Ulziykh. Fabaceae

Origins:

From the Greek *spongias* "sponge" and *karpos* "fruit."

Spongiola J.J. Wood & A.L. Lamb Orchidaceae

Origins:

Greek *spongias* "sponge."

Spongiosperma Zarucchi Apocynaceae

Origins:

From the Greek *spongias* "sponge" and *sperma* "seed."

Spongiosyndesmus Gilli Umbelliferae

Origins:

Greek *spongias* and *syndesmos* "that which binds together."

Sponia Commerson ex Decaisne Ulmaceae

Origins:

Presumably after the French physician Jacob (Jacques) Spon, 1647-1685 (see Pritzel), traveler. Among his numerous publications are *Histoire de la ville e de l'estat de Geneve.* Lyon 1682, *Ignotorum atque obscurorum quorundam deorum arae; notis ... illustratae.* Lugduni 1676, *Observations sur les fièvres et les fébrifuges.* Lyon 1687, *Recherche des antiquités et curiosités de la Ville de Lyon. Avec un mémoire des principaux antiquaires & curieux de l'Europe.* Lyon 1676, *Voyage d'Italie, de Dalmatie, de*

Grece et du Levant. Lyon 1678-1680 and *Tractatus novi de potu caphè, de Chinensium thè, et de chocolata.* [Edition issued under Spon's pseudonym of P. Sylvestre Dufour] Parisiis 1685, with Jean Huguetan wrote *Voyage d'Italie curieux et nouveau.* Lyon, Thomas Amaulry 1681. See Sylvestre Dufour (Philippe) (1622-1687) [pseud. of Jacob Spon], Libellus primus sub titulo: *Jacobi Sponii Bevanda asiatica,* etc. [Lugduni] 1705; A. Mollière, *Une Famille médicale lyonnaise au XVIIe siècle. Charles et Jacob Spon.* 1905; Joseph Decaisne (1807-1882), in *Nouvelles Annales du Muséum d'Histoire Naturelle.* 3: 498. 1834; Arthur D. Chapman, ed., *Australian Plant Name Index.* 2709. Canberra 1991; [Librairie Paul Jammes — Paris], Cabinets de Curiosités. Collections. Collectionneurs. [Items no. 255, 332-336.] 1998.

Sporadanthus F. Muell. Restionaceae

Origins:
Greek *sporaden* "scatteredly," *sporadikos* "scattered" and *anthos* "flower."

Sporichloe Piler Gramineae

Origins:
Greek *spora, sporos* "seed, spore" and *chloe, chloa* "grass."

Sporobolus R. Br. Gramineae

Origins:
Greek *spora, sporos* "seed, spore" and *ballo, bolis, bolos* "casting," *boleo, bollein* "to throw," in reference to the dropping and the dispersion of the seeds; see Robert Brown (1773-1858), *Prodromus florae Novae Hollandiae et Insulae van-Diemen.* 169. London 1810.

Species/Vernacular Names:
S. actinocladus (F. Muell.) F. Muell.

English: ray grass

S. africanus (Poir.) Robyns & Tournay (*Sporobolus capensis* (Willd.) Kunth)

English: rat's tail dropseed, rat's tailgrass, rush grass, tough dropseed, smutgrass, African dropseed, South African rat-tail grass, Parramatta grass

South Africa: rotstert fynsaadgras, saadgras, taaipol, taaipolfynsaadgras, vleigras, matshiki

S. airoides (Torrey) Torrey

English: alkali sacaton

S. carolii Mez

English: Yakka grass, fairy grass

S. consimilis Fresen. (*Sporobolus robustus* sensu Chippind., non Kunth)

German: schilfgras

S. contractus A. Hitchc.

English: spike dropseed

S. cryptandrus (Torrey) A. Gray

English: sand dropseed

S. festivus Hochst.

Yoruba: irun awo

S. fimbriatus (Trin.) Nees (*Sporobolus fimbriatus* (Trin.) Nees var. *latifolius* Stent)

English: common dropseed

Southern Africa: blousaadgras, blousygras, gewone fynsaadgras, grootsoetgras, grootvleigras, soetvleigras; matolo-a-maholo (Sotho)

S. flexuosus (Vasey) Rydb.

English: Mesa dropseed

S. indicus (L.) R. Br.

English: smutgrass, Indian dropseed

China: shu wei su

S. mitchellii (Trin.) S.T. Blake (*Vilfa mitchellii* Trin.)

English: short rat-tail grass

S. nebulosus Hack.

German: nebelgras

S. pyramidalis Beauv.

English: cat's tail dropseed, cat's tail grass, narrow-plumed dropseed

Southern Africa: katstertfynsaadgras, smalpluimfynsaadgras, taaipol, vleigras; mixikijane (Tsonga)

Yoruba: motisan, sekogbona, ida odo

S. vaginiflorus (A. Gray) Alph. Wood

English: poverty grass

S. virginicus (L.) Kunth

English: seashore rushgrass, beach dropseed, salt couch, sand-and-Mud couch, coast rat-tail grass, Virginian dropseed

Hawaii: 'aki'aki, 'aki mahiki, mahikihiki, manienie, manienie 'aki'aki, manienie mahikihiki, manienie maoli

Spraguea Torrey Portulacaceae

Origins:
After the American Isaac Sprague, 1811-1895, botanical and zoological illustrator; see John H. Barnhart, *Biographical Notes upon Botanists.* 3: 311. 1965; T.W. Bossert, *Bio-*

graphical Dictionary of Botanists Represented in the Hunt Institute Portrait Collection. 378. 1972; Ethelyn Maria Tucker, Catalogue of the Library of the Arnold Arboretum of Harvard University. Cambridge, Massachusetts 1917-1933.

Spragueanella Balle Loranthaceae

Origins:

For the Scottish (b. Edinburgh) botanist Thomas Archibald Sprague, 1877-1958 (d. Chelthenham, Glos.), from 1900 to 1945 at Kew, authority on Loranthaceae, plant collector and traveler (Venezuela, Canary Islands, Colombia), 1903 Fellow of the Linnean Society, husband of the British botanist and bibliographer at Kew Mary Letitia Sprague (née Green) (1886-1978). His writings include "The botanical name of tara." Bull. Misc. Inform. 1931: 91-96. 1931, "The herbal of Leonhart Fuchs." J. Linn. Soc., Bot. 48: [545]-642. 1931, "The herbal of Valerius Cordus." J. Linn. Soc., Bot. 52: [1]-113. 1939 and "Preliminary report on the botany of Captain Dowding's Columbian Expedition, 1898-1899." Trans. Bot. Soc. Edinburgh. 22: 425-436. 1905. See Ray Desmond, Dictionary of British & Irish Botanists and Horticulturists. 647. 1994; John H. Barnhart, Biographical Notes upon Botanists. 3: 311. 1965; R. Zander, F. Encke, G. Buchheim and S. Seybold, Handwörterbuch der Pflanzennamen. 14. Aufl. Stuttgart 1993; Stafleu and Cowan, Taxonomic Literature. 5: 798-802. 1985; T.W. Bossert, Biographical Dictionary of Botanists Represented in the Hunt Institute Portrait Collection. 378. 1972; Hermia Newman Clokie, Account of the Herbaria of the Department of Botany in the University of Oxford. 247. Oxford 1964; Ethelyn Maria Tucker, Catalogue of the Library of the Arnold Arboretum of Harvard University. Cambridge, Massachusetts 1917-1933; Ida Kaplan Langman, A Selected Guide to the Literature on the Flowering Plants of Mexico. Philadelphia 1964.

Sprekelia Heister Amaryllidaceae (Liliaceae)

Origins:

For Johann Heinrich von Sprekelsen (or Sprekel), 1691-1764, botanist, sent a specimen to the author; see F. Boerner & G. Kunkel, Taschenwörterbuch der botanischen Pflanzennamen. 4. Aufl. 172. 1989; Helmut Genaust, Etymologisches Wörterbuch der botanischen Pflanzennamen. 606. 1996; Georg Christian Wittstein, Etymologisch-botanisches Handwörterbuch. 835. 1852.

Species/Vernacular Names:

S. formosissima (L.) Herbert

English: Jacobean lily

Sprengelia J.A. Schultes Sterculiaceae (Byttneriaceae)

Origins:

For the German (b. Boldekow) botanist Kurt Polycarp Joachim Sprengel, 1766-1833 (d. Halle), physician, M.D. Halle 1787, noted for his botanical work in plant-cell formation, professor of medicine and botany at the University of Halle, botanical collector. His works include Antiquitatum botanicarum. Lipsiae 1798, Historia rei herbaria. Amsterdam 1807, Geschichte der Botanik. Altenburg and Leipzig 1817-1818, Index plantarum quae in horto botanico halensi anno 1807 viguerunt. [Halle] [1807] and Flora halensis. Halae [Halle] 1832. He was son of the theologian Joachim Friedrich Sprengel, nephew of the German botanist Christian Konrad Sprengel (1750-1816) and father of the German botanist Anton Sprengel (1803-1851); see Guenter B. Risse, in Dictionary of Scientific Biography 12: 591-592. 1981; Lawrence J. King, in Dictionary of Scientific Biography 12: 587-591. 1981; R. Zander, F. Encke, G. Buchheim and S. Seybold, Handwörterbuch der Pflanzennamen. 14. Aufl. Stuttgart 1993; Stafleu and Cowan, Taxonomic Literature. 5: 804-814. 1985; John H. Barnhart, Biographical Notes upon Botanists. 3: 312. 1965; E. Bretschneider, History of European Botanical Discoveries in China. Leipzig 1981; Ida Kaplan Langman, A Selected Guide to the Literature on the Flowering Plants of Mexico. 715. University of Pennsylvania Press, Philadelphia 1964; A. Lasègue, Musée botanique de Benjamin Delessert. Paris 1845; S. Lenley et al., Catalog of the Manuscript and Archival Collections and Index to the Correspondence of John Torrey. Library of the New York Botanical Garden. 469. 1973; Günther Schmid, Chamisso als Naturforscher. Eine Bibliographie. Leipzig 1942; E.M. Tucker, Catalogue of the Library of the Arnold Arboretum of Harvard University. Cambridge, Massachusetts 1917-1933; Edward Lee Greene, Landmarks of Botanical History. Edited by Frank N. Egerton. Stanford University Press, Stanford, California 1983; G. Schmid, Goethe und die Naturwissenschaften. Halle 1940; J. A. Schultes, Observ. Bot. 134. 1809; Giulio Giorello & Agnese Grieco, a cura di, Goethe scienziato. Einaudi Editore, Torino 1998; Mariella Azzarello Di Misa, ed., Il Fondo Antico della Biblioteca dell'Orto Botanico di Palermo. 260-261. Regione Siciliana, Palermo 1988.

Sprengelia J.E. Smith Epacridaceae

Origins:

After the German botanist Christian Konrad Sprengel, 1750-1816 (Berlin), studied theology, teacher, naturalist, wrote Das entdeckte Geheimnis der Natur im Bau und in der Befruchtung der Blumen. [First edition of this work on the sexuality of plants, which influenced Darwin's later work on evolution and botany.] Berlin 1793. He was uncle of Kurt Polycarp Joachim Sprengel (1766-1833); see John H. Barnhart, *Biographical Notes upon Botanists*. 3: 312. 1965; J.E. Smith, in *Kongl. Vetenskaps Academiens Nya Handlingar*. 15: 260. Stockholm 1794; Guenter B. Risse, in *Dictionary of Scientific Biography* 12: 591-592. 1981; Lawrence J. King, in *Dictionary of Scientific Biography* 12: 587-591. 1981; T.W. Bossert, *Biographical Dictionary of Botanists Represented in the Hunt Institute Portrait Collection*. 379. 1972; Jonas C. Dryander, *Catalogus bibliothecae historico-naturalis Josephi Banks*. London 1800; Stafleu and Cowan, *Taxonomic Literature*. 5: 804-814. Utrecht 1985.

Species/Vernacular Names:

S. distichophylla (Rodway) W.M. Curtis

English: alpine swamp-heath

S. incarnata Smith

English: pink swamp-heath

S. monticola (DC.) Druce

English: mountain swamp-heath

S. sprengelioides (R. Br.) Druce (*Ponceletia sprengelioides* R. Br.)

English: white swamp-heath

Springalia Andrews Epacridaceae

Origins:

Misspelling for *Sprengelia* Smith; see Arthur D. Chapman, ed., *Australian Plant Name Index*. 2713. Canberra 1991.

Sprucea Benth. Rubiaceae

Origins:

Dedicated to the English (b. Ganthorpe, near Malton) botanist Richard Spruce, 1817-1893 (Yorkshire, Coneysthorpe, Castle Howard, near Malton), naturalist, traveler, cartographer, botanical explorer, plant collector, bryologist, from June 1849 to 1864 in South America (Venezuela, Brazil, Peru, Ecuador, Bolivia, on the Amazon and Andes), collected cinchona plants and seeds. His writings include "Palmae amazonicae." *J. Linn. Soc., Bot.* 11(50-51): 65-183. 1869 and *Hepaticae of the Amazon and of the Andes of Peru and Ecuador*. London [1884-]1885; see R.G.C. Desmond, in *Dictionary of Scientific Biography* 12: 594. 1981; Alfred Russell Wallace, ed., *Notes of a Botanist on the Amazon & Andes ...* during the years 1849-1864 by Richard Spruce. London 1908; Stafleu and Cowan, *Taxonomic Literature*. 5: 816-820. 1985; J.H. Barnhart, *Biographical Notes upon Botanists*. 3: 312. 1965; T.W. Bossert, *Biographical Dictionary of Botanists Represented in the Hunt Institute Portrait Collection*. 379. 1972; G. Murray, *History of the Collections Contained in the Natural History Departments of the British Museum*. 1: 184. London 1904; E.M. Tucker, *Catalogue of the Library of the Arnold Arboretum of Harvard University*. 1917-1933; H.N. Clokie, *Account of the Herbaria of the Department of Botany in the University of Oxford*. 247. Oxford 1964; S. Lenley et al., *Catalog of the Manuscript and Archival Collections and Index to the Correspondence of John Torrey*. Library of the New York Botanical Garden. 381. 1973; Charles Lyte, *The Plant Hunters*. London 1983; Henri Pittier, *Manual de las Plantas Usuales de Venezuela* y su Suplemento. Caracas 1978; August Weberbauer (1871-1948), *Die Pflanzenwelt der peruanischen Andes in ihren Grundzügen dargestellt*. 14-15. Leipzig 1911; R. Desmond, *The European Discovery of the Indian Flora*. Oxford 1992; Ray Desmond, *Dictionary of British & Irish Botanists and Horticulturists*. 647-648. London 1994.

Spruceanthus Sleumer Tiliaceae

Origins:

For the British botanist Richard Spruce, 1817-1893, traveler, botanical explorer, plant collector, from 1849 to 1864 in South America (Venezuela, Brazil, Peru, Ecuador, Bolivia, on the Amazon and Andes).

Sprucella Pierre Sapotaceae

Origins:

Dedicated to the British botanist Richard Spruce, 1817-1893.

Sprucina Nied. Malpighiaceae

Origins:

After the British botanist Richard Spruce, 1817-1893.

Spuriodaucus C. Norman Umbelliferae

Origins:

Latin *spurius, i* "a bastard, false, spurious" and *daucus, i* "a plant of the carrot kind," genus *Daucus* L.

Spuriopimpinella Kitag. Umbelliferae

Origins:

From the Latin *spurius, i* "a bastard, false, spurious" plus *Pimpinella*.

Spyridium Fenzl Rhamnaceae

Origins:

Greek *spyridion* "a little basket," *spyris* "a basket," referring to the flowerheads surrounded by bracts or to the form of the calyx; see S.L. Endlicher et al., *Enumeratio plantarum quas in Novae Hollandiae* ... collegit C. de Hügel. 24, in nota. Wien 1837.

Species/Vernacular Names:

S. bifidum (F. Muell.) Benth. (*Trymalium bifidum* F. Muell.)

English: forked spyridium

S. cinereum Wakef.

English: slender spyridium

S. coactilifolium Reissek (*Cryptandra coactilifolia* (Reissek) F. Muell.) (Latin *coactus* "felted")

English: butterfly spyridium, felted spiridium

S. halmaturinum (F. Muell.) Benth. (*Cryptandra halmaturina* (F. Muell.) F. Muell.; *Trymalium halmaturinum* F. Muell.)

English: Kangaroo Island spyridium

S. parvifolium (Hook.) F. Muell. (*Pomaderris parvifolia* Hook.; *Cryptandra parvifolia* (Hook.) Hook.f.; *Trymalium parvifolium* (Hook.) F. Muell. ex Reissek)

Australia: dusty miller, Australian dusty miller

Squamellaria Becc. Rubiaceae

Origins:

From the Latin *squama, ae* "a scale."

Squamopappus Jansen, Harriman & Urbatsch Asteraceae

Origins:

Latin *squama, ae* "a scale" and *pappus* "the woolly hairy seed of certain plants."

Staavia Dahl Bruniaceae

Origins:

After one Martin Staaf, a correspondent of Linnaeus; see W.P.U. Jackson, *Origins and Meanings of Names of South African Plant Genera.* 168. Rondebosch 1990.

Stachyandra Leroy ex Radcl.-Sm. Euphorbiaceae

Origins:

Greek *stachys* "a spike, an ear of grain, ear of corn" and *aner, andros* "man, male, stamen," see also *Androstachys* Prain.

Stachyanthus DC. Asteraceae

Origins:

Greek *stachys* and *anthos* "flower."

Stachyanthus Engl. Icacinaceae

Origins:

Greek *stachys* "a spike, an ear of grain, ear of corn" and *anthos* "flower."

Stachyanthus Engl. Orchidaceae

Origins:

From the Greek *stachys* and *anthos* "flower."

Stachyarrhena Hook.f. Rubiaceae

Origins:

From the Greek *stachys* "a spike, an ear of grain, ear of corn" and *arrhen* "male."

Stachycarpus (Endl.) Tieghem Podocarpaceae

Origins:

Greek *stachys* and *karpos* "fruit"; see Philippe Édouard Léon van Tieghem, in *Bulletin de la Société Botanique de France.* 38: 163, 173. 1891.

Stachycephalum Sch. Bip. ex Bentham Asteraceae

Origins:

Greek *stachys* "a spike, an ear of grain, ear of corn" and *kephale* "head."

Stachydeoma Small Labiatae

Origins:
Greek *stachys* plus the genus *Hedeoma* Pers.

Stachygynandrum P. Beauv. Selaginellaceae

Origins:
From the Greek *stachys* "a spike," *gyne* "female" and *aner, andros* "male."

Stachyococcus Standley Rubiaceae

Origins:
From the Greek *stachys* and *kokkos* "berry."

Stachyophorbe (Martius) Liebm. Palmae

Origins:
From the Greek *stachys* and *phorbe* "food."

Stachyopsis Popov & Vved. Labiatae

Origins:
From the Greek *stachys* and *opsis* "appearance," genus *Stachys* L.

Species/Vernacular Names:
S. lamiiflora (Ruprecht) Popov & Vved.
English: heart-leaf falsebetony
China: xin ye jia shui su
S. marrubioides (Regel) Ikonn.-Gal.
English: hairy falsebetony
China: duo mao jia shui su
S. oblongata (Schrenk ex Fischer & C.A. Meyer) Popov & Vved.
English: oblong falsebetony
China: jia shui su

Stachyothyrsus Harms Caesalpiniaceae

Origins:
From the Greek *stachys* "a spike, an ear of grain, ear of corn" and *thyrsos* "a panicle."

Stachyphrynium K. Schumann Marantaceae

Origins:
From the Greek *stachys* and *phrynos* "a frog," *phrynion* for a plant, called also *poterion*, referring to the habitat; see Engler, *Pflanzenr.* IV. 48(Heft 11): 45. 8 Jul. 1902.

Stachyphyllum Tieghem Eremolepidaceae

Origins:
From the Greek *stachys* "a spike, ear of wheat" and *phyllon* "a leaf."

Stachys L. Labiatae

Origins:
Greek *stachys* "a spike, an ear of grain, ear of corn," Latin *stachys, yos* "a plant, horsemint," referring to the inflorescence; see Carl Linnaeus, *Species Plantarum.* 580. 1753 and *Genera Plantarum.* Ed. 5. 253. 1754; Giovanni Semerano, *Le origini della cultura europea.* Dizionario della lingua Greca. 2(1): 276. Leo S. Olschki Editore, Firenze 1994.

Species/Vernacular Names:
S. adulterina Hemsley
English: few-hair betony
China: shao mao gan lu zi
S. arvensis (L.) L. (*Glechoma arvensis* L.)
English: stagger weed, field woundwort, fieldnettle betony
China: tian ye shui su
S. baicalensis Fischer ex Bentham
English: Baikal betony, Chinese fieldnettle
China: mao shui su, shui su
S. byzantina K. Koch (*Stachys lanata* Jacq.)
English: lamb's tongue, lamb's tail, lamb's ears, woolly betony, woolly stachys
S. chinensis Bunge ex Bentham
English: Chinese betony
China: hua shui su
S. geobombicis C.Y. Wu
English: earth-silkworm betony
China: di can
S. germanica L.
English: downy woundwort
S. hyssopoides Burch. ex Benth.
Southern Africa: pienksalie; selaoane (Sotho)

S. japonica Miquel

English: Japanese betony

China: shui su

S. lanata Jacquin

English: lanate betony

China: mian mao shui su

S. melissaefolia Bentham

English: melissa-leaf betony

China: duo zhi shui su

S. oblongifolia Wallich ex Bentham

English: oblong-leaf betony

China: zheng tong cai

S. officinalis (L.) Trev. (*Stachys betonica* Benth.; *Betonica officinalis* L.)

English: bishop's wort, wood betony, betony, woundwort

Arabic: batuniqa, shatra

S. palustris L.

English: marsh betony, marshy betony

China: zhao sheng shui su

S. pseudophlomis C.Y. Wu

English: narrow-teeth betony

China: xia chi shui su

S. pycnantha Benth.

English: short-spiked hedge nettle

S. sieboldii Miquel

English: Siebold betony, Chinese artichoke

China: gan lu zi, kan lu tsu, cao shi can, tsao shih tsan, ti tsan

S. sieboldii Miquel var. *sieboldii* (*Stachys affinis* Bunge; *Stachys tuberifera* Naudin)

English: Chinese artichoke, Japanese artichoke, knotroot, chorogi, artichoke betony

French: crosnes du Japon, crosnes

China: gan lu zi, tsao shih tsan, ti tsan, kan lu tsu

S. sylvatica L.

English: white-spot betony

China: lin di shui su

Stachystemon Planchon Euphorbiaceae

Origins:

Greek *stachys* "a spike" and *stemon* "a thread, stamen," referring to the united stamens; see Jules Émile Planchon (1823-1888), *Hooker's London Journal of Botany*. 4: 471. 1845.

Stachytarpheta Vahl Verbenaceae

Origins:

Greek *stachys* and *tarphys* "thick," *tarpheios* "dense," referring to the flower spikes; see Martin Vahl (1749-1804), *Enumeratio Plantarum*. 1: 205. 1804; Maria Helena Farelli, *Plantas que curam e cortam feitiços*. Rio de Janeiro 1988; W.G. Herter & S.J. Rambo, "Nas pegadas dos naturalistas Sellow e Saint-Hilaire." *Revista Sudamericana de Botánica*. 10, 3: 61-98. 1953; Pierre Fatumbi Verger, *Ewé: The Use of Plants in Yoruba Society*. São Paulo 1995; Celia Blanco, *Santeria Yoruba*. Caracas 1995; William W. Megenney, *A Bahian Heritage*. University of North Carolina at Chapel Hill 1978.

Species/Vernacular Names:

S. angustifolia (Mill.) Vahl

Yoruba: ogan akuko, ogangan, igbole

S. cayennensis (Rich.) Vahl

English: Brazilian tea

Japan: nagabo-sô, honaga-sô

Yoruba: iru eko, pasaloke

S. dichotoma (Ruíz & Pavón) Vahl (*Verbena dichotoma* Ruíz & Pavón; *Stachytarpheta australis* Moldenke)

Hawaii: owi, oi

S. jamaicensis (L.) Vahl (*Stachytarpheta indica* Vahl; *Verbena jamaicensis* L.)

English: blue snakeweed, common snakeweed, Jamaica vervain, jamaica false-valerian, devil's coach whip, Brazilian tea

Japan: futo-bo-naga-bo-sô

China: jia ma bian

The Philippines: kandi-kandilaan, bolo-moros, albaka

Malaya: selaseh dandi, selaseh hutan

Hawaii: owi, oi

Yoruba: opapa, opapara, panipani, akitipa, iru alangba, iru amore, aagba, agogo igun

S. mutabilis (Jacq.) Vahl (*Verbena mutabilis* Jacq.)

English: pink snakeweed

S. urticifolia Sims (*Cymburus urticifolia* Salisb.)

English: snakeweed, blue rat's tail

Japan: ho-naga-sô

Stachyurus Siebold & Zucc. Stachyuraceae

Origins:

Greek *stachys* "a spike" and *oura* "a tail," referring to the shape of the racemes.

Stackhousia J.E. Smith Stackhousiaceae (Rosidae, Celastrales)

Origins:

For the British (b. Cornwall) botanist John Stackhouse, 1742-1819 (d. Bath, Somerset), algologist, botanical artist, 1795 Fellow of the Linnean Society of London, collected algae. His works include *Nereis britannica.* Bathoniae [Bath] and Londini [1795-] 1801, *Extracts from Bruce's Travels in Abyssinia.* Bath 1815 and *Illustrationes Theophrasti.* Oxonii [Oxford] 1811, editor of *Theophrastii Eresii de Historia plantarum* libri decem. Oxonii 1813-1814; see Jonas C. Dryander, *Catalogus bibliothecae historico-naturalis Josephi Banks.* London 1800; J.H. Barnhart, *Biographical Notes upon Botanists.* 3: 313. 1965; H.N. Clokie, *Account of the Herbaria of the Department of Botany in the University of Oxford.* 247. Oxford 1964; Lady Pleasance Smith, ed., *Memoir and Correspondence of ... Sir J.E. Smith.* London 1832; A. Lasègue, *Musée botanique de Benjamin Delessert.* Paris 1845; Blanche Elizabeth Edith Henrey (1906-1983), *British Botanical and Horticultural Literature before 1800.* Oxford 1975; J.E. Smith, *Transactions of the Linnean Society of London. Botany.* 4: 218. 1798; T.W. Bossert, *Biographical Dictionary of Botanists Represented in the Hunt Institute Portrait Collection.* 380. 1972.

Species/Vernacular Names:

S. monogyna Labill. (*Stackhousia linariifolia* Cunn.; *Stackhousia pubescens* A. Rich.)

Australia: candles, creamy stackhousia, creamy candles, native mignonette

S. muricata Lindley (*Stackhousia elata* F. Muell.; *Stackhousia occidentalis* Domin)

Australia: western stackhousia

S. pulvinaris F. Muell.

Australia: alpine stackhousia

S. scoparia Benth.

Australia: leafless stackhousia

S. spathulata Sieber ex Sprengel (*Tripterococcus spathulatus* F. Muell.)

Australia: coast stackhousia

Stadiochilus Ros.M. Sm. Zingiberaceae

Origins:

From the Greek *stadios* "firm, standing fast and firm, fixed" and *cheilos* "lip."

Stadmannia Lam. Sapindaceae

Origins:

After a German botanist and traveler, M. Stadman; see Jean Baptiste Antoine Pierre de Monnet de Lamarck (1744-1829), *Tableau encyclopédique et méthodique* des trois règnes de la nature. Botanique. 2: 443, t. 312. Paris 1793.

Species/Vernacular Names:

S. oppositifolia (Lam.) Poir. subsp. *rhodesica* Exell (*Stadmannia sideroxylon* DC.; *Melanodiscus venulosus* Bullock ex Dale & Greenway)

English: Bourbon ironwood, silky plum

Mauritius: bois de fer, fer

South Africa: sypruim

Stagmaria Jack Anacardiaceae

Origins:

Greek *stagma, stagmatos* "that which drips, perfume, aromatic oil."

Stahelia Jonker Gentianaceae

Origins:

After the Swiss botanist Gerold Stahel, 1887-1955, traveler; see J.H. Barnhart, *Biographical Notes upon Botanists.* 3: 314. 1965; Ida Kaplan Langman, *A Selected Guide to the Literature on the Flowering Plants of Mexico.* 716. 1964; S. Lenley et al., *Catalog of the Manuscript and Archival Collections and Index to the Correspondence of John Torrey.* Library of the New York Botanical Garden. 382. 1973; E.M. Tucker, *Catalogue of the Library of the Arnold Arboretum of Harvard University.* 1917-1933.

Stahlia Bello Caesalpiniaceae

Origins:

After the Puerto Rican physician Augustin Stahl, 1842-1917, botanist. His works include "El tortugo amarillo de Puerto Rico." *Anales Soc. Esp. de Hist. Nat.* 4: 19-40. 1875 and *Estudios sobre la flora de Puerto Rico.* 6 panfletos. Porto Rico 1883-1888; see E.M. Tucker, *Catalogue of the Library of the Arnold Arboretum of Harvard University.* 1917-1933.

Stahlianthus Kuntze Zingiberaceae

Origins:

In honor of Helene Kuntze (*née* von Stahl), wife of the German botanist Otto Kuntze (1843-1907); see Thomas A. Zanoni, "Otto Kuntze, botanist. I. Biography, bibliography and travels." *Brittonia*. 32(4): 551-571. 1980.

Staintoniella Hara Brassicaceae

Origins:

For Adam Stainton, author of *Forests of Nepal*. London 1972, with Oleg Polunin wrote *Flowers of the Himalaya*. Delhi 1988.

Stalkya Garay Orchidaceae

Origins:

See Leslie Andrew Garay, *Bot. Mus. Leafl*. 28: 371. 25 Jun 1982 [Dec. 1980]; for the English (Devon) orchidologist Galfrid Clement Keyworth Dunsterville, 1905-1988 (Caracas), collected orchids in Venezuela, painted and drew Venezuelan orchids, author of *Venezuelan Orchids*. Caracas 1987, with Leslie Garay wrote *Venezuelan Orchids Illustrated*. London 1959-1976.

Standleya Brade Rubiaceae

Origins:

For the American botanist Paul Carpenter Standley, 1884-1963, plant collector and botanical explorer (Central and Western North America, Costa Rica, Guatemala), lichenologist; see J.H. Barnhart, *Biographical Notes upon Botanists*. 3: 315. 1965; Ida Kaplan Langman, *A Selected Guide to the Literature on the Flowering Plants of Mexico*. Philadelphia 1964; S. Lenley et al., *Catalog of the Manuscript and Archival Collections and Index to the Correspondence of John Torrey*. Library of the New York Botanical Garden. 382. 1973; T.W. Bossert, *Biographical Dictionary of Botanists Represented in the Hunt Institute Portrait Collection*. 380. 1972; Joseph Ewan, *Rocky Mountain Naturalists*. The University of Denver Press 1950; R. Zander, F. Encke, G. Buchheim and S. Seybold, *Handwörterbuch der Pflanzennamen*. 14. Aufl. 1993; E.M. Tucker, *Catalogue of the Library of the Arnold Arboretum of Harvard University*. 1917-1933; J. Ewan, ed., *A Short History of Botany in the United States*. 1969; Stafleu and Cowan, *Taxonomic Literature*. 5: 831-837. 1985; Irving William Knobloch, compil.,

"A preliminary verified list of plant collectors in Mexico." *Phytologia Memoirs*. VI. 1983.

Standleyacanthus Leonard Acanthaceae

Origins:

After the American botanist Paul Carpenter Standley, 1884-1963.

Standleyanthus R. King & H. Robinson Asteraceae

Origins:

After the American botanist Paul Carpenter Standley, 1884-1963.

Stanfieldiella Brenan Commelinaceae

Origins:

For the British botanist Dennis Percival Stanfield, 1903-1971, taxonomist, 1926-1950 Nigerian Civil Service, District Officer, traveler and plant collector (Northern and Southern Nigeria), a Fellow of the Linnean Society (1959); see F.N. Hepper and Fiona Neate, *Plant Collectors in West Africa*. 76. 1971.

Stangea Graebner Valerianaceae

Origins:

In honor of the family of the wife of Graebner.

Stangeria T. Moore Stangeriaceae

Origins:

For the British physician William Stanger, 1811-1854 (d. Durban, South Africa), geologist, visited Australia, 1841 naturalist on the Niger River Expedition of Capt. Henry Dundas Trotter (1802-1859), collected in South Nigeria, from 1843 in South Africa, 1845-1854 Surveyor-General in Natal; see Arthur Hugh Garfit Alston (1902-1958), *Ferns and Fern-allies of West Tropical Africa*. 42(6): 15. London 1959; P. Vorster and Elsa Vorster, "*Stangeria eriopus*." in *Excelsa*. 4: 79-89. 1974; S. Crowther and J.C. Taylor, *Niger Expedition of 1857-1859*. 1859; J.H. Barnhart, *Biographical Notes upon Botanists*. 3: 316. 1965; Mary Gunn and Leslie Edward W. Codd, *Botanical Exploration of Southern*

Africa. 331-332. Cape Town 1981; W.J. Hooker, *Niger Flora.* London 1849; William Allen (1793-1864) and Thomas Richard Heywood Thomson, *A Narrative of the Expedition ... to the River Niger, in 1841.* Under the command of Capt. H.D. Trotter. London 1848; J.F. Schön & S. Crowther, *Journals of the Expedition up the Niger in 1841.* London 1848; F.N. Hepper and Fiona Neate, *Plant Collectors in West Africa.* 76. 1971; James Britten (1846-1924) and George E. Simonds Boulger (1853-1922), *A Biographical Index of Deceased British and Irish Botanists.* London 1931; F. Boerner & G. Kunkel, *Taschenwörterbuch der botanischen Pflanzennamen.* 4. Aufl. 173. 1989; H. Genaust, *Etymologisches Wörterbuch der botanischen Pflanzennamen.* 608. [b. 1812] 1996.

Species/Vernacular Names:
S. eriopus (Kunze) Baillard (*Lomaria eriopus* Kunze; *Stangeria paradoxa* Moore) (from the Greek *erion* "wool" and *pous* "foot")

Southern Africa: Natal grass cycad; umNcuma (cone only), umFingwani (Xhosa)

Stanhopea Frost Orchidaceae

Origins:
For the English (b. London) Right Honorable Philip Henry (Stanhope), 4th Earl Stanhope, 1781-1855 (d. Kent), politician, 1807 Fellow of the Royal Society, from 1829 to 1837 President of the Medico-Botanical Society of London. Among his writings are *A Discourse on Medical Botany.* London 1854, *A Letter ... on the Corn Laws.* London 1826 and *A Letter to the Owners and Occupiers of Sheep Farms.* London 1828; see Edward Binns, *The Anatomy of Sleep ...* With annotations and additions by ... Earl Stanhope. London 1845.

Stanhopeastrum Reichb.f. Orchidaceae

Origins:
Resembling *Stanhopea.*

Stanleya Nutt. Brassicaceae

Origins:
For the English ornithologist Edward Smith Stanley, 1775-1851, 13th Earl of Derby; see Henry Salt (1780-1827, d. Alexandria), *A Voyage to Abyssinia and Travels into ... that Country ... in ... 1809 and 1810.* (Appendix. Zoology, with descriptions of the Birds by J. Latham, and additional remarks on them by Lord Stanley, afterwards 13th Earl of Derby. Botany, list of new and rare plants by R. Brown) London 1814; John Edward Gray (1800-1875), *Gleanings from the Menagerie and Aviary at Knowsley Hall.* Knowsley 1846 and *Hoofed Quadrupeds.* Knowsley 1850.

Stanleyella Rydb. Brassicaceae

Origins:
The diminutive of *Stanleya.*

Stapelia L. Asclepiadaceae

Origins:
After the Dutch physician Jan Bode van Stapel (Johannes Bodaeus Stapelius), botanist, d. *circa* 1636, author of *Disputatio medica inauguralis de variolis et morbillis,* etc. Lugduni Batavorum [Leyden] 1625, editor of Theophrastus's work on plants; see *Theophrastii Eresii De historia plantarum libri decem, graece et latine, ...* latinam *Gazae* versionem nova interpretatione ad margines: ... item rariorum plantarum iconibus illustravit *Joannes Bodaeus a Stapel,* medicus Amstelodamensis, etc. Amstelodami 1644; *Caii Plinii Secundi Historiae naturalis libri XXXVII,* etc. (vol. 5-9. Cum selectis ... *Bodaei* ... notis et excursibus) 1831; Edward Lee Greene, *Landmarks of Botanical History.* Edited by Frank N. Egerton. 459-460. 1983; Carl Linnaeus, *Species Plantarum.* 217. 1753 and *Genera Plantarum.* Ed. 5. 102. 1754; Gilbert Westacott Reynolds, *The Aloes of South Africa.* 3, 73, 75. Balkema, Rotterdam 1982.

Species/Vernacular Names:
S. gigantea N.E. Br. (*Stapelia nobilis* N.E. Br.)

English: giant stapelia, Zulu-giant, giant toad plant

Japan: chii-oo-saiku

S. grandiflora Mass. (*Stapelia ambigua* Mass. var. *ambigua*; *Stapelia ambigua* Mass. var. *fulva* Sweet; *Stapelia desmetiana* N.E. Br.; *Stapelia desmetiana* N.E. Br. var. *apicalis* N.E. Br.; *Stapelia desmetiana* N.E. Br. var. *desmetiana*; *Stapelia desmetiana* N.E. Br. var. *fergusoniae* R.A. Dyer; *Stapelia desmetiana* N.E. Br. var. *pallida* N.E. Br.; *Stapelia flavirostris* N.E. Br.; *Stapelia senilis* N.E. Br.)

English: carrion flower

S. variegata L. (*Orbea variegata* (L.) Haw.)

English: carrion flower, toad cactus, toad plant, starfish cactus

Japan: giyu-u-kaku

Stapelianthus Choux ex A.C. White & B. Sloane Asclepiadaceae

Origins:
The genus *Stapelia* L. and *anthos* "flower."

Stapeliopsis Choux Asclepiadaceae

Origins:
Resembling *Stapelia* L.

Stapeliopsis Pillans Asclepiadaceae

Origins:
The genus *Stapelia* L. and *opsis* "resemblance, resembling."

Stapfia Burtt Davy Gramineae

Origins:
After the Austrian botanist Otto Stapf, 1857-1933 (d. Innsbruck), traveler, Wien 1882-1889 assistant with Anton Joseph Kerner von Marilaun (1831-1898), 1900-1922 Keeper of the Herbarium of the Royal Botanic Gardens, Kew, 1908-1916 botanical secretary of the Linnean Society, 1922-1933 editor of the *Botanical Magazine*, contributor to Daniel Oliver (1830-1916), *Flora of Tropical Africa* (Apocynaceae, Verbenaceae, Myristicaceae, Gramineae, etc.), contributor to Harvey and Sonder, *Flora Capensis* (Apocynaceae, Laurineae, Proteaceae, etc.), contributed to Hooker's *Icones Plantarum*. Among his numerous and valuable publications are *On the Flora of Mount Kinabalu in North Borneo.* London 1894 and *The Aconites of India.* Calcutta 1905. See John H. Barnhart, *Biographical Notes upon Botanists.* 3: 317. 1965; Mia C. Karsten, *The Old Company's Garden at the Cape and Its Superintendents*: involving an historical account of early Cape botany. Cape Town 1951; James Edgar Dandy (1903-1976), in *The Journal of Botany.* 69: 54. (Feb.) 1931; Ida Kaplan Langman, *A Selected Guide to the Literature on the Flowering Plants of Mexico.* Philadelphia 1964; S. Lenley et al., *Catalog of the Manuscript and Archival Collections and Index to the Correspondence of John Torrey.* Library of the New York Botanical Garden. 382. 1973; T.W. Bossert, *Biographical Dictionary of Botanists Represented in the Hunt Institute Portrait Collection.* 380. 1972; E.M. Tucker, *Catalogue of the Library of the Arnold Arboretum of Harvard University.* 1917-1933; Mea Allan, *The Hookers of Kew.* London 1967; Ray Desmond, *Dictionary of British & Irish Botanists and Horticulturists.* 650. 1994; Leonard Huxley, *Life and Letters of Sir Joseph Dalton Hooker.* London 1918; Stafleu and Cowan, *Taxonomic Literature.* 5: 839-843. 1985; R.

Zander, F. Encke, G. Buchheim and S. Seybold, *Handwörterbuch der Pflanzennamen.* 14. Aufl. 1993; Emil Bretschneider (1833-1901), *History of European Botanical Discoveries in China.* [Reprint of the original edition, St. Petersburg 1898.] Leipzig 1981.

Stapfiella Gilg Turneraceae

Origins:
After the Austrian botanist Otto Stapf, 1857-1933.

Stapfiola Kuntze Gramineae

Origins:
After the Austrian botanist Otto Stapf, 1857-1933.

Stapfiophyton Li Melastomataceae

Origins:
After the Austrian botanist Otto Stapf, 1857-1933.

Staphidiastrum Naudin Melastomataceae

Origins:
The genus *Staphidium* Naudin and *astrum* "incomplete resemblance."

Staphidium Naudin Melastomataceae

Origins:
Greek *staphis, staphidos, astaphis*, for a plant, *staphis agria*, stavesacre, licebane, a species of *Delphinium*, Latin *staphis, idis*.

Staphylea L. Staphyleaceae

Origins:
Greek *staphyle* "bunch of grapes," referring to the inflorescence.

Species/Vernacular Names:
S. bolanderi A. Gray
English: Sierra bladdernut
S. pinnata L.
English: bladdernut

Staphylosyce Hook.f. Cucurbitaceae

Origins:

Greek *staphyle* "bunch of grapes," *sikyos* "wild cucumber, gourd," *sykon* "fig."

Statice L. Plumbaginaceae

Origins:

Latin *statice, es* "an herb of an astringent quality" (Plinius), Greek *statike*; see Carl Linnaeus, *Species Plantarum*. 421. 1753 and *Genera Plantarum*. Ed. 5. 193. 1754.

Staudtia Warb. Myristicaceae

Origins:

After the German botanist and botanical collector Alois Staudt, d. 1897 (Johann-Albrechtshöhe [Kumba], Cameroon), collected in Togo Rep. and West Cameroon, with Georg August Zenker (1855-1922) in East Cameroon; see G.W.J. Mildbraed (1879-1954), *Notizbl. Bot. Gart. Berl.* 8: 317-324. 1923; H. Walter, "Afrikanische pflanzen in Hamburg." in *Mitt. Geog. Gesell. Hamburg.* 56: 92-93. Hamburg 1965; Réné Letouzey (1918-1989), "Les botanistes au Cameroun." in *Flore du Cameroun.* 7: 1-110. Paris 1968; Frank Nigel Hepper, "Botanical collectors in West Africa, except French territories, since 1860." in *Comptes Rendus de l'Association pour l'étude taxonomique de la flore d'Afrique*, (A.E.T.F.A.T.). 69-75. Lisbon 1962; Anthonius Josephus Maria Leeuwenberg, "Isotypes of which holotypes were destroyed in Berlin." in *Webbia.* 19: 861-863. 1965; F.N. Hepper and Fiona Neate, *Plant Collectors in West Africa.* 76. 1971.

Species/Vernacular Names:

S. sp.

Nigeria: bopae, bopai, bope, metange, ekop, hikob

S. stipitata Warb.

French: arbre à pagayes, arbre à pagaies

Cameroon: bambele, oho, ovos, bopana, bope, bope bambale, ekop, hikob, ikop, malanga, mbondjo, m'bonda, mbonda, mbounde, mboune, ongom

Central Africa: molanga, oropa

Congo: mbu, menga menga

Gabon: mbassina, m'boum, m'boun, mbona, mbonn, mboun, mbounde, nbounde, mougoubi, mogoubi, mooum, mugubi, muguvi, mulanga, niové, ohohé, bobe, ndjolz, ngakombo, ngobye, ngubo, ninegone, niobe, niogo, niohe, niohue, niove, nikoubi, nkubi, nyowe, ogobe, ogoweni, olangi, todo, ungubu

Yoruba: afo, oropa, erubabaseje

Nigeria: oropa, oropupa, oropo, amuje, abala, icha, itshi, itchi, nja, umaza, itsha, iyipokoyo, ekop, metange, niove, niovi, ukpufem-obo, ukpufemme, mbundi, bopae, hikob, bope-bambali, babasheje, afo, amoje; oropa, amoje (Yoruba); umaza (Edo); abala (Ijaw); iyipokoyo (Efik); ichala (Igbo)

Zaire: somve menge, sunzu menga, sofi menga, susu menga, adjua, kamashi, bokole, bokolombe, bokolofe, bope, bososa, bufuki, gobe, kafi, ngubi, mbu, mohanga, molanga, mvingi, m'soma, onanga, wanga zanga

Stauntonia DC. Lardizabalaceae

Origins:

For the English (b. Galway) naturalist Sir George Leonard Staunton, 1737-1801 (d. London), diplomat, plant collector in China, physician, studied medicine in Montpellier (M.D. 1758), 1762-1779 medical officer in West Indies, 1781-1784 Madras, 1785 Baronet, Fellow of the Royal Society of London (1787) and of the Linnean Society (1789), 1792 accompanied Lord Macartney (1737-1806) on his mission to the Emperor of China, published *An Authentic Account of an Embassy from the King of Great Britain to the Emperor of China* ... Taken chiefly from the papers of ... the Earl of Macartney. London 1797. He was father of Sir George Thomas Staunton (1781-1859, d. London); see Joseph François Charpentier-Cossigny de Palma, *Voyage à Canton ... à la Chine ...* suivi d'observations sur le voyage à la Chine de Lord Macartney (rédigé par Sir G. Staunton). Paris, an VII [1798-1799]; Sir George Thomas Staunton, *Memoir of the Life and Family of Sir George Leonard Staunton.* 1823; E.H.M. Cox, *Plant-Hunting in China. A history of botanical exploration in China and the Tibetan marches.* London 1945; Emil Bretschneider, *History of European Botanical Discoveries in China.* Leipzig 1981; John H. Barnhart, *Biographical Notes upon Botanists.* 3: 318. 1965; Alice Margaret Coats, *The Quest for Plants. A History of the Horticultural Explorers.* London 1969; A. Lasègue, *Musée botanique de Benjamin Delessert.* Paris 1845; Ethelyn Maria Tucker, *Catalogue of the Library of the Arnold Arboretum of Harvard University.* Cambridge, Massachusetts 1917-1933; G. Murray, *History of the Collections Contained in the Natural History Departments of the British Museum.* London 1904.

Species/Vernacular Names:

S. hexaphylla Decne.

English: Japanese Staunton-vine

Japan: mube

China: ye mu gua

Stauracanthus Link Fabaceae

Origins:

Greek *stauros* "a cross" and *akantha*, *akanthes* "thorn, prickle."

Stauranthera Benth. Gesneriaceae

Origins:

From the Greek *stauros* "a cross" and *anthera* "anther," referring to the anthers, coherent and forming a cross.

Stauranthus Liebm. Rutaceae

Origins:

From the Greek *stauros* and *anthos* "flower."

Staurites Reichb.f. Orchidaceae

Origins:

From the Greek *stauros* "a cross, upright pale," an allusion to the position of the lateral lobes; see also *Phalaenopsis* Blume.

Staurochilus Ridley ex Pfitzer Orchidaceae

Origins:

Greek *stauros* "a cross" and *cheilos* "lip," in some species the lip is cruciform.

Staurochlamys Baker Asteraceae

Origins:

Greek *stauros* and *chlamys*, *chlamydos* "cloak, mantle."

Stauroglottis Schauer Orchidaceae

Origins:

From the Greek *stauros* and *glotta* "tongue," referring to the lateral lobes of the lip; see also *Phalaenopsis* Blume.

Staurogyne Wallich Acanthaceae

Origins:

From the Greek *stauros* "cross" and *gyne* "a woman, female," referring to the stigma; see Nathaniel Wallich

(1786-1854), *Plantae Asiaticae rariores*. 2: 80, t. 186. London 1831.

Staurogynopsis Mangenot & Ake Assi Acanthaceae

Origins:

Resembling the genus *Staurogyne* Wall.

Staurophragma Fischer & C.A. Meyer Scrophulariaceae

Origins:

Greek *stauros* and *phragma* "a partition, compartment, wall, fence."

Stauropsis Reichb.f. Orchidaceae

Origins:

Greek *stauros* "cross" and *opsis* "appearance," alluding to the shape of the flowers, to the cross-shaped lip.

Staurospermum Thonn. Rubiaceae

Origins:

Greek *stauros* "cross, upright pale" and *sperma* "seed"; see also *Mitracarpus* Zucc. ex Schultes & Schultes f.

Staurostigma Scheidw. Araceae

Origins:

Greek *stauros* "cross" and *stigma* "stigma"; see also *Asterostigma* Fischer & C.A. Meyer.

Staurothylax Griff. Euphorbiaceae

Origins:

From the Greek *stauros* "cross" and *thylakos* "bag, sack, pouch."

Stawellia F. Muell. Anthericaceae (Liliaceae)

Origins:

After the Irish lawyer Sir William Foster Stawell, 1815-1889, emigrated to Australia 1842, from 1857 to 1886 Chief Justice of Victoria, Australia, President Philosophical Institute (afterwards Royal Society of Victoria), supervised the arrangements for the ill-fated Burke and Wills expedition (the Victoria Exploring Expedition, 1860); see *The Political*

Sentiments of W.F. Stawell. Melbourne [1856]; Ferdinand von Mueller, *Fragmenta Phytographiae Australiae.* 7: 85. (Apr.) 1870; Jonathan Wantrup, *Australian Rare Books, 1788-1900.* 234-239. Hordern House, Sydney 1987.

Stayneria L. Bolus Aizoaceae

Origins:

For the South African (b. Cedara, Natal) horticulturist Frank J. Stayner, 1907-1981, district officer in Natal, a succulent plant enthusiast and specialist, 1935-1946 assistant Superintendent of Parks in the Port Elizabeth Parks Department, 1959-1969 Curator of the Karoo Botanic Gardens at Worcester (South Africa), a collector of succulents; see Mary Gunn and Leslie E. Codd, *Botanical Exploration of Southern Africa.* 333. Cape Town 1981; M.B. Bayer, "In memoriam: Frank J. Stayner (1907-1981)." *Veld & Flora.* 68: 27. 1982; I.H. Vegter, *Index Herbariorum.* Part II (6), *Collectors S.* Regnum Vegetabile vol. 114. 1986; J. Scott, "News from the gardens. Karoo Garden — 60th anniversary celebrations." *Veld & Flora.* 68: 31. 1982; Gordon Douglas Rowley, *A History of Succulent Plants.* Strawberry Press, Mill Valley, California 1997.

Stebbinsia Lipsch. Asteraceae

Origins:

Named for the American geneticist George Ledyard Stebbins, Jr., b. 1906, evolutionist, botanist, studied biology at Harvard.

Stebbinsoseris K.L. Chambers Asteraceae

Origins:

For the American geneticist George Ledyard Stebbins, Jr., b. 1906, evolutionist, botanist, studied biology at Harvard, established the Department of Genetics at the University of California, professor of genetics. Among his writings are "Critical notes on *Lactuca* and related genera." *J. Bot.* 75: 12-18. 1937, "Notes on *Lactuca* in western North America." *Madroño.* 5(4): 123-126. 1939, "Types in polypoids: Their classification and significance." *Advances in Genetics.* 1: 403-429. 1947, *Variation and Evolution in Plants.* Columbia University Press, New York, NY 1950, "A new classification of the tribe Cichorieae, family Compositae." *Madroño.* 12(3): 65-81. 1953 and "A preliminary list of the vascular plants found on the Northern California Coast Range Preserve, Nature Conservancy." *Ecological Studies Leaflet* No. 14. The Nature Conservancy, Washington, DC. 1968, with E.F. Paddock wrote "The *Solanum nigrum* complex in Pacific North America." *Madroño.* 10: 70-80. 1949.

See Vassiliki Betty Smocovitis, "G. Ledyard Stebbins Jr. and the Evolutionary Synthesis (1924-1950)." *America Journal of Botany.* 84(12): 1625-1637. 1997; Bentley Glass, ed., *The Roving Naturalist. Travel Letters of Theodosius Dobzhansky.* Philadelphia 1980; P. Raven, "Plant systematics 1947-1972." *Annals of the Missouri Botanical Garden.* 61: 166-178. 1974; Ida Kaplan Langman, *A Selected Guide to the Literature on the Flowering Plants of Mexico.* 720. Philadelphia 1964; John H. Barnhart, *Biographical Notes upon Botanists.* 3: 319. 1965; T.W. Bossert, *Biographical Dictionary of Botanists Represented in the Hunt Institute Portrait Collection.* 381. 1972; Ethelyn Maria Tucker, *Catalogue of the Library of the Arnold Arboretum of Harvard University.* Cambridge, Massachusetts 1917-1933; J. Ewan, ed., *A Short History of Botany in the United States.* New York and London 1969.

Species/Vernacular Names:

S. decipiens (Chambers) Chambers (*Microseris decipiens* Chambers)

English: Santa Cruz microseris

Steenisia Bakh.f. Rubiaceae

Origins:

Named for the Dutch botanist Cornelis Gijsbert Gerrit Jan van Steenis, 1901-1986, traveler and plant collector, from 1927 to 1946 botanist Buitenzorg (Bogor), professor of tropical botany, from 1962 to 1972 Director of the Leyden Rijksherbarium, editor and joint author of *Flora malesiana,* from 1931 to 1950 editor of the *Bulletin du Jardin Botanique de Buitenzorg* (with Dirk Fok van Slooten, 1891-1953, and Marinus Anton Donk, 1908-1972), 1961-1971 editor of *Blumea.* Among his most valuable writings are *Malayan Bignoniaceae.* Arnhem 1927 and *The Styracaceae of Netherlands India.* Buitenzorg (Archipel Drukkerij) 1932; see John H. Barnhart, *Biographical Notes upon Botanists.* 3: 320. 1965; Maria Johann van Steenis-Kruseman, *Malaysian Plant Collectors and Collections.* in *Flora Malesiana.* 1(1): 1-639; T.W. Bossert, *Biographical Dictionary of Botanists Represented in the Hunt Institute Portrait Collection.* 381. 1972; S. Lenley et al., *Catalog of the Manuscript and Archival Collections and Index to the Correspondence of John Torrey.* Library of the New York Botanical Garden. 383. 1973; Ethelyn Maria Tucker, *Catalogue of the Library of the Arnold Arboretum of Harvard University.* Cambridge, Massachusetts 1917-1933; R. Zander, F. Encke, G. Buchheim and S. Seybold, *Handwörterbuch der Pflanzennamen.* 14. Aufl. Stuttgart 1993.

Steenisioblechnum Hennipman Blechnaceae (Blechnoideae)

Origins:

The Dutch botanist Cornelis Gijsbert Gerrit Jan van Steenis, 1901-1986; see Elbert Hennipman (1937-), in *Blumea*. 30: 17. 1984.

Steetzia Sonder Asteraceae

Origins:

For the German botanist Joachim Steetz, 1804-1862, physician, M.D. Würzburg 1829, studied pharmacy, orchidologist, at the Hamburg Botanical Garden, wrote *Die Familie der Tremandreen*. Hamburg 1853, contributed to Wilhelm Carl Hartwig Peters (1815-1883), *Naturwissenschaftliche Reise nach Mossambique*. Berlin (Georg Reimer) [1861-] 1862-1864; see J.H. Barnhart, *Biographical Notes upon Botanists*. 3: 320. 1965; Otto Wilhelm Sonder, in *Linnaea*. 25: 450. 1853; R. Zander, F. Encke, G. Buchheim and S. Seybold, *Handwörterbuch der Pflanzennamen*. 14. Aufl. 1993; Johann G.C. Lehmann, *Plantae Preissianae*. Hamburgi 1848; Berthold Carl Seeman (1825-1871), *The Botany of the Voyage of HMS Herald ... 1845-51*. London 1854; E.M. Tucker, *Catalogue of the Library of the Arnold Arboretum of Harvard University*. 1917-1933.

Stefanoffia H. Wolff Umbelliferae

Origins:

For the Bulgarian botanist Boris Stefanov (Stefanoff), 1894-1979, professor of botany; see John H. Barnhart, *Biographical Notes upon Botanists*. 3: 320. 1965; E.M. Tucker, *Catalogue of the Library of the Arnold Arboretum of Harvard University*. 1917-1933; R. Zander, F. Encke, G. Buchheim and S. Seybold, *Handwörterbuch der Pflanzennamen*. 14. Aufl. Stuttgart 1993.

Stegania R. Br. Blechnaceae

Origins:

From the Greek *steganos* "covered, secret," *stegane* "a covering," *stege* "roof," *stego, stegein* "to cover closely, shelter, protect, to hold water, cover and conceal"; see Robert Brown, *Prodromus florae Novae Hollandiae et Insulae van-Diemen*. 152. London 1810.

Steganotaenia Hochst. Umbelliferae

Origins:

Greek *steganos* "covered, secret" (*stegane* "a covering," *stege* "roof") and *tainia* "fillet," Latin *taenia, ae* "band, ribbon."

Species/Vernacular Names:

S. araliacea Hochst. (*Peucedanum fraxinifolium* Hiern ex Oliver; *Peucedanum araliaceum* (Hochst.) Benth. & Hook.f. ex Vatke; *Peucedanum araliaceum* (Hochst.) Bentham & Hook.f. ex Vatke var. *fraxinifolium* (Hiern ex Oliv.) Engler; *Peucedanum fraxinifolium* Hiern ex Oliv. var. *haemanthum* Welw. ex Hiern)

English: carrot tree, pop-gun tree

Southern Africa: geelwortelboom; morobolo (Tswana: western Transvaal, northern Cape, Botswana); muBama, muBojana, muKetu, muPombohlove, muSante, muShoriwondo, muSumfu, muSodzambudzi (Shona)

Mozambique: psukula (Nhungo, Tete province)

Zimbabwe: popgun tree

Angola: mupandalolo (Namibe district)

Kenya: bisugu (Northern Frontier province)

Malawi: mpoloni (Chinyana, Chikwawa district; see R. Sutherland Rattray, *Some Folk-Lore Stories and Songs in Chinyanja*. London 1907)

Nigeria: hano, raken giwa (Hausa)

Steganotropis Lehm. Fabaceae

Origins:

From the Greek *steganos* "covered, secret" and *tropis, tropidos* "a keel"; see also *Centrosema* (DC.) Benth.

Steganthera Perkins Monimiaceae

Origins:

Greek *steganos* "covered, secret" and *anthera* "anther"; see Janet Russell Perkins (1853-1933), in *Botanische Jahrbücher*. 25: 557, 564. (Sep.) 1898.

Steganthus Knobl. Oleaceae

Origins:

From the Greek *steganos* "covered, secret" and *anthos* "flower."

Stegastrum Tieghem Loranthaceae

Origins:

Greek *stegane* "covering," *stege* "roof" and the Latin substantival suffix *-astrum*, indicating inferiority or incomplete resemblance; see also *Lepeostegeres* Blume, from the Greek *lepos* "bark, husk, scale" and *stegeres* "roofed."

Stegnocarpus Torrey & A. Gray Boraginaceae

Origins:

Greek *stegnos* "covered, sheltered" and *karpos* "fruit."

Stegnogramma Blume Thelypteridaceae

Origins:

Greek *stegnos* and *gramma* "a letter, line, character."

Stegnosperma Benth. Stegnospermataceae (Phytolaccaceae)

Origins:

Greek *stegnos* and *sperma* "seed," possibly referring to the aril covering the seeds.

Stegolepis Klotzsch ex Koern. Rapateaceae

Origins:

From the Greek *stege*, *stegos* "roof, shelter, cover" and *lepis* "scale."

Stegosia Lour. Gramineae

Origins:

Greek *stege*, *stegos* "roof, shelter, cover."

Steigeria Müll. Arg. Euphorbiaceae

Origins:

After the Swiss botanist Jakob Robert Steiger, 1801-1862, physician; see John H. Barnhart, *Biographical Notes upon Botanists*. 3: 321. 1965; T.W. Bossert, *Biographical Dictionary of Botanists Represented in the Hunt Institute Portrait Collection*. 381. 1972.

Steinchisma Raf. Gramineae

Origins:

See Constantine S. Rafinesque, *Seringe Bull. Bot.* 1: 220. 1830; E.D. Merrill, *Index rafinesquianus*. 76. 1949.

Steinheilia Decne. Asclepiadaceae

Origins:

For the Alsatian botanist Adolphe (Adolph) Steinheil, 1810-1839, traveler, surgeon, pharmacist; see J.H. Barnhart, *Biographical Notes upon Botanists*. 3: 322. 1965; A. Lasègue, *Musée botanique de Benjamin Delessert*. Paris 1845; R. Zander, F. Encke, G. Buchheim and S. Seybold, *Handwörterbuch der Pflanzennamen*. 14. Aufl. 1993; Stafleu and Cowan, *Taxonomic Literature*. 5: 877-878. Utrecht 1985.

Steinmannia Philippi Alliaceae (Liliaceae)

Origins:

Named after the German geologist Johann Heinrich Conrad Gottfried Gustav Steinmann, 1856-1929, palaeontologist; see J.H. Barnhart, *Biographical Notes upon Botanists*. 3: 323. 1965; Stafleu and Cowan, *Taxonomic Literature*. 5: 879-880 1985.

Steirachne Ekman Gramineae

Origins:

Greek *steiros* "barren, sterile" and *achne* "chaff, glume."

Steiractinia S.F. Blake Asteraceae

Origins:

From the Greek *steiros* and *aktin* "ray."

Steirodiscus Less. Asteraceae

Origins:

Greek *steiros* and *diskos* "a disc," referring to the sterile disc florets.

Steiroglossa DC. Asteraceae

Origins:

From the Greek *steiros* "barren, sterile" and *glossa* "tongue"; see A.P. de Candolle, *Prodromus*. 6: 38. 1838.

Steironema Raf. Primulaceae

Origins:
Greek *steiros* "barren, sterile" and *nema* "filament, thread," referring to the staminodes; see C.S. Rafinesque, *Ann. Gén. Sci. Phys.* 7: 193. 1820; E.D. Merrill, *Index rafinesquianus.* 187. 1949.

Steiropteris (C. Christensen) Pic. Serm. Thelypteridaceae

Origins:
Greek *steiros* and *pteris* "fern"; see Rodolfo E.G. Pichi Sermolli, "Fragmenta Pteridologiae — IV." in *Webbia.* 28: 445-477. 30 Dec. 1973.

Steirosanchezia Lindau Acanthaceae

Origins:
Greek *steiros* "sterile" plus the genus *Sanchezia* Ruíz & Pavón.

Steirotis Raf. Loranthaceae

Origins:
Greek *steiros* "sterile," *steirotes* "sterility"; see C.S. Rafinesque, *Ann. Gén. Sci. Phys.* 6: 79. 1820; E.D. Merrill, *Index rafinesquianus.* 113. 1949.

Stekhovia Vriese Goodeniaceae

Origins:
After the Dutch botanist Jacobus Herman (Hermanus) Schuurmans Stekhoven, 1792-1855, naturalist, Curator of the Leyden Botanical Garden, author of *Kruidkundig handboek.* Amsterdam 1815-1818 and *Kruikundig kunstwoordenboek.* Leyden 1825; see H. Veendorp and L.G.M. Baas Becking, *Hortus Academicus Lugduno-Batavus 1587-1937. The development of the gardens of Leyden University.* Harlemi [Haarlem]. 1938; Willem Hendrik de Vriese (1806-1862), in *Natuurkundige Verhandelingen van de Hollandsche Maatshappij der Wetenschappen te Haarlem.* Ser. 2, 10: 166. Amsterdam 1854.

Stelechanteria Thouars ex Baillon Euphorbiaceae

Origins:
Possibly from the Greek *stelechos* "branch, trunk, stem" and *anthera* "anther," Latin *stela, ae* "a pillar, column."

Stelechantha Bremek. Rubiaceae

Origins:
Greek *stelechos* "branch, the crown of the root, stool, trunk, stem," *stele* "a pillar, column, trunk, central part of stem" and *anthos* "flower."

Stelechocarpus Hook.f. & Thomson Annonaceae

Origins:
Greek *stelechos* and *karpos* "fruit."

Species/Vernacular Names:
S. burahol (Blume) Hook.f. & Thomson
English: keppel fruit

Steleocodon Gilli Asteraceae

Origins:
From the Greek *steleon, stelea* "haft, shaft, manubrium," *steleos, stelea* "of an axe" and *kodon* "a bell."

Steleostemma Schltr. Asclepiadaceae

Origins:
Greek *steleon, stelea* "haft, shaft," *steleos, stelea* "of an axe" and *stemma* "garland, crown."

Stelephuros Adans. Gramineae

Origins:
From the Latin name *stelephuros*, used by Plinius for a plant, perhaps Ravenna sugar cane, a species of *Saccharum*; Greek *stelephouros*, ancient name applied by Theophrastus (*HP.* 7.11.2) to the haresfoot plantain, a species of *Plantago.*

Stelestylis Drude Cyclanthaceae

Origins:
Greek *stele* "a pillar, column" and *stylos* "style," fruit spadix not screw-like.

Steliopsis Swartz Orchidaceae

Origins:
Resembling *Stelis* Sw.

Stelis Swartz Orchidaceae

Origins:
Greek *stelis*, *stelidos* "mistletoe," also these epiphytes orchids grow on trees.

Species/Vernacular Names:
S. sp.
Spanish: cucharita blanca, cucharita negra

Stellaria L. Caryophyllaceae

Origins:
Latin *stella*, *ae* "a star," *stellaris*, *e* "starry," the flowers are star-shaped; see Carl Linnaeus, *Species Plantarum*. 421. 1753 and *Genera Plantarum*. Ed. 5. 193. 1754.

Species/Vernacular Names:
S. angustifolia Hook.
English: swamp starwort
S. filiformis (Benth.) Mattf. (*Drymaria filiformis* Benth.)
Australia: thread starwort
S. holostea L.
English: stitchwort, greater stitchwort, adder's meat
S. longifolia Willd.
English: long-leaved starwort
S. media (L.) Villars (*Alsine media* L.)
English: bindweed, chickweed, chickweed starwort, common chickweed, satin flower, starweed, starwort, stitchwort, tonguegrass, white bird's eye, winterweed
Arabic: meshit
Japan: ko-hakobe
Okinawa: minna
China: fan lu
Southern Africa: gewone sterremuur, mier, muggiesgras, muur, sterremier, sterremuur; qoqobala (Sotho)
S. multiflora Hook.
English: rayless starwort
S. neglecta Weihe
English: greater chickweed
S. nitens Nutt.
English: shining chickweed
S. obtusa Engelm.
English: obtuse stellaria
S. pungens Brongn.
English: prickly starwort

Stellariopsis (Baillon) Rydb. Rosaceae

Origins:
Stellaris, *e* "starry" and *opsis* "aspect, resemblance."

Stellera L. Thymelaeaceae

Origins:
For the German (b. Windsheim) botanist Georg Wilhelm Steller (Stöller), 1709-1746 (Siberia), physician, naturalist, traveler and explorer, plant collector in Russia and Alaska, geographer, 1738-1742 on the second Bering Expedition. His major work is *Beschreibung von dem Lande Kamtschatka*. Frankfurt, Leipzig 1774; see Sten Lindroth, in *Dictionary of Scientific Biography* 13: 28-29. 1981; John H. Barnhart, *Biographical Notes upon Botanists*. 3: 323. 1965; Jonas C. Dryander, *Catalogus bibliothecae historico-naturalis Josephi Banks*. London 1800; A. Lasègue, *Musée botanique de Benjamin Delessert*. Paris 1845; Peter [Pyotr] Simon Pallas (1741-1811), *A Naturalist in Russia. Letters from Peter Simon Pallas to Thomas Pennant*. Edited by Carol Urness. Minneapolis [1967].

Stellera Turcz. Gentianaceae

Origins:
After the German botanist Georg Wilhelm Steller (Stöller), 1709-1746, physician, naturalist, traveler and explorer, plant collector in Russia, 1738-1742 on the second Bering Expedition; see J.H. Barnhart, *Biographical Notes upon Botanists*. 3: 323. 1965.

Stelleropsis Pobed. Thymelaeaceae

Origins:
Resembling *Stellera* L.

Stelligera A.J. Scott Chenopodiaceae

Origins:
Latin *stelliger*, *gera*, *gerum* (*stella-gero*) "star-bearing, starry," referring to the fruits bearing spines; A.J. Scott, *Feddes Repertorium*. 89: 114. (Jun.) 1978.

Species/Vernacular Names:
S. endecaspinis A.J. Scott (*Kochia stelligera* (F. Muell.) Benth.; *Bassia stelligera* (F. Muell.) F. Muell.)
English: star-fruit bassia, star copperburr, starred bluebush

Stellilabium Schltr. Orchidaceae

Origins:

Latin *stella, ae* "a star" and *labium, ii* "a lip," referring to the nature of the lip.

Stellorchis Thouars Orchidaceae

Origins:

Latin *stella* and Greek *orchis* "orchid," referring the position of the perianth segments.

Stellorkis Thouars Orchidaceae

Origins:

Latin *stella* "a star" and Greek *orchis* "orchid."

Stellularia Benth. Scrophulariaceae

Origins:

From Latin *stellula, ae* "a little star."

Stelmacrypton Baillon Asclepiadaceae

Origins:

Greek *stelma, stelmatos* "a girdle, belt" and *krypto* "to hide."

Species/Vernacular Names:

S. khasianum (Kurz) Baillon

English: common stelmacrypton

China: xu yao teng

Stelmagonum Baillon Asclepiadaceae

Origins:

Greek *stelma* "a girdle, belt" and *gony* "a knee, bend, joint," *gonia* "an angle."

Stelmanis Raf. Rubiaceae

Origins:

Greek *anisos* "unequal" and *stelma, stelmatos* "a girdle, crown, garland, wreath," or from *Anistelma* Raf.? See C.S. Rafinesque, *Aut. Bot.* 13, 69. 1840; E.D. Merrill, *Index rafinesquianus.* 225, 227. 1949.

Stelmation Fourn. Asclepiadaceae

Origins:

Greek *stelma, stemma* "a wreath," *stemmatias* "one who wears a wreath"; see also *Metastelma* R. Br.

Stelmatocodon Schltr. Asclepiadaceae

Origins:

Greek *stelma, stelmatos* "a wreath" and *kodon* "bell."

Stelmatocrypton Baillon Asclepiadaceae

Origins:

Greek *stelma, stelmatos* "a wreath" and *krypto* "to hide."

Stelmotis Raf. Rubiaceae

Origins:

Greek *stelma, stelmatos* "a wreath" and *ous, otos* "an ear," see also *Hedyotis* L. (Greek *hedys* "sweet" and *ous, otos,* possibly referring to the scented leaves and flowers); see Carl Linnaeus, *Species Plantarum.* 101. 1753 and *Genera Plantarum.* Ed. 5. 44. 1754); C.S. Rafinesque, *New Fl. N. Am.* 2: 21. 1836 [1837] and 4: 101. 1836 [1838].

Stemaria Schott Pteridaceae

Origins:

"Adern zu beiden Seiten der mittleren Hauptrippe des Blattes" (Theophr.), in Helmut Genaust, *Etymologisches Wörterbuch der botanischen Pflanzennamen.* 609. Basel 1996.

Stemmacantha Cass. Asteraceae

Origins:

Greek *stemma, stemmatos* "garland, crown" and *akantha* "thorn."

Stemmadenia Bentham Apocynaceae

Origins:

Greek *stemma, stemmatos* and *aden* "a gland," referring to the style.

Stemmatella Wedd. ex Bentham Asteraceae

Origins:
Greek *stemma*, *stemmatos* "garland, crown," the diminutive.

Stemmatium Phil. Alliaceae (Liliaceae)

Origins:
Greek *stemma*, *stemmatos* "garland, chaplet."

Stemmatodaphne Gamble Lauraceae

Origins:
Greek *stemma*, *stemmatos* plus *Daphne*; see also *Alseodaphne* Nees.

Stemmatophyllum Tieghem Loranthaceae

Origins:
From the Greek *stemma*, *stemmatos* and *phyllon* "leaf."

Stemmatospermum P. Beauv. Gramineae

Origins:
From the Greek *stemma*, *stemmatos* and *sperma* "a seed," referring to the seeds.

Stemodia L. Scrophulariaceae

Origins:
Greek *stemon* "stamen" and *dis* "two, double," referring to the anthers; see C. Linnaeus, *Systema Naturae*. Ed. 10. 1091, 1118, 1374. 1759.

Stemodiopsis Engl. Scrophulariaceae

Origins:
Resembling *Stemodia* L.

Stemona Lour. Stemonaceae

Origins:
Greek *stemon* "stamen," referring to the protruding and foliaceous stamens; see João (Joannes) de Loureiro (1717-1791), *Flora cochinchinensis*. 401, 404. [Lisboa] 1790.

Species/Vernacular Names:
S. tuberosa Lour.

English: tuber stemona
China: bai bu, pai pu
Vietnam: bach bo
Malaya: kemili hutan

Stemonocoleus Harms Caesalpiniaceae

Origins:
Greek *stemon* "pillar, stamen" and *koleos* "sheath," referring to the united filaments of the stamens.

Species/Vernacular Names:
S. micranthus Harms
Nigeria: oruro; erhaneybeni (Edo); nre (Igbo)
Congo: edoumba, odoumba
Gabon: pongo
Cameroon: gondou, esson, gondu
Ivory Coast: ahianana, akianana

Stemonoporus Thwaites Dipterocarpaceae

Origins:
Greek *stemon* "pillar, stamen" and *poros* "opening, pore."

Stemonurus Blume Icacinaceae

Origins:
Greek *stemon* and *oura* "a tail," referring to the tufted stamens; see Karl Ludwig von Blume, *Bijdragen tot de flora van Nederlandsch Indië*. 648. Batavia (Jan.) 1826.

Stenachaenium Benth. Asteraceae

Origins:
From the Greek *stenos* "narrow" and *achaenium* or *achenium* "achene."

Stenactis Cass. Asteraceae

Origins:
Greek *stenos* "narrow" and *aktis*, *aktin* "a ray."

Stenadenium Pax Euphorbiaceae

Origins:
Greek *stenos* and *aden* "gland," see also *Monadenium* Pax.

Stenandriopsis S. Moore Acanthaceae

Origins:

Resembling *Stenandrium* Nees.

Stenandrium Nees Acanthaceae

Origins:

From the Greek *stenos* "narrow" and *aner, andros* "man, male, stamen, anther."

Stenanona Standley Annonaceae

Origins:

Greek *stenos* "narrow" plus *Annona*.

Stenanthella Rydb. Melanthiaceae (Liliaceae)

Origins:

Referring to the genus *Stenanthium* (A. Gray) Kunth, the diminutive.

Stenanthemum Reissek Rhamnaceae

Origins:

Greek *stenos* and *anthos* "flower"; see Siegfried Reissek (or Reisseck) (1819-1871), in *Linnaea*. 29: 295. (Sep.) 1858.

Stenanthera Engl. & Diels Annonaceae

Origins:

Greek *stenos* and *anthera* "anther."

Stenanthera R. Br. Epacridaceae

Origins:

From the Greek *stenos* and *anthera* "anther"; see Robert Brown, *Prodromus florae Novae Hollandiae et Insulae van-Diemen*. 538. London 1810.

Stenanthium (A. Gray) Kunth Melanthiaceae (Liliaceae)

Origins:

Greek *stenos* "narrow" and *anthos* "flower," referring to the perianth segments and panicles.

Stenanthus Oerst. ex Hanst. Gesneriaceae

Origins:

From the Greek *stenos* and *anthos* "flower."

Stenaria Raf. Rubiaceae

Origins:

See C.S. Rafinesque, *Ann. Gén. Sci. Phys.* 5: 226. 1820; E.D. Merrill, *Index rafinesquianus*. 227. 1949.

Stenia Lindl. Orchidaceae

Origins:

Greek *stenos* "narrow," referring to the form of the pollinia, long and narrow.

Stenocactus (K. Schumann) A.W. Hill Cactaceae

Origins:

From the Greek *stenos* plus *cactus*.

Stenocalyx O. Berg Myrtaceae

Origins:

Greek *stenos* "narrow" and *kalyx* "calyx."

Stenocalyx Turcz. Malpighiaceae

Origins:

Greek *stenos* "narrow" and *kalyx* "calyx."

Stenocarpha S.F. Blake Asteraceae

Origins:

Greek *stenos* and *karphos* "chip of straw, chip of wood."

Stenocarpus R. Br. Proteaceae

Origins:

Greek *stenos* "narrow" and *karpos* "a fruit"; see Robert Brown, *Transactions of the Linnean Society of London*. 201. London (Mar.) 1810.

Species/Vernacular Names:

S. sinuatus (Loudon) Endl.

English: firewheel tree, wheel of fire

Stenocephalum Sch. Bip. Asteraceae

Origins:
Greek *stenos* and *kephale* "head."

Stenocereus (A. Berger) Riccobono Cactaceae

Origins:
Greek *stenos* "narrow" with *Cereus*.

Stenochasma Miq. Moraceae

Origins:
From the Greek *stenos* "narrow" and *chasme* "gaping, yawning."

Stenochilus R. Br. Myoporaceae

Origins:
Greek *stenos* "narrow" and *cheilos* "a lip, margin"; see Robert Brown, *Prodromus florae Novae Hollandiae et Insulae van-Diemen*. 517. London 1810.

Stenochlaena J. Sm. Blechnaceae (Stenochlaenaceae, Stenochlaenoideae)

Origins:
Greek *stenos* and *chlaena, chlaenion* "a cloak, blanket," referring to the sporophyll; see J. Smith, *Hooker's Journal of Botany*. 3: 149. 1841.

Species/Vernacular Names:
S. palustris (Burm.) Bedd.

English: climbing fern

The Philippines: diliman, hagnaya, lanas

Malaya: akar paku, lembiding, paku mesin, paku miding, paku ranu, lamiding

S. tenuifolia (Desv.) T. Moore

English: climbing fern, bracken

Stenochloa Nutt. Gramineae

Origins:
From the Greek *stenos* "narrow" and *chloe, chloa* "grass."

Stenocline DC. Asteraceae

Origins:
From the Greek *stenos* and *kline* "a bed, couch, receptacle."

Stenocoelium Ledeb. Umbelliferae

Origins:
From the Greek *stenos* "narrow" and *koilos* "hollow."

Stenocoryne Lindley Orchidaceae

Origins:
Greek *stenos* "narrow" and *koryne* "a club," referring to the spur formed by the foot of the column, to the pollen masses (*caudiculis 2, glandulisque totidem ovalibus*).

Stenodiscus Reissek Rhamnaceae

Origins:
From the Greek *stenos* "narrow" and *diskos* "a disc"; see Siegfried Reissek, in *Linnaea*. 29: 295. (Sep.) 1858.

Stenodon Naudin Melastomataceae

Origins:
From the Greek *stenos* and *odous, odontos* "a tooth."

Stenodraba O.E. Schulz Brassicaceae

Origins:
From the Greek *stenos* "narrow" and the genus *Draba* L.

Stenodrepanum Harms Caesalpiniaceae

Origins:
Greek *stenos* plus *drepanon, drapanon, drepane* "scythe, sickle."

Stenofestuca (Honda) Nakai Gramineae

Origins:
From the Greek *stenos* "narrow, close" with *Festuca*.

Stenofilix Nakai Grammitidaceae

Origins:

From the Greek *stenos* "narrow, close" and Latin *filix* "fern."

Stenogastra Hanst. Gesneriaceae

Origins:

From the Greek *stenos* "narrow" and *gaster* "abdomen, belly, paunch."

Stenoglossum Kunth Orchidaceae

Origins:

Greek *stenos* "narrow" and *glossa* "tongue," the shape of the free part of the lip.

Stenoglottis Lindley Orchidaceae

Origins:

From the Greek *stenos* and *glotta* "tongue," referring to the divisions of the lip, trilobed in the apical third and with the side lobes often fimbriate.

Stenogonum Nutt. Polygonaceae

Origins:

Greek *stenos* "narrow" and *gony* "a knee, bend, joint," *gonia* "an angle."

Stenogyne Benth. Labiatae

Origins:

From the Greek *stenos* "narrow" and *gyne* "a woman, female, pistil."

Species/Vernacular Names:

S. rotundifolia A. Gray

Hawaii: pua'ainaka

S. rugosa Benth.

Hawaii: ma'ohi'ohi

S. scrophularioides Benth. (*Stenogyne biflora* (Sherff) St. John; *Stenogyne hirsutula* St. John; *Stenogyne nelsonii* Benth.; *Stenogyne scandens* Sherff; *Stenogyne sororia* Sherff)

Hawaii: mohihi

Stenolepia Alderw. Dryopteridaceae (Aspleniaceae)

Origins:

From the Greek *stenos* "narrow" and *lepis* "scale."

Stenolirion Baker Amaryllidaceae (Liliaceae)

Origins:

Greek *stenos* "narrow" and *leirion* "a lily."

Stenolobium D. Don Bignoniaceae

Origins:

From the Greek *stenos* "narrow" and *lobos* "pod," *lobion* "a little pod."

Stenolobus C. Presl Davalliaceae

Origins:

From the Greek *stenos* "narrow" and *lobos* "pod, lobe," possibly referring to the pouch-shaped indusia.

Stenoloma Fée Dennstaedtiaceae (Lindsaeoideae)

Origins:

From the Greek *stenos* and *loma* "border, margin, fringe, edge," an allusion to the narrow indusia; see Antoine Laurent Apollinaire Fée (1789-1874), *Mémoires sur la famille des Fougères*. Genera Filicum. 5: 330. 1850-1852.

Stenomeria Turcz. Asclepiadaceae

Origins:

From the Greek *stenos* "narrow" and *meris* "a portion, part."

Stenomeris Planchon Dioscoreaceae

Origins:

From the Greek *stenos* and *meris* "a portion, part."

Stenomesson Herb. Amaryllidaceae (Liliaceae)

Origins:

Greek *stenos* "narrow" and *messon, messos, mesos* "middle, in the middle," referring to the shape of the perianth.

Species/Vernacular Names:
S. sp.
Peru: china sullu sullu, pichi huay

Stenonema Hook. Brassicaceae

Origins:

From the Greek *stenos* and *nema* "filament, thread."

Stenonia Baill. Euphorbiaceae

Origins:

From the Greek *stenos* "narrow," referring to the flowers; see also *Cleistanthus* Hook.f. ex Planch.

Stenonia Didr. Euphorbiaceae

Origins:

Greek *stenos* "narrow," see also *Ditaxis* Vahl ex A. Juss.

Stenoniella Kuntze Euphorbiaceae

Origins:

The diminutive of *Stenonia*, see also *Cleistanthus* Hook.f. ex Planchon.

Stenopadus S.F. Blake Asteraceae

Origins:

Pachycaul shrub and trees, possibly from the Greek *stenopous, stenopodus* "narrow-footed," Greek *pados, pedos* used by Theophrastus for a species of *Prunus*.

Stenopetalum R. Br. ex DC. Brassicaceae

Origins:

From the Greek *stenos* and *petalum* or *petalon* "petal."

Species/Vernacular Names:
S. nutans F. Muell.
English: stinking thread-petal
S. sphaerocarpum F. Muell.
English: wiry thread-petal, pea thread-petal

Stenophalium A. Anderb. Asteraceae

Origins:

Presumably from the Greek *stenos* "narrow, scanty" and *phalos* "shining, bright, white," see also *Stenocline* DC.

Stenophragma Celak. Brassicaceae

Origins:

From the Greek *stenos* and *phragma* "a partition, compartment, wall, fence."

Stenophyllus Raf. Cyperaceae

Origins:

Greek *stenos* "narrow" and *phyllon* "leaf"; see C.S. Rafinesque, *Neogenyton, or Indication of Sixty-Six New Genera of Plants of North America*. 4. 1825; E.D. Merrill, *Index rafinesquianus*. 79. 1949.

Stenopolen Rafinesque Orchidaceae

Origins:

Possibly from the Greek *stenos* "narrow" and Latin *pollen, inis* "fine dust, pollen," Spanish *polen* "pollen," referring probably to the shape of the pollinia; see Constantine Samuel Rafinesque, *Flora Telluriana*. 4: 49. 1836 [1838]; Elmer Drew Merrill, *Index rafinesquianus*. 104. Massachusetts, USA 1949; see also *Stenia* Lindl.

Stenops B. Nord. Asteraceae

Origins:

Greek *stenos* and *opsis* "aspect, resemblance."

Stenoptera C. Presl Orchidaceae

Origins:

From the Greek *stenos* "narrow" and *pteron* "wing," the narrow petals.

Stenorrhynchos Rich. ex Sprengel Orchidaceae

Origins:
The narrow snout, Greek *stenos* and *rhynchos* "horn, beak, snout," referring to the narrow and slender rostellum and blade-like viscidium.

Stenoschista Bremek. Acanthaceae

Origins:
From the Greek *stenos* with *schistos* "cut," *schizo* "to divide."

Stenosemia C. Presl Dryopteridaceae (Aspleniaceae)

Origins:
From the Greek *stenos* plus *sema* "standard, a sign, mark," referring to the fertile fronds.

Stenoseris Shih Asteraceae

Origins:
Greek *stenos* "narrow" and *seris, seridos* "chicory, lettuce."

Stenosiphanthus A. Samp. Bignoniaceae

Origins:
From the Greek *stenos* "narrow," *siphon* "a tube" and *anthos* "flower."

Stenosiphon Spach Onagraceae

Origins:
Greek *stenos* and *siphon* "a tube."

Stenosiphonium Nees Acanthaceae

Origins:
Greek *stenos* "narrow" and *siphon* "a tube."

Stenosolen (Müll. Arg.) Markgraf Apocynaceae

Origins:
Greek *stenos* "narrow" and *solen* "a tube."

Stenosolenium Turcz. Boraginaceae

Origins:
From the Greek *stenos* and *solen* "a tube."

Species/Vernacular Names:
S. saxatile (Pallas) Turcz.
English: cliff stenosolenium
China: zi tong cao

Stenospermation Schott Araceae

Origins:
Greek *stenos* "narrow" and *sperma* "a seed," *spermation* "a little seed"; see D.H. Nicolson, "Derivation of aroid generic names." *Aroideana*. 10: 15-25. 1988.

Stenospermum Sweet ex Heynhold Myrtaceae

Origins:
Greek *stenos* and *sperma* "a seed."

Stenostachys Turcz. Gramineae

Origins:
From the Greek *stenos* "narrow" and *stachys* "a spike."

Stenostelma Schltr. Asclepiadaceae

Origins:
From the Greek *stenos* and *stelma, stelmatos* "a girdle, belt."

Stenostephanus Nees Acanthaceae

Origins:
From the Greek *stenos* "narrow" and *stephanos* "a crown."

Stenostomum C.F. Gaertn. Rubiaceae

Origins:
Greek *stenos* "narrow" and *stoma* "mouth."

Stenotaenia Boiss. Umbelliferae

Origins:
Greek *stenos* "narrow" and *tainia* "a band, ribbon."

Stenotaphrum Trin. Gramineae

Origins:

Greek *stenos* and *taphros* "a trench, ditch," in the raceme axis there are some depressions or cavities.

Species/Vernacular Names:

S. dimidiatum (L.) Brongn.

Rodrigues Island: chiendent bourrique

S. secundatum (Walter) Kuntze (*Ischaemum secundatum* Walter; *Stenotaphrum americanum* Schrank)

English: buffalo grass, coastal buffalo grass, buffalo quick grass, Cape quick grass, carpet grass, coarse couch grass, coarse quick grass, couch grass, mission grass, quick grass, ramsammy grass, seaside quick grass, seaside quick, Saint Augustine grass

Southern Africa: Augustinus gras, Augustinus kweek, buffelgras, buffelskweek, Cape kweek, coast kweek, grove kweek, growwekweek, kweekgras, lidjieskweek, olifantskweek, rivierkweek, strand-buffelskweek; marotlo-a-mafubelu (Sotho); umtombo (Zulu)

Hawaii: 'aki'aki haole, manienie 'aki'aki, manienie 'aki'aki haole, manienie mahikihiki

Stenothyrsus C.B. Clarke Acanthaceae

Origins:

From the Greek *stenos* "narrow" and *thyrsos* "a panicle."

Stenotopsis Rydb. Asteraceae

Origins:
Resembling *Stenotus*.

Stenotus Nutt. Asteraceae

Origins:

Greek *stenos* "narrow" and *ous, otos* "an ear."

Species/Vernacular Names:
S. lanuginosus A. Gray

English: woolly stenotus

Stephanachne Keng Gramineae

Origins:

Greek *stephein* "to crown," *stephanos* "a crown, garland, that which surrounds or encompasses" and *achne* "chaff, glume," Latin *stephanos* is the name of several plants (Plinius).

Stephanandra Siebold & Zucc. Rosaceae

Origins:

Greek *stephanos* "a crown" and *aner, andros* "man, stamen," the stamens as a crown around the capsule.

Stephania Lour. Menispermaceae

Origins:

Greek *stephein* "to crown," *stephanos* "a crown," referring to the arrangement of the stamens or to the nature of the seeds; the genus *Stephania* Willd. is dedicated to the German botanist Christian Friedrich Stephan, 1757-1814, physician, professor of botany and chemistry at Moscow, from 1811 Director of the Forestry Institute St. Petersburg, author of *Icones plantarum mosquensium*. Mosquae 1795, see John H. Barnhart, *Biographical Notes upon Botanists*. 3: 324. 1965.

Species/Vernacular Names:
S. abyssinica (Dillon & A. Rich.) Walp.

Yoruba: gbejedi, gbejegi

S. japonica (Thunberg) Miers

China: qian jin teng

S. japonica (Thunberg) Miers var. *japonica* (*Menispermum japonicum* Thunberg)

Japan: hasu-no-ha-kazura

Okinawa: yama-kanda

Stephanium Schreb. Rubiaceae

Origins:

Greek *stephanos* "a crown, wreath," *stephanion*, the diminutive.

Stephanocaryum Popov Boraginaceae

Origins:

Greek *stephanos* "a crown" and *karyon* "nut."

Stephanocereus A. Berger Cactaceae

Origins:

Greek *stephanos* "a crown" with *Cereus*.

Stephanochilus Cosson & Durieu ex Maire Asteraceae

Origins:

From the Greek *stephanos* and *cheilos* "lip."

Stephanococcus Bremek. Rubiaceae

Origins:

From the Greek *stephanos* "a crown" and *kokkos* "berry."

Stephanocoma Less. Asteraceae

Origins:

Greek *stephanos* "a crown" and *kome* "hair, hair of the head," referring to the crown-like pappus.

Stephanodaphne Baillon Thymelaeaceae

Origins:

Greek *stephanos* and the genus *Daphne* L.

Stephanodoria E. Greene Asteraceae

Origins:

Possibly from the Greek *stephanos* "a crown" and *dorea, doron* "a gift."

Stephanolepis S. Moore Asteraceae

Origins:

From the Greek *stephanos* and *lepis* "a scale."

Stephanoluma Baill. Sapotaceae

Origins:

Greek *stephanos* "a crown, garland" plus *luma*.

Stephanomeria Nutt. Asteraceae

Origins:

Greek *stephanos* "a crown" and *meris* "part."

Species/Vernacular Names:

S. blairii Munz & I.M. Johnston

English: Blair's munzothamnus

S. pauciflora (Nutt.) Nelson

English: wire lettuce

S. tenuifolia (Torrey) H.M. Hall

English: wire lettuce

Stephanopholis S.F. Blake Asteraceae

Origins:

From the Greek *stephanos* "a crown" and *pholis, pholidos* "scale, horny scale."

Stephanophysum Pohl Acanthaceae

Origins:

From the Greek *stephanos* and *physa* "bladder."

Stephanopodium Poeppig Dichapetalaceae

Origins:

From the Greek *stephanos* "a crown" and *podion* "little foot," presumably referring to the inflorescences on the petioles.

Stephanorossia Chiov. Umbelliferae

Origins:

For the Italian botanist Stefano Rossi, 1851-1898, author of *Studi sulla flora ossolana*, Domodossola 1883; see E.M. Tucker, *Catalogue of the Library of the Arnold Arboretum of Harvard University*. 1917-1933.

Stephanostachys (Klotzsch) Klotzsch ex O.E. Schulz Palmae

Origins:

From the Greek *stephanos* "garland, crown" and *stachys* "spike."

Stephanostegia Baillon Apocynaceae

Origins:

From the Greek *stephanos* and *stege, stegos* "roof, cover."

Stephanostema K. Schumann Apocynaceae

Origins:
From the Greek *stephanos* "a crown" and *stema* "stamen, penis."

Stephanotella E. Fourn. Asclepiadaceae

Origins:
Greek *stephanos* "a crown," the diminutive, referring to the genus *Stephanotis* Thouars.

Stephanothelys Garay Orchidaceae

Origins:
From the Greek *stephanos* and *thelys* "feminine, female."

Stephanotis Thouars Asclepiadaceae

Origins:
Stephanotis is a name of the myrtle, Greek *stephanos* "crown," Latin *stephanos* is the name of several plants (Plinius), Latin *stephanites* or *stephanitis* is a kind of vine, which winds about in the shape of garlands (Plinius); see F. Boerner & G. Kunkel, *Taschenwörterbuch der botanischen Pflanzennamen*. 4. Aufl. 174. 1989; Helmut Genaust, *Etymologisches Wörterbuch der botanischen Pflanzennamen*. 610. Basel 1996; Ernest Weekley, *An Etymological Dictionary of Modern English*. 2: 1415. 1967.

Species/Vernacular Names:
S. floribunda (R. Br.) Brongn. (*Stephanotis jasminoides* hort.)
English: Madagascar jasmine, bridal wreath, chaplet flower, waxflower
Spanish: floradora

Stephanotrichum Naudin Melastomataceae

Origins:
From the Greek *stephanos* "a crown" and *thrix, trichos* "hair," referring to the nature of the plants.

Stephegyne Korth. Rubiaceae

Origins:
Greek *stephos* "a garland, crown" and *gyne* "woman, ovary," see also *Mitragyna* Korth.

Steptorhamphus Bunge Asteraceae

Origins:
Greek *steptos* "crowned" and *rhamphos* "a beak."

Sterculia L. Sterculiaceae

Origins:
Sterculius (and also Stercutus, Sterculus, Sterculinius and Sterces were all epithets of Saturnus or Picumnus; see Q. Serenus Sammonicus, 3rd century a.C., author of a *liber medicinalis*) was the Roman god of privies, a Roman god who presided over cultivation and manuring; from the Latin *stercus, oris* "dung, manure," it refers to the unpleasant smell of the flowers and leaves of some species.

Species/Vernacular Names:
S. sp.
Nigeria: bongele, kukuki
E. Africa: mgua, mhosia
Uganda: mutumbwi, muvule
Gabon: obachou
Malaya: kepayang hantu, rengkong
S. africana (Lour.) Fiori (*Sterculia guerichii* K. Schum.; *Sterculia tomentosa* sensu Sim; *Sterculia triphaca* R. Br.; *Triphaca africana* Lour.)
English: mopopoja tree
Southern Africa: muKukubuyu, muNera, muNgoza, muRere (Shona); mopopoja (Subya: Botswana, eastern Caprivi); mokokobuyu (Kololo: Barotseland); omumbambahako (Herero)
Nigeria: kukuki
S. apeibophylla Ducke
Bolivia: sujo
S. apetala (Jacq.) Karsten (*Sterculia carthaginensis* Cav.)
English: Panama tree
Tropical America: bellota
Bolivia: sujo
S. bequaertii De Wild.
Zaire: elalombe, otutu, kausa
S. foetida L.
English: Indian almond, Java olive, great sterculia, hazel sterculia, horse almond
French: bois puant
Malaya: kelumpang jari, kelupang
Sri Lanka: telambu, telembu, kadutenga, kaduteynga, pinari
India: jangal beddam

S. macrophylla Vent.

English: broad-leaf sterculia

Malaya: kelumpang, milian, berlubuh bukit

S. monosperma Vent. (*Sterculia nobilis* Sm.)

English: noble battle tree

Japan: pinpon-no-ki

S. murex Hemsley (from the Latin *murex* "prickly")

English: lowveld chestnut

Southern Africa: laeveldkastaiing; umBhaba (Swazi)

S. oblonga Martius

English: yellow sterculia

Nigeria: okoko, kokoniko, ako-girishi

Yoruba: orodo, kokoniko, okonko

Ivory Coast: assasodau, azodo

Cameroon: bongele, eyon, lom, mvan

Tropical Africa: ekonge

S. quinqueloba (Garcke) K. Schum. (*Sterculia zastrowiana* Engl.; *Cola quinqueloba* Garcke)

English: five-lobed sterculia

Southern Africa: muGoza, muNgoza, murere, muTedza, muTondoshure (Shona)

S. rhinopetala K. Schumann

English: brown sterculia

Cameroon: nkanna, nkanga, nkannang, nkanang, bulembo

Ghana: wawabima

Yoruba: aye, orodo, otutu, oro

Nigeria: orodo, aye, bongele, ogbo, ogiokoko; aye, orodo (Yoruba); ogiokoko (Edo); abang abuga (Boki)

Ivory Coast: lotofa

S. rogersii N.E. Br. (after the South African (b. England, Dorset) botanist, missionary, Archdeacon of Pietersburg (Transvaal), Frederick Arundel Rogers, 1876-1944 (d. London), son of the Rev. William Moyle Rogers (1835-1920, d. Hants.), botanical collector in South Africa, Belgian Congo and Rhodesia, author of *Provisional List of Flowering Plants and Ferns Found in the Divisions of Albany and Bathurst.* Grahamstown 1909; see J.H. Barnhart, *Biographical Notes upon Botanists.* 3: 171. 1965; Mary Gunn and Leslie E. Codd, *Botanical Exploration of Southern Africa.* 298-299. Cape Town 1981; T.W. Bossert, *Biographical Dictionary of Botanists Represented in the Hunt Institute Portrait Collection.* 336. 1972; Alain White (1880-1951) and Boyd Lincoln Sloane (1886-1955), *The Stapelieae.* Pasadena 1937; Ray Desmond, *Dictionary of British & Irish Botanists and Horticulturists.* 592. London 1994; Stafleu and Cowan, *Taxonomic Literature.* 4: 858. 1983)

English: common star-chestnut, succulent chestnut

Southern Africa: gewone sterkastaiing, bolstamkastaiing, koedoeklapper, Ulumbu tree; iNkuphenkuphe, uLumbu, inKuphenkuphe, umuMbu (Zulu); uLumba (Swazi); samani (Tsonga); mokakata (Mangwato dialect, Botswana)

S. rubiginosa Vent.

English: rusty sterculia

Malaya: bayor betina, dedamak hitam, kelumpang bukit, kelumpang gajah, kelunting, sakelat, sakhlat, seburu, unting unting besar, tangisan burong

S. setigera Del.

Nigeria: kukuki (Hausa); kokongiga (Nupe); bo'boli (Fula); sugubo (Kanuri); shadarat ad damn (Shuwa Arab); ose awere (Yoruba)

W. Africa: korofogo, koronfogo

S. striata St. Hil. & Naudin

Bolivia: sujo

S. subviolacea K. Schumann

Congo: naboubouk

Cameroon: efok oswé, yebolo

Central Africa: pomboli

S. tragacantha Lindley

Central Africa: dototo, efok, fembe

Zaire: diolofo

Cameroon: efok afum, mboli, popoko

Ivory Coast: poré-poré

Gabon: ezelfou

Yoruba: owun, omorun, alawefan, iwanranwanran, wanran-wanran, ikakaale, ilakaale ogun, okaagbo

Nigeria: oloko, omorin, owun, okagbo, woran-woran, oporipor, omu, udo-gohoho, eyong, yebel, wonron-wonron, paupau, pio, ndo-toto, etude, lengui, lengue, botopia, pori-pori, apompom, aloyefun, tutumbo, kukukin-rafi, lakale, rofo-rofo; kukukin rafi (Hausa); nyichi kuso (Nupe); alawefun (Yoruba); oporipor (Edo); apompor (Kwale); oloko (Igbo); kemuan (Boki); udot eto (Ibibio)

S. urens Roxb.

India: gulu, kulu, kulru, gular, gulli, kalauri

S. villosa Roxb.

India: gulkandar, massu, osha, gudgudala, udial, udar, udalla

Stereocarpus (Pierre) Hallier f. Theaceae

Origins:

Greek *stereos* "solid, firm, tight" and *karpos* "fruit."

Stereocaryum Burret Myrtaceae

Origins:
From the Greek *stereos* and *karyon* "nut."

Stereochilus Lindley Orchidaceae

Origins:
Greek *stereos* "solid, firm, tight" and *cheilos* "lip," the shape and the nature of the lip.

Stereochlaena Hackel Gramineae

Origins:
From the Greek *stereos* "solid, firm, tight" and *chlaena*, *chlaenion* "a cloak, covering."

Stereoderma Blume Oleaceae

Origins:
From the Greek *stereos* and *derma*, *dermatos* "skin."

Stereosandra Blume Orchidaceae

Origins:
Greek *stereos* "solid, firm, tight" and *aner*, *andros* "male, a man, stamen," referring to the texture of the anthers.

Species/Vernacular Names:
S. javanica Blume (*Stereosandra koidzumiana* Ohwi; *Stereosandra liukiuensis* Tuyama)
Japan: Iriomote-muyô-ran

Stereospermum Cham. Bignoniaceae

Origins:
From the Greek *stereos* plus *sperma* "seed."

Species/Vernacular Names:
S. acuminatissimum K. Schum.
Nigeria: osuobon, akoko-igbo; eru yeye (Yoruba); oshuobon (Edo)
Yoruba: eru iyeje
S. fimbriatum (Wallich ex G. Don) A. DC.
English: snake tree
Malaya: chachah, chichah, lempoyan
S. kunthianum Cham.
English: pink jacaranda
Mali: mogoiri, mogokolo, shimugo

Southern Africa: muKuku (Shona)
Nigeria: ayagdo, sansami, ajade, jiri, golombi, turken-doki; sansami (Hausa); golombi (Fula); golombi (Kanuri); umana tumba (Tiv); ayada (Yoruba)
Yoruba: ajade, ayada, rijarija
S. personatum (Hassk.) Chatterji (*Stereospermum caudatum* Miq.; *Stereospermum chelonoides* sensu (Roxb.) DC.; *Stereospermum tetragonum* A. DC.)
English: yellow snake tree

Stereoxylon Ruíz & Pavón Grossulariaceae (Escalloniaceae)

Origins:
Greek *stereos* "solid, firm, tight" and *xylon* "wood."

Sterigmanthe Klotzsch & Garcke Euphorbiaceae

Origins:
Greek *sterigmos* "a setting firmly," *sterizo*, *sterizein* "to stand fast, to fix," *sterigma*, *sterigmatos* "a support, foundation" and *anthos* "flower."

Sterigmapetalum Kuhlm. Rhizophoraceae

Origins:
Greek *sterigma*, *sterigmatos* "a support, foundation" and *petalon* "petal."

Sterigmostemum M. Bieb. Brassicaceae

Origins:
From the Greek *sterigma*, *sterigmatos* "a support, foundation" and *stemon* "stamen."

Steriphoma Sprengel Capparidaceae (Capparaceae)

Origins:
Probably from the Greek *steriphos* "barren, unfruitful."

Sternbergia Waldstein & Kitaibel Amaryllidaceae (Liliaceae)

Origins:
Dedicated to the Austrian (b. Bohemia, Prague) botanist and paleobotanist Kaspar (Caspar) Maria Reichsgraf von Stern-

berg, 1761-1838 (d. Radnice), geologist, clergyman, devoted to the study of Carboniferous phytopaleontology, a founder of the Bohemian National Museum (Böhmische Nationalmuseum) in Prague and from 1822 to 1838 President of the Museum, author of *Versuch einer geognostisch-botanischen Darstellung der Flora der Vorwelt*. Regensburg [1820-] 1825-1838. See Robert P. Beckinsale & Jan Krejci, in *Dictionary of Scientific Biography* 13: 43-44. 1981; J.W. von Goethe, *Briefwechsel zwischen Goethe und K. Graf von Sternberg. (1820-1832)*. Wien 1866; Giulio Giorello & Agnese Grieco, a cura di, *Goethe scienziato*. Einaudi Editore, Torino 1998; Günther Schmid, *Chamisso als Naturforscher. Eine Bibliographie*. Leipzig 1942; R. Zander, F. Encke, G. Buchheim and S. Seybold, *Handwörterbuch der Pflanzennamen*. 14. Aufl. 786. Stuttgart 1993; G.C. Wittstein, *Etymologisch-botanisches Handwörterbuch*. 844. Ansbach 1852; Agnes Arber, *Goethe's Botany. The Metamorphosis of Plants* (1790) and Tobler's *Ode to Nature* (1782). Waltham 1946; J.H. Barnhart, *Biographical Notes upon Botanists*. 3: 326. 1965; Carolus Clusius (Charles de l'Écluse), 1526-1609, *Rariorum plantarum historia*. Antwerpen 1601; T.W. Bossert, *Biographical Dictionary of Botanists Represented in the Hunt Institute Portrait Collection*. 382. 1972; A. Lasègue, *Musée botanique de Benjamin Delessert*. 333. Paris 1845; G. Schmid, *Goethe und die Naturwissenschaften*. Halle 1940; J.D. Milner, *Catalogue of Portraits of Botanists Exhibited in the Museums of the Royal Botanic Gardens*. Royal Botanic Gardens, Kew, London 1906.

Sterrhymenia Griseb. Solanaceae

Origins:

Greek *sterros* "firm, stiff, solid" and *hymen* "a membrane"; see also *Sclerophylax* Miers.

Sterropetalum N.E. Br. Aizoaceae

Origins:

Greek *sterros* "firm, stiff, solid" and *petalon* "a petal, leaf."

Stethoma Raf. Acanthaceae

Origins:

Greek *stethos* "tumor, breast, swelling"; see C.S. Rafinesque, *Flora Telluriana*. 4: 61. 1836 [1838]; Elmer D. Merrill, *Index rafinesquianus*. 224. 1949.

Stetsonia Britton & Rose Cactaceae

Origins:

For Francis Lynde Stetson, of New York, an American specialist in Cactaceae.

Steudelago O. Kuntze Rubiaceae

Origins:

For the German botanist Ernst Gottlieb von Steudel, 1783-1856, physician, botanical collector, December 1825 with Christan Ferdinand Hochstetter (1787-1860) founded Der Botanische Reiseverein Esslingen; see J.H. Barnhart, *Biographical Notes upon Botanists*. 3: 326. 1965; H.N. Clokie, *Account of the Herbaria of the Department of Botany in the University of Oxford*. 249, 280. Oxford 1964; A. Lasègue, *Musée botanique de Benjamin Delessert*. Paris 1845; E.M. Tucker, *Catalogue of the Library of the Arnold Arboretum of Harvard University*. 1917-1933; Ida Kaplan Langman, *A Selected Guide to the Literature on the Flowering Plants of Mexico*. 722. University of Pennsylvania Press, Philadelphia 1964; Mariella Azzarello Di Misa, ed., *Il Fondo Antico della Biblioteca dell'Orto Botanico di Palermo*. 261. Palermo 1988; S. Lenley et al., *Catalog of the Manuscript and Archival Collections and Index to the Correspondence of John Torrey*. Library of the New York Botanical Garden. 470. 1973; Leonard Huxley, *Life and Letters of Sir Joseph Dalton Hooker*. London 1918; R. Zander, F. Encke, G. Buchheim and S. Seybold, *Handwörterbuch der Pflanzennamen*. 14. Aufl. Stuttgart 1993.

Steudelella Honda Gramineae

Origins:

For the German botanist Ernst Gottlieb von Steudel, 1783-1856.

Steudnera Koch Araceae

Origins:

After the German naturalist Hermann Steudner, 1832-1863, plant collector in Sudan and tropical North Africa; see Georg August Schweinfurth, 1836-1925, *Beitrag zur Flora Aethiopiens*. Berlin 1867 and *Novae species aethiopicae*. [Wien 1868].

Species/Vernacular Names:

S. virosa (Kunth) Prain

India: bish kachu

Stevenia Adams ex Fischer Brassicaceae

Origins:

For the Finnish plant collector Christian von Steven, 1781-1863, botanist, physician, traveler (southern Russia and central Europe), M.D. St. Petersburg 1798, 1812 in Crimea established the State Nikita Botanical Garden; see R. Zander, F. Encke, G. Buchheim and S. Seybold, *Hand-*

wörterbuch der Pflanzennamen. 14. Aufl. Stuttgart 1993; J.H. Barnhart, *Biographical Notes upon Botanists*. 3: 327. 1965.

Steveniella Schlechter Orchidaceae

Origins:

For the Finnish plant collector Christian von Steven, 1781-1863, botanist, physician, discoverer of the genus, traveler (southern Russia and central Europe), M.D. St. Petersburg 1798, 1812 in Crimea established the State Nikita Botanical Garden; see J.H. Barnhart, *Biographical Notes upon Botanists*. 3: 327. 1965; R. Zander, F. Encke, G. Buchheim and S. Seybold, *Handwörterbuch der Pflanzennamen*. 14. Aufl. Stuttgart 1993; T.W. Bossert, *Biographical Dictionary of Botanists Represented in the Hunt Institute Portrait Collection*. 383. 1972; H.N. Clokie, *Account of the Herbaria of the Department of Botany in the University of Oxford*. 249. Oxford 1964; A. Lasègue, *Musée botanique de Benjamin Delessert*. Paris 1845; E.M. Tucker, *Catalogue of the Library of the Arnold Arboretum of Harvard University*. 1917-1933; Günther Schmid, *Chamisso als Naturforscher*. Eine Bibliographie. Leipzig 1942; Stafleu and Cowan, *Taxonomic Literature*. 5: 911-915. Utrecht 1985.

Stevenorchis Wankow & Kraenzlin Orchidaceae

Origins:

For the plant collector Christian von Steven, 1781-1863.

Stevensia Poit. Rubiaceae

Origins:

For Edward Stevens, an American author.

Stevia Cav. Asteraceae

Origins:

After the Spanish botanist and physician Pedro (Petrus) Jaime (Jago, Jacobus) Esteve (Stevius), d. 1566, author of *Nicandri ... Theriaca*. P.J.S. interprete. Valentiae 1552; see F. Boerner & G. Kunkel, *Taschenwörterbuch der botanischen Pflanzennamen*. 4. Aufl. 174. 1989; Georg Christian Wittstein (1810-1887), *Etymologisch-botanisches Handwörterbuch*. 844. Ansbach 1852.

Species/Vernacular Names:

S. sp.

Spanish: orégano cimarrón

Steviopsis R. King & H. Robinson Asteraceae

Origins:

Resembling *Stevia*.

Stewartia L. Theaceae

Origins:

See also *Stuartia* L.; dedicated to John Stuart, 3rd Earl of Bute, 1713-1792, botanist, a patron of science as well as of literature and art, studied at Eton, statesman, 1737 he was elected one of the representative peers of Scotland, Knight of the Thistle, one of the lords of the bedchamber to the Prince of Wales, Prime Minister 1762-1763, plantsman, introduced many new species to Kew, author of *Botanical Tables*, containing the different families of British plants. [London 1785?]; see Dawson Turner, *Extracts from the Literary and Scientific Correspondence of R. Richardson, of Bierly, Yorkshire: Illustrative of the State and Progress of Botany*. Yarmouth 1835; W. Darlington, *Memorials of John Bartram and Humphry Marshall*. Philadelphia 1849; T.W. Bossert, *Biographical Dictionary of Botanists Represented in the Hunt Institute Portrait Collection*. 1972; Blanche Henrey, *British Botanical and Horticultural Literature before 1800*. Oxford 1975; M. Hadfield et al., *British Gardeners: A Biographical Dictionary*. London 1980; George Taylor, in *Dictionary of Scientific Biography* 1: 88. 1981; James A.L. Fraser, *John Stuart, Earl of Bute*. 1912; J.H. Barnhart, *Biographical Notes upon Botanists*. 1: 291. 1965; Jonas C. Dryander, *Catalogus bibliothecae historico-naturalis Josephi Banks*. London 1800; Frans Antonie Stafleu (1921-1997), *Linnaeus and the Linnaeans. The spreading of their ideas in systematic botany, 1735-1789*. 231. ["the botanophilic Lord Bute..."] Utrecht 1971.

Stewartiella E. Nasir Umbelliferae

Origins:

For the American botanist Ralph Randles Stewart, b. 1890, traveler, botanical collector in many parts of Pakistan and Kashmir (Mussoorie, in India, Ladak, Frontier Regions, Punjab, Chamba, Kashmir), Principal Emeritus of Gordon College (Rawalpindi), Research Associate at the Herbarium (University of Michigan, Ann Arbor, USA), compiled *An Annotated Catalogue of the Vascular Plants of West Pakistan and Kashmir*. [in *Flora of West Pakistan*. Editors: E. Nasir and S.I. Ali] Karachi 1972; see T.W. Bossert, *Biographical Dictionary of Botanists Represented in the Hunt Institute Portrait Collection*. 383. 1972; J.H. Barnhart, *Bio-*

graphical Notes upon Botanists. 3: 329. 1965; S. Lenley et al., *Catalog of the Manuscript and Archival Collections and Index to the Correspondence of John Torrey*. Library of the New York Botanical Garden. 388. 1973.

Steyermarkia Standley Rubiaceae

Origins:

Dedicated to the American botanist Julian Alfred Steyermark, 1909-1988, botanical explorer, from 1937 to 1958 with the Field Museum of Natural History, from 1959 to 1984 at the Instituto Botanico Caracas, specialist in Rubiaceae. Among his writings are "New Species of Rubiaceae from Peru collected by John Wurdack." *Bol. Soc. Venez. Ci. Nat.* 25: 232-244. 1964, *Spring Flora of Missouri*. St. Louis and Chicago 1940, *Behind the Scenes*. St. Louis, Mo. 1984 and "Rubiaceae." in Bassett Maguire, J.J. Wurdack and collaborators, "The Botany of the Guayana Highland-part V, VI, VII, IX." in *Mem. New York Bot. Gard.* 10(5): 186-278. 1964, 12(3): 178-285. 1965, 17(1): 230-439. 1967 and 23: 227-832. 1972, with Otto Huber wrote *Flora del Avila*. Caracas 1978, with Paul C. Standley curator of the *Flora of Guatemala*. [Fieldiana, Botany. volume 24, etc.] Published by Chicago Natural History Museum; see J.A. Steyermark & collaborators, "Botanical Exploration in Venezuela." in *Fieldiana, Bot.* 1951-1952; T.W. Bossert, *Biographical Dictionary of Botanists Represented in the Hunt Institute Portrait Collection*. 383. 1972; J.H. Barnhart, *Biographical Notes upon Botanists*. 3: 330. 1965; S. Lenley et al., *Catalog of the Manuscript and Archival Collections and Index to the Correspondence of John Torrey*. Library of the New York Botanical Garden. 388. 1973; Laurence J. Dorr, "In Memoriam. John J. Wurdack, 1921-1998." in *Plant Science Bulletin*. 44(2): 41. Summer 1998; R. Zander, F. Encke, G. Buchheim and S. Seybold, *Handwörterbuch der Pflanzennamen*. 14. Aufl. Stuttgart 1993; Stafleu and Cowan, *Taxonomic Literature*. 5: 952-956. Utrecht 1985.

Steyermarkina R.M. King and H. Robinson Asteraceae

Origins:

For the American botanist Julian Alfred Steyermark, 1909-1988.

Steyermarkochloa Davidse & R.P. Ellis Gramineae

Origins:

For the American botanist Julian Alfred Steyermark, 1909-1988.

Stibasia C. Presl Marattiaceae

Origins:

Greek *stibas, stibados* "bed of straw, rushes, or leaves," *stibasis* "building up, laying."

Stiburus Stapf Gramineae

Origins:

Possibly from the Latin *stibi, is* and *stibium, ii* "antimony, black antimony"; or from Greek *stibi* and *oura* "a tail," or *stibe* "stipa."

Species/Vernacular Names:

S. alopecuroides (Hack.) Stapf

English: Pongwa grass

South Africa: koperdraadgras

Stichianthus Valeton Rubiaceae

Origins:

Greek *stichos* "a row, line" and *anthos* "flower."

Stichoneuron Hook.f. Stemonaceae (Croomiaceae)

Origins:

Greek *stichos* "a row, line" and *neuron* "nerve."

Stichorkis Thouars Orchidaceae

Origins:

From the Greek *stichos* "a row, line" plus *orchis* "orchid," indicating the inflorescence.

Stictocardia Hallier f. Convolvulaceae

Origins:

From the Greek *stiktos* "spotted, punctured" and *kardia* "heart," referring to the spotted leaves.

Species/Vernacular Names:

S. beraviensis (Vatke) Hallier f.

Yoruba: namunamu, digbaro, abeesun digbaro

S. tiliifolia (Desr.) Hallier f.

English: common stictocardia

China: xian ye teng

Stictophyllorchis Carnevali & Dodson Orchidaceae

Origins:

The genus *Stictophyllum* and *orchis* "orchid."

Stictophyllum Dodson & M.V. Chase Orchidaceae

Origins:

From the Greek *stiktos* "spotted, punctured" and *phyllon* "leaf."

Stigmanthus Lour. Rubiaceae

Origins:

Greek *stigma*, *stigmatos* "stigma, mark, spot, tattoo-mark" and *anthos* "flower."

Stigmaphyllon A. Juss. Malpighiaceae

Origins:

Greek *stigma*, *stigmatos* "stigma, mark" and *phyllon* "leaf."

Stigmatanthus Roemer & Schultes Rubiaceae

Origins:

From the Greek *stigma*, *stigmatos* and *anthos* "flower."

Stigmatella Eig Brassicaceae

Origins:

Greek *stigma*, *stigmatos* "stigma, mark, spot," the diminutive.

Stigmatocarpum L. Bolus Aizoaceae

Origins:

From the Greek *stigma*, *stigmatos* "stigma, mark, spot, tattoo-mark" and *karpos* "fruit."

Stigmatodactylus Maxim. ex Makino Orchidaceae

Origins:

Greek *stigma*, *stigmatos* and *daktylos* "a finger," a dactyliform projection below the stigma.

Stigmatopteris C. Chr. Dryopteridaceae (Aspleniaceae)

Origins:

From the Greek *stigma*, *stigmatos* and *pteris* "fern."

Stigmatorhynchus Schltr. Asclepiadaceae

Origins:

From the Greek *stigma*, *stigmatos* plus *rhynchos* "horn, beak, snout."

Stigmatorthos M. Chase & D. Bennett Orchidaceae

Origins:

Greek *stigma*, *stigmatos* with *orthos* "upright, straight."

Stigmatosema Garay Orchidaceae

Origins:

Greek *stigma*, *stigmatos* "stigma, mark" and *sema* "standard, a sign, mark."

Stilaginella Tul. Euphorbiaceae

Origins:

The diminutive of the genus *Stilago* L.

Stilbanthus Hook.f. Amaranthaceae

Origins:

Greek *stilbe* "lamp, mirror," *stilbo*, *stilbein* "to shine, glitter, glisten" and *anthos* "flower."

Stilbe Bergius Stilbaceae (Verbenaceae)

Origins:

Greek *stilbe* "lamp, mirror."

Stilbocarpa (Hook.f.) Decne. & Planchon Araliaceae

Origins:

From the Greek *stilbein* "to shine, glitter, glisten" and *karpos* "fruit."

Stillingfleetia Bojer Euphorbiaceae

Origins:

After the British botanist Benjamin Stillingfleet, 1702-1771 (London).

Stillingia Garden ex L. Euphorbiaceae

Origins:

After the British botanist Benjamin Stillingfleet, 1702-1771 (London). His works include *Miscellaneous Tracts Relating to Natural History, Husbandry, and Physic*. Translated from the Latin, with notes by Benj. Stillingfleet. London 1759; see *Memoir and Correspondence of ... Sir J.E. Smith ...* Edited by Lady Pleasance Smith. London 1832; Ray Desmond, *Dictionary of British & Irish Botanists and Horticulturists*. 656. London 1994; R. Pulteney, *Historical and Biographical Sketches of the Progress of Botany in England*. London 1790; Blanche Henrey, *British Botanical and Horticultural Literature before 1800*. Oxford 1975; J.H. Barnhart, *Biographical Notes upon Botanists*. 3: 331. Boston 1965; T.W. Bossert, *Biographical Dictionary of Botanists Represented in the Hunt Institute Portrait Collection*. 384. 1972; E.M. Tucker, *Catalogue of the Library of the Arnold Arboretum of Harvard University*. 1917-1933; J.D. Milner, *Catalogue of Portraits of Botanists Exhibited in the Museums of the Royal Botanic Gardens*. Royal Botanic Gardens, Kew, London 1906; Georg Christian Wittstein, *Etymologisch-botanisches Handwörterbuch*. 846. Ansbach 1852; F. Boerner & G. Kunkel, *Taschenwörterbuch der botanischen Pflanzennamen*. 4. Aufl. 174. 1989.

Stilpnogyne DC. Asteraceae

Origins:

Greek *stilpnos* "shining, glittering" and *gyne* "woman, female."

Stilpnolepis Kraschen. Asteraceae

Origins:

From the Greek *stilpnos* "shining, glittering" and *lepis* "scale."

Stilpnopappus Martius ex DC. Asteraceae

Origins:

Greek *stilpnos* and *pappos* "fluff, downy appendage."

Stilpnophleum Nevski Gramineae

Origins:

Greek *stilpnos* "shining, glittering" and the genus *Phleum* L.

Stilpnophyllum (Endl.) Drury Moraceae

Origins:

Greek *stilpnos* plus *phyllon* "leaf."

Stilpnophyllum Hook.f. Rubiaceae

Origins:

From the Greek *stilpnos* "shining, glittering" and *phyllon* "leaf."

Stilpnophyton Less. Asteraceae

Origins:

From the Greek *stilpnos* with *phyton* "plant."

Stilpnophytum Less. Asteraceae

Origins:

From the Greek *stilpnos* "shining, glittering" and *phyton* "plant," referring to the involucre.

Stimegas Raf. Orchidaceae

Origins:

Greek *stigma* "stigma" and *megas* "large," a dilated column with a trilobate stigma; see C.S. Rafinesque, *Flora Telluriana*. 4: 45. 1836 [1838]; E.D. Merrill, *Index rafinesquianus*. 101, 103, 104. 1949.

Stimenes Raf. Solanaceae

Origins:

See C.S. Rafinesque, *Flora Telluriana*. 3: 76. 1836 [1837]; E.D. Merrill, *Index rafinesquianus*. 213. 1949.

Stimomphis Raf. Solanaceae

Origins:

See C.S. Rafinesque, *Flora Telluriana*. 3: 76. 1836 [1837]; E.D. Merrill, *Index rafinesquianus*. 213. 1949.

Stimoryne Raf. Solanaceae

Origins:
See Constantine Samuel Rafinesque (1783-1840), *Flora Telluriana*. 3: 76. 1836 [1837]; E.D. Merrill, *Index rafinesquianus*. 213. 1949.

Stimpsonia C. Wright ex A. Gray Primulaceae

Origins:
Named after the American conchologist William Stimpson, 1832-1872, Charles Wright's companion, author of *The Crustacea and Echinodermata of the Pacific Shores of North America*. Riverside 1857, of *Report on the Crustacea, Brachyura and Anomura collected by the North Pacific Exploring Expedition, 1853-1856*. 1907, co-author of *Check Lists of the Shells of North America*. Prepared by ... W. Stimpson [and others] [Smithsonian Miscellaneous Collections. vol. 2.] 1860; see Alfred Goldsborough Mayer, *Biographical Memoir of William Stimpson, 1832-1872*. [City of Washington. National Academy of Sciences. Biographical Memoirs. vol. 8.] Washington 1919.

Species/Vernacular Names:
S. chamaedryoides Wright ex A. Gray
English: common stimpsonia
China: jia po po na

Stipa L. Gramineae

Origins:
Greek *stype, styppe* "tow," in some species the awns are feathery; Latin *stuppa, stipa* "tow, oakum, hards"; see Carl Linnaeus, *Species Plantarum*. 78. 1753 and *Genera Plantarum*. Ed. 5. 34. 1754; Helmut Genaust, *Etymologisches Wörterbuch der botanischen Pflanzennamen*. 611. Basel 1996.

Species/Vernacular Names:
S. arundinacea Hook.f.
English: New Zealand wind grass, pheasant's tail grass
S. avenacea L.
English: black oat grass
S. drummondii Steudel (*Stipa luehmanii* Reader; *Stipa horrifolia* J. Black)
English: cottony spear-grass
S. eremofila Reader (*Stipa fusca* C.E. Hubb.; *Stipa variegata* Summerh. & C.E. Hubb.; *Stipa dura* J. Black)
English: desert spear-gras
S. ichu (Ruíz & Pav.) Kunth
Latin America: ichu

S. macalpinei Reader (*Stipa compressa* R. Br. var. *lachnocolea* Benth.; *Stipa lachnocolea* (Benth.) Hughes; *Stipa setacea* R. Br. var. *latifolia* Benth.; *Stipa scelerata* Behr. ex J. Black)
English: annual spear-grass, one year grass
S. muelleri Tate
English: wiry spear-grass
S. nitida Summerh. & C.E. Hubb. (*Stipa scabra* Lindley var. *pallida* Reader)
Australia: Balcarra grass
S. pennata L.
English: European feather grass
S. pulchra Hitchc.
English: purple needlegrass
S. splendens Trin.
English: chee grass
S. tenacissima L.
English: esparto grass, esparto, Algerian grass, alfa, halfa

Stipagrostis Nees Gramineae

Origins:
Greek *stype* "tow," *Stipa* and *Agrostis*.

Species/Vernacular Names:
S. brevifolia (Nees) De Winter (*Aristida brevifolia* (Nees) Steud.)
English: Twabushman grass
South Africa: bloudakgras, boesmangras, bossiesgras, ghagras, grootboesmangras, langbeentoagras, langbeentwaagras, twaagras, twaboesmangras
S. ciliata (Desf.) De Winter var. *capensis* (Trin. & Rupr.) De Winter (*Aristida ciliata* sensu Desf., non Steud. & Hochst. ex Steud.; *Aristida ciliata* Desf. var. *capensis* Trin. & Rupr.; *Aristida ciliata* Desf. var. *pectinata* Henr.; *Aristida ciliata* Desf. var. *tricholaena* Hack.; *Aristida ciliata* Desf. var. *villosa* Hack.)
English: large bushman grass
South Africa: langbeenboesmangras
S. uniplumis (Licht.) De Winter var. *uniplumis* (*Aristida uniplumis* Licht. var. *pearsonii* Henr.; *Aristida uniplumis* Licht. var. *uniplumis*)
English: bushman grass
South Africa: beesgras, blinksaadgras
South West Africa: blinkhaar-federgras
S. zeyheri (Nees) De Winter subsp. *zeyheri* (*Aristida capensis* Thunb. var. *canescens* Trin. & Rupr.)
Southern Africa: blinkaargras; bohlanya-ba-pere (Sotho)

Stipavena Vierh. Gramineae

Origins:
Greek *stype* "tow," *Stipa* plus *Avena*.

Stipecoma Müll. Arg. Apocynaceae

Origins:
Greek *stype* "tow" and *kome* "hair, hair of the head."

Stiptanthus (Benth.) Briq. Labiatae

Origins:
Greek *steibo, stibo* "to tread firmly," *stiptos* "trodden down, firm, solid" and *anthos* "flower."

Stipularia Delpino Ranunculaceae

Origins:
From the Latin *stipula, ae* "a stalk, blade, straw, stem"; see also *Thalictrum* L.

Stipularia P. Beauv. Rubiaceae

Origins:
Latin *stipula, ae* "a stalk, blade, straw, stem."

Stipulicida Michaux Caryophyllaceae

Origins:
From the Latin *stipula* "a stalk, blade, straw, stem," *stipulus* "firm."

Stirlingia Endl. Proteaceae

Origins:
For Sir James Stirling, 1791-1865, from 1828 to 1839 first Lieut.-Governor of W. Australia, 1862 Admiral, married Ellen Mangles (cousin of Capt. James Mangles); see Stephan F. Ladislaus Endlicher, *Genera Plantarum.* 339. (Dec.) 1837; Joseph Henry Maiden (1859-1925), *Sir Joseph Banks, the Father of Australia.* Sydney and London 1909; *Journals of Several Expeditions Made in Western Australia* ... under the sanction of the Governor Sir J. Sterling. 1833; Malcolm John Leggoe Uren, *Land Looking West. The Story of Governor James Stirling in Western Australia.* London 1948; James A. Baines, *Australian Plant Genera.* An Etymological Dictionary of Australian Plant Genera. 355. N.S.W. 1981; Francis Aubie Sharr, *Western Australian Plant Names and Their Meanings.* 66. 1996.

Stixis Lour. Capparidaceae (Capparaceae)

Origins:
From the Greek *stizo, stizein* "to prick," *stixis* "a marking with a pointed instrument, puncture."

Stizolobium P. Browne Fabaceae

Origins:
Greek *stizo* "to prick" and *lobion* diminutive of *lobos* "a pod, lobe, capsule," referring to the stinging hairs of the pods of some species.

Stizolophus Cass. Asteraceae

Origins:
Greek *stizo* "to prick" and *lophos* "a crest."

Stizophyllum Miers Bignoniaceae

Origins:
Greek *stizo* "to prick" and *phyllon* "leaf."

Stoebe L. Asteraceae

Origins:
Greek *steibein, stibo* "to tread firmly," *stoibe* "thorny burnet, a species of *Poterium*," Latin *stoebe, es* for a plant, called also *pheos* (Plinius).

Species/Vernacular Names:
S. spiralis Less. (*Stoebe filaginea* (DC.) Sch.Bip.; *Stoebe spiralis* Less. var. *flavescens* (DC.) Harv.)
South Africa: slangbos
S. vulgaris Levyns
English: bankrupt bush
Southern Africa: asbossie, bankrotbossie, slangbos, slanghoutjie; sehalahala (Sotho); hanya (Inyanga)

Stoibrax Raf. Umbelliferae

Origins:
See Constantine Samuel Rafinesque, *The Good Book.* 52. 1840; Elmer D. Merrill, *Index rafinesquianus.* 183. 1949.

Stokesia L'Hérit. Asteraceae

Origins:

For the British (b. Chesterfield, Derbyshire) physician Jonathan Stokes, 1755-1831 (d. Chesterfield), botanist, M.D. Edinburgh 1782, a student of the Scottish botanist and physician John Hope (1725-1786) and friend of Linnaeus, wrote *A Botanical Materia Medica*. London 1812, *Botanical Commentaries*. London 1830 and *Dissertatio inauguralis de aere dephlogisticato*, etc. Edinburgi 1782, edited William Withering (1741-1799), *A Botanical Arrangement of British Plants*. Birmingham and London 1787; see J.H. Barnhart, *Biographical Notes upon Botanists*. 3: 333. 1965; S. Lenley et al., *Catalog of the Manuscript and Archival Collections and Index to the Correspondence of John Torrey*. Library of the New York Botanical Garden. 470. 1973; *Memoir and Correspondence of ... Sir J.E. Smith ...* Edited by Lady Pleasance Smith. London 1832; Blanche Elizabeth Edith Henrey (1906-1983), *British Botanical and Horticultural Literature before 1800*. Oxford 1975; R. Zander, F. Encke, G. Buchheim and S. Seybold, *Handwörterbuch der Pflanzennamen*. 14. Aufl. 1993; E.M. Tucker, *Catalogue of the Library of the Arnold Arboretum of Harvard University*. Cambridge, Massachusetts 1917-1933; Ray Desmond, *Dictionary of British & Irish Botanists and Horticulturists*. 658. 1994.

Stokoeanthus E.G.H. Oliver Ericaceae

Origins:

For the South African (Yorkshire-born, England) plant collector Thomas Pearson Stokoe, 1868-1959 (d. Cape Town), to South Africa in 1911, mountaineer, watercolor artist; see M. Gunn and L.E. Codd, *Botanical Exploration of Southern Africa*. 336. Cape Town 1981.

Stollaea Schlechter Cunoniaceae

Origins:

After the German botanist Emil Stolle, 1868-1940, bryologist and lichenologist, gardener; see J.H. Barnhart, *Biographical Notes upon Botanists*. 3: 334. 1965.

Stolzia Schltr. Orchidaceae

Origins:

After the German missionary Adolf Ferdinand Stolz, 1871-1917, plant and orchid collector (Nyasaland, Africa), merchant; see A.J.M. Leeuwenberg, "Isotypes of which holotypes were destroyed in Berlin." *Webbia*. 19(2): 863. 1965; Richard Neuhauss, *Deutsch Neu-Guinea*. Berlin 1911;

Alain White (1880-1951) and Boyd Lincoln Sloane (1886-1955), *The Stapelieae*. Pasadena 1937.

Stomandra Standley Rubiaceae

Origins:

Greek *stoma* "mouth, opening" and *aner, andros* "male, a man, stamen."

Stomarrhena DC. Epacridaceae

Origins:

Greek *stoma* "opening" and *arrhen* "male, anther."

Stomatanthes R.M. King & H. Robinson Asteraceae

Origins:

From the Greek *stoma, stomatos* "mouth, opening" and *anthos* "flower."

Stomatium Schwantes Aizoaceae

Origins:

Greek *stomation*, the diminutive of *stoma, stomatos* "mouth, opening," referring to the pairs of toothed leaves.

Stomatocalyx Müll. Arg. Euphorbiaceae

Origins:

From the Greek *stoma, stomatos* "mouth, opening" and calyx.

Stomatochaeta (S.F. Blake) Maguire & Wurdack Asteraceae

Origins:

From the Greek *stoma, stomatos* plus *chaite* "a bristle, foliage."

Stomatostemma N.E. Br. Asclepiadaceae

Origins:

Greek *stoma, stomatos* with *stemma, stemmatos* "garland, crown," an allusion to the corona at the mouth of the corolla tube.

Stomoisia Raf. Lentibulariaceae

Origins:

See E.D. Merrill, *Index rafinesquianus*. The plant names published by C.S. Rafinesque, etc. 222. Jamaica Plain, Massachusetts, USA 1949; C.S. Rafinesque, *Flora Telluriana*. 4: 108. 1836 [1838].

Storthocalyx Radlk. Sapindaceae

Origins:

Greek *storthe* "point, spike" and *kalyx* "calyx."

Stracheya Bentham Fabaceae

Origins:

For the British (b. Somerset) naturalist Lieutenant-general Richard Strachey, 1817-1908 (d. Hampstead, London), Royal (Bengal) Engineers, traveler, plant collector in India, 1854 Fellow of the Royal Society and in 1859 of the Linnean Society, meteorologist, geological collector, interested in botany, 1848 with James Edward Winterbottom (1803-1854) in northwest Himalaya and Tibet; see J.H. Barnhart, *Biographical Notes upon Botanists*. 3: 337. 1965; Ray Desmond, *Dictionary of British & Irish Botanists and Horticulturists*. 660. 1994; H.N. Clokie, *Account of the Herbaria of the Department of Botany in the University of Oxford*. 250. Oxford 1964; E. Bretschneider, *History of European Botanical Discoveries in China*. Leipzig 1981; E.M. Tucker, *Catalogue of the Library of the Arnold Arboretum of Harvard University*. Cambridge, Massachusetts 1917-1933; Isaac Henry Burkill (1870-1965), *Chapters on the History of Botany in India*. Delhi 1965; Leonard Huxley, *Life and Letters of Sir Joseph Dalton Hooker*. London 1918; Joseph Ewan, *Rocky Mountain Naturalists*. The University of Denver Press 1950.

Stramentopappus H. Robinson & V.A. Funk Asteraceae

Origins:

Latin *stramen, inis* "straw, litter," *stramentum* "straw, litter" and *pappus, i* "pappus, the woolly hairy seed."

Stramonium Mill. Solanaceae

Origins:

Perhaps from the Latin *stramus* "= *solanum*" plus *s'ramen* "straw"; see H. Genaust, *Etymologisches Wörterbuch der botanischen Pflanzennamen*. 613. Basel 1996; Georg Christian Wittstein, *Etymologisch-botanisches Handwörterbuch*. 847. 1852; F. Boerner & G. Kunkel, *Taschenwörterbuch der botanischen Pflanzennamen*. 4. Aufl. 370. 1989; Ernest Weekley, *An Etymological Dictionary of Modern English*. 2: 1426. New York 1967; C.T. Onions, *The Oxford Dictionary of English Etymology*. Oxford University Press 1966.

Strangea Meissner Proteaceae

Origins:

After the British naturalist Frederick Strange, d. 1854 (murdered on Percy Island, Queensland, Australia), collector of natural history, with the explorer Charles Sturt (1795-1869), 1848-1849 in New Zealand; see C.F. Meissner, in *Hooker's Journal of Botany & Kew Garden Miscellany*. 7: 66. (Mar.) 1855.

Strangweja Bertol. Hyacinthaceae (Liliaceae)

Origins:

Named in honor of the English botanist William Thomas Horner Fox-Strangways, 1795-1865 (d. Dorset), 4th Earl of Ilchester, diplomat, 1821 Fellow of the Royal Society and of the Linnean Society, owner of a garden at Abbotsbury, Dorset; see A. Bertoloni, "Descrizione di un nuovo genere, et di una nuova specie di pianta gigliacea." *Memorie di Matematica e di Fisica della Società Italiana delle Scienze residente in Modena* 21, Parte Fisica: 1-4. 1937; K. Persson & P. Wendelbo, "*Bellevalia hyacinthoides*, a new name for *Strangweia spicata* (Liliaceae)." *Botaniska Notiser*. 132: 65-70. 1979; E. Bretschneider, *History of European Botanical Discoveries in China*. 1981; H. Genaust, *Etymologisches Wörterbuch der botanischen Pflanzennamen*. 613. [for William Thomas Homer Fox-Strangways] 1996.

Stranvaesia Lindley Rosaceae

Origins:

For the English botanist William Thomas Horner Fox-Strangways, 1795-1865 (d. Dorset), 4th Earl of Ilchester, diplomat, 1821 Fellow of the Royal Society and of the Linnean Society, owner of a garden at Abbotsbury, Dorset.

Strasburgeria Baillon Strasburgeriaceae (Ochnaceae)

Origins:

For the German (b. Warsaw, Poland) botanist Eduard (Edward) Adolf Strasburger, 1844-1912 (Poppelsdorf, Germany), professor of botany; see J.H. Barnhart, *Biographical Notes upon Botanists*. 3: 338. 1965; T.W. Bossert, *Biographical Dictionary of Botanists Represented in the Hunt Institute Portrait Collection*. 386. 1972; E.M. Tucker, *Catalogue of the Library of the Arnold Arboretum of Harvard University*. Cambridge, Massachusetts 1917-1933; Gloria Robinson, in *Dictionary of Scientific Biography* 13: 87-90. 1981; Stafleu and Cowan, *Taxonomic Literature*. 6: 31-40. Utrecht 1986.

Strateuma Raf. Orchidaceae

Origins:

Greek *strateuma, atos* "armament, army, military company, host," referring to the name of *Orchis strateumatica* L.; see C.S. Rafinesque, *Flora Telluriana*. 2: 89. 1836 [1837]; E.D. Merrill, *Index rafinesquianus*. 105. 1949

Strateuma Salisbury Orchidaceae

Origins:

Greek *strateuma* "campaign, army, military company, armament," from the name of the type species *Orchis militaris*.

Stratiotes L. Hydrocharitaceae

Origins:

Greek *stratiotes* "soldier, water-lettuce," Latin *stratiotes, ae* for a water-plant, either the aloe-leaved water-soldier (a species of *Stratiotes*), or the great duck-weed applied by Plinius to a kind of *Pistia*.

Species/Vernacular Names:
S. aloides L.

English: water soldier

Straussia A. Gray Rubiaceae

Origins:

Named for the German physician Lorenz (Laurentius) Strauss, 1633-1687 (see Pritzel, 307). His writings include *Disputatio physico-medica de potu Coffi*. Giessae 1666 and *Exercitatio physica de ovo galli*. Gissae 1669; see Sylvestre Dufour (Philippe) (1622-1687) [pseud. of Jacob Spon], Libellus primus sub titulo: *Jacobi Sponii Bevanda asiatica*, etc. [Lugduni] 1705.

Straussiella Haussknecht Brassicaceae

Origins:

After the German plant collector Theodor Strauss, 1859-1911, merchant, traveler, in Iran; see J.H. Barnhart, *Biographical Notes upon Botanists*. 3: 338. 1965.

Streblacanthus Kuntze Acanthaceae

Origins:

Greek *streblos* "twisted, crooked" and the genus *Acanthus* L., Greek *akantha* "thorn."

Streblochaete Hochst. ex Pilger Gramineae

Origins:

From the Greek *streblos* "twisted, crooked" and *chaite* "a bristle."

Streblorrhiza Endl. Fabaceae

Origins:

From the Greek *streblos* and *rhiza* "a root."

Streblosa Korth. Rubiaceae

Origins:
Greek *streblos* "twisted, crooked."

Streblosiopsis Valeton Rubiaceae

Origins:
Resembling *Streblosa* Korth.

Streblus Lour. Moraceae

Origins:

Possibly from the Latin *strebula, ae* and *strebula (stribula), orum*, the flesh about the haunches, possibly referring to the base of the fruit; or from the Greek *streblos* "twisted, crooked," indicating the twisted branches or trunk of the type species.

Species/Vernacular Names:
S. asper Lour. (*Trophis aspera* Retz.; *Trophis aculeata* Roth; *Trophis cochinchinensis* Poir.; *Epicarpurus orientalis* Blume; *Calius lactescens* Blanco; *Streblus lactescens*

Blume; *Achymus pallens* Soland. ex Blume; *Cudrania crenata* C.H. Wright; *Vanieria crenata* (C.H. Wright) Chun; *Diplothorax tonkinensis* Gagnep.)

English: Siamese rough bush, crooked rough bush, sand paper tree

Malaya: kesinai, kesinah, serinai

The Philippines: calios, aludig, ampas, bagtak

Sri Lanka: géta-netul, patpiray, pirasu

Nepal: kakshi

India: sakhotaka, shakhotaka, shaorha, sahora, sahor, sahuda, sheora, sehora, sihora dahya, daheya, dahya, siora, sahra, kharota, kavati, karvati, karera, kharaoli, khorua, paruka, paruva, parava, tintapparuya, paraya, prayam, pirayam, piraayamaram, pira, piray, kurripla, baranki, kharanchi-bol, barinki, baranika, barinika, barivenkachettu, karchanua, rusa, cheroot-pathi, akhor moranu, poi, pakki, pukki, kakkabedi, sitanike, kuttippirai, pasuna, vittil, mitlemare, mitligade, ponalige, punje, serphang, jindi, dieng-soh-khyrdang

S. brunonianus F. Muell. (*Morus brunoniana* Endl.; *Pseudomorus brunoniana* (Endl.) Bureau)

Australia: white handlewood, whalebone tree, prickly fig, axehandle wood, ragwood, waddywood, grey handlewood

S. pendulinus (Endl.) F. Muell. (*Morus pendulina* Endl.; *Morus brunoniana* Endl.; *Pseudomorus brunoniana* (Endl.) Bureau; *Pseudomorus pendulina* (Endl.) Stearn; *Pseudomorus sandwicensis* Degener; *Streblus brunonianus* (Endl.) F. Muell.; *Streblus sandwicensis* (Degener) St. John)

Hawaii: a'ia'i

S. taxoides (Heyne) Kurz (*Trophis taxoides* Heyne ap. Roth; *Trophis taxiformis* Spreng.; *Trophis spinosa* Roxb.; *Epicarpurus timorensis* Decne; *Epicarpurus spinosus* (Roxb.) Wight; *Taxotrophis roxburghii* Blume; *Phyllochlamys spinosa* (Roxb.) Bur.; *Streblus microphyllus* Kurz; *Streblus taxoides* Kurz var. *microphylla* Kurz; *Phyllochlamys wallichii* King ap. Hook.f.; *Phyllochlamys taxoides* (Heyne) Koorders; *Phyllochlamys taxoides* var. *parvifolia* Merr.; *Phyllochlamys tridentata* Gagnep.; *Taxotrophis poilanei* Gagnep.)

English: fig lime

Malaya: merlimau, limau limau

Sri Lanka: gongotu

Streleskia Hook.f. Campanulaceae

Origins:

After the Polish explorer Sir Paul Edmund de Strzelecki, Count, 1797-1873 (d. London), scientist, 1853 a Fellow of the Royal Society, 1869 K.C.M.G., traveler, author of *Physical Description of New South Wales and Van Diemen's Land*. London 1845 and *Gold and Silver: A Supplement to S.'s Physical Description of New South Wales and Van Diemen's Land*. London 1856; see Geoffrey Rawson, *The Count. A Life of Sir Paul Edmund Strzelecki*. London and Melbourne 1954; E. Mroczkowski, *W sprawie Edmunda Strzeleckiego*, etc. New York [1935]; Percival Serle, *Dictionary of Australian Biography*. Vol. II: 377-378; Arthur D. Chapman, ed., *Australian Plant Name Index*. 2749. Canberra 1991; J.D. Hooker, *Hooker's Journal of Botany*. 6: 266. 1847; Douglas Pike, ed., *Australian Dictionary of Biography*. 2: 494-495. Melbourne 1967; Ray Desmond, *Dictionary of British & Irish Botanists and Horticulturists*. 662. [b. 1796] London 1994; Jonathan Wantrup, *Australian Rare Books, 1788-1900*. Sydney 1987.

Strelitzia Banks ex W. Aiton Strelitziaceae (Musaceae, Strelitzioideae)

Origins:

Dedicated to H.M. Queen, Charlotte Sophia of Mecklenburg-Strelitz, who in 1761 became Queen to George III of England and bore him 15 children, she was a pupil of Rev. J. Lightfoot; see W. Aiton, *Hort. Kew*. ed. 1. 1: 285. 7 Aug-1 Oct 1789; Blanche Henrey, *British Botanical and Horticultural Literature before 1800*. Oxford 1975; Ray Desmond, *Dictionary of British & Irish Botanists and Horticulturists*. 142-143. 1994.

Species/Vernacular Names:

S. alba (L.f.) Skeels (*Heliconia alba* L.f.; *Strelitzia augusta* Wright p.p.)

English: wild banana, bird-of-paradise flower, crane flower

Japan: shiro-gokuraku-cho-bana

South Africa: wild mealie, piesang, witpiesang

S. caudata R.A. Dyer

English: Transvaal wild banana, wild banana

Southern Africa: wildepiesang; inKamanga (Swazi)

S. nicolai Regel & Körn. (the specific name after the Czar Nicholas I (1796-1855) of Russia)

English: Natal wild banana, wild banana, Natal strelitzia

Southern Africa: Natalse wildepiesang, wildepiesang; isiGude, iGceba, iNkamanga (Zulu); muChema, muTsundi (Shona); iKhamanga (Xhosa)

S. reginae Banks ex Dryander

English: bird-of-paradise, crane flower, ship on fire

Japan: gokuraku-cho-bana

South Africa: geel piesang

Strempelia A. Richard ex DC. Rubiaceae

Origins:

For the German botanist Johannes Carl Friedrich Strempel, 1800-1872, physician; see J.H. Barnhart, *Biographical Notes upon Botanists*. 3: 339. 1965; R. Zander, F. Encke, G. Buchheim and S. Seybold, *Handwörterbuch der Pflanzennamen*. 14. Aufl. Stuttgart 1993.

Strempeliopsis Benth. Rubiaceae

Origins:

Resembling *Strempelia*.

Strephium Schrader ex Nees Gramineae

Origins:

From the Greek *strepho, strephein* "to twist."

Strephonema Hook.f. Combretaceae

Origins:

Greek *strepho, strephein* "to twist" and *nema* "thread," referring to the stamens.

Species/Vernacular Names:

S. gilletii De Wild.

Zaire: pese bikolo

S. sericeum Hook.f.

Gabon: andong

Strepsiloba Raf. Mimosaceae

Origins:

See E.D. Merrill, *Index rafinesquianus*. 148. 1949; C.S. Rafinesque, *Sylva Telluriana*. 119. 1838.

Strepsilobus Raf. Mimosaceae

Origins:

Greek *strepho* "to twist, turn, bend," *strepsis* "a turning round" and *lobos* "capsule, pod, lobe"; see E.D. Merrill (1876-1956), *Index rafinesquianus*. 148. 1949; C.S. Rafinesque, *Sylva Telluriana*. 117, 118. 1838.

Strepsimela Raf. Loranthaceae

Origins:

From the Greek *strepsis* "a turning round" and *melos* "a limb, part, member"; see C.S. Rafinesque, *Sylva Telluriana*. 159. 1838; see also *Helixanthera* Lour.

Streptachne R. Br. Gramineae

Origins:

From the Greek *streptos* "twisted" and *achne* "chaff, glume."

Streptanthella Rydb. Brassicaceae

Origins:

The diminutive of the generic name *Streptanthus* Nutt.

Streptanthera Sweet Iridaceae

Origins:

From the Greek *streptos* "twisted" and *anthera* "anther."

Streptanthus Nutt. Brassicaceae

Origins:

From the Greek *streptos* plus *anthos* "flower," referring to the petals.

Species/Vernacular Names:

S. batrachopus J. Morrison

English: Tamalpais jewelflower (Mount Tamalpais)

S. bernardinus (E. Greene) Parish

English: Laguna Mountains jewelflower

S. callistus J. Morrison

English: Mount Hamilton jewelflower

S. campestris S. Watson

English: southern jewelflower

S. diversifolius S. Watson

English: varied-leaved jewelflower

S. drepanoides Kruckeb. & J. Morrison

English: sickle-fruit jewelflower

S. farnsworthianus J. Howell

English: Farnsworth's jewelflower

S. fenestratus (E. Greene) J. Howell

English: Tehipite Valley jewelflower

S. glandulosus Hook.

English: jewelflower

S. hispidus A. Gray

English: Mount Diablo jewelflower

S. niger E. Greene

English: Tiburon jewelflower (Tiburon Peninsula)

S. oliganthus Rollins

English: Masonic Mountain jewelflower

S. polygaloides A. Gray

English: milkwort jewelflower

S. tortuosus Kellogg

English: mountain jewelflower

Streptia Döll Gramineae

Origins:

From the Greek *streptos* "twisted," see also *Streptogyna* P. Beauv.

Streptilon Raf. Rosaceae

Origins:

See E.D. Merrill, *Index rafinesquianus*. 141. 1949; C.S. Rafinesque, *Autikon botanikon*. Icones plantarum select. nov. vel rariorum, etc. 173. Philadelphia 1840; Greek *streptos* "twisted" and *ptilon* "feather, wing."

Streptima Raf. Frankeniaceae

Origins:

See E.D. Merrill, *Index rafinesquianus*. 169. 1949; C.S. Rafinesque, *Flora Telluriana*. 2: 93. 1836 [1837]; Arthur D. Chapman, ed., *Australian Plant Name Index*. 2749. Canberra 1991.

Streptocalyx Beer Bromeliaceae

Origins:

Greek *streptos* "twisted, pliant, turned" and *kalyx* "calyx"; see J.G. Beer, *Die Familie der Bromeliaceen*. Vienna 1857.

Streptocarpus Lindley Gesneriaceae

Origins:

Greek *streptos* with *karpos* "fruit," referring to the nature of the capsules, twisted in narrow spirals.

Species/Vernacular Names:

S. rexii (Hook.) Lindl. (*Didymocarpus rexii* Hook.)

Southern Africa: umFazi onengxolo (Xhosa)

Streptocaulon Wight & Arnott Asclepiadaceae

Origins:

From the Greek *streptos* "twisted, pliant, turned" and *kaulos* "stem, stalk."

Species/Vernacular Names:

S. juventas (Loureiro) Merrill (*Streptocaulon griffithii* Hook.f.)

English: infant streptocaulon, Griffith streptocaulon

China: ma lian an

Vietnam: ha thu o trang, sung bo

Streptochaeta Schrader ex Nees Gramineae

Origins:

From the Greek *streptos* "twisted" and *chaite* "a bristle."

Streptoglossa Steetz ex F. Muell. Asteraceae

Origins:

From the Greek *streptos* and *glossa* "tongue," referring to the bracts.

Species/Vernacular Names:

S. adscendens (Benth.) Dunlop (*Pterigeron adscendens* Benth.)

English: desert daisy

S. liatroides (Turcz.) Dunlop (*Erigeron liatroides* Turcz.; *Pterigeron liatroides* (Turcz.) Benth.; *Pluchea ligulata* F. Muell.)

English: Wertaloona daisy

Streptogyna P. Beauv. Gramineae

Origins:

From the Greek *streptos* "twisted" and *gyne* "female, ovary," referring to the nature of the ovary.

Species/Vernacular Names:

S. crinita P. Beauv.

Yoruba: apala odo

Congo: mana mango, manamango

Streptolirion Edgew. Commelinaceae

Origins:
Greek *streptos* and *leirion* "a lily," referring to the nature of the inflorescence.

Streptoloma Bunge Brassicaceae

Origins:
Greek *streptos* "twisted" and *loma* "border, margin, fringe."

Streptolophus D.K. Hughes Gramineae

Origins:
From the Greek *streptos* "twisted" and *lophos* "a crest."

Streptomanes K. Schumann Asclepiadaceae (Periplocaceae)

Origins:
Greek *streptos* "twisted" and *manes* "a cup, a kind of cup."

Streptopetalum Hochst. Turneraceae

Origins:
From the Greek *streptos* and *petalon* "a petal."

Streptopus Michx. Convallariaceae (Liliaceae)

Origins:
Greek *streptos* "twisted" and *pous*, *podos* "foot," referring to the peduncles.

Streptosiphon Mildbr. Acanthaceae

Origins:
From the Greek *streptos* "twisted" and *siphon* "tube, pipe."

Streptosolen Miers Solanaceae

Origins:
From the Greek *streptos* and *solen* "pipe."

Streptostachys Desv. Gramineae

Origins:
From the Greek *streptos* plus *stachys* "ear of corn, spike."

Streptostigma Regel Solanaceae

Origins:
From the Greek *streptos* "twisted, bent" and *stigma* "stigma."

Streptothamnus F. Muell. Flacourtiaceae

Origins:
Greek *streptos* "twisted" and *thamnos* "bush, shrub," referring to the habit; see Ferdinand von Mueller, *Fragmenta Phytographiae Australiae*. 3: 27. (May) 1862.

Streptotrachelus Greenman Apocynaceae

Origins:
From the Greek *streptos* and *trachelos* "a neck."

Striga Lour. Scrophulariaceae

Origins:
Referring to the roughness and habit of the plants, Latin *striga* (*strix*, *strigis*) "a witch, hag," Latin *striga* "a furrow, streak, swath, a row of grain," Latin *stria, ae* "a channel, furrow," Akkadian *sir'u, sirihu* "a furrow"; see João de Loureiro, *Flora cochinchinensis*. 17, 22. [Lisboa] 1790; Giovanni Semerano, *Le origini della cultura europea. Dizionario della lingua Latina e di voci moderne*. 2(2): 576, 577. 1994; H. Genaust, *Etymologisches Wörterbuch der botanischen Pflanzennamen*. 614. 1996; William T. Stearn, *Botanical Latin*. 506. ["Striga: striga, a straight rigid close-pressed rather short bristle-like hair."] 1993.

Species/Vernacular Names:
S. bilabiata (Thunb.) Kuntze (*Buchnera bilabiata* Thunb.; *Striga thunbergii* Benth.)

English: witchweed

S. elegans Benth.

English: large mealie-witchweed, Matabele flower, witchweed

Southern Africa: groot mielierooiblom, kopseerblommetjie, mieliegif, rooiblom, rooibossie, vuurbossie; umNake (Xhosa); lethibela (Sotho); seona (South Sotho)

S. forbesii Benth.

English: giant mealie-witchweed, witchweed

South Africa: rooiblom

S. gesnerioides (Willd.) Vatke ex Engl. (*Buchnera gesnerioides* Willd.; *Striga chloroleuca* Dinter; *Striga orobanchoides* Benth.)

English: purple witchweed, tobacco witchweed

Southern Africa: rooiblommetjie, bloublom; iGalo (Xhosa)

S. hermontheca Bentham

English: purple witchweed, striga

East Africa: hayongo, kayongo, kituha, ekeyongo, emoto

S. lutea Lour. (*Buchnera asiatica* L.; *Striga asiatica* (L.) Kuntze)

English: red witchweed, common mealie-witchweed, mealie witchweed, witchweed, buri, striga, isona weed, Matabele flower, mealie poison, scarlet lobelia

China: du jiao gan

East Africa: akarebwa omwe, ekeyongo, emoto

Southern Africa: gewone mielierooiblom, kopseerblommetjie, mieliegif, rooiblom, rooiblombossie, vuurbossie; bisi (Shona); isona (Zulu), seona (South Sotho)

Yoruba: irokodu, oloyin

Strigina Engl. Scrophulariaceae

Origins:
Referring to the genus *Striga* Lour.

Strigosella Boiss. Brassicaceae

Origins:
Latin *strigosus* "covered with strigae, with stiff bristles."

Striolaria Ducke Rubiaceae

Origins:
Latin *strio* "to hollow out, groove, flute, striate," *striolatus* "with fine lines, striped."

Strobilacanthus Griseb. Acanthaceae

Origins:
Greek *strobilos* "a cone" and the genus *Acanthus* L., *akantha* "thorn."

Strobilanthes Blume Acanthaceae

Origins:
Greek *strobilos* and *anthos* "a flower," leaves and bracts enclose the flowers.

Species/Vernacular Names:
S. anisophylla (Lodd.) T. Anderson

English: goldfussia

S. atropurpureus Nees

English: Mexican petunia

S. cusia (Nees) Kuntze (*Baphicacanthus cusia* (Nees) Bremek.)

English: Assam indigo

China: malan

S. dyerianus Mast. (*Perilepta dyeriana* (Mast.) Bremek.)

English: Persian shield

Japan: ura-murasaki

S. isophyllus (Nees) Anderson (*Goldfussia isophylla* Nees)

English: bedding conehead, goldfussia

Strobilanthopsis S. Moore Acanthaceae

Origins:
Resembling *Strobilanthes* Blume.

Strobilocarpus Klotzsch Grubbiaceae (Dilleniidae, Ericales)

Origins:
From the Greek *strobilos* "a cone" and *karpos* "a fruit."

Strobilopanax R. Viguier Araliaceae

Origins:
Greek *strobilos* "a cone" plus *Panax*.

Strobilopsis Hilliard & B.L. Burtt Scrophulariaceae

Origins:
Greek *strobilos* "a cone" and *opsis* "aspect, resemblance."

Strobilorhachis Klotzsch Acanthaceae

Origins:
Greek *strobilos* "a cone, round ball" plus *rhachis* "rachis, axis, midrib of a leaf."

Strobopetalum N.E. Br. Asclepiadaceae

Origins:

Greek *strobos* "a whirling round," *strobeo* "to spin a top spin and *petalon* "leaf, petal."

Stromanthe Sonder Marantaceae

Origins:

Greek *stroma* "a bed" and *anthos* "flower," referring to the form of the inflorescence.

Stromatocactus Karw. ex Rümpler Cactaceae

Origins:

From the Greek *stroma, stromatos* "a bed, mattress, pavement" plus *Cactus*.

Stromatopteris Mett. Stromatopteridaceae (Gleicheniaceae, Stromatopteridoideae)

Origins:

Greek *stroma, stromatos* and *pteris* "fern"; frond is continuous and there is no clear distinction between stem and leaf.

Strombocactus Britton & Rose Cactaceae

Origins:

Greek *strombos* "a spinning-top, turban" and *cactus*.

Strombocarpa (Benth.) A. Gray Mimosaceae

Origins:

From the Greek *strombos* and *karpos* "fruit."

Strombosia Blume Olacaceae

Origins:

Greek *strombos* "a spinning-top, turban," Latin *strombus, i* "a kind of spiral snail," in reference to the nature of the disc; see Y. Tailfer, *La Forêt dense d'Afrique centrale*. CTA, Ede/Wageningen 1989; J. Vivien & J.J. Faure, *Arbres des Forêts denses d'Afrique Centrale*. Agence de Coopération Culturelle et Technique. Paris 1985.

Species/Vernacular Names:

S. grandifolia Hook.f. ex Benth.

Ghana: afina

Zaire: nghila; booko (Basankusu); botaka (Lokundu); bowlisangu, gwarzabiri (Azande); dengele (Babua); ebelakalaka, okwelekwelesi (Turumbu); etaka (Mai-Ndombe lake); kambukozi (Kiombe); moa tembo (= the tree that kills the elephant) (Maniéma); muhika (Kinyaruanda); tshipulupulu (Tshiluba)

Gabon: ezipt

Cameroon: edjip, epinedo, mbang mbazoa

Congo: nghila

Nigeria: arememila, poyi; aramemila (Edo); itako pupa (Yoruba)

S. javanica Blume

Malaya: dali dali, dedali, bayam badak

S. pustulata Oliv. (*Strombosia glaucescens* Engl.)

Zaire: botaka (Lokundu); dalakadaba (Tshofa); dandala, n'zugambakala, tshipulupulu tshitoke (Tshiluba); dzuwe (Basankusu); ikelenge (Mbelo); ikwanga mbongo (Turumbu); mpaniate (lake Mai-Ndombe)

Gabon: mbazork, mbeza, mouguiba

Cameroon: adjip, bongongo, komondip, mbang mbazoa, nkemelo

Nigeria: atako, itako, poyi, ubelu, ubaelu, otingbo, arembadi, odogbo; ubelu (Edo); itako (Yoruba)

Ghana: afina

Ivory Coast: azoubé, fognan, heilé, poé

Yoruba: itako, otingbo, odogbo

S. scheffleri Engl.

Cameroon: epinedo, mbazoa

S. zenkeri Engl.

Cameroon: epinedo, mbazoa, akolembole

Strombosiopsis Engl. Olacaceae

Origins:

Resembling the genus *Strombosia* Blume.

Species/Vernacular Names:

S. tetrandra Engl.

Zaire: tip; bogwane (Kisangani); botaka (Lokundu); fifulula (Tshofa); ebelakabelaka, obwelekwelesi (Turumbu); epopo, puika (Kiswahili); etaka (lake Mai-Ndombe); tshipulupulu tshifike (Tshiluba)

Gabon: egipt, engomegom

Cameroon: bazoa, bosiko, edoup, oyen, mandjoc, edipmbazoa, pindé

Congo: tip

Strongylocalyx Blume Myrtaceae

Origins:
Greek *strongylos* "round" and *kalyx* "calyx."

Strongylocaryum Burret Palmae

Origins:
From the Greek *strongylos*, *stroggylos* "round" and *karyon* "nut," see also *Ptychosperma* Labill.

Strongylodon Vogel Fabaceae

Origins:
Greek *strongylos* "round" and *odous, odontos* "a tooth," referring to the teeth of the calyx; see *Linnaea*. 10: 585. 1836.

Species/Vernacular Names:
S. macrobotrys A. Gray
English: jade vine
The Philippines: tayabak, bayou
S. ruber Vogel
Hawaii: nuku 'i'iwi, ka'i'iwi, nuku

Strongylosperma Less. Asteraceae

Origins:
Greek *strongylos* "round" and *sperma* "seed."

Strophacanthus Lindau Acanthaceae

Origins:
From the Greek *strophe* "twist, turning," *strephein* "to twist" and *akantha* "a thorn."

Strophanthus DC. Apocynaceae

Origins:
Greek *strophe* "twist, turning," *strophos* "twisted cord or band," *strephein* "to twist" and *anthos* "a flower," referring to the twisted petal appendages or tails or corolla lobes; see M.A. Marchi, *Dizionario tecnico-etimologico-filologico*. Milano 1828-1829; Manlio Cortelazzo & Paolo Zolli, *Dizionario etimologico della lingua italiana*. 5: 1288. Zanichelli, Bologna 1988; B. Migliorini, *Parole d'autore (onomaturgia)*. Firenze 1975.

Species/Vernacular Names:
S. boivini Baillon (*Roupellina boivini* (Baillon) Pichon)
Madagascar: kabocala, befe, lalondo, tambio, hiba, saritangena
S. caudatus (L.) Kurz
English: ovate-sepal strophanthus
China: luan e yang jiao niu
S. divaricatus (Loureiro) Hooker & Arnott
English: divaricate strophanthus
China: yang jiao niu, yang liao ao
Vietnam: thuoc ban, sung de
S. gratus (Wallich & Hooker) Baillon
English: climbing oleander, spiny-flower strophanthus
China: xuan hau yang jiao niu
S. hispidus DC.
English: Transvaal strophanthus
China: jian du yang jiao niu
Yoruba: isa, isa gidi, isa ogbugu, isa giri, isa gere, oro, iwase dudu, sagere
S. preussii Engl. & Pax
Yoruba: iya funfun, isa wewe, isa kekere
S. sarmentosus DC.
English: arrow-poison strophanthus
China: xi fei yang jiao niu
Nigeria: isha wewe, isha kere
Yoruba: irewu, ilagba omode, isa kekere, lagba omode
Mali: mangana, kuna, kunankale
S. speciosus (Ward & Harv.) Reber (*Christya speciosa* Ward & Harv.; *Strophanthus capensis* A. DC.)
English: common poison rope, corkscrew-flower
Southern Africa: gewone giftou, osdoring; amaSebele, umHlazazane (Zulu); umKhukhumeza (Xhosa)
S. wallichii DC.
English: Wallich strophanthus
China: yun nan yang jiao niu

Strophioblachia Boerl. Euphorbiaceae

Origins:
Greek *strophe* "twist, turning" and the genus *Blachia* Baill.

Strophiodiscus Choux Sapindaceae

Origins:
Greek *strophos* "twisted band or cord, woman's girdle," *strophion* "a band" and *diskos* "a disc"; see also *Plagios-*

cyphus Radlk., Greek *plagios* "oblique" and *skyphos* "a cup, goblet, a jug."

Strophocactus Britton & Rose Cactaceae

Origins:
From the Greek *strophos* plus *cactus*, see also *Selenicereus* (A. Berger) Britton & Rose.

Strophocaulos Small Convolvulaceae

Origins:
From the Greek *strophe* "twist, turning," *strophos* "twisted cord or band" and *kaulos* "stem," see also *Convolvulus* L.

Strophocereus Fric & Kreuz. Cactaceae

Origins:
From the Greek *strophos* plus *Cereus*, see also *Selenicereus* (A. Berger) Britton & Rose.

Stropholirion Torrey Alliaceae (Liliaceae)

Origins:
Greek *strophos* and *leirion* "a lily," see also *Dichelostemma* Kunth.

Strophostyles Elliott Fabaceae

Origins:
Greek *strophe* "twist, turning" and *stylos* "pillar, style."

Strumaria Jacq. ex Willd. Amaryllidaceae (Liliaceae)

Origins:
Latin *struma, ae* "a scrofulous tumor."

Species/Vernacular Names:
S. hardyana D. & U. Müller-Doblies (after Dave Hardy of the Botanical Research Institute, Pretoria, who made the type gatherings of *Strumaria hardyana*, in October 1978, and its second collector)

English: strumaria

Struthanthus Martius Loranthaceae

Origins:
Greek *strouthion* "a small bird, small sparrow," *strouthos* "a sparrow, ostrich" and *anthos* "flower."

Struthiola L. Thymelaeaceae

Origins:
The Greek name for a plant used by Theophrastus (*stroutheion* "a kind of quince"), from *strouthion* "a small bird, small sparrow," *strouthos* "a sparrow, ostrich"; Latin *struthion, ii* for Plinius was a plant, soapwort, *Saponaria officinalis* L.; Latin *turdus*, Irish *truid*, Akkadian *taradu* and Hebrew *tarad* "to hunt, to push."

Struthiolopsis E. Phillips Thymelaeaceae

Origins:
Resembling *Struthiola* L.

Struthiopteris Scop. Blechnaceae

Origins:
From *strouthion* "a small bird, small sparrow" plus *pteris* "fern."

Struthiopteris Willd. Dryopteridaceae (Woodsiaceae)

Origins:
Strouthion "a small bird, small sparrow" and *pteris* "fern."

Struthopteris Bernh. Osmundaceae

Origins:
From the Greek *strouthos* "a sparrow, ostrich" and *pteris* "fern."

Strychnopsis Baillon Menispermaceae

Origins:
Resembling the genus *Strychnos* L.

Strychnos L. Strychnaceae (Loganiaceae)

Origins:

Greek name for many and different poisonous plants, from *strychnon*, *strychnos* "acrid, bitter"; Latin *strychnos* or *trychnos* for a kind of nightshade; see Carl Linnaeus, *Species Plantarum*. 189. 1753 and *Genera Plantarum*. Ed. 5. 86. 1754; Helmut Genaust, *Etymologisches Wörterbuch der botanischen Pflanzennamen*. 616. Basel 1996; B. Migliorini, *Parole d'autore (onomaturgia)*. Firenze 1975; A. Bonavilla, *Dizionario etimologico di tutti i vocaboli usati nelle scienze, arti e mestieri, che traggono origine dal greco*. Milano 1819-1821; Ernest Weekley, *An Etymological Dictionary of Modern English*. 2: 1432. New York 1967.

Species/Vernacular Names:

S. angustiflora Bentham

English: narrow-flower poison nut

China: niu yan ma qian

S. cocculoides Bak. (*Strychnos schumanniana* Gilg)

Southern Africa: suurklapper; mogorugorwana (Kwena dialect, Botswana); mohoruhoru (Tawana dialect, Ngamiland); morapa (North Sotho); mohoruhorwana, mohoruhoru (Kololo, Barotseland); umi (Mbukushu); uguni (Deiriku); omukwakwa (Ovambo); mutamba, muOno (Shona)

S. decussata (Pappe) Gilg (*Atherstonea decussata* Pappe; *Strychnos atherstonei* Harv.) (genus *Atherstonea* named for the English (b. Nottingham) botanist William Guybon Atherstone, 1814-1898 (d. Grahamstown, C.P., South Africa), physician and plant collector, geologist, naturalist; see Mary Gunn and Leslie E. Codd, *Botanical Exploration of Southern Africa*. 82-83. Cape Town 1981)

English: Cape teak, Chaka's wood, panda's walking stick, king's tree

Southern Africa: Kaapse kiaat, Kaapse kiaathout; umPathawenkosi, umPhathawenkosi-emhlophe, umKhombazulu, umGangele, umLahlankosi, umPhathankosi, umPhathankosi omhlophe, umKhangala, umHlamahlala, iNama (Zulu); umHlamahlala, umHlamalala, umKhangele (Xhosa)

S. henningsii Gilg (after the German mycologist and amateur poet Paul Christoph Hennings, 1841-1908, cryptogamist, professor, author of *Botanische Wanderungen durch die Umgebung Kiel's* ... Kiel 1879, from 1874 to 1880 with Eichler at the Botanical Garden Kiel; see J.H. Barnhart, *Biographical Notes upon Botanists*. 2: 158. 1965; Stafleu and Cowan, *Taxonomic Literature*. 2: 157-159. 1979; R. Zander, F. Encke, G. Buchheim and S. Seybold, *Handwörterbuch der Pflanzennamen*. 14. Aufl. Stuttgart 1993; T.W. Bossert, *Biographical Dictionary of Botanists Represented in the Hunt Institute Portrait Collection*. 1972; S. Lenley et al., *Catalog of the Manuscript and Archival Collections and Index to the Correspondence of John Torrey*. Library of the New York Botanical Garden. 220. 1973; Ida

Kaplan Langman, *A Selected Guide to the Literature on the Flowering Plants of Mexico*. University of Pennsylvania Press, Philadelphia 1964; Elmer Drew Merrill, in *Contr. U.S. Natl. Herb*. 30(1): 148. 1947; August Weberbauer, *Die Pflanzenwelt der peruanischen Andes in ihren Grundzügen dargestellt*. Leipzig 1911)

English: Natal teak, red bitterberry, coffee-bean strychnos, coffee hard pear, panda's walking stick

Southern Africa: rooibitterbessie, koffiehardpeer; umQalothi, uManana, umNono, umDunye (Zulu); umNonono (Xhosa); umNonono (Swazi)

S. ignatii Bergius (*Strychnos hainanensis* Merrill & Chun)

English: St. Ignatius's bean, Hainan poison nut, Ignatius bean

India: pipita, kayap-pankottai, kayapan-kottai, papita

China: lu song guo, lu sung kuo

S. innocua Del.

Southern Africa: dull-leaved mukwakwa; muDo, mujuni, muKwakwa (Shona)

Nigeria: kokirmo (Hausa)

N. Rhodesia: mukolo

Mali: gongoroni, kulegan

S. madagascariensis Poir. (*Strychnos innocua* Del. subsp. *dysophylla* (Benth.) Verdoorn; *Strychnos gerrardii* N.E. Br.; *Strychnos dysophylla* Benth.; *Strychnos innocua* Del. subsp. *gerrardii* (N.E. Br.) Verdoorn)

English: black monkey orange, kaffir-orange, spineless monkey orange

Southern Africa: swartklapper, shiny-leaved mukwakwa, gulagula, mkwakwa, klapper, botterklapper; umKwakwa, umNconjwa, umGuluguhla, umGuluguza (Zulu); umGulugula (Xhosa); anKwakwa (Thonga or Tsonga); iHlala (Swazi); nkwakwa (Tsonga); mookgwane (North Sotho)

S. mellodora S. Moore

English: forest strychnos

Southern Africa: chiTonga (Shona)

S. minor Dennst. (*Strychnos colubrina* L.)

English: snakewood

India: kuchila lata, kuchilalata, kajar wel, kajarwel, naagamushadi, nagamusti, nagamusadi, kongu-kandira, konsukandira, tansupaum, tansoopaum, madura kanjiram, modira-kanni, modira-caniram, goagarilakri, goagari lakei, kanal, taral, devakadu

S. mitis S. Moore

English: yellow bitterberry, pitted-leaf bitterberry

Southern Africa: geelbitterbessie; tambabunga, muTambawebungu, muTambungu (Shona); uManana, umHlamahlala, uMahlabekufeni, umNono, umNgqungquthi, umPhathankosi (Zulu); iBholo, umNqonqodi, umNgqun-quti, umNqunquti (Xhosa); umPhathankosi (Swazi)

S. nitida G. Don (*Strychnos cheliensis* Hu)

English: south Yunnan poison nut

China: mao zhu ma qian

S. nux-vomica L.

English: strychnine, nux-vomica tree, poison nut, nux-vomica, snakewood, strychnine tree, strychnine plant, Quaker button

French: noix-vomique

India: kuchla, kuchala, kuchila, kupilu, kulaka, ruchila, kachila, rajra, karaskara, kasarkana mara, karya-ruku, kajra, kar, kara, kora, kanjira, kanjirai, kanniram, kanjiram, kariram, nirmali, hemmush, mushti, mushti-vittulu, muchidi, vishna-mushti, vishnamushti, vishtindu, yetti-kottai, yetti, etti, itti, chibbige, chipita, dirghapatra, geradruma, kakasphurja, marakatindu, bailewa, thalkesur, jahar, hub-ul-jarab, jharkatachura, jharakatachura, kagodi, ittangi, kosila, kagophale, kagphala, khaboung

Tibetan: bya-phur-leb, ldum-stag

Nepal: nirmali

China: ma qian zi, fan mu pieh, ma chien

Vietnam: cu chi, ma tien, co ben kho

S. ovata A.W. Hill (*Strychnos confertiflora* Merrill & Chun)

English: denseflower poison nut

China: mi hua ma qian

S. potatorum L.f. (*Strychnos stuhlmannii* Gilg) (Latin *potator, oris* "drinker")

English: black bitterberry, clearing nut, water filter tree, Kataka nuts, clearing nut tree

Southern Africa: swartbitterbessie; ntsupa (Tsonga or Thonga)

India: kuchla, nelmal, neimal, nirmali, tattamaram, tetta, tetankottai, tetan-kottai, tetan-kotai, tettran, tettian, tetranparal, tettamparel, kotaku, kataka, katakamu, katakam, katakami, kattakami-chettu, ambu-prasada, ambu-prasad, ingini, indupa, induga, indupachettu, chilladabeeja, chillachettu, chillaginjalu, gajrah, jalada, chali-mara, chel-beey

S. pungens Soler. (*Strychnos occidentalis* Soler.)

English: monkey orange, kaffir orange

Southern Africa: klapper, botterklapper (word based on the butter-like pulp), blouklapper, bobbejaanklapper, kierieklapper, grootklapper; mohwahwa (Kwena dialect, Botswana); mokgwaha (North Sotho); iHlala, muNgono, muTamba (Shona); mukubudu, muramba (Venda); mutu (Mbukushu); matu (Sambui); omuoni (Ovambo)

S. spinosa Lam. (*Brehmia spinosa* (Lam.) Harv. ex DC.; *Strychnos spinosa* subsp. *lokua* (A. Rich.) E.A. Bruce)

English: monkey orange, spiny monkey orange, green monkey orange, kaffir orange, Natal orange

Mali: kankoro, gongoroba

Nigeria: kokiya (Hausa); kumbija (Fula); manvovogi (Nupe); amaku (Tiv); angboroko, atako (Yoruba)

Yoruba: robo, eekannase adiye, ata oro

N. Rhodesia: mutunga, mutungi

Southern Africa: groenklapper, klapper, msala, wildekalabasboom; umSala, umHlala (Swazi); muGono, iHlala, muMberi, muTamba, muZhumi, muZhume (Shona); muramba, muramba-khomu (Venda); muyimbili (Kololo, Barotseland); nsala (Tsonga); mogorogoro (Western Transvaal, northern Cape, Botswana); umHlala, umHlalawehlathi, umHlalakolontshe, umHlalakolotsho (Zulu); umHlala, umHleli (Xhosa)

S. umbellata (Loureiro) Merrill (*Cissus umbellata* Loureiro)

English: umbella-flowered poison nut

China: san hua ma qian

S. usambarensis Gilg (*Strychnos micans* S. Moore)

English: blue bitterberry, little monkey orange, Dingaan's wood

Southern Africa: bloubitterbessie; mukangala (Venda); chiKwakwani, Tamba we hlanzi (Shona); umDlamalala, uNdlunye (Xhosa); umPhathawenkosi-onmyama, iNdlunye, uNdlunye, umPhathankosi-lemnyama (Zulu)

Stryphnodendron Martius Mimosaceae

Origins:

Greek *stryphnos* "astringent" and *dendron* "tree," referring to the taste of the bark.

Species/Vernacular Names:

S. adstringens (Martius) Coville

Brazil: barbatimao

Strzeleckya F. Muell. Rutaceae

Origins:

After the Polish explorer Sir Paul Edmund de Strzelecki, Count, 1797-1873 (d. London), scientist, 1853 Fellow of the Royal Society, 1869 K.C.M.G., traveler, author of *Physical Description of New South Wales and Van Diemen's Land*. London 1845 and *Gold and Silver: A Supplement to S.'s Physical Description of New South Wales and Van Diemen's Land*. London 1856; see Geoffrey Rawson, *The Count. A Life of Sir P.E. Strzelecki*. London and Melbourne 1954; E. Mroczkowski, *W sprawie Edmunda Strzeleckiego*, etc. New York [1935]; Percival Serle, *Dictionary of Austra-*

lian Biography. Vol. II, 377-378; F.J.H. von Mueller, Hooker's Journal of Botany & Kew Garden Miscellany. 9: 308. (Oct.) 1857; Ray Desmond, Dictionary of British & Irish Botanists and Horticulturists. 662. [b. 1796] 1994; Arthur D. Chapman, ed., Australian Plant Name Index. 2749. 1991.

Stuartia L'Hérit. Theaceae

Origins:

Named for John Stuart, 3rd Earl of Bute, b. Edinburgh 1713-d. London 1792, botanist, a patron of science as well as of literature and art, studied at Eton, statesman, 1737 he was elected one of the representative peers of Scotland, Knight of the Thistle, one of the lords of the bedchamber to the Prince of Wales, Prime Minister 1762-1763, plantsman, introduced many new species to Kew, author of Botanical Tables, containing the different familys of British plants. [London 1785?]; see Dawson Turner, Extracts from the Literary and Scientific Correspondence of R. Richardson, of Bierly, Yorkshire: Illustrative of the State and Progress of Botany. Yarmouth 1835; W. Darlington, Memorials of John Bartram and Humphry Marshall. Philadelphia 1849; James A.L. Fraser, John Stuart, Earl of Bute. 1912; T.W. Bossert, Biographical Dictionary of Botanists Represented in the Hunt Institute Portrait Collection. 1972; Blanche Henrey, British Botanical and Horticultural Literature before 1800. 2: 242-243. Oxford 1975; M. Hadfield et al., British Gardeners: A Biographical Dictionary. London 1980; George Taylor, in Dictionary of Scientific Biography 1: 88. 1981; J.H. Barnhart, Biographical Notes upon Botanists. 1: 291. 1965; Jonas C. Dryander, Catalogus bibliothecae historico-naturalis Josephi Banks. London 1800; Frans A. Stafleu, Linnaeus and the Linnaeans. The spreading of their ideas in systematic botany, 1735-1789. 231. Utrecht 1971.

Stuartina Sonder Asteraceae

Origins:

After the botanist and gardener Charles Stuart, 1802-1877 (d. New South Wales), trained as a nurseryman in England, travelled and collected in New South Wales, South Australia and Tasmania, botanical collector for Mueller, William Henry Harvey (1811-1866), Otto Wilhelm Sonder (1812-1881) and Ronald Campbell Gunn (1808-1881); see C. Daley, "Charles Stuart, an early Australian botanist." Victorian Nat. 52: 106-110, 132-137 and 154-157. 1935; I.H. Vegter, Index Herbariorum. Part II (6), Collectors S. Regnum Vegetabile vol. 114. 1986; O.W. Sonder, Linnaea. 25: 521. (Jun.) 1853; N. Hall, Botanists of the Eucalypts. Melbourne 1978 and Supplement 1980.

Species/Vernacular Names:

S. hamata Philipson

English: hooked cudweed

S. muelleri Sonder (Gnephosis rotundifolia Diels)

English: Wertaloona daisy, spoon cudweed

Stuckertia Kuntze Asclepiadaceae

Origins:

For the Swiss (b. Basel) botanist Theodor (Teodoro) Juan Vicente Stuckert, 1852-1932, pharmacist, in Argentina (Cordoba); see J.H. Barnhart, Biographical Notes upon Botanists. 3: 342. 1965; T.W. Bossert, Biographical Dictionary of Botanists Represented in the Hunt Institute Portrait Collection. 387. 1972; E.M. Tucker, Catalogue of the Library of the Arnold Arboretum of Harvard University. Cambridge, Massachusetts 1917-1933; Stafleu and Cowan, Taxonomic Literature. 6: 55-56. Utrecht 1986.

Stuckertiella Beauverd Asteraceae

Origins:

For the Swiss botanist Theodor (Teodoro) Juan Vicente Stuckert, 1852-1932.

Stuebelia Pax Capparidaceae (Capparaceae)

Origins:

For the German plant collector Moritz Alphons Stübel, 1835-1904, explorer and geologist, naturalist, traveler (Egypt, South and Central America, Europe), made volcanological studies. His writings include Die Vulkanberge von Ecuador ... Berlin 1897 and Die Vulkanberge von Colombia ... Dresden 1906; see J.H. Barnhart, Biographical Notes upon Botanists. 3: 343. 1965; August Weberbauer (1871-1948), Die Pflanzenwelt der peruanischen Andes in ihren Grundzügen dargestellt. Leipzig 1911.

Stuhlmannia Taubert Caesalpiniaceae

Origins:

After the German naturalist Franz Ludwig Stuhlmann, 1863-1928, traveler (East Africa and Dutch East Indies), botanical collector in East Africa, Director of the biological-agricultural institute at Amani 1903-1908, author of Mit Emin Pasha ins Herz von Afrika. Berlin 1894, Beiträge zur Kenntnis der Tsetsefliege (Glossina fusca und Glossina tachinoides). Berlin 1907 and Beiträge zur Kulturgeschichte von Ostafrika. Berlin 1909; see J.H. Barnhart, Biographical *Notes upon Botanists*. 3: 343. 1965; E.M. Tucker, *Catalogue of the Library of the Arnold Arboretum of Harvard University*. Cambridge, Massachusetts 1917-1933; Alain White and Boyd Lincoln Sloane, *The Stapelieae*. Pasadena 1937.

Stultitia E. Phillips Asclepiadaceae

Origins:
Latin *stultitia* "folly, foolishness, simplicity, silliness, extravagance."

Stupa Asch. Gramineae

Origins:
Greek *stype, styppe* "tow," Latin *stuppa, stupa* "tow, oakum, hards," see also *Stipa* L.

Sturmia C.F. Gaertner Rubiaceae

Origins:
Referring possibly to the German engraver Jacob (Jakob) Sturm, 1771-1848, botanical artist.

Sturmia Hoppe Gramineae

Origins:
Dedicated to the German engraver Jacob Sturm, 1771-1848, botanical artist, the son of the Nürnberg engraver Johann Georg Sturm; see J.H. Barnhart, *Biographical Notes upon Botanists*. 3: 343. 1965; *Deutschlands Flora*. Nürnberg [1796-] 1798-1862.

Sturmia Reichb.f. Orchidaceae

Origins:
Dedicated to the German engraver Jacob (Jakob) Sturm, 1771-1848, botanical artist, naturalist. Among his works *Catalog der Kaefer-Sammlung von J.S.* Nürnberg 1843, *Catalog meiner Insecten-Sammlung*. Nürnberg 1826, *Deutschlands Fauna*. Nürnberg 1805 and *Icones Coleopterorum Germaniae*. Berlin 1877, he was the son of the Nürnberg engraver Johann Georg Sturm; see J.H. Barnhart, *Biographical Notes upon Botanists*. 3: 343. 1965; *Deutschlands Flora*. Nürnberg [1796-] 1798-1862; T.W. Bossert, *Biographical Dictionary of Botanists Represented in the Hunt Institute Portrait Collection*. 387. 1972; E.M. Tucker, *Catalogue of the Library of the Arnold Arboretum of Harvard*

University. Cambridge, Massachusetts 1917-1933; Günther Schmid, *Chamisso als Naturforscher*. Eine Bibliographie. Leipzig 1942; Stafleu and Cowan, *Taxonomic Literature*. 6: 65-72. Utrecht 1986.

Sturtia R. Br. Malvaceae

Origins:
For the British soldier Charles Sturt, 1795-1869 (Glos., England), explorer of the inland of Australia, plant collector, Surveyor General to South Australia, Colonial Treasurer of South Australia, Colonial Secretary 1849-1851, friend of Robert Brown, 1833 Fellow of the Linnean Society. Among his publications are *Two Expeditions into the Interior of Southern Australia During the Years 1828, 1829, 1830 and 1831*. London 1833 and *Narrative of an Expedition into Central Australia, Performed ... During the Years 1844, 1845 and 1846*. London 1849; see Jonathan Wantrup, *Australian Rare Books, 1788-1900*. Sydney 1987; J.H. Barnhart, *Biographical Notes upon Botanists*. 3: 343. 1965; E. Beale, *Sturt, the Chipped Idol*. Sydney 1979; Dennis John Carr and S.G.M. Carr, eds., *People and Plants in Australia*. 1981; John H.L. Cumpston, *Charles Sturt. His Life and Journeys of Exploration*. Melbourne 1951; Mrs. Napier George Sturt, *Life of Charles Sturt*. London 1899; T.W. Bossert, *Biographical Dictionary of Botanists Represented in the Hunt Institute Portrait Collection*. 387. 1972; E.M. Tucker, *Catalogue of the Library of the Arnold Arboretum of Harvard University*. Cambridge, Massachusetts 1917-1933; Douglas Pike, ed., *Australian Dictionary of Biography*. 2: 495-498. Melbourne 1967; A. Lasègue, *Musée botanique de Benjamin Delessert*. Paris 1845.

Styasasia S. Moore Acanthaceae

Origins:
An anagram of *Asystasia*.

Stygnanthe Hanst. Gesneriaceae

Origins:
From the Greek *stygnos* "hated, gloomy, hateful" and *anthos* "flower."

Stylagrostis Mez Gramineae

Origins:
Greek *stylos* "a column" and *agrostis, agrostidos* "grass, weed, couch grass"; see also *Calamagrostis* Adans.

Stylanthus Reichb.f. & Zoll. Euphorbiaceae

Origins:

From the Greek *stylos* "a column" and *anthos* "flower."

Stylapterus A. Juss. Penaeaceae

Origins:

Greek *stylos* "a column," *a* "negative, without" and *pteron* "wing," referring to the winged style of *Penaea* L.

Stylarthropus Baillon Acanthaceae

Origins:

From the Greek *stylos* "a column," *arthron* "a joint" and *pous* "foot."

Stylidium Sw. ex Willd. Stylidiaceae (Asteridae, Asterales)

Origins:

Greek *stylos* "a column, pillar," *stylidion* "a small pillar," in these plants the stamens and the style are united; see Arthur D. Chapman, ed., *Australian Plant Name Index*. 2752-2768. Canberra 1991.

Species/Vernacular Names:

S. beaugleholei J.H. Willis

English: Beaugleholes trigger-plant

S. calcaratum R. Br. (*Stylidium calcaratum* R. Br. var. *ecorne* F. Muell. ex R. Erickson & J.H. Willis)

English: book trigger-plant

S. inundatum R. Br. (*Stylidium brachyphyllum* Sonder)

English: hundreds and thousands

S. tepperanum (F. Muell.) Mildbr. (*Candollea tepperiana* F. Muell.)

English: Kangaroo Island trigger-plant

Stylidium Swartz Stylidiaceae (Asteridae, Asterales)

Origins:

From the Greek *stylidion* "a small pillar"; see Arthur D. Chapman, ed., *Australian Plant Name Index*. 2752. Canberra 1991.

Stylisma Raf. Convolvulaceae

Origins:

From the Greek *stylos* "a column, pillar"; see C.S. Rafinesque, *Am. Monthly Mag. Crit. Rev.* 3: 101. 1818 and 4: 191. 1818, *Jour. Phys. Chim. Hist. Nat.* 89: 258. 1819, *Ann. Gén. Sci. Phys.* 8: 270, 272. 1821, *Neogenyton, or Indication of Sixty-Six New Genera of Plants of North America.* 2. 1825, *New Fl. N. Am.* 4: 54. 1836 [1838], *Flora Telluriana.* 4: 83. 1836 [1838] and *The Good Book.* 47. 1840; E.D. Merrill, *Index rafinesquianus.* 200. 1949.

Stylites Amstutz Isoetaceae

Origins:

Greek *stylites* "standing or dwelling on a pillar," possibly referring to the elongate stems.

Stylobasium Desf. Surianaceae (Stylobasiaceae, Rosidae, Rosales)

Origins:

Greek *stylos* and *basis* "base, pedestal," referring to the gynoecium, the style rises from a globular base; see R.L. Desfontaines, in *Mémoires du Muséum d'Histoire Naturelle.* 5: 37. 1819.

Styloceras Kunth ex Juss. Buxaceae (Stylocerataceae)

Origins:

Greek *stylos* "a column" and *keras* "a horn."

Stylochaeton Lepr. Araceae

Origins:

Greek *stylos* and *chiton* "a tunic, covering"; see D.H. Nicolson, "Derivation of aroid generic names." *Aroideana.* 10: 15-25. 1988.

Species/Vernacular Names:

S. sp.

Yoruba: kuo

Stylochiton Schott Araceae

Origins:

Greek *stylos* "a column" and *chiton* "a tunic, covering."

Stylocline Nutt. Asteraceae

Origins:
Greek *stylos* plus *kline* "a bed, receptacle," female column, referring to the long receptacle.

Species/Vernacular Names:
S. citroleum Morefield
English: oil nest straw

S. gnaphaloides Nutt.
English: everlasting nest straw

S. masonii Morefield
English: Mason nest straw

S. micropoides A. Gray
English: desert nest straw

S. psilocarphoides M. Peck
English: Peck nest straw

S. sonorensis Wiggins
English: Mesquite nest straw

Styloconus Baillon Haemodoraceae

Origins:
Greek *stylos* "a column" and *konos* "a cone"; see H.E. Baillon, *Histoire des Plantes.* 13: 28, 75. (Oct.) 1894.

Stylocoryna Cavanilles Rubiaceae

Origins:
Greek *stylos* "a column" and *koryne* "a club"; see Antonio José Cavanilles (1745-1804), *Icones Plantarum.* 4: 45. 1798; Arthur D. Chapman, ed., *Australian Plant Name Index.* 2768. Canberra 1991.

Stylocoryne Wight & Arn. Rubiaceae

Origins:
Greek *stylos* and *koryne* "a club."

Stylodiscus Benn. Euphorbiaceae

Origins:
From the Greek *stylos* "a column" and *diskos* "a disc."

Stylodon Raf. Verbenaceae

Origins:
Greek *stylos* "a column" and *odous, odontos* "tooth"; see C.S. Rafinesque, *Neogenyton, or Indication of Sixty-Six New Genera of Plants of North America.* 2. 1825; E.D. Merrill, *Index rafinesquianus.* 205. 1949.

Styloglossum Breda Orchidaceae

Origins:
From the Greek *stylos* "a column" and *glossa* "a tongue," lip adnate to the column.

Stylogyne A. DC. Myrsinaceae

Origins:
From the Greek *stylos* "a column, style" and *gyne* "a woman, female."

Stylolepis Lehm. Asteraceae

Origins:
From the Greek *stylos* "a column, style" and *lepis* "scale."

Styloma O.F. Cook Palmae

Origins:
From the Greek *styloma, stylomatos* "support."

Stylomecon G. Taylor Papaveraceae

Origins:
From the Greek *stylos* with *mekon* "poppy," referring to the pistil with a distinct style.

Stylophorum Nutt. Papaveraceae

Origins:
From the Greek *stylos* "a column" and *phoros* "bearing, carrying," referring to the columnar style.

Stylophyllum Britton & Rose Crassulaceae

Origins:
From the Greek *stylos* "a column, style" and *phyllon* "leaf."

Stylosanthes Sw. Fabaceae

Origins:
Greek *stylos* and *anthos* "a flower," the flowers have a long style; see Olof Peter Swartz (1760-1818), *Nova genera et species plantarum* seu *Prodromus*. 108. Stockholm, Uppsala & Åbo 1788.

Species/Vernacular Names:
S. erecta P. Beauv. (*Stylosanthes guineensis* Schum.)

English: Nigerian stylo

S. fruticosa (Retz.) Alston (*Stylosanthes bojeri* Vogel; *Stylosanthes mucronata* Willd., nom. illeg.; *Arachis fruticosa* Retz.)

English: wild lucerne

Yoruba: ekun debe, eramora

S. guianensis (Aublet) Sw.

English: Brazilian stylo, Brazilian lucerne

S. hamata (L.) Taubert (*Hedysarum hamatum* L.)

English: Caribbean stylo

S. humilis Kunth (*Stylosanthes figueroae* Mohl.)

English: Townsville stylo

Stylosiphonia Brandegee Rubiaceae

Origins:
From the Greek *stylos* "style, column" and *siphon* "a tube."

Stylotrichium Mattf. Asteraceae

Origins:
From the Greek *stylos* "style, column" and *thrix, trichos* "hair."

Stylurus Salisb. ex J. Knight Proteaceae

Origins:
From the Greek *stylos* "style, column" and *oura* "tail"; see [R.A. Salisbury] in Joseph Knight, *On the Cultivation of the Plants Belonging to the Natural Order of Proteeae*. 115. London (Dec.) 1809.

Stypandra R. Br. Phormiaceae (Liliaceae)

Origins:
From the Greek *stype* "tow, flax-fiber" and *aner, andros* "male, a man, anther," referring to the woolly stamens; see Robert Brown, *Prodromus florae Novae Hollandiae et Insulae van-Diemen*. 278. London 1810.

Species/Vernacular Names:
S. glauca R. Br. (*Arthropodium glaucum* (R. Br.) Sprengel; *Stypandra imbricata* R. Br.; *Arthropodium imbricatum* (R. Br.) Sprengel)

English: nodding blue lily, candyup poison, blind grass

Styphelia Smith Epacridaceae

Origins:
Greek *styphelos* "sour, hard, astringent, rough," referring to the leaves.

Species/Vernacular Names:
S. adscendens R. Br.

English: golden heath

S. exarrhena (F. Muell.) F. Muell. (*Leucopogon exarrhenus* F. Muell.; *Leucopogon intermedius* Cheel; *Leucopogon hirtellus* F. Muell. ex Benth.; *Styphelia pusilliflora* F. Muell.; *Styphelia hirtella* (F. Muell. ex Benth.) F. Muell. ex Tate)

English: desert styphelia, beard-heath styphelia

S. tameiameiae (Cham. & Schltdl.) F. Muell. (*Cyathodes tameiameiae* Cham. & Schltdl.; *Cyathodes banksii* Gaud.; *Cyathodes douglasii* A. Gray; *Styphelia douglasii* (A. Gray) F. Muell. ex Skottsb.; *Styphelia grayana* Rock)

Hawaii: pukiawe, 'a'ali'i mahu, kanehoa, kawa'u, maiele, maieli, puakeawe, puakiawe, pukeawe, pupukiawe

S. triflora Andrews

English: pink fivecorner

S. tubiflora Smith

English: narrow fivecorner, red fivecorners

S. viridis Andrews

English: green fivecorners

Styphnolobium Schott ex Endl. Fabaceae

Origins:
From the Greek *styphnos, stryphnos, styphelos* "sour, harsh, rough, astringent" and *lobos* "pod, capsule, lobe."

Styponema Salisb. Liliaceae

Origins:

Greek *stype* "tow, flax-fiber" and *nema* "thread, filament."

Styppeiochloa de Winter Gramineae

Origins:

Greek *styppeion* "hemp, tow" and *chloe, chloa* "grass."

Styrax L. Styracaceae

Origins:

Latin *styrax* or *storax, acis* for a tree and also the resinous gum of that tree, storax; the classical Greek name used by Theophrastus (*HP.* 9.7.3) and Dioscorides, *styrax, styrakos,* a corruption of the Arabic or Semitic name *assthirak* for *Styrax officinalis* L.

Species/Vernacular Names:

S. sp.

Mexico: nocuana-yala

S. americanum Lam.

English: mock orange, American snowbell

S. argenteus C. Presl

Mexico: capulín, chicamay, chucamay, chilacuate, estoraque, hoja de jabón, ruin

S. argenteus C. Presl var. *ramirezii* (Greenman) Gonsoulin

Mexico: chilacuate

S. benzoin Dryander

English: Sumatra snowbell, styrax

Malaya: kemenyan, kemian, gum benjamin

China: an xi xiang, an hsi hsiang

S. japonicus Siebold & Zuccarini (*Styrax japonica* var. *kotoensis* (Hay.) Masam. & Suzuki; *Styrax japonica* var. *iriomotensis* Masam.)

English: Japanese snowbell, snowbell tree

Japan: ego-no-ki

China: ye mo li

S. obassia Siebold & Zucc.

English: fragrant snowbell, big leaf storax

Japan: hakuunboku

China: yu ling hua

S. officinalis L.

English: drug snowbell, storax, styrax

S. officinalis L. var. *redivivus* (Torrey) H. Howard

English: snowdrop bush

S. ovatus (Ruíz & Pav.) A. DC. (*Foveolaria ovata* Ruíz & Pav.; *Strigilia ovata* (Ruíz & Pav.) DC.; *Tremanthus ovata* (Ruíz & Pav.) Persoon)

Peru: incienso macho

S. tessmannii Perkins (*Styrax longifolius* Standley)

Peru: utsupa cacao

S. weberbaueri Perkins

Peru: incienso macho

Styrosinia Raf. Gesneriaceae

Origins:

See C.S. Rafinesque, *Flora Telluriana.* 2: 95. 1836 [1837].

Suaeda Forssk. ex Scopoli Chenopodiaceae

Origins:

Suiwed mullah, suivada, suwayd, suweid, suwed, suda, sauda, sawad, souïda "salty," Arabic names for *Suaeda baccata* or *Suaeda aegyptiaca.*

Species/Vernacular Names:

S. spp.

Mexico: guixi cete, quixi cete

S. aegyptiaca (Hasselq.) Zoh. (*Chenopodium aegyptiacum* Hasselq.; *Suaeda baccata* Forssk. ex J. Gmelin)

English: sea blite

S. californica S. Watson

English: California sea blite

S. esteroa W. Ferren & S. Whitmore

English: estuary sea blite

S. maritima (L.) Dumort. (*Chenopodium maritimum* L.)

Japan: hama-matsu-na

S. moquinii (Torrey) E. Greene

English: bush seepweed

S. taxifolia (Standley) Standley

English: woolly sea blite

S. vera Forssk. ex J. Gmelin (*Suaeda fruticosa* (L.) Forssk.)

English: inkbush, shrubby sea blite, sea rosemary

South Africa: brakbos, brakganna, inkbos

Arabic: hatab-shâmi, hatab-suweidi

Portuguese: barrilha

Suardia Schrank Gramineae

Origins:

For Paulus Suardus, author of *Thesaurus Aromatariorum*. Venetiis 1504 [1510?].

Suarezia Dodson Orchidaceae

Origins:

After Carola Lindberg de Súarez, orchid collector and painter from Ecuador; see Hubert Mayr, *Orchid Names and Their Meanings*. Vaduz 1998.

Suberanthus Borhidi & Fernández Rubiaceae

Origins:

Latin *suber, eris* "the cork-oak, cork-tree, a thick bark" and Greek *anthos* "flower"; Latin *supparus* "a linen garment, a topsail, a small sail on the foremast," Akkadian *sabaru* "to bend," Hebrew *safa* "to turn oneself."

Sublimia Comm. ex Mart. Palmae

Origins:

Latin *sublimo* "to raise," *sublime* "lofty, eminent," *sublimis* "lofty, elevated."

Submatucana Backeb. Cactaceae

Origins:

Referring to *Matucana* Britton & Rose.

Subpilocereus Backeb. Cactaceae

Origins:

Referring to *Pilocereus*.

Subularia L. Brassicaceae

Origins:

Latin *subula, ae* "an awl, a small weapon," *subulo, onis* "a kind of hart with pointed horns," presumably referring to the leaves.

Species/Vernacular Names:

S. aquatica L.

English: awlwort

Subulatopuntia Fric & Schelle Cactaceae

Origins:

From the Latin *subula, ae* "an awl, a small weapon" plus *Opuntia*.

Succisa Haller Dipsacaceae

Origins:

Latin *succido* "cut from below, cut off," referring to the rhizome.

Species/Vernacular Names:

S. pratensis L.

English: devil's bit, blue buttons

Succisella Beck Dipsacaceae

Origins:

The diminutive of the genus *Succisa* Haller.

Succowia Medikus Brassicaceae

Origins:

For the German botanist Georg Adolph Suckow, 1751-1813, physician, naturalist, professor of natural sciences, son of the German naturalist Laurenz (Lorenz) Johann Daniel Suckow (Succow) (1722-1801); see J.H. Barnhart, *Biographical Notes upon Botanists*. 3: 344. 1965; Jonas C. Dryander, *Catalogus bibliothecae historico-naturalis Josephi Banks*. London 1796-1800; E.M. Tucker, *Catalogue of the Library of the Arnold Arboretum of Harvard University*. Cambridge, Massachusetts 1917-1933; R. Zander, F. Encke, G. Buchheim and S. Seybold, *Handwörterbuch der Pflanzennamen*. 14. Aufl. Stuttgart 1993.

Suckleya A. Gray Chenopodiaceae

Origins:

For George Suckley, 1830-1869, surgeon, naturalist, explorer; see Joseph Ewan, *Rocky Mountain Naturalists*. 243, 317. The University of Denver Press 1950.

Sucomorus Raf. Moraceae

Origins:

Based on *Sycomorus*; see C.S. Rafinesque, in *Sylva Telluriana*. 5. 1838; E.D. Merrill, *Index rafinesquianus*. 112. 1949.

Suessenguthia Merxm. Acanthaceae

Origins:

Dedicated to the German botanist Karl Suessenguth, 1893-1955, a student of Karl Immanuel E. Goebel (1855-1932) at München, from 1927 to 1955 professor of botany at the University of München, Curator of the Botanische Staatssammlung 1927-1955. His works include *Neue Ziele der Botanik*. München and Berlin 1938, *Amarantaceae of Southeastern Polynesia*. Honolulu, Hawaii 1936 and *Untersuchungen über Variationsbewegungen von Blättern*. Jena 1922, contributed to H.G.A. Engler & K.A.E. Prantl, *Die Natürlichen Pflanzenfamilien* ed. 2 (Rhamnaceae, Vitaceae and Leeaceae); see J.H. Barnhart, *Biographical Notes upon Botanists*. 3: 345. 1965; T.W. Bossert, *Biographical Dictionary of Botanists Represented in the Hunt Institute Portrait Collection*. 388. 1972; Ida Kaplan Langman, *A Selected Guide to the Literature on the Flowering Plants of Mexico*. University of Pennsylvania Press, Philadelphia 1964; R. Zander, F. Encke, G. Buchheim and S. Seybold, *Handwörterbuch der Pflanzennamen*. 14. Aufl. Stuttgart 1993.

Suessenguthiella Friedrich Molluginaceae

Origins:

Dedicated to the German botanist Karl Suessenguth, 1893-1955, a student of Karl Immanuel E. Goebel (1855-1932) at München, from 1927 to 1955 professor of botany at the University of München, Curator of the Botanische Staatssammlung 1927-1955; see J.H. Barnhart, *Biographical Notes upon Botanists*. 3: 345. 1965.

Suitenia Stokes Meliaceae

Origins:

See *Swietenia* Jacq., named after the Dutch botanist and physician Gerard L.B. van Swieten, 1700-1772.

Suitramia Reichb. Melastomataceae

Origins:

In some of his writings Carl Friedrich Philipp von Martius used the pseudonym *Suitram*, an anagram of his name.

Sukeon Raf. Moraceae

Origins:

Greek *sykon* "fig"; see C.S. Rafinesque, in *Sylva Telluriana*. 58. 1838; E.D. Merrill, *Index rafinesquianus*. 112. 1949.

Suksdorfia A. Gray Saxifragaceae

Origins:

After the American (b. Holstein, Germany) botanist Wilhelm Nikolaus Suksdorf, 1850-1932 (killed by a train), plant collector (Iowa, California, Montana, Oregon and Washington), worked with Asa Gray. His writings include *Flora washingtonensis*. [White Salmon, Washington 1892] and "Untersuchungen in der Gattung *Amsinckia*." *Werdenda*. 1: 47-113. 1931; see J.H. Barnhart, *Biographical Notes upon Botanists*. 3: 346. 1965; R. Zander, F. Encke, G. Buchheim and S. Seybold, *Handwörterbuch der Pflanzennamen*. 14. Aufl. 1993; T.W. Bossert, *Biographical Dictionary of Botanists Represented in the Hunt Institute Portrait Collection*. 388. 1972; S. Lenley et al., *Catalog of the Manuscript and Archival Collections and Index to the Correspondence of John Torrey*. Library of the New York Botanical Garden. 360, 398. 1973; E.M. Tucker, *Catalogue of the Library of the Arnold Arboretum of Harvard University*. Cambridge, Massachusetts 1917-1933; Joseph Ewan, *Rocky Mountain Naturalists*. 1950.

Sulcorebutia Backeb. Cactaceae

Origins:

Latin *sulcus* "a furrow, a ditch" plus *Rebutia*.

Sullivania F. Muell. Orchidaceae

Origins:

Possibly after the Australian botanist David Sullivan, 1836-1895 (d. Ararat, Victoria, Australia), in 1884 a Fellow of the Linnean Society, correspondent of F. von Mueller.

Sullivantia Torrey & A. Gray Saxifragaceae

Origins:

Dedicated to the American botanist William Starling Sullivant, 1803-1873, businessman, bryologist. His writings include *A Catalogue of Plants, Native or Naturalized, in the Vicinity of Columbus, Ohio*. Columbus 1840 and *The Musci and Hepaticae of the Northern United States*. Cambridge 1848; see J.H. Barnhart, *Biographical Notes upon Botanists*. 3: 346. 1965; Stafleu and Cowan, *Taxonomic Literature*. 6: 87-91. 1986; William Jay Youmans, ed., *Pioneers of Science in America*. 111-118. New York 1896; J. Ewan, ed., *A Short History of Botany in the United States*. 1969; H.N. Clokie, *Account of the Herbaria of the Department of Botany in the University of Oxford*. 250. Oxford 1964; T.W. Bossert, *Biographical Dictionary of Botanists Represented in the Hunt Institute Portrait Collection*. 388. 1972; E.M. Tucker, *Catalogue of the Library of the Arnold*

Arboretum of Harvard University. 1917-1933; S. Lenley et al., *Catalog of the Manuscript and Archival Collections and Index to the Correspondence of John Torrey.* Library of the New York Botanical Garden. 1973; A.D. Rodgers, *William Starling Sullivant.* 1940; Samuel Wood Geiser, *Naturalists of the Frontier.* Dallas 1948.

Sulpitia Raf. Orchidaceae

Origins:

According to Rafinesque named for a nymph; see C.S. Rafinesque, *Flora Telluriana.* 4: 37. 1836 [1838].

Sumachium Raf. Anacardiaceae

Origins:

From the Arabic *summaq*; see Constantine Samuel Rafinesque, *Am. Monthly Mag. Crit. Rev.* 2: 265. 1818; E.D. Merrill, *Index rafinesquianus.* 158. 1949; Ernest Weekley, *An Etymological Dictionary of Modern English.* 2: 1444. New York 1967; G.B. Pellegrini, *Gli arabismi nelle lingue neolatine con speciale riguardo all'Italia.* 119, 195. Brescia 1972.

Sumacus Raf. Anacardiaceae

Origins:

Arabic *summaq*; see Constantine Samuel Rafinesque, *Flora Telluriana.* 3: 56. 1836 [1837] and *Autikon botanikon.* Icones plantarum select. nov. vel rariorum, etc. 83. Philadelphia 1840; E.D. Merrill, *Index rafinesquianus.* 158-159. 1949.

Sumatroscirpus Oteng-Yeboah Cyperaceae

Origins:
Scirpus from Sumatra.

Sumbavia Baillon Euphorbiaceae

Origins:

For the island Sumbavia in Indonesia, see M.P. Nayar, *Meaning of Indian Flowering Plant Names.* 335. 1985.

Sumbaviopsis J.J. Sm. Euphorbiaceae

Origins:

Resembling Sumbavia, see M.P. Nayar, *Meaning of Indian Flowering Plant Names.* 335. 1985.

Summerhayesia P.J. Cribb Orchidaceae

Origins:

For the British (b. Somerset) botanist Victor Samuel Summerhayes, 1897-1974, orchidologist, traveler, from 1924 to 1964 in charge of the Kew Orchid Herbarium, author of *An Enumeration of the Angiosperms of the Seychelles Archipelago.* London 1931 and *Wild Orchids of Britain.* London 1951, he was father of the English plant collector Gerald Victor Summerhayes (b. 1928); see J.H. Barnhart, *Biographical Notes upon Botanists.* 3: 347. Boston 1965; R. Zander, F. Encke, G. Buchheim and S. Seybold, *Handwörterbuch der Pflanzennamen.* 14. Aufl. 1993; T.W. Bossert, *Biographical Dictionary of Botanists Represented in the Hunt Institute Portrait Collection.* 388. 1972; H.N. Clokie, *Account of the Herbaria of the Department of Botany in the University of Oxford.* 251. Oxford 1964; Elmer Drew Merrill, in *Contr. U.S. Natl. Herb.* 30(1): 291. 1947; F.N. Hepper and Fiona Neate, *Plant Collectors in West Africa.* 77. 1971.

Sunaptea Griff. Dipterocarpaceae

Origins:

Presumably from the Greek *syn* "with, together" and *apto* "to fasten," *synapsis* "a binding, union."

Sundacarpus (Buchholz & N.E. Gray) C.N. Page Podocarpaceae

Origins:

Sunda Islands, Malay Archipelago, between the Indian Ocean and the Java Sea.

Sunipia Buch.-Ham. ex Lindley Orchidaceae

Origins:

From a Nepalese or Indian/Himalayan vernacular name, *sunipiang.*

Suregada Roxburgh ex Rottler Euphorbiaceae

Origins:

From *suregada*, a Telugu name (from India), *soora gade*, for *Suregada multiflora* (Juss.) Baill.; see H. Genaust, *Etymologisches Wörterbuch der botanischen Pflanzennamen.* 621. Basel 1996; M.P. Nayar, *Meaning of Indian Flowering Plant Names.* 336. Dehra Dun 1985.

Species/Vernacular Names:

S. africana (Sond.) Kuntze (*Gelonium africanum* (Sond.) Müll. Arg.; *Suregada ceratophora* Baill.; *Ceratophorus africanus* Sond.)

English: common Canary berry, isitubi tree

Southern Africa: gewone Kanariebessie; umHlebezi-omhlope, umMemezi-omhlope, iThambo lempaka (= wild-cat bone), iThambolempisi (Zulu); isiThubi (Zulu, Xhosa); umTiyankawu (Xhosa)

S. glomerulata (Blume) Baill.

English: false lime

Malaya: limau limau, merlimau, limau hantu, lima, melima, ruas puas, penawa puteh

S. procera (Prain) Croizat (*Gelonium procerum* Prain)

English: forest Canary berry, tall suregada

South Africa: boskanariebessie

S. zanzibariensis Baill. (*Gelonium serratum* Pax & K. Hoffm.; *Gelonium zanzibariense* (Baill.) Müll. Arg.)

English: sand Canary berry, woodland suregada, Zanzibar suregada

Southern Africa: sandkanariebessie; iKhushwane-elikulu, umDlankawu (Zulu)

Suriana L. Surianaceae (Rosidae, Rosales)

Origins:

After the French physician Joseph Donat Surian, d. 1691, pharmacist, plant collector, collaborator of the French botanist and artist Charles Plumier (1646-1704), traveler in the West Indies, accompanied Plumier to the Caribbean (West Indies) sent by Louis XIV, in 1689 and 1690 visited Martinique, Guadeloupe and Haiti; see *Insignium et rariorum plantarum semina, ex insulis americanis recentes allata, offerentur & communicantur à Josepho Donato de Surian ... in Nicolas Lémery (1645-1715), Traité universel des drogues simples.* Paris 1698; J.H. Barnhart, *Biographical Notes upon Botanists.* 3: 348. 1965; H.N. Clokie, *Account of the Herbaria of the Department of Botany in the University of Oxford.* 251. Oxford 1964; A. Lasègue, *Musée botanique de Benjamin Delessert.* Paris 1845; M.P. Nayar, *Meaning of Indian Flowering Plant Names.* 336. [Suriana is a name of a Dutch colony in South America] 1985; Carl Linnaeus (Carl von Linnaeus) (1707-1778), *Species Plantarum.* 284. 1753 and *Genera Plantarum.* Ed. 5. 137. 1754.

Species/Vernacular Names:

S. maritima L.

Rodrigues Island: bois matelot

Surwala M. Roemer Meliaceae

Origins:

The anagram of *Walsura* Roxb.

Susanna E. Phillips Asteraceae

Origins:

For Mrs. Susan Phillips, *née* Kriel, second wife to the South African botanist Edwin Percy Phillips (1884-1967), E.P. Phillips wrote "The thorn pears (*Scolopia* spp.)." in *Bothalia.* 1: 83-86. 1922; see J.H. Barnhart, *Biographical Notes upon Botanists.* 3: 81. 1965; T.W. Bossert, *Biographical Dictionary of Botanists Represented in the Hunt Institute Portrait Collection.* 309. 1972; R. Zander, F. Encke, G. Buchheim and S. Seybold, *Handwörterbuch der Pflanzennamen.* 14. Aufl. Stuttgart 1993; Mary Gunn and Leslie E. Codd, *Botanical Exploration of Southern Africa.* Cape Town 1981; I.C. Hedge and J.M. Lamond, *Index of Collectors in the Edinburgh Herbarium.* Edinburgh 1970; Mia C. Karsten, *The Old Company's Garden at the Cape and Its Superintendents*: involving an historical account of early Cape botany. Cape Town 1951; Alain White and Boyd Lincoln Sloane, *The Stapelieae.* Pasadena 1937; A.P. Backer, D.J.B. Killick and D. Edwards, "A plant ecological bibliography and thesaurus for Southern Africa up to 1975." *Memoirs of the Botanical Survey of South Africa*, no. 52. 1986.

Susum Blume Flagellariaceae (Hanguanaceae)

Origins:

Probably from the Latin *susum, sursum* "upwards, on high."

Sutera A.W. Roth Scrophulariaceae

Origins:

After the Swiss botanist Johann Rudolf Suter, 1766-1827, Dr. med. 1794, physician, 1820-1827 professor of philosophy and Greek at Berne, author of *Helvetiens Flora.* Zürich 1802; see J.H. Barnhart, *Biographical Notes upon Botanists.* 3: 348. 1965; E.M. Tucker, *Catalogue of the Library of the Arnold Arboretum of Harvard University.* 1917-1933; Georg Christian Wittstein, *Etymologisch-botanisches Handwörterbuch.* 855. 1852; R. Zander, F. Encke, G. Buchheim and S. Seybold, *Handwörterbuch der Pflanzennamen.* 14. Aufl. 787. Stuttgart 1993.

Species/Vernacular Names:

S. microphylla (L.f.) Hiern (*Manulea microphylla* L.f.)

Southern Africa: iGwanishe lenyoka

Suteria DC. Rubiaceae

Origins:

Possibly for the Swiss physician Johann Rudolf Suter, 1766-1827; see J.H. Barnhart, *Biographical Notes upon Botanists.* 3: 348. 1965.

Sutherlandia R. Br. Fabaceae

Origins:

For the Scottish (b. Edinburgh) botanist James Sutherland, *circa* 1639-1719 (d. Edinburgh), 1699 King's Botanist for Scotland, Superintendent of the Royal Botanical Garden and professor of botany at Edinburgh, author of *Hortus medicus edinburgensis.* Edinburgh 1683, sent plants to James Petiver (1658-1718); see R. Pulteney, *Historical and Biographical Sketches of the Progress of Botany in England.* London 1790; John Claudius Loudon (1783-1843), *Arboretum et fruticetum britannicum.* London 1838; J.H. Barnhart, *Biographical Notes upon Botanists.* 3: 349. 1965; A.D. Boney, *The Lost Gardens of Glasgow University.* London 1988; B.D. Schrire and S. Andrews, "*Sutherlandia*: Part I." *The Plantsman.* 14(2): 65-69. September 1992; E.M. Tucker, *Catalogue of the Library of the Arnold Arboretum of Harvard University.* 1917-1933; Jonas C. Dryander, *Catalogus bibliothecae historico-naturalis Josephi Banks.* 1796-1800; M. Hadfield et al., *British Gardeners: A Biographical Dictionary.* London 1980; B. Henrey, *British Botanical and Horticultural Literature before 1800.* 1975; James Britten, *The Sloane Herbarium,* revised and edited by J.E. Dandy. 1958; F. Boerner & G. Kunkel, *Taschenwörterbuch der botanischen Pflanzennamen.* 4. Aufl. 175. 1989; H.R. Fletcher and W.H. Brown, *Royal Botanic Garden Edinburgh, 1670-1970.* Edinburgh 1970; Dawson Turner, *Extracts from the Literary and Scientific Correspondence of R. Richardson, of Bierly, Yorkshire: Illustrative of the State and Progress of Botany* [Edited by D. Turner. Extracted from the memoir of the Richardson family, by Mrs. D. Richardson] Yarmouth 1835; Arthur D. Chapman, ed., *Australian Plant Name Index.* 2784. 1991; Francis Wall Oliver (1864-1951), ed., *Makers of British Botany.* 281-282. Cambridge 1913; Robert Morison (1620-1683), *Plantarum historiae universalis oxoniensis.* Pars tertia, post auctoris mortem expleta et absoluta a Jacobo Bobartio [Jacobo Bobartio the younger, *circa* 1640-1719]. Oxonii, e Theatro Sheldoniano 1699; James Petiver, *Museii Petiveriani.* London 1695-1703.

Species/Vernacular Names:

S. frutescens (L.) R. Br. (*Colutea frutescens* L.)

English: cancer bush, bladder senna, duck plant, balloon pea

South Africa: kankerbos, cancer bush, kipkippers, kippiebos, kalkoenbelletjie, hoenderbel, aandjies, gansies keur

Sutrina Lindley Orchidaceae

Origins:

Latin *sutrinus, a, um* "belonging to a cobbler," *sutrina, ae* "a shoe-maker's shop, cobbler's stall," referring to the gland.

Suttonia A. Rich. Myrsinaceae

Origins:

After Rev. Charles Sutton, 1756-1846 (d. Norwich); see Ray Desmond, *Dictionary of British & Irish Botanists and Horticulturists.* 665. London 1994.

Suzukia Kudô Labiatae

Origins:

Named after the Japanese botanist Shigeyoshi (Sigeyosi) Suzuki, 1894-1937, plant collector on Iheya Island, Okinawa, and Ishigaki, in 1935 companion to Genkei Masamune (b. 1899) in Iriomote; see G. Masamune, "On the origin of the flora of Iriomote Island." *Bot. & Zool.* (Tokyo) 5: 63-64. 1937.

Species/Vernacular Names:

S. shikikunensis Kudô (*Glechoma shikikunensis* (Kudô) Masamune)

English: Taiwan suzukia

China: tai qian cao

Svenkoeltzia Burns-Balogh Orchidaceae

Origins:

For Sven Koeltz, born Nov. 27th, 1941, at Dresden/Germany, German bookseller and publisher, studied economics; see Pamela Burns-Balogh, *A Reference Guide to Orchidology.* Königstein 1989.

Svensonia Moldenke Verbenaceae

Origins:

For the American botanist Henry Knute (Knut) Svenson, 1897-1986, traveler, botanical explorer and collector, an authority on *Eleocharis* (Cyperaceae); see J.H. Barnhart, *Biographical Notes upon Botanists.* 3: 350. 1965; T.W. Bossert, *Biographical Dictionary of Botanists Represented in the Hunt Institute Portrait Collection.* 389. 1972; S. Lenley et al., *Catalog of the Manuscript and Archival Collections and Index to the Correspondence of John Torrey.* Library of the New York Botanical Garden. 398. 1973; Ida

Kaplan Langman, *A Selected Guide to the Literature on the Flowering Plants of Mexico.* University of Pennsylvania Press, Philadelphia 1964.

Svitramia Cham. Melastomataceae

Origins:

Suitramia, see Chamisso, *Linnaea.* 9: 445. 1835 (prim.) [1834].

Swainsona Salisb. Fabaceae

Origins:

For the British (b. Lancs.) botanist Isaac Swainson, 1746-1812 (d. Middx.), gardener, M.D. 1785, owner of a private botanic garden at Twickenham. He was a cousin of the British zoologist and naturalist William Swainson (1789-1855, d. Wellington, New Zealand); see Henry C. Andrews (1794-1830), *The Botanist's Repository.* 348. London 1804; John Claudius Loudon (1783-1843), *Arboretum et fruticetum britannicum.* London 1838; Ray Desmond, *Dictionary of British & Irish Botanists and Horticulturists.* 666. London 1994; F. Boerner & G. Kunkel, *Taschenwörterbuch der botanischen Pflanzennamen.* 4. Aufl. 175. 1989; Arthur D. Chapman, ed., *Australian Plant Name Index.* 2784-2793. 1991; R. Zander, F. Encke, G. Buchheim and S. Seybold, *Handwörterbuch der Pflanzennamen.* 14. Aufl. 535. 1993; James A. Baines, *Australian Plant Genera.* An Etymological Dictionary of Australian Plant Genera. 359-360. 1981; Francis Aubie Sharr, *Western Australian Plant Names and Their Meanings.* 67. University of Western Australia Press, Nedlands, Western Australia 1996; Nora F. McMillan, in *Dictionary of Scientific Biography* 13: 167-168. 1981.

Species/Vernacular Names:

S. galegifolia (Andrews) R. Br. (*Swainsona coronillifolia* Salisb.; *Vicia galegifolia* Andrews)

English: darling pea, swanflower, wintersweet pea

S. lessertiifolia DC.

English: bog pea, coast swainson pea, Darling pea, poison pea, poison vetch

S. procumbens (F. Muell.) F. Muell. (*Cyclogyne procumbens* F. Muell.; *Swainsona procumbens* (F. Muell.) F. Muell. var. *parvifolia* J. Black)

Australia: Broughton pea, Tatiara pea, pretty swainson pea

Swallenia Söderstr. & H.F. Decker Gramineae

Origins:

For the American botanist Jason Richard Swallen, 1903-1991, agrostologist, with USDA Bureau of Plant Industry. Among his works are "The grass genus *Schizachne*." *J.* *Wash. Acad. Sci.* 18: 203-206. 1928, "The grass genus *Amphibromus*." *Amer. J. Bot.* 18: 411-415. 1931, "The grass genus *Gouinia*." *Amer. J. Bot.* 22: 31-41. 1935, "Three new grasses from Mexico and Chile." *J. Wash. Acad. Sci.* 26: 207-209. 1936 and "Gramineae." in *Flora of Panama. Ann. Missouri Bot. Gard.* 30: 104-280. 1943; see J.H. Barnhart, *Biographical Notes upon Botanists.* 3: 351. 1965; Irving William Knobloch, compil., "A preliminary verified list of plant collectors in Mexico." *Phytologia Memoirs.* VI. 1983; T.W. Bossert, *Biographical Dictionary of Botanists Represented in the Hunt Institute Portrait Collection.* 390. 1972; E.M. Tucker, *Catalogue of the Library of the Arnold Arboretum of Harvard University.* 1917-1933; Ida Kaplan Langman, *A Selected Guide to the Literature on the Flowering Plants of Mexico.* University of Pennsylvania Press, Philadelphia 1964.

Species/Vernacular Names:

S. alexandrae (Swallen) Söderstrom & Decker

English: Eureka Valley dune grass

Swallenochloa F.A. McClure Gramineae

Origins:

For the American botanist Jason Richard Swallen, 1903-1991.

Swartsia J.F. Gmel. Solanaceae

Origins:

After the Swedish botanist Olof Peter Swartz, 1760-1818.

Swartzia Schreber Caesalpiniaceae

Origins:

Named after the Swedish botanist Olof Peter Swartz (Svarts, Svartz, Swarts, Swarz), 1760-1818, traveler (West Indies and northeast South America), physician, he published *Nova genera et species plantarum* seu *Prodromus descriptionum vegetabilium maximam partem incognitorum.* Stockholm, Uppsala & Åbo 1788, *Icones plantarum incognitarum quas in Indiae occidentali...* Erlangae 1794-1800 and *Flora Indiae occidentalis.* Erlangae 1797-1806; see J.H. Barnhart, *Biographical Notes upon Botanists.* 3: 351. 1965; T.W. Bossert, *Biographical Dictionary of Botanists Represented in the Hunt Institute Portrait Collection.* 390. 1972; E.M. Tucker, *Catalogue of the Library of the Arnold Arboretum of Harvard University.* 1917-1933; A. Lasègue, *Musée botanique de Benjamin Delessert.* Paris 1845; Georg Christian Wittstein, *Etymologisch-botanisches Handwörterbuch.* 855. 1852; Frederico Carlos Hoehne, M. Kuhlmann and Oswaldo Handro, *O jardim botânico de São*

Paulo. 1941; R. Zander, F. Encke, G. Buchheim and S. Seybold, *Handwörterbuch der Pflanzennamen*. 14. Aufl. 787. 1993; Emil Bretschneider, *History of European Botanical Discoveries in China*. [Reprint of the original edition 1898.] Leipzig 1981; R.E.G. Pichi Sermolli, "Names and types of fern genera — 2. Angiopteridaceae, Marattiaceae, Danaeaceae, Kaulfussianaceae, Matoniaceae, Parkeriaceae." *Webbia*. 12(2): 339-373. 1957; Ida Kaplan Langman, *A Selected Guide to the Literature on the Flowering Plants of Mexico*. University of Pennsylvania Press, Philadelphia 1964; Mariella Azzarello Di Misa, a cura di, *Il Fondo Antico della Biblioteca dell'Orto Botanico di Palermo*. 262. Regione Siciliana, Palermo 1988.

Species/Vernacular Names:

S. sp.

Suriname: aliana-oeu, boco, bois de fer, ijzerhout, ijzerhart, gewone ijzerhart, ironwood, isrihati, jaroroballi, jesriharti, jetiki boroballi hororadikoro, gewone jizerhart, kapoekoeroe japoekoetjare, kharemero-jaroro, larakusana, panacoco, zwarte parelhout, siribidanni, toepoera apoekoetja, wajewoe, wepeteno tamoene, yzerhart

Guyana (British Guiana): bania, ebony, serebadan, bastard yararu

French Guiana: bois à fleches, paddehout, pagaai-hout, panacoco, sikki-sikki-danna, sikki-sikki-danni, tounou, yakalele

Panama: cornudo, cutaro

Peru: chisnan

Brazil: jacaranda rosa, mira-pishuna, muiracutaca, pacapeua, pacora de macaco, parachutaca, pitaica

Mexico: naranjito

S. fistuloides Harms

Nigeria: udoghogho, udegueghor, ezi; udoghogho (Edo); akite (Igbo)

Zaire: ngakala

Gabon: awong, eluku, oken

Cameroon: ndina, njibo, elekwo, eleku

Ivory Coast: boto, bobo

Central Africa: m'guessa

S. madagascariensis Desv. (*Tounatea madagascariensis* (Desv.) Kuntze)

English: snake bean tree, ironheart tree

W. Africa: samakara, samagara

Southern Africa: moshakashela (Mbukushu); muCherechedzi, muCherechese, muCherekese, muChezeze, muChiakacha, muPondo, muShakashele, muTeherenge, muTseketsa, muTshakatsha, muWengerezi (Shona)

North Nigeria: bayama, bozo zage, gamma fada, guazkiya

Nigeria: bayama, bogo-zage, gama-fada, gwanja-kusa, gwaskiya; bayama (Hausa); hil ighgom (Tiv); yawolawogi (Nupe)

Zaire: kabi, n'daale, ndale, pampi

East Africa: kasanda

N. Rhodesia: mutete

Zambia: mushakashela, ndale, bayama, bozo zage, gamma fada, guazkiya, Mpingo

Togo: subando

S. polyphylla DC. (*Swartzia acuminata* Willdenow ex Vogel; *Swartzia opacifolia* J.F. Macbride)

English: black paddle wood

Brazil: pitaſca-da-terra-firme, maracutaca, paracutaca, paracutaca-da-terra-firme, pitaſca, pracuſba, tachi-pequeno, muiracutaca

Suriname: blakka kabisie, boegoe-boegoe, bois pagaies, jaroroballi, kharemero-jaroro, larakusana, black paddlewood, parakusan, zwart parelhout, sepietoena, toepoera apoekoetja

Sweetia Sprengel Fabaceae

Origins:

For the British (b. Devon) botanist Robert Sweet, 1783-1835 (Chelsea, London), nurseryman, foreman, horticulturist, naturalist, ornithologist, 1812 Fellow of the Linnean Society, half-brother of James Sweet; see J.H. Barnhart, *Biographical Notes upon Botanists*. 3: 352. 1965; Emil Bretschneider, *History of European Botanical Discoveries in China*. Leipzig 1981; Ida Kaplan Langman, *A Selected Guide to the Literature on the Flowering Plants of Mexico*. University of Pennsylvania Press 1964; Mariella Azzarello Di Misa, ed., *Il Fondo Antico della Biblioteca dell'Orto Botanico di Palermo*. 264. Regione Siciliana, Palermo 1988; S. Lenley et al., *Catalog of the Manuscript and Archival Collections and Index to the Correspondence of John Torrey*. Library of the New York Botanical Garden. 470. 1973; E.M. Tucker, *Catalogue of the Library of the Arnold Arboretum of Harvard University*. 1917-1933; A. Lasègue, *Musée botanique de Benjamin Delessert*. Paris 1845; H.R. Fletcher, *Story of the Royal Horticultural Society, 1804-1968*. Oxford 1969; Alain White and Boyd Lincoln Sloane, *The Stapelieae*. Pasadena 1937; Stafleu and Cowan, *Taxonomic Literature*. 6: 122-125. 1986; Ray Desmond, *Dictionary of British & Irish Botanists and Horticulturists*. 667. 1994; R. Zander, F. Encke, G. Buchheim and S. Seybold, *Handwörterbuch der Pflanzennamen*. 14. Aufl. Stuttgart 1993; N. Hall, *Botanists of the Eucalypts*. Melbourne 1978 and Supplement 1980.

Sweetiopsis Chodat & Hasskarl Fabaceae

Origins:

Resembling *Sweetia*, for the British botanist Robert Sweet, 1783-1835.

Swertia L. Gentianaceae

Origins:

For the Dutch herbalist Emanuel (Emmanuel) Swert (Sweerts, Sweert), 1552-1612, florist, cultivator of bulbs, author of *Florilegium*. Frankfurt 1612; see Jean Théodore de Bry (1561-1623), *Florilegium Novum*. Oppenheim 1611; Pierre Vallet (*circa* 1575-*circa* 1657), *Le Jardin du Roy tres chrestien Henry IV.* [Paris] 1608; Mariella Azzarello Di Misa, ed., *Il Fondo Antico della Biblioteca dell'Orto Botanico di Palermo*. 262. Palermo 1988; Carl Linnaeus, *Species Plantarum*. 226. 1753 and *Genera Plantarum*. Ed. 5. 107. 1754; G.C. Wittstein, *Etymologisch-botanisches Handwörterbuch*. 855. 1852; F. Boerner & G. Kunkel, *Taschenwörterbuch der botanischen Pflanzennamen*. 4. Aufl. 175. Berlin & Hamburg 1989.

Species/Vernacular Names:

S. angustifolia Buch.-Ham. ex D. Don.
China: xia ye zhang ya cai

S. angustifolia Buch.-Ham. ex D. Don. var. *pulchella* (D. Don) Burkill (*Swertia affinis* C.B. Clarke)
English: beautiful swertia
China: mei li zhang ya cai

S. bimaculata (Siebold & Zuccarini) Hook.f. & Thomson (*Swertia platyphylla* Merrill)
English: twospotted swertia
China: zhang ya cai

S. cincta Burkill (*Swertia alboviolacea* H. Léveillé)
English: surrounded swertia
China: xi nan zhang ya cai

S. delavayi Franchet
English: Delavay swertia
China: li jiang zhang ya cai

S. diluta (Turcz.) Bentham & Hook.f.
English: diluted swertia
China: bei fang zhang ya cai

S. erythrosticta Maxim.
English: redspotted swertia
China: hong zhi zhang ya cai

S. fastigiata Pursh
English: clustered green-gentian

S. macrosperma (C.B. Clarke) C.B. Clarke

English: bigseeded swertia
China: da zi zhang ya cai

S. neglecta (H.M. Hall) Jepson
English: pine green-gentian

S. perennis L. (*Swertia manshurica* (Komarov) Kitagawa)
English: marsh felwort
China: bei wen dai zhang ya cai

S. pseudochinensis H. Hara
English: false Chinese swertia
China: liu mao zhang ya cai

S. pubescens Franchet
English: pubescent swertia, hairy swertia
China: mao zhang ya cai

S. punicea Hemsley
English: scarlet swertia
China: zi hong zhang ya cai

S. radiata (Kellogg) Kuntze
English: monument plant

S. tibetica Batalin
English: Tibet swertia
China: da yao zhang ya cai

S. wolfgangiana Grüning
English: Wolfgang swertia
China: hua bei zhang ya cai

S. yunnanensis Burkill
English: Yunnan swertia
China: yun nan zhang ya cai

Swertopsis Makino Gentianaceae

Origins:
Resembling *Swertia* L.

Swietenia Jacquin Meliaceae

Origins:

For the Dutch (b. Leiden) botanist and physician Gerard L.B. van Swieten, 1700-1772 (d. Vienna), *Protomedicus*, studied medicine (M.D. 1725) and pharmacy at Louvain and Leyden/Leiden under Herman Boerhaave, 1745 physician to the Empress Maria Theresa of Austria, 1754 a founder of the Vienna University Botanical Garden, founder of the first Vienna medical school and Director of the Medical Faculty of the University of Vienna. His works include Gerardi van Swieten *Commentaria in Hermanni Boerhaave aphorismos de cognoscendis et curandis morbis*. Lugduni Batavorum 1745-1772 and *Description abrégé des mala-*

dies qui regnent le plus communément dans les armées, avec la méthode de les traiter. Amsterdam 1761; see Johann Georg Hasenöhrl (1729-1796), *Historia medica morbi epidemici sive febris petechialis* quae ab anno 1757, etc. [8vo, the account of the Vienna plague of 1758-1759.] Vienna 1760; Anton Störck (1731-1803), *Annus medicus quo sistuntur observationes circa morbos acutos et chronicos.* [Two vols, 8vo, records of dissections.] Vienna 1760, 1761; Peter W. van der Pas, in *Dictionary of Scientific Biography* 13: 181-183. 1981; G.C. Wittstein, *Etymologisch-botanisches Handwörterbuch.* 855. 1852; Garrison and Morton, *Medical Bibliography.* 2152, 2200, 4488. 1961.

Species/Vernacular Names:

S. humilis Zucc.

English: Mexican mahogany

S. macrophylla King

English: Honduras mahogany, big-leaf mahogany, broadleafed mahogany

Peru: aguano, caoba, mahogani, pasich, tuxw

Japan: ôba-mahogani

S. mahagoni (L.) Jacq.

English: mahogany, mahogany tree, West Indian mahogany, Spanish mahogany, Madeira redwood, Cuba mahogany

Japan: mahogani

India: mahagoni

Swinglea Merrill Rutaceae

Origins:

After the American botanist Walter Tennyson Swingle, 1871-1952, a specialist in *Citrus* and Rutaceae; see J.H. Barnhart, *Biographical Notes upon Botanists.* 3: 352. 1965; T.W. Bossert, *Biographical Dictionary of Botanists Represented in the Hunt Institute Portrait Collection.* 391. 1972; Ida Kaplan Langman, *A Selected Guide to the Literature on the Flowering Plants of Mexico.* Philadelphia 1964; S. Lenley et al., *Catalog of the Manuscript and Archival Collections and Index to the Correspondence of John Torrey.* Library of the New York Botanical Garden. 399. 1973; E.M. Tucker, *Catalogue of the Library of the Arnold Arboretum of Harvard University.* 1917-1933; J. Ewan, ed., *A Short History of Botany in the United States.* New York and London 1969; R. Zander, F. Encke, G. Buchheim and S. Seybold, *Handwörterbuch der Pflanzennamen.* 14. Aufl. 1993.

Swintonia Griffith Anacardiaceae

Origins:

After the British writer George Swinton, 1780-1854, translator, 1802 East India Company, 1827 Chief Secretary in Bengal, (*circa* 1840) a senior official of East India Company in Bengal, plant collector for N. Wallich, friend of William Griffith (1810-1845); see Abu Talib Ibn Muhammad Khan, Isfahani. [Masnavi.] (Poem in praise of Miss Julia Burrell.) With an English translation by G. Swinton. London [1802?].

Species/Vernacular Names:

S. sp.

Malaya: rengas, mempenai, bermi, mepauh, balau

Swynnertonia S. Moore Asclepiadaceae

Origins:

For Charles Francis Massey Swynnerton, b. India 1877-d. Tanganyika/Tanzania 1938 (in an air crash), farmer, botanical collector in Rhodesia and Mozambique, 1907 Fellow of the Linnean Society, naturalist; see Mary Gunn and Leslie E. Codd, *Botanical Exploration of Southern Africa.* 339. A.A. Balkema Cape Town 1981; Ray Desmond, *Dictionary of British & Irish Botanists and Horticulturists.* 668. London 1994.

Syagrus Mart. Palmae

Origins:

Latin *syagrus, i* used by Plinius for a kind of palm-tree; possibly from the Greek *sys* "pig" and *agrios, agrimaios* (*agra*) "wild," referring to the habitat.

Species/Vernacular Names:

S. botryophora (Mart.) Becc.

Brazil: pati, patioba

S. cearensis Noblick

Brazil: catolé, coco catolé

S. cocoides Mart.

Brazil: jatá, jatá uva, piririma, pererina, uaperena

S. comosa (Mart.) Mart.

Brazil: catolé, babão, palmito amargoso, coco babão, guariroba do campo, jerivá

S. coronata (Mart.) Beccari

Brazil: coqueiro cabeçudo, licuri, licurizeiro, ouricuri

S. duartei Glassman

Brazil: coqueirinho, coco da lapa, coquinho da serra

S. flaxuosa (Mart.) Beccari

Brazil: acumã, acuman, akumá, coco do campo, coco babão, coco de vaqueiro, acumão, ariri, coqueiro do campo, palmito do campo, coco da serra, coco de vassoura, coqueiro acumão

S. glaucescens Glaz. ex Beccari

Brazil: coco da pedra, palmeirinha azul

S. graminifolia (Drude) Beccari

Brazil: coquinho do cerrado, acumã rasteiro

S. harleyi Glassman

Brazil: coco de raposa

S. inajai (Spruce) Beccari

Brazil: pupunharana, pupunha brava, pirima, piririma, pupunha de porco, buruarana, jararana

S. macrocarpa Barb. Rodr.

Brazil: maria rosa, mari rosa, mariroba, gerivá, guariroba

S. x matafome (Bondar) Glassman (fome = Latin *fames* meaning hunger)

Brazil: ariroba, licuri de boi, licuri mata fome, licuriroba, mata-fome

S. microphylla Burret

Brazil: coquinho, ariri

S. oleracea (Mart.) Beccari

Brazil: gueiroba, guariroba, coqueiro guariroba, gariroba, pati, palmito amargoso, catolé, coco babão, paty amargoso, coco amargoso, coqueiro amargoso, coco de quaresma

S. petrea (Mart.) Beccari

Brazil: coco de vassoura, ariri, acuman rasteiro, guriri, hariri

S. picrophylla Barb. Rodr.

Brazil: coco de quaresma, coco de quarta, licuri

S. pleioclada Burret

Brazil: coqueirinho, palmeirinha

S. pseudococos (Raddi) Glassman

Brazil: coco amargoso, piririma, coco verde, gariroba, guariroba, palha branca, palmito amargoso, paty amargoso

S. romanzoffiana (Cham.) Glassm. (*Arecastrum romanzoffianum* (Cham.) Beccari)

English: queen palm

Japan: giri-ba-yashi

Brazil: coqueiro, coqueiro datil, coqueiro de Santa Catarina, coqueiro juvena, coqueiro pindaba, coqueiro pindoba, pindó, coco babão, baba de boi, baba de bio, coco de catarro, coco de cachorro, coco de sapo, gerivá, jerivá, yarivá

Argentina: coquito

Portuguese: coqueiro de jardim

S. ruschiana (Bondar) Glassman

Brazil: coco de pedra

S. sancona H. Karst.

Brazil: jaciarana, açairana

S. schizophylla (Martius) Glassman

Brazil: coco babão, licurioba, aricuriroba, coco caboclo

S. vagans (Bondar) A.D. Hawkes

Brazil: ariri, licurioba, licuriroba, pindoba, licuriroba das caatingas

S. werdermannii Burret

Brazil: coco de vassoura, coco de raposa, coco de peneira

Sychinium Desv. Moraceae

Origins:
Greek *sykon* "fig."

Sycios Medik. Cucurbitaceae

Origins:
Greek *sikyos* "wild cucumber, gourd," Latin *sycion agron* "a plant, called also *cucumis anguinus*"; see also *Sicyos* L.

Sycocarpus Britton Meliaceae

Origins:
From the Greek *sykon* "fig" and *karpos* "fruit."

Sycodendron Rojas Acosta Moraceae

Origins:
Greek *sykon* "fig" and *dendron* "tree."

Sycomorphe Miq. Moraceae

Origins:
From the Greek *sykon* "fig" and *morphe* "a form, shape"; see also *Ficus* L.

Sycomorus Gasp. Moraceae

Origins:
From the Greek *sykon* "fig" and *moron* "mulberry," Latin *sycaminus* or *sycaminos, sycaminon, sycomorus* "a mulberry tree"; see also *Ficus* L.

Sycophila Welw. ex Tieghem Loranthaceae

Origins:
From the Greek *sykon* "fig" and *philos* "lover, loving."

Sycopsis Oliver Hamamelidaceae

Origins:

Greek *sykon* "fig" and *opsis* "appearance."

Syena Schreb. Mayacaceae (Commelinidae, Commelinales)

Origins:

For Arnoldus Syen, 1640-1678, author of Disputatio ... *de hydrope ascite*. Lugduni Batavorum 1659; see Pritzel, *Thesaurus*. 311; Hendrik Adriaan van Rheede tot Draakestein (1637-1691), *Hortus Indicus Malabaricus*. [... notis adauxit et commentariis illustravit A.S.] Amstelodami [Amsterdam] 1678, etc.

Sylitra E. Meyer Fabaceae

Origins:

From a Greek name used by Dioscorides for a kind of *Glycyrrhiza*.

Sylvalismis Thouars Orchidaceae

Origins:

Latin *silva, sylva* "forest, wood" and the genus *Alisma*, water plantain, referring to the leaves.

Symbasiandra Steud. Gramineae

Origins:

Greek *symbasis* "juncture" and *aner, andros* "man, stamen."

Symbegonia Warb. Begoniaceae

Origins:

From the Greek *syn* "with, together" and the genus *Begonia*, referring to the united perianth segments of the female flowers.

Symbolanthus G. Don f. Gentianaceae

Origins:

Greek *symbolos, symbolon* "a sign, a mark" and *anthos* "flower."

Symethus Raf. Convolvulaceae

Origins:

Simeto, a Sicilian river; see Constantine Samuel Rafinesque, *Flora Telluriana*. 4: 83. 1836 [1838]; E.D. Merrill, *Index rafinesquianus*. 200. 1949.

Symingtonia Steenis Hamamelidaceae

Origins:

After the Scottish (b. Edinburgh) botanist Colin Fraser Symington, 1905-1943 (d. Nigeria), plant collector, 1940 Fellow of the Linnean Society, Forest botanist at the Forest Research Institute of Kepong (Malayan Forestry Service) and in South Nigeria, botanical assistant to Frederick William Foxworthy (1877-1950). His works include *Forester's Manual of Dipterocarps*. Kuala Lumpur 1941, "A simple recipe for poison." *Malayan Forester*. 10: 25-27. 1941 and "The Flora of Gunong Tapis in Pahang: with notes on the altitudinal zonation of the forests of the Malay Peninsula." *Jour. Malay. Branch Roy. Asiatic Soc.* 14: 333-365. 1936; see J.H. Barnhart, *Biographical Notes upon Botanists*. 3: 353. 1965; F.N. Hepper and Fiona Neate, *Plant Collectors in West Africa*. 77. 1971; T.W. Bossert, *Biographical Dictionary of Botanists Represented in the Hunt Institute Portrait Collection*. 391. 1972; Ray Desmond, *Dictionary of British & Irish Botanists and Horticulturists*. 669. London 1994.

Species/Vernacular Names:

S. populnea (R. Br. ex Griff.) Steenis

English: Malayan aspen

Malaya: tiga sagi, drok, grok

Sympagis (Nees) Bremek. Acanthaceae

Origins:

From the Greek *sympages* "joined together," referring to the basal membrane joining the stamens; see F. Boerner & G. Kunkel, *Taschenwörterbuch der botanischen Pflanzennamen*. 4. Aufl. 175. 1989; G.C. Wittstein, *Etymologisch-botanisches Handwörterbuch*. 856. 1852.

Sympegma Bunge Chenopodiaceae

Origins:

Greek *sympegma* "framework, superstructure," *pegma, pegmatos* "congealed, fixed, a framework, that which makes to curdle."

Sympetalandra Stapf Caesalpiniaceae

Origins:
Greek *syn* "with, together," *petalon* "leaf, petal" and *aner, andros* "man, stamen," Latin *sympetalus* "gamopetalous, having united petals."

Sympetaleia A. Gray Loasaceae

Origins:
From the Greek *syn* "with, together" and *petalon* "leaf, petal."

Symphiandra Steud. Campanulaceae

Origins:
Greek *symphyes* "born with one, congenital, grown together, coalescing" and *aner, andros* "male, stamen"; see also *Symphyandra* A. DC.

Symphionema R. Br. Proteaceae

Origins:
Greek *symphyes* "grown together, coalescing" and *nema* "thread, filament"; see R. Brown, *Transactions of the Linnean Society of London. Botany.* 10: 157. 1810.

Symphonia L.f. Guttiferae

Origins:
Greek *syn* plus *phone* "voice, sound," Latin *symphonia* "concord, harmony, symphony," referring to the stamens, united.

Species/Vernacular Names:
S. globulifera L.f.

English: hog plum, chewstick, doctor gum, sambo gum

Central Africa: agberigbedi, beu, bon

Yoruba: agberigbede

Nigeria: ossol, arkwani, arquani, osoe, ogolo, okilolo, agberigbede, agberigbedi, ovien-edun-eze, ovien edun eze, umochoneze, manil; agben (Yoruba); ovien edun eze (Edo); okilolo (Ijaw)

Ivory Coast: arquane, aruane, beu

Gabon: osodou, osongho, ossol, monianga, ossol

Cameroon: gambé, mokoa, onié, sisako, nom onié à fleur rouge

Congo: niang-nianga, sondzo

Zaire: bolaka, bolongo, bombiko, dake, mbiko, botete, bulungu, ulundu, ilungu, nsongia

Symphorema Roxburgh Verbenaceae (Symphoremataceae)

Origins:
From the Greek *symphorema* "brought together, compound," *symphoreo* "to bring together, collect," referring to the flowers or to the capitate cymes within the involucre.

Species/Vernacular Names:
S. involucratum Roxburgh

English: involucrate symphorema

China: liu bao teng

Symphoricarpos Duhamel Caprifoliaceae

Origins:
Greek *symphoreo* "to bring together, collect" and *karpos* "fruit," referring to the clustered fruits.

Species/Vernacular Names:
S. albus (L.) S.F. Blake

English: snowberry

S. longiflorus A. Gray

English: fragrant snowberry

S. mollis Nutt.

English: creeping snowberry, trip vine

S. orbiculatus Moench

English: coral berry

Symphostemon Hiern Labiatae

Origins:
From the Greek *symphoreo* and *stemon* "stamen, pillar."

Symphyandra A. DC. Campanulaceae

Origins:
Greek *symphyes* "grown together, coalescing" (*syn* and *phyo, phyein* "to grow") and *aner, andros* "male, man, stamen, anther," referring to the joined anthers.

Symphyglossum Schltr. Orchidaceae

Origins:

Greek *symphyes* and *glossa* "tongue," the lip is adnate to the column.

Symphyllarion Gagnepain Rubiaceae

Origins:

From the Greek *syn* "together" and *phyllarion*, the diminutive of *phyllon* "leaf."

Symphyllia Baillon Euphorbiaceae

Origins:

Greek *syn* "together" and *phyllon* "leaf," alluding to the nature of the leaves.

Symphyllocarpus Maxim. Asteraceae

Origins:

From the Greek *syn* "together," *phyllon* "leaf" and *karpos* "fruit."

Symphyllophyton Gilg Gentianaceae

Origins:

Greek *syn* "together," *phyllon* "leaf" and *phyton* "plant."

Symphyobasis K. Krause Goodeniaceae

Origins:

Greek *symphyo, symphyein* "to grow together" plus *basis* "base, pedestal," the ovaries are adnate to the corolla.

Symphyochaeta (DC.) Skottsb. Asteraceae

Origins:

Greek *symphyes* "born with one, congenital, grown together, coalescing" (*syn* "together" and *phyo* "to grow") and *chaite* "a bristle."

Symphyochlamys Gürke Malvaceae

Origins:

Greek *symphyo, symphyein* "to grow together" and *chlamys, chlamydos* "cloak, mantle."

Symphyogyne Burret Palmae

Origins:

Greek *symphyo, symphyein* "to grow together" with *gyne* "woman, female."

Symphyoloma C.A. Meyer Umbelliferae

Origins:

Greek *symphyo, symphyein* and *loma* "border, margin, fringe."

Symphyomera Hook.f. Asteraceae

Origins:

From the Greek *symphyes* "born with one, coalescing" and *meros* "part."

Symphyomyrtus Schauer Myrtaceae

Origins:

Greek *symphyes* "born with one, congenital, grown together, coalescing," *symphyo, symphyein* "to grow together" and *Myrtus*.

Symphyonema Sprengel Proteaceae

Origins:

Orthographic variant of *Symphionema*.

Symphyopappus Turcz. Asteraceae

Origins:

Greek *symphyes* "born with one, congenital, grown together, coalescing" and *pappos* "fluff, downy appendage."

Symphyopetalon Drumm. ex Harvey Rutaceae

Origins:

Greek *symphyes* "grown together, coalescing" and *petalon* "petal, leaf."

Symphyosepalum Hand.-Mazz. Orchidaceae

Origins:

Greek *symphyo, symphyein* "to grow together" and Latin *sepalum* "sepal," the sepals are connate.

Symphyostemon Miers ex Klatt Iridaceae

Origins:
From the Greek *symphyes* "born with one, congenital, grown together, coalescing," *symphyo*, *symphyein* "grow together" and *stemon* "stamen, pillar."

Symphysia C. Presl Ericaceae

Origins:
Greek *symphysis* "growing together, natural junction," *symphyein* "to grow together."

Symphysicarpus Hassk. Asclepiadaceae

Origins:
Greek *symphysis* "growing together, natural junction" plus *karpos* "fruit."

Symphytonema Schltr. Asclepiadaceae (Periplocaceae)

Origins:
Greek *symphyo*, *symphyein* "to grow together," *phyton* "plant" and *nema* "filament, thread," Latin *symphyton* for wallwort, comfrey, boneset, for a plant called also *helenion* (Plinius).

Symphytosiphon Harms Meliaceae

Origins:
Greek *symphyo* "to grow together," *phyton* "plant" and *siphon* "tube."

Symphytum L. Boraginaceae

Origins:
Growing together, *symphyo* and *phyton*, Greek *symphyton* used by Dioscorides for a plant, comfrey, a species of *Symphytum*, or for low pine, a kind of *Coris*; Latin *symphyton, i* applied by Plinius to the wallwort, comfrey, boneset or to a plant, called also *helenion*.

Species/Vernacular Names:
S. officinale L.
English: comfrey, healing herb, wound knit, bone knit
China: ju he cao

Sympieza Lichtenst. ex Roemer & Schultes Ericaceae

Origins:
Greek *sympiezo* "to squeeze together, to press," referring to the flowers or to the stamens adhering to the corolla tube.

Symplectochilus Lindau Acanthaceae

Origins:
Greek *syn* "with, together" and *plektos* "twisted, plaited" and *cheilos* "lip."

Symplocarpus Salisb. ex W. Barton Araceae

Origins:
Greek *symploke* "combination, connection," *symplokos* "entwined, interwoven" and *karpos* "fruit," referring to the ovaries coalescing into a compound fruit; see D.H. Nicolson, "Derivation of aroid generic names." *Aroideana.* 10: 15-25. 1988.

Species/Vernacular Names:
S. foetidus (L.) W. Barton
English: skunk cabbage

Symplococarpon Airy Shaw Theaceae

Origins:
Greek *symploke* "combination, connexion, connection," *symplokos* "entwined, interwoven" (*syn* "with, together" and *pleko*, *plekein* "to twist, twine, tie, enfold") and *karpos* "fruit."

Symplocos Jacq. Symplocaceae

Origins:
Greek *symploke* "combination," *symplokos* "entwined, interwoven" (*syn* "with, together" and *pleko* "to twist, twine, tie, enfold"), referring to the stamens (united at the base) or to the ovary.

Species/Vernacular Names:
S. cochinchinensis (Loureiro) S. Moore
English: sweetleaf, Cochinchina sweetleaf
Sri Lanka: bobu, bombu
China: yue nan shan fan

S. laurina (Retzius) Wall.

English: laurel sweetleaf

Malaya: kendong

S. lucida (Thunberg) Siebold & Zuccarini (*Symplocos japonica* A. DC.; *Dicalix lucida* (Thunberg) Hara)

Japan: kuro-ki

China: guang liang shan fan

S. paniculata (Thunb.) Miquel (*Symplocos crataegoides* Buch.-Ham. ex D. Don)

English: sapphire berry sweetleaf, Asiatic sweetleaf

China: bai tan

S. pendula Wight var. *hirtistylis* (C.B. Clarke) Nooteboom (*Symplocos sonoharai* Koidz.; *Symplocos confusa* Brand; *Cordyloblaste confusa* (Brand) Ridley; *Bobua confusa* (Brand) Kaneh. & Sasaki)

English: confuse sweetleaf

Japan: miyama-shiro-bai

China: nan ling shan fan

S. racemosa Roxburgh

English: lodh bark, racemose sweetleaf

China: zhu zi shu

Malaya: melilin, mempatu

S. theiformis (L.f.) Oken

English: Bogota tea

Synadena Raf. Orchidaceae

Origins:

Greek *syn* plus *aden* "gland," referring to the bifid gland at the apex of the column; see C.S. Rafinesque, *Flora Telluriana*. 4: 9. 1836 [1838].

Synadenium Boissier Euphorbiaceae

Origins:

Greek *syn* "with, together, united" and *aden* "gland," it refers to the united gland enclosing and surrounding the flower.

Species/Vernacular Names:

S. cupulare (Boiss.) L.C. Wheeler (*Synadenium arborescens* Boiss.; *Euphorbia cupularis* Boiss.; *Euphorbia arborescens* E. Mey.)

English: dead-man's tree, crying tree

Southern Africa: dooiemansboom; umDlebe, umDletshane, umZilanyoni (Zulu); mulamba-noni (= not eaten by birds), muswoswo (Venda)

S. grantii Hook.f. non Jex-Blake

English: the African milk-bush

Synallodia Raf. Gentianaceae

Origins:

See C.S. Rafinesque, *New Fl. N. Am.* 4: 93. 1836 [1838].

Synandra Nutt. Labiatae

Origins:

Greek *syn* "united" and *aner, andros* "stamen, male, man, anther"; see also *Aphelandra* R. Br.

Synandra Schrad. Acanthaceae

Origins:

Greek *syn* "together, united" and *aner, andros* "stamen, male, anther."

Synandrina Standley & L.O. Williams Flacourtiaceae

Origins:

Greek *syn* "together, united" and *aner, andros* "stamen, male," the diminutive.

Synandrodaphne Gilg Thymelaeaceae

Origins:

From the Greek *syn* "with, together, united," *aner, andros* "stamen, male, man, anther" plus *Daphne*.

Synandrogyne Buchet Araceae

Origins:

Greek *syn* plus *androgynos* "man-woman, hermaphrodite," Latin *androgynus* "having male and female flowers separate but on the same inflorescence"; see D.H. Nicolson, "Derivation of aroid generic names." *Aroideana*. 10: 15-25. 1988.

Synandropus A.C. Sm. Menispermaceae

Origins:

From the Greek *syn* "with, together, united," *aner, andros* "stamen, male, man, anther" and *pous, podos* "a foot."

Synandrospadix Engl. Araceae

Origins:

Greek *syn* "with, together," *aner, andros* "stamen, man" and *spadix* "a palm frond, palm branch, spadix"; see D.H. Nicolson, "Derivation of aroid generic names." *Aroideana*. 10: 15-25. 1988; F. Boerner & G. Kunkel, *Taschenwörterbuch der botanischen Pflanzennamen*. 4. Aufl. 176. Berlin & Hamburg 1989.

Synantherias Schott Araceae

Origins:

Greek *syn* "with, together, united" and *anthera* "anther."

Synanthes Burns-Balogh, H. Robinson & M.S. Foster Orchidaceae

Origins:

Greek *syn* "united" and *anthos* "flower."

Synaphe Dulac Gramineae

Origins:

Greek *synaphe* "union, connession," *syn* with *apto, aptein* "to fasten."

Synaphea R. Br. Proteaceae

Origins:

Greek *synaphe* "union, connession," referring to the membrane connecting the filament of the upper sterile anther to the stigma; see Robert Brown, *Transactions of the Linnean Society of London. Botany*. 10: 155. 1810; James A. Baines, *Australian Plant Genera. An Etymological Dictionary of Australian Plant Genera*. 361. Chipping Norton, N.S.W. 1981; Francis Aubie Sharr, *Western Australian Plant Names and Their Meanings*. 67. University of Western Australia Press 1996.

Synaphlebium J. Sm. Dennstaedtiaceae

Origins:

From the Greek *synaphe* "union" and *phleps, phlebos* "vein."

Synapsis Griseb. Bignoniaceae

Origins:

From the Greek *synapsis* "a binding, union."

Synaptantha Hook.f. Rubiaceae

Origins:

From the Greek *syn* "with, together," *aptein* "to fasten" and *anthos* "flower," referring to the filaments.

Synaptea Kurz Dipterocarpaceae

Origins:

From the Greek *syn* "with, together" and *apto* "to fasten."

Synaptolepis Oliver Thymelaeaceae

Origins:

Greek *synapto* "join together" and *lepis* "scale," referring to the disk.

Synaptophyllum N.E. Br. Aizoaceae

Origins:

From the Greek *synapto* "join together" and *phyllon* "leaf."

Synardisia (Mez) Lundell Myrsinaceae

Origins:

Greek *syn* "with, together" plus *Ardisia* Sw.

Synarrhena F. Muell. Actinidiaceae

Origins:

Greek *syn* plus *arrhen* "male, anther," see Arthur D. Chapman, ed., *Australian Plant Name Index*. 2797. Canberra 1991.

Synarrhena Fischer & C.A. Meyer Sapotaceae

Origins:

From the Greek *syn* "with, together" and *arrhen* "male, anther."

Synaspisma Endl. Euphorbiaceae

Origins:

Greek *synaspismos* "holding of the shields together," *syn* "together" and *aspis* "a shield."

Synassa Lindley Orchidaceae

Origins:

Greek *syn* and *asson* "nearer," referring to the petals and dorsal sepals (*sepala superiora et petala agglutinata*), see also *Sauroglossum* Lindl.

Synastemon F. Muell. Euphorbiaceae

Origins:

From the Greek *syn* with *stemon* "pillar, stamen," see also *Sauropus* Blume.

Syncalathium Lipsch. Asteraceae

Origins:

From the Greek *syn* "united, together" and *kalathos* "a basket."

Syncarpha DC. Asteraceae

Origins:

From the Greek *syn* and *karphos* "chip of straw, chip of wood."

Species/Vernacular Names:

S. argentea (Thunb.) B. Nord. (*Helichrysum argenteum* (Thunb.) Thunb.)

English: pink everlasting

S. striata (Thunb.) B. Nord. (*Helichrysum striatum* (Thunb.) Thunb.)

English: everlasting

Syncarpia Ten. Myrtaceae

Origins:

Greek *syn* and *karpos* "fruit," indicating the carpels, united; see Michele Tenore (1780-1861), *Index Seminum in Horto Botanico Neapolitano Collectorum*. 1839.

Species/Vernacular Names:

S. glomulifera (Sm.) Niedenzu

English: turpentine wood

Syncephalantha Bartling Asteraceae

Origins:

Greek *syn* "with, together," *kephale* "head" and *anthos* "flower"; see also *Dyssodia* Cav.

Syncephalum DC. Asteraceae

Origins:

Greek *syn* "with, together" and *kephale* "head."

Synchoriste Baillon Acanthaceae

Origins:

Greek *syn* and *choristes* "one who separates," *synchoreo* "come together, meet"; see also *Lasiocladus* Bojer ex Nees.

Synclisia Benth. Menispermaceae

Origins:

From the Greek *synkleisis* "shutting up, closing up, safe storage."

Syncolostemon E. Meyer Labiatae

Origins:

Greek *syn* "united," *kolos* "stunted, clipped" and *stemon* "pillar, stamen," possibly referring to the lower pair of filaments.

Syncretocarpus S.F. Blake Asteraceae

Origins:

From the Greek *synkretismos* "union" and *karpos* "fruit."

Syndesmanthus Klotzsch Ericaceae

Origins:

From the Greek *syndesmos* "that which binds together" and *anthos* "flower."

Syndesmis Wall. Anacardiaceae

Origins:

Greek *syndesmos* "fastening, conjunction," see also *Gluta* L.

Syndesmon (Hoffmanns. ex Endl.) Britton Ranunculaceae

Origins:
From the Greek *syndesmos* "that which binds together."

Syndiclis Hook.f. Lauraceae

Origins:
From the Greek *syn* plus *diklis* "double-folding," referring to the nature of the stamens.

Syndyophyllum K. Schumann & Lauterb. Euphorbiaceae

Origins:
Greek *syndyo* "in pairs, two together" and *phyllon* "leaf."

Synechanthus H.A. Wendland Palmae

Origins:
Greek *syneches* "continuous, holding together" and *anthos* "flower," referring to the arrangement of the inflorescence. The flowers are in groups of several in linear arrangement along the rachillae.

Species/Vernacular Names:
S. warscewiczianus H.A. Wendland

Brazil: palmeira mexicana

Synedrella Gaertner Asteraceae

Origins:
Greek *synedrion* (*syn* "with, together" and *hedra* "a seat, chair") "a sitting together, council," referring to the clustered flowers.

Species/Vernacular Names:
S. nodiflora (L.) Gaertner (*Verbesina nodiflora* L.)

English: nodeweed

Japan: fushi-zaki-sô

Yoruba: apa iwofa, aluganbi, apawofa, arasan, ogbugbo, iba igbo

Congo: ndongu

Synedrellopsis Hieron. & Kuntze Asteraceae

Origins:
Resembling *Sinedrella* Gaertner.

Syneilesis Maxim. Asteraceae

Origins:
From the Greek *syneilesis* "rolling up, synthesis, rolling oneself up."

Synelcosciadium Boiss. Umbelliferae

Origins:
Greek *synelko* "draw together, help to pull" and *skiadion* "umbel, parasol."

Synema Dulac Euphorbiaceae

Origins:
Greek *syn* "together" and *nema* "thread, filament."

Synepilaena Baillon Gesneriaceae

Origins:
From the Greek *syn* "together," *epi* "upon" and *chlaena, laina* "cloak, blanket," Latin *laena* "a cloak, mantle."

Synexemia Raf. Euphorbiaceae

Origins:
From the Greek *synexomai* "sit together"; see C.S. Rafinesque, *Neogenyton, or Indication of Sixty-Six New Genera of Plants of North America*. 2. 1825 and *Flora Telluriana*. 4: 117. 1836 [1838]; E.D. Merrill, *Index rafinesquianus*. 154, 156. 1949.

Syngonanthus Ruhland Eriocaulaceae

Origins:
Greek *syngonos* "congenital, natural, native, of one's country" and *anthos* "flower."

Syngonium Schott Araceae

Origins:
From the Greek *syn* "together" and *gone* "womb, ovary," referring to the united ovaries; see D.H. Nicolson, "Derivation of aroid generic names." *Aroideana*. 10: 15-25. 1988.

Syngramma J. Sm. Pteridaceae (Adiantaceae)

Origins:
From the Greek *syn* "together" and *gramma, grammatos* "a letter, line, character, thread," *syngramma* "written paper."

Syngrammatopsis Alston Pteridaceae (Adiantaceae)

Origins:
Resembling *syn* "together" and *gramma, grammatos* "a letter, line, character, thread."

Synima Radlk. Sapindaceae

Origins:
From the Greek *syn* "with" and *heima* "clothing, cloak," *syneimi* "be joined with, assemble, gather," possibly referring to the covering on the seeds.

Synisoon Baillon Rubiaceae

Origins:
Presumably from the Greek *synisoomai* "to be identical, made identical."

Synmeria Nimmo Orchidaceae

Origins:
From the Greek *syn* "together" and *meris* "part," referring to the petals connate with the sepal.

Synnema Benth. Acanthaceae

Origins:
From the Greek *syn* plus *nema* "thread, filament," referring to the united stamens.

Synnotia Sweet Iridaceae

Origins:
Dedicated to the Irish (b. Ballymoyer, Co. Armagh, Northern Ireland) Captain Walter Synnot, 1773-1851 (d. Launceston, Tasmania), plant collector at the Cape of Good Hope. He sent many bulbs to Robert Sweet, 1820 to South Africa, 1836 Tasmania; see Mary Gunn and L.E. Codd, *Botanical Exploration of Southern Africa*. 340. [*sphalm.* "Synnetia"] Cape Town 1981; Ray Desmond, *Dictionary of British & Irish Botanists and Horticulturists*. 669. 1994.

Species/Vernacular Names:
S. villosa (Burm.f.) N.E. Br. (*Gladiolus villosus* Burm.f.)
South Africa: ferweelblom

Synochlamys Fée Pteridaceae (Adiantaceae)

Origins:
Greek *syn* "united, together" and *chlamys* "cloak, mantle."

Synoecia Miq. Moraceae

Origins:
Latin *synoecium* "a room where several persons dwell together," Greek *synoikia* "a joint lodging"; see also *Ficus* L.

Synoplectris Raf. Orchidaceae

Origins:
Greek *syn* with *plektron* "a spur, cock's spur," from the shape of the spur, the connivent lateral sepals form a spur; see C.S. Rafinesque, *Flora Telluriana*. 2: 89, 90. 1836 [1837].

Synoplectrodia Lazarides Gramineae

Origins:
Relationships with *Plectrachne* Henr. and *Triodia* R. Br.; see M. Lazarides, "New taxa of tropical Australian grasses (Poaceae)." in *Nuytsia*. 5(2): 273-303. 1984; Arthur D. Chapman, ed., *Australian Plant Name Index*. 2795. Canberra 1991.

Synosma Raf. ex Britton & Brown Asteraceae

Origins:
Greek *syn* plus *osme* "smell, odor, perfume"; see C.S. Rafinesque, *Loud. Gard. Mag.* 8: 247. 1832 and *New Fl. N. Am.* 4: 78, 79. 1836 [1838]; E.D. Merrill, *Index rafinesquianus*. 236, 242. 1949.

Synostemon F. Muell. Euphorbiaceae

Origins:
From the Greek *syn* "united, together" and *stemon* "stamen," see also *Sauropus* Blume.

Synotis (C.B. Clarke) C. Jeffrey & Y.L. Chen Asteraceae

Origins:
Greek *syn* "united, together" and *ous, otos* "an ear."

Synotoma (G. Don f.) R. Schulz Campanulaceae

Origins:
Greek *syn* "united" and *tome, tomos, temno* "division, section, to slice," *syntomos* "cut short, abridged"; see also *Physoplexis* (Endl.) Schur, from the Greek *physa* "a bladder" and *plexis* "plaiting, weaving," referring to the divisions of the corolla, basally swollen.

Synoum A. Juss. Meliaceae

Origins:
Greek *syn* "united" and *oon* "egg," referring to the ovules, seeds united by a common aril.

Species/Vernacular Names:
S. glandulosum (Sm.) A. Juss.
English: bastard rosewood, scented rosewood

Synptera Llanos Orchidaceae

Origins:
Greek *syn* "united" and *pteron* "wing," *synpteroomai* "get wings together," referring to the base of the lateral sepals.

Synsepalum (A. DC.) Daniell Sapotaceae

Origins:
From the Greek *syn* "united, together" and Latin *sepalum*, referring to the nature of the sepals.

Species/Vernacular Names:
S. dulcificum (Schum. & Thonn.) Daniell
English: miraculous berry
Nigeria: abayunkun
Yoruba: agbayun, agbaro aaya
S. stipulatum (Radlk.) Engl.
English: blacksmiths' charcoal wood,
Nigeria: azimomo, ogeromo (Edo); udalaenwe (Igbo)
S. subcordatum De Wild.

Zaire: bonge

Synsiphon Regel Colchicaceae (Liliaceae)

Origins:
From the Greek *syn* "united" and *siphon* "tube."

Synstemon Botsch. Brassicaceae

Origins:
Greek *syn* "united, together" and *stemon* "stamen."

Synstemonanthus Botsch. Brassicaceae

Origins:
The genus *Synstemon* Botsch. and *anthos* "flower."

Syntherisma Walter Gramineae

Origins:
Greek *syn* and *therizo* "to mow, to cut the hair," *syntherizo* "reap together"; see Thomas Walter (*circa* 1740-1789), *Flora Caroliniana*. 76. 1788.

Synthyris Benth. Scrophulariaceae

Origins:
Greek *syn* "together" and *thyris* "a little door, small door," referring to the valves of the capsule.

Species/Vernacular Names:
S. reniformis (Douglas) Benth.
English: snow queen

Syntriandrium Engl. Menispermaceae

Origins:
Greek *syn* "together," *treis, tris* "three" and *aner, andros* "man, male, stamen."

Syntrichopappus A. Gray Asteraceae

Origins:
Fused bristly pappus, from the Greek *syn* "together," *thrix, trichos* "hair" and *pappos* "fluff, downy appendage."

Species/Vernacular Names:
S. lemmonii (A. Gray) A. Gray
English: Lemmon's syntrichopappus

Syntrinema H. Pfeiffer Cyperaceae

Origins:
Greek *syn* "together," *treis*, *tris* "three" and *nema* "filament, thread."

Synurus Iljin Asteraceae

Origins:
From the Greek *syn* "together" and *oura* "tail."

Syrenia Andrz. ex Besser Brassicaceae

Origins:
Latin *Siren* and Greek *Seiren* "a Siren, a sea nymph," Greek *syrinx* "a pipe, tube," *Syrinx* "a nymph of Arcadia"; see Giovanni Semerano, *Le origini della cultura europea. Dizionari Etimologici. Basi semitiche delle lingue indeuropee. Dizionario della lingua Greca.* 2(1): 257, 281. Leo S. Olschki Editore, Firenze 1994; Helmut Genaust, *Etymologisches Wörterbuch der botanischen Pflanzennamen.* 624. Basel 1996.

Syrenopsis Jaub. & Spach Brassicaceae

Origins:
Resembling *Syrenia*.

Syringa L. Oleaceae

Origins:
Greek *syrinx* "a pipe, tube," referring to the stems; see G. Semerano, *Le origini della cultura europea. Dizionari Etimologici. Basi semitiche delle lingue indeuropee. Dizionario della lingua Greca.* 2(1): 281. 1994; H. Genaust, *Etymologisches Wörterbuch der botanischen Pflanzennamen.* 625. 1996.

Species/Vernacular Names:
S. oblata Lindley
English: early lilac
China: zi ding xiang
S. pinnatifolia Hemsl.

English: pinnate-leaf lilac
China: yu ye ding xiang
S. pubescens Turcz.
English: hairy lilac
China: qiao ling hua
S. villosa Vahl
English: late lilac
China: hong ding xiang
S. yunnanensis Franchet
English: Yunnan lilac
China: yunnan ding xiang

Syringantha Standl. Rubiaceae

Origins:
Greek *syrinx*, *syrigx* "a pipe, tube" and *anthos* "flower."

Syringidium Lindau Acanthaceae

Origins:
Referring to a *syrinx* "a pipe, tube."

Syringodea Hook.f. Iridaceae

Origins:
Like a *syrinx* "a pipe, tube."

Syringodium Kûtz. Cymodoceaceae

Origins:
From the Greek *syrinx* "a pipe, tube," referring to the leaves.

Species/Vernacular Names:
S. isoëtifolium (Aschers.) Dandy (*Cymodocea isoetifolia* Aschers.)
English: Manatee grass
Japan: bôba-ama-mo, shio-nira

Syrrheonema Miers Menispermaceae

Origins:
Greek *rheo* "to flow," *syrrheo* "flow together" and *nema* "filament, thread."

Systeloglossum Schltr. Orchidaceae

Origins:

Greek *systellein* "to draw together" and *glossa* "tongue," indicating the margins of the lip.

Systenotheca Reveal & Hardham Polygonaceae

Origins:

Tapering case, *stenos* "narrow" and *theke* "a box, case," referring to the involucre; see also *Centrostegia* A. Gray ex Benth.

Species/Vernacular Names:

S. vortriedei (Brandegee) Rev. & Hardham

English: Vortriede's spineflower

Syzygiopsis Ducke Sapotaceae

Origins:

Resembling *Syzygium*, Greek *syzygos* "coupled, joined, jointed."

Syzygium Gaertner Myrtaceae

Origins:

Greek *syzygos* "coupled, joined, jointed," from *syn* "together" and *zygon, zygos* "yoke"; the allusion is to the paired branches and leaves or to the calyptrate petals.

Species/Vernacular Names:

S. aqueum (Burm.f.) Alston (*Eugenia aquea* Burm.f.; *Eugenia sylvestris* (non Wight) Moon; *Jambosa aquea* DC.; *Eugenia grandis* Wight; *Syzygium montanum* Thw.)

English: water rose-apple, water apple

Malaya: jambu chili, jambu ayer

S. aromaticum (L.) Merrill & Perry (*Caryophyllus aromaticus* L.; *Myrtus caryophyllus* Spreng.; *Eugenia aromatica* (L.) Baillon; *Jambosa caryophyllus* (Spreng.) Niedz., non *Jambosa aromatica* Miq.; *Eugenia caryophyllus* (Spreng.) Bullock & Harrison)

English: clove, cloves, Zanzibar red head, clove tree

Arabic: qoronfel

French: clou de girofle, giroflier, girofle

Madagascar: jirofo, jorofo, karafoy, girofle

Sri Lanka: karabu neti, karambu

China: ding xiang, ting hsiang, ting tzu hsiang

Central America: clavo, clavo de olor

Mexico: guiña xtilla cica guechi-guiba, guiña, castilla cica queechi-quiba

S. assimile Thw. subsp. *assimile* (*Eugenia assimilis* Duthie)

Sri Lanka: damba

S. balfourii (Baker) Guého & A.J. Scott

Rodrigues Island: bois clou

S. brazzavillense Aubrév. & Pellegr.

Congo: mukizu, nkumbi

S. caryophyllatum (L.) Alston (*Eugenia corymbosa* Lam.; *Syzygium corymbosum* (Lam.) DC.; *Myrtus caryophyllatus* L.)

Sri Lanka: dan, rin-dan

S. congolense Verm. ex Amsh.

Zaire: alika, kotoko

Gabon: ebobogo, etom

S. cordatum Hochst. (*Eugenia cordata* (Hochst.) Laws.)

English: water berry, umdoni tree, water tree, waterwood

Southern Africa: umdoni, waterbessie; umDoni, umJoni (Zulu); umJomi, umSwi (Xhosa); umCozi (Swazi); muhlwa, muthwa, onDoni (Tsonga or Thonga); montlho, motlho (North Sotho); mutu (Venda); muKute, muWototo (Shona)

N. Rhodesia: musombo, musombu

S. cordifolium Walp. subsp. *cordifolium* (*Eugenia cordifolia* Wight; *Calyptranthes cordifolia* Moon; *Eugenia androsaemoides* (non DC.) Beddome)

Sri Lanka: walijambu

S. cumini (L.) Skeels (*Myrtus cumini* L.; *Eugenia cumini* (L.) Druce; *Eugenia jambolana* Lam.; *Jambolifera pedunculata* Gaertn.; *Calyptranthes cumini* Moon; *Syzygium jambolanum* DC.)

English: jambolan, Java plum, jambolan plum, black plum

Madagascar: rotra, robahaza, rotrambazaha, rotravazaha, varotra, jamblon

India: jambuh, naval, nakam, jaman, jambhul, perin-njara

Nepal: jamun

Sri Lanka: madan, mahadan, naval, perunaval

Malaya: jambolan, jiwat, salam, kerian duat

The Philippines: duhat, duat-nasi, lomboi, lomboy, longboi, dunghoi

S. fergusoni Gamble subsp. *fergusoni* (*Eugenia fergusoni* Trimen subsp. *fergusoni*)

Sri Lanka: wel karabu (= wild clove)

S. firmum Thw. (*Eugenia grandis* Wight; *Syzygium montanum* Thw.)

Sri Lanka: waljambu

S. gardneri Thw. (*Eugenia gardneri* Beddome)

Sri Lanka: damba, nir-nawal

S. gerrardii (Harvey ex Hook.f.) Burtt Davy (*Acmena gerrardii* Harvey ex Hook.f.; *Eugenia gerrardii* (Harvey ex Hook.f.) Sim) (after the British William Tyrer Gerrard (d. 1866, Madagascar), botanical collector in Natal and Madagascar in the 1860s, collected with Mark Johnston McKen (1823-1872); see Mary Gunn and Leslie E. Codd, *Botanical Exploration of Southern Africa.* 165-166. Cape Town 1981)

English: forest water berry, forest umdoni, bush waterwood

Southern Africa: boswaterhout, waterhoutboom; motlhatlhanya (North Sotho); umDoniwehlati, umDoni-wehlathi, umDunwana, isiFesane, umDlumuthwa (= bushman food), umDunywana (Zulu); umJomi-wehlathi, umJome-wehlathi, uManzani, uMansane (Xhosa); mutahwi, mutuwi (Venda)

S. grandis (Wight) Walp. (*Eugenia grandis* Wight)

English: sea apple

Malaya: jambu laut, jambu ayer laut, jambu jembah, jembah, kerian acheh, kerian ayer, ubah

S. guineense (Willd.) DC. (*Calyptranthes guineensis* Willd.)

English: water pear, white umdoni

French: kissa d'eau

Southern Africa: waterpeer; umDoni-wamanzi, umDonivungu (Zulu); muthwa (Tsonga); mosane, mmako (Western Transvaal, northern Cape, Botswana); motoya (Kololo); muboa, mubowa, mukute, muKuti dombo, muKuti bango, muRgwi, muSiribango (Shona)

N. Rhodesia: mafuwu

Mali: kokisa

S. guineense (Willd.) DC. var. *guineense* (see R.W.J. Keay, *Trees of Nigeria.* 76-78. Oxford Science Publications. Clarendon Press, Oxford 1989)

Nigeria: adere, isinren, malmo, asafra, keska, kurumibi, busu; malmo (Hausa); asurahi (Fula); mho (Tiv); adere (Yoruba)

Yoruba: aderee, ori ira, isinren

Zaire: alika, kotoko

Gabon: ebobogo, etom

S. guineense (Willd.) DC. var. *macrocarpum* Engl.

Nigeria: malmo (Hausa); asurahi (Fula); mho (Tiv); adere (Yoruba)

Congo: omwaya, omwaga

Mali: kisa

S. huillense (Hiern) Engl.

Southern Africa: muKundangwedzi (Shona)

N. Rhodesia: chimomomo

S. jambos (L.) Alston (*Eugenia jambos* L.; *Eugenia vulgaris* (DC.) Baill.; *Jambosa jambos* (L.) Millsp.; *Jambosa vulgaris* DC.; *Myrtus jambos* Kunth)

English: rose apple

India: gulab jamun

Sri Lanka: veli jambu, seenijambu

Japan: futo-momo

Malaya: jambu mawar, jambu ayer mawar

Hawaii: 'ohi'a loke

S. makul Gaertn. (*Eugenia sylvestris* Moon ex Wight; *Syzygium sylvestre* Thw.)

Sri Lanka: alu-bo

S. malaccense (L.) Merrill & Perry (*Eugenia malaccensis* L.; *Eugenia macrophylla* Lam.; *Jambosa malaccensis* DC.; *Jambosa purpurascens* DC.; *Eugenia purpurea* Roxb.; *Jambosa domestica* Blume; *Calophyllus malaccensis* Stokes)

English: Malay apple, mountain apple

Sri Lanka: peria-jambu, jambu

Japan: mare-atsupuru

Malaya: jambu bol, jambu keling, jambu merah, jambu

Hawaii: 'ohi'a 'ai, 'ohi'a, 'ohi'a 'ai ke'oke'o, 'ohi'a hakea, 'ohi'a kea, 'ohi'a 'ula

S. neesianum Arn. (*Eugenia neesiana* Wight)

Sri Lanka: panu-kera

S. oleosum (F. Muell.) B. Hyland

English: blue lilly pilly

S. operculatum (Roxb.) Niedz. (*Eugenia operculatum* Roxb.; *Syzygium nervosum* DC.; *Calyptranthes caryophyllifolia* Moon; *Cleistocalyx operculatus* (Roxb.) Merr. & Perry)

Nepal: kyaman

Sri Lanka: bata-damba, diya-damba, kobo-mal

S. owariense (P. Beauv.) Benth.

Nigeria: ori-odo (Yoruba)

S. pondoense Engl.

English: Pondo waterwood

South Africa: Pondowaterhout

S. pycnanthum Merrill & Perry (*Eugenia densiflora* (Bl.) Miq.)

English: wild rose, wild rose apple

Malaya: kelat jambu

S. rowlandii Sprague

Cameroon: bibolo afum, bilolo afun, esosi

Congo: eguim

Ivory Coast: guessiguie ako

Nigeria: asafra; ori (Yoruba)

S. rubicundum Wight & Arnott (*Eugenia rubicunda* Wight; *Syzygium lissophyllum* Thw.; *Eugenia lyssophylla* Beddome)

Sri Lanka: pinibaru, maha kuretiye, karaw, damba

S. samarangense (Bl.) Merrill & Perry (*Eugenia javanica* Lam.; *Myrtus samarangensis* Bl.; *Jambosa samarangensis* DC.)

English: Java apple, wax apple

Japan: ô-futo-momo, renbu

Malaya: jambu ayer, jambu ayer rhio

Sri Lanka: pini jambu

S. sandwicensis (A. Gray) Nied. (*Eugenia sandwicensis* A. Gray; *Syzygium oahuense* Degener & E.A. Ludw.)

Hawaii: 'ohi'a ha, ha, kauokahiki, pa'ihi, pa'ihi'ihi

S. staudtii (Engl.) Mildbr.

Zaire: lungu

Gabon: etom

Cameroon: wé

S. umbrosum Thw. (*Eugenia umbrosa* (non Berg) Beddome; *Eugenia subavenis* (non Berg) Duthie)

Sri Lanka: vali damba, hin damba, naval

S. zeylanicum DC. (*Eugenia spicata* Lam.; *Eugenia zeylanica* Wight; *Acmena zeylanica* Thw.; *Myrtus zeylanica* L.)

Sri Lanka: yakul maran, marung, mariangi, maranda

T

Tabacum Gilib. Solanaceae

Origins:
Tobacco, see *Nicotiana* L.

Tabacus Moench Solanaceae

Origins:
Tobacco, see *Nicotiana* L.

Tabebuia Gomes ex DC. Bignoniaceae

Origins:
Tabebuia, tabebuya or *taiaveruia*, native Brazilian names.

Species/Vernacular Names:
T. sp.

Peru: paliperro, coda, ipe, lapacho, papelillo caspi, tahuari amarillo, tahuari negro, tamuratuira

Brazil: ipê-roxo, ipê-amarelo, ipê prêto, ipeúva roxa, peúva, ipê tabaco, peúva roxa, ipê mirim

Paraguay: lapacho, Tajy pytã

Mexico: palo cortés, palo serrano

T. argentea Britt. (*Tecoma argentea* Bureau & K. Schum.)

English: tree of gold, silver trumpet tree, Paraguayan silver trumpet tree

T. billbergii (Bureau & Schumann) Standley subsp. *ampla* A. Gentry (*Tecoma billbergii* Bureau & Schumann)

Peru: coraliba

T. capitata (Bureau & Schumann) Sandwith (*Tecoma capitata* Bureau & Schumann)

Peru: asta venado amarillo, tahuari, tahuari colorado

T. chrysantha (Jacq.) Nicholson

Bolivia: tajibo amarillo

Mexico: amapa, amapa prieta, lombricillo, verdecillo, hahuaché

T. dubia (C. Wright) Britt. ex Seib. (*Tabebuia crassifolia* Britt.)

Tropical America: roble negro, cucharillo

T. haemantha DC.

Tropical America: roble cimarrón

T. heptaphylla (Vell.) Toledo

Bolivia: tajibo

T. heterophylla Britt. (*Tabebuia triphylla* DC.)

English: pink trumpet tree, white cedar, pink manjack, pink cedar

Tropical America: roble blanco

T. impetiginosa (C. Martius ex DC.) Standley

Bolivia: tajibo, tajibo morado

Peru: guayacan

Paraguay: Tajy pytã

T. incana A. Gentry

Peru: tahuari, tarota

T. lapacho (Schumann) Sandw.

Bolivia: tajibo

T. nodosa (Griseb.) Griseb.

Latin America: Martin Gil, palo cruz, uinaj, toroguatay

Peru: caspi, palo cruz

T. ochracea (Cham.) Standley subsp. *ochracea* (*Tabebuia heteropoda* (A. DC.) Sandwith; *Tecoma grandiceps* Kränzl.; *Tecoma heteropoda* A. DC.; *Tecoma ochracea* Chamisso)

Peru: tahuari, tahuari rojo, papelillo caspi, guayacán

T. pallida (Lindl.) Miers

Tropical America: white cedar, Cuban pink trumpet tree, Cuban pink trumpet

T. rosea (Bertol.) DC. (*Tabebuia mexicana* (Martius ex DC.) Hemsl.; *Tabebuia punctatissima* (Kränzl.) Standl.)

English: trumpet tree

Tropical America: pink poui, rosy trumpet tree, roble morado, matilisguate

Peru: amapá, apamate, maquilishuat, palo blanco, roble blanco, roble de savana, roble maguiligua

T. roseo-alba (Ridley) Sandw.

Bolivia: tajibo

T. serratifolia (Vahl) Nichols.

Tropical America: yellow poui, guayacán, guayacán polvillo, curarire, asta de venado, chonta

Bolivia: tajibo

Peru: araguaney, chonta, pau de arco

Taberna Miers Apocynaceae

Origins:

After the German herbalist Jakob Theodor or Jacobus Theodorus von Bergzaben, 1520-1590; see also *Tabernaemontana* L.

Tabernaemontana L. Apocynaceae

Origins:

For the German herbalist Jakob Theodor (Jacobus Theodorus) von Bergzaben (Latinized Tabernaemontanus, "tavern in the mountain"), 1520-1590, he was the personal physician to the Count of the Palatine at Heidelberg, Germany; he published *Eicones plantarum* ... quae partim Germania sponte producit; partim ab exteris regionibus allata in Germania plantantur ... Francoforti ad Maenium [Frankfurt] 1590 and *Neuw Kreuterbuch* ... Franckfurt am Mayn 1588-1591; see John Gerard (1545-1612), *Herball*, or Generall Historie of Plantes. London 1597; Eleanour Sinclair Rohde, *The Old English Herbals*. London 1922; Asher Rare Books & Antiquariaat Forum, *Catalogue Natural History*. item no. 149. The Netherlands 1998.

Species/Vernacular Names:

T. sp.

Peru: lobo sanango

Mexico: mahuma, mani, mañi, torotexis, yaga nichi

T. angolensis Stapf (*Conopharyngia stapfiana* (Britten) Stapf; *Tabernaemontana stapfiana* Britten)

English: soccer ball fruit

Southern Africa: muchenga, muChanga, muChesa (Shona)

T. divaricata (L.) R. Brown ex Roemer & Schultes (*Tabernaemontana coronaria* (Jacq.) Willd.)

English: moon beam, crepe jasmine, crepe gardenia, pinwheel flower, East Indian rosebay, East India rosebay, Adam's apple, Nero's crown, coffee rose, broad-leaved rosebay, crape gardenia, crape jasmine, Ceylon jasmine

Japan: Indo-sôkei

China: gou ya hua

Malaya: susun kelapa, susoh ayam

The Philippines: pandakaking-tsina, rosa de hielo

T. elegans Stapf (*Conopharyngia elegans* (Stapf) Stapf)

English: toad tree

Southern Africa: paddaboom; umKhadlu, umKhalu, umKhahlwana, umKahlwana, umKahlwane, iNomfi (Zulu); kahlwane (Tsonga); muhatu (Venda); ruChene (Shona)

T. macrocalyx Müll. Arg.

Brazil (Amazonas): akia hi, asokoma hi, akiama hanaki, axokama hi

T. mauritiana Poiret

La Réunion Island: bois de lait

T. orientalis var. *angustisepala* Benth. (*Ervatamia angustisepala* (Benth.) Domin)

English: banana bush, windmill bush

T. pachysiphon Stapf (*Tabernaemontana holstii* Schum.; *Conopharyngia holstii* (K. Schum.) Stapf)

Nigeria: dodo (Yoruba); ibu (Edo); pete-pete (Igbo)

Yoruba: dodo, abo dodo

T. pandacaqui Lam.

China: ping mai gou ya hua

The Philippines: pandakaki-puti, busbusilak, halibutbut

T. ventricosa Hochst. ex A. DC. (*Conopharyngia ventricosa* (Hochst. ex DC.) Stapf; *Conopharyngia usambarensis* Stapf)

English: forest toad tree, nomfi tree, fever tree, swollen-on-one-side tabernaemontana

Southern Africa: bospaddaboom, koorsboom; ruChene (Shona); uKhamamasane, inKamamasana, uNomfi (= birdlime), iNomfi-yehlathi, umCikimanzi, umKhahlu (Zulu)

Tabernanthe Baillon Apocynaceae

Origins:

Latin *taberna* "hut, stall" or the genus *Tabernaemontana* plus Greek *anthos* "flower."

Species/Vernacular Names:

T. iboga Baillon

Congo: liboka

Tabernaria Raf. Apocynaceae

Origins:

Based on the genus *Tabernaemontana*; see C.S. Rafinesque, *Flora Telluriana*. 1: 17, 86. 1837.

Tacazzea Decne. Asclepiadaceae (Periplocaceae)

Origins:

An Ethiopian river, Takazze, between Semien and Tigre; see J.B. Gillett, "W.G. Schimper's botanical collecting localities in Ethiopia." *Kew Bulletin*. 27(1): 127. 1972.

Species/Vernacular Names:

T. apiculata Oliv. (*Tacazzea apiculata* Oliv. var. *benedicta* Scott Elliot; *Tacazzea apiculata* Oliv. var. *glabra* K. Schum.; *Tacazzea bagshawei* S. Moore; *Tacazzea bagshawei* S. Moore var. *occidentalis* Norman; *Tacazzea barteri* Baill.; *Tacazzea brazzaeana* Baill.; *Tacazzea kirkii* N.E. Br.; *Tacazzea laxiflora* Engl.; *Tacazzea nigritana* N.E. Br.; *Tacazzea stipularis* N.E. Br.; *Tacazzea thollonii* Baill.; *Tacazzea verticillata* K. Schum.; *Tacazzea welwitschii* Baill.)

Southern Africa: isimondane (Zulu)

Tacca Forst. & Forst.f. Taccaceae (Liliidae, Haemodorales)

Origins:

From *taka laoet*, the Indonesian name for *Tacca leontopetaloides*.

Species/Vernacular Names:

T. sp.

English: bat flower

T. leontopetaloides (L.) Kuntze

English: Polynesian arrowroot, East Indian arrowroot, Tahiti arrowroot

Hawaii: pia

Yoruba: olopapaniraga

Taccarum Brongn. ex Schott Araceae

Origins:

From *Tacca* plus the genus *Arum* L.; see D.H. Nicolson, "Derivation of aroid generic names." *Aroideana*. 10: 15-25. 1988.

Tachia Aublet Gentianaceae

Origins:

Presumably from the Greek *tachys* "rapid, sudden," referring to the growth.

Species/Vernacular Names:

T. guianensis Aublet

Brazil: tingua aba, caferona, jacaré-açu, jacurmarua, falso café, guassia, guassia do Pará, guassia amargosa, guassia do Amazonas, tupurapós, tuperaba

Peru: etosima-ey, sacha café

Tachiadenus Griseb. Gentianaceae

Origins:

Presumably from the genus *Tachia* Aubl. with the Greek *aden* "gland."

Tachibota Aubl. Chrysobalanaceae

Origins:

Presumably from the Greek *tachys* "rapid, sudden" and *boton* "beast, grazing beasts," referring to the nature of the plants, also used as fodder.

Tachigali Aublet Caesalpiniaceae

Origins:

Vernacular name for this ant-loving genus, see Luíz Caldas Tibiriçá, *Dicionário Guarani-Português*. 159. [vernacular name *tahîi* for ant.] Traço Editora, Liberdade 1989; Luíz Caldas Tibiriçá, *Dicionário Tupi-Português*. Traço Editora, Liberdade 1984; Timothy J. Killeen, Emilia Garcfa E. and Stephan G. Beck, eds., *Guía de Arboles de Bolivia*. 418. Herbario Nacional de Bolivia and Missouri Botanical Garden 1993; some suggest from the Greek *tachys* "rapid" and *gala, galaktos* "milk," fodder for cattle and goats.

Tachigalia Juss. Caesalpiniaceae

Origins:

See *Tachigali* Aublet.

Tacinga Britton & Rose Cactaceae

Origins:

An anagram of the Portuguese *caatinga* or *catinga* "scrub."

Tacoanthus Baillon Acanthaceae

Origins:

Possibly from the Greek *tachos* "speed" or Spanish *taco* "plug" and *anthos* "flower."

Taeckholmia Boulos Asteraceae

Origins:

For the Swedish botanist Vivi Täckholm (*née* Laurent), 1898-1978, in Egypt, botanical collector, professor of botany at the University of Cairo (Faculty of Science). His writings include *Bibliographical Notes to the Flora of Egypt*. [Stockholm 1932] and *Student's Flora of Egypt*. Cairo 1974, with her husband Gunnar Vilhelm Täckholm (1891-1933) and Mohammed Drar wrote *Flora of Egypt*. Cairo 1941-1969; see T.W. Bossert, *Biographical Dictionary of Botanists Represented in the Hunt Institute Portrait*

Collection. 392. 1972; R. Zander, F. Encke, G. Buchheim and S. Seybold, *Handwörterbuch der Pflanzennamen*. 14. Aufl. Stuttgart 1993; Stafleu and Cowan, *Taxonomic Literature*. 6: 147-149. 1986; Loutfy Boulos and M. Nabil el-Hadidi, *The Weed Flora of Egypt*. The American University in Cairo Press. Cairo 1984; John H. Barnhart, *Biographical Notes upon Botanists*. 3: 354. 1965.

Taeniandra Bremek. Acanthaceae

Origins:

From the Greek *tainia* "a band, ribbon" and *aner, andros* "man, male."

Taenianthera Burret Palmae

Origins:

Greek *tainia* "a band, ribbon" and *anthera* "anther."

Taeniatherum Nevski Gramineae

Origins:

Greek *tainia* "a band, ribbon" and *ather* "an awn," the awn of the lemma is flat, alluding to the ribboned awn.

Species/Vernacular Names:

T. caput-medusae (L.) Nevski (*Elymus caput-medusae* L.)

English: Medusa's head

Taenidia (Torrey & A. Gray) Drude Umbelliferae

Origins:

From *tainidion* "small band," the diminutive of the Greek *tainia* "a band, fillet."

Taeniochlaena Hook.f. Connaraceae

Origins:

From the Greek *tainia* and *chlaena, chlaenion* "a cloak."

Taeniophyllum Blume Orchidaceae

Origins:

From the Greek *tainia* "a fillet," Latin *taenia, ae* "band, ribbon" and *phyllon* "leaf," these epiphytic plants are leafless,

the roots form a low mound; see Karl Ludwig von Blume, *Bijdragen tot de flora van Nederlandsch Indië*. 335, t. 70. Batavia (Sep.-Dec.) 1825.

Species/Vernacular Names:

T. glandulosum Blume (*Taeniophyllum aphyllum* (Makino) Makino)

Japan: kumo-ran

Taeniopleurum J.M. Coulter & Rose Umbelliferae

Origins:

From the Greek *tainia* "a fillet" and *pleura, pleuro, pleuron* "side, rib."

Taeniopsis J. Sm. Vittariaceae (Adiantaceae)

Origins:

Greek *tainia* "a fillet" and *opsis* "resemblance."

Taeniopteris Hook. Vittariaceae (Adiantaceae)

Origins:

From the Greek *tainia* "a fillet" and *pteris* "fern."

Taeniorrhiza Summerh. Orchidaceae

Origins:

From the Greek *tainia* "a fillet" and *rhiza* "root," roots like ribbons.

Taeniosapium Müll. Arg. Euphorbiaceae

Origins:

Greek *tainia* "a fillet" and the genus *Sapium* P. Browne.

Taenitis Willdenow ex Schkuhr Pteridaceae (Adiantaceae, Taenitidaceae)

Origins:

From the Greek *tainia* "a fillet, ribbon," referring to the linear sori.

Tagetes L. Asteraceae

Origins:

Named after Tages, an Etruscan deity, grandson of Jupiter; see Cicero, *De Divinatione*, ii. 23; Carl Linnaeus, *Species*

Plantarum. 887. 1753 and *Genera Plantarum*. Ed. 5. 378. 1754.

Species/Vernacular Names:

T. sp.

Peru: cravo de defunto, supi quehua, kita huacatai

Mexico: guia nagati, lipa-caca-no

Yoruba: eru inabo

T. erecta L.

English: marigold, African marigold, Aztec marigold, big marigold, French marigold

Peru: aya sisa, rosario, rosa sisa

Mexico: cempasóchil, cempoal, cempoalxochitl (= flor de veinte pétalos = flower with twenty petals), mcuilxochitl, cimpual, flor de muerto, picoa, musá, nulibé, tiringuini, xkanlol, guie nagati

Guatemala: flor de muerto, ix tupuj

India: genda, makhmal, gul-jafari, guljajari, rojiacha-pul, banti, zanduga, turukkasamandi

Japan: senju-giku

China: wan shou ju

The Philippines: amarillo, ahito

Indonesia: marigold Afrika

Bali: bungan mitir, gemitir

T. filifolia Lagasca

English: Irish lace

Spanish: anicillo, tuna anís

Mexico: anisillo, curucumín, flor de Santa María, Santa María, hierba anís, pericón, periquillo, manzanilla

T. lucida Cav.

English: sweet mace, sweet-scented marigold, sweet-scented Mexican marigold

Mexico: pericón, yauhtli (= niebla, nube = cloud), guie laga zaa, quie laga zaa

T. minuta L.

English: khaki bush, khaki weed, master-John-Henry, muster-John-Henry, Mexican marigold, stinking Roger, tall khaki weed, stinkweed

Peru: huacatai, huacatay, huatacay

East Africa: ang'we, anyach, bhangi, mubangi, muvangi, nyanjaga, nyanjagra, omotioku, omubazi gwemhazi

Southern Africa: Africander bossie, kakiebos, kleinafrikander, langkakiebos, stinkbos, stinkkhakibos, Transvaalsekakiebos; jeremane (Sotho); mbanje (Ndebele); mbanje (Shona)

Hawaii: 'okole'oi'oi

T. patula L.

English: French marigold

Latin America: flor de muerto

Mexico: clemole, clemolitos, flor de muerto, iscoque, pastora, pastoral, pastorcita, tlemole

Japan: kô-ô-sô

China: kong que cao

T. tenuifolia Cav.

English: signet marigold, striped Mexican marigold

Taguaria Raf. Loranthaceae

Origins:

See Constantine Samuel Rafinesque, *Sylva Telluriana*. 125. 1838; E.D. Merrill, *Index rafinesquianus*. 113-114. 1949.

Tahitia Burret Tiliaceae

Origins:

From Tahiti.

Tainia Blume Orchidaceae

Origins:

Greek *tainia* "a fillet," Latin *taenia, ae* "band, ribbon," referring to the leaves or to the sepals and petals, or to the elevated keels on the lip; see Karl Ludwig von Blume, *Bijdragen tot de flora van Nederlandsch Indië*. 354. Batavia 1825.

Species/Vernacular Names:

T. laxiflora Makino

Japan: hime-token-ran

Tainionema Schltr. Asclepiadaceae

Origins:

From the Greek *tainia* plus *nema* "filament, thread."

Tainiopsis Schltr. Orchidaceae

Origins:

The orchid genus *Tainia* Blume plus *opsis* "resembling."

Taiwania Hayata Taxodiaceae

Origins:

From Taiwan.

Takhtajania Baranova & J.-F. Leroy Winteraceae (Takhtajaniaceae)

Origins:

For the Armenian botanist Armen Leonovich Takhtajan, b. 1910, plant taxonomist, Komarov Botanical Institute; see T.W. Bossert, *Biographical Dictionary of Botanists Represented in the Hunt Institute Portrait Collection.* 393. 1972; Ida Kaplan Langman, *A Selected Guide to the Literature on the Flowering Plants of Mexico.* 729. University of Pennsylvania Press, Philadelphia 1964; Stafleu and Cowan, *Taxonomic Literature.* 6: 155. 1986.

Takhtajaniantha Nazarova Asteraceae

Origins:

For the Armenian botanist Armen Leonovich Takhtajan, b. 1910, plant taxonomist.

Takhtajanianthus De Asteraceae

Origins:

For the Armenian botanist Armen Leonovich Takhtajan, b. 1910, plant taxonomist.

Talasium Spreng. Gramineae

Origins:

Presumably from the Greek *talasios, talaseios* "of wool-spinning," *talasia* "wool-spinning."

Talauma Juss. Magnoliaceae

Origins:

Perhaps a native name for a West Indian species, or from the Greek *talao, tlao* "hold out," referring to the stamens and pistils; see David Hunt, ed., *Magnolias and Their Allies.* Proceedings of an International Symposium, Royal Holloway, University of London, Egham, Surrey, U.K., 12-13 April 1996. International Dendrology Society and The Magnolia Society. 1998.

Species/Vernacular Names:
T. sp.

Latin America: granadilla

Talbotia Balf. Velloziaceae

Origins:

After the English mathematician William Henry Fox Talbot, 1800-1877 (Wiltshire), linguist, botanist and plant collector in Ionian Islands, in 1831 a Fellow of the Linnean Society, 1831 Fellow of the Royal Society, translator of Assyrian texts, a pioneer of photography. His writings include *English Etymologies.* London 1847, *Sun Pictures in Scotland.* London 1845, *On the Eastern Origin of the Name and Worship of Dionysus.* [London 1866] and *The Pencil of Nature.* London 1844. He was the son of Lady Elisabeth Theresa Fox-Strangways. See *Memoir and Correspondence of ... Sir J.E. Smith ...* Edited by Lady Pleasance Smith. London 1832; Reese V. Jenkins, in *Dictionary of Scientific Biography* (Editor in Chief Charles Coulston Gillispie.) 13: 237-239. New York 1981.

Talbotia S. Moore Acanthaceae

Origins:

For the British plant collectors in Nigeria Percy Amaury Talbot (1877-1945) and his wife, Dorothy A. Talbot (1871-1916, d. Degama, Nigeria); see Percy Amaury Talbot, *In the Shadow of the Bush.* London 1912; Dorothy Amaury Talbot, *Women's Mysteries of a Primitive People. The Ibibios of Southern Nigeria.* London 1915; I.H. Vegter, *Index Herbariorum.* Part II (7), *Collectors T-Z.* Regnum Vegetabile vol. 117. 1988; F.N. Hepper and Fiona Neate, *Plant Collectors in West Africa.* 77-78. 1971.

Talbotiella Bak.f. Caesalpiniaceae

Origins:

For the botanists Percy Amaury Talbot (1877-1945) and his wife, Dorothy Amaury Talbot (1871-1916, Nigeria), plant collectors in Nigeria, their collection made between 1909-1912 includes over 1,000 species and varieties, 1912-1913 a further collection was made in the Eket District; 1904-1905 P.A. Talbot and Captain G.B. Gosling collected from Ibi to Lake Chad via Bauchi. See Alfred Barton Rendle (1865-1938) et al., "Catalogue of the Plants collected by Mr. and Mrs. P.A. Talbot in the Oban District of South Nigeria." *British Museum Trustees, Natural History.* London 1913; Edmund Gilbert Baker (1864-1949) (and others), "Plants from the Eket District, S. Nigeria." in *Journ. Bot.* 1: 25. 1914; J. Lanjouw and F.A. Stafleu, *Index Herbariorum.* Part II, *Collectors A-D.* Regnum Vegetabile vol. 2. 1954.

Talbotiopsis L.B. Sm. Velloziaceae

Origins:

Resembling the genus *Talbotia* Balf.

Talinaria Brandegee Portulacaceae

Origins:
Referring to the genus *Talinum*.

Talinella Baill. Portulacaceae

Origins:
The diminutive of the genus *Talinum*.

Talinopsis A. Gray Portulacaceae

Origins:
Resembling the genus *Talinum*.

Talinum Adans. Portulacaceae

Origins:
Derivation of the name is obscure, perhaps from the Greek *thaleia* "full of bloom, blooming, luxuriant," *thalia* "bloom," or from *tali* a native African (Senegal) name for one species, *Erythrophleum guineense* G. Don.

Species/Vernacular Names:
T. sp.

Spanish: lengua de vaca

Yoruba: gbure osun

T. paniculatum (Jacq.) Gaertn. (*Portulaca paniculata* Jacq.)

English: jewels-of-Opar, fame flower

Japan: haze-ran

Okinawa: ichijûrû

Vietnam: tho cao ly sam, man sam dam

T. triangulare (Jacq.) Willd.

Yoruba: alawere, gbure, ajigborere

Talisia Aubl. Sapindaceae

Origins:
A native name in Guiana, or from the Greek *thaleo* "to grow green, flourish," referring to the leaves.

Species/Vernacular Names:
T. cerasina (Bentham) Radlkofer (*Sapindus cerasinus* Bentham; *Sapindus oblongus* Bentham)

Peru: juapina, pitomba

T. peruviana Standley

Peru: siuca sanango, sinca sanango

Talisiopsis Radlk. Sapindaceae

Origins:
Referring to the genus *Talisia* Aublet.

Talpinaria Karsten Orchidaceae

Origins:
Latin *talpa* "mole," *talpinus* "mole-like," referring to the lobes of the lip.

Tamaricaria Qaiser & Ali Tamaricaceae

Origins:
Referring to *Tamarix* L. and *Myricaria*.

Tamarindus L. Caesalpiniaceae

Origins:
From the Arabic *tamar* "date" and *hindi* "Indian"; see Carl Linnaeus, *Species Plantarum*. 34. 1753 and *Genera Plantarum*. Ed. 5. 20. 1754.

Species/Vernacular Names:
T. indica L.

English: tamarind, tamarind tree

Arabic: tamr hindi

W. Africa: domi, tombi

Yoruba: ajagbon

Nigeria: ezi, icheko-oyinbo, tamsugu, udeguegor, ajagbon, tsamiya; tsamiya (Hausa); jatami (Fula); tamsugu (Kanuri); tamr hindi (Shuwa Arabic); darachi (Nupe); ajagbon (Yoruba); icheku oyibo (Igbo)

Tanzania: nsisi, kirarahe, kitune

Madagascar: kily, dilo, diro, fampivalanana, monte, madilo, voamadilo, tamarinier

Southern Africa: muSeka, muSika (Shona)

Congo: tomi

Mali: kataanga, tombi

Mexico: pachuhuk, tamarindo

The Philippines: tamarindo, salamagi, salomagi, salumagi, salunagi, sambag, sambagi, salomangue, sambak, sambalagi, kambalagi, kalamagi, asam, sampalok

Japan: tama-rindo

Malaya: asam, asam jawa, chelagi

Nepal: imli

China: suan jiao, an mi lo

India: tintiri, tintidi, tintili, tintil, tentul, tentuli, tintrani, tintrini, amlavrasksha, amilam, amlam, amla, amli, ambli, anbli, amlika, amlica, ambia, imli, imbli, tamar-i-hind, chinch, chintz, chinta-pandu, puliyan, puli, puliyam-palam, puliyam-pazham, hunasehannu, hunisay, siyambula, siyembela, magi

Tamarix L. Tamaricaceae

Origins:

Latin *tamarix, icis* (A. Cornelius Celsus, L. Junius Moderatus Columella) for a tamarisk, tamarisk-shrub, called also *tamarice, es* (Plinius) and *tamariscus* (Palladius Rutilius Taurus Aemilianus), *Tamarici, orum* "a people of *Hispania Tarracconensis*, on the river Tamaris" (Plinius, Pomponius Mela); see Carl Linnaeus, *Species Plantarum*. 270. 1753 and *Genera Plantarum*. Ed. 5. 131. 1754; Manlio Cortelazzo & Paolo Zolli, *Dizionario etimologico della lingua italiana*. 5: 1310, 1311. Zanichelli, Bologna 1988; Ernest Weekley, *An Etymological Dictionary of Modern English*. 2: 1469. New York 1967.

Species/Vernacular Names:

T. aphylla (L.) Karsten (*Thuja aphylla* L.)

English: Athel pine, Athel tamarix, Athel tree, Athel

North East Africa: tarfa, athl

Arabic: etel

T. chinensis Lour. (*Tamarix elegans* Spach)

English: Chinese tree, Chinese tamarisk

China: cheng liu, chi ai

T. gallica L. (*Tamarix anglica* Webb)

English: French tree, manna plant, English tree

India: jhavuca, koan, rukh, leinya, ghazlei, pilchi, lei, lai, jhau, yelta, rgelta, jhau

Arabic: tarfa

T. hispida Willd.

English: Kashgar tree

T. juniperina Bunge

English: tamarisk

Japan: sakki-gyo-ryû (ryu = liu = *Salix*)

T. mannifera Ehrenb.

English: manna tamarisk

T. usneoides E. Meyer ex Bunge (*Tamarix austro-africana* Schinz; *Tamarix articulata* Harv.) (the specific name means "like *Usnea*," a genus of lichens)

English: tamarisk

Southern Africa: abiekwasgeelhout, dawee, daweep, dabee, abiqua (= robbers); omunguati (Herero); daweb (Nama: southern southwest Africa)

Namibia: dawip (dawep), tamarisk, abikwa tree; omungwati (Herero); daweb (Nama/Damara)

Tamaulipa R.M. King & H. Robinson Asteraceae

Origins:
Tamaulipas, Mexico.

Tamilnadia Tirveng. & Sastre Rubiaceae

Origins:
From Tamil Nadu State, India.

Tamus L. Dioscoreaceae

Origins:
Latin *taminia uva* "a kind of wild grape"; the vine on which it grew was called *tamnus*; see Helmut Genaust, *Etymologisches Wörterbuch der botanischen Pflanzennamen*. 628. Basel 1996.

Species/Vernacular Names:
T. communis L.

English: black bryony, murraim berries, edible bryony, yam

Portuguese: norça

Tanacetum L. Asteraceae

Origins:
Greek *athanasia* "immortality," the Medieval Latin *tanazita*; see Carl Linnaeus, *Species Plantarum*. 843. 1753 and *Genera Plantarum*. Ed. 5. 366. 1754; H. Genaust, *Etymologisches Wörterbuch der botanischen Pflanzennamen*. 628-629. 1996.

Species/Vernacular Names:
T. camphoratum Less.

English: dune tansy

T. parthenium (L.) Sch.Bip. (*Matricaria parthenium* L.; *Chrysanthemum parthenium* (L.) Benth.)

English: feverfew

Latin America: altamiza, artemisa, chusita, margarita, Santa María

T. vulgare L.

English: tansy

Tanaecium Sw. Bignoniaceae

Origins:

Probably from the Greek *tanaekes* "with long point, tall," referring to the appearance of the plants.

Tanakaea Franchet & Sav. Saxifragaceae

Origins:

For the Japanese botanist Yoshio Tanaka, 1838-1916, entomologist, author of *Useful Plants of Japan*. Tokyo 1895; see John H. Barnhart, *Biographical Notes upon Botanists*. 3: 359. Boston 1965; T.W. Bossert, *Biographical Dictionary of Botanists Represented in the Hunt Institute Portrait Collection*. 393. 1972; E.M. Tucker, *Catalogue of the Library of the Arnold Arboretum of Harvard University*. 1917-1933; R. Zander, F. Encke, G. Buchheim and S. Seybold, *Handwörterbuch der Pflanzennamen*. 14. Aufl. 788. Stuttgart 1993; F. Boerner & G. Kunkel, *Taschenwörterbuch der botanischen Pflanzennamen*. 4. Aufl. 176. Berlin & Hamburg 1989.

Tanaosolen N.E. Br. Iridaceae

Origins:

From the Greek *tanaos* "tall" and *solen* "a tube."

Tandonia Baill. Euphorbiaceae

Origins:

From *Tannodia* Baillon; see Baillon, in *Adansonia*. 1: 184. Feb 1861.

Tankervillia Link Orchidaceae

Origins:

In honor of the English plant collector Lady Tankerville, d. 1836, wife of Charles, Earl of Tankerville.

Tannodia Baillon Euphorbiaceae

Origins:

An anagram from *Tandonia* Baill., substitute name for *Tandonia* Baillon; dedicated to the French botanist Christian Horace Bénédict Alfred Moquin-Tandon, 1804-1863, naturalist, from 1834 to 1853 Director of the Botanic Garden of Toulouse, one of the founders of the Société Botanique de France, professor of botany at the Faculté de Médecine at Paris, expert on the Provençal language, contributed to A.P. de Candolle, *Prodromus* (Phytolaccaceae, Salsolaceae, Basellaceae, Amaranthaceae). His publications include *Histoire naturelle des Mollusques ... de France*. Paris 1855, *Eléments de tératologie végetale*. Paris 1841, *Lettres inédites de Moquin-Tandon à Auguste de Saint-Hilaire*. Clermont l'Hérault 1893, *Eléments de Zoologie médicale*. Paris 1860 [1859], *Carya Magalonensis, ou Noyer de Maguelonne*. Montpellier 1844 and *Eléments de botanique médicale*. Paris 1861; see John H. Barnhart, *Biographical Notes upon Botanists*. 2: 510. 1965; Stafleu and Cowan, *Taxonomic Literature*. 3: 573-575. 1981; Ida Kaplan Langman, *A Selected Guide to the Literature on the Flowering Plants of Mexico*. Philadelphia 1964; R. Zander, F. Encke, G. Buchheim and S. Seybold, *Handwörterbuch der Pflanzennamen*. 14. Aufl. Stuttgart 1993; Ethelyn Maria Tucker, *Catalogue of the Library of the Arnold Arboretum of Harvard University*. Cambridge, Massachusetts 1917-1933; A. Lasègue, *Musée botanique de Benjamin Delessert*. Paris 1845; T.W. Bossert, *Biographical Dictionary of Botanists Represented in the Hunt Institute Portrait Collection*. 273. 1972; Elmer Drew Merrill, *Contr. U.S. Natl. Herb*. 30(1): 220. 1947; H. Baillon, in *Adansonia*. 1: 251. Apr. 1861 and 5: 149-175. 1865; H. Genaust, *Etymologisches Wörterbuch der botanischen Pflanzennamen*. 629. [from the Greek *thamnos* "shrub"] 1996.

Species/Vernacular Names:

T. swynnertonii (S. Moore) Prain (*Croton swynnertonii* S. Moore)

English: tannodia

Southern Africa: muNyabotabota (Shona)

Tanquana H.E.K. Hartmann & Liede Aizoaceae

Origins:
From the Tanqua Karoo.

Tanulepis Balf.f. Asclepiadaceae (Periplocaceae)

Origins:
Greek *tanyo* or *teino* "stretch, stretched" and *lepis* "scale."

Species/Vernacular Names:

T. sphenophylla Balf.f.

Rodrigues Island: liane à cornes

Tapeinanthus Boiss. ex Benth. Labiatae

Origins:

From the Greek *tapeinos* "low" and *anthos* "flower"; see also *Thuspeinanta* T. Durand.

Tapeinanthus Herbert Amaryllidaceae (Liliaceae)

Origins:

Greek *tapeinos* "low" and *anthos* "flower"; see also *Braxireon* Raf.

Tapeinia Comm. ex Juss. Iridaceae

Origins:

From the Greek *tapeinos* "low, humble, depressed."

Tapeinidium (Presl) C. Chr. Dennstaedtiaceae (Lindsaeoideae)

Origins:

Greek *tapeinos* "low, humble, depressed."

Species/Vernacular Names:

T. pinnatum (Cav.) C. Chr. (*Davallia pinnata* Cav.)

Japan: goza-dake-shida (= Gozadake fern)

Tapeinochilos Miq. Zingiberaceae (Costaceae)

Origins:

Greek *tapeinos* "low, low-lying, depressed, base" and *cheilos* "lip," referring to the short labellum.

Tapeinoglossum Schltr. Orchidaceae

Origins:

From the Greek *tapeinos* and *glossa* "tongue," referring to the size of the lip.

Tapeinosperma Hook.f. Myrsinaceae

Origins:

Greek *tapeinos* plus *sperma* "seed," the seeds are small.

Tapeinostelma Schltr. Asclepiadaceae

Origins:

Greek *tapeinos* with *stelma, stelmatos* "a girdle, belt, crown, garland, wreath"; see also *Brachystelma* R. Br.

Tapeinostemon Benth. Gentianaceae

Origins:

From the Greek *tapeinos* and *stemon* "stamen, thread."

Tapeinotes DC. Gesneriaceae

Origins:

Greek *tapeinotes* "lowness, dejection."

Tapesia C.F. Gaertner Rubiaceae

Origins:

From the Greek *tapes* "carpet, rug."

Taphrospermum C.A. Meyer Brassicaceae

Origins:

Greek *taphros* "a trench, ditch" and *sperma* "seed."

Tapina Mart. Gesneriaceae

Origins:

From the Greek *tapeinos* "low, humble, depressed, base."

Tapinanthus (Blume) Reichb. Loranthaceae

Origins:

Greek *tapeinos* "low, humble, depressed" and *anthos* "a flower," possibly referring to the size of the flowers.

Species/Vernacular Names:

T. sp.

Yoruba: afomo, ose, etu, afomo olo boruje, ojele

T. oleifolius (Wendl.) Danser (*Lichtensteinia oleifolia* Wendl.; *Loranthus meyeri* Presl; *Loranthus meyeri* Presl var. *inachabensis* Engl.; *Loranthus namaquensis* Harv.; *Loranthus namaquensis* Harv. var. *ligustrifolius* Engl.; *Loranthus oleifolius* (Wendl.) Cham. & Schltdl.; *Loranthus oleifolius* (Wendl.) Cham. & Schltdl. var. *luteus* Neusser;

Loranthus speciosus F.G. Dietr.; *Tapinanthus namaquensis* (Harv.) Tieghem)

English: mistletoe

Namibia: vuurhoutjies (= matches), voëlent (Afrikaans); otjiraura (Herero)

German: streichholzpflanze

T. kraussianus (Meisn.) Tieghem subsp. *kraussianus* (*Loranthus kraussianus* Meisn.; *Loranthus bulawayensis* Engl.; *Tapinanthus kraussianus* (Meisn.) Tieghem)

English: mistletoe

Southern Africa: gomarangwa, gomarara (Shona)

T. terminaliae (Engl. & Gilg) Danser (*Loranthus terminaliae* Engl. & Gilg; *Loranthus villosiflorus* Engl.)

English: mistletoe

Tapinopentas Bremek. Rubiaceae

Origins:

Greek *tapeinos* "low, humble, depressed" and the genus *Pentas* Benth.

Tapinostemma Tieghem Loranthaceae

Origins:

Greek *tapeinos* "low, humble, depressed" and *stemma*, *stemmatos* "garland, crown."

Tapiphyllum Robyns Rubiaceae

Origins:

Latin *tapetum, i* and *tapete, is*; Greek *tapes, tapetos* "rug" and *phyllon* "a leaf," referring to the nature of the plant, the leaves are tomentose or villose; see J.R. Tennant, in *Kew Bulletin*. 19(2): 280-283. 1965.

Species/Vernacular Names:

T. parvifolium (Sond.) Robyns (*Vangueria parvifolia* Sond.)

English: mountain medlar, small velvet-leaf, small wild medlar

Southern Africa: bergmispel, kleinmispel; monyunwana (Western Transvaal, northern Cape, Botswana)

Tapirira Aubl. Anacardiaceae

Origins:

A vernacular name.

Species/Vernacular Names:

T. sp.

Peru: nucñu baras

T. guianensis Aublet (*Tapirira myriantha* Triana & Planchon)

Peru: isa paritsi, jemeco, joy-ey, nucñu varas, pau pombo, itil

Tapiscia Oliver Staphyleaceae (Tapisciaceae)

Origins:

The anagram of the generic name *Pistacia*.

Taplinia N.S. Lander Asteraceae

Origins:

The name honors Theodore Ernest Holmes Taplin, botanist at the Western Australian Herbarium, first collector of this genus; see N.S. Lander, "*Taplinia*, a new genus of Asteraceae (Inuleae) from Western Australia." *Nuytsia*. 7(1): 37-42. 1989.

Tapogamea Raf. Rubiaceae

Origins:

See C.S. Rafinesque, *Sylva Telluriana*. 147. 1838; for *Tapogomea* Aublet (1775).

Tapura Aubl. Dichapetalaceae

Origins:

Origins obscure, possibly from some vernacular African or South American name.

Species/Vernacular Names:

T. fischeri Engl. (*Tapura fischeri* Engl. var. *pubescens* Verdc. & Torre) (for the German physician Gustav Adolf Fischer, 1848-1886, botanical collector in Masailand, East Africa)

English: leaf-berry tree

Nigeria: asasa-igbo

South Africa: blaarbessieboom, tapura; umPengende, uMahlosane (Zulu)

T. guianensis Aublet

Guadeloupe: bois côte noir

Tara Molina Caesalpiniaceae

Origins:

A vernacular Amerindian name for *Caesalpinia spinosa*.

Tarachia C. Presl Aspleniaceae

Origins:

From the Greek *tarachos, tarache, tarche* "disorder."

Taraktogenos Hassk. Flacourtiaceae

Origins:

From the Greek *taraktos* "disturbed" and *genos* "people, nation, race, kind"; see also *Hydnocarpus* Gaertner.

Taramea Raf. Rubiaceae

Origins:

For *Faramea* Aublet (1775); see C.S. Rafinesque, *Ann. Gén. Sci. Phys.* 6: 68. 1820.

Taraxacum G.H. Weber ex F.H. Wiggers Asteraceae

Origins:

Persian *tarashqum, tarkhashqun, talkh chakok* "bitter herb"; Arabic *tarahshaqun, tarakhshagog*; Greek *tarasso* "to trouble, to trouble the mind, disturb, confound"; see M.A. Marchi, *Dizionario tecnico-etimologico-filologico.* Milano 1828-1829; R. Zander, F. Encke, G. Buchheim and S. Seybold, *Handwörterbuch der Pflanzennamen.* 14. Aufl. 539, 580. 1993; Manlio Cortelazzo & Paolo Zolli, *Dizionario etimologico della lingua italiana.* 5: 1314. 1988; H. Genaust, *Etymologisches Wörterbuch der botanischen Pflanzennamen.* 629-630. 1996; G.C. Wittstein, *Etymologisch-botanisches Handwörterbuch.* 865. 1852; Ernest Weekley, *An Etymological Dictionary of Modern English.* 2: 1473. 1967; C.T. Onions, *The Oxford Dictionary of English Etymology.* 1966.

Species/Vernacular Names:

T. californicum Munz & I.M. Johnston

English: California dandelion

T. glaucanthum (Ledeb.) DC. (*Taraxacum bicorne* Dahlst.)

English: Russian dandelion

Russia: kok-saghyz

T. megalorhizon (Forssk.) Hand.-Mazz.

Western Asia: krim-saghyz

T. officinale Weber ex Wiggers (*Leontodon taraxacum* L.)

English: common dandelion, dandelion, lion's tooth, Irish daisy

Italian: tarassaco

Spanish: achicoria, amargon, diente de león

Latin America: amragon, achicoria, botón de oro, diente de león, lechuguilla

Mexico: nocuana gueta, nocuana queeta

South Africa: perdeblom, platdissel, irwabe lenyoka

Japan: seiyo-tanpo-po, tanpupu

China: pu gong ying, pu kung ying, chiang nou tsao, chin tsan tsao

Hawaii: laulele, lauhele

T. serotinum (Waldst. & Kitaibel) Poir. (*Leontodon serotinus* Waldst. & Kitaibel)

English: dandelion

South Africa: platdissel

Taraxia (Nutt.) Raimann Onagraceae

Origins:

From the Greek *taraxe* "disorder, confusion."

Tarchonanthus L. Asteraceae

Origins:

Tarchon or *tarkon* is the Arabian name for *Artemisia dracunculus, anthos* "flower," in allusion to the resemblance of its flower-heads to those of that plant, see Helmut Genaust, *Etymologisches Wörterbuch der botanischen Pflanzennamen.* 630. 1996; according to some authors the generic name from the Greek *tarchos* "funeral rites, a solemn funeral," *tarchyo, tarchuo* "to bury solemnly."

Species/Vernacular Names:

T. camphoratus L. (*Tarchonanthus abyssinicus* Sch.Bip.; *Tarchonanthus litakunensis* DC.; *Tarchonanthus camphoratus* L. var. *litakunensis* (DC.) Harv.; *Tarchonanthus minor* Less.)

English: wild camphor bush, wild sage, wild cotton, sagewood, camphor tree, camphor wood, camphor bush, camphor bush tree, camphor bush wood, African fleabane

Southern Africa: siri, siriehout, wildekanferbos, kanferboom, kanferbos, kanferhout, kapokboom, wildesalie, salie, saliehout, basterolien, vaalbos, bastervaalbos, bergvaalbos, kleinvaalbos, olienvaalbos, veldvaalbos, waaibos, witbos; qoboqobo (Swati); isiDuli selindle, mathola (Xhosa); isiDuli-sehlathi, iGqeba-elimhlophe (Zulu); mofahlana (South Sotho); sefahla (North Sotho); mohatlha, mohata (Tswana:

western Transvaal, northern Cape, Botswana); moologa (Venda); omuteatupa (Herero); umgebe, umNgebe (Shona); umnqebe (Ndebele)

T. trilobus DC. var. *galpinii* (Hutch. & Phill.) J. Paiva (*Tarchonanthus galpinii* Hutch. & Phill.) (the name of the variety after Ernest Edward Galpin, 1858-1941, banker and plant collector, author of "The native timbers of the Springbok Flats." *Mem. Bot. Surv. S. Afr.* 7: 25. 1924)

English: isimemela tree

Southern Africa: isiMemela (Zulu)

T. trilobus DC. var. *trilobus*

English: three-lobed tarchonanthus, broad-leaved camphor bush, trident camphor bush

Southern Africa: breëblaarkanferbos, drietandkanferbos; iGqeba lamatshe (= the iGqeba of the rocks), isiMemela (Zulu); isiDuli sehlathi (Xhosa)

Tarenna Gaertner Rubiaceae

Origins:
Tarana is a Sinhalese name, the type-species was collected in Sri Lanka (Ceylon).

Species/Vernacular Names:
T. borbonica (E.G. & A. Henderson) Verdc.

La Réunion Island: bois de pintade

Mauritius: bois de rat

T. depauperata Hutch.

English: whitebark tarenna

T. junodii (Schinz) Brem. (*Chomelia junodii* Schinz) (for the Swiss missionary and botanical collector Rev. Henri Junod who botanized in Portuguese East Africa. He published "Considérations sur la flore sud-africaine." *Bull. Soc. Bot. Genève.* sér. 2, 18: 316. 1926, and with H. Schinz, "Pflanzenwelt der Delagoa Bay." *Bull. Herb. Boiss.* 7: 870-892. 1898, *loc. cit.* sér. 2, 3: 653-662. 1903, *Mém. Herb. Boiss.* 10: 25-79. 1900; see Henri Junod, *The Life of a South African Tribe (the Thonga).* London, Neuchâtel [printed] 1912, *Les Ba-Ronga.* Étude ethnographique sur les indigènes de la Baie de Delagoa. Neuchâtel 1898, *Manuel de conversation et Dictionnaire Ronga-Portugais-Français-Anglais.* Lausanne 1896 and *Grammaire Ronga.* Lausanne 1896; John Wesley Haley, *Life in Mozambique and South Africa.* Free Methodist Publishing House, Chicago 1926; D.M. Goodfellow, *Principles of Economic Sociology. The Economics of Primitive Life as Illustrated from the Bantu Peoples of South and East Africa.* London 1939; António Augusto Pereira Cabral, *Vocabulario português, shironga, shitsua, guitonga, shishope, shisena, shinhungue, shishuabo, kikua, shi-yao e kissuahili.* Lourenço Marques 1924; Henri Philippe Junod, *NwaMpfundla-NwaSisana.*

(The Romance of the Hare). [Written in the new Tsonga Orthography] Pretoria 1940; Henri A. Junod, *Moeurs et Coutumes des Bantous.* Paris 1936)

English: climbing tarenna

Southern Africa: klimbasterbruidsbos; amaBamba (Zulu)

T. littoralis (Hiern) Bridson (*Enterospermum littorale* Hiern)

English: dune butterspoon bush

South Africa: duinebotterlepelbos

T. mollissima (Hook. & Arn.) Robins.

English: hairy tarenna

T. pavettoides (Harv.) Sim subsp. *pavettoides* (*Kraussia pavettoides* Harv.)

English: false bride's bush, tarenna, pavetta-like tarenna

Southern Africa: basterbruidsbos; uMuthi-caShaka (Zulu)

T. supra-axillaris (Hemsl.) Brem. subsp. *barbertonensis* (Brem.) Bridson (*Tarenna barbertonensis* (Brem.) Brem.; Stylocoryne *barbertonensis* Brem.)

English: narrow-leaved false bride's bush, Barberton tarenna

Southern Africa: smalblaar; basterbruidsbos; ukuKula-ketelo-lehlathi (Zulu); mufhanza (Venda)

T. tsangii Merr.

English: Hainan tarenna

T. zeylanica Gaertn.

English: Ceylon tarenna

T. zimbabwensis Bridson (*Enterospermum rhodesiacum* Brem.)

English: Rhodesian tarenna

Southern Africa: muPakamabke, muPakamabgwe (Shona)

Tarennoidea Tirveng. & Sastre Rubiaceae

Origins:
Resembling *Tarenna*.

Tarpheta Raf. Verbenaceae

Origins:
To replace *Stachytarpheta* Vahl; see Constantine Samuel Rafinesque, *Flora Telluriana.* 2: 103. 1836 [1837].

Tarphochlamys Bremek. Acanthaceae

Origins:
From the Greek *tarphos* "thicket" and *chlamys* "cloak, mantle."

Tarrietia Blume Sterculiaceae

Origins:
Latinized Javanese name for the type species.

Tartonia Raf. Thymelaeaceae

Origins:
See Constantine Samuel Rafinesque, *Autikon botanikon. Icones plantarum select. nov. vel rariorum*, etc. 146. Philadelphia 1840; E.D. Merrill, *Index rafinesquianus*. 172. 1949.

Tasmannia R. Br. ex DC. Winteraceae

Origins:
After the Dutch navigator and explorer Abel Janszoon Tasman, 1603?-d. probably before Oct. 22, 1659, certainly before Feb. 5, 1661, entered the service of the Dutch East India Company in 1632 or 1633, a member of the Council of Justice of Batavia, in 1639 sailed under Commander Mathijs Hendrickszoon Quast on an expedition in search of the "islands of gold and silver" (in the seas east of Japan), between 1642-43 first voyage in the service of the Dutch East India Company (leaving Batavia [modern Jakarta] on August 14, 1642, with two ships, the *Heemskerk* and *Zeehaen*). He was chosen by the Governor-General of the Dutch East Indies, Anthony van Diemen, to command the voyage for the exploration of the Southern Hemisphere. (Following instructions based on a memoir by Frans Jacobszoon Visscher, his chief pilot, he was instructed to explore the Indian Ocean from west to east, south of the ordinary trade route, and, proceeding eastward into the Pacific.) Discovered Tasmania, New Zealand, Tonga, and the Fiji Islands. In 1644 the council of the company decided to send him on a new expedition to the "South Land" with instructions to establish relationship with New Guinea, the "great known South Land" (Western Australia), Van Diemen's Land, and the "unknown South Land," in 1647 commanded a trading fleet to Siam (now Thailand) and in the following year he commanded a war fleet against the Spaniards in the Philippines. See G.H. Kenihan, *The Journal of Abel Jansz Tasman*. Australian Heritage Press, Adelaide [n.d.]; P.A. Tiele, *Mémoire Bibliographique sur les Journaux des Navigateurs Néerlandais* ... la plupart en la possession de Frederick Muller à Amsterdam. Amsterdam 1867; Frederick Muller, *Four Catalogues. [Les Indes Orientales: Catalogue de Livres sur les Possessions Néerlandaises aux Indes*; etc.] Amsterdam 1882; James Burney, *A Chronological History of the Discoveries in the South Sea or Pacific Ocean*. London 1803-1817; John Landwehr, *VOC: A Bibliography of Publications Relating to the Dutch East India Company,* *1602-1800*. Ed. Peter van der Krogt. HES Publisher, Utrecht 1991.

Tatea F. Muell. Labiatae (Verbenaceae)

Origins:
For the British (b. Northumberland) geologist and botanist Ralph Tate, 1840-1901 (d. Adelaide, Australia), science teacher, traveler, plant collector, from 1875 to 1901 professor of Natural Science at Adelaide University, founder of Belfast Naturalists Field Club, 1883 Fellow of the Linnean Society. Among his writings are *Flora belfastiensis. The plants around Belfast*, etc. Belfast 1863, *A Class-Book of Geology, Physical and Historical*. London 1872, *A Plain and Easy Account of the Land and Fresh-Water Mollusks of Great Britain*. London 1866, *A Handbook of the Flora of Extratropical South Australia*. Adelaide 1890 and *Rudimentary Treatise on Geology*. London 1871, nephew of George Tate (1805-1871); see F. von Mueller and R. Tate, *List of Plants Collected during Mr. Tietken's Expedition into Central Australia*. 1889, in William H. Tietkens (1844-1933), *Journal of the Central Australian Exploring Expedition, 1889, under Command of W.H. Tietkens*, etc. Adelaide 1891; Jonathan Wantrup, *Australian Rare Books, 1788-1900*. Sydney 1987; J.H. Barnhart, *Biographical Notes upon Botanists*. 3: 361. 1965; H.N. Clokie, *Account of the Herbaria of the Department of Botany in the University of Oxford*. 251. Oxford 1964; E.M. Tucker, *Catalogue of the Library of the Arnold Arboretum of Harvard University*. Cambridge, Massachusetts 1917-1933; Dennis John Carr and S.G.M. Carr, eds., *People and Plants in Australia*. 1981.

Tatea Seem. Rubiaceae

Origins:
For the British naturalist Ralph Tate, 1840-1901, South Australia geologist and botanist, from 1875 to 1901 professor of Natural Science at Adelaide University, founder of Belfast Naturalist Field Club, nephew of George Tate (1805-1866); see F. von Mueller and R. Tate, *List of Plants Collected during Mr. Tietkens' Expedition into Central Australia*. 1889, in William H. Tietkens (1844-1933), *Journal of the Central Australian Exploring Expedition, 1889, under Command of W.H. Tietkens*, etc. Adelaide 1891; J.H. Barnhart, *Biographical Notes upon Botanists*. 3: 361. 1965; Jonathan Wantrup, *Australian Rare Books, 1788-1900*. Sydney 1987.

Tateanthus Gleason Melastomataceae

Origins:

Named for the American (b. London) naturalist George Henry Hamilton Tate, 1894-1953 (d. Morristown, New Jersey), zoologist, plant and botanical collector in South America (Venezuela, Bolivia, Peru), 1922 Field Assistant American Museum of Natural History, author of *Mammals of Eastern Asia*. New York 1947, with Thomas D. Carter wrote *Mammals of the Pacific World*. 1945; see Douglas C. McMurtrie, *A Bibliography of Morristown Imprints 1798-1820*. [from the *Proceedings of the New Jersey Historical Society*, April 1936] Newark 1936; J.H. Barnhart, *Biographical Notes upon Botanists*. 3: 361. 1965; E.M. Tucker, *Catalogue of the Library of the Arnold Arboretum of Harvard University*. Cambridge, Massachusetts 1917-1933.

Tatina Raf. Sapotaceae (Aquifoliaceae)

Origins:

See Constantine Samuel Rafinesque, *Autikon botanikon. Icones plantarum select. nov. vel rariorum, etc.* 75. Philadelphia 1840; E.D. Merrill, *Index rafinesquianus*. 160. 1949.

Taubertia Schumann Menispermaceae

Origins:

For the German botanist Paul Hermann Wilhelm Taubert, 1862-1897, traveler and plant collector (in Brazil); see H. Barnhart, *Biographical Notes upon Botanists*. 3: 362. 1965; Ida Kaplan Langman, *A Selected Guide to the Literature on the Flowering Plants of Mexico*. University of Pennsylvania Press, Philadelphia 1964; R. Zander, F. Encke, G. Buchheim and S. Seybold, *Handwörterbuch der Pflanzennamen*. 14. Aufl. Stuttgart 1993.

Tauroceras Britton & Rose Mimosaceae

Origins:

From the Greek *tauros* "bull" and *keras* "a horn."

Taurostalix Reichb.f. Orchidaceae

Origins:

Greek *tauros* "bull" and *stalix, stalikos* "a stake," indicating the anterior teeth of the column.

Tauscheria Fischer ex DC. Brassicaceae

Origins:

Named for a collector in western Russia, M. Tauscher; see Stafleu and Cowan, *Taxonomic Literature*. 6: 184. 1986.

Tauschia Schltdl. Umbelliferae

Origins:

For the Czech botanist Ignaz Friedrich Tausch, 1793-1848, naturalist, botanical collector; see J.H. Barnhart, *Biographical Notes upon Botanists*. 3: 362. 1965; Ida Kaplan Langman, *A Selected Guide to the Literature on the Flowering Plants of Mexico*. University of Pennsylvania Press, Philadelphia 1964; A. Lasègue, *Musée botanique de Benjamin Delessert*. Paris 1845; E.M. Tucker, *Catalogue of the Library of the Arnold Arboretum of Harvard University*. 1917-1933; Stafleu and Cowan, *Taxonomic Literature*. 6: 182-184. Utrecht 1986; R. Zander, F. Encke, G. Buchheim and S. Seybold, *Handwörterbuch der Pflanzennamen*. 14. Aufl. Stuttgart 1993; N. Hall, *Botanists of the Eucalypts*. Melbourne 1978 and Supplement 1980.

Species/Vernacular Names:

T. glauca (J. Coulter & Rose) Mathias & Constance
English: glaucous tauschia

T. howellii (J. Coulter & Rose) J.F. Macbr.
English: Howell's tauschia

Tavaresia Welw. Asclepiadaceae

Origins:

After the Portuguese naturalist Joaquim (Joachim) da Silva Tavares, 1866-1931, clergyman, entomologist, traveler; see J.H. Barnhart, *Biographical Notes upon Botanists*. 3: 362. 1965; T.W. Bossert, *Biographical Dictionary of Botanists Represented in the Hunt Institute Portrait Collection*. 395. 1972.

Taveunia Burret Palmae

Origins:

A modification of Taveuni, Fiji Islands.

Taxillus Tieghem Loranthaceae

Origins:

Latin *taxillus, i* "a small die," or the diminutive of the genus *Taxus*.

Species/Vernacular Names:

T. yadoriki (Maxim.) Danser (*Loranthus yadoriki* Sieb. ex Maxim.)

Japan: ôba-yadori-gi (= large-leaved *Viscum*)

Okinawa: hanmuchi

Taxodium Richard Taxodiaceae (Cupressaceae)

Origins:

The genus *Taxus* L. and the Greek *eidos* "resemblance," referring to the leaves.

Species/Vernacular Names:

T. distichum (L.) Richard (*Cupressus disticha* Richard)

English: bald cypress, swamp cypress, southern cypress

Japan: raku-u-sho (= deciduous feather)

T. mucronatum Ten.

English: Mexican swamp cypress

Mexico: sabino, bachil, ahuehuetl, quetzalhuehuetl (meaning sabino precioso), ahuehuete (meaning tambor de agua, viejo del agua)

Taxotrophis Blume Moraceae

Origins:

From the Greek *taxo* "to put in order" and *trophe* "food."

Taxus L. Taxaceae

Origins:

The Latin name *taxus, i* for the yew tree (Plinius); see Carl Linnaeus, *Species Plantarum*. 1040. 1753 and *Genera Plantarum*. Ed. 5. 462. 1754.

Species/Vernacular Names:

T. baccata L.

English: English yew, yew

Nepal: barma salla

T. brevifolia Nutt.

English: Californian yew

T. canadensis Marshall

English: American yew

T. cuspidata Siebold & Zucc.

English: Japanese yew

China: zi shan

T. mairei (Lemée & Lév.) Liu

English: Chinese yew

T. sumatrana (Miq.) Laubenf.

English: Sumatra yew

Tayloriophyton M.P. Nayar Melastomataceae

Origins:

For the British (b. Edinburgh) botanist George Taylor, b. 1904, traveler and botanical collector, 1956-1971 Director of the Royal Botanic Gardens, Kew; see J.H. Barnhart, *Biographical Notes upon Botanists*. 3: 363. 196; T.W. Bossert, *Biographical Dictionary of Botanists Represented in the Hunt Institute Portrait Collection*. 395. 1972; H.R. Fletcher, *Story of the Royal Horticultural Society, 1804-1968*. Oxford 1969; H.N. Clokie, *Account of the Herbaria of the Department of Botany in the University of Oxford*. 252. Oxford 1964; R. Zander, F. Encke, G. Buchheim and S. Seybold, *Handwörterbuch der Pflanzennamen*. 14. Aufl. Stuttgart 1993; Mary Gunn and Leslie E. Codd, *Botanical Exploration of Southern Africa*. 341. Cape Town 1981; M. Hadfield et al., *British Gardeners: A Biographical Dictionary*. London 1980.

Tchihatchewia Boissier Brassicaceae

Origins:

After the Russian botanist Pierre de Tchihatcheff (Petr Aleksandrovich Tchichatscheff, Chikhachef, Tschihatcheff), 1812-1890, traveler, geographer; see John H. Barnhart, *Biographical Notes upon Botanists*. 3: 365. 196; T.W. Bossert, *Biographical Dictionary of Botanists Represented in the Hunt Institute Portrait Collection*. 396. 1972; E.M. Tucker, *Catalogue of the Library of the Arnold Arboretum of Harvard University*. 1917-1933.

Teclea Delile Rutaceae

Origins:

After St. Takla Hemanout, a legendary protagonist of the history of the Coptic Church; he was a son of an Ethiopian Orthodox priest; the Coptic Church recognizes him as a saint.

Species/Vernacular Names:

T. sp.

Uganda: nzo

East Africa: kwati

T. gerrardii Verdoorn (after William Tyrer Gerrard, d. 1866 (d. Foulpointe, Madagascar), English botanical collector in Natal and Madagascar in the 1860s, collected with Mark

Johnston McKen (1823-1872), naturalist and traveler; see Mary Gunn and Leslie E. Codd, *Botanical Exploration of Southern Africa*. 165-166. Cape Town 1981; J. Lanjouw and F.A. Stafleu, *Index Herbariorum*. Part II (2), *Collectors E-H*. Regnum Vegetabile vol. 9. 1957)

English: Zulu cherry-orange

Southern Africa: Zoeloekersielemoen, umboza; uMozane (Zulu, Xhosa); iNzanyana, umBoza (Xhosa); umBoza, umBozane (Zulu)

T. natalensis (Sond.) Engl. (*Toddalia natalensis* Sond.)

English: Natal cherry-orange, bastard white ironwood

Southern Africa: Natalkersielemoen; uMozane (Zulu); umBoza, iNzanyana (Xhosa)

Tecleopsis Hoyle & Leakey Rutaceae

Origins:
Resembling the genus *Teclea* Delile.

Tecoma Juss. Bignoniaceae

Origins:
The name *Tecoma* originates from the Mexican name *tecomaxochitl*, which is given by the local people to all those plants with flowers that are tubular, or trumpet-like or ship-like, etc.

Species/Vernacular Names:
T. arequipensis (Sprague) Sandwith (*Stenolobium arequipense* Sprague)

Peru: cahuato, pichus

T. capensis (Thunb.) Lindley (*Bignonia capensis* Thunb.; *Tecoma krebsii* Klotzsch; *Tecoma petersii* Klotzsch; *Tecomaria capensis* (Thunb.) Spach)

English: Cape honeysuckle, Cape trumpet flower, kaffir honeysuckle, tecoma

Southern Africa: trompetters; iCakatha (Xhosa); uChacha, uGcangca, uMunyane (Zulu); luvhengwa-mmbwa (= hated by dogs) (Venda)

Japan: hime-nôzen-kazura

T. fulva (Cav.) G. Don (*Bignonia fulva* Cavanilles; *Bignonia meyeniana* Schauer; *Stenolobium fulvum* (Cav.) Sprague; *Tecomaria fulvum* (Cav.) Seemann)

Peru: chuvé

T. garrocha Hieron.

Argentina: guaran colorado, garrocha

T. rosifolia Kunth (*Stenolobium huancabambae* Kränzl.; *Tecomaria roseifolia* (Kunth) Seemann)

Peru: fresno, fresnillo hada

T. sambucifolia Kunth (*Stenolobium sambucifolium* (Kunth) Seemann)

Peru: huaranhua, huaruma, huarumo, huaranhui, huaranhuai, huaranhuay, huarauma, huaraumo

T. stans (L.) Kunth (*Bignonia stans* L.; *Stenolobium stans* (L.) Seem.; *Tecoma incisa* Sweet; *Tecoma stans* var. *angustatum* Rehd.; *Tecoma stans* var. *apiifolia* hort. ex DC.)

English: yellow elder, yellow bells, shrubby trumpet flower

Central America: timboco, barreto, chacté, flor amarilla, San Andrés, timboque, trompeta, tronadora

Mexico: batilimi, matilimi, borla de San Pedro, hierba de San Pedro, corneta amarilla, flor de San Pedro, gloria, guiabiche, guie-bacaná, guie-biche, hierba de San Nicolás, hoja de baño, ixontli, mazorca, miñona, caballito, retama, San Pedro, trompeta, tronador, tronadora, tulasúchil, xochimitl, palo de arco, lluvia de oro

Peru: huaranhua

Bolivia: marangaya, guaranguay, sau-sau, árbol canario

N. Rhodesia: kapasa

Tecomanthe Baillon Bignoniaceae

Origins:
The generic name *Tecoma* and *anthos* "flower."

Tecomaria Spach Bignoniaceae

Origins:
From the genus *Tecoma*, of the same family.

Tecomella Seemann Bignoniaceae

Origins:
The diminutive of the genus *Tecoma*.

Tecophilaea Bertero ex Colla Tecophilaeaceae (Liliaceae)

Origins:
For Tecofila Billotti-Colla, an Italian botanical illustrator, daughter of the Italian botanist Luigi A. Colla; see F. Boerner & G. Kunkel, *Taschenwörterbuch der botanischen Pflanzennamen*. 4. Aufl. 177. Berlin & Hamburg 1989.

Tectaria Cavanilles Dryopteridaceae

Origins:
Latin *tectum, i* "a covering" and the adjectival suffix *-aria*, referring to the indusia of some species.

Species/Vernacular Names:
T. coriandrifolia (Swartz) L. Underwood (*Aspidium coriandrifolium* Swartz)
English: hairy halberd fern
T. heracleifolia (Willdenow) L. Underwood (*Aspidium heracleifolium* Willdenow)
English: broad halberd fern

Tectaridium Copel. Dryopteridaceae (Aspleniaceae)

Origins:
Referring to the genus *Tectaria* Cav.

Tecticornia Hook.f. Chenopodiaceae

Origins:
Latin *tectum* "a covering, roof" and *cornu* "a horn," possibly referring to a relationship to the genus *Salicornia*, or presumably alluding to the outer margin of the bracts; see P.G. Wilson, "A taxonomic revision of the genus *Tecticornia* (Chenopodiaceae)." in *Nuytsia*. 1(3): 277-288. 1972.

Species/Vernacular Names:
T. verrucosa Paul G. Wilson
Australia: samphire

Tectiphiala H.E. Moore Palmae

Origins:
From the Latin *tectum, i* "a covering" and Greek *phiale* "a vial."

Tectona L.f. Labiatae (Verbenaceae)

Origins:
Greek *tekton, onos* "carpenter, worker in wood"; Portuguese *teca*, Malayalam and Tamil *tekka* or *theku, tekku* for *Tectona grandis* L.f.

Species/Vernacular Names:
T. grandis L.f.

English: teak tree, teak, Indian oak, common teakwood, common teak
Mexico: teca, teka
Southeast Asia: saigun, saga, teka
India: saka, segan, sagwan, peddateku, tega, tegu, tekku, tekku-maram, tekkoo, tekka, sal, saj, ky-won, jati, jaddi, ching-jagu
Japan: chiku-no-ki
China: you mu
Malaya: jati

Teesdalia R. Brown Brassicaceae

Origins:
After the British botanist Robert Teesdale, *circa* 1740-1804 (d. Hammersmith, London), gardener and seedsman, friend of J.E. Smith, 1788 Fellow of the Linnean Society, contributed to James Sowerby and J.E. Smith, *English Botany*, son of Robert Teesdale (d. 1773); see Dawson Turner and Lewis Weston Dillwyn (1778-1855), *The Botanist's Guide Through England and Wales*. 663-665. London 1805; J.H. Barnhart, *Biographical Notes upon Botanists*. 3: 365. 1965; [Robert Teesdale] "Plantae Eboracenses." *Trans. Linn. Soc.* 2: 103-125. (May) 1794 and "A Supplement to the Plantae Eboracenses." *ib.* 5: 36-95. (Feb.) 1800; Warren R. Dawson, *The Banks Letters, a Calendar of the Manuscript Correspondence of Sir Joseph Banks*. London 1958; E.M. Tucker, *Catalogue of the Library of the Arnold Arboretum of Harvard University*. Cambridge, Massachusetts 1917-1933; Jonas C. Dryander, *Catalogus bibliothecae historico-naturalis Josephi Banks*. London 1796-1800; Blanche Henrey, *No Ordinary Gardener — Thomas Knowlton, 1691-1781*. Edited by A.O. Chater. British Museum (Natural History). London 1986; F. Boerner & G. Kunkel, *Taschenwörterbuch der botanischen Pflanzennamen*. 4. Aufl. 177. Berlin & Hamburg 1989.

Species/Vernacular Names:
T. nudicaulis (L.) R. Br.
English: shepherd's cress

Teesdaliopsis (Willkomm) Rothmaler Brassicaceae

Origins:
Resembling *Teesdalia*; see *Repert. Spec. Nov. Regni Veg.* 49: 178. 1940.

Teganium Schmidel Solanaceae

Origins:

From *teganion*, the diminutive of the Greek *tegane, teganon, tagenon* "saucepan."

Tegularia Reinw. Dryopteridaceae (Aspleniaceae)

Origins:

Latin *tegula, tegulae* "tiles, roof-tiles."

Teichmeyera Scopoli Lecythidaceae

Origins:

After the German botanist Hermann Friedrich Teichmeyer, 1685-1744, physician; see John H. Barnhart, *Biographical Notes upon Botanists*. 3: 365. 1965; T. W. Bossert, *Biographical Dictionary of Botanists Represented in the Hunt Institute Portrait Collection*. 396. Boston, Massachusetts 1972; Jonas C. Dryander, *Catalogus bibliothecae historico-naturalis Josephi Banks*. London 1796-1800.

Teijsmanniodendron Koorders Verbenaceae

Origins:

Named for the Dutch botanist Johannes Elias Teijsmann (Teysmann), 1809-1882, traveler, gardener and botanical explorer, plant collector, 1831-1869 Curator of the Buitenzorg (Bogor) Botanic Gardens; see John H. Barnhart, *Biographical Notes upon Botanists*. 3: 365. 1965; T. W. Bossert, *Biographical Dictionary of Botanists Represented in the Hunt Institute Portrait Collection*. 397. Boston, Massachusetts 1972; E.M. Tucker, *Catalogue of the Library of the Arnold Arboretum of Harvard University*. Cambridge, Massachusetts 1917-1933]; Stafleu and Cowan, *Taxonomic Literature*. 6: 201-204. 1986; R. Zander, F. Encke, G. Buchheim and S. Seybold, *Handwörterbuch der Pflanzennamen*. 14. Aufl. Stuttgart 1993.

Teinosolen Hook.f. Rubiaceae

Origins:

Greek *tainia* "fillet, a ribbon" or *teino* "stretch, stretched" and *solen* "a tube"; see also *Heterophyllaea* Hook.

Teinostachyum Munro Gramineae

Origins:

Greek *tainia* "fillet, a ribbon" or *teino* "stretch, stretched" and *stachys* "spike," referring to the long spikes; see also *Schizostachyum* Nees.

Telanthera R. Br. Amaranthaceae

Origins:

Greek *tele* "far, distant, at a distance" and *anthera* "anther"; see also *Alternanthera* Forssk.

Telanthophora H. Robinson & Brettell Asteraceae

Origins:

From the Greek *tele* "far, distant, at a distance," *anthos* "flower" and *phoreo* "to bear."

Teleozoma R. Br. Pteridaceae (Parkeriaceae)

Origins:

Greek *teleios, teleos* "complete, perfect" and *zoma* "a belt, dress," probably referring to the sporangia.

Telephium L. Caryophyllaceae (Molluginaceae)

Origins:

Greek *telephion* for a species of *Andrachne*, Latin *telephion, ii* "a kind of herb resembling purslane" (Plinius).

Telesonix Raf. Saxifragaceae

Origins:

See Constantine S. Rafinesque, *Flora Telluriana*. 2: 69. 1836 [1837]; E.D. Merrill, *Index rafinesquianus*. 135. 1949.

Telfairia Hook. Cucurbitaceae

Origins:

For the Irish (b. Belfast, County Antrim) surgeon Charles Telfair, 1778-1833 (d. Mauritius, Port Louis), a keen amateur naturalist, botanist, plant collector, arrived at Mauritius in 1810. He created a garden at Bois Chéri (Moka), he was appointed Superintendent of Pamplemousses Gardens (to help John White and John Newman, the Directors), he was the first President of the Société d'Histoire Naturelle de l'Ile Maurice (founded in 1829) (later became the Royal Society of Arts and Sciences of Mauritius), owner of Bois Chéri, Bel Ombre, Beau Manguier and Bon Espoir. His writings include *Some Account of the State of Slavery in Mauritius*, since the British occupation in 1810. Mauritius 1830, *An Account of the Conquest of the Island of Bourbon* [by an Officer of the Expedition.] London 1811 and "Rodrigues in 1809." *Revue Retrospective de l'Ile Maurice.* 6: 311-312. 1955, husband of Annabella Telfair *née* Chamberlain (d. 1832, Port Louis, Mauritius); see Nathaniel Wallich, *Plantae Asiaticae rariores.* 2: 78-79. London 1832; Ernest Nelmes and William Cuthbertson, *Curtis's Botanical Magazine Dedications, 1827-1927.* 15-16. [1931]; M. Hadfield et al., *British Gardeners: A Biographical Dictionary.* London 1980; Jules Desjardins, *Notice historique sur Charles Telfair, Esq.,* Fondateur et Président de la Société d'Histoire Naturelle de l'Ile Maurice. Port Louis, Ile Maurice 1836; Guy Rouillard and Joseph Guého, *Le Jardin des Pamplemousses 1729-1979. Histoire et Botanique.* Les Pailles 1983; W. J. Hooker, *Bot. Mag.* 2751-52. 1 Jul 1827; F. Boerner & G. Kunkel, *Taschenwörterbuch der botanischen Pflanzennamen.* 4. Aufl. 177. 1989; Ray Desmond, *Dictionary of British & Irish Botanists and Horticulturists.* 675. London 1994.

Species/Vernacular Names:
T. occidentalis Hook.f.

Tropical Africa: krobonko

Yoruba: ila iroko

T. pedata (Smith) Hook.

English: oyster nut

Tropical Africa: kweme

Mozambique: cungo, djiungo, macungo-Jicungo, castanhas de Inhambane

Teline Medik. Fabaceae

Origins:
Greek *teline* = *kytisos* for woody legumes, for a shrubby kind of clover much valued by the ancients, perhaps the shrubby snail-clover; Plinius, Hippocrates in *De Natura Muliebri* 93 and Theophrastus in *HP.* 4.16.5. (Loeb Classical Library 1916) presumably applied to *Medicago arborea* L.

Teliostachya Nees Acanthaceae

Origins:
From the Greek *teleios, teleos* "complete, perfect" and *stachys* "spike."

Telipogon Mutis ex Kunth Orchidaceae

Origins:
Greek *thelys* "feminine, female" or *telos* "end, extremity" and *pogon* "beard," referring to the style or to the apex of the pilose column.

Tellima R. Br. Saxifragaceae

Origins:
An anagram of *Mitella* L.

Species/Vernacular Names:
T. grandiflora (Pursh) Douglas ex Lindley

English: fringe cups

Telmatophila Martius ex Baker Asteraceae

Origins:
Greek *telma, telmatos* "marsh" and *philos* "lover, loving."

Telminostelma Fourn. Asclepiadaceae

Origins:
From the Greek *telmis, telminos* "mud, slime" and *stelma* "a girdle, belt, crown."

Telmissa Fenzl Crassulaceae

Origins:
From the Greek *telma, telmatos* "marsh," *telmis* "mud, slime," referring to the habitat.

Telogyne Baillon Euphorbiaceae

Origins:
From the Greek *tele* "far, distant" and *gyne* "female, woman."

Telopea R. Br. Proteaceae

Origins:
Greek *tele* "far, distant" and *ops, opos* "aspect, resemblance, appearance, sight," *telopos* "seen from afar," referring to the conspicuous inflorescences.

Species/Vernacular Names:
T. speciosissima (Sm.) R. Br.

Australia: waratah

Telosma Coville Asclepiadaceae

Origins:
From the Greek *tele* "far, distant" and *osme* "smell, odor, perfume."

Species/Vernacular Names:

T. cordata (Burm.f.) Merrill (*Asclepias cordata* Burm.f.; *Telosma odoratissima* (Lour.) Coville)

English: cordate telosma

Japan: ya-rai-sô

China: ye lai xiang

T. pallida (Roxb.) Craib

English: Taiwan telosma

China: tai wan ye lai xiang

T. procumbens (Blanco) Merrill (*Telosma cathayensis* Merrill)

English: south China telosma, creeping telosma

China: wo jing ye lai xiang

Teloxys Moq. Chenopodiaceae

Origins:

Greek *tele* and *oxys* "sharp-pointed, sharp, acid, sour."

Temminckia Vriese Goodeniaceae

Origins:

After the Dutch zoologist Coenraad Jacob Temminck, 1778-1858, ornithologist, Director of the Royal Netherlands Zoological Museum at Leiden, professor of botany and zoology. Among his works are *Coup-d'oeil général sur les possessions néerlandaises dans l'Inde archipélagique*. Leide 1847-1849, *Histoire naturelle générale des Pigeons ... avec figures ... peintes par Pauline de Courcelles (le texte par F. Prevost)*. Paris 1808, *Manuel d'Ornithologie*, ou tableau systématique des oiseaux qui se trouvent en Europe. Amsterdam 1815 and *Monographie de Mammalogie*. Paris, Leiden 1827-1841, with Guillaume Michel Jérôme Meiffren, Baron Langier (or Laugier) de Chartrouse, wrote *Nouveau recueil de planches coloriées d'oiseaux*, pour servir de suite et de complément aux Planches enluminées de Buffon ... édition de 1770. Paris [1822-]1838; see J.H. Barnhart, *Biographical Notes upon Botanists*. 3: 366. 1965; E.M. Tucker, *Catalogue of the Library of the Arnold Arboretum of Harvard University*. Cambridge, Massachusetts 1917-1933.

Temmodaphne Kosterm. Lauraceae

Origins:

From the Greek *temno* "division, section, to slice" and *daphne* "bay laurel."

Temnadenia Miers Apocynaceae

Origins:

Greek *temno* "division, section, to slice" and *aden* "gland."

Temnocalyx Robyns Rubiaceae

Origins:

From the Greek *temno* plus *kalyx* "calyx."

Temnopteryx Hook.f. Rubiaceae

Origins:

From the Greek *temno* with *pteryx, pteron* "wing."

Templetonia R. Br. Fabaceae

Origins:

After the Irish (b. Belfast, County Antrim) botanist John Templeton, 1766-1825 (d. Malone, Belfast), naturalist, founding member of Belfast Natural History Society, contributed to James Sowerby (1757-1822) and J.E. Smith, *English Botany*, to L.W. Dillwyn, *British Confervae*. London [1802-] 1809. Father of the Irish surgeon and plant collector Robert Templeton (1802-1892); see J.H. Barnhart, *Biographical Notes upon Botanists*. 3: 366. 1965; John Claudius Loudon, *Arboretum et fruticetum britannicum*. London 1838; John Templeton, *Eulogium on Henry William Tennent*. Belfast 1815; H.N. Clokie, *Account of the Herbaria of the Department of Botany in the University of Oxford*. 252. 1964; F. Boerner & G. Kunkel, *Taschenwörterbuch der botanischen Pflanzennamen*. 4. Aufl. 177. 1989; Ray Desmond, *Dictionary of British & Irish Botanists and Horticulturists*. 676. 1994.

Species/Vernacular Names:

T. aculeata (F. Muell.) Benth. (*Bossiaea aculeata* F. Muell.)

Australia: spiny mallee-pea

T. egena (F. Muell.) Benth. (*Daviesia egena* F. Muell.)

Australia: desert broombush, round broombush, round templetonia

T. retusa (Vent.) R. Br. (*Rafnia retusa* Vent.)

Australia: coral bush, cockies tongue, common templetonia, red templetonia, flame bush, parrot bush, bullock bush, cocky's beaks

Tenagocharis Hochst. Alismataceae (Limnocharitaceae)

Origins:

From the Greek *tenagos* "a shoal shallow" and *charis* "delight, grace, beauty," referring to the habitat.

Tenaris E. Meyer Asclepiadaceae

Origins:

Probably from the Greek *teino* "to stretch, extend," referring to the extended lobes of the corolla.

Tengia Chun Gesneriaceae

Origins:

For the Chinese botanist Teng-Shi-Wei, plant collector; see Stafleu and Cowan, *Taxonomic Literature*. 6: 209. 1986.

Tenicroa Raf. Hyacinthaceae (Liliaceae)

Origins:

See C.S. Rafinesque, *Flora Telluriana*. 3: 52. 1836 [1837]; Elmer D. Merrill, *Index rafinesquianus*. 94. 1949.

Tenorea Gasp. Moraceae

Origins:

For the Italian botanist Michele Tenore, 1780-1861. His writings include *Catalogo delle piante che si coltivano nel Regio Orto Botanico di Napoli*. Napoli 1845 and *Index Seminum in Horto Botanico Neapolitano Collectorum*. 1839. Uncle of the Italian botanist Vincenzo Tenore (1825-1886); see Stafleu and Cowan, *Taxonomic Literature*. 6: 212-219. 1986; Mariella Azzarello Di Misa, ed., *Il Fondo Antico della Biblioteca dell'Orto Botanico di Palermo*. 265-268. Palermo 1988; J.H. Barnhart, *Biographical Notes upon Botanists*. 3: 367. 1965; R. Zander, F. Encke, G. Buchheim and S. Seybold, *Handwörterbuch der Pflanzennamen*. 14. Aufl. 789. 1993; A. Lasègue, *Musée botanique de Benjamin Delessert*. Paris 1845; T.W. Bossert, *Biographical Dictionary of Botanists Represented in the Hunt Institute Portrait Collection*. 396. 1972; Ethelyn Maria Tucker, *Catalogue of the Library of the Arnold Arboretum of Harvard University*. Cambridge, Massachusetts 1917-1933; Ida Kaplan Langman, *A Selected Guide to the Literature on the Flowering Plants of Mexico*. University of Pennsylvania Press, Philadelphia 1964; H.N. Clokie, *Account of the Herbaria of the Department of Botany in the University of Oxford*. 252. Oxford 1964.

Tenorea Raf. Rutaceae

Origins:

For the Italian botanist Michele Tenore, 1780-1861; see C.S. Rafinesque, in *Specchio delle scienze, o giornale enciclopedico di Sicilia etc.* 1: 192, 193. Palermo 1814; E.D. Merrill (1876-1956), *Index rafinesquianus*. The plant names published by C.S. Rafinesque, etc. 150. Jamaica Plain, Massachusetts, USA 1949.

Tephrocactus Lem. Cactaceae

Origins:

Greek *tephros, tephra* "ashen, ash-colored" plus *cactus*.

Tephroseris (Reichb.) Reichb. Asteraceae

Origins:

From the Greek *tephros, tephra* "ashen, ash-colored" and *seris, seridos* "chicory, lettuce."

Species/Vernacular Names:

T. integrifolia (L.) Holub (*Senecio campestris* (Retz.) DC.)

English: fleawort

China: kou she tsao

Tephrosia Pers. Fabaceae

Origins:

Greek *tephros, tephra* "ashen, ash-colored," referring to the leaves of many species, to the gray pubescence on the leaves; see Pierre Fatumbi Verger, *Ewé: The Use of Plants in Yoruba Society*. São Paulo 1995; Celia Blanco, *Santeria Yoruba*. Caracas 1995; Maria Helena Farelli, *Plantas que curam e cortam feitiços*. Rio de Janeiro 1988; William W. Megenney, *A Bahian Heritage*. University of North Carolina at Chapel Hill 1978.

Species/Vernacular Names:

T. sp.

Yoruba: roro odan

T. bracteolata Guill. & Perr.

Yoruba: akororo, roro, roro funfun, iroro

T. candida DC.(*Cracca candida* (DC.) Kuntze)

English: white tephrosia

India: lashtia

Congo: mbaha

T. elegans Schumach.

Yoruba: amati, eyofo

T. glomeruliflora Meissn.

English: pink tephrosia

T. linearis (Willd.) Pers.

Yoruba: weje

T. platycarpa Guill. & Perr.

Yoruba: aboroyo, roro, ororo

T. pondoensis (Codd) Schrire (*Mundulea pondoensis* Codd)

English: Pondo poison pea

South Africa: Pondogifertjie

T. purpurea (L.) Pers. (*Galega piscatoria* Aiton; *Tephrosia piscatoria* (Aiton) Pers.; *Cracca purpurea* L.)

English: fish-poison-tree, Vogel tephrosia

India: sharapunkha, bannilgach, sarphunkha, sarphonka, phanike, kolinnil, mollukkay

Japan: nanban-kusa-fuji

Hawaii: 'auhuhu, ahuhu, 'auhola, hola

T. sinapou (Buc'hoz) A. Chev. (*Tephrosia toxicaria* (Sw.) Pers.; *Galega toxicaria* Sw.; *Galega sinapou* Buc'hoz; *Cracca toxicaria* (Sw.) Kuntze)

Tropical America: yarroconalli

Guyana (Brit. Guiana): yawrukunan

Peru: barbasco, cube, cube ordinario, huasca barbasco, kumu, kkumu, motuy kube, muyuy cube, tingui de cayenne, tirana barbasco

T. sphaerospora F. Muell.

Australia: Mulga trefoil

T. virginiana (L.) Pers.

English: catgut, goat's rue, rabbit's pea

T. vogelii Hook. (*Cracca vogelii* (Hook.f.) Kuntze)

English: fish-poison-tree, Vogel tephrosia

East Africa: mtunungu

Nigeria: were, majimfa

Yoruba: igun, lakuta, agba odo, oro beja, were, ifo

N. Rhodesia: wusungu

Teramnus P. Browne Fabaceae

Origins:

Greek *teramna, teremna, teramnon, teremnon* "a roof, room, home, chamber, house," Latin *teramon* or *teramum* for a plant growing near Philippi (Plinius), Greek *teramon* "soft, tender," Latin *trabs, trabis, trabes* "a timber, roof, beam, a tree."

Species/Vernacular Names:

T. labialis (L.f.) Sprengel (*Glycine labialis* L.f.)

English: blue wiss

Yoruba: adagbudu

Teratophyllum Mett. ex Kuhn Lomariopsidaceae (Aspleniaceae)

Origins:

From the Greek *teras, teratos* "a wonder, marvel, monster" and *phyllon* "leaf."

Terebinthina Kuntze Scrophulariaceae

Origins:

Pertaining to turpentine, from Greek *terebinthos*, the name of *Pistacia terebinthus*; see Manlio Cortelazzo & Paolo Zolli, *Dizionario etimologico della lingua italiana*. 5: 1329. Zanichelli, Bologna 1988; Ernest Weekley, *An Etymological Dictionary of Modern English*. 2: 1487. New York 1967.

Terebraria O. Kuntze Rubiaceae

Origins:

Latin *terebra, ae* "a borer, gimlet."

Teremis Raf. Solanaceae

Origins:

See Elmer D. Merrill (1876-1956), *Index rafinesquianus*. 213. 1949; Constantine S. Rafinesque, *Sylva Telluriana*. 53, 54. 1838.

Terminalia L. Combretaceae

Origins:

Latin *terminus* "end," referring to the leaves borne in whorls close to the ends of the shoots, branchlets and branches; *Terminus* was also the name of the Roman god of boundaries and frontiers, and *Terminalia* was the festival (celebrated on 23rd February).

Species/Vernacular Names:

T. spp.

English: Indian almond

Mexico: membrillo, almendro, canshán, cortés amarillo, sombrerete, volador

Nigeria: baushe, edo

T. alata Roth

English: Indian laurel

Nepal: asna, saj

T. amazonica (Gmelin) Exell

English: Indian almond

Bolivia: verdolago blanco, verdolago colorado, verdolago amarillo, mara macho, membrillo, almendro, canshán, cortés amarillo, sombrerete, volador

Mexico: almendro, canolté, cashán, cortés amarillo, guayabo, k'anzaan, tepesúchil

T. arjuna (Roxb.) Wight & Arnott (*Pentaptera arjuna* Roxb.)

English: arjuna myrobalan

India: arjuna, arjun, arjunasadara, arjuna-sadra, arjan, raktarjuna, shardul sanmadat, sajadan, sadado, kukubha, marudu, maruthu, vellamarutu, vellamarda, vellai marudamaram, billimatti, tormatti, holematti, tellamaddi, tellamadoi, yermaddi, kahu, maochettu

Sri Lanka: kumbulu, kumbuk, marutu

T. arostrata Ewart & O. Davies

English: nutwood

T. avicennioides Guill. & Perr.

French: badamier duveteux

Nigeria: baushe (Hausa); idi (Yoruba); bo'deyi (Fula); kumanda (Kanuri); kuwah (Tiv)

W. Africa: woloke

T. bellirica (Gaertn.) Roxb. (*Myrobalanus bellirica* Gaertner)

English: Siamese terminalia, bastard myrobalan

Malaya: uji, jelawei, jelawai, mentalun

India: akkam, anilaghnaka, aksha, behedan, behesa, behara, behaira, behada, behda, beda, bherdha, bherda, balra, bahudda, bahira, baheri, bhaira, bhairah, bayrah, bahera, baheda, bhera, birha, bharla, buhura, bohera, boyra, buhuru, balela, bhuta-vasah, berang, sagona, karshapalah, kaligrvamah, vibhitaka, bibithak, vibhitaki, vipitakaha, yella, goting, yel, sagwan, tandra, tusham, tandi, hulluch

Tibetan: barura

Nepal: barro

Sri Lanka: bulu, tanti, ahdan koddai

T. bentzoë (L.) L.f.

La Réunion Island: benjoin, benjoin-pays

Mauritius: bois benzoin

Rodrigues Island: bois charron

T. bentzoë (L.) L.f. subsp. *rodriguesensis* Wickens

Rodrigues Island: bois benjoin, bois charron

T. brachystemma Welw. ex Hiern (*Terminalia baumii* Engl. & Diels) (Greek *brachys* "short" and *stemma, stemmatos* "garland, crown")

South Africa: Kalahari sand terminalia

T. calamansanai (Blanco) Rolfe

Malaya: Kedah tree, mentalun, mentalun batu, batalong

T. catappa L. (*Terminalia latifolia* Blanco non Sw.; *Terminalia mauritania* Blanco non Lam.; *Myrobalanus catappa* (L.) Kuntze) (from the Malabar name katapang)

English: Indian almond, tropical almond, myrobalan, sea almond, Barbados almond, wild almond

Arabic: bedam

Peru: almendra, almendro de Indias, castaña

Mexico: almendro, almendro de Tehuantepec, almendrón, nocuana-huenaa

Japan: momo-tama-na, koba-teishi, kudadi-ishi, unmagii

Malaya: ketapang, lingtak, lingkak

The Philippines: talisay, dalinsi, banilak, almendras

India: jangli badam, bangla badam, deshi badam, desi badam, badami mara, hatbadam, ingudi, badami, tapasataruvu, grahadruma, nattuvadumai, natubadamu, nattubadam, amandi-maram, katappa, adamaram

Sri Lanka: kottamba, kottan

Hawaii: false kamani, kamani haole, kamani 'ula

Nigeria: akumtan, bokomewa-bakala, fonem-baf-mefkof, jombe-ja-mokala

T. chebula Retz.

English: myrobalan, chebulic myrobalan, black myrobalan, ink nut, ink nut tree, Indian gall-nut, gallnut, medicinal terminalia

India: haritaki, hari-taki, haimayathi, karitaki, harara, har, pile-har, pile-hara, bal-har, zangihar, kalehar, hana, harrar, hurh, halela, harir, hirda, harda, hora, kadukkai, katukka, hilikha, pathya, suddha, vayastha, amritha, jivanthi, imachi, zard halela, himaja, kabuli-harda

China: ho tzu, he zi, ho li le

Tibetan: arura

Nepal: barro

Sri Lanka: aralu, kadukkay

T. cunninghamii C. Gardner

Australia: pindang quandong

T. glaucescens Planchon ex Benth.

Nigeria: edo, idi-ogan, idiodan, idi, baushe, kano, karno; baushe (Hausa); idi odan (Yoruba); edo (Igbo)

Yoruba: idi, idi odan, idijo

T. grandiflora Benth. (*Myrobalanus grandiflora* (Benth.) Kuntze)

English: plumwood, nutwood

T. ivorensis A. Chev.

Nigeria: awunsin, awenwen, ipepe, owewe, ebi, egoynlukan, eghoin-nebi, ogoyn-odan, idigbo, didigbo, framire, framiri, nkondi, nkomi, kekange, kekangi, okpoha, obiri, ubiri; idigbo (Yoruba); eghoin-nekwi (Edo); ubiri (Ijaw); awunshin (Igbo); afia (Efik); kekange (Boki)

Yoruba: idigbo, idiigbo, afara dudu

Cameroon: lidia, ngal-ruidou

Ivory Coast: framire

T. laxiflora Engl.

Nigeria: idi odan (Yoruba); zindi, farin baushe (Hausa)

T. macroptera Guill. & Perr.

French: badamier sessile

Nigeria: kandari, kwumanda, ponha, poba; kwandari (Hausa); orin idi odan, ponla (Yoruba)

Yoruba: poba

W. Africa: wolofira, woloba

T. mollis Laws.

Nigeria: baushen giwa (Hausa)

T. oblonga (Ruíz & Pav.) Steudel

Bolivia: verdolago, verdolago amarillo, manicillo

Mexico: guayabo, guayabo volador, sombrerete, volador

T. phanerophlebia Engl. & Diels (Greek *phaneros* "evident, visible, distinct" and *phleps, phlebos* "vein")

English: Lebombo cluster-leaf

Southern Africa: Lebombotrosblaar; amaNgwe-amnyama, amaNgwe-omphofu, amaNgwe (Zulu)

T. prunioides Laws. (*Terminalia porphyrocarpa* Schinz; *Terminalia rautanenii* Schinz; *Terminalia petersii* Engl.; *Terminalia benguellensis* Welw. ex Hiern; *Combretum holstii* Engl.)

English: Lowveld cluster-leaf, Lowveld terminalia, purple-pod terminalia

Southern Africa: sterkbos, hardekoolboom, vaalboom, deurmekaarbos, deurmekaar; xaxandzawu, nshashantsawu (Tsonga); ivikani (Ndebele); muchanana (Shona); mutororo (Mbukushu); mochara (Subya and Kololo); mochiara, motsiyara (Tswana: Western Transvaal, northern Cape, Botswana); mutsiara (North Sotho)

T. sericea Burch. ex DC.

English: silver cluster-leaf, Transvaal silver leaf, silver terminalia, silver tree, Transvaal silver tree, wild quince, silky terminalia, assegai wood, sand yellow wood

Southern Africa: vaalboom, sandvaalboom, sandvaalbos, vaalbos, geelhout, geelhoutboom, sandgeelhout, sandsalie, silwerboom, bloubos, bosvaalbos; mangwe (Ndebele); muHonono, chiJuju, iKonono, namasimba, muSoso, muSusu, muTabvu (Shona); umKhonono, amangwe-amhlope, aMangwe amhlophe, uMangwe, amaNgwe-amhlophe (Zulu); umHonono (Swazi); nkonola (Tsonga); mogonono (Western Transvaal, northern Cape, Botswana); mushosho (Mbukushu: Okavango Swamps and Western Caprivi); mokuba (Malete dialect, Botswana); mogo-nono, moxonono (North Sotho); mususu (Venda); mususu (Kalanga: Northern Botswana); mususu (Shona); omuseasetu (Herero)

N. Rhodesia: mufunji

T. superba Engler & Diels

Zaire: limba

Congo: mulomba, limba, mulimba

Gabon: akom

Central Africa: akom, yasa, n'ganga, offram

Yoruba: afara, afa

Nigeria: afa, afara, afi-eto, afram, offram, bokome, aaha, akom, edo, ega-en, eghoin, eghoin-nofua, egonni, end, fraki, frake, gbarada, jombe, nkon, epion, moukonia, nfako, umwomon-ron, mikoma, nkombi; eji (Nupe); afara (Yoruba); eghoin-nofwa (Edo); egonni (Itsekiri); unwonron (Urhobo); gbarada (Ijaw); edo (Igbo); afia eto (Efik)

Cameroon: ende, akom, lande, ngolu

Ivory Coast: frake

T. triflora (Griseb.) Lillo

Bolivia: lanza amarilla, lanza

T. zeylanica van Heurck & Müll. Arg. (*Terminalia parviflora* Thw.; *Terminalia chebula* Retz. var. *parviflora* Clarke)

Sri Lanka: hampalanda

Terminaliopsis Danguy Combretaceae

Origins:
Resembling *Terminalia*.

Terminthia Bernh. Anacardiaceae

Origins:
From the Greek *terminthos* "terebinth."

Terminthodia Ridley Rutaceae

Origins:
Resembling *terminthos* "terebinth."

Ternatea Miller Fabaceae

Origins:
Of the island of Ternate, Moluccas.

Terniola Tul. Podostemaceae

Origins:
Latin *ternio, ternionis* "the number three."

Terniopsis Chao Podostemaceae

Origins:

Resembling Latin *ternio, ternionis* "the number three."

Ternstroemia Mutis ex L.f. Theaceae

Origins:

Named for the Swedish clergyman Christopher Tärnström (Ternström), 1703-1746 (d. Indochina, Poeloe Candor, Pulo Condor, Poulo Condor or Palicandre), student of Linnaeus, traveler in China and Southeast Asia, chaplain in the Swedish East-Indian Company, teacher, botanist, author of *De Alandia*. Upsaliae 1739, collected plants for Linnaeus; see John Hendley Barnhart, *Biographical Notes upon Botanists*. 3: 360. 1965; A. Lasègue, *Musée botanique de Benjamin Delessert*. Paris 1845; Frans A. Stafleu, *Linnaeus and the Linnaeans*. The spreading of their ideas in systematic botany, 1735-1789. 148. Utrecht 1971.

Species/Vernacular Names:

T. cherryi (Bailey) Bailey & C. White (*Garcinia cherryi* Bailey) (after F.J. Cherry, original collector, Queensland, Cook District, Coen, or for Thomas Cherry, professor of agriculture, University of Melbourne)

Australia: Cherry's mangosteen, Cherry beech

T. gymnanthera (Wight & Arn.) Sprague (*Cleyera gymnanthera* Wight & Arn.) (Greek *gymnos* "naked" and *anthera* "anther")

Japan: mok-koku, iiku, iku

China: bai hua guo

Ternstroemiopsis Urban Theaceae

Origins:
Resembling *Ternstroemia*.

Tersonia Moq. Gyrostemonaceae

Origins:
From the Latin *tersus, a, um* "wiped off, clean, neat," referring to the appearance of the leafless stems.

Species/Vernacular Names:
T. brevipes Moq.
English: button creeper

Tessarandra Miers Oleaceae

Origins:

From the Greek *tessares, tessara* "four" and *aner, andros* "male, stamen."

Tessaria Ruíz & Pav. Asteraceae

Origins:

For the physician Ludovico Tessari, author of *Materia medica*. Venet. 1752, in 1772 he wrote a letter to Albrecht von Haller.

Tessmannia Harms Caesalpiniaceae

Origins:

Dedicated to the German ethnographer Günther (Guenther) Tessmann, explorer, plant collector (western tropical Africa, Spanish Guinea, Peru). Among his writings are *Die Bubi von Fernando Poo*. Herausgegeben von Prof. Dr. O. Reche. Hagen & Darmstadt 1923, *Menschen Ohne Gott*. Stuttgart 1928 and *Die Pangwe*. ... Ergebnisse der Lübecker Pangwe-Expedition 1907-1909 und früherer Forschungen 1904-1907. Berlin 1913; see John Hendley Barnhart, *Biographical Notes upon Botanists*. 3: 369. Boston 1965; Réné Letouzey, "Les botanistes au Cameroun." in *Flore du Cameroun*. 7: 58. Paris 1968; F.N. Hepper and Fiona Neate, *Plant Collectors in West Africa*. 78-79. Utrecht 1971; A.J.M. Leeuwenberg, "Isotypes of which holotypes were destroyed in Berlin." in *Webbia*. 19(2): 863. 1965; Y. Tailfer, *La Forêt dense d'Afrique centrale*. CTA, Ede/Wageningen 1989; J. Vivien & J.J. Faure, *Arbres des Forêts denses d'Afrique Centrale*. Agence de Coopération Culturelle et Technique. Paris 1985.

Species/Vernacular Names:
T. africana Harms
Congo: pamiel
Gabon: nkaga
Cameroon: eba, nkaga, paka, nom esingang
Zaire: waka, wamba, ngare

Tessmanniacanthus Mildbr. Melastomataceae

Origins:
Named for the German ethnographer Günther Tessmann, explorer and plant collector in Africa and Peru.

Tessmannianthus Markgraf Melastomataceae

Origins:

For the German ethnographer Günther Tessmann, explorer and plant collector in Africa and Peru; see John H. Barnhart, *Biographical Notes upon Botanists*. 3: 369. 1965.

Tessmanniodoxa Burret Palmae

Origins:

For the German ethnographer Günther Tessmann, explorer and plant collector in Africa and Peru.

Tessmanniophoenix Burret Palmae

Origins:

For the German ethnographer Günther Tessmann, explorer and plant collector in Africa and Peru.

Testudipes Markgraf Apocynaceae

Origins:

From the Latin *testudo, inis* "a tortoise" and *pes, pedis* "foot," referring to the fruit.

Testulea Pellegrin Ochnaceae

Origins:

After the French (b. Caen) Colonial Administrator (*Administrateur des Colonies*) Georges Marie Patrice Charles Le Testu, 1877-1967 (d. Caen), traveler, explorer, plant collector in West Africa, 1900-1902 Dahomey, 1904-1906 Mozambique, 1907-1934 Gabon and Ubangi-Shari, 1935 Botanic Garden at Caen; see Auguste Jean Baptiste Chevalier (1873-1956), *Flore vivante de l'Afrique Occidentale Française*. 1: xxvii-xxx. Paris 1938; N. Hallé, in *Adansonia*. sér. 2, 7: 263-273. 1967; F.N. Hepper and F. Neate, *Plant Collectors in West Africa*. 50. 1971; H. Lecomte, *Les Bois de la Forêt d'Analamazaotra*. Madagascar 1922; François Pellegrin (1881-1965), *La Flore du Mayombe* d'après les Récoltes de M. Georges Le Testu. [in *Mémoires de la Société Linnéenne de Normandie*. XXVI volume. Two parts.] Caen 1924-1928.

Species/Vernacular Names:

T. gabonensis Pellegrin

Gabon: aké, akéwé, inogongou, isombe, izombe, izombé, zombé

Cameroon: rôné, n'gron, n'rone, rone

Congo: ngwaki, wassa-wassa

Tetraberlinia (Harms) Hauman Caesalpiniaceae

Origins:

From the Greek *tetra* "four" plus *Berlinia*.

Species/Vernacular Names:

T. bifoliolata (Harms) Hauman

Cameroon: ekop, ribi, kaka, ekop ribi

Gabon: eko, kogo, mukuru, parola

Tetracanthus A. Rich. Asteraceae

Origins:

From the Greek *tetra* "four" and *akantha* "thorn."

Tetracarpaea Benth. Anisophylleaceae (Rosidae, Rosales)

Origins:

Greek *tetra* "four" and *karpos* "a fruit."

Tetracarpaea Hook. Grossulariaceae (Escalloniaceae)

Origins:

From the Greek *tetra* "four" and *karpos* "a fruit," four carpels.

Tetracarpidium Pax Euphorbiaceae

Origins:

Greek *tetra* "four" and *karpos* "a fruit," Latin *carpidium* "carpel."

Species/Vernacular Names:

T. conophorum (Müll. Arg.) Hutch. & Dalz.

English: conophor nut, awusa nut

Tetracarpus Post & Kuntze Grossulariaceae (Escalloniaceae)

Origins:

From the Greek *tetra* "four" and *karpos* "a fruit."

Tetracentron Oliver Trochodendraceae (Tetracentraceae)

Origins:
Greek *tetra, tetras* "four" and *kentron* "a spur, prickle," referring to the fruit.

Tetracera L. Dilleniaceae

Origins:
From the Greek *tetra* "four" and *keras* "a horn," an allusion to the curved fruits; see Carl Linnaeus, *Species Plantarum*. 533. 1753 and *Genera Plantarum*. Ed. 5. 737. 1754.

Species/Vernacular Names:
T. sp.
Yoruba: ahon ekun, itakun opon
T. alnifolia Willd.
Sierra Leone: ndo paneneh
Congo: mukazu, moukazou
T. potatoria Afzel. ex F. Don
French: liane à eau
Yoruba: owere, atoyipo
Congo: ndili

Tetrachaete Chiov. Gramineae

Origins:
From the Greek *tetra* "four" and *chaite* "a bristle."

Tetracheilos Lehm. Mimosaceae

Origins:
Greek *tetra* "four" and *cheilos* "lip"; see Johann Georg Christian Lehmann, *Plantae Preissianae ...* Plantarum quas in Australasia occidentali et meridionali-occidentali annis 1838-41 collegit L. Preiss. 2: 368. 1848.

Tetrachne Nees Gramineae

Origins:
From the Greek *tetra* "four" and *achne* "chaff, glume."

Tetrachondra Petrie ex Oliver Labiatae (Tetrachondraceae)

Origins:
From the Greek *tetra* "four" and *chondros* "cartilage, gristle."

Tetrachyron Schltdl. Asteraceae

Origins:
From the Greek *tetra* "four" and *achyron* "chaff, husk."

Tetraclea A. Gray Labiatae

Origins:
From the Greek *tetra* plus *kleis* "lock, key, bar, bolt."

Tetraclinis Mast. Cupressaceae

Origins:
Greek *tetra* "four" and *kline* "a bed, couch," the scale leaves are in whorls of four; see J. Templado, "El Araar, *Tetraclinis articulata* (Vahl) en las Sierra de Cartagena." *Bol. Estac. Centr. Ecol.* 3(5): 43-56. 1975; Martin F. Gardner and Stephen L. Jury, "*Tetraclinis articulata*." in *The Plantsman*. 15(1): 54-59. June 1993.

Species/Vernacular Names:
T. articulata (Vahl) Masters
Morocco/S. Spain: Araar, arar, alerce, thuya

Tetraclis Hiern Ebenaceae

Origins:
From the Greek *tetra* "four" and *kleis* "lock, key, bar."

Tetracme Bunge Brassicaceae

Origins:
From the Greek *tetra* "four" and *akme* "the highest point," referring to the fruit.

Tetracmidion Korsh. Brassicaceae

Origins:
From the Greek *tetra* "four" and *akme* "the highest point"; see also *Tetracme* Bunge.

Tetracoccus Engelm. ex C. Parry Euphorbiaceae

Origins:
Greek *tetra* "four" and *kokkos* "berry," alluding to the four-lobed ovary.

Species/Vernacular Names:
T. dioicus C. Parry

English: Parry's tetracoccus

T. ilicifolius Cov. & Gilman

English: holly-leaved tetracoccus

Tetracoilanthus Rappa & Camarrone Aizoaceae

Origins:

From the Greek *tetra* "four," *koilos* "hollow" and *anthos* "flower."

Tetractinostigma Hassk. Euphorbiaceae

Origins:

From the Greek *tetra* "four," *aktin* "ray" and *stigma* "stigma."

Tetradenia Benth. Labiatae

Origins:

From the Greek *tetra, tetras* "four" and *aden* "a gland."

Species/Vernacular Names:

T. riparia (Hochst.) Codd (*Iboza bainesii* N.E. Br.; *Iboza galpinii* N.E. Br.; *Iboza riparia* (Hochst.) N.E. Br.; *Iboza multiflora* (Benth.) E.A. Bruce; *Moschosma riparium* Hochst.)

English: ginger bush

Southern Africa: chaRogwa, chiRurwe, chororgwe, sharorwe (Shona)

East Africa: fukufuku (Nyika; see J.H. Patterson, *In the Grip of the Nyika.* London 1909); honwa (Marakwet); lilaaku (Lunyore); mshunshu (Baroka Distr.); mwache (Pare); mwaraka (Meru)

Tetradiclis Steven ex M. Bieb. Zygophyllaceae

Origins:

From the Greek *tetra* "four" and *diklis* "double-folding."

Tetradium Dulac Crassulaceae

Origins:

Greek *tetras* "four," Latin *tetradium, ii* "the number four, quaternion, tetrad."

Tetradium Lour. Rutaceae

Origins:

Greek *tetras* "four," referring to the flowers; Greek *tetradion* is a guard of soldiers, normally consisting of four men; Latin *tetradium, ii* "the number four, quaternion, tetrad."

Species/Vernacular Names:

T. sp.

Malaya: inggek burong, tinggek burong, setenggek burong, pauh pauh, pepauh

Tetradoxa C.Y. Wu Adoxaceae

Origins:

Greek *tetras, tetra* plus *Adoxa* L., Greek *a* "lacking, without, negative" and *doxa* "glory, repute," referring to the flowers.

Tetradyas Danser Loranthaceae

Origins:

Possibly from the Greek *tetras, tetrados* "four," referring to the flower.

Tetradymia DC. Asteraceae

Origins:

From the Greek *tetradymos* "fourfold," referring to the four-flowered heads of some species.

Species/Vernacular Names:

T. argyraea Munz & Roos

English: striped horsebrush

T. spinosa Hook. & Arn.

English: cotton-thorn

Tetraedrocarpus O. Schwarz Boraginaceae

Origins:

Greek *hedra* "seat, chair," *tetraedros* "having four faces" and *karpos* "fruit."

Tetraeugenia Merr. Myrtaceae

Origins:

From the Greek *tetras* "four" plus *Eugenia*.

Tetragamestus Reichb.f. Orchidaceae

Origins:
From the Greek *tetras* "four, fourfold" and *gamos* "marriage, stigma, female part," referring to the shape of the stigma, square.

Tetragastris Gaertner Burseraceae

Origins:
Greek *tetras, tetra* with *gaster* "belly, paunch," referring to the form of the fruits.

Species/Vernacular Names:
T. balsamifera (Sw.) Oken (*Hedwigia balsamifera* Sw.)
Spanish: azucarero, azucarero del monte, azucarero de montaña, copal,
French: bois cochon, bois à barrique, bois à flambeaux, bois d'encens, bois de gommier rouge, bois de sucrier, gommier de montagne

Tetraglochidion K. Schumann Euphorbiaceae

Origins:
Greek *tetras, tetra* "four" and *glochis, glochin* "a point, the point of an arrow, a projecting point," *Glochidion* Forst. & Forst.f.

Tetraglochidium Bremek. Acanthaceae

Origins:
Greek *tetras* "four" and *glochis, glochin, glochinos* "a point, the point of an arrow, a projecting point."

Tetraglochin Poeppig Rosaceae

Origins:
Greek *tetras, tetra* "four" and *glochis, glochin* "a point, the point of an arrow, a projecting point."

Tetraglossa Bedd. Euphorbiaceae

Origins:
From the Greek *tetras, tetra* plus *glossa* "a tongue."

Tetragompha Bremek. Acanthaceae

Origins:
From the Greek *tetras, tetra* "four" and *gomphos* "a nail, pin, peg, club."

Tetragonella Miq. Aizoaceae

Origins:
The diminutive of *Tetragonia* L.

Tetragonia L. Aizoaceae

Origins:
Greek *tetras* with *gonia* "an angle, corner," in reference to the shape of the fruit of some species; see Carl Linnaeus, *Species Plantarum*. 480. 1753 and *Genera Plantarum*. Ed. 5. 215. 1754.

Species/Vernacular Names:
T. echinata Ait.
South Africa: klappiesbrak
T. calycina Fenzl (*Tetragonia karasmontana* Dinter ex Adamson; *Tetragonia macroptera* Pax)
South Africa: klappiesbrak
T. reduplicata Welw. ex Oliv. (*Tetragonia arbusculoides* Engl.)
South Africa: klappiesbrak
T. tetragonioides (Pallas) Kuntze (*Tetragonia expansa* Murr.; *Demidovia tetragonioides* Pallas)
English: New Zealand spinach, New Zealand ice-plant, Warrigal cabbage
Maori names: rengamutu, kokihi
Rodrigues Island: épinard
Australia: Warragul cabbage, Warrigal cabbage
Japan: tsuru-na

Tetragonocalamus Nakai Gramineae

Origins:
From the Greek *tetras* "four," *gonia* "an angle" and *kalamos* "a reed, cane."

Tetragonolobus Scop. Fabaceae

Origins:
From the Greek *tetras* "four," *gonia* "an angle" and *lobos* "pod," referring to the fruits.

Tetragonotheca L. Asteraceae

Origins:
From the Greek *tetras* "four," *gonia* "an angle" and *theke* "a box, case."

Tetragyne Miq. Pandaceae

Origins:
Greek *tetras* "four" and *gyne* "woman, female," referring to the pistils.

Tetralepis Steudel Cyperaceae

Origins:
From the Greek *tetras* "four" and *lepis* "scale."

Tetralix Griseb. Tiliaceae

Origins:
Greek *tetraelix* "four times wound round," *tetralix* was mentioned by Theophrastus (*HP.* 6.4.4) for a plant of the thistle kind, Latin *tetralix, icis* applied by Plinius to a plant, *erice*, a species of heath; see H. Genaust, *Etymologisches Wörterbuch der botanischen Pflanzennamen.* 637. Basel 1996.

Tetralobus A. DC. Lentibulariaceae

Origins:
From the Greek *tetras* "four" and *lobos* "pod," referring to the fruits.

Tetralocularia O'Don. Convolvulaceae

Origins:
From the Greek *tetras* "four" and Latin *loculus* "a little place, coffin, compartment."

Tetralopha Hook.f. Rubiaceae

Origins:
From the Greek *tetras* "four" and *lophos* "a crest."

Tetrameles R. Br. Datiscaceae (Tetramelaceae)

Origins:
Greek *tetras, tetra* "four" and *melos* "a limb, part, member," referring to the calyx lobes; see Dixon Denham (1786-1828), Hugh Clapperton (1788-1827) and Walter Oudney (1790-1824), *Narrative of Travels and Discoveries in Northern and Central Africa. 1822-1824.* [Botany by R. Brown.] 230. London 1826.

Tetrameranthus R.E. Fries Annonaceae

Origins:
From the Greek *tetras, tetra* "four," *meros* "part" and *anthos* "flower."

Tetramerista Miq. Tetrameristaceae

Origins:
From the Greek *tetras, tetra* "four" and *meris* "a portion, part."

Tetramerium Gaertner Rubiaceae

Origins:
Greek *tetras, tetra* "four" and *meris* "a portion, part."

Tetramerium Nees Acanthaceae

Origins:
From the Greek *tetras, tetra* "four" and *meris* "a portion, part."

Tetramicra Lindley Orchidaceae

Origins:
From the Greek *tetras, tetra* "fourfold, four" and *mikros* "small," referring to the four-locular anther.

Tetramolopium Nees Asteraceae

Origins:
Greek *tetras* "four" and *molops* "a stripe," referring to the nerved achenes.

Species/Vernacular Names:
T. capillare (Gaud.) St. John (*Senecio capillaris* Gaud.; *Luteidiscus capillaris* (Gaud.) St. John; *Tetramolopium bennettii* Sherff)
Hawaii: pamakani

Tetranema Benth. Scrophulariaceae

Origins:
Greek *tetras* "four" and *nema* "thread, filament," referring to the stamens.

Tetraneuris Greene Asteraceae

Origins:

From the Greek *tetras* "four" and *neuron* "nerve."

Tetranthera Jacq. Lauraceae

Origins:

Greek *tetras* "four" and *anthera* "anther."

Tetranthus Sw. Asteraceae

Origins:

From the Greek *tetras* "four" and *anthos* "flower."

Tetrapanax (K. Koch) K. Koch Araliaceae

Origins:

Greek *tetra* "four" and the genus *Panax*, referring to the four seeds or to the flowers, in four.

Species/Vernacular Names:

T. papyrifer (Hook.) K. Koch (*Aralia papyrifera* Hook.; *Fatsia papyrifera* (Hook.) Miq. ex Witte)

English: rice-paper plant, Chinese rice-paper plant, rice-paper tree, pith-paper tree

Japan: kami-yatsu-de, tsû-datsu-boku

China: tong cao, tung to mu, tung tsao

Tetrapathaea (DC.) Reichb. Passifloraceae

Origins:

See H.E. Connor and E. Edgar, "Name changes in the indigenous New Zealand flora, 1960-1986 and Nomina Nova IV, 1983-1986." *New Zealand Journal of Botany.* Vol. 25: 115-170. 1987; P.S. Green, *Kew Bull.* 26: 539-558. 1972; W.J.J.O. de Wilde, in *Blumea.* 22: 37-50. 1974.

Tetrapeltis Lindley Orchidaceae

Origins:

Greek *tetra* plus *peltis* "a shield," referring to the pollinia.

Tetraperone Urb. Asteraceae

Origins:

From the Greek *tetra* "four" and *perone* "rivet, buckle."

Tetrapetalum Miq. Annonaceae

Origins:

From the Greek *tetra* "four" and *petalon* "petal, leaf."

Tetraphylax (G. Don) Vriese Goodeniaceae

Origins:

From the Greek *tetra* and *phylax, phylakos* "a guardian, protector, defender."

Tetraphyllaster Gilg Melastomataceae

Origins:

From the Greek *tetra* "four," *phyllon* "leaf" and *aster, astron* "a star, the stars."

Tetraphyllum Griffith ex C.B. Clarke Gesneriaceae

Origins:

Greek *tetra* and *phyllon* "leaf," the leaves whorled at the apex.

Tetraphysa Schltr. Asclepiadaceae

Origins:

From the Greek *tetra* "four" and *physa* "bladder."

Tetrapilus Lour. Oleaceae

Origins:

From the Greek *tetra* "four" and *pilos* "hat, cap, felt cap, hair."

Tetraplandra Baillon Euphorbiaceae

Origins:

Greek *tetraploos, tetraplous* "fourfold" and *aner, andros* "male, stamen."

Tetraplasandra A. Gray Araliaceae

Origins:

Greek *tetraplasios* "fourfold" and *aner, andros* "male, stamen," referring to the number of stamens.

Species/Vernacular Names:

T. gymnocarpa (Hillebr.) Sherff (*Pterotropia gymnocarpa* Hillebr.; *Dipanax gymnocarpa* (Hillebr.) A. Heller; *Heptapleurum gymnocarpa* (Hillebr.) Drake)

Hawaii: 'ohe'ohe

T. havaiensis A. Gray

Hawaii: 'ohe

T. kavaiensis (H. Mann) Sherff (*Tetraplasandra micrantha* Sherff; *Tetraplasandra turbans* Sherff; *Heptapleurum kavaiense* H. Mann; *Agalma kavaiense* (H. Mann) Seem.)

Hawaii: 'ohe'ohe

T. oahuensis (A. Gray) Harms (*Tetraplasandra meiandra* (Hillebr.) Harms; *Tetraplasandra lydgatei* Hillebr.; *Tetraplasandra kaalae* (Hillebr.) Harms; *Tetraplasandra bisattenuata* Sherff)

Hawaii: 'ohe mauka

T. waimeae Wawra (*Tetraplasandra waimeae* (Wawra) A. Heller)

Hawaii: 'ohe kiko'ola

Tetraplasia Rehder Rubiaceae

Origins:

From the Greek *tetraplasios* "fourfold."

Tetrapleura Benth. Fabaceae

Origins:

From the Greek *tetras* and *pleura*, *pleuro*, *pleuron* "side, rib," referring to the fruit.

Species/Vernacular Names:

T. sp.

Zaire: munyenye, munienie

T. tetraptera (Schum.) Taubert (*Adenanthera tetraptera* Schum.; *Tetrapleura thonningii* Benth., nom. illeg.)

Cameroon: akpa, dawo, essanga sanga, essesse, esesié, kpwa, djaga

Central Africa: angulu

Congo: badiok, ezibil, kiaka, eyaka

Gabon: cocso, enkagouma, mbaghesa, ogagoume, ouogueso, nkouarsa

Ghana: esem

Ivory Coast: belfou, esse-hesse, esehese, tchiekboue, tické, m'bekréhia

Yoruba: aridan, aidan

Nigeria: aqua, apapa, aridan, aidan, aridan, ayidan, ebuk, bokomake, dawo, kalangon-daji, esisang, esekeseki, mfe, iminiyi, igirihimi, ighimiaki, ighimiakia, kombolo, osakirisa, osshosha, osshogisha, sassas, sekokbagu, ushosho; ikoho (Nupe); aridan (Yoruba); ighimiakia (Edo); ighirehimi (Ishan); imiminje (Etsako); apapa (Ijaw); oshosho (Igbo); edeminang (Efik); ebuk (Boki)

Togo: prekese

Tetrapodenia Gleason Malpighiaceae

Origins:

From the Greek *tetras* "four" and *pous*, *podos* "foot."

Tetrapogon Desf. Gramineae

Origins:

From the Greek *tetras*, *tetra* "four" and *pogon* "beard," referring to the tufts of hairs.

Tetrapollinia Maguire & B.M. Boom Gentianaceae

Origins:

From the Greek *tetras* "four" and *pollinia*.

Tetrapoma Turcz. ex Fischer & C.A. Meyer Brassicaceae

Origins:

From the Greek *tetras* and *poma*, *pomatos* "a lid, cover."

Tetrapora Schauer Myrtaceae

Origins:

Greek *tetras* "four" and *poros* "opening, pore."

Tetraptera Phil. Malvaceae

Origins:

From the Greek *tetrapteros* "four-winged."

Tetrapteris Cav. Malpighiaceae

Origins:

See *Tetrapterys* Cav.

Tetrapterocarpon Humbert Caesalpiniaceae

Origins:

From the Greek *tetras, tetra* "four," *pteron* "wing" and *karpos* "fruit," referring to the fruits.

Tetrapterys Cav. Malpighiaceae

Origins:

Greek *tetras, tetra* "four" and *pteron* "wing," referring to the winged samaras.

Tetrardisia Mez Myrsinaceae

Origins:

From the Greek *tetras, tetra* "four" plus *Ardisia*, the genus is 4-merous.

Tetraria P. Beauv. Cyperaceae

Origins:

From the Greek *tetra* "four," referring to the style branches or to the four stamens.

Species/Vernacular Names:
T. capillaris (F. Muell.) J. Black (*Chaetospora capillaris* F. Muell.; *Elynanthus capillaceus* Benth.; *Machaerina capillacea* (Benth.) Koyama; *Cladium capillaceum* (Benth.) C.B. Clarke; *Heleocharis halmaturina* J. Black; *Tetraria halmaturina* (J. Black) J. Black)

English: hair sedge, bristle twig-rush

Tetrariopsis C.B. Clarke Cyperaceae

Origins:
Resembling *Tetraria* P. Beauv.

Tetrarrhena R. Br. Gramineae

Origins:

From the Greek *tetra* "four" and *arrhen* "male," the spp. have four stamens.

Species/Vernacular Names:
T. distichophylla (Labill.) R. Br. (*Ehrharta distichophylla* Labill.)

English: hairy rice-grass

Tetraselago Junell Scrophulariaceae

Origins:

Greek *tetras* "four" and the genus *Selago*.

Tetrasida Ulbr. Malvaceae

Origins:

From the Greek *tetras* "four" plus *Sida* L.

Tetrasiphon Urban Celastraceae

Origins:

From the Greek *tetras* "four" and *siphon* "tube."

Tetraspidium Baker Scrophulariaceae

Origins:

Greek *tetras* and *aspidion* "a small round shield," *aspidos, aspis* "a shield."

Tetraspora Miq. Myrtaceae

Origins:

Orthographic variant of *Tetrapora* Schauer; see Arthur D. Chapman, ed., *Australian Plant Name Index.* 2827. Canberra 1991.

Tetrastemma Diels ex H. Winkl. Annonaceae

Origins:

From the Greek *tetras* "four" and *stemma, stemmatos* "a garland, crown."

Tetrastemon Hooker & Arnott Myrtaceae

Origins:

From the Greek *tetras* "four" and *stemon* "a stamen."

Tetrastichella Pichon Bignoniaceae

Origins:
From the Greek *tetras* "four" and *stichos* "a row, line."

Tetrastigma (Miq.) Planchon Vitaceae

Origins:
Greek *tetras* "four" and *stigma*, referring to the four-lobed stigma.

Species/Vernacular Names:
T. formosana (Hemsl.) Nakai (*Vitis formosana* Hemsl.)
Japan: mitsuba-binbô-kazura
T. obtectum (Wall. ex Lawson) Planchon
China: zou you cao
T. voinieranum (Baltet) Pierre ex Gagnep. (*Cissus voinierana* (Baltet) Viala; *Vitis voinierana* Baltet)
English: chestnut vine, lizard plant

Tetrastigma K. Schumann Rubiaceae

Origins:
Greek *tetras* "four" plus *stigma*.

Tetrastylidium Engl. Olacaceae

Origins:
From the Greek *tetras* "four" and *stylos* "a column," *stylidion* "a small pillar."

Tetrastylis Barb. Rodr. Passifloraceae

Origins:
From the Greek *tetras* "four" and *stylos* "a column."

Tetrasynandra Perkins Monimiaceae

Origins:
Greek *tetras* "four," *syn* "together, united" and *aner, andros* "man, stamen," four stamens close together.

Tetrataenium (DC.) Manden. Umbelliferae

Origins:
From the Greek *tetras* "four" and *tainia* "fillet, a ribbon."

Tetrataxis Hook.f. Lythraceae

Origins:
From the Greek *tetras* "four" and *taxis* "a series, order, arrangement."

Tetrateleia Arwidsson Capparidaceae (Capparaceae)

Origins:
Greek *tetras* "four" and *telos, teleos* "an end, accomplishment, consummation," or *thele* "nipple," *thelion* "a little nipple or teat."

Tetratelia Sonder Capparidaceae (Capparaceae)

Origins:
Greek *tetras* "four" and *telos, teleos* "an end, accomplishment, consummation," or *thele* "nipple," *thelion* "a little nipple or teat."

Tetrathalamus Lauterb. Winteraceae

Origins:
Greek *tetras* "four" and *thalamos* "the base of the flower"; see also *Zygogynum* Baill.

Tetratheca Smith Tremandraceae

Origins:
Greek *tetras* "four" and *theke* "a box, case," referring to the four anther-loculi, the anthers are often four-lobed or four-celled.

Species/Vernacular Names:
T. halmaturina J. Black (*Tetratheca ericifolia* Smith var. *aphylla* Tate)
Australia: curly pink-bells

Tetrathylacium Poeppig Flacourtiaceae

Origins:
From the Greek *tetras* "four" and *thylakos* "bag, sack, pouch."

Tetrathyrium Benth. Hamamelidaceae

Origins:
From the Greek *tetras* "four" and *athyros, athyron* "without door, open."

Tetraulacium Turcz. Scrophulariaceae

Origins:
From the Greek *tetras* "four" and *aulax, aulakos* "a furrow."

Tetrazygia Rich. ex DC. Melastomataceae

Origins:
From the Greek *tetras* plus *zygon, zygos* "yoke."

Tetrazygiopsis Borhidi Melastomataceae

Origins:
Resembling *Tetrazygia* Rich. ex DC.

Tetroncium Willd. Juncaginaceae

Origins:
Greek *tetras* "four" and *onkinos* "hook, hooked," *onkos* "tumor, tubercle."

Tetrorchidiopsis Rauschert Euphorbiaceae

Origins:
Resembling *Tetrorchidium* Poepp.

Tetrorchidium Poeppig Euphorbiaceae

Origins:
Greek *tetras* "four" and *orchis, orchidos* "a testicle," referring to the anthers.

Species/Vernacular Names:
T. sp.
Liberia: plor-plor
T. didymostemon (Baillon) Pax & K. Hoffm. (*Tetrorchidium minus* (Prain) Pax & K. Hoffmann)
Nigeria: ofun-oke (Yoruba); iheni (Edo)
Yoruba: ofun oke
Ivory Coast: oulogbaoue, amene, bredoue
Congo: onlili, oulili

Teucridium Hook.f. Labiatae (Verbenaceae)

Origins:
Referring to *Teucrium*.

Teucrium L. Labiatae

Origins:
Greek *teukrion*, possibly for Teucer (Teukros) the founder of the town of Salamis in Cyprus; Latin *teucrion, ii* for a plant, the germander, *Teucrium chamaedrys* L., the herb spleenwort, *Teucrium flavium* L.; see Helmut Genaust, *Etymologisches Wörterbuch der botanischen Pflanzennamen.* 638. Basel 1996

Species/Vernacular Names:
T. bidentatum Hemsley
English: two-dentate germander
China: er chi xiang ke ke
T. canadense L. (*Teucrium littorale* E. Bickn.)
English: American germander, wood sage
T. chamaedrys L. (*Teucrium officinale* Lam.)
English: wall germander, ground oak, common germander
Arabic: bellout el-ard
T. corymbosum R. Br.
English: forest germander
T. fruticans L.
English: tree germander
T. japonicum Willdenow
English: Japanese germander
China: shi hua xiang ke ke
T. marum L.
English: cat thyme
T. occidentale A. Gray
English: hairy germander
T. pernyi Franchet
English: Perny germander
China: lu shang xiang ke ke
T. quadrifarium Buch.-Ham. ex D. Don
English: fourfile germander
China: tie zhou cao
T. racemosum R. Br. (*Teucrium racemosum* R. Br. var. *triflorum* J. Black)
English: grey germander
T. scorodonia L.
English: wood germander, wood sage
T. sessiliflorum Benth.
English: camel bush
T. simplex Vaniot
English: simplex germander
China: exiang ke ke

T. viscidum Blume

English: viscid germander

China: xue jian chou, shan huo xiang

Teuscheria Garay Orchidaceae

Origins:

For the German botanist Heinrich (Henry) Teuscher, 1891-1984, gardener, dendrologist, orchidologist, landscape architect, 1922 to the United States, from 1936 to 1972 Curator of the Botanical Garden of Montreal, orchid-explorer in Ecuador. Among his writings are *Window-Box Gardening*. New York 1956 and "Two species of the genus *Galeandra*." *Amer. Orch. Soc. Bull.* 30: 802-806. 1961, with Rudolph Adler wrote *The Soil and Its Fertility*. New York [1960]; see John H. Barnhart, *Biographical Notes upon Botanists*. 3: 369. 1965; R. Zander, F. Encke, G. Buchheim and S. Seybold, *Handwörterbuch der Pflanzennamen*. 14. Aufl. Stuttgart 1993; T.W. Bossert, *Biographical Dictionary of Botanists Represented in the Hunt Institute Portrait Collection*. 397. 1972; Ethelyn Maria Tucker, *Catalogue of the Library of the Arnold Arboretum of Harvard University*. Cambridge, Massachusetts 1917-1933; Ida Kaplan Langman, *A Selected Guide to the Literature on the Flowering Plants of Mexico*. 733. University of Pennsylvania Press, Philadelphia 1964; S. Lenley et al., *Catalog of the Manuscript and Archival Collections and Index to the Correspondence of John Torrey*. Library of the New York Botanical Garden. 408. 1973.

Textoria Miq. Araliaceae

Origins:

Latin *textor, textoris* "a weaver."

Teysmannia Reichb. & Zoll. Palmae

Origins:

Named for the Dutch botanist Johannes Elias Teijsmann (Teysmann), 1809-1882, traveler, gardener and botanical explorer, plant collector, 1831-1869 Curator of the Buitenzorg (Bogor) Botanic Gardens; see J.H. Barnhart, *Biographical Notes upon Botanists*. 3: 365. Boston 1965; R. Zander, F. Encke, G. Buchheim and S. Seybold, *Handwörterbuch der Pflanzennamen*. 14. Aufl. 1993; T. W. Bossert, *Biographical Dictionary of Botanists Represented in the Hunt Institute Portrait Collection*. 397. Boston, Massachusetts 1972; E.M. Tucker, *Catalogue of the Library of the Arnold Arboretum of Harvard University*. Cambridge, Massachusetts 1917-1933]; Stafleu and Cowan, *Taxonomic*

Literature. 6: 201-204. 1986; Hazel Cropper, Natalie W. Uhl and John Dransfield, "Index to *Principes*, Volume 1-40." *Principes*. 41(3S): 3-132. 1997.

Teyssmannia Reichb. & Zoll. Palmae

Origins:

Named for the Dutch botanist Johannes Elias Teijsmann (Teysmann), 1809-1882; see *Teysmannia*.

Thalassia Banks ex C. Koenig Hydrocharitaceae (Thalassiaceae)

Origins:

Greek *thalassa, thalatta* "the sea," submerged marine aquatics; Akkadian *apsu* "sea, deep water," *sala'u* "to besprinkle"; see G. Semerano, *Le origini della cultura europea. Dizionari Etimologici. Basi semitiche delle lingue indeuropee. Dizionario della lingua Greca*. 2(1): 113. Leo S. Olschki Editore, Firenze 1994.

Species/Vernacular Names:

T. hemprichii (Solms) Asch. (*Schizotheca hemprichii* Ehrenb.) (named for the plant collector F.W. Hemprich, companion to Christian Gottfried Ehrenberg (1795-1876) on Red Sea coral reefs)

English: turtle grass

Japan: Ryûkyû-suga-mo

Thalassodendron Hartog Cymodoceaceae

Origins:

Greek *thalassa, thalatta* "the sea" and *dendron* "tree," branched like a tree and aquatic.

Thalia L. Marantaceae

Origins:

Named for the German physician Johannes Thal (Thalius), 1542/1543-1583, botanist, *Stadtphysikus* in Stolberg am Harz, author of *Sylva Hercynia*. Francofurto ad Moenum [Frankfurt a.M.] 1588; see John H. Barnhart, *Biographical Notes upon Botanists*. 3: 369. 1965; H. Genaust, *Etymologisches Wörterbuch der botanischen Pflanzennamen*. 638. 1996; Jonas C. Dryander, *Catalogus bibliothecae historico-naturalis Josephi Banks*. London 1796-1800; G.C. Wittstein, *Etymologisch-botanisches Handwörterbuch*. 874. Ansbach 1852; E.M. Tucker, *Catalogue of the Library of the Arnold Arboretum of Harvard University*. Cambridge,

Massachusetts 1917-1933; Richard J. Durling, *A Catalogue of Sixteenth Century Printed Books in the National Library of Medicine*. 806. Bethesda, Maryland 1967; Carl Linnaeus, *Species Plantarum*. 1193. 1753 and *Genera Plantarum*. Ed. 5. 3. 1754.

Thalictrum L. Ranunculaceae

Origins:

Taliktron is a name used by Dioscorides for a plant with coriander-like leaves, Latin *thalictrum* or *thalitruum* applied by Plinius to a plant, meadow-rue; see H. Genaust, *Etymologisches Wörterbuch der botanischen Pflanzennamen*. 638. 1996.

Species/Vernacular Names:

T. flavum L.

English: yellow meadow rue

Thalysia Kuntze Gramineae

Origins:

Greek *thalysia* "offerings of the first-fruits, made to Artemis," *thalos, thallos* "scion, child."

Thaminophyllum Harvey Asteraceae

Origins:

From the Greek *thaminos* "crowded" and *phyllon* "leaf."

Thamnea Sol. ex Brongn. Bruniaceae

Origins:

From the Greek *thamnos* "shrub, bush."

Thamnocalamus Munro Gramineae

Origins:

Greek *thamnos* "shrub, bush" and *kalamos* "reed," referring to the nature of the plant.

Species/Vernacular Names:

T. spathaceus (Franchet) Söderstr.

English: umbrella bamboo

T. tessellatus (Nees) Söderstrom & Ellis (*Arundinaria tessellata* (Nees) Munro) (Latin *tessellatus, a, um* "checkered, tesselated, of small square stones")

English: berg bamboo, miniature bamboo

South Africa: bergbamboes

Thamnocharis W.T. Wang Gesneriaceae

Origins:

From the Greek *thamnos* "shrub, bush" and *charis* "delicacy, delight, grace, beauty."

Thamnochortus Bergius Restionaceae

Origins:

Greek *thamnos* "shrub, bush" and *chortos* "green herbage, grass"; see Peter Jonas Bergius (1730-1790), *Descriptiones plantarum ex Capite Bonae Spei*. 353. Stockholmiae 1767.

Species/Vernacular Names:

T. fruticosus Berg.

English: Cape grass, love grass

Southern Africa: isiNama (Xhosa)

T. insignis Mast.

South Africa: dekriet

Thamnojusticia Mildbr. Acanthaceae

Origins:

From the Greek *thamnos* "shrub, bush" plus the genus *Justicia* L.

Thamnopteris (C. Presl) C. Presl Aspleniaceae

Origins:

From the Greek *thamnos* and *pteris* "fern"; see C. Presl (1794-1852), *Epimeliae botanicae*. Pragae 1849 [reprinted from *Abhandlungen der Königlichen Böhmischen Gesellschaft der Wissenschaften*. 1851].

Thamnosciadium Hartvig Umbelliferae

Origins:

From the Greek *thamnos* "shrub, bush" and *skias, skiados* "a canopy, umbel."

Thamnoseris F. Philippi Asteraceae

Origins:

Greek *thamnos* "shrub, bush" and *seris, seridos* "chicory, lettuce."

Thamnosma Torrey & Frémont Rutaceae

Origins:
Greek *thamnos* with *osme* "smell, odor, perfume."

Species/Vernacular Names:
T. montana Torrey & Frémont
English: turpentine-broom

Thamnus Klotzsch Ericaceae

Origins:
From the Greek *thamnos* "a shrub, bush."

Thapsia L. Umbelliferae

Origins:
Greek *thapsos, thapsia* "a plant or wood used for dying yellow," Latin *thapsia* "a poisonous shrub" (Plinius); see Helmut Genaust, *Etymologisches Wörterbuch der botanischen Pflanzennamen.* 639. Basel 1996.

Species/Vernacular Names:
T. garganica L.
English: Spanish turpeth root, drias plant, smooth thapsia
Arabic: diryes, derias, bounefaa

Tharpia Britton & Rose Caesalpiniaceae

Origins:
Dedicated to the American botanist Benjamin Carroll Tharp, 1885-1964, plant physiologist, professor of biology and botany. His works include "Texas parasitic fungi." *Mycologia.* 9(2): 105-124. 1917 and *The Vegetation of Texas.* Houston 1939, with Fred A. Barkley wrote "The genus *Ruellia* in Texas." *Amer. Midl. Nat.* 42: 1-86. 1949; see Ida Kaplan Langman, *A Selected Guide to the Literature on the Flowering Plants of Mexico.* 733. Philadelphia 1964; E.M. Tucker, *Catalogue of the Library of the Arnold Arboretum of Harvard University.* Cambridge, Massachusetts 1917-1933; Irving William Knobloch, compil., "A preliminary verified list of plant collectors in Mexico." *Phytologia Memoirs.* VI. 1983.

Thaumasianthes Danser Loranthaceae

Origins:
Greek *thaumasia* "wonder," *thaumazo, thaumazein* "to wonder, marvel at a thing," *thauma, thaumatos* "a wonder, marvel" and *anthos* "flower."

Thaumastochloa C.E. Hubb. Gramineae

Origins:
From the Greek *thauma* "a wonder, marvel," *thaumastos* "wonderful, marvellous, strange" and *chloe, chloa* "grass," referring to the nature of the plant.

Species/Vernacular Names:
T. cochinchinensis (Lour.) C.E. Hubb. (*Phleum cochinchinense* Lour.)
Japan: hime-ushi-no-shippei

Thaumatocaryon Baillon Boraginaceae

Origins:
From the Greek *thauma, thaumatos* and *karyon* "nut."

Thaumatococcus Benth. Marantaceae

Origins:
Greek *thauma, thaumatos* "a wonder" and *kokkos* "berry."

Species/Vernacular Names:
T. daniellii (Benn.) Benth.
Yoruba: katemfe, ketenfe, kekerenfe

Thaumatophyllum Schott Araceae

Origins:
Greek *thauma, thaumatos* with *phyllon* "leaf"; see D.H. Nicolson, "Derivation of aroid generic names." *Aroideana.* 10: 15-25. 1988.

Thayera Copeland Polypodiaceae

Origins:
For Alfred Thayer, a professor at the University of Texas, USA; see Stafleu and Cowan, *Taxonomic Literature.* 6: 236. 1986.

Thea L. Theaceae

Origins:
Chinese (Amoy) *t'e*, Dutch *thee*, Mandarin *ch'a* (and *cha-ye* are the leaves of tea); see H. Genaust, *Etymologisches Wörterbuch der botanischen Pflanzennamen.* 639. 1996 Ernest Weekley, *An Etymological Dictionary of Modern English.* 2: 1480. New York 1967; Manlio Cortelazzo &

Paolo Zolli, *Dizionario etimologico della lingua italiana.* 5: 1319. Bologna 1988.

Theaphyla Raf. Theaceae

Origins:
For *Thea* L.= *Camellia* L.; see C.S. Rafinesque, *Flora Telluriana.* 1: 17. 1836 [1837].

Thecacoris A. Juss. Euphorbiaceae

Origins:
Greek *theke* "a box, case, envelope" and *koris* "a bug."

Species/Vernacular Names:
T. sp.
Congo: monon, munon

Thecagonum Babu Rubiaceae

Origins:
From the Greek *theke* "a box, case, envelope" and *gonia* "angle."

Thecanthes Wikstrom Thymelaeaceae

Origins:
Derived from the Greek *theke* "a box, case, envelope, sac" and *anthos* "flower," referring to the sac-like concave receptacle and attached bracts enclosing the flowers; see B.L. Rye, "A revision of Western Australian Thymelaeaceae." *Nuytsia.* 6(2): 129-278. 1988.

Thecocarpus Boiss. Umbelliferae

Origins:
From the Greek *theke* and *karpos* "fruit," referring to the mericarps.

Thecophyllum E.F. André Bromeliaceae

Origins:
From the Greek *theke* "a box, case" and *phyllon* "leaf."

Thecopus Seidenf. Orchidaceae

Origins:
From the Greek *theke* "a box, case, envelope" and *pous, podos* "a foot," indicating the base of the column.

Thecostele Reichb.f. Orchidaceae

Origins:
From the Greek *theke* "a receptacle" and *stele* "a pillar, column, trunk," referring to the gynostemium.

Theilera E. Phillips Campanulaceae

Origins:
For Sir Arnold Theiler, 1867-1936 (d. London), veterinarian, botanist; see Mary Gunn and Leslie E. Codd, *Botanical Exploration of Southern Africa.* 342-343. Cape Town 1981.

Thelasis Blume Orchidaceae

Origins:
Greek *thele* "nipple," presumably referring to the shape of the rostellum, emarginate and attenuate; see Karl Ludwig von Blume, *Bijdragen tot de flora van Nederlandsch Indië.* 385, t. 75. Batavia (Sep.-Dec.) 1825.

Thelecarpus Tieghem Loranthaceae

Origins:
From the Greek *thele* "nipple" and *karpos* "fruit."

Thelechitonia Cuatrec. Asteraceae

Origins:
From the Greek *thele* and *chiton* "a tunic, covering."

Thelepaepale Bremek. Acanthaceae

Origins:
Greek *thele* "nipple" and *pale* "pollen, any fine dust, ashes," *paipale* "the finest flour or meal," referring to the nipple-like spinules on the pollen grains.

Thelepogon Roth ex Roemer & Schultes Gramineae

Origins:
Greek *thele* "nipple" and *pogon* "beard," possibly referring to the stamens.

Thelesperma Less. Asteraceae

Origins:
Greek *thele* and *sperma* "seed," referring to the rough achenes.

Thelethylax C. Cusset Podostemaceae

Origins:
From the Greek *thele* and *thylakos* "bag, sack, pouch."

Theligonum L. Rubiaceae (Theligonaceae)

Origins:
Greek *thelys, theleia, thelia* "female" and *gonos* "progeny, seed, offspring," it was supposed to give the power of producing female offspring; Latin *thelygonon* for a species of the plant *phyllum* (Plinius).

Thelionema R.J.F. Henderson Phormiaceae (Liliaceae)

Origins:
From the Greek *thele* "nipple," *thelion* "a little nipple or teat" and *nema* "thread," referring to the covering of the staminal filaments.

Species/Vernacular Names:
T. caespitosum (R. Br.) R. Henderson (*Stypandra caespitosa* R. Br.)
English: tufted blue lily

Thelipteris Adans. Pteridaceae (Adiantaceae)

Origins:
From the Greek *thelys* "female" and *pteris* "fern."

Thellungia Stapf Gramineae

Origins:
After the Swiss botanist Albert Thellung, 1881-1928, at the Botanical Museum of the University of Zürich. Among his numerous publications are "*Lepidium*-Studien." *Bull. Herb. Boissier.* ser. 2. 4(7): [695]-716. 1904, *Die Gattung Lepidium* (L.) R. Br. Zürich 1906 and *La Flore adventice de Montpellier.* Cherbourg 1912; see John H. Barnhart, *Biographical Notes upon Botanists.* 3: 371. 1965; T.W. Bossert, *Biographical Dictionary of Botanists Represented in the Hunt Institute Portrait Collection.* 398. 1972; E.M. Tucker, *Catalogue of the Library of the Arnold Arboretum of Harvard University.* Cambridge, Massachusetts 1917-1933; Ida Kaplan Langman, *A Selected Guide to the Literature on the Flowering Plants of Mexico.* 733-734. Philadelphia 1964; Stafleu and Cowan, *Taxonomic Literature.* 6: 242-245. Utrecht 1986.

Species/Vernacular Names:
T. advena Stapf
English: coolibah grass

Thellungiella O.E. Schulz Brassicaceae

Origins:
For the Swiss botanist Albert Thellung, 1881-1928.

Thelocactus (K. Schumann) Britton & Rose Cactaceae

Origins:
Greek *thele* "teat, nipple" plus *Cactus.*

Thelocephala Y. Ito Cactaceae

Origins:
From the Greek *thele* "teat, nipple" and *kephale* "head."

Thelomastus Fric Cactaceae

Origins:
Greek *thele* "teat, nipple" and *mastos* "a breast," referring to the areolae; see also *Thelocactus* (K. Schum.) Britton & Rose.

Thelychiton Endl. Orchidaceae

Origins:
Greek *thelys* "feminine, female" and *chiton* "a tunic, covering, dress," referring to the column.

Thelycrania (Dumort.) Fourr. Cornaceae

Origins:
From the Greek *thelys* plus *kraneia* for cornelian cherry, a species of *Cornus.*

Thelymitra Forster & Forster f. Orchidaceae

Origins:

Greek *thelys* "feminine, female" and *mitra* "a turban, hat," referring to the hood-like appendage of the column.

Species/Vernacular Names:

T. antennifera Hook.f. (*Macdonaldia antennifera* Lindley)

English: rabbit ears, lemon orchid

T. canaliculata R. Br. (*Thelymitra azurea* R. Rogers)

English: azure sun-orchid

T. ixioides Sw.

English: spotted sun-orchid, dotted sun-orchid

T. mucida Fitzg. (Latin *mucidus* "moldy, musty")

English: plum orchid

T. nuda R. Br. (*Thelymitra megcalyptra* Fitzg.; *Thelymitra aristata* Lindley var. *megcalyptra* (Fitzg.) Nicholls ex J. Black)

English: scented sun-orchid, plain sun-orchid

T. venosa R. Br.

English: veined sun-orchid

Thelypodiopsis Rydb. Brassicaceae

Origins:

Resembling *Thelypodium* Endl.

Thelypodium Endl. Brassicaceae

Origins:

Greek *thelys* "feminine, female" and *pous, podos* "a foot," *podion* "a little foot," fruit generally stalked above the receptacle.

Species/Vernacular Names:

T. brachycarpum Torrey

English: short-podded thelypodium

T. stenopetalum S. Watson

English: slender-petaled thelypodium

Thelypteris Schmidel Thelypteridaceae

Origins:

Greek *thelypteris, thelys* "female" and *pteris* "fern," name used by Theophrastus (*HP*. 9.18.8) and Dioscorides for a fern, bracken; Latin *thelypteris, is,* Plinius applied to the female plant of the *filix* (fern polypody).

Species/Vernacular Names:

T. palustris Schott

English: marsh fern

Thelyschista Garay Orchidaceae

Origins:

From the Greek *thelys* "female" and *schistos* "cut," *schizo, schizein* "to divide."

Themeda Forssk. Gramineae

Origins:

From *thaemed*, an Arabic name for a grass; see Pehr (Peter) Forsskål (1732-1763), *Flora aegyptiaco-arabica*. 178. Copenhagen 1775.

Species/Vernacular Names:

T. triandra Forssk. (*Themeda polygama* J.F. Gmel.; *Themeda forskalii* Hack.; *Themeda australis* (R. Br.) Stapf; *Themeda triandra* Forssk. var. *burchellii* (Hack.) Stapf; *Themeda triandra* Forssk. var. *hispida* (Nees) Stapf; *Themeda triandra* Forssk. var. *imberbis* (Retz.) A. Camus; *Themeda triandra* Forssk. var. *trachyspathea* Goossens; *Themeda triandra* Forssk. var. *vulgaris* auct. non Hack.; *Anthistiria australis* R. Br.; *Anthistiria forskalii* Kunth)

English: angle grass, blue grass, kangaroo grass, red grass, red oat grass

Arabic: thaemed, themed, alaf

Southern Africa: rooigras, angelgras, asgras, blougras, hoëveldrooigras, platgras, rooiangel, rooihawergras, rooisaadgras, soetgras, swartangel; insinde (Zulu), makgulu (Tswana); seboko (Sotho); mSinde (Xhosa)

The Philippines: bagokbok, panau, samsamong, taau, usimau

Theobroma L. Sterculiaceae

Origins:

From the Greek *theos* "a god" and *broma* "food"; see Ernest Weekley, *An Etymological Dictionary of Modern English*. 1: 230, 325. New York 1967.

Species/Vernacular Names:

T. spp.

Peru: huappanaja, huata, mamo

T. angustifolium DC.

English: monkey cacao

Mexico: cacao

T. bicolor Bonpl.

English: tiger cacao

Mexico: patashte, cacao blanco, balam-té, cacao, cacao malacayo, pataste, papatle, pataxte

Peru: bacao, cacao de anta, macambo, machin tsau, majambu, maraca, mojambu, cacao blanco

Brazil: cacaueiro do Peru, cupuaçurana

T. cacao L.

English: cacao tree, chocolate tree, cacao bean, cocoa bean

Mexico: cacaotl, cacahuacuahuitl, cacaocuáhuitl, cacauatzaua, cacaotero, cacao, kakau, caco, cágau, kahau, biziáa, piziaa, bizoya, pizoya, yaga bizoya, yaga bisoya, yaga-pi-zija, haa, cajecua, chudechú, ma-micha-moya, ma-mu-guía, mo-chá, yau

Bolivia: chocolate, cacao criollo, cacao

Paraguay: kupuá, kupuasú

Brazil: cupuassú, cupuaçu, cupuhi, cacau, cacaueiro

Peru: bakau, cacahua, cacahua caspi, cacahuillo, cacao, cacao arisco, cacao común, cacao silvestre, cacao uchpa cacao, canga, ccahua, ccarhua, chaxon runxan, chepere, cocoa, cumala, cupuassú, cupuhi, dsohuero, kagka, kavarolli, kimituki, macambo, musena, muse-na, nucan, quemitoqui, sarguiminiqui, sarhuiminiqui, sariguieminike, sariyeminiki, sariyeminiqui, turampi, turanqui, turanti

The Philippines: cacao, kakaw

T. cacao subsp. *cacao* fo. *leiocarpum* (Bernoulli) Ducke (*Theobroma leiocarpa* Bernoulli; *Theobroma cacao* subsp. *leiocarpum* (Bernoulli) Cuatrecasas)

Mexico: cacao calabacillo

T. grandiflorum (Sprengel) Schumann

Brazil: cupuaçu, cupuaçueiro

Peru: cupuassú

T. sinuosum Ruíz & Pav. ex Huber (*Theobroma tessmannii* Mildbr.)

Peru: cacao del monte, paco cacao

T. speciosum Willd. ex Sprengel (*Theobroma quinquenervia* Bernoulli)

Bolivia: chocolatillo

Peru: macambillo, cacao, macambo, majambo, fohtuuro

Brazil: cacaueirof, cacauí

T. subincanum C. Martius (*Theobroma ferruginea* Bernoulli; *Cacao sylvestris* Aublet)

Bolivia: chocolatillo

Peru: uchpa cacao, cacahuillo, cacao ceniza, cumala, mapanahã, paco cacao, shiña, tsiña, yurac cacao, cacao blanco, uchpa cacao xoveedyo

Brazil: cupuaizeiro, cupuaí

Theodorea Barb. Rodr. Orchidaceae

Origins:
For the Brazilian Theodoro Machado Freire Pereira da Silva, Minister of Public Works and promoter of botanical explorations.

Theophrasta L. Theophrastaceae

Origins:
For Theophrastus, a Greek philosopher of Eresus, a disciple of Plato and Aristotle, see [John Lemprière, 1765?-1824], *Lemprière's Classical Dictionary* of Proper Names mentioned in Ancient Authors. Third edition. 622. London and New York 1984; J.B. McDiarmid, in *Dictionary of Scientific Biography* 13: 328-334. 1981.

Therefon Raf. Saxifragaceae

Origins:
Perhaps from the Greek *theros* "summer," *thero* "heat, warm" and *phonos* "murder"; see C.S. Rafinesque, *The Good Book*. 46. Philadelphia 1840; E.D. Merrill, *Index rafinesquianus*. 135. 1949.

Thereianthus G.J. Lewis Iridaceae

Origins:
Greek *thereios, theros* "summer, in summer, summer-time" (*thero* "to warm, heat, make hot") and *anthos* "flower."

Theriophonum Blume Araceae

Origins:
Greek *therion* "wild animal, beast" and *phonos* "murder"; see D.H. Nicolson, "Derivation of aroid generic names." *Aroideana*. 10: 15-25. 1988.

Thermopsis R. Br. Fabaceae

Origins:
Like a lupin, from the Greek *thermos* "lupin" and *opsis* "like," referring to the yellow flower heads.

Species/Vernacular Names:
T. fabacea (Pall.) DC.

China: huang hua

T. macrophylla Hook. & Arn. (*Thermopsis gracilis* Howell)

English: golden pea, false lupine

T. villosa (Walter) Fern. & B.G. Schubert (*Sophora villosa* Walter; *Thermopsis caroliniana* M. Curtis)

English: Carolina lupin

Therocistus Holub Cistaceae

Origins:
From the Greek *theros* "summer" plus *Cistus*.

Therofon Raf. Saxifragaceae

Origins:
Who knows? Perhaps from the Greek *theros* "summer," *thero* "heat, warm" and *phonos* "murder"; see C.S. Rafinesque, *New Fl. N. Am.* 4: 66. 1836 [1838] and *Autikon botanikon*. Icones plantarum select. nov. vel rariorum, etc. 15. Philadelphia 1840; E.D. Merrill, *Index rafinesquianus*. 135. 1949.

Therogeron DC. Asteraceae

Origins:
Possibly from the Greek *theros* "summer," *thero* "heat, warm" and *geron* "an old man."

Therophon Rydb. Saxifragaceae

Origins:
Possibly from the Greek *theros* "summer, summer-fruits," *thero* "heat, warm" and *phonos* "murder, slaughter," see also *Terofon* Raf and *Terefon* Raf.

Theropogon Maxim. Convallariaceae (Liliaceae)

Origins:
Greek *theros* "summer" and *pogon* "beard," referring to the summer flowering and to the tufted habit.

Therorhodion (Maxim.) Small Ericaceae

Origins:
From the Greek *thero* "heat, warm" and *rhodon* "rose."

Thesidium Sonder Santalaceae

Origins:
Referring to *Thesium*.

Thesium L. Santalaceae

Origins:
Latin *thesion* and *thesium, ii*, an ancient name for a species of *Linaria*, toad flax, used by Plinius. According to some authors the genus was named after Theseus, hero of Attic legend, son of Aegeus, king of Athens, and Aethra, daughter of Pittheus, king of Troezen (in Argolis), or of the sea god, Poseidon, and Aethra; Greek *theseion* "the temple of Theseus"; see Carl Linnaeus, *Species Plantarum*. 207. 1753 and *Genera Plantarum*. Ed. 5. 97. 1754; H. Genaust, *Etymologisches Wörterbuch der botanischen Pflanzennamen*. 640-641. 1996; F. Boerner & G. Kunkel, *Taschenwörterbuch der botanischen Pflanzennamen*. 4. Aufl. 378. 1989.

Species/Vernacular Names:
T. australe R. Br.

English: Austral toadflax, toadflax

T. chinense Turcz.

Japan: kana-biki-sô

T. humifusum DC.

English: bastard toadflax

T. namaquense Schltr.

English: Namaqua thesium, poison bush

South Africa: poison bush, gifbossie, Namakwa-thesium

T. racemosum Bernh.

Southern Africa: bohoho (Sotho)

Thespesia Soland. ex Corrêa Malvaceae

Origins:
Greek *thespesios* "divine, wonderful," referring to *Thespesia populnea*, a sacred plant in Tahiti; see also [about the 24 "Canoe Plants" of Ancient Hawai`i] http://hawaii-nation.org/nation/canoe/canoe.html.

Species/Vernacular Names:
T. acutiloba (Bak.f.) Exell & Mendonça (*Thespesia populnea* (L.) Corrêa var. *acutiloba* Bak.f.)

English: wild tulip tree, tulip tree

Southern Africa: wildetulpboom, thespesia; iBhicongo, iPhuphume (Zulu)

T. populnea (L.) Corrêa (*Hibiscus populneus* L.)

English: portia tree, false rosewood, bhendi tree, Pacific rosewood, seaside mahoe, cork tree

Rodrigues Island: mahoe, Ste. Marie

Japan: sakishima-hamabô, shima-aoi, toyûma

Malaya: baru, baru baru, baru laut, bebaru, waru

Hawaii: milo

India: arasi, bhendi tree, puvarasu, gangarevi, gangaravi, gangareni, parisha, paraspital, parash, parsipu, kallal

Thespesiopsis Exell & Hillc. Malvaceae

Origins:
Resembling *Thespesia* Sol. ex Corrêa.

Thespidium F. Muell. Asteraceae

Origins:
Resembling the genus *Thespis* DC.

Thespis DC. Asteraceae

Origins:
Probably from the Greek *thespis* "divine, awful, wondrous," *thespesios* "divine, wonderful."

Thevetia L. Apocynaceae

Origins:
After the French monk F. André Thevet, 1502-1592, plant collector and traveler in Brazil and Guiana. His writings include *Cosmographie du Levant*. Lyon 1554, *Les singularitez de la France Antarctique, autrement nommée Amérique, et de plusieurs terres et isles découvertes de notre temps*. Paris 1558 (Italian translation: *Historia dell'India America, detta altramente Francia antartica...* Vinegia 1561) and *Les vrais portraits et vies des hommes illustres grecs, latins et payens, anciens et modernes ...* Paris 1584; see J. and R. Parmentier, *Discours de la navigation de Jean et Raoul Parmentier de Dieppe. Voyage à Sumatra en 1529. Description de l'isle de Sainct-Dominigo*. Publié par M.C. Schefer. 1883; C. von Posadowsky-Wehner, *Jean Parmentier, 1494-1529*. Leben und Werk. 1937; H. Genaust, *Etymologisches Wörterbuch der botanischen Pflanzennamen*. 641. [d. 1590] 1996.

Species/Vernacular Names:
T. spp.

Mexico: becuaa, pecuaa, yaba becuaa

T. ahouai (L.) A. DC.

English: broadleaf thevetia

China: kuo ye zhu tao

T. peruviana (Persoon) K. Schumann (*Thevetia neriifolia* A. Jussieu ex Steudel; *Cerbera peruviana* Persoon; *Cascabela thevetia* (L.) Lippold)

English: yellow oleander, be-still tree, lucky nut, lucky bean, trumpet flower

Spanish: narciso amarillo

Yoruba: olomi ojo, sopa sopa

Peru: bellaquillo, ischacapa, maichill, maichil, siática, ahoahai mirim, ahoui guacu, árbol de Panamá, bellaco caspi, cruceta real, jorro-jorro, lechero, louro rosa, pájaro bobo, suche

Guatemala: chirrto

Brazil: jorro-jorro, chapéu de Napoleão, cega-olho

India: hapusha, kanai, marang kanaili, gohai phul, pila kaner, pivala-kanher, kolka-phul, kolkaphul, ashvaha, pachaiyalari, pachchaarali, pachchai-alari, pachchaganeru, pachcha-ganneru, sherani, thivati

Japan: kibana-kyô-chiku-tô

China: huang hua jia zhu tao

Vietnam: thong thien, cay dau tay

Malaya: bunga Jepun, Jepun, jitong, zetun

The Philippines: campanilla

Hawaii: nohomalie

Thibaudia Ruíz & Pav. Ericaceae

Origins:
For the French botanist Thibaud de Chauvalon or Jean Baptiste Thibault de Chanvalon, 1725-1788.

Species/Vernacular Names:
T. sp.

Peru: macha-macha

Thicuania Raf. Orchidaceae

Origins:
A vernacular name in Burma; see C.S. Rafinesque, *Flora Telluriana*. 4: 47. 1836 [1838]; Richard Evans Schultes and Arthur Stanley Pease, *Generic Names of Orchids. Their Origin and Meaning*. 305. Academic Press, New York and London 1963; E.D. Merrill, *Index rafinesquianus*. 105. 1949.

Thiebautia Colla Orchidaceae

Origins:

For the French botanist Arsenne (Arsène) Thiébaut-de-Berneaud (Thiébaud de Berneaud), 1777-1850, agronomist, Secretary of the Linnean Society of Paris (Société Linnéenne de Paris). His writings include *Mémoire sur la culture des Dahlies*. Paris 1812, *Éloge historique de C.S. Sonnini de Manoncourt*. Paris 1812, *Mémoire sur le cactus Opuntia*. Paris 1813, *Manuel théorique et pratique du Vigneron français*. Paris 1836 and *Voyage à l'Isle d'Elbe*. Paris 1808; see John H. Barnhart, *Biographical Notes upon Botanists*. 3: 372. 1965; S. Lenley et al., *Catalog of the Manuscript and Archival Collections and Index to the Correspondence of John Torrey*. Library of the New York Botanical Garden. 470. 1973; G. Schmid, *Goethe und die Naturwissenschaften*. Halle 1940; E.M. Tucker, *Catalogue of the Library of the Arnold Arboretum of Harvard University*. Cambridge, Massachusetts 1917-1933; Giulio Giorello & Agnese Grieco, a cura di, *Goethe scienziato*. Einaudi Editore, Torino 1998.

Thieleodoxa Chamisso Rubiaceae

Origins:

For the German botanist Friedrich Leopold Thiele, d. 1841.

Thilachium Lour. Capparidaceae (Capparaceae)

Origins:

From Greek *thylakos* "bag, sack, pouch," referred to the calyx splits in half.

Species/Vernacular Names:

T. africanum Lour.

English: cucumber bush, African thilachium

Southern Africa: komkommerbos; isiKonze (Zulu); nkobva (Eastern Transvaal)

East Africa: mdudu, mtunguru

Thiloa Eichler Combretaceae

Origins:

Dedicated to the German botanist Johann Friedrich Thilo Irmisch, 1816-1879, school teacher. His writings include *Beiträge zur Biologie und Morphologie der Orchideen*. Leipzig 1853 *Ueber einige Fumariaceen*. Halle 1862 and *Ueber Papaver trilobum Wallroth. Ein Beitrag zur Naturgeschichte der gattung Papaver*. Halle 1865, joint author of *Handbuch der Physiologischen Botanik* in Verbindung mit A. de Bary, T. Irmisch und J. Sachs herausgegeben von Wilhelm Friedrich Benedict Hofmeister. Leipzig 1867, studied with Diederich F.L. von Schlechtendal (1794-1866); see John H. Barnhart, *Biographical Notes upon Botanists*. 2: 233. 1965.

Thinogeton Benth. Solanaceae

Origins:

Greek *thinoo* "to fill, choke with sand," *this, thin* "a heap, the sand-heaps on the beach, sand-banks' and *geiton* "a neighbor"; see also *Exodeconus* Raf.

Thinopyrum Á. Löve Gramineae

Origins:

Greek *thinoo* "to fill, choke with sand," *this, thin* "a heap, the sand-heaps on the beach, sand-banks" and *pyros* "grain, wheat," a salt-tolerant cereal.

Thiseltonia Hemsley Asteraceae

Origins:

Dedicated to the botanist G.H. Thiselton-Dyer, plant collector in Western Australia; see *Hooker's Icon. Pl.* 2781. Jan. 1905; I.H. Vegter, *Index Herbariorum*. Part II (7), *Collectors T-Z*. Regnum Vegetabile vol. 117. 1988.

Thladiantha Bunge Cucurbitaceae

Origins:

Greek *thlao, thlaein* "to break, to crush, pound," *thladias* "eunuch" and *anthos* "flower," referring to the suppression of stamens.

Species/Vernacular Names:

T. dubia Bunge

China: wang kua, tu kua, chih pao

Thlaspeocarpa C.A. Sm. Brassicaceae

Origins:

Greek *thlastes* "a crusher," *thlastos* "crushed," *thlao* "to break, to crush, pound," *thlaspi* "a kind of cress" and *karpos* "fruit."

Thlaspi L. Brassicaceae

Origins:

From *thlaspi, thlaspis*, a Greek name for a cress (*thlao, thlaein* "to break, to crush, pound"); Latin *thlaspi*, used by A. Cornelius Celsus and Plinius for a cress; see Carl Linnaeus, *Species Plantarum*. 645. 1753 and *Genera Plantarum*. Ed. 5. 292. 1754.

Species/Vernacular Names:

T. arvense L.

English: fanweed, field penny cress, French weed, Mithridate mustard, penny cress, stinkweed

Tibetan: bre-ga

China: xi ming, hsi ming, ta chi (= large shepherd's burse)

T. californicum S. Watson

English: Kneeland Prairie penny cress

T. montanum L.

English: mountain penny cress

T. perfoliatum L.

English: penny cress

Thlaspiceras F.K. Mey. Brassicaceae

Origins:

From *thlaspi, thlaspis* and *keras* "horn."

Thollonia Baillon Icacinaceae

Origins:

For the French François-Romain Thollon, 1855-1896 (d. Libreville, Gabon), plant collector (Ivory Coast, southern Nigeria, Congo and Gabon); see F.N. Hepper and Fiona Neate, *Plant Collectors in West Africa*. 79. 1971; John H. Barnhart, *Biographical Notes upon Botanists*. Boston 1965.

Thomasia J. Gay Sterculiaceae

Origins:

After the Thomases, a Swiss family of botanists and botanical collectors, Pierre (Peter) Thomas (fl. 1750) and his brother Abram (Abraham) Thomas (1740-1822) and the latter's sons Philippe Thomas (d. 1831), Louis Thomas (1784-1823) and (Abraham Louis) Emmanuel (Emanuel) Thomas (1788-1859); Emmanuel Thomas was the author of *Catalogue des plantes de Sardaigne*, qui se vendent chez Emmanuel Thomas a Bex. 1841; see Stafleu and Cowan, *Taxonomic Literature*. 6: 271-272. Utrecht 1986; Susan Paust, "Taxonomic studies in *Thomasia* and *Lasiopetalum* (Sterculiaceae)." in *Nuytsia*. 1(4): 348-366. 1974; John H.

Barnhart, *Biographical Notes upon Botanists*. 3: 374. 1965; Mariella Azzarello Di Misa, ed., *Il Fondo Antico della Biblioteca dell'Orto Botanico di Palermo*. 269. Regione Siciliana, Palermo 1988.

Species/Vernacular Names:

T. petalocalyx F. Muell.

Australia: paper-flower

Thompsonella Britton & Rose Crassulaceae

Origins:

For the American botanist Charles Henry Thompson, 1870-1931; see John H. Barnhart, *Biographical Notes upon Botanists*. 3: 376. Boston 1965; E.M. Tucker, *Catalogue of the Library of the Arnold Arboretum of Harvard University*. Cambridge, Massachusetts 1917-1933; Ida Kaplan Langman, *A Selected Guide to the Literature on the Flowering Plants of Mexico*. Philadelphia 1964; Irving William Knobloch, compiled by, "A preliminary verified list of plant collectors in Mexico." *Phytologia Memoirs*. VI. Plainfield, N.J. 1983.

Thompsonia R. Br. Passifloraceae

Origins:

After the British (b. Northumberland) botanist John Vaughan Thompson, 1779-1847 (d. Sydney, New South Wales, Australia), studied medicine and surgery, naturalist, interested in marine biology, surgeon (to His Majesty's 37th Regiment of Foot), zoologist, 1799-1809 West Indies and Guiana, 1810 Fellow of the Linnean Society of London, corresponded with Alexander MacLeay (1767-1848), 1812-1816 Madagascar and Mauritius, made studies of marine invertebrates. Among his writings are *A Catalogue of the Exotic Plants Cultivated in the Mauritius ...* to which are added the English and French names... Compiled under the auspices of R.T. Farquhar Esq. Governor of Mauritius. Mauritius 1816, *Zoological Researches and Illustrations*. Cork, Ireland 1828-1834 and *A Catalogue of Plants Growing in the Vicinity of Berwick upon Tweed*. London 1807; see John H. Barnhart, *Biographical Notes upon Botanists*. 3: 377. Boston 1965; E.M. Tucker, *Catalogue of the Library of the Arnold Arboretum of Harvard University*. Cambridge, Massachusetts 1917-1933; Alwyne Wheeler, in *Dictionary of Scientific Biography* 13: 353-356. 1981.

Thomsonia Wallich Araceae

Origins:

For the British (b. Edinburgh) physician Anthony Todd Thomson, 1778-1849 (d. Ealing, Middx.), botanist, 1812

Fellow of the Linnean Society; see John H. Barnhart, *Biographical Notes upon Botanists*. 3: 378. 1965; D.H. Nicolson, "Derivation of aroid generic names." *Aroideana*. 10: 15-25. 1988; Ray Desmond, *Dictionary of British & Irish Botanists and Horticulturists*. 681. London 1994.

Thonnera De Wild. Annonaceae

Origins:

After the Austrian botanist Franz Thonner, 1863-1928, naturalist, traveler (Belgian Congo); see John H. Barnhart, *Biographical Notes upon Botanists*. 3: 379. 1965; E.M. Tucker, *Catalogue of the Library of the Arnold Arboretum of Harvard University*. 1917-1933; R. Zander, F. Encke, G. Buchheim and S. Seybold, *Handwörterbuch der Pflanzennamen*. 14. Aufl. Stuttgart 1993.

Thonningia Vahl Balanophoraceae

Origins:

After the Danish botanist and traveler Peter Thonning, 1775-1848, Councillor of State, plant collector, between 1799-1803 he was with the Danish Ole Haaslund Smith (d. 1802) on a botanical expedition to Danish Guinea, now in Ghana; see Heinrich Christian Friedrich Schumacher (1757-1830), *Beskrivelse af Guineiske Planter* som ere fundne af Danske Botanikere isaer af Etatsraad Thonning. [Copenhagen 1828-29]; Martin H. Vahl (1749-1804), *M. Vahlii ... Enumeratio Plantarum*. Hauniae (& Lipsiae) [1804-] 1805-1806; Carl Frederik Albert Christensen (1872-1942), *Den danske Botaniks Historie med tilhørende Bibliografi*. Copenhagen 1924-1926 and *Den danske botaniske litteratur 1880-1911*. Kopenhagen 1913; C.D. Adams, "Activities of Danish botanists in Guinea, 1783-1850." in *Transactions of the Historical Society of Ghana*. 3: 30-46. 1957; R. Zander, F. Encke, G. Buchheim and S. Seybold, *Handwörterbuch der Pflanzennamen*. 14. Aufl. Stuttgart 1993; Frank N. Hepper, *The West African Herbaria of Isert & Thonning*. Kew 1976; F.N. Hepper and Fiona Neate, *Plant Collectors in West Africa*. 41, 80. 1971; Joseph Vallot (1854-1925), "Études sur la flore du Sénégal." in *Bull. Soc. Bot. de France*. 29: 168-238. Paris 1882; R.W.J. Keay, "Botanical collectors in West Africa prior to 1860." in *Comptes Rendus A.E.T.F.A.T.* 55-68. Lisbon 1962; John H. Barnhart, *Biographical Notes upon Botanists*. Boston 1965; Georg Nørregaard, *Danish Settlements in West Africa 1658-1850*. Boston 1966; A. Lasègue, *Musée botanique de Benjamin Delessert*. Paris 1845.

Species/Vernacular Names:

T. sanguinea Vahl

Yoruba: ade ile, adele, oyaile, oyale

Congo: lilanda

Thoracocarpus Harling Cyclanthaceae

Origins:

From the Greek *thorax, thorakos* "a breast-plate, cuirasse, the chest, outer wall" and *karpos* "fruit."

Thoracosperma Klotzsch Ericaceae

Origins:

Greek *thorax, thorakos* and *sperma* "seed," a well-protected seed.

Thoracostachyum Kurz Cyperaceae

Origins:

Greek *thorax, thorakos* plus *stachys* "spike."

Thorea Rouy Gramineae

Origins:

After the French naturalist Jean Thore, 1762-1823, physician; see John H. Barnhart, *Biographical Notes upon Botanists*. 3: 380. 1965; R. Zander, F. Encke, G. Buchheim and S. Seybold, *Handwörterbuch der Pflanzennamen*. 14. Aufl. Stuttgart 1993.

Thoreldora Pierre Rutaceae

Origins:

After the French botanist Clovis Thorel, 1833-1911, physician, plant collector; see John H. Barnhart, *Biographical Notes upon Botanists*. 3: 380. 1965; Emil Bretschneider, *History of European Botanical Discoveries in China*. Leipzig 1981; E.M. Tucker, *Catalogue of the Library of the Arnold Arboretum of Harvard University*. 1917-1933.

Thorelia Gagnepain Asteraceae

Origins:

After the French botanist Clovis Thorel, 1833-1911, physician, plant collector; see John H. Barnhart, *Biographical Notes upon Botanists*. 3: 380. 1965; Emil Bretschneider, *History of European Botanical Discoveries in China*. Leipzig 1981; E.M. Tucker, *Catalogue of the Library of the Arnold Arboretum of Harvard University*. 1917-1933.

Thorella Briquet Umbelliferae

Origins:

After the French naturalist Jean Thore, 1762-1823, physician; see John H. Barnhart, *Biographical Notes upon Botanists*. 3: 380. 1965; R. Zander, F. Encke, G. Buchheim and S. Seybold, *Handwörterbuch der Pflanzennamen*. 14. Aufl. Stuttgart 1993.

Thoreochloa J. Holub Gramineae

Origins:

After the French naturalist Jean Thore, 1762-1823, physician; see John H. Barnhart, *Biographical Notes upon Botanists*. 3: 380. 1965; R. Zander, F. Encke, G. Buchheim and S. Seybold, *Handwörterbuch der Pflanzennamen*. 14. Aufl. Stuttgart 1993.

Thornbera Rydberg Fabaceae

Origins:

After the American botanist John James Thornber, 1872-1962; see John H. Barnhart, *Biographical Notes upon Botanists*. 3: 380. 1965; T.W. Bossert, *Biographical Dictionary of Botanists Represented in the Hunt Institute Portrait Collection*. 400. 1972; E.M. Tucker, *Catalogue of the Library of the Arnold Arboretum of Harvard University*. Cambridge, Massachusetts 1917-1933.

Thorncroftia N.E. Br. Labiatae

Origins:

For the British (b. Kent) botanist George Thorncroft, 1857-1934 (d. Transvaal, Barberton), plant collector, merchant; see Mary Gunn and Leslie E. Codd, *Botanical Exploration of Southern Africa*. 346-347. Cape Town 1981.

Thorntonia Reichenbach Malvaceae

Origins:

After the British botanist John James Thornber, *circa* 1768-1837 (d. London), physician, pupil of Thomas Martyn (1735-1825); see John H. Barnhart, *Biographical Notes upon Botanists*. 3: 380. 1965; T.W. Bossert, *Biographical Dictionary of Botanists Represented in the Hunt Institute Portrait Collection*. 400. 1972; E.M. Tucker, *Catalogue of the Library of the Arnold Arboretum of Harvard University*. Cambridge, Massachusetts 1917-1933; Blanche Henrey, *British Botanical and Horticultural Literature before 1800*.

1975; A. Lasègue, *Musée botanique de Benjamin Delessert*. Paris 1845; William Munk, *The Roll of the Royal College of Physicians of London*. London 1878; Alain White and Boyd Lincoln Sloane, *The Stapelieae*. Pasadena 1937; Ray Desmond, *Dictionary of British & Irish Botanists and Horticulturists*. 684. 1994; M. Hadfield et al., *British Gardeners: A Biographical Dictionary*. London 1980.

Thorvaldsenia Liebmann Orchidaceae

Origins:

For the Danish sculptor Bertel Thorvaldsen (Thorwaldsen), 1770-1844 (d. Copenhagen), traveler (Malta, Sicily and Napoli), 1797 Rome, author of *Triumph of Alexander* (in the Palazzo del Quirinale, Rome) and *Giasone*; see P. Kragelund & M. Nykjër, *Thorvaldsen. L'ambiente, l'influsso, il mito*. L'Erma di Bretschneider. Roma; Bertel Thorvaldsen, *Bas reliefs*. Frankfurt am Main, s.d.

Thouarsia Kuntze Gramineae

Origins:

After the French botanist Louis-Marie Aubert Aubert du Petit-Thouars, 1758-1831, traveler and plant collector.

Thouarsiora Homolle ex Arènes Rubiaceae

Origins:

Named after the French botanist Louis-Marie Aubert Aubert du Petit-Thouars, 1758-1831, traveler and plant collector. Among his publications are *Mélanges de botanique et de voyages* ... Paris 1811, *Plantes des îles de l'Afrique australe* ... Paris [1804], *Genera nova madagascariensia*. [Paris 1806], *Histoire particulière des plantes orchidées* ... Paris 1822 and *Histoire des végétaux recueillis dans les isles australes d'Afrique* ... Paris 1806 [-1808]; see Mariella Azzarello Di Misa, a cura di, *Il Fondo Antico della Biblioteca dell'Orto Botanico di Palermo*. 98. Palermo 1988; T.W. Bossert, *Biographical Dictionary of Botanists Represented in the Hunt Institute Portrait Collection*. 109. 1972; Ida Kaplan Langman, *A Selected Guide to the Literature on the Flowering Plants of Mexico*. Philadelphia 1964; Stafleu and Cowan, *Taxonomic Literature*. 1: 704-706. 1976.

Thouinia Poiteau Sapindaceae

Origins:

After the French (b. Paris) horticulturist André Thouin, 1747-1824 (d. Paris), botanist, traveler and botanical collector, a student of Bernard de Jussieu and Buffon, friend of Chrétien-Guillaume de Lamoignon de Malesherbes (1721-1794) and Rousseau, collaborated with A.P. de Candolle and Desfontaines (1750-1833), gardener at the Jardin des Plantes in Paris (Jardin du Roi), a member of the Academy of Sciences. Among his works are *Monographie des greffes*. [4to, first edition, he describes 16 different grafts.] [Paris] 1821, *Histoire d'une nouvelle espèce d'arbre fruitier*. [Paris 1813] and *Voyage dans le Belgique, la Hollande, et l'Italie ... rédigé sur le journal autographe ... par le Baron Trouvé*. Paris 1841, with Réné L. Desfontaines wrote *Rapport fait à la Classe des sciences physique et mathématiques de l'Institut*. [Paris 1801?]. Son of Jean André Thouin (1745-1768), brother of Jean Thouin (1756-1821) and Gabriel Thouin (1747-1829); see J.H. Barnhart, *Biographical Notes upon Botanists*. 3: 381. 1965; P. Jovet & M. Mallet, in *Dictionary of Scientific Biography* 13: 390. 1981; Royal Society, *Catalogue of Scientific Papers*. V. 983-984; R. Zander, F. Encke, G. Buchheim and S. Seybold, *Handwörterbuch der Pflanzennamen*. 14. Aufl. 1993; Frans A. Stafleu, *Linnaeus and the Linnaeans*. The spreading of their ideas in systematic botany, 1735-1789. Utrecht 1971; Emil Bretschneider (1833-1901), *History of European Botanical Discoveries in China*. Leipzig 1981; Georges Léopold Chrétien Frédéric Dagobert Cuvier (1769-1832), *Recueil des éloges historiques* lus dans les séances publiques de l'Institut royal de France. Strasbourg & Paris 1819-1827; T.W. Bossert, *Biographical Dictionary of Botanists Represented in the Hunt Institute Portrait Collection*. 400. 1972; E.M. Tucker, *Catalogue of the Library of the Arnold Arboretum of Harvard University*. Cambridge, Massachusetts 1917-1933; Jonas C. Dryander, *Catalogus bibliothecae historico-naturalis Josephi Banks*. London 1796-1800; A. Lasègue, *Musée botanique de Benjamin Delessert*. Paris 1845.

Thouinidium Radlk. Sapindaceae

Origins:

After the French horticulturist André Thouin, 1747-1824, a student of Bernard de Jussieu and Buffon, friend of Malesherbes and Rousseau, collaborated with A.P. de Candolle and Desfontaines, botanist, gardener at the Jardin des Plantes in Paris, a member of the Academy of Sciences, traveler and botanical collector; see J.H. Barnhart, *Biographical Notes upon Botanists*. 3: 381. 1965; Royal Society, *Catalogue of Scientific Papers*. V. 983-984; P. Jovet & M. Mallet, in *Dictionary of Scientific Biography* 13: 390. 1981.

Thozetia F. Muell. ex Benth. Asclepiadaceae

Origins:

For the Australian (b. near Lyons, France) plant collector Anthelme Thozet, *circa* 1826-1878 (d. Rockhampton,

Queensland), gardener, farmer, 1867 Fellow of the Linnean Society, sent plants to F. von Mueller, author of *Notes on Some of the Roots, Tubers, Bulbs, and Fruits, Used as Vegetable Food by the Aboriginals of Northern Queensland, Australia*. Rockhampton 1866; see Ray Desmond, *Dictionary of British & Irish Botanists and Horticulturists*. 684. 1994; John Alexander Ferguson, *Bibliography of Australia. 1851-1900*. Sydney and London 1969; J.H. Barnhart, *Biographical Notes upon Botanists*. 3: 381. 1965; I.H. Vegter, *Index Herbariorum*. Part II (7), *Collectors T-Z*. Regnum Vegetabile vol. 117. 1988; N. Hall, *Botanists of the Eucalypts*. Melbourne 1978 and Supplement 1980.

Thrasya Kunth Gramineae

Origins:

From the Greek *thrasys* "bold."

Thrasyopsis Parodi Gramineae

Origins:

Resembling *Thrasya*.

Thraulococcus Radlk. Sapindaceae

Origins:

From the Greek *thraulos* "frangible, brittle" and *kokkos* "a berry"; see also *Lepisanthes* Blume.

Threlkeldia R. Br. Chenopodiaceae

Origins:

For the British (b. Cumberland) botanist Caleb Threlkeld, 1676-1728 (d. Dublin), clergyman, physician, M.D. Edinburgh 1713, author of *Synopsis stirpium hibernicarum*. Dublin 1726; see J.H. Barnhart, *Biographical Notes upon Botanists*. 3: 381. 1965; R. Pulteney, *Historical and Biographical Sketches of the Progress of Botany in England*. 2: 196-201. London 1790; Edward Lee Greene, *Landmarks of Botanical History*. Edited by Frank N. Egerton. 971. Stanford, California 1983; E.M. Tucker, *Catalogue of the Library of the Arnold Arboretum of Harvard University*. Cambridge, Massachusetts 1917-1933; John C. Loudon, *An Encyclopaedia of Gardening*. 282. London 1878; Jonas C. Dryander, *Catalogus bibliothecae historico-naturalis Josephi Banks*. London 1796-1800; Blanche Henrey, *British Botanical and Horticultural Literature before 1800*. 1975; Robert Brown (1773-1858), *Prodromus florae Novae Hollandiae*. 409. London 1810.

Species/Vernacular Names:

T. diffusa R. Br.

Australia: coast bonefruit

T. inchoata (J. Black) J. Black (*Bassia inchoata* J. Black) (Latin *inchoatus, a, um* "unfinished, imperfect, rudimentary")

Australia: tall bonefruit

Thrinax Sw. Palmae

Origins:

Greek *thrinax, thrinakos* "a fan," referring to the form of the leaves.

Species/Vernacular Names:

T. morrisii H.A. Wendland

Brazil: palmeira vassoura

T. parviflora Sw.

English: thatch palm

Brazil: palmeira leque

T. radiata Loddiges ex Schult. & Schult.f.

Brazil: palmeira leque brilhante

Thrincoma O.F. Cook Palmae

Origins:

Greek *thrinax, thrinakos* "a fan" and *kome* "hair"; see also *Coccothrinax* Sargent.

Thringis O.F. Cook Palmae

Origins:

Perhaps from the Greek *thringion, thrinkion* "wall, row"; see also *Coccothrinax* Sargent.

Thrixanthocereus Backeb. Cactaceae

Origins:

From the Greek *thrix, trichos* "hair," *anthos* "flower" plus *Cereus.*

Thrixgyne Keng Gramineae

Origins:

From the Greek *thrix, trichos* "hair" and *gyne* "female, woman."

Thrixspermum Loureiro Orchidaceae

Origins:

Greek *thrix, trichos* "hair" and *sperma* "seed," the hair-like seeds are long and thin; see B.A. Lewis and P.J. Cribb, *Orchids of the Solomon Islands and Bougainville.* Royal Botanic Gardens, Kew 1991.

Species/Vernacular Names:

T. fantasticum L.O. Williams (*Thrixspermum neglectum* Fukuyama)

Japan: ha-gakure-naga-me-ran

T. platystachys (Bailey) Schltr.

Bougainville Island: nuo-wacki

Thryallis L. Malpighiaceae

Origins:

An old Greek name used by Theophrastus.

Thryallis Martius Malpighiaceae

Origins:

An old Greek name used by Theophrastus for *Verbascum, thryallis, thryallidos* "a wick, a plant used for making wicks," *thryon* "a rush, reed," Latin *juncus*; Latin *thryallis, idis* is the name of two different plants, a plant called also *lychnitis* (Plinius), and an ear-shaped plant (Plinius); see Helmut Genaust, *Etymologisches Wörterbuch der botanischen Pflanzennamen.* 642. Basel 1996.

Thryothamnus Philippi Verbenaceae

Origins:

Greek *thryon* "a rush, reed" and *thamnos* "shrub, bush."

Thryptomene Endl. Myrtaceae

Origins:

Greek *thrypto* "to break in pieces, weaken," *thryptikos* "easily broken, soft, delicate," referring to the low stature and dimension of the species.

Species/Vernacular Names:

T. calycina (Lindley) Stapf (*Baeckea calycina* Lindley; *Paryphantha mitchelliana* Schauer; *Thryptomene mitchelliana* (Schauer) F. Muell.)

Australia: Grampians thryptomene

T. micrantha Hook.f. (*Thryptomene miqueliana* F. Muell.)

Australia: ribbed thryptomene, heather bush

Thuarea Persoon Gramineae

Origins:

Named after the French botanist Louis-Marie Aubert Aubert du Petit-Thouars, 1758-1831, traveler and plant collector. Among his publications are *Mélanges de botanique et de voyages* ... Paris 1811, *Plantes des îles de l'Afrique australe* ... Paris [1804], *Genera nova madagascariensia*. [Paris 1806], *Histoire particulière des plantes orchidées* ... Paris 1822 and *Histoire des végétaux recueillis dans les isles australes d'Afrique* ... Paris 1806 [-1808]; see Mariella Azzarello Di Misa, a cura di, *Il Fondo Antico della Biblioteca dell'Orto Botanico di Palermo*. 98. Palermo 1988; T.W. Bossert, *Biographical Dictionary of Botanists Represented in the Hunt Institute Portrait Collection*. 109. 1972; Ida Kaplan Langman, *A Selected Guide to the Literature on the Flowering Plants of Mexico*. Philadelphia 1964.

Species/Vernacular Names:

T. involuta (G. Forst.) Roem. & Schult.

English: beach grass

Japan: Kuroiwa-zasa (Hisashi Kuroiwa was a plant collector in the Ryukyus, author of "A list of phanerogams collected in the southern part of Isl. Okinawa, one of the Loochoo chain." *Bot. Mag. Tokyo*. 14: 109-112, 122-126, 139-143. 1900)

Thuinia Raf. Oleaceae

Origins:

After the French horticulturist André Thouin, 1747-1824; see Constantine Samuel Rafinesque, *Sylva Telluriana*. 137. 1838.

Thuja L. Cupressaceae

Origins:

Greek *thyia*, for a kind of resinous tree or a juniper (Theophr.), *thyo, thyein* "to sacrifice, to burn a victim"; Latin *thya* or *thyia* is the Greek name for the citrus-tree (Plinius), adj. *thyinus, a, um* "made of the citrus-tree"; see E. Zaccaria, *L'elemento iberico nella lingua italiana*. Bologna 1927; M. Cortelazzo & P. Zolli, *Dizionario etimologico della lingua italiana*. 5: 1383. Bologna 1988; E. Zaccaria, *Raccolta di voci affatto sconosciute o mal note ai lessicografi ed ai filologi*. Marradi 1919; G. Semerano, *Le origini della cultura europea*. Dizionari Etimologici.

Dizionario della lingua Greca. 2(1): 121. Firenze 1994; Helmut Genaust, *Etymologisches Wörterbuch der botanischen Pflanzennamen*. 643. Basel 1996; E. Weekley, *An Etymological Dictionary of Modern English*. 2: 1502. New York 1967.

Species/Vernacular Names:

T. occidentalis L.

English: white cedar, American arbor-vitae

T. orientalis L.f.

English: arbor vitae, oriental arbor vitae

Japan: konote-gashiwa (= palm of a child's hand)

Okinawa: futauonoti

China: ce bai ye, po, pien po, ai po, tse po

T. plicata Donn ex D. Don

English: western red cedar, western arbor-vitae, giant arbor-vitae

T. standishii (Gordon) Carrière

English: Japanese arbor-vitae

Thujopsis Siebold & Zucc. ex Endl. Cupressaceae

Origins:

Resembling *Thuja*.

Species/Vernacular Names:

T. dolabrata (L.f.) Siebold & Zucc.

Japan: hiba

Thunbergia Retzius Acanthaceae (Thunbergiaceae)

Origins:

The generic name honors the Swedish botanist and physician Carl Peter Thunberg, 1743-1828 (Tunaberg, near Uppsala, Sweden), 1772 Dr. med. at Uppsala, a pupil of Carl Linnaeus, plant collector, explorer, traveler, naturalist, professor of botany and medicine at Uppsala, to the Cape and interior with the British gardener and plant collector Francis Masson (1741-1805). Among his numerous writings are *Flora japonica*. Lipsiae 1784, *Voyage en Afrique et en Asie, principalement au Japon, pendant les années 1770-1779*. Paris 1794, *Icones plantarum japonicarum*. Upsaliae [Uppsala] 1794-1805 and *Prodromus plantarum Capensium: quas in promontorio Bonae Spei Africes, annis 1772-1775 collegit...* Upsaliae 1794-1800; see Thorgny Ossian Bolivar Napoleon Krok (1834-1921), *Bibliotheca botanica suecana*. 705-716. Stockholm, Uppsala 1925; Mia C. Karsten,

The Old Company's Garden at the Cape and Its Superintendents: involving an historical account of early Cape botany. Cape Town 1951; M.C. Karsten, "Carl Peter Thunberg. An Early Investigator of Cape Botany." *Journal of South Africa Botany*. 5: 1-27 and 87-155. 1939; Gunnar Eriksson, in *Dictionary of Scientific Biography* 13: 391-392. 1981; J.H. Barnhart, *Biographical Notes upon Botanists*. 3: 382. 1965; Stafleu and Cowan, *Taxonomic Literature*. 6: 306-334. 1986; Gilbert Westacott Reynolds (1895-1967), *The Aloes of South Africa*. Balkema, Rotterdam 1982; H. Genaust, *Etymologisches Wörterbuch der botanischen Pflanzennamen*. 643. [d. 1822] 1996; R. Zander, F. Encke, G. Buchheim and S. Seybold, *Handwörterbuch der Pflanzennamen*. 14. Aufl. 790. 1993; Hans Oscar Juel (1863-1931), *Plantae Thunbergianae*. Uppsala 1918; Blanche Henrey, *British Botanical and Horticultural Literature before 1800*. 1975; T.W. Bossert, *Biographical Dictionary of Botanists Represented in the Hunt Institute Portrait Collection*. 400. 1972; E.M. Tucker, *Catalogue of the Library of the Arnold Arboretum of Harvard University*. Cambridge, Massachusetts 1917-1933; Günther Schmid, *Chamisso als Naturforscher*. Eine Bibliographie. Leipzig 1942; Gordon Douglas Rowley, *A History of Succulent Plants*. Strawberry Press, Mill Valley, California 1997; H.N. Clokie, *Account of the Herbaria of the Department of Botany in the University of Oxford*. 254. Oxford 1964; Emil Bretschneider, *History of European Botanical Discoveries in China*. Leipzig 1981; Jonas C. Dryander, *Catalogus bibliothecae historico-naturalis Josephi Banks*. London 1796-1800; A. Lasègue, *Musée botanique de Benjamin Delessert*. Paris 1845; F.A. Stafleu, *Linnaeus and the Linnaeans*. The spreading of their ideas in systematic botany, 1735-1789. 1971; Mary Gunn and Leslie E. Codd, *Botanical Exploration of Southern Africa*. 347-350. Cape Town 1981; S. Lenley et al., *Catalog of the Manuscript and Archival Collections and Index to the Correspondence of John Torrey*. Library of the New York Botanical Garden. 470. 1973; Mariella Azzarello Di Misa, a cura di, *Il Fondo Antico della Biblioteca dell'Orto Botanico di Palermo*. 270. Palermo 1988; Alain White and Boyd Lincoln Sloane, *The Stapelieae*. Pasadena 1937; F.N. Hepper and Fiona Neate, *Plant Collectors in West Africa*. 80. 1971; John Hutchinson (1884-1972), *A Botanist in Southern Africa*. London 1946.

Species/Vernacular Names:

T. alata Bojer ex Sims

English: black-eyed Susan, black-eyed Susan vine

T. atriplicifolia E. Mey. ex Nees (*Thunbergia aspera* Nees; *Thunbergia bachmannii* Lindau; *Thunbergia cordibracteolata* C.B. Cl.; *Thunbergia galpinii* Lindau; *Thunbergia hirtistyla* C.B. Cl.; *Thunbergia xanthotricha* Lindau)

English: Natal primrose, Cape primrose

Southern Africa: isiphondo (Zulu)

T. capensis Retz.

English: Cape primrose

Southern Africa: iYeza lehashe (Xhosa)

T. erecta (Benth.) Anderson (*Meyenia erecta* Benth.)

English: king's mantle, bush clock vine

T. fragrans Roxb.

English: white thunbergia, sweet clock vine

Japan: yahazu-kazura

T. grandiflora Roxb.

English: blue trumpet vine, clock vine, Bengal clock vine, sky vine, skyflower, blue skyflower, Bengal trumpet

Vietnam: bong bao, day bong xanh, madia

China: tong gu xiao

T. laurifolia Lindl.

English: purple allamanda

T. natalensis Hook.

English: Natal blue bell

Bali: bungan pelung-pelung (pelung = blue)

Thunbergianthus Engler Scrophulariaceae

Origins:

For the Swedish botanist and physician Carl Peter Thunberg, 1743-1828, explorer, a student of Carl Linnaeus, professor of botany and medicine at Uppsala, plant collector; see J.H. Barnhart, *Biographical Notes upon Botanists*. 3: 382. 1965; Gunnar Eriksson, in *Dictionary of Scientific Biography* 13: 391-392. 1981.

Thunbergiella H. Wolff Umbelliferae

Origins:

For the Swedish botanist and physician Carl Peter Thunberg, 1743-1828, explorer, a student of Carl Linnaeus, professor of botany and medicine at Uppsala, plant collector. His works include *De Gardenia*. Upsaliae [1780], *De Protea*. Upsaliae [1781], *Oxalis*. Upsaliae [Uppsala] [1781] and *Iris*. Upsaliae [1782]; see J.H. Barnhart, *Biographical Notes upon Botanists*. 3: 382. 1965; Gunnar Eriksson, in *Dictionary of Scientific Biography* (Editor in Chief Charles Coulston Gillispie.) 13: 391-392. New York 1981.

Thunbgeria Montin Rubiaceae

Origins:

For the Swedish botanist and physician Carl Peter Thunberg, 1743-1828.

Thunia Reichb.f. Orchidaceae

Origins:

Named for Franz Anton Graf von Thun und Hohenstein, 1786-1873; see Franz de Paula Philipp Anton von Thun-Hohenstein, *Vorschläge zur Reorganisirung des öffentlichen Baudienstes in Oesterreich.* Prag 1861; Ignacy Daszynski, *Socyalna demokracya wobec rządów hr. Thuna.* Kraków 1898; Joseph Burgerstein, *Franz Anton Graf von Thun-Hohenstein.* Biographische Skizze. Wien 1871; Legis Glueckselig, *Denkwürdigkeiten des Grafenhauses Thun-Hohenstein, grössentheils aus Familien-Archiven geschöpft.* Prag 1867; F. Boerner & G. Kunkel, *Taschenwörterbuch der botanischen Pflanzennamen.* 4. Aufl. 179. 1989.

Thuranthos C.H. Wright Hyacinthaceae (Liliaceae)

Origins:

Latin *tus* (*thus*), *turis* "incense, frankincense," Greek *thyos* "incense, an offering" and *anthos* "flower," *thyo*, *thyein* "to sacrifice, to burn a victim," Sumerian *du* "to burn"; Akkadian *asû*, *wasû*, *usû* "to rise, to go out," *susû* "to expel, to make leave."

Thurberia A. Gray Malvaceae

Origins:

After the American botanist George Thurber, 1821-1890.

Thurberia Bentham Gramineae

Origins:

After George Thurber, 1821-1890, American botanist, naturalist, 1850-1853 Mexican-U.S. Boundary Survey, chemist; see Joseph Ewan, *Rocky Mountain Naturalists.* 64, 321. The University of Denver Press 1950; John H. Barnhart, *Biographical Notes upon Botanists.* 3: 382. 1965; T.W. Bossert, *Biographical Dictionary of Botanists Represented in the Hunt Institute Portrait Collection.* 400. 1972; E.M. Tucker, *Catalogue of the Library of the Arnold Arboretum of Harvard University.* Cambridge, Massachusetts 1917-1933; J. Ewan, ed., *A Short History of Botany in the United States.* 45. New York and London 1969; S. Lenley et al., *Catalog of the Manuscript and Archival Collections and Index to the Correspondence of John Torrey.* Library of the New York Botanical Garden. 1973; Stafleu and Cowan, *Taxonomic Literature.* 6: 334-335. 1986; Irving William Knobloch, compiled by, "A preliminary verified list of plant collectors in Mexico." *Phytologia Memoirs.* VI. Plainfield, N.J. 1983.

Thurnia Hook.f. Thurniaceae (Commelinidae, Juncales)

Origins:

For the British plant collector Sir Everard Ferdinand Im Thurn, *circa* 1852-1932 (d. Prestonpans, E. Lothian), colonial administrator, 1904-1910 Governor of the Fiji Islands, traveler; see John H. Barnhart, *Biographical Notes upon Botanists.* Boston 1965; E.M. Tucker, *Catalogue of the Library of the Arnold Arboretum of Harvard University.* Cambridge, Massachusetts 1917-1933; Ray Desmond, *Dictionary of British & Irish Botanists and Horticulturists.* 684-685. London 1994.

Thurya Boissier & Balansa Caryophyllaceae

Origins:

For the Swiss naturalist Jean Marc Antoine Thury, 1822-1905, teacher; see John H. Barnhart, *Biographical Notes upon Botanists.* 3: 382. 1965; T.W. Bossert, *Biographical Dictionary of Botanists Represented in the Hunt Institute Portrait Collection.* 401. 1972.

Thuspeinanta T. Durand Labiatae

Origins:

An anagram of *Tapeinanthus* Boiss. ex Benth.

Thyella Raf. Convolvulaceae

Origins:

See Constantine Samuel Rafinesque, *Flora Telluriana.* 4: 84. 1836 [1838]; Elmer D. Merrill, *Index rafinesquianus.* 200. 1949.

Thylacanthus Tul. Caesalpiniaceae

Origins:

From the Greek *thylakos* "a bag, sack, pouch" and *anthos* "flower," referring to the flowers.

Thylachium DC. Capparidaceae (Capparaceae)

Origins:

Orthographic variant of *Thilachium* Lour., from the Greek *thylakos* "bag, sack, pouch," referring to the fact that the calyx splits in half.

Thylacis Gagnep. Orchidaceae

Origins:
Greek *thylakos* "a bag, sack, pouch," referring to the shape of the lip, saccate.

Thylacodraba (Nábelek) O.E. Schulz Brassicaceae

Origins:
From the Greek *thylakos* "a bag, sack, pouch" and the genus *Draba* L.

Thylacophora Ridley Zingiberaceae

Origins:
From the Greek *thylakos* plus *phoreo* "to bear."

Thylacopteris Kunze ex J. Sm. Polypodiaceae

Origins:
From the Greek *thylakos* "a bag, sack, pouch" and *pteris* "a fern."

Thylacospermum Fenzl Caryophyllaceae

Origins:
From the Greek *thylakos* and *sperma* "a seed."

Thymbra L. Labiatae

Origins:
From the Latin and Greek *thymbra, ae* applied by Plinius, Dioscorides and Theophrastus to a plant, savory, a species of *Satureia*.

Thymelaea Miller Thymelaeaceae

Origins:
Thymelaia, the Greek name for a plant, *Daphne gnidium* L., *thymos* "thyme" and *elaia* "olive," referring to the foliage and fruit; Latin *thymelaea, ae* for a plant, the flax-leaved daphne, *Daphne gnidium* L.; see Manlio Cortelazzo & Paolo Zolli, *Dizionario etimologico della lingua italiana*. 5: 1339. Zanichelli, Bologna 1988.

Thymocarpus Nicolson, Steyerm. & Sivad. Marantaceae

Origins:
From the Greek *thymos* "thyme" and *karpos* "fruit."

Thymophylla Lagasca Asteraceae

Origins:
Greek *thymos* "thyme" and *phyllon* "leaf."

Species/Vernacular Names:
T. tenuiloba (DC.) Small
English: Dahlberg daisy

Thymopsis Benth. Asteraceae

Origins:
From the Greek *thymos* "thyme" and *opsis* "like, resembling."

Thymus L. Labiatae

Origins:
The ancient Greek name used by Dioscorides probably for this plant, *thymon* or *thymos* (*thymiao* "perfume"); Latin *thymum, i* was used for *Thymus vulgaris* L. and *Satureia capitata* L. (Plinius, Vergilius, M.T. Quintilianus); see Ernest Weekley, *An Etymological Dictionary of Modern English*. 2: 1504. 1967; M. Cortelazzo & P. Zolli, *Dizionario etimologico della lingua italiana*. 5: 1339. Zanichelli, Bologna 1988; N. Tommaseo & B. Bellini, *Dizionario della lingua italiana*. Torino 1865-1879; Carl Linnaeus, *Species Plantarum*. 590. 1753.

Species/Vernacular Names:
T. amurensis Klokov
English: Amur thyme
China: hei long jiang bai li xiang
T. caespititius Brot.
English: mountain thyme
T. disjunctus Klokov
English: long-tooth thyme
China: chang chi bai li xiang
T. herbabarona Lois.
English: caraway thyme
T. inaequalis Klokov
English: oblique-leaf thyme

China: xie ye bai li xiang

T. maschallianus Willdenow

English: Marschall thyme

China: yi zhu bai li xiang

T. mongolicus (Ronniger) Ronniger

English: Mongolian thyme

China: bai li xiang

T. nervulosus Klokov

English: distinct-vein thyme

China: xian mai bai li xiang

T. quinquecostatus Celak.

English: five-ribbed thyme

China: di jiao

T. serpyllum L.

English: wild thyme

T. vulgaris L.

English: common thyme, garden thyme, thyme

Arabic: za'ater

South Africa: tiemie

China: she xiang cao

Thyrasperma N.E. Br. Aizoaceae

Origins:
From the Greek *thyra* "a door, entrance" and *sperma* "seed"; see also *Hymenogyne* Haw.

Thyridachne C.E. Hubb. Gramineae

Origins:
From the Greek *thyridos* "a window, small door" and *achne* "chaff, glume."

Thyridocalyx Bremek. Rubiaceae

Origins:
From the Greek *thyridos* and *kalyx* "calyx."

Thyridolepis S.T. Blake Gramineae

Origins:
From the Greek *thyridos* "a window, small door," *thyra* "a door, entrance" and *lepis* "scale," referring to a depression in the glume, the scales are transparent.

Species/Vernacular Names:
T. mitchelliana (Nees) S.T. Blake (*Neurachne mitchelliana* Nees)

Australia: window mulga-grass, mulga-grass

Thyridostachyum Nees Gramineae

Origins:
From the Greek *thyridos* "a window, small door" and *stachys* "spike."

Thyrocarpus Hance Boraginaceae

Origins:
From the Greek *thyra* "a door, entrance" and *karpos* "fruit."

Species/Vernacular Names:
T. glochidiatus Maxim.

English: curved-tooth thyrocarpus

China: wan chi dun guo cao

T. sampsonii Hance

English: Sampson thyrocarpus

China: dun guo cao

Thyrsacanthus Nees Acanthaceae

Origins:
Greek *thyrsos* "a panicle, a thyrse" and *akantha* "thorn," genus *Acanthus* L., referring to the arrangement of the flowers.

Thyrsanthella Pichon Apocynaceae

Origins:
From the Greek *thyrsos* "a panicle, a thyrse" and *anthos* "flower."

Thyrsanthemum Pichon Commelinaceae

Origins:
From the Greek *thyrsos* and *anthemon* "flower."

Thyrsanthera Pierre ex Gagnepain Euphorbiaceae

Origins:
From the Greek *thyrsos* plus *anthera* "anther."

Thyrsia Stapf Gramineae

Origins:
From the Greek *thyrsos* "a panicle, a thyrse"; see also *Phacelurus* Griseb.

Thyrsodium Salzm. ex Benth. Anacardiaceae

Origins:
From the Greek *thyrsos* "a panicle, a thyrse."

Thyrsopteris Kunze Dicksoniaceae (Thyrsopteridoideae)

Origins:
From the Greek *thyrsos* "a panicle, a thyrse" and *pteris* "fern."

Thyrsosalacia Loes. Celastraceae

Origins:
Greek *thyrsos* "a panicle, a thyrse" plus *Salacia* L.

Thyrsostachys Gamble Gramineae

Origins:
From the Greek *thyrsos* with *stachys* "spike."

Thysanachne C. Presl Gramineae

Origins:
Greek *thysanos* "a fringe, tassel" and *achne* "chaff, glume."

Thysanella A. Gray Polygonaceae

Origins:
The diminutive of the Greek *thysanos* "a fringe, tassel."

Thysanobotrya Alderw. Cyatheaceae

Origins:
Greek *thysanos* "a fringe, tassel" and *botrys* "cluster, a bunch of grapes, a cluster of grapes."

Thysanocarpus Hook. Brassicaceae

Origins:
From the Greek *thysanos* "a fringe, tassel" and *karpos* "fruit," with winged fruit; wing can be entire or lobed.

Species/Vernacular Names:
T. conchuliferus E. Greene
California: Santa Cruz Island fringepod (Santa Cruz Island)

Thysanochilus Falc. Orchidaceae

Origins:
From the Greek *thysanos* and *cheilos* "lip," indicating the fringed lip; see also *Eulophia* R. Br., Greek *eu* "well, good" and *lophos* "a crest, plume," in reference to the crests on the labellum of some species.

Thysanoglossa Porto & Brade Orchidaceae

Origins:
From the Greek *thysanos* "a fringe, tassel" and *glossa* "tongue," descriptive of the margin of the lateral lobes of the lip.

Thysanolaena Nees Gramineae

Origins:
Greek *thysanos* "a fringe, tassel" and *chlaena, chlaenion* "a cloak," referring to the inflorescence.

Species/Vernacular Names:
T. latifolia (Hornem.) Honda
English: tiger grass
The Philippines: tambu, bugubui, buybui, eagadu, gatbo, lasa, tagadeu, tagisa, talankaran

Thysanosoria Gepp Lomariopsidaceae (Aspleniaceae)

Origins:
From the Greek *thysanos* "a fringe, tassel" and *soros* "heap, mound, a spore case."

Thysanospermum Champ. ex Bentham Rubiaceae

Origins:
From the Greek *thysanos* "a fringe, tassel" and *sperma* "seed."

Thysanostemon Maguire Guttiferae

Origins:

Greek *thysanos* "a fringe, tassel" and *stemon* "stamen."

Thysanostigma J.B. Imlay Acanthaceae

Origins:

From the Greek *thysanos* "a fringe, tassel" and *stigma* "stigma."

Thysanothus Poir. Anthericaceae (Liliaceae)

Origins:

Orthographic variant of *Thysanotus* R. Br.

Thysanotus R. Br. Anthericaceae (Liliaceae)

Origins:

Greek *thysanotos* "fringed," *thysanos* "a fringe, tassel," referring to the inner perianth segments; see Robert Brown (1773-1858), *Prodromus florae Novae Hollandiae*. 282. London (Mar.) 1810.

Species/Vernacular Names:

T. baueri R. Br. (*Chlamysporum baueri* (R. Br.) Kuntze; *Thysanotus humilis* F. Muell.)

English: mallee fringe-lily

T. dichotomus (Labill.) R. Br. (*Chlamysporum dichotomum* (Labill.) Kuntze; *Thysanotus divaricatus* R. Br.; *Ornithogalum dichotomum* Labill.)

English: branching fringe-lily

T. juncifolius (Salisb.) J.H. Willis & Court (*Chlamysporum juncifolium* Salisb.)

English: rush fringe-lily

T. patersonii R. Br.

English: twining fringe-lily

T. tuberosus R. Br. (*Chlamysporum tuberosum* (R. Br.) Kuntze; *Thysanotus elatior* R. Br.)

English: common fringe-lily, fringed violet

Thysanurus O. Hoffm. Asteraceae

Origins:

Greek *thysanos* "a fringe, tassel" and *oura* "tail."

Thysanus Lour. Connaraceae

Origins:

From the Greek *thysanos* "a fringe, tassel."

Tianschaniella Fedtsch. ex Popov Boraginaceae

Origins:

Tien Shan mountain systems, Turkestan, central Asia.

Tiarella L. Saxifragaceae

Origins:

The diminutive of the Greek *tiara* "a Persian diadem, a small crown," Latin *tiara* or *tiaras* "the head-dress of the Oriental, a turban, tiara," an allusion to the shape of the fruits.

Species/Vernacular Names:

T. cordifolia L.

English: coolwort, foamflower

T. trifoliata L.

English: lace-flower, foamflower

Tiarocarpus Rech.f. Asteraceae

Origins:

From the Greek *tiara* "a tiara, a small crown" and *karpos* "fruit."

Tiarrhena (Maxim.) Nakai Gramineae

Origins:

Greek *tiara* "a small crown" and *arrhen* "male."

Tibestina Maire Asteraceae

Origins:

Tibesti Massif, the highest mountain group of the Sahara and of western equatorial Africa.

Tibetia (Ali) H.P. Tsui Fabaceae

Origins:

Tibet, central Asia, south China, and north of India, Nepal, Sikkim and Bhutan.

Tibouchina Aublet Melastomataceae

Origins:
From a native name for the plants in Guiana.

Species/Vernacular Names:
T. elegans (Naudin) Cogn.

English: bristlecup glory bush

T. urvilleana (DC.) Cogn. (*Tibouchina semidecandra* hort. non (DC.) Cogn.; *Tibouchina grandiflora* hort.; *Lasiandra urvilleana* DC.; *Pleroma grandiflora* hort.; *Pleroma splendens* hort.)

English: glory bush, lasiandra, princess flower, purple glory tree, Brazilian spider flower

Malaya: Brazilian sendudok

Portuguese: aranhas

Tibouchinopsis Markgr. Melastomataceae

Origins:
Resembling *Tibouchina* Aubl.

Ticodendron Gómez-Laurito & Gómez P. Ticodendraceae (Hamamelidae, Fagales)

Origins:
Possibly from the Greek *teichos, tichos* "wall, fortified, walled" and *dendron* "tree"; see Rafaël Govaerts & David G. Frodin, *World Checklist and Bibliography of Fagales*. Royal Botanic Gardens, Kew 1996.

Ticoglossum Lucas Rodriguez ex Halbinger Orchidaceae

Origins:
From *tico*, a Central American Spanish vernacular word given to the Costa Ricans, plus the genus *Odontoglossum*.

Ticorea Aublet Rutaceae

Origins:
A native name in Guiana.

Tidestromia Standley Amaranthaceae

Origins:
After the Swedish-born American botanist Ivar Frederick Tidestrom (Tidestrøm), 1864-1956, to the United States 1880, assistant to Edward Lee Greene (1843-1915), botanical collector, author of *Flora of Utah and Nevada*. Wash-ington 1925; see J.H. Barnhart, *Biographical Notes upon Botanists*. 3: 383. 1965; R. Zander, F. Encke, G. Buchheim and S. Seybold, *Handwörterbuch der Pflanzennamen*. 14. Aufl. Stuttgart 1993; T.W. Bossert, *Biographical Dictionary of Botanists Represented in the Hunt Institute Portrait Collection*. 401. 1972; E.M. Tucker, *Catalogue of the Library of the Arnold Arboretum of Harvard University*. Cambridge, Massachusetts 1917-1933; J. Ewan, ed., *A Short History of Botany in the United States*. 101. New York and London 1969; Ida Kaplan Langman, *A Selected Guide to the Literature on the Flowering Plants of Mexico*. Philadelphia 1964; Joseph Ewan, *Rocky Mountain Naturalists*. 321-322. [b. 1865] The University of Denver Press 1950.

Tieghemella Pierre Sapotaceae

Origins:
The generic name honors the French (b. Bailleul) botanist Philippe Édouard Léon van Tieghem, 1839-1914 (d. Paris), professor of botany, author of *Recherches sur la structure des Aroïdées*. Paris 1867, *Traité de Botanique*. Paris 1881-1884 and *Eléments de botanique ...* Paris 1886; see John H. Barnhart, *Biographical Notes upon Botanists*. 3: 425. 1965; A. Nougarède, in *Dictionary of Scientific Biography* 13: 405-406. 1981; T.W. Bossert, *Biographical Dictionary of Botanists Represented in the Hunt Institute Portrait Collection*. Boston, Massachusetts 1972; R. Zander, F. Encke, G. Buchheim and S. Seybold, *Handwörterbuch der Pflanzennamen*. 14. Aufl. 793. Stuttgart 1993; Ida Kaplan Langman, *A Selected Guide to the Literature on the Flowering Plants of Mexico*. Philadelphia 1964; E.M. Tucker, *Catalogue of the Library of the Arnold Arboretum of Harvard University*. Cambridge, Massachusetts 1917-1933.

Species/Vernacular Names:
T. africana Pierre

Central Africa: nduka, okala, douka, n'duka, duka

Cameroon: nom adjap elang

Congo: n'duka

Gabon: okola

T. heckelii Pierre ex A. Chev. (*Mimusops heckelii* (A. Chev.) Hutch. & Dalz.)

English: cherry mahogany

Tropical Africa: makoré, bacu, baku

Nigeria: aghanokpe (Edo)

Tieghemia Balle Loranthaceae

Origins:
For the French botanist Philippe Édouard Léon van Tieghem, 1839-1914, professor of botany. His works

include *Recherches sur la structure des Aroïdées*. Paris 1867, *Traité de Botanique*. Paris 1881-1884 and *Eléments de botanique ...* Paris 1886; see J.H. Barnhart, *Biographical Notes upon Botanists*. 3: 425. 1965; A. Nougarède, in *Dictionary of Scientific Biography* 13: 405-406. 1981.

Tieghemopanax R. Viguier Araliaceae

Origins:

After the French botanist Philippe Édouard Léon van Tieghem, 1839-1914, author of *Recherches sur la structure des Aroïdées*. Paris 1867 and *Eléments de botanique ...* Paris 1886; see J.H. Barnhart, *Biographical Notes upon Botanists*. 3: 425. 1965; A. Nougarède, in *Dictionary of Scientific Biography* 13: 405-406. 1981.

Tiglium Klotzsch Euphorbiaceae

Origins:

See Helmut Genaust, *Etymologisches Wörterbuch der botanischen Pflanzennamen*. 644-645. Basel 1996; R. Zander, F. Encke, G. Buchheim and S. Seybold, *Handwörterbuch der Pflanzennamen*. 14. Aufl. 666. Stuttgart 1993; Georg Christian Wittstein, *Etymologisch-botanisches Handwörterbuch*. 881. Ansbach 1852; F. Boerner & G. Kunkel, *Taschenwörterbuch der botanischen Pflanzennamen*. 4. Aufl. 380. Berlin & Hamburg 1989; Manlio Cortelazzo & Paolo Zolli, *Dizionario etimologico della lingua italiana*. 5: 1338. Zanichelli, Bologna 1988.

Tigridia Juss. Iridaceae

Origins:

Latin *tigris* "a tiger," referring to the flowers and the spots in the tube.

Species/Vernacular Names:

T. pavonia (L.f.) Ker Gawler
English: peacock tiger flower

Tilia L. Tiliaceae

Origins:

From the Latin name for the linden or lime tree; see M. Cortelazzo & P. Zolli, *Dizionario etimologico della lingua italiana*. 5: 1338. Zanichelli, Bologna 1988; Helmut Genaust, *Etymologisches Wörterbuch der botanischen Pflanzennamen*. 645. 1996; R. Zander, F. Encke, G. Buchheim and S. Seybold, *Handwörterbuch der Pflanzennamen*. 14. Aufl. 666. 1993; F. Boerner & G. Kunkel, *Taschenwörterbuch der botanischen Pflanzennamen*. 4. Aufl. 380. 1989.

Species/Vernacular Names:

T. americana L.
English: American lime, basswood, American basswood, whitewood

T. caroliniana Mill.
English: Carolina basswood

T. chinensis Maxim.
China: tuan

T. cordata Miller
English: small-leaved lime

T. x *europaea* L.
English: common lime

T. heterophylla Vent.
English: white basswood

T. japonica (Miq.) Simonkai
English: Japanese lime

T. mandshurica Rupr. & Maxim.
English: Manchurian linden

T. mexicana Schltdl.
English: Mexican basswood

T. miqueliana Maxim.
China: pu ti shu

T. mongolica Maxim.
English: Mongolian lime

T. platyphyllos Scop.
English: broad-leaved lime, large-leaved lime

T. tomentosa Moench
English: silver lime

Tiliacora Colebr. Menispermaceae

Origins:

The native name in Bengal for *Tiliacora racemosa* Colebr.

Species/Vernacular Names:

T. racemosa Colebr.
India: tiliakoru, karwant, karrauth, rangoe

Tilingia Regel & Tiling Umbelliferae

Origins:

For Heinrich Sylvester Theodor Tiling (Tilling), 1818-1871, plant collector, physician and botanist; see J.H. Barnhart, *Biographical Notes upon Botanists*. 3: 384. 1965; H.N. Clokie, *Account of the Herbaria of the Department of Botany in the University of Oxford*. 254. Oxford 1964; E.M. Tucker, *Catalogue of the Library of the Arnold Arboretum of Harvard University.* Cambridge, Massachusetts 1917-1933.

Tillaea L. Crassulaceae

Origins:

For the Italian botanist Michelangelo Tilli (Michael Angelus Tillius), 1655-1740, scientist, professor of botany, a Fellow of the Royal Society of London, from 1685 to 1740 Praefectus of the Botanical Garden of the University of Pisa, Italy, author of *Catalogus plantarum horti Pisani.* Florentiae 1723. A member of the family, Giovanni Lorenzo Tilli, published an *Enumeratio stirpium in horto academico Pisano viventium.* Pisa 1806-1807 and 1810. See L. Amadei, "Note sull'*Herbarium Horti Pisani:* l'origine delle collezioni." *Museol. Scient.* 4(1-2): 119-129. 1987; Fabio Garbari (b. 1937), L. Tomasi Tongiorgi and A. Tosi, *Giardino dei Semplici: L'Orto Botanico di Pisa dal XVI al XX secolo.* Pisa 1991; F.A. Stafleu, "Die Geschichte der Herbarien." *Bot. Jahrb. Syst.* 108(2/3): 155-166. 1987; Carl Linnaeus, *Species Plantarum.* 128. 1753 and *Genera Plantarum.* Ed. 5. 62. 1754; Mariella Azzarello Di Misa, a cura di, *Il Fondo Antico della Biblioteca dell'Orto Botanico di Palermo.* 270. Palermo 1988; Gordon Douglas Rowley, *A History of Succulent Plants.* Strawberry Press, Mill Valley, California 1997; Mary Gunn and Leslie E. Codd, *Botanical Exploration of Southern Africa.* 62. Cape Town 1981.

Tillaeastrum Britton Crassulaceae

Origins:

Referring to the genus *Tillaea* L.

Tillandsia L. Bromeliaceae

Origins:

Named in honor of the Swedish physician Elias Erici Tillands (Tillander, Tillandz, Til-Landz, Til-Lands), 1640-1693, botanist, Dr. med. Leiden 1670, professor of medicine at Åbo, wrote *Disputatio de atrophia.* Lugduni Batavorum [Leyden] 1670, *Catalogus plantarum.* Aboae 1673 and Eliae Til-Landz ... *Icones novae.* Aboae 1683; see J.H. Barnhart, *Biographical Notes upon Botanists.* 3: 384. 1965; Carl Linnaeus, *Species Plantarum.* 286. 1753 and *Genera Plantarum.* Ed. 5. 138. 1754; Helmut Genaust, *Etymologisches Wörterbuch der botanischen Pflanzennamen.* 645-646. Basel 1996; Jonas C. Dryander, *Catalogus bibliothecae historico-naturalis Josephi Banks.* London 1796-1800; Georg Christian Wittstein, *Etymologisch-botanisches Handwörterbuch.* 882. Ansbach 1852; F. Boerner & G. Kunkel, *Taschenwörterbuch der botanischen Pflanzennamen.* 4. Aufl. 179. 1989.

Species/Vernacular Names:

T. usneoides (L.) L. (the plant resembles *Usnea,* a lichen)

English: Spanish moss, Florida moss, old man's beard, grandfather's whiskers, air plant

Spanish: barba de capuchino

Peru: ccacca suncja, cotataura, inti suncja, millmahina, saccropa, salvaje, salvajina, ucushpa hueclan

Tillospermum Salisbury Myrtaceae

Origins:

Greek *tillo* "pluck, pick so as to extract fiber" and *sperma* "a seed"; see *Monthly Rev.* ser. 2. 75: 74. 1814.

Timandra Klotzsch Euphorbiaceae

Origins:

Greek *timaios* "highly prized," *timao* "honor" and *aner, andros* "male, stamen."

Timonius DC. Rubiaceae

Origins:

Timon is an Amboina plant name; in the Malay–Indonesian language *timun* means a gourd or a pumpkin. Ambon Island is located 11 km off the southwestern coast of the island of Seram; Ambon, formerly Amboina or Amboyna, island and *kotamadya* (city) in Maluku Tengah *kabupaten* (Central Moluccas regency) of Maluku *propinsi* (province), Indonesia.

Tina Schultes Sapindaceae

Origins:

Possibly from the Greek and Latin *tina* "wine-vessel."

Tinaea Garzia Gramineae

Origins:

Possibly from the Latin *tinea, ae* "a gnawing worm, moth."

Tinantia Scheidw. Commelinaceae

Origins:

For François Auguste Tinant, 1803-1853, botanist, author of *Flore luxembourgeoise.* Luxembourg 1836; see J.H.

Barnhart, *Biographical Notes upon Botanists*. 3: 385. 1965; F. Boerner & G. Kunkel, *Taschenwörterbuch der botanischen Pflanzennamen*. 4. Aufl. 179. Berlin & Hamburg 1989; Ethelyn Maria Tucker, *Catalogue of the Library of the Arnold Arboretum of Harvard University*. Cambridge, Massachusetts 1917-1933.

Tinea Bivona Orchidaceae

Origins:

After the Italian botanist Vincenzo (Vincentius) Tineo, 1791-1856, professor of botany, from 1814 to 1856 Director of the Botanical Garden of Palermo. His works include *Catalogus plantarum Horti Regii Panormitani*. Panormi 1827, *Plantarum rariorum Siciliae minus cognitarum* pugillus primus. Palermo 1817 and *Plantarum rariorum Siciliae minus cognitarum* ... Panormi 1846, he was son of Giuseppe Tineo (1757-1812); see John H. Barnhart, *Biographical Notes upon Botanists*. 3: 386. 1965; T.W. Bossert, *Biographical Dictionary of Botanists Represented in the Hunt Institute Portrait Collection*. 402. 1972; A. Lasègue, *Musée botanique de Benjamin Delessert*. 314. Paris 1845; Ethelyn Maria Tucker, *Catalogue of the Library of the Arnold Arboretum of Harvard University*. Cambridge, Massachusetts 1917-1933; F. Tornabene, "Elogio accademico del cav. Vincenzo Tineo." *Atti Accad. Gioenia Sci. Nat*. Catania 1856; Giuseppe M. Mira, *Bibliografia Siciliana*. 2: 411. Palermo 1881; R. Zander, F. Encke, G. Buchheim and S. Seybold, *Handwörterbuch der Pflanzennamen*. 14. Aufl. 790. Stuttgart 1993; Mariella Azzarello Di Misa, a cura di, *Il Fondo Antico della Biblioteca dell'Orto Botanico di Palermo*. 271. Palermo 1988.

Tinospora Miers Menispermaceae

Origins:

Possibly from the Latin *tinus* "a plant, laurustinus" and Greek *sporos, spora* "a seed."

Species/Vernacular Names:

T. crispa (L.) Hook.f. & Thomson (*Tinospora rumphii* Boerl.)

The Philippines: makabuhay, paliaban, panauan, taganagtagua

Tintinabulum Rydb. Polemoniaceae

Origins:

Latin *tintinnabulum* "a bell, signal-bell," *tintinno* "to ring, to jingle."

Tintinnabularia Woodson Apocynaceae

Origins:

From the Latin *tintinnabulum* "a bell, signal-bell."

Tipuana (Benth.) Benth. Fabaceae

Origins:

From the South American name, *tipu*.

Species/Vernacular Names:

T. tipu (Benth.) Kuntze

English: pride of Bolivia

Latin America: tipu

Peru: tipa

Tipularia Nutt. Orchidaceae

Origins:

Latin *tippula, tipula, ae* "an insect that runs swiftly over the water, the water-spider, water-spinner," possibly referring to the corm or to the shape of the flower, an allusion to the shape of the spur on the lip.

Tiquilia Pers. Boraginaceae (Ehretiaceae)

Origins:

Vernacular or native South American name.

Tiquiliopsis A. Heller Boraginaceae (Ehretiaceae)

Origins:
Resembling *Tiquilia*.

Tirucalia Raf. Euphorbiaceae

Origins:

The Malabar names, *tiru-malu, tirukalli, kalli*, see *Euphorbia tirucalli* L.; see C.S. Rafinesque, *Flora Telluriana*. 4: 112. 1836 [1838] and *Autikon botanikon*. Icones plantarum select. nov. vel rariorum, etc. 90. Philadelphia 1840; E.D. Merrill, *Index rafinesquianus*. 156. 1949.

Tischleria Schwantes Aizoaceae

Origins:

For the German botanist Georg Friedrich Leopold Tischler, 1878-1955, traveler (Java, Ceylon, East Africa), plant collector, cytologist, professor of botany; see J.H. Barnhart, *Biographical Notes upon Botanists*. 3: 386. 1965; T.W. Bossert, *Biographical Dictionary of Botanists Represented in the Hunt Institute Portrait Collection*. 402. 1972; Ida Kaplan Langman, *A Selected Guide to the Literature on the Flowering Plants of Mexico*. Philadelphia 1964; E.M. Tucker, *Catalogue of the Library of the Arnold Arboretum of Harvard University*. Cambridge, Massachusetts 1917-1933; Stafleu and Cowan, *Taxonomic Literature*. 6: 366-368. 1986.

Tisonia Baillon Flacourtiaceae

Origins:

For the French botanist Eugène Édouard Augustin Tison, b. 1842; see J.H. Barnhart, *Biographical Notes upon Botanists*. 3: 387. 1965; E.M. Tucker, *Catalogue of the Library of the Arnold Arboretum of Harvard University*. Cambridge, Massachusetts 1917-1933.

Tisserantia Humbert Asteraceae

Origins:

For the French botanist Charles Tisserant, 1886-1962, traveler, explorer and plant collector (Angola, Equatorial Africa, Central African Republic, Mozambique); see I.H. Vegter, *Index Herbariorum*. Part II (7), *Collectors T-Z*. Regnum Vegetabile vol. 117.

Tisserantiella Mimeur Gramineae

Origins:

For the French botanist Charles Tisserant, 1886-1962, traveler, explorer and plant collector (Angola, Equatorial Africa, Central African Republic, Mozambique); see I.H. Vegter, *Index Herbariorum*. Part II (7), *Collectors T-Z*. Regnum Vegetabile vol. 117.

Tisserantiodoxa Aubrév. & Pellegrin Sapotaceae

Origins:

For the French botanist Charles Tisserant, 1886-1962, traveler, explorer and plant collector (Angola, Equatorial Africa, Central African Republic, Mozambique); see I.H. Vegter, *Index Herbariorum*. Part II (7), *Collectors T-Z*. Regnum Vegetabile vol. 117.

Tisserantodendron Sillans Bignoniaceae

Origins:

For the French botanist Charles Tisserant, 1886-1962, traveler, explorer and plant collector (Angola, Equatorial Africa, Central African Republic, Mozambique); see I.H. Vegter, *Index Herbariorum*. Part II (7), *Collectors T-Z*. Regnum Vegetabile vol. 117.

Titania Endlicher Orchidaceae

Origins:

Named after Titania, wife of Oberon; genus closely related to *Oberonia* Lindl.

Titanopsis Schwantes Aizoaceae

Origins:

From the Greek *titan* "sun, the Sun-god" and *opsis* "like," referring to the flowers; see H. Genaust, *Etymologisches Wörterbuch der botanischen Pflanzennamen*. 647. [from the Greek *titanos* "lime, gypsum, chalk"] Basel 1996; William T. Stearn, *Stearn's Dictionary of Plant Names for Gardeners*. 296. Cassell, London 1993; F. Boerner & G. Kunkel, *Taschenwörterbuch der botanischen Pflanzennamen*. 4. Aufl. 179. Berlin & Hamburg 1989.

Titanotrichum Soler. Gesneriaceae

Origins:

Greek *titanos* "lime, gypsum, chalk" and *thrix, trichos* "hair," referring to the hairs of the indumentum.

Species/Vernacular Names:

T. oldhamii (Hemsley) Soler.

Japan: matsumura-sô

Tithonia Desf. ex Juss. Asteraceae

Origins:

After Tithonus, a young man who was a favorite and companion of Aurora, the goddess of dawn.

Species/Vernacular Names:

T. sp.

Yoruba: gbawobo

T. diversifolia (Hemsley) A. Gray (*Mirasolia diversifolia* Hemsl.)

English: Mexican sunflower, tree marigold

T. rotundifolia (Mill.) Blake (*Tagetes rotundifolia* Mill.)

English: red sunflower

Tithymaloides Ortega Euphorbiaceae

Origins:

Tithymalos, tithymallos, tithymallon are Greek names for spurge and *euphorbia*; Latin *tithymalis, idis, tithymalus* and *tithymallus* for a plant with a milk-like sap, spurge and sea-spurge (Plinius); see H. Genaust, *Etymologisches Wörter-buch der botanischen Pflanzennamen*. 647. 1996.

Tithymalopsis Klotzsch & Garcke Euphorbiaceae

Origins:

Resembling *Tithymalus*.

Tithymalus Gaertner Euphorbiaceae

Origins:

Greek name for spurge and euphorbia, for a plant with a milk-like sap; Latin *tithymalus, tithymallus* "a plant with a milk-like sap," a female species, called *tithymalis, idis* "sea spurge"; see Edward Lee Greene, *Landmarks of Botanical History*. Edited by Frank N. Egerton. Stanford, California 1983.

Tithymalus Hill Euphorbiaceae

Origins:

From*tithymalos, tithymallos, tithymallon*, Greek names for spurge and euphorbia, for a plant with a milk-like sap; Latin *tithymalus, tithymallus* "a plant with a milk-like sap," a female species, called *tithymalis, idis* "sea spurge."

Tittmannia Brongn. Bruniaceae

Origins:

For the German botanist Johann August Tittmann, 1774-1840, physician, agronomist, Dr. med. 1801 Leipzig, published *Ueber den Embryo des Saamenkorns*. Dresden 1817 and *Keimung der Pflanzen*. Dresden 1821; see J.H. Barn-hart, *Biographical Notes upon Botanists*. 3: 387. 1965.

Tmesipteris Bernh. Psilotaceae (Tmesipteridaceae)

Origins:

Greek *tmesis* "cutting" and *pteris* "fern," referring to the fronds.

Tobagoa Urban Rubiaceae

Origins:

From Tobago.

Tococa Aublet Melastomataceae

Origins:

From a vernacular name, French Guiana, the Galibis called *Tococa guianensis* Aubl. *tococo*.

Species/Vernacular Names:

T. glandulosa Gleason

Peru: yaco mullaca

T. guianensis Aublet

Peru: sacha mullaca

T. juruensis Pilger

Peru: maraño

T. lasiostyla Cogniaux

Peru: sacha mullaca, yacu mullaca

Tocoyena Aublet Rubiaceae

Origins:

From a vernacular name.

Species/Vernacular Names:

T. formosa (Cham. & Schltdl.) Schumann

Bolivia: bicito, bicito de la pampa, tutumillo

Todaroa A. Richard & Galeotti Orchidaceae

Origins:

Probably dedicated to the Italian botanist Agostino Todaro, 1818-1892, author of *Orchideae siculae*. Panormi 1842; see J.H. Barnhart, *Biographical Notes upon Botanists*. 3: 388. 1965; Gordon Douglas Rowley, *A History of Succulent Plants*. Strawberry Press, Mill Valley, California 1997.

Todaroa Parlatore Umbelliferae

Origins:

Dedicated to the Italian botanist Agostino Todaro, 1818-1892, from 1856 to 1892 Director of the Palermo Botanical Garden. Among his writings are *Orchideae siculae*. Panormi 1842, *Piante vive del Real Orto Botanico di Palermo disponibili per baratti nell'anno 1857*. Palermo 1856, *Hortus Botanicus Panormitanus*. Palermo 1876-1891, *Nuovi generi e nuove specie coltivate nel R. Orto Botanico di Palermo*. Palermo 1858-1861 and *Relazione sulla coltura dei cotoni in Italia seguita da una monografia del genere Gossypium*. Roma 1877-1878; see P.A. Fryxell and C. Earle Smith, Jr., "The contribution of Agostino Todaro to the *Gossypium* nomenclature." *Taxon*. 21(1): 139-145. (Feb.) 1972; D. Lanza, "Agostino Todaro." *Malpighia*. 6(1): 120-132. 1892; J.H. Barnhart, *Biographical Notes upon Botanists*. 3: 388. 1965; T.W. Bossert, *Biographical Dictionary of Botanists Represented in the Hunt Institute Portrait Collection*. 402. 1972; H.N. Clokie, *Account of the Herbaria of the Department of Botany in the University of Oxford*. 255. Oxford 1964; Ida Kaplan Langman, *A Selected Guide to the Literature on the Flowering Plants of Mexico*. Philadelphia 1964; E.M. Tucker, *Catalogue of the Library of the Arnold Arboretum of Harvard University*. Cambridge, Massachusetts 1917-1933; Alain White and Boyd Lincoln Sloane, *The Stapelieae*. Pasadena 1937; Giuseppe M. Mira, *Bibliografia Siciliana*. 2: 414-415. Palermo 1881; Gordon Douglas Rowley, *A History of Succulent Plants*. Mill Valley, California 1997; R. Zander, F. Encke, G. Buchheim and S. Seybold, *Handwörterbuch der Pflanzennamen*. 14. Aufl. Stuttgart 1993.

Toddalia Juss. Rutaceae

Origins:

Kaki-toddali or *koka-toddali* is a Malayalam name for *Toddalia asiatica* (L.) Lam.

Species/Vernacular Names:

T. asiatica (L.) Lam. (*Toddalia aculeata* Pers.)

English: Lopez root

Southern Africa: iGado, ruKato, Nyachara (Shona)

Rodrigues Island: bambara, patte de poule à piquants

Japan: saru-kake-mikan, sara-kachâ

China: fei long zhang zue

Toddaliopsis Engl. Rutaceae

Origins:

The name means "like *Toddalia*" or "resembling the genus Toddalia," *kaki-toddali* or *koka-toddali* is a Malayalam name for *Toddalia asiatica* (L.) Lam.

Species/Vernacular Names:

T. bremekampii Verdoorn (the specific name after the Dutch botanist Cornelis Eliza Bertus Bremekamp, 1888-1984, professor at the University of Pretoria. His works include "A revision of the South African species of *Pavetta*." in *Ann. Transv. Mus.* 13: 182-213. 1929, "A monograph of the genus *Pavetta* L." in *Feddes Repert.* 37: 1-208. 1934, and *The African Species of Oldenlandia*. Amsterdam 1952)

English: wild mandarin

Southern Africa: wildenartjie; uMozane, iNtane (Zulu)

Toddavaddia Kuntze Oxalidaceae

Origins:

A vernacular name.

Todea Willdenow ex Bernh. Osmundaceae

Origins:

After the German botanist Heinrich Julius Tode, 1733-1797, clergyman, cryptogamist, author of *Fungi mecklenburgensis selecti*. Luneburgi [Lüneburg] 1790-1791; see J.H. Barnhart, *Biographical Notes upon Botanists*. 3: 388. 1965; Jonas C. Dryander, *Catalogus bibliothecae historico-naturalis Josephi Banks*. London 1796-1800.

Species/Vernacular Names:

T. barbara (L.) T. Moore (*Acrostichum barbarum* L.; *Osmunda barbara* (L.) Thunb.; *Todea africana* Willd.)

English: king fern, crape fern

Toechima Radlk. Sapindaceae

Origins:

Greek *toichos* "a wall" and *ima* "lowermost" or *heima* "clothing," possibly referring to the fruits; see J.A. Baines, *Australian Plant Genera. An Etymological Dictionary of Australian Plant Genera*. 375-376. Chipping Norton, N.S.W. 1981.

Species/Vernacular Names:

T. erythrocarpum (F. Muell.) Radlk. (*Cupania erythrocarpa* F. Muell.; *Ratonia nugentii* Bailey)

English: pink tamarind, foambark

T. lanceolatum C. White

English: pear tamarind, pink foambark

Tofieldia Hudson Melanthiaceae (Liliaceae)

Origins:

For the British (b. Yorkshire) botanist Thomas Tofield, 1730-1779 (d. Yorks.), correspondent of W. Hudson; see R.

Zander, F. Encke, G. Buchheim and S. Seybold, *Handwörterbuch der Pflanzennamen*. 14. Aufl. 548. 1993; F. Boerner & G. Kunkel, *Taschenwörterbuch der botanischen Pflanzennamen*. 4. Aufl. 179. 1989; Helmut Genaust, *Etymologisches Wörterbuch der botanischen Pflanzennamen*. 647. [d. 1793] Basel 1996; Ray Desmond, *Dictionary of British & Irish Botanists and Horticulturists*. 687. London 1994.

Species/Vernacular Names:
T. pusilla (Michaux) Pers.
English: Scotch asphodel

Tolmachevia Á. Löve & D. Löve Crassulaceae

Origins:
For the Russian botanist Alexsandr (Alexandr) Innokent'evich (Innokentevich) Tolmachew (Tolmacev, Tolmatchev), 1903-1979; see Stafleu and Cowan, *Taxonomic Literature*. 6: 386-387. Utrecht 1986.

Tolmiea Torrey & A. Gray Saxifragaceae

Origins:
After the British (b. Inverness) botanist William Fraser Tolmie, 1812-1886 (d. Victoria, B.C., Canada), physician, plant collector, fur trader, a pupil of W.J. Hooker, with G.M. Dawson wrote *Comparative Vocabularies of the Indian Tribes of British Columbia*. 1884; see J.H. Barnhart, *Biographical Notes upon Botanists*. 3: 389. 1965; F. Boerner & G. Kunkel, *Taschenwörterbuch der botanischen Pflanzennamen*. 4. Aufl. 179. 1989; T.W. Bossert, *Biographical Dictionary of Botanists Represented in the Hunt Institute Portrait Collection*. 403. Boston, Massachusetts 1972; H.N. Clokie, *Account of the Herbaria of the Department of Botany in the University of Oxford*. 255. Oxford 1964; Howard Atwood Kelly and Walter Lincoln Burrage, *Dictionary of American Medical Biography*. New York 1928; Wilson Duff, *The Indian History of British Columbia*. Anthropology in British Columbia. Memoir no. 5. Royal British Columbia Museum 1969; William Fraser Tolmie, *The Journals of William Fraser Tolmie: physician and fur trader*. Mitchell Press Limited, Vancouver 1963; R. Tanghe (sous la direction de), *Bibliographie des Bibliographies Canadiennes*, Bibliography of Canadian Bibliographies. T. U. of T. Press 1960; G.P.V. Akrigg & Helen B. Akrigg, *British Columbia Place Names*. Sono Nis Press, Victoria 1986; John T. Walbran, *British Columbia Coast Names, 1592-1906. To Which are Added a Few Names in Adjacent United States Territory, Their Origin and History*. First edition. Ottawa: Government Printing Bureau, 1909.

Species/Vernacular Names:
T. menziesii (Hook.) Torrey & A. Gray
English: pig-a-back plant, pickaback plant, piggyback plant

Tolpis Adans. Asteraceae

Origins:
Origin unknown, possibly from the Greek *tolype* "a ball of wool, lump," referring to the fruiting capitula; some suggest from *Crepis*.

Species/Vernacular Names:
T. barbata (L.) Gaertner (*Crepis barbata* L.; *Tolpis umbellata* Bertol.)
English: yellow hawkweed, tolpis
T. capensis (L.) Sch.Bip. (*Hieracium capense* L.)
English: yellow hawkweed
Southern Africa: fukuthoane (Sotho)
T. macrorhiza (Lowe) Lowe
English: Madeira hawkweed
Portuguese: leituga

Tolumnia Raf. Orchidaceae

Origins:
Origin unknown, according to Rafinesque, named for a nymph; Tolumnius was a king of the Veientes or a Rutulian soothsayer mentioned by Vergilius; see Constantine Samuel Rafinesque, *Flora Telluriana*. 2: 101. Philadelphia 1836 [1838].

Tolypanthus (Blume) Reichb. Loranthaceae

Origins:
From the Greek *tolype* "a ball of wool, lump" and *anthos* "flower."

Tomanthera Raf. Scrophulariaceae

Origins:
See C.S. Rafinesque, *New Fl. N. Am*. 2: 65. Philadelphia 1836 [1837] and *The Good Book*. 45. 1840; E.D. Merrill, *Index rafinesquianus*. 219. 1949.

Tomex Forssk. Salvadoraceae

Origins:
From the Latin *thomix, thomex, tomex, tomix* "a cord, string, line, thread."

Tomilix Raf. Scrophulariaceae

Origins:

See C.S. Rafinesque, *New Fl. N. Am.* 2: 72. Philadelphia 1836 [1837]; E.D. Merrill, *Index rafinesquianus.* 219. 1949.

Tomista Raf. Asteraceae

Origins:

See C.S. Rafinesque, *New Fl. N. Am.* 4: 73. Philadelphia 1836 [1838]; E.D. Merrill, *Index rafinesquianus.* 242. 1949.

Tommasinia A. Bertolini Umbelliferae

Origins:

For the Austro-Hungarian-Italian botanist Muzio Giuseppe Spirito de Tommasini (Mutius Joseph Spiritus, Ritter von), 1794-1879, magistrate, Podestà del Comune (mayor of Trieste); see John H. Barnhart, *Biographical Notes upon Botanists.* 3: 390. 1965; T.W. Bossert, *Biographical Dictionary of Botanists Represented in the Hunt Institute Portrait Collection.* 403. 1972; H.N. Clokie, *Account of the Herbaria of the Department of Botany in the University of Oxford.* 255. Oxford 1964; E.M. Tucker, *Catalogue of the Library of the Arnold Arboretum of Harvard University.* Cambridge, Massachusetts 1917-1933; Antoine Lasègue, *Musée botanique de M. Benjamin Delessert.* 1845; Renato Mezzena, "Il Civico Orto Botanico di Trieste." in Francesco Maria Raimondo, ed., *Orti Botanici, Giardini Alpini, Arboreti Italiani.* 321-325. Palermo 1992.

Tomodon Raf. Amaryllidaceae

Origins:

See C.S. Rafinesque, *Flora Telluriana.* 4: 22. Philadelphia 1836 [1838]; E.D. Merrill, *Index rafinesquianus.* 98. 1949.

Tomotris Raf. Orchidaceae

Origins:

Greek *tomos* "division, section, slice" and *treis* "three," probably referring to the lip or to the inflorescence; see C.S. Rafinesque, *Flora Telluriana.* 2: 89. Philadelphia 1836 [1837].

Tonduzia Pittier Apocynaceae

Origins:

For the Swiss botanist Adolphe Tonduz, 1862-1921, plant collector, traveler and botanical explorer (Costa Rica and Guatemala); see J.H. Barnhart, *Biographical Notes upon Botanists.* 3: 390. 1965; E.M. Tucker, *Catalogue of the Library of the Arnold Arboretum of Harvard University.* Cambridge, Massachusetts 1917-1933.

Tonestus A. Nelson Asteraceae

Origins:

Anagram of *Stenotus.*

Species/Vernacular Names:

T. lyallii (A. Gray) Nelson

English: Lyall's haplopappus

Tonina Aublet Eriocaulaceae

Origins:

A vernacular name, French Guiana.

Toona (Endlicher) M. Roemer Meliaceae

Origins:

From the words *toon* or *tunna*, Sanskrit for *Toona ciliata* M. Roemer; Hindi *tun*; see Jennifer M. Edmonds & Martin Staniforth, "*Toona sinensis.* Meliaceae." in *Curtis's Botanical Magazine.* 15(3): 186-193. August 1998.

Species/Vernacular Names:

T. ciliata M. Roemer (*Cedrela toona* Roxburgh ex Rottl. & Willd.; *Cedrela australis* F. Muell.; *Toona australis* (F. Muell.) Harms)

English: Burma cedar, cedrela, Indian mahogany, Indian mahogany tree, red cedar, red toon, toon, toon tree, tun tree, Australian red cedar, Australian cedar

Australia: (aboriginal names:) polai, woolia, mamin, mugurpul, woota

French: bois de Chittagong, bois de toon

Burma: Moulmein cedar, latsai, mai yum, thitkado, toon, yedama

Indochina: Xoan moc

Java: Soeren poeti, soerken-meira

Philippines Islands: red cedar

India: khusing, deodari, devadari, cuveraca, kuberaka, kachla, tun, tuni, tuna, toona, tunni, tunna, tunumaram, aranamaram, sevvagil, lim, maha limbo, mahalimbu, maha nim, mahlun, nandi-vraksha, nandibriksha, nandi-chettu, kempu-gandhagiri, kooruk, thit-ka-du; drawi, chiti-sirin (Punjab); gorianim (East Bengal); lud, tun, tuni (Bengal); maha-limbu (West Bengal); devdari, huruk, katangia, kooveruka, kurpodol, lim, maha-nim, mahalim, mahalum, mahanim, mathagiri vembu, vedi-vembu, noge, toona, tunna (East India); henduri poma, poma (Assam); malayapoo-toon-marum, tood (Deccan); santhana vembu, thevatharam, tood (Madras); todu, tundu (Bombay)

Nepal: tooni, toon, tun, tuni, labshi

South Africa: Indian mahogany, toonboom

Canary Islands: tunda

T. sinensis (A. Juss.) M. Roem. (*Cedrela sinensis* A. Juss.)

Japan: chan-chin (= fragrant *Camellia*)

China: ch'un

Tordyliopsis DC. Umbelliferae

Origins:
Resembling the genus *Tordylium* L.

Tordylium L. Umbelliferae

Origins:
From *tordylon, tordylion* ancient Greek names for an umbelliferous plant, hartwort; Latin *tordylion* or *tordylon*; according to some the seed of the plant *seselis*, according to others, a plant, hartwort.

Species/Vernacular Names:
T. apulum L.
English: hartwort

Torenia L. Scrophulariaceae

Origins:
Named for the Swedish clergyman Reverend Olof Torén, 1718-1753, traveler, botanist and plant collector, ship's chaplain with the Swedish East India Company at Surat, India (1750-1752) and in China (1748-1749); see J.H. Barnhart, *Biographical Notes upon Botanists.* 3: 391. 1965; Pehr Osbeck (1723-1805), *Dagbok öfwer en Ostindisk Resa åren 1750, 1751, 1752 ...* Stockholm 1757, English translation by Johann Reinhold Forster (1729-1798), (in English) *A Voyage to China and the East Indies,* by P.O., ... together with a voyage to Suratte, by Olof Toreen, etc. London 1771; Carl Linnaeus, *Species Plantarum.* 619. 1753 and *Genera Plantarum.* Ed. 5. 270. 1754; Emil Bretschneider (1833-1901), *History of European Botanical Discoveries in China.* [Reprint of the original edition 1898.] Leipzig 1981; A. Lasègue, *Musée botanique de Benjamin Delessert.* Paris 1845; E.M. Tucker, *Catalogue of the Library of the Arnold Arboretum of Harvard University.* Cambridge, Massachusetts 1917-1933.

Species/Vernacular Names:
T. asiatica L.
Hawaii: ola'a beauty, naioola'a
Malaya: kelelawat
T. fournieri Linden ex Fourn.
English: wishbone flower, bluewings
T. thouarsii (Cham. & Schltdl.) Kuntze
Yoruba: alofo odo, alofohun, reku reku

Toresia Pers. Gramineae

Origins:
Orthographic variant of *Torresia* Ruíz & Pav.

Torfasadis Raf. Euphorbiaceae

Origins:
See C.S. Rafinesque, *Flora Telluriana.* 4: 112. 1836 [1838]; E.D. Merrill, *Index rafinesquianus.* 157. 1949.

Torgesia Bornm. Gramineae

Origins:
After the German botanist Karl Emil Wilhelm Torges, 1831-1917, military physician, agrostologist; see J.H. Barnhart, *Biographical Notes upon Botanists.* 3: 391. 1965; Stafleu and Cowan, *Taxonomic Literature.* 6: 395-396. 1986; T.W. Bossert, *Biographical Dictionary of Botanists Represented in the Hunt Institute Portrait Collection.* 403. 1972.

Torilis Adans. Umbelliferae

Origins:
A meaningless name or possibly from *toreo* "to bore through, to pierce," referring to the prickled fruit.

Species/Vernacular Names:
T. japonica (Houtt.) DC. (*Caucalia japonica* Houtt.)
English: hedge parsley
Japan: yabu-jirami, umagusa
China: he shi
T. nodosa (L.) Gaertner (*Tordylium nodosum* L.; *Caucalis nodosa* (L.) Scop.)
English: knotted hedge-parsley, knotted parsley, hedge parsley

Tornabenea Parl. Umbelliferae

Origins:

After the Italian botanist Francesco Tornabene, 1813-1897, Benedictine monk, from 1850 professor of botany at the University of Catania (Sicily), Director of the Botanical Garden at Catania. His publications include *Ricerche bibliografiche sulle opere botaniche del secolo decimoquinto*. Catania 1840, *Catalogo ragionato delle edizioni del secolo XV° de' manoscritti* ... compilato dal bibliotecario ... F. Tornabene. Catania 1846, *Flora Aetnea* seu descriptio plantarum in Monte Aetna nascentium. Catinae 1889-1892 and *Discorso e descrizione per la solenne cerimonia nel porsi la prima pietra alla fondazione del R. Orto Botanico in Catania*. Catania 1858. See V. Giacomini, "Un secolo di vita dell'Orto Botanico dell'Università di Catania (1858-1958)." *Boll. Ist. Bot. Univ. Catania*. Vol. 2. Catania; J.H. Barnhart, *Biographical Notes upon Botanists*. 3: 392. Boston 1965; T.W. Bossert, *Biographical Dictionary of Botanists Represented in the Hunt Institute Portrait Collection*. 403. 1972; E.M. Tucker, *Catalogue of the Library of the Arnold Arboretum of Harvard University*. Cambridge, Massachusetts 1917-1933; Giuseppe M. Mira, *Bibliografia Siciliana*. 2: 424-425. Palermo 1881; Mariella Azzarello Di Misa, a cura di, *Il Fondo Antico della Biblioteca dell'Orto Botanico di Palermo*. 271. Palermo 1988.

Torralbasia Krug & Urban Celastraceae

Origins:

For the Cuban botanist José Ildefonso Torralbas, 1842-1903, agronomist, Director of the Botanical Museum, Habana, wrote *Discurso ... pronunciado en el acto de recibir la investidura de Doctor en Medicina y Cirujia de la Universidad Central de Venezuela*. Carácas 1875 and *Estudio teratológico de un caso de exencefalia*. Carácas 1876, with Manuel Gómez de la Maza y Jimenez (1867-1916) published *Flórula fanerogámica del Jardin botanico y del Vedado* ... Habana 1895; see J.H. Barnhart, *Biographical Notes upon Botanists*. Boston 1965; Johannes Wilbert, *Identificacion etno-linguistica de las tribus indigenas del Occidente de Venezuela*. Caracas, La Salle 1961; E.M. Tucker, *Catalogue of the Library of the Arnold Arboretum of Harvard University*. Cambridge, Massachusetts 1917-1933.

Torrenticola Domin Podostemaceae

Origins:

From the Latin *torrens, torrentis* "a torrent," referring to the habitat.

Torreya Arnott Taxaceae

Origins:

After the North American (b. New York) botanist John Torrey, 1796-1873 (d. New York), physician, chemist, professor of chemistry 1824-1827 (West Point) and 1827-1855 (New York). His writings include *Botany of New York*. Albany 1853 and *Plantae Frémontianae*. Washington and New York 1853, co-author with Asa Gray (1810-1888) of *A Flora of North America*. New York, London, Paris 1838-1843; see A. Hunter Dupree, in *Dictionary of Scientific Biography* 13: 432-433. 1981; J.H. Barnhart, *Biographical Notes upon Botanists*. 3: 392. 1965; United States. *Pacific Railroad Survey*. [Includes *Descriptions of the General Botanical Collections* by J. Torrey.] Washington 1856; Joseph Ewan, *Rocky Mountain Naturalists*. The University of Denver Press 1950; Jeannette Elizabeth Graustein, *Thomas Nuttall, Naturalist. Explorations in America, 1808-1841*. Cambridge, Harvard University Press 1967; J. Ewan, ed., *A Short History of Botany in the United States*. New York and London 1969; T.W. Bossert, *Biographical Dictionary of Botanists Represented in the Hunt Institute Portrait Collection*. 404. 1972; H.N. Clokie, *Account of the Herbaria of the Department of Botany in the University of Oxford*. 255. Oxford 1964; William Jay Youmans, ed., *Pioneers of Science in America*. 327-335. New York 1896; E.M. Tucker, *Catalogue of the Library of the Arnold Arboretum of Harvard University*. Cambridge, Massachusetts 1917-1933; Ida Kaplan Langman, *A Selected Guide to the Literature on the Flowering Plants of Mexico*. 742. 1964; Antoine Lasègue, *Musée botanique de M. Benjamin Delessert*. 1845; S. Lenley et al., *Catalog of the Manuscript and Archival Collections and Index to the Correspondence of John Torrey*. Library of the New York Botanical Garden. 1973; Stafleu and Cowan, *Taxonomic Literature*. 6: 401-408. Utrecht 1986; Mariella Azzarello Di Misa, ed., *Il Fondo Antico della Biblioteca dell'Orto Botanico di Palermo*. 272. Palermo 1988; R. Zander, F. Encke, G. Buchheim and S. Seybold, *Handwörterbuch der Pflanzennamen*. 14. Aufl. 790. Stuttgart 1993; Howard Atwood Kelly and Walter Lincoln Burrage, *Dictionary of American Medical Biography*. New York 1928; Georg Christian Wittstein, *Etymologisch-botanisches Handwörterbuch*. 885. Ansbach 1852; Irving William Knobloch, compiled by, "A preliminary verified list of plant collectors in Mexico." *Phytologia Memoirs*. VI. Plainfield, N.J. 1983; William Darlington, *Reliquiae Baldwinianae*. Philadelphia 1843; J.D. Milner, *Catalogue of Portraits of Botanists Exhibited in the Museums of the Royal Botanic Gardens*. Royal Botanic Gardens, Kew, London 1906; Ernest Nelmes and William Cuthbertson, *Curtis's Botanical Magazine Dedications, 1827-1927*. 91-92. [1931]; D.B. Tyler, *The Wilkes Expedition*: The First United States Exploring Expedition (1838-1842). Philadelphia 1968.

Species/Vernacular Names:
T. californica Torrey (*Torreya myristica* Hooker; *Tumion californicum* (Torrey) Greene)

English: California torreya, California nutmeg

T. grandis Fort. ex Lindl.

English: Chinese nutmeg-yew, Torreya nut

China: fei zi

T. nucifera (L.) Siebold & Zucc.

English: kaya nut

China: fei

T. taxifolia Arnott

English: stinking cedar, stinking yew

Torreyochloa Church Gramineae

Origins:
After the North American botanist John Torrey, 1796-1873.

Torularia O.E. Schulz Brassicaceae

Origins:
Latin *torulus, i* "a tuft of hair, the sap-wood, alburnum of a tree," *torus* "a little elevation, thickness of trees."

Torulinium Desv. Cyperaceae

Origins:
Latin *torulus, i* "tuft of hair, sap-wood, alburnum of a tree," *torus* "a little elevation, thickness of trees."

Species/Vernacular Names:
T. sp.
Hawaii: mau'u, kaluhaluha

Tosagris P. Beauv. Gramineae

Origins:
Greek *tosos* "so many" and *agrios* "living in the fields, wild, savage," *agros* "field," Latin *ager, agri* "a field, country, a measure of length."

Toubaouate Aubrév. & Pellegrin Caesalpiniaceae

Origins:
From an African vernacular name, Guineo-Congolian forests.

Species/Vernacular Names:
T. brevipaniculata (J. Léonard) Aubr. & Pellegr.

Gabon: andoung, imboumba, ngangui, ngal-ngang

Cameroon: zing, evele abai, ekop zing

Ivory Coast: toubaouate, touone

Touchardia Gaudich. Urticaceae

Origins:
Name derivation unknown.

Species/Vernacular Names:
T. sp.
Hawaii: olona

Toumeya Britton & Rose Cactaceae

Origins:
For the American botanist James William Toumey, 1865-1932; see Gordon Douglas Rowley, *A History of Succulent Plants*. Strawberry Press, Mill Valley, California 1997; J.H. Barnhart, *Biographical Notes upon Botanists*. 3: 393. 1965; T.W. Bossert, *Biographical Dictionary of Botanists Represented in the Hunt Institute Portrait Collection*. 404. 1972; E.M. Tucker, *Catalogue of the Library of the Arnold Arboretum of Harvard University*. Cambridge, Massachusetts 1917-1933; Ida Kaplan Langman, *A Selected Guide to the Literature on the Flowering Plants of Mexico*. Philadelphia 1964.

Tournefortia L. Boraginaceae

Origins:
For the French (b. Aix-en-Provence) botanist Joseph Pitton de Tournefort, 1656-1708 (d. Paris), physician, naturalist, professor of medicine and botany, studied under Pierre Magnol (1638-1715) at the University of Montpellier, friend of Charles Plumier (1646-1704) and Pierre Joseph Garidel (1658-1737), traveler with Claude Aubriet (1665?-1742) and Andreas v. Gundelsheimer (1668-1715) in the Levant. He is the father of the generic concept (first to define genus and genera). Among his many works are *Elémens de botanique*. Paris 1694, *Institutiones rei herbariae*. Parisiis 1700, *Materia medica*. London 1708 and *Relation d'un voyage du Levant*. Paris 1717; see René Louiche dit Desfontaines (1750-1833), *Choix de Plantes du Corollaire des Instituts de Tournefort*, publiées d'après son herbier, et gravées sur les dessins originaux d'Aubriet. Paris 1808; Johann Christian Daniel von Schreber (1739-1810), *Icones et descriptiones*

plantarum minus cognitarum. Halae [Halle a.S.] 1765; Jean F. Leroy, *Dictionary of Scientific Biography* 13: 442-444. 1981; J.H. Barnhart, *Biographical Notes upon Botanists.* 3: 394. 1965; D.O. Wijnands, *The botany of the Commelins.* Rotterdam 1983; H. Daudin, *De Linné à Jussieu.* Paris 1926; T.W. Bossert, *Biographical Dictionary of Botanists Represented in the Hunt Institute Portrait Collection.* 404. 1972; James Britten, *The Sloane Herbarium,* revised and edited by J.E. Dandy. 1958; H.N. Clokie, *Account of the Herbaria of the Department of Botany in the University of Oxford.* 255. Oxford 1964; E.M. Tucker, *Catalogue of the Library of the Arnold Arboretum of Harvard University.* Cambridge, Massachusetts 1917-1933; Mariella Azzarello Di Misa, ed., *Il Fondo Antico della Biblioteca dell'Orto Botanico di Palermo.* 272-273. Palermo 1988; Emil Bretschneider, *History of European Botanical Discoveries in China.* Leipzig 1981; Stafleu and Cowan, *Taxonomic Literature.* 6: 412-415. Utrecht 1986; Frans A. Stafleu, *Linnaeus and the Linnaeans.* The spreading of their ideas in systematic botany, 1735-1789. 1971; Edward Lee Greene, *Landmarks of Botanical History.* Edited by Frank N. Egerton. Stanford, California 1983; A. Lasègue, *Musée botanique de Benjamin Delessert.* Paris 1845; Blanche Henrey, *No Ordinary Gardener — Thomas Knowlton, 1691-1781.* Edited by A.O. Chater. British Museum (Natural History). London 1986.

Species/Vernacular Names:
T. sp.
Peru: ckapichinya
T. acuminata DC.
La Réunion Island: bois de Laurent-Martin
T. argentea L.f. (*Argusia argentea* L.f.; *Messerschmidia argentea* (L.f.) I.M. Johnst.)
English: velvet-leaf tree, tree heliotrope
Japan: monpa-no-ki, hama-suki
China: yin mao shu
Rodrigues Island: veloutier
T. montana Loureiro
English: montane tournefortia
China: zi dan

Tournefortiopsis Rusby Rubiaceae

Origins:
For the French botanist Joseph Pitton de Tournefort, 1656-1708 (Paris); see J.H. Barnhart, *Biographical Notes upon Botanists.* 3: 394. 1965.

Tournesol Adans. Euphorbiaceae

Origins:
French *tournesol,* see also *Chrozophora* A. Juss., from the Greek *chrozo* "to dye, tinge, stain" and *phoros* "bearing," source of turn-sole dye.

Tournesolia Scop. Euphorbiaceae

Origins:
French *tournesol, bezetta rubra,* see also *Chrozophora* A. Juss., Greek *chrozo* and *phoros* "bearing," source of turn-sole dye.

Tourneuxia Cosson Asteraceae

Origins:
For the French botanist Aristide-Horace Letourneux, 1820-1890, magistrate, malacologist, 1851-1876 and 1881-1890 in Algeria, 1876-1881 Alexandria. Among his writings are *Étude botanique sur la Kabylie du Jurjura.* Paris 1871 and *Catalogue des Mollusques terrestres et fluviatiles recueillis dans le Département de la Vendée.* Paris 1869, with Jules René Bourguignat wrote *Prodrome de la malacologie terrestre et fluviatile de la Tunisie.* [France–Ministère de l'Instruction Publique, etc. Exploration scientifique de la Tunisie, etc.] 1887; see J.H. Barnhart, *Biographical Notes upon Botanists.* 2: 373. 1965; Stafleu and Cowan, *Taxonomic Literature.* 2: 858-859. 1979; T.W. Bossert, *Biographical Dictionary of Botanists Represented in the Hunt Institute Portrait Collection.* 236. 1972; H.N. Clokie, *Account of the Herbaria of the Department of Botany in the University of Oxford.* 199. Oxford 1964; Ethelyn Maria Tucker, *Catalogue of the Library of the Arnold Arboretum of Harvard University.* Cambridge, Massachusetts 1917-1933; Ernest Saint-Charles Cosson (1819-1889), *Compendium florae atlanticae.* Paris 1881-1887 and *Répertoire alphabétique des principales localités mentionnées dans le Compendium et le Conspectus florae atlanticae ... Par E. Cosson ... avec le concours de MM. L. Kralik, A. Letourneux, etc.* 1882; Michel Charles Durieu de Maisonneuve (1796-1878), *Exploration scientifique de l'Algérie pendant les années 1840, 1841, 1842* publiée par ordre du gouvernement et avec le concours d'une commission académique. Sciences physiques. Botanique par MM. Bory de St.-Vincent et Durieu de Maisonneuve membres de la Commission scientifique d'Algérie. Paris 1846-1855[-1869]. The genus is also dedicated to Henri René Letourneux de la Perraudière (1831-1861).

Tourrettia Fougeroux Bignoniaceae

Origins:

After the French naturalist Marc Antoine Louis Claret de Latourrette (Tourrette, de la Tourette) (Fleurieu de), 1729-1793, botanist, author of *Voyage au Mont-Pilat* dans la province du Lyonnois. Avignon 1770, with Abbé (Jean-F.) François Rozier (Rosier) (1734-1793) wrote *Démonstrations élémentaires de botanique*, à l'usage de l'École royale vétérinaire. Lyon 1766 and *Partie des figures*. Lyon 1796; see J.H. Barnhart, *Biographical Notes upon Botanists*. 2: 349. 1965; Stafleu and Cowan, *Taxonomic Literature*. 2: 764-766. 1979; E.M. Tucker, *Catalogue of the Library of the Arnold Arboretum of Harvard University*. Cambridge, Massachusetts 1917-1933; Jonas C. Dryander, *Catalogus bibliothecae historico-naturalis Josephi Banks*. London 1796-1800; S. Lenley et al., *Catalog of the Manuscript and Archival Collections and Index to the Correspondence of John Torrey*. Library of the New York Botanical Garden. 462. 1973.

Species/Vernacular Names:

T. lappacea (L'Hérit.) Willd.

Peru: acoquilla

Toussaintia Boutique Annonaceae

Origins:

For L. Toussaint, in Central Africa; see Stafleu and Cowan, *Taxonomic Literature*. 6: 417. 1986.

Tovaria Ruíz & Pavón Tovariaceae (Dilleniidae, Capparidales)

Origins:

After the Spanish physician and botanist Simón de Tovar. His writings include *De compositorum medicamentorum examine*. Antverpiae 1586, *Examen i Censura del modo di averiguar las altura de las tierras, por la altura de la Estrella del Norte*, tomada con la Ballestillá. Sevilla 1595 and *Hispalensium pharmacopoliorum recognitio*. [in Joannes Duboys, *Joannis Du Boys methodus miscendi et conficiendi medicamenta diligenter recognita et expurgata*. Hagae Comitis] 1640, he sent plants to Clusius; see Richard J. Durling, *A Catalogue of Sixteenth Century Printed Books in the National Library of Medicine*. Bethesda, Maryland 1967. Some suggest an origin from the town of Tovar, W. Venezuela.

Tovomitidium Ducke Guttiferae

Origins:

Resembling *Tovomita* Aublet.

Tovomitopsis Planch. & Triana Guttiferae

Origins:

Resembling *Tovomita* Aublet.

Townsendia Hook. Asteraceae

Origins:

After the American amateur botanist David Townsend, 1787-1858, see F. Boerner & G. Kunkel, *Taschenwörterbuch der botanischen Pflanzennamen*. 4. Aufl. 180. Berlin & Hamburg 1989 and Stafleu and Cowan, *Taxonomic Literature*. 6: 420. Utrecht 1986; according to Joseph Ewan (see his *Rocky Mountain Naturalists*. 323. The University of Denver Press 1950) the genus was named for John Kirk Townsend (1809-1851), American naturalist and plant collector.

Species/Vernacular Names:

T. parryi Eaton

English: Parry's townsendia

Townsonia Cheeseman Orchidaceae

Origins:

Named for the New Zealand (b. Liverpool, England) chemist William Townson, 1850-1926 (d. Thames, New Zealand), plant collector in the northwestern side of the South Island, explorer of a portion of southwestern Nelson, discoverer of *Townsonia deflexa* Cheesem., sent specimens to Thomas Frederic Cheeseman (Director of the Auckland Museum, New Zealand); see J.H. Barnhart, *Biographical Notes upon Botanists*. 3: 395. Boston 1965; T.F. Cheeseman (1846-1923), *Manual of the New Zealand Flora*. 691. Wellington 1906.

Toxanthera Hook.f. Cucurbitaceae

Origins:

From the Greek *toxon* "a bow" and *anthera* "anther"; see also *Kedrostis* Medik.

Toxanthes Turcz. Asteraceae

Origins:

Greek *toxon* "a bow" and *anthos* "flower," referring to the corolla tube.

Species/Vernacular Names:

T. muelleri (Sonder) Benth. (*Anthocerastes muelleri* Sonder)

English: common bow-flower

T. perpusillus Turcz.

English: tiny bow-flower

Toxanthus Benth. Asteraceae

Origins:
Orthographic variant of *Toxanthes* Turcz.

Toxicodendron Miller Anacardiaceae

Origins:
Latin *toxicum* for a poison in which arrows were dipped, *toxicon* for a kind of laudanum (Plinius); Greek *toxikos* "for the bow, belonging to the bow, to archery" (*toxikon pharmakon* "poison for smearing arrows with," Aristotle, *Mirabilia*; *toxikon*, orig. a poison in which the arrows were dipped, hence a poison) and *dendron* "tree."

Species/Vernacular Names:
T. diversilobum (Torrey & A. Gray) E. Greene

English: western poison oak

Toxicodendrum Thunb. Euphorbiaceae

Origins:
From the Greek *toxikos* "for the bow, belonging to the bow, to archery" (*toxikon pharmakon* "poison for smearing arrows with," Aristotle, *Mirabilia*; *toxikon*, orig. a poison in which the arrows were dipped, hence a poison) and *dendron* "tree."

Toxocarpus Wight & Arnott Asclepiadaceae

Origins:
From the Greek *toxon* "bow" and *karpos* "fruit," referring to the follicles.

Species/Vernacular Names:
T. fuscus Tsiang

English: rusty-hair toxocarpus

China: guang hua gong guo teng

T. hainanensis Tsiang

English: Hainan toxocarpus

China: hai nan gong guo teng

T. himalensis Falconer ex Hook.f.

English: Himalayan toxocarpus

China: xi zang gong guo teng

T. villosus (Blume) Decaisne

English: hairy toxocarpus

China: mao gong guo teng

T. wightianus Hook. & Arn. (*Toxocarpus ovalifolius* Tsiang)

English: Wight toxocarpus, oval-leaf toxocarpus

China: gong guo teng

Toxophoenix Schott Palmae

Origins:
From the Greek *toxon* "bow" and *Phoenix*.

Toxopteris Trevis. Pteridaceae (Adiantaceae)

Origins:
From the Greek *toxon* "bow" and *pteris* "a fern."

Toxostigma A. Rich. Boraginaceae

Origins:
From the Greek *toxon* with *stigma*.

Toxylon Raf. Moraceae

Origins:
See also *Ioxylon, Joxylon, Toxylon, Toxylus*, etc.; see C.S. Rafinesque, in *Am. Monthly Mag. Crit. Rev.* 2: 118 1817 and 4: 188. 1818; E.D. Merrill, *Index rafinesquianus*. 111, 112. 1949.

Toxylus Raf. Moraceae

Origins:
See Constantine S. Rafinesque, *New Fl. N. Am.* 3: 42. 1836 [1838]; Elmer D. Merrill, *Index rafinesquianus*. 111, 112. 1949.

Tozzettia Savi Gramineae

Origins:
For the Italian botanist Ottaviano Targioni Tozzetti, 1755-1829, physician, professor of botany, 1801-1829 Director of the Botanical Garden of Florence. He was father of the

Italian botanist Antonio Targioni Tozzetti (1785-1856, 1829-1856 Director of the Botanical Garden of Florence) and son of the Italian naturalist and physician Giovanni Targioni Tozzetti (b. Florence 1712-d. Florence 1783, 1737-1746 Director of the Botanical Garden of Florence); see Francesco Rodolico, in *Dictionary of Scientific Biography* 13: 257-258. 1981; O. Mattirolo, *Cenni cronologici sugli Orti Botanici di Firenze.* Firenze 1899; J.H. Barnhart, *Biographical Notes upon Botanists.* 3: 360. 1965; Stafleu and Cowan, *Taxonomic Literature.* 6: 167-171. 1986; T. W. Bossert, *Biographical Dictionary of Botanists Represented in the Hunt Institute Portrait Collection.* 394. Boston, Massachusetts 1972; E.M. Tucker, *Catalogue of the Library of the Arnold Arboretum of Harvard University.* Cambridge, Massachusetts 1917-1933; R. Zander, F. Encke, G. Buchheim and S. Seybold, *Handwörterbuch der Pflanzennamen.* 14. Aufl. Stuttgart 1993; Mariella Azzarello Di Misa, ed., *Il Fondo Antico della Biblioteca dell'Orto Botanico di Palermo.* 264-265. Palermo 1988.

Tozzia L. Scrophulariaceae

Origins:
After the Italian physician Luca Tozzi, 1633-1717, botanist, wrote *Virtù del Caffè.* 1716; see Pritzel, 9433 (b. 1638); H. Genaust, *Etymologisches Wörterbuch der botanischen Pflanzennamen.* 649. 1996; Georg Christian Wittstein, *Etymologisch-botanisches Handwörterbuch.* 887. Ansbach 1852; Dale M.J. Mueller, in *Dictionary of Scientific Biography* 13: 449. (For the Italian botanist Don Bruno Tozzi, [b. Florence 1656-d. Vallombrosa 1743], a friend of Pier Antonio Micheli) 1981.

Trachelanthus Kunze Boraginaceae

Origins:
From the Greek *trachelos* "a neck" and *anthos* "flower."

Tracheliopsis Buser Campanulaceae

Origins:
Resembling *Trachelium* L., from the Greek *trachelos* "a neck" and *opsis* "aspect, resemblance, appearance"; see also *Campanula* L.

Trachelium L. Campanulaceae

Origins:
Greek *trachelos* "a neck," supposed to be effective against diseases of the trachea; Helmut Genaust, *Etymologisches Wörterbuch der botanischen Pflanzennamen.* 649. Basel

1996; Georg Christian Wittstein, *Etymologisch-botanisches Handwörterbuch.* 887. Ansbach 1852.

Species/Vernacular Names:
T. caeruleum L.
English: throatwort

Trachelosiphon Schltr. Orchidaceae

Origins:
Greek *trachelos* "a neck" and *siphon* "tube," probably referring to the style or to the basally connate sepals; see also *Eurystyles* Wawra.

Trachelospermum Lemaire Apocynaceae

Origins:
Greek *trachelos* "a neck" and *sperma* "a seed."

Species/Vernacular Names:
T. asiaticum (Siebold & Zuccarini) Nakai (*Trachelospermum foetidum* (Matsum. & Nakai) Nakai; *Trachelospermum gracilipes* Hook.f.)
English: Taiwan star jasmine, slender-stalk star jasmine
China: ya zhou luo shi
T. axillare Hook.f.
English: purple-flower star jasmine
China: zi hua luo shi
T. bodinieri (H. Léveillé) Woodson (*Trachelospermum cathayanum* Schneider)
English: Bodinier star jasmine, Cathayan star jasmine
China: gui zhou luo shi
T. brevistylum Hand.-Mazz.
English: short-style star jasmine
China: duan zhu luo shi
T. dunnii (H. Léveillé) H. Léveillé (*Trachelospermum tenax* Tsiang)
English: Dunn star jasmine, firmbark star jasmine
China: xiu mao luo shi
T. jasminoides (Lindl.) Lem. (*Rhynchospermum jasminoides* Lindley)
English: Confederate jasmine, star jasmine, Chinese star jasmine, diversifolious star jasmine, Chinese ivy
China: luo shit eng, luo shi, lo shih, nai tung (= enduring the winter), wan lan

Trachoma Garay Orchidaceae

Origins:

Greek *trachoma* "that which is made rough, roughness," *trachys* "rough, shaggy."

Trachomitum Woodson Apocynaceae

Origins:

Greek *trachoma* "roughness" and *mitos* "a thread, web, string," kendyr fiber used for sails and nets.

Trachyandra Kunth Asphodelaceae (Liliaceae)

Origins:

From the Greek *trachys* "rough" and *aner, andros* "male," the filaments are hairy or scabrid.

Trachycalymma (K. Schumann) Bullock Asclepiadaceae

Origins:

Greek *trachys* plus *kalymma* "a covering."

Trachycarpus H.A. Wendl. Palmae

Origins:

Greek *trachys* with *karpos* "fruit," referring to the rough surface of the fruits.

Species/Vernacular Names:

T. fortunei (Hooker) H.A. Wendland (*Chamaerops excelsus* Mart.)

English: hemp palm, fan palm, windmill palm, Chusan palm, Chinese windmill palm

Brazil: palmeira moinho de vento, palmeira moinho de vento da China

Japan: shuro, tojuro

Okinawa: chigu, suru

China: zong lu zi, tsung lu, ping lu

T. wagnerianus Becc.

Japan: tô-juro

Trachycaryon Klotzsch Euphorbiaceae

Origins:

From the Greek *trachys* "rough" and *karyon* "nut."

Trachydium Lindley Umbelliferae

Origins:

Greek *trachys* "rough."

Trachylobium Hayne Caesalpiniaceae

Origins:

From the Greek *trachys* "rough" and *lobos* "pod, capsule, lobe."

Trachymene Rudge Umbelliferae

Origins:

The derivation of the name is obscure, probably from the Greek *trachys* and *hymen* "a membrane" or *mene* "moon," referring to the fruits; see James A. Baines, *Australian Plant Genera. An Etymological Dictionary of Australian Plant Genera*. 377. Chipping Norton, N.S.W. 1981; Francis Aubie Sharr, *Western Australian Plant Names and Their Meanings*. 69. University of Western Australia Press 1996.

Species/Vernacular Names:

T. coerulea Graham

English: blue lace flower

T. cyanopetala (F. Muell.) Benth. (*Dimetopia cyanopetala* F. Muell.; *Didiscus cyanopetalus* (F. Muell.) F. Muell.)

English: purple trachymene, purple parsnip

Trachynia Link Gramineae

Origins:

From the Greek *trachys* "rough"; see Johann Heinrich Friedrich Link (1767-1851), *Enumeratio plantarum horti regii berolinensis altera*. 1: 42. Berolini [Berlin] 1827.

Trachynotia Michx. Gramineae

Origins:

From the Greek *trachys* "rough" and *notos* "back."

Trachyozus Reichb. Gramineae

Origins:

From the Greek *trachys* and *ozos* "branch, knot"; see also *Trachys* Pers.

Trachyphrynium Benth. Marantaceae

Origins:
From the Greek *trachys* "rough" and *Phrynium*.

Trachyphrynium K. Schum. Marantaceae

Origins:
Greek *trachys* "rough" plus *Phrynium* Willd.

Trachyphytum Nutt. Loasaceae

Origins:
From the Greek *trachys* "rough" and *phyton* "a plant."

Trachypoa Bubani Gramineae

Origins:
From the Greek *trachys* and *poa* "grass, pasture grass."

Trachypogon Nees Gramineae

Origins:
From the Greek *trachys* "rough" and *pogon* "a beard," the fertile spikelet has a plumose awn.

Species/Vernacular Names:
T. spicatus (L.f.) Kuntze (*Trachypogon capensis* (Thunb.) Trin.)
English: giant spear grass, grey tussock grass
Southern Africa: bokbaardgras, reuse pylgras, steekgras; isitube (Zulu); selokana (Sotho)

Trachypremnon Lindig Cyatheaceae

Origins:
From the Greek *trachys* "rough" and *premnon* "the stump of a tree, trunk, stem, basis, root."

Trachypteris E.F. André ex H. Christ Pteridaceae (Adiantaceae)

Origins:
From the Greek *trachys* "rough" and *pteris* "a fern."

Trachyrhizum (Schltr.) Brieger Orchidaceae

Origins:
From the Greek *trachys* with *rhiza* "root."

Trachys Pers. Gramineae

Origins:
From the Greek *trachys* "rough."

Trachysciadium Ecklon & Zeyher Umbelliferae

Origins:
Greek *trachys* "rough" and *skiadion, skiadeion* "umbel, parasol."

Trachyspermum Link Umbelliferae

Origins:
From the Greek *trachys* "rough" and *sperma* "seed."

Trachystachys A. Dietr. Gramineae

Origins:
From the Greek *trachys* "rough" and *stachys* "spike"; see also *Trachys* Pers.

Trachystemon D. Don Boraginaceae

Origins:
Greek *trachys* "rough" and *stemon* "stamen."

Trachystigma C.B. Clarke Gesneriaceae

Origins:
From the Greek *trachys* "rough" and *stigma* "stigma."

Trachystoma O.E. Schulz Brassicaceae

Origins:
From the Greek *trachys* plus *stoma* "mouth."

Trachystylis S.T. Blake Cyperaceae

Origins:
From the Greek *trachys* "rough" and *stylos* "style."

Tracyina S.F. Blake Asteraceae

Origins:
For the California botanist Joseph Prince Tracy, 1879-1953, plant collector; see J.H. Barnhart, *Biographical Notes upon Botanists*. 3: 396. 1965.

Species/Vernacular Names:
T. rostrata S.F. Blake
English: beaked tracyina

Tradescantella Small Commelinaceae

Origins:
Referring to *Tradescantia*.

Tradescantia L. Commelinaceae

Origins:
For the British (b. Meopham, Kent) naturalist and botanist John Tradescant, 1608-1662 (d. South Lambeth, Surrey), traveler, plant collector in Virginia, from 1638 gardener to Queen Henrietta Maria (Keeper of His Majesty's Gardens at Oatlands), author of *Musaeum Tradescantianum*. London 1656, and for his father John Tradescant, *circa* 1570/1575-1638 (d. South Lambeth, Surrey), traveler and botanical collector, friend of John Parkinson (1567-1650), gardener to Robert Cecil (Earl of Salisbury) and to Sir Wotton, 1618 went to Russia (the first western European botanist), 1630 Keeper of His Majesty's Gardens at Oatlands; see Mea Allan, *The Tradescants. Their Plants, Gardens and Museum. 1570-1662*. London 1964; Prudence Leith-Ross, *The John Tradescants. Gardeners to the Rose and Lily Queen*. London 1984; John H. Barnhart, *Biographical Notes upon Botanists*. 3: 397. 1965; Richard Pulteney, *Historical and Biographical Sketches of the Progress of Botany in England*. 1: 175-179. London 1790; A. Lasègue, *Musée botanique de Benjamin Delessert*. Paris 1845; Ray Desmond, *Dictionary of British & Irish Botanists and Horticulturists*. 689. (genus dedicated to John Tradescant, c. 1570-1638) 1994; Carl Linnaeus, *Species Plantarum*. 288. 1753 and *Genera Plantarum*. Ed. 5. 139. 1754; Joseph Ewan, in *Dictionary of Scientific Biography* (Editor in Chief Charles Coulston Gillispie.) 13: 449-451. New York 1981; J. Tradescant, *Musaeum Tradescantianum: or, A Collection of Rarities Preserved at South Lambeth Near London.*

(Reprint prepared by R.T. Gunther for the Opening of the Old Ashmolean Museum for the Lewis Ewans Collection.) Oxford 1925; T. W. Bossert, *Biographical Dictionary of Botanists Represented in the Hunt Institute Portrait Collection*. 405. Boston, Massachusetts 1972; E.M. Tucker, *Catalogue of the Library of the Arnold Arboretum of Harvard University*. Cambridge, Massachusetts 1917-1933; Jonas C. Dryander, *Catalogus bibliothecae historico-naturalis Josephi Banks*. London 1796-1800; Robert William Theodore Gunther (1869-1940), *Early British Botanists and their Gardens*. Oxford 1922; R.T. Gunther, *Early Science in Cambridge*. Oxford 1937; M. Hadfield et al., *British Gardeners: A Biographical Dictionary*. London 1980.

Species/Vernacular Names:
T. albiflora Kunth
English: green wandering Jew
T. cerinthoides Kunth
English: flowering inch plant
T. spathacea Sw. (*Rhoeo discolor* (L'Hérit.) Hance; *Rhoeo spathacea* (Sw.) Stearn)
English: oyster plant, boat lily, cradle lily, Moses in his cradle, Moses on a raft, Moses in the bulrushes, man in a boat, men in a boat, three men in a boat, purple leaved spiderwort
Japan: fuiri-murasaki-omoto
China: bang lan ye
The Philippines: bangka-bangkaan
T. zebrina hort. ex Bosse (*Zebrina pendula* Schnizl.)
English: inch plant, wandering Jew
Japan: Hakata-kara-kusa
China: diao zhu mei

Traganopsis Maire & Wilczek Chenopodiaceae

Origins:
The genus *Traganum* Delile and *opsis* "aspect, resemblance, appearance."

Traganthus Klotzsch Euphorbiaceae

Origins:
Greek *tragos* "a goat" and *anthos* "flower," referring to the smell of the flowers; Latin *traganthes, is* for a species of the plant *artemisia*.

Traganum Delile Chenopodiaceae

Origins:

Greek *tragos* "a goat," referring to a plant of which goats are fond; Latin *traganus, i* "a suckling pig dressed in a particular way."

Species/Vernacular Names:

T. nudatum Del.

Arabic: dhemran

French: tragan dénudé

Tragia L. Euphorbiaceae

Origins:

Named after the German botanist Hieronymus (Jerome) Bock (*latine* Tragus), 1498-1554, physician and teacher, herbalist, author of *New Kreütter Buch*. Strassburg 1539, *Teütsche Speiszkammer*. [Small 4to, second edition.] Strasbourg 1555 and *De stirpium, maxime earum, quae in Germania nostra nascuntur*. [Argentorati 1552], with Brunfels and Fuchs one of the German fathers of botany; see Stafleu and Cowan, *Taxonomic Literature*. 1: 243-245. 1976; Otto Brunfels (1488-1534), *Herbarum vivae eicones* ad naturae imitationem summa cum diligentia et artificio effigiatae. Argentorati 1530-1540; Ida Kaplan Langman, *A Selected Guide to the Literature on the Flowering Plants of Mexico*. Philadelphia 1964; J.H. Barnhart, *Biographical Notes upon Botanists*. 1: 206. 1965; T. W. Bossert, *Biographical Dictionary of Botanists Represented in the Hunt Institute Portrait Collection*. 42. Boston, Massachusetts 1972; Garrison and Morton, *Medical Bibliography*. 1806 and 1803. 1961; Carl Linnaeus, *Species Plantarum*. 980. 1753 and *Genera Plantarum*. Ed. 5. 421. 1754; Mariella Azzarello Di Misa, a cura di, *Il Fondo Antico della Biblioteca dell'Orto Botanico di Palermo*. 44. Regione Siciliana, Palermo 1988; Jerry Stannard, in *Dictionary of Scientific Biography* 2: 218-220. 1981; Edward Lee Greene, *Landmarks of Botanical History*. Edited by Frank N. Egerton. Stanford, California 1983. Some suggest the genus was named from the Greek *tragos* "a billy-goat, he-goat," referring to hairs on the margins of the leaves.

Species/Vernacular Names:

T. sp.

Yoruba: obo

T. benthamii Baker

Yoruba: esisi funfun

T. involucrata L.

India: sengel sing

T. ramosa Torrey

English: noseburn

T. tenuifolia Benth.

Sierra Leone: morlinyényé

Tragiella Pax & Hoffmann Euphorbiaceae

Origins:

The diminutive of the genus *Tragia* L., after the German botanist Hieronymus (Jerome) Bock (*latine* Tragus), 1498-1554.

Tragiola Small & Pennell Scrophulariaceae

Origins:

The anagram of the generic name *Gratiola* L.

Tragiopsis Karsten Euphorbiaceae

Origins:

Resembling the genus *Tragia* L., after the German botanist Hieronymus (Jerome) Bock (*latine* Tragus), 1498-1554.

Tragiopsis Pomel Umbelliferae

Origins:

Resembling the genus *Tragium* Spreng.

Tragium Spreng. Umbelliferae

Origins:

Latin *tragion* "a plant called goatwort," Dioscorides used Greek *tragion* for a plant smelling like a goat, stinking tutsan, a species of *Hypericum*, or for pimpinel, a kind of *Pimpinella* L.

Tragopogon L. Asteraceae

Origins:

Greek *tragopogon, tragos* "he-goat, goat" and *pogon* "a beard," referring to the hairy or silky pappus; see Carl Linnaeus, *Species Plantarum*. 789. 1753 and *Genera Plantarum*. Ed. 5. 346. 1754.

Species/Vernacular Names:

T. porrifolius L.

English: oyster plant, purple goat's beard, salsify, vegetable oyster, wild salsify, purple flowered salsify, Jerusalem star

South Africa: bokbaard, persbokbaard, sydissel

Hawaii: kalapi

T. pratensis L.

English: goat's beard, shepherd's clock, Johnny-go-to-bed-at-noon, Jack-go-to-bed-at-noon, meadow goat's beard, meadow salsify

Tragoselinum Hall. Umbelliferae

Origins:

From the Greek *tragos* "goat" and *selinon* "parsley, celery."

Tragus Haller Gramineae

Origins:

For the German botanist Hieronymus Bock (called Tragus, *bock* being German for goat), 1498-1554, physician, teacher, author of *New Kreütter Buch*. Strassburg 1537, *Teütsche Speiszkammer.* [Small 4to, second edition.] Strasbourg 1555 and *De stirpium, maxime earum, quae in Germania nostra nascuntur.* [Strassburg 1552], with Brunfels and Fuchs one of the German fathers of botany; see Otto Brunfels (1488-1534), *Herbarum vivae eicones* ad naturae imitationem summa cum diligentia et artificio effigiatae. Argentorati 1530-1540; Garrison and Morton, *Medical Bibliography.* 1806 and 1803. 1961; Edward Lee Greene, *Landmarks of Botanical History.* Edited by Frank N. Egerton. Stanford, California 1983; J.H. Barnhart, *Biographical Notes upon Botanists.* 1: 206. 1965; Mariella Azzarello Di Misa, a cura di, *Il Fondo Antico della Biblioteca dell'Orto Botanico di Palermo.* 44. Regione Siciliana, Palermo 1988; Jerry Stannard, in *Dictionary of Scientific Biography* 2: 218-220. 1981.

Species/Vernacular Names:

T. berteronianus Schult.

English: burgrass, carrot-seed grass, small carrot-seed grass, spiked carrot-seed grass, Bertero goatgrass

South Africa: digaarwortelsaadgras, gewone wortelsaadgras, haasgras, hassklits, kleinwortelsaadgras, klitsgras, kousgras, kousklits, luisgras, lysgras, raasklits, wolgras, wolklits, wortelsaadgras

T. koelerioides Aschers.

English: creeping carrot-seed grass, cushion grass, perennial carrot-seed grass, goat's-beard grass

South Africa: kophaargras, kruipgras, kruipwortelsaadgras, kwaggakweek, olifantskweek, roletjiesgras, rooiwortelsaadgras, langbeenwortelsaadgras, meerjarige, wortelsaadgras, vetmakergras, wortelsaadgras

T. racemosus (L.) All.

English: carrot-seed grass, carrot grass, large carrot-seed grass, stalked bristle grass, stalked carrot-seed grass

Southern Africa: grootwortelsaadgras, losaarwortelsaadgras, wortelsaadgras, klitsgras, kousgras, luisgras, lysgras; bore-ba-ntjia (Sotho)

Trailliaedoxa W.W. Smith & Forrest Rubiaceae

Origins:

After the daughter of the British botanist George William Traill (b. Kirkwall, Orkney 1836–d. Edinburgh 1897) and wife of the botanical explorer and traveler George Forrest (1873-1932); see Stafleu and Cowan, *Taxonomic Literature.* 6: 430-431. 1986; Ray Desmond, *Dictionary of British & Irish Botanists and Horticulturists.* 254, 690. London 1994.

Transcaucasia M. Hiroe Umbelliferae

Origins:

A region of southeastern Russia between the Caucasus mountains on the north and Iran and Turkey in Asia on the south.

Trapa L. Trapaceae (Rosidae, Myrtales)

Origins:

See Helmut Genaust, *Etymologisches Wörterbuch der botanischen Pflanzennamen.* 651-652. Basel 1996.

Species/Vernacular Names:

T. bicornis Osbeck

English: ling nut, water chestnut

China: ling

T. bispinosa Roxb.

English: singhara nut

China: chi shih, ling, shiu li

T. natans L.

English: water caltrops, horn nut, Jesuit's nut, water chestnut

China: chi shih, ling, ling chio, lao ling, shiu li

Trapella Oliver Pedaliaceae (Trapellaceae)

Origins:
Diminutive of the genus *Trapa* L., referring to a superficial resemblance.

Trasera Raf. Gentianaceae

Origins:
Dedicated to the Scottish traveler John Fraser, 1750-1811 (Chelsea, London), see also *Frasera* Walter.

Trattinnickia Willdenow Burseraceae

Origins:
After the Austrian botanist Leopold von Trattinnick (Trattinick, Trattinnik), 1764-1849, naturalist. Among his works are *Fungi austriaci*. Wien [1804-] 1805 [-1806] and *Thesaurus botanicus*. Viennae [1805-] 1819; see J.H. Barnhart, *Biographical Notes upon Botanists*. 3: 398. 1965; G. Schmid, *Goethe und die Naturwissenschaften*. Halle 1940; T. W. Bossert, *Biographical Dictionary of Botanists Represented in the Hunt Institute Portrait Collection*. 405. Boston, Massachusetts 1972; E.M. Tucker, *Catalogue of the Library of the Arnold Arboretum of Harvard University*. Cambridge, Massachusetts 1917-1933; H.N. Clokie, *Account of the Herbaria of the Department of Botany in the University of Oxford*. 256. Oxford 1964; Jonas C. Dryander, *Catalogus bibliothecae historico-naturalis Josephi Banks*. London 1796-1800; Mariella Azzarello Di Misa, a cura di, *Il Fondo Antico della Biblioteca dell'Orto Botanico di Palermo*. 273-274. Regione Siciliana, Palermo 1988; R. Zander, F. Encke, G. Buchheim and S. Seybold, *Handwörterbuch der Pflanzennamen*. 14. Aufl. 1993.

Traubia Moldenke Amaryllidaceae (Liliaceae)

Origins:
After the American botanist Hamilton Paul Traub, 1890-1983, horticulturist; see J.H. Barnhart, *Biographical Notes upon Botanists*. 3: 398. 1965; T. W. Bossert, *Biographical Dictionary of Botanists Represented in the Hunt Institute Portrait Collection*. 405. Boston, Massachusetts 1972; Ida Kaplan Langman, *A Selected Guide to the Literature on the Flowering Plants of Mexico*. Philadelphia 1964.

Traunsteinera Reichb. Orchidaceae

Origins:
Named after the Austrian apothecary Josef Traunsteiner, 1798-1850, botanist, author of *Monographie der Weiden von Tirol und Vorarlberg*. Innsbruck 1842; see J.H. Barn-

hart, *Biographical Notes upon Botanists*. 3: 398. 1965; H. Genaust, *Etymologisches Wörterbuch der botanischen Pflanzennamen*. 652. 1996; Ethelyn Maria Tucker, *Catalogue of the Library of the Arnold Arboretum of Harvard University*. Cambridge, Massachusetts 1917-1933.

Trautvetteria Fischer & C.A. Meyer Ranunculaceae

Origins:
After the Russian botanist Ernst Rudolf von Trautvetter, 1809-1889, professor of botany, Director of the Botanical Garden in St. Petersburg. Among his many works are *Elenchus stirpium anno 1880 in isthmo caucasico lectarum*. Petropoli [St. Petersburg] 1881 and *Contributio ad floram Turcomaniae*. Petropoli 1885; see John H. Barnhart, *Biographical Notes upon Botanists*. 3: 398. 1965; R. Zander, F. Encke, G. Buchheim and S. Seybold, *Handwörterbuch der Pflanzennamen*. 14. Aufl. 1993; Stafleu and Cowan, *Taxonomic Literature*. 6: 443-452. 1986; T. W. Bossert, *Biographical Dictionary of Botanists Represented in the Hunt Institute Portrait Collection*. 405. Boston, Massachusetts 1972; A. Lasègue, *Musée botanique de Benjamin Delessert*. 341. Paris 1845; E.M. Tucker, *Catalogue of the Library of the Arnold Arboretum of Harvard University*. Cambridge, Massachusetts 1917-1933; Ida Kaplan Langman, *A Selected Guide to the Literature on the Flowering Plants of Mexico*. Philadelphia 1964; Emil Bretschneider, *History of European Botanical Discoveries in China*. Leipzig 1981; Günther Schmid, *Chamisso als Naturforscher. Eine Bibliographie*. Leipzig 1942.

Traversia Hook.f. Asteraceae

Origins:
For the British (b. County Limerick) botanist William Thomas Locke Travers, 1819-1903 (d. Wellington, New Zealand), magistrate, in 1863 Fellow of the Linnean Society; see R. Glenn, *The Botanical Explorers of New Zealand*. Wellington 1950; Ray Desmond, *Dictionary of British & Irish Botanists and Horticulturists*. 690. London 1994.

Traxara Raf. Boraginaceae

Origins:
See Elmer D. Merrill (1876-1956), *Index rafinesquianus*. 204. 1949; Constantine S. Rafinesque, *Flora Telluriana*. 4: 85. Philadelphia 1836 [1838].

Traxilum Raf. Boraginaceae

Origins:

See C.S. Rafinesque, *Sylva Telluriana*. 42. 1838; E.D. Merrill, *Index rafinesquianus*. 204. 1949.

Traxisuke Raf. Moraceae

Origins:

See C.S. Rafinesque, *Sylva Telluriana*. 58. 1838; E.D. Merrill, *Index rafinesquianus*. 112. 1949.

Treculia Decne. ex Trécul Moraceae

Origins:

After the French botanist Auguste Adolphe Lucien Trécul, 1818-1896, traveler in North America between 1848 and 1850, author of *Recherches sur la structure et le développement du Nuphar lutea*. Paris 1845 and "Du suc propre dans les feuilles des Aloes." *Ann. Sci. Nat.* sér. 5, 14: 80-90. 1872; see J.H. Barnhart, *Biographical Notes upon Botanists*. 3: 399. Boston 1965; S. Lenley et al., *Catalog of the Manuscript and Archival Collections and Index to the Correspondence of John Torrey*. Library of the New York Botanical Garden. 1973; E.M. Tucker, *Catalogue of the Library of the Arnold Arboretum of Harvard University*. Cambridge, Massachusetts 1917-1933; Joseph Ewan, *Rocky Mountain Naturalists*. The University of Denver Press 1950; R. Zander, F. Encke, G. Buchheim and S. Seybold, *Handwörterbuch der Pflanzennamen*. 14. Aufl. 791. 1993.

Species/Vernacular Names:

T. africana Decne. ex Trécul (*Myriopeltis edulis* Welw. ex Hook.f.) (see Y. Tailfer, *La Forêt dense d'Afrique centrale*. CTA, Ede/Wageningen 1989; J. Vivien & J.J. Faure, *Arbres des Forêts denses d'Afrique Centrale*. Agence de Coopération Culturelle et Technique. Paris 1985)

English: bread tree, African breadfruit

French: arbre à pain d'Afrique

Angola: isa quente (Portuguese); disanha (Kimbundu); musanha, mutáli (Kioko)

Zaire: baimi, boimbo, bolimbo, bombumbo, lesaia, mobumbo (Lingala); banga (Likimi); bobimbo (Mai-Ndombe lake); disaya, muzania (Tshiluba); m'bimbo, m'bimbu, mbondi (Kitetela); mondonga (Gombe); mvaye (Kaniama); njani (Mayumbe); n'zaga (Kisantu); n'tui (Matubi); ombimbo (Turumbu); owimbu (Bashobwa)

Malawi: mjaya, njayi (the fruits, Chichewa); lyaja, maja(ja) (Yao)

Congo: toum, bleblendou

Cameroon: boembe, bongo, bwembi, etoup, pusa, zilo

Central African Republic: zilo

Yoruba: afon

Nigeria: akim, afon, ifon, bopembe, bwembe, etub, ediang, ize, izenagan, ije, okwa, etube, aji, ukwa, yukugu, okorinye, izaquenti; afon (Yoruba); ize (Edo); ukwa mbu, nneukwa, ukwa odo-owa, ukwa isienwe, oke ukwa (Igbo); ediang (Efik)

Ivory Coast: blébendou, bléblendou, gloutué, nlandié

T. obovoides N.E. Br. (*Treculia staudtii* Engl.)

Zaire: para (Mayumbe)

Congo: ompoh

Nigeria: ogini (Ekoi); oken (Boki)

Treichelia Vatke Campanulaceae

Origins:

For the German botanist Alexander Johann August Treichel, 1837-1901; see J.H. Barnhart, *Biographical Notes upon Botanists*. 3: 399. Boston 1965; E.M. Tucker, *Catalogue of the Library of the Arnold Arboretum of Harvard University*. Cambridge, Massachusetts 1917-1933.

Treleasea Rose Commelinaceae

Origins:

For the American (b. Mount Vernon, New York) botanist William Trelease, 1857-1945 (d. Urbana, Illinois), professor of botany, Director of the Missouri Botanical Garden, specialist in genera *Yucca* and *Agave*, plant collector; see J.H. Barnhart, *Biographical Notes upon Botanists*. 3: 399. 1965; Joseph Ewan, in *Dictionary of Scientific Biography* 13: 456. 1981; J.T. Buchholz, "William Trelease 1857-1945." *Science*. n.s. 101: 192-193. 1945; R. Zander, F. Encke, G. Buchheim and S. Seybold, *Handwörterbuch der Pflanzennamen*. 14. Aufl. 1993; Stafleu and Cowan, *Taxonomic Literature*. 6: 458-468. Utrecht 1986; S. Lenley et al., *Catalog of the Manuscript and Archival Collections and Index to the Correspondence of John Torrey*. Library of the New York Botanical Garden. 1973; Ida Kaplan Langman, *A Selected Guide to the Literature on the Flowering Plants of Mexico*. Philadelphia 1964; E.M. Tucker, *Catalogue of the Library of the Arnold Arboretum of Harvard University*. Cambridge, Massachusetts 1917-1933; Joseph Ewan, *Rocky Mountain Naturalists*. The University of Denver Press 1950; T.W. Bossert, *Biographical Dictionary of Botanists Represented in the Hunt Institute Portrait Collection*. 406. 1972; Irving William Knobloch, compil., "A preliminary verified list of plant collectors in Mexico." *Phytologia Memoirs*. VI. Plainfield, N.J. 1983.

Trema Lour. Ulmaceae

Origins:

Greek *trema* "hole, aperture," in reference to the pitted stone of the fruit; see João de Loureiro (1717-1791), *Flora cochinchinensis*. 2: 539, 562. [Lisboa] 1790.

Species/Vernacular Names:

T. spp.

Nigeria: afe, afefe, afoforo, affi, ehu-ogo, ukpafiami, ayinyin

T. cannabina Lour. (*Trema virgata* (Planch.) Blume; *Sponia virgata* Planch.)

English: smooth trema, lesser trema

Japan: koba-fungi, kiri-enoki, usuba-urajiro-enoki

T. guineensis (Schum. & Thonn.) Ficalho

Southern Africa: muButibuti, muGubvura, siGungoro, muFeteti, muPutiputi, muTiputi (Shona)

Congo: osossi, ossossi, ombumba, yayaka, muyayaka

Zaire: baumba, besesu, bukengi (Katalinga); esesu, inmesu, limesu (Turumbu); kadiabahote (Kaniama); kpewadi, leioso, likefu (Katabira or Kitabira); mangosa-bolebo, memvo (Tshiluba); mudiankutshi (Kasai); mubesu o mubessu (Tshofa); mukelekele (Kinande); muhelehele, muafu, muhengi (Kimbumba); mushowshow (Kihunde); nkingi (Kipakompe); nsengensenge (Mayombe); pewa (Azande); umunyereso (Shangungu); wesesu (Eala)

Cameroon: echen, evekoulou, fafa

Angola: musamba-samba, mugudisosola, kipukulo (Kimbundu); maiapa, qipexi, kituense (Umbundu); omuiakalenga (Lunyaneka); nsengue-nsengue (Kikongo); tjimonomono (Kioko); cabra (S. Tomé)

Ivory Coast: adaschia, foué

Nigeria: ayinyin

T. lamarckiana (Roem. & Schult.) Blume

English: West Indian trema

T. micrantha (L.) Blume

English: Florida trema, Jamaican nettle tree, white capulin

Peru: yana caspi, atadyo

Tropical America: guacimilla, capulincillo, pala de cabra, memiso, white capulin

Honduras: capulin negro

Guatemala: capulii

Mexico: yaco de cuero

Brazil: grandiúva, crindiúva, coatidiba, periquiteira, orindeúva, orinduíba, Orindiba, gurindiba, gurindiva, pau de pólvora, candiúba, taleira, motamba, pessegueiro de capoeira, seriúva

Argentina: afata colorada, fruta paloma, palo pólvora, moroti

Costa Rica: capulin, jucó, vara blanca

Venezuela: carraspero, masaquila

Colombia: venaco, berraco, raspador, majaguá colorado

Cuba: guacimella, capuli, cimarrón

Ecuador: tortolero, muchi chilón

T. orientalis (L.) Blume (*Celtis orientalis* L.; *Trema bracteolata* (Hochst.) Blume; *Trema guineensis* (Schumach. & Thonn.) Ficalho; *Trema tomentosa* (Roxb.) H. Hara)

English: pigeonwood, Rhodesian elm, trema, gunpowder tree, charcoal tree, rough trema, glaucous trema, tree peach, woolly cedar, Indian nettle tree

Ethiopia: alele, talalaä, halele, hudu-farda (Orominya); bal-ambal-dured (Somalinya)

Ghana: aisie

Ivory Coast: aissien, alakra-bogbena, assissien, assessian

Burkina: assissien, assessian

Yoruba: afee, afoforo, afoforo afe, afere, amokole, ayinyin, aferi, afeeri

Nigeria: afe, afo-aforo, afoforo-afe, ayinyin; telemukwu (Igbo); afere (Yoruba); ehuogo (Edo)

Southern Africa: hophout; umBalalaqane (Swazi); mutiputi (Shona); umCebekhazana, isiKhwelamfene, uBathini, nbantini, umBokhangabokhanga, umBhangabhanga, iFamu, iPhubane, umVangazi, umSekeseke, umDindwa (Zulu); aPhubani, mpukupuku (Thonga or Tsonga); modutu (North Sotho); umVangazi, uPhakane, umVumvu, umBengele (Xhosa); makuru-kuru (Venda)

Malawi: mpefu (Chichewa, Nyanja); musyasya (Sukwa); mpesi (Tonga); msakasa(ka) (Tumbuka, Henga); majanja-juni, mjajajuni, tesa, yesa (Yao)

Tanzania: mpehe (Zaramo)

Kenya: musakala (Kakamega and Tiriki; see Walter H. Sangree, *Age, Prayer and Politics in Tiriki, Kenya*. London 1966)

India: jhupon (Assam), jivanti, chenkolam

Sri Lanka: gedumba

Malaya: menarong, mendarong, bendarong, marong, mengkirai, mengkira

The Philippines: hanagdóng

Mascarène Islands: andrèze, bois d'andrèze

T. tomentosa (Roxb.) H. Hara (*Celtis tomentosa* Roxb.)

Australia: peach leaf poison bush, native peach

Tremacanthus S. Moore Acanthaceae

Origins:

From the Greek *trema* "hole, aperture" and *akantha* "thorn."

Tremacron Craib Gesneriaceae

Origins:

Greek *trema* "aperture" and *makron* "large," referring to the dehiscence of the anthers.

Tremandra R. Br. ex DC. Tremandraceae

Origins:

Greek *trema* "hole, aperture" and *aner, andros* "man, anther," basifixed anthers with single apical pore or small slit.

Tremastelma Raf. Dipsacaceae

Origins:

Greek *trema* "hole, aperture" and *stelma, stelmatos* "a girdle, belt, crown, garland, wreath"; see Constantine Samuel Rafinesque, *Flora Telluriana.* 4: 96. Philadelphia 1836 [1838].

Trematocarpus A. Zahlbr. Campanulaceae

Origins:

Greek *trematos, trema* "hole, aperture" and *karpos* "fruit."

Trematolobelia A. Zahlbr. ex Rock Campanulaceae

Origins:

Greek *trematos, trema* "hole, aperture" plus the genus *Lobelia* L., referring to the capsules; see also *Trematocarpus* Zahlbr.

Species/Vernacular Names:

T. kauaiensis (Rock) Skottsb. (*Trematolobelia macrostachys* (Hook. & Arnott) A. Zahlbr. var. *kauaiensis* Rock)

Hawaii: koli'i

T. macrostachys (Hook. & Arnott) A. Zahlbr. (*Lobelia macrostachys* Hook. & Arnott; *Delissea macrostachys* (Hook. & Arnott) C. Presl; *Dortmanna macrostachys* (Hook. & Arnott) Kuntze; *Trematocarpus macrostachys* (Hook. & Arnott) A. Zahlbr.; *Trematolobelia auriculata* St. John)

Hawaii: koli'i

Trematosperma Urb. Icacinaceae

Origins:

From the Greek *trematos, trema* plus *sperma* "seed."

Trembleya DC. Melastomataceae

Origins:

In honor of the Swiss (b. Geneva) botanist Abraham Trembley, 1710-1784 (d. near Geneva), zoologist and naturalist, scientist and philosopher, the father of experimental zoology, best known for his studies of the freshwater hydra, mainly the green hydra *Chlorohydra viridissima*. He did his most important biological works while serving as a private tutor to distinguished families near The Hague (in the early 1740s); he was the first to describe multiplication and colony formation in Protozoa. Among his writings are *Instructions d'un père à ses enfants, sur la nature et sur la religion*. Genève 1775 and *Mémoires pour servir à l'histoire d'un genre de polypes d'eau douce à bras en forme de cornes*. Leiden 1744; see John Randal Baker, *Abraham Trembley of Geneva*, etc. London 1952; John R. Baker, in *Dictionary of Scientific Biography* 13: 457-458. 1981; John Turberville Needham, *Nouvelles découvertes faites avec le microscope. Avec un mémoire sur les Polypes à bouquet, et sur ceux en entonnoir, par A. Trembley*. Leiden 1747.

Tremotis Raf. Moraceae

Origins:

See C.S. Rafinesque, *Sylva Telluriana.* 59. 1838; E.D. Merrill, *Index rafinesquianus.* 112. 1949.

Tremularia Fabr. Gramineae

Origins:

From the Latin *tremulus, a, um* "shaking, trembling"; see also *Briza* L.

Trepocarpus Nutt. ex DC. Umbelliferae

Origins:

From the Greek *trepo, trope* "a turning" and *karpos* "fruit," referring to the fruit.

Tresanthera Karsten Rubiaceae

Origins:

Probably from the Greek *tresis* "a perforation" and *anthera* "anther."

Tretorhiza Adanson Gentianaceae

Origins:

Greek *tretos* "perforated, with a hole in it" and *rhiza* "a root."

Treubania Tieghem Loranthaceae

Origins:

After the Dutch botanist Melchior Treub, 1851-1910 (d. St. Raphael, France), traveler and botanical collector, from 1873 to 1880 collaborated in the botanical institute at Leiden with professor of botany Willem Frederik Reinier Suringar (1832-1898), from 1880 Director (he succeeded Rudolf Herman Christiaan Carel Scheffer, 1844-1880) of the Buitenzorg Botanical Gardens (now Bogor, West Java). Among his scientific works are "Notice sur la nouvelle flore de Krakatau." *Ann. Jard. Bot. Buitenzorg.* Leide [Leiden] 1888 and "Observations sur les Loranthacées." *Ann. Jard. Bot. Buitenzorg.* Leide [Leiden] 1881, from 1881 to 1910 editor of the *Annales du Jardin Botanique de Buitenzorg.* See Friedrich August Ferdinand Christian Went (1863-1935), "In Memoriam." *Ann. Jard. Bot. Buitenzorg.* 9: i-xxxii. 1911; C.G.G.J. van Steenis, in *Dictionary of Scientific Biography* 13: 458-460. 1981; G.J. Symons, ed., *The Eruption of Krakatoa and Subsequent Phenomena.* Report of the Krakatoa Committee of the Royal Society. London 1888; J.H. Barnhart, *Biographical Notes upon Botanists.* 3: 592. 1965; H.N. Clokie, *Account of the Herbaria of the Department of Botany in the University of Oxford.* 406. Oxford 1964; E.M. Tucker, *Catalogue of the Library of the Arnold Arboretum of Harvard University.* Cambridge, Massachusetts 1917-1933; P.M.W. Dakkus, *An Alphabetical List of Plants Cultivated in the Botanic Gardens, Buitenzorg.* Buitenzorg/Java, Dutch-East Indies 1930; Stafleu and Cowan, *Taxonomic Literature.* 6: 469-474. Utrecht 1986; Gordon Douglas Rowley, *A History of Succulent Plants.* Strawberry Press, Mill Valley, California 1997.

Treubella Pierre Sapotaceae

Origins:

After the Dutch botanist Melchior Treub, 1851-1910.

Treubella Tieghem Sapotaceae

Origins:

After the Dutch botanist Melchior Treub, 1851-1910, from 1873 to 1880 collaborated with Willem Frederik Reinier Suringar (1832-1898), from 1880 Director of the Buitenzorg Botanical Gardens.

Treutlera Hook.f. Asclepiadaceae

Origins:

For the British (b. India) physician William John Treutler, 1841-1915 (d. Sussex), plant collector in Sikkim, 1868 Fellow of the Linnean Society.

Trevesia Vis. Araliaceae

Origins:

Named after Enrichetta Treves de' Bonfigli and her family, 19th century, Italian patrons of botany and supporters of botanical research.

Species/Vernacular Names:

T. burckii Boerl. (*Trevesia cheirantha* (C.B. Clarke) Kuntze; *Trevesia sanderi* hort.; *Trevesia palmata* ver. *cheirantha* C.B. Clarke)

English: ghost's foot

Malaya: daun tapak badak, daun tapak rimau, daun tapak hantu, mati sedangor

Trevia L. Euphorbiaceae

Origins:

For the German botanist Christoph Jakob Trew, 1695-1769, see also *Trewia* L.

Trevirana Willd. Gesneriaceae

Origins:

After the German (b. Bremen) botanist Ludolph Christian Treviranus, 1779-1864 (d. Bonn), physician, professor of botany and of natural history, brother of the German biologist and physician Gottfried Reinhold Treviranus (b. Bremen 1776–d. Bremen 1837); see John H. Barnhart, *Biographical Notes upon Botanists.* 3: 400. 1965; P. Smit, in *Dictionary of Scientific Biography* 13: 460-463. 1981; Mariella Azzarello Di Misa, ed., *Il Fondo Antico della Biblioteca dell'Orto Botanico di Palermo.* 274. Regione Siciliana, Palermo 1988; T.W. Bossert, *Biographical Dictionary of*

Botanists Represented in the Hunt Institute Portrait Collection. 406. 1972; H.N. Clokie, Account of the Herbaria of the Department of Botany in the University of Oxford. 256. Oxford 1964; E.M. Tucker, Catalogue of the Library of the Arnold Arboretum of Harvard University. Cambridge, Massachusetts 1917-1933; Antoine Lasègue, Musée botanique de M. Benjamin Delessert. 1845; G. Schmid, Goethe und die Naturwissenschaften. Halle 1940.

Trevirania Heynh. Rubiaceae

Origins:

After the German botanist and physician Ludolph Christian Treviranus, 1779-1864.

Trevoria F. Lehmann Orchidaceae

Origins:

Dedicated to Sir James John Trevor Lawrence, 1831-1913 (Surrey), orchidist and orchid grower, 1854-1864 Indian Medical Service, President of the Royal Horticultural Society, father of Sir William Matthew Trevor Lawrence (1870-1934); see M. Hadfield et al., British Gardeners: A Biographical Dictionary. London 1980; D.G. Crawford, A History of the Indian Medical Service, 1600-1913. London 1914; Marcus Bourne Huish, Catalogue of the Collection of Japanese Works of Art formed between ... 1869 and 1894 by Sir Trevor Lawrence. London 1895; Ernest Nelmes and William Cuthbertson, Curtis's Botanical Magazine Dedications, 1827-1927. 235-236. [1931]; Ray Desmond, Dictionary of British & Irish Botanists and Horticulturists. 417. [President, Cactus and Succulent Society, 1924-1928] London 1994.

Trewia L. Euphorbiaceae

Origins:

For the German botanist Christoph Jakob Trew, 1695-1769, physician, traveler, correspondent of the botanical artist Georg Dionysius Ehret (1708-1770), author of Plantae selectae, quarum imagines ad exemplaria naturalia Londini in hortis Curiosorum nutrita manu artificiosa doctaque pinxit Georgius Dionysius Ehret ... [Norimbergae] 1750-1773; see John H. Barnhart, Biographical Notes upon Botanists. 3: 400. 1965; J.D. Milner, Catalogue of Portraits of Botanists Exhibited in the Museums of the Royal Botanic Gardens. Royal Botanic Gardens, Kew, London 1906; R. Zander, F. Encke, G. Buchheim and S. Seybold, Handwörterbuch der Pflanzennamen. 14. Aufl. Stuttgart 1993; T.W. Bossert, Biographical Dictionary of Botanists Represented in the Hunt Institute Portrait Collection. 406. 1972;

Jonas C. Dryander, Catalogus bibliothecae historico-naturalis Josephi Banks. London 1800; Blanche Elizabeth Edith Henrey (1906-1983), British Botanical and Horticultural Literature before 1800. Oxford 1975; Gordon Douglas Rowley, A History of Succulent Plants. Strawberry Press, Mill Valley, California 1997.

Species/Vernacular Names:
T. nudiflora L.
Nepal: gurel

Triachyrum Hochst. Gramineae

Origins:

From the Greek treis, tria "three" and achyron "chaff, husk"; see also Sporobolus R. Br.

Triactina Hook.f. & Thomson Crassulaceae

Origins:

From the Greek treis, tria "three" and aktin "ray."

Triadenia Spach Guttiferae

Origins:

From the Greek treis, tria "three" and aden "gland."

Triadenum Raf. Guttiferae

Origins:

Greek treis, tria "three" and aden "gland"; see Constantine Samuel Rafinesque, Med. Repos. II. 5: 352. 1808, Am. Monthly Mag. Crit. Rev. 2: 267. 1818, Jour. Phys. Chim. Hist. Nat. 89: 261. 1819 and Flora Telluriana. 3: 78. 1936 [1837]; Elmer D. Merrill, Index rafinesquianus. 168-169. 1949.

Triadica Lour. Euphorbiaceae

Origins:

Greek treis, tria "three," referring to the calyx, ovary and fruit.

Triadodaphne Kosterm. Lauraceae

Origins:

From the Greek triadikos "of three" plus Daphne.

Triaena Kunth Gramineae

Origins:
Greek *triaina* "trident," referring to the flowers.

Triaenacanthus Nees Acanthaceae

Origins:
Greek *triaina* "trident, three-pronged fork" and *akantha* "thorn."

Triaenophora (Hook.f.) Soler. Scrophulariaceae

Origins:
Greek *triaina* "trident" and *phoros* "bearing, carrying."

Triainolepis Hook.f. Rubiaceae

Origins:
From the Greek *triaina* "trident" and *lepis* "scale."

Trianaea Planchon & Linden Solanaceae

Origins:
For the Colombian (b. Zipaquirá) botanist José Jerónimo (or Gerónimo) Triana, 1834-1890 (d. Paris), traveler, plant collector, botanical explorer, journalist. His writings include *Nuevos jeneros i especies de plantas para la flora Neo-Granadina*. Bogotá 1854 and *Prodromus florae Novogranatensis*. Paris 1862-1867, pupil of Francisco Javier Matis (the last survivor of the Mutis botanical expedition); see Enrique Pérez Arbeláez, in *Dictionary of Scientific Biography* 13: 463-464. 1981; J.H. Barnhart, *Biographical Notes upon Botanists*. 3: 400. 1965; T. W. Bossert, *Biographical Dictionary of Botanists Represented in the Hunt Institute Portrait Collection*. 406. Boston, Massachusetts 1972; M. Colmeiro y Penido, *La Botánica y los Botánicos de la Peninsula Hispano-Lusitana*. Madrid 1858; Ida Kaplan Langman, *A Selected Guide to the Literature on the Flowering Plants of Mexico*. Philadelphia 1964; E.M. Tucker, *Catalogue of the Library of the Arnold Arboretum of Harvard University*. Cambridge, Massachusetts 1917-1933; R. Zander, F. Encke, G. Buchheim and S. Seybold, *Handwörterbuch der Pflanzennamen*. 14. Aufl. Stuttgart 1993.

Trianaeopiper Trelease Piperaceae

Origins:
For the Colombian botanist José Jerónimo Triana, 1834-1890.

Triandrophora O. Schwarz Capparidaceae (Capparaceae)

Origins:
From the Greek *treis, tria* "three," *aner, andros* "man, stamen" and *phoros* "bearing, carrying."

Trianoptiles Fenzl ex Endl. Cyperaceae

Origins:
Greek *triaina* "trident" and *ptilon* "feather, wing."

Trianosperma (Torrey & A. Gray) Mart. Cucurbitaceae

Origins:
From the Greek *triaina* "trident" and *sperma* "a seed."

Trianthema L. Aizoaceae

Origins:
Greek *treis, tria* "three" and *anthemon* "flower," the flowers generally grouped in threes; see Carl Linnaeus, *Species Plantarum*. 223. 1753 and *Genera Plantarum*. Ed. 5. 105. 1754.

Species/Vernacular Names:
T. portulacastrum L.

Yoruba: alowonjeja funfun, alowo njeja pupa, akisan, akisa, atankale, afakale

T. triquetra Willd. (*Trianthema crystallina* sensu J. Black)

Australia: red spinach, small hogweed

Trianthera Wettst. Scrophulariaceae

Origins:
Greek *treis, tria* "three" and *anthera* "anther."

Trianthium Desv. Gramineae

Origins:
From the Greek *treis, tria* and *anthos* "flower," referring to the flowers.

Triaristella Brieger Orchidaceae

Origins:
Latin *tri* and *arista* "the awn or the beard of grain," *aristella* "small awn," referring to the shape of the sepals; see also *Trisetella* Luer.

Triaristellina Rauschert Orchidaceae

Origins:
The diminutive of *Triaristella*; see also *Trisetella* Luer.

Triarrhena (Maxim.) Nakai Gramineae

Origins:
From the Greek *treis*, *tria* "three" and *arrhen* "male"; see also *Miscanthus* Andersson.

Triarthron Baill. Loranthaceae

Origins:
From the Greek *treis*, *tria* plus *arthron* "a joint."

Trias Lindley Orchidaceae

Origins:
Greek *treis*, *tria* "three," referring to the arrangement of the floral envelopes.

Triaspis Burchell Malpighiaceae

Origins:
From the Greek *treis*, *tris* "three" and *aspis* "a shield," referring to the winged fruits.

Species/Vernacular Names:
T. stipulata Oliv.
Yoruba: abebe odan

Triathera Desv. Gramineae

Origins:
From the Greek *treis*, *tria* "three" and *ather* "stalk, barb."

Triatherus Raf. Gramineae

Origins:
Greek *treis*, *tria* and *ather* "stalk, barb"; see C.S. Rafinesque, in *Am. Monthly Mag. Crit. Rev.* 3: 99. 1818; E.D. Merrill, *Index rafinesquianus*. 76. 1949.

Triavenopsis Candargy Gramineae

Origins:
Greek *treis*, *tria* with *Avena*.

Tribeles Philippi Grossulariaceae (Tribelaceae, Escalloniaceae)

Origins:
Greek *tribeles* "three-pointed."

Triblemma (J. Sm.) Ching Dryopteridaceae (Aspleniaceae, Woodsiaceae)

Origins:
Greek *treis*, *tria* with *blemma* "the look, glance, the eye"; see also *Diplazium* Sw.

Tribolium Desv. Gramineae

Origins:
From the Greek *tribolos*, a name of various prickly plants.

Tribonanthes Endl. Haemodoraceae

Origins:
Greek *tribon* "cloak, a thredbare cloak" and *anthos* "flower," alluding to the indumentum of the flowers of most species, the filament appendages rise above the anthers.

Tribrachia Lindley Orchidaceae

Origins:
Greek *tribrachys* "a poetical foot consisting of three short syllables," Latin *tribrachys* "a tribrach," referring to the three sepals.

Tribrachya Korth. Rubiaceae

Origins:
Greek *treis*, *tria* "three" and *brachys* "short."

Tribroma O.F. Cook Sterculiaceae

Origins:
From the Greek *treis*, *tria* "three" and *broma* "food," referring to the fruit; see also *Theobroma* L.

Tribulocarpus S. Moore Aizoaceae (Tetragoniaceae)

Origins:
Greek *tribolos* "a threshing-machine, caltrop, a prickly plant" and *karpos* "fruit," Latin *tribulus* or *tribolos*.

Tribulopis R. Br. Zygophyllaceae

Origins:

Resembling *Tribulus*; see Charles Sturt (1795-1869), *Narrative of an Expedition into Central Australia*. 2. App. 70. London 1849.

Tribulopsis F. Muell. Zygophyllaceae

Origins:

Orthographic variant, see *Tribulopis* R. Br.; see Ferdinand von Mueller, *Fragmenta Phytographiae Australiae*. 1: 47. (Jul.) 1858; Arthur D. Chapman, ed., *Australian Plant Name Index*. 2881. Canberra 1991.

Tribulopsis R. Br. Zygophyllaceae

Origins:

See *Tribulopis* R. Br.

Tribulus L. Zygophyllaceae

Origins:

Latin *tribulus*, *tribolos*, *i* "a caltrop"; Greek *tribolos*, *treis*, *tria* "three" and *bolos* "a point," the fruit resembling a club-shaped crest with sharp spines (each fruit breaking into five triangular-shaped segments, each with two large spines at the tip and several smaller spines); see Carl Linnaeus, *Species Plantarum*. 386. 1753 and *Genera Plantarum*. Ed. 5. 183. 1754.

Species/Vernacular Names:

T. cistoides L. (*Kallstroemia cistoides* (L.) Endl.)

Hawaii: nohu, nohunohu

Mexico: guechi bioba roo, quechi pioba, quechi quiqui pioba

T. terrestris L.

English: burnut, devil's thorn, cat-head, puncture vine, goat head, puncture vine, caltrop

Arabic: hasak, hasaka

India: gokshru

Tibetan: gzema

China: ci ji li, sha yuan chi li, chi li, tu chi li, pai chi li, tzu

East Africa: esuguru, eziguru, kungu, mbigiri, mbiliwili, okuro, shokolo

Tanzania: ndaghandaghu

South Africa: devil's thorn, tshetlo, dubbeltjie, common dubbeltjie, gewone dubbeltjie, rankdubbeltjie, duwweltjie, duiweltjie, platdubbeltjie, volstruisdubbeltjie, volstruisdoring, drie-doring dubbeltjie, dubbeltjie doring, volstruisdubbeltjie

Yoruba: dagunro, dagunro nla

T. zeyheri Sond. subsp. *zeyheri*

English: devil's thorn, devil-thorn weed

Namibia: dubbeltjie, dubbeltjie doring, platdubbeltjie, duiweltjies (= little devils) (Afrikaans); ohongwe (= thorn), ongarayohongwe (= flowers) Herero; nhonho (Kwangali); shosho (Mbukushu, Gciriku, Shambyu); oshosholo (Kwanyama); ondjoho (Ndonga)

German: morgenstern

Tricalysia A. Rich. ex DC. Rubiaceae

Origins:

The generic name is based on the Greek *treis*, *tria* "three" and *kalyx*, *kalykos* "calyx," referring to the true calyx and two epicalyces (or often double), or to the bracteoles beneath the flower.

Species/Vernacular Names:

T. africana (Sim) Robbrecht (*Diplospora africana* Sim)

Southern Africa: isiCeza (Xhosa)

T. capensis (Meisn. ex Hochst.) Sim var. *capensis* (*Bunburya capensis* Meisn. ex Hochst.)

English: Cape coffee, kaffir coffee

Southern Africa: Kaapse koffie, kafferkoffie; iBhicongo, inDulwane (Zulu); umPonyane, iNdulwane, isiCeza (Xhosa); tshituku-tsha-dake (Venda)

T. deightonii Brenan

Sierra Leone: kpoli wuli

T. junodii (Schinz) Brenan var. *junodii* (*Tricalysia allenii* (Stapf) Brenan var. *australis* (Schweick.) Brenan)

South Africa: jackal-coffee

T. lanceolata (Sond.) Burtt Davy (*Kraussia lanceolata* Sond.)

English: jackal-coffee, wild coffee

Southern Africa: jakkalskoffie, wildekoffie, tolbalie; iSanwane, umDleza, uQaganja wentaba, umsanyana, umCakasho, isAnyana (Zulu); umPhonyana, umDlesa, isiXeza (Xhosa)

T. macrophylla K. Schum.

Nigeria: uviame (Edo)

T. okelensis Hiern

Yoruba: ataparaja

T. sonderiana Hiern (*Kraussia coriacea* Sond.) (in honor of the German botanist and pharmacist Otto Wilhelm Sonder, 1812-1881, plant collector, co-author with W.H.

Harvey of the first three volumes of *Flora Capensis*, author of "Beiträge zur Flora von Süd-Afrika." *Linnaea*. 23: 1-138. 1850; see Michel Gandoger (1850-1926), "L'herbier africain de Sonder." *Bull. Soc. bot. France*. 60: 414-422, 445-462. 1913; J.H. Barnhart, *Biographical Notes upon Botanists*. 3: 303. 1965; R. Zander, F. Encke, G. Buchheim and S. Seybold, *Handwörterbuch der Pflanzennamen*. 14. Aufl. 1993; Stafleu and Cowan, *Taxonomic Literature*. 5: 741-743. 1985; Dennis John Carr and S.G.M. Carr, eds., *People and Plants in Australia*. 1981; Mary Gunn and Leslie E. Codd, *Botanical Exploration of Southern Africa*. 328-329. Cape Town 1981)

English: coast coffee, Sonder's tricalysia

Southern Africa: kuskoffie; iNkweza, inKweza, isiGilamthonjane, uTholombe-wehlathi (Zulu)

Tricardia Torrey Hydrophyllaceae

Origins:
Greek *treis*, *tria* "three" and *kardia* "heart," referring to the calyx lobes, outer three-cordate.

Species/Vernacular Names:
T. watsonii S. Watson
English: three hearts

Tricarium Lour. Euphorbiaceae

Origins:
Greek *treis*, *tria* "three" and *karyon* "nut," referring to the fruit.

Tricarpha Longpre Asteraceae

Origins:
Greek *treis*, *tria* "three" and *karphos* "chip of straw, chip of wood, mote, splinter, nail."

Tricera Schreb. Buxaceae

Origins:
Greek *treis*, *tria* "three" and *keras* "horn," referring to the fruits.

Triceratella Brenan Commelinaceae

Origins:
Greek *treis*, *tria* "three" and *keras* "horn," the diminutive.

Triceratia A. Rich. Cucurbitaceae

Origins:
From the Greek *treis*, *tria* and *keras* "horn."

Triceratorhynchus Summerh. Orchidaceae

Origins:
Greek *treis*, *tria* "three," *keras*, *keratos* "horn" and *rhynchos* "horn, beak, snout," referring to the rostellum.

Tricerma Liebm. Celastraceae

Origins:
From the Greek *treis*, *tria* "three" and *kerma* "a small coin, morsel."

Triceros Griff. Cucurbitaceae

Origins:
From the Greek *treis*, *tria* "three" and *keras* "horn," referring to the fruit; see also *Gomphogyne* Griff.

Trichacanthus Zoll. Acanthaceae

Origins:
Greek *thrix*, *trichos* "hair" plus *acanthus*.

Trichachne Nees Gramineae

Origins:
From the Greek *thrix*, *trichos* "hair" and *achne* "chaff, glume."

Trichadenia Thwaites Flacourtiaceae

Origins:
From the Greek *thrix*, *trichos* "hair" and *aden* "gland."

Trichaeta P. Beauv. Gramineae

Origins:
From the Greek *treis*, *tria* "three" and *chaite* "a bristle," referring to the glume; see also *Trisetaria* Forssk.

Trichantha Hook. Gesneriaceae

Origins:

Greek *thrix*, *trichos* "hair" and *anthos* "flower," referring to the corolla.

Trichantha Karst. & Triana Convolvulaceae

Origins:

Greek *thrix*, *trichos* "hair" and *anthos* "flower."

Trichanthemis Regel & Schmalh. Asteraceae

Origins:

From the Greek *thrix*, *trichos* "hair" and *anthemis, anthemon* "flower."

Trichanthera Kunth Acanthaceae

Origins:

From the Greek *thrix*, *trichos* "hair" and *anthera* "anther."

Trichanthodium Sonder & F. Muell. Asteraceae

Origins:

Greek *thrix*, *trichos* "hair" and *anthos* "flower."

Trichelostylis T. Lestib. Cyperaceae

Origins:

Greek *treis*, *tria* "three," *chele* "a horse's hoof" and *stylos* "a pillar, style," *tricheilos* "three-lipped."

Tricherostigma Boiss. Euphorbiaceae

Origins:

Presumably from the Greek *tricha*, *triche* "in three parts, in three ways," *tricheiros* "three-handed" and *stigma* "stigma."

Trichilia P. Browne Meliaceae

Origins:

Greek *tricha*, *triche* "three parts," *tricheilos* "three-lipped," referring to the ovary and fruits; see Patrick Browne (1720-1790), *The Civil and Natural History of Jamaica*. 278. London 1756.

Species/Vernacular Names:

T. sp.

English: bastard cedar

Ivory Coast: asamoiake, konangbri, mutigbanaye rasso

Cameroon: ebangbemva

Nigeria: dissoko, ebangbemva, irere, urere

Yoruba: ajigbagbo

Belize: cedrillo, si-sin

Salvador: barredero, canelillo, canjuro, jocotillo, ojo de muñeca, pimientillo

Puerto Rico: cabo de hacha, caracolillo, go ita, guaita, guayavacon, palo de anastasio

Brazil: catigua graudo, catigua miudo, guamichava

Argentina: catigua puita, fe de gozo, guamiri, mangacitara, palo anis, tape rigua

Colombia: manglesito, yayo, yayo colorado

Nicaragua: matapiojo

Peru: chiajap, chuchuhuasi, lechuza caspi

Mexico: bola de tejon (Hidalgo); mi-cha-hi (Chinanteca l., Oaxaca); cabo de hacha (Oaxaca); cauache (Sinaloa); chobenché (Maya l., Yucatan); estribillo, cucharillo (Tamaulipas); estribillo (southeast San Luis Potosí); garrapatilla (Colima; see Richard D. Reynolds, *The Ancient Art of Colima, Mexico*. Walnut Creek, Squibob Press 1993); garbancillo (Sonora)

Cuba: guaban, jubaban, siguraya

T. dregeana Sonder (*Trichilia dregei* E. Meyer; *Trichilia dregei* var. *oblonga* C. DC.; *Trichilia chirindensis* Swynnerton & Bak.f.; *Trichilia splendida* A. Chev.) (after the German botanist and botanical collector (Johann Franz) Jean François Drège, 1794-1881, botanical explorer, traveler. His writings include *Zwei pflanzengeografische Dokumente*. Leipzig [1843-1844] and *Catalogus plantarum exsiccatarum Africae australioris, quas emturis offert*. 1837-1840. He was the brother of Carl Friedrich Drège (1791-1867); see George Arnott Walker Arnott (1799-1868), "Notes on some South African plants." *Hook., J.Bot.* 3: 147-156. 1841; John H. Barnhart, *Biographical Notes upon Botanists*. 1: 471. 1965; Peter MacOwan, "Personalia of botanical collectors at the Cape." *Trans. S. Afr. Philos. Soc.* 4(1): xlix-l. 1884-1886; J.H. Verduyn den Boer, *Botanists at the Cape*. 55-58. Cape Town and Stellenbosch 1929; John Hutchinson (1884-1972), *A Botanist in Southern Africa*. 642. London 1946; R. Zander, F. Encke, G. Buchheim and S. Seybold, *Handwörterbuch der Pflanzennamen*. 14. Aufl. 704. 1993; T.W. Bossert, *Biographical Dictionary of Botanists Represented in the Hunt Institute Portrait Collection*. 107. 1972; A. White and B.L. Sloane, *The Stapelieae*. Pasadena 1937; Gordon Douglas Rowley, *A History of Succulent Plants*. California 1997; Mary Gunn and Leslie

E. Codd, *Botanical Exploration of Southern Africa*. 145, 317, 382, 386, 389. Cape Town 1981)

English: forest mahogany, Cape mahogany, white mahogany, thunder tree, Christmas bells, red ash

Ivory Coast: ouagon, aribanda

Uganda: sekoba

Southern Africa: bosrooiessenhout, rooiessenhout, basteressenhout; umKhuhlu (Xhosa); muKuhlu (Shona); umKhuhlu, uMathunzini, uMantunzini (Zulu); nkuhlu (Tsonga); mmaba (North Sotho); mutuhu, mutshikili (Venda)

Zaire: ivundi, soko

T. emetica Vahl (*Trichilia roka* Chiov. nom. illegit.)

English: Natal mahogany, Cape mahogany, mafurra tallow, mafoureira tallow

French: cèdre d'Afrique

Mali: warakatiga, sulafinzan

Tropical Africa: adyanya peso, mtshigizi

Yoruba: papa, papa odan, karuntan

Nigeria: gwandar kusa, jan saye, irere, gwanja-kusa, jansaye, jansayi, tanuba, ebangbemva, dissoko, jakan

Eastern Africa: mafoureire

Zambia: msikiri, mushikishi, musikili

Uganda: sekoba

Southern Africa: rooiessenhout; umKhuhlu, uMathunzini, iGwolo (Zulu); ankhulu, nkuhlu (Thonga or Tsonga); umKhuhlu (Swazi); mosikiri (Western Transvaal, northern Cape, Botswana); mmaba (North Sotho); mutshikili, mutuhu (Venda); muChenya, muChichiri, muKuhlu, muSikili, muSikivi, muTshitshivi, muTsikiri (Shona)

T. emetica Vahl subsp. *suberosa* J.J. De Wilde

Nigeria: jan saiwa, jan saye, gwanja kusa, goron talaka (Hausa); asapa (Yoruba)

T. gilletii De Wild.

Cameroon: eyon

Gabon: eyou

T. martineaui Aubrév. & Pellegr.

Nigeria: ogiovaloa; ogiovalo (Edo)

Ivory Coast: mietandabo

Uganda: musuga

T. megalantha Harms

Yoruba: itabira

T. monadelpha (Thonn.) J.J. De Wilde (*Limonia monadelpha* Thonn.; *Trichilia heudelotii* Planch. ex Oliv.)

Yoruba: rere, ako rere, owatete, awuya, rereigbo

Nigeria: akoledo, olomi, ovalo, agbigbon, ovien-urhen, okawrere; akika, ako ere olomi (Yoruba); oyallo (Edo); ovien urhaen (Sobo); agbigben (Jerki)

Sierra Leone: njawae

Congo: kinkuma, muheho

T. patens Oliv.

Sierra Leone: fok lobae

T. prieuriana A. Juss.

Yoruba: awe, odofin igbo, oloja ebana, eyin eye, odofin oko

Nigeria: irere, urere, awe, igogo, odofin-igbo; awe, enyiade, odapingbo, olajebane (Yoruba); atantan, eghogho, igogo (Edo)

Ghana: aboya, kakadikro

W. Africa: dilifu

T. retusa Oliv.

Nigeria: owo (Aguleri); nyamenyok (Boki)

T. tessmannii Harms (*Trichilia lanata* A. Chev.)

Yoruba: iseko, isinko, aje, alaje

Nigeria: aje, iseko, ogiovalo, ogiovallo; ogiovalo (Edo); ubenwenwe (Igbo)

Cameroon: lobonda, lebonda, nom oswe, mayimbonagbanga

Ivory Coast: aribanda, ndia, bohia, dandi

Congo: tobla

T. welwitschii C. DC.

Cameroon: ebangbemva oswe, mayimbo

Ivory Coast: banaye

Nigeria: akika, olomi oyallo

Trichinium R. Br. Amaranthaceae

Origins:
Greek *trichinos* "of hair," Latin *trichinus* "slight, meager, poor," referring to the inflorescence; see Robert Brown, *Prodromus florae Novae Hollandiae*. 414. London 1810.

Trichiocarpa (Hook.) J. Sm. Dryopteridaceae

Origins:
From *trichion*, the diminutive of the Greek *thrix*, *trichos* "hair" and *karpos* "fruit."

Trichiogramme Kuhn Pteridaceae (Adiantaceae)

Origins:
From *trichion*, the diminutive of *thrix*, *trichos* "hair" and *gramma* "a letter, line, character."

Trichipteris C. Presl Cyatheaceae

Origins:

Greek *thrix*, *trichos* "hair" and *pteris* "a fern," referring to the indusium.

Trichloris E. Fourn. ex Benth. Gramineae

Origins:

From the Greek *tri-* "three" plus *Chloris* Swartz., referring to the awns.

Trichobasis Turcz. Myrtaceae

Origins:

From the Greek *thrix*, *trichos* "hair" and *basis* "base, pedestal."

Trichocalyx Balf.f. Acanthaceae

Origins:

From the Greek *thrix*, *trichos* "hair" and *kalyx* "calyx."

Trichocalyx Schauer Myrtaceae

Origins:

Greek *thrix*, *trichos* "hair" and *kalyx* "calyx," see also *Calytrix* Labill.

Trichocarya Miq. Chrysobalanaceae

Origins:

From the Greek *thrix*, *trichos* "hair" and *karyon* "nut."

Trichocaulon N.E. Br. Asclepiadaceae

Origins:

From the Greek *thrix*, *trichos* "hair" and *kaulos* "stem."

Trichocentrum Poepp. & Endl. Orchidaceae

Origins:

Greek *thrix*, *trichos* "hair" and *kentron* "a spur, prickle," referring to the spur of the labellum.

Trichocereus (A. Berger) Riccob. Cactaceae

Origins:

From the Greek *thrix*, *trichos* "hair" plus *cereus*.

Trichoceros Kunth Orchidaceae

Origins:

Greek *thrix*, *trichos* "hair" and *keras* "horn," referring to the processes from the column.

Trichochilus Ames Orchidaceae

Origins:

From the Greek *thrix*, *trichos* and *cheilos* "lip," referring to the nature of the mid-lobe of the column.

Trichochiton Komarov Brassicaceae

Origins:

From the Greek *thrix*, *trichos* "hair" and *chiton* "a tunic, covering."

Trichochlaena Kuntze Gramineae

Origins:

From the Greek *thrix*, *trichos* plus *chlaena*, *chlaenion* "a cloak"; see also *Tricholaena* Schrad.

Trichochloa DC. Gramineae

Origins:

From the Greek *thrix*, *trichos* "hair" and *chloe*, *chloa* "grass."

Trichocladus Pers. Hamamelidaceae

Origins:

From the Greek *thrix*, *trichos* "hair" and *klados* "branch."

Species/Vernacular Names:

T. crinitus (Thunb.) Pers. (*Dahlia crinita* Thunb.; *Trichocladus peltatus* Meisn.; *Trichocladus vittatus* Meisn.)

English: black hazel

Southern Africa: swarthaselaar, onderbos; isiTha, uBisana (Zulu); iThambo (= bone) (Xhosa)

T. ellipticus Eckl. & Zeyh. subsp. *ellipticus*

English: white hazel, Natal hazel

Southern Africa: onderbos, withaselaar; umGqonci (Zulu); umVa wenyathi, umGqonci (Xhosa)

T. grandiflorus Oliv.

English: green hazel, green witch-hazel, honeysuckle tree, wild peach

Southern Africa: groenhaselaar; rooihout; uGabavu (Zulu)

Trichocline Cass. Asteraceae

Origins:
From the Greek *thrix, trichos* "hair" and *kline* "a bed, couch."

Trichocoronis A. Gray Asteraceae

Origins:
Greek *thrix, trichos* "hair" and *koronis, koronidos* "garland, wreath," an allusion to the pappus.

Trichocoryne S.F. Blake Asteraceae

Origins:
From the Greek *thrix, trichos* and *koryne* "a club."

Trichocyamos Yakovlev Fabaceae

Origins:
Greek *thrix, trichos* "hair" and *kyamos* "a bean."

Trichocyclus Dulac Dryopteridaceae (Aspleniaceae, Woodsiaceae)

Origins:
Greek *thrix, trichos* "hair" and *kyklos* "a circle, ring."

Trichodesma R. Br. Boraginaceae

Origins:
Greek *thrix, trichos* "hair" and *desmos* "a bond, band," referring to the anthers.

Species/Vernacular Names:
T. calycosum Collett & Hemsley

English: large-calyx trichodesma

China: mao shu cao

T. physaloides (Fenzl) A. DC. (*Friedrichsthalia physaloides* Fenzl)

English: chocolate bells

South Africa: slangkop

T. zeylanicum (Burm.f.) R. Br. (*Borago zeylanica* Burm.f.; *Trichodesma zeylanicum* (Burm.f.) R. Br. var. *latisepalum* F. Muell. ex Benth.)

English: late weed, cattle bush, water bush, camel bush, Kuenstler bush

East Africa: eileili, machacha, magundulu, nyalak-dede

Trichodiadema Schwantes Aizoaceae

Origins:
From the Greek *thrix, trichos* "hair" and *diadema, diadematos* "band, fillet."

Trichodiclida Cerv. Gramineae

Origins:
From the Greek *thrix, trichos* "hair" and *diklis, diklidos* "double-folding, double-folding door."

Trichodon Benth. Gramineae

Origins:
From the Greek *thrix, trichos* "hair" and *odous, odontos* "tooth."

Trichodrymonia Oerst. Gesneriaceae

Origins:
From the Greek *thrix, trichos* "hair" plus *Drymonia* Mart.

Trichodypsis Baillon Palmae

Origins:
From the Greek *thrix, trichos* "hair" plus *Dypsis* Noronha ex Mart.

Trichogalium Fourr. Rubiaceae

Origins:
From the Greek *thrix, trichos* "hair" plus *Galium* L.

Trichoglottis Blume Orchidaceae

Origins:
Greek *thrix, trichos* "hair" and *glotta* "tongue," the dactylate process of the lip is often hairy or pubescent.

Species/Vernacular Names:
T. luchuensis (Rolfe) Garay & Sweet (*Stauropsis luchuensis* Rolfe; *Vandopsis luchuensis* (Rolfe) Schltr.; *Staurochilus luchuensis* (Rolfe) Fukuyama)

Japan: Iriomote-ran

T. papuana Schltr. (see B.A. Lewis and P.J. Cribb, *Orchids of the Solomon Islands and Bougainville.* Royal Botanic Gardens, Kew 1991)

Bougainville Island: marapauna

Trichogonia (DC.) Gardner Asteraceae

Origins:
From the Greek *thrix, trichos* and *gonia* "an angle."

Trichogoniopsis R.M. King & H. Robinson Asteraceae

Origins:
Resembling *Trichogonia.*

Trichogyne Less. Asteraceae

Origins:
Greek *thrix, trichos* "hair" and *gyne* "woman, female," referring to the female flowers.

Tricholaena Schrader Gramineae

Origins:
From the Greek *thrix, trichos* "hair" and *chlaena, chlaenion* "a cloak."

Species/Vernacular Names:
T. monachne (Trin.) Stapf & C.E. Hubb. (*Tricholaena monachne* (Trin.) Stapf & C.E. Hubb. var. *annua* J.G. Anders.)

English: blue-seed tricholaena

South Africa: blousaadgras, blousaadtricholaena

Tricholaser Gilli Umbelliferae

Origins:
Latin *laser, lasar, eris* for the plant *laserpitium* or for the juice of the plant *laserpitium,* assafoetida.

Tricholepidium Ching Polypodiaceae

Origins:
From the Greek *thrix, trichos* plus *lepidion* "a little scale," *lepis, lepidos* "scale."

Tricholepis DC. Asteraceae

Origins:
Greek *thrix, trichos* "hair" and *lepis, lepidos* "scale."

Tricholobus Blume Connaraceae

Origins:
From the Greek *thrix, trichos* "hair" and *lobos* "pod, lobe."

Trichomanes Hill Aspleniaceae

Origins:
Ancient Greek name applied by Theophrastus and Dioscorides to a fern, said to be derived from *thrix* "hair, bristle" and *manos* "soft, thin, porous"; Latin *trichomanes,* for a plant resembling *adiantum* (Plinius); see H. Genaust, *Etymologisches Wörterbuch der botanischen Pflanzennamen.* 653-654. Basel 1996.

Trichomanes L. Hymenophyllaceae

Origins:
Greek *trichomanes,* ancient name applied by Theophrastus (*HP.* 7.14.1) and Dioscorides to a fern, bristle or kidney fern; Latin *trichomanes,* is used by Plinius for a plant resembling *adiantum*; see Carl Linnaeus, *Species Plantarum.* 1097. 1753 and *Genera Plantarum.* Ed. 5. 485. 1754; H. Genaust, *Etymologisches Wörterbuch der botanischen Pflanzennamen.* 653-654. Basel 1996.

Species/Vernacular Names:
T. auriculatum Blume (*Vandenboschia auriculata* (Blume) Copel.)

Japan: tsuru-hora-goke (= vine *Trichomanes*)

T. proliferum Blume (*Trichomanes minutum* Blume)

Japan: uchiwa-goke (= fan moss)

T. speciosum Willd.

English: Killarney fern

Trichonema Ker Gawler Iridaceae

Origins:
From the Greek *thrix, trichos* "hair" and *nema* "filament, thread."

Trichoneura Andersson Gramineae

Origins:

From the Greek *thrix*, *trichos* and *neuron* "nerve."

Species/Vernacular Names:

T. grandiglumis (Nees) Ekman var. *grandiglumis* (*Crossotropis grandiglumis* (Nees) Rendle)

English: rolling grass, tumble weed

Southern Africa: rolgras, waaigras, warrelgras; joang-batsela (Sotho)

Trichoneuron Ching Aspleniaceae (Woodsiaceae, Tectarioideae)

Origins:

From the Greek *thrix*, *trichos* "hair" and *neuron* "nerve."

Trichoon Roth Gramineae

Origins:

From the Greek *thrix*, *trichos* "hair" and *oon* "egg."

Trichopetalum Lindley Anthericaceae (Liliaceae)

Origins:

From the Greek *thrix*, *trichos* "hair" and *petalon* "petal."

Trichophorum Pers. Cyperaceae

Origins:

From the Greek *thrix*, *trichos* "hair" and *phoros* "bearing, carrying."

Trichopilia Lindley Orchidaceae

Origins:

Greek *thrix*, *trichos* "hair" and *pilos* "hat, cap, felt cap," referring to the margins of the clinandrium.

Trichopodium C. Presl Fabaceae

Origins:

Greek *thrix*, *trichos* and *pous*, *podos* "a foot," *podion* "a little foot."

Trichopodium Lindley Dioscoreaceae (Trichopodaceae)

Origins:

Greek *thrix*, *trichos* "hair" and *pous*, *podos* "a foot," *podion* "a little foot," see also *Trichopus* Gaertn.

Trichopteria Nees Gramineae

Origins:

Greek *thrix*, *trichos* "hair" and *pteron* "wing," referring to the fruits; see also *Trichopteryx* Nees.

Trichopteryx Nees Gramineae

Origins:

From the Greek *thrix*, *trichos* "hair" and *pteron* "wing," referring to the fruits.

Trichoptilium A. Gray Asteraceae

Origins:

Feathery bristle, Greek *thrix*, *trichos* "hair" and *ptilon* "feather, wing," referring to the pappus.

Trichopus Gaertner Dioscoreaceae (Trichopodaceae)

Origins:

From the Greek *thrix*, *trichos* "hair" and *pous*, *podos* "a foot."

Trichopyrum Á. Löve Gramineae

Origins:

From the Greek *thrix*, *trichos* and *pyros* "grain, wheat."

Trichosacme Zucc. Asclepiadaceae

Origins:

From the Greek *thrix*, *trichos* "hair" and *akme* "the highest point, the top, summit."

Trichosalpinx Luer Orchidaceae

Origins:

Greek *thrix*, *trichos* and *salpinx* "a trumpet, tube," referring to the ribs and margins of the trumpet-shaped sheats of the secondary stems.

Trichosanchezia Mildbr. Acanthaceae

Origins:
From the Greek *thrix, trichos* "hair" plus *Sanchezia* Ruíz & Pavón.

Trichosandra Decne. Asclepiadaceae

Origins:
Greek *thrix, trichos* "hair" and *aner, andros* "male, a man, anther, stamen."

Species/Vernacular Names:
T. borbonica Decne.
Rodrigues Island: liane à cornes

Trichosantha Steud. Gramineae

Origins:
From the Greek *thrix, trichos* plus *anthos* "a flower."

Trichosanthes L. Cucurbitaceae

Origins:
Greek *thrix, trichos* and *anthos* "a flower," referring to the fringed corolla; see Carl Linnaeus, *Species Plantarum.* 1008. 1753 and *Genera Plantarum.* Ed. 5. 439. 1754.

Species/Vernacular Names:
T. cucumerina L. (*Trichosanthes colubrina* Jacq.f.; *Trichosanthes anguina* L.)
English: serpent cucumber, club gourd, snake gourd, viper's gourd, serpent gourd
Malaya: ketola ular
T. kirilowii Maxim.
English: Mongolian snake gourd
China: gua lou, kua lou
T. ovigera Bl. (*Trichosanthes cucumeroides* Maxim. ex Franch. & Savat.)
English: snake gourd
China: wang gua
T. tricuspidata Lour. (*Trichosanthes bracteata* (Lam.) Voigt; *Modecca bracteata* Lam.; *Trichosanthes palmata* Roxb. non L.)
English: snake gourd
Japan: ô-karasu-uri
India: marang
Malaya: ketola ular, akar ketomenan, suntoh

Trichoschoenus J. Raynal Cyperaceae

Origins:
From the Greek *thrix, trichos* "hair" and *schoinos* "rush, reed, cord."

Trichoscypha Hook.f. Anacardiaceae

Origins:
From the Greek *thrix, trichos* and *skyphos* "a cup, goblet," referring to the disc.

Species/Vernacular Names:
T. acuminata Engler
Gabon: amvut
Congo: seng
Cameroon: ngoyo, ndoi, amvut, abut, ekong, levboda
Central Africa: avbambe, abambe

Trichosia Blume Orchidaceae

Origins:
From the Greek *thrix, trichos* "hair," *trichosis* "growth of hair"; see also *Eria* Lindl.

Trichosiphum Schott & Endl. Sterculiaceae

Origins:
From the Greek *thrix, trichos* "hair" and *siphon* "tube."

Trichosma Lindley Orchidaceae

Origins:
Greek *thrix, trichos* "hair," referring to the lamellae on the lip; some suggest from *trichos* plus *osme* "smell, odor, perfume"; see also *Eria* Lindl.

Trichosorus Liebm. Lophosoriaceae (Dicksoniaceae)

Origins:
From the Greek *thrix, trichos* "hair" and *soros* "a heap"; see also *Lophosoria* Presl.

Trichospermum Blume Tiliaceae

Origins:
From the Greek *thrix, trichos* "hair" and *sperma* "seed."

Trichospira Kunth Asteraceae

Origins:
From the Greek *thrix, trichos* and *speira* "a spiral."

Trichosporum D. Don Gesneriaceae

Origins:
From the Greek *thrix, trichos* "hair" and *sporos, spora* "a seed."

Trichostachys Hook.f. Rubiaceae

Origins:
From the Greek *thrix, trichos* "hair" and *stachys* "a spike."

Trichostegia Turcz. Asteraceae

Origins:
Greek *thrix, trichos* with *stege, stegos* "roof, cover"; see Porphir Kiril N.S. Turczaninow (1796-1863), *Bulletin de la Société Impériale des Naturalistes de Moscou.* 24(2): 81. 1851.

Trichostelma Baillon Asclepiadaceae

Origins:
Greek *thrix, trichos* "hair" and *stelma, stelmatos* "a girdle, belt, crown, garland, wreath."

Trichostema L. Labiatae

Origins:
Greek *thrix, trichos* "hair" and *stema* "stamen."

Species/Vernacular Names:
T. lanatum Benth.
English: woolly bluecurls, bluecurls
T. lanceolatum Benth.
English: vinegar weed
T. laxum A. Gray
English: turpentine weed
T. micranthum A. Gray
English: small-flowered bluecurls
T. ovatum Curran
English: San Joaquin bluecurls

T. rubisepalum Elmer
English: Hernandez bluecurls

Trichostephania Tardieu Connaraceae

Origins:
Greek *thrix, trichos* "hair" and *stephanos* "a crown"; see also *Ellipanthus* Hook.f.

Trichostephanus Gilg Flacourtiaceae

Origins:
Greek *thrix, trichos* and *stephanos* "a crown."

Trichosterigma Klotzsch & Garcke Euphorbiaceae

Origins:
Greek *thrix, trichos* and *sterigmos* "a setting firmly," *sterizo, sterizein* "to stand fast, to fix," *sterigma, sterigmatos* "a support, foundation."

Trichostigma A. Rich. Phytolaccaceae

Origins:
From the Greek *thrix, trichos* "hair" and *stigma* "a stigma."

Trichostomanthemum Domin Apocynaceae

Origins:
From the Greek *thrix, trichos* "hair," *stoma* "mouth" and *anthemon* "flower."

Trichotaenia Yamaz. Scrophulariaceae

Origins:
Greek *thrix, trichos* "hair" and *tainia* "fillet, a ribbon, band."

Trichothalamus Sprengel Rosaceae

Origins:
Greek *thrix, trichos* "hair" and *thalamos* "the base of the flower."

Trichotosia Blume Orchidaceae

Origins:
Greek *thrix, trichos* "hair" and *toxon* "a bow," referring to the shape of the hairy sepals, or from *trichotos* "furnished with hair, hairy," indicating the very hairy epiphitic plants; see Richard Evans Schultes and Arthur Stanley Pease, *Generic Names of Orchids. Their Origin and Meaning.* 308. Academic Press, New York and London 1963; Hubert Mayr, *Orchid Names and Their Meanings.* 185. Vaduz 1998.

Trichovaselia Tieghem Ochnaceae

Origins:
Greek *thrix, trichos* "hair" plus the anagram of the genus *Elvasia* DC.

Trichuriella Bennet Amaranthaceae

Origins:
The diminutive of the genus *Trichurus.*

Trichurus C.C. Towns. Amaranthaceae

Origins:
Greek *thrix, trichos* "hair" and *oura* "tail."

Tricliceras Thonn. ex DC. Turneraceae

Origins:
From the Greek *treis, tria* "three," *kleis* "lock, key, bar, bolt" and *keras* "a horn."

Triclisia Benth. Menispermaceae

Origins:
From the Greek *treis, tria* "three" and *kleis* "lock, key, bar."

Species/Vernacular Names:
T. subcordata Oliv.
Yoruba: alugbonron, sona gburu

Tricochilus Ames Orchidaceae

Origins:
From the Greek *thrix, trichos* and *cheilos* "lip."

Tricomariopsis Dubard Malpighiaceae

Origins:
Resembling *Tricomaria* Gillies ex Hook. & Arn.

Tricondylus Salisbury ex J. Knight Proteaceae

Origins:
Greek *treis, tria* "three" and *kondylos* "prominence, knuckle, knot"; see Arthur D. Chapman, ed., *Australian Plant Name Index.* 2898. Canberra 1991.

Tricoryne R. Br. Anthericaceae (Liliaceae)

Origins:
Greek *treis* "three" and *koryne* "a club," referring to the fruits, club shaped and separated into three mericarps at maturity; see Robert Brown (1773-1858), *Prodromus florae Novae Hollandiae.* 278. London (Mar.) 1810; James A. Baines, *Australian Plant Genera. An Etymological Dictionary of Australian Plant Genera.* 380. Chipping Norton, N.S.W. 1981.

Species/Vernacular Names:
T. elatior R. Br. (*Tricoryne scabra* R. Br.)
English: yellow rush-lily, yellow autumn lily

Tricostularia Nees ex Lehm. Cyperaceae

Origins:
Latin *tri-* "three" and *costula* "little rib," *costa, ae* "a rib, wall, side," referring to the ribbed nuts, see also *Costularia* C.B. Clarke.

Species/Vernacular Names:
T. pauciflora (F. Muell.) Benth. (*Lepidosperma pauciflorum* F. Muell.)
English: needle bog-rush

Tricuspidaria Ruíz & Pavón Elaeocarpaceae

Origins:
From the Latin *tricuspis* "having three points, three-pointed."

Tricuspis P. Beauv. Gramineae

Origins:
Latin *tricuspis* "having three points, three-pointed, three-tined"; see also *Tridens* Roemer & Schultes.

Tricycla Cav. Nyctaginaceae

Origins:
Greek *treis, tria* "three" and *kyklos* "a circle, ring."

Tricyclandra Keraudren Cucurbitaceae

Origins:
From the Greek *treis, tria* "three," *kyklos* "a circle, ring" and *aner, andros* "male, a man, stamen."

Tricyrtis Wallich Liliaceae (Convallariaceae)

Origins:
Greek *treis, tria* and *kyrtos* "curved, arched, humped," referring to the three outer petals; see Brian Mathew, "A review of the genus *Tricyrtis*." in *The Plantsman*. 6(4): 193-224. March 1985.

Species/Vernacular Names:
T. hirta (Thunb.) Hook.
English: Japanese toad lily

Tridactyle Schltr. Orchidaceae

Origins:
From Greek *treis* "three" and *daktylos* "a finger," indicating the more or less trilobed lip.

Species/Vernacular Names:
T. gentilii (De Wild.) Schltr. (*Angraecum gentilii* De Wild.) (the species honors Ambroise Gentil, 1842-1929, first discoverer of the taxon in the forests of Haut Lomami in Zaire)

Tridactylina (DC.) Sch. Bip. Asteraceae

Origins:
The diminutive of the Greek *treis* "three" and *daktylos* "a finger."

Tridax L. Asteraceae

Origins:
Ancient name for another plant, from the Greek *tridaknos* "thrice-bitten, eaten-in-three-bites"; Latin *thridax, acis* and Greek *thridax* "a kind of wild lettuce"; Latin *tridacna* "a kind of oysters"; see Carl Linnaeus, *Species Plantarum*. 900. 1753 and *Genera Plantarum*. Ed. 5. 382. 1754; Helmut

Genaust, *Etymologisches Wörterbuch der botanischen Pflanzennamen*. 655. 1996.

Species/Vernacular Names:
T. procumbens L.
English: daisy, tridax daisy, coat buttons, wild daisy
South Africa: aster
Yoruba: iyo esin, aaragba
Japan: kôtobuki-giku
Malaya: rumput kanching baju
Central America: hierba del toro, cadillo, chisaaca, curagusano, hierba de San Juan, romerillo, San Juan del monte

Tridens Roemer & Schultes Gramineae

Origins:
Latin *tridens, tridentis* "a trident, three-pronged," referring to the lemma tip.

Species/Vernacular Names:
T. muticus (Torrey) Nash
English: slim tridens

Tridentea Haw. Asclepiadaceae

Origins:
Latin *tridens, tridentis* "a trident, three-pronged."

Tridesmis Lour. Euphorbiaceae

Origins:
Greek *treis* "three" and *desmos* "a bond," referring to the seeds.

Tridesmostemon Engl. Sapotaceae

Origins:
Greek *treis, tris* "three," *desmos* "a bond" and *stemon* "stamen."

Species/Vernacular Names:
T. claessensii De Wild.
Central Africa: ekouekoue, babama, tuba
T. omphalocarpoides Engl.
Central Africa: ekouekoue, babama, tuba
Cameroon: babama, tuba
Gabon: ekouekoue

Tridimeris Baillon Annonaceae

Origins:

Greek *treis, tris* "three" and *dimeres* "bipartite."

Tridynamia Gagnepain Convolvulaceae

Origins:

Greek *dynamis* "power, a store," *tridynamos* "of three powers."

Species/Vernacular Names:

T. sinensis (Hemsley) Staples var. *sinensis* (*Porana sinensis* Hemsley)

English: Chinese porana

China: da guo san chi teng (yuan bian zhong)

Trieenia Hilliard Scrophulariaceae

Origins:

For Elsie Elizabeth Esterhuysen, b. 1912, botanist at the Bolus Herbarium, botanical collector, wrote "Regeneration after clearing at Kirstenbosch." *Journ. S. Afr. Bot.* 2: 177-185. 1936.

Trientalis L. Primulaceae

Origins:

Latin *trientalis, e* "that contains a third of a foot."

Species/Vernacular Names:

T. arctica Hook.

English: Arctic starflower, northern starflower

T. europaea Hook.

English: chickweed wintergreen, European starflower

China: qi ban lian

Trifidacanthus Merr. Fabaceae

Origins:

Latin *trifidus* "three-cleft, triangular" and Greek *akantha* "thorn."

Trifoliada Rojas Acosta Oxalidaceae

Origins:

Latin *tres, treis, tris, tri-* "three" and *folium, ii* "a leaf."

Trifolium L. Fabaceae

Origins:

Latin *trifolium* "three-leaved grass, trefoil" (Plinius); see Carl Linnaeus, *Species Plantarum*. 764. 1753 and *Genera Plantarum*. Ed. 5. 337. 1754.

Species/Vernacular Names:

T. alexandrinum L.

English: berseem clover, Egyptian clover

T. alpestre L.

English: mountain zigzag clover, purple globe clover

T. amabile Kunth

English: Aztec clover

T. ambiguum M. Bieb.

English: Kura clover

T. amoenum E. Greene

English: showy Indian clover

T. angustifolium L.

English: white clover, narrow-leaved clover

T. arvense L.

English: old-field clover, rabbit-foot clover, stone clover

T. aureum Pollich (*Trifolium strepens* Crantz, nom. illeg.)

English: hop clover

T. burchellianum Ser. subsp. *johnstonii* (Oliver) J.B. Gillett (*Trifolium basileianum* Chiov.; *Trifolium burchellianum* Ser. var. *johnstonii* (Oliver) J.B. Gillett; *Trifolium johnstonii* Oliver)

English: Uganda clover

T. campestre Schreber

English: large hop clover, hop clover, low clover, low hop clover, yellow clover

South Africa: geelklawer

T. carolinianum Michaux

English: Carolina clover

T. cernuum Brot. (*Trifolium perrymondii* Gren. & Godron)

English: drooping-flowered clover, drooping-flower clover, nodding clover, drooping clover

T. dubium Sibth. (*Trifolium filiforme* L. var. *dubium* (Sibth.) Fiori; *Trifolium minus* Smith; *Trifolium parviflorum* Bunge ex Nyman, nom. illeg.; *Trifolium procumbens* L., nom. ambig.)

English: small hop clover, cow hop clover, little hop clover, Irish shamrock, least hop clover, shamrock, suckling clover, yellow clover, yellow suckling clover

South Africa: kleingeelklawer

T. fragiferum L. (*Trifolium bonannii* J.S. Presl & C. Presl; *Trifolium fragiferum* L. subsp. *bonannii* (J.S. Presl & C. Presl) Soják; *Trifolium neglectum* C. Meyer)

English: strawberry clover

T. glomeratum L.

English: cluster clover, ball clover

T. hirsutum All. (*Trifolium hispidum* Desf.)

English: rose clover

T. hybridum L.

English: Alsike clover (from Alsike near Uppsala)

T. incarnatum L.

English: crimson clover, Italian clover

T. lappaceum L.

English: lappa clover, burdock clover, bristly clover, burr-clover

T. medium L.

English: zigzag clover

T. michelianum Savi

English: bigflower clover

T. nigrescens Viv.

English: ball clover

T. pannonicum Jacq.

English: Hungarian clover

T. polymorphum Poiret (*Trifolium amphianthum* Torrey & A. Gray; *Trifolium megalanthum* Steudel)

English: peanut clover

T. pratense L.

English: red clover

Japan: murasaki-tsume-kusa, ake-tsume-kusa

China: hong che zhou cao

Mexico: guie tuu xtilla, quie too castilla

T. reflexum L.

English: buffalo clover

T. repens L.

English: Dutch clover, white Dutch clover, ladino clover, white clover

South Africa: Brabantse klawer, geelklawer, witteklawer

Japan: shiro-tsume-kusa

China: san xiao cao

T. resupinatum L. (= bent back, referring to the corolla)

English: Persian clover, reversed clover, Shaftal clover

Arabic: qort

T. retusum L.

English: teasel clover

T. scabrum L.

English: rough clover

T. semipilosum Fresen.

English: Kenya clover, Kenya wild white clover

T. stellatum L.

English: star clover

T. stoloniferum Muhlenb. ex Eaton

English: running buffalo clover

T. subterraneum L.

English: sub clover, subterranean clover

T. tomentosum L.

English: woolly clover, woolly-headed clover

T. variegatum Nutt.

English: whitetip clover

T. vesiculosum Savi

English: arrowleaf clover

Trifurcia Herbert Iridaceae

Origins:
Latin *trifurcium, ii* "anything of a three-forked shape."

Trigastrotheca F. Muell. Molluginaceae

Origins:
Greek *treis* "three," *gaster* "belly, paunch" and *theke* "case"; see *Hooker's Journal of Botany & Kew Garden Miscellany.* 9: 16. (Jan.) London 1857.

Triglochin L. Juncaginaceae

Origins:
Greek *treis* plus *glochin, glochis* "a point, projecting point," *triglochis, triglochinos* "three-barbed," the capsules or carpels of some species have three angles or points; see Carl Linnaeus, *Species Plantarum.* 338. 1753 and *Genera Plantarum.* Ed. 5. 157. 1754.

Species/Vernacular Names:
T. bulbosa L. (*Triglochin compacta* Adamson; *Triglochin elongata* Buchen.; *Triglochin tenuifolia* Adamson)

English: arrowgrass

Southern Africa: iNqoba yomgxobhozo (Xhosa)

T. concinna Burtt Davy

English: arrowgrass

T. maritima L.

English: seaside arrowgrass

T. palustris L.

English: marsh arrowgrass

T. striata Ruíz & Pav.

English: three-ribbed arrowgrass

Triglossum Roemer & Schultes Gramineae

Origins:
Greek *treis* "three" and *glossa* "a tongue."

Trigonachras Radlk. Sapindaceae

Origins:
Greek *trigonos* (*tri*- "three," *gonia* "angle") and *achras* "a kind of wild pear," a three-angled pear.

Species/Vernacular Names:
T. acuta (Hiern) Radlk.

English: cardinal tree

Malaya: serengkan, tangisong burong

Trigonanthe (Schltr.) Brieger Orchidaceae

Origins:
From the Greek *trigonos* and *anthos* "flower."

Trigonella L. Fabaceae

Origins:
Diminutive of the Latin *trigonum, i* "a triangle," *trigonus, a, um* "three-cornered," Greek *trigonos*, referring to the appearance of the corolla of *Trigonella foenum-graecum* L.; see Carl Linnaeus, *Species Plantarum*. 776. 1753 and *Genera Plantarum*. Ed. 5. 338. 1754.

Species/Vernacular Names:
T. foenum-graecum L.

English: fenugreek, Greek clover, Greek hay

Arabic: helba, hulba

Tibetan: shu-mo-za

China: hu lu ba, hu lu pa, ku tou

T. hamosa L.

English: Egyptian fenugreek

Arabic: eshb el-malik

T. laciniata L.

English: jagged fenugreek

Trigonia Aublet Trigoniaceae (Rosidae, Polygalales)

Origins:
Greek *trigonos* (*tri* –"three" and *gonia* "angle"), the fruit is a septicidal capsule.

Trigoniastrum Miq. Trigoniaceae

Origins:
Referring to the genus *Trigonia* Aubl. plus the Latin substantival suffix *astrum*, the fruit is a three-winged samara.

Trigonidium Lindley Orchidaceae

Origins:
Greek *trigonos* and *eidos*, *oides* "resemblance," referring to the nature of sepals, stigma and viscidium, indicating the triangular form of the floral parts.

Trigoniodendron E.F. Guim. & J.R. Miguel Trigoniaceae

Origins:
Greek *trigonos* and *dendron* "tree."

Trigonobalanus Forman Fagaceae

Origins:
Greek *trigonos* "three-cornered, triangular" and *balanos* "acorn," the fruits are conspicuously three-winged; see L.L. Forman, in *Kew Bulletin*. 17(3): 381-396. 1964; D.F. Cutler, in *Kew Bulletin*. 17(3): 401-409. 1964.

Species/Vernacular Names:
T. sp.

English: trig oak

Trigonocapnos Schltr. Fumariaceae (Papaveraceae)

Origins:
Greek *trigonos* "three-cornered, triangular" and *kapnos* "smoke."

Trigonocaryum Trautv. Boraginaceae

Origins:
From the Greek *trigonos* and *karyon* "nut."

Trigonochlamys Hook.f. Burseraceae

Origins:

From the Greek *trigonos* plus *chlamys, chlamydos* "cloak, mantle."

Trigonophyllum (Prantl) Pic. Serm. Hymenophyllaceae

Origins:

Greek *trigonos* "three-cornered, triangular" and *phyllon* "leaf"; see R.E.G. Pichi Sermolli, "Fragmenta Pteridologiae–VI." in *Webbia*. 31(1): 237-259. 1977.

Trigonopleura Hook.f. Euphorbiaceae

Origins:

Greek *trigonos* and *pleura, pleuro, pleuron* "rib."

Trigonopyren Bremek. Rubiaceae

Origins:

From the Greek *trigonos* with *pyren* "a kernel, a fruit stone."

Trigonosciadium Boiss. Umbelliferae

Origins:

Greek *trigonos* "three-cornered, triangular" and *skiadion, skiadeion* "umbel, parasol."

Trigonospermum Less. Asteraceae

Origins:

From the Greek *trigonos* and *sperma* "seed."

Trigonospora Holttum Thelypteridaceae

Origins:

From the Greek *trigonos* "three-cornered, triangular" and *spora* "seed, spore, offspring"; see also *Cyclosorus* Link.

Trigonostemon Blume Euphorbiaceae

Origins:

Greek *trigonos* "three-cornered, triangular" and *stemon* "thread, stamen"; see Karl Ludwig von Blume (1796-1862),

Bijdragen tot de flora van Nederlandsch Indië. 600. Batavia 1825-1826.

Trigonotheca Sch. Bip. Asteraceae

Origins:

Greek *trigonos* "three-cornered, triangular" and *theke* "case."

Trigonotis Steven Boraginaceae

Origins:

Latin *trigonum, i* "a triangle," *trigonus, a, um* "three-cornered, triangular," referring to the fruits; Greek *trigonos* and *ous, otos* "an ear."

Species/Vernacular Names:

T. cavaleriei (H. Léveillé) Handel-Mazzetti

English: Cavalerie trigonotis

China: xi nan fu di cai

T. compressa I.M. Johnston

English: narrow-leaf trigonotis

China: xia ye fu di cai

T. floribunda I.M. Johnston

English: manyflowers trigonotis

China: duo hua fu di cai

T. laxa I.M. Johnston

English: Nanchuan trigonotis

China: nan chuan fu di cai

T. macrophylla Vaniot

English: large-leaf trigonotis

China: da ye fu di cai

T. omeiensis Matsuda

English: Omei Mountain trigonotis

China: e mei fu di cai

T. peduncularis (Trevir.) Benth. ex Baker & Moore (*Myosotis peduncularis* Trevir.)

English: pedunculate trigonotis

Japan: kiuri-gusa, tabirako

China: fu di cai, chi chang tsao

Trigynaea Schltdl. Annonaceae

Origins:

Greek *treis, tria* "three" and *gyne* "a woman, female, pistil," referring to the number of pistils.

Trigyneia Reichb. Annonaceae

Origins:
Greek *treis*, *tria* "three" and *gyne* "a woman, female, pistil"; see also *Trigynaea* Schltdl.

Trigynia Jacq.-Fél. Melastomataceae

Origins:
From the Greek *tri-*, *treis* and *gyne* "a woman, female, pistil."

Trikalis Raf. Chenopodiaceae

Origins:
See E.D. Merrill, *Index rafinesquianus*. The plant names published by C.S. Rafinesque, etc. 118. Jamaica Plain, Massachusetts, USA 1949; Constantine Samuel Rafinesque, *Flora Telluriana*. 3: 47. 1836 [1837].

Trikeraia Bor Gramineae

Origins:
Greek *treis*, *tria*, *tri-* "three" and *keras* "horn," referring to the nature of the spikelets.

Trilepidea Tieghem Loranthaceae

Origins:
From the Greek *treis*, *tria* and *lepis*, *lepidos* "a scale."

Trilepis Nees Cyperaceae

Origins:
Greek *treis* plus *lepis*, *lepidos* "a scale."

Trilepisium Thouars Moraceae

Origins:
Greek *treis* "three" and *lepis* "scale," in reference to the scales among the stamen and the pistil.

Species/Vernacular Names:
T. madagascariense DC. (*Bosqueia angolensis* (Welw.) Ficalho; *Bosqueia phoberos* Baill.)

Southern Africa: muphopha-madi (= exudes water in drops), muvhadela-phanga (Venda)

Angola: muguengue ia muxito (Kimbundu); okakuana, muholo (Umbundu); nsekenhe, munkenhe (Kikongo)

Zaire: kid; agbute (Azande); bofonge (Lokundu); bonge, bonke, boonge (Turumbu); bopongi (Tumba); kalanda kakunze (makunze, kakunze = red, in allusion to the latex) (Tshiluba); mukobakoba mukunze (Bakuba); mulamba, mupelenge (Tshofa); mungagnya, sacagne (Mayumbe); nsuga, n'zungwa (Bambesa); sehenie (Kiombe); the seeds are: mangeria, muscagna (Mayumbe) and tofiolio (Turumbu)

Cameroon: arasalia, dingbimi, fongi, osombo, osomzo, pongi

Congo: kid

Yoruba: apongbe, gangaran, saworo, ikobe

Nigeria: koko eram, gangaran, saworo (Yoruba; see S.O. Biobaku, *The Origin of the Yoruba*. Lagos 1955; S.O. Biobaku, ed., *Sources of Yoruba History*. Oxford 1973; Rev. Samuel Johnson, *The History of the Yorubas*. Lagos 1921; William Bascom, *Ifa Divination. Communication between Gods and Men in West Africa*. Bloomington 1969; Pierre Fatumbi Verger, *Ewé: The Use of Plants in Yoruba Society*. São Paulo 1995; Celia Blanco, *Santeria Yoruba*. Caracas 1995; William W. Megenney, *A Bahian Heritage*. University of North Carolina at Chapel Hill 1978); ukputu (Edo); igobe (Itsekiri); otukhurhu (Urhobo); binu (Ijwa); oze (Igbo); katep nkim (Boki)

Ethiopia: juya, shoro (Orominya)

Ivory Coast: béhio, daocou, nantié, n'kuihia paa

Trilepta Raf. Verbenaceae

Origins:
Greek *leptos* "slender, small"; see E.D. Merrill, *Index rafinesquianus*. The plant names published by C.S. Rafinesque, etc. 205. Jamaica Plain, Massachusetts, USA 1949; Constantine Samuel Rafinesque, *Sylva Telluriana*. 81. Philadelphia 1838.

Triliena Raf. Solanaceae

Origins:
See E.D. Merrill, *Index rafinesquianus*. The plant names published by C.S. Rafinesque, etc. 213. Jamaica Plain, Massachusetts, USA 1949; Constantine Samuel Rafinesque (1783-1840), *Sylva Telluriana*. 54. Philadelphia 1838.

Trilisa (Cass.) Cass. Asteraceae

Origins:
Probably an anagram of the genus *Liatris* Gaertner ex Schreber.

Trillidium Kunth Trilliaceae (Liliaceae)

Origins:
See *Trillium* L.

Trillium L. Trilliaceae (Liliaceae)

Origins:
Latin *tri-* "triple, three"; see F. Boerner & G. Kunkel, *Taschenwörterbuch der botanischen Pflanzennamen*. 4. Aufl. 181. 1989; Georg Christian Wittstein, *Etymologisch-botanisches Handwörterbuch*. 898f. Ansbach 1852; Helmut Genaust, *Etymologisches Wörterbuch der botanischen Pflanzennamen*. 657. Basel 1996.

Species/Vernacular Names:
T. chloropetalum (Torrey) Hovell
English: giant trillium

T. erectum L.
English: birthroot, stinking Benjamin

T. ovatum Pursh
English: white trillium, western trillium

T. rivale S. Watson
English: Brook wakerobin

Trilobachne Schenck ex Henrard Gramineae

Origins:
Greek *treis, tria* "three," *lobos* "pod, lobe" and *achne* "chaff, glume," referring to the lobed glumes.

Trilobulina Raf. Lentibulariaceae

Origins:
From the Greek *treis, tria* "three" and *lobos* "pod, lobe," the diminutive; see Constantine Samuel Rafinesque, *Flora Telluriana*. 4: 110. 1836 [1838]; E.D. Merrill, *Index rafinesquianus*. 222. Jamaica Plain, Massachusetts, USA 1949.

Trilomisa Raf. Begoniaceae

Origins:
See Constantine Samuel Rafinesque, *Flora Telluriana*. 2: 91. 1836 [1837]; E.D. Merrill, *Index rafinesquianus*. 171. 1949.

Trilopus Mitch. Hamamelidaceae

Origins:
Possibly from the Greek *treis, tria* "three" and *lopos* "a covering, husk, bark, peel, hide."

Trimenia Seemann Trimeniaceae

Origins:
After the British (b. London) botanist Henry Trimen, 1843-1896 (d. Peradeniya), 1866 Fellow of the Linnean Society, from 1879 to 1896 Director of the Peradeniya Botanic Gardens in Sri Lanka, 1888 Fellow of the Royal Society, plant collector. Among his many works are *Hortus zeylanicus*. Colombo 1888 and *A Hand-Book to the Flora of Ceylon*. London 1893-1931, editor of the *Journal of Botany*, with Robert Bentley (1821-1893) wrote *Medicinal Plants*. London [1875]-1880. He was brother of the British (b. London) entomologist Roland Trimen (1840-1916, d. Epsom, Surrey); see J.H. Barnhart, *Biographical Notes upon Botanists*. 3: 401. 1965; R. Zander, F. Encke, G. Buchheim and S. Seybold, *Handwörterbuch der Pflanzennamen*. 14. Aufl. Stuttgart 1993; B. Henrey, *British Botanical and Horticultural Literature before 1800*. Oxford 1975; Andrew Thomas Gage, *A History of the Linnean Society of London*. London 1938; T. W. Bossert, *Biographical Dictionary of Botanists Represented in the Hunt Institute Portrait Collection*. 406. Boston, Massachusetts 1972; H.N. Clokie, *Account of the Herbaria of the Department of Botany in the University of Oxford*. 256. Oxford 1964; E.M. Tucker, *Catalogue of the Library of the Arnold Arboretum of Harvard University*. Cambridge, Massachusetts 1917-1933; Isaac Henry Burkill (1870-1965), *Chapters on the History of Botany in India*. Delhi 1965; Ernest Nelmes and William Cuthbertson, *Curtis's Botanical Magazine Dedications, 1827-1927*. [1931]; H. Dolezal, in *Portug. Acta Biol*. 6: 257-323. 1959 and 7: 324-551. 1961; Miles Hadfield et al., *British Gardeners: A Biographical Dictionary*. London 1980; Ray Desmond, *Dictionary of British & Irish Botanists and Horticulturists*. 692. London 1994; R. Desmond, *The European Discovery of the Indian Flora*. Oxford 1992; Mary Gunn and Leslie Edward W. Codd, *Botanical Exploration of Southern Africa*. 351-352. Cape Town 1981; Stafleu and Cowan, *Taxonomic Literature*. 6: 489-492. Utrecht 1986.

Trimeranthus Karst. Melastomataceae

Origins:
Greek *treis, tria, tri-* "three," *meros* "part" and *anthos* "flower."

Trimeria Harvey Flacourtiaceae

Origins:
Greek *trimereia* "division into three parts," referring to the flowers.

Species/Vernacular Names:
T. grandifolia (Hochst.) Warb. (*Trimeria alnifolia* (Hook.) Harv.; *Trimeria rotundifolia* (Hochst.) Gilg; *Monospora grandifolia* Hochst.; *Monospora rotundifolia* Hochst.)

English: wild mulberry, big ears, big-leafed trimeria, elephant's ears

Southern Africa: wildemoerbei, grootblaarysterhout, lindeboom, olifants oor; iDlebenlendlovu (Zulu); iDlebenlendlovu, iTabatane, iGqabi, iliTye, iXabelo, iGqabela, umNqabane (Xhosa); isiCandamashane, maHlebe (Swazi); mufhanza, muhasha-phande (Venda)

T. trinervis Harv.

English: small-leaved wild mulberry, trimeria, three-nerved trimeria

Southern Africa: fynblaarwildemoerbei; iDlebenlendlovu (Zulu); iKomanci (Xhosa)

Trimeris C. Presl Campanulaceae

Origins:
Greek *trimeres* "tripartite, threefold."

Trimeriza Lindl. Aristolochiaceae

Origins:
Greek *trimerizo* "divide into three parts."

Trimerocalyx (Murb.) Murb. Scrophulariaceae

Origins:
Greek *treis, tria* "three," *meris* "a portion, part" and *kalyx* "calyx."

Trimezia Salisb. ex Herbert Iridaceae

Origins:
Greek *treis, tria* "three" and *meizon* "greater," *megas* "big, large, great," the outer perianth segments are bigger than the inner.

Trimorpha Cass. Asteraceae

Origins:
Greek *treis, tria* "three," *morphe* "a form, shape," referring to the flowers.

Species/Vernacular Names:
T. acris (L.) S.F. Gray var. *debilis* (A. Gray) G. Nesom
English: northern daisy

Trimorphandra Brongn. & Gris Dilleniaceae

Origins:
Greek *treis, tria* "three," *morphe* "a form, shape" and *aner, andros* "male, stamen."

Trineuria Presl Fabaceae

Origins:
From the Greek *treis, tria* "three" and *neuron* "nerve."

Trineuron Hook.f. Asteraceae

Origins:
From the Greek *treis, tria* "three" and *neuron* "nerve."

Trinia Hoffm. Umbelliferae

Origins:
For the German botanist Carl (Karl) Bernhard Freiherr von Trinius, 1778-1844, physician, traveler, agrostologist, wrote *Fundamenta Agrostographiae.* Viennae 1820; see Stafleu and Cowan, *Taxonomic Literature.* 6: 493-497. 1986; John H. Barnhart, *Biographical Notes upon Botanists.* 3: 401. 1965; T. W. Bossert, *Biographical Dictionary of Botanists Represented in the Hunt Institute Portrait Collection.* 406. Boston, Massachusetts 1972; Ida Kaplan Langman, *A Selected Guide to the Literature on the Flowering Plants of Mexico.* Philadelphia 1964; Antoine Lasègue, *Musée botanique de M. Benjamin Delessert.* 1845; S. Lenley et al., *Catalog of the Manuscript and Archival Collections and Index to the Correspondence of John Torrey.* Library of the New York Botanical Garden. 1973; G. Schmid, *Goethe und die Naturwissenschaften.* Halle 1940; Mariella Azzarello Di Misa, ed., *Il Fondo Antico della Biblioteca dell'Orto Botanico di Palermo.* 274-277. Palermo 1988; Günther Schmid, *Chamisso als Naturforscher. Eine Bibliographie.* Leipzig 1942; R. Zander, F. Encke, G. Buchheim and S. Seybold, *Handwörterbuch der Pflanzennamen.* 14. Aufl. 791. Stuttgart 1993; F. Boerner & G. Kunkel, *Taschenwörterbuch der botanischen Pflanzennamen.* 4. Aufl. 181. Berlin & Hamburg 1989; E.M. Tucker, *Catalogue of the Library of*

the Arnold Arboretum of Harvard University. Cambridge, Massachusetts 1917-1933; G.C. Wittstein, *Etymologisch-botanisches Handwörterbuch*. 899. Ansbach 1852; Helmut Genaust, *Etymologisches Wörterbuch der botanischen Pflanzennamen*. 657. [d. 1841] Basel 1996.

Species/Vernacular Names:
T. glauca (L.) Dumort.
English: honewort

Triniella Calest. Umbelliferae

Origins:
The diminutive of *Trinia* Hoffm.

Triniochloa Hitchc. Gramineae

Origins:
For the German botanist Carl Bernhard Trinius, 1778-1844.

Triniusa Steudel Gramineae

Origins:
For the German botanist Carl Bernhard Trinius, 1778-1844.

Triodanis Raf. Campanulaceae

Origins:
From the Greek *treis, tria* "three" and *odous, odontos* "tooth," referring to the divided calyx; see Edward Lee Greene (1843-1915), *Manual of the Botany of the Region of San Francisco Bay*. 230. San Francisco 1894; C.S. Rafinesque, *New Fl. N. Am.* 4: 67. 1836 [1838] and *Autikon botanikon*. Icones plantarum select. nov. vel rariorum, etc. 16. Philadelphia 1840; E.D. Merrill, *Index rafinesquianus*. 231. 1949.

Species/Vernacular Names:
T. biflora (Ruíz & Pavón) Greene (*Campanula biflora* Ruíz & Pavón; *Specularia biflora* (Ruíz & Pavón) Fisch. & C.A. Mey.; *Triodanis perfoliata* (L.) Nieuwl. var. *biflora* (Ruíz & Pavón) Bradley)
English: Venus' looking glass

Triodia R. Br. Gramineae

Origins:
Greek *treis, tria* "three" and *odous, odontos* "tooth," referring to the divided lemma; see Robert Brown, *Prodromus florae Novae Hollandiae et Insulae van-Diemen*. 182. London 1810;

M. Lazarides, "New taxa of tropical Australian grasses (Poaceae)." in *Nuytsia*. 5(2): 273-303. 1984.

Species/Vernacular Names:
T. basedowii E. Pritzel
English: lobed spinifex, hard spinifex, spinifex
T. longiceps J. Black
English: black spinifex

Triodoglossum Bullock Asclepiadaceae (Periplocaceae)

Origins:
Greek *treis, tria* "three," *odous, odontos* "tooth" and *glossa* "tongue."

Triodon Baumg. Gramineae

Origins:
From the Greek *treis, tria* and *odous, odontos* "tooth," see *Triodia* R. Br.

Triodon DC. Rubiaceae

Origins:
Greek *treis, tria* "three" and *odous, odontos* "tooth," see also *Diodia* L.

Triolena Naudin Melastomataceae

Origins:
Greek *treis, tria* "three" and *olene* "an arm, the elbow, bundle" or *chlaena, laina* "cloak, blanket," referring to the three thread-like appendages on the stamens, the fruit is a three-winged capsule; see F. Boerner & G. Kunkel, *Taschenwörterbuch der botanischen Pflanzennamen*. 4. Aufl. 181. 1989; H. Genaust, *Etymologisches Wörterbuch der botanischen Pflanzennamen*. 657. 1996.

Triomma Hook.f. Burseraceae

Origins:
Greek *tri-* and *omma, ommatos* "eye, light."

Trioncinia (F. Muell.) Veldkamp Asteraceae

Origins:
From the Greek *treis, tria* "three" and *onkinos* "hook, hooked," *onkos* "tumor, tubercle."

Triopteris L. Malpighiaceae

Origins:
See *Triopterys* L.

Triopterys L. Malpighiaceae

Origins:
From the Greek *treis*, *tria* "three" and *pteron* "wing," an allusion to the winged samaras.

Triorchis Agosti Orchidaceae

Origins:
From the Greek *treis*, *tria* plus *orchis*, *orchidos* "a testicle, an orchid," referring to the tubers.

Triorchos Small & Nash Orchidaceae

Origins:
From the Greek *triorches*, *triorchos* "with three testicles, buzzard," indicating the lip.

Triosteum L. Caprifoliaceae

Origins:
From the Greek *treis*, *tria* "three" and *osteon* "bone," referring to the fruit.

Species/Vernacular Names:
T. perfoliatum L.
English: fever root, tinker's weed

Tripetalanthus A. Chev. Caesalpiniaceae

Origins:
Greek *treis*, *tria* "three," *petalon* "petal" and *anthos* "flower."

Tripetaleia Siebold & Zucc. Ericaceae

Origins:
From the Greek *treis*, *tria* and *petalon* "petal," referring to the corolla.

Tripetalum K. Schumann Guttiferae

Origins:
From the Greek *treis*, *tria* "three" and *petalon* "petal."

Triphasia Lour. Rutaceae

Origins:
Greek *triphasios* "triple, threefold" (*tri-* and *phasis* "an appearance, apparition"), referring to the parts of the flowers; see João de Loureiro, 1717-1791, *Flora cochinchinensis: sistens plantas in regno Cochinchina nascentes*. 1: 96, 152. Ulyssipone [Lisboa] 1790.

Species/Vernacular Names:
T. trifolia (Burm.f.) P. Wilson (*Triphasia aurantiola* Lour.)
English: lime berry, myrtle lime
Malaya: limau kiah, limau kaya, limau kikir, limau kelingket
The Philippines: limoncito, kalamansito, limonsitong-kastila, suang-kastila, sua-sua, dayap, kamalitos, tagimunao

Triphelia R. Brown ex Endlicher Myrtaceae

Origins:
Presumably from the Greek *tryphaleia*, *triphaleia* "helmet."

Triphlebia Stapf Gramineae

Origins:
From the Greek *treis*, *tria* and *phleps*, *phlebos* "vein."

Triphlebium Baker Aspleniaceae

Origins:
From the Greek *treis*, *tria* and *phleps*, *phlebos* "vein."

Triphora Nutt. Orchidaceae

Origins:
Greek *treis*, *tria* "three" and *phoros* "bearing, carrying," referring to the flowers.

Triphylleion Suess. Umbelliferae

Origins:
Greek *treis*, *tria* "three" and *phylleion* "leafage, green stuff, small herbs, radish-tops."

Triphyophyllum Airy Shaw Dioncophyllaceae (Dilleniidae, Nepenthales)

Origins:
From the Greek *triphyes* "of threefold nature, threefold, with three stems, of threefold form" and *phyllon* "leaf."

Triphysaria Fischer & C. Meyer Scrophulariaceae

Origins:
Greek *tri-* "three-" and *physa* "bladder," referring to the pouches of the lower lip.

Species/Vernacular Names:
T. eriantha (Benth.) Chuang & Heckard
English: butter-and-eggs, Johnny-tuck
T. floribunda (Benth.) Chuang & Heckard
English: San Francisco owl's-clover

Triplachne Link Gramineae

Origins:
From the Greek *triploos* "triple, threefold" and *achne* "chaff, glume."

Tripladenia D. Don Convallariaceae (Liliaceae)

Origins:
From the Greek *triploos* "triple, threefold" and *aden* "gland," referring to the basal appendages of the perianth; see *Proceedings of the Linnean Society of London*. 1: 46. 1839.

Triplandra Raf. Euphorbiaceae

Origins:
Greek *triploos* "triple, threefold" and *aner, andros* "man, male," referring to the stamens; see C.S. Rafinesque, *Sylva Telluriana*. 62, 63. 1838.

Triplarina Raf. Myrtaceae

Origins:
Latin *triplaris* "triple, threefold," referring to the flowers; see Constantine Samuel Rafinesque (1783-1840), *Sylva Telluriana*. 104. Philadelphia 1838; Arthur D. Chapman, ed., *Australian Plant Name Index*. 2909. ["Combination made by indirect reference to the basionym."] Canberra 1991; E.D. Merrill, *Index rafinesquianus*. The plant names published by C.S. Rafinesque, etc. 174. Jamaica Plain, Massachusetts, USA 1949.

Triplaris Loefl. ex L. Polygonaceae

Origins:
Latin *triplaris* "triple, threefold," referring to the calyx, stamens and styles.

Triplasandra Seem. Araliaceae

Origins:
Greek *triplasios* "thrice as many, triple" and *aner, andros* "man, stamen"; see also *Tetraplasandra* A. Gray.

Triplasis P. Beauv. Gramineae

Origins:
Latin *triplasius* and Greek *triplasios* "threefold, thrice as many, triple," referring to the arrangement of the flowers.

Triplathera (Endl.) Lindley Gramineae

Origins:
From the Greek *triploos* "triple, threefold" and *ather* "stalk, barb."

Triplectrum Wight & Arn. Melastomataceae

Origins:
Greek *treis, tria* "three" and *plektron* "a spur, cock's spur."

Tripleura Lindl. Orchidaceae

Origins:
From the Greek *treis, tria* "three" and *pleura, pleuro, pleuron* "rib," referring to the keeled sepals.

Tripleurospermum Sch. Bip. Asteraceae

Origins:
From the Greek *treis, tria* "three," *pleura, pleuro, pleuron* "rib" and *sperma* "seed," referring to the achenes.

Triplisomeris Aubrév. & Pellegrin Caesalpiniaceae

Origins:
Latin *triplus* "triple," Greek *isos* "equal" and *meris* "a part, portion."

Triplocephalum O. Hoffm. Asteraceae

Origins:

Greek *triploos* "triple, threefold" and *kephale* "head."

Triplochiton K. Schumann Sterculiaceae

Origins:

From the Greek *triploos* "triple, threefold" and *chiton* "a tunic, covering," referring to the flowers; see Y. Tailfer, *La Forêt dense d'Afrique centrale*. CTA, Ede/Wageningen 1989; J. Vivien & J.J. Faure, *Arbres des Forêts denses d'Afrique Centrale*. Agence de Coopération Culturelle et Technique. Paris 1985.

Species/Vernacular Names:

T. scleroxylon K. Schumann

English: whitewood

Yoruba: arere, igioro

Nigeria: arere, asigka, ayous, digiten, eba-lebe, egin-fifen, ewowo, nkom, obeche, okp, opobo, samba, uvori, lomangeni, di, dee, nwelechi, wawa, eruku; arere (Yoruba); dihitin (Ijaw); obeche (Edo); askikga (Etsako); ewowo (Urhobo); egin-fifen (Itsekiri); uvori (Engenni); okpobo (Igbo)

Central Africa: mbado, bado

Ghana: wawa

Ivory Coast: samba

Cameroon: ngo, ayos, ayous, mbado

Congo: eguess

Gabon: ayous

Triplochlamys Ulbr. Malvaceae

Origins:

Greek *triploos* "triple, threefold" and *chlamys, chlamydos* "cloak, mantle."

Triplolepis Turcz. Asclepiadaceae

Origins:

From the Greek *triploos* "triple, threefold" and *lepis* "scale."

Triplopetalum Nyarady Brassicaceae

Origins:

From the Greek *triploos* and *petalon* "petal, leaf."

Triplophyllum Holttum Dryopteridaceae

Origins:

Greek *triploos* and *phyllon* "leaf."

Triplopogon Bor Gramineae

Origins:

From the Greek *triploos* plus *pogon* "beard."

Triplorhiza Ehrh. Orchidaceae

Origins:

From the Greek *triploos* with *rhiza* "root," the digitate tuber; see also *Pseudorchis* Séguier.

Triplostegia Wallich ex DC. Valerianaceae (Triplostegiaceae)

Origins:

From the Greek *triploos* "triple, threefold" and *stege, stegos* "roof, cover."

Triplotaxis Hutch. Asteraceae

Origins:

From the Greek *triploos* and *taxis* "a series, order, arrangement."

Tripodandra Baillon Menispermaceae

Origins:

Greek *tripodeios, tripodios* "three-footed" and *aner, andros* "man, male organs."

Tripodanthera M. Roemer Cucurbitaceae

Origins:

From the Greek *tripodeios, tripodios* "three-footed" and *anthera* "anther."

Tripodanthus (Eichler) Tieghem Loranthaceae

Origins:

Greek *tripodeios, tripodios* and *anthos* "flower."

Tripodion Medikus Fabaceae

Origins:
Greek *tripodeios*, *tripodios* "three-footed," *podion* "a little foot."

Tripodium Medikus Fabaceae

Origins:
Greek *tripodeios*, *tripodios* "three-footed," *podion* "a little foot."

Tripogandra Raf. Commelinaceae

Origins:
Greek *treis*, *tria* "three," *pogon* "beard" and *aner, andros* "man, stamen," referring to the three longer stamens; see E.D. Merrill, *Index rafinesquianus*. The plant names published by C.S. Rafinesque, etc. 84. Jamaica Plain, Massachusetts, USA 1949; C.S. Rafinesque, *Flora Telluriana*. 2: 16. 1836 [1837].

Tripogon Roemer & Schultes Gramineae

Origins:
Greek *treis* "three" and *pogon* "beard," indicating the lemma.

Species/Vernacular Names:
T. loliiformis (F. Muell.) C.E. Hubb. (*Festuca loliiformis* F. Muell.; *Diplachne loliiformis* (F. Muell.) F. Muell. ex Benth.)

English: five-minute grass, rye beetlegrass, eight-day grass

T. major Hook.f.

Yoruba: irungbon efon

Tripolium Nees Asteraceae

Origins:
Latin *tripolium* for a plant growing on cliffs (Plinius), Greek *tripolion*, a genus of halophyte.

Species/Vernacular Names:
T. vulgare Besler ex Nees

English: sea aster

Tripsacum L. Gramineae

Origins:
Origin obscure, probably from Greek *tripsis* "rubbing, friction, durability" and *psakas* "a grain, any small piece broken off"; see C. Linnaeus, *Systema Naturae*. Ed. 10. 1261. (May-June) 1759; Helmut Genaust, *Etymologisches Wörterbuch der botanischen Pflanzennamen*. 658-659. Basel 1996.

Species/Vernacular Names:
T. dactyloides (L.) L.
English: gama grass
T. fasciculatum Trin. ex Asch.
English: Guatemala grass

Tripteris Less. Asteraceae

Origins:
Greek *treis*, *tria* "three" and *pteron* "wing"; see *Linnaea*. 6: 95. 1831.

Tripterocalyx (Torrey) Hook. Nyctaginaceae

Origins:
From the Greek *tripteros* "having three wings" and *kalyx* "calyx."

Tripterococcus Endl. Stackhousiaceae

Origins:
From the Greek *tripteros* and *kokkos* "a berry," referring to the cocci.

Tripterodendron Radlk. Sapindaceae

Origins:
From the Greek *tripteros* plus *dendron* "tree."

Tripterospermum Blume Gentianaceae

Origins:
Greek *treis*, *tria* "three," *pteron* "wing" and *sperma* "seed," referring to the triquetrous seeds, often winged.

Species/Vernacular Names:
T. coeruleum (Hand.-Mazz.) H. Smith
China: yan yuan shuang hu die

Tripterygium Hook.f. Celastraceae

Origins:

Greek *treis*, *tria* "three" and *pteron* "wing," *pteryx* "wing, pterygos* "small wing," an allusion to the winged fruits.

Triptilion Ruíz & Pavón Asteraceae

Origins:

Greek *treis*, *tria* "three" and *ptilon* "feather," referring to the divisions of the pappus.

Triptilodiscus Turcz. Asteraceae

Origins:

Greek *treis*, *tria* "three," *ptilon* "feather" and *diskos* "a disc," referring to the plumose pappus; see Porphir Kiril N.S. Turczaninow (1796-1863), *Bulletin de la Société Impériale des Naturalistes de Moscou.* 24(2): 66. 1851.

Species/Vernacular Names:

T. pygmaeus Turcz. (*Dimorpholepis australis* A. Gray; *Helipterum australe* (A. Gray) Druce; *Helipterum dimorpholepis* Benth.; *Duttonia sessiliceps* F. Muell.)

English: common sunray, pygmy daisy

Triraphis R. Br. Gramineae

Origins:

Greek *treis* "three" and *rhaphis*, *rhaphidos* "a needle," referring to the lemmas, to the awns of the flowering glume; Robert Brown, *Prodromus florae Novae Hollandiae et Insulae van-Diemen.* 185. London (Mar.) 1810.

Species/Vernacular Names:

T. mollis R. Br.

English: purple heads, purple plume grass, needle grass

Trirhaphis F. Muell. Gramineae

Origins:

Orthographic variant, see *Triraphis* R. Br.

Trirostellum Z.P. Wang & Xie Cucurbitaceae

Origins:

From *tri-* "three" and Latin *rostellum, i* "a little beak, a small snout."

Trisanthus Lour. Umbelliferae

Origins:

From the Greek *treis* "three" and *anthos* "flower."

Triscenia Griseb. Gramineae

Origins:

Greek *triskenes* "three-legged."

Trisciadia Hook.f. Rubiaceae

Origins:

Greek *treis*, *tria* "three" and *skias*, *skiados* "a canopy, umbel."

Trisciadium Phil. Umbelliferae

Origins:

From the Greek *treis*, *tria* "three" and *skias*, *skiados* "a canopy, umbel," *skiadion*, *skiadeion* "umbel, parasol."

Triscyphus Taubert ex Warm. Burmanniaceae

Origins:

From the Greek *treis*, *tria* "three" and *skyphos* "a cup, goblet, a jug."

Trisema Hook.f. Dilleniaceae

Origins:

Greek *treis*, *tria* "three" and *sema* "standard, a sign, mark."

Trisepalum C.B. Clarke Gesneriaceae

Origins:

From *treis*, *tria* and *sepalum* "sepal."

Trisetaria Forssk. Gramineae

Origins:

From *tri-* "three" plus Latin *saeta* (*seta*), *ae* "a bristle, hair," genus *Setaria* P. Beauv.

Trisetarium Poir. Gramineae

Origins:

From *tri-* plus Latin *saeta* (*seta*), *ae* "a bristle, hair," see also *Trisetum* Pers.

Trisetella Luer Orchidaceae

Origins:
A diminutive, from *tri-* "three" and *saeta, ae* "a bristle, hair," an allusion to the tails of the sepals.

Trisetobromus Nevski Gramineae

Origins:
From *tri-* "three," *saeta, ae* "a bristle, hair" plus *Bromus* L.

Trisetum Pers. Gramineae

Origins:
From *tri-* "three" and Latin *saeta* (*seta*), *ae* "a bristle, hair," referring to the three awns of the lemmas.

Species/Vernacular Names:
T. glomeratum (Kunth) Trin. (*Koeleria glomerata* Kunth)
Hawaii: pili uka, he'upueo, mountain pili
T. pensylvanicum (L.) Beauv.
English: swamp oat

Trisiola Raf. Gramineae

Origins:
Referring to *Uniola*; see E.D. Merrill, *Index rafinesquianus.* 77. 1949; C.S. Rafinesque, *Neogenyton, or Indication of Sixty-Six New Genera of Plants of North America.* 4. 1825, *Am. Monthly Mag. Crit. Rev.* 2: 175. 1817 and 4: 188, 190. 1819, *Jour. Phys. Chim. Hist. Nat.* 89: 104. 1819 and *Florula ludoviciana.* 144. New York 1817.

Trisiola Raf. Gramineae

Origins:
See E.D. Merrill, *Index rafinesquianus.* 77. 1949; C.S. Rafinesque, *Neogenyton, or Indication of Sixty-Six New Genera of Plants of North America.* 4. 1825.

Trismeria Fée Pteridaceae (Adiantaceae)

Origins:
From *treis* "three" and *meros* "part."

Trispermium Hill Selaginellaceae

Origins:
From Greek *treis* "three" and *sperma* "seed," referring to the fruit.

Tristachya Nees Gramineae

Origins:
From the Greek *treis* and *stachys* "spike."

Species/Vernacular Names:
T. leucothrix Nees (*Apochaete hispida* (L.f.) J.B. Phipps; *Tristachya hispida* (L.f.) K. Schum.)
English: hairy trident grass
South Africa: harige drieblomgras, rooisaadgras

Tristagma Poepp. Alliaceae (Liliaceae)

Origins:
Greek *treis* "three" and *stagma* "something which drips, that which drips," referring to the ovary and the nectar-producing pores.

Tristania R. Br. Myrtaceae

Origins:
After the French botanist Jules Marie Claude Comte de Tristan, 1776-1861, erudite naturalist. His works include *Mémoire sur la situation botanique de l'Orléanaise.* Orléans 1810 and *Mémoire sur les organes caulinares des Asperges.* Orléans 1813; see I.H. Vegter, *Index Herbariorum.* Part II (7), *Collectors T-Z.* Regnum Vegetabile vol. 117. 1988; J.H. Barnhart, *Biographical Notes upon Botanists.* 3: 401. 1965; A. Lasègue, *Musée botanique de Benjamin Delessert.* Paris 1845; Stafleu and Cowan, *Taxonomic Literature.* 6: 497-499. 1986; F. Boerner & G. Kunkel, *Taschenwörterbuch der botanischen Pflanzennamen.* 4. Aufl. 182. 1989.

Species/Vernacular Names:
T. sp.
Malaya: selunchor, lidah mara, belian sipi, palan, kepayang burong
T. neriifolia (Sims) R. Br.
English: water gum

Tristaniopsis Brongn. & Gris Myrtaceae

Origins:
Resembling *Tristania* R. Br.

Tristeca P. Beauv. Psilotaceae

Origins:
From Greek *treis* "three" and *theke* "case, box."

Tristegis Nees Gramineae

Origins:

From the Greek *treis, tria* "three" and *stege, stegos* "roof, cover."

Tristellateia Thouars Malpighiaceae

Origins:

Latin *tri-* "three" and *stellatus, a, um* "starry, star shaped," referring to the three star-like samaras; see Louis-Marie Aubert Aubert du Petit-Thouars, 1758-1831, *Genera nova madagascariensia*. 14. [Paris 1806].

Species/Vernacular Names:

T. australasiae A. Rich. (*Tristellateia australis* A. Rich.)

Japan: Kôshun-kazura

The Philippines: bagnít

Tristemma A.L. Juss. Melastomataceae

Origins:

Greek *treis, tria* and *stemma, stemmatos* "a garland, crown," referring to the hypanthium; see A.L. de Jussieu, *Genera Plantarum*. 329. (Aug.) 1789.

Species/Vernacular Names:

T. hirtum P. Beauv.

Yoruba: apiko

T. littorale Benth.

Yoruba: ajalugborogan

Tristemon Scheele Cucurbitaceae

Origins:

From the Greek *treis, tria* plus *stemon* "a stamen."

Tristemonanthus Loes. Celastraceae

Origins:

Greek *treis, tria* "three," *stemon* "a stamen" and *anthos* "flower."

Tristicha Thouars Podostemaceae (Tristichaceae)

Origins:

Greek *tristichos* "three-row," referring to the arrangement of the leaves; see Louis-Marie Aubert Aubert du Petit-Thouars, *Genera nova madagascariensia*. 3. [Paris 1806].

Tristichocalyx F. Muell. Menispermaceae

Origins:

Greek *tristichos* and *kalyx* "a calyx"; see Ferdinand von Mueller, *Fragmenta Phytographiae Australiae*. 4: 27. (Oct.) 1863.

Tristiropsis Radlk. Sapindaceae

Origins:

Resembling *Tristira* Radlk.; see Arthur D. Chapman, ed., *Australian Plant Name Index*. 2914. Canberra 1991.

Tristylium Turcz. Theaceae

Origins:

From the Greek *treis, tria* "three" and *stylos* "a pillar, style."

Tritaxis Baill. Euphorbiaceae

Origins:

Greek *treis* "three" and *taxis* "a series, order, arrangement," referring to the whorled stamens; see also *Trigonostemon* Blume.

Tritelandra Raf. Orchidaceae

Origins:

Greek *treis* "three," *teleios, teleion, teleia, teleos* "complete, perfect" and *aner, andros* "man, stamen," referring to the fertile stamens; see E.D. Merrill, *Index rafinesquianus*. 105. 1949.

Triteleia Douglas ex Lindley Alliaceae (Liliaceae)

Origins:

Greek *treis* "three" and *teleios, teleos* "complete, perfect," referring to the ovary and all parts of the flower.

Species/Vernacular Names:

T. clementina Hoover

English: San Clemente Island triteleia

T. hyacinthina (Lindley) E. Greene

English: white brodiaea

T. laxa Benth.

English: Ithuriel's spear

T. peduncularis Lindley

English: long-rayed brodiaea

Triteleiopsis Hoover Alliaceae (Liliaceae)

Origins:
Resembling *Triteleia* Douglas ex Lindl.

Trithecanthera Tieghem Loranthaceae

Origins:
From the Greek *treis* "three," *theke* "case, box" and *anthera* "anther."

Trithrinax Martius Palmae

Origins:
Greek *treis* "three" and *thrinax* "fan, trident," referring to the leaves.

Species/Vernacular Names:
T. brasiliensis Martius

Brazil: buriti-palito, buriti, carandaí, caraná, carandá, palma de escoba

Trithuria Hook.f. Hydatellaceae

Origins:
Greek *treis* "three" and *thyra* "a door, entrance," referring to the valves and to the dehiscence of the fruits.

Species/Vernacular Names:
T. submersa Hook.f. (*Juncella submersa* (Hook.f.) Hieron.; *Trithuria occidentalis* Benth.; *Juncella occidentalis* (Benth.) Hieron.)

English: trithuria, juncella

Triticum L. Gramineae

Origins:
Latin *triticum, i* for wheat, *tero, -is, trivi, tritum, terere* "to grind, to wear away, to waste," Greek *teiro, teirein* "distress, weaken," see Plinius, L. Junius Moderatus Columella, Plautus, Cicero; Carl Linnaeus, *Species Plantarum*. 85. 1753 and *Genera Plantarum*. Ed. 5. 37. 1754; Giovanni Semerano, *Le origini della cultura europea*. Dizionario della lingua Latina e di voci moderne. 2(2): 588, 594. Firenze 1994.

Species/Vernacular Names:
T. aestivum L. (*Triticum vulgare* Villars; *Triticum sativum* Lam.)

English: common bread, volunteer wheat, wheat, common wheat

South Africa: opslagkoring

Japan: ko-mugi

Okinawa: umamuji

Tibetan: gro

China: xiao mai, hsiao mai, lai

Tritomopterys (A. Juss. ex Endl.) Nied. Malpighiaceae

Origins:
Greek *treis* "three," *tome, tomos, temno* "division, section, to slice" and *pteron* "wing."

Tritonia Ker Gawler Iridaceae

Origins:
Greek *triton* "a vane, a weathercock, weathervane," in allusion to the variable direction of the stamens in different species; Triton was the legendary son of Poseidon and Amphitrite, or of Neptune and Salacia, of semi-human form, represented as human to the waist and dolphin below; see *Curtis's Botanical Magazine*. 581. 1802.

Species/Vernacular Names:
T. crispa (L.f.) Ker Gawl. var. *crispa*

South Africa: agretjie, vlieërtjie

T. crocata (L.) Ker Gawl. (*Tritonia hyalina* Bak.)

South Africa: kalkoentjie, orange tritonia, Mossel Bay tritonia, rooikalkoentjie, rooikalossie

T. deusta (Ait.) Ker Gawl. subsp. *deusta* (Latin *deustus, a, um* "burned, scorched," part. *deuro, urere, ussi, ustum* "to burn down, to destroy")

South Africa: kalkoentjie

T. lineata (Salisb.) Ker Gawl. (*Gladiolus lineatus* Salisb.)

English: lined tritonia

T. lineata (Salisb.) Ker Gawl. var. *lineata* (*Tritonia flavida* Schltr.; *Tritonia kraussii* Bak.)

South Africa: bergkatjietee

T. pallida (Ait.) Ker Gawl. subsp. *pallida*

South Africa: katjietee

T. securigera (Ait.) Ker Gawl. (from the Latin *securiger, era, erum* (*securis gero*) "axe-bearing," *securis, is* "an axe, hatchet")

South Africa: kalkoentjie, tritonia

T. squalida (Ait.) Ker Gawl. (*Ixia squalida* Sol.)

South Africa: kalkoentjie, Mosselbaaikalkoentjie

Tritoniopsis L. Bolus Iridaceae

Origins:
Resembling *Tritonia*.

Species/Vernacular Names:
T. antholyza (Poir.) Goldbl. (*Anapalina nervosa* (Thunb.) G.J. Lewis)

South Africa: rooipypie, karkarblom

T. pulchra (Bak.) Goldbl. (*Anapalina pulchra* (Bak.) N.E. Br)

South Africa: rooipypie

Triumfetta L. Tiliaceae

Origins:
For the Italian botanist Giovanni Battista Trionfetti, 1658-1708, author of *Observationes de ortu ac vegetatione Plantarum cum novarum stirpium historia iconibus illustrata*. Romae 1685 and *Syllabus plantarum horto medico Romanae Sapientiae*. 1688; see Marcello Malpighi (1628-1694), *Opera posthuma*, cura Faustini Gavinelli, edita objectionibus circa plantas Johannis Baptistae Triumfetti. Venetiis 1698; Carl Linnaeus, *Species Plantarum*. 444. 1753 and *Genera Plantarum*. Ed. 5. 203. 1754; Mariella Azzarello Di Misa, ed., *Il Fondo Antico della Biblioteca dell'Orto Botanico di Palermo*. 277. Regione Siciliana, Palermo 1988.

Species/Vernacular Names:
T. spp.

Mexico: yechi y ruz

T. angolensis Sprague & Hutch.

South Africa: maagbossie, rolbossie, spinnekopbossie, tolbossie, waaierbossie

T. bartramia L.

The Philippines: kulut-kulutan, kullukullot, kolokolot, kolotang-bilog, kulutan, pallopallot, pallot-pallot, kupot, maropoto, balangot, bulagon, daupang

China: huang hua di tao hua

T. cordifolia A. Rich.

Yoruba: esura, akeriri, amaramo, orisemayin

Congo: mponga, mpunga, mpounga

T. pilosa Roth var. *effusa* (E. Meyer ex Harvey) Wild (*Triumfetta effusa* E. Meyer ex Harvey)

South Africa: klitsbossie

T. pilosa Roth var. *tomentosa* Szyszyl. ex Sprague & Hutch.

English: burs

South Africa: klitse

T. rhomboidea Jacq. (*Waltheria fauriei* H. Lév.)

English: Chinese bur

South Africa: klitsbossie

Congo: mpunga, binkampula

Yoruba: ilasa oku, ilasa omode, ako bolo bolo, bokoo pupa, eepafo

Malaya: chempadang

T. welwitschii Mast. var. *hirsuta* (Sprague & Hutch.) Wild (*Triumfetta hirsuta* Sprague & Hutch.)

South Africa: klitsbossie

Triumfettoides Rauschert Tiliaceae

Origins:
Resembling the genus *Triumfetta*.

Triunia L. Johnson & B.G. Briggs Proteaceae

Origins:
Latin *tri-* "three" and *unus* "one," referring to the perianth; see Lawrence Alexander Sidney Johnson (1925-) and Barbara Gillian Briggs (1934-), in *Botanical Journal of the Linnean Society*. 70: 175. (Sep.) 1975.

Triunila Raf. Gramineae

Origins:
Referring to *Uniola* L.; see E.D. Merrill (1876-1956), *Index rafinesquianus*. 77. 1949; C.S. Rafinesque, *Neogenyton, or Indication of Sixty-Six New Genera of Plants of North America*. 4. 1825.

Triuranthera Backer Melastomataceae

Origins:
From the Greek *treis* "three," *oura* "tail" and *anthera* "anther."

Triurocodon Schltr. Burmanniaceae

Origins:
From the Greek *treis* "three," *oura* "tail" and *kodon* "a bell."

Trivalvaria (Miq.) Miq. Annonaceae

Origins:
Latin *tri-* "three" and *valva, ae* "a valve, a folding-door," referring to the petals.

Trixago Raf. Labiatae

Origins:
Latin *trixago* or *trissago* for a plant, called also *chamaedrys*, germander (Plinius et al.); see E.D. Merrill (1876-1956), *Index rafinesquianus*. 210. 1949; C.S. Rafinesque, *Flora Telluriana*. 3: 85. 1836 [1837].

Trixanthera Raf. Acanthaceae

Origins:
Referring to *Trichanthera*; see E.D. Merrill, *Index rafinesquianus*. 224. 1949; C.S. Rafinesque, *Sylva Telluriana*. 146. 1838.

Trixapias Raf. Lentibulariaceae

Origins:
See E.D. Merrill, *Index rafinesquianus*. 222. 1949; C.S. Rafinesque, *Flora Telluriana*. 4: 108. 1836 [1838].

Trixis P. Browne Asteraceae

Origins:
Greek *trixos, trissos* "threefold," Latin *triplex*, referring to the angled fruit or to the outer corolla lip.

Trixostis Raf. Gramineae

Origins:
See Elmer D. Merrill (1876-1956), *Index rafinesquianus*. 77. 1949; C.S. Rafinesque, *Seringe Bull. Bot*. 1: 221. 1830.

Trizeuxis Lindley Orchidaceae

Origins:
Greek *treis, tri-* "three" and *zeuxis* "yoking, union," triple hook, referring to the cohesion of the sepals.

Trochera Rich. Gramineae

Origins:
Greek *trochos* "a wheel, a round ball, anything round or circular," *trocheros* "running, tripping."

Trochetia DC. Sterculiaceae

Origins:
Greek *trochos* "a wheel," referring to the flowers.

Trochetiopsis Marais Sterculiaceae

Origins:
Resembling *Trochetia* DC.; see Sue Brodie, Martin Cheek & Martin Staniforth, "*Trochetiopsis ebenus*." in *Curtis's Botanical Magazine*. 15(1): 27-36. 1998; W. Marais, "*Trochetiopsis* (Sterculiaceae), a new genus from St. Helena." *Kew Bulletin*. 36: 645-646. 1981.

Species/Vernacular Names:
T. ebenus Cronk
English: ebony

Trochilocactus Lindinger Cactaceae

Origins:
Greek *trochos* "a wheel, a round ball, anything round or circular" plus *cactus*.

Trochisandra Bedd. Celastraceae

Origins:
From the Greek *trochos* "a wheel, any thing circular," *trochis* "a runner" and *aner, andros* "man, stamen."

Trochiscanthes W.D.J. Koch Umbelliferae

Origins:
Greek *trochiskos* "a small wheel" and *anthos* "flower."

Trochiscus O.E. Schulz Brassicaceae

Origins:
Latin *trochiscus* and Greek *trochiskos* "a small wheel, a small circle, a small globe or ball."

Trochocarpa R. Br. Epacridaceae

Origins:

Greek *trochos* "a wheel, a round ball" and *karpos* "fruit," referring to the cells of the fruits; see Robert Brown, *Prodromus Florae Novae Hollandiae*. 548. 1810.

Trochocodon Candargy Campanulaceae

Origins:

From the Greek *trochos* and *kodon* "a bell."

Trochodendron Siebold & Zucc. Trochodendraceae (Hamamelidae, Trochodendrales)

Origins:

Greek *trochos* "a wheel" and *dendron* "tree," referring to the leaves and stamens.

Species/Vernacular Names:

T. aralioides Siebold & Zucc.

Japan: yama-guruma (= mountain wheel)

Trochomeria Hook.f. Cucurbitaceae

Origins:

Greek *trochos* and *meris* "a portion, part."

Trochomeriopsis Cogn. Cucurbitaceae

Origins:

Resembling *Trochomeria* Hook.f.

Trochopteris Gardner Schizaeaceae

Origins:

From the Greek *trochos* "a wheel" and *pteris* "a fern."

Troglophyton Hilliard & B.L. Burtt Asteraceae

Origins:

From the Greek *trogle* "cave-dwelling, hole, hollow" and *phyton* "plant," Greek *trogo* "to gnaw, chew," Akkadian *taraku* "to beat."

Trogostolon Copel. Davalliaceae

Origins:

From the Greek *trogo* "to gnaw, chew" and *stolos* "appendage, excrescence," Latin *stolo, stolonis* "a shoot, twig, branch."

Trollius L. Ranunculaceae

Origins:

From *trollblume* a Swiss German name for *Trollius europaeus*, Latinized by C. Gesner in 1555 as *trollius flos* "rounded flower"; see Carl Linnaeus, *Species Plantarum*. 556. 1753 and *Genera Plantarum*. Ed. 5. 243. 1754; Ernest Weekley, *An Etymological Dictionary of Modern English*. 2: 1544-1545. New York 1967.

Tromotriche Haw. Asclepiadaceae

Origins:

Greek *tremo* "to tremble, quake," *tromos* "trembling, fear" and *trichos* "hair."

Tropaeastrum Mabb. Tropaeolaceae

Origins:

See *Trophaeastrum* Sparre.

Tropaeolum L. Tropaeolaceae

Origins:

The leaf suggesting a shield, and the flower a helmet, from the Greek *tropaion* "trophy," Latin *tropaeum* or *trophaeum, i* "a trophy"; see Carl Linnaeus, *Species Plantarum*. 345. 1753 and *Genera Plantarum*. Ed. 5. 162. 1754; C.T. Onions, *The Oxford Dictionary of English Etymology*. Oxford University Press 1966; Ernest Weekley, *An Etymological Dictionary of Modern English*. 2: 1545. New York 1967; Helmut Genaust, *Etymologisches Wörterbuch der botanischen Pflanzennamen*. 661. Basel 1996; National Research Council, *Lost Crops of the Incas: Little-Known Plants of the Andes with Promise for Worldwide Cultivation*. National Academy Press, Washington, D.C. 1989.

Species/Vernacular Names:

T. sp.

Peru: pató, capuchinha

T. crenatiflorum Hook.f.

Peru: mashua-mashua

T. dipetalum Ruíz & Pav. (*Chymocarpus stipulaceus* Klatt ex Otto)

Andes: monte massua

T. majus L.

English: garden nasturtium, Indian cress, tall nasturtium, nasturtium

Mexico: capuchina, mastuerzo, cuitziquiendas, curutziti, pelonmexixquilitl, pelonchili

Peru: mastuerzo, ticsau

South Africa: kapkappertjies, kappertjie

Japan: nasutachûmu, kin-renka (kin = gold), nôzen-haren

China: han lian hua, chin lien hua

Hawaii: pohe haole

T. minus L.

Andes: yucyushpa-malán

T. peregrinum L. (*Tropaeolum aduncum* J.E. Smith)

English: Canary creeper

Mexico: pajaritos, palonillas

Peru: añu, añu-añu, chichirique, hualpa-hualpa, huallpa-huallpa, malla, osañu, otolo-kcora, pajarillos, pajarillos amarillos, pajarito, vilcu, quita añu

T. seemannii Buchenau

Peru: kita añu

T. smithii DC.

Peru: pajarito

T. speciosum Poeppig & Endl.

English: flame nasturtium, Scottish flame flower

T. tuberosum Ruíz & Pav.

English: mashua, anu

Peru: añu, anyu, osañu, apiña mama, isaño, mashua, massua, mayua, tuna, tuna mashua, occe-añu, yana-añu, puca añu, yurac añu, ckello añu, sapallu añu, checche añu, muru añu

Quechua names: mashua, añu, apiñu, apiña-mama, yanaoca

Aymara names: isau, issanu, kkayacha

Colombia: puel

Spanish: mashua, majua, mafua, mauja, maxua, mashuar, añú, anyú, cubios, navios, navo, isaño, isañu, apilla, ysaño

Tropalanthe S. Moore Sapotaceae

Origins:

From the Greek *tropalis* "bundle, bunch" and *anthos* "flower"; see also *Pycnandra* Benth.

Trophaeastrum Sparre Tropaeolaceae

Origins:

Greek *tropaion* "trophy" and *astrum*, the suffix indicating incomplete resemblance.

Trophianthus Scheidw. Orchidaceae

Origins:

Greek *trophis* "large, huge, swollen" and *anthos* "flower," referring to the showy flowers.

Trophis P. Browne Moraceae

Origins:

Greek *trophis* "well-fed, stout, large, huge."

Species/Vernacular Names:

T. caucana (Pittier) C.C. Berg (*Olmedia aspera* Ruíz & Pav.; *Olmedia caucana* Pittier; *Olmedia poeppigiana* Klotzsch, nom. illeg.)

Peru: huacha huasca, llanchama, lanchmana, minshi-pata, minchipata, muichipata, palo amarillo

Trophisomia Rojas Acosta Moraceae

Origins:

From the Greek *trophis* "well-fed, stout, large, huge" and *soma* "body."

Tropidia Lindley Orchidaceae

Origins:

Greek *tropis, tropidos* "a keel," referring to the shape of the labellum; see *Edward's Botanical Register*. 19 (Oct.) 1833.

Species/Vernacular Names:

T. calcarata Ames (*Tropidia somai* Hayata)

Japan: akô-nettai-ran

T. disticha Schltr. (see B.A. Lewis and P.J. Cribb, *Orchids of the Solomon Islands and Bougainville*. Royal Botanic Gardens, Kew 1991)

The Solomon Islands: tukurutukuru, puni, oi'oi

Tropidocarpum Hook. Brassicaceae

Origins:

From the Greek *tropis, tropidos* "a keel" and *karpos* "fruit," keeled fruit.

Species/Vernacular Names:
T. capparideum E. Greene
English: caper-fruited tropidocarpum

Tropidopetalum Turcz. Anacardiaceae

Origins:
Greek *tropis, tropidos* and *petalon* "petal."

Tropilis Raf. Orchidaceae

Origins:
According to Rafinesque "keel-lip," etymology not clear; see E.D. Merrill (1876-1956), *Index rafinesquianus*. 105. 1949; C.S. Rafinesque, *Flora Telluriana*. 2: 95. Philadelphia 1836 [1837].

Trouettia Pierre ex Baillon Sapotaceae

Origins:
For the French pharmacist Édouard Trouette, born 1855, wrote *De l'introduction et de l'acclimatation des Quinquinas à l'Ile de la Réunion*. Paris 1879; see E.M. Tucker, *Catalogue of the Library of the Arnold Arboretum of Harvard University*. Cambridge, Massachusetts 1917-1933; Stafleu and Cowan, *Taxonomic Literature*. 6: 509. Utrecht 1986.

Trozelia Raf. Solanaceae

Origins:
After the Swedish clergyman Clas Blechert (Claes Bliechert) Trozelius, 1719-1794, economist, botanist, professor of economy, with Edward Ulrich von Hausswolff wrote *Korta anmerkningar vid Svio-Göthernas fordna* etc. Stockholm [1752]; see John H. Barnhart, *Biographical Notes upon Botanists*. 3: 403. Boston 1965; *Dissertatio academica de generatione ac nutritione Arborum*, quam ... sub praesidio ... C.B. Trozelii ... defert C. Revigin. Londini Gothorum [1768]; T. W. Bossert, *Biographical Dictionary of Botanists Represented in the Hunt Institute Portrait Collection*. 407. Boston, Massachusetts 1972; E.D. Merrill, *Index rafinesquianus*. 211, 213. 1949; C.S. Rafinesque, *Sylva Telluriana*. 54. 1838; Jonas C. Dryander, *Catalogus bibliothecae historico-naturalis Josephi Banks*. London 1796-1800.

Trudelia Garay Orchidaceae

Origins:
For Nikolaus Trudel, Swiss orchid enthusiast and orchid photographer; see Hubert Mayr, *Orchid Names and Their Meanings*. 186. Vaduz 1998.

Trungboa Rauschert Scrophulariaceae

Origins:
A place in Southeast Asia.

Trybliocalyx Lindau Acanthaceae

Origins:
Latin *tryblium* "a plate, salver," Greek *tryblion* "a cup, bowl" and *kalyx* "a calyx."

Trychinolepis B.L. Robinson Asteraceae

Origins:
From the Greek *trychinos* "ragged, made of rags" and *lepis* "scale."

Trymalium Fenzl Rhamnaceae

Origins:
Greek *tryma, trymalia* "a hole, perforation," referring to the slits at the summit of the fruit.

Species/Vernacular Names:
T. wayi F. Muell. & Tate
English: grey trymalium

Trymatococcus Poeppig & Endl. Moraceae

Origins:
From the Greek *tryma, trymalia, trymatos* "a hole" and *kokkos* "berry."

Tryonella Pichi Sermolli Pteridaceae (Adiantaceae)

Origins:
See Rodolfo E.G. Pichi Sermolli, "Fragmenta Pteridologiae — V." in *Webbia*. 29(1): 1-16. Feb. 1975

[1974] and *Authors of Scientific Names in Pteridophyta.* 56, 76. Royal Botanic Gardens, Kew 1996; F.N. Hepper and Fiona Neate, *Plant Collectors in West Africa.* 81. 1971; J.H. Barnhart, *Biographical Notes upon Botanists.* Boston 1965.

Tryphane Reichb. Caryophyllaceae

Origins:
Probably from the Greek *tryphao* "to pride oneself, to live sumptuously," *tryphema* "luxuries, the object in which one takes pride or pleasure."

Tryphia Lindley Orchidaceae

Origins:
Greek *tryphe* "softness, tenderness," referring to the leaves; see also *Holothrix* Rich. ex Lindl.

Tryphostemma Harvey Passifloraceae

Origins:
Greek *tryphos* "a piece, fragment" and *stemma* "a crown," referring to the corona rays.

Tryptomene Endl. ex F. Muell. Myrtaceae

Origins:
Orthographic variant of *Thryptomene* Endl.; see Arthur D. Chapman, ed., *Australian Plant Name Index.* 2922. Canberra 1991.

Tryptomene Walp. Myrtaceae

Origins:
See the genus *Thryptomene* Endl.

Tryssophyton Wurdack Melastomataceae

Origins:
From the Greek *tryssos* "friable" and *phyton* "a plant."

Tsaiorchis Tang & Wang Orchidaceae

Origins:
After Hse Tao Tsai, 1911-1981, botanist and plant collector (Yunnan).

Tsavo Jarmol. Salicaceae

Origins:
Tsavo river and Tsavo National Park, Kenya.

Tsingya Capuron Sapindaceae

Origins:
A place in Madagascar, Tsingy de Namoroka or de Bemaraha; see Werner Rauh (1913-1997), *Succulent and Xerophytic Plants of Madagascar.* 1: 81. 1995. Strawberry Press, Mill Valley, California 1998.

Tsotria Raf. Orchidaceae

Origins:
For *Isotria* Raf.; see C.S. Rafinesque, *Am. Monthly Mag. Crit. Rev.* 2: 173, 174. 1817.

Tsuga (Antoine) Carrière Pinaceae

Origins:
From the Japanese name; see Helmut Genaust, *Etymologisches Wörterbuch der botanischen Pflanzennamen.* 661. 1996.

Species/Vernacular Names:
T. canadensis (L.) Carrière
English: Canada hemlock, eastern hemlock, white hemlock
T. caroliniana Engelm.
English: Carolina hemlock
T. dumosa (D. Don) Eichler
English: Himalayan hemlock
Nepal: tengre salla, thigre salla
Bhutan: ba shing
T. heterophylla (Raf.) Sarg.
English: western hemlock, Alaskan pine
T. sieboldii Carrière
English: Japanese hemlock

Tsusiophyllum Maxim. Ericaceae

Origins:
A Japanese name, *Rhododendron* L. is *Tsutsuji Zoku*; see Jisaburo Ohwi, *Flora of Japan.* 699. [*Rhododendron tsusiophyllum* Sugimoto (*Rhododendron tanakae* (Maxim.) Ohwi, non Hayata; *Tsusiophyllum tanakae* Maxim.)] Edited

by Frederick G. Meyer and Egbert H. Walker. Smithsonian Institution, Washington, D.C. 1965.

Tuberaria (Dunal) Spach Cistaceae

Origins:
From the Latin *tuber, eris* "hump, bump, swelling, truffle."

Tuberolabium Yamamoto Orchidaceae

Origins:
Latin *tuber, eris* "hump, bump, swelling" and *labium, ii* "a lip," referring to the appearance and shape of the lip.

Tuberostyles Benth. Asteraceae

Origins:
Latin *tuber, eris* and Greek *stylos* "a pillar, style"; see *Tuberostylis.*

Tuberostylis Steetz Asteraceae

Origins:
Latin *tuber, eris* "hump, swelling" and Greek *stylos* "a pillar, style."

Tubiflora J.F. Gmel. Acanthaceae

Origins:
Latin *tubus, i* "pipe, tube" and *flos, floris* "a flower, blossom."

Tubilabium J.J. Smith Orchidaceae

Origins:
From the Latin *tubus, i* "pipe, tube" and *labium, ii* "a lip," referring to the margins of the lip.

Tubocapsicum Makino Solanaceae

Origins:
From the Latin *tubus, i* "pipe, tube" plus *Capsicum.*

Species/Vernacular Names:
T. anomalum (Franchet & Sav.) Makino

English: Japanese tubocapsicum

China: long zhu

Tuckermannia Nuttall Asteraceae

Origins:
For the American botanist Edward Tuckerman, 1817-1886, traveler, lichenologist; see J.H. Barnhart, *Biographical Notes upon Botanists.* 3: 407. 1965; E.M. Tucker, *Catalogue of the Library of the Arnold Arboretum of Harvard University.* Cambridge, Massachusetts 1917-1933; Antoine Lasègue, *Musée botanique de M. Benjamin Delessert.* 1845; R. Zander, F. Encke, G. Buchheim and S. Seybold, *Handwörterbuch der Pflanzennamen.* 14. Aufl. Stuttgart 1993; Jeannette Elizabeth Graustein, *Thomas Nuttall, Naturalist. Explorations in America, 1808-1841.* Cambridge, Harvard University Press 1967; T. W. Bossert, *Biographical Dictionary of Botanists Represented in the Hunt Institute Portrait Collection.* 408. Boston, Massachusetts 1972; S. Lenley et al., *Catalog of the Manuscript and Archival Collections and Index to the Correspondence of John Torrey.* Library of the New York Botanical Garden. 1973; Joseph Ewan, ed., *A Short History of Botany in the United States.* New York and London 1969; D.C. Haskell, *The United States Exploring Expedition 1838-1842 and Its Publications 1844-1874.* New York 1942; D.B. Tyler, *The Wilkes Expedition*: The First United States Exploring Expedition (1838-1842). Philadelphia 1968; Stafleu and Cowan, *Taxonomic Literature.* 6: 523-527. 1986 and 7: 295-297. 1988.

Tuctoria Reeder Gramineae

Origins:
An anagram of *Orcuttia.*

Species/Vernacular Names:
T. greenei (Vasey) Reeder

English: Greene's tuctoria

T. mucronata (Crampton) Reeder

English: Crampton's tuctoria

Tuerckheimocharis Urban Scrophulariaceae

Origins:
For the German plant collector Hans von Türckheim, 1853-1920, traveler, in Guatemala and Santo Domingo; see T. W. Bossert, *Biographical Dictionary of Botanists Represented in the Hunt Institute Portrait Collection.* 408. Boston, Massachusetts 1972.

Tula Adans. Solanaceae

Origins:

In Mexico, Hidalgo and Tamaulipas States.

Tulakenia Raf. Asteraceae

Origins:

See E.D. Merrill, *Index rafinesquianus*. 242. 1949; C.S. Rafinesque, *Flora Telluriana*. 4: 120. Philadelphia 1836 [1838].

Tulasnea Naudin Melastomataceae

Origins:

For the French botanist Louis-René (Edmond, Étienne) Tulasne, 1815-1885 (d. Hyères, France), scientist, brother of the French botanist and physician Charles Tulasne (1816-1884); see J.H. Barnhart, *Biographical Notes upon Botanists*. 3: 407. 1965; E.M. Tucker, *Catalogue of the Library of the Arnold Arboretum of Harvard University*. Cambridge, Massachusetts 1917-1933; Stafleu and Cowan, *Taxonomic Literature*. 6: 529-535. 1986; G. Viennot-Bourgin, in *Dictionary of Scientific Biography* 13: 489-490. 1981; Ida Kaplan Langman, *A Selected Guide to the Literature on the Flowering Plants of Mexico*. Philadelphia 1964; Antoine Lasègue, *Musée botanique de M. Benjamin Delessert*. 1845.

Tulasnea Wight Podostemaceae

Origins:

For the French botanist Louis-René (Edmond, Étienne) Tulasne, 1815-1885.

Tulasneantha P. Royen Podostemaceae

Origins:

For the French botanist Louis-René (Edmond, Étienne) Tulasne, 1815-1885.

Tulbaghia L. Alliaceae (Liliaceae)

Origins:

The genus was named to honor the Dutch (bapt. Utrecht, Holland) Ryk (Rijk) Tulbagh, 1699-1771 (d. Cape Town, South Africa), Dutch Governor of the Cape of the Good Hope 1751-1771, sent plant specimens, bulbs and seeds to Linnaeus; see Carolus Linnaeus, *Correspondence between Carl von Linné and C. Rijk Tulbagh*, Governor of the Dutch Colony. London 1918; Peter MacOwan, "Personalia of botanical collectors at the Cape." *Trans. S. Afr. Philos. Soc.* 4(1): xxxiv. 1884-1886; Mary Gunn and Leslie Edward W. Codd, *Botanical Exploration of Southern Africa*. 55-60, 83, 352. Cape Town 1981; F. Boerner & G. Kunkel, *Taschenwörterbuch der botanischen Pflanzennamen*. 4. Aufl. 182. Berlin & Hamburg 1989.

Species/Vernacular Names:

T. alliacea L.f.

Southern Africa: iVimba 'mpunzi (Xhosa)

T. simmleri Beauv. (*Tulbaghia fragrans* Verdoorn)

English: sweet garlic, pink agapanthus

T. violacea Harvey (*Tulbaghia cepacea* L.f.)

English: wild garlic

South Africa: wild garlic, wildeknoffel, wilde knoflok

Tulexis Raf. Orchidaceae

Origins:

Greek *tylos* "wart," referring to the appearance of petals and sepals; see E.D. Merrill, *Index rafinesquianus*. 105. 1949; C.S. Rafinesque, *Flora Telluriana*. 4: 42. Philadelphia 1836 [1838].

Tulipa L. Liliaceae

Origins:

See H. Genaust, *Etymologisches Wörterbuch der botanischen Pflanzennamen*. 663-664. 1996; Edward Lee Greene, *Landmarks of Botanical History*. Edited by Frank N. Egerton. 759, 769-771, 790, 907, 1017. Stanford 1983; E. Weekley, *An Etymological Dictionary of Modern English*. 2: 1552. New York 1967; M. Cortelazzo & P. Zolli, *Dizionario etimologico della lingua italiana*. 5: 1383. Zanichelli, Bologna 1988; B. Migliorini, *Lingua e cultura*. Roma 1948; Augier (Ogier) Ghislain de Busbecq (1522-1592), A.-Gislenii Busbequii *omnia quae extant*. Lugd.-Batavor., Elzevir. 1633, *Epistolae ... ad Rudolphum II*. [written at Paris from 1582 to 1585, and edited by Jan Baptista Houwaert] [Brussels] 1630 and *Itinera Constantinopolitanum et Amasianum ab A.G.B*. Antverpiae 1581; [Auger Gislain de Busbecq] *Travels into Turkey*. London 1744; Jacques-Charles Brunet, *Manuel du libraire et de l'amateur de livres*. 1: 1417. Paris 1861; Jane Quinby, compiled by, *Catalogue of Botanical Books in the Collection of Rachel McMasters Miller Hunt*. 1: 134. [Reprint edition.] Pittsburgh, Pennsylvania 1958.

Species/Vernacular Names:

T. edulis (Miq.) Bak.

English: tulip

China: guang ci gu

Tulisma Raf. Gesneriaceae

Origins:

See Elmer D. Merrill, *Index rafinesquianus*. 221. 1949; C.S. Rafinesque, *Flora Telluriana*. 2: 98. Philadelphia 1836 [1837].

Tulocarpa Raf. Euphorbiaceae

Origins:

See E.D. Merrill, *Index rafinesquianus*. 157. 1949; C.S. Rafinesque, *Flora Telluriana*. 4: 115. Philadelphia 1836 [1838].

Tuloclinia Raf. Asteraceae

Origins:

See E.D. Merrill, *Index rafinesquianus*. 242. 1949; C.S. Rafinesque, *Flora Telluriana*. 4: 119. Philadelphia 1836 [1838].

Tulophos Raf. Alliaceae (Liliaceae)

Origins:

See E.D. Merrill, *Index rafinesquianus*. 96. 1949; C.S. Rafinesque, *Flora Telluriana*. 3: 71. Philadelphia 1836 [1837].

Tulorima Raf. Saxifragaceae

Origins:

See E.D. Merrill, *Index rafinesquianus*. 135. 1949; C.S. Rafinesque, *Flora Telluriana*. 2: 69. Philadelphia 1836 [1837].

Tulosuke Raf. Moraceae

Origins:

See Elmer D. Merrill, *Index rafinesquianus*. 112. 1949; Constantine Samuel Rafinesque, *Sylva Telluriana*. 58. 1838.

Tulotis Raf. Orchidaceae

Origins:

Greek *tylos* "callus, knob," *tylotos* "knobbed," indicating the lip; see E.D. Merrill, *Index rafinesquianus*. 105. 1949;

C.S. Rafinesque, *Flora Telluriana*. 2: 37. Philadelphia 1836 [1837].

Tulotropis Raf. Convolvulaceae

Origins:

See E.D. Merrill, *Index rafinesquianus*. 201. 1949; C.S. Rafinesque, *Flora Telluriana*. 4: 74. Philadelphia 1836 [1838].

Tumalis Raf. Euphorbiaceae

Origins:

See E.D. Merrill, *Index rafinesquianus*. 157. 1949; C.S. Rafinesque, *Flora Telluriana*. 4: 114. Philadelphia 1836 [1838].

Tumelaia Raf. Thymelaeaceae

Origins:

Referring to *Thymelaea*; see C.S. Rafinesque, *Flora Telluriana*. 4: 105. 1836 [1838]; Elmer D. Merrill, *Index rafinesquianus*. 172. 1949.

Tumidinodus H.W. Li Gesneriaceae

Origins:

Latin *tumidus, a, um* "swollen, swelling" and *nodus* "knot."

Tunas Lunell Cactaceae

Origins:

Tuna is a vernacular name for *Opuntia* spp.

Tunica Ludwig Caryophyllaceae

Origins:

From the Latin *tunica, ae* "skin, coating, membrane, husk," referring to the bracts filling the calyx.

Tunica Mert. & W.D.J. Koch Caryophyllaceae

Origins:

Latin *tunica, ae* "skin, coating, membrane, husk."

Tupa G. Don Campanulaceae

Origins:

From a vernacular Chilean name.

Tupeia Cham. & Schltdl. Loranthaceae

Origins:

From the Maori name *tupia* for a plant; see H.E. Connor and E. Edgar, "Name changes in the indigenous New Zealand flora, 1960-1986 and Nomina Nova IV, 1983-1986." *New Zealand Journal of Botany.* Vol. 25: 115-170. 1987; H.H. Allan, *Fl. New Z.* 1: 412-416. 1961.

Tupidanthus Hook.f. & Thomson Araliaceae

Origins:

Greek *tupis* "a mallet" and *anthos* "flower," referring to the shape of the flower buds.

Tupistra Ker Gawler Convallariaceae (Liliaceae)

Origins:

From the Greek *tupis, tupas* "a mallet, hammer," referring to the form of the stigma.

Turaniphytum Poljakov Asteraceae

Origins:

Turanian: A person of Ural–Altaic stock; Persian *Turan* for a country north of the Oxus River (the ancient name for the Amu Darya, a river of central Asia).

Turbina Raf. Convolvulaceae

Origins:

Latin *turbo, turbinis* "a whirlwind, a spinning-top, a spinning object," referring to the flowers; see Elmer D. Merrill, *Index rafinesquianus.* The plant names published by C.S. Rafinesque, etc. 201. 1949; Constantine Samuel Rafinesque (1783-1840), *Flora Telluriana.* 4: 81. Philadelphia 1836 [1838]; R. Gordon Wasson, "Notes on the present status of Ololiuhqui and the other hallucinogens of Mexico." from *Botanical Museum Leaflets, Harvard University.* Vol. 20(6): 161-212. Nov. 1963; Blas Pablo Reko, *Mitobotánica Zapoteca.* [Appended by an analysis of "Lienzo de Santiago Guevea"] Tacubaya 1945.

Species/Vernacular Names:

T. corymbosa (L.) Raf.

Mexico: flor de la Virgen, xtabentún, ololiuhqui (meaning el redondito), coatlxoxouhqui (meaning culebra verde), coaxihuitl (= yerba de la culebra), cuetzpallin (meaning lagartija)

T. oblongata (E. Mey. ex Choisy) A. Meeuse (*Ipomoea lambtoniana* Rendle; *Ipomoea oblongata* E. Mey. ex Choisy; *Ipomoea randii* Rendle)

Southern Africa: mothokho (Sotho)

T. oenotheroides (L.f.) A. Meeuse (*Ipomoea argyreoides* Choisy; *Ipomoea barrettii* Rendle; *Convolvulus oenotheroides* Choisy)

South Africa: nuwejaarsblom

Turbinicarpus (Backeb.) F. Buxb. & Backeb. Cactaceae

Origins:

Latin *turbo, turbinis* "a whirlwind, a spinning-top, a spinning object" and *carpus* "fruit," Greek *karpos* "fruit."

Turczaninovia DC. Asteraceae

Origins:

For the Russian botanist Porphir Kiril Nicolai Stepanowitsch Turczaninow, 1796-1863 (or 1864), traveler, botanical explorer, administrator; see J.H. Barnhart, *Biographical Notes upon Botanists.* 3: 408. 1965; Emil Bretschneider (1833-1901), *History of European Botanical Discoveries in China.* [Reprint of the original edition 1898.] Leipzig 1981; H.N. Clokie, *Account of the Herbaria of the Department of Botany in the University of Oxford.* 257. Oxford 1964; E.M. Tucker, *Catalogue of the Library of the Arnold Arboretum of Harvard University.* Cambridge, Massachusetts 1917-1933; Ida Kaplan Langman, *A Selected Guide to the Literature on the Flowering Plants of Mexico.* Philadelphia 1964; Antoine Lasègue, *Musée botanique de M. Benjamin Delessert.* 1845; Stafleu and Cowan, *Taxonomic Literature.* 6: 537-541. 1986; R. Zander, F. Encke, G. Buchheim and S. Seybold, *Handwörterbuch der Pflanzennamen.* 14. Aufl. Stuttgart 1993; N. Hall, *Botanists of the Eucalypts.* Melbourne 1978 and Supplement 1980.

Turczaninoviella Koso-Pol. Umbelliferae

Origins:

For the Russian botanist Porphir Kiril Nicolai Stepanowitsch Turczaninow, 1796-1863 (or 1864).

Turgenia Hoffm. Umbelliferae

Origins:

Latin *turgeo* "to swell," referring to the fruit.

Turgeniopsis Boiss. Umbelliferae

Origins:
Resembling *Turgenia*.

Turia Forssk. ex J.F. Gmelin Cucurbitaceae

Origins:

From *turia*, an Arabic vernacular name; see Pehr (Peter) Forsskål (1732-1763), *Flora aegyptiaco-arabica*. 165. Copenhagen 1775.

Turnera L. Turneraceae

Origins:

For the English (b. Morpeth, Northumberland) botanist Rev. William Turner, *circa* 1508-1568 (d. London), physician, herbalist, naturalist, zoologist, clergyman, dean of Wells cathedral, traveled in Europe, friend of Conrad von Gesner (1516-1565). His writings include *Libellus de Re Herbaria Novus*. Londini [London] 1538, *The names of herbes in Greke, Latin, Englishe Duche*, etc. London 1548 and *A New Herball*. London 1551; see Charles Webster, in *Dictionary of Scientific Biography* 13: 501-503. New York 1981; Ray Desmond, *Dictionary of British & Irish Botanists and Horticulturists*. 697. 1994; R.D. Jones, *William Turner, Tudor Nat., Physician and Divine*. 1988; W.A. Cooke, *First Records of British Flowering Plants*. London 1900; C.E. Raven, *English Naturalists from Neckam to Ray*. Cambridge 1947; Blanche Henrey, *British Botanical and Horticultural Literature before 1800*. Oxford 1975; Robert William Theodore Gunther (1869-1940), *Early British Botanists and their Gardens*. Oxford 1922; R. Pulteney, *Historical and Biographical Sketches of the Progress of Botany in England*. London 1790.

Species/Vernacular Names:

T. diffusa Willd.

Central America: damiana, chac-mixib, hierba de la pastora, mejorana, misibcoc

T. ulmifolia L.

English: West Indian holly, sage rose, yellow alder

Malaya: lidah kuching

Turpethum Raf. Convolvulaceae

Origins:

See C.S. Rafinesque, *Flora Telluriana*. 4: 71. 1836 [1838]; E.D. Merrill (1876-1956), *Index rafinesquianus*. The plant names published by C.S. Rafinesque, etc. 201. Jamaica Plain, Massachusetts, USA 1949.

Turpinia Ventenat Staphyleaceae

Origins:

After the French (b. Vire) botanical artist Pierre Jean François Turpin, 1775-1840 (d. Paris), naturalist and botanist, from 1794 to 1802 traveled in Haiti, Tortuga and the United States, author of *Observations sur la famille des Cactées*. Paris 1830; see François Pierre Chaumeton (1775-1819), *Flore médicale*. Paris 1828-1832; Jules Paul Benjamin Delessert (1773-1847), *Icones selectae Plantarum*. Paris 1820-1846; J.H. Barnhart, *Biographical Notes upon Botanists*. 3: 410. 1965; R. Zander, F. Encke, G. Buchheim and S. Seybold, *Handwörterbuch der Pflanzennamen*. 14. Aufl. Stuttgart 1993; M. Hocquette, in *Dictionary of Scientific Biography* 13: 506-507. 1981; Ida Kaplan Langman, *A Selected Guide to the Literature on the Flowering Plants of Mexico*. Philadelphia 1964; Antoine Lasègue, *Musée botanique de M. Benjamin Delessert*. 1845; E.M. Tucker, *Catalogue of the Library of the Arnold Arboretum of Harvard University*. Cambridge, Massachusetts 1917-1933; G. Schmid, *Goethe und die Naturwissenschaften*. Halle 1940; Giulio Giorello & Agnese Grieco, a cura di, *Goethe scienziato*. Einaudi Editore, Torino 1998.

Species/Vernacular Names:

T. ternata Nakai

Japan: shôben-no-ki, jiibuta

Turraea L. Meliaceae

Origins:

Origin not very clear, possibly after the Italian botanist and physician Antonio Turra (1730-1796), mineralogist, author of *Istoria del arbore della China*. Livorno 1764 and *Florae italicae prodromus*. Vicetiae [Vicenza] 1780; see C. Linnaeus, *Mantissa Plantarum*. 2: 150, 237. 1771; J.H. Barnhart, *Biographical Notes upon Botanists*. 3: 410. 1965; T. W. Bossert, *Biographical Dictionary of Botanists Represented in the Hunt Institute Portrait Collection*. 409. Boston, Massachusetts 1972; G. Schmid, *Goethe und die Naturwissenschaften*. Halle 1940; Jonas C. Dryander, *Catalogus bibliothecae historico-naturalis Josephi Banks*. London 1796-1800; Helmut Genaust, *Etymologisches Wörterbuch der botanischen Pflanzennamen*. 665. 1996; A. Lasègue, *Musée botanique de Benjamin Delessert*. Paris

1845; R. Zander, F. Encke, G. Buchheim and S. Seybold, *Handwörterbuch der Pflanzennamen.* 14. Aufl. 792. 1993; James A. Baines, *Australian Plant Genera.* An Etymological Dictionary of Australian Plant Genera. 383-384. [named for Giorgio dalla Torre (1607-1688), professor of botany at Padua. Among his writings are *Catalogus plantarum horti patavini novo incremento locupletior.* Padova 1660 and *Historia plantarum.* Padova 1685] Chipping Norton, N.S.W. 1981.

Species/Vernacular Names:

T. spp.

East Africa: mwoza nyama (Digo); gal-l (Iraqw); ol-nyerima (Masai)

T. casimiriana Harms

La Réunion Island: bois de Quivi, Quivi, petit Quivi, ti bois de quivi

T. floribunda Hochst. (*Turraea heterophylla* sensu Sond.)

English: wild honeysuckle tree, many-flowered turraea

Malawi: chikwisimbi (Chichewa, Nyanja)

Southern Africa: uMadlozane, uMadlozana, umLulama, uBhugulo (Zulu); umHlatholana, umLahlana (Xhosa)

Uganda: pogoliech (Acholi); muhojole (Lunyuli)

Tanzania: msonganya (Shambaa)

T. laciniata (Balf.f.) Harms

Rodrigues Island: bois balais

T. nilotica Kotschy e Peyr. (*Turraea randii* Bak.f.; *Turraea tubulifera* C. DC.)

English: small mahogany

Eastern Africa: mkobala (Gogo)

Southern Africa: tshigombo (Venda)

Malawi: msindira, msindila, mkulabala (Chichewa and Nyanja and Yao)

Zimbabwe: chiPindura, muZaramanga, muZaramanya (Shona); muSusi an Singongo (Lozi; see E.Colson and Max Gluckman, ed., *Seven Tribes of British Central Africa.* [The tribes being the Lozi, Tonga, Bemba, Ngoni, Nyakyusa, Yao and Shona.] London 1951; Wilfred Whiteley (and J. Slaski), *Bemba and Related Peoples of Northern Rhodesia* (and) *Peoples of the Lower Luapula Valley.* London 1951); muZaramanga (Karanga; see Rev. Francisque Marconnés, *A Grammar of Central Karanga.* Witwatersrand 1931)

T. obtusifolia Hochst. (*Turraea oblancifolia* Brem.; *Turraea obtusifolia* Hochst. var. *microphylla* C. DC.; *Turraea obtusifolia* Hochst. var. *matopensis* Bak.f.)

English: small honeysuckle tree, lesser honeysuckle tree

Southern Africa: kleinkamperfoelieboom, kleinkanferfoelieboom (Afrikaans); amehlo, inkosikasi (= the Queen's Eyes, the fruits and seeds are very showy), inKunzi, inKunzi ebomvana, amaZulu (Zulu)

T. ovata Cav.

La Réunion Island: gros bois de quivi

T. robusta Guerke (*Turraea volkensii* Guerke)

Tanzania: mzikoziko (Zaramo)

Uganda: kivunambasa (Lusoga); omukarakare (Lukiga)

Eastern Africa: dwayu (Usambara)

T. vogelii Hook.f.

Yoruba: aha omode, asa omode, omimi

T. zambesica Sprague & Hutch.

Southern Africa: Zambesi turraea; motulu (Kololo, Barotseland)

Turraeanthus Baillon Meliaceae

Origins:

The genus *Turraea* and *anthos* "flower," very likely for the Italian botanist and physician Antonio Turra, 1730-1796.

Species/Vernacular Names:

T. sp.

Ghana: apaya

Liberia: blimah pu

T. africanus (C. DC.) Pellegrin

W. Africa: aks

Nigeria: avodire, avodiri

Ghana: apaya, kwadwuma, suinguisu, wansawa

Cameroon: asama, engang

Ivory Coast: avodire, agboui, avodiré, hague, hakue

Zaire: esasaw, kisanda

Congo: avodire, kilembe lengoba

Turricula J.F. Macbr. Hydrophyllaceae

Origins:

From the Latin *turricula, ae* "a little tower, a turret," *turris, is* "a tower."

Species/Vernacular Names:

T. parryi (A. Gray) J.F. Macbr.

English: poodle-dog bush

Turrigera Decne. Asclepiadaceae

Origins:

Latin *turriger, gera, gerum* (*turris gero*) "turret-bearing, turreted," *turrigera, ae* "turret-crowned," an epithet of Cybele.

Turrillia A.C. Smith Proteaceae

Origins:

For the British (b. Woodstock, Oxfordshire) botanist William Bertram Turrill, 1890-1961 (d. Kew, Surrey), from 1916 to 1918 with the Royal Army Medical Corps in Macedonia, 1925 Fellow of the Linnean Society, 1958 Fellow of the Royal Society, at Oxford with George Claridge Druce (1850-1932), collaborated with Eric Marsden Marsden-Jones (1887-1960), from 1946 to 1957 Keeper of the herbarium and library at Kew. His publications include "A contribution to the flora of Fiji." *J. Linn. Soc., Bot.* 43: 15-39. 1915, *Joseph Dalton Hooker*, botanist, explorer, and administrator. Canadian Book Club 1963 and *The Plant-Life of the Balkan Peninsula*. Oxford 1929; see J.H. Barnhart, *Biographical Notes upon Botanists.* 3: 410. 1965; T. W. Bossert, *Biographical Dictionary of Botanists Represented in the Hunt Institute Portrait Collection.* 409. Boston, Massachusetts 1972; H.N. Clokie, *Account of the Herbaria of the Department of Botany in the University of Oxford.* 257. Oxford 1964; Stafleu and Cowan, *Taxonomic Literature.* 6: 552-553. 1986; E.M. Tucker, *Catalogue of the Library of the Arnold Arboretum of Harvard University.* Cambridge, Massachusetts 1917-1933; R. Zander, F. Encke, G. Buchheim and S. Seybold, *Handwörterbuch der Pflanzennamen.* 14. Aufl. 792. Stuttgart 1993; Ray Desmond, *Dictionary of British & Irish Botanists and Horticulturists.* 697-698. London 1994.

Turrita Wallr. Brassicaceae

Origins:

Latin *turrita, ae* "tower-crowned, tower-shaped, high."

Turritis L. Brassicaceae

Origins:

Latin *turris* "a tower," *turrita, ae* "tower-crowned, tower-shaped," referring to the pyramidal appearance of the plant; see Carl Linnaeus, *Species Plantarum.* 666. 1753 and *Genera Plantarum.* Ed. 5. 298. 1754.

Tussaca Rafinesque Orchidaceae

Origins:

For the French botanist François Richard de Tussac, 1751-1837, traveler and plant collector (Martinique, Haiti and Jamaica), author of *Flore des Antilles.* Paris 1808-1827 [1828] and *Cri des colons contre un ouvrage de M. l'évêque et sénateur Grégoire*, ayant pour titre "De la Littérature des Nègres," etc. Paris 1810; see J.H. Barnhart, *Biographical*

Notes upon Botanists. 3: 410. 1965; A. Lasègue, *Musée botanique de Benjamin Delessert.* Paris 1845; Ethelyn Maria Tucker, *Catalogue of the Library of the Arnold Arboretum of Harvard University.* Cambridge, Massachusetts 1917-1933; R. Zander, F. Encke, G. Buchheim and S. Seybold, *Handwörterbuch der Pflanzennamen.* 14. Aufl. 1993; E.D. Merrill, *Index rafinesquianus.* The plant names published by C.S. Rafinesque, etc. 105. Jamaica Plain, Massachusetts, USA 1949; C.S. Rafinesque, *Jour. Phys. Chim. Hist. Nat.* 89: 261. 1819 and *Flora Telluriana.* 2: 87. 1836 [1837].

Tussacia Benth. Gesneriaceae

Origins:

For the French botanist François Richard de Tussac, 1751-1837.

Tussacia Raf. ex Desv. Orchidaceae

Origins:

For the French botanist François Richard de Tussac, 1751-1837.

Tussilago L. Asteraceae

Origins:

Latin *tussilago, inis* "the herb colt's foot" (Plinius).

Species/Vernacular Names:

T. farfara L.

English: colt's foot

China: kuan dong hua, kuan tung

Tutcheria S.T. Dunn Theaceae

Origins:

For the British (b. Bristol) botanist William James Tutcher, 1867-1920 (d. Hong Kong), plant collector (Chinese plants), gardener, 1904 Fellow of the Linnean Society, Kew gardener, 1910-1920 Superintendent of the Botanical and Forestry Department Hong Kong (Director of the Botanical Garden of Hong Kong), with Stephen Troyte Dunn (1868-1938) wrote *Flora of Kwangtung and Hongkong (China)*, etc. [Royal Botanic Gardens. Bulletin of Miscellaneous Information. Additional Series X.] London 1912; see J.H. Barnhart, *Biographical Notes upon Botanists.* 3: 410. 1965; Emil Bretschneider, *History of European Botanical Discov-*

eries in China. Leipzig 1981; E.M. Tucker, *Catalogue of the Library of the Arnold Arboretum of Harvard University.* Cambridge, Massachusetts 1917-1933; Elmer D. Merrill and Egbert H. Walker, *A Bibliography of Eastern Asiatic Botany.* 506. 1938; Ray Desmond, *Dictionary of British & Irish Botanists and Horticulturists.* 222, 698. London 1994.

Tuxtla Villaseñor & Strother Asteraceae

Origins:
Tuxtla Gutiérrez, the capital of Chiapas state, southern Mexico.

Tweedia Hooker & Arnott Asclepiadaceae

Origins:
After the Scottish (b. Lanarkshire) botanist John (or James) Tweedie, 1775-1862 (d. Santa Catalina, Buenos Aires, Argentina), botanical explorer (Uruguay, Argentina), gardener at the Royal Botanic Garden, Edinburgh, traveler and plant collector, sent seeds and plants to W.J. Hooker; see A. Lasègue, *Musée botanique de Benjamin Delessert.* 486-487. Paris 1845; J.H. Barnhart, *Biographical Notes upon Botanists.* 3: 411. 1965; A. Castellanos, in *Holmbergia.* 4(8): 3-14. 1945; E.C. Nelson & Eileen May McCracken (1920-1988), *The Brightest Jewel: A History of the National Botanic Gardens, Glasnevin, Dublin.* Kilkenny 1987; H.N. Clokie, *Account of the Herbaria of the Department of Botany in the University of Oxford.* 257. Oxford 1964; H.R. Fletcher and W.H. Brown, *Royal Botanic Garden Edinburgh, 1670-1970.* Edinburgh 1970; Gordon Douglas Rowley, *A History of Succulent Plants.* Strawberry Press, Mill Valley, California 1997; Ray Desmond, *Dictionary of British & Irish Botanists and Horticulturists.* 698. 1994; Alice Margaret Coats, *The Quest for Plants. A History of the Horticultural Explorers.* London 1969.

Tylanthera C. Hansen Melastomataceae

Origins:
Greek *tylos* "lump, knob" and *anthera* "anther."

Tylecarpus Engl. Icacinaceae

Origins:
Greek *tylos, tyle* "lump, knob, any swelling" and *karpos* "fruit"; see also *Medusanthera* Seem.

Tylecodon Tölken Crassulaceae

Origins:
Anagram of *Cotyledon,* in which the genus *Tylecodon* was formerly placed.

Species/Vernacular Names:
T. ellaphieae E.J. van Jaarsveld (in honor of the South African (b. Pretoria) botanical artist Johanna Ellaphie Ward-Hilhorst (née Hilhorst), 1920-1994 (d. Cape Town), interested in the genus *Pelargonium,* illustrator of Johannes Jacobus Adriaan Van der Walt (b. 1938) & Pieter Johannes Vorster (b. 1945), *Pelargoniums of Southern Africa.* Cape Town 1981, and of Deirdré Anne Snijman (b. 1949), *A Revision of the genus Haemanthus L.* (Amaryllidaceae). Claremont 1984; in 1990 she was awarded of the Royal Horticultural Society's Gold Medal for her paintings of *Haemanthus;* see Gordon Douglas Rowley, *A History of Succulent Plants.* Mill Valley, California 1997; Mary Gunn and Leslie Edward W. Codd, *Botanical Exploration of Southern Africa.* 371. Cape Town 1981)

T. kritzingeri E.J. van Jaarsveld (the species was named after Mr. Kobus Kritzinger of the Cape Department of Nature Conservation, one of the collectors of the species in September 1981)

T. paniculatus (L.f.) Tölken (*Cotyledon paniculata* L.f.; *Cotyledon fascicularis* Ait.; *Cotyledon tardiflora* Bonpl.)

South Africa: botterboom, baadjiebos, bandjiebos, botterbaanboom, botterblom, botterbos, plakkie

English: butter tree, butter bush

T. reticulatus (L.f.) Tölken subsp. *reticulatus* (*Cotyledon parvula* Burch.; *Cotyledon reticulata* L.)

South Africa: oukoe, oukoeicotyledon, poppiebos

T. ventricosus (Burm.f.) Tölken (*Cotyledon ventricosa* Burm.f.; *Cotyledon ventricosa* Burm.f. var. *alpina* Harv.)

South Africa: klipnenta, klipnentacotyledon, krimpsiekte, neenta, nentabossie, nanta, nenta

T. wallichii (Harv.) Tölken subsp. *wallichii* (*Cotyledon wallichii* Harv.)

English: Wallich cotyledon

South Africa: Wallich cotyledon, bandjiesbos, c'nenta, kandelaarbos, karkeibos, kokerbos, komi-ganna, krimpsiektebos, nenta, poppebos

Tyleropappus Greenm. Asteraceae

Origins:
Greek *tyleros* "callous" and *pappos* "fluff, downy appendage."

Tylocarya Nelmes Cyperaceae

Origins:
From the Greek *tylos* "lump, knob, any swelling" and *karyon* "nut"; see also *Fimbristylis* Vahl.

Tylochilus Nees Orchidaceae

Origins:
Greek *tylos* and *cheilos* "a lip, margin," referring to the warty lip.

Tylodontia Griseb. Asclepiadaceae

Origins:
Greek *tylos, tyle* "lump, knob, any swelling" and *odous, odontos* "tooth."

Tyloglossa Hochst. Acanthaceae

Origins:
From the Greek *tylos, tyle* plus *glossa* "tongue."

Tylopetalum Barneby & Krukoff Menispermaceae

Origins:
From the Greek *tylos, tyle* with *petalon* "a petal, leaf."

Tylophora R. Br. Asclepiadaceae

Origins:
From the Greek *tylos* "lump, knob" and *phoros* "bearer," referring to the pollen masses or to the tubercled corona; see R. Brown, *Prodromus Florae Novae Hollandiae*. 460. 1810.

Species/Vernacular Names:
T. arenicola Merrill
English: sandhill tylophora
China: hu xu wa er teng
T. augustiniana (Hemsley) Craib
English: Yichang tylophora
China: yi chang wa er teng
T. chingtungensis Tsiang & P.T. Li
English: Chingtung tylophora
China: xian mai wa er teng
T. floribunda Miquel
English: manyflowered tylophora
China: duo hua wa er teng, wa er teng
T. glabra Costantin (*Tylophora longipedicellata* Tsiang & P.T. Li; *Tylophora renchangii* Tsiang)
English: long-pedicel tylophora, Renchang tylophora
China: chang geng wa er teng
T. hui Tsiang
English: Hu tylophora
China: jian shui wa er teng
T. indica (Burm.f.) Merr.
India: antomul, anantamul, damni, nach-churuppam, nanja-murichchaan
T. japonica Miq. (*Tylophora liukiuensis* Matsum.)
Japan: tokiwa-kamome-zuru
T. kerrii Craib
English: Kerr tylophora
China: ren shen wa er teng
T. koi Merrill
English: Ko tylophora
China: tong tian lian
T. leptantha Tsiang
English: broad-inflorescence tylophora
China: guang hua wa er teng
T. nana C.K. Schneider
English: dwarf tylophora
China: wen chuan wa er teng
T. ovata (Lindley) Hooker ex Steudel
English: ovate tylophora
China: wa er teng, san shii liu dang
T. rotundifolia Buch.-Ham. ex Wight (*Tylophora trichophylla* Tsiang)
English: hairy-leaf tylophora
China: yuan ye wa er teng
T. secamonoides Tsiang
English: secamone-like tylophora
China: she den cao
T. silvestris Tsiang
English: woods tylophora
China: gui zhou wa er teng
T. sylvatica Decne.
Yoruba: afeje kosun
T. yunnanensis Schlechter

English: Yunnan tylophora

China: yun nan wa er teng

Tylophoropsis N.E. Br. Asclepiadaceae

Origins:

Resembling *Tylophora* R. Br.

Tylopsacas Leeuwenb. Gesneriaceae

Origins:

Greek *tylos* "lump, knob" and *psakas*, *psekas* "a grain, crumb, a small drop, any small piece rubbed or broken off."

Tylosema (Schweinf.) Torre & Hillcoat Caesalpiniaceae

Origins:

From the Greek *tylos* and *sema* "standard, a sign, mark," referring to the large seeds; see also *Bauhinia*.

Species/Vernacular Names:

T. fassoglense (Kotschy ex Schweinf.) Torre & Hillc. (*Bauhinia cissoides* Oliv.; *Bauhinia fassoglensis* Kotschy ex Schweinf.; *Bauhinia kirkii* Oliv.; *Bauhinia welwitschii* Oliv.), "Apapa"

English: yellow creeping bauhinia, camel's foot

Zimbabwe: muMbova, muRama, muTewa (Shona)

Tylosepalum Kurz ex Teijsm. & Binn. Euphorbiaceae

Origins:

From the Greek *tylos* "lump, knob" and Latin *sepalum* "a sepal"; see also *Trigonostemon* Blume.

Tylosperma Botsch. Rosaceae

Origins:

Greek *tylos* "lump, knob" and *sperma* "seed."

Tylosperma Leeuwenb. Gesneriaceae

Origins:

From the Greek *tylos* and *sperma* "seed"; see also *Tylopsacas* Leeuwenb.

Tylostemon Engl. Lauraceae

Origins:

From the Greek *tylos* "lump, knob" and *stemon* "stamen."

Tylostigma Schltr. Orchidaceae

Origins:

From the Greek *tylos* and *stigma* "stigma, a spot, mark," referring to the shape of the stigma.

Tylostylis Blume Orchidaceae

Origins:

From the Greek *tylos* "lump, knob" and *stylos* "a pillar, style," descriptive of the foot of the column.

Tylothrasya Döll Gramineae

Origins:

From the Greek *tylos* plus the genus *Thrasya* Kunth.

Tynanthus Miers Bignoniaceae

Origins:

See *Tynnanthus* Miers.

Tynnanthus Miers Bignoniaceae

Origins:

Greek *tynnos* "so small" and *anthos* "flower."

Typha L. Thyphaceae

Origins:

Typhe, tiphe applied by Theophrastus, Dioscorides and Aristotle (*Historia animalium*) to a kind of grass or straw or other aquatic plants used for stuffing beds and bolsters, Aristophanes (*The Acharnians*) and Claudius Aelianus (*De natura animalium*) applied to a species of water bug or beetle; Latin *tiphe, es* for a kind of grain, Peter's corn, one-grained wheat (Plinius); see Carl Linnaeus, *Species Plantarum*. 971. 1753 and *Genera Plantarum*. Ed. 5. 418. *1754*; Carmen Aguilera, *Flora y fauna Mexicana*. Mitología y tradiciones. 152-153. México s.d. [1985]; Blas Pablo Reko, *Mitobotánica Zapoteca*. [Appended by an analysis of "Lienzo de Santiago Guevea"] Tacubaya 1945; M. Cortelazzo & P. Zolli, *Dizionario etimologico della lingua italiana*. 5: 1338. Bologna 1988; G. Semerano, *Le origini della cultura europea*. Dizionari Etimologici. Dizionario della lingua Greca. 2(1): 293. Leo S. Olschki Editore, Firenze 1994; H. Genaust, *Etymologisches Wörterbuch der botanischen Pflanzennamen*. 666. 1996.

Species/Vernacular Names:

T. spp.

Mexico: tule, pop, tollin, beecho, peecho, taa beecho, taa peecho, coba beecho, coba peecho, coba gana, coba yaguemma, coba yaquemma

T. angustifolia L.

English: cat tail, narrow-leaved cat tail, soft flag, lesser bulrush, bullrush, bulrush, narrow-leaved reedmace, lesser reedmace, small reedmace

China: puhuang

The Philippines: balangot, kaidked, lampakanai, tubal-tubal, homai-homai

Maori names: raupo, hune (= pappus of the seeds)

T. capensis (Rohrb.) N.E. Br. (*Typha latifolia* L. subsp. *capensis* Rohrb.)

English: bullrush, bulrush, cat's tail, common bullrush, common cat tail, Cossack asparagus, nail-rod, poker plant, reedmace

Southern Africa: gewone papkuil, matjiesgoed, palmiet, papies, papkuil; ibhuma (Zulu); mositla (Sotho); umkhanzi, inGcongolo (Xhosa)

T. domingensis (Pers.) Steud. (*Typha angustata* Bory & Chaub.; *Typha brownii* Kunth; *Typha basedowii* Graebner)

English: cat tail, bulrush, southern cat tail, reed-mace, small bulrush

Arabic: berdi, bardi

Australia: narrow-leaved cumbungi, narrow-leaved bulrush

Japan: hime-gama

Okinawa: kama, hama

China: xiang pu

Yoruba: ewu egungun

T. latifolia L.

English: bullrush, cat's tail, cat tail, nailrod, common cat's tail, great reedmace, broad-leaved cat tail

China: xiang pu

T. orientalis C. Presl

English: broad-leaved bulrush, bulrush

Australia: broad-leaved cumbungi

China: hsiang pu

Typhoides Moench Gramineae

Origins:

Greek *typhe, tiphe* and *eidos, oides* "resemblance."

Typhonium Schott Araceae

Origins:

See Helmut Genaust, *Etymologisches Wörterbuch der botanischen Pflanzennamen*. 666. Basel 1996; D.H. Nicolson, "Derivation of aroid generic names." *Aroideana*. 10: 15-25. 1988; Greek *typhonios, typhonion* applied by Dioscorides (3.26) to a kind of lavandula, *typhonia* is the plant *stoichas*.

Species/Vernacular Names:

T. divaricatum (L.) Decne. (*Arum divaricatum* L.)

China: li tou jian

Japan: Ryûkyû-hange

Okinawa: uwen-chunujû

Malaya: birak kechil

T. giganteum Engl.

China: tu chio lien

Typhonodorum Schott Araceae

Origins:

Greek *typhos* "a furious whirlwind, typhoon, stormy wind, a waterspout" and *doron* "a gift." In Greek legend Typhon or Typhaon was a hundred-headed monster buried by Zeus in Tartarus under Mt. Aetna; see Helmut Genaust, *Etymologisches Wörterbuch der botanischen Pflanzennamen*. 666. 1996; D.H. Nicolson, "Derivation of aroid generic names." *Aroideana*. 10: 15-25. 1988.

Typtomene F. Muell. Myrtaceae

Origins:

Orthographic variant, see *Thryptomene* Endl.

Tyria Klotzsch ex Endl. Euphorbiaceae

Origins:

Probably from the Greek *tyros* "cheese," *tyreia* "a making of cheese," or dedicated to Tyre, the ancient town of Syria, the famous maritime and commercial city of the Phoenicians.

Tyrimnus (Cass.) Cass. Asteraceae

Origins:

Probably from the Greek *tyros* "cheese," *tyreia* "a making of cheese," or dedicated to Tyre, the ancient town of Syria, the famous maritime and commercial city of the Phoenicians.

Tysonia Bolus Boraginaceae

Origins:

After the South African (b. Port Royal, Jamaica) botanist William Tyson, 1851-1920 (d. Grahamstown, C.P., South Africa), c. 1874 to South Africa, plant collector and tutor, a school teacher, 1896 Fellow of the Linnean Society, collected marine algae; see E.P. Phillips, "A brief historical sketch of the development of botanical science in South Africa and the contribution of South Africa to botany." *S. Afr. J. Sc.* 27: 53. 1930; A.W. Bayer, *S. Afr. J. Sc.* 67: 407. 1971; I.H. Vegter, *Index Herbariorum.* Part II (7), *Collectors T-Z.* Regnum Vegetabile vol. 117. 1988; M. Gunn and L.E. Codd, *Botanical Exploration of Southern Africa.* 353. Cape Town 1981.

Tysonia F. Muell. Asteraceae

Origins:

Named after Isaac Tyson, c. 1859-1942, plant collector in Western Australia and Queensland; see F.J.H. von Mueller, *The Chemist and Druggist of Australasia.* 11: 215. 1896; I.H. Vegter, *Index Herbariorum.* Part II (7), *Collectors T-Z.* Regnum Vegetabile vol. 117. 1988; James A. Baines, *Australian Plant Genera. An Etymological Dictionary of Australian Plant Genera.* 385. N.S.W. 1981.

Tytthostemma Nevski Caryophyllaceae

Origins:

From the Greek *tytthos* "little, young" and *stemma* "crown, wreath, garland"; see also *Stellaria* L.

U

Uapaca Baillon Euphorbiaceae (Uapacaceae)

Origins:

A Madagascan vernacular name; see Werner Rauh, *Succulent and Xerophytic Plants of Madagascar.* 1: 42, 43, 92. 1995 and 2: 287. 1998; Helmut Genaust, *Etymologisches Wörterbuch der botanischen Pflanzennamen.* 667. Basel 1996.

Species/Vernacular Names:

U. sp.

Gabon: asam, moussanfi, moussanfi benga, ntombo, n'tyombi, tchombi, okess, ozombi

Ivory Coast: rikio, somon

Nigeria: ashiha

U. esculenta A. Chev. ex Aubrév. & Leandri

Nigeria: asisa, gagara (Yoruba)

Ivory Coast: borikio, kebi, namby

Zaire: mole, bossenge

U. guineensis Müll. Arg.

English: sugar plum, red cedar

Cameroon: asam, assam, banka, sengui, lissamba, lesambo, bossambi

Central Africa: dobo

Congo: n'samvi, n'sanvi

Gabon: assam, ossambi, yambi

Ivory Coast: elehoha, nan, rikio, kebi

Liberia: beyor, kindi

Yoruba: ajegbe, ujobe, yeye, yere, emido, abo emido

Nigeria: abo-emido, abo amido, abo, abua-agom, achi, ajele, bosambi, cosomon, berong, ile, kafafogo, obia, goramfi, nderiti, okon, ovia, oreng, oyen, sam, shem, rikio, ogbom, ubia, yere, likamba; abo-emido (Yoruba); obia (Igbo); okon (Ijaw); umpwenek (Edo); oriang (Ekoi); obubit nkpenek (Ibibio)

Sudan: somo

Zaire: bossenghe, kisamfi, samfi, makala, molengo, ndzara

U. heudelotii Baillon

W. Africa: ko somon

Cameroon: asam, ebam

Ivory Coast: kosomo, rikio

Yoruba: yere, yeye, akun, abo akun, abo emido, opon, opon atakun

Nigeria: onyen, ighen, ile, otehor; senchi (Nupe); akun, yeye (Yoruba); oyen (Edo); ighen (Itsekiri); ile (Ijaw); otehor (Urhobo); ibia-ile (Igbo); odang (Boki)

Zaire: bossenge

U. kirkiana Müll. Arg.

English: wild loquat

S. Central Africa: mahobo-hobo, mduku

Southern Africa: mahobohobo; muHobohobo, muJanje, muShugu, muShuku, muSuko, muTsangidze, muZhanje (Shona)

Zaire: makoom-makooba

Zambia: mohobohobo, musuku, msuku

U. nitida Müll. Arg.

Southern Africa: narrow-leaved mahobohobo; muKonokono, muKono u muZhanzhi, muSangidze, muTongoro, muTsangidze (Shona)

U. paludosa Aubrév. & Leandri

Nigeria: gagara (Yoruba)

U. sansibarica Pax

Southern Africa: lesser mahobohobo; muShenzi, muTongoro (Shona)

U. staudtii Pax

French: chêne d'Afrique

Cameroon: assam, bosambi, lissamba, rikio, sam

Tropical Africa: bihambi

Nigeria: onyen, akun, rikio, gagara, ashiha, ajele, likamba, shem; akun (Yoruba); onye (Edo); obubit nkpenek (Ibibio); nkpenek (Efik); oda (Boki)

U. togoensis Pax

Nigeria: kafafago (Hausa); bakurehi (Fula); senchi (Nupe); shase (Tiv); ajegbe (Yoruba); onyen (Edo); obia (Igbo); odang (Boki); goramfi (Kanuri)

Ivory Coast: rikio des montagnes, somon

W. Africa: somon, lisogho

Ucriana Willdenow Rubiaceae

Origins:

For the Italian botanist Father Bernardino da Ucria (Bernardinus ab Ucria) (Michelangelo Aurifici), 1739-1796, clergyman; see J.H. Barnhart, *Biographical Notes upon Botanists.* 1: 91. 1965; Ethelyn Maria Tucker, *Catalogue of the Library of the Arnold Arboretum of Harvard University.* Cambridge, Massachusetts 1917-1933; Mariella Azzarello Di Misa, a cura di, *Il Fondo Antico della Biblioteca dell'Orto Botanico di Palermo.* 37. Soprintendenza per i Beni Culturali e Ambientali. Sezione per i Beni Bibliografici. Regione Siciliana, Palermo 1988; Giuseppe M. Mira, *Bibliografia Siciliana.* 1: 97-98. Palermo 1881; R. Zander, F. Encke, G. Buchheim and S. Seybold, *Handwörterbuch der Pflanzennamen.* 14. Aufl. Stuttgart 1993.

Udora Nutt. Hydrocharitaceae

Origins:

Greek *hydor* "water," submerged aquatics, grown in ponds.

Udoza Raf. Hydrocharitaceae

Origins:

Based on *Udora* Nutt.; see C.S. Rafinesque, *Am. Monthly Mag. Crit. Rev.* 4: 195. 1819.

Uechtritzia Freyn Asteraceae

Origins:

After the German botanist Rudolf Karl Friedrich von Uechtritz, 1838-1886, son of the German entomologist and botanist Maximilian Friedrich Sigismund Freiherr von Uechtritz (1785-1851); see J.H. Barnhart, *Biographical Notes upon Botanists.* 3: 413. 1965; Stafleu and Cowan, *Taxonomic Literature.* 6: 562-564. 1986; R. Zander, F. Encke, G. Buchheim and S. Seybold, *Handwörterbuch der Pflanzennamen.* 14. Aufl. 1993; T. W. Bossert, *Biographical Dictionary of Botanists Represented in the Hunt Institute Portrait Collection.* 410. Boston, Massachusetts 1972; E.M. Tucker, *Catalogue of the Library of the Arnold Arboretum of Harvard University.* Cambridge, Massachusetts 1917-1933; Günther Schmid, *Chamisso als Naturforscher.* Eine Bibliographie. 263. Leipzig 1942.

Ugni Turcz. Myrtaceae

Origins:

A Chilean vernacular name; see F. Boerner & G. Kunkel, *Taschenwörterbuch der botanischen Pflanzennamen.* 4. Aufl. 183. Berlin & Hamburg 1989; Helmut Genaust, *Etymologisches Wörterbuch der botanischen Pflanzennamen.* 667. Basel 1996; Juan Ortiz Garmendia, *Plantas silvestres Chilenas de frutos comestibiles por el hombre.* 17. Museo

de la Serena, Chile 1969; National Research Council, *Lost Crops of the Incas: Little-Known Plants of the Andes with Promise for Worldwide Cultivation.* 219, 220. National Academy Press, Washington, D.C. 1989.

Species/Vernacular Names:

U. molinae Turcz.

English: Chilean guava, strawberry myrtle, Chilean cranberry

Chile: murta, murtilla, uñi

Uittienia Steenis Caesalpiniaceae

Origins:

After the Dutch botanist Hendrik Uittien, 1898-1944; see J.H. Barnhart, *Biographical Notes upon Botanists.* 3: 414. 1965; Stafleu and Cowan, *Taxonomic Literature.* 6: 568-571. Utrecht 1986.

Uladendron Marc.-Berti Malvaceae

Origins:

Greek *oulos* "tough, compact, crinkled" and *dendron* "tree."

Ulantha Hooker Orchidaceae

Origins:

From the Greek *oulos* "woolly, crisp, crinkled" and *anthos* "flower," descriptive of the floral parts.

Ulbrichia Urban Malvaceae

Origins:

After the German botanist Oskar Eberhard Ulbrich, 1879-1952; see J.H. Barnhart, *Biographical Notes upon Botanists.* 3: 414. 1965; Ida Kaplan Langman, *A Selected Guide to the Literature on the Flowering Plants of Mexico.* Philadelphia 1964; T. W. Bossert, *Biographical Dictionary of Botanists Represented in the Hunt Institute Portrait Collection.* 410. Boston, Massachusetts 1972; E.M. Tucker, *Catalogue of the Library of the Arnold Arboretum of Harvard University.* Cambridge, Massachusetts 1917-1933; Stafleu and Cowan, *Taxonomic Literature.* 6: 571-578. Utrecht 1986; R. Zander, F. Encke, G. Buchheim and S. Seybold, *Handwörterbuch der Pflanzennamen.* 14. Aufl. Stuttgart 1993.

Uldinia J.M. Black Umbelliferae

Origins:

From *uldilnga gabi*, native name of the Ooldea Soak, South Australia; see Arthur D. Chapman, ed., *Australian Plant Name Index*. 2926. Canberra 1991 James A. Baines, *Australian Plant Genera. An Etymological Dictionary of Australian Plant Genera.* 385. Chipping Norton, N.S.W. 1981; Francis Aubie Sharr, *Western Australian Plant Names and Their Meanings.* 72. University of Western Australia Press, Nedlands, Western Australia 1996.

Species/Vernacular Names:

U. ceratocarpa (W. Fitzg.) N. Burb. (*Hydrocotyle ceratocarpa* W. Fitzg.; *Hydrocotyle mercurialis* (J. Black) Hiroe; *Uldinia mercurialis* J. Black; *Maidenia acroptera* Domin; *Maidenia ceratocarpa* (W. Fitzg.) Domin; *Dominia acroptera* (Domin) Fedde)

English: creeping carrot

Uleanthus Harms Fabaceae

Origins:

For the German botanist Ernst Heinrich Georg Ule, 1854-1915, plant collector, botanical explorer in the Brazilian highlands and in Venezuela (Alta Guayana venezolana, Mniang, Caroní, Cuquenán), from 1913 to 1914 assistant at the Botanical Museum Berlin. Among his many works are *Die Vegetation des Roraima.* Leipzig und Berlin 1914, "Bericht über den Verlauf der zweiten Expedition in das Gebiet des Amazonenstromes in den Jahren 1908 bis 1912." *Notizbl. d. königl. bot. Gart. u. Mus.-Berlin.* 6: 78. 1914, "Die Pflanzenformationen des Amazonas-gebietes. II." *Bot. Jahrb. Syst.* 40(3): 398-432. 1908, "II. Beiträge zur Flora der Hylaea nach den Sammlungen von Ule's Amazonas-Expedition." *Verh. Bot. Vereins Prov. Brandenburg.* 48(2): 116-222. 1907 and "Über eine neue Gattung der Capparidaceen." *Ber. deut. bot. Ges.* 26a(3): [220]-224. pl. 2. 1908; see J.H. Barnhart, *Biographical Notes upon Botanists.* 3: 414. 1965; Stafleu and Cowan, *Taxonomic Literature.* 6: 578-583. 1986; R. Zander, F. Encke, G. Buchheim and S. Seybold, *Handwörterbuch der Pflanzennamen.* 14. Aufl. 792. 1993; T. W. Bossert, *Biographical Dictionary of Botanists Represented in the Hunt Institute Portrait Collection.* 410. Boston, Massachusetts 1972; H.N. Clokie, *Account of the Herbaria of the Department of Botany in the University of Oxford.* 258. Oxford 1964; Ethelyn Maria Tucker, *Catalogue of the Library of the Arnold Arboretum of Harvard University.* Cambridge, Massachusetts 1917-1933; Henri François Pittier, *Manual de las Plantas Usuales de Venezuela* y su Suplemento. Caracas 1971; H. Genaust, *Etymologisches Wörterbuch der botanischen Pflanzennamen.* 667. 1996; August Weberbauer (1871-1948), *Die Pflanzenwelt der peruanischen Andes in ihren Grundzügen dargestellt.* Leipzig 1911; Ida Kaplan Langman, *A Selected Guide to the Literature on the Flowering Plants of Mexico.* Philadelphia 1964.

Ulearum Engl. Araceae

Origins:

After the German botanist and explorer Ernst Heinrich Georg Ule, 1854-1915, plus the genus *Arum*; see A. Engler, "Ulearum Engl. nov. gen." *Botanische Jahrbücher für Systematik, Pflanzengeschichte und Pflanzengeographie.* 37: 95-96. 1905 [1906]; D.H. Nicolson, "Derivation of aroid generic names." *Aroideana.* 10: 15-25. 1988.

Uleiorchis Hoehne Orchidaceae

Origins:

Named for the German botanist and botanical explorer Ernst Heinrich Georg Ule, 1854-1915. He collected the type material of the genus.

Uleodendron Rauschert Moraceae

Origins:

After the German botanist and botanical explorer Ernst Heinrich Georg Ule, 1854-1915, and *dendron* "tree."

Uleophytum Hieron. Asteraceae

Origins:

After the German botanist and botanical explorer of the Amazon Ernst Heinrich Georg Ule, 1854-1915, and *phyton* "plant."

Ulex L. Fabaceae

Origins:

Ulex, ulicis, the ancient Latin name used by Plinius for a kind of heather or a shrub resembling rosemary; see Carl Linnaeus, *Species Plantarum.* 741. 1753 and *Genera Plantarum.* Ed. 5. 329. 1754.

Species/Vernacular Names:

U. europaeus L.

English: furze, gorse, Irish gorse, whin

Portuguese: carqueja, tojo

U. gallii Planchon

English: dwarf gorse

U. minor Roth (*Ulex nanus* T. Forster ex Symons)

English: dwarf furze, dwarf gorse

Ulleria Bremek. Acanthaceae

Origins:

See *Ruellia* L.

Ulloa Pers. Solanaceae

Origins:

See *Juanulloa* Ruíz & Pav.

Ullucus Caldas Basellaceae

Origins:

A native name from Latin America, Peru; see Henri François Pittier (1857-1950), *Manual de las Plantas Usuales de Venezuela* y su Suplemento. Caracas 1971; National Research Council, *Lost Crops of the Incas: Little-Known Plants of the Andes with Promise for Worldwide Cultivation*. National Academy Press, Washington, D.C. 1989.

Species/Vernacular Names:

U. tuberosus Caldas

English: ulluco, melloco

Peru: olloco, papa lisa, tuna ullush, uljucu, ullucu, oca quina

Venezuela: ruba, timbós, mucuchí, migurí, tiquiño

Quechua name: ullucu

Aymara names: ulluma, ullucu

Spanish: melloco, olluco, ulluco, rubas, rubia, ruba, tiquiño, timbós, mucuchi, michuri, michiruí migurí, camarones de tierra, ruhuas, hubas, chuguas, chigua, papa lisas, lisas, olloco, ulluca, ulluma, papa lisa

Ulmus L. Ulmaceae

Origins:

The classical Latin name for an elm-tree; Akkadian *u'ulum* "to bind," *u'iltu* "bond"; see Carl Linnaeus, *Species Plantarum*. 225. 1753 and *Genera Plantarum*. Ed. 5. 106. 1754; G. Semerano, *Le origini della cultura europea*. Dizionario della lingua Latina e di voci moderne. 2(2): 598. 1994; H. Genaust, *Etymologisches Wörterbuch der botanischen Pflanzennamen*. 667. 1996; Ernest Weekley, *An Etymological*

Dictionary of Modern English. 1: 502. Dover Publications, New York 1967; Manlio Cortelazzo & Paolo Zolli, *Dizionario etimologico della lingua italiana*. 4: 828. Zanichelli, Bologna 1985; Salvatore Battaglia, *Grande dizionario della lingua italiana*. XI: 882. UTET, Torino 1981.

Species/Vernacular Names:

U. alata Michaux

English: red elm, winged elm, wahoo elm, small-leaved elm

U. americana L.

English: white elm, American elm, water elm

U. angustifolia (Weston) Weston

English: Goodyer's elm

U. carpinifolia Rupp ex Suckow

English: European field elm, smooth-leaved elm, elm

Arabic: neshem, gharghar

U. crassifolia Nutt.

English: cedar elm

U. glabra Hudson

English: wych elm, Scotch elm

U. x *hollandica* Miller

English: Dutch elm

U. japonica (Rehder) Sarg.

English: Japanese elm

U. macrocarpa Hance

China: wu i, pien, chou wu i

U. minor Miller

English: smooth-leaved elm, English elm

U. parvifolia Jacq.

English: Chinese elm, lacebark, leather-leaf elm

Japan: aki-nire

China: lang yu pi, lang yu

U. plotii Druce

English: lock elm

U. procera Salisb.

English: English elm

U. pumila L.

English: Siberian elm, dwarf elm, Chinese elm

Italian: olmo siberiano

Japan: no-nire

Tibetan: yo-bog

China: yu dai pi

U. rubra Muhl.

English: red elm, slippery elm

U. serotina Sarg.

English: red elm, September elm

U. thomasii Sarg.

English: cork elm, hickory elm, rock elm

U. x *vegeta* (Loudon) Ley

English: Huntingdon elm

U. villosa Brandis ex Gamble

English: cherry bark elm, marn elm

Ulostoma G. Don Gentianaceae

Origins:
From the Greek *oulos* "woolly, crisp, crinkled" and *stoma* "mouth."

Ulticona Raf. Solanaceae

Origins:
See E.D. Merrill (1876-1956), *Index rafinesquianus*. The plant names published by C.S. Rafinesque, etc. 213. Jamaica Plain, Massachusetts, USA 1949; C.S. Rafinesque, *Sylva Telluriana*. 55. Philadelphia 1838.

Ultragossypium Roberty Malvaceae

Origins:
Latin *ultra* "over, more, besides, beyond" plus *Gossypium*.

Umbellularia (Nees) Nutt. Lauraceae

Origins:
Latin *umbella, ae* "a sunshade, parasol," dim. of *umbra, ae* "shade, shadow," referring to the inflorescence, umbel in upper axils.

Species/Vernacular Names:
U. californica (Hook. & Arn.) Nutt.

English: California laurel, California olive, California bay, pepperwood

Umbilicus DC. Crassulaceae

Origins:
From the Latin *umbilicus, i* "the middle, the navel," referring to the leaves.

Species/Vernacular Names:
U. rupestris (Salisb.) Dandy

English: navelwort, penny pies, wall pennywort, pennywort

Umtiza Sim Caesalpiniaceae

Origins:
From the native name.

Species/Vernacular Names:
U. listeriana Sim (after Joseph Storr Lister, a former Conservator of Forests in the Cape, author of *Report on the Extent, Value and Administration of the Forests of the Transkei and East Griqualand*. Cape Town 1893 and of *Practical Hints on Tree-Planting in the Cape Colony*. W.A. Richards & Sons, Cape Town 1884)

Southern Africa: omtisa, umtiza; umThiza (Xhosa)

Unanuea Ruíz & Pav. ex Pennell Scrophulariaceae

Origins:
For the Peruvian (b. Arica, Peru/Chile) scientist José Hipólito Unanue, 1755/1758-1833 (d. Lima), physician, naturalist, professor of medicine, botanist; see J.H. Barnhart, *Biographical Notes upon Botanists*. 3: 415. 1965; M. Colmeiro y Penido, *La Botánica y los Botánicos de la Peninsula Hispano-Lusitana*. Madrid 1858; Thomas F. Glick, in *Dictionary of Scientific Biography* 13: 541. 1981.

Uncaria Burch. Pedaliaceae

Origins:
Latin *uncus, i* "a hook," referring to the capsules; see also *Harpagophytum* DC. ex Meissner.

Uncaria Schreber Rubiaceae (Naucleaceae)

Origins:
Latin *uncus, i* "a hook," referring to the petioles and inflorescence axes.

Species/Vernacular Names:
U. gambir (Hunter) Roxb.

English: pale catechu

Tibetan: gadur khyung-sder dkar-po, khyung-sder, byi-tsher khyung-sder smug-po

Malaya: gambir

China: kou teng, tiao teng

Uncarina (Baillon) Stapf Pedaliaceae

Origins:
The diminutive from the Latin *uncus* "a hook, hooked, crooked," the fruits with many rows of long spines, more

or less hooked; see J.J. Lavranos, "Two new taxa in *Uncarina* (Pedaliaceae, Madagascar)." *Haseltonia*. 3: 83-88. 1995.

Species/Vernacular Names:
U. decaryi Humbert (for M. Raymond Decary, c. 1890-1973, plant collector in Madagascar. He collected about 19,000 specimens of vascular plants, fungi and bryophytes, see Raymond Decary and Ravoanarivo, "La médication antirabique chez les Antandroy." in *Bull. Acad. Malg.* 9: 17. 1926; Werner Rauh, *Succulent and Xerophytic Plants of Madagascar*. Strawberry Press, Mill Valley, California 1995-1998)

Madagascar: farehitra, faretsy, faretra, farehitsa, farahitso (South of Madagascar)

Uncariopsis Karst. Rubiaceae

Origins:
Resembling *Uncaria* Schreb.

Uncifera Lindley Orchidaceae

Origins:
From the Latin *uncus, i* "a hook, barb" and *fero, fers, tuli, latum, ferre* "to bear, carry," referring to the pollen masses or to the shape of the spur on the lip.

Uncinaria Reichb. Rubiaceae

Origins:
From the Latin *uncinus, i* "a hook"; see also *Uncaria* Schreb.

Uncinia Pers. Cyperaceae

Origins:
Latin *uncinus, i* "a hook," indicating the awns.

Species/Vernacular Names:
U. sp.
Hawaii: mau'u, kaluhaluha

Ungeria Schott & Endlicher Sterculiaceae

Origins:
After the Austrian (b. near Leutschach, Austria) botanist Franz Joseph Andreas Nicolaus Unger, 1800-1870 (d. Graz,

Austria), Dr. med. 1827 Wien, professor of botany and zoology, Director of the Botanical Garden of the Johanneum (Graz), professor of plant anatomy and physiology. Among his works are *Synopsis plantarum fossilium*. Lipsiae 1845, (written with Stephan L. Endlicher) *Grundzüge der Botanik*. Wien 1843, *Botanische Briefe*. Wien 1852 and *Genera et species plantarum fossilium*. Vindobonae [Wien] 1850; see G. Haberlandt, ed., *Briefwechsel zwischen Franz Unger und Stephen Endlicher*. Berlin 1899; J.H. Barnhart, *Biographical Notes upon Botanists*. 3: 416. 1965; Stafleu and Cowan, *Taxonomic Literature*. 6: 594-602. 1986; Robert Olby, in *Dictionary of Scientific Biography* 13: 542-543. 1981; A. Reyer, *Leben und Werke des Naturhistorikers Dr. Franz Unger*. Graz 1871; T. W. Bossert, *Biographical Dictionary of Botanists Represented in the Hunt Institute Portrait Collection*. 411. Boston, Massachusetts 1972; A. Lasègue, *Musée botanique de Benjamin Delessert*. Paris 1845; Mariella Azzarello Di Misa, ed., *Il Fondo Antico della Biblioteca dell'Orto Botanico di Palermo*. 278. Palermo 1988; Ethelyn Maria Tucker, *Catalogue of the Library of the Arnold Arboretum of Harvard University*. Cambridge, Massachusetts 1917-1933; Bentley Glass et al., eds., *Forerunners of Darwin: 1745-1859*. Edited by ... O. Temkin, William Strauss, Jr. First edition. John Hopkins Press, Baltimore 1959.

Ungernia Bunge Amaryllidaceae (Liliaceae)

Origins:
For Baron Franz Ungern-Sternberg, 1808-1885, German botanist and physician, 1866 Dr. med. Dorpat, published *Versuch einer Systematik der Salicornieen*. Dorpat 1866 and *Salicorniearum Synopsis*. Florentinae 1876; see Maximilian Leidesdorf, *Trattato delle malattie mentali*. Torino 1878; J.H. Barnhart, *Biographical Notes upon Botanists*. 3: 416. 1965; T. W. Bossert, *Biographical Dictionary of Botanists Represented in the Hunt Institute Portrait Collection*. 411. 1972; O. Mattirolo, *Cronistoria dell'Orto Botanico della Regia Università di Torino*. in *Studi sulla vegetazione nel Piemonte* pubblicati a ricordo del II Centenario della fondazione dell'Orto Botanico della R. Università di Torino. Torino 1929.

Ungnadia Endl. Sapindaceae

Origins:
For Christian Samuel Baron von Ungnad, d. 1804; see F. Boerner & G. Kunkel, *Taschenwörterbuch der botanischen Pflanzennamen*. 4. Aufl. 183. 1989; Georg Christian Wittstein, *Etymologisch-botanisches Handwörterbuch*. 913. 1852; H. Genaust, *Etymologisches Wörterbuch der botanischen Pflanzennamen*. 669. 1996.

Species/Vernacular Names:
U. speciosa Endl.
English: Mexican buck-eye

Ungula Barlow Loranthaceae

Origins:
From the Latin *ungula, ae* "a hoof, claw."

Ungulipetalum Moldenke Menispermaceae

Origins:
Latin *ungula, ae* "a hoof, claw" and *petalum* "petal."

Unigenes E. Wimmer Campanulaceae

Origins:
Latin *unigena, ae* "only, born of one parent, of the same family."

Uniola L. Gramineae

Origins:
From *uniola, ae* for a plant, otherwise unknown, diminutive of the Latin *unio, unionis* "unity, a single large pearl, a kind of single onion, oneness," referring to the glumes; see Carl Linnaeus, *Species Plantarum*. 71. 1753 and *Genera Plantarum*. Ed. 5. 32. 1754; Helmut Genaust, *Etymologisches Wörterbuch der botanischen Pflanzennamen*. 670. Basel 1996.

Unona L.f. Annonaceae

Origins:
From the Latin *uno, are* "to make one, to join" or formed from *Anona* or *Annona*, an allusion to the stamens; see C. Linnaeus (filius), *Supplementum Plantarum*. 44, 270. 1782.

Unonopsis R.E. Fries Annonaceae

Origins:
Resembling *Unona*.

Unxia L.f. Asteraceae

Origins:
From the Latin *Unxia, ae* "the goddess of anointing."

Upopion Raf. Umbelliferae

Origins:
See C.S. Rafinesque, *New Fl. N. Am.* 4: 29. 1836 [1838] and *The Good Book*. 46, 55. 1840; Elmer D. Merrill, *Index rafinesquianus*. 183-184. 1949.

Upudalia Raf. Acanthaceae

Origins:
See C.S. Rafinesque, *Flora Telluriana*. 4: 66. 1836 [1838]; Elmer D. Merrill, *Index rafinesquianus*. 224. 1949.

Upuntia Raf. Cactaceae

Origins:
Referring to *Opuntia*; see C.S. Rafinesque, *New Fl. N. Am.* 3: 94. 1836 [1838]; Elmer D. Merrill, *Index rafinesquianus*. 171. 1949.

Urachne Trin. Gramineae

Origins:
Greek *oura* "tail" and *achne* "chaff, glume"; see Carl Bernhard von Trinius (1778-1844), *Fundamenta Agrostographiae*. 109. Viennae 1820.

Uralepis Nutt. Gramineae

Origins:
Greek *oura* "tail" and *lepis* "scale"; see Arthur D. Chapman, ed., *Australian Plant Name Index*. 2929. [" Comments: Spelling corrected by C.S. Rafinesque in *Amer. Monthly Mag. Crit. Rev.* 4: 190. 1819"] Canberra 1991; E.D. Merrill, *Index rafinesquianus*. 77. 1949.

Uralepsis Nutt. Gramineae

Origins:
See *Uralepis*.

Urandra Thwaites Icacinaceae

Origins:
From the Greek *oura* "tail" and *aner, andros* "man, male"; see also *Stemonurus* Blume.

Uranthera Naudin Melastomataceae

Origins:

From the Greek *oura* "tail" and *anthera* "anther"; see also *Acisanthera* P. Browne.

Uranthera Pax. & K. Hoffm. Euphorbiaceae

Origins:

From the Greek *oura* "tail" and *anthera* "anther"; see also *Phyllanthus* L.

Uranthoecium Stapf Gramineae

Origins:

Greek *ouranos* "sky, roof, vaulted roof" and *thekion* a diminutive of *theke* "case, capsule, box," referring to the shape of the seed-box, see James A. Baines, *Australian Plant Genera. An Etymological Dictionary of Australian Plant Genera.* 385-386. 1981; some suggest from *oura* "tail," *anthos* "flower" and *oikeo, ikeo* "to inhabit."

Species/Vernacular Names:

U. truncatum (Maiden & Betche) Stapf (*Rottboellia truncata* Maiden & Betche)

English: flat stem grass

Uraria Desv. Fabaceae

Origins:

Greek *oura* "a tail," referring to the shape of the bracts or to the inflorescences; see James A. Baines, *Australian Plant Genera. An Etymological Dictionary of Australian Plant Genera.* 386. 1981.

Species/Vernacular Names:

U. crinita (L.) Desv. ex DC. (*Hedysarum crinitum* L.)

India: dieng-kha-riu

Japan: fuji-bô-gusa

Malaya: ekor kuching, pua acoraging, serengan, ekor asu

U. lagopodioides (L.) DC. (*Hedysarum lagopodioides* L.)

India: prasniparni, prisniparni, prishniparni, atigupta, atiguha, pitvan, pithvan, pithavana, chakulia, golak chakulia, orila, dowla, dowala, davala, debra, kolku-ponna, kola-ponna, kolaponna, ranaganja, nabiyalbone, anghri-parnika, bilai kata

Japan: yaeyama-fuji-bô-gusa

China: hu li wei

U. picta (Jacq.) Desv. ex DC. (*Hedysarum pictum* Jacq.)

India: dabra, sankarjata, krishniparni, prasniparni, prisniparni, chitraparni, pilavan, pithavan, pilo samarveo, sittirappaladai, ishworojota, deterdane, hansia dafar, bannapada, bir ete teod

Japan: hosoba-fuji-bô-gusa

Malaya: ekor kuching

Nigeria: alupayida

Yoruba: alupayida, alupayida funfun, alupayida osanyin, apada

Urariopsis Schindler Fabaceae

Origins:

Resembling *Uraria* Desv.

Urbananthus R.M. King & H. Robinson Asteraceae

Origins:

The generic name honors the German botanist Ignatz Urban, 1848-1931, Assistant Director of the Botanical Garden and Museum at Berlin-Dahlem. Among his most valuable writings are *Morphologie der Gattung Bauhinia.* 1885, *Geschichte des Königlichen Botanischen Museums zu Berlin-Dahlem (1815-1913). Nebst Aufzählung seiner Sammlungen.* Dresden 1916 and *Zur Flora Südamerikas.* Halle a.S. 1882; see J.H. Barnhart, *Biographical Notes upon Botanists.* 3: 417. 1965; Stafleu and Cowan, *Taxonomic Literature.* 6: 606-619. 1986; T. W. Bossert, *Biographical Dictionary of Botanists Represented in the Hunt Institute Portrait Collection.* 411. Boston, Massachusetts 1972; Ida Kaplan Langman, *A Selected Guide to the Literature on the Flowering Plants of Mexico.* Philadelphia 1964; S. Lenley et al., *Catalog of the Manuscript and Archival Collections and Index to the Correspondence of John Torrey.* Library of the New York Botanical Garden. 1973; E.M. Tucker, *Catalogue of the Library of the Arnold Arboretum of Harvard University.* Cambridge, Massachusetts 1917-1933; R. Zander, F. Encke, G. Buchheim and S. Seybold, *Handwörterbuch der Pflanzennamen.* 14. Aufl. Stuttgart 1993.

Urbanella Pierre Sapotaceae

Origins:

After the German botanist Ignatz Urban, 1848-1931, from 1878 to 1883 Assistant Director of the Botanical Garden and Museum at Berlin-Dahlem. Among his most valuable writings are *Prodromus einer Monographie der Gattung Medicago* L. Berlin 1873 and *Papayaceae africanae.* Leipzig [1892]; see J.H. Barnhart, *Biographical Notes upon Botanists.* 3: 417. 1965.

Urbania Philippi Verbenaceae

Origins:

After the German botanist Ignatz Urban, 1848-1931, from 1889 professor Botanical Garden and Museum at Berlin-Dahlem. Among his most valuable writings are *Plantae novae Antillanae*. Berlin 1895 and *Zur Flora Südamerikas*. Halle a.S. 1882, edited *Symbolae Antillanae* seu fundamenta florae Indiae occidentalis. Berolini etc. 1898-1928; see J.H. Barnhart, *Biographical Notes upon Botanists*. 3: 417. 1965.

Urbanodendron Mez Lauraceae

Origins:

After the German botanist Ignatz Urban, 1848-1931, Assistant Director of the Botanical Garden and Museum at Berlin-Dahlem, author of *Plantae novae americanae imprimis Glaziovianae*. 1897; see J.H. Barnhart, *Biographical Notes upon Botanists*. 3: 417. 1965.

Urbanodoxa Muschler Brassicaceae

Origins:

After the German botanist Ignatz Urban, 1848-1931.

Urbanoguarea Harms Meliaceae

Origins:

After the German botanist Ignatz Urban (1848-1931) plus the genus *Guarea* L.

Urbanolophium Melchior Bignoniaceae

Origins:

After the German botanist Ignatz Urban, 1848-1931.

Urbanosciadium H. Wolff Umbelliferae

Origins:

After the German botanist Ignatz Urban, 1848-1931.

Urbinella Greenman Asteraceae

Origins:

After the Mexican botanist Manuel Urbina y Altamirano, 1843-1906, physician, zoologist, from 1881 professor of botany at the Museo Nacional. Among his writings are *Catalogo de plantas mexicanas*. Mexico 1897 and *Plantas comestibiles de los antiquos mexicanos*. Mexico 1904; see

J.H. Barnhart, *Biographical Notes upon Botanists*. 3: 417. 1965; Irving William Knobloch, compiled by, "A preliminary verified list of plant collectors in Mexico." *Phytologia Memoirs*. VI. Plainfield, N.J. 1983; T. W. Bossert, *Biographical Dictionary of Botanists Represented in the Hunt Institute Portrait Collection*. 411. Boston, Massachusetts 1972; Ida Kaplan Langman, *A Selected Guide to the Literature on the Flowering Plants of Mexico*. Philadelphia 1964; E.M. Tucker, *Catalogue of the Library of the Arnold Arboretum of Harvard University*. Cambridge, Massachusetts 1917-1933.

Urbinia Britton & Rose Crassulaceae

Origins:

After the Mexican botanist Manuel Urbina y Altamirano, 1843-1906; see *Bull. New York Bot. Gard.* 3(9): 11. 1903.

Urceola Roxburgh Apocynaceae

Origins:

Latin *urceolus, i* "a small cup or pitcher, urn, a little water-pot," *urceus, i* "a pitcher, ewer."

Species/Vernacular Names:
U. esculenta Benth. ex Hook.f.
Burma (Myanmar): kyetpaung
U. micrantha (Wallich ex G. Don) D.J. Middleton
China: du zhong teng
U. rosea (Hook. & Arn.) D.J. Middleton
China: suan ye jiao teng

Urceolaria Willd. ex Cothen. Rubiaceae

Origins:

Latin *urceolaris, e* "of or belonging to pitchers," *urceus* "a pitcher, ewer."

Urceolina Reichb. Amaryllidaceae (Liliaceae)

Origins:

Latin *urceus* "a pitcher, ewer," *urceolus* "a small cup, little pitcher, urn, a little water-pot," referring to the shape of the perianth.

Urechites Müll. Arg. Apocynaceae

Origins:

Greek *oura* "a tail" plus *Echites*, referring to the appendages of the anthers.

Urelytrum Hack. Gramineae

Origins:

Greek *oura* "a tail" and *elytron* (*elyo* "to wind") "a sheath, a cover."

Urena L. Malvaceae

Origins:

Urena is a Malabar vernacular name used by van Rheede in *Hortus Indicus Malabaricus*; see Carl Linnaeus, *Species Plantarum*. 692. 1753 and *Genera Plantarum*. Ed. 5. 309. 1754.

Species/Vernacular Names:

U. lobata L.

English: Congo jute, cousin mahoe, burweed

Nigeria: ramaniya, ka-fi-rama, uwar-mangani, okeriri, bolobolo

Yoruba: ilasa omode, ilasa agborin, ilasa oyibo, akeri, ake iri, ake riri, bolobolo

Japan: ôba-bonten-ka (bon-ten = Buddha's heaven)

India: wanbhendi, bachata, bachita, bonokra, ben-ochra, ban okhra, ottatti, ottututti, otte, lotloti, kunjia, bhidi-janelet, mota-behedi-janelet, tapkote, vanabendha, vanabhenda, wagdau bhendi, vana-bhenda, van bhendi, peddabenda, rantupkada, valta epala, udiram, uram, uran, unga, lapetua, pithia, bilokapasiva

Malaya: kelulut, pulut pulut

The Philippines: kulutkulutan, kulit, kulut, kulukulut, kulokullot, kulotan, malopolo, mangkit, palisin, dalupang, supang, saligut, baranggot, barangot, anonongkot, puriket

Urera Gaudich. Urticaceae

Origins:

Latin *uro, -is, ussi, ustum, urere* "to burn, burn up, to sting," referring to the powerful stinging hairs; see Charles Gaudichaud-Beaupré (1789-1854), [Botany of the Voyage.] *Voyage autour du Monde ... sur ... l'Uranie et la Physicienne, pendant ... 1817-1820*. 496. Paris 1826 [-1830].

Species/Vernacular Names:

U. sp.

Nigeria: ela

Mexico: laa guii, laa qui

U. baccifera (L.) Wedd.

Paraguay: pyno guasu

Panama: ñenge

U. cordifolia Engl.

Yoruba: esigala, esagbonrin, esisi agbonrin, jagbonrin

U. glabra (Hook. & Arnott) Wedd. (*Procris glabra* Hook. & Arnott; *Urera konaensis* St. John; *Urera sandwicensis* Wedd.)

Hawaii: opuhe, hopue, hona

U. kaalae Wawra

Hawaii: opuhe

U. repens (Wedd.) Rendle

Yoruba: omi odan

U. trinervis (Hochst. apud Krauss) Friis & Immelman (*Urera cameroonensis* Wedd.; *Urera woodii* N.E. Br.)

English: climbing nettle

South Africa: rankbrandnetel

Urginea Steinh. Hyacinthaceae (Liliaceae)

Origins:

From Beni Urgin, the Afro-Arab tribe in Algeria; see W.T. Stearn, "Mediterranean and Indian species of *Drimia* (Liliaceae): A nomenclature survey with special reference to the medicinal squill, *D. maritima* (syn. *Urginea maritima*)." in *Annales Musei Goulandris*. 4: 199-210. 1978.

Species/Vernacular Names:

U. altissima (L.f.) Bak.

South Africa: jeukbol, maermanbol, maermanui, slangkop

U. macrocentra Bak. (*Urginea lilacina* Bak.)

English: poison bulb, snake's head

Southern Africa: bergtulp, maerman, Natal slangkop, slangkop; injobo (Xhosa); injobo (Zulu)

U. maritima (L.) Baker

English: sea onion, squill

Italian: scilla marittima

Tunisia: ansal

Arabic: 'onsul, 'onsel, 'onsal, 'onseil

French: scille

U. pusilla (Jacq.) Bak.

South Africa: mountain slangkop, bergslangkop, slangkop

U. sanguinea Schinz (*Urginea burkei* Bak.)

Southern Africa: krimpsiekteblaar, krimpsiekte-ui, slangkop, rooislangkop, red slangkop, Transvaal slangkop, Transvaalse slangkop, Burke's slangkop; sekaname (Tswana)

Urgineopsis Compton Hyacinthaceae (Liliaceae)

Origins:
Resembling *Urginea* Steinh.

Uribea Dugand & Romero Fabaceae

Origins:
After the Colombian clergyman Antonio Lorenzo Uribe-Uribe, 1900-1980, botanist, in 1916 entered the Company of Jesus, from 1958 to 1980 Curator at the Instituto de Ciencias Naturales, author of *Flora de Antoquia.* Medellin 1940; see T. W. Bossert, *Biographical Dictionary of Botanists Represented in the Hunt Institute Portrait Collection.* 411. Boston, Massachusetts 1972.

Urinaria Medik. Euphorbiaceae

Origins:
Possibly from the Latin *urina, ae* "urine," Greek *ouron,* or Greek *oura* "a tail."

Urnularia Stapf Apocynaceae

Origins:
Latin *urnula, ae* "a little urn, a water-urn," *urna, ae* "urn, water-pot."

Urobotrya Stapf Opiliaceae

Origins:
From the Greek *oura* "a tail" and *botrys* "cluster, a bunch of grapes, a cluster of grapes."

Urocarpidium Ulbr. Malvaceae

Origins:
From the Greek *oura* "a tail" and *karpos* "fruit," Latin *carpidium* "carpel."

Urocarpus J. Drumm. ex Harvey Rutaceae

Origins:
Greek *oura* "a tail" and *karpos* "fruit," the fruits are beaked.

Urochlaena Nees Gramineae

Origins:
From the Greek *oura* "a tail" and *chlaena, chlaenion* "cloak, blanket."

Urochloa P. Beauv. Gramineae

Origins:
Greek *oura* "a tail" and *chloe, chloa* "grass," referring to the awns.

Species/Vernacular Names:
U. brachyura (Hackel) Stapf
English: urochloa
South Africa: urochloa

U. mosambicensis (Hackel) Dandy (*Urochloa pullulans* Stapf; *Urochloa rhodesiensis* Stent)
English: common urochloa, buffalo grass, bushveld herringbone grass, bushveld signal grass, herring-bone grass, white buffalo grass
South Africa: gewone urochloa, bosveldbeesgras, buffelsgras, kruipsinjaalgras, witbuffelgras

U. panicoides P. Beauv. (*Urochloa ruschii* sensu Chippind., non Pilg.)
English: annual signal grass, garden grass, garden urochloa, herringbone grass, kuri-millet, poke
Southern Africa: tuin-urochloa, beesgras, eenjarige sinjaalgras, kurimanna, tuingras; bore-ba-ntjia (Sotho)

U. trichopus (Hochst.) Stapf (*Urochloa engleri* Pilger)
English: roundseed urochloa, Gonya grass
South Africa: rondesaadurochloa, witgras

Urochondra C.E. Hubb. Gramineae

Origins:
Greek *oura* "a tail" and *chondros* "wheat, grain of wheat, big, a grain," referring to the nature of the spikelets.

Urodesmium Naudin Melastomataceae

Origins:
From the Greek *oura* "a tail" and *desmos* "a bond"; see also *Pachyloma* DC.

Urodon Turcz. Fabaceae

Origins:

Greek *oura* "a tail" and *odous, odontos* "tooth"; see Porphir Kiril N.S. Turczaninow, *Bulletin de la Société Impériale des Naturalistes de Moscou*. 22(2): 16. 1849.

Urogentias Gilg & Gilg-Ben. Gentianaceae

Origins:

From the Greek *oura* "a tail" plus *Gentiana*, corolla with long processes.

Urolepis (DC.) R.M. King & H. Robinson Asteraceae

Origins:

From the Greek *oura* "a tail" and *lepis* "scale."

Uromorus Bureau Moraceae

Origins:

From the Greek *oura* "a tail" plus *Morus*.

Uromyrtus Burret Myrtaceae

Origins:

Greek *oura* "a tail" plus *Myrtus, myrtos* "myrtle."

Uropappus Nutt. Asteraceae

Origins:

Tailed pappus, from the Greek *oura* "a tail" and *pappos* "fluff, downy appendage," referring to the scales.

Species/Vernacular Names:
U. lindleyi (DC.) Nutt.
English: silver puffs

Uropedium Lindley Orchidaceae

Origins:

Greek *oura* "a tail" and the genus *Cypripedium*, referring to the shape of the lip; see also *Phragmipedium* Rolfe.

Urophyllum Jack ex Wallich Rubiaceae

Origins:

From the Greek *oura* "a tail" and *phyllon* "leaf."

Urophysa Ulbr. Ranunculaceae

Origins:

Greek *oura* "a tail" and *physa* "bladder."

Uroskinnera Lindley Scrophulariaceae

Origins:

For the British (b. Northumberland) naturalist George Ure Skinner, 1804-1867 (Panama), merchant, orchidologist, ornithologist, traveler, 1866 Fellow of the Linnean Society, collected orchids (for James Bateman, 1811/1812-1897), sent plants to W.J. Hooker; see J.H. Barnhart, *Biographical Notes upon Botanists*. 3: 285. Boston 1965; Ray Desmond, *Dictionary of British & Irish Botanists and Horticulturists*. 631. London 1994; Merle A. Reinikka, *A History of the Orchid*. Timber Press 1996; M. Hadfield et al., *British Gardeners: A Biographical Dictionary*. London 1980; Ron Rigby, "James Bateman, 1811-1897." *The Orchid Review*. 106(1222): 214-218. 1998.

Urospatha Schott Araceae

Origins:

From the Greek *oura* "a tail" and *spatha* "spathe"; see D.H. Nicolson, "Derivation of aroid generic names." *Aroideana*. 10: 15-25. 1988.

Urospathella Bunting Araceae

Origins:

Diminutive of the genus *Spatha* Schott.

Urospermum Scop. Asteraceae

Origins:

Greek *oura* "a tail" and *sperma* "a seed," referring to the tail-like beak of the seeds.

Species/Vernacular Names:
U. picroides (L.) F.W. Schmidt (*Tragopogon picroides* L.; *Arnopogon picroides* (L.) Willd.) (resembling *Picris*, Asteraceae)

English: false hawkbit, urospermum

Arabic: salis

Urostachya (Lindley) Brieger Orchidaceae

Origins:
Greek oura "a tail" and stachys "spike"; see also Eria Lindl.

Urostachys (E. Pritz.) Herter Lycopodiaceae

Origins:
From the Greek oura "a tail" and stachys "spike."

Urostemon B. Nord. Asteraceae

Origins:
From the Greek oura "a tail" and stemon "stamen."

Urostephanus B.L. Robinson & Greenman Asclepiadaceae

Origins:
From the Greek oura "a tail" and stephanos "a crown."

Urostigma Gasp. Moraceae

Origins:
From the Greek oura "a tail" and stigma "a stigma."

Urotheca Gilg Melastomataceae

Origins:
From the Greek oura "a tail" and theke "case, box."

Ursinia Gaertner Asteraceae

Origins:
Latin ursus "bear," Greek arktos "bear, the north"; see H. Genaust, Etymologisches Wörterbuch der botanischen Pflanzennamen. 671. Basel 1996; Joseph Gaertner (1732-1791), De fructibus et seminibus plantarum. Stuttgart, Tübingen 1788-1791 [-1792]; Francis Aubie Sharr, Western Australian Plant Names and Their Meanings. 72. University of Western Australia Press 1996; some suggest the genus was named for the German botanical author Johann

Heinrichus Ursinus of Regensburg, 1608-1667, theologian, author of Arboretum Biblicum. Norimbergae 1663 and De Ecclesiarum Germanicarum origine et progressu. Norimbergae 1664, see F. Boerner & G. Kunkel, Taschenwörterbuch der botanischen Pflanzennamen. 4. Aufl. 183. 1989.

Species/Vernacular Names:
U. abrotanifolia (R. Br.) Spreng. (Sphenogyne abrotanifolia R. Br.)

South Africa: fynkruie, lammetjieskruie

U. chrysanthemoides (Less.) Harv. (Sphenogyne chrysanthemoides Less.; Ursinia chrysanthemoides (Less.) Harv. var. chrysanthemoides; Ursinia chrysanthemoides (Less.) Harv. var. geyeri (L. Bol. & Hall) Prassler; Ursinia geyeri L. Bol. & Hall)

English: coral ursinia

U. nana DC. subsp. nana (Ursinia abyssinica Sch.Bip. ex Walp.; Ursinia affinis Harv.; Ursinia annua Less. ex Harv.; Ursinia engleriana Muschl.; Ursinia indecora DC.; Ursinia matricariifolia Dinter; Ursinia schinzii Dinter)

English: dwarf ursinia, yellow Margaret

Southern Africa: dwerg-ursinia, geelmargriet; sehalikane (Sotho)

Ursiniopsis E. Phillips Asteraceae

Origins:
Resembling Ursinia Gaertn.

Urtica L. Urticaceae

Origins:
Urtica is the Latin name for the nettle (Plinius), wanting etymology, possibly from Latin uro, -is, ussi, ustum, urere "to burn, burn up, to sting" and Akkadian essatu, issatu "fire"; see Carl Linnaeus, Species Plantarum. 983. 1753 and Genera Plantarum. Ed. 5. 423. 1754; Manlio Cortelazzo & Paolo Zolli, Dizionario etimologico della lingua italiana. 4: 848. Zanichelli, Bologna 1985; Giovanni Semerano, Le origini della cultura europea. Dizionario della lingua Latina e di voci moderne. 2(2): 601. Leo S. Olschki Editore, Firenze 1994; Salvatore Battaglia, Grande dizionario della lingua italiana. XII: 162-164. Torino 1984; H. Genaust, Etymologisches Wörterbuch der botanischen Pflanzennamen. 671. 1996; F. Boerner and G. Kunkel, Taschenwörterbuch der botanischen Pflanzennamen. 4. Aufl. 183. Berlin and Hamburg 1989.

Species/Vernacular Names:
U. dioica L.

English: nettle, tall nettle, stinging nettle, common nettle, common stinging nettle, great stinging nettle, large stinging nettle, perennial stinging nettle, stinger, Swedish hemp

Guatemala: ortiga, chichicaste

Mexico: guechi bidoo

Italian: ortica comune, ortica maschia

India: bichu, bichhu booti, bichhua, chichru, chicru

Southern Africa: brandnetel, brandneukel, brandneuker, gewone brandnetel; bobatsi (Sotho); umbabazane (Swati); umbabazane (Xhosa)

U. ferox Forst.f.

English: tree nettle

Maori names: ongaonga

U. incisa Poir. (*Urtica lucifuga* Hook.)

English: native nettle, scrub nettle, stinging nettle

U. parviflora Roxb.

India: anachoriyanom, paharah-bichuti, berain, shishona, bichu, bichhu, kaniyali, seusni

U. pilulifera L.

English: Roman nettle

Italian: ortica romana

Arabic: horriqua, horreiq, qorreis

U. urens L.

English: nettle, annual stinging nettle, burning nettle, bush nettle, small nettle, dwarf nettle, bush stinging nettle, dwarf stinging nettle, nettle, small stinging nettle, stinging nettle, lesser stinging nettle, dog nettle

Paraguay: ortiga blanca, pyno-i

Arabic: horreiq, qorreis

Southern Africa: bosbrandnetel, bosbrandnekel, brandblare, brandnetel, brandneukel, brandneuker, klein brandnetel, perde brandnetel; bubati (Pedi); umbabazane (Zulu); umbabazane (Xhosa)

Urticastrum Fabr. Urticaceae

Origins:
Referring to the genus *Urtica* L.

Uruparia Raf. Rubiaceae

Origins:
Based on the genus *Ourouparia* Aublet (1775); see E.D. Merrill, *Index rafinesquianus*. 227. 1949; C.S. Rafinesque, *Sylva Telluriana*. 148. 1838.

Urvillea Kunth Sapindaceae

Origins:
Named for the French traveler and explorer Jules Sébastian César Dumont d'Urville, 1790-1842, plant collector, a member of the Linnean Society and the Société de Géographie, he took part in the voyage of the *Coquille* (commanded by L.I. Duperrey), from 1825 commander of the *Astrolabe* (former *Coquille*). His writings include *Enumeratio Plantarum quas in insulis archipelagi aut littoribus Ponti-Euxinii, annis 1819 et 1820, collegit atque detexit J. Dumont D'Urville*. Parisiis 1822, *Voyage au Pôle Sud and dans l'Océanie sur les corvettes L'Astrolabe et La Zélée*. Paris 1841-1846 and *Voyage de Découvertes autour du Monde ... sur la corvette L'Astrolabe pendant les Années 1826-1829*. Paris 1832-1848; see E.S. Dodge, *Islands and Empire: Western Impact on the Pacific and East Asia*. Minneapolis 1976; Gaston Meissas, *Les grands voyageurs de notre siècle*. Paris 1889; J.H. Barnhart, *Biographical Notes upon Botanists*. 1: 480. 1965; T. W. Bossert, *Biographical Dictionary of Botanists Represented in the Hunt Institute Portrait Collection*. 109. Boston, Massachusetts 1972; R. Glenn, *The Botanical Explorers of New Zealand*. Wellington 1950; John Dunmore, *Who's Who in Pacific Navigation*. University of Hawaii Press, Honolulu 1991.

Utahia Britton & Rose Cactaceae

Origins:
Utah, USA.

Utleria Beddome ex Benth. Asclepiadaceae

Origins:
Uttallari "jungle beauty," which is a Tamil name for *Utleria salicifolia* Bedd.

Utricularia L. Lentibulariaceae

Origins:
Latin *utriculus, i* "a small bottle, leather bag, little bladder," diminutive of *uterus, i* "the womb, matrix," referring to the inflated insect-trapping bladders; see Carl Linnaeus, *Species Plantarum*. 18. 1753 and *Genera Plantarum*. Ed. 5. 11. 1754.

Species/Vernacular Names:
U. australis R. Br. (*Utricularia japonica* Makino)

Japan: tanuki-mo

U. gibba L. (*Utricularia exoleta* R. Br.) (Latin *exoletus, a, um* "dying away, mature, grown up")

Japan: mikawa-tanuki-mo

U. livida E. Mey. (*Utricularia transrugosa* Stapf)

English: bladderwort

South Africa: blaaskruid

U. menziesii R. Br.

English: redcoats

U. minor L.

English: lesser bladderwort

U. stellaris L.f. (*Utricularia inflexa* Forssk. var. *stellaris* (L.f.) P. Tayl.)

English: bladderwort, star bladderwort

Southern Africa: sterblasiekruid; tlamana-sa-metsi (Sotho)

U. tenella R. Br. (*Polypompholyx tenella* (R. Br.) Lehm.)

English: pink bladderwort, pink fan

U. vulgaris L.

English: greater bladderwort, bladderwort, common bladderwort

Uva Kuntze Annonaceae

Origins:
Latin *uva, ae* "cluster, cluster of grapes, a grape"; see also *Uvaria* L.

Uvaria L. Annonaceae

Origins:
Latin *uva* "cluster, cluster of grapes, bunch of grapes," referring to the fruits; see Carl Linnaeus, *Species Plantarum*. 536. 1753 and *Genera Plantarum*. Ed. 5. 140. 1754.

Species/Vernacular Names:
U. afzelii Scott-Elliot

Yoruba: gbogbonse

U. acuminata Oliver var. *cartocarpa* (Diels) Cavaco & Keraudren

Madagascar: sena sena, sena sena vavy

U. caffra E. Mey. ex Sond.

English: small cluster pear

Southern Africa: kleintrospeer; iNkonjane, uMavumba, uMazwenda-omnyama (Zulu); iDwaba (Xhosa)

U. chamae P. Beauv.

English: finger root

Nigeria: iru-ugu, akossa, akisan

Yoruba: agako, eeruiju, eeruju, eerugbo

U. lucida Benth. subsp. *virens* (N.E. Br.) Verdc. (*Uvaria gazensis* Bak.f.; *Uvaria virens* N.E. Br.)

English: large cluster pear

Southern Africa: groottrospeer; mavumba, uMavumba (Zulu)

U. scheffleri Diels

Tanzania: nsilimbu

Uvariastrum Engl. Annonaceae

Origins:
Incompletely resembling the genus *Uvaria* L.

Uvariella Ridl. Annonaceae

Origins:
The diminutive of the genus *Uvaria* L.

Uvariodendron (Engl. & Diels) R.E. Fries Annonaceae

Origins:
From the genus *Uvaria* and the Greek *dendron* "tree."

Species/Vernacular Names:
U. angustifolium (Engl. & Diels) R.E. Fries

Nigeria: igbere (Yoruba)

Uvariopsis Engl. & Diels Annonaceae

Origins:
Like *Uvaria*, from the genus *Uvaria* and the Greek *opsis* "aspect, resemblance, appearance."

Species/Vernacular Names:
U. dioica (Diels) Robyns & Ghesq.

Nigeria: akossa; akosa (Edo)

Uvedalia R. Br. Scrophulariaceae

Origins:
After the English (b. Westminster, London) clergyman Rev. Robert Uvedale, 1642-1722 (d. Enfield, Middx.), schoolmaster, Rector (Orpington, Kent), had a garden of exotic plants at Enfield; see R. Brown, *Prodromus Florae Novae*

Hollandiae. 440. 1810; L. Plukenet, *Almagestum botanicum*. Londini 1696; Robert Morison (1620-1683), *Plantarum historiae universalis oxoniensis*. Oxonii, e Theatro Sheldoniano 1699; James Britten, *The Sloane Herbarium*, revised and edited by J.E. Dandy. London 1958; R. Pulteney, *Historical and Biographical Sketches of the Progress of Botany in England*. 2: 30. London 1790; Dawson Turner, *Extracts from the Literary and Scientific Correspondence of R. Richardson, of Bierly, Yorkshire: Illustrative of the State and Progress of Botany* [Edited by D. Turner. — Extracted from the memoir of the Richardson family, by Mrs. D. Richardson] Yarmouth 1835; John Claudius Loudon, *Arboretum et fruticetum britannicum*. London 1838; Ray Desmond, *Dictionary of British & Irish Botanists and Horticulturists*. 701. London 1994; Francis Aubie Sharr, *Western Australian Plant Names and Their Meanings*. 214. 1996.

Uvularia L. Colchicaceae (Convallariaceae, Liliaceae, Uvulariaceae)

Origins:

Latin *uva, ae* "cluster, cluster of grapes," *uvula* "the soft palate"; see Helmut Genaust, *Etymologisches Wörterbuch der botanischen Pflanzennamen*. 673. 1996.

V

Vaccaria Wolf Caryophyllaceae

Origins:
Latin *vacca, ae* "a cow," fodder, pastures, referring to the milk production of cows, the roots are galactogogue, the plant was regarded as good fodder; see H. Genaust, *Etymologisches Wörterbuch der botanischen Pflanzennamen.* 673. 1996.

Species/Vernacular Names:
V. hispanica (Miller) Rauschert (*Vaccaria pyramidata* Medikus; *Saponaria vaccaria* L.; *Vaccaria vulgaris* Host)
English: cockle, cow cockle, cow herb, cow soapwort, dairy pink
China: wang pu liu hsing
Arabic: foul el-'arab
South Africa: akkerkoeikruid

Vacciniopsis Rusby Ericaceae

Origins:
Resembling *Vaccinium* L.

Vaccinium L. Ericaceae

Origins:
Vaccinium (Plinius), a Latin name for the blueberry, whortleberry, a corruption of the Greek *hyakinthos* "the hyacinth, purple, dark red," Bakinthos; Akkadian *bakkitu* "professional mourner, paid mourner," *bakku* "lachrymose," *bakum, bikitu* "lamented"; see G. Semerano, *Le origini della cultura europea.* Dizionari Etimologici. Dizionario della lingua Greca. 2(1): 298. Leo S. Olschki Editore, Firenze 1994; Pietro Bubani, *Flora Virgiliana.* 117-121. Bologna 1978; Giovanni Semerano, *Le origini della cultura europea.* Dizionario della lingua Latina e di voci moderne. 2(2): 603. Firenze 1994; H. Genaust, *Etymologisches Wörterbuch der botanischen Pflanzennamen.* 673. 1996; G.C. Wittstein, *Etymologisch-botanisches Handwörterbuch.* 916. 1852.

Species/Vernacular Names:
V. spp.
Mexico: yaga bidina
V. arboreum Marshall
English: farkleberry
V. arctostaphylos L.
English: Broussa tea
V. bracteatum Thunb.
English: sea bilberry, Asiatic bilberry
China: nan zhu zi
Japan: shashan-bo
Malaya: mempadang, kelampadang
V. caespitosum Michaux
English: dwarf bilberry
V. calycinum Sm. (*Metagonia calycina* (Sm.) Nutt.; *Vaccinium calycinum* var. *grandifolium* Fosb.; *Vaccinium fauriei* H. Lév.)
Hawaii: 'ohelo, 'ohelo kau la'au
V. deliciosum Piper
English: cascade bilberry
V. dentatum Sm. (*Metagonia penduliflora* (Gaud.) Nutt.)
Hawaii: 'ohelo
V. macrocarpon Aiton
English: cranberry
V. membranaceum Hook.
English: thinleaf huckleberry
V. mortinia Benth.
Latin America: mortiña
V. myrtillus L.
English: bilberry, whortleberry, blaeberry
V. ovalifolium Sm.
English: mathers
V. ovatum Pursh
English: California huckleberry
V. oxycoccos L.
English: cranberry
V. padifolium Sm.
English: Madeiran whortleberry, Madeira bilberry
Portuguese: uveira da serra
V. parvifolium Smith
English: red huckleberry
V. reticulatum Sm. (*Vaccinium berberidifolium* (A. Gray) Skottsb.)
Hawaii: 'ohelo, 'ohelo 'ai, 'ohelo berry

V. scoparium Cov.

English: littleleaf huckleberry

V. stramineum L.

English: deerberry

V. vitis-idaea L.

English: cowberry, lingberry, foxberry, cranberry

China: yue ju ye

Vachellia Wight & Arnott Mimosaceae

Origins:

After the English (b. Littleport, Cambridgeshire) Rev. George Harvey Vachell, b. 1799, chaplain to the Hon. East India Company at Macao from 1825 to 1836, collected plants in China; see Emil Bretschneider, *History of European Botanical Discoveries in China.* Leipzig 1981.

Vagaria Herbert Amaryllidaceae (Liliaceae)

Origins:.

Latin *vagus* "roaming, wandering," the specimen was of uncertain origin.

Vaginopteris Nakai Vittariaceae (Adiantaceae)

Origins:

Latin *vagina, ae* "sheath, scabbard, the covering, husk" and *pteris* "fern, a species of fern"; see also *Vaginularia* Fée.

Vaginularia Fée Vittariaceae (Adiantaceae)

Origins:

From the Latin *vagina, ae* "sheath, scabbard, the covering, husk."

Vahlia Thunberg Saxifragaceae (Vahliaceae)

Origins:

For the Norwegian-born Danish botanist Martin Vahl, 1749-1804, traveler, studied under Johann Zoëga (1742-1788) and Linnaeus, professor of botany. Among his most valuable writings are *Symbolae botanicae.* Hauniae 1790-1794 and *Eclogae americanae.* Hauniae 1796, father of the Danish botanist Jens Laurentius (Lorenz) Moestue Vahl (1796-1854); see J.H. Barnhart, *Biographical Notes upon Botanists.* 3: 419. 1965; Stafleu and Cowan, *Taxonomic Literature.* 6: 628-631. 1986; R. Zander, F. Encke, G. Buchheim and S. Seybold, *Handwörterbuch der Pflanzennamen.* 14. Aufl. 1993; C.F.A. Christensen, *Den danske Botaniks Historie*

med tilhørende Bibliografi. Copenhagen 1924-1926; Ida Kaplan Langman, *A Selected Guide to the Literature on the Flowering Plants of Mexico.* Philadelphia 1964; Mariella Azzarello Di Misa, ed., *Il Fondo Antico della Biblioteca dell'Orto Botanico di Palermo.* 279. Regione Siciliana, Palermo 1988; Blanche Elizabeth Edith Henrey (1906-1983), *British Botanical and Horticultural Literature before 1800.* Oxford 1975; Jonas C. Dryander, *Catalogus bibliothecae historico-naturalis Josephi Banks.* London 1796-1800; A. Lasègue, *Musée botanique de Benjamin Delessert.* Paris 1845; T. W. Bossert, *Biographical Dictionary of Botanists Represented in the Hunt Institute Portrait Collection.* 412. Boston, Massachusetts 1972; E.M. Tucker, *Catalogue of the Library of the Arnold Arboretum of Harvard University.* Cambridge, Massachusetts 1917-1933.

Vahlodea Fries Gramineae

Origins:

For the Danish botanist Jens Laurentius (Lorenz) Moestue Vahl (1796-1854), plant collector, traveler, librarian, son of Martin Vahl (1749-1804); see Paul Gaimard, *Voyages de la Commission Scientifique du Nord, en Scandinavie, en Laponie, au Spitzberg et aux Feröe, pendant les années 1838, 1839 et 1840, sur la Corvette La Recherche*, commandée par M. Fabvre ... Géographie physique, Géographie botanique, Botanique et Physiologie, etc. Paris [1842-1848]; J.H. Barnhart, *Biographical Notes upon Botanists.* 3: 419. 1965; Stafleu and Cowan, *Taxonomic Literature.* 6: 628. Utrecht 1986; R. Zander, F. Encke, G. Buchheim and S. Seybold, *Handwörterbuch der Pflanzennamen.* 14. Aufl. 1993; C.F.A. Christensen, *Den danske Botaniks Historie med tilhørende Bibliografi.* Copenhagen 1924-1926; A. Lasègue, *Musée botanique de Benjamin Delessert.* Paris 1845.

Vailia Rusby Asclepiadaceae

Origins:

After the American botanist Anna Murray Vail, 1863-1955; see J.H. Barnhart, *Biographical Notes upon Botanists.* 3: 419. 1965; Ida Kaplan Langman, *A Selected Guide to the Literature on the Flowering Plants of Mexico.* Philadelphia 1964; S. Lenley et al., *Catalog of the Manuscript and Archival Collections and Index to the Correspondence of John Torrey.* Library of the New York Botanical Garden. 1973; E.M. Tucker, *Catalogue of the Library of the Arnold Arboretum of Harvard University.* Cambridge, Massachusetts 1917-1933; Thomas Morong (1827-1894), "An enumeration of the plants collected by Dr. Thomas Morong in Paraguay, 1888-1890." *Ann. New York Acad.* vol. 7, no. 2. 1892-1893 [by T. Morong and N.L. Britton, with the assistance of Miss A.M. Vail].

Vaillanta Raf. Rubiaceae

Origins:

For *Valantia* L. = *Vaillantia* Gled.; see E.D. Merrill, *Index rafinesquianus*. 227. 1949; C.S. Rafinesque, in *Specchio delle scienze*, o giornale enciclopedico di Sicilia etc. 1: 87. Palermo 1814.

Vaillantia Hoffm. Rubiaceae

Origins:

See *Valantia* L.

Valantia L. Rubiaceae

Origins:

For the French (b. Vigny, Val d'Oise) botanist Sébastien Vaillant, 1669-1722 (d. Paris), surgeon, author of *Botanicon parisiense*. Leiden 1723; see Stafleu and Cowan, *Taxonomic Literature*. 6: 634-636. Utrecht 1986; Mariella Azzarello Di Misa, ed., *Il Fondo Antico della Biblioteca dell'Orto Botanico di Palermo*. 279-280. Palermo 1988; P. Jovet & J. Mallet, in *Dictionary of Scientific Biography* (Editor in Chief Charles Coulston Gillispie.) 13: 553-554. New York 1981; J.H. Barnhart, *Biographical Notes upon Botanists*. 3: 419. 1965; G.C. Wittstein, *Etymologisch-botanisches Handwörterbuch*. 916. 1852; Frans A. Stafleu, *Linnaeus and the Linnaeans*. Utrecht 1971; T. W. Bossert, *Biographical Dictionary of Botanists Represented in the Hunt Institute Portrait Collection*. 413. Boston, Massachusetts 1972; H.N. Clokie, *Account of the Herbaria of the Department of Botany in the University of Oxford*. 258. Oxford 1964; Edm. and D.S. Berkeley, *John Clayton, Pioneer of American Botany*. Chapel Hill 1963; Edward Lee Greene, *Landmarks of Botanical History*. Edited by Frank N. Egerton. 964, 1035. Stanford, California 1983; Jonas C. Dryander, *Catalogus bibliothecae historico-naturalis Josephi Banks*. London 1796-1800; Antoine Lasègue, *Musée botanique de M. Benjamin Delessert*. 1845; E.M. Tucker, *Catalogue of the Library of the Arnold Arboretum of Harvard University*. Cambridge, Massachusetts 1917-1933; James Britten, *The Sloane Herbarium*, revised and edited by J.E. Dandy. 1958.

Valdivia C. Gay ex J. Remy Grossulariaceae (Escalloniaceae)

Origins:

Valdivia, Chile; Pedro de Valdivia (1500?-1554), Spanish conqueror of Chile.

Valentiana Raf. Acanthaceae (Thunbergiaceae)

Origins:

After the English (b. Arley Castle, Staffs.) botanist George Annesley, 1769-1844, 2nd Earl of Mountnorris, Viscount Valentia, traveler and plant collector (India and the Middle East), 1796 Fellow of the Linnean Society and the Royal Society, correspondent of Sir J.E. Smith and Sir J. Banks. His writings include *Short Instructions for Collecting Shells*. [1815] and *Voyages and Travels to India, Ceylon, the Red Sea, Abyssinia, and Egypt*. 1802, 1803, 1804, 1805 and 1806. London 1809; see Ray Desmond, *Dictionary of British & Irish Botanists and Horticulturists*. 17. London 1994; R. Desmond, *The European Discovery of the Indian Flora*. Oxford 1992; *Proceedings in the Court of King's Bench*, ex parte George Viscount Valentia [to recover his infant son from the custody of Lady Valentia] Kidderminster 1799; Frederick Robinson, R.N., *Refutation of Lieutenant Wellsted's Attack upon Lord Valentia's ... Work upon the Red Sea*. London 1842; E.D. Merrill (1876-1956), *Index rafinesquianus*. 228. 1949; C.S. Rafinesque, in *Specchio delle scienze*, o giornale enciclopedico di Sicilia etc. 1: 87. Palermo 1814; Henry C. Andrews (1794-1830), *The Botanist's Repository*. London 1801.

Valentiniella Speg. Boraginaceae

Origins:

The diminutive of *Valentina* Speg.

Valeriana L. Valerianaceae

Origins:

A plant from Valeria, ancient name of a province of low Pannonia, or from the personal name Valerius; see F. Boerner & G. Kunkel, *Taschenwörterbuch der botanischen Pflanzennamen*. 4. Aufl. 183. Berlin & Hamburg 1989; V. Bertoldi, "Il nome botanico *Valeriana*." *Archivum romanicum*. X: 210-216. Genève & Firenze 1926; Helmut Genaust, *Etymologisches Wörterbuch der botanischen Pflanzennamen*. 674. Basel 1996; Manlio Cortelazzo & Paolo Zolli, *Dizionario etimologico della lingua italiana*. 5: 1410. 1988; Ernest Weekley, *An Etymological Dictionary of Modern English*. 2: 1578. Dover Publications, New York 1967; [John Lemprière, 1765?-1824], *Lemprière's Classical Dictionary* of Proper Names mentioned in Ancient Authors. Third edition. 442. London and New York 1984.

Species/Vernacular Names:

V. sp.

Peru: ancu-ancu, huaina-curi

V. officinalis L.

English: common valerian, garden heliotrope, valerian

Italian: valeriana

China: xie cao

Valerianella Miller Valerianaceae

Origins:
Diminutive of the genus *Valeriana*.

Species/Vernacular Names:
V. discoidea (L.) Lois.

English: lesser corn salad

V. eriocarpa Desv.

English: Italian corn salad, narrow-fruit corn salad

V. locusta (L.) Laterrade (*Valerianella locusta* var. *olitoria* L.; *Valerianella olitoria* (L.) Pollich)

English: common corn salad, lamb's lettuce, corn salad

V. muricata (Steven ex Roemer & Schultes) Baxter & Wooster (*Fedia muricata* Steven ex Roemer & Schultes; *Valerianella truncata* Betcke; *Valerianella truncata* Betcke var. *muricata* (Steven) Boiss.)

English: narrow-fruit corn salad

Valetonia T. Durand Icacinaceae

Origins:
For the Dutch botanist Theodoric Valeton, 1855-1929, bacteriologist, scientist, traveler; see R. Zander, F. Encke, G. Buchheim and S. Seybold, *Handwörterbuch der Pflanzennamen*. 14. Aufl. 1993; Stafleu and Cowan, *Taxonomic Literature*. 6: 649-651. 1986.

Valisneria Scop. Hydrocharitaceae (Vallisneriaceae)

Origins:
After the Italian physician Antonio Vallisnieri (Vallisneri), 1661-1730 (Padua), see also *Vallisneria* L.

Vallariopsis Woodson Apocynaceae

Origins:
Resembling the genus *Vallaris* Burm.f.

Vallaris Burm.f. Apocynaceae

Origins:
Latin *vallaris, e* "belonging to a rampart," *vallum, i* "a wall, rampart," *vallus* "a stake in a palisade, a stake," referring to the nature of these plants, used for fences.

Species/Vernacular Names:
V. indecora (Baillon) Tsiang & P.T. Li

English: giant vallaris

China: da niu zi hua

V. solanacea (Roth) Kuntze

English: nightshade-like vallaris

China: niu zi hua

Vallaris Raf. Euphorbiaceae

Origins:
See Constantine Samuel Rafinesque (1783-1840), *Flora Telluriana*. 4: 114. Philadelphia 1836 [1838]; E.D. Merrill (1876-1956), *Index rafinesquianus*. The plant names published by C.S. Rafinesque, etc. 157. Jamaica Plain, Massachusetts, USA 1949.

Vallea Mutis ex L.f. Elaeocarpaceae

Origins:
Named for Felice Valle, d. 1747, botanist, wrote a *Florula Corsicae*.

Species/Vernacular Names:
V. stipularis L.f.

Peru: achacapuli, cugur, cunhur, chillunmay, gellccoy, gorgor, gorgosh, olla-olla, quellccoy, sacha capuli, úlas

Vallesia Ruíz & Pav. Apocynaceae

Origins:
After the Spanish physician Francisco de Vallés (Franciscus Vallesius), 1524-1592. His writings include *Tratado de las aguas destiladas, pesos, y medidas de que los boticarios deven usar*. Madrid 1592, De iis quae scripta sunt physice in libris sacris, sive *de sacra philosophia*. Lugduni 1592 and *Methodus medendi*. Venetiis 1589; see Richard J. Durling, comp., *A Catalogue of Sixteenth Century Printed Books in the National Library of Medicine*. Bethesda, Maryland 1967; E. Ortega and B. Marcos González, *Francisco de Valles*. 1914; M. Colmeiro y Penido, *La Botánica y los Botánicos de la Peninsula Hispano-Lusitana*. Madrid 1858.

Vallifilix Thouars Schizaeaceae

Origins:

From the Latin *valles* or *vallis* "a valley" and *filix* "fern."

Vallisneria L. Hydrocharitaceae (Vallisneriaceae)

Origins:

After the Italian (b. Lucca province) physician Antonio Vallisnieri (or Vallisneri), 1661-1730 (d. Padua), biologist, naturalist, botanist, a pupil of Marcello Malpighi (1628-1694) at the University of Bologna, 1700-1710 professor of practical medicine at the University of Padua and from 1710 to 1730 professor of theoretical medicine, 1705 member of the Royal Society of London, 1718 knighted by the Duke Rinaldo I of Modena, showed that all parasitic insects of plants derive from eggs (to demonstrate the fallacy of spontaneous generation). His writings include *Dialoghi sopra la curiosa origine di molti insetti*. Venezia 1700, *Istoria del camaleonte Affricano e di varj animali d'Italia*. [4to, first edition, a monograph on the chameleon, salamander and other animals.] Venezia 1715 and *Istoria della generazione dell'uomo, e degli animali*. Venezia 1721. See *Opere fisico-mediche stampate e manoscritte del ... Antonio Vallisneri* raccolte da Antonio suo figliuolo. Venezia 1733; Mario Sabia, *Le opere di Antonio Vallisneri. Bibliografia ragionata*. Rimini 1996; P. Masat Lucchetta, *Antonio Vallisnieri medico naturalista*. Venezia 1984; G.A. Porcia, *Notizie della vita e degli studi del cavalier Antonio Vallisnieri*. Bologna 1986; Giuseppe Montalenti, in *Dictionary of Scientific Biography* 13: 562-565. New York 1981; Donald Culross Peattie (1898-1964), *De natuur ontsloten in de werken van Marcello Malpighi, Jan Swammerdam*, etc. Den Haag 1937; Jan Swammerdam (1637-1680), *Historia Insectorum Generalis* ... Ex Belgica Latinam fecit Henricus Christianus Henninius. Editio Nova. [The third Latin edition; originally published in Dutch in 1669.] Leyden 1733 and *The Book of Nature*; or the History of Insects ... With the life of the author, by Herman Boerhaave. [First edition in English, first published under the title *Biblia Naturae* in a Dutch/Latin edition in 1737-38.] London 1758; Garrison and Morton, *Medical Bibliography*. 302. New York 1961; Alexander B. Adams, *Eternal Quest. The Story of the Great Naturalists*. New York 1969; Howard B. Adelmann, *Marcello Malpighi and the Evolution of Embryology*. Ithaca 1966; Carl Linnaeus, *Species Plantarum*. 1015. 1753 and *Genera Plantarum*. Ed. 5. 446. 1754; Abraham Schierbeek, *J. Swammerdam, His Life and Work*. Amsterdam 1967; G.C. Wittstein, *Etymologisch-botanisches Handwörterbuch*. 917. 1852; K. Mägdefrau, *Geschichte der Botanik. Leben und Leistung großer Forscher*. 2. Aufl. Stuttgart 1992; Asher Rare Books & Antiquariaat Forum, *Catalogue Natural History*. item no. 148. The Netherlands 1998.

Species/Vernacular Names:

V. aethiopica Fenzl

English: eel grass, tape grass

V. americana Michx.

English: flumine Mississippi

V. spiralis L. (*Vallisneria nana* R. Br.; *Vallisneria gigantea* Graebner)

English: eel grass, tape grass, ribbonweed, eelweed

The Philippines: sintas-sintasan, sabotan-buaia, balaiba, lanteng, mariu-bariu

China: ku cao, ku tsao

Vallota Salisb. ex Herbert Amaryllidaceae

Origins:

After the French botanist Antoine Vallot, 1594-1671, physician.

Valteta Raf. Solanaceae

Origins:

See Constantine Samuel Rafinesque, *Sylva Telluriana*. 53. 1838; E.D. Merrill, *Index rafinesquianus*. The plant names published by C.S. Rafinesque, etc. 211, 213. Jamaica Plain, Massachusetts, USA 1949; Julian M.H. Shaw, "*Iochroma* — a review." *The New Plantsman*. 5(3): 154-192. 1998.

Valvanthera C.T. White Hernandiaceae

Origins:

From the Latin *valva* "a valve, a folding-door" and *anthera* "anther," indicating the nature of the anthers; see Cyril Tenison White (1890-1950), *Proceedings of the Royal Society of Queensland*. 47: 76. (Feb.) 1936.

Valvaria Seringe Ranunculaceae

Origins:

Latin *valva* "a valve, a folding-door."

Van-royena Aubréville Sapotaceae

Origins:

For the Dutch botanist Adriaan (Adrian) van Royen, 1704-1779, Dr. med. Leyden 1728, professor of botany and medicine, 1730-1754 Director of Botanic Garden at Leyden, friend of Linnaeus, correspondent of A. v. Haller, physician, author of Dissertatio ... *de anatome et oeconomia plantarum*. Lugduni Batavorum [Leyden] [1728], and *Florae Leydensis Prodromus*, exhibens plantas quae in horto academico Lugduno-Batavo aluntur. Lugduni Batavorum 1740, and *Ericetum africanum*. 40 tab. absque descriptione. (J. v. d. Spyk fecit. J.P. Catell del.). Opus ineditum. Lugdunum Batavorum circa 1760 (Illustrations of 40 unnamed species of *Erica*, without text or title page, in the library of the Royal Botanic Gardens, Kew); see Martin Wilhelm Schwencke (1707-1785), *Novae plantae Schwenckia dictae a celeberrimo Linnaeo in Gen. plant.* ed. VI. p. 567 ex celeb. Davidiis van Rooijen Charact. mss. 1761 communicata brevis descriptio et delineatio cum notis characteristicis. Hagae Comitum [The Hague], typ. van Karnebeek. 1766; H. Veendorp and L.G.M. Baas Becking, *Hortus Academicus Lugduno-Batavus 1587-1937*. The development of the gardens of Leyden University. Harlemi [Haarlem] 1938; D.O. Wijnands, *The Botany of the Commelins*. Rotterdam 1983; John H. Barnhart, *Biographical Notes upon Botanists*. 3: 187. 1965; Mariella Azzarello Di Misa, a cura di, *Il Fondo Antico della Biblioteca dell'Orto Botanico di Palermo*. 238. Regione Siciliana, Palermo 1988. Or named for David van Royen, 1727-1799, Dutch botanist and physician, Adriaan van Royen was his uncle.

Vanasushava Mukherjee & Constance Umbelliferae

Origins:

Indicating its habitat in the forest, from the Sanskrit words for *forest* and *Carum*, respectively; see Prasanta K. Mukherjee & Lincoln Constance, in *Kew Bulletin*. 29(3): 593-596. 1974.

Vancouveria C. Morren & Decne. Berberidaceae

Origins:

After the British explorer Captain George Vancouver, Royal Navy, 1757-1798, navigator, with Captain James Cook on his second and third voyages, 1791-1795 survey of the Pacific coast of North America (with Archibald Menzies on the Voyage of the *Discovery* and *Chatham*, commanded by William Broughton), 1794 promoted Captain, wrote *A Voyage of Discovery to the North Pacific Ocean and round the World*. [Completed by his brother John, aided by Peter John Puget, 1765-1822.] London 1798; see Archibald Menzies, *Menzies' Journal of Vancouver's Voyage, April to October, 1792*. Edited, with ... notes, by C.F. Newcombe ... and a biographical note by J. Forsyth. Victoria 1923 [Archives of British Columbia. Memoir no. 5]; G. Godwin, *Vancouver:*

A Life 1757-1798. London 1930; B. Anderson, *Surveyor of the Sea: The Life and Voyages of George Vancouver*. Seattle 1960; E. Bell, ed., "The log of the *Chatham* (by Peter Puget)." *Honolulu Mercury*. 1, 4. 1929; W. Kaye Lamb, ed., *The Voyage of George Vancouver 1791-1795*. London 1984; John Dunmore, *Who's Who in Pacific Navigation*. 254-256. Honolulu 1991; John T. Walbran, *British Columbia Coast Names, 1592-1906. To Which are Added a Few Names in Adjacent United States Territory, Their Origin and History*. First edition. Ottawa: Government Printing Bureau, 1909; G.P.V. Akrigg & Helen B. Akrigg, *British Columbia Place Names*. Sono Nis Press, Victoria 1986; Jonathan Wantrup, *Australian Rare Books, 1788-1900*. Sydney 1987.

Species/Vernacular Names:

V. chrysantha E. Greene

California: Siskiyou inside-out-flower

V. hexandra (Hook.) Morren & Decne.

English: inside-out-flower

V. planipetala Calloni

English: redwood ivy

Vanda W. Jones ex R. Br. Orchidaceae

Origins:

From a Sanskrit word applied to *Vanda tessellata* (Roxb.) G. Don or *Vanda roxburghii* R. Br.

Species/Vernacular Names:

V. bensoni Batem.

Thailand: saam poi chom poo

V. brunnea Reichb.f.

Thailand: saam poi nok

V. coerulea Griff.

Thailand: fa mui

V. coerulescens Griff.

Thailand: fa mui noi

V. lamellata Lindl. (*Vanda amiensis* Masam. & Segawa)

Japan: Kôtô-hishui-ran (named for Kôtôshô Island, Taiwan)

V. sanderiana Reichb.f.

The Philippines: waling-waling

Vandasia Domin Fabaceae

Origins:

Named after the Bohemian botanist Karel Vandas, 1861-1923; see also *Vandasina* Rauschert.

Vandasina Rauschert Fabaceae

Origins:

After the Bohemian botanist Karel Vandas, 1861-1923 (Serbian Macedonia), traveler and plant collector, assistant at the National Museum, Praha, school teacher, wrote *Beiträge zur Kenntniss der Flora Bulgariens*. Praha 1889 and *Reliquiae Formánekianae*. Brunae [Brno] 1909; see J.H. Barnhart, *Biographical Notes upon Botanists*. 3: 477. 1965; R. Zander, F. Encke, G. Buchheim and S. Seybold, *Handwörterbuch der Pflanzennamen*. 14. Aufl. Stuttgart 1993; Ethelyn Maria Tucker, *Catalogue of the Library of the Arnold Arboretum of Harvard University*. Cambridge, Massachusetts 1917-1933.

Vandellia P. Browne ex L. Scrophulariaceae

Origins:

For the Italian botanist Domenico Vandelli, 1735-1816, physician, traveler, professor of chemistry and natural history at the University of Coimbra, from 1773 to 1791 Director of the Coimbra Botanical Garden. Among his writings are D.V. *Viridarium Grisley lusitanicum*. Olisipone 1789, *Dell'acqua di Brandola dissertazione*. Modena 1763, *Diccionario dos termos technicos de historia natural*. Coimbra 1788, *Florae lusitanicae et brasiliensis specimen*. Conimbricae [Coimbra] 1788, D. Vandellii *Epistola de holothurio et testudine coriacea*, etc. Patavii 1761, D. Vandellii *Fasciculus plantarum*, cum novis generibus, et speciebus. Olisipone 1771 and *Dissertatio de arbore draconis seu Dracaena*. Olisipone [Lisboa] 1768; see J.H. Barnhart, *Biographical Notes upon Botanists*. 3: 422. 1965; R. Zander, F. Encke, G. Buchheim and S. Seybold, *Handwörterbuch der Pflanzennamen*. 14. Aufl. Stuttgart 1993; H. Dolezal, in *Portug. Acta Biol*. 6: 257-323. 1959 and 7: 324-551. 1961; M. Colmeiro y Penido, *La Botánica y los Botánicos de la Peninsula Hispano-Lusitana*. Madrid 1858; Ida Kaplan Langman, *A Selected Guide to the Literature on the Flowering Plants of Mexico*. Philadelphia 1964; Antoine Lasègue, *Musée botanique de M. Benjamin Delessert*. 1845; Jonas C. Dryander, *Catalogus bibliothecae historico-naturalis Josephi Banks*. London 1796-1800; Ethelyn (Daliaette) Maria Tucker, *Catalogue of the Library of the Arnold Arboretum of Harvard University*. Cambridge, Massachusetts 1917-1933.

Vandenboschia Copel. Hymenophyllaceae

Origins:

For the Dutch physician Roelof Benjamin van den Bosch, 1810-1862, botanist, author of *Hymenophyllaceae javanicae*; sive descriptio Hymenophyllacearum archipelagi indici, iconibus illustrata. Amsterdam 1861; see Arthur D.

Chapman, ed., *Australian Plant Name Index*. 2946. Canberra 1991; Edwin Bingham Copeland (1873-1964), *Philippine Journal of Science*. 67: 51. 1938; J.H. Barnhart, *Biographical Notes upon Botanists*. 1: 225. 1965; T.W. Bossert, *Biographical Dictionary of Botanists Represented in the Hunt Institute Portrait Collection*. 46. 1972; Elmer Drew Merrill, *Contr. U.S. Natl. Herb*. 30(1): 70. 1947.

Vandera Raf. Euphorbiaceae

Origins:

See Constantine Samuel Rafinesque (1783-1840), *Autikon botanikon*. Icones plantarum select. nov. vel rariorum, etc. 49. Philadelphia 1840; E.D. Merrill (1876-1956), *Index rafinesquianus*. The plant names published by C.S. Rafinesque, etc. 157. Jamaica Plain, Massachusetts, USA 1949.

Vanderystia De Wild. Sapotaceae

Origins:

To honor the Belgian botanist Hyacinthe Julien Robert Vanderyst, 1860-1934, missionary in the Congo, agronomist, traveler and plant collector. Among his works are *Prodrome des maladies cryptogamiques belges*. Bruxelles 1904-1905 and *Jardin agrostologique de Kisanti*. Bruxelles 1928; see Stafleu and Cowan, *Taxonomic Literature*. 6: 662-663. Utrecht 1986.

Vandopsis Pfitzer Orchidaceae

Origins:

Resembling *Vanda* Jones ex R. Br.

Species/Vernacular Names:

V. parishii Schltr.

Thailand: lin krabue

Vanessa Raf. Rubiaceae

Origins:

See Constantine Samuel Rafinesque, *Flora Telluriana*. 3: 57. 1836 [1837]; E.D. Merrill, *Index rafinesquianus*. The plant names published by C.S. Rafinesque, etc. 227. Jamaica Plain, Massachusetts, USA 1949.

Vangueria Comm. ex Juss. Rubiaceae

Origins:

From the local Madagascan names *voa vanguer, vavangue* for *Vangueria madagascariensis* Gmel.

Species/Vernacular Names:

V. apiculata K. Schum. (*Vangueria longicalyx* Robyns)

English: tangle-flowered wild medlar

Southern Africa: mudenza, mundziru, muNemsva, muNjiro, muNzwiro (Shona)

Tanzania: ngundai, mgholoma

V. cyanescens Robyns

English: bush medlar

Southern Africa: bosmispel; umViyo, umViyo-wehlathi (Zulu)

V. esculenta S. Moore

English: wild medlar, Chirinda medlar

Southern Africa: mispel, edible vangueria; muGarampin-doro (Shona)

V. infausta Burch. subsp. *infausta* (*Vangueria tomentosa* Hochst.)

English: wild medlar

Southern Africa: wildemispel, mispel, grootmispel; umViyo, umVili, umVilo, umThulwa, isAntulutshwana (Zulu); muNjiro, nyakamwenje, munziro, ziRuwombe, muTufa, muZiringombe, muZozo (Shona); mmilo (North Sotho); mmilo, mothwanye (Ngwaketse dialect, Botswana); mpfilwa (Tsonga); iMandulu (Swazi); umVilo (Xhosa); muzwilu (Venda); omundjenja (Herero)

V. lasioclados K. Schum.

South Africa: shaggy-branched vangueria

V. madagascariensis J. Gmelin (*Vavanga chinensis* Rohr; *Vavanga edulis* Vahl; *Vangueria edulis* Lam.)

English: tamarind of the Indies, Spanish tamarind

Mascarene Islands: vavangue (La Réunion, Mauritius, Rodrigues)

V. randii S. Moore subsp. *chartacea* (Robyns) Verdc. (*Vangueria chartacea* Robyns) (after the British botanist and apothecary Isaac Rand, d. 1743, gardener, 1719 Fellow of the Royal Society, Curator of the Chelsea Physic Garden 1724-1743, author of *Index plantarum officinalium* ... Londini [London] 1730 and *Horti medici Chelseiani* ... Londini 1739; see John Martyn (1699-1768), *Historia plantarum rariorum*. Londini 1728; J.H. Barnhart, *Biographical Notes upon Botanists*. 3: 127. 1965; H. Field, *Memoirs of the Botanic Garden of Chelsea* belonging to the Society of Apothecaries of London. London 1878 [R.H. Semple, *Memoirs of the Botanic Garden at Chelsea* ... by the Late Henry Field. London 1878]; F.D. Drewitt, *The Romance of the Apothecaries' Garden at Chelsea*. London 1924; Blanche Elizabeth Edith Henrey, *British Botanical and Horticultural Literature before 1800*. Oxford 1975; G. Murray, *History of the Collections Contained in the Natural History Departments of the British Museum*. 1: 176. London 1904; D.E. Allen, *The Botanists*. Winchester 1986)

English: Natal bush medlar

Southern Africa: Natalbosmispel; umViyo-wehlathi (Zulu)

V. spinosa Roxb.

India: pindituka, pinda, muyuna, moyna, atu, alu, mullakare, manak-karai, manakkarai, visikilamu, pundrika, bangariki-lakri, chircholi, madandriksh, peddamaoga, veliki, vedankike, segagadda

Malaya: false thorn-randa, duri timun tahil, duri timbang tahil

Vangueriella Verdc. Rubiaceae

Origins:

Diminutive of the genus *Vangueria*.

Vangueriopsis Robyns Rubiaceae

Origins:

From *Vangueria* and the Greek *-opsis*, resembling the genus *Vangueria*.

Species/Vernacular Names:

V. lanciflora (Hiern) Robyns ex Good (*Canthium lanciflorum* Hiern; *Canthium platyphyllum* Hiern; *Vangueria lateritia* Dinter)

English: Rhodesian wild medlar

Southern Africa: muTofu, muTufu (Shona)

N. Rhodesia: musole

V. nigerica Robyns

Yoruba: erelu

Vanheerdia L. Bolus ex Hartmann Aizoaceae

Origins:

For Mr. P. van Heerde from Springbok, plant collector.

Vanhouttea Lemaire Gesneriaceae

Origins:

After the Belgian botanist Louis Benoît Van Houtte, 1810-1876, horticulturist, traveler and plant collector in Brazil, Honduras, Guatemala, with the Belgian botanist Charles François Antoine Morren (1807-1858) in 1839 founded *L'Horticulteur belge*, general editor of *Flore des serres*. 1845-1880, author of *Hortus Vanhoutteanus*. Gand [Gent] 1845-1846; see J.H. Barnhart, *Biographical Notes upon Botanists*. 3: 424. 1965; T.W. Bossert, *Biographical Dictionary of Botanists Represented in the Hunt Institute Portrait Collection.* 414. 1972; E.M. Tucker, *Catalogue of the Library of the Arnold Arboretum of Harvard University.* Cambridge, Massachusetts 1917-1933; R. Zander, F. Encke, G. Buchheim and S. Seybold, *Handwörterbuch der Pflanzennamen.* 14. Aufl. Stuttgart 1993; Stafleu and Cowan, *Taxonomic Literature.* 6: 670-675. 1986.

Vanilla Miller Orchidaceae

Origins:

Spanish *vaina* and *vainilla*, referring to the cylindrical and sheath-like pod, Latin *vagina* "a sheath," the fruits appear like little pods; see Ernest Weekley, *An Etymological Dictionary of Modern English.* 2: 1580. New York 1967; Manlio Cortelazzo & Paolo Zolli, *Dizionario etimologico della lingua italiana.* 5: 1412. 1988; F. D'Alberti di Villanuova, *Dizionario universale, critico, enciclopedico della lingua italiana.* Lucca 1797-1805.

Species/Vernacular Names:
V. madagascariensis Rolfe

Madagascar: amalo

V. planifolia Andr.

English: vanilla

Italian: vaniglia, vainiglia

French: vanille

Madagascar: vanila, vanilina, vanille

Mexico: vainilla, siisbik (= aromático, oloroso), tlilxochitl (= flor negra)

Vanillophorum Necker Orchidaceae

Origins:

From the Spanish *vaina* and *vainilla* and Greek *phoros* "bearer," descriptive of the shape of the fruits.

Vanillosmopsis Sch. Bip. Asteraceae

Origins:

From the genus *Vanilla*, *osme* "smell, odor, perfume" and *opsis* "like, resembling."

Vaniotia H. Lév. Gesneriaceae

Origins:

For the French botanist Eugène Vaniot, d. 1913, clergyman, Jesuit; see J.H. Barnhart, *Biographical Notes upon Botanists.* 3: 424. 1965.

Vanwykia Wiens Loranthaceae

Origins:

The genus was named for the South African botanist Pieter van Wyck, born 1931, ecologist, head of the Department of Research and Communication of the National Parks Board of South Africa, biologist, plant collector, author of *Trees of the Kruger National Park.* Cape Town 1972-1974 and *Field Guide to the Trees of the Kruger National Park.* Cape Town 1984; see Mary Gunn and Leslie E. Codd, *Botanical Exploration of Southern Africa.* 360. Cape Town 1981.

Species/Vernacular Names:
V. remota (Baker & Sprague) Wiens (*Loranthus remotus* Baker & Sprague)

English: mistletoe

Vanzijlia L. Bolus Aizoaceae

Origins:

After the South African Mrs. Dorothy Constantia van Zijl, 1886-1938.

Vargasiella C. Schweinf. Orchidaceae

Origins:

Named for the Peruvian botanist Julio César Vargas Calderón, 1907-1960, plant collector (Peruvian Andes), professor of botany at the University of Cuzco, discoverer of many species of orchids. His writings include "La flora de la región descubierta por la expedición de *The Viking Fund.*" *Revista Univ.* (Cuzco) 84: 1-19. 1943 and "Orchids of Machupichu." *Amer. Orchid Soc. Bull.* 34(11): 960-966. 1965.

Varilla A. Gray Asteraceae

Origins:
Spanish *varilla* "small rod, wand, flexible twig, rib."

Varinga Raf. Euphorbiaceae

Origins:
See Constantine S. Rafinesque, *Sylva Telluriana.* 58. 1838; E.D. Merrill, *Index rafinesquianus.* The plant names published by C.S. Rafinesque, etc. 112. Jamaica Plain, Massachusetts, USA 1949.

Vaseya Thurber Gramineae

Origins:

For the American (b. Scarborough, Yorks.) botanist George Vasey, 1822-1893 (d. Washington, USA), agrostologist, physician, in 1868 and 1869 explorer with J.W. Powell in Colorado, Curator of the United States National Herbarium, botanized in the Rockies, editor of *The American Entomologist and Botanist*, with J.W. Chickering, E. Foreman, Wm.H. Seaman and L.F. Ward wrote *Flora columbiana*. Washington 1876. His works include *The Grasses of the United States*. Washington 1883, *Grasses of the South*. Washington 1887, *Grasses of the Southwest*. Washington 1890-1891 and *Grasses of the Pacific Slope*. Washington 1892-1893, father of George Richard Vasey; see John H. Barnhart, *Biographical Notes upon Botanists*. 3: 427. 1965; Ida Kaplan Langman, *A Selected Guide to the Literature on the Flowering Plants of Mexico*. Philadelphia 1964; S. Lenley et al., *Catalog of the Manuscript and Archival Collections and Index to the Correspondence of John Torrey*. Library of the New York Botanical Garden. 1973; Ethelyn Maria Tucker, *Catalogue of the Library of the Arnold Arboretum of Harvard University*. Cambridge, Massachusetts 1917-1933; Howard Atwood Kelly and Walter Lincoln Burrage, *Dictionary of American Medical Biography*. New York 1928; J. Ewan, ed., *A Short History of Botany in the United States*. 45. New York and London 1969; Joseph Ewan, *Rocky Mountain Naturalists*. 327. The University of Denver Press 1950; Ray Desmond, *Dictionary of British & Irish Botanists and Horticulturists*. 703. 1994; William Marriott Canby (1831-1904) & Joseph Nelson Rose (1862-1928), "George Vasey: A biographical sketch." *Bot. Gaz.* 18: 170-183. 1893; Frederick Vernon Coville (1867-1937), "Death of Dr. George Vasey." *Bull. Torr. Bot. Club.* 20: 218-220. 1893; W.R. Maxon, "Frederick Vernon Coville." *Science*. 85: 280-281. 1937; J.W. Harshberger, *The Botanists of Philadelphia and Their Work*. Philadelphia 1899; Irving William Knobloch, compiled by, "A preliminary verified list of plant collectors in Mexico." *Phytologia Memoirs*. VI. Plainfield, N.J. 1983.

Vaseyanthus Cogn. Cucurbitaceae

Origins:

For the American botanist George Vasey, 1822-1893, agrostologist, physician, explorer in Colorado, Curator of the United States National Herbarium.

Vaseyochloa Hitchc. Gramineae

Origins:

After the American botanist George Vasey, 1822-1893, agrostologist, physician, author of *The Grasses of the United States*. Washington 1883, *Grasses of the South*. Washington 1887, *Grasses of the Southwest*. Washington 1890-1891, *Grasses of the Pacific Slope*. Washington 1892-1893 and *Monograph of the Grasses of the United States and British America*. Washington 1892; see Joseph Ewan, *Rocky Mountain Naturalists*. The University of Denver Press 1950.

Vasquesiella Dodson Orchidaceae

Origins:

Named for the Bolivian botanist Vasquez.

Vateria L. Dipterocarpaceae

Origins:

For the German physician Abraham Vater, 1684-1751, botanist. His writings include *Catalogus plantarum inprimis exoticarum horti Academici Wittenbergensis*. Wittenbergae 1721-1724 and *Catalogus variorum exoticorum ... quae in Museo suo ... possidet A.V.* Wittembergae 1726; see J.H.S. Formey, *Éloges des académiciens de Berlin, II*. Berlin 1757; Henry Guerlac, in *Dictionary of Scientific Biography* 6: 169-175. 1981; Bruno von Freyberg, in *Dictionary of Scientific Biography* 8: 146-148. 1981; Pietro Franceschini, in *Dictionary of Scientific Biography* 10: 266-268. 1981; G.C. Wittstein, *Etymologisch-botanisches Handwörterbuch*. 919. Ansbach 1852; Garrison and Morton, *Medical Bibliography*. By Leslie T. Morton. 976. New York 1961.

Species/Vernacular Names:
V. indica L.

English: white dammar, piney varnish

Vateriopsis F. Heim Dipterocarpaceae

Origins:
Resembling *Vateria* L.

Vatica L. Dipterocarpaceae

Origins:

Latin *vates, is* (*vatis*) "a foreteller, prophet"; *vatica herba*, a plant called also *Apollinaria*, see Pseudo Apuleius Barbarus, *Herbarium* 74.

Vatovaea Chiovenda Fabaceae

Origins:

After the Italian botanist Aristocle Vatova, born 1897, plant collector (Ethiopia and Somalia), author of *Compendio della flora e fauna del Mare Adriatico presso Rovigno.* Venezia 1928.

Vatricania Backeb. Cactaceae

Origins:

For Louis Vatrican, author of "Monaco's Exotic Garden." *Cact. Succ. J. Amer.* 38: 39-42. 1966; see Gordon Douglas Rowley, *A History of Succulent Plants.* 294. Strawberry Press, Mill Valley, California 1997.

Vauanthes Haw. Crassulaceae

Origins:

Greek *vau* "V" and *anthos* "flower," referring to the markings on the petals.

Vaughania S. Moore Fabaceae

Origins:

After the British (b. Northumberland) botanist John Vaughan Thompson, 1779-1847 (d. Sydney, New South Wales, Australia), studied medicine and surgery, naturalist, interested in marine biology, surgeon (to His Majesty's 37th Regiment of Foot), zoologist, 1799-1809 West Indies and Guiana, 1810 Fellow of the Linnean Society of London, corresponded with Alexander MacLeay (1767-1848), 1812-1816 Madagascar and Mauritius, made studies of marine invertebrates. Among his writings are *A Catalogue of the Exotic Plants Cultivated in the Mauritius ... to which are* added the English and French names... Compiled under the auspices of R.T. Farquhar Esq. Governor of Mauritius. Mauritius 1816, *Zoological Researches and Illustrations.* Cork, Ireland 1828-1834 and *A Catalogue of Plants Growing in the Vicinity of Berwick upon Tweed.* London 1807; see John H. Barnhart, *Biographical Notes upon Botanists.* 3: 377. Boston 1965; E.M. Tucker, *Catalogue of the Library of the Arnold Arboretum of Harvard University.* Cambridge, Massachusetts 1917-1933.

Vaupelia Brand Boraginaceae

Origins:

After the German botanist Friedrich Karl Johann Vaupel, 1876-1927, from 1899 to 1910 traveled in Mexico and the West Indies, India and Australia, Curator of the Botanical Museum Berlin-Dahlem, founder and editor of the *Zeitschrift für Sukkulentenkunde.* Berlin 1923-1928, editor of *Monatschrift für Kakteenkunde.* 1911-1920, with Karl

Moritz Schumann (1851-1904) and Robert Louis August Maximilian Guerke (Gürke) (1854-1911) wrote *Blühende Kakteen.* Among his works are *Cactaceae andinae.* 1913 and *Iridaceae africanae novae.* 1912; see John H. Barnhart (1871-1949), *Biographical Notes upon Botanists.* 3: 428. 1965; Ida Kaplan Langman, *A Selected Guide to the Literature on the Flowering Plants of Mexico.* Philadelphia 1964; Irving William Knobloch, compil., "A preliminary verified list of plant collectors in Mexico." *Phytologia Memoirs.* VI. 1983; Ethelyn Maria Tucker, *Catalogue of the Library of the Arnold Arboretum of Harvard University.* Cambridge, Massachusetts 1917-1933; Gordon Douglas Rowley, *A History of Succulent Plants.* Strawberry Press, Mill Valley, California 1997; R. Zander, F. Encke, G. Buchheim and S. Seybold, *Handwörterbuch der Pflanzennamen.* 14. Aufl. Stuttgart 1993.

Vauquelinia Corrêa ex Humb. & Bonpl. Rosaceae

Origins:

After the French (b. Normandy) pharmacist Louis Nicolas (Nicolas Louis) Vauquelin, 1763-1829 (d. Normandy), chemist and professor of chemistry at the Collège de France and at the Muséum d'Histoire Naturelle, wrote *Expériences sur les sèves des végétaux.* Paris [1798-1799]; see J.H. Barnhart, *Biographical Notes upon Botanists.* 3: 428. 1965; W.A. Smeaton, in *Dictionary of Scientific Biography* 13: 596-598. 1981; Mariella Azzarello Di Misa, a cura di, *Il Fondo Antico della Biblioteca dell'Orto Botanico di Palermo.* 280. Palermo 1988; T. W. Bossert, *Biographical Dictionary of Botanists Represented in the Hunt Institute Portrait Collection.* 415. Boston, Massachusetts 1972; Jonas C. Dryander, *Catalogus bibliothecae historico-naturalis Josephi Banks.* London 1796-1800; Ethelyn Maria Tucker, *Catalogue of the Library of the Arnold Arboretum of Harvard University.* Cambridge, Massachusetts 1917-1933.

Vausagesia Baillon Ochnaceae

Origins:

See *Sauvagesia* L.

Vavaea Benth. Meliaceae

Origins:

From the South Sea island of Vavau or Babo, Tonga (Friendly Isles); see Arthur D. Chapman, ed., *Australian Plant Name Index.* 2946. Canberra 1991; James A. Baines, *Australian Plant Genera. An Etymological Dictionary of*

Australian Plant Genera. 388. Chipping Norton, N.S.W. 1981.

Vavanga Rohr Rubiaceae

Origins:

From the local Madagascan names *voa vanguer, vavangue* for *Vangueria madagascariensis* Gmel.

Vavara Benoist Acanthaceae

Origins:

A vernacular name.

Vavilovia Fedorov Fabaceae

Origins:

For the Russian (b. Moscow) botanist Nikolaj (Nikolay, Nikolai) Ivanovich Vavilov, 1887-1943 (d. Saratov), scientist, teacher, phytogeographer, agronomist, explorer in Iran, 1929-1940 traveled and collected economic plants, he was the author of *Studies on the Origin of Cultivated Plants.* Leningrad 1926 and *The Role of Central Asia in the Origin of Cultivated Plants.* Leningrad 1931, brother of the physicist and President of the Soviet Academy of Sciences Sergey Ivanovich Vavilov (1891-1951); see Mark B. Adams, in *Dictionary of Scientific Biography* 15: 505-513. 1981; David Joravsky, in *Dictionary of Scientific Biography* 15: 425-427. 1981; J. Dorfman, in *Dictionary of Scientific Biography* 13: 598-599. 1981; John H. Barnhart, *Biographical Notes upon Botanists.* 3: 428. Boston 1965; T. W. Bossert, *Biographical Dictionary of Botanists Represented in the Hunt Institute Portrait Collection.* 416. Boston, Massachusetts 1972; Ida Kaplan Langman, *A Selected Guide to the Literature on the Flowering Plants of Mexico.* Philadelphia 1964; S. Lenley et al., *Catalog of the Manuscript and Archival Collections and Index to the Correspondence of John Torrey.* Library of the New York Botanical Garden. 1973; J. Ewan, ed., *A Short History of Botany in the United States.* 87. New York and London 1969; Bentley Glass, ed., *The Roving Naturalist. Travel Letters of Theodosius Dobzhansky.* Philadelphia 1980; R. Zander, F. Encke, G. Buchheim and S. Seybold, *Handwörterbuch der Pflanzennamen.* 14. Aufl. Stuttgart 1993.

Veatchia A. Gray Anacardiaceae

Origins:

For John Allan Veatch, plant collector; see Irving William Knobloch, compil., "A preliminary verified list of plant collectors in Mexico." *Phytologia Memoirs.* VI. 1983; Ida Kaplan Langman, *A Selected Guide to the Literature on the*

Flowering Plants of Mexico. Philadelphia 1964; George E. Lindsay, *Notes Concerning the Botanical Explorers and Exploration of Lower California, Mexico.* San Francisco 1955; Ira L. Wiggins, *Flora of Baja California.* Stanford, California 1980.

Veconcibea (Müll. Arg.) Pax & K. Hoffm. Euphorbiaceae

Origins:

An anagram of the generic name *Conceveiba* Aublet.

Veeresia Monach. & Moldenke Sterculiaceae

Origins:

An anagram of the generic name *Reevesia* Lindley.

Veitchia H.A. Wendland Palmae

Origins:

Named for the nurseryman and horticulturist James Veitch (1815-1869, d. Chelsea, London) of Exeter (Devon), England, 1862 Fellow of the Linnean Society, son of James Veitch (1792-1863), father of the British horticulturists Sir Harry James Veitch (1840-1924) and John Gould Veitch (1839-1870); see James H. Veitch, *Hortus Veitchii, A History of the Rise and Progress of the Nurseries of Messrs. James Veitch and Sons,* etc. London 1906; Miles Hadfield et al., *British Gardeners: A Biographical Dictionary.* London 1980; R. Zander, F. Encke, G. Buchheim and S. Seybold, *Handwörterbuch der Pflanzennamen.* 14. Aufl. 727, 794. Stuttgart 1993; Stafleu and Cowan, *Taxonomic Literature.* 6: 688-690. 1986; Ethelyn Maria Tucker, *Catalogue of the Library of the Arnold Arboretum of Harvard University.* Cambridge, Massachusetts 1917-1933; John H. Barnhart, *Biographical Notes upon Botanists.* 3: 429. Boston 1965; T. W. Bossert, *Biographical Dictionary of Botanists Represented in the Hunt Institute Portrait Collection.* 416. Boston, Massachusetts 1972; Ray Desmond, *Dictionary of British & Irish Botanists and Horticulturists.* 703-704. [the genus was named for James Veitch (1815-1869) and John Gould Veitch] London 1994; Euan Hillhouse Methven Cox, *Plant-Hunting in China. A History of Botanical Exploration in China and the Tibetan Marches.* London 1945; Alice Margaret Coats, *The Quest for Plants. A History of the Horticultural Explorers.* London 1969; Shirley Heriz-Smith, "Western collectors on eastern islands." *Hortus.* 7: 83-92. 1988; Shirley Heriz-Smith, "William Purdom (1880-1921). A Westmorland Planthunter in China." *Hortus.* 38: 49-62. 1996; H.R. Fletcher, *Story of the Royal Horticultural Society, 1804-1968.* Oxford 1969; Merle A. Reinikka, *A*

History of the Orchid. Timber Press 1996; Ernest Nelmes and William Cuthbertson, *Curtis's Botanical Magazine Dedications, 1827-1927.* [1931]; Michael S. Tyler-Whittle, *The Plant Hunters.* Philadelphia 1970; Emil Bretschneider, *History of European Botanical Discoveries in China.* Leipzig 1981.

Species/Vernacular Names:

V. joannis H.A. Wendland

Brazil: palmeira véitia

V. merrillii (Beccari) H.E. Moore (*Adonidia merrillii* Becc.)

English: Manila palm, Christmas palm

Brazil: palmeira de Manila

The Philippines: bunga de jolo, bunga de China, oring-oring

V. montgomeryana H.E. Moore

Brazil: palmeira véitia

Velezia L. Caryophyllaceae

Origins:

After a C. Velez, friend of the botanist Loefling; see James C. Hickman, ed., *The Jepson Manual: Higher Plants of California.* 497. University of California Press, Berkeley 1993.

Vella L. Brassicaceae

Origins:

Latin *vela, ae* the Gallic name for the plant *erysimon* (Plinius); see Helmut Genaust, *Etymologisches Wörterbuch der botanischen Pflanzennamen.* 676. Basel 1996; Carl Linnaeus, *Species Plantarum.* 641. 1753 and *Genera Plantarum.* Ed. 5. 289. 1754; Arthur D. Chapman, ed., *Australian Plant Name Index.* 2946. Canberra 1991.

Velleia Smith Goodeniaceae

Origins:

After the English botanist Thomas Velley, 1748/1749-1806 (d. Reading, Berks.), algologist, 1792 Fellow of the Linnean Society, in 1799 Lieutenant-Colonel of the Oxfordshire Militia, correspondent of Sir Thomas Gery Cullum (1741-1831), John Stackhouse (1742-1819), Sir James Edward Smith (1759-1828) and Richard Relhan (1754-1823), friend of Dawson Turner, author of *Coloured Figures of Marine Plants.* Bathoniae [Bath] 1795; see Helen Blackler, "The herbarium of Thomas Velley (1748-1806). *The North Western Naturalist.* vol. 13. 1938; J.H. Barnhart, *Biographical Notes upon Botanists.* 3: 429. 1965; I.H. Vegter, *Index Herbariorum.* Part II (7), *Collectors T-Z.* Regnum Vegetabile vol. 117. 1988; Jonas C. Dryander, *Catalogus bibliothecae historico-naturalis Josephi Banks.* London 1796-1800; Blanche Henrey, *British Botanical and Horticultural Literature before 1800.* Oxford 1975; Ray Desmond, *Dictionary of British & Irish Botanists and Horticulturists.* 704. London 1994; *Memoir and Correspondence of ... Sir J.E. Smith ...* Edited by Lady Pleasance Smith. London 1832.

Species/Vernacular Names:

V. arguta R. Br. (*Antherostylis calcarata* C. Gardner)

English: spur velleia

V. connata F. Muell. (*Velleia helmsii* K. Krause)

Australia: cup velleia

V. cycnopotamica F. Muell.

English: Swan River velleia (Swan River, Western Australia)

V. discophora F. Muell.

Australia: cabbage poison

V. macrocalyx Vriese (*Velleia prostrata* Ewart & L. Kerr)

Australia: pale poison

V. montana Hook.f.

Australia: mountain velleia

V. panduriformis A. Cunn. ex Benth.

Australia: Pindan poison

V. paradoxa R. Br.

English: spur velleia

V. parvisepta Carolin

English: smooth velleia

V. perfoliata R. Br.

English: Hawkesbury velleia

V. pubescens R. Br.

English: hairy velleia

V. spathulata R. Br.

English: northern velleia

Vellereophyton Hilliard & B.L. Burtt Asteraceae

Origins:

From the Latin *vellus, velleris* "a fleece, wool, down" and Greek *phython* "a plant."

Species/Vernacular Names:

V. dealbatum (Thunb.) Hilliard & B.L. Burtt (*Gnaphalium dealbatum* Thunb.; *Gnaphalium candidissimum* Lam.)

English: white cudweed

Velleya Roemer & Schultes Goodeniaceae

Origins:

An orthographic variant, genus dedicated to the English botanist Thomas Velley, 1748/49-1806 (Reading, Berks).

Vellosiella Baillon Scrophulariaceae

Origins:

Named to honor the Brazilian (b. San José, Minas Gerais) clergyman José Mariano de Conceição Vellozo (Velloso, Veloso), 1742-1811 (d. Rio de Janeiro), botanist, author of *Flora fluminensis*. Flumine Januario [Rio de Janeiro] 1825; see Thomas F. Glick, in *Dictionary of Scientific Biography* 13: 601-602. 1981; Stafleu and Cowan, *Taxonomic Literature*. 6: 696-699. 1986; R. Zander, F. Encke, G. Buchheim and S. Seybold, *Handwörterbuch der Pflanzennamen*. 14. Aufl. 794. 1993; Antoine Lasègue, *Musée botanique de M. Benjamin Delessert*. 1845; Mariella Azzarello Di Misa, ed., *Il Fondo Antico della Biblioteca dell'Orto Botanico di Palermo*. 281. Palermo 1988; Jacques-Charles Brunet, *Manuel du libraire et de l'amateur de livres*. 2: 1294. Paris 1861; Pritzel, *Thesaurus*. no. 10696; Ethelyn Maria Tucker, *Catalogue of the Library of the Arnold Arboretum of Harvard University*. Cambridge, Massachusetts 1917-1933; H. Genaust, *Etymologisches Wörterbuch der botanischen Pflanzennamen*. 676. 1996.

Vellozia Vandelli Velloziaceae

Origins:

After the Portuguese botanist Joaquim Velloso de Miranda, 1733-1815, plant collector in Brazil, correspondent of the Italian botanist Domenico Vandelli (1735-1816); see Mariella Azzarello Di Misa, ed., *Il Fondo Antico della Biblioteca dell'Orto Botanico di Palermo*. 281. Palermo 1988; Jacques-Charles Brunet, *Manuel du libraire et de l'amateur de livres*. 2: 1294. Paris 1861; Pritzel, *Thesaurus*. no. 10696; Stafleu and Cowan, *Taxonomic Literature*. 6: 698. 1986; R. Zander, F. Encke, G. Buchheim and S. Seybold, *Handwörterbuch der Pflanzennamen*. 14. Aufl. 794. Stuttgart 1993; Helmut Genaust, *Etymologisches Wörterbuch der botanischen Pflanzennamen*. 676. Basel 1996.

Veltheimia Gled. Hyacinthaceae (Liliaceae)

Origins:

The genus was named after August Ferdinand von Veltheim, 1741-1801, a German patron of botany, author of *Conjectures sur l'Urne de Barberini appartenant au Duc de Portland*. Helmstedt 1801 and *Ueber der Herren Werner und Karsten Reformen in der Mineralogie*. Helmstedt 1793; see

Hans Baumgärtel, in *Dictionary of Scientific Biography* 14: 60-61. 1981.

Species/Vernacular Names:

V. bracteata Harv. ex Bak. (*Veltheimia undulata* Moench; *Veltheimia viridifolia* Jacq.)

English: forest lily

South Africa: sandui, sandlelie

V. capensis (L.) DC. (*Veltheimia deasii* Barnes; *Veltheimia glauca* (Aiton) Jacq.; *Veltheimia roodeae* Phill.)

South Africa: sandlelie, kwarobe

Velvitsia Hiern Scrophulariaceae

Origins:

After the Austrian explorer Friedrich (Frederick) Martin Joseph Welwitsch, 1806-1872 (London), traveler, physician, botanist, plant collector, M.D. Vienna 1836, zoologist, from 1839 to 1853 Director of the Botanical Garden of Lisbon, 1865 Fellow of the Linnean Society. Among his works are "On the botany of Benguela, Mossamedes, etc., in Western Africa." *J. Linn. Soc. (Bot.)* 5:182-187. 1861 (Including the first description of *Welwitschia* under the name *Tumboa*) and *Apontamentos phytogeographicos sobre a flora da provincia de Angola na Africa equinocial*. Lisboa 1858 [1859], co-author (with Frederick Currey, 1819-1881) of *Fungi angolenses*. London 1868; see William Philip Hiern (1839-1925), *Catalogue of the African Plants Collected by F. Welwitsch in 1853-61*. London 1896-1901; F.N. Hepper and Fiona Neate, *Plant Collectors in West Africa*. 85. 1971; John H. Barnhart, *Biographical Notes upon Botanists*. 3: 476. 1965; Anthonius Josephus Maria Leeuwenberg, "Isotypes of which holotypes were destroyed in Berlin." *Webbia*. 19: 861-863. 1965; M. Colmeiro y Penido, *La Botánica y los Botánicos de la Peninsula Hispano-Lusitana*. Madrid 1858; Antoine Lasègue, *Musée botanique de M. Benjamin Delessert*. 1845; T. W. Bossert, *Biographical Dictionary of Botanists Represented in the Hunt Institute Portrait Collection*. 431. Boston, Massachusetts 1972; Ethelyn Maria Tucker, *Catalogue of the Library of the Arnold Arboretum of Harvard University*. Cambridge, Massachusetts 1917-1933; H.N. Clokie, *Account of the Herbaria of the Department of Botany in the University of Oxford*. 263. Oxford 1964; H. Dolezal, in *Portug. Acta Biol*. 6: 257-323. 1959 and 7: 324-551. 1961.

Vendredia Baillon Asteraceae

Origins:

"In a little time I began to speak to him, and teach him to speak to me; and, first, I made him know his name should be Friday, which was the day I saved his life, and I called him so for the memory of the time..." (Daniel Defoe, *The Life and Adventures of Robinson Crusoe*, chapter XIV); *vendredi* is the French for Friday, the syn. of *Vendredia* Baill. is *Robinsonia* DC.

Venegasia DC. Asteraceae

Origins:
After the Mexican writer Miguel P. Venegas, 1680-1764, missionary to California, author of *Noticia de la California*. Madrid 1757 [English translation: *Natural and Civil History of California*. London 1759] and *Vida y virtudes del V.P.J.B. Zappa*, ... Sacada de la que escrivió el padre M.V. ... y ordenada por otro padre (F.X. Fluviá). Barcelona 1754; see Irving William Knobloch, compil., "A preliminary verified list of plant collectors in Mexico." *Phytologia Memoirs*. VI. 1983; Ira L. Wiggins, *Flora of Baja California*. Stanford, California 1980 and "Botanical investigations in Baja California, Mexico." *Plant Sci. Bull*. 9(1): 1-6. 1963.

Species/Vernacular Names:
V. carpesioides DC.
English: canyon sunflower

Venidium Lessing Asteraceae

Origins:
Latin *vena* "a blood-vessel, vein," referring to the ribbed fruits; see F. Boerner & G. Kunkel, *Taschenwörterbuch der botanischen Pflanzennamen*. 4. Aufl. 184. Berlin & Hamburg 1989.

Species/Vernacular Names:
V. fastuosum (Jacq.) Stapf
English: monarch-of-the-veld, Cape daisy

Ventenata G.L. Koeler Gramineae

Origins:
For the French botanist Étienne Pierre Ventenat, 1757-1808, clergyman, librarian, brother of Louis Ventenat (1765-1794). His writings include *Monographie du genre Tilleul*. Paris 1802, *Choix de plantes*. Paris 1803 [-1808] and *Jardin de la Malmaison*. Paris 1803-1804 [-1805]; see G. Cuvier, *Éloge historique de M. Ventenat*. Paris 1809; Mariella Azzarello Di Misa, a cura di, *Il Fondo Antico della Biblioteca dell'Orto Botanico di Palermo*. 281. Regione Siciliana, Palermo 1988; Yves Laissus, in *Dictionary of Scientific Biography* 3: 172-173. 1981; Ida Kaplan Langman, *A Selected Guide to the Literature on the Flowering Plants of Mexico*. Philadelphia 1964; Antoine Lasègue, *Musée botanique de M. Benjamin Delessert*. 1845; J.H. Barnhart, *Biographical*

Notes upon Botanists. 3: 430. 1965; Emil Bretschneider, *History of European Botanical Discoveries in China*. Leipzig 1981; Jonas C. Dryander, *Catalogus bibliothecae historico-naturalis Josephi Banks*. London 1796-1800; Ethelyn Maria Tucker, *Catalogue of the Library of the Arnold Arboretum of Harvard University*. Cambridge, Massachusetts 1917-1933; R. Zander, F. Encke, G. Buchheim and S. Seybold, *Handwörterbuch der Pflanzennamen*. 14. Aufl. Stuttgart 1993.

Ventenatia Cav. Epacridaceae

Origins:
For the French botanist Étienne Pierre Ventenat, 1757-1808, clergyman, librarian; see J.H. Barnhart, *Biographical Notes upon Botanists*. 3: 430. 1965.

Ventenatia Smith Stylidiaceae

Origins:
For the French botanist Étienne Pierre Ventenat, 1757-1808, clergyman, librarian; see J.H. Barnhart, *Biographical Notes upon Botanists*. 3: 430. 1965.

Ventenatia Tratt. Euphorbiaceae

Origins:
Named for the French botanist Étienne Pierre Ventenat, 1757-1808, clergyman, librarian; see G. Cuvier, *Éloge historique de M. Ventenat*. Paris 1809; J.H. Barnhart, *Biographical Notes upon Botanists*. 3: 430. 1965.

Ventilago Gaertner Rhamnaceae

Origins:
Latin *ventilo, avi, atum* "to swing, to move, to fan," *ventus* "wind," an allusion to the winged fruits.

Ventricularia Garay Orchidaceae

Origins:
Latin *ventriculus* "the belly," diminutive of *venter, ventris* "the belly," referring to the lip spur.

Veprecella Naudin Melastomataceae

Origins:
Latin *veprecula, ae* "a little thorn-bush, a small brier-bush," diminutive of *vepres, vepris, veper, is* "a bramble-bush."

Vepris Comm. ex A. Juss. Rutaceae

Origins:

Latin *vepres, is* "a bramble, thorny bush, spiny shrub," the generic name is not clear because the genus is unarmed; many species live in scrub and thorn thicket.

Species/Vernacular Names:

V. sp.

East Africa: mlunguschigiti

V. carringtoniana Mendonça (the specific name after the professor Joao S. Carrington da Costa)

English: coastal white ironwood, coastal white ironwood

Southern Africa: kuswitysterhout; uMozane (Zulu)

V. lanceolata (Lam.) G. Don (*Toddalia lanceolata* Lam.; *Vepris undulata* (Thunb.) Verdoorn & C.A. Sm., nom. illegit.)

English: white ironwood

Rodrigues Island: bois de patte de poule, rampoule, bois trois feuilles

Southern Africa: witysterhout; uMozane, iSutha (Zulu); umZane (Zulu, Xhosa); iMotane (Swazi); muhondwa (Venda)

V. louisii G. Gilbert (for J. Louis, botanist of INEAC, professor at Gembloux, France)

Zaire: yolowe, tshitempa, kolomasumu

V. madagascariensis (Baillon) H. Perrier

Madagascar: anisety, laposintiala, maingitsangeala, tolongoala

V. reflexa Verdoorn

English: bushveld white ironwood, drooping-leaved vepris

Southern Africa: bosveld witysterhout; uMosane (Zulu)

Veratrilla Baillon ex Franchet Gentianaceae

Origins:

Latin *veratrum, i* used by Plinius et al. for a plant, hellebore.

Species/Vernacular Names:

V. baillonii Franchet

English: Baillon veratrilla

China: huang qin jiao

Veratrum L. Melanthiaceae (Liliaceae)

Origins:

Latin *veratrum, i* used by Plinius et al. for a plant, hellebore; see M. Cortelazzo & P. Zolli, *Dizionario etimologico della lingua italiana.* 5: 1425. Bologna 1988; Helmut Genaust, *Etymologisches Wörterbuch der botanischen Pflanzennamen.* 677-678. 1996; F. Boerner & G. Kunkel, *Taschenwörterbuch der botanischen Pflanzennamen.* 4. Aufl. 184. 1989.

Species/Vernacular Names:

V. album L.

English: white ellebore

China: li lu, shan tsung

V. fimbriatum A. Gray

English: fringed false ellebore

V. insolitum Jepson

English: Siskiyou false ellebore

V. nigrum L.

English: false hellebore

China: li hu, li lu, shan tsung

Verbascum L. Scrophulariaceae

Origins:

Ancient Latin name for the plant used by Plinius, *verbascum, i*, mullein; see Carl Linnaeus, *Species Plantarum.* 177. 1753 and *Genera Plantarum.* Ed. 5. 83. 1754; Manlio Cortelazzo & Paolo Zolli, *Dizionario etimologico della lingua italiana.* 5: 1425. 1988.

Species/Vernacular Names:

V. blattaria L.

English: moth mullein

V. chaixii Vill.

English: nettle-leaved mullein

V. lychnitis L.

English: white mullein

V. nigrum L.

English: dark mullein, black mullein

V. phlomoides L.

English: clasping mullein, mullein

V. phoeniceum L.

English: purple mullein

V. speciosum Schrader

English: showy mullein

V. thapsus L. (Greek *thapsia*, Latin *thapsia* or *thapsos* for a poisonous shrub, *Thapsia asclepium* L.)

English: Aaron's rod, hag-taper, torches, great mullein, woolly mullein, flannel plant, velvet plant, common mullein

V. virgatum Stokes

English: Aaron's rod, twiggy mullein, green mullein, virgate mullein, wand mullein, purple-stamen mullein

Southern Africa: setla-bocha (Sotho)

Verbena L. Verbenaceae

Origins:

Verbena, ae, the Latin name for a plant sacred to the Romans, "sacred boughs, etc."; Akkadian *erum, arum, harum* "branch," *banu* "to engender, to produce, to create: said of deity," *binu* "son"; usually in plur., Latin *verbenae, arum* "foliage, leaves, twigs and branches sacred boughs, laurel, olive, myrtle, etc.," *verber, eris* "a lash, whip, scourge, rod," *verbero, avi, atum, are* "to lash, whip, scourge, beat"; see Carl Linnaeus, *Species Plantarum.* 18. 1753 and *Genera Plantarum*. Ed. 5. 12. 1754; Ernest Weekley, *An Etymological Dictionary of Modern English*. 2: 1588, 1591. Dover Publications, New York 1967; Giovanni Semerano, *Le origini della cultura europea*. Dizionario della lingua Latina e di voci moderne. 2(2): 608. Leo S. Olschki Editore, Firenze 1994; Manlio Cortelazzo & Paolo Zolli, *Dizionario etimologico della lingua italiana*. 5: 1425. 1988.

Species/Vernacular Names:

V. sp.

Peru: azulina, huallkjapaya, huayñu-huayta, maycha

V. bipinnatifida Nutt.

English: Dakota vervain

V. bonariensis L. (*Verbena trichotoma* Moench; *Verbena quadrangularis* Vell.) (named for the city of Buenos Aires, Argentina)

English: purple top, purple top verbena, wild verbena, weed verbena, South American vervain, tall verbena, tall vervain

Peru: poleo

South Africa: blouwaterbossie, wilde verbena

Japan: tachi-ba-bena

V. bracteata Lagasca & Rodriguez

English: creeping vervain, bracted vervain, prostrate vervain, wild verbena

V. brasiliensis Vell.

English: Brazilian vervain

V. canadensis (L.) Britton

English: rose vervain, clump verbena, creeping vervain, rose verbena

V. californica Mold.

English: California vervain

V. x *enegelmannii* Mold.

English: Engelmann's vervain

V. hastata L.

English: American blue vervain, wild hyssop, blue verbena, simpler's joy

V. hispida Ruíz & Pavón

English: hairy vervain, hairy verbena

V. x *hybrida* Groenl. & Ruempl.

English: common garden verbena, florist's verbena

V. lasiostachys Link

English: vervain

V. litoralis Kunth

English: seashore vervain

Central America: verbena, chachalbe, chichavac, cotacám, dorí, verbena fina

Peru: verbena, verbena del campo, wirwina, yapo

Japan: hime-kuma-tsuzura

Hawaii: owi, oi, ha'uoi, ha'uowi

V. macdougalii A.A. Heller

English: Macdougal's vervain

V. menthifolia Benth.

English: mint vervain

V. monacensis Mold.

English: Munich vervain

V. officinalis L.

English: vervain, common vervain, common verbena, European verbena, verbain, wild verbena, European vervain, pigeon's grass, holy herb

Southern Africa: Europese verbena; seona-se-seholo (South Sotho)

Japan: kuma-tsuzura

China: ma bian cao, ma pien tsao

Arabic: tronjia, ben nout

V. rigida Spreng. (*Verbena venosa* Gillies & Hook.; *Verbena scaberrima* Cham.)

English: veined vervain, large-veined vervain

Japan: bijo-zakura

V. runyonii Mold.

English: Rio Grande vervain

V. simplex Lehm.

English: narrow-leaved vervain, narrow-leaf vervain

V. stricta Vent.

English: hoary vervain, woolly verbena

V. supina L.

English: procumbent vervain, trailing verbena

V. tenera Spreng.

English: thin-leaf verbena

V. tenuisecta Briq. (*Verbena tenera* auct. non Spreng.; *Glandularia tenuisecta* (Briq.) Small)

English: moss vervain, fine-leaved verbena, moss verbena, tuber vervain, wild verbena, Maynes pest

South Africa: fynblaarverbena, wilde verbena

V. urticifolia L.

English: white vervain, white verbena

V. venosa Gill. & Hook.

English: veined vervain, verbena

Southern Africa: morod (Sotho)

V. xutha Lehm.

English: gulf vervain

Verbenoxylum Tronc. Verbenaceae

Origins:
The genus *Verbena* and *xylon* "wood."

Verbesina L. Asteraceae

Origins:
Resembling the leaves of the genus *Verbena*; see Carl Linnaeus, *Species Plantarum*. 901. 1753 and *Genera Plantarum*. Ed. 5. 384. 1754.

Species/Vernacular Names:
V. alternifolia (L.) Britt. ex C. Mohr
English: wingstem, yellow ironweed
V. dissita A. Gray
English: crownbeard
V. encelioides (Cav.) A. Gray var. *encelioides*
English: butter daisy, golden crownbeard, wild sunflower
South Africa: wilde sonneblom
V. encelioides (Cav.) A. Gray ssp. *exauriculata* (Robinson & Greenman) J. Coleman
English: golden crownbeard
V. helianthoides Michaux
English: crownbeard
V. virginica L.
English: frostweed

Verlotia E. Fournier Asclepiadaceae

Origins:
For the French botanist Pierre Bernard Lazare Verlot, 1836-1897, horticulturist; see J.H. Barnhart, *Biographical Notes upon Botanists*. 3: 431. 1965; Ethelyn Maria Tucker, *Catalogue of the Library of the Arnold Arboretum of Harvard University*. Cambridge, Massachusetts 1917-1933; R. Zander, F. Encke, G. Buchheim and S. Seybold, *Handwörterbuch der Pflanzennamen*. 14. Aufl. Stuttgart 1993.

Vermeulenia Löve & D. Löve Orchidaceae

Origins:
For the Dutch botanist Pieter Vermeulen, 1899-1981, orchidologist. Among his writings are "The vanished stamens." in *Amer. Orchid Soc. Bull.* 22: 437-440. 1953 and "The system of the Orchidales." *Acta Botan. Neerlandica.* 15: 224-253. 1966; see T. W. Bossert, *Biographical Dictionary of Botanists Represented in the Hunt Institute Portrait Collection*. 418. Boston, Massachusetts 1972; Helga Dietrich, *Bibliographia Orchidacearum*. 2: 185. Jena 1981; R. Zander, F. Encke, G. Buchheim and S. Seybold, *Handwörterbuch der Pflanzennamen*. 14. Aufl. Stuttgart 1993.

Vermifrux J.B. Gillett Fabaceae

Origins:
Latin *vermis* "a worm" and *frux, frugis* or *fruges, um* "legumes, fruits of the earth."

Vernicia Lour. Euphorbiaceae

Origins:
Latin *vernix, icis* "varnish, paint," it is possible to make varnish from the oily seeds; see João (Joannes) de Loureiro, 1717-1791, *Flora cochinchinensis*: sistens plantas in regno Cochinchina nascentes. 2: 541, 586. Ulyssipone [Lisboa] 1790.

Vernonanthera H. Robinson Asteraceae

Origins:
The genus *Vernonia* and Greek *anthera*.

Vernonia Schreber Asteraceae

Origins:
In honor of the English botanist William Vernon, 1666/1667-c. 1715, bryologist, 1702 Fellow of the Royal Society, traveler and plant collector in North America, Maryland 1698, with David Krieg (c. 1670-1710); see I.H. Vegter, *Index Herbariorum*. Part II (7), *Collectors T-Z*. Regnum

Vegetabile vol. 117. 1988; Charles Earle Raven, *English Naturalists from Neckam to Ray*. Cambridge 1947; R. Pulteney, *Historical and Biographical Sketches of the Progress of Botany in England*. 2: 57-58. London 1790; Dawson Turner (1775-1858), *Extracts from the Literary and Scientific Correspondence of R. Richardson, of Bierly, Yorkshire: Illustrative of the State and Progress of Botany* [Edited by D. Turner. Extracted from the memoir of the Richardson family, by Mrs. D. Richardson] Yarmouth 1835; James Britten, *The Sloane Herbarium*, revised and edited by J.E. Dandy. 1958; Georg Christian Wittstein, *Etymologisch-botanisches Handwörterbuch*. Ansbach 1852; H. Genaust, *Etymologisches Wörterbuch der botanischen Pflanzennamen*. 679. [genus named for English plant collector in North America William Vernon, d. 1711] 1996; James Petiver, *Museii Petiveriani*. London 1695-1703.

Species/Vernacular Names:

V. sp.

Peru: shuca

Nigeria: domashi, ewo-jedi, bopolopolo, burzu, sabulum-mata, shekani, ewuro, onimagungun, shapo, shiwaka, olubo, ujuju, bagashi

V. adoensis Sch.Bip. ex Walp. (*Vernonia shirensis* Oliv. & Hiern)

English: ironweed

Yoruba: ero oko, ewuro oko, ewuro odan, orubu

V. ambigua Kotschy & Peyr.

Yoruba: tabale, oorungo

V. amygdalina Del. (*Vernonia randii* S. Moore) ("like an almond," from the Greek *amygdale* "almond")

Southern Africa: monqo (Yei); sikakadzi, sikavakadzi (Shona)

Nigeria: ewuro-oko, ewuro, burzu, sabulum-mata, shiwaka, olubo

Yoruba: ewuro jije, ewuro gidi, ewuro oko, ewuro, pako, orin

Congo: matulu, munkankali

V. arborea Buch.-Ham.

English: tree vernonia

Malaya: medang gambong, merambong

V. baldwinii Torr.

English: western ironweed

V. capensis (Houtt.) Druce (*Vernonia pinifolia* (Lam.) Less.)

English: wild heliotrope

V. cinerascens Sch.Bip. (*Vernonia luederitziana* O. Hoffm.; *Vernonia porta-taurinae* Dinter ex Merxm.; *Vernonia squarrosa* Dinter ex Merxm.)

English: ash-grey vernonia

V. cinerea (L.) Less.

English: ashy colored ironweed

The Philippines: tagulinaw, tagulinai, bulak-manok, agasmoro, magmansi, sagit, kolong-kugon

Malaya: bujang samalam, ekor kuda, sebagi, sembong hutan, rumput sepagi, susor daun, tahi babi, tambak bukit, tombak bukit

Yoruba: jedi jedi, bojure, oorungo

V. colorata (Willd.) Drake subsp. *colorata* (*Vernonia senegalensis* (Pers.) Less.; *Decaneurum senegalense* (Less.) DC.; *Eupatorium coloratum* Willd.)

English: lowveld bitter-tea, colored vernonia, star-flowered tree vernonia

Southern Africa: laeveldbittertee, laeveldbloutee, vernonia; iBozane (Zulu); nyamepo (Shona)

Yoruba: ewuro oko

Mali: ko safina, ko safune

V. conferta Benth.

Nigeria: orungo, shapo, agori, oluworoworo, bagashi, domashi, oruwenni; sapo (Yoruba); oriweni (Edo); ubol (Igbo)

V. escuilenta Hemsl.

English: edible ironweed

V. fastigiata Oliv. & Hiern (*Vernonia schinzii* O. Hoffm.)

Southern Africa: blouteebossie, langbeenbossie; lehlanye (Pedi)

V. flexuosa Sims

English: zig-zag vernonia

V. frondosa Oliv. & Hiern

Nigeria: akisa (Yoruba)

V. lindheimeri A. Gray & Engelm.

English: woolly ironweed

V. marginata (Torr.) Raf.

English: Plains ironweed

V. myriantha Hook.f. (*Vernonia ampla* O. Hoffm.; *Vernonia podocoma* Sch.Bip. ex Vatke; *Vernonia stipulacea* Klatt)

English: blue bitter tea, poison tree vernonia

Southern Africa: bloubittertee, bosbloutee; uHluhlunga (Zulu); muDambasese, penembe (Shona)

V. nantcianensis (Pamp.) Hand.- Mazz.

English: Nanchang ironweed

V. natalensis Sch.Bip. ex Vatke

Southern Africa: tshiavonika, muTshiwonika, chiWonika (Shona)

V. staehelinoides Harv.

South Africa: blouteebossie

Vernoniopsis Humbert Asteraceae

Origins:
Resembling the genus *Vernonia*.

Veronica L. Scrophulariaceae

Origins:
A derivation from vetonica/vettonica/betonica (Plinius) or named in honor of Saint Veronica; see H. Genaust, *Etymologisches Wörterbuch der botanischen Pflanzennamen*. 679-680. Basel 1996; Carl Linnaeus, *Species Plantarum*. 9. 1753 and *Genera Plantarum*. Ed. 5. 10. 1754; Ernest Weekley, *An Etymological Dictionary of Modern English*. 2: 1590. New York 1967; Manlio Cortelazzo & Paolo Zolli, *Dizionario etimologico della lingua italiana*. 5: 1428. Zanichelli, Bologna 1988; P. Sella, *Glossario latino emiliano*. Città del Vaticano 1937; G.P. Bergantini, *Voci italiane d'autori approvati dalla Crusca*, nel Vocabolario d'essa non registrate ... Venezia 1745.

Species/Vernacular Names:
V. agrestis L.

English: field speedwell

V. americana (Raf.) Schwein.

English: brooklime, America brooklime

V. anagallis-aquatica L. (*Veronica aquatica* Bernh.; *Veronica salina* Schür)

English: water speedwell, blue speedwell, pink speedwell, long-leaved water speedwell, water pimpernel

Arabic: habaq

Nepal: tite

Southern Africa: ereprys, waterereprys; moqhoboqhobo-o-monyenyane (Sotho)

V. arvensis L. (*Veronica hawaiensis* H. Lév.)

English: corn speedwell

V. beccabunga L.

English: brooklime, European brooklime

V. catenata Pennell

English: water speedwell, chain speedwell

V. chamaedrys L.

English: angel's eyes, bird's eye, God's eye, germander speedwell

V. copelandii Eastw.

English: Copeland's speedwell

V. cusickii A. Gray

English: Cusick's speedwell

V. fruticans Jacq.

English: speedwell

V. incana L.

English: silver speedwell

V. officinalis L.

English: common speedwell, gypsyweed

V. parnkalliana J. Black

English: Port Lincoln speedwell

V. peregrina L.

English: purslane speedwell, necklace weed

China: jie gu xian tao

V. perfoliata R. Br.

English: digger's speedwell

V. persica Poiret

English: bird's-eye speedwell, field speedwell, creeping speedwell, Persian speedwell, Buxbaum's speedwell

Japan: ô-inu-no-fuguri

V. plebeia R. Br.

English: trailing speedwell, creeping speedwell, common speedwell

V. scutellata L.

English: marsh speedwell

V. serpyllifolia L.

English: thyme-leaved speedwell

V. wormskjoldii Roemer & Schultes

English: American alpine speedwell

Veronicastrum Heister ex Fabricius Scrophulariaceae

Origins:
Incompletely resembling the genus *Veronica* L.

Species/Vernacular Names:
V. virginicum (L.) O.A. Farw. (*Leptandra virginica* (L.) Nutt.; *Veronica virginica* L.)

English: Culver's root, blackroot, bowman's root

Verreauxia Benth. Goodeniaceae

Origins:
Named for the French naturalist Pierre Jules Verreaux, 1807-1873 (d. Paris), ornithologist, botanist, plant collector in Australia (Tasmania, New South Wales and Moreton Bay) and in South Africa, with his brother Jean Baptiste Édouard Verreaux wrote *L'Océanie en estampes, ou description géographique et historique de toutes les îles du grand océan*

et du continent de la Nouvelle Hollande. Paris et London 1832; see James Thomson, *Arcana Naturae*, ou recueil d'Histoire Naturelle. Paris 1859; I.H. Vegter, *Index Herbariorum*. Part II (7), *Collectors T-Z*. Regnum Vegetabile vol. 117. 1988; Mary Gunn and Leslie E. Codd, *Botanical Exploration of Southern Africa*. 128, 361-362. Cape Town 1981; Antoine Lasègue, *Musée botanique de M. Benjamin Delessert*. Paris 1845.

Verrucifera N.E. Br. Aizoaceae

Origins:

Latin *verruca* "a steep place, wart, excrescence" and *fero, fers, tuli, latum, ferre* "to bear, carry."

Verrucularia A. Juss. Malpighiaceae

Origins:

Latin *verruca* "wart, excrescence," the diminutive *verrucula, ae* "a little eminence, a small wart."

Verrucularina Rauschert Malpighiaceae

Origins:

The diminutive of *Verrucularia*.

Verschaffeltia H.A. Wendland Palmae

Origins:

After the Belgian horticulturist Ambroise Colette Alexandre Verschaffelt, 1825-1886, founder and publisher of *L'Illustration horticole*. 1854-1868/69(1870), author of *Nouvelle iconographie des Camellias*. Gand [Gent] 1848-1860; see J.H. Barnhart, *Biographical Notes upon Botanists*. 3: 432. 1965; Ethelyn Maria Tucker, *Catalogue of the Library of the Arnold Arboretum of Harvard University*. Cambridge, Massachusetts 1917-1933; H.R. Fletcher, *Story of the Royal Horticultural Society, 1804-1968*. Oxford 1969; R. Zander, F. Encke, G. Buchheim and S. Seybold, *Handwörterbuch der Pflanzennamen*. 14. Aufl. Stuttgart 1993.

Species/Vernacular Names:
V. splendida H.A. Wendland
Brazil: palmeira esplêndida

Verticordia DC. Myrtaceae

Origins:

Verticordia, ae "the turner of hearts," an epithet of the goddess Venus, who was supposed to restrain maidens from unchastity; Latin *verto, is, ti, sum, ere* "to turn" and *cor,*

cordis "the heart"; see A.S. George, "New taxa, combinations and typifications in *Verticordia* (Myrtaceae: Chamelaucieae)." *Nuytsia*. 7(3): 231-394. 1991.

Species/Vernacular Names:
V. wilhelmii F. Muell. (*Homoranthus wilhelmii* (F. Muell.) Cheel)
Australia: eastern feather-flower

Verulamia DC. ex Poir. Rubiaceae

Origins:
A vernacular name.

Vescisepalum (J.J. Sm.) Garay, Hamer & Siegerist Orchidaceae

Origins:

Latin *vescus, a, um* "small, little, thin, wretched, eating away" and *sepalum* "sepal."

Veseyochloa J.B. Phipps Gramineae

Origins:

For the British entomologist Leslie Desmond Edward Foster Vesey-Fitzgerald, c. 1910-1974 (d. Nairobi, Kenya), ecologist, conservationist, plant collector, see "Central African Grasslands." *J. Ecol*. 243-274. 1963.

Vesicarex Steyerm. Cyperaceae

Origins:
Latin *vesica* "a bladder, purse" plus *Carex*.

Vesicaria Adans. Brassicaceae

Origins:

Latin *vesicaria* applied by Plinius to a plant that cures pain in the bladder, bladder-wort, *vesica* "a bladder," referring to the pods.

Vesiculina Raf. Lentibulariaceae

Origins:

Latin *vesicula* "a little blister, vesicle," the diminutive of *vesica* "a blister"; see Constantine Samuel Rafinesque,

Flora Telluriana. 4: 109. Philadelphia 1836 [1838]; Elmer D. Merrill, *Index rafinesquianus*. The plant names published by C.S. Rafinesque, etc. 222. Jamaica Plain, Massachusetts, USA 1949.

Vesselowskya Pampanini Cunoniaceae

Origins:

For the Russian scientist E. Veselovsky (Vesselowsky), of Saratov; see Renato Pampanini (1875-1949), in *Annali di Botanica*. 2: 93, t. 6. Rome 1905.

Vestia Willdenow Solanaceae

Origins:

For the Austrian physician Lorenz Chrysanth von Vest, 1776-1840, chemist and botanist, 1798 Dr. med. at Freiburg, professor of botany and chemistry at the Johanneum, Graz. Among his writings the *Manuale botanicum*. Klagenfurt 1805; see H.N. Clokie, *Account of the Herbaria of the Department of Botany in the University of Oxford*. 259. Oxford 1964; J.H. Barnhart, *Biographical Notes upon Botanists*. 3: 433. 1965; H. Genaust, *Etymologisches Wörterbuch der botanischen Pflanzennamen*. 681. [1775-1846] 1996; R. Zander, F. Encke, G. Buchheim and S. Seybold, *Handwörterbuch der Pflanzennamen*. 14. Aufl. 794. 1993.

Vetiveria Bory Gramineae

Origins:

Malayalam and Tamil names of the plant *Vetiveria zizanioides* (L.) Nash, in Malayalam *veti* "cut" and *ver* "root," referring to the method of propagation; see M.P. Nayar, *Meaning of Indian Flowering Plant Names*. 360. Dehra Dun 1985; Helmut Genaust, *Etymologisches Wörterbuch der botanischen Pflanzennamen*. 682. 1996; Arthur D. Chapman, ed., *Australian Plant Name Index*. 2958-2959. [Lemaire ex Cassini] Canberra 1991.

Species/Vernacular Names:

V. arguta (Steud.) C.E. Hubbard

Rodrigues Island: vetiver

V. zizanioides (L.) Nash

English: vetiver, sevendara grass, khus-khus grass, cuscus grass

India: khushkhus, cuscus, khas, khas bena, khaskhas, khasakhasa, khas-khas, khus, khus-khus, reshira, usheera, ushira, sugandhimula, panni, amranalam, mudivala, bena, bhanavalo, valo, vala, kuruveeru, kuruvaeru, veeranam, vettiveellu, vetti-vellu, vetti-veru, vettiveru, vettiveeru, vettiver, vattiveeru, vattiveru, laamanche, lavanchi, kaadu karidappasajje hallu, ramaccham, ramachham

Sri Lanka: sevendara, vettiver

The Philippines: moras, amoora, amoras, anias de moras, giron, ilib, mora, moro, narawastu, raiz de moras, rimodas, rimora, rimoras, tres-moras

Malaya: rumput wangi, narawastu, kus kus

Vexatorella Rourke Proteaceae

Origins:

Latin *vexator, oris* "a troubler, abuser," *vexo, as, avi, atum, are* "to shake, to injure, to damage, to maltreat," an allusion to the problems created by the new genus; Akkadian *mehsu, wehsu* "stroke, attack," *mahasu, wahasu* "to hit, to strike, to affect, to hurt, to smash, to knock down, to weave."

Vexibia Raf. Fabaceae

Origins:

See Constantine Samuel Rafinesque, *Neogenyton, or Indication of Sixty-Six New Genera of Plants of North America*. 3. 1825 and *New Fl. N. Am.* 2: 53. 1836; Elmer D. Merrill, *Index rafinesquianus*. The plant names published by C.S. Rafinesque, etc. 147, 149. Jamaica Plain, Massachusetts, USA 1949.

Vexillabium F. Maekawa Orchidaceae

Origins:

From the Latin *vexillum, i* "a flag, standard, banner" and *labium, ii* "a lip," referring to the apical portion of the lip.

Species/Vernacular Names:

V. yakushimense (Yamamoto) Maekawa (*Anoectochilus yakushimensis* Yamamoto; *Pristiglottis yakushimensis* (Yamamoto) Masam.) (the island of Yaku-shima south of Japan's main southern island, Kyushu)

Japan: Yakushima-hime-ari-dôshi-ran (Yakushima = Yaku Island)

Vexillaria Raf. Fabaceae

Origins:

See Constantine S. Rafinesque, *Am. Monthly Mag. Crit. Rev.* 2: 268. 1818 and *Jour. Phys. Chim. Hist. Nat.* 89: 261. 1819; Elmer D. Merrill, *Index rafinesquianus*. 149. Jamaica Plain, Massachusetts, USA 1949.

Vexillifera Ducke Fabaceae

Origins:

From the Latin *vexillifer, fera, ferum* "standard-bearing."

Vibexia Raf. ex Jacks. Fabaceae

Origins:

See Elmer D. Merrill, *Index rafinesquianus*. The plant names published by C.S. Rafinesque, etc. 147, 149. Massachusetts, USA 1949.

Vibones Raf. Polygonaceae

Origins:

See Constantine S. Rafinesque, *Flora Telluriana*. 3: 45. 1836 [1837]; Elmer D. Merrill, *Index rafinesquianus*. 118. Jamaica Plain, Massachusetts, USA 1949.

Viburnum L. Caprifoliaceae (Adoxaceae)

Origins:

The Latin name, *viburnum, i* for the wayfaring tree: *Viburnum lantana* L.; see Carl Linnaeus, *Species Plantarum*. 267. 1753 and *Genera Plantarum*. Ed. 5. 129. 1754; Pietro Bubani, *Flora Virgiliana*. 124-126. [Ristampa dell'edizione di Bologna 1870] Bologna 1978; Ernest Weekley, *An Etymological Dictionary of Modern English*. 2: 1593. Dover Publications, New York 1967; John Gerard (Gerarde) (1545-1612), *Herball*, or Generall Historie of Plantes Gathered by John Gerard ... London 1597.

Species/Vernacular Names:

V. acerifolium L.

English: dockmackie, maple-leaved viburnum, arrow wood, possum haw

V. cassinoides L.

English: withe-rod, Appalachian tea, swamp haw, teaberry, wild raisin

V. dentatum L. (*Viburnum canbyi* Sargent; *Viburnum scabrellum* (Torrey & A. Gray) Chapm.) (after William Marriott Canby, 1831-1904, Delaware botanist, traveler and plant collector, editor of the autobiography of August Fendler (1813-1883), merchant, banker, accompanied Henry Villard's Northern Transcontinental Survey with C.S. Sargent; see Joseph Ewan, *Rocky Mountain Naturalists*. The University of Denver Press 1950; Joseph William Blankinship (1862-1938), "A century of botanical exploration in Montana, 1805-1905: collectors, herbaria and bibliography." in *Montana Agric. Coll. Sci. Studies Bot.* 1: 1-31.

1904; J.W. Harshberger, *The Botanists of Philadelphia and Their Work*. Philadelphia 1899)

English: arrow wood, southern arrow wood

V. dilatatum Thunb.

China: chia mi, hsi mi

V. japonicum (Thunb.) Spreng. (*Cornus japonica* Thunb.; *Viburnum fusiforme* Nakai; *Viburnum macrophyllum* Bl.)

Japan: hakusan-boku, meishigii

V. lantana L.

English: wayfaring tree, twistwood

V. lantanoides Michx. (*Viburnum alnifolium* Marsh.)

English: hobble bush, witch hobble, witch hopple, tanglefoot, tangle-legs, trip-toe, White Mountain dogwood, American wayfaring tree, devil's shoestrings, dogberry, dog hobble, moosewood, moose bush, mooseberry

V. lentago L.

English: sheepberry, nannyberry, black haw, cowberry, nanny plum, sweetberry, tea plant, wild raisin, sweet viburnum

V. molle Michx.

English: poison haw, black alder

V. nudum L.

English: smooth withe-rod, nannyberry haw, possum haw, swamp haw, black alder, naked alder

V. odoratissimum Ker.-Gawl.

English: sweet viburnum

V. opulus L.

English: guelder rose, European cranberry bush, crampbark, cranberry bush, whitten tree, snowball tree, snowball plant

China: hsueh chiu

V. pauciflorum Raf.

English: mooseberry

V. prunifolium L.

English: black haw, sweet haw, sheepberry, nannyberry, stagbush

V. rafinesquianum Schult.

English: downy-leaved arrow wood, downy arrow wood

V. recognitum Fern.

English: arrow wood

V. rufidulum Raf.

English: southern black haw, blue haw, rusty nannyberry

V. suspensum Lindl. (*Viburnum sandankwa* Hassk.)

Japan: gome-ju, gumuru

V. tinus L.

English: laurustinus

V. trilobum Marsh. (*Viburnum americanum* sensu Dipp.; *Viburnum opulus* var. *americanum* Ait.)

English: American cranberry bush, highbush cranberry, cranberry, cranberry bush, cranberry tree, tree cranberry, crampbark, grouseberry, squawbush, summerberry, pimbina

V. wrightii Miq.

English: leatherleaf

Vicatia DC. Umbelliferae

Origins:

After the Swiss botanist Philippe Rodolphe Vicat, 1720-1783, physician; see Jonas C. Dryander, *Catalogus bibliothecae historico-naturalis Josephi Banks*. London 1796-1800; Ethelyn Maria Tucker, *Catalogue of the Library of the Arnold Arboretum of Harvard University*. Cambridge, Massachusetts 1917-1933; Stafleu and Cowan, *Taxonomic Literature*. 6: 724. 1986.

Vicia L. Fabaceae

Origins:

Latin *vicia, ae* "a vetch" (Plinius, Marcus Porcius Cato, L. Junius Moderatus Columella, Marcus Terentius Varro, Palladius Rutilius Taurus Aemilianus and Vergilius), possibly related to *vinco, -is, vici, victum, vincere* "to be victorious" ("herba vicia, id est victorialis," Isidorus or Saint Isidore, Archbishop of Seville); see Carl Linnaeus, *Species Plantarum*. 734. 1753 and *Genera Plantarum*. Ed. 5. 327. 1754; Manlio Cortelazzo & Paolo Zolli, *Dizionario etimologico della lingua italiana*. 5: 1416. 1988; Giovanni Semerano, *Le origini della cultura europea*. Dizionario della lingua Latina e di voci moderne. 2(2): 613. Leo S. Olschki Editore, Firenze 1994; P. Sella, *Glossario latino emiliano*. Città del Vaticano 1937; N. Tommaseo & B. Bellini, *Dizionario della lingua italiana*. Torino 1865-1879; Ernest Weekley, *An Etymological Dictionary of Modern English*. 2: 1593. New York 1967; Arthur D. Chapman, ed., *Australian Plant Name Index*. 2959-2960. Canberra 1991; Shri S.P. Ambasta, ed., *The Useful Plants of India*. Council of Scientific and Industrial Research, New Delhi 1986.

Species/Vernacular Names:

V. americana Willd. (*Vicia americana* Muhlenb. ex Willd. var. *oregana* (Nutt.) Nelson; *Vicia oregana* Nutt.)

English: American vetch

V. articulata Hornem. (*Vicia monanthos* (L.) Desf., nom. illeg.; *Ervum monanthos* L.)

English: single-flowered vetch

V. benghalensis L. (*Vicia albicans* Lowe, nom. illeg.; *Vicia atropurpurea* Desf.; *Vicia loweana* Steudel; *Vicia micrantha* R. Lowe, nom. illeg.)

English: purple vetch, narrow-leaved purple vetch, wild purple vetch

South Africa: perswieke, smalblaar perswieke, wilde-ertjie

V. caroliniana Walter

English: Carolina vetch

V. cracca L.

English: tufted vetch, bird vetch, Canada pea, boreal vetch

China: shan ye wan dou

V. cracca L. subsp. *tenuifolia* (Roth) Gaudin (*Vicia brachytropis* Karelin & Kir.; *Vicia tenuifolia* Roth)

English: bramble vetch, fine-leaved vetch

V. ervilia (L.) Willd. (*Ervum ervilia* L.)

English: bitter vetch, ervil

V. faba L. (*Faba vulgaris* Moench)

English: broad bean, horse bean, English bean, European bean, field bean, fava bean, Windsor bean

Peru: haba, lipta

Japan: soṛa-mame, tômami

Tibetan: monsran

China: can dou, tsan tou

India: bakla, anhuri, kala matar, katun, kadu huralikayee, bakla sem, chas tang, chastang raiun, kabli bakla, mattzrewari, raj-rawan, nakshan, hende matar

Arabic: foul, foul hashadi

V. faba L. var. *equina* Pers. (*Vicia faba* L. var. *minor* Peterm., nom. illeg.)

English: horsebean

V. faba L. var. *minuta* (hort. ex Alef.) Mansf. (*Faba vulgaris* Moench var. *minor* Harz; *Faba vulgaris* Moench var. *minuta* hort. ex Alef.; *Vicia faba* L. var. *minor* (Harz) Beck, nom. illeg.)

English: tickbean

V. hirsuta (L.) S.F. Gray (*Ervum hirsutum* L.; *Ervum terronii* Ten.; *Vicia hirsuta* (L.) Gray var. *terronii* (Ten.) Burnat)

English: tiny vetch, hairy tare, tiny purple vetch, hairy vetch

South Africa: klein perswieke, ringelwieke, wiekies, wilde-ertjie

Japan: suzume-no-endô, garasa-mami

China: xiao chao cai, hsiao chao tsai, chiao yao, yao che

India: jhunjhuni ankari, jhanjhaniya, mun-muna, masuri, masur chana, musur chana, chirinji, chirinji arxa, birbut, tiririte

V. lathyroides L.

English: spring vetch

V. lutea L. (*Vicia laevigata* Smith; *Vicia lutea* L. var. *laevigata* (Smith) Boiss.)

English: yellow vetch

V. menziesii Spreng.

English: Hawaiian vetch

V. monantha Retz. (*Vicia biflora* Desf.; *Vicia calcarata* Desf.)

English: bard vetch, spurred vetch, square-stemmed vetch, one-flower vetch, Syrian vetch

Arabic: kharig

V. narbonensis L. (*Vicia serratifolia* Jacq. f. *integrifolia* Beck)

English: Narbonne vetch

V. nigricans Hook. & Arn. subsp. *gigantea* (*Vicia gigantea* Hook.)

English: large vetch, Sitka vetch

China: wei

V. orobus DC.

English: bitter vetch

V. pannonica Crantz

English: Hungarian vetch

V. pisiformis L.

English: pea vetch

V. sativa L.

English: spring vetch, tare, common vetch, broad-leaved purple vetch, vetch, wild vetch

Arabic: gilban, djelban

South Africa: breëblaarperswieke, breëblaarwilde-ertjie, gewone wieke, wieke, wilde-ertjie

Japan: karasu-no-endô, garasa-mami

India: ankra, ankari, akra, akta, rothi, choni, chatri-matri, matra, chirinji, chirinji arxa

V. sativa L. subsp. *nigra* (L.) Ehrh. (*Vicia angustifolia* L.; *Vicia angustifolia* L. var. *segetalis* (Thuill.) W. Koch; *Vicia pilosa* M. Bieb.; *Vicia sativa* L. var. *angustifolia* L.; *Vicia sativa* L. var. *nigra* L.; *Vicia segetalis* Thuill.)

English: blackpod vetch, narrow leaved vetch, common vetch

India: ankra, ankari

V. sativa L. subsp. *sativa* (*Vicia leucosperma* Moench; *Vicia sativa* L. var. *leucosperma* (Moench) Ser.; *Vicia sativa* L. var. *obovata* Ser.)

English: common vetch, tare, spring vetch

V. sepium L.

English: bush vetch, hedge vetch

V. tetrasperma (L.) Schreber (*Ervum tetraspermum* L.; *Vicia gemella* Crantz)

English: fourseed vetch, sparrow vetch, slender vetch

Japan: kasuma-gusa, garasa-mami

V. villosa Roth subsp. *varia* (Host) Corbière (*Vicia dasycarpa* Ten.; *Vicia varia* Host; *Vicia villosa* Roth subsp. *dasycarpa* (Ten.) Cavill.)

English: winter vetch, woollypod vetch

V. villosa Roth subsp. *villosa*

English: hairy vetch, Russian vetch, large Russian vetch, winter vetch

South Africa: harige wieke

Vicilla Schur Fabaceae

Origins:
The diminutive of *Vicia*.

Vicioides Moench Fabaceae

Origins:
Resembling *Vicia*.

Victoria Lindley Nymphaeaceae (Euryalaceae)

Origins:
Named in honor of Her Majesty Queen Victoria, 1819-1901.

Species/Vernacular Names:
V. amazonica (Poeppig) Sowerby

English: giant waterlily

Peru: atun sisa

Victorinia Léon Euphorbiaceae

Origins:
After the Canadian botanist Frère Marie-Victorin (Joseph Louis Conrad Kirouac), 1885-1944, clergyman, 1920-1944 University of Montreal, traveler, botanical explorer. Among his writings *La flore du Temiscouata*. Québec 1916; see John H. Barnhart, *Biographical Notes upon Botanists*. 2: 295. 1965; Stafleu and Cowan, *Taxonomic Literature*. 3: 293-295. Utrecht 1981; R. Zander, F. Encke, G. Buchheim and S. Seybold, *Handwörterbuch der Pflanzennamen*. 14. Aufl. Stuttgart 1993; I.C. Hedge and J.M. Lamond, *Index of Collectors in the Edinburgh Herbarium*. Edinburgh 1970.

Vidalia Fern.-Vill. Guttiferae

Origins:

For the Spanish botanist Sebastian Vidal y Soler, 1842-1889, traveler and plant collector; see John H. Barnhart, *Biographical Notes upon Botanists.* 3: 435. 1965; R. Zander, F. Encke, G. Buchheim and S. Seybold, *Handwörterbuch der Pflanzennamen.* 14. Aufl. Stuttgart 1993; Ethelyn Maria Tucker, *Catalogue of the Library of the Arnold Arboretum of Harvard University.* Cambridge, Massachusetts 1917-1933.

Vieillardorchis Kraenzlin Orchidaceae

Origins:

For the French naval surgeon Eugène Vieillard, 1819-1896, botanist, plant collector in New Caledonia *circa* 1857-1869 and Tahiti, collected at the Cape, from 1871 to 1895 Director of the Botanical Garden of Caen, wrote *Plantes de la Nouvelle-Calédonie.* Caen 1865, co-author with Dr. Émile Deplanche (1824-1875) of *Essais sur la Nouvelle-Calédonie.* Paris 1863; see John H. Barnhart, *Biographical Notes upon Botanists.* 3: 435. 1965; M.E. and H.S. McKee, *Histoire et Nature.* 17-18: 49-68. 1981; R. Zander, F. Encke, G. Buchheim and S. Seybold, *Handwörterbuch der Pflanzennamen.* 14. Aufl. Stuttgart 1993; Mary Gunn and Leslie E. Codd, *Botanical Exploration of Southern Africa.* 362. Cape Town 1981; Ethelyn Maria Tucker, *Catalogue of the Library of the Arnold Arboretum of Harvard University.* Cambridge, Massachusetts 1917-1933; Elmer Drew Merrill, *Contr. U.S. Natl. Herb.* 30(1): 306. 1947; E.D. Merrill, in *Bernice P. Bishop Mus. Bull.* 144: 186. 1937.

Viereckia R.M. King & H. Robinson Asteraceae

Origins:

For the German botanist H.W. Viereck (d. 1945), plant collector in Mexico (for Haage's nursery in Erfurt), pupil of Hugo Baum (1867-1950); see Ida Kaplan Langman, *A Selected Guide to the Literature on the Flowering Plants of Mexico.* 779. Philadelphia 1964; Irving William Knobloch, compil., "A preliminary verified list of plant collectors in Mexico." *Phytologia Memoirs.* VI. 1983; Helia Bravo-Hollis & Hernando Sánchez-Mejorada R., *Las Cactáceas de México.* Universidad Nacional Autónoma de México, México 1991; Gordon Douglas Rowley, *A History of Succulent Plants.* 356-357, 385. 1997.

Vietnamosasa T.Q. Nguyen Gramineae

Origins:

From Vietnam plus genus *Sasa*.

Vietsenia C. Hansen Melastomataceae

Origins:

A genus from Vietnam.

Vigna Savi Fabaceae

Origins:

After the Italian botanist Domenico Vigna, d. 1647, professor of botany and Director of the Botanical Garden of Pisa (in 1614, 1616-1617 and 1632-1634), author of *Animadversiones, sive observationes in libro de Historia, et de Causis Plantarum Theophrasti.* Pisis 1625; see L. Amodei, "Note sull'*Herbarium Horti Pisani*: l'origine delle collezioni." *Museol. Scient.* 4(1-2): 119-129. 1987; P.E. Tomei and Carlo Del Prete, "The Botanical Garden of the University of Pisa." *Herbarist.* 49: 47-71. Concord, Massachusetts 1983.

Species/Vernacular Names:

V. sp.

Yoruba: awuje were

V. aconitifolia (Jacq.) Maréchal

English: matbean, mothbean

V. adenantha (G. Meyer) Marechal et al. (*Phaseolus adenanthus* G. Meyer)

Japan: kochô-ingen

V. angularis (Willd.) Ohwi & H. Ohashi (*Phaseolus angularis* W. Wight)

English: adzuki bean, aduki bean

Japan: azuki, akamâmii, azicha

China: chi xiao dou

V. angularis (Willd.) Ohwi & H. Ohashi var. *angularis* (*Dolichos angularis* Willd.; *Phaseolus angularis* (Willd.) W. Wight)

English: adzuki-bean, aduki bean

V. angustifoliolata Verdc. (*Vigna stenophylla* (Harv.) Burtt Davy; *Vigna triloba* Walp. var. *stenophylla* Harv.)

English: wild sweet pea

South Africa: wilde ertjie

V. caracalla (L.) Verdc. (*Phaseolus caracalla* L.)

English: Bertoni bean, snail flower, corkscrew flower, snail bean

V. hosei (Craib) Backer (*Dolichos hosei* Craib)

English: Sarawak bean

V. lanceolata Benth. (*Vigna lanceolata* Benth. var. *latifolia* C. White)

Australia: maloga bean, native bean, yam

V. luteola (Jacq.) Benth. (*Dolichos luteolus* Jacq.; *Dolichos niloticus* Del.; *Dolichos repens* L.; *Vigna nilotica* (Del.) Hook.f.; *Vigna repens* (L.) Kuntze, nom. illeg.)

English: yellowish bean

Japan: nagaba-hama-sasage

V. marina (Burm.) Merr. (*Dolichos luteus* Sw.; *Phaseolus marinus* Burm.; *Scytalis retusa* E. Meyer; *Vigna lutea* (Sw.) A. Gray; *Vigna retusa* (E. Meyer) Walp.)

English: beach pea

Japan: hama-sasage

Hawaii: mohihihi, lemuomakili, nanea, nenea, 'okolemak-ili, puhili, puhilihili, pulihilihi, wahine 'oma'o

V. minima (Roxb.) Ohwi & H. Ohashi (*Phaseolus minimus* Roxb.)

Japan: koba-no-tsuru-azuki, hime-azuki, hime-tsuru-azuki, hosoba-tsuru-azuki

V. mungo (L.) Hepper (*Phaseolus mungo* L.)

English: black gram

China: lu tou

Australia: komin (by the Rockhampton Aborigines), kadolo (by the Cleveland Bay aborigines)

V. racemosa (G. Don) Hutch. & Dalziel

Yoruba: eree igbo

V. radiata (L.) R. Wilczek

English: mung bean, green gram, golden gram

Japan: ke-tsuru-azuki

China: lu dou, chih hsiao tou, hung tou

Tanzania: ndotodoto

V. radiata (L.) R. Wilczek var. *radiata* (*Phaseolus aureus* Roxb.; *Phaseolus radiatus* L.)

English: golden gram, green gram, mungbean

V. subterranea (L.) Verdc. (*Glycine subterranea* L.; *Voandzeia subterranea* (L.) Thouars ex DC.)

English: groundbean, Bambarra groundnut, Bambara groundnut

French: pois de terre

N. Nigeria: gujiya, gurujiya

Yoruba: epa roro, epa orubu, epa lorubu, epa ruburubu, epa olojukan, epa boro, epa oboro, epa okuta, paruru, epakun

West Africa: depi (depo = squat)

Congo: nionzi, voandzou

V. umbellata (Thunb.) Ohwi & H. Ohashi (*Dolichos umbel-latus* Thunb.; *Phaseolus calcaratus* Roxb.; *Vigna calcarata* (Roxb.) Kurz)

English: ricebean

Japan: tsuru-azuki

V. unguiculata (L.) Walp. subsp. *cylindrica* (L.) Verdc. (*Dolichos biflorus* L.; *Dolichos catiang* L.; *Dolichos cat-jang* Burm.f.; *Phaseolus cylindricus* L.; *Vigna catjang* (Burm.f.) Walp.; *Vigna cylindrica* (L.) Skeels)

English: catjang, Jerusalem pea, marble pea

V. unguiculata (L.) Walp. subsp. *sesquipedalis* (L.) Verdc. (*Dolichos sesquipedalis* L.; *Vigna sesquipedalis* (L.) Fruw.; *Vigna sinensis* (L.) Savi ex Hassk. subsp. *sesquipedalis* (L.) Eselt.)

English: asparagus bean, yard-long bean

Japan: juroku sasage

The Philippines: sitaw, karakala, kibal, hamtak, otong, anta, batong, kauat, laston

V. unguiculata (L.) Walp. subsp. *unguiculata* (*Dolichos sinensis* L.; *Dolichos unguiculatus* L.; *Phaseolus unguicu-latus* (L.) Piper; *Vigna sinensis* (L.) Savi ex Hassk.)

English: black-eyed pea, cowpea, crowder pea, southern pea, horse gram, cherry bean

India: kulatthah, mutira, kollu, kulthi, kulith

Peru: chiclayo

Japan: sasage, hachigawa

China: jiang dou

Tanzania: soloko, msambura

Yoruba: eree ahun, ewa, ewe, ewa funfun, ewa dudu, ewa erewe

V. vexillata (L.) A. Rich.

English: zombi-pea, wild-cowpea, wild sweetpea

South Africa: wilde-akkerboontjie, wilde-ertjie

Japan: aka-sasage

Vignopsis De Wild. Fabaceae

Origins:
Resembling *Vigna*.

Viguiera Kunth Asteraceae

Origins:
After the French botanist L.G. Alexandre Viguier, 1790-1867, physician, author of *Histoire naturelle, médicale et économique, des pavots et des argémones*. Montpellier 1814; see R. Zander, F. Encke, G. Buchheim and S. Seybold, *Handwörterbuch der Pflanzennamen*. 14. Aufl. Stuttgart 1993.

Species/Vernacular Names:
V. laciniata A. Gray
English: San Diego County viguiera

Viguierella A. Camus Asteraceae

Origins:

For the French botanist René Viguier, 1880-1931, bryologist and specialist in Araliaceae, from 1919 to 1931 professor of botany at the University and Director of the Municipal Botanical Garden at Caen, author of *Recherches sur le genre Grewia*. 1917 and *Sur les Araliacées du groupe des Polyscias*. 1905; see J.H. Barnhart, *Biographical Notes upon Botanists*. 3: 436. 1965; Ethelyn Maria Tucker, *Catalogue of the Library of the Arnold Arboretum of Harvard University*. Cambridge, Massachusetts 1917-1933; Elmer Drew Merrill, in *Contr. U.S. Natl. Herb.* 30(1): 307. 1947 and in *Bernice P. Bishop Mus. Bull.* 144: 186. 1937.

Vilfagrostis Döll Gramineae

Origins:

The genera *Vilfa* and *Agrostis*, Greek *agrostis, agrostidos* "grass, weed, couch grass"; see also *Eragrostis* Wolf.

Villadia Rose Crassulaceae

Origins:

For the Mexican scientist Dr. Manuel M. Villada; see Ida Kaplan Langman, *A Selected Guide to the Literature on the Flowering Plants of Mexico*. Philadelphia 1964; Irving William Knobloch, compil., "A preliminary verified list of plant collectors in Mexico." *Phytologia Memoirs*. VI. 1983.

Villaresia Ruíz & Pav. Icacinaceae

Origins:

For Mattias Villares (Matthias Villarez), a monk who had a botanic garden (in Chile); see William T. Stearn, *Stearn's Dictionary of Plant Names for Gardeners*. 308. Cassell, London 1993; A.W. Smith, *A Gardener's Book of Plant Names*. 368. New York 1963.

Villaresiopsis Sleumer Icacinaceae

Origins:
Resembling *Villaresia*.

Villarsia Ventenat Menyanthaceae

Origins:
After the French botanist Dominique Villars (Villar until 1785), 1745-1814, physician, a friend of Dominique Chaix (1730-1799), from 1803 professor of botany and medicine at the University of Strasbourg. Among his writings are *Histoire des plantes de Dauphiné*. Grenoble, Lyon and Paris 1786-1789 and *Catalogue méthodique des plantes du jardin de l'École de medicine de Strasbourg*. Strasbourg 1807; see John H. Barnhart (1871-1949), *Biographical Notes upon Botanists*. 3: 437. 1965; Stafleu and Cowan, *Taxonomic Literature*. 6: 739-741. 1986; Antoine Lasègue, *Musée botanique de M. Benjamin Delessert*. 1845; S. Lenley et al., *Catalog of the Manuscript and Archival Collections and Index to the Correspondence of John Torrey*. Library of the New York Botanical Garden. 1973; T. W. Bossert, *Biographical Dictionary of Botanists Represented in the Hunt Institute Portrait Collection*. 419. Boston, Massachusetts 1972; Jonas C. Dryander, *Catalogus bibliothecae historiconaturalis Josephi Banks*. London 1796-1800; Ethelyn Maria Tucker, *Catalogue of the Library of the Arnold Arboretum of Harvard University*. Cambridge, Massachusetts 1917-1933; R. Zander, F. Encke, G. Buchheim and S. Seybold, *Handwörterbuch der Pflanzennamen*. 14. Aufl. 794. 1993.

Species/Vernacular Names:
V. reniformis R. Br. (*Limnanthemum stygium* J. Black; *Nymphoides stygia* (J. Black) H. Eichler)
English: running marsh-flower

Villocuspis (A. DC.) Aubrév. & Pellegrin Sapotaceae

Origins:
Latin *villus* "shaggy hair, tuft of hair" and *cuspis, idis* "the pointed end."

Vilmorinia DC. Fabaceae

Origins:

For the French botanist Philippe Victor Lévêque de Vilmorin; see Pritzel, *Thesaurus*. page 332. [b. 1746-d. 1804] 1871-1877; Diana M. Simpkins, in *Dictionary of Scientific Biography* 14: 33-34. 1981; R. Zander, F. Encke, G. Buchheim and S. Seybold, *Handwörterbuch der Pflanzennamen*. 14. Aufl. 795. 1993; Stafleu and Cowan, *Taxonomic Literature*. 6: 741-748. 1986.

Vilobia Strother Asteraceae

Origins:
Bolivia.

Viminaria J.E. Smith Fabaceae

Origins:

Latin *vimen, viminis* "a twig, pliant twig," Akkadian *hamamu* "to gather, to gather to oneself," *humumu* "to collect," "specus virgis ac vimine densus", Ovidius; see G. Semerano, *Le origini della cultura europea*. Dizionario della lingua Latina e di voci moderne. 2(2): 613. Leo S. Olschki Editore, Firenze 1994.

Species/Vernacular Names:

V. juncea (Schrader) Hoffsgg. (*Sophora juncea* Schrader & Wendl.; *Daviesia denudata* Vent.; *Viminaria denudata* (Vent.) Smith)

English: golden spray

Vinca L. Apocynaceae

Origins:

Contracted from the Latin *vincapervinca* and *vicapervica, ae* (Plinius), *vica pervica* (Pseudo Apuleius Barbarus), perhaps related to *vincio, -is, vinxi, vinctum, vincire* "to bind, to wind about," or to *vinco, is, vici, vistum, vincere* "to overcome, to be victorious," *pervinco, vici, victum, ere* "to conquer completely, to achieve," *pervicax, acis* "determined, obstinate"; see Carl Linnaeus, *Species Plantarum*. 209. 1753 and *Genera Plantarum*. Ed. 5. 98. 1754; E. Weekley, *An Etymological Dictionary of Modern English*. 2: 1072. 1967; G. Semerano, *Le origini della cultura europea*. Dizionario della lingua Latina e di voci moderne. 2(2): 613. 1994.

Species/Vernacular Names:

V. major L.

English: band plant, blue buttons, blue periwinkle, greater periwinkle, periwinkle

Spanish: vinca

South Africa: gewone opklim, maagdepalm

China: man chang chun hua

V. minor L.

English: common periwinkle, lesser periwinkle, myrtle, running myrtle

China: hua ye man chang chun hua

Vincentella Pierre Sapotaceae

Origins:

After a Monsieur Vincent.

Vincentia Gaudich. Cyperaceae

Origins:

After the French naturalist Jean Baptiste Georges Geneviève Marcellin Bory de Saint-Vincent, 1778-1846, microscopist, traveler, geographer and explorer, with Baudin in Australia 1801, on the Isle de Bourbon and Canary Islands 1801-1802, took part in the voyage of the *Coquille* commanded by Louis-Isidor Duperrey (1786-1865), author of *Voyage dans les quatres principales îles des mers d'Afrique*. Paris 1804, *Sur les Grenadilles ou Passionaires*. Bruxelles 1819 and *Histoire des hydrophytes, ou plantes agames des eaux*. Paris 1829, joint author of *Dictionnaire classique d'histoire naturelle*. Paris 1822-1831; see P. Romieux, *Les carnets de Bory de Saint-Vincent 1813-1815*. Paris 1934; Stafleu and Cowan, *Taxonomic Literature*. 1: 284-286. 1976; Frans A. Stafleu and Erik A. Mennega, *Taxonomic Literature. Supplement II*. 362-364. 1993; R. Zander, F. Encke, G. Buchheim and S. Seybold, *Handwörterbuch der Pflanzennamen*. 14. Aufl. Stuttgart 1993; John H. Barnhart, *Biographical Notes upon Botanists*. 1: 224. 1965; J. Lanjouw and F.A. Stafleu, *Index Herbariorum*. Part II, *Collectors A-D*. Regnum Vegetabile vol. 2. 1954; T. W. Bossert, *Biographical Dictionary of Botanists Represented in the Hunt Institute Portrait Collection*. 46. 1972; M. Colmeiro y Penido, *La Botánica y los Botánicos de la Peninsula Hispano-Lusitana*. Madrid 1858; Ida Kaplan Langman, *A Selected Guide to the Literature on the Flowering Plants of Mexico*. Philadelphia 1964; Antoine Lasègue, *Musée botanique de M. Benjamin Delessert*. 1845; S. Lenley et al., *Catalog of the Manuscript and Archival Collections and Index to the Correspondence of John Torrey*. Library of the New York Botanical Garden. 1973; Günther Schmid, *Chamisso als Naturforscher*. Eine Bibliographie. Leipzig 1942.

Vincetoxicopsis Costantin Asclepiadaceae

Origins:

Resembling the genus *Vincetoxicum* Wolf.

Vincetoxicum Wolf Asclepiadaceae

Origins:

Latin *vinco* "to conquer, overcome" and *toxicum* "poison," an antidote to snake-bite?

Species/Vernacular Names:

V. nigrum Moench

English: black swallowort

Vintenatia Cav. Epacridaceae

Origins:

Orthographic variant, see *Ventenatia* Cav.

Viola L. Violaceae

Origins:

Latin *viola, ae* and Greek *ion* "violet," Akkadian *asu* and Hebrew *js'* "to grow, to turn, to come from," Akkadian *isum* "plant"; see Carl Linnaeus, *Species Plantarum*. 933. 1753 and *Genera Plantarum*. Ed. 5. 402. 1754; Giovanni Semerano, *Le origini della cultura europea*. Dizionario della lingua Latina e di voci moderne. 2(2): 614. 1994; Manlio Cortelazzo & Paolo Zolli, *Dizionario etimologico della lingua italiana*. 5: 1440. 1988; Giovanni Semerano, *Le origini della cultura europea*. Dizionario della lingua Greca. 2(1): 126. Firenze 1994; G. Volpi, "Le falsificazioni di Francesco Redi nel Vocabolario della Crusca." in *Atti della R. Accademia della Crusca per la lingua d'Italia*. 33-136. 1915-1916.

Species/Vernacular Names:

V. adunca Smith

English: western dog violet, hooked-spur violet

V. arvensis Murray

English: field pansy

V. beckwithii Torrey & A. Gray (for Edwin Griffin Beckwith, 1838-1881, explorer, see Joseph Ewan, *Rocky Mountain Naturalists*. The University of Denver Press 1950)

English: Great Basin violet

V. blanda Willd.

English: Willdenow violet, woodland white violet, sweet white violet

V. canadensis L.

English: Canada violet, tall white violet

V. canina L.

English: dog violet, heath violet

V. chamissoniana Ging.

Hawaii: pamakani, 'olopu

V. conspersa Reichb.

English: American dog violet

V. cornuta L.

English: horned violet, viola, bedding pansy

V. douglasii Steudel

English: Douglas violet

V. fimbriatula Sm.

English: fringed violet, rattlesnake violet, northern downy violet

V. flettii Piper

English: rock violet, olympic violet

V. glabella Nutt.

English: stream violet

V. hederacea Labill. (*Erpetion hederaceum* (Labill.) Don)

English: ivy-leaved violet, native violet, Australian violet, trailing violet

V. japonica Langsd.

English: Japanese violet

China: li tou cao

V. kauaensis A. Gray

Hawaii: nani Wai'ale'ale, kalili, liliwai, pohe hiwa

V. lobata Benth.

English: yellow wood violet, pine violet

V. lutea Huds.

English: mountain pansy

V. ocellata Torrey & A. Gray

English: western heart's ease

V. odorata L.

English: English violet, florist's violet, garden violet, sweet viola, violet, violet tea, sweet violet, common violet

Peru: laram ajjiri, violeta

South Africa: viooltjie

Arabic: banaf sag

V. palustris L.

English: marsh violet

V. patrinii DC.

China: hua to cao, tzu hua ti ting

V. pedunculata Torrey & A. Gray

English: Johnny-jump-up

V. pinnata L.

China: hu chin tsao

V. praemorsa Douglas

English: Astoria violet

V. sempervirens E. Greene

English: evergreen violet

V. tomentosa M. Baker & J. Clausen

English: woolly violet

V. tricolor L.

English: herb trinity, European wild pansy, wild pansy, field pansy, heart's ease, heartease, Johnny-jump-up, miniature pansy, pansy, love-in-idleness, pink of my John

Arabic: belesfendj

Peru: juacallicma, pensamiento, trinitaria

V. tuberifera Franch.

English: tuberous violet

V. verecunda A. Gray

English: common violet

China: xiao du yao

V. yedoensis Makino

English: wild Chinese violet, Tokyo violet

China: zi hua di ding

Viorna Reichb. Ranunculaceae

Origins:

Viorne, the French word for *Viburnum*; see C.T. Onions, *The Oxford Dictionary of English Etymology*. Oxford University Press 1966; G.C. Wittstein, *Etymologisch-botanisches Handwörterbuch*. 924. 1852; F. Boerner & G. Kunkel, *Taschenwörterbuch der botanischen Pflanzennamen*. 4. Aufl. 391. Berlin & Hamburg 1989; H. Genaust, *Etymologisches Wörterbuch der botanischen Pflanzennamen*. 685. 1996; F. Boerner, *Taschenwörterbuch der botanischen Pflanzennamen*. 2. Aufl. 351. Berlin & Hamburg 1966.

Viraya Gaudich. Asteraceae

Origins:

For the French (b. Haute-Marne) naturalist Jules-Joseph Virey, 1775-1846 (d. Paris), physician, M.D. Paris 1814, pharmacist, from 1794 to 1813 military pharmacist. Among his numerous publications are *Histoire naturelle des médicamens*. Paris 1820 and *Philosophie de l'histoire naturelle*. Paris 1835; see John H. Barnhart, *Biographical Notes upon Botanists*. 3: 439. Boston 1965; E.M. Tucker, *Catalogue of the Library of the Arnold Arboretum of Harvard University*. Cambridge, Massachusetts 1917-1933; Alex Berman, in *Dictionary of Scientific Biography* 14: 44-45. [Julien-Joseph?] 1981.

Virecta L.f. Rubiaceae

Origins:

Latin *virectum, i* "a place overgrown with grass, a green place," *vireo, -es, -ui, -ere* "to be green, to be verdant."

Virecta Smith Rubiaceae

Origins:

Latin *virectum, i* "a place overgrown with grass, a green place"; see also *Virectaria* Bremek.

Virectaria Bremek. Rubiaceae

Origins:

Latin *virectum, i* "a place overgrown with grass, a green place, greenness."

Vireya Raf. Gesneriaceae

Origins:

Named for the French naturalist Jules-Joseph Virey, 1775-1846 (Paris), physician, M.D. Paris 1814, pharmacist, from 1794 to 1813 military pharmacist. Among his numerous publications are *Histoire naturelle des médicamens*. Paris 1820 and *Philosophie de l'histoire naturelle*. Paris 1835; see J.H. Barnhart, *Biographical Notes upon Botanists*. 3: 439. 1965; E.M. Tucker, *Catalogue of the Library of the Arnold Arboretum of Harvard University*. Cambridge, Massachusetts 1917-1933; Alex Berman, in *Dictionary of Scientific Biography* 14: 44-45. 1981; Constantine Samuel Rafinesque (1783-1840), in *Specchio delle scienze, o giornale enciclopedico di Sicilia* etc. 1: 194. Palermo 1819; E.D. Merrill, *Index rafinesquianus*. The plant names published by C.S. Rafinesque, etc. 221. 1949.

Virgilia Poiret Fabaceae

Origins:

The generic name after the Latin poet and farmer Publius Vergilius Maro (70-19 BC.); see Craig Kallendorf, *A Bibliography of Venetian Editions of Virgil, 1470-1599*. Olschki, Firenze 1991.

Species/Vernacular Names:

V. divaricata Adamson (*Virgilia capensis* sensu Pole Evans, non Lam.)

South Africa: keurboom (keur = choice, or pick; keurboom = pick of the trees)

V. oroboides (Bergius) T.M. Salter subsp. *oroboides* (*Hypocalyptus capensis* (L.) Thunb.; *Podalyria capensis* (L.) Willd.; *Sophora capensis* L.; *Sophora oroboides* Berg.; *Virgilia capensis* (L.) Lam.; *Virgilia oroboides* (Berg.) Salter; *Virgilia capensis* Lam.) (the specific name means "resembling *Orobus* L.," a genus of plants now included in *Lathyrus* L.)

English: blossom tree, snowdrop tree

South Africa: keurboom, Amaquasboom, keur, wildekeur

Virgulaster Semple Asteraceae

Origins:

From the Latin *virgula* "a small rod, wand" plus *Aster* L.

Virgulus Raf. Asteraceae

Origins:

Latin *virgula* "a small rod, wand"; see Constantine Samuel Rafinesque (1783-1840), *Flora Telluriana*. 2: 46. Philadelphia 1836 [1837].

Viridivia J.H. Hemsley & Verdcourt Passifloraceae

Origins:

Latin *viridis, e* "green" and *via* "way, path," the Latin translation of the name Greenway. Genus dedicated to the South African (b. Transvaal) botanist Percy (Peter) James Greenway, 1897-1980, from 1927 to 1950 East African Agricultural Research Station (Amani), 1928 Fellow of the Linnean Society, 1950-1958 botanist of the East African Herbarium in Nairobi, systematic botanist, plant collector with Colin Graham Trapnell and John P. Micklethwait Brenan (1917-1985) in Northern and Southern Rhodesia and Nyasaland, 1970-1971 President of the Kew Guild, author of *A Swahili-Botanical-English Dictionary of Plant Names*. Dar es Salaam 1940 and "The Pawpaw or Papaya." *E.A. Agri. Journ.* 13: 228-233. Nairobi 1948, co-author (with Ivan Robert Dale, 1904-1963) of *Kenya Trees & Shrubs*. Nairobi 1961, editor of Jessie Williamson's *Useful Plants of Nyasaland*. Zomba, Nyasaland 1955 and of the *East African Agricultural Journal*; see Ray Desmond, *Dictionary of British & Irish Botanists and Horticulturists*. 295. 1994.

Virola Aublet Myristicaceae

Origins:

A native name for *Virola sebifera* Aublet, this tree is called *dayapa* and *virola* by the Galibis, French Guiana.

Viscainoa Greene Zygophyllaceae

Origins:

Baja California Sur, El Vizcaíno, Norte de Baja California Sur, Municipio de Mulegé; Sebastièn Vizcaíno was a merchant who sailed the southern California coast in 1602.

Viscaria Roehl. Caryophyllaceae

Origins:

Latin *viscus* "birdlime," referring to the nature of the stems.

Viscoides Jacq. Rubiaceae

Origins:

Resembling *viscus* "birdlime."

Viscum L. Viscaceae

Origins:

Latin *viscus, i* "bird-lime, clammy, the mistletoe," the berry is sticky; Akkadian *wasu* "to go out, to come from, going out"; Hebrew *jasi* "come forth, descended," *jasa* "to spread out"; see Carl Linnaeus, *Species Plantarum*. 1023. 1753 and *Genera Plantarum*. Ed. 5. 448. 1754.

Species/Vernacular Names:

V. album L.

English: European mistletoe, mistletoe

Central America: muerdago, liga, ligamatapalo, matapalo, nigüita

V. anceps E. Mey. ex Sprague (*Aspidixia anceps* E. Mey.)

English: mistletoe

V. continuum E. Mey. ex Sprague

English: mistletoe

South Africa: voëlent, mistel

V. obscurum Thunb. (*Aspidixia bivalvis* Tieghem; *Viscum bivalve* (Tieghem) Engl.; *Viscum brevifolium* (Harv.) Engl.; *Viscum obscurum* Thunb. var. *brevifolium* Harv.; *Viscum obscurum* Thunb. var. *longiflorum* Harv.)

English: mistletoe, bird-lime

Southern Africa: voëlent, mistel, martak, voëllym, ligji-estee; inDembu, isiSende (Xhosa)

V. rotundifolium L.f. (*Viscum bosciae-foetidae* Dinter; *Viscum glaucum* Eckl. & Zeyh.; *Viscum macowanii* Engl.; *Viscum thymifolium* Presl; *Viscum tricostatum* E. Mey. ex Harv.; *Viscum ziziphi-mucronati* Dinter)

English: mistletoe

Visiania DC. Oleaceae

Origins:

After the Italian botanist Roberto de Visiani, 1800-1878, physician.

Visiania Gasp. Moraceae

Origins:

After the Italian botanist Roberto de Visiani, 1800-1878, physician, explorer of Dalmatia, from 1836 to 1878 Director of the Botanical Garden of Padova. Among his many works are *L'Orto Botanico di Padova nell'anno MDCCCXLII*. Padova 1842, *Di alcune piante storiche del Giardino di Padova. Cenni critici*. Padova 1857, *Stirpium dalmaticarum specimen*. Patavii [Padova] 1826 and *Flora dalmatica*. Lipsiae 1842-1852, co-author with Pier Andrea Saccardo (1845-1920) of *Catalogo delle piante vascolari del Veneto*. Venezia 1869; see J.H. Barnhart, *Biographical Notes upon Botanists*. 1: 449. 1965; Mariella Azzarello Di Misa, a cura di, *Il Fondo Antico della Biblioteca dell'Orto Botanico di Palermo*. 282. Palermo 1988; T. W. Bossert, *Biographical Dictionary of Botanists Represented in the Hunt Institute Portrait Collection*. 420. Boston, Massachusetts 1972; H.N. Clokie, *Account of the Herbaria of the Department of Botany in the University of Oxford*. 259. Oxford 1964; Ida Kaplan Langman, *A Selected Guide to the Literature on the Flowering Plants of Mexico*. 784. Philadelphia 1964; Antoine Lasègue, *Musée botanique de M. Benjamin Delessert*. 1845; E.M. Tucker, *Catalogue of the Library of the Arnold Arboretum of Harvard University*. Cambridge, Massachusetts 1917-1933; R. Zander, F. Encke, G. Buchheim and S. Seybold, *Handwörterbuch der Pflanzennamen*. 14. Aufl. Stuttgart 1993; Edward Lee Greene, *Landmarks of Botanical History*. Edited by Frank N. Egerton. 727, 729-730. Stanford, California 1983.

Vismia Vand. Guttiferae

Origins:
Named for M. de Visme, a Portuguese merchant.

Species/Vernacular Names:
V. sp.

Peru: huaca shambu, loro micuna, mandor, roble, yana pichirina

V. guineensis (L.) Choisy

Nigeria: huda tukunya (Hausa); ovitue (Edo); oke oturu (Igbo)

Brazil (Amazonas): siiriama sihi

Vismianthus Mildbr. Connaraceae

Origins:
The genus *Vismia* and *anthos* "flower."

Visnaga Miller Umbelliferae

Origins:
See *Ammi visnaga* (L.) Lam. and also Helmut Genaust, *Etymologisches Wörterbuch der botanischen Pflanzennamen*. 687-688. Basel 1996; R. Zander, F. Encke, G. Buchheim and S. Seybold, *Handwörterbuch der Pflanzennamen*.

14. Aufl. 670. Stuttgart 1993; Georg Christian Wittstein, *Etymologisch-botanisches Handwörterbuch*. 107, 925. Ansbach 1852.

Visnea L.f. Theaceae

Origins:
For a Portuguese botanist, Giraldo Visne; see Helmut Genaust, *Etymologisches Wörterbuch der botanischen Pflanzennamen*. 688. 1996; F. Boerner and G. Kunkel, *Taschenwörterbuch der botanischen Pflanzennamen*. 4. Aufl. 185. Berlin and Hamburg 1989; Georg Christian Wittstein, *Etymologisch-botanisches Handwörterbuch*. 925. Ansbach 1852.

Species/Vernacular Names:
V. mocanera L.f.

English: mocan

Vitaliana Sesler Primulaceae

Origins:
After the Italian naturalist Vitaliano Donati, 1713-1763, studied at Padova (Padua), physician, traveler and botanical collector, biologist, apothecary at Venezia, from 1749 to 1762 Praefectus of the Botanical Garden of Turin, author of *Della storia naturale marina dell'Adriatico*. [Edited by G. Rubbi-Carli.] Venezia 1750; see A. Ceruti, "L'Orto Botanico di Torino." *Agricoltura*. 7. 1963; O. Mattirolo, *Cronistoria dell'Orto Botanico della Regia Università di Torino*. in *Studi sulla vegetazione nel Piemonte* pubblicati a ricordo del II Centenario della fondazione dell'Orto Botanico della R. Università di Torino. Torino 1929; Karl Alfred von Zittel (1839-1904), *History of Geology and Palaeontology* to the end of the nineteenth century. 34. London 1901; John H. Barnhart, *Biographical Notes upon Botanists*. 1: 464. 1965; T. W. Bossert, *Biographical Dictionary of Botanists Represented in the Hunt Institute Portrait Collection*. 105. Boston, Massachusetts 1972; A. Lasègue, *Musée botanique de Benjamin Delessert*. Paris 1845; H. Genaust, *Etymologisches Wörterbuch der botanischen Pflanzennamen*. 688. [1717-1762] 1996; F. Boerner & G. Kunkel, *Taschenwörterbuch der botanischen Pflanzennamen*. 4. Aufl. 185. 1989; G.C. Wittstein, *Etymologisch-botanisches Handwörterbuch*. 925. 1852.

Vitellaria C.F. Gaertner Sapotaceae

Origins:
Latin *vitellus, i* "the yolk of an egg" and the adjectival suffix *-aria* indicating connection or possession, referring to the fruits.

Species/Vernacular Names:

V. paradoxa C.F. Gaertn. (*Butyrospermum parkii* (G. Don) Kotschy; *Butyrospermum paradoxum* (C.F. Gaertn.) Hepper)

Nigeria: kadanya (Hausa); kareje (Fula); chammal (Tiv); emi, emi-emi, emi-gidi (Yoruba); osisi (Igbo)

Mali: si, yiri, sie

Vitellariopsis Baillon ex Dubard Sapotaceae

Origins:

Resembling (*opsis*) the genus *Vitellaria*.

Species/Vernacular Names:

V. dispar (N.E. Br.) Aubrév. (*Austromimusops dispar* (N.E. Br.) A. Meeuse; *Mimusops dispar* N.E. Br.)

English: Tugela bush milkwood, Tugela milkwood

South Africa: Tugelabosmelkhout (Tugela river in Natal)

V. marginata (N.E. Br.) Aubrév. (*Austromimusops marginata* (N.E. Br.) A. Meeuse; *Mimusops marginata* N.E. Br.; *Mimusops natalensis* Schinz; *Mimusops schinzii* Engl.; *Inhambanella natalensis* (Schinz) Dubard)

English: Natal bush milkwood, flat-cap milkwood

Southern Africa: Natalbosmelkhout; amaSethole, umNqambomabele, umPhumbulu (Zulu); umBumbulu, umThunzi (Xhosa)

Vitex L. Labiatae (Verbenaceae)

Origins:

Latin name used by Plinius for the chaste-tree, Abraham's balm, *Vitex agnus-castus* or a similar shrub; probably derived from the Latin *vieo, -es, -etum, -ere* "to plait, to tie up, to twine"; see Carl Linnaeus, *Species Plantarum*. 638. 1753 and *Genera Plantarum*. Ed. 5. 285. 1754.

Species/Vernacular Names:

V. sp.

Malaya: leban, haleban, halban

V. agnus-castus L.

English: chaste tree, hemp tree, monk's pepper tree, sage tree, Indian spice, wild pepper, lilac chaste tree

Arabic: ghar, kherwa'

India: panjangusht, athlac, ranukabija, shambhaluka-bija

V. altissima L.f.

India: balage, banalgay, balgay, nemiliadogu, maila, mayilai, mayila, myrole, bulgi, ahoi, ashoi, jharua, selongphang, naviladi, tin-patte

V. amboniensis Guerke (the species was collected in Amboni, tropical east Africa)

English: large-fruited vitex, plum fingerleaf, Amboni vitex

Southern Africa: pruimvingerblaar; umBendula, umPhenduka (Zulu); mupfumbu-pfumbu (Venda)

V. burmensis Moldenke (*Vitex lanceifolia* S.C. Huang)

English: Burma (or Myanmar) chaste tree, lance-leaf chaste tree

China: chang ye jing

V. canescens Kurz

English: grey-hairy chaste tree

China: hui mao mu jing

V. chrysocarpa Planch. ex Benth.

Nigeria: dinyar rafi (Hausa); oori-eta (Yoruba)

V. ciliata (Pierre) Pellegrin

Gabon: angona

V. congolensis De Wild. & T. Dur.

Zaire: nembule, dufulu

V. doniana Sweet

Nigeria: dinya (Hausa); oori-nla (Yoruba); galbihi (Fula); ucha koro (Igbo)

Yoruba: ori, ori nla, ori odan

Mali: korofin, koro ba, koriyerefo

V. ferruginea Schum. & Thonn.

Nigeria: oori-eta (Yoruba); ibang (Edo); koronta (Igbo); ogi (Efik)

Yoruba: ori eta

Congo: mufilu, moufilou

V. gamosepala Griff.

English: glabrous yellow vitex

Malaya: galang dapur, leban pachat, leban pelandok

V. grandifolia Guerke

Ivory Coast: lokoubo

Gabon: nom angona

Yoruba: ori

Nigeria: ambro, efotaliamba, efol, nsum, lilenge, lilengi, orabia, ori, ori-nla, okurutu, ovuruburu, oraku, ogikhimi; orabia (Igbo); oori (Yoruba); ogikhimi (Edo)

Cameroon: evoula, mvule, fulu

V. harveyana H. Pearson (*Vitex geminata* H. Pearson; *Vitex schlechteri* Guerke) (named to honor the Irish botanist W.H. Harvey, 1811-1866; see Sir Charles James Fox Bunbury (1809-1886), *Botanical fragments*. Part 5: 205-206. London 1883; B. Seemann, "William Henry Harvey." *J. Bot., Lond.* 4: 236-238. 1866; J.H. Barnhart, *Biographical Notes upon Botanists*. 2: 135. 1965; Robert Lloyd Praeger (1865-1953), "William Henry Harvey." in Francis Wall Oliver (1864-1951), ed., *Makers of British Botany*. 204-224. Cambridge 1913; D.A. Webb, "William Henry Harvey, 1811-1866, and the tradition of systematic botany." *Hermathena*. 103: 32-45. 1966; E. Charles Nelson, "William Henry Harvey: A portrait of the artist as a young man." *Curtis's Botanical Magazine*. 13(1): 36-41. February 1996; Michael E. Mitchell, in *Dictionary of Scientific Biography* 6: 162-163. 1981; Mary Gunn and Leslie E. Codd, *Botanical Exploration of Southern Africa*. 179-181. Cape Town 1981; Stafleu and Cowan, *Taxonomic Literature*. 2: 70-90. 1979)

English: three-fingerleaf

Southern Africa: drievingerblaar, umbendula; umBhendula, umBendula (Zulu)

V. lucens T. Kirk

New Zealand: pururi (Maori name)

V. megapotamica (Spreng.) Moldenke

Brazil: tarumã, tapinhoan, tarumã preta, maria preta, tarumã de montevidéu, azeitona do mato, azeitona brava, azeitona da terra, sombra de touro, tarumã romã

V. mombassae Vatke (*Vitex flavescens* Rolfe)

English: wild cherry, smelly-berry vitex

N. Rhodesia: mumbomba

Southern Africa: pypsteel, vaalbos, poerabessie; umDuli, umLuthu (Zulu); mokwele (North Sotho); muChunkule (Shona)

V. negundo L.

English: horse shoe vitex, negundo chaste tree, five-leaved chaste tree

China: huang ching, huang jing

Malaya: lagundi, lenggundi, lemuning, peninchang

The Philippines: lagundi, dangla

India: nirgundi, nirgandi, nirkkundi, nila-nirgundi, nirgunda, nisinda, nishinda, nisind, mewri, katri, nigad, nigudi, nagoda, nagaol, sephalika, svetasurasa, vaavili, tellavaaviti, tellavavili, vavili, nalla-vavili, nochchi, notchi, cara-nosi, kari-nocci, bem-nosi, ven-nocci, venmochi, nocci, vellai-nocohi, vellai-noch-chi, vellanocchi, shamalic, samalu, sambhalu, sawbhalu, shambalu, shivari, lakkigida, nakkilu, nekki, beyguna, begundia, banna, torbanna, marwan, marwa, maura, mawa, shwari, shiwali, shiwari, pasutia, aggla chita, vrikshaha, chinduvaram, sindhuvaram, sindhuvaruma, indrani, bile-nekki

V. obovata E. Meyer

English: white vitex, Kei fingerleaf, Cape vitex

South Africa: Keivingerblaar

V. oxycuspis Bak.

Nigeria: ojikanaba (Edo)

V. pachyphylla Bak.

Gabon: angona

V. parviflora Juss.

The Philippines: molabe, molave, molauin, molaue, sagad, sagat, amaraun, hamoraon, hamuraon, adagauon, hamolauen, hamyaon, marauin, mulaon, murain, salingkapa, molauebatu tugas, tugas-abgauon, tugas-lanhan

V. patula E.A. Bruce

South Africa: the spreading vitex

V. payos (Lour.) Merr.

English: chocolate berry

Southern Africa: muBalebale, muBindoyo, muHubva, chiKubai, muKubvu, muTentera, muTsubvu (Shona)

V. peduncularis Wallich ex Schauer

English: longspike chaste tree

China: chang xu jing

V. pierreana Dop

English: Pierre chaste tree

China: ying ge mu

V. pooara Corbishley (the specific name has an African origin)

Southern Africa: stinkbessie; mphuru (North Sotho)

V. quinata (Lour.) F.N. Williams (*Cornutia quinata* Lour.)

English: orange-barked vitex, fiveleaf chaste tree

Japan: ô-ninjin-boku

China: shan mu jing

Malaya: merboh

V. rehmannii Guerke (after the botanist Dr. Anton(i) Rehmann (Rehman), 1840-1917, geographer, traveler, plant collector in South Africa (1875-1877 and 1879-1880); see H.N. Dixon and A. Gepp, "Rehmann's South African mosses." *Kew Bull.* 1923: 193-238. 1923; John H. Barnhart, *Biographical Notes upon Botanists*. 3: 139. Boston 1965; Stafleu and Cowan, *Taxonomic Literature*. 4: 655-656. 1983; G. Murray, *History of the Collections Contained in the Natural History Departments of the British Museum*. 1: 176. 1904; T.W. Bossert, *Biographical Dictionary of Botanists Represented in the Hunt Institute Portrait Collection*. 328. 1972; H.N. Clokie, *Account of the Herbaria of the Department of Botany in the University of Oxford*. 230. 1964; Mary Gunn and Leslie E. Codd, *Botanical Exploration of Southern Africa*. 292-294. 1981; R. Zander, F. Encke, G. Buchheim and S. Seybold, *Handwörterbuch der Pflanzennamen*. 14. Aufl. 1993)

English: pipe-stem tree

Southern Africa: pypsteelboom, vaalbos, pypsteel; umDuli, umLuthu (Zulu); mokwele (North Sotho)

V. rivularis Guerke

Nigeria: boron, ori, ububan; oori (Yoruba); ububan (Edo)

Yoruba: ububan

V. rotundifolia L.f. (*Vitex trifolia* L. var. *simplicifolia* Cham.; *Vitex ovata* Thunb.)

English: beach vitex

Japan: hama-gô

Hawaii: kolokolo kahakai, hinahina kolo, manawanawa, mawanawana, pohinahina, polinalina

V. sampsonii Hance

English: Sampson chaste tree

China: guang dong mu jing

V. simplicifolia Oliv.

Nigeria: dinyar biri (Hausa); bummehi (Fula)

Mali: kotoni, korofon, koronafan, koronifin, koro yafan

V. thyrsiflora Baker

Yoruba: aanu, alaanu, aayanrin

V. trifolia L.

English: common blue vitex, shrub chaste tree, threeleaf chaste tree, Indian wild pepper, Indian privet, seashore vites

Japan: mitsu-ba-hama-gô, hôgâgii

China: man jing, man jing zi, man ching

Vietnam: mac nim, man kinh

Malaya: lenggundi, lagundi, legundi, lemuning, muning, demundi

India: surasa, valuru, pani-ki-sanbhalu, panikisanbhalu, paniki-shumbala, sufed-sanbhalu, panisamalu, pani-samalu, lagondi, nirgundi, jalanirgundi, begundia, lingur, indrani, niruvavili, vaavili chiruvaavili, shiruvavili, nirnos-chi, nirnocchi, nirnochchi, nirnochi, sirunochi, shirunoch-chi, nira-lakki-gida, lakki, nekki nochi, karinochi, sindhuka, nichinda

V. tripinnata (Loureiro) Merrill

English: tripinnate chaste tree

China: yue nan mu jing

V. vestita Wallich ex Schauer

English: yellow-hairy chaste tree

China: huang mao mu jing

V. wilmsii Guerke var. *reflexa* (H. Pearson) Pieper (*Vitex reflexa* Pearson) (after the German (b. Münster) apothecary and botanical collector Friedrich Wilms, Jr., 1848-1919, traveler, author of "Ein botanischer Ausflug ins Boeren-land." *Verh. Bot. Ver. Prov. Brandenb.* 1898; see J.H. Barn-hart (1871-1949), *Biographical Notes upon Botanists*. 3: 502. 1965; Mary Gunn and Leslie E. Codd, *Botanical Exploration of Southern Africa*. 378. Cape Town 1981)

English: hairy vitex, hairy fingerleaf, Wilm's vitex

Southern Africa: harige vingerblaar; umLuthu (Zulu); ama-Khosikati (Swazi)

V. zeyheri Sond. (for the German botanical collector Carl Ludwig Philipp Zeyher, 1799-1858)

Southern Africa: mokwele; mokwele (Tswana: western Transvaal, northern Cape, Botswana)

Viticella Mitch. Hydrophyllaceae

Origins:
From the Latin *viticella, ae* for a plant, otherwise unknown.

Viticipremna H.J. Lam. Labiatae (Verbenaceae)

Origins:
Latin *vitex, viticis* plus the genus *Premna* L.; some suggest from the Latin *viticella, ae* "a plant" or *viticula* "a little vine."

Vitiphoenix Becc. Palmae

Origins:
Viti Levu, Fiji Islands, plus *Phoenix*.

Vitis L. Vitaceae

Origins:
The Latin name for the grapevine, Latin *vieo, -es, -etum, -ere* "to bend, plait, weave," Akkadian *ebitu* "to be tied, girt"; see Carl Linnaeus, *Species Plantarum*. 202. 1753 and *Genera Plantarum*. Ed. 5. 95. 1754; Giovanni Semerano, *Le origini della cultura europea*. Dizionario della lingua Latina e di voci moderne. 2(2): 613, 616. Leo S. Olschki Editore, Firenze 1994.

Species/Vernacular Names:
V. acerifolia Raf.

English: bush grape

V. aestivalis Michx.

English: summer grape

V. amurensis Rupr.

English: Amur grape

China: shan teng teng yang

V. arizonica Engelm.

English: canyon grape

V. bicolor Le Conte

English: blue grape

V. californica Benth.

English: California wildgrape

V. coignetiae Raf.

English: crimson glory vine

V. flexuosa Thunb.

Japan: sankaku-tsuru

China: ge lei zhi, chien sui lei, chang chun teng

V. girdiana Munson

English: desert wildgrape

V. labrusca L.

English: fox grape, skunk grape

V. labruscana L.H. Bailey

English: labruscan vineyard grape

Japan: Amerika-budô

V. linecumii Buckl.

English: post-oak grape

V. mustangensis Buckl.

English: mustang grape

V. palmata Vahl

English: red grape, cat grape

V. riparia Michx.

English: frost grape, riverbank grape

V. rotundifolia Michx.

English: muscadine grape, bullace, fox grape

V. rupestris Scheele

English: bush grape, sand grape

V. thunbergii Siebold & Zuccarini (*Vitis ficifolia* Bunge)

Japan: ebi-zuru, Ryûkyû ganebu, kanibu

V. vinifera L.

English: common grape vine, wine grape, cultivated grape, European grape, grapes, grape

Mexico: bicholi yaha castilla

India: angur, angura, angurphal, onguro, drakh, draksha, draksha-pondu, drakshai, drakshya, drakshi, drakya, mudraka, dakh, dakha, drakhyaluta, darakh, mundiri, gostani, gostoni, gostanidraksha, kismis, kishmish, sougi, manakka, mridirka

Japan: budô

China: pu tao

French: vigne

Arabic, anba, dalia, anab

V. vulpina L.

English: frost grape, chicken grape

Vittadinia A. Richard Asteraceae

Origins:

After the Italian botanist and physician Carlo Vittadini, 1800-1865, mycologist, author of *Tentamen mycologicum.*

Mediolani [1826] and of *Descrizione dei funghi mangerecci più comuni ... e de' velenosi.* Milano [1832-]1835; see S. Garovaglio, *Notizie sulla vita e sugli scritti del dott. Carlo Vittadini.* Milano 1867; Mariella Azzarello Di Misa, a cura di, *Il Fondo Antico della Biblioteca dell'Orto Botanico di Palermo.* 283. Regione Siciliana, Palermo 1988; I.H. Vegter, *Index Herbariorum.* Part II (7), *Collectors T-Z.* Regnum Vegetabile vol. 117. 1988; J.H. Barnhart, *Biographical Notes upon Botanists.* 3: 440. 1965; F. Boerner & G. Kunkel, *Taschenwörterbuch der botanischen Pflanzennamen.* 4. Aufl. 185. 1989; T. W. Bossert, *Biographical Dictionary of Botanists Represented in the Hunt Institute Portrait Collection.* 420. Boston, Massachusetts 1972; Antoine Lasègue, *Musée botanique de M. Benjamin Delessert.* 1845.

Species/Vernacular Names:

V. blackii N. Burb.

Australia: Western New Holland daisy

V. cuneata DC.

Australia: woolly New Holland daisy, fuzzweed

V. gracilis (Hook.f.) N. Burb. (*Eurybiopsis gracilis* Hook.f.; *Vittadinia triloba* (Gaudich.) DC. var. *lanuginosa* J. Black)

Australia: woolly New Holland daisy

V. megacephala (F. Muell. ex Benth.) J. Black (*Vittadinia australis* A. Rich. var. *megacephala* F. Muell. ex Benth.)

Australia: giant New Holland daisy, fuzzweed

Vittaria Smith Vittariaceae (Adiantaceae)

Origins:

Latin *vitta, ae* (*vieo, -es, -etum, -ere* "to bend, twist together"; Akkadian *ebetu, ebitu, ebiu* "to be tied, girt") "a band, fillet, ribbon," linear fronds with linear sori.

Species/Vernacular Names:

V. flexuosa Fée (*Vittaria japonica* Miquel)

Japan: shishi-ran

V. lineata (L.) Smith (*Pteris lineata* L.)

English: shoestring fern

V. zosterifolia Willd.

Japan: shima-shishi-ran (= island *Vittaria*)

Viviania Cavanilles Geraniaceae (Vivianiaceae)

Origins:

Named after the Italian botanist Domenico (Dominicus) Viviani, 1772-1840, physician, professor of botany and natural history, 1803 founder of the Genova Botanical Garden and Director 1803-1837. Among his writings are *Flora Libycae specimen*. Genuae [Genova] 1824 and *I funghi d'Italia*. Genova 1834 [-1838]; see John H. Barnhart, *Biographical Notes upon Botanists*. 3: 440. 1965; Rodolfo Emilio Giuseppe Pichi Sermolli, "L'Orto Botanico di Genova." in *Agricoltura*. 12(4): 87-90. Roma 1963; Mariella Azzarello Di Misa, a cura di, *Il Fondo Antico della Biblioteca dell'Orto Botanico di Palermo*. 284. Regione Siciliana, Palermo 1988; T. W. Bossert, *Biographical Dictionary of Botanists Represented in the Hunt Institute Portrait Collection*. 420. Boston, Massachusetts 1972; Antoine Lasègue, *Musée botanique de M. Benjamin Delessert*. 1845; E.M. Tucker, *Catalogue of the Library of the Arnold Arboretum of Harvard University*. Cambridge, Massachusetts 1917-1933; R. Zander, F. Encke, G. Buchheim and S. Seybold, *Handwörterbuch der Pflanzennamen*. 14. Aufl. Stuttgart 1993.

Viviania Colla Rubiaceae

Origins:

After the Italian botanist Domenico Viviani, 1772-1840.

Viviania Raf. Rubiaceae

Origins:

After the Italian botanist Domenico Viviani, 1772-1840; see Constantine Samuel Rafinesque (1783-1840), *Specchio delle scienze, o giornale enciclopedico di Sicilia* etc. 1: 117. Palermo 1814; E.D. Merrill (1876-1956), *Index rafinesquianus*. The plant names published by C.S. Rafinesque, etc. 227. Jamaica Plain, Massachusetts, USA 1949.

Vlamingia Vriese Violaceae

Origins:

After the Dutch navigator Willem de Vlamingh (Vlaming) (= William the Fleming), 300 years ago arrived off the coast of Western Australia and explored and named Rottnest Island (= rats' nest) and the Swan River; however, de Vlamingh was not the first Dutchman to reach the coast of Australia, Willem Janszoon claimed this honor when he explored the area around Cape York in the *Duyfken* in 1606; the first Dutchman to land on the Western Australian coast was Dirk Hartog who arrived there in a ship, *Eendracht*, in 1616; see Arthur D. Chapman, ed., *Australian Plant Name Index*. 2976-2977. Canberra 1991; James A. Baines, *Australian Plant Genera. An Etymological Dictionary of Australian Plant Genera*. 394. Chipping Norton, N.S.W. 1981.

Voacanga Thouars Apocynaceae

Origins:

From the vernacular Madagascan name.

Species/Vernacular Names:
V. spp.

Nigeria: ako-dodo, dodo, akanta

V. africana Stapf

Nigeria: ako dodo (Yoruba); pete-pete (Igbo)

Yoruba: dodo nla, ako dodo, ajik(un)efun, farajoyan

China: fei zhou ma ling guo

V. thouarsii Roemer & Schultes (*Voacanga dregei* E. Meyer; *Annularia natalensis* Hochst.; *Piptolaena dregei* (E. Meyer) A. DC.) (for the French botanist Louis-Marie Aubert Aubert du Petit-Thouars, 1758-1831, traveler, author of *Plantes des îles de l'Afrique australe ...* Paris [1804] and *Histoire des végétaux recueillis dans les isles australes d'Afrique ...* Paris 1806-1808)

English: wild frangipani

Nigeria: kokiyar biri (Hausa)

Southern Africa: wilde frangipani, voacanga; iNomfi (meaning bird-lime), uNomfi, umKhadlu, umHlambamanzi, uNokhahlu, umKhahlu, umThondisa (Zulu); umThofu, umThomfi, umTomvi (Xhosa)

Voandzeia Thouars Fabaceae

Origins:
An African vernacular name.

Voanioala J. Dransfield Palmae

Origins:
From the Malagasy local name for the palm, the Forest Coconut, *voa-nio-ala*; see J. Dransfield, "*Voanioala* (Arecoideae: Cocoeae: Butiinae), a new palm genus from Madagascar." in *Kew Bulletin*. 44(2): 191-198. 1989; F.N. Hepper, ed., *Plant Hunting for Kew*. Royal Botanic Gardens, Kew, London 1989.

Species/Vernacular Names:
V. gerardii J. Dransfield

English: forest coconut

Voatamalo Capuron ex Bosser Euphorbiaceae

Origins:
From the vernacular Madagascan name.

Vochy Aublet Vochysiaceae (Rosidae, Polygales)

Origins:

From the native name in tropical America, French Guiana; see also *Vochysia* Aublet.

Vochysia Aublet Vochysiaceae

Origins:

From the native name in tropical America.

Species/Vernacular Names:

V. hondurensis Sprague

Tropical America: quaruba

Vogelia Lam. Plumbaginaceae

Origins:

According to M.P. Nayar (see *Meaning of Indian Flowering Plant Names*. 362. Dehra Dun 1985) the genus was named for the German (b. Berlin) botanist and plant collector Julius Rudolph Theodor Vogel (1812-1841, d. F. Po) who accompanied Henry Dundas Trotter (1802-1859) on the Niger expedition.

Vogelocassia Britton Caesalpiniaceae

Origins:

To honor the German botanist Julius Rudolph Theodor Vogel, 1812-1841, traveler, explorer and plant collector, co-Director of the Botanical Garden in Bonn, from 1841 to 1842 with John Ansell and Moffat on Capt. Allen's Niger Expedition, author of *Generis Cassia synopsis*. Berolini [1837], brother of the German explorer and plant collector Eduard Vogel (1829-1856); see William Allen (1793-1864) and Thomas Richard Heywood Thomson, *A Narrative of the Expedition ... to the River Niger, in 1841*. London 1848; R. Zander, F. Encke, G. Buchheim and S. Seybold, *Handwörterbuch der Pflanzennamen*. 14. Aufl. Stuttgart 1993; Joseph Vallot (1854-1925), "Études sur la flore du Sénégal." in *Bull. Soc. Bot. de France*. 29: 168-238. Paris 1882; R.W.J. Keay, "Botanical collectors in West Africa prior to 1860." in *Comptes Rendus A.E.T.F.A.T.* 55-68. Lisbon 1962; Ludolf Christian Treviranus (1779-1864), in *Linnaea*. 16: 533-560. 1842; J.F. Schön & S. Crowther, *Journals of the Expedition up the Niger in 1841*. London 1848; E.M. Tucker, *Catalogue of the Library of the Arnold Arboretum of Harvard University*. Cambridge, Massachusetts 1917-1933; F.N. Hepper and Fiona Neate, *Plant Collectors in*

West Africa. 83. 1971; Günther Schmid, *Chamisso als Naturforscher*. Eine Bibliographie. 173. Leipzig 1942.

Voharanga Costantin & Bois Asclepiadaceae

Origins:

From the vernacular Madagascan name.

Vohemaria Buchenau Asclepiadaceae

Origins:

From the vernacular Madagascan name.

Voigtia Klotzsch Rubiaceae

Origins:

For the German botanist Friedrich Siegmund Voigt, 1781-1850, physician, 1802 Dr. med. in Jena, professor of medicine, Director of the Botanical Garden at Jena University. His works include *Conspectus tractatus de plantis hybridis*. Jenae 1802, *System der Botanik*. Jena 1808 and *Catalogus plantarum* quae in hortis ducalibus botanici jenensi et belvederensi coluntur. Jenae 1812; see G. Schmid, *Goethe und die Naturwissenschaften*. Halle 1940.

Volkameria L. Labiatae (Verbenaceae)

Origins:

Probably dedicated to the German botanist Johann Georg Volckamer or Volcamer, the Younger (Joannes Georgius Volcamerus, filius), 1662-1744, author of *Flora Noribergensis*. Noribergae 1700, Dissertatio ... *de lethargo*. Altdorffii [1684], *Epistola de Stomacho*. Altdorii [sic] Noricorum 1682 and *Opobalsami orientalis in Theriaces confectionem Romae revocati examen*, etc. Norimbergae 1644; see Antonio Colmenero de Ledesma, *Chocolata Inda*. Opusculum de qualitate & natura chocolatae ... in Latinum traslatum. Norimbergae 1644; Marcus Aurelius Severinus (Marco Aurelio Severino) (1580-1656), *Zootomia Democritaea id est Anatome generalis totius animantium opificii*. Noribergae 1645; Carl Linnaeus, *Species Plantarum*. 637. 1753 and *Genera Plantarum*. Ed. 5. 284. 1754; Blanche Henrey, *No Ordinary Gardener — Thomas Knowlton, 1691-1781*. Edited by A.O. Chater. British Museum (Natural History). London 1986; Mariella Azzarello Di Misa, a cura di, *Il Fondo Antico della Biblioteca dell'Orto Botanico di Palermo*. 284-285. Palermo 1988; Gilbert Westacott Reynolds (1895-1967), *The Aloes of South Africa*. 79-80. Balkema, Rotterdam 1982.

Volkensia O. Hoffmann Asteraceae

Origins:

After the German botanist Georg Ludwig August Volkens, 1855-1917, in 1884-1885 traveler in the Egyptian-Arab Desert, collaborator of A. Engler and Simon Schwendener (1829-1919), explorer. Among his works are *Exkursionen am Kilima-Ndjaro*. Berlin 1895, *Der Kilimandscharo*. Berlin 1897, *Die Vegetation der Karolinen*. Leipzig 1901, *Die Botanische Zentralstelle für die Kolonien*. Berlin 1907 and *Beiträge zur Flora von Mikronesien*. Leipzig und Berlin 1914; see John H. Barnhart, *Biographical Notes upon Botanists*. 3: 443. 1965; Sir Harry Hamilton Johnston (1858-1927), *Der Kilima-Ndjaro*. Leipzig 1886; Alex Johnston, *The Life and Letters of Sir Harry Johnston*. London 1929; T. W. Bossert, *Biographical Dictionary of Botanists Represented in the Hunt Institute Portrait Collection*. 421. Boston, Massachusetts 1972; E.M. Tucker, *Catalogue of the Library of the Arnold Arboretum of Harvard University*. Cambridge, Massachusetts 1917-1933; Elmer Drew Merrill, *Contr. U.S. Natl. Herb.* 30(1): 308. 1947.

Volkensiella H. Wolff Umbelliferae

Origins:

For the German botanist Georg Ludwig August Volkens, 1855-1917; see John H. Barnhart, *Biographical Notes upon Botanists*. 3: 443. 1965.

Volkensinia Schinz Amaranthaceae

Origins:

Named after the German botanist Georg Ludwig August Volkens, 1855-1917, explorer, collaborator of A. Engler and Simon Schwendener (1829-1919); see John H. Barnhart, *Biographical Notes upon Botanists*. 3: 443. 1965.

Volkensiophyton Lindau Acanthaceae

Origins:

After the German botanist Georg Ludwig August Volkens, 1855-1917, explorer. Among his works are *Die Nutzpflanzen Togos*. Leipzig 1909-1910, *Über die Karolinen-Insel Yap*. Berlin 1901 and *Die Flora der aegyptisch-arabischen Wüste*. Berlin 1887; see John H. Barnhart, *Biographical Notes upon Botanists*. 3: 443. 1965.

Volutaria Cass. Asteraceae

Origins:

Latin *voluta, ae* "a volute, spiral scroll," *voluto* "to roll, to twist."

Volvulopsis Roberty Convolvulaceae

Origins:

Latin *volvola, ae* applied by Hieronymus to the plant *convolvulus*; see also *Evolvulus* L.

Volvulus Medik. Convolvulaceae

Origins:

Latin *volvola* applied by Hieronymus to the plant *convolvulus*, Latin *volvo, volvi, volutum* "to roll, turn round."

Vonitra Beccari Palmae

Origins:

Based on Malagasy plant name, or from the Latin *nitrum* "soda, natron"; see John Dransfield and Henk J. Beentje, *The Palms of Madagascar*. Royal Botanic Gardens, Kew 1995; Natalie W. Uhl & John Dransfield, *Genera Palmarum*. 344-346. Allen Press, Lawrence, Kansas 1987; O. Beccari, *Palme del Madagascar*. Firenze 1914.

Species/Vernacular Names:

V. fibrosa (C. Wright) Beccari

English: vonitra palm, piassava

Vonroemeria J.J. Smith Orchidaceae

Origins:

Named after L.S.A.M. von Roemer, plant collector and member of the second Lorentz Expedition to Dutch New Guinea in 1909. His works include *Over de Familie van Koningin Elisabeth van Engeland*. (Overdruk uit de De Malanger.) 1936 and *Historical Sketches* [of Dutch Medicine]. An Introduction to the Fourth Congress of the Far Eastern Association of Tropical Medicine ... translated ... by Duncan MacColl, G. Da Silva, and others. Batavia 1921.

Vossia Wallich & Griffith Gramineae

Origins:

For the German poet Johann Heinrich Voss, 1751-1826, remembered chiefly for his translations of Homer, or the Dutch Gerhard Johann Voss (Vossius) (1577-1649), Dutch Humanist theologian, one of the foremost scholars of the Dutch Republic's "Golden Age" and father of the distinguished classical scholar Isaac Vossius (1618-1689); see Frans A. Stafleu and Richard S. Cowan, *Taxonomic Literature*. 6: 784. [genus named for the German poet J.G. Voss]

1986; W.P.U. Jackson, *Origins and Meanings of Names of South African Plant Genera*. 37. [genus named for the Belgian botanist L. Voss, 17th cent.] Rondebosch 1990.

Species/Vernacular Names:
V. cuspidata (Roxb.) Griffith

English: hippo-grass, floating grass

South Africa: um-soof reed

Vossianthus Kuntze Tiliaceae

Origins:
For the German gardener Andreas Voss, 1857-1924, author of *Vilmorin's Blumengärtnerei*. (with the collaboration of the German botanist and gardener August Siebert, 1854-1923) Berlin 1894-1896; see John H. Barnhart, *Biographical Notes upon Botanists*. 3: 444. 1965; E.M. Tucker, *Catalogue of the Library of the Arnold Arboretum of Harvard University*. Cambridge, Massachusetts 1917-1933; R. Zander, F. Encke, G. Buchheim and S. Seybold, *Handwörterbuch der Pflanzennamen*. 14. Aufl. Stuttgart 1993.

Votomita Aubl. Melastomataceae

Origins:
A vernacular name.

Vouacapoua Aublet Caesalpiniaceae

Origins:
A native name for *Vouacapoua americana* Aublet. This tree is called *vouacapou* by the Galibis, French Guiana.

Vouarana Aubl. Sapindaceae

Origins:
A native name in French Guiana.

Vouay Aubl. Palmae

Origins:
A native name, French Guiana.

Voyria Aubl. Gentianaceae

Origins:
A vernacular name in French Guiana.

Voyriella Miq. Gentianaceae

Origins:
The diminutive of the genus *Voyria* Aubl.

Vriesea Lindley Bromeliaceae

Origins:
For the Dutch botanist Willem Hendrik de Vriese, 1806-1862, physician, 1834-1845 professor of botany at Amsterdam and from 1845 to 1862 at Leyden (Leiden), traveler in the Dutch East Indies. His writings include *Specimen academicum medicum*. Lugduni Batavorum [Leyden] 1830, *Novae species Cycadearum Africae australis*. [1838], *Hortus Spaarn-Bergensis*. Enumeratio stirpium quas, in Villa Spaarn-Berg prope Harlemum, alit Adr. van der Hoop, ... Amstelodami 1839 and *Plantarum javanicarum*. [1844], co-author of *Plantae Indiae Batavae orientalis*. Lugduni Batavorum 1856; see J.H. Barnhart, *Biographical Notes upon Botanists*. 3: 445. 1965; F.A. Stafleu and R.S. Cowan, *Taxonomic Literature*. 6: 792-799. 1986; T.W. Bossert, *Biographical Dictionary of Botanists Represented in the Hunt Institute Portrait Collection*. 422. Boston, Massachusetts 1972; R. Zander, F. Encke, G. Buchheim and S. Seybold, *Handwörterbuch der Pflanzennamen*. 14. Aufl. 702. 1993; E.M. Tucker, *Catalogue of the Library of the Arnold Arboretum of Harvard University*. Cambridge, Massachusetts 1917-1933; Ida Kaplan Langman, *A Selected Guide to the Literature on the Flowering Plants of Mexico*. 786. 1964; Antoine Lasègue, *Musée botanique de M. Benjamin Delessert*. 1845.

Species/Vernacular Names:
V. carinata Wawra

English: lobster claws

V. x *mariae* André

English: painted feather

V. splendens (Brongn.) Lem.

English: flaming sword

Vriesia Lindley Bromeliaceae

Origins:
See *Vriesea* Lindl.

Vrydagzynea Blume Orchidaceae

Origins:
After the Dutch pharmacologist Theodore Daniel Vrydag Zynen; see Arthur D. Chapman, ed., *Australian Plant Name*

Index. 2977. Canberra 1991; Mark Alwin Clements, in *Australian Orchid Research*. 1: 1-160. 1989; Carl (Karl) Ludwig Blume, *Flora Javae et insularum adjacentium nova series*. 60. Leiden, Amsterdam 1858[-1859]; Pritzel, *Thesaurus*. no. 9888.

Species/Vernacular Names:
V. salomonensis Schltr. (see B.A. Lewis and P.J. Cribb, *Orchids of the Solomon Islands and Bougainville*. Royal Botanic Gardens, Kew 1991)

The Solomon Islands: waingongi

Vulpia C.C. Gmelin Gramineae

Origins:
Latin *vulpes, is* "a fox," indicating the appearance of the plants; according to Helmut Genaust (see *Etymologisches Wörterbuch der botanischen Pflanzennamen*. 691. 1996) the genus was named after the German chemist Johann Samuel Vulpius, 1760-1840 (or 1846), pharmacist and amateur botanist of Baden, Germany.

Species/Vernacular Names:
V. bromoides (L.) S.F. Gray (*Festuca bromoides* L.; *Festuca sciuroides* Roth; *Vulpia sciuroides* (Roth) Roth ex C. Gmelin)

English: squirrel tail fescue, rat's tail fescue, silver grass, brome fescue

V. myuros (L.) C.C. Gmelin (*Festuca myuros* L.)

English: rat's tail fescue, silver grass

South Africa: langbaardswenkgras, wildegars

Vulpiella (J. A. Battandier & L.C. Trabut) P.-A. Burollet Gramineae

Origins:
The diminutive of the genus *Vulpia*.

Vvedenskya Korovin Umbelliferae

Origins:
For Aleksei Ivanovich Vvedensky, 1898-1972, botanist; see Patricia K. Holmgren, Noel H. Holmgren and Lisa C. Barnett, eds., *Index Herbariorum. Part I: The Herbaria of the World*. 342. Eighth edition. New York Botanical Garden, New York 1990.

Vvedenskyella Botsch. Brassicaceae

Origins:
For Aleksei Ivanovich Vvedensky, 1898-1972, botanist; see Patricia K. Holmgren, Noel H. Holmgren and Lisa C. Barnett, eds., *Index Herbariorum. Part I: The Herbaria of the World*. 342. Eighth edition. New York Botanical Garden, New York 1990.

W

Wachendorfia Burm. Haemodoraceae

Origins:

The genus commemorates the name of the Dutch botanist and physician Evert Jacob (Everardus Jacobus) van Wachendorff, 1702-1758, professor of medicine, botany and chemistry at Utrecht, one of the first Directors of the Botanic Gardens of Utrecht, author of *Oratio botanico-medica de plantis immensitatis intellectus divini testibus locupletissimis*. Trajecti ad Rhenum [Utrecht] 1743 and *Horti Ultrajectini index*. Trajecti ad Rhenum 1747; see J.H. Barnhart, *Biographical Notes upon Botanists*. 3: 446. 1965; Jonas C. Dryander, *Catalogus bibliothecae historico-naturalis Josephi Banks*. London 1796-1800; F.A. Stafleu, *Linnaeus and the Linnaeans*. Utrecht 1971; F. Boerner & G. Kunkel, *Taschenwörterbuch der botanischen Pflanzennamen*. 4. Aufl. 185f. Berlin & Hamburg 1989.

Species/Vernacular Names:

W. brachyandra W.F. Barker (Greek *brachys* "short" and *aner, andros* "male, a man, stamen," with short anthers)

South Africa: rooikanol

W. paniculata Burm.

South Africa: rooikanol, spinnekopblom

W. thyrsiflora Burm.

South Africa: rooikanol

Wadea Raf. Solanaceae

Origins:

For the Irish botanist Walter Wade, c. 1740/1760-1825 (d. Dublin), physician, professor of botany for the Dublin Society, founder and Director of the Dublin Society's Botanical Garden at Glasnevin, 1811 Fellow of the Royal Society. Among his writings are *Catalogue of Plants*. Dublin 1802 and *Plantae rariores in Hibernia*. Dublin 1804; see John H. Barnhart, *Biographical Notes upon Botanists*. 3: 447. 1965; Constantine Samuel Rafinesque, *Sylva Telluriana*. 56. Philadelphia 1838; E. Charles Nelson and E.M. McCracken, *The Brightest Jewel: A History of the National Botanic Gardens, Glasnevin, Dublin*. Kilkenny 1987; T. W. Bossert, *Biographical Dictionary of Botanists Represented in the Hunt Institute Portrait Collection*. 422. Boston, Massachusetts 1972; [Sir J.E. Smith], *Memoir and Correspondence of ... Sir J.E. Smith ... Edited by Lady Pleasance Smith*. London 1832; Jonas C. Dryander, *Catalogus bibliothecae*

historico-naturalis Josephi Banks. London 1796-1800; Blanche Henrey, *British Botanical and Horticultural Literature before 1800*. Oxford 1975; E.M. Tucker, *Catalogue of the Library of the Arnold Arboretum of Harvard University*. Cambridge, Massachusetts 1917-1933; E. Charles Nelson, "National Botanic Gardens, Glasnevin, Retrospect and Prospect." *Curtis's Botanical Magazine*. 12(4): 181-185. November 1995.

Wagatea Dalzell Caesalpiniaceae

Origins:

A local Marathi name derived from Sanskrit *vaghanti, vag* "tiger," referring to the plant, a thorny scrambler.

Wahlenbergia Blume Rubiaceae

Origins:

For the Swedish botanist Georg (Göran) Wahlenberg, 1780-1851 (Uppsala).

Wahlenbergia Schrader ex Roth Campanulaceae

Origins:

For the Swedish botanist Georg (Göran) Wahlenberg, 1780-1851 (d. Uppsala), studied under Thunberg, plant geographer, physician, Dr. med. 1806, traveler, naturalist, professor of botany and medicine (succeeding C.P. Thunberg 1828-1829) at Uppsala, friend of Carl Ludwig von Willdenow (1765-1812) and Christian Leopold von Buch (1774-1853). Among his main scientific works are *Flora lapponica*. Berolini [Berlin] 1812, *Flora upsaliensis*. Upsaliae 1820, *Flora suecica*. Upsaliae 1824-1826 and *Tractatio anatomica*. Upsaliae [1806]; see Gunnar Eriksson, in *Dictionary of Scientific Biography* 14: 116-117. 1981; Elias M. Fries (1794-1878), *Novitiae Florae suecicae. Edit. altera, auctior et in formam commentarii in cel. Wahlenbergii* floram suecicam redacta. Lund (Berling) 1828; Thorgny Ossian Bolivar Napoleon Krok (1834-1921), *Bibliotheca botanica suecana*. 741-745. Stockholm, Uppsala 1925; J.H. Barnhart, *Biographical Notes upon Botanists*. 3: 619. 1965; R. Zander, F. Encke, G. Buchheim and S. Seybold, *Handwörterbuch der Pflanzennamen*. 14. Aufl. 796. 1993; Mariella Azzarello Di Misa, a cura di, *Il Fondo Antico della*

Biblioteca dell'Orto Botanico di Palermo. 285. Regione Siciliana, Palermo 1988; T. W. Bossert, *Biographical Dictionary of Botanists Represented in the Hunt Institute Portrait Collection*. 423. Boston, Massachusetts 1972; Antoine Lasègue, *Musée botanique de M. Benjamin Delessert*. 1845; E.M. Tucker, *Catalogue of the Library of the Arnold Arboretum of Harvard University*. Cambridge, Massachusetts 1917-1933.

Species/Vernacular Names:
W. gracilis (Forster f.) A. DC. (*Campanula gracilis* (Forster f.) A. DC.; *Campanula quadrifida* R. Br.; *Wahlenbergia quadrifida* (R. Br.) A. DC.; *Wahlenbergia sieberi* A. DC.; *Wahlenbergia marginata* (Thunb.) A. DC.; *Wahlenbergia trichogyna* Stearn)

English: sprawling bluebell, Australian bluebell, gentian rock-bell

Japan: hina-gikyô

China: hsi yeh sha shen

W. hederacea (L.) Reichb.

English: ivy-leaved bellflower

W. krebsii Cham. subsp. *krebsii* (*Wahlenbergia zeyheri* Buek)

English: fairy bluebell, fairy bellflower

W. lobelioides (L.f.) Link

English: bellflower, hare bell

W. procumbens (Thunb.) A. DC.

English: wild violet

W. rivularis Diels (*Wahlenbergia tysonii* Zahlb.)

English: bellflower

Southern Africa: iPipiyo (Xhosa)

W. stellarioides Cham. & Schltdl.

Southern Africa: iPepilana (little pepper) (Xhosa)

Wailesia Lindley Orchidaceae

Origins:
For the British naturalist George Wailes, c. 1802-1882, cultivated and studied the orchids; see Ray Desmond, *Dictionary of British & Irish Botanists and Horticulturists*. 709. London 1994.

Waitzia J.C. Wendland Asteraceae

Origins:
The name *Waitzia* commemorates Karl Friedrich Waitz, 1774-1848, privy councillor of the Duchy of Saxe-Altenburg. His writings include *Beschreibung der Gattung und*

Arten der Heiden nebst einer Anweisung zur zweckmässigen Kultur derselben. Ein Handbuch für Botaniker, Gärtner und Gartenfreunde. Leipzig und Altenburg 1809 and *Romanzen und Balladen der Deutschen*. Altenburg 1799-1800; see Johann Christoph Wendland (1755-1828), *Collectio plantarum*. 2: 13, t. 42. 1808; Helmut Genaust, *Etymologisches Wörterbuch der botanischen Pflanzennamen*. 691. Basel 1996; G.C. Wittstein, *Etymologisch-botanisches Handwörterbuch*. 929. 1852; F. Boerner & G. Kunkel, *Taschenwörterbuch der botanischen Pflanzennamen*. 4. Aufl. 186. 1989; A.D. Chapman, ed., *Australian Plant Name Index*. 2980-2981. Canberra 1991; James A. Baines, *Australian Plant Genera. An Etymological Dictionary of Australian Plant Genera*. 394-395. N.S.W. 1981; F.A. Sharr, *Western Australian Plant Names and Their Meanings*. 75. University of Western Australia Press 1996; Paul G. Wilson, "The classification of the genus *Waitzia* Wendl. (Asteraceae: Gnaphalieae)." *Nuytsia*. 8(3): 461-477. September 1992.

Species/Vernacular Names:
W. acuminata Steetz

English: orange immortelle

W. citrina (Benth.) Steetz (*Leptorhynchos citrinus* Benth.)

English: pale immortelle

Walafrida E. Meyer Scrophulariaceae

Origins:
After the German monk Walahfrid (Walafridus) Strabo, c. 809-849, poet and politician, theologian, abbot of Reichenau in Baden, botanist, author of *Hortulus*. Wien 1510, for the first English translation see *Hortulus or The Little Garden*. Wembley Hill, Middlesex 1924; see also Werner Näf and Mathäus Gabathuler, *Walahfrid Strabo Hortulus vom Gartenbau*. St. Gallen 1942, reprint of the Latin editions of 1510 and 1512 with German translation; see Edward Lee Greene, *Landmarks of Botanical History*. Edited by Frank N. Egerton. 445-446. Stanford, California 1983.

Walcottia F. Muell. Verbenaceae

Origins:
For Pemberton Walcott (d. 1883), plant collector in Australia in 1861; see Sir Ferdinand J.H. von Mueller (1825-1896), *Fragmenta phytographiae Australiae*. 1: 241. Melbourne 1859; Francis Aubie Sharr, *Western Australian Plant Names and Their Meanings*. 218. University of Western Australia Press, Nedlands, Western Australia 1996; Ray Desmond, *Dictionary of British & Irish Botanists and Horticulturists*. 710.

London 1994; A.D. Chapman, ed., *Australian Plant Name Index*. 2981. Canberra 1991.

Waldheimia Karelin & Kirilov Asteraceae

Origins:

After the German paleobiologist (Johann) Gotthelf (Friedrich) Fischer von Waldheim, 1771-1853, entomologist, in 1805 founder of the Société Impériale des Naturalistes de Moscou. His works include *Spicilegium entomographiae Rossicae*. [Moscow 1844], *Oryctographie du Gouvernement de Moscou*. Moscou 1837, *Entomographia Imperii Russici*. Mosquae 1820-1828 and *Bibliographia Palaeonthologica animalium systematica*. Mosquae 1834; see J.H. Barnhart, *Biographical Notes upon Botanists*. 1: 544. 1965; T. W. Bossert, *Biographical Dictionary of Botanists Represented in the Hunt Institute Portrait Collection*. 126. Boston, Massachusetts 1972; Günther Schmid, *Chamisso als Naturforscher*. Eine Bibliographie. Leipzig 1942; F.A. Stafleu and Richard S. Cowan, *Taxonomic Literature*. 1: 840. 1976.

Waldsteinia Willdenow Rosaceae

Origins:

Named for the Austrian botanist Franz de Paula Adam von Waldstein-Wartemburg, 1759-1823, soldier in the Austrian Army, a friend of Paul (Pál) Kitaibel (1757-1817); see Francisci comitis Waldstein, ... et Pauli Kitaibel ... *Descriptiones et icones plantarum rariorum Hungariae*. Viennae [1799] 1802-1812; John H. Barnhart, *Biographical Notes upon Botanists*. 3: 451. 1965; Antoine Lasègue, *Musée botanique de M. Benjamin Delessert*. Paris 1845; E.M. Tucker, *Catalogue of the Library of the Arnold Arboretum of Harvard University*. Cambridge, Massachusetts 1917-1933; R. Zander, F. Encke, G. Buchheim and S. Seybold, *Handwörterbuch der Pflanzennamen*. 14. Aufl. 796. Stuttgart 1993.

Walidda (A. DC.) Pichon Apocynaceae

Origins:

An Arabic name.

Walkeria A. Chev. Sapotaceae

Origins:

To commemorate A. Walker, clergyman, plant collector in Africa.

Wallacea Spruce ex Hooker Ochnaceae

Origins:

To honor the British (b. Usk, Monmouthshire, Wales) botanist Alfred Russel Wallace, 1823-1913 (d. Broadstone, Dorset, England), naturalist, biologist, explorer and zoologist, co-discoverer of the theory of evolution, plant collector with Henry Walter Bates (1825-1892) in South America and in the Malay Archipelago (Indonesia and Malaysia), travelled with Bates on the Amazon 1848-1850, 1872 Fellow of the Linnean Society, 1893 elected to the Royal Society. Among his most valuable writings are *The Malay Archipelago*. New York 1869, *Palm Trees of the Amazon and Their Uses*. London 1853, *My Life: A Record of Events and Opinions*. London 1905, "On the River Negro." *Journal of the Royal Geographical Society*. XXIII 212-217. 1853 and *Darwinism: An Exposition of the Theory of Selection with Some of Its Applications*. London and New York 1889; see H.W. Bates, *The Naturalist on the River Amazons*. A record of adventures, habits of animals, sketches of Brazilian and Indian life, and aspects of nature under the Equator, during eleven years of travel. London 1863; John H. Barnhart, *Biographical Notes upon Botanists*. 3: 453. 1965; Joseph Ewan, *Rocky Mountain Naturalists*. The University of Denver Press 1950; B. Glass et al., eds., *Forerunners of Darwin: 1745-1859*. Baltimore 1959; H. Lewis McKinney, in *Dictionary of Scientific Biography* (Editor in Chief Charles Coulston Gillispie.) 14: 133-140. New York 1981; F.A. Stafleu and R.S. Cowan, *Taxonomic Literature*. 7: 33-36. 1988; S. Lenley et al., *Catalog of the Manuscript and Archival Collections and Index to the Correspondence of John Torrey*. Library of the New York Botanical Garden. 426. 1973; E.M. Tucker, *Catalogue of the Library of the Arnold Arboretum of Harvard University*. Cambridge, Massachusetts 1917-1933; Mea Allan, *The Hookers of Kew*. London 1967; Andrew Thomas Gage, *A History of the Linnean Society of London*. London 1938; Alexander B. Adams, *Eternal Quest. The Story of the Great Naturalists*. New York 1969; [Don Conner Fine Books & Jeff Weber Rare Books], Catalogue 41: *Charles Darwin and His Circle* including: Alfred Russel Wallace, Thomas Huxley and Charles Lyell. Mostly from the Library of Eric T. Pengelley. Sacramento and Glendale, California 1996; R. Zander, F. Encke, G. Buchheim and S. Seybold, *Handwörterbuch der Pflanzennamen*. 14. Aufl. Stuttgart 1993; Leonard Huxley, *Life and Letters of Sir Joseph Dalton Hooker*. London 1918; [Maggs Bros Ltd], Catalogue 1260: *Huxley*. London 1998.

Wallaceodendron Koorders Mimosaceae

Origins:

Named after the British botanist Alfred Russel Wallace (1823-1913), explorer and zoologist, plant collector with

H.W. Bates in South America and in the Malay Archipelago. Among his most valuable writings are *A Narrative of Travels on the Amazon and Rio Negro*. London 1853, *Tropical Nature*. London 1878 and *Island Life*. London 1880; see J.H. Barnhart, *Biographical Notes upon Botanists*. 3: 453. 1965.

Wallenia Swartz Myrsinaceae

Origins:

After the Irish botanist Matthew Wallen, resident in Jamaica, with P. Browne; see Ray Desmond, *Dictionary of British & Irish Botanists and Horticulturists*. 713. London 1994.

Walleniella P. Wilson Myrsinaceae

Origins:

The diminutive of *Wallenia*.

Walleria J. Kirk Tecophilaeaceae (Liliaceae)

Origins:

For the British (b. London) botanist Rev. Horace Waller, 1833-1896 (d. Hants.), missionary in Central Africa, plant collector (Mozambique); see Ray Desmond, *Dictionary of British & Irish Botanists and Horticulturists*. 713. 1994.

Wallichia Reinw. ex Blume Rubiaceae

Origins:

For the Danish physician and botanist Nathaniel Wallich (originally Nathan Wulff or Wolff), 1786-1854, botanical collector.

Wallichia Roxburgh Palmae

Origins:

For the Danish (b. Copenhagen) physician Nathaniel Wallich (originally Nathan Wulff or Wolff), 1786-1854 (d. London), botanist and botanical collector (India, Malaya, Cape, Nepal), pupil of Vahl at Copenhagen, in 1807 went out to India as a surgeon, 1809 with William Roxburgh (1751-1815) at Calcutta, 1813 with the Hon. East India Company, 1815-1846 Superintendent of the Calcutta Botanic Garden (like his predecessor W. Roxburgh), 1818 Fellow of the Linnean Society, 1820-1822 plant collector in Nepal, 1829 Fellow of the Royal Society, 1833 visited Assam, correspondent of the naturalist and plant collector John Reeves (1774-1856). Among his most valuable writings are *Tentamen Florae Nepalensis*. Calcutta and Serampore 1824-1826, *Descriptions of Some Rare Indian Plants*. [Asiatic

Researches 1820] 1820 and *Plantae Asiaticae rariores*. London [1829-] 1830-1832. Father of George Charles Wallich (b. Calcutta 1815-d. Marylebone, London 1899); see J.H. Barnhart, *Biographical Notes upon Botanists*. 3: 454. 1965; I.H. Vegter, *Index Herbariorum*. Part II (7), *Collectors T-Z*. Regnum Vegetabile vol. 117. 1988; R. Zander, F. Encke, G. Buchheim and S. Seybold, *Handwörterbuch der Pflanzennamen*. 14. Aufl. 796. 1993; C.F.A. Christensen, *Den danske Botaniks Historie med tilhørende Bibliografi*. Copenhagen 1924-1926; I.H. Vegter, *Index Herbariorum*. Part II (5), *Collectors N-R*. Regnum Vegetabile vol. 109. 1983; Mary Gunn and Leslie E. Codd, *Botanical Exploration of Southern Africa*. 369-370. 1981; K. Biswas, ed., *The Original Correspondence of Sir Joseph Banks Relating to the Foundation of the Royal Botanic Garden, Calcutta* and *The Summary of the 150th Anniversary Volume of the Royal Botanic Garden, Calcutta*. Calcutta 1950; Isaac Henry Burkill, *Chapters on the History of Botany in India*. Delhi 1965; Andrew Thomas Gage, *A History of the Linnean Society of London*. London 1938; K. Lemmon, *Golden Age of Plant Hunters*. London. 1968. D.G. Crawford, *A History of the Indian Medical Service, 1600-1913*. London 1914; E.M. Tucker, *Catalogue of the Library of the Arnold Arboretum of Harvard University*. Cambridge, Massachusetts 1917-1933; Antoine Lasègue, *Musée botanique de M. Benjamin Delessert*. 1845; E. Bretschneider, *History of European Botanical Discoveries in China*. Leipzig 1981; R. Desmond, *The European Discovery of the Indian Flora*. Oxford 1992; T. W. Bossert, *Biographical Dictionary of Botanists Represented in the Hunt Institute Portrait Collection*. 425. Boston, Massachusetts 1972; Daniel Merriman, in *Dictionary of Scientific Biography* 14: 145-146. 1981; Leonard Huxley, *Life and Letters of Sir Joseph Dalton Hooker*. London 1918; J.D. Milner, *Catalogue of Portraits of Botanists Exhibited in the Museums of the Royal Botanic Gardens*. Royal Botanic Gardens, Kew, London 1906; [Sir J.E. Smith], *Memoir and Correspondence of ... Sir J.E. Smith ... Edited by Lady Pleasance Smith*. London 1832; M. Archer, *Natural History Drawings in the India Office Library*. London 1962; James Britten and George E. Simonds Boulger, *A Biographical Index of Deceased British and Irish Botanists*. London 1931.

Species/Vernacular Names:

W. disticha T. Anderson

Brazil: palmeira rabo de peixe

Wallrothia Spreng. Umbelliferae

Origins:

For the German botanist Carl Friedrich Wilhelm Wallroth, 1792-1856, physician; see J.H. Barnhart, *Biographical Notes upon Botanists*. 3: 455. 1965; E.M. Tucker, *Catalogue of the Library of the Arnold Arboretum of Harvard University*.

Cambridge, Massachusetts 1917-1933; R. Zander, F. Encke, G. Buchheim and S. Seybold, *Handwörterbuch der Pflanzennamen*. 14. Aufl. 1993.

Walpersia Harvey Fabaceae

Origins:

For the German botanist Wilhelm (Guilelmo, Guilielmo) Gerhard (Gerardo) Walpers, 1816-1853 (death by suicide), plant collector. His works include *Animadversiones criticae in Leguminosas Capenses herbarii Regii Berolinensis*. Halis [Halle a.S.] 1839 and *Annales botanices systematicae*. Lipsiae [Leipzig] 1848-1871; see J.H. Barnhart, *Biographical Notes upon Botanists*. 3: 455. 1965; M. Azzarello Di Misa, a cura di, *Il Fondo Antico della Biblioteca dell'Orto Botanico di Palermo*. 285-287. Palermo 1988; Antoine Lasègue, *Musée botanique de M. Benjamin Delessert*. 1845; R. Zander, F. Encke, G. Buchheim and S. Seybold, *Handwörterbuch der Pflanzennamen*. 14. Aufl. 1993; E.M. Tucker, *Catalogue of the Library of the Arnold Arboretum of Harvard University*. Cambridge, Massachusetts 1917-1933.

Walsura Roxb. Meliaceae

Origins:

A Telugu name for *Walsura trifolia* (A. Juss.) Harms.

Waltheria L. Sterculiaceae

Origins:

After the German botanist Augustin Friedrich Walther, 1688-1746, physician, professor of pathology, author of *Designatio plantarum quas Hortus A.F. Waltheri ... complectitur. Accedunt novae Plantarum icones*. Lipsiae [Leipzig] 1735 and *Plantarum exoticarum indigenarumque index tripartitus*. Lipsiae 1732; see John Hendley Barnhart, *Biographical Notes upon Botanists*. 3: 456. 1965; Carl Linnaeus, *Species Plantarum*. 673. 1753 and *Genera Plantarum*. Ed. 5. 304. 1754; Mariella Azzarello Di Misa, a cura di, *Il Fondo Antico della Biblioteca dell'Orto Botanico di Palermo*. 287. Regione Siciliana, Palermo 1988; T. W. Bossert, *Biographical Dictionary of Botanists Represented in the Hunt Institute Portrait Collection*. 425. Boston, Massachusetts 1972; Jonas C. Dryander, *Catalogus bibliothecae historico-naturalis Josephi Banks*. London 1796-1800; E.M. Tucker, *Catalogue of the Library of the Arnold Arboretum of Harvard University*. Cambridge, Massachusetts 1917-1933.

Species/Vernacular Names:

W. indica L. (*Waltheria americana* L.; *Waltheria americana* L. var. *indica* (L.) K. Schumann; *Waltheria americana* L. var. *subspicata* (L.) K. Schumann; *Waltheria pyrolaefolia* A. Gray)

Southern Africa: meidebossie; isikolanjiba (Ndebele); lexutasela (Sotho); seretwane (Tswana); simbongana (Shangaan)

Yoruba: erokosunkasi, ekuru, olorun kunmi lefun, wara wara odan, opa emere, korikodi, opa abiku, agamago

Hawaii: 'uhaloa, 'ala'ala pu loa, hala 'uhaloa, hi'aloa, kanakaloa

Waluewa Regel Orchidaceae

Origins:

For Graf P.A. Walujew (Valuev), Minister of Crown Lands, patron of Regel.

Wangenheimia Dietrich Araliaceae

Origins:

For the German forester Friedrich Adam Julius von Wangenheim, 1749-1800, soldier; see John Hendley Barnhart, *Biographical Notes upon Botanists*. 3: 457. 1965; T. W. Bossert, *Biographical Dictionary of Botanists Represented in the Hunt Institute Portrait Collection*. 425. Boston, Massachusetts 1972; William Darlington (1782-1863), *Reliquiae Baldwinianae*. Philadelphia 1843; R. Zander, F. Encke, G. Buchheim and S. Seybold, *Handwörterbuch der Pflanzennamen*. 14. Aufl. Stuttgart 1993; E.M. Tucker, *Catalogue of the Library of the Arnold Arboretum of Harvard University*. Cambridge, Massachusetts 1917-1933; Jonas C. Dryander, *Catalogus bibliothecae historico-naturalis Josephi Banks*. London 1796-1800; D.H. Nicolson, "Derivation of aroid generic names." *Aroideana*. 10: 15-25. 1988.

Wangenheimia Moench Gramineae

Origins:

For the German forester Friedrich Adam Julius von Wangenheim, 1749-1800, soldier; see John Hendley Barnhart, *Biographical Notes upon Botanists*. 3: 457. 1965.

Wangerinia Franz Caryophyllaceae

Origins:

For the German botanist Walther Leonhard Wangerin, 1884-1938, naturalist, professor of botany, assistant to A. Engler; see John Hendley Barnhart, *Biographical Notes upon Botanists*. 3: 457. 1965; T. W. Bossert, *Biographical Dictionary of Botanists Represented in the Hunt Institute Portrait Collection.* 425. Boston, Massachusetts 1972; E.M. Tucker, *Catalogue of the Library of the Arnold Arboretum of Harvard University.* Cambridge, Massachusetts 1917-1933; R. Zander, F. Encke, G. Buchheim and S. Seybold, *Handwörterbuch der Pflanzennamen.* 14. Aufl. Stuttgart 1993; Werner Burau, in *Dictionary of Scientific Biography* 14: 158-159. [about the German mathematician Albert Wangerin, 1844-1933, d. Halle] 1981.

Warburgia Engler Canellaceae

Origins:

Dedicated to the German botanist and traveler Otto Warburg, 1859-1938. His works include *Die Pflanzenwelt.* Leipzig und Wien 1913-1922, *Die Muskatnuss.* Leipzig 1897 and *Monographie der Myristicaceen.* Halle 1897, editor and publisher of many composite works and author of *Plantae Hellwigianae.* Flora von Kaiser Wilhelms-Land ... Leipzig 1894 (collector Franz Carl Hellwig, 1861-1889) and *Die Kulturpflanzen Usambaras* ... Berlin 1894; see John Hendley Barnhart, *Biographical Notes upon Botanists.* 3: 458. 1965; T. W. Bossert, *Biographical Dictionary of Botanists Represented in the Hunt Institute Portrait Collection.* 426. Boston, Massachusetts 1972; E.M. Tucker, *Catalogue of the Library of the Arnold Arboretum of Harvard University.* Cambridge, Massachusetts 1917-1933; R. Zander, F. Encke, G. Buchheim and S. Seybold, *Handwörterbuch der Pflanzennamen.* 14. Aufl. 797. Stuttgart 1993.

Species/Vernacular Names:

W. salutaris (Bertol.f.) Chiov. (*Warburgia breyeri* Pott; *Warburgia ugandensis* Sprague; *Chibaca salutaris* Bertol.f.)

English: fever tree, pepper-bark tree, pepper root

Southern Africa: peperbasboom, sterkbos, koorsboom; isiBaha, isiBhaha, amaZwecehlabayo (Zulu); mulanga (Venda)

Zanzibar: karambaki (only the bark)

Warburgina Eig Rubiaceae

Origins:

For the German botanist Otto Warburg, 1859-1938; see J.H. Barnhart, *Biographical Notes upon Botanists.* 3: 458. 1965.

Warburtonia F. Muell. Dilleniaceae

Origins:

For the Australian (British-born) soldier Major Peter Egerton Warburton, 1813-1889 (d. Adelaide, South Australia),

explorer, plant collector in Western and South Australia, Commissioner of Police, 1872 led expedition from Alice Springs (to Roebourne, W.A.), sent specimens to F. Mueller, wrote *Journey across the Western Interior of Australia.* Edited by H.W. Bates, Esq. Assistant Secretary of the Royal Geographical Society. London 1875; see Sir Ferdinand J.H. von Mueller, *Fragmenta phytographiae Australiae.* 1: 229. Melbourne 1859; H. Lewis McKinney, in *Dictionary of Scientific Biography* 1: 502. 1981; Ray Desmond, *Dictionary of British & Irish Botanists and Horticulturists.* 716. 1994; Jonathan Wantrup, *Australian Rare Books, 1788-1900.* Sydney 1987.

Wardaster Small Asteraceae

Origins:

For the British (b. Manchester) plant collector Francis (Frank) Kingdon Ward, 1885-1958 (d. London), 1907 Shanghai, 1909-1956 collected in China, Burma, Tibet and Thailand, 1920 Fellow of the Linnean Society, son of Harry Marshall Ward (1854-1906, d. Devon); see T. W. Bossert, *Biographical Dictionary of Botanists Represented in the Hunt Institute Portrait Collection.* 212. 1972; M. Hadfield et al., *British Gardeners: A Biographical Dictionary.* London 1980; R. Desmond, *The European Discovery of the Indian Flora.* Oxford 1992; Charles Lyte, *The Plant Hunters.* 151-165. London 1983; E.H.M. Cox, *Plant-Hunting in China. A History of Botanical Exploration in China and the Tibetan Marches.* 180-191. 1945; S. Lenley et al., *Catalog of the Manuscript and Archival Collections and Index to the Correspondence of John Torrey.* 251. 1973; Alice Margaret Coats, *The Quest for Plants. A History of the Horticultural Explorers.* 128-131. 1969; H.R. Fletcher and W.H. Brown, *Royal Botanic Garden Edinburgh, 1670-1970.* 245. Edinburgh 1970; Richard Eyde and Keith Ferguson, "The Little Dogwood of Frank Kingdon Ward." *Curtis's Botanical Magazine.* 6(2): 74-83. May 1989; H.R. Fletcher, *Story of the Royal Horticultural Society, 1804-1968.* Oxford 1969; R. Zander, F. Encke, G. Buchheim and S. Seybold, *Handwörterbuch der Pflanzennamen.* 14. Aufl. Stuttgart 1993; J.H. Barnhart, *Biographical Notes upon Botanists.* 3: 458. 1965.

Warionia Bentham & Cosson Asteraceae

Origins:

For the French physician Jean Pierre Adrien Warion, 1837-1880, botanist, traveler (Italy, Algeria, Morocco, North Africa), with Jean Louis Kralik (1813-1892), Aristide-Horace Letourneux (1820-1890) and Victor Constant Reboud (1821-1889) contributed to Ernest Saint-Charles Cosson (1819-1889), *Répertoire alphabétique des principales localités mentionnées dans le Compendium et le Conspectus florae atlanticae.* Paris 1882, wrote *Herborisations dans les Pyrénées-orientales en 1878 et 1879.* Perpignan 1880; see J.H. Barnhart, *Biographical Notes upon Botanists.* 3: 460. 1965; E. Walter, *Le docteur Warion,* botaniste lorrain et algérien (1837-1880). Metz 1935; E.M. Tucker, *Catalogue of the Library of the Arnold Arboretum of Harvard University.* Cambridge, Massachusetts 1917-1933.

Warmingia Reichb.f. Orchidaceae

Origins:

For the Danish (b. Mandø) botanist Johannes Eugenius Bülow Warming, 1841-1924 (Copenhagen), the founder of plant ecology, plant geographer, taxonomist, morphologist, traveler, explorer and plant collector, student of the vegetation of tropical South America (Brazil, West Indies, Venezuela) and of the Arctic flora (Greenland and Norway), 1863-1866 in Lagoa Santa (Minas Gerais, Brazil), 1885-1911 Director of the Botanical Garden at Copenhagen, professor of botany at Stockholm and Copenhagen. Among his numerous publications are *Die Blüthe der Compositen.* Bonn 1876, *Lagoa Santa.* Copenhagen 1892, *Plantesamfund.* Copenhagen 1895 [English transl., *Oecology of Plants. An Introduction to the Study of Plant-Communities.* Oxford 1909] and *Om Caryophyllaceernes Blomster.* Copenhagen 1890, secretary to the Danish zoologist P.W. Lund; see J.H. Barnhart, *Biographical Notes upon Botanists.* 3: 460. 1965; D. Müller, in *Dictionary of Scientific Biography* 14: 181-182. 1981; F.A. Stafleu and R.S. Cowan, *Taxonomic Literature.* 7: 71-81. 1988; C.F.A. Christensen, *Den danske Botaniks Historie med tilhørende Bibliografi.* Copenhagen 1924-1926; T. W. Bossert, *Biographical Dictionary of Botanists Represented in the Hunt Institute Portrait Collection.* 426. Boston, Massachusetts 1972; S. Lenley et al., *Catalog of the Manuscript and Archival Collections and Index to the Correspondence of John Torrey.* Library of the New York Botanical Garden. 427. 1973; E.M. Tucker, *Catalogue of the Library of the Arnold Arboretum of Harvard University.* Cambridge, Massachusetts 1917-1933; C.J.É. Morren, *Correspondance botanique.* Liège 1884.

Warneckea Gilg Melastomataceae

Origins:

Named for the German Otto Warnecke, c. 1899 plant collector and gardener in Togo, 1903 to East Africa; see Frank Nigel Hepper (1929-), "Botanical collectors in West Africa, except French territories, since 1860." in *Comptes Rendus de l'Association pour l'étude taxonomique de la flore*

d'Afrique, (A.E.T.F.A.T.). 69-75. Lisbon 1962; F.N. Hepper and Fiona Neate, *Plant Collectors in West Africa.* 84. 1971.

Warrea Lindley Orchidaceae

Origins:

After the English plant collector Frederick Warre, fl. 1820s, collected orchids in Brazil, sent plants to Lindley.

Warreella Schltr. Orchidaceae

Origins:

The diminutive of *Warrea.*

Warreopsis Garay Orchidaceae

Origins:

Resembling *Warrea.*

Warscewiczella Reichb.f. Orchidaceae

Origins:

For Joseph (Józef) von Rawicz Warszewicz (Warscewicz), 1812-1866.

Warszewiczia Klotzsch Rubiaceae

Origins:

For Joseph (Józef) von Rawicz Warszewicz (Josef von Warscewicz), 1812-1866, gardener, botanist, plant collector (Guatemala, St. Thomas, Central and South America); see J.H. Barnhart, *Biographical Notes upon Botanists.* 3: 461. 1965; Merle A. Reinikka, *A History of the Orchid.* Timber Press 1996; R. Zander, F. Encke, G. Buchheim and S. Seybold, *Handwörterbuch der Pflanzennamen.* 14. Aufl. 797. Stuttgart 1993; T. W. Bossert, *Biographical Dictionary of Botanists Represented in the Hunt Institute Portrait Collection.* 426. Boston, Massachusetts 1972; Irving William Knobloch, compil., "A preliminary verified list of plant collectors in Mexico." *Phytologia Memoirs.* VI. 1983.

Wartmannia Müll. Arg. Euphorbiaceae

Origins:

For the Swiss naturalist Friedrich Bernhard Wartmann, 1830-1902, a friend of Carl Wilhelm von Nägeli (1817-1891), professor of natural history, 1868-1902 President of the St. Gallen naturwissenschaftliche Gesellschaft, Director of the St. Gallen natural history museum. Among his publications are *Beiträge zur Anatomie und Entwicklungsgeschichte der Algengattung Lemanea*. St. Gallen 1854, *Beiträge zur St. Gallischen Volksbotanik*. St. Gallen 1861 and *Zweite, stark vermehrte und total umgearbeitete Auflage*. (Separatabdruck aus den Verhandlungen der St. Gallischen naturwissenschaftliche Gesellschaft.) St. Gallen 1874; see John H. Barnhart, *Biographical Notes upon Botanists*. 3: 461. 1965; Ethelyn Maria Tucker, *Catalogue of the Library of the Arnold Arboretum of Harvard University*. Cambridge, Massachusetts 1917-1933; R. Zander, F. Encke, G. Buchheim and S. Seybold, *Handwörterbuch der Pflanzennamen*. 14. Aufl. Stuttgart 1993.

Wasabia Matsum. Brassicaceae

Origins:
A vernacular name for one species, *Wasabia wasabi* (Siebold) Makino.

Species/Vernacular Names:
W. wasabi (Siebold) Makino

English: Japanese horseradish, wasabi

Japan: wasabi

Washingtonia H.A. Wendland Palmae

Origins:
For George Washington, 1732-1799, first President of the United States; see A.S.W. Rosenbach, *1776 Americana. A Catalogue of Autograph Letters and Documents Relating to the Declaration of Independence and the Revolutionary War*. Philadelphia 1926 and *The History of America in Documents*. Philadelphia and New York 1949-1950.

Species/Vernacular Names:
W. filifera (André) H.A. Wendl.

English: fountain palm, thread palm, Californian Washington palm, cotton palm, desert fan palm, Mexican fan palm, California fan palm, northern washingtonia, Californian washingtonia, cabbage trees

Brazil: washingtonia de saia, palmeira de saia da Califórnia

Japan: Washington-yashi

W. robusta H.A. Wendland

English: southern washingtonia, Mexican washingtonia

Brazil: washingtonia do sul, palmeira de saia, palmeira de leque do México

Washingtonia Raf. Umbelliferae

Origins:
See Constantine Samuel Rafinesque (1783-1840), *Am. Monthly Mag. Crit. Rev.* 2: 176. 1818; E.D. Merrill (1876-1956), *Index rafinesquianus*. The plant names published by C.S. Rafinesque, etc. 184. Jamaica Plain, Massachusetts, USA 1949.

Waterhousia B. Hyland Myrtaceae

Origins:
See Arthur D. Chapman, ed., *Australian Plant Name Index*. 2982. Canberra 1991; Bernard Patrick Matthew Hyland, in *Australian Journal of Botany*. Supplementary Series No. 9: 138. 1983.

Watsonamra Kuntze Rubiaceae

Origins:
Dedicated to the American (b. Connecticut) botanist Sereno Watson, 1826-1892 (Cambridge, Massachusetts), attended Yale College, traveler, plant collector, assisted William Whitman Bailey in botanical collections, in March 1868 official botanist of Clarence King's Geological Exploration of the fortieth parallel, Assistant with Asa Gray at the Gray Herbarium and later Curator, in 1889 elected to the National Academy of Sciences. His most significant works are *Botany of California*. Cambridge, Massachusetts 1876, 1880 and *Bibliographical Index to North American Botany ... Part I, Polypetalae*. Washington, D.C. 1878; see *United States Geological Expolration* [sic] of the fortieth parallel. Botany. by Sereno Watson. Washington 1871 and *List of Plants Collected in Nevada and Utah 1867-1869*. Sereno Watson, collector. [Washington, D.C. 1871]; Elizabeth Noble Shor, in *Dictionary of Scientific Biography* 14: 192-193. 1981; John H. Barnhart, *Biographical Notes upon Botanists*. 3: 465. 1965; William H. Brewer, *Biographical Memoirs*. National Academy of Sciences. 5: 267-290. 1903; T. W. Bossert, *Biographical Dictionary of Botanists Represented in the Hunt Institute Portrait Collection*. 427. Boston, Massachusetts 1972; Ida Kaplan Langman, *A Selected Guide to the Literature on the Flowering Plants of Mexico*. 1964; R. Zander, F. Encke, G. Buchheim and S. Seybold, *Handwörterbuch der Pflanzennamen*. 14. Aufl. Stuttgart 1993; S. Lenley et al., *Catalog of the Manuscript and Archival Collections and Index to the Correspondence of John Torrey*. Library of the New York Botanical Garden. 1973; Ethelyn Maria Tucker, *Catalogue of the Library of the Arnold Arboretum of Harvard University*. Cambridge, Massachusetts 1917-1933.

Watsonia Miller Iridaceae

Origins:

After the English (b. London) physician Sir William Watson, 1715-1787 (London), apothecary, naturalist and botanist, physicist, scientist, Fellow of the Royal Society 1741, an adherent of Franklin's theory, knighted 1786, active in the Royal Society. Among his writings are *Experiments and Observations Tending to Illustrate the Nature and Properties of Electricity*. London 1745 and *Observations upon the Effects of Lightning*, with an account of the apparatus proposed to prevent its mischiefs to buildings ... being answers to certain questions proposed by Mr. Calandrini. London 1764; see Richard Pulteney (1730-1801), *Geschichte der Botanik bis auf die neueren Zeiten mit besonderer Rücksicht auf England*. II: 479-512. Leipzig 1798; R. Pulteney, *Historical and Biographical Sketches of the Progress of Botany in England*. 2: 295-340. London 1790; J.H. Barnhart, *Biographical Notes upon Botanists*. 3: 465. 1965; William Munk, *The Roll of the Royal College of Physicians of London*. 348-349. London 1878; J.L. Heilbron, in *Dictionary of Scientific Biography* 14: 193-196. 1981; T. W. Bossert, *Biographical Dictionary of Botanists Represented in the Hunt Institute Portrait Collection*. 428. Boston, Massachusetts 1972; Jonas C. Dryander, *Catalogus bibliothecae historico-naturalis Josephi Banks*. London 1796-1800; Ethelyn Maria Tucker, *Catalogue of the Library of the Arnold Arboretum of Harvard University*. Cambridge, Massachusetts 1917-1933.

Species/Vernacular Names:

W. aletroides (Burm.f.) Ker Gawler

South Africa: kanolpypie, pypie, rooikanolpypie, rooi Afrikaner

W. angusta Ker Gawler

South Africa: rooikanolpypie, patryskos

W. borbonica (Pourr.) Goldbl. (*Watsonia ardernei* Sander)

English: bugle lily

Portuguese: hastes de S. José

W. borbonica (Pourr.) Goldbl. subsp. *borbonica* (*Watsonia cooperi* (Bak.) L. Bol.; *Watsonia pyramidata* (Andr.) Stapf; *Watsonia ardernei* Sander) (the species was named after Gaston de Bourbon or Gaston d'Orléans, 1608-1660, a patron of botany, or to honor all the Bourbon kings of France; see Abel Brunyer, *Hortus Regius Blesensis*. Paris, Antoine Vitré 1653)

English: bugle lily

Portuguese: hastes de S. José

South Africa: suurkanol, witkanol

W. fourcadei J.W. Mathews & L. Bolus (*Watsonia ryderae* L. Bol.; *Watsonia stanfordiae* L. Bol.; *Watsonia stohrii* L. Bol.)

South Africa: suurkanol

W. hysterantha J.W. Mathews & L. Bol.

South Africa: rooipypie, suikerkan

W. x longifolia J.W. Mathews & L. Bol.

South Africa: pypie, watsonia

W. marginata (L.f.) Ker Gawl.

South Africa: kanolpypie, rooikanolpypie

W. meriana (L.) Miller (*Watsonia leipoldtii* L. Bol.; *Watsonia vivipara* J.W. Mathews & L. Bol.; *Antholyza meriana* L.) (the specific name after a German-Dutch painter of plants and insects, Maria Sybilla Merian afterwards Graff, 1647-1717, author of *Dissertatio de generatione et metamorphosibus insectorum Surinamensium*. The Hague 1726. See W. Blunt, *The Art of Botanical Illustration*. 127-29. London 1950; W. Blunt and S. Raphael, *The Illustrated Herbal*. London 1979; Bertus Aafjes, *Maria Sybilla Merian*. [First edition, 8vo.] Amsterdam 1952; Leonard de Vries, *De exotische kunst van Maria Sybilla Merian*. Amsterdam 1984; S. Sitwell and W. Blunt, *Great Flower Books 1700-1900: A Bibliographical Record of Two Centuries of Finely Illustrated Flower Books*. 5-6, 18. London 1956)

South Africa: rooikanol, rooikanolpypie, lakpypie, waspypie, knolpypie, suurknol, waspypie

W. pillansii L. Bolus (*Watsonia archbelliae* L. Bol.; *Watsonia beatricis* J.W. Mathews & L. Bol.; *Watsonia masoniae* L. Bol.; *Watsonia priorii* L. Bolus; *Watsonia socium* J.W. Mathews & L. Bolus) (after the South African (b. Rosebank, Cape Town) botanist Neville Stuart Pillans, 1884-1964 (d. Plumstead, Cape Town), from 1918 at the Bolus Herbarium. His writings include "The genus *Phylica* Linn." *J. S. Afr. Bot.* 8: 1-164. 1942 and "Destruction of indigenous vegetation by burning on the Cape Peninsula." *S. Afr. J. Sci.* 21: 348-350. 1924; see John H. Barnhart, *Biographical Notes upon Botanists*. 3: 86. 1965; T.W. Bossert, *Biographical Dictionary of Botanists Represented in the Hunt Institute Portrait Collection*. 311. 1972; Mary Gunn and Leslie Edward W. Codd, *Botanical Exploration of Southern Africa*. 281-282. Cape Town 1981; Gordon Douglas Rowley, *A History of Succulent Plants*. Mill Valley, California 1997; John Hutchinson (1884-1972), *A Botanist in Southern Africa*. 53-54. London 1946; A. White and B.L. Sloane, *The Stapelieae*. Pasadena 1937)

South Africa: pypie

Wattakaka Hassk. Asclepiadaceae

Origins:

A popular name, *watta kakadodi*.

Weatherbya Copeland Polypodiaceae

Origins:

After the American botanist Charles Alfred Weatherby, 1875-1949, pteridologist; see J.H. Barnhart, *Biographical Notes upon Botanists*. 3: 467. 1965; T. W. Bossert, *Biographical Dictionary of Botanists Represented in the Hunt Institute Portrait Collection*. 428. Boston, Massachusetts 1972; Ida Kaplan Langman, *A Selected Guide to the Literature on the Flowering Plants of Mexico*. Philadelphia 1964; Ethelyn Maria Tucker, *Catalogue of the Library of the Arnold Arboretum of Harvard University*. Cambridge, Massachusetts 1917-1933; R. Zander, F. Encke, G. Buchheim and S. Seybold, *Handwörterbuch der Pflanzennamen*. 14. Aufl. Stuttgart 1993; S. Lenley et al., *Catalog of the Manuscript and Archival Collections and Index to the Correspondence of John Torrey*. Library of the New York Botanical Garden. 429. 1973.

Webera Cramer Rubiaceae

Origins:

For the German physician Georg Heinrich Weber, 1752-1828, botanist.

Webera J.F. Gmelin Melastomataceae

Origins:

For the German physician Georg Heinrich Weber, 1752-1828, botanist, professor of medicine, father of the German botanist Friedrich Weber (1781-1823); see J.H. Barnhart, *Biographical Notes upon Botanists*. 3: 468. 1965; Jonas C. Dryander, *Catalogus bibliothecae historico-naturalis Josephi Banks*. London 1796-1800; R. Zander, F. Encke, G. Buchheim and S. Seybold, *Handwörterbuch der Pflanzennamen*. 14. Aufl. 1993.

Webera Schreber Rubiaceae

Origins:

For the German physician Georg Heinrich Weber, 1752-1828, botanist.

Weberbauera Gilg & Muschler Brassicaceae

Origins:

After the German botanist August Weberbauer, 1871-1948 (Lima), explorer, plant collector in Peru, Director of the Botanical Garden and Agricultural Station at Victoria, W.

Cameroon, professor of botany and pharmaceutical botany, Director of the Parque zoológico y botánico de Lima, Peru. His works include *Plantas tóxicas que sirven para la pesca en el Perú*. [Lima 1933], *Die Pflanzenwelt der peruanischen Andes in ihren Grundzügen dargestellt*. Leipzig 1911, "Die Vegetationsgliederung des nördlichen Peru um 5° südl. Br." *Bot. Jahrb. Syst*. 50(Suppl.): 72-94. 1914 and *Phytogeography of the Peruvian Andes*, in James Francis Macbride (1892-1976), *Flora of Peru*. pt. 1. 1936 [Field Museum of Natural History. Botanical Series. vol. 13.]; see John H. Barnhart, *Biographical Notes upon Botanists*. 3: 469. 1965; Réné Letouzey, "Les botanistes au Cameroun." in *Flore du Cameroun*. 7: 60. Paris 1968; T. W. Bossert, *Biographical Dictionary of Botanists Represented in the Hunt Institute Portrait Collection*. 429. Boston, Massachusetts 1972; Ethelyn Maria Tucker, *Catalogue of the Library of the Arnold Arboretum of Harvard University*. Cambridge, Massachusetts 1917-1933; F.N. Hepper and Fiona Neate, *Plant Collectors in West Africa*. 84. 1971; T. Harper Goodspeed, *Plant Hunters in the Andes*. University of California Press, Berkeley and Los Angeles 1961.

Weberbauerella Ulbr. Fabaceae

Origins:

Dedicated to the German botanist August Weberbauer, 1871-1948 (Lima); see John H. Barnhart, *Biographical Notes upon Botanists*. 3: 469. 1965.

Weberbauerocereus Backeb. Cactaceae

Origins:

For the German botanist August Weberbauer, 1871-1948 (Lima); see John H. Barnhart, *Biographical Notes upon Botanists*. 3: 469. 1965; Réné Letouzey, "Les botanistes au Cameroun." in *Flore du Cameroun*. 7: 60. Paris 1968.

Weberiopuntia Fric Cactaceae

Origins:

After the French military surgeon Frédéric Albert Constantin Weber, 1830-1903, botanist, specialist in Cactaceae.

Weberocereus Britton & Rose Cactaceae

Origins:

After the French military surgeon Frédéric Albert Constantin Weber, 1830-1903 (Paris), botanist, specialist in Cactaceae, 1864-1867 in Mexico; see Gordon Douglas Rowley, *A History of Succulent Plants*. Strawberry Press, Mill Valley, California 1997; R. Zander, F. Encke, G. Buchheim and S. Seybold, *Handwörterbuch der Pflanzennamen*. 14. Aufl.

798. Stuttgart 1993; John H. Barnhart, *Biographical Notes upon Botanists*. 3: 468. Boston 1965; Ethelyn Maria Tucker, *Catalogue of the Library of the Arnold Arboretum of Harvard University*. Cambridge, Massachusetts 1917-1933; Frans A. Stafleu and Richard S. Cowan, *Taxonomic Literature*. 7: 124-125. Utrecht 1988.

Websteria S.H. Wright Cyperaceae

Origins:

For the American botanist G.W. Webster, 1833-1914, farmer.

Weddellina Tulasne Podostemaceae (Tristichaceae)

Origins:

Named for the British (b. Glos.) botanist Hugh Algernon Weddell, 1819-1877 (Poitiers, France), physician, Dr. med. Paris 1841, traveler and plant collector (Brazil, Bolivia, Peru), 1859 Fellow of the Linnean Society of London, associated with A. Jussieu, 1843-1848 with François Louis Nompar de Caumat de Laporte Castelnau (1810-1880) to S. America. Among his main works are *Histoire naturelle des Quinquinas* ou monographie du genre *Cinchona*. Paris 1849, *Chloris andina*. Paris 1855-1857 [-1861], *Monographie de la famille des Urticacées*. Paris 1856 [-1857] and *Voyage dans le Nord de la Bolivie* et dans les parties voisines du Pérou ou visite du district aurifère de Tipuani. Paris, Londres 1853; see John H. Barnhart, *Biographical Notes upon Botanists*. 3: 470. Boston 1965; Ethelyn Maria Tucker, *Catalogue of the Library of the Arnold Arboretum of Harvard University*. Cambridge, Massachusetts 1917-1933; R. Desmond, *The European Discovery of the Indian Flora*. Oxford 1992; T. W. Bossert, *Biographical Dictionary of Botanists Represented in the Hunt Institute Portrait Collection*. 429. Boston, Massachusetts 1972; Ida Kaplan Langman, *A Selected Guide to the Literature on the Flowering Plants of Mexico*. Philadelphia 1964; Antoine Lasègue, *Musée botanique de M. Benjamin Delessert*. 1845; R. Zander, F. Encke, G. Buchheim and S. Seybold, *Handwörterbuch der Pflanzennamen*. 14. Aufl. Stuttgart 1993; August Weberbauer (1871-1948), *Die Pflanzenwelt der peruanischen Andes in ihren Grundzügen dargestellt*. Leipzig 1911.

Wedelia Jacquin Asteraceae

Origins:

For the German (b. Golssen) physician Georg Wolfgang Wedel (Georgius Wolfgangus Wedelius), 1645-1721 (Jena), botanist, a pupil of G. Rolfink, M.D. at Jena 1669, defender of astrology and alchemy, professor of medicine at Jena, physician at Gotha, author of *Opiologia*. Jenae 1674 and *Pharmacia in artis formam redacta, experimentis, observationibus et discursu perpetuo illustrata*. Jenae 1677; see Johann Heinrich Zedler, ed., *Grosses vollständiges Universal-Lexicon*. Graz 1964; Lynn Thorndike, *A History of Magic and Experimental Science*. New York 1958; F. Boerner & G. Kunkel, *Taschenwörterbuch der botanischen Pflanzennamen*. 4. Aufl. 186. 1989; Karl Hufbauer, in *Dictionary of Scientific Biography* (Editor in Chief Charles Coulston Gillispie.) 14: 212-213. New York 1981.

Species/Vernacular Names:

W. sp.

Peru: botoncillo

W. glauca (Ort.) S.F. Blake (*Pascalia glauca* Ortega)

English: pascalia weed, pascalia

Latin America: sunchillo

W. trilobata (L.) A. Hitchc.

English: Singapore daisy

Indonesia: serunai rambat

Balinese name: bungan padang lumut-lumut (lumut = moss)

Wedelia Loefl. Nyctaginaceae

Origins:

After the German physician Georg Wolfgang Wedel (Georgius Wolfgangus Wedelius), 1645-1721 (Jena).

Wedeliella Cockerell Nyctaginaceae

Origins:

For the German physician Georg Wolfgang Wedel (Georgius Wolfgangus Wedelius), 1645-1721 (Jena), botanist.

Wehlia F. Muell. Myrtaceae

Origins:

After Dr. Eduard Wehl, 1823-1876, physician, married Ferdinand Mueller's sister (Clara); see Ferdinand von Mueller, *Fragmenta phytographiae Australiae*. 10: 22. Melbourne 1876; James A. Baines, *Australian Plant Genera. An Etymological Dictionary of Australian Plant Genera*. Chipping Norton, N.S.W. 1981; Francis Aubie Sharr, *Western Australian Plant Names and Their Meanings*. University of Western Australia Press, Nedlands, Western Australia 1996.

Weigela Thunberg Caprifoliaceae

Origins:

After the German (b. Stralsund) botanist Christian Ehren-fried von Weigel, 1748-1831 (d. Greifswald), chemist, physician; see John H. Barnhart, *Biographical Notes upon Botanists*. 3: 471. 1965; Jonas C. Dryander, *Catalogus bibliothecae historico-naturalis Josephi Banks*. London 1796-1800; Karl Hufbauer, in *Dictionary of Scientific Biography* 14: 224-225. 1981; Ethelyn Maria Tucker, *Catalogue of the Library of the Arnold Arboretum of Harvard University*. Cambridge, Massachusetts 1917-1933; Mariella Azzarello Di Misa, a cura di, *Il Fondo Antico della Biblioteca dell'Orto Botanico di Palermo*. 288. Regione Siciliana, Palermo 1988; R. Zander, F. Encke, G. Buchheim and S. Seybold, *Handwörterbuch der Pflanzennamen*. 14. Aufl. 798. Stuttgart 1993; H. Genaust, *Etymologisches Wörterbuch der botanischen Pflanzennamen*. 692. [b. 1746] Basel 1996.

Species/Vernacular Names:

W. florida (Bunge) A. DC.

English: weigelia

W. japonica Thunberg

China: yang lu

Weigeltia A. DC. Myrsinaceae

Origins:

For the German botanist Christoph Weigelt, d. 1828, physician, plant collector (Suriname); see John H. Barnhart, *Biographical Notes upon Botanists*. 3: 471. 1965; Antoine Lasègue, *Musée botanique de M. Benjamin Delessert*. 1845.

Weihea Sprengel Rhizophoraceae

Origins:

For the German botanist Carl Ernst August Weihe, 1779-1834, physician, batologist; see J.H. Barnhart, *Biographical Notes upon Botanists*. 3: 471. 1965; Ethelyn Maria Tucker, *Catalogue of the Library of the Arnold Arboretum of Harvard University*. Cambridge, Massachusetts 1917-1933; R. Zander, F. Encke, G. Buchheim and S. Seybold, *Handwörterbuch der Pflanzennamen*. 14. Aufl. Stuttgart 1993.

Weingartia Werderm. Cactaceae

Origins:

After Wilhelm Weingart, 1856-1936, amateur botanist, expert on succulents and cacti; see Gordon Douglas Rowley, *A History of Succulent Plants*. Strawberry Press, Mill Valley, California 1997.

Weinmannia L. Cunoniaceae

Origins:

For the German apothecary Johann Wilhelm Weinmann, 1683-1741, botanist, author of *Phytanthoza iconographia*. Ratisbonae [Regensburg] [1734-] 1737; see C. Linnaeus, *Systema Naturae*. 2: 997, 1005, 1367. 1759; T. W. Bossert, *Biographical Dictionary of Botanists Represented in the Hunt Institute Portrait Collection*. 431. Boston, Massachusetts 1972; Gilbert Westacott Reynolds (1895-1967), *The Aloes of South Africa*. Balkema, Rotterdam 1982; Jonas C. Dryander, *Catalogus bibliothecae historico-naturalis Josephi Banks*. 3: 62. London 1796-1800; Ethelyn Maria Tucker, *Catalogue of the Library of the Arnold Arboretum of Harvard University*. Cambridge, Massachusetts 1917-1933; Mariella Azzarello Di Misa, a cura di, *Il Fondo Antico della Biblioteca dell'Orto Botanico di Palermo*. 288-289. Palermo 1988; Blanche Elizabeth Edith Henrey (1906-1983), *British Botanical and Horticultural Literature before 1800*. Oxford 1975.

Species/Vernacular Names:

W. racemosa L.f.

New Zealand: kamahi, tauhero, tawhero, towai (Maori names)

W. sylvicola Sol. ex A. Cunn.

New Zealand: tauhero, tawhero

W. trichosperma Cav.

Latin America: tineo, maden

Weldenia Schult.f. Commelinaceae

Origins:

For the German botanist Franz Ludwig Freiherr von Welden, 1782-1853, soldier, horticulturist; see J.H. Barnhart, *Biographical Notes upon Botanists*. 3: 475. 1965; Ethelyn Maria Tucker, *Catalogue of the Library of the Arnold Arboretum of Harvard University*. Cambridge, Massachusetts 1917-1933; H.N. Clokie, *Account of the Herbaria of the Department of Botany in the University of Oxford*. 262-263. Oxford 1964; Antoine Lasègue, *Musée botanique de M. Benjamin Delessert*. 1845.

Welwitschia Hook.f. Welwitschiaceae

Origins:

For the Austrian (b. near Klagenfurt) explorer Friedrich (Frederick) Martin Joseph Welwitsch (Welvich), 1806-1872 (d. London), M.D. Vienna 1836, zoologist, physician, botanist, traveler and plant collector (Angola, southwest Africa, Portugal), from 1839 to 1853 Director of the Botanical Garden of Lisbon, 1865 Fellow of the Linnean Society.

Among his works are "On the botany of Benguela, Mossamedes, etc., in Western Africa." *J. Linn. Soc. (Bot.)* 5:182-187. 1861 (Including the first description of *Welwitschia* under the name *Tumboa*) and *Apontamentos phytogeographicos sobre a flora da provincia de Angola na Africa equinocial*. Lisboa 1858 [1859], co-author (with Frederick Currey, 1819-1881) of *Fungi angolenses*. London 1868; see Ray Desmond, *Dictionary of British & Irish Botanists and Horticulturists*. 729. 1994; Joseph Vallot (1854-1925), "Études sur la flore du Sénégal." in *Bull. Soc. Bot. de France*. 29: 195-196. Paris 1882; R. Zander, F. Encke, G. Buchheim and S. Seybold, *Handwörterbuch der Pflanzennamen*. 14. Aufl. 798f. 1993; Frans A. Stafleu and Richard S. Cowan, *Taxonomic Literature*. 7: 174-178. 1988; William Philip Hiern (1839-1925), *Catalogue of the African Plants Collected by F. Welwitsch in 1853-61*. London 1896-1901; Helmut Dolezal (1922-1981), in *Portug. Acta Biol*. 6(3-4): 257-323. 1959 and 7(1-3): 324-551. 1961; F.N. Hepper and Fiona Neate, *Plant Collectors in West Africa*. 85. 1971; Gordon Douglas Rowley, *A History of Succulent Plants*. Mill Valley, California 1997; J.H. Barnhart, *Biographical Notes upon Botanists*. 3: 476. 1965; Anthonius Josephus Maria Leeuwenberg, "Isotypes of which holotypes were destroyed in Berlin." *Webbia*. 19: 861-863. 1965; M. Colmeiro y Penido, *La Botánica y los Botánicos de la Peninsula Hispano-Lusitana*. Madrid 1858; Antoine Lasègue, *Musée botanique de M. Benjamin Delessert*. 1845; T. W. Bossert, *Biographical Dictionary of Botanists Represented in the Hunt Institute Portrait Collection*. 431. Boston, Massachusetts 1972; Ethelyn Maria Tucker, *Catalogue of the Library of the Arnold Arboretum of Harvard University*. Cambridge, Massachusetts 1917-1933; H.N. Clokie, *Account of the Herbaria of the Department of Botany in the University of Oxford*. 263. Oxford 1964; Ernest Nelmes and William Cuthbertson, *Curtis's Botanical Magazine Dedications, 1827-1927*. 146. [1931]; Alain White and Boyd Lincoln Sloane, *The Stapelieae*. Pasadena 1937; Merle A. Reinikka, *A History of the Orchid*. Timber Press 1996.

Species/Vernacular Names:

W. mirabilis Hook.f. (*Welwitschia bainesii* (Hook.f.) Carr.; *Tumboa bainesii* Hook.f.; *Tumboa strobilifera* Hook.f.)

English: welwitschia

Namibia: tumboa (from the Angolan name N'Tumbo)

Welwitschia Reichb. Polemoniaceae

Origins:

After the Austrian explorer Friedrich (Frederick) Martin Joseph Welwitsch, 1806-1872 (London).

Welwitschiella O. Hoffmann Asteraceae

Origins:

Named after the Austrian explorer Friedrich (Frederick) Martin Joseph Welwitsch, 1806-1872 (London), doctor of medicine, botanist, 1836 Dr. med. Vienna, zoologist, in 1865 a Fellow of the Linnean Society of London, plant collector; see John H. Barnhart, *Biographical Notes upon Botanists*. 3: 476. 1965.

Wenderothia Schltdl. Fabaceae

Origins:

For the German botanist Georg Wilhelm Franz Wenderoth, 1774-1861, M.D. 1801 Marburg, professor of medicine, professor of botany and pharmacology, founded of the Marburg Botanical Garden. His writings include *Flora hassiaca*. Cassel 1846 and *Der Pflanzengarten der Universität Marburg*. Marburg 1850; see John H. Barnhart, *Biographical Notes upon Botanists*. 3: 476. 1965; Ethelyn Maria Tucker, *Catalogue of the Library of the Arnold Arboretum of Harvard University*. Cambridge, Massachusetts 1917-1933; Ida Kaplan Langman, *A Selected Guide to the Literature on the Flowering Plants of Mexico*. Philadelphia 1964; R. Zander, F. Encke, G. Buchheim and S. Seybold, *Handwörterbuch der Pflanzennamen*. 14. Aufl. Stuttgart 1993.

Wendlandia Bartling ex DC. Rubiaceae

Origins:

To honor the German botanist Johann Christoph Wendland, 1755-1828, gardener, author of *Hortus herrenhusanus*. Hannoverae 1798-1801 and *Ericarum icones et descriptiones*. Hannover [1798-] 1804-1823; see John H. Barnhart, *Biographical Notes upon Botanists*. 3: 476. 1965; James A. Baines, *Australian Plant Genera. An Etymological Dictionary of Australian Plant Genera*. 396. [genus named for Johann Christian Wendland] NSW 1981; I.H. Vegter, *Index Herbariorum*. Part II (7), *Collectors T-Z*. Regnum Vegetabile vol. 117. 1988; Mariella Azzarello Di Misa, a cura di, *Il Fondo Antico della Biblioteca dell'Orto Botanico di Palermo*. 289. Palermo 1988; Jonas C. Dryander, *Catalogus bibliothecae historico-naturalis Josephi Banks*. London 1796-1800; E.M. Tucker, *Catalogue of the Library of the Arnold Arboretum of Harvard University*. Cambridge, Massachusetts 1917-1933; R. Zander, F. Encke, G. Buchheim and S. Seybold, *Handwörterbuch der Pflanzennamen*. 14. Aufl. Stuttgart 1993.

Species/Vernacular Names:
W. aberrans How.

English: Kwangsi wendlandia

Wendlandiella Dammer Palmae

Origins:

For the German botanist Hermann Wendland, 1825-1903, taxonomist, gardener, palm botanist and collector, Director of the Royal Botanical Gardens at Herrenhausen near Hannover from 1879 until his death, travelled in Guatemala and Central America, author of *Index palmarum*. Hannoverae 1854, with the German botanist Gustav Mann (1836-1916) wrote "On the palms of Western Tropical Africa." *Trans. Linn. Soc.* 24: 421-439. (Nov.) 1864, with Carl Georg Oscar Drude (1852-1933) wrote "Ueber *Grisebachia*." in *Nachtr. Kön. Ges. Wiss. Georg-Augusts-Univ.* 1875: 54-60 and "Palmae australasicae." *Linnaea*. 39: 153-238. 1875. He was the son of the German botanist and gardener Heinrich Ludolph Wendland (1791/1792-1869); see John H. Barnhart, *Biographical Notes upon Botanists*. 3: 476. 1965; T. W. Bossert, *Biographical Dictionary of Botanists Represented in the Hunt Institute Portrait Collection*. 431. Boston, Massachusetts 1972; Ethelyn Maria Tucker, *Catalogue of the Library of the Arnold Arboretum of Harvard University*. Cambridge, Massachusetts 1917-1933; Ida Kaplan Langman, *A Selected Guide to the Literature on the Flowering Plants of Mexico*. Philadelphia 1964; R. Zander, F. Encke, G. Buchheim and S. Seybold, *Handwörterbuch der Pflanzennamen*. 14. Aufl. 1993.

Wendtia Meyen Geraniaceae (Ledocarpaceae)

Origins:

To commemorate Captain W. Wendt, the commander of the ship during Meyen's trip around the world; see the German (Baltic German, Prussian, b. Tilsit) botanist Franz Julius Ferdinand Meyen, 1804-1840 (d. Berlin), *Observationes botanicas* in itinere circum terram institutas. Opus posthumum, sociorum Academiae curis suppletum. Vratislaviae et Bonnae [Breslau and Bonn] 1843 and *Reise um die Erde ausgeführt auf dem Königlich Preussischen Seehandlungs-Schiffe Prinzess Louise*, commandirt von Capitain W. Wendt, in den Jahren 1830, 1831 und 1832. Berlin 1834-1843; Hans Querner, in *Dictionary of Scientific Biography* 9: 344-345. 1981.

Wercklea Pittier & Standley Malvaceae

Origins:

For the French botanist Karl (Carl) Wercklé, 1860-1924, horticulturist, traveler (Costa Rica); see John H. Barnhart, *Biographical Notes upon Botanists*. 3: 477. 1965; E.M.

Tucker, *Catalogue of the Library of the Arnold Arboretum of Harvard University*. Cambridge, Massachusetts 1917-1933; R. Zander, F. Encke, G. Buchheim and S. Seybold, *Handwörterbuch der Pflanzennamen*. 14. Aufl. 799. Stuttgart 1993.

Werckleocereus Britton & Rose Cactaceae

Origins:

For the French botanist Karl Wercklé, 1860-1924, horticulturist, traveler (Costa Rica).

Werdermannia O.E. Schulz Brassicaceae

Origins:

Named for the German (b. Berlin) botanist Erich Werdermann, 1892-1959 (Bremen), specialist in Cactaceae and fungi, plant geographer, traveler (Latin America, South Africa, Namibia, Chile, Bolivia, Brazil, Mexico), explorer, plant collector, anatomist, professor of botany, plant physiologist; see John H. Barnhart, *Biographical Notes upon Botanists*. 3: 477. 1965; Mary Gunn and Leslie E. Codd, *Botanical Exploration of Southern Africa*. 374. Cape Town 1981; T.W. Bossert, *Biographical Dictionary of Botanists Represented in the Hunt Institute Portrait Collection*. 431. 1972; Ida Kaplan Langman, *A Selected Guide to the Literature on the Flowering Plants of Mexico*. 798-799. Philadelphia 1964; Gordon Douglas Rowley, *A History of Succulent Plants*. Strawberry Press, Mill Valley, California 1997; R. Zander, F. Encke, G. Buchheim and S. Seybold, *Handwörterbuch der Pflanzennamen*. 14. Aufl. Stuttgart 1993; Irving William Knobloch, compil., "A preliminary verified list of plant collectors in Mexico." *Phytologia Memoirs*. VI. 1983.

Werneria Kunth Asteraceae

Origins:

For the German (b. Osiecznica, Poland) geologist Abraham Gottlob Werner, 1749-1817 (d. Dresden), professor of mineralogy at Freiburg, gave his name to the Wernerian or Neptunian theory of deposition; see Alexander Ospovat, in *Dictionary of Scientific Biography* 14: 256-264. 1981; Alexander B. Adams, *Eternal Quest. The Story of the Great Naturalists*. [b. 1750] New York 1969.

Species/Vernacular Names:

W. sp.

Peru: varita de San José

Bolivia: ak'ana, akjana

Wernhamia S. Moore Rubiaceae

Origins:

For the British (b. Hatcham, London) botanist Herbert Fuller Wernham, 1879-1941 (d. Southend, Essex), Department of Botany of the British Museum (Natural History) (succeeded James Britten, 1846-1924), a specialist in tropical Rubiaceae; see J.H. Barnhart, *Biographical Notes upon Botanists*. 3: 624. 1965; Ida Kaplan Langman, *A Selected Guide to the Literature on the Flowering Plants of Mexico*. Philadelphia 1964; Ray Desmond, *Dictionary of British & Irish Botanists and Horticulturists*. 730. 1994; E.M. Tucker, *Catalogue of the Library of the Arnold Arboretum of Harvard University*. Cambridge, Massachusetts 1917-1933.

Species/Vernacular Names:

W. boliviensis S. Moore

Bolivia: jotavio

Westeringia Dum.-Cours. Labiatae

Origins:

Orthographic variant, see *Westringia* Smith.

Westia Vahl Caesalpiniaceae

Origins:

After the Danish botanist Hans West, 1758-1811, philologist, plant collector and traveler; see John H. Barnhart, *Biographical Notes upon Botanists*. 3: 479. 1965; R. Zander, F. Encke, G. Buchheim and S. Seybold, *Handwörterbuch der Pflanzennamen*. 14. Aufl. Stuttgart 1993; Carl Frederik Albert Christensen (1872-1942), *Den danske Botaniks Historie med tilhørende Bibliografi*. Copenhagen 1924-1926.

Westonia Sprengel Fabaceae

Origins:

After the British Richard Weston, 1733-1806 (d. Leicester); see John H. Barnhart, *Biographical Notes upon Botanists*. 3: 481. 1965; R. Zander, F. Encke, G. Buchheim and S. Seybold, *Handwörterbuch der Pflanzennamen*. 14. Aufl. Stuttgart 1993; Jonas C. Dryander, *Catalogus bibliothecae historico-naturalis Josephi Banks*. London 1796-1800; Blanche Elizabeth Edith Henrey, *British Botanical and Horticultural Literature before 1800*. Oxford 1975; E.M. Tucker, *Catalogue of the Library of the Arnold Arboretum of Harvard University*. Cambridge, Massachusetts 1917-1933; Dawson Turner and Lewis Weston Dillwyn, *The Botanist's Guide Through England and Wales*. London 1805;

Arthur D. Chapman, ed., *Australian Plant Name Index*. 2985. Canberra 1991.

Westoniella Cuatrecasas Asteraceae

Origins:

For the American botanist Dr. Arthur Stewart Weston, born 1932, student of Costa Rica flora (*los páramos*); see Karl Weidmann, *Páramos venezolanos*. Caracas 1991; Volkmar Vareschi (1906-1991), *Flora de Los Páramos de Venezuela*. Universidad de Los Andes. Merida–Venezuela 1970.

Westringia J.E. Smith Labiatae

Origins:

After the Swedish physician Johan Peter Westring, 1753-1833, Dr. med. 1780 Uppsala, botanist, lichenologist, student of C. Linnaeus, from 1794 physician to the King of Sweden, author of *Svenska lafvarnas färghistoria*. Stockholm 1805 [-1809]; see John H. Barnhart, *Biographical Notes upon Botanists*. 3: 481. 1965; Jonas C. Dryander, *Catalogus bibliothecae historico-naturalis Josephi Banks*. London 1796-1800; H. Genaust, *Etymologisches Wörterbuch der botanischen Pflanzennamen*. 692. [d. 1813] 1996; F. Boerner & G. Kunkel, *Taschenwörterbuch der botanischen Pflanzennamen*. 4. Aufl. 186. 1989; G. Christian Wittstein, *Etymologisch-botanisches Handwörterbuch*. 933. Ansbach 1852.

Species/Vernacular Names:

W. dampieri R. Br. (*Westringia cinerea* R. Br.; *Westringia grevillina* F. Muell.)

English: shore westringia

W. eremicola Cunn. ex Benth.

English: slender westringia, slender western rosemary

W. rigida R. Br.

English: stiff westringia, stiff western rosemary

Wetria Baillon Euphorbiaceae

Origins:

An anagram of *Trewia* L., for the German botanist Christoph Jakob Trew, 1695-1769, physician, traveler, correspondent of the botanical artist Georg Dionysius Ehret (1708-1770), author of *Plantae selectae, quarum imagines ad exemplaria naturalia Londini in hortis Curiosorum nutrita manu artificiosa doctaque pinxit Georgius Dionysius Ehret ... [Norimbergae]* 1750-1773; see John H. Barnhart, *Biographical Notes upon Botanists*. 3: 400. 1965; T.W. Bossert, *Biographical Dictionary of Botanists Represented in the*

Hunt Institute Portrait Collection. 406. 1972; Jonas C. Dryander, *Catalogus bibliothecae historico-naturalis Josephi Banks.* London 1800; J.D. Milner, *Catalogue of Portraits of Botanists Exhibited in the Museums of the Royal Botanic Gardens.* Royal Botanic Gardens, Kew, London 1906; Blanche Elizabeth Edith Henrey (1906-1983), *British Botanical and Horticultural Literature before 1800.* Oxford 1975; Gordon Douglas Rowley, *A History of Succulent Plants.* California 1997; R. Zander, F. Encke, G. Buchheim and S. Seybold, *Handwörterbuch der Pflanzennamen.* 14. Aufl. Stuttgart 1993.

Wetriaria Pax Euphorbiaceae

Origins:

For the German botanist Christoph Jakob Trew, 1695-1769.

Wettinella O.F. Cook & Doyle Palmae

Origins:

Takes its name from King Frederick August of Saxony of the house of Wettin.

Wettinia Poepp. Palmae

Origins:

Takes its name from King Frederick August of Saxony of the house of Wettin.

Wettiniicarpus Burret Palmae

Origins:

From King Frederick August of Saxony of the house of Wettin.

Wettsteinia Petrak Asteraceae

Origins:

For the Austrian botanist Richard Wettstein von Westersheim, 1863-1931, traveler, plant collector, phylogenist, father of the Austrian botanist and plant physiologist Fritz (Friedrich) Wettstein von Westersheim (1895-1945); see T.W. Bossert, *Biographical Dictionary of Botanists Represented in the Hunt Institute Portrait Collection.* 433. 1972; Ida Kaplan Langman, *A Selected Guide to the Literature on the Flowering Plants of Mexico.* Philadelphia 1964; E.M. Tucker, *Catalogue of the Library of the Arnold Arboretum of Harvard University.* Cambridge, Massachusetts 1917-

1933; R. Zander, F. Encke, G. Buchheim and S. Seybold, *Handwörterbuch der Pflanzennamen.* 14. Aufl. 1993; Stafleu and Cowan, *Taxonomic Literature.* 7: 219-235. 1988.

Whipplea Torrey Hydrangeaceae (Philadelphaceae)

Origins:

Named for Lieutenant Amiel Weeks Whipple, 1816-1863, Commander, Pacific Railroad Expedition of 1853-1854, see *A Pathfinder in the Southwest. The Itinerary of Lieutenant Amiel Weeks Whipple* during his explorations for a railway route from Port Smith to Los Angeles in the years 1853 & 1854. Edited and annotated by Grant Foreman. [American Exploration and Travel. vol. 6.] University of Oklahoma Press 1941; Joseph Ewan, *Rocky Mountain Naturalists.* The University of Denver Press 1950.

Species/Vernacular Names:
W. modesta Torrey

English: modesty

Spanish: yerba de selva

Whiteheadia Harvey Hyacinthaceae (Liliaceae)

Origins:

After the Anglican missionary Rev. Henry Whitehead, 1817-1884 (St. Helena), clergyman, 1855-1861 plant collector in South Africa (Namaqualand and the Tulbagh area), sent plants to William H. Harvey, 1861 to St. Helena, fern collector for Karl Wilhelm Ludwig Pappe (1803-1862) and Sir Rawson William Rawson (1812-1899), sent ferns to Kew; see M. Gunn and Leslie E. Codd, *Botanical Exploration of Southern Africa.* 375-376. Cape Town 1981; Pappe and Rawson, *Synopsis filicum Africae australis.* Cape Town 1858; J.C. Melliss, *St. Helena.* London 1875.

Species/Vernacular Names:
W. bifolia (Jacq.) Bak.

South Africa: bobbejaanskoen

Whiteochloa C.E. Hubbard Gramineae

Origins:

For the Australian botanist Cyril Tenison White, 1890-1950, plant collector, from 1917 to 1950 Government Botanist of Queensland, (maternal) grandson of Frederick Manson Bailey (1827-1915). Among his publications are *Contributions to the*

Queensland Flora. Brisbane 1921-1946, *A Contribution to Our Knowledge of the Flora of Papua* (British New Guinea). Brisbane 1922, "The Bailey family and its place in the botanical history of Australia." *J. Hist. Soc. Qld.* 3: 362-383. 1944, and "Ligneous plants collected in New Caledonia by C.T. White in 1923." *J. Arnold Arb.* 7: 74-103. 1926. John Frederick Bailey wrote "Introduction of economic plants into Queensland." *Proc. R. Soc. Queensland.* 22: 77-102. 1910; Frederick Manson Bailey was the author of *Catalogue of the Indigenous and Naturalized Plants of Queensland*. Brisbane 1890, *The Queensland Flora*. Brisbane 1899-1905, and *Comprehensive Catalogue of Queensland Plants both Indigenous and Naturalised*. Brisbane [1913]; see Charles Edward Hubbard (1900-1980), *Proceedings of the Royal Society of Queensland*. 62: 111. (Aug.) 1952; J.H. Barnhart, *Biographical Notes upon Botanists*. 3: 485 and 1: 104. 1965; Dennis John Carr (1915-) and S.G.M. Carr (1912-1988), eds., *People and Plants in Australia*. 1981; J. Lanjouw and F.A. Stafleu, *Index Herbariorum*. Part II, *Collectors A-D*. Regnum Vegetabile vol. 2. 1954; E.M. Tucker, *Catalogue of the Library of the Arnold Arboretum of Harvard University*. Cambridge, Massachusetts 1917-1933; S. Lenley et al., *Catalog of the Manuscript and Archival Collections and Index to the Correspondence of John Torrey*. Library of the New York Botanical Garden. 432. 1973; N. Hall, *Botanists of the Eucalypts*. Melbourne 1978 and Supplement 1980.

Whiteodendron Steenis Myrtaceae

Origins:

For the Australian botanist and plant collector Cyril Tenison White, 1890-1950, 1917-1950 Government Botanist of Queensland. His works include *Contributions to the Queensland Flora*. Brisbane 1921-1946, *A Contribution to Our Knowledge of the Flora of Papua* (British New Guinea). Brisbane 1922, "The Bailey family and its place in the botanical history of Australia." *J. Hist. Soc. Qld.* 3: 362-383. 1944; see John H. Barnhart, *Biographical Notes upon Botanists*. 3: 485. 1965.

Whitesloanea Chiovenda Asclepiadaceae

Origins:

After Boyd Lincoln Sloane (1886-1955, d. California) and the American botanist Alain Campbell White (1880-1951, d. South Carolina), authors of *The Stapelieae*. Pasadena 1933 [Ed. 2, Feb. 1937]; see John H. Barnhart, *Biographical Notes upon Botanists*. 3: 485 and 3: 287. 1965; Larry W. Mitich, "Stapeliads and Euphorbias — The White and Sloane Story." *Haseltonia*. 2: 61-78. 1994; Gordon Douglas Rowley, *A History of Succulent Plants*. Strawberry Press, Mill Valley, California 1997; Gilbert Westacott Reynolds

(1895-1967), *The Aloes of South Africa*. Balkema, Rotterdam 1982; R. Zander, F. Encke, G. Buchheim and S. Seybold, *Handwörterbuch der Pflanzennamen*. 14. Aufl. 781, 799. Stuttgart 1993.

Whitfieldia Hook. Acanthaceae

Origins:

For the British Thomas Whitfield (Withfield), plant collector in Sierra Leone and Gambia (for Lord Derby); see F.N. Hepper and Fiona Neate, *Plant Collectors in West Africa*. 85. 1971; John H. Barnhart, *Biographical Notes upon Botanists*. Boston 1965; Ray Desmond, *Dictionary of British & Irish Botanists and Horticulturists*. 736. London 1994; F. Boerner & G. Kunkel, *Taschenwörterbuch der botanischen Pflanzennamen*. 4. Aufl. 186. Berlin & Hamburg 1989.

Whitfordia Elmer Fabaceae

Origins:

After the American botanist Harry Nichols Whitford, 1872-1941.

Whitfordiodendron Elmer Fabaceae

Origins:

After the American botanist Harry Nichols Whitford, 1872-1941, plant collector, forester, author of *The Forests of the Flathead Valley*, Montana. 1905 and *The Forests of the Philippines*. Manila 1911; see John H. Barnhart, *Biographical Notes upon Botanists*. 3: 488. 1965; Ida Kaplan Langman, *A Selected Guide to the Literature on the Flowering Plants of Mexico*. Philadelphia 1964; E.M. Tucker, *Catalogue of the Library of the Arnold Arboretum of Harvard University*. Cambridge, Massachusetts 1917-1933.

Species/Vernacular Names:

W. atropurpureum (Wallich) Merr. (*Millettia atropurpurea* (Wallich) Benth.; *Pongamia atropurpurea* Wallich)

English: purple millettia

Malaya: tulang diang, tulang dain, jenaris, jenerek

Whitneya A. Gray Asteraceae

Origins:

For the American (b. Northampton, Massachusetts) geologist Josiah Dwight Whitney, 1819-1896 (d. Lake Sunapee, New Hampshire), scientist, chemist, studied at Yale, in 1840 joined the New Hampshire Geological Survey, professor at

Iowa University, 1860 State geologist of California, professor at Harvard in 1865, explorer. Among his publications are *The Auriferous Gravels of the Sierra Nevada of California*. Cambridge, Massachusetts 1880, *The Metallic Wealth of the United States*. Philadelphia 1854 and *Report of a Geological Survey of the Upper Missouri Lead Region*. Albany 1862. He was the brother of the philologist William Dwight Whitney (1827-1894); see Gerald D. Nash, in *Dictionary of Scientific Biography* 14: 315-316. 1981; Edwin T. Brewster, *Life and Letters of Josiah Dwight Whitney*. Boston 1909; John Wells Foster (1815-1873) and J.D. Whitney, *Report on the Geology and Topography of a Portion of the Lake Superior Land District*, in the State of Michigan. 1850, 1851; Joseph Ewan, *Rocky Mountain Naturalists*. The University of Denver Press 1950.

Whyanbeelia Airy Shaw & B. Hyland Euphorbiaceae

Origins:

"QUEENSLAND. Cook District: Timber Reserve 55 Whyanbeel, lat. 16°20'S., long. 145°20'E., rain forest, alt. 150 m., 3 July 1974 ..."; see *Kew Bulletin*. 31: 375. (Aug.) 1976; Arthur D. Chapman, ed., *Australian Plant Name Index*. 2989. Canberra 1991.

Wibelia Bernhardi Davalliaceae

Origins:

For the German physician August Wilhelm Eberhard Christoph Wibel, 1775-1813, botanist; see John H. Barnhart, *Biographical Notes upon Botanists*. 3: 490. 1965; R. Zander, F. Encke, G. Buchheim and S. Seybold, *Handwörterbuch der Pflanzennamen*. 14. Aufl. Stuttgart 1993.

Wiborgia Thunberg Fabaceae

Origins:

The genus was named after the Danish botanist Erik Nissen Viborg, 1759-1822, student of M. Vahl, professor of botany at the University of Copenhagen and Director of the Botanical Garden, he published *Om Pilevaands Afbarkning og Plantning*. Copenhagen 1821, and, by Theodor Holm afterwards Holmskjold (1732-1794), *Beata ruris otia Fungis Danicis* ... Volumen II. post obitum auctoris editum, curante E. Viborg. [1799]; see John H. Barnhart, *Biographical Notes upon Botanists*. 3: 434. 1965; Carl Frederik Albert Christensen (1872-1942), *Den danske Botaniks Historie med tilhørende Bibliografi*. Copenhagen 1924-1926; T.W. Bossert, *Biographical Dictionary of Botanists Represented in the Hunt Institute Portrait Collection*. 418. 1972; Jonas C.

Dryander, *Catalogus bibliothecae historico-naturalis Josephi Banks*. London 1800; E.M. Tucker, *Catalogue of the Library of the Arnold Arboretum of Harvard University*. Cambridge, Massachusetts 1917-1933.

Species/Vernacular Names:

W. sericea Thunb. (*Jacksonago sericea* (Thunb.) Kuntze; *Loethainia sericea* (Thunb.) Heynhold; *Wiborgia lanceolata* E. Mey.)

South Africa: wiborgia

Wichuraea Nees ex Reisseck Rhamnaceae

Origins:

After the German botanist Max Ernst Wichura, 1817-1866, magistrate, traveler, from 1859 to 1863 participated in the Prussian Eastern Asiatic Expedition. Among his works *Aus vier Welttheilen. Ein Reise-Tagebuch in Briefen*. Breslau, Leipzig 1868 and *Die Bastardbefruchtung im Pflanzenreich*. Breslau 1865, with Ferdinand Julius Cohn (1828-1898) wrote *Über Stephanosphaera pluvialis*. [1858]; see John H. Barnhart, *Biographical Notes upon Botanists*. 3: 490. 1965; Emil Bretschneider, *History of European Botanical Discoveries in China*. Leipzig 1981; E.M. Tucker, *Catalogue of the Library of the Arnold Arboretum of Harvard University*. Cambridge, Massachusetts 1917-1933.

Wickstroemia Endlicher Thymelaeaceae

Origins:

For the Swedish botanist Johan (Johann) Emanuel Wikström, 1789-1856, physician.

Species/Vernacular Names:

W. indica (L.) C. Meyer

China: liao ge wang

Widdringtonia Endlicher Cupressaceae

Origins:

For the Commander (of the Royal Navy) Samuel Edward Widdrington (formerly Cook), 1787-1856 (d. Felton, Northumberland), traveler in Spain, author of *Sketches in Spain during ... 1829-1832, containing notices ... of the ... Natural History*. London 1834 and *Spain & the Spaniards, in 1843*. 2 vols. London 1844 (The Appendix to Vol. 1 contains notices of the Geology, Forests and Ornithology), in 1840 took name of Widdrington; see Ray Desmond, *Dictionary of British & Irish Botanists and Horticulturists*. 738. London 1994.

Species/Vernacular Names:

W. cedarbergensis Marsh (*Widdringtonia juniperoides* sensu Dur. & Schinz, non Endl.) (after Cedarberg, in the Clanwilliam district, north of Cape Town, South Africa)

English: Clanwilliam cedar, cedar tree, Cape cedar, Clanwilliam cypress

South Africa: sederboom, sederhout

W. nodiflora (L.) Powrie (*Widdringtonia cupressoides* (L.) Endl.; *Widdringtonia dracomontana* Stapf; *Widdringtonia whytei* Rendle; *Thuja cupressoides* L.)

English: mountain cypress, Cape cypress, Mlanji cedar, Milanje cedar

Southern Africa: bergsipres, sapreehout, bergsapree, sapree wood, bobbejaankers, sipresboom, dupresknop; mapande, muSheza (Shona); uNwele-lwe-entaba (Zulu); thaululo (Venda)

W. schwarzii (Marloth) Mast. (the species was named after E. Schwarz)

English: Willowmore cedar

South Africa: Willowmore cedar (this species grows in the Baviaanskloof and Kouga Mountains in the Willowmore district, South Africa)

Wiedemannia Fischer & C.A. Meyer Labiatae

Origins:

For Edward Wiedemann, d. 1844, physician, plant collector (Asia Minor and Armenia); see Stafleu and Cowan, *Taxonomic Literature*. 7: 261. 1988.

Wiegmannia Meyen Rubiaceae

Origins:

After the German pharmacist Arend Joachim Friedrich Wiegmann, 1770-1853, naturalist, father of the German zoologist Arend Friedrich August Wiegmann (1802-1841); see J.H. Barnhart, *Biographical Notes upon Botanists*. 3: 492. 1965; Herbert F. Roberts, *Plant Hybridization before Mendel*. Princeton 1929.

Wielandia Baillon Euphorbiaceae

Origins:

After J.F. Wieland, 1804-1872.

Wierzbickia Reichb. Caryophyllaceae

Origins:

For the botanist Piotr Pawlus (Peter) Wierzbicki (Petrus Paulus Wirzbicki), 1794-1847, surgeon, entomologist, plant collector; see J.H. Barnhart, *Biographical Notes upon Botanists*. 3: 492. 1965; H.N. Clokie, *Account of the Herbaria of the Department of Botany in the University of Oxford*. 264. Oxford 1964.

Wiesneria Micheli Alismataceae

Origins:

For the Austrian botanist Julius Ritter von Wiesner, 1838-1916, traveler (India, Dutch East Indies, North America), 1909 ennobled, professor of plant anatomy and physiology; see J.H. Barnhart, *Biographical Notes upon Botanists*. 3: 492. 1965; T.W. Bossert, *Biographical Dictionary of Botanists Represented in the Hunt Institute Portrait Collection*. 435. 1972; Richard Biebl, in *Dictionary of Scientific Biography* 14: 349-350. 1981; Ethelyn Maria Tucker, *Catalogue of the Library of the Arnold Arboretum of Harvard University*. Cambridge, Massachusetts 1917-1933; Stafleu and Cowan, *Taxonomic Literature*. 7: 268-272. 1988.

Wigandia Kunth Hydrophyllaceae

Origins:

After Johannes Wigand, 1523-1587, Bishop of Pomerania, "Preussens erster Botaniker", writer on Prussian plants, author of *Vera historia de succino borussico; de alce borussica, et de herbis in Borussia nascentibus*. Jenae 1590; see Kurt Wein, in *Sudhoffs Arch. Gesch. Med. Naturw.* 35(3-4): 160-205. 1942; Francis Aubie Sharr, *Western Australian Plant Names and Their Meanings*. 75. University of Western Australia Press, Nedlands, Western Australia 1996. According to some authors the genus was named for the German physician Justus Heinrich Wigand, 1769-1817; see G.C. Wittstein, *Etymologisch-botanisches Handwörterbuch*. 934. Ansbach 1852; F. Boerner & G. Kunkel, *Taschenwörterbuch der botanischen Pflanzennamen*. 4. Aufl. 186. 1989; H. Genaust, *Etymologisches Wörterbuch der botanischen Pflanzennamen*. 692. Basel 1996.

Species/Vernacular Names:

W. caracasana Kunth

English: wigandia, Caracas big-leaf

Wigginsia Duncan M. Porter Cactaceae

Origins:

For the American botanist Ira Loren Wiggins, 1899-1987, professor of botany, botanical explorer and traveler, plant collector, professor emeritus of biological sciences at Stanford University, author of *Flora of Baja California*. Stanford, California 1980; see J.H. Barnhart, *Biographical*

Notes upon Botanists. 3: 493. 1965; T.W. Bossert, *Biographical Dictionary of Botanists Represented in the Hunt Institute Portrait Collection.* 435. 1972; Ida Kaplan Langman, *A Selected Guide to the Literature on the Flowering Plants of Mexico.* Philadelphia 1964; Irving William Knobloch, compil., "A preliminary verified list of plant collectors in Mexico." *Phytologia Memoirs.* VI. 1983.

Wightia Wallich Scrophulariaceae

Origins:

For the British (b. East Lothian) botanist Robert Wight, 1796-1872 (d. Berks.), East India Company, surgeon, traveler, plant collector; see Nathaniel Wallich (1786-1854), *Plantae Asiaticae Rariores.* London 1830-1832; J.H. Barnhart, *Biographical Notes upon Botanists.* 3: 493. 1965; T.W. Bossert, *Biographical Dictionary of Botanists Represented in the Hunt Institute Portrait Collection.* 435. 1972; J.D. Milner, *Catalogue of Portraits of Botanists Exhibited in the Museums of the Royal Botanic Gardens.* Royal Botanic Gardens, Kew, London 1906; E.M. Tucker, *Catalogue of the Library of the Arnold Arboretum of Harvard University.* Cambridge, Massachusetts 1917-1933; S. Lenley et al., *Catalog of the Manuscript and Archival Collections and Index to the Correspondence of John Torrey.* Library of the New York Botanical Garden. 1973; Ernest Nelmes and William Cuthbertson, *Curtis's Botanical Magazine Dedications, 1827-1927.* 143-144. [1931]; H.N. Clokie, *Account of the Herbaria of the Department of Botany in the University of Oxford.* 265. Oxford 1964; Antoine Lasègue, *Musée botanique de M. Benjamin Delessert.* 1845; R. Zander, F. Encke, G. Buchheim and S. Seybold, *Handwörterbuch der Pflanzennamen.* 14. Aufl. Stuttgart 1993; M. Archer, *Natural History Drawings in the India Office Library.* London 1962; Isaac Henry Burkill, *Chapters on the History of Botany in India.* Delhi 1965; Ray Desmond, *Dictionary of British & Irish Botanists and Horticulturists.* London 1994; R. Desmond, *The European Discovery of the Indian Flora.* Oxford 1992; D.G. Crawford, *A History of the Indian Medical Service, 1600-1913.* London 1914; Leonard Huxley, *Life and Letters of Sir Joseph Dalton Hooker.* London 1918; Alain White and Boyd Lincoln Sloane, *The Stapelieae.* Pasadena 1937; Merle A. Reinikka, *A History of the Orchid.* Timber Press 1996.

Wikstroemia Endlicher Thymelaeaceae

Origins:

For the Swedish botanist Johan (Johann) Emanuel Wikström, 1789-1856, physician, from 1818 to 1856 professor at the Bergius Garden, author of *Dissertatio de Daphne.* Upsaliae [Uppsala] [1817] and *Conspectus litteraturae*

botanicae in Suecia. Holmiae [Stockholm] 1831, editor of Olof Peter Swartz (1760-1818), *Adnotationes botanicae quas reliquit ... post mortem auctoris collectae ... atque notis et praefatione instructae a J.E. Wikstrom, accedit Biographia Swartzii, auctoribus C. Sprengel et C.A. Agardh* [a bibliography]. Holmiae 1829; see J.H. Barnhart, *Biographical Notes upon Botanists.* 3: 494. 1965; Mariella Azzarello Di Misa, a cura di, *Il Fondo Antico della Biblioteca dell'Orto Botanico di Palermo.* 291. Palermo 1988; Ida Kaplan Langman, *A Selected Guide to the Literature on the Flowering Plants of Mexico.* Philadelphia 1964; Antoine Lasègue, *Musée botanique de M. Benjamin Delessert.* 1845; Ethelyn Maria Tucker, *Catalogue of the Library of the Arnold Arboretum of Harvard University.* Cambridge, Massachusetts 1917-1933; S. Lenley et al., *Catalog of the Manuscript and Archival Collections and Index to the Correspondence of John Torrey.* Library of the New York Botanical Garden. 434. 1973; R. Zander, F. Encke, G. Buchheim and S. Seybold, *Handwörterbuch der Pflanzennamen.* 14. Aufl. 800. 1993; G. Schmid, *Goethe und die Naturwissenschaften.* Halle 1940; Giulio Giorello & Agnese Grieco, a cura di, *Goethe scienziato.* Einaudi Editore, Torino 1998.

Species/Vernacular Names:

W. sp.

Hawaii: 'akia, kauhi

W. oahuensis (A. Gray) Rock (*Diplomorpha elongata* (A. Gray) A. Heller)

Hawaii: aoaoa

W. retusa A. Gray

Japan: ao-ganpi, kajigii

W. uva-ursi A. Gray (*Diplomorpha uva-ursi* (A. Gray) A. Heller)

Hawaii: aoaoa

Wikstroemia Schrader Theaceae

Origins:

For the Swedish botanist Johan Emanuel Wikström, 1789-1856.

Wilbrandia Silva Manso Cucurbitaceae

Origins:

For the German (b. Clarholz) botanist Johann Bernhard Wilbrand, 1779-1846 (d. Giessen), physician; see John H. Barnhart, *Biographical Notes upon Botanists.* 3: 494. 1965; Guenter B. Risse, in *Dictionary of Scientific Biography* 14:

351-352. 1981; G. Schmid, *Goethe und die Naturwissenschaften*. Halle 1940.

Wildemaniodoxa Aubrév. & Pellegr. Sapotaceae

Origins:

For the Belgian botanist Émile Auguste Joseph De Wildeman, 1866-1947, a specialist in the Congolese flora; see J.H. Barnhart, *Biographical Notes upon Botanists*. 1: 450. 1965; Ida Kaplan Langman, *A Selected Guide to the Literature on the Flowering Plants of Mexico*. University of Pennsylvania Press, Philadelphia 1964; Frans A. Stafleu and Richard S. Cowan, *Taxonomic Literature*. 1: 639-643. Utrecht 1976.

Wilibald-schmidtia Conrad Gramineae

Origins:

For the German (b. Bohemia) botanist Franz Wilibald Schmidt, 1764-1796.

Wilibalda Roth Gramineae

Origins:

For the German (b. Bohemia) botanist Franz Wilibald Schmidt, 1764-1796, professor of botany, physician; see John H. Barnhart, *Biographical Notes upon Botanists*. 3: 231. 1965; Jonas C. Dryander, *Catalogus bibliothecae historico-naturalis Josephi Banks*. London 1800; E.M. Tucker, *Catalogue of the Library of the Arnold Arboretum of Harvard University*. Cambridge, Massachusetts 1917-1933; R. Zander, F. Encke, G. Buchheim and S. Seybold, *Handwörterbuch der Pflanzennamen*. 14. Aufl. Stuttgart 1993.

Wilkesia A. Gray Asteraceae

Origins:

For Charles Wilkes, 1798-1877, American naval officer, traveler, in 1818 joined the US Navy, studied hydrography, from 1838 to 1842 explorer in the Pacific Ocean (explored the South Pacific islands and the Antarctic continent), in 1864 was court-martialed for disobedience, author of *Narrative of the United States Exploring Expedition*. During the years ... Philadelphia 1845; see D.M. Henderson, *The Hidden Coasts: A Biography of Admiral Charles Wilkes*. New York 1953; G.A. Doumani, ed., *Antarctic Bibliography*. Washington, Library of Congress 1965-1979; John H. Barnhart, *Biographical Notes upon Botanists*. 3: 496. 1965; W. Bixby, *The Forgotten Voyage of Charles Wilkes*. New York 1966; D.C. Haskell, *The United States Exploring Expedition 1838-1842 and Its Publications 1844-1874*.

New York 1942; W.J. Morgan et al., eds., *Autobiography of Rear Admiral Charles Wilkes U.S. Navy 1798-1877*. Washington, D.C. 1978; D.B. Tyler, *The Wilkes Expedition: The First United States Exploring Expedition (1838-1842)*. Philadelphia 1968; Edmund Fanning, *Voyages round the World*; with selected sketches of voyages to the south seas, north and south Pacific oceans, China, etc., ... 1792 and 1832, ... N.Y. 1833; John Dunmore, *Who's Who in Pacific Navigation*. 265-267. Honolulu 1991; T.W. Bossert, *Biographical Dictionary of Botanists Represented in the Hunt Institute Portrait Collection*. 436. 1972; J. Ewan, ed., *A Short History of Botany in the United States*. New York and London 1969; E.M. Tucker, *Catalogue of the Library of the Arnold Arboretum of Harvard University*. Cambridge, Massachusetts 1917-1933; S. Lenley et al., *Catalog of the Manuscript and Archival Collections and Index to the Correspondence of John Torrey*. Library of the New York Botanical Garden. 472. 1973; Stafleu and Cowan, *Taxonomic Literature*. 7: 295-297. 1988.

Species/Vernacular Names:

W. gymnoxiphium A. Gray (*Argyroxiphium gymnoxiphium* (A. Gray) D. Keck)

Hawaii: iliau

Wilkia F. Muell. Brassicaceae

Origins:

Orthographic variant of *Wilckia* Scopoli; see Arthur D. Chapman, ed., *Australian Plant Name Index*. 2991. Canberra 1991.

Wilkiea F. Muell. Monimiaceae

Origins:

After the Australian David Elliott Wilkie, M.D., physician to the Melbourne Hospital, vice-president of the Philosophical Institute of Victoria (later the Royal Society of Victoria). His writings include *On the Expected Water Supply of the City of Melbourne from Yan Yean*. Melbourne 1855, *The Failure of the Yan Yean Reservoir*. Melbourne 1855 and *On the Delivery on the Right Side*. Melbourne 1863; see F. Mueller, in *Transactions and Proceedings of the Philosophical Institute of Victoria*. 2: 64. (Sept.) 1858.

Willbleibia Herter Gramineae

Origins:

For the German botanist Heinrich Moriz Willkomm, 1821-1895, explorer, traveler, naturalist, 1874-1892 professor of botany and Director of the Botanical Garden of the University of Prague; see J.H. Barnhart, *Biographical Notes upon*

Botanists. 3: 501. 1965; R. Zander, F. Encke, G. Buchheim and S. Seybold, *Handwörterbuch der Pflanzennamen.* 14. Aufl. 1993; Stafleu and Cowan, *Taxonomic Literature.* 7: 336-346. 1988; E.M. Tucker, *Catalogue of the Library of the Arnold Arboretum of Harvard University.* Cambridge, Massachusetts 1917-1933; T.W. Bossert, *Biographical Dictionary of Botanists Represented in the Hunt Institute Portrait Collection.* 438. 1972.

Willdenowia J.F. Gmelin Rubiaceae

Origins:

For the German (b. Berlin) botanist Karl (Carl) Ludwig Willdenow, 1765-1812 (d. Berlin), physician, naturalist, professor of botany, Director of the Berlin Botanical Garden, friend of D.F.L. von Schlechtendal; see J.H. Barnhart, *Biographical Notes upon Botanists.* 3: 497. 1965; Jerome J. Bylebyl, in *Dictionary of Scientific Biography* 14: 386-388. New York 1981; R. Zander, F. Encke, G. Buchheim and S. Seybold, *Handwörterbuch der Pflanzennamen.* 14. Aufl. 1993; Stafleu and Cowan, *Taxonomic Literature.* 7: 298-305. 1988; Emil Bretschneider (1833-1901), *History of European Botanical Discoveries in China.* [Reprint of the original edition, St. Petersburg 1898.] Leipzig 1981; Jonas C. Dryander, *Catalogus bibliothecae historico-naturalis Josephi Banks.* London 1796-1800; E.M. Tucker, *Catalogue of the Library of the Arnold Arboretum of Harvard University.* Cambridge, Massachusetts 1917-1933; Ida Kaplan Langman, *A Selected Guide to the Literature on the Flowering Plants of Mexico.* Philadelphia 1964; T.W. Bossert, *Biographical Dictionary of Botanists Represented in the Hunt Institute Portrait Collection.* 437. 1972; S. Lenley et al., *Catalog of the Manuscript and Archival Collections and Index to the Correspondence of John Torrey.* Library of the New York Botanical Garden. 435. 1973; H.N. Clokie, *Account of the Herbaria of the Department of Botany in the University of Oxford.* 265. Oxford 1964; A. Lasègue, *Musée botanique de M. Benjamin Delessert.* 1845; Mariella Azzarello Di Misa, a cura di, *Il Fondo Antico della Biblioteca dell'Orto Botanico di Palermo.* 293-294. Soprintendenza per i Beni Culturali e Ambientali. Sezione per i Beni Bibliografici. Regione Siciliana, Palermo 1988; William Darlington (1782-1863), *Reliquiae Baldwinianae.* Philadelphia 1843; G. Schmid, *Goethe und die Naturwissenschaften.* Halle 1940; Günther Schmid, *Chamisso als Naturforscher.* Eine Bibliographie. Leipzig 1942.

Willdenowia Thunberg Restionaceae

Origins:

For the German (b. Berlin) botanist Karl (Carl) Ludwig Willdenow, 1765-1812 (d. Berlin).

Willemetia Necker ex Cass. Asteraceae

Origins:

For the French botanist Pierre Remi (Remy) Willemet, 1735-1807, pharmacist, father of the French botanist and physician Pierre Rémi François de Paule Willemet (1762-1790); see R. Zander, F. Encke, G. Buchheim and S. Seybold, *Handwörterbuch der Pflanzennamen.* 14. Aufl. 800. 1993; Stafleu and Cowan, *Taxonomic Literature.* 7: 312-314. 1988; F. Boerner & G. Kunkel, *Taschenwörterbuch der botanischen Pflanzennamen.* 4. Aufl. 187. 1989; Helmut Genaust, *Etymologisches Wörterbuch der botanischen Pflanzennamen.* 693. 1996; John H. Barnhart, *Biographical Notes upon Botanists.* 3: 497. 1965; Jonas C. Dryander, *Catalogus bibliothecae historico-naturalis Josephi Banks.* London 1796-1800; E.M. Tucker, *Catalogue of the Library of the Arnold Arboretum of Harvard University.* Cambridge, Massachusetts 1917-1933; G.C. Wittstein (1810-1887), *Etymologisch-botanisches Handwörterbuch.* 934. Ansbach 1852.

Williamsia Merrill Rubiaceae

Origins:

Dedicated to the American botanist Robert Statham Williams, 1859-1945, miner, bryologist, explorer, scientist, traveler, rider for the Pony Express, botanical collector; see T.W. Bossert, *Biographical Dictionary of Botanists Represented in the Hunt Institute Portrait Collection.* 437. 1972; S. Lenley et al., *Catalog of the Manuscript and Archival Collections and Index to the Correspondence of John Torrey.* Library of the New York Botanical Garden. 436-437. 1973; Joseph Ewan, *Rocky Mountain Naturalists.* The University of Denver Press 1950; Joseph William Blankinship (1862-1938), "A century of botanical exploration in Montana, 1805-1905: collectors, herbaria and bibliography." in *Montana Agric. Coll. Sci. Studies Bot.* 1: 1-31. 1904.

Willisia Warming Podostemaceae

Origins:

For the British (b. Cheshire) botanist John Christopher Willis, 1868-1958 (d. Montreux, Switzerland), traveler, Director of the Royal Botanic Gardens of Ceylon (Peradeniya, 1896-1911), Director of the Botanic Garden Rio de Janeiro (1912-1915), Fellow of the Linnean Society 1897, Fellow of the Royal Society in 1919; see Ida Kaplan Langman, *A Selected Guide to the Literature on the Flowering Plants of Mexico.* Philadelphia 1964; E.M. Tucker, *Catalogue of the Library of the Arnold Arboretum of Harvard University.* Cambridge, Massachusetts 1917-1933; Isaac Henry Burkill, *Chapters on the History of Botany in India.* Delhi 1965; Ray Desmond, *Dictionary of British & Irish Botanists and*

Horticulturists. 744. London 1994; R. Desmond, *The European Discovery of the Indian Flora.* Oxford 1992; Stafleu and Cowan, *Taxonomic Literature.* 7: 331-335. 1988.

Willkommia Hackel Gramineae

Origins:

For the German botanist Heinrich Moriz Willkomm, 1821-1895, explorer, traveler, naturalist, 1874-1892 professor of botany and Director of the Botanical Garden of the University of Prague; see J.H. Barnhart, *Biographical Notes upon Botanists.* 3: 501. 1965; R. Zander, F. Encke, G. Buchheim and S. Seybold, *Handwörterbuch der Pflanzennamen.* 14. Aufl. 1993; Stafleu and Cowan, *Taxonomic Literature.* 7: 336-346. 1988; E.M. Tucker, *Catalogue of the Library of the Arnold Arboretum of Harvard University.* Cambridge, Massachusetts 1917-1933; T.W. Bossert, *Biographical Dictionary of Botanists Represented in the Hunt Institute Portrait Collection.* 438. 1972; H. Dolezal, *Portug. Acta Biol.* 6: 257-323. 1959 and 7: 324-551. 1961.

Willughbeia Roxb. Apocynaceae

Origins:

For the British (b. Middleton, Warwickshire) botanist Francis Willughby (Willoughby), 1635-1672 (d. Middleton), naturalist, Fellow of the Royal Society in 1663, traveler and plant collector. His works include *De historia piscium* libri quatuor ... totum opus recognovit ... supplevit, librum etiam primum et secundum integros adjecit J. Raius. [Folio, first edition, one of the most important English illustrated books of the seventeenth century.] Ozonii 1686 and *Ornithologiae* libri tres ... Totum opus recognovit, digessit, supplevit J. Raius. Londini 1676, a friend of John Ray and Sir Philip Skippon; see Mary A. Welch, in *Dictionary of Scientific Biography* 14: 412-414. 1981; Alexander B. Adams, *Eternal Quest. The Story of the Great Naturalists.* New York 1969; C.E. Raven, *English Naturalists from Neckam to Ray.* Cambridge 1947; James Britten, *The Sloane Herbarium,* revised and edited by J.E. Dandy. 1958; W. Blunt, *The Art of Botanical Illustration.* 68. London 1950; C.E. Raven, *John Ray Naturalist, His Life and Works.* 368. Cambridge 1942; Roger Gaskell Rare Books, *Catalogue 24,* item no. 55. September 1998; Georg Christian Wittstein, *Etymologisch-botanisches Handwörterbuch.* 935. 1852; Helmut Genaust, *Etymologisches Wörterbuch der botanischen Pflanzennamen.* 693. Basel 1996.

Species/Vernacular Names:

W. coriacea Wallich

English: Borneo rubber

Willughbeiopsis Rauschert Apocynaceae

Origins:

Resembling *Willughbeia.*

Wilsonia R. Br. Convolvulaceae

Origins:

After the British botanist John Wilson, 1696-1751 (d. Kendal, Westmorland), baker and (possibly) shoemaker, author of *A Synopsis of British Plants,* in Mr. Ray's method. Newcastle upon Tyne 1744; see John H. Barnhart, *Biographical Notes upon Botanists.* 3: 504. 1965; R. Brown, *Prodromus Florae Novae Hollandiae.* 490. 1810; Jonas C. Dryander, *Catalogus bibliothecae historico-naturalis Josephi Banks.* London 1796-1800; R. Pulteney, *Historical and Biographical Sketches of the Progress of Botany in England.* 2: 264-269. London 1790; Blanche Henrey, *British Botanical and Horticultural Literature before 1800.* Oxford 1975; E.M. Tucker, *Catalogue of the Library of the Arnold Arboretum of Harvard University.* Cambridge, Massachusetts 1917-1933; James A. Baines, *Australian Plant Genera. An Etymological Dictionary of Australian Plant Genera.* 397-398. Chipping Norton, N.S.W. 1981; Francis Aubie Sharr, *Western Australian Plant Names and Their Meanings.* 75. University of Western Australia Press, Nedlands, Western Australia 1996.

Species/Vernacular Names:

W. backhousei Hook.f.

English: narrow-leaved wilsonia

W. humilis Hook.f. (*Frankenia cymbifolia* Hook.)

English: silky wilsonia

Wilsonia Raf. Epacridaceae

Origins:

For the British (b. Lancs.) bryologist William Wilson, 1799-1871 (d. near Warrington, Lancs.); see J.H. Barnhart, *Biographical Notes upon Botanists.* 3: 505. 1965; H.N. Clokie, *Account of the Herbaria of the Department of Botany in the University of Oxford.* 266. Oxford 1964; T. W. Bossert, *Biographical Dictionary of Botanists Represented in the Hunt Institute Portrait Collection.* 439. Boston, Massachusetts 1972; S. Lenley et al., *Catalog of the Manuscript and Archival Collections and Index to the Correspondence of John Torrey.* Library of the New York Botanical Garden. 438. 1973; G. Murray, *History of the Collections Contained in the Natural History Departments of the British Museum.* London 1904; Ray Desmond, *Dictionary of British & Irish Botanists and Horticulturists.* 748. London 1994; Constantine Samuel

Rafinesque (1783-1840), in *Specchio delle scienze*, o giornale enciclopedico di Sicilia etc. 1: 157. Palermo 1814; E.D. Merrill (1876-1956), *Index rafinesquianus*. The plant names published by C.S. Rafinesque, etc. 186. Jamaica Plain, Massachusetts, USA 1949.

Wimmeria Schltdl. & Chamisso Celastraceae

Origins:

For the German botanist Christian Friedrich (Heinrich) Wimmer, 1803-1868, naturalist; see John H. Barnhart, *Biographical Notes upon Botanists*. 3: 505. 1965; R. Zander, F. Encke, G. Buchheim and S. Seybold, *Handwörterbuch der Pflanzennamen*. 14. Aufl. Stuttgart 1993; Ethelyn Maria Tucker, *Catalogue of the Library of the Arnold Arboretum of Harvard University*. Cambridge, Massachusetts 1917-1933; G. Schmid, *Goethe und die Naturwissenschaften*. Halle 1940; Stafleu and Cowan, *Taxonomic Literature*. 7: 359-362. 1988; Mariella Azzarello Di Misa, a cura di, *Il Fondo Antico della Biblioteca dell'Orto Botanico di Palermo*. 294. Soprintendenza per i Beni Culturali e Ambientali. Sezione per i Beni Bibliografici. Regione Siciliana, Palermo 1988.

Winchia A. DC. Apocynaceae

Origins:

For the British (b. Middx.) botanist Nathaniel John Winch, 1768-1838 (d. Northumberland), traveler, Fellow of the Linnean Society 1803; see John H. Barnhart, *Biographical Notes upon Botanists*. 3: 506. 1965; Ethelyn Maria Tucker, *Catalogue of the Library of the Arnold Arboretum of Harvard University*. Cambridge, Massachusetts 1917-1933; J.D. Milner, *Catalogue of Portraits of Botanists Exhibited in the Museums of the Royal Botanic Gardens*. Royal Botanic Gardens, Kew, London 1906; Ray Desmond, *Dictionary of British & Irish Botanists and Horticulturists*. 749. 1994; T. W. Bossert, *Biographical Dictionary of Botanists Represented in the Hunt Institute Portrait Collection*. 439. Boston, Massachusetts 1972; Andrew Thomas Gage, *A History of the Linnean Society of London*. London 1938.

Windsoria Nuttall Gramineae

Origins:

For the British (b. Yorks.) botanist John Windsor, 1787-1868 (d. Manchester), surgeon, Fellow of the Linnean Society 1814, a friend of Nuttall; see John H. Barnhart, *Biographical Notes upon Botanists*. 3: 506. 1965; Ethelyn Maria Tucker, *Catalogue of the Library of the Arnold Arboretum of Harvard University*. Cambridge, Massachusetts 1917-1933; Ray Desmond, *Dictionary of British & Irish Botanists*

and Horticulturists*. 749. 1994; H.N. Clokie, *Account of the Herbaria of the Department of Botany in the University of Oxford*. 266. Oxford 1964; Jeannette Elizabeth Graustein, *Thomas Nuttall, Naturalist. Explorations in America, 1808-1841*. Cambridge, Harvard University Press 1967.

Windsorina Gleason Rapateaceae

Origins:

For the British royal family, the House of Windsor.

Winklerella Engl. Podostemaceae

Origins:

For the German botanist Hans Karl Albert Winkler, 1877-1945, plant collector, traveler; see John H. Barnhart, *Biographical Notes upon Botanists*. 3: 507. Boston 1965; F.N. Hepper and Fiona Neate, *Plant Collectors in West Africa*. 86. [genera *Winklera* Regel, *Winklera* P. & K., *Winkleria* Reichb. and *Winklerella* Engler dedicated to the German botanist Hubert J.P. Winkler, 1875-1941] Utrecht 1971; S. Lenley et al., *Catalog of the Manuscript and Archival Collections and Index to the Correspondence of John Torrey*. Library of the New York Botanical Garden. 439. 1973; E.M. Tucker, *Catalogue of the Library of the Arnold Arboretum of Harvard University*. Cambridge, Massachusetts 1917-1933; Stafleu and Cowan, *Taxonomic Literature*. 7: 371-373. 1988.

Wintera Forster f. Winteraceae

Origins:

To honor John Winter, traveled with Sir Francis Drake (c. 1540-1596, he was buried at sea) on his first voyage to Virginia in 1577, secretly commissioned by Queen Elizabeth I; it was an expedition against the Spanish colonies on the Pacific seabord of America; he was vice-admiral of Sir Francis Drake's voyage to Tierra del Fuego in 1578. See E.F. Benson, *Sir Francis Drake*. London 1927; H. Suzanne Maxwell and Martin F. Gardner, "The quest for Chilean green treasure: some notable British collectors before 1940." *The New Plantsman*. 4(4): 195-214. December 1997; H.E. Connor and E. Edgar, "Name changes in the indigenous New Zealand flora, 1960-1986 and Nomina Nova IV, 1983-1986." *New Zealand Journal of Botany*. Vol. 25: 115-170. 1987.

Wintera J.A. Murray Winteraceae

Origins:

To honor John Winter, traveled with Sir Francis Drake on his first voyage to Virginia in 1577.

Winterana L. Canellaceae (Winteraceae, Magnoliidae, Illiciales)

Origins:

To honor John Winter, traveled with Sir Francis Drake on his first voyage to Virginia in 1577; see Arthur D. Chapman, ed., *Australian Plant Name Index*. 2991. Canberra 1991; Linnaeus, *Syst. Nat.* ed. 10. 1041, 1045, 1370. 7 Jun 1759.

Winteria F. Ritter Cactaceae

Origins:

After Hildegard Winter, 1893-1975, she was the sister of the German cactus hunter and explorer Friedrich Ritter (1898-1989); see Gordon Douglas Rowley, *A History of Succulent Plants*. 379-380, 386. Strawberry Press, Mill Valley, California 1997.

Winterocereus Backeb. Cactaceae

Origins:

After Hildegard Winter, 1893-1975; see Gordon Douglas Rowley, *A History of Succulent Plants*. Strawberry Press, Mill Valley, California 1997.

Wislizenia Engelmann Capparidaceae

Origins:

Named for the German-born physician Friedrich (Frederick) Adolph(us) Wislizenus, 1810-1889, traveler, writer, naturalist, explorer, plant collector in the southwest United States, author of "Memoir of a tour of Northern México in 1846 and 1847." *Misc. Publ. 26. U.S. Senate*. Washington 1848; see Lyman David Benson, "The publication date of Wislizenus's memoir of a tour to northern Mexico in 1846 and 1847." *Cactus & Succ. Journ.* 46: 74. 1974 and 47: 40-43. 1975; J.H. Barnhart, *Biographical Notes upon Botanists*. 3: 509. 1965; Joseph Ewan, *Rocky Mountain Naturalists*. The University of Denver Press 1950; S. Lenley et al., *Catalog of the Manuscript and Archival Collections and Index to the Correspondence of John Torrey*. Library of the New York Botanical Garden. 472. 1973; E.M. Tucker, *Catalogue of the Library of the Arnold Arboretum of Harvard University*. Cambridge, Massachusetts 1917-1933; Howard Atwood Kelly and Walter Lincoln Burrage, *Dictionary of American Medical Biography*. New York 1928; Gordon Douglas Rowley, *A History of Succulent Plants*. Mill Valley, California 1997; Ida Kaplan Langman, *A Selected Guide to the Literature on the Flowering Plants of Mexico*. 808. Philadelphia 1964; Irving William Knobloch, compil., "A preliminary verified list of plant collectors in Mexico." *Phytologia*

Memoirs. VI. 1983; G.J. Englemann, "Frederick Adolphus Wislizenus." *Trans. St. Louis Acad. Sci.* 5: 464-468. 1892; Samuel Wood Geiser, *Naturalists of the Frontier*. Dallas 1948.

Species/Vernacular Names:

W. refracta Engelm.

English: jackass clover

Wissadula Medikus Malvaceae

Origins:

From an African vernacular name; see Helmut Genaust, *Etymologisches Wörterbuch der botanischen Pflanzennamen*. 693. 1996.

Species/Vernacular Names:

W. amplissima (L.) R. Fries

Yoruba: ifin, ewe ifin, lagbo lagbo funfun

Wistaria Sprengel Fabaceae

Origins:

An orthographic variant, see *Wisteria* Nutt.

Wisteria Nuttall Fabaceae

Origins:

For the American (b. Philadelphia, Pennsylvania) physician Caspar Wistar, 1761-1818 (d. Philadelphia), studied anatomy under John Hunter, professor of chemistry, 1787 elected to the American Philosophical Society, from 1792 professor of anatomy at the University of Pennsylvania, abolitionist, President of the Pennsylvania Abolition Society. His writings include (his graduation thesis) *De animo demisso*. 1786 and *A System of Anatomy for the Use of Students of Medicine*. Philadelphia 1811; see David Hosack, *A Tribute to the Memory of the Late C. Wistar, M.D.* New York 1818; John Golder, *Life of the Honourable W. Tilghman*, late Chief Justice ... of Pennsylvania. Compiled from the Eulogies of two distinguished Members of the Philadelphia Bar, etc. (An eulogium in commemoration of Doctor C. Wistar) Philadelphia 1829; B(enjamin) F(ranklin) Morris, *Life of Thomas Morris: Pioneer and long a Legislator of Ohio and U.S. Senator from 1833 to 1839*. Edited by His Son, B. F. Morris. Cincinnati 1856; C. Edwards Lester, *Life and Public Services of Charles Sumner*. (Printed and Bound by John F. Trow & Son for the) United States Publishing Company, New York 1874; Whitfield J. Bell, Jr., in *Dictionary of Scientific Biography* 14: 456-457. 1981; Peter Valder, *Wisterias*. Balmain, NSW, Florilegium 1995;

Thomas Nuttall (1786-1859), *The Genera of North American Plants*, and catalogue of the species, to the year 1817. 2: 115. Philadelphia 1818.

Species/Vernacular Names:

W. floribunda (Willd.) DC.

English: Japanese wisteria

Japan: fuji

W. frutescens (L.) Poiret

English: American wisteria

W. macrostachys (Torr. & A. Gray) Nutt.

English: Kentucky wisteria

W. sinensis (Sims) Sweet

English: Chinese wisteria, wisteria

China: zi teng, tzu teng, chao tu teng

W. venusta Rehd. & Wils.

English: silky wisteria

Withania Pauquy Solanaceae

Origins:

According to many authors the genus was named possibly (misspelling included!) after the English (b. Minster Acres, Northumberland) paleobotanist Henry Thomas Maire Witham (né Silvertop), 1779-1844 (d. Lartington Hall, Yorkshire), geologist, author of *Observations on Fossil Vegetables*. Edinburgh, London 1831 and *The Internal Structure of Fossile Vegetables*. Edinburgh, London 1833, in 1829 founded the Natural History Society of Northumberland, Durham, and Newcastle upon Tyne; see J.H. Barnhart, *Biographical Notes upon Botanists*. 3: 509. 1965; Albert G. Long, in *Dictionary of Scientific Biography* 14: 462-463. 1981; Francis Wall Oliver (1864-1951), ed., *Makers of British Botany*. Cambridge 1913; Arthur D. Chapman, ed., *Australian Plant Name Index*. 2992. Canberra 1991; Stafleu and Cowan, *Taxonomic Literature*. 4: 114-115. 1983.

Species/Vernacular Names:

W. coagulans (Stocks) Dunal

English: vegetable rennet

W. somnifera (L.) Dunal (*Physalis somnifera* L.; *Withania microphysalis* Suess.; *Withania kansuensis* Kuang & A.M. Lu)

English: poisonous gooseberry, winter cherry, clustered winter cherry, clustered withania, Kansu withania

Arabic: morgan, simm frakh, babu, ben nour, semm el-far, sharma

India: ashvagandha

Tibetan: a-sho gandha dman-pa

China: shui qie

Southern Africa: bitterappelliefie, geneesblaar, geneesblaarbossie, meidjieblaar, spanjoolbossie, stuipbossie, vernietbossie, vuilsiektebossie; bofepha (Sotho); ubuvuma (Xhosa); umuvimba (Zulu); vimhepe (Swati)

Witheringia L'Hérit. Solanaceae

Origins:

For the British (b. Wellington, Shropshire) physician and botanist William Withering, 1741-1799 (d. Birmingham), mineralogist, lichenologist, member of the Lunar Society of Birmingham, in 1785 a Fellow of the Royal Society, 1791 Fellow of the Linnean Society, friend of Erasmus Darwin (1731-1802). His writings include *A Botanical Arrangement*. London 1776, *Analyse chimica da agoa das Caldas da Rainha. A Chemical Analysis of the Water at Caldas da Rainha*. Lisboa 1795 and *An Account of the Scarlet Fever and Sore Throat*, or scarlatina anginosa; particularly as it appeared at Birmingham in the year 1778. London 1779; see T. Whitmore Peck and K. Douglas Wilkinson, *William Withering of Birmingham*. Bristol, London 1950; R. Zander, F. Encke, G. Buchheim and S. Seybold, *Handwörterbuch der Pflanzennamen*. 14. Aufl. 1993; William F. Bynum, in *Dictionary of Scientific Biography* 14: 463-465. 1981; John H. Barnhart, *Biographical Notes upon Botanists*. 3: 510. 1965; T. W. Bossert, *Biographical Dictionary of Botanists Represented in the Hunt Institute Portrait Collection*. 440. Boston, Massachusetts 1972; Jonas C. Dryander, *Catalogus bibliothecae historico-naturalis Josephi Banks*. London 1796-1800; E.M. Tucker, *Catalogue of the Library of the Arnold Arboretum of Harvard University*. Cambridge, Massachusetts 1917-1933; Andrew Thomas Gage, *A History of the Linnean Society of London*. London 1938; Blanche Henrey, *British Botanical and Horticultural Literature before 1800*. Oxford 1975; Garrison and Morton, *Medical Bibliography*. By Leslie T. Morton. 1836, 5079. New York 1961.

Witsenia Thunb. Iridaceae

Origins:

For Nicolaas Witsen, 1641-1717, a Dutch naturalist, Director of the Dutch East India Company; see Gordon Douglas Rowley, *A History of Succulent Plants*. 232. Strawberry Press, Mill Valley, California 1997; Mary Gunn and Leslie E. Codd, *Botanical Exploration of Southern Africa*. 38-42, 119. Cape Town 1981.

Wittia K. Schumann Cactaceae

Origins:

For N.H. Witt, plant collector in S. America.

Wittiocactus Rauschert Cactaceae

Origins:

For N.H. Witt, plant collector in S. America; see Helmut Genaust, *Etymologisches Wörterbuch der botanischen Pflanzennamen*. 693. Basel 1996.

Wittmackanthus Kuntze Rubiaceae

Origins:

After the German botanist Marx Carl Ludwig Wittmack, 1839-1929; see John H. Barnhart, *Biographical Notes upon Botanists*. 3: 511. 1965; E.M. Tucker, *Catalogue of the Library of the Arnold Arboretum of Harvard University*. Cambridge, Massachusetts 1917-1933; T.W. Bossert, *Biographical Dictionary of Botanists Represented in the Hunt Institute Portrait Collection*. 460. 1972; Ida Kaplan Langman, *A Selected Guide to the Literature on the Flowering Plants of Mexico*. 809. University of Pennsylvania Press, Philadelphia 1964; R. Zander, F. Encke, G. Buchheim and S. Seybold, *Handwörterbuch der Pflanzennamen*. 14. Aufl. Stuttgart 1993.

Wittmackia Mez Bromeliaceae

Origins:

After the German botanist Marx Carl Ludwig Wittmack, 1839-1929; see Stafleu and Cowan, *Taxonomic Literature*. 7: 404-406. 1988.

Wittrockia Lindm. Bromeliaceae

Origins:

After the Swedish botanist Veit Brecher Wittrock, 1839-1914; see Stafleu and Cowan, *Taxonomic Literature*. 7: 406-412. 1988; see John H. Barnhart, *Biographical Notes upon Botanists*. 3: 511. 1965S. Lenley et al., *Catalog of the Manuscript and Archival Collections and Index to the Correspondence of John Torrey*. Library of the New York Botanical Garden. 440. 1973; E.M. Tucker, *Catalogue of the Library of the Arnold Arboretum of Harvard University*. Cambridge, Massachusetts 1917-1933; Helmut Genaust, *Etymologisches Wörterbuch der botanischen Pflanzennamen*. 693. 1996; T.W. Bossert, *Biographical Dictionary of Botanists Represented in the Hunt Institute Portrait Collection*. 440. 1972; Ida Kaplan Langman, *A Selected Guide to the Literature on the Flowering Plants of Mexico*. 809. University of Pennsylvania Press, Philadelphia 1964.

Wittsteinia F. Muell. Alseuosmiaceae

Origins:

After the German pharmacist Georg Christian Wittstein, 1810-1887, teacher, wrote *Etymologisch-botanisches Handwörterbuch*. Ansbach 1852; see John H. Barnhart, *Biographical Notes upon Botanists*. 3: 511. 1965; E.M. Tucker, *Catalogue of the Library of the Arnold Arboretum of Harvard University*. Cambridge, Massachusetts 1917-1933; Ferdinand von Mueller, *Fragmenta Phytographiae Australiae*. 2: 136. 1861.

Wodyetia Irvine Palmae

Origins:

A vernacular name, from the aboriginal people of the Bathurst Bay area, Australia; see A. Irvine, "*Wodyetia*, a new Arecoid genus from Australia." *Principes*. 27(4): 158-167. 1983; Arthur D. Chapman, ed., *Australian Plant Name Index*. 2992. 1991.

Woikoia Baehni Sapotaceae

Origins:

A vernacular name.

Wokoia Baehni Sapotaceae

Origins:

A vernacular name.

Wolffia Horkel ex Schleiden Lemnaceae

Origins:

After the German physician Johann Friedrich Wolff, 1778-1806, Dr. med. Altorf 1801, botanist, author of Dissertatio inauguralis de *Lemna*. Altorfii et Norimbergiae 1801 and *Icones Cimicum* descriptionibus illustratae. Erlangae 1800-1811; see John H. Barnhart, *Biographical Notes upon Botanists*. 3: 514. 1965; Helmut Genaust, *Etymologisches Wörterbuch der botanischen Pflanzennamen*. 693. Basel 1996; F. Boerner & G. Kunkel, *Taschenwörterbuch der botanischen Pflanzennamen*. 4. Aufl. 187. Berlin & Hamburg 1989; James A. Baines, *Australian Plant Genera. An Etymological Dictionary of Australian Plant Genera*. 398. [genus named after Nathan Matthias von Wolff, 1724-1784] Chipping Norton, N.S.W. 1981; Francis Aubie Sharr, *Western Australian Plant Names and Their Meanings*. 75. University of Western Australia Press, Nedlands, Western Australia 1996.

Species/Vernacular Names:

W. angusta Landolt

English: water meal

W. arhiza (L.) Wimmer

English: duckweed, water meal

South Africa: eendekroos

W. brasiliensis Wedd.

English: water meal

W. columbiana Karst.

English: common wolffia

W. microscopica (Griffith) Kurz

English: water meal

Wolffiella (Hegelm.) Hegelm. Lemnaceae

Origins:

The diminutive of the genus *Wolffia*, for the German physician Johann Friedrich Wolff, 1778-1806.

Wolffiopsis Hartog & van der Plas Lemnaceae

Origins:

Resembling *Wolffia*, after the German physician Johann Friedrich Wolff, 1778-1806.

Wollemia W.G. Jones, K. Hill & J.M. Allen Araucariaceae

Origins:

A name for Wollemi pine, *Wollemia nobilis* Jones, Hill & Allen; the Wollemi pine was recently discovered by NPWS naturalists in a rainforest gorge and on rock ledges in Wollemi National Park, just 200 km from Sydney.

Woodfordia R.A. Salisbury Lythraceae

Origins:

Presumably for the English gardener E. John (or James? see Stafleu and Cowan, *Taxonomic Literature*. 7: 442. 1988) Alexander Woodford (fl. 1790s), owner of a garden at Belmont house, Vauxhall, London. Despite statements by some authors this genus was not named after the British physician James Woodforde, 1771-1837, botanist, Dr. med. Edinburgh 1825, Fellow of the Linnean Society 1826, author of *A Catalogue of the Indigenous Phenogamic Plants Growing in the Neighborhood of Edinburgh*; and of certain species of the class Cryptogamia: with reference to their localities. Edinburgh 1824; see John H. Barnhart, *Biographical Notes upon Botanists*. 3: 517. 1965; Ray Desmond, *Dictionary of British & Irish Botanists and Horticulturists*. 754. London 1994.

Species/Vernacular Names:

W. fruticosa (L.) Kurz

India: dhataki

Nepal: dhayaro

Woodia R. Brown Dryopteridaceae (Aspleniaceae, Woodsiaceae)

Origins:

See *Woodsia* R.Br.

Woodia Schlechter Asclepiadaceae

Origins:

After the Natal (b. England, Notts.) botanist John Medley Wood, 1827-1915 (d. Durban), 1852 to Durban, from 1882 to 1903 Curator of the Natal Botanic Garden and Director of the Natal Herbarium (1903-1913). His writings include *An Analytical Key to the Natural Orders and Genera of Natal Indigenous Plants*. Durban 1888, "Poisonous plants." *Natal Mercury*. 1894, "Indigenous food plants." *Rep. Colon. Herb. (Natal)* 1900: 12-24. 1901, *Handbook to the Flora of Natal*. Cape Town 1907 and *Catalogue of Plants in Natal Botanic Gardens*. Durban 1890, with Maurice Smethurst Evans (1854-1920) wrote *Natal Plants*. Durban 1898-1912; see Alain White and Boyd Lincoln Sloane, *The Stapelieae*. Pasadena 1937; Ray Desmond, *Dictionary of British & Irish Botanists and Horticulturists*. 753. 1994; J.H. Barnhart, *Biographical Notes upon Botanists*. 3: 516. 1965; T.W. Bossert, *Biographical Dictionary of Botanists Represented in the Hunt Institute Portrait Collection*. 442. 1972; E.M. Tucker, *Catalogue of the Library of the Arnold Arboretum of Harvard University*. 1917-1933; Mary Gunn and Leslie E. Codd, *Botanical Exploration of Southern Africa*. 379-381. Cape Town 1981; Ernest Nelmes and William Cuthbertson, *Curtis's Botanical Magazine Dedications, 1827-1927*. 338-340. [1931]; Gordon Douglas Rowley, *A History of Succulent Plants*. Mill Valley, California 1997; R. Zander, F. Encke, G. Buchheim and S. Seybold, *Handwörterbuch der Pflanzennamen*. 14. Aufl. Stuttgart 1993.

Woodiella Merrill Annonaceae

Origins:

Named for D.D. Wood, a forester in Borneo; see Frans A. Stafleu and Richard S. Cowan, *Taxonomic Literature*. 7: 439. Utrecht 1988.

Woodiellantha Rauschert Annonaceae

Origins:

For D.D. Wood, a forester in Borneo.

Woodrowia Stapf Gramineae

Origins:

To commemorate the Kew gardener George Marshall Woodrow, 1846-1911, Director of Botanical Survey of Western India 1893-1899; see Isaac Henry Burkill, *Chapters on the History of Botany in India*. Delhi 1965; Ray Desmond, *Dictionary of British & Irish Botanists and Horticulturists*. 755. 1994; M.P. Nayar, *Meaning of Indian Flowering Plant Names*. 365. Dehra Dun 1985.

Woodsia R. Brown Dryopteridaceae (Aspleniaceae, Woodsiaceae)

Origins:

After the English (b. London) architect Joseph Woods, 1776-1864 (Lewes, Sussex), botanist, Fellow of the Linnean Society 1807, author of *The Tourist's Flora*. London 1850 and *A Synopsis of the British Species of Rosa*. London 1818; see Dawson Turner and Lewis Weston Dillwyn (1778-1855), *The Botanist's Guide Through England and Wales*. London 1805; H.N. Clokie, *Account of the Herbaria of the Department of Botany in the University of Oxford*. 267. Oxford 1964; Mariella Azzarello Di Misa, a cura di, *Il Fondo Antico della Biblioteca dell'Orto Botanico di Palermo*. 294. Palermo 1988; R. Zander, F. Encke, G. Buchheim and S. Seybold, *Handwörterbuch der Pflanzennamen*. 14. Aufl. 802. Stuttgart 1993; John H. Barnhart, *Biographical Notes upon Botanists*. 3: 518. 1965; Robert Brown, *Prodromus Florae Novae Hollandiae*. 158. 1810; E.M. Tucker, *Catalogue of the Library of the Arnold Arboretum of Harvard University*. Cambridge, Massachusetts 1917-1933.

Woodsonia L.H. Bailey Palmae

Origins:

After the American botanist Robert Woodson, Jr., 1904-1963 (d. heart attack), traveler, Herbarium Missouri Botanical Garden; see John H. Barnhart, *Biographical Notes upon Botanists*. 3: 518. 1965; T.W. Bossert, *Biographical Dictionary of Botanists Represented in the Hunt Institute Portrait Collection*. 442. 1972; Ida Kaplan Langman, *A Selected Guide to the Literature on the Flowering Plants of Mexico*. 811-812. University of Pennsylvania Press, Philadelphia 1964; E.M. Tucker, *Catalogue of the Library of the Arnold Arboretum of Harvard University*. Cambridge, Massachusetts 1917-1933; Frans A. Stafleu and Richard S. Cowan, *Taxonomic Literature*. 7: 446-448. Utrecht 1988; R. Zander, F. Encke, G. Buchheim and S. Seybold, *Handwörterbuch der Pflanzennamen*. 14. Aufl. Stuttgart 1993.

Woodwardia J.E. Smith Blechnaceae (Blechnoideae)

Origins:

After the English (b. Huntingdon) botanist Thomas Jenkinson Woodward, 1745-1820 (d. Diss, Norfolk), 1789 a Fellow of the Linnean Society, contributor to Philip Miller (1691-1771), *The Gardener's and Botanist's Dictionary ... newly arranged ...* by Thomas Martyn. London 1807, with Samuel Goodenough (1743-1827) wrote *Observations on the British Fuci*. London 1797; see John H. Barnhart, *Biographical Notes upon Botanists*. 3: 518. Boston 1965; Jonas C. Dryander, *Catalogus bibliothecae historico-naturalis Josephi Banks*. London 1796-1800; Ray Desmond, *Dictionary of British & Irish Botanists and Horticulturists*. 756. 1994; Blanche Elizabeth Edith Henrey (1906-1983), *British Botanical and Horticultural Literature before 1800*. Oxford 1975; Helmut Genaust, *Etymologisches Wörterbuch der botanischen Pflanzennamen*. 693. [d. 1826] 1996; F. Boerner & G. Kunkel, *Taschenwörterbuch der botanischen Pflanzennamen*. 4. Aufl. 187. 1989.

Species/Vernacular Names:

W. areolata (L.) T. Moore (*Acrostichum areolatum* L.; *Lorinseria areolata* (L.) C. Presl)

English: chain fern

W. harlandii Hook. (named for the English (b. Yorks.) physician William Aurelius Harland, d. 1857 in Hong Kong, M.D. Edinburgh 1845, Government Surgeon at Hong Kong, in China before 1847, early plant collector and botanical explorer in Hong Kong, a friend of the botanist and investigator of the Chinese flora Henry Fletcher Hance (1827-1886), in 1847 H. published a treatise on Chinese anatomy and physiology, in 1850 an article on Chinese manufacture of magnetic needles and vermilion; see G.A.C. Herklots, *The Hong Kong Countryside*. 164. Hong Kong 1965; Emil Bretschneider (1833-1901), *History of European Botanical Discoveries in China*. [Reprint of the original edition 1898.] Leipzig 1981; Ray Desmond, *Dictionary of British & Irish Botanists and Horticulturists*. 318. 1994)

Japan: Ogimi-shida

W. japonica (L.f.) Sm.

China: kou chi

W. radicans (L.) Sm.

China: kuang chung, hei kou chi

W. virginica (L.) Smith (*Blechnum virginicum* L.; *Anchistea virginica* (L.) C. Presl)

English: Virginia chain fern

Wooleya L. Bolus Aizoaceae

Origins:

For a Major C.H.F. Wooley.

Woollsia F. Muell. Epacridaceae

Origins:

After the Australian (British-born, Winchester, Hants.) Rev. William Woolls, 1814-1893 (d. Burwood near Sydney, New South Wales), botanist, journalist, 1832 emigrated to Australia (New South Wales), 1865 Fellow of the Linnean Society, corresponded with F. von Mueller, a friend of Rev. James Walker (1794-1854), ordained 1873. His works include *A Contribution to the Flora of Australia.* Sydney 1867, *The Plants of the New South Wales.* Sydney 1885, *Progress of Botanical Discovery in Australia.* Sydney 1869 and *Plants Indigenous and Naturalised in the Neighborhood of Sydney.* Sydney 1891; see John H. Barnhart, *Biographical Notes upon Botanists.* 3: 520. 1965; Dennis John Carr and Stella Grace Maisie Carr, eds., *People and Plants in Australia.* 1981; M.M.H. Thompson, *William Woolls: A Man of Parramatta.* 1986; Ferdinand von Mueller, *Fragmenta Phytographiae Australiae.* 8: 52, 55. Melbourne 1873; T. W. Bossert, *Biographical Dictionary of Botanists Represented in the Hunt Institute Portrait Collection.* 442. Boston, Massachusetts 1972; Ethelyn Maria Tucker, *Catalogue of the Library of the Arnold Arboretum of Harvard University.* Cambridge, Massachusetts 1917-1933; N. Hall, *Botanists of the Eucalypts.* Melbourne 1978 and Supplement 1980; James A. Baines, *Australian Plant Genera. An Etymological Dictionary of Australian Plant Genera.* 398-399. N.S.W. 1981.

Species/Vernacular Names:

W. pungens (Cav.) F. Muell.

English: woollsia

Wootonella Standley Asteraceae

Origins:

After the American botanist Elmer Ottis Wooton, 1865-1945, with U.S. Department of Agriculture, traveler and plant collector, professor of biology; see John H. Barnhart, *Biographical Notes upon Botanists.* Boston 1965; Joseph Ewan, *Rocky Mountain Naturalists.* The University of Denver Press 1950; T.W. Bossert, *Biographical Dictionary of Botanists Represented in the Hunt Institute Portrait Collection.* 442. 1972; Ida Kaplan Langman, *A Selected Guide to the Literature on the Flowering Plants of Mexico.* 812.

Philadelphia 1964; Ethelyn Maria Tucker, *Catalogue of the Library of the Arnold Arboretum of Harvard University.* Cambridge, Massachusetts 1917-1933; Frans A. Stafleu and Richard S. Cowan, *Taxonomic Literature.* 7: 453-455. 1988; R. Zander, F. Encke, G. Buchheim and S. Seybold, *Handwörterbuch der Pflanzennamen.* 14. Aufl. Stuttgart 1993; Irving William Knobloch, compil., "A preliminary verified list of plant collectors in Mexico." *Phytologia Memoirs.* VI. Plainfield, N.J. 1983.

Wootonia Greene Asteraceae

Origins:

After the American botanist Elmer Ottis Wooton, 1865-1945.

Worsleya (Traub) Traub Amaryllidaceae (Liliaceae)

Origins:

For the British (b. London) horticulturist Arthington Worsley, 1861-1944, civil engineer, traveler and plant collector in Americas (Central and South), specialist in *Amaryllis*; see J.H. Barnhart, *Biographical Notes upon Botanists.* 3: 521. 1965; Ray Desmond, *Dictionary of British & Irish Botanists and Horticulturists.* 758. 1994; Gordon Douglas Rowley, *A History of Succulent Plants.* Strawberry Press, Mill Valley, California 1997; R. Zander, F. Encke, G. Buchheim and S. Seybold, *Handwörterbuch der Pflanzennamen.* 14. Aufl. 802. Stuttgart 1993.

Species/Vernacular Names:

W. rayneri (Hook.) Traub

English: blue amaryllis

Wrightea Roxburgh Palmae

Origins:

For the British botanist and physician William Wright, 1735-1819 (Edinburgh); see John H. Barnhart, *Biographical Notes upon Botanists.* 3: 524. 1965.

Wrightia R. Br. Apocynaceae

Origins:

The generic name honors the British (b. Perthshire) botanist and physician William Wright, 1735-1819 (d. Edinburgh), traveler, 1778 Fellow of the Royal Society, plant collector in Jamaica, Physician-General of Jamaica, discovered *Cinchona jamaicensis*, author of many botanical publications on Jamaican plants, he wrote "Description of the Jesuits bark tree of Jamaica and the Caribbees." *Phil. Trans. R. Soc.* London 67: 504-506. 1777 and "Descriptions and use of the Cabbage-bark tree of Jamaica." *id.* 67: 507-512. 1777. See *Memoir of ... W. Wright with Extracts from His Correspondence and a Selection of his Papers on Medical and Botanical Subjects.* Edinburgh & London 1828; John H. Barnhart, *Biographical Notes upon Botanists.* 3: 524. Boston 1965; R. Zander, F. Encke, G. Buchheim and S. Seybold, *Handwörterbuch der Pflanzennamen.* 14. Aufl. 802. [1740-1827] Stuttgart 1993; Robert Brown, *Prodromus Florae Novae Hollandiae.* 467. 1810 and "On the Asclepiadeae." *Memoirs of the Wernerian Natural History Society.* 1: 73. Edinburgh 1811; Stafleu and Cowan, *Taxonomic Literature.* 7: 469-470. 1988; H.N. Clokie, *Account of the Herbaria of the Department of Botany in the University of Oxford.* 268. Oxford 1964; Jonas C. Dryander, *Catalogus bibliothecae historico-naturalis Josephi Banks.* London 1796-1800; Ethelyn Maria Tucker, *Catalogue of the Library of the Arnold Arboretum of Harvard University.* Cambridge, Massachusetts 1917-1933; Ray Desmond, *Dictionary of British & Irish Botanists and Horticulturists.* 760. London 1994.

Species/Vernacular Names:

W. arborea (Dennst.) Mabberley (*Wrightia tomentosa* (Roxb.) Roemer & Schultes)

English: tomentose wrightia

China: yan mu

W. coccinea (Loddiges) Sims

English: scarlet wrightia

China: yun nan dao diao bi

W. laevis Hook.f.

English: smooth wrightia

China: lan shu

W. natalensis Stapf

English: saddle pod

Southern Africa: saalpeultjieboom; umPengende, umBengende, umTshatshali (Zulu)

W. pubescens R. Br. (*Wrightia kwangtungensis* Tsiang)

English: common wrightia

China: dao diao bi

W. sikkimensis Gamble

English: Sikkim wrightia

China: ge pu

W. tinctoria (Rottler) R. Br.

English: pala indigo plant

Wrixonia F. Muell. Labiatae

Origins:

For the Australian Sir Henry John Wrixon, 1839-1913, barrister-at-law, politician and patron of science, Member of Parliament for Belfast, Major of Melbourne, 1870 Solicitor-General of Victoria, 1886 Attorney-General of Victoria (Australia). Among his writings are *Democracy in Australia.* Melbourne 1868, *The Condition & Prospects of Australia as Compared with Older Lands.* Melbourne 1869, *Political Progress. Its Tendency and Limit.* Melbourne 1870, *Self-Government in Victoria.* Melbourne, Sydney and Adelaide 1876 and *Socialism: Being Notes on a Political Tour.* London 1896; see Percival Serle, *Dictionary of Australian Biography.* Vol. II, 512-513; Ferdinand von Mueller, *Fragmenta Phytographiae Australiae.* 10: 18. Melbourne 1876; Francis Aubie Sharr, *Western Australian Plant Names and Their Meanings.* 75. University of Western Australia Press, Nedlands, Western Australia 1996; James A. Baines, *Australian Plant Genera. An Etymological Dictionary of Australian Plant Genera.* 399. Chipping Norton, N.S.W. 1981.

Wulfenia Jacq. Scrophulariaceae

Origins:

For the Austrian botanist Franz Xavier (Xaver, see Pritzel and Genaust) Freiherr von Wulfen, 1728-1805, lichenologist, Jesuit, naturalist, botanical collector; see J.H. Barnhart, *Biographical Notes upon Botanists.* 3: 525. 1965; H. Genaust, *Etymologisches Wörterbuch der botanischen Pflanzennamen.* 693. 1996; R. Zander, F. Encke, G. Buchheim and S. Seybold, *Handwörterbuch der Pflanzennamen.* 14. Aufl. 802. Stuttgart 1993; Ethelyn Maria Tucker, *Catalogue of the Library of the Arnold Arboretum of Harvard University.* Cambridge, Massachusetts 1917-1933; T.W. Bossert, *Biographical Dictionary of Botanists Represented in the Hunt Institute Portrait Collection.* 444. 1972; Jonas C. Dryander, *Catalogus bibliothecae historico-naturalis Josephi Banks.* London 1796-1800; Antoine Lasègue, *Musée botanique de M. Benjamin Delessert.* 1845; Frans A. Stafleu, *Linnaeus and the Linnaeans.* 193. Utrecht 1971.

Wulfeniopsis D.Y. Hong Scrophulariaceae

Origins:
Resembling *Wulfenia.*

Wulffia Necker ex Cassini Asteraceae

Origins:

For the German physician Johann Christoph Wulff, d. 1767, botanist; see Jonas C. Dryander, *Catalogus bibliothecae historico-naturalis Josephi Banks.* London 1796-1800; Ethelyn Maria Tucker, *Catalogue of the Library of the Arnold Arboretum of Harvard University.* Cambridge, Massachusetts

1917-1933; Stafleu and Cowan, *Taxonomic Literature*. 7: 477-478. Utrecht 1988.

Wullschlaegelia Reichb.f. Orchidaceae

Origins:

For the German plant collector (in Jamaica) Henrich Rudolph Wullschlaegel (Wullschlägel), 1805-1864, teacher and missionary, author of *Deutsch-Negerenglisches Wörterbuch*. Löbau 1856 and *Lebensbilder aus der Geschichte der Brüdermission*. Stuttgart 1843-1848.

Wunschmannia Urban Bignoniaceae

Origins:

After the German botanist Ernst Wunschmann, b. 1848, naturalist.

Wurdackanthus Maguire Gentianaceae

Origins:

Dedicated to the American botanist John Julius Wurdack, 1921-1998, plant collector; see Julien Alfred Steyermark (1909-1988), "New Species of Rubiaceae from Peru collected by John Wurdack." *Bol. Soc. Venez. Ci. Nat.* 25: 232-244. 1964; Bassett Maguire (1904-1991), "Guttiferae." in B. Maguire, J.J. Wurdack and collaborators, "The Botany of the Guayana Highland-part IV(2)." *Mem. New York Bot. Gard.* 10(4): 21-32. 1961, "Rapateaceae." in B. Maguire, J.J. Wurdack and collaborators, "The Botany of the Guayana Highland-part VI." *Mem. New York Bot. Gard.* 12(3): 69-102. 1965; Bassett Maguire and Y.-C. Hung, "Styracaceae." in B. Maguire, J.J. Wurdack and collaborators, "The Botany of the Guayana Highland-part X." *Mem. New York Bot. Gard.* 29: 204-223. 1978; Bassett Maguire, J.A. Steyermark and D.G. Frodin, "Araliaceae." in B. Maguire, J.J. Wurdack and collaborators, "The Botany of the Guayana Highland-part XII." *Mem. New York Bot. Gard.* 38: 46-84. 1984; Lois Brako and James Lee Zarucchi, *Catalogue of the Flowering Plants and Gymnosperms of Peru*. 1249-1250. Missouri Botanical Garden, St. Louis, Missouri 1993; Laurence J. Dorr, "In Memoriam. John J. Wurdack, 1921-1998." in *Plant Science Bulletin*. 44(2): 41. Summer 1998.

Wurdackia Moldenke Eriocaulaceae

Origins:

Dedicated to the American botanist John Julius Wurdack, 1921-1998, plant collector; see Lois Brako and James Lee

Zarucchi, *Catalogue of the Flowering Plants and Gymnosperms of Peru*. 1229-1231. St. Louis, Missouri 1993; Laurence J. Dorr, "In Memoriam. John J. Wurdack, 1921-1998." in *Plant Science Bulletin*. 44(2): 41. Summer 1998.

Wurmbaea Steudel Colchicaceae (Liliaceae)

Origins:

Orthographic variant, see *Wurmbea* Thunb.

Wurmbea Thunb. Colchicaceae (Liliaceae)

Origins:

The genus was named by C.P. Thunberg after Friedrich von Wurmb, d. 1781 (or 1783, see Genaust), botanist, plant collector, Dutch colonial administrator, merchant and Secretary of the Academy of Sciences, Batavia (Dutch Netherlands); see Baron von Wurmb, *Briefe des Herrn von Wurmb und des Herrn Baron von Wollzogen auf ihren Reisen nach Afrika und Ostindien in den Jahren 1774 bis 1792*. Gotha 1794, and *Merkwürdigkeiten aus Ostindien, die Länder-VolkerKunde und Naturgeschichte betreffend ... Herausgegeben vom Major von Wurmb*. Gotha 1797; see F. Boerner & G. Kunkel, *Taschenwörterbuch der botanischen Pflanzennamen*. 4. Aufl. 187. Berlin & Hamburg 1989; Mary Gunn and Leslie E. Codd, *Botanical Exploration of Southern Africa*. 382. 1981; H. Genaust, *Etymologisches Wörterbuch der botanischen Pflanzennamen*. 693. 1996; R. Zander, F. Encke, G. Buchheim and S. Seybold, *Handwörterbuch der Pflanzennamen*. 14. Aufl. 802. 1993; Arthur D. Chapman, ed., *Australian Plant Name Index*. 2995-2996. Canberra 1991.

Species/Vernacular Names:

W. dioica (R. Br.) F. Muell. (*Anguillaria dioica* R. Br.)

English: early Nancy

W. recurva B. Nord.

South Africa: swartkoppie

W. spicata (Burm.f.) Dur. & Schinz var. *spicata*

South Africa: swartkoppie

W. spicata (Burm.f.) Dur. & Schinz var. *ustulata* (B. Nord.) B. Nord. (*Wurmbea conferta* N.E. Br.; *Wurmbea ustulata* B. Nord.)

South Africa: swartkoppie

Wycliffea Ewart & A. Petrie Caryophyllaceae

Origins:

Wycliffe, A.J.E., Australia; see Arthur D. Chapman, ed., *Australian Plant Name Index*. 2996. Canberra 1991.

Wyethia Nuttall Asteraceae

Origins:

For the American explorer Nathaniel Jarvis Wyeth, 1802-1856, plant collector for Nuttall, fur trader, led two expeditions to the Pacific, see *The Correspondence and Journals of Captain N.J. Wyeth, 1831-1836. A Record of Two Expeditions for the Occupation of the Oregon Country.* Edited by F.G. Young. [Oregon Historical Society. Sources of the History of Oregon. vol. 1. pt. 3-6.] 1899; Joseph Ewan, *Rocky Mountain Naturalists.* The University of Denver Press 1950; J.W. Harshberger, *The Botanists of Philadelphia and Their Work.* Philadelphia 1899; F. Boerner & G. Kunkel, *Taschenwörterbuch der botanischen Pflanzennamen.* 4. Aufl. 187. Berlin & Hamburg 1989.

Species/Vernacular Names:

W. elata H.M. Hall

English: Hall's wyethia

W. longicaulis A. Gray

English: Humboldt County wyethia

W. reticulata E. Greene

California: El Dorado County mule ears

X

Xamesike Raf. Euphorbiaceae

Origins:
Possibly derived from *chamaesyce*, see Constantine Samuel Rafinesque (1783-1840), *Flora Telluriana*. 4: 115. Philadelphia 1836 [1838]; E.D. Merrill (1876-1956), *Index rafinesquianus*. The plant names published by C.S. Rafinesque, etc. 157. Jamaica Plain, Massachusetts, USA 1949.

Xamesuke Raf. Euphorbiaceae

Origins:
Possibly derived from *chamaesyce*, see Constantine Samuel Rafinesque, *Aut. Bot.* 138. 1840; E.D. Merrill, *Index rafinesquianus*. 157. 1949.

Xananthes Raf. Lentibulariaceae

Origins:
See Constantine Samuel Rafinesque, *Flora Telluriana*. 4: 108. Philadelphia 1836 [1838]; E.D. Merrill, *Index rafinesquianus*. 222. 1949.

Xanthanthos St.-Lag. Gramineae

Origins:
Greek *xanthos* "yellow" and *anthos* "flower"; see also *Anthoxanthum* L.

Xantheranthemum Lindau Acanthaceae

Origins:
Greek *xanthos* "yellow" and the genus *Eranthemum*.

Xanthisma DC. Asteraceae

Origins:
From the Greek *xanthisma* "that which is dyed yellow, dyed," referring to the flowers.

Xanthium L. Asteraceae

Origins:
Greek *xanthion* "a plant used for dyeing yellow"; see Carl Linnaeus, *Species Plantarum*. 987. 1753 and *Genera Plantarum*. Ed. 5. 424. 1754; H. Genaust, *Etymologisches Wörterbuch der botanischen Pflanzennamen*. 693-694. 1996.

Species/Vernacular Names:
X. occidentale Bertol. (*Xanthium pungens* Wallr.)
Australia: Noogoora burr, clotburr, cockleburr
X. spinosum L.
English: burweed, spiny burweed, clotbur, spiny clotbur, cocklebur, dagger cocklebur, spiny cocklebur, dagger-weed, Bathurst bush, Bathurst burr
Arabic: shobbeit
Peru: allco-quisca, allcco-quisca, amor seco, espina de perro, Juan Alonso, yerba de Alonso, yerba de Juan Alonso
Bolivia: amor seco, clonqui, anu ch'ap'i, allqu k'iska
Southern Africa: boetebossie, boeteklits, pinotiebossie, speldebossie, vaalboetebossie, vernietbossie; hlaba-hlabane (Sotho); iligcume (Zulu); lepero (Tswana)
X. strumarium L. (*Xanthium natalense* Widder; *Xanthium californicum* Greene; *Xanthium italicum* Moretti; *Xanthium pensylvanicum* Wallr.; *Xanthium saccharatum* Wallr.) (with cushion-like swellings, Latin *struma* "a tumor, struma")
English: broad cocklebur, burweed, clotbur, cocklebur, large cocklebur
Arabic: shabka
Southern Africa: groot boetebossie, kankerroos; hlaba-hlabane (Sotho)
Japan: o-namomi, nan-mumi
Malaya: anjang
China: hsi erh, tsang erh, chuan erh
Tibetan: byi-tsher khyung-sder smug-po, byi-tsher
Hawaii: kikania

Xanthobrychis Galushko Fabaceae

Origins:
Greek *xanthos* "yellow" and *bryche* "gnashing, bellowing," *bryko, brycho* "bite, gobble, eat greedily"; see also *Onobrychis* Mill.

Xanthocephalum Willd. Asteraceae

Origins:
Greek *xanthos* "yellow" and *kephale* "head," an allusion to the color of the capitula or flower-heads.

Xanthoceras Bunge Sapindaceae

Origins:
Greek *xanthos* "yellow" and *keras* "a horn," referring to the growths between the petals.

Species/Vernacular Names:
X. sorbifolium Bunge
China: wen kuang kuo

Xanthocercis Baillon Fabaceae

Origins:
Yellow *Cercis*, from the Greek *xanthos* "yellow" and the genus *Cercis*, Greek *kerkis*, *kerkidos* "a measuring rod, shuttle."

Species/Vernacular Names:
X. zambesiaca (Baker) Dumaz-le-Grand (*Pseudocadia zambesiaca* (Baker) Harms; *Sophora zambesiaca* Beker)
English: Nyala tree
Southern Africa: hoenderspoor, nhlaru, Njalaboom; nhlahu (Thonga); motha (Mangwato dialect, Botswana); muChetuchetu, muSharo, muTsha (Shona)

Xanthochrysum Turcz. Asteraceae

Origins:
Greek *xanthos* "yellow" and *chrysos* "gold, golden," genus *Helichrysum* Miller; see Porphir Kiril N.S. Turczaninow, *Bulletin de la Société Impériale des Naturalistes de Moscou.* 24(1): 199. 1851.

Xanthochymus Roxb. Guttiferae

Origins:
Greek *xanthos* "yellow" and *chymos* "juice."

Xanthogalum Avé-Lall. Umbelliferae

Origins:
From the Greek *xanthos* "yellow" and *gala*, *galaktos* "milk."

Xanthomyrtus Diels Myrtaceae

Origins:
From the Greek *xanthos* plus *Myrtus*.

Xanthonanthus St.-Lag. Gramineae

Origins:
A word play, an anagram of the generic name *Anthoxanthum* L., Greek *xanthos* "yellow" and *anthos* "flower"; see also *Xanthanthos* St.-Lag.

Xanthopappus C. Winkler Asteraceae

Origins:
From the Greek *xanthos* "yellow" and *pappos* "fluff, downy appendage."

Xanthophyllum Roxb. Xanthophyllaceae (Polygalaceae)

Origins:
Greek *xanthos* "yellow" and *phyllon* "leaf"; see William Roxburgh (1751-1815), *Plants of the Coast of Coromandel.* 3: 81. London 1795-1820.

Xanthophytopsis Pitard Rubiaceae

Origins:
Resembling *Xanthophytum* Reinw. ex Blume.

Xanthophytum Reinw. ex Blume Rubiaceae

Origins:
From the Greek *xanthos* "yellow" and *phyton* "plant."

Xanthorhiza Marshall Ranunculaceae

Origins:
From the Greek *xanthos* "yellow" and *rhiza* "root."

Xanthorrhoea Smith Xanthorrhoeaceae (Liliaceae)

Origins:
Greek *xanthos* "yellow" and *rheo* "to flow," referring to a yellow resinous gum extracted from the stem; see J.E. Smith, in *Transactions of the Linnean Society of London. Botany.* 4: 219. (May) 1798; Arthur D. Chapman, ed., *Australian Plant Name Index.* 2997-3001. Canberra 1991.

Species/Vernacular Names:

X. australis R. Br.

English: Austral grass-tree

X. quadrangulata F. Muell.

English: Mount Lofty grass-tree

Xanthosia Rudge Umbelliferae

Origins:

Greek *xanthos* "yellow," referring to the colors of the hairs; see E. Rudge, in *Transactions of the Linnean Society of London. Botany.* 10: 301. (Sept.) 1811; Arthur D. Chapman, ed., *Australian Plant Name Index.* 3001-3003. Canberra 1991.

Species/Vernacular Names:

X. dissecta Hook.f. (*Xanthosia leiophylla* F. Muell. ex Klatt)

English: creeping carrot

Xanthosoma Schott Araceae

Origins:

From the Greek *xanthos* with *soma* "a body," referring to the stigma or to the yellow inner tissues; see D.H. Nicolson, "Derivation of aroid generic names." *Aroideana.* 10: 15-25. 1988.

Species/Vernacular Names:

X. sp.

Peru: hualusa, bore, cacabo, cocoyam, mairina, ñuá, papa china, sacha papa, taioba, uncucha, uncuchu

Mexico: guia bu niza, quije pu niza

Yoruba: koko arira oja

X. sagittifolium (L.) Schott

English: tannia, tania

Peru: malanga, mairino, mangará, mangareto, pituca, taiazes, taro, yautia, yantia

Brazil: taioba

Japan: taro-imo, yabane-imo

Okinawa: seiban-imo

X. violaceum Schott

English: blue tannia, blue taro

Peru: daledale

Brazil: taioba

Xanthostachya Bremek. Acanthaceae

Origins:

Greek *xanthos* "yellow" and *stachys* "spike."

Xanthostemon F. Muell. Myrtaceae

Origins:

Greek *xanthos* "yellow" and *stemon* "stamen"; in *Hooker's Journal of Botany & Kew Garden Miscellany.* 9: 17. 1857.

Species/Vernacular Names:

X. oppositifolius Bailey

Australia: penda

Xanthoxalis Small Oxalidaceae

Origins:

From the Greek *xanthos* "yellow" and the genus *Oxalis* L., *oxys* "acid, sour, sharp."

Xanthoxylon Sprengel Rutaceae

Origins:

An orthographic variant, see *Zanthoxylum* L.

Xanthoxylum Miller Rutaceae

Origins:

An orthographic variant, see *Zanthoxylum* L.

Xantolis Raf. Sapotaceae

Origins:

See C.S. Rafinesque, *Sylva Telluriana.* 36. Philadelphia 1838; E.D. Merrill, *Index rafinesquianus.* The plant names published by C.S. Rafinesque, etc. 188. Jamaica Plain, Massachusetts, USA 1949.

Species/Vernacular Names:

X. longispinosa (Merrill) H.S. Lo

English: longspine xantolis

China: qiong ci lan

X. stenosepala (Hu) P. Royen

English: narrow sepal xantolis

China: dian ci lan

X. shweliensis (W.W. Smith) P. Royen

English: Shweli xantolis

China: rui li ci lan

Xantorrhoea Diels Xanthorrhoeaceae

Origins:

An orthographic variant, see *Xanthorrhoea* L.

Xaritonia Raf. Orchidaceae

Origins:

Named for a nymph, according to Rafinesque; see C.S. Rafinesque, *Flora Telluriana*. 4: 9. 1836 [1838]; E.D. Merrill, *Index rafinesquianus*. 105. Jamaica Plain, Massachusetts, USA 1949.

Xeilyathum Raf. Orchidaceae

Origins:

According to Rafinesque from the Greek *cheilos* "lip"; see C.S. Rafinesque, *Flora Telluriana*. 2: 62. 1836 [1837]; E.D. Merrill, *Index rafinesquianus*. 105. Jamaica Plain, Massachusetts, USA 1949.

Xenacanthus Bremek. Acanthaceae

Origins:

From the Greek *xenos* "a host, foreigner, alien, stranger, guest" and the genus *Acanthus* L.

Xenikophyton Garay Orchidaceae

Origins:

From the Greek *xenikos*, "strange, foreign, peculiar to a stranger" and *phyton* "plant."

Xenochloa Roemer & Schultes Gramineae

Origins:

Greek *xenos* "a host, foreigner, alien, stranger, guest" and *chloe, chloa* "grass."

Xenodendron K. Schum. & Lauterb. Myrtaceae

Origins:

Greek *xenos* "foreigner, alien, stranger, guest" and *dendron* "tree."

Xenophya Schott Araceae

Origins:

Greek *xenophyes* "strange of shape or nature," *phye* "shape, nature, growth," *phyo* "to grow"; see D.H. Nicolson, "Derivation of aroid generic names." *Aroideana*. 10: 15-25. 1988.

Xenostegia D.F. Austin & Staples Convolvulaceae

Origins:

From the Greek *xenos* "foreigner, alien, stranger" and *stege, stegos* "roof, cover."

Xeraea Kuntze Amaranthaceae

Origins:

Greek *xeros* "dry"; see Otto Kuntze (1843-1907), *Revisio generum Plantarum*. 2: 545. (Nov.) 1891.

Xeranthemum L. Asteraceae

Origins:

Greek *xeros* "dry" and *anthemon, anthos* "flower," referring to the everlasting flowers; see Carl Linnaeus, *Species Plantarum*. 857. 1753 and *Genera Plantarum*. Ed. 5. 369. 1754.

Xeroaloysia Tronc. Verbenaceae

Origins:

From the Greek *xeros* "dry" plus the genus *Aloysia* Juss.

Xerocarpa (G. Don) Spach Goodeniaceae

Origins:

Greek *xeros* "dry" and *karpos* "fruit."

Xerocarpa H.J. Lam Labiatae (Verbenaceae)

Origins:

From the Greek *xeros* "dry" and *karpos* "fruit."

Xerocarpus Guill. & Perr. Fabaceae

Origins:

Greek *xeros* "dry" and *karpos* "fruit."

Xerocassia Britton & Rose Caesalpiniaceae

Origins:
Greek *xeros* "dry" plus the genus *Cassia* L.

Xerochlamys Baker Sarcolaenaceae

Origins:
From the Greek *xeros* "dry" plus *chlamys, chlamydos* "cloak, mantle."

Xerochloa R. Br. Gramineae

Origins:
Greek *xeros* "dry" and *chloe, chloa* "grass"; see Robert Brown, *Prodromus Florae Novae Hollandiae*. 196. 1810.

Xerocladia Harvey Mimosaceae

Origins:
From the Greek *xeros* "dry" and *klados* "a branch."

Xerococcus Oersted Rubiaceae

Origins:
From the Greek *xeros* "dry" and *kokkos* "a berry."

Xeroderris Roberty Fabaceae

Origins:
From the Greek *xeros* "dry" and the genus *Derris*.

Species/Vernacular Names:
X. stuhlmannii (Taub.) Mendonça & Sousa (*Deguelia stuhlmannii* Taub.; *Xeroderris chevalier* (Dunn) Roberty; *Derris stuhlmannii* (Taub.) Harms; *Ostryoderris stuhlmannii* (Taub.) Dunn ex Harms) (for the German botanical collector in East Africa Franz Ludwig Stuhlmann, 1863-1928, Director of the Biological-agricultural institute at Amani 1903-1908, author of *Mit Emin Pasha ins Herz von Afrika*. Berlin 1894 and *Beiträge zur Kulturgeschichte von Ostafrika*. Berlin 1909; see J.H. Barnhart, *Biographical Notes upon Botanists*. 3: 343. 1965; E.M. Tucker, *Catalogue of the Library of the Arnold Arboretum of Harvard University*. Cambridge, Massachusetts 1917-1933; Alain White and Boyd Lincoln Sloane, *The Stapelieae*. Pasadena 1937)

Nigeria: durbi (Hausa)

Yoruba: erumaki

Southern Africa: muzamalowa; muzamalowa (Kololo); murumanyama (Shona)

W. Africa: kalajege, kalajige

Xerodraba Skottsb. Brassicaceae

Origins:
From the Greek *xeros* "dry" plus the genus *Draba* L.

Xerolekia A. Anderb. Asteraceae

Origins:
Greek *xeros* "dry" plus the genus *Telekia* Baumg.

Xerolirion A.S. George Xanthorrhoeaceae (Lomandraceae)

Origins:
Greek *xeros* "dry" plus *leirion* "a lily"; see Alexander Segger George (b. 1939), in *Flora of Australia*. 46: 229. 1986.

Xeromphis Raf. Rubiaceae

Origins:
Or a meaningless name, or possibly from the Greek *xeros* "dry" and *omphalos* "umbilicus"; see Constantine S. Rafinesque, *Sylva Telluriana*. 21. Philadelphia 1838; E.D. Merrill, *Index rafinesquianus*. 227. 1949.

Xeronema Brongn. & Gris Phormiaceae (Liliaceae)

Origins:
Greek *xeros* "dry" and *nema* "thread, filament," referring to the dry and persistent filaments.

Species/Vernacular Names:
X. callistemon W. Oliver

English: Poor Knights Island lily

Xerophyllum Michaux Melanthiaceae (Liliaceae)

Origins:
Greek *xeros* "dry" and *phyllon* "leaf," an allusion to the dry, tough and grass-like leaves.

Species/Vernacular Names:
X. tenax (Pursh) Nutt.
English: bear grass

Xerophyta Juss. Velloziaceae

Origins:
From the Greek *xeros* "dry" and *phyton* "plant."

Xeroplana Briq. Stilbaceae (Verbenaceae)

Origins:
Greek *xeros* "dry" and *planos* "a wandering, roaming."

Xerorchis Schltr. Orchidaceae

Origins:
Greek *xeros* "dry" plus *orchis* "orchid," descriptive of the habitat, xerophytic.

Xerosicyos Humbert Cucurbitaceae

Origins:
From the Greek *xeros* "dry" and *sikyos* "wild cucumber, gourd."

Xerosiphon Turcz. Amaranthaceae

Origins:
Greek *xeros* "dry" and *siphon* "tube."

Xerosollya Turcz. Pittosporaceae

Origins:
From the Greek *xeros* plus the genus *Sollya* Lindley; see Porphir Kiril N.S. Turczaninow (1796-1863), *Bulletin de la Société Impériale des Naturalistes de Moscou.* 27(2): 362. 1854.

Xerospermum Blume Sapindaceae

Origins:
From the Greek *xeros* "dry" and *sperma* "a seed," referring to the nature of the seeds.

Species/Vernacular Names:
X. sp.
Malaya: nasi dingi, buah sakor
X. noronhianum Blume
Malaya: kikir buntal, gigi buntal, rambutan pachat

Xerosphaera Soják Fabaceae

Origins:
From the Greek *xeros* "dry" and *sphaira* "sphere, ball, globe."

Xerospiraea Henrickson Rosaceae

Origins:
From the Greek *xeros* "dry" plus the genus *Spiraea* L.

Xerotecoma J. Gómes Bignoniaceae

Origins:
Greek *xeros* "dry" plus the genus *Tecoma* Juss.

Xerotes R. Br. Xanthorrhoeaceae (Lomandraceae)

Origins:
From the Greek *xerotes* "dryness, drought, thirst," *xeros* "dry"; see R. Brown, *Prodromus Florae Novae Hollandiae.* 259. 1810; Arthur D. Chapman, ed., *Australian Plant Name Index.* 3005-3012. Canberra 1991.

Xerothamnella C.T. White Acanthaceae

Origins:
Greek *xeros* "dry" and *thamnos* "shrub, bush"; see Cyril Tenison White, in *Proceedings of the Royal Society of Queensland.* 55: 72. (Feb.) 1944.

Xerotia Oliver Caryophyllaceae (Illecebraceae)

Origins:
Greek *xerotes* "dryness, drought, thirst."

Ximenia L. Olacaceae

Origins:

The genus was named in honor of the Spanish monk and botanist Francisco Ximenez, a native of Luna in the Kingdom of Aragon, in 1605 he came to New Spain, in 1612 (25th February) he became a lay brother of the Convento de Santo Domingo de Mexico; he edited and translated, by the Spanish physician Francisco Hernandez (1514-1587), *Quatro libros* de la naturaleza y virtudes de las plantas y animales que estan recevidos en el uso de *Medicina en la Nueva España* ... Traduzido, y aumentados muchos simples, y Compuestos y otro muchos secretos curativos, por Fr. *Francisco Ximenez* ... Mexico 1615, and *Rerum Medicarum Novae Hispaniae Thesaurus.* Romae 1651; see Francisco Guerra, *Bibliografia de la Materia Medica Mexicana.* 319 (under Ximenez, 203 fol.) Mexico City 1950; Mario Sartor, "Libri dell'altro mondo." *La Bibliofilia.* Anno XCLX. 1: 1-37. 1997; Carl Linnaeus, *Species Plantarum.* 1193. 1753 and *Genera Plantarum.* Ed. 5. 500. 1754.

Species/Vernacular Names:

X. americana L. (*Ximenia exarmata* F. Muell.)

English: blue sour plum, American hog plum, tallow wood, yellow plum, hog plum, monkey plum, tallow nut, wild oliver, wild lime, spiny mountain

French: citron de mer, pommiere de Cithère

Malaya: bedarah laut, rukam laut

Nigeria: tsada, tswada, ego, igo; tsada (Hausa); chabbuli (Fula); anomadze (Tiv); igo (Yoruba)

Yoruba: igo, ego

Mali: tonkè, sinkè, sene, kagban, gban

Florida/British West Indies: hog plum, tallow wood, mountain plum, false sandalwood, wild olive

Latin America: cimelillo, limoncilla, manzanillo, ciruelo; xkuk-ché (Yucatan, Maya); pepe nance (El Salvador); chocomico (Nicaragua); limoncillo (Colombia); yanà, jia manzanilla, ciruelo cimarron, ciruelillo (Cuba); manzanilla (Guatemala); manzanilla Honduras); albarillo del campo (Argentina)

X. americana L. var. *americana*

Southern Africa: blousuurpruim, blue sourplum; umThunduluka-omncane, umKholotshwana (Zulu)

X. americana L. var. *microphylla* Welw. ex Oliv. (*Ximenia rogersii* Burtt Davy)

English: sour plum

Southern Africa: suurpruim, sour plum; uKolotshane (Zulu); mohambia (Yei, Ngamiland); umThunduluka (Swazi: Swaziland and eastern Transvaal); ntsengele (eastern Transvaal); morotologana (Tswana: western Transvaal, northern Cape, Botswana); morotologa (Tawana dialect, Ngamiland); morotologa, motShidi, hwele, moSidi-wa-

serotologane (north Sotho: north and northeast Transvaal); mutanzwa (Venda: Soutpansberg, northern Transvaal); omuninga (Herero: central southwest Africa); osipeke (northern southwest Africa)

X. caffra Sond. var. *caffra*

English: Natal sour plum, sour plum, sour plum of the bushveld

Tanzania: loma, mloma, ngoromoko

Southern Africa: Natalsuurpruim, suurpruim; umThunduluka-obomvu, umThunduluka, umThundulukwa, umGwenya (Zulu); anTshunduluka (Thonga); umThunduluka (Swazi: Swaziland and eastern Transvaal); muSanza, muTengeni, muTengeno, iTsengeni, muTunguru (Shona); iTsengeni (Ndebele: central and southern Transvaal); morotologa, morokolo (Tswana: western Transvaal, northern Cape, Botswana); morotonoga (Tawana dialect, Ngamiland); moretologa-kgomo (Malete dialect, Botswana); mosidi (north Sotho: north and northeast Transvaal); mutanzwa (Venda: Soutpansberg, northern Transvaal); omumbeke (Herero: central southwest Africa); mohambia (Yei, Ngamiland)

N. Rhodesia: musongosongo

Ximeniopsis Alain Olacaceae

Origins:

Resembling the genus *Ximenia* L.

Xiphagrostis Coville Gramineae

Origins:

From the Greek *xiphos* "a sword" plus *agrostis, agrostidos* "grass, weed, couch grass."

Xiphidium Aublet Haemodoraceae

Origins:

From the Greek *xiphos* "a sword, dagger," the diminutive *xiphidion*.

Xiphion Miller Iridaceae

Origins:

Theophrastus (*HP.* 6.8.1) and Dioscorides used *xiphion* for a species of *Gladiolus*, corn-flag, from the Greek *xiphos* "sword," referring to the sword-like leaves.

Xiphium Mill. Iridaceae

Origins:
See *Xiphion*.

Xiphizusa Reichb.f. Orchidaceae

Origins:
Greek *xiphizein, xiphizo* "dance the sword-dance, to dance with arms extended as if holding a sword," referring to the nature of the lip.

Xiphochaeta Poeppig Asteraceae

Origins:
Greek *xiphos* "sword" and *chaite* "a bristle."

Xiphophyllum Ehrh. Orchidaceae

Origins:
Lanceolate leaves, Greek *xiphos* "sword" and *phyllon* "leaf," referring to the type *Serapias xiphophyllum*.

Xiphopteris Kaulf. Grammitidaceae

Origins:
From the Greek *xiphos* "sword" and *pteris* "a fern," sword fern.

Xiphosium Griffith Orchidaceae

Origins:
From the Greek *xiphos* "sword," descriptive of the bracts of the scape.

Xiphotheca Ecklon & Zeyher Fabaceae

Origins:
Greek *xiphos* "sword" and *theke* "case."

Xolisma Raf. Ericaceae

Origins:
See C.S. Rafinesque, in *Am. Monthly Mag. Crit. Rev.* 4: 193. 1819; E.D. Merrill, *Index rafinesquianus*. 185-186. 1949.

Xuris Raf. Xyridaceae (Commelinidae, Commelinales)

Origins:
An orthographic variant, see *Xyris* L.; see Constantine Samuel Rafinesque, *Flora Telluriana*. 2: 14. 1836 [1837]; E.D. Merrill, *Index rafinesquianus*. 82. 1949.

Xylanche Beck Orobanchaceae (Scrophulariaceae)

Origins:
Greek *xylon* "wood" and *ancho* "to bind, to strangle," *anchein* "to strangle," *anche* "poison," *anchi* "near, close by," Akkadian *hanaqu* "to strangle, to constrict, to compress, suffocation," Latin *angina, ango* "to press tight."

Xylanthemum Tzvelev Asteraceae

Origins:
From the Greek *xylon* "wood" and *anthemon* "flower."

Xylia Benth. Mimosaceae

Origins:
From the Greek *xylon* "wood," the timber is very hard.

Species/Vernacular Names:
X. sp.

Ghana: kotoprepre

X. ghesquieri Robyns

Zaire: kariaria (from ku lia = to cry)

X. torreana Brenan (after the Portuguese botanist Antonio Rocha da Torre, born 1904, plant collector in Mozambique and student of that flora)

English: sand ash

South Africa: xylia, sandessenhout

X. xylocarpa (Roxb.) Theob.

English: ironwood

The Philippines: acle

India: eruvalu, jamba, jambe, panga, suria, tirawa, trul, trumalla, yerul, boja, konda tangedu, kongora

Thailand: mai deng

Burma: cam ne, mai sa-lan, pkhay, praing, pran, pyin, pyingado

Indochina: camxe, cam xé, chau kram, deng, sokram

Xylinabariopsis Pit. Apocynaceae

Origins:
Resembling the related genus *Xylinabaria* Pierre.

Xylobium Lindley Orchidaceae

Origins:
Greek *xylon* "wood, log" and *bios* "life," epiphytic orchids growing on trees.

Xylocalyx Balf.f. Scrophulariaceae

Origins:
From the Greek *xylon* "wood" and *kalyx* "a calyx."

Xylocarpus J. König Meliaceae

Origins:
From the Greek *xylon* "wood" and *karpos* "a fruit," alluding to the woody fruits.

Xylochlamys Domin Loranthaceae

Origins:
From the Greek *xylon* "wood" and *chlamys, chlamydos* "cloak, mantle."

Xylococcus Nutt. Ericaceae

Origins:
Greek *xylon* "wood" and *kokkos* "berry"; see Thomas Nuttall (1786-1859), in *Transactions of the American Philosophical Society*. ser. 2. 8: 258. 1843.

Xylococcus R. Br. ex Britt. & S. Moore Euphorbiaceae

Origins:
Greek *xylon* "wood" and *kokkos* "berry."

Xylolobus Kuntze Mimosaceae

Origins:
From the Greek *xylon* "wood" and *lobos* "a pod, capsule"; see also *Xylia* Benth.

Xylomelum Sm. Proteaceae

Origins:
Greek *xylon* "wood" and *melon* "an apple," referring to the woody and pear-shaped fruits; see J.E. Smith, in *Transactions of the Linnean Society of London. Botany*. 4: 214. 1798.

Xylonagra J.Donn. Sm. & Rose Onagraceae

Origins:
From the Greek *xylon* "wood" and the genus *Onagra* Mill., Dioscorides (4.117) and Galenus used *onagra, onagron* for oleander.

Xylonymus Kalkman ex Ding Hou Celastraceae

Origins:
From the Greek *xylon* "wood" plus the genus *Euonymus* L.

Xyloolaena Baillon Sarcolaenaceae

Origins:
From the Greek *xylon* "wood" and *chlaena* "a cloak, blanket."

Xylophacos Rydb. Fabaceae

Origins:
From the Greek *xylon* "wood" and *phakos* "a lentil."

Xylophragma Sprague Bignoniaceae

Origins:
From the Greek *xylon* "wood" and *phragma* "a partition, compartment, wall, fence."

Xylophylla L. Euphorbiaceae

Origins:
From the Greek *xylon* "wood" and *phyllon* "a leaf"; see also *Phyllanthus* L.

Xylophyllos Kuntze Santalaceae

Origins:
Greek *xylon* "wood" and *phyllon* "a leaf"; see Otto Kuntze (1843-1907), in *Revisio generum Plantarum*. 2: 589. (Nov.) 1891.

Xylopia L. Annonaceae

Origins:

Greek *xylon* "wood" and *ope* "opening, hole," some suggest from *pikros*, *pikron* "bitter, pungent"; see C. Linnaeus, *Systema Naturae*. 2: 1241, 1250, 1378. 1759; Helmut Genaust, *Etymologisches Wörterbuch der botanischen Pflanzennamen*. 695-696. 1996; F. Boerner & G. Kunkel, *Taschenwörterbuch der botanischen Pflanzennamen*. 4. Aufl. 187. Berlin & Hamburg 1989.

Species/Vernacular Names:

X. sp.

Peru: anona, mataro grande, sabina, sabino, espintana negra, pinsha caspi

The Philippines: white lanutan

X. aethiopica (Dunal) A. Rich.

English: negro pepper, Guinea pepper

French: poivrier de Guinée, poivre de singe

Togo: ssosi, tso

Central Africa: bosange, bossanghe, nsanghe, diluluka, inkala, kassana, kwa, makwa, likungu, mukala, mukula, mukuba, ndunga, okala, akwi, kimba, mbare, okala, bolopharan, halédé, simingui

Congo: kani

Sierra Leone: hewe

Nigeria: aghako, atta, akadu, eru, uda, sesedo, unien, kanifig, fondi, olorin, kimba, arunje, erinje; erunji (Yoruba); unien (Edo); kimba (Hausa); kimbare (Fulani); kyimba (Arabic); tsunfyanya (Nupe); uda (Igbo); atta (Efik); kenya (Boki); ata (Ibibio)

Yoruba: eeru, eerunje, olorin

Zaire: bikuwe, bossanghe, bossanghi, inkala, kuwa, makuwa, mokwa, mossanghe, mossanki, mukwa, ndunga

Liberia: deo

Cameroon: ebongo mbonji, ikola hindi, kimba, okolo

Ivory Coast: efomu, fonde, poivrier de Guinée

Gabon: kanifing, ogana, okala

X. aromatica (Lamarck) C. Martius (*Uvaria aromatica* Lamarck)

Peru: mataro, omechuai caspi

X. calophylla R.E. Fries

Bolivia: piraquina blanca

X. chrysophylla Louis ex Boutique

Central Africa: bompaie bo fufow, bonsanghe, nsanghe, ishanga, lole mutoke, lukangwa, lusangi, mbomente, ndunga, ontshengi

X. ferruginea Hook.f. & Thomson

Malaya: stilted antoi, antoi jangkang, jangkang bukit, jangkang betina, jangkang merah, jangkang paya, medang kenanga, pisang pisang jari, jari ayam

X. gilbertii Boutique

Central Africa: bompaie bo fufow, likungu, lole mutoke, yanga

X. hypolampra Mildbr.

Central Africa: abies, lukangua, biez, moinzou, ndong-éli, moley, sangui

Cameroon: nom akwi, sangui, moley

Congo: biez

Gabon: ndong-eli

X. ligustrifolia Humb. & Bonpl. ex Dunal

Bolivia: piraquina

X. odoratissima Welw. ex Oliv. (*Xylopia antunesii* Engl. & Diels)

English: fragrant bitterwood, savanna bitterwood

Southern Africa: savanna bitterhout; mogohlo (north Sotho); muvhula-vhusiku (Venda); muTsitsi (Shona)

X. parviflora Spruce (*Xylopia neglecta* (Kuntze) R.E. Fries, nom. illeg.)

Nigeria: sesedo (Yoruba); aghako (Edo); kimba (Hausa)

Yoruba: sesedo

South America: espintana de varillal

X. peruviana R.E. Fries

Peru: pichi varilla

Bolivia: tetemetsijoni

South America: espintana

X. phloidora Mildbr.

Central Africa: bendjo, bolonda, bompaie bo fufow, tshiena muye mutoke

X. polyantha R.E. Fries

Bolivia: xahuiria

X. quintasii Engl. & Diels

Liberia: gbay, gbay-dee

Ivory Coast: elo

Central Africa: mossangi, mvouma, nséli, bombaya, mvoma, omviala, elo

Cameroon: mfoenga, njie, otouan, otungui, pene

Nigeria: opalifon, aghako, ovonien, ofun, awonka; opalifon (Yoruba); ovunien (Edo); udaofia (Igbo); bolonge (Boki)

X. rubescens Oliv.

Central Africa: bana, bobom, bofundu, bonsanghe, m'bambana, nsanghe, lukungu li lowe, mogunda, mponde, ndunga, koumba, ntoleng, odzibi, odiobozam, odjobi, odzobi, dandou, ntouleng

Nigeria: ofunoke (Yoruba); unien-eze (Edo); atarabang (Efik)

X. sericea A. St.-Hil.

Bolivia: piraquina negra, pequi

X. staudtii Engler

Cameroon: ntouleng, mbanbana, odjobi

Ivory Coast: fondé

Central Africa: bosanghe, nsanghe, dungamufiki, ndunga mufike, iole mukunze, lupusu, ishenge, niombe, ntsua, adjobi, odjwé, adouébé, efomou, fondé

Gabon: ntsua

X. villosa Chipp

Nigeria: unien; palufon dudu (Yoruba); aghako (Edo); uda (Igbo)

Xylopiastrum Roberty Annonaceae

Origins:
Resembling *Xylopia* L.

Xylopicron Adans. Annonaceae

Origins:
From the Greek *xylon* "wood" and *pikros*, *pikron* "bitter, pungent"; see also *Xylopia* L.

Xylopicrum P. Browne Annonaceae

Origins:
From the Greek *xylon* "wood" and *pikros*, *pikron* "bitter, pungent"; see also *Xylopia* L.

Xylopleurum Spach Onagraceae

Origins:
From the Greek *xylon* "wood" and *pleura*, *pleuro*, *pleuron* "rib."

Xylorhiza Nutt. Asteraceae

Origins:
From the Greek *xylon* "wood" and *rhiza* "root," woody root.

Species/Vernacular Names:
X. cognata (H.M. hall) T.J. Watson
English: Mecca-aster

X. orcuttii (Vasey & Rose) E. Greene
English: Orcutt's woody-aster

X. tortifolia (Torrey & A. Gray) E. Greene var. *tortifolia*
English: Mojave-aster

Xylosma Forster f. Flacourtiaceae

Origins:
Fragrant wood, from the Greek *xylon* "wood" and *osme* "smell, odor, perfume"; see Johann Georg Adam Forster (1754-1794), *Florulae insularum australium prodromus.* 72. (Oct.-Nov.) 1786.

Species/Vernacular Names:
X. sp.
English: wild lime

X. congestum (Lour.) Merrill (*Xylosma senticosum* Hance; *Xylosma racemosum* (Miq.) Siebold & Zucc.; *Myroxylon senticosum* (Hance) Warb.)
Japan: kusudo-ige
China: tung ching

X. hawaiiense Seem. (*Xylosma hillebrandii* Wawra; *Drypetes forbesii* Sherff)
Hawaii: maua, a'e

Xylosterculia Kosterm. Sterculiaceae

Origins:
From the Greek *xylon* "wood" plus the genus *Sterculia* L.

Xylothamia G.L. Nesom, Y.B. Suh, D.R. Morgan & B.B. Simpson Asteraceae

Origins:
Greek *xylon* "wood" plus the genus *Euthamia* (Nutt.) Cass.

Xylotheca Hochst. Flacourtiaceae

Origins:
Greek *xylon* "wood" and *theke* "case," *xylotheke* "woodhouse," the fruit is a woody capsule.

Species/Vernacular Names:
X. kraussiana Hochst. (*Oncoba kraussiana* (Hochst.) Planch.; *Xylotheca kraussiana* Hochst. var. *glabrifolia* Wild; *Xylotheca kotzei* Phill.) (for the German (b. Stuttgart) botanist and traveler Christian Ferdinand Friedrich von

Krauss, 1812-1890 (d. Stuttgart), zoologist, scientist, botanical explorer, from 1838 to 1840 plant collector in South Africa; see Philipp Bruch (1781-1847), Gottlieb W. Bischoff and Johann Bernhard Wilhelm Lindenberg (1781-1851), "*Musci et Hepatici* Kraussiani." *Flora*. 29: 132-136. 1846; Christian Ferdinand Hochstetter (1787-1860), "Pflanzen des Cap- und Natal-landes, gesammelt und zusammengestelt von Dr. Ferdinand Krauss." *Flora*. 28: 337-344, 753-764. 1845 and 29: 113-129, 129-138, 209-219. 1846; J. Hutchinson, *A Botanist in Southern Africa*. London 1946; Mary Gunn and Leslie E. Codd, *Botanical Exploration of Southern Africa*. 210-212. Cape Town 1981; P. MacOwan, *Trans. S. Afr. Philos. Soc.* 4: xxx-liii. 1887; J. Lanjouw and F.A. Stafleu, *Index Herbariorum*. 2: 386. Utrecht 1972; H.N. Clokie, *Account of the Herbaria of the Department of Botany in the University of Oxford*. 195. 1964; O.H. Spohr, *Ferdinand Krauss: Travel Journal / Cape to Zululand*. Observations by a collector and naturalist 1838-40. Cape Town 1973; John H. Barnhart, *Biographical Notes upon Botanists*. 2: 319. Boston 1965; T.W. Bossert, *Biographical Dictionary of Botanists Represented in the Hunt Institute Portrait Collection*. 219. 1972; Stafleu and Cowan, *Taxonomic Literature*. 2: 668. 1979)

English: African dog-rose, umbalekani tree

Southern Africa: Afrikaanse hondsroos; umDubu (Xhosa); umBhalekani, umBalekani, umNaminami, uNamnyani, uMavuthwa-emfuleni, isiNkankanka, isiShwashwa (Zulu)

Xylothermia Greene Fabaceae

Origins:

Probably from the Greek *xylon* "wood" and *thermos* "lupin," or *therme, therma* "heat," referring to the habitat.

Xymalos Baillon Monimiaceae (Trimeniaceae)

Origins:

The generic name is the anagram of *Xylosma* (from the Greek *xylon* "wood" and *osme* "smell, odor, perfume"), the genus in which the tree was first placed and described; Helmut Genaust, in *Etymologisches Wörterbuch der botanischen Pflanzennamen*. 696. Basel 1996, suggests an origin from the Greek *oxymalon, oxymelon, oxys* "sharp-pointed, sharp, acid, sour" and *melon* "an apple."

Species/Vernacular Names:
X. monospora (Harvey) Baillon (*Xymalos monospora* Harvey)

English: lemonwood, wild lemon, one-seeded xymalos

Southern Africa: lemoenhout, borriehout; umHlwehlwe, iThotshe, umHlungwane (Zulu); nyagazani, muNyamazani,

muVeti (Shona); uVethe, umVethi (Xhosa); isiPhisimakhata (Swazi)

Xyochlaena Stapf Gramineae

Origins:

Greek *xyo* "to scrape, to make smooth, to carve wood" and *chlaena, chlaenion* "cloak, blanket"; see also *Tricholaena* Schrader.

Xyridanthe Lindley Asteraceae

Origins:

From *xyris*, a Greek name for an aromatic plant, so called from the razor-like leaves, *xyron* "a razor" and *anthos* "flower," Latin *xyris, idis* applied by Plinius to a wild iris.

Xyridopsis Welw. ex B. Nord. Asteraceae

Origins:

Resembling *xyris*.

Xyris L. Xyridaceae

Origins:

Xyris, Greek name for an aromatic plant, a species of *Iris*, so called from its razor-like leaves, *xyron* "a razor," Latin *xyris, idis* applied by Plinius to a wild iris; see Carl Linnaeus, *Species Plantarum*. 42. 1753 and *Genera Plantarum*. Ed. 5. 25. 1754.

Species/Vernacular Names:
X. sp.
English: yellow-eyed grass
X. capensis Thunb.
English: yellow-eyed grass
X. operculata Labill.
English: tall yellow-eye

Xyropteris K.U. Kramer Dennstaedtiaceae (Lindsaeoideae)

Origins:

Greek *xyo* "to scrape, to make smooth, to carve wood, *xyron* "a razor" and *pteris* "a fern."

Xyropterix Airy Shaw Dennstaedtiaceae (Lindsaeoideae)

Origins:
See *Xyropteris*.

Xysmalobium R. Br. Asclepiadaceae

Origins:
Greek *xysme*, *xysma* "shavings, particles, a scar, fragment" and *lobos* "lobe, division," *lobion* dim. from *lobos*, referring to the crown or corona.

Species/Vernacular Names:
X. heudelotianum Decne.

Yoruba: disoke

Xystidium Trin. Gramineae

Origins:
Perhaps from the Greek *xyston* "a spear, dart, pole" or from *xystis*, *idos* "a rich and soft robe, a garment."

Xystrolobus Gagnep. Hydrocharitaceae

Origins:
Greek *xystron* "an instrument for scraping" and *lobos* "lobe, division."

Y

Yabea Koso-Pol. Umbelliferae

Origins:
For the Japanese botanist Yoshitaka (Yoshitada) Yabe, 1876-1931, professor of botany at Tokyo. Among his writings are "Revisio Umbelliferarum Japonicarum." in *Journal of the College of Science*. Univ. of Tokyo 1902 and "Umbelliferae Koreae Uchiyamanae." *Botanical Magazine*. 17: 105-108. Tokyo 1903; see J.H. Barnhart, *Biographical Notes upon Botanists*. 3: 527. 1965; Ethelyn Maria Tucker, *Catalogue of the Library of the Arnold Arboretum of Harvard University*. Cambridge, Massachusetts 1917-1933; Elmer D. Merrill and Egbert H. Walker, *A Bibliography of Eastern Asiatic Botany*. 537-538. The Arnold Arboretum of Harvard University, Jamaica Plain, Massachusetts, USA 1938.

Yakirra Lazarides & R.D. Webster Gramineae

Origins:
The aboriginal (in Australia) name for a species of the genus; see Michael Lazarides (b. 1928) and Robert D. Webster (b. 1950), in *Brunonia*. 7: 292. (Mar.) 1985.

Species/Vernacular Names:
Y. australiensis (Domin) Lazarides & R. Webster (*Panicum australiense* Domin; *Panicum pauciflorum* R. Br. var. *fastigiatum* Benth.; *Ichnanthus australiensis* (Domin) Hughes)

English: bunch panic

Yamala Raf. Saxifragaceae

Origins:
See E.D. Merrill, *Index rafinesquianus*. The plant names published by C.S. Rafinesque, etc. 135. Jamaica Plain, Massachusetts, USA 1949; C.S. Rafinesque, *Flora Telluriana*. 2: 75. 1836 [1837].

Yangapa Raf. Rubiaceae

Origins:
See E.D. Merrill, *Index rafinesquianus*. The plant names published by C.S. Rafinesque, etc. 227. Jamaica Plain, Massachusetts, USA 1949; C.S. Rafinesque, *Sylva Telluriana*. 20. 1838.

Ygramela Raf. Scrophulariaceae

Origins:
See E.D. Merrill, *Index rafinesquianus*. 217, 219. 1949; C.S. Rafinesque, *Flora Telluriana*. 4: 57. 1836 [1838].

Yoania Maxim. Orchidaceae

Origins:
For the Japanese botanist Wudogawa Yoan, botanical artist, physician, sent many drawings to the Siebold Herbarium (St. Petersburg); see Richard Evans Schultes and Arthur Stanley Pease, *Generic Names of Orchids. Their Origin and Meaning*. 323. Academic Press, New York and London 1963.

Yolanda Hoehne Orchidaceae

Origins:
Named for Yolanda, daughter of the Brazilian botanist Frederico Carlos Hoehne (1882-1959).

Youngia Cass. Asteraceae

Origins:
According to Cassini the genus was named after two Englishmen, a famous poet and writer (possibly Edward Young, 1684-1765) and a physician (presumably Thomas Young, 1773-1829); according to other authors the generic name honors Charles, James and Peter Young, nurserymen at Epson, Surrey, during the early 19th century. Edward Young obtained in 1708 a law fellowship in All Soul's College, in 1727 entered orders and was named one of the royal chaplains, Rector of Welwyn in Hertfordshire. Among his many works are *The Complaint: or, Night-Thoughts on Life, Death, & Immortality*. 1742, *The Centaur not Fabulous*. 1755, and *Resignation*. London 1762. The Egyptologist Thomas Young, physician and physicist, was a member of the Society of Friends, in 1794 elected Fellow of the Royal Society and of the Linnean Society, M.D. Göttingen 1796, from 1801 till 1804 professor of natural philosophy in the Royal Institution, in 1817 founded the Egyptian Society, in 1818 appointed secretary to the board of longitude, he was a frequent contributor to the *Quarterly Review* and to the *Encyclopaedia Britannica*. His principal writings are

Hieroglyphics; collected by the Egyptian Society, arranged by T. Young. London 1823-1828, *Rudiments of a Dictionary of the Ancient Egyptian Language, in the Enchorial Characters.* 1830 and *A Course of Lectures on Natural Philosophy and the Mechanical Arts.* London 1807. See Alexandre Henri Gabriel Comte de Cassini (1781-1832), in *Annales des Sciences Naturelles.* ser. 1. 23: 88. 1831; Alexander Wood and Frank Oldham, *Thomas Young, Natural Philosopher 1773-1829.* Cambridge 1954; Hudson Gurney, *Memoir of the Life of Thomas Young.* London 1831.

Species/Vernacular Names:
Y. japonica (L.) DC. (*Crepis japonica* (L.) Benth.; *Prenanthes japonica* L.)

English: native hawk's bird, oriental hawk's bird

Japan: oni-tabirako, tuinufisa

China: huang an cai

Ypsilandra Franchet Melanthiaceae (Liliaceae)

Origins:
From the Greek letter Y and *aner, andros* "male, a man, stamen."

Ypsilopus Summerh. Orchidaceae

Origins:
Greek letter Y and *pous* "a foot," indicating the Y-shaped stipe of the pollinarium, from the stipe attaching the pollinia to the viscidium.

Yuca Raf. Agavaceae

Origins:
For *Yucca*, see Elmer D. Merrill, *Index rafinesquianus.* The plant names published by C.S. Rafinesque, etc. 96. Jamaica Plain, Massachusetts, USA 1949; Constantine S. Rafinesque, *Flora Telluriana.* 1: 73. 1836 [1837]; vernacular *yuca* for *Manihot esculenta.*

Yucca L. Agavaceae

Origins:
The vernacular name in Peru or in the Carib for manihot or cassava, misapplied to these plants, *yuca* for *Manihot esculenta*; see Carl Linnaeus, *Species Plantarum.* 319. 1753 and *Genera Plantarum.* Ed. 5. 150. 1754; F. D'Alberti di Villanuova, *Dizionario universale, critico, enciclopedico*

della lingua italiana. Lucca 1797-1805; G. Friederici, *Amerikanisches Wörterbuch.* Hamburg 1947; Gordon Douglas Rowley, *A History of Succulent Plants.* Strawberry Press, Mill Valley, California 1997.

Species/Vernacular Names:
Y. aloifolia L.

English: Spanish bayonet, dagger plant, yucca, aloe yucca

Bolivia: yuca

Japan: chimo-ran

South Africa: yucca

Y. baccata Torrey (*Yucca circinata* Bak.)

English: Spanish bayonet, banana yucca, blue yucca

Y. brevifolia Engelm. (*Yucca arborescens* (Torr.) Trel.)

English: Joshua tree

Y. carnerosana (Trel.) McKelv.

English: Spanish dagger

Y. elata Engelm.

English: soap tree, soap weed

Spanish: palmella

Y. filamentosa L. (*Yucca concava* Haw.)

English: spoonleaf yucca, Adam's needle, needle palm

Y. filifera Chabaud (*Yucca flaccida* Haw.; *Yucca meldensis* Engelm.; *Yucca puberula* Haw. non Torr.)

English: weak-leaved yucca

Japan: ito-ran

Y. gloriosa L.

English: Spanish dagger, Roman candle, palm lily, Adam's needle

Y. recurvifolia Salisb.

Japan: kamigayo-ran

Y. reverchonii Trel.

English: San Angelo yucca

Y. rupicola Scheele

English: twisted leaf yucca

Y. smalliana Fern.

English: Adam's needle, bear grass

Y. schidigera Roezl ex Ortgies

English: Mohave yucca

Spanish: palmilla

Y. treculeana Carr. (*Yucca canaliculata* Hook.)

English: Spanish dagger

Spanish: palma pita

Y. whipplei Torrey

English: Our Lord's candle

Yucea Raf. Agavaceae

Origins:

For *Yucca*, see E.D. Merrill, *Index rafinesquianus*. 96. 1949; Constantine S. Rafinesque, *Flora Telluriana*. 1: 16. 1836 [1837].

Yunckeria Lundell Myrsinaceae

Origins:

Named after the American botanist Truman George Yuncker, 1891-1964, professor of biology and botany, plant collector in Ecuador. His works include *Revision of the North American and West Indian Species of Cuscuta*. University of Illinois 1921, *Anatomy of Hawaiian Peperomias*. Honolulu, Hawaii 1933, *Flora of the Aguan Valley*. [Chicago] 1940, *The Flora of Niue Island*. Honolulu, Hawaii 1943 and *Plants of the Manua Islands*. Honolulu, Hawaii 1945, with the American botanist William Trelease (1857-1945) wrote *The Piperaceae of Northern South America*. Urbana 1950; see J.H. Barnhart, *Biographical Notes upon Botanists*. 3: 535. 1965; R. Zander, F. Encke, G. Buchheim and S. Seybold, *Handwörterbuch der Pflanzennamen*. 14. Aufl. Stuttgart 1993; T.W. Bossert, *Biographical Dictionary of Botanists Represented in the Hunt Institute Portrait Collection*. 447. 1972; Ida Kaplan Langman, *A Selected Guide to the Literature on the Flowering Plants of Mexico*. Philadelphia 1964; Gordon Douglas Rowley, *A History of Succulent Plants*. Strawberry Press, Mill Valley, California 1997.

Yunnanea Hu Theaceae

Origins:

Yunnan, China.

Yushania Keng f. Gramineae

Origins:

Of Yushan, Taiwan.

Yuyba (Barb. Rodr.) L.H. Bailey Palmae

Origins:

A vernacular name.

Yvesia A. Camus Gramineae

Origins:

Dedicated to the French botanist Alfred Marie Augustine Saint-Yves, 1855-1933, soldier, agrostologist; see John H. Barnhart, *Biographical Notes upon Botanists*. 3: 201. 1965; T.W. Bossert, *Biographical Dictionary of Botanists Represented in the Hunt Institute Portrait Collection*. 1972; Ida Kaplan Langman, *A Selected Guide to the Literature on the Flowering Plants of Mexico*. 664-665. Philadelphia 1964; R. Zander, F. Encke, G. Buchheim and S. Seybold, *Handwörterbuch der Pflanzennamen*. 14. Aufl. Stuttgart 1993.

Z

Zabelia (Rehder) Makino Caprifoliaceae

Origins:
After the German botanist Hermann Zabel, 1832-1912, forester, from 1854 to 1860 Assistant at the Botanical Garden and Museum, Greifswald, he was the author of *Synoptische Tabellen*. Münden 1872 and *Die strauchigen Spiräen der deutschen Gärten*. Berlin 1893; see J.H. Barnhart, *Biographical Notes upon Botanists*. 3: 535. 1965; Ethelyn Maria Tucker, *Catalogue of the Library of the Arnold Arboretum of Harvard University*. Cambridge, Massachusetts 1917-1933; R. Zander, F. Encke, G. Buchheim and S. Seybold, *Handwörterbuch der Pflanzennamen*. 14. Aufl. Stuttgart 1993.

Zacateza Bullock Asclepiadaceae

Origins:
A vernacular name; see Louise C. Schoenhals, *A Spanish-English Glossary of Mexican Flora and Fauna*. Hidalgo, México, D.F. 1988.

Zacintha Mill. Asteraceae

Origins:
An island in the Ionian Sea, Zante or Zacyntho.

Zaczatea Baillon Asclepiadaceae

Origins:
An anagram of *Zacateza*.

Zahlbrucknera Reichb. Saxifragaceae

Origins:
For the Austrian botanist Johann Baptist Zahlbruckner, 1782-1851, mineralogist; see John H. Barnhart, *Biographical Notes upon Botanists*. 3: 536. 1965; R. Zander, F. Encke, G. Buchheim and S. Seybold, *Handwörterbuch der Pflanzennamen*. 14. Aufl. Stuttgart 1993; Stafleu and Cowan, *Taxonomic Literature*. 7: 510-511. 1988.

Zalacca Blume Palmae

Origins:
See *Salacca*.

Zalaccella Becc. Palmae

Origins:
The diminutive of *Zalacca*, see also *Calamus* L.

Zaleja Burm.f. Aizoaceae

Origins:
See *Zaleya* Burm.f.

Zaleya Burm.f. Aizoaceae

Origins:
Presumably from the Greek *zaleia*, ancient name applied by Dioscorides to the *daphne Alexandreia*; or named from the corruption of an Indian name, *vallai-sharunai*; see James A. Baines, *Australian Plant Genera. An Etymological Dictionary of Australian Plant Genera*. 402. Chipping Norton, N.S.W. 1981.

Species/Vernacular Names:
Z. galericulata (Melville) H. Eichler (*Trianthema galericulata* Melville)
English: hogweed
Z. pentandra (L.) Jeffrey (*Trianthema pentandra* L.)
English: African purslane, zaleya
South Africa: muisvygie

Zalitea Raf. Euphorbiaceae

Origins:
See C.S. Rafinesque, *New Fl. N. Am.* 4: 98, 99. 1836 [1838]; Elmer D. Merrill (1876-1956), *Index rafinesquianus. The plant names published by C.S. Rafinesque, etc.* 157. Jamaica Plain, Massachusetts, USA 1949.

Zalmaria B.D. Jacks. Rubiaceae

Origins:

See Elmer D. Merrill (1876-1956), *Index rafinesquianus*. The plant names published by C.S. Rafinesque, etc. 227. Jamaica Plain, Massachusetts, USA 1949; see *Zamaria* Raf.

Zaluzania Pers. Asteraceae

Origins:

For the Bohemian botanist Adam Zaluziansky à Zaluzian [Zaluziansky ze Zaluzian, Zaluziansky von Zaluzian], 1558-1613, physician; see F. Boerner & G. Kunkel, *Taschenwörterbuch der botanischen Pflanzennamen*. 4. Aufl. 188. Berlin and Hamburg 1989.

Zaluzianskia Necker Marsileaceae

Origins:

After the Bohemian botanist Adam Zaluziansky à Zaluzian [Zaluziansky ze Zaluzian], 1558-1613, physician at Prague, lecturer and administrator at Charles University in Prague, author of *Methodi herbariae libri tres*. Pragae 1592 and *Animadversionum medicarum in Galenum et Avicennam libri vii. ...* item synopsis legitimae institutionis medicinae, etc. Francofurti 1604.

Zaluzianskya F.W. Schmidt Scrophulariaceae

Origins:

Named for the Bohemian botanist Adam Zaluziansky à Zaluzian [Zaluziansky ze Zaluzian, Zaluzansky ze Zaluzan], 1558-1613 (d. Prague), physician at Prague, lecturer and administrator at Charles University in Prague, author of *Methodi herbariae libri tres*. Pragae 1592 and *Animadversionum medicarum in Galenum et Avicennam libri vii. ...* item synopsis legitimae institutionis medicinae, etc. Francofurti 1604; see O. Hilliard and B.L. Burtt, "*Zaluzianskya* (Scrophulariaceae) in South Eastern Africa and the correct application of the names *Z. capensis* and *Z. maritima*." *Notes Roy. Bot. Gard. Edinb.* 41: 1-43. 1983; F. Boerner & G. Kunkel, *Taschenwörterbuch der botanischen Pflanzennamen*. 4. Aufl. 188. Berlin & Hamburg 1989; Vera Eisnerova, in *Dictionary of Scientific Biography* 14: 586. 1981.

Species/Vernacular Names:

Z. divaricata (Thunb.) Walp. (*Manulea divaricata* Thunb.)

English: spreading night-phlox

Z. maritima (L.f.) Walp. (*Zaluzianskya lychnidea* (D. Don) Walp.; *Zaluzianskya maritima* (L.f.) Walp. var. *fragrantissima* Hiern)

English: drumsticks

Zamaria Raf. Rubiaceae

Origins:

See C.S. Rafinesque, *Ann. Gén. Sci. Phys.* 6: 85. 1820; Elmer D. Merrill (1876-1956), *Index rafinesquianus*. The plant names published by C.S. Rafinesque, etc. 227. Jamaica Plain, Massachusetts, USA 1949.

Zamia L. Zamiaceae

Origins:

Possibly from Latin *zamia* or *samia, ae* "hurt, loss, damage, detrimentum," or from *azaniae nuces* (Greek *azano* "to dry up"), used by Plinius for a kind of pine-cone or pine-nuts; see C. Linnaeus, *Species Plantarum*. 2: 1659. 1763; H. Genaust, *Etymologisches Wörterbuch der botanischen Pflanzennamen*. 697. 1996; F. Boerner & G. Kunkel, *Taschenwörterbuch der botanischen Pflanzennamen*. 4. Aufl. 188. Berlin & Hamburg 1989; W.P.U. Jackson, *Origins and Meanings of Names of South African Plant Genera*. 38-39. Rondebosch 1990.

Species/Vernacular Names:

Z. integrifolia L.f. (*Zamia floridana* DC.; *Zamia silvicola* Small; *Zamia umbrosa* Small)

English: Florida arrowroot, coontie, conti hateka (Seminole)

Zamioculcas Schott Araceae

Origins:

From *Zamia* plus *Colocasia* or *Culcasia*, Araceae; see D.H. Nicolson, "Derivation of aroid generic names." *Aroideana*. 10: 15-25. 1988.

Zamzela Raf. Chrysobalanaceae

Origins:

See C.S. Rafinesque, *Sylva Telluriana*. 90. 1838; Elmer D. Merrill, *Index rafinesquianus*. The plant names published by C.S. Rafinesque, etc. 142. Jamaica Plain, Massachusetts, USA 1949.

Zanha Hiern Sapindaceae

Origins:

Meaning uncertain, genus perhaps dedicated to the German plant collector K.H. Zahn; see Réné Letouzey, "Les botanistes au Cameroun." in *Flore du Cameroun*. 7: 1-110. Paris 1968; F.N. Hepper and Fiona Neate, *Plant Collectors in West Africa*. 87. 1971.

Species/Vernacular Names:

Z. africana (Radlk.) Exell (*Dialiopsis africana* Radlk.)

English: velvet-fruited zanha

Southern Africa: muChenyua, muChenyura, muChenyuwa (Shona)

Z. golungensis Hiern

English: smooth-fruited zanha

Southern Africa: muGarayezhe (Shona)

Zaire: ndingo, penzi

Zanichellia Roth Zannichelliaceae

Origins:

An orthographic variant, see *Zannichellia* L.

Zannichellia L. Zannichelliaceae

Origins:

After the Italian botanist Giovanni Gerolamo (Gian Girolamo) Zannichelli (Zanichelli), 1662-1729, physician, pharmacist. Among his works are *Istoria delle piante* che nascono ne' lidi intorno a Venezia opera postuma di Gian-Girolamo Zanichelli ... Venezia 1735 and *Considerazioni ... intorno ad una pioggia di terra caduta nel Golfo di Venezia, e sopra l'incendio del Vesuvio*. Venezia 1737; see John H. Barnhart, *Biographical Notes upon Botanists*. 3: 537. 1965; Carl Linnaeus, *Species Plantarum*. 969. 1753 and *Genera Plantarum*. Ed. 5. 416. 1754; T.W. Bossert, *Biographical Dictionary of Botanists Represented in the Hunt Institute Portrait Collection*. 448. Boston, Massachusetts 1972; H.N. Clokie, *Account of the Herbaria of the Department of Botany in the University of Oxford*. 268. Oxford 1964; Jonas C. Dryander, *Catalogus bibliothecae historico-naturalis Josephi Banks*. London 1796-1800; Antoine Lasègue, *Musée botanique de M. Benjamin Delessert*. 12. 1845; Ethelyn Maria Tucker, *Catalogue of the Library of the Arnold Arboretum of Harvard University*. Cambridge, Massachusetts 1917-1933.

Species/Vernacular Names:

Z. palustris L.

English: horned-pondweed

Zanonia L. Cucurbitaceae

Origins:

After the Italian botanist Giacomo Zanoni, 1615-1682, author of *Rariorum stirpium Historia*. Bologna 1742 (first published as *Istoria botanica*. Bologna 1675); see Carl Linnaeus, *Species Plantarum*. 1028. 1753 and *Genera Plantarum*. Ed. 5. 454. 1754; R. Zander, F. Encke, G. Buchheim and S. Seybold, *Handwörterbuch der Pflanzennamen*. 14. Aufl. 671. Stuttgart 1993; Mariella Azzarello Di Misa, a cura di, *Il Fondo Antico della Biblioteca dell'Orto Botanico di Palermo*. 295. Palermo 1988.

Zantedeschia Sprengel Araceae

Origins:

After the Italian botanist Francesco Zantedeschi, 1798-1873, professor of physics at Padua, he wrote *Dell'influenza dei raggi solari rifratti dai vetri colorati sulla vegetazione delle piante e germinazione de' semi*. Venezia 1843 and *Della elettricità degli stami e pistilli delle piante esplorata all'atto della fecondazione e di una nuova classificazione delle linfe o succhi Vegetabili*, etc. Padova 1853; see Kurt Polycarp Joachim Sprengel (1766-1833), in *Systema Vegetabilium*. 3: 756, 765. (Jan.-Jun.) 1826; H. Genaust, *Etymologisches Wörterbuch der botanischen Pflanzennamen*. 697. [1773-1846] Basel 1996; D.H. Nicolson, "Derivation of aroid generic names." *Aroideana*. 10: 15-25. 1988; W.P.U. Jackson, *Origins and Meanings of Names of South African Plant Genera*. 39. [genus named for Francesco Zantedeschi, 1773-1846] Rondebosch 1990; R.K. Brummitt and C.E. Powell, *Authors of Plant Names*. 724. [dates: 1773-1846, Giovanni Zantedeschi] Royal Botanic Gardens, Kew 1992; Francis Aubie Sharr, *Western Australian Plant Names and Their Meanings*. 77. [for Giovanni Zantedeschi, 1773-1806, or Francesco Zantedeschi, 1797-1873] University of Western Australia Press, Nedlands, Western Australia 1996; William T. Stearn, *Stearn's Dictionary of Plant Names for Gardeners*. 314. [named for Francesco Zantedeschi, b. 1797] Cassell, London 1993.

Species/Vernacular Names:

Z. aethiopica (L.) Sprengel (*Calla aethiopica* L.; *Richardia africana* Kunth)

English: calla, calla lily, white calla lily, common calla lily, common calla, florist's calla, garden calla, immaculate aroid lily, Egyptian lily, Jack-in-the-pulpit, lily-of-the-Nile, pig lily, trumpet lily, arum lily, white arum lily, pig lily

Japan: oranda-kaiu

The Philippines: immaculate aroid lily

Southern Africa: aronskelk, Hottentotsblare, Hottentotsbrood, varkblaar, varkblom, varklelie, varkoor, varkore, varkwortel, ystervarkwortel; intebe (Zulu); mahalalitoe (Sotho); inYibiba (Xhosa)

Z. albomaculata (Hook.) Baill. subsp. *albomaculata* (*Zantedeschia angustiloba* (Schott) Engl.; *Zantedeschia hastata* (Hook.) Engl.; *Zantedeschia melanoleuca* (Hook.f.) Engl.; *Zantedeschia oculata* (Lindl.) Engl.; *Zantedeschia tropicalis* (N.E. Br.) Letty)

English: calla lily

South Africa: kleinvarkblom

Z. pentlandii (Watson) Wittm.

South Africa: geelvarkblom

Z. rehmannii Engl. (*Richardia rehmannii* (Engl.) N.E. Br.)

South Africa: purple arum

Zanthorhiza L'Hérit. Ranunculaceae

Origins:
See *Xanthorhiza.*

Zanthoxylum L. Rutaceae

Origins:
Greek *xanthos* "yellow" and *xylon* "wood," a yellow dye is sometimes contained in the roots of some species; see Carl Linnaeus, *Species Plantarum.* 270. 1753 and *Genera Plantarum.* Ed. 5. 130. 1754.

Species/Vernacular Names:
Z. sp.

English: sand forest knobwood, sea ash, doctor's club, wild lime, Cuban yellowwood, peppermint wood, prickly ash

China: wan chiao, chu chiao (= pig pepper), kou chiao, chin chiao, ti chiao

French West Indies: espin de bobo, épineux blanc, épineux jaune, frêne piquant, bois jaune

Hawaii: a'e, manele, hea'e

Panama: alcabe, ruda

Argentina: coco de Cordoba, cococho, mamica de candela negra, manduvina, sacha limon, sauce hediondo, tembetary

Mexico: colima, alacran; cola de alacran (Tuxtla Gutierrez); palo mulato (Puebla; see Ana María Huerta Jaramillo, *El jardin de Cal. Antonio de la Cal y Bracho, la botánica y las ciencias de la salud en Puebla, 1766-1833.* Puebla 1996); pipima (Izmala, Sinaloa); rabo de lagarto (Chiapas; Tehuantepec, Oaxaca)

Peru: alcanfor sacha, shapallejo

Cuba: ayua, ayua amarilla, ayua blanca, ayuda, ayuda varia, bayua, bayua lisa, bayuda, espino, cafe limon, mate arbol, tomeguin, zorillo

Venezuela: calabori

Brazil: espinho de vintem, tembetaru, tembetary

Puerto Rico: carubio, espino

Jamaica: Caesarwood, yellow Hercules, licca tree, saven tree, suarra tree, prickly yellow

Santo Domingo: Santo Domingo harewood, male pine

Trinidad: mapourita, nogal

Costa Rica: lagartillo

Salvador: salitrero

Suriname: geel steckelboom

Gabon: lomvogo, nungo, olonvogo

Cameroon: nungo

Southern Africa: umNungwane (Zulu)

Uganda: mubajangabo, ntaleuerunga

Z. ailanthoides Siebold & Zucc.

Japan: karasu-zan-shô, angi

China: tang tzu, shih chu yu, yueh chiao

Z. americanum Mill.

English: toothache tree

Z. capense (Thunb.) Harv. (*Fagara capensis* Thunb.; *Fagara magaliesmontana* Engl.; *Fagara armata* Thunb.; *Fagara multifoliolata* Engl.; *Zanthoxylum thunbergii* DC. var. *obtusifolia* Harv.)

English: small knobwood, woodland knobwood, wild cardamon, knob thorn

South Africa: kleinperdepram, paardepram, knophout, wildekardamon, lemoenhout, lemoendoring, umnunquazi; muDzanganyia (Shona); umNungumabele, umLungumabele (Xhosa); umNungwane, umNungwane omncane, umNungumabele, amaBelentombi, amaBelezintshingezi (Zulu); umNungwane (Swazi); monokomabele (Hebron dialect, central Transvaal); monokwane, senoko-maropa (North Sotho); murandela, munungu (Venda)

Z. caribaeum Lamarck (*Fagara caribaea* (Lamarck) Krug & Urban)

French: bois piquant

Puerto Rico: espinillo rubial, espino rubial

Mexico: sinanché (Maya language, Yucatan); zorrillo (Sinaloa)

Z. clava-herculis L.

English: Hercules-club, sea ash, pepperwood

Z. davyi (Verdoorn) Waterm. (*Fagara davyi* Verdoorn; *Zanthoxylum thunbergii* DC. var. *grandifolia* Harv.)

English: knobwood, forest knobwood

South Africa: perdepram, knophout, perdepramboom; munungu (Venda); umNungumabele, umNungwane omkhulu, umNungwane (Zulu); umLungamabele, umNungumabele (Xhosa)

Z. dipetalum H. Mann (*Connarus kavaiensis* H. Mann; *Fagara waianensis* Degener & Skottsb.)

Hawaii: kawa'u, kawa'u kua kuku kapa

Z. fagara (L.) Sargent (*Fagara pterota* L.; *Schinus fagara* L.)

English: wild lime

Honduras: chincho

Mexico: garabatillo; palo mulato (Jalisco); tenaza (Guadalcazar); rabo de lagarto (Tuxtepec, Oaxaca; see Aurora Chimal Hernández et al., *Las plantas medicinales y su uso tradicional en el ejido "Paraiso," Municipio de Tuxtepec, Oaxaca.* México 1993); alacran (Chiapas); colima (Nuevo Leon, Tamaulipas); limoncillo (Sinaloa); tankaché (Yucatan); uña de gato (Tamaulipas); uolé, xik-ché (Maya language, Yucatan); yichasmis (Tzotzil language, San Cristobal); huipuy (Huasteca language, San Luis Potosí); naranjillo (southeast San Luis Potosí)

Colombia: uña de gato

Puerto Rico: espino rubial

Z. gilletii (De Wild.) Waterman (*Fagara macrophylla* (Oliv.) Engl.; *Fagara tessmannii* Engl.; *Fagara melanorchis* Hoyle) (see Y. Tailfer, *La Forêt dense d'Afrique centrale.* CTA, Ede/Wageningen 1989; J. Vivien & J.J. Faure, *Arbres des Forêts denses d'Afrique Centrale.* Agence de Coopération Culturelle et Technique. Paris 1985)

Gabon: olonvongo, nongo, olon

Nigeria: ata igbo (Yoruba); okor (Edo); uko (Igbo)

Cameroon: akombo, longue, wongo, bongo, elongo, bolongo, olon, eyeyongo, timba eyidi, timba, eyidi, ledjon

Ivory Coast: gbessi, bahe, pahe, mene hane, n'pahe, g'pon, hanwgo, hanwogo, kengue, kingue

Ghana: ahenkyen, okuo, yea

Zaire: ditsende-isende, nungu-tsende, endongo, mapanga, empende

Z. heitzii (Aubr. & Pellegrin) Waterm.

Congo: m'banza

Gabon: nongo, olon

Cameroon: bolongo, bongo, timba-eyidi

Zaire: lisumba, mbanza

Z. heterophyllum (Lam.) Smith

Rodrigues Island: bambara

La Réunion Island: bois de poivre, poivrier, bois de poivrier, poivrier des hauts, poivrier mal de dents, bois blanc rouge, bois de catafaille noir, bois de rat

Z. juniperinum Poeppig (*Fagara acreana* K. Krause; *Fagara juniperina* (Poeppig) Engler; *Fagara procera* (J.D. Smith) Engler; *Zanthoxylum acreanum* (K. Krause) J.F. Macbride; *Zanthoxylum procerum* J.D. Smith)

English: mountain rooder bush, wild rooder

Mexico: abrojo, palo de ropa; lagarto (Pichucalco); carricillo, rabo de lagarto, ruda (Tehuantepec, Oaxaca); limoncillo (Oaxaca); lomo lagarto de hoja grande (northeast Chiapas)

Belize: cedrillo

Z. kellermanii P. Wilson (*Fagara kellermannii* (P. Wilson) Engler)

Honduras: cedro espino, lagarto amarillo

Belize: prickly yellow, yellow prickle

Guatemala: ceibillo, lagarto

Z. lemairei (De Wild.) Waterman (*Fagara lemairei* De Wild.)

Nigeria: ukhankhan (Edo)

Z. leprieurii Guill. & Perr. (*Fagara leprieurii* (Guill. & Perr.) Engl.) (after the French botanist F.M.R. Leprieur, 1799-1869, a dispenser in the French Navy, between 1824-29 in Senegambia, in 1829 returned to France and began his *Flora* continued by [Jean Baptiste] Antoine Guillemin (1796-1842), George Samuel Perrottet (1793-1870) and Achille Richard (1794-1852) as *Florae Senegambiae tentamen.* Paris (Treuttel et Wurtz), London 1830-1833; see Joseph Vallot (1854-1925), "Études sur la flore du Sénégal." in *Bull. Soc. Bot. de France.* 29: 168-238. Paris 1882; R.W.J. Keay, "Botanical collectors in West Africa prior to 1860." in *Comptes Rendus A.E.T.F.A.T.* 55-68. Lisbon 1962; F.N. Hepper and Fiona Neate, *Plant Collectors in West Africa.* 49. 1971; G. Murray, *History of the Collections Contained in the Natural History Departments of the British Museum.* London 1904; John H. Barnhart, *Biographical Notes upon Botanists.* Boston 1965)

English: sand knobwood

Nigeria: fasa kwari (Hausa); fasakorihi (Fula); ata (Yoruba); ughanghan (Edo); atufio (Etsako); atako (Itsekiri); ujo (Urhobo); korokumo (Ijaw)

South Africa: sandperdepram, sand knobwood; umNungwane (Zulu)

Z. paniculatum Balf.f.

Rodrigues Island: bambara, bois pasner

Z. piperitum DC.

English: Japan pepper, Japanese pepper, Japanese prickly ash

Japan: sansho

China: shu chiao, chuan chiao, nan chiao, chiao mu (meaning pepper eyes)

Z. planispinum Sieb. & Zucc.

China: zhu ye jiao

Z. rhetsa (Roxb.) DC. (*Fagara rhetsa* Roxb.; *Fagara budrunga* Roxb.; *Tipalia limonella* Dennst.; *Zanthoxylum limonella* (Dennst.) Alston; *Zanthoxylum budrunga* (Roxb.) DC.)

English: Indian ivy-rue

Malaya: hantar duri, hantu duri

Sri Lanka: katu-kina, rhetsu

Z. rhoifolium Lamarck (*Fagara rhoifolia* (Lamarck) Engler; *Zanthoxylum obscurum* var. *ruizianum* Engler; *Zanthoxylum rhoifolium* var. *sessilifolium* Engler; *Zanthoxylum ruizianum* (Engler) J.F. Macbride)

Brazil: mamica de porca, tamanqueira, mamica de cadela, teta de cadela, teta de porca, maminha de porca, juva, juvevê, jubebê, espinho de vintém, tembetaru, tambetari, tambatarão, tamanqueira, tamanqueira da terra firme, tinguaciba guarita, tembatery

Z. schinifolium Sieb. & Zucc.

English: anise pepper

Japan: inu-zan-shô

China: yai chiao, yeh chiao

Z. senegalense DC.

Yoruba: ata, ata dudu

Z. simulans Hance (*Zanthoxylum bungei* Planch.)

English: Szechuan pepper

China: ye hua jiao ye, hua chiao, chin chiao, ta chiao

Z. thomense (Engl.) A. Chev. ex Waterman

Zaire: angi, lesebule, olon, olonvogo, emonge

Z. viride (A. Chev.) Waterman

Yoruba: eegun

Z. zanthoxyloides (Lam.) Zepernick & Timler (*Fagara zanthoxyloides* Lam.)

Nigeria: fasa kwari (Hausa); fasakorihi (Fula); ata (Yoruba); ughanghan (Edo); atufio (Etsako); atako (Itsekiri); ujo (Urhobo); korokumo (Ijaw)

Congo: wankare

Mali: wuo, wo, nkologo

Zapania Lam. Verbenaceae

Origins:

Orthographic variant of *Zappania* Scop.

Zapoteca H.M. Hern. Mimosaceae

Origins:

The Zapotecs and zapoteca language, Oaxaca, Mexico; see Laura Nader, *Harmony Ideology: Justice and Control in a Zapotec Mountain Village.* Stanford 1990; Joseph W.

Whitecotton, *The Zapotecs: Princes, Priests, and Peasants.* Norman, University of Oklahoma 1977; Blas Pablo Reko, *Mitobotánica Zapoteca.* [Appended by an analysis of "Lienzo de Santiago Guevea"] Tacubaya 1945.

Zauschneria Presl Onagraceae

Origins:

For Johann Baptist(a) Josef Zauschner, 1737-1799.

Zea L. Gramineae

Origins:

Greek *zeia, zea*, name used for a kind of cereal, probably a coarse barley or a fodder for horses; Akkadian *se'u* "grain corn," Sanskrit *yava* "barley," Latin *zea, ae* for spelt, *Triticum spelta* L., or for rosemary; see Carl Linnaeus, *Species Plantarum.* 971. 1753 and *Genera Plantarum.* Ed. 5. 419. 1754; Carmen Aguilera, *Flora y fauna Mexicana.* Mitología y tradiciones. 148-152. México [1985]; H. Genaust, *Etymologisches Wörterbuch der botanischen Pflanzennamen.* 697-698. Basel 1996.

Species/Vernacular Names:

Z. sp.

Peru: hahua sara, parhuay, paru sara

Z. mays L. (*Zea curagua* Molina)

English: Arab wheat, corn, corn of Mecca, Indian corn, maize, mealie, Turkish grain, sweet corn

French: maïs, blé d'Inde, blé de Turquie

Italian: mais

Peru: maiz, chaxu, chinqui, chroop, misha, muti, sara, sha, shinki, trigo de Indias, wawati, xequi, xofiiro, yovato

Mexico: maíz, ixim, cintl, xooba, xoopa, yooba, xupac, nite

Paraguay: guejna

Tunisia: ktania

Yoruba: agbado, igbado, erinigbado, eginrin agbado, erinka, elepee, ijeere, oka, yangan

Southern Africa: kiepiemielie, mielie, milie, miljie, soetmielie, springmielie, Turksch koorn, Turksche taruw; chaake (Sotho); godi (Shoan); ilimone (Zulu); lefela (Pedi); mavhele (Venda); umbila (Swati); umnuli (Ndebele)

Tanzania: elubaeg

India: cholam, yavanala, janar, jagung, bhutta, bottah, bhuththe, makka, makai, maka, makaibonda, mokkajanna, makkasholam, makka-cholam, makka-zonnalu, makkajonna, mukka-jauri, mekkejola, goinjol, munwairingu

The Philippines: mais, gahilang, igi, mait, mañgi, mangi, tigi

Japan: tô-morokoshi

Okinawa: gusun-tojin

China: yu mi xu, yu shu shu, yu kao liang

Vietnam: ngo, bap, ho bo, ma khau ly

South Laos: (people Nya Hön) plää hlii, plää hlii duan (meaning corn with small grains), plää hlii maat (meaning corn with large grains)

Malaya: jagong

Zebrina Schnizl. Commelinaceae

Origins:

From native West African name, Portuguese *zebra*.

Zederachia Heist. ex Fabr. Meliaceae

Origins:

See *Melia azedarach* L.

Zederbauera H.P. Fuchs Brassicaceae

Origins:

Named for the Austrian botanist Emmerich Zederbauer, 1877-1950, professor of botany. Among his works are *Exkursion in die niederösterreichischen Alpen und in das Donautal.* Wien 1905 and *Geschlechtliche und ungeschlechtliche Fortpflanzung von Ceratium hirundinella.* 1904, with the Austrian botanical collector and naturalist Arnold Penther (1865-1931) published *Ergebnisse einer naturwissenschaftlichen Reise zum Erdschias-Dag (Kleinasien).* Vienna 1905-1906; see S. Lenley et al., *Catalog of the Manuscript and Archival Collections and Index to the Correspondence of John Torrey.* Library of the New York Botanical Garden. 444. 1973; Ethelyn (Daliaette) Maria Tucker, *Catalogue of the Library of the Arnold Arboretum of Harvard University.* Cambridge, Massachusetts 1917-1933.

Zehneria Endl. Cucurbitaceae

Origins:

Possibly after the botanical artist Jos Zehner, author of some plates in H.W. Schott and S.F.L. Endlicher, *Meletemata botanica.* Vindobonae [Wien] 1832, in Heinrich Wilhelm Schott (1794-1865), *Rutaceae.* Vindobonae [Wien] 1834, and in S.F.L. Endlicher, *Flora posoniensis.* Posen [Poznan] 1830 and *Iconographia generum plantarum.* Wien [1837-] 1838 [1841-]; or Nicol. Zehner, author of the plates of

Skizzen oesterreichischer Ranunkelen sectionis Allophanes. Wien 1852, by H.W. Schott.

Species/Vernacular Names:

Z. scabra (L.f.) Sond. subsp. *scabra* (*Bryonia scabra* L.f.; *Melothria cordata* (Thunb.) Cogn.; *Melothria punctata* Cogn.)

South Africa: Davidjeswortel, Dawetjieswortel, Dawidjieswortel

Z. thwaitesii (Schweinf.) C. Jeffrey (*Melothria thwaitesii* Schweinf.)

Mozambique: nacaraca-muancane

Zehntnerella Britton & Rose Cactaceae

Origins:

For the German naturalist Leo Zehntner, in Brazil worked for J.N. Rose; see Gordon Douglas Rowley, *A History of Succulent Plants.* 1997.

Zeia Lunell Gramineae

Origins:

Greek *zeia, zea,* name used by Dioscorides for one-seeded wheat, a species of *Triticum,* and by Theophrastus (*HP.* 8.9.2) for rice-wheat.

Zeiba Raf. Bombacaceae

Origins:

For *Ceiba* L.; see C.S. Rafinesque, *Flora Telluriana.* 1: 29. 1836 [1837].

Zelkova Spach Ulmaceae

Origins:

From the Caucasian name; see R. Zander, F. Encke, G. Buchheim and S. Seybold, *Handwörterbuch der Pflanzennamen.* 14. Aufl. 575. Stuttgart 1993.

Species/Vernacular Names:

Z. carpinifolia (Pall.) Koch

English: Caucasian elm

Z. schneiderana Hand.-Mazz.

English: Chinese zelkova

Z. serrata (Thunb.) Mak. (*Ulmus keakii* Sieb.)

English: Japanese zelkova, saw leaf zelkova

Japan: keyaki

China: chu liu

Zelonops Raf. Palmae

Origins:

See C.S. Rafinesque, *Flora Telluriana*. 2: 102. 1836 [1837]; E.D. Merrill (1876-1956), *Index rafinesquianus*. The plant names published by C.S. Rafinesque, etc. 80. Jamaica Plain, Massachusetts, USA 1949; Natalie W. Uhl & John Dransfield, *Genera Palmarum*. 215. Allen Press, Lawrence, Kansas 1987.

Zenkerella Taubert Caesalpiniaceae

Origins:

After the German botanist and plant collector Georg August Zenker (1855-1922), 1896-1916 professional plant collector for Woermann Trading Co. in East and West Cameroon, with the German botanist and botanical collector Alois Staudt in East Cameroon; see G.W.J. Mildbraed (1879-1954), *Notizbl. Bot. Gart. Berl.* 8: 317-324. 1923; Réné Letouzey (1918-1989), "Les botanistes au Cameroun." in *Flore du Cameroun*. 7: 1-110. Paris 1968; Frank Nigel Hepper, "Botanical collectors in West Africa, except French territories, since 1860." in *Comptes Rendus de l'Association pour l'étude taxonomique de la flore d'Afrique*, (A.E.T.F.A.T.). 69-75. Lisbon 1962; F.N. Hepper and Fiona Neate, *Plant Collectors in West Africa*. 87. 1971; Anthonius Josephus Maria Leeuwenberg, "Isotypes of which holotypes were destroyed in Berlin." *Webbia*. 19: 861-863. 1965; H. Walter, "Afrikanische pflanzen in Hamburg." in *Mitt. Geog. Gesell. Hamburg*. 56: 92-93. Hamburg 1965.

Zenkeria Reichb. Bignoniaceae

Origins:

After the German botanist Jonathan Karl Zenker, 1799-1837; see John H. Barnhart, *Biographical Notes upon Botanists*. 3: 539. 1965.

Zenkeria Trinius Gramineae

Origins:

After the German botanist Jonathan Karl Zenker, 1799-1837, studied theology, natural history and medicine, botanical collector, 1825 Dr. med. at Jena, professor of natural history, 1836 professor in the medical faculty. His works include *Die Pflanzen*. Eisenach 1830 and *Plantae indicae*.

Jena 1835-1837, with David Nathaniel Friedrich Dietrich (1799-1888) wrote *Musci thuringici*. Jenae 1821; see John H. Barnhart, *Biographical Notes upon Botanists*. 3: 539. 1965; H.N. Clokie, *Account of the Herbaria of the Department of Botany in the University of Oxford*. 268. Oxford 1964; G. Schmid, *Goethe und die Naturwissenschaften*. Halle 1940; Giulio Giorello & Agnese Grieco, a cura di, *Goethe scienziato*. Einaudi Editore, Torino 1998.

Zenkerina Engler Acanthaceae

Origins:

After the German botanist Georg August Zenker, 1855-1922, 1896-1916 professional plant collector for Woermann Trading Co. in East and West Cameroon.

Zenobia D. Don Ericaceae

Origins:

Greek and Latin Zenobia, Queen of Palmyra, wife of Odaenethus, defeated by Aurelian, taken prisoner on the capture of Palmyra (273) and carried to Rome.

Zeocriton Wolf Gramineae

Origins:

Greek *zeia*, *zea* "a kind of cereal" and *krithe* "barley-corn, barley."

Zephyra D. Don Tecophilaeaceae (Liliaceae)

Origins:

From the Greek *zephyros* "the west wind," referring to the flowers.

Zephyranthella (Pax) Pax Amaryllidaceae (Liliaceae)

Origins:

Diminutive of the genus *Zephyranthes* Herbert.

Zephyranthes Herbert Amaryllidaceae (Liliaceae)

Origins:

Greek *zephyros* "the west wind" and *anthos* "flower."

Species/Vernacular Names:

Z. atamasco (L.) Herbert

English: zephyr lily

Z. candida (Lindl.) Herb. (*Amaryllis candida* Herbert)

English: autumn zephyr lily

Japan: tama-sudare

China: gan feng cao

Z. carinata Herbert (*Zephyranthes grandiflora* Lindl.)

English: rose-pink zephyr lily, rain lily

Japan: safuran-modoki

Okinawa: zikujiku, gushii

China: sai fan hong hua

Zerumbet J.C. Wendl. Zingiberaceae

Origins:

See *Zingiber zerumbet* (L.) Sm. (*Amomum zerumbet* L.) and *Alpinia zerumbet* (Pers.) B.L. Burtt & Rosemary M. Sm.; see H. Genaust, *Etymologisches Wörterbuch der botanischen Pflanzennamen*. 699. Basel 1996.

Zetagyne Ridley Orchidaceae

Origins:

From the Greek letter *zeta* and *gyne* "female, woman, pistil," referring to the shape of the column.

Zeugandra P.H. Davis Campanulaceae

Origins:

Greek *zeugos* "a yoke, chariot" and *aner, andros* "male, a man, male organs."

Zeugites P. Browne Gramineae

Origins:

Greek *zeugites* "yoked, joined in pairs," Latin *zeugites, ae* for Plinius a kind of reed.

Zeuktophyllum N.E. Br. Aizoaceae

Origins:

From the Greek *zeuktos* "yoked, joined, fastened" and *phyllon* "leaf."

Zeuxanthe Ridley Rubiaceae

Origins:

From the Greek *zeuxis* "yoking, fastening" and *anthos* "flower."

Zeuxine Lindley Orchidaceae

Origins:

Greek *zeuxis* "yoking, fastening," the column and the lip (labellum) are partially united.

Species/Vernacular Names:

Z. strateumatica (L.) Schltr. (*Orchis strateumatica* L.; *Spiranthes strateumatica* (L.) Lindl.; *Zeuxine sulcata* Lindl.; *Zeuxine rupicola* Fuk.) (Greek *strateuma, atos* "armament, army, military company, host")

Japan: Yakushima-hime-ari-dôshi-ran (Yakushima = Yaku Island)

Zexmenia La Llave & Lex. Asteraceae

Origins:

An anagram of the surname of Francisco Ximenez.

Zeyhera Mart. Bignoniaceae

Origins:

See *Zeyheria* Martius.

Zeyherella (Pierre ex Engler) Aubréville & Pellegrin Sapotaceae

Origins:

For the German (b. Hesse) botanist and botanical collector Carl (Karl) Ludwig Philipp Zeyher, 1799-1858 (Cape Town, died of smallpox), traveler with Franz Wilhelm Sieber (1789-1844), with Christian Friedrich Ecklon (1795-1868) published *Enumeratio plantarum Africae Australis* 1835-1837, collected in South Africa with John Burke, from 1849 botanist at the Botanical Garden at the Cape; see William Jackson Hooker (1785-1865), *London J. Bot*. 2: 163-165. 1843, 5: 242. 1846; *Kew J. Bot*. 2: 61-62. 1850; Gordon Douglas Rowley, *A History of Succulent Plants*. 1997; *Catalogue of the Books, Manuscripts, Maps and Drawings in the British Museum (Natural History)*. 2: 504 and 5: 2389. Weinheim 1964; Mary Gunn and Leslie E. Codd, *Botanical Exploration of Southern Africa*. 383-387, 388-395. Cape Town 1981; J.H. Barnhart, *Biographical Notes upon Botanists*. 3:

540. 1965; C.F. Ecklon, "Nachricht über die von Ecklon und Zeyher unternommenen Reisen und deren Ausbeute in botanischer Hinsicht." *Linnaea*. 8: 390-400. 1833; Karl Boriwog Presl (1794-1852), *Botanische Bemerkungen*. Prague 1844; H.N. Clokie, *Account of the Herbaria of the Department of Botany in the University of Oxford*. Oxford 1964; Antoine Lasègue (1793-1873), *Musée botanique de M. Benjamin Delessert*. Paris, Leipzig 1845; Ethelyn (Daliaette) Maria Tucker, *Catalogue of the Library of the Arnold Arboretum of Harvard University*. Cambridge, Massachusetts 1917-1933; Alain White (1880-1951) and Boyd Lincoln Sloane (1886-1955), *The Stapelieae*. Pasadena 1937; R. Zander, F. Encke, G. Buchheim and S. Seybold, *Handwörterbuch der Pflanzennamen*. 14. Aufl. Stuttgart 1993; Stafleu and Cowan, *Taxonomic Literature*. 7: 534-535. Utrecht 1988.

Zeyheria Martius Bignoniaceae

Origins:
Named for the German botanist Johann Michael Zeyher, 1770-1843, horticulturist, traveler, plant collector, from 1806 to 1843 Director of the garden of the grand-duke of Baden at Schwetzingen, author of *Verzeichniss der Gewaechse in dem Grossherzoglichen Garten zu Schwetzingen*. Mannheim 1819; see J.H. Barnhart, *Biographical Notes upon Botanists*. 3: 540. 1965; Antoine Lasègue, *Musée botanique de M. Benjamin Delessert*. Paris, Leipzig 1845; Ethelyn Maria Tucker, *Catalogue of the Library of the Arnold Arboretum of Harvard University*. Cambridge, Massachusetts 1917-1933; Stafleu and Cowan, *Taxonomic Literature*. 7: 535-536. 1988.

Zeylanidium (Tul.) Engl. Podostemaceae

Origins:
Of Ceylon (now Sri Lanka).

Zieria Smith Rutaceae

Origins:
After the Polish botanist John Zier, d. 1796 (London), assistant to Ehrhart and William Curtis, 1788 a Fellow of the Linnean Society, contributor to James (Jacobus) J. Dickson, *J. Dickson fasciculus Plantarum Cryptogamicarum Britanniae*. London 1785-1801; see J. Britten, "John Zier, F.L.S." *J. Bot.* (London). 24. 1886; J.E. Smith, in *Transactions of the Linnean Society of London. Botany*. 4: 216. 1798; Arthur D. Chapman, ed., *Australian Plant Name Index*. 3023-3025. Canberra 1991.

Species/Vernacular Names:
Z. arborescens Sims

English: stinkwood

Z. veronica (F. Muell.) Benth. (*Boronia veronicea* F. Muell.)

English: pink zieria

Zieridium Baillon Rutaceae

Origins:
Resembling the genus *Zieria* Sm.

Zigadenus Michaux Melanthiaceae (Liliaceae)

Origins:
Greek *zygon* "a yoke" and *aden* "a gland," referring to the floral glands.

Species/Vernacular Names:
Z. densus (Desr.) Fern.

English: crow poison

Z. glaucus Nutt.

English: white camash

Z. nuttallii A. Gray

English: death camash, death camas

Zigmaloba Raf. Mimosaceae

Origins:
Greek *sigma* "S-shaped" and *lobos* "a pod, lobe"; see C.S. Rafinesque, *Sylva Telluriana*. 120. 1838.

Zilla Forssk. Brassicaceae

Origins:
An Arabic vernacular name; see Pehr (Peter) Forsskål (1732-1763), *Flora aegyptiaco-arabica*. 121. Copenhagen 1775.

Species/Vernacular Names:
Z. spinosa (L.) Prantl

Arabic: zilla, zillae

Zimmermannia Pax Euphorbiaceae

Origins:
Named after the German professor and plant collector Philipp William (Wilhelm) Albrecht Zimmermann, 1860-1931,

naturalist, traveler, botanist and assistant Director of the Biological-Agronomical Institute at Amani in Tanzania, he published *Beiträge zur Morphologie und Physiologie der Pflanzenzelle* ... Tübingen 1893 and *Die Cucurbitaceen*. Jena 1922; see John H. Barnhart, *Biographical Notes upon Botanists*. 3: 541. 1965; T. W. Bossert, *Biographical Dictionary of Botanists Represented in the Hunt Institute Portrait Collection*. 449. Boston, Massachusetts 1972; Ethelyn Maria Tucker, *Catalogue of the Library of the Arnold Arboretum of Harvard University*. Cambridge, Massachusetts 1917-1933.

Zimmermanniopsis Radcl.-Sm. Euphorbiaceae

Origins:
Resembling *Zimmermannia* Pax.

Zingeriopsis Prob. Gramineae

Origins:
Resembling *Zingeria* P.A. Smirn.

Zingiber Boehmer Zingiberaceae

Origins:
Greek *zingiberis* for ginger or an Arabian spice (Dioscorides, Galenus, Soranus and Oribasius), late Latin *gingiber*, ancient Indian *srngavera*, Malayalam *inchiver* (*inchi* "root"), Latin *zingiberi* or *zimbiperi*, also *zingiber* "ginger"; see Manlio Cortelazzo & Paolo Zolli, *Dizionario etimologico della lingua italiana*. 5: 1465. Zanichelli, Bologna 1988; E. Weekley, *An Etymological Dictionary of Modern English*. 1: 639. Dover Publications, New York 1967.

Species/Vernacular Names:
Z. spp.
English: ginger
South Africa: gemmer

Z. mioga (Thunb.) Roscoe (*Amomum mioga* Thunb.)
English: Japanese ginger, mioga ginger
Japan: myôga
China: rang he, jang ho

Z. officinale Roscoe (*Amomum zingiber* L.; *Zingiber cholmondeleyi* (Bailey) Schumann)
English: common ginger, Canton ginger, stem ginger, Queensland ginger, ginger, East Indian ginger
Arabic: zingibil, zenjabil

French: gingembre
Tanzania: tangauzi
Yoruba: atale
Congo: tangawissi
India: ardraka, adrak, ada, ala, alen, allam, adu, adi, inji, ardrakamu, ardhrakam, varam, katu-patram, andrakam, srangavera, sringa-beram, nagara, nagaram, visoushada, maha-oushadam, mahaushada, sonth, sho-ont, sunta, sunt, sonti, shukhu, chukku, chukka, hashi-shunti, vona-shunti, soonti, inguru
Vietnam: gung, khuong
Tibetan: sga, bca'-sga, sman-sga, dong-gra
The Philippines: luya, laya, baseng, agat
Japan: shôga
Okinawa: soga
China: sheng jiang, sheng chiang, gan jiang, kan chiang, chiang
Brazil (Amazonas): amata kiki

Z. purpureum Roscoe
English: Bengal ginger, cassumar ginger

Z. zerumbet (L.) Sm. (*Amomum zerumbet* L.)
English: shampoo ginger, wild ginger
Japan: hana-shôga
Hawaii: 'awapuhi, 'awapuhi kuahiwi, 'opuhi

Zinnia L. Asteraceae

Origins:
For the German botanist Johann Gottfried Zinn, 1727-1759, physician, professor of botany, was Haller's favorite pupil, professor of medicine and anatomy, Director of the Botanical Gardens at Göttingen, botanical collector, author of *Observationes quaedam botanicae*. Gottingae 1753, *Descriptio anatomica oculi humani iconibus illustrata*. [correctly described "fibrae radiatae"] Gottingae 1755 and *Catalogus plantarum horti academici et agri gottingensis*. Gottingae 1757; see John H. Barnhart, *Biographical Notes upon Botanists*. 3: 542. 1965; C. Linnaeus, *Systema Naturae*. 1189, 1221 and 1377. 1759; Jonas C. Dryander, *Catalogus bibliothecae historico-naturalis Josephi Banks*. London 1796-1800; T. W. Bossert, *Biographical Dictionary of Botanists Represented in the Hunt Institute Portrait Collection*. 450. Boston, Massachusetts 1972; R. Zander, F. Encke, G. Buchheim and S. Seybold, *Handwörterbuch der Pflanzennamen*. 14. Aufl. 803. Stuttgart 1993; Garrison and Morton, *Medical Bibliography*. By Leslie T. Morton. 1484. New York 1961; Ethelyn Maria Tucker, *Catalogue of the Library of the Arnold Arboretum of Harvard University*. Cambridge, Massachusetts 1917-1933.

Species/Vernacular Names:

Z. elegans Jacq. (*Zinnia gracillima* hort.)

English: youth and old age

Japan: hyaku-nichi-sô

Z. peruviana (L.) L. (*Zinnia multiflora* L.; *Zinnia pauciflora* L.)

English: kaffor daisy, redstar zinnia, wild zinnia

Southern Africa: Engelsmannetjies, fluitjiesbossie, Jakobre-gop, regopjakob, wilde Jakobregop; lipii (Sotho)

Zizania L. Gramineae

Origins:

Greek *zizanion*, ancient name for a wild grain, a weed that grows in wheat, Latin *zizania, orum* "darnel, cockle, tares"; see Manlio Cortelazzo & Paolo Zolli, *Dizionario etimologico della lingua italiana.* 5: 1468. Zanichelli, Bologna 1988; Giovanni Semerano, *Le origini della cultura europea. Dizionario della lingua Latina e di voci moderne.* 2(2): 620. Leo S. Olschki Editore, Firenze 1994; Giovanni Semerano, *Le origini della cultura europea.* Dizionari Etimologici. Basi semitiche delle lingue indeuropee. Dizionario della lingua Greca. Leo S. Olschki Editore, Firenze 1994.

Species/Vernacular Names:
Z. aquatica L.

English: annual wild rice, water rice, Canadian wild rice, Manchurian water rice, wild oat, Tuscarora rice

China: ku

Z. latifolia (Griseb.) Stapf (*Hydropyrum latifolium* Griseb. ex Ledeb.)

English: broad-leaved rice grass, Manchurian wild rice, water rice, shattering wild rice

Japan: ma-komo

Okinawa: ma-komu

China: jiao bai, ku, chiao tsao, chiang tsao

Zizaniopsis Döll & Asch. Gramineae

Origins:

Resembling the genus *Zizania* L.

Ziziphus Miller Rhamnaceae

Origins:

From the Persian *zizfum* or *zizafun*; from the Arabian *zizouf*, the name for *Ziziphus lotus* Lam.; Latin *ziziphus, zizyphus*, the jujube-tree (Plinius and Columella), *ziziphum, zizyphum*, the fruit jujube (Plinius); the Greek *zizouphon, zizyphon* for the tree, and *zizoula* for the fruit; see Giovanni Semerano, *Le origini della cultura europea.* Dizionario della lingua Latina e di voci moderne. 2(2): 620. Leo S. Olschki Editore, Firenze 1994.

Species/Vernacular Names:
Z. sp.

The Philippines: aligamen, alinau, alisalanga, baliknit, biaa, biga, bigaa, dagaau, dalipatsan, diaan, diklab, diklap, dir'an, diraan, duklap, dukulab, labalaba, ligaa, ligamen, limiggien, maglangka, malipaga, mangaluas, marautong, matang-hipon, pirokalau, salindangat

Neth. Indies: bedaroh

Tibetan: chi-pa kha

China: chung ssu tsao, hsien tsao, ku tsao, chueh i

Z. abyssinica Hochst.

Nigeria: magariyar kura

N. Rhodesia: mupundukaina

Z. jujuba Mill. (*Ziziphus sativa* Gaertner; *Ziziphus vulgaris* Lam.)

English: jujube, common jujube, Chinese date, Chinese jujube

China: suan tsao, suan zao ren, tsao, pei tsao, nan tsao, da zao

Malaya: bedara china, jujube

Nigeria: ekanesi, ekanse-adie, magariya, kusuru, kusulu

Z. mauritiana Lam. (*Ziziphus jujuba* (L.) Lam. not Mill.)

English: common jujube, Chinese jujube, Indian jujube, Chinese date, cottony jujube

French: jujubier de l'Inde, massonnier, jujubier

Venezuela: azufaifo, ponsigue, yuyubo

Mexico: ciruela gobernadora; quetembilla (Yucatan); jujube (Parras, Coahuila)

Burma: mahkaw, mak-hkaw, zi, zidaw

Vietnam: tao, tao chua

Thailand: phoot mai

Japan: inu-natsu-me

Nepal: bayar

India: bayr, ber, beri, bier, bir, jum-janum, ringa, yellande; budree, dedoari-janum, kool, kul, narikeleekool (Bengal); caroukouva-marom, elandar-pazham, iharberi, ilandampa-jam, ilenden-marom (South India); ilangi, ilantai, ilantha (West India); koli kukhunda, nabig, nazuc, regi, renga, rengha, unab, vadari (East India); jhar-beri (N.E. India); reyghoo, yelchi (Bombay)

Malaya: bidara, bedara, bedara China, apple Siam

Mali: ndomon, ndomonon, tomboron

Nigeria: kurna; magariya (Hausa); jali (Fula); kusulu (Kanuri)

Southern Africa: muChecheni, muSawu, muSau, muTjejele (Shona)

Mauritius: masson

Z. mistol Griseb.

Argentina: mistol

Z. mucronata Willd. subsp. *mucronata*

English: buffalo thorn, Cape thorn, shiny leaf, wait-a-bit, wait-a-bit tree

French: jujubier à griffes

Nigeria: magariya-kura; magariyar kura (Hausa)

Yoruba: eekannase adie

Togo: mangu, pangbainga, sansanyebui

Mali: jabi folo, suruku tomboron

Tropical Africa: omukaru

Southern Africa: blinkblaar, blinkblaar-wag-'n-bietjie (meaning shiny-leaved-wait-a-bit), haak-en-steek-wag-'n-bietjie, buffelsdoring, bokgalo; umLahlabantu (Swazi); umPhafa, umLahlankosi (meaning that buries the chief), isiLahla, umKhobonga (Zulu); muChecheni, chiNanga, muPakwe, muPasamala, muTsotsomva (Shona); umPhafa (Xhosa); mphasamhala (Tsonga or Thonga); moksalo (Pedi); mokgalo (Tswana: western Transvaal, northern Cape, Botswana); mokhalo (Sotho); mokgalo, moonaona (North Sotho); mukhalu, mutshetshete (Venda); monganga (Kololo, Barotseland); umpafa (Ndebele); omukaru (Herero); omukekete (Ovambo: northern southwest Africa); aros (Nama: southern southwest Africa)

Z. oenoplia (L.) Mill. (Greek *oinos* "wine" and *pleios* "many")

English: squirrel's jujube, littlefruit jujube

Malaya: duri sakah, kuku lang

Z. rivularis Codd

English: false buffalo-thorn, ziziphus of the stream

South Africa: vals-wag-'n-bietjie, river blinkblaar

Z. spina-christi (L.) Desf.

English: Christ's thorn, crown of thorns

French: jujubier vert

Arabic: sidr, zegzeg, zefzouf, ardj, ourdj, nabq

Nigeria: kurna, ekanse-adiye, kurunwa

North Africa: kabk, negba

Mali: bandomon, tomboron

Z. spina-christi (L.) Desf. var. *spina-christi*

Nigeria: kurna (Hausa); kurnahi (Fula); korna (Kanuri); karno (Shuwa Arabic); eekanannase-adie

Z. spinosus Hu

English: spiny Chinese date

China: suanzaoren, suan zau ren

Ziziphus Adanson Rhamnaceae

Origins:
Orthographic variant of *Ziziphus* Miller.

Zoisia J. Black Gramineae

Origins:
An orthographic variant, see *Zoysia* Willd.

Zollernia Wied-Neuwied & Nees Fabaceae

Origins:
In honor of Friderico Guilelmo III, a member of the Hohenzollern family; see Gwilym Lewis & Clive Foster, "*Zollernia splendens*." in *Curtis's Botanical Magazine*. 14(4): 194-197. 1997.

Zollingeria Kurz Sapindaceae

Origins:
For the Swiss botanist Heinrich Zollinger, 1818-1859, school-teacher, studied with Alph. de Candolle, plant collector and botanical explorer in Java and Sumatra. Among his writings are *Observationes phytographicae*. [Batavia 1844-1846], *Over de soorten van Rottlera*. Batavia 1856, *Observationes botanicae novae*. [Batavia 1857] and *Sur la végétation autour des cratères volcaniques de l'île de Java*. [Genève] 1858, see also in *Verhandelingen der Natuurkundige Vereen. in Nederlandsche Indië*. 1: 19. 1856 [Acta Soc. Regiae Sci. Indo-Neerl. 1(4): 19. post 1 Sep. 1856]; see John H. Barnhart, *Biographical Notes upon Botanists*. 3: 543. 1965; T. W. Bossert, *Biographical Dictionary of Botanists Represented in the Hunt Institute Portrait Collection*. 450. Boston, Massachusetts 1972; Antoine Lasègue, *Musée botanique de M. Benjamin Delessert*. Paris, Leipzig 1845; Ethelyn Maria Tucker, *Catalogue of the Library of the Arnold Arboretum of Harvard University*. Cambridge, Massachusetts 1917-1933; R. Zander, F. Encke, G. Buchheim and S. Seybold, *Handwörterbuch der Pflanzennamen*. 14. Aufl. Stuttgart 1993.

Zombia L.H. Bailey Palmae

Origins:
A common name.

Zomicarpa Schott Araceae

Origins:
Greek *zoma* "a belt, dress" and *karpos* "fruit"; see D.H. Nicolson, "Derivation of aroid generic names." *Aroideana.* 10: 15-25. 1988.

Zomicarpella N.E. Br. Araceae

Origins:
Diminutive of the genus *Zomicarpa* Schott.

Zonanthemis Greene Asteraceae

Origins:
From the Greek *zone* "a belt, armor, girdle" and *anthos, anthemis, anthemon* "flower"; see also *Hemizonia* DC.

Zonanthus Griseb. Gentianaceae

Origins:
Greek *zone* "a belt, armor, girdle" and *anthos* "flower."

Zonotriche (C.E. Hubb.) J.B. Phipps Gramineae

Origins:
From the Greek *zone* "a belt, armor, girdle" and *thrix, trichos* "hair."

Zoophora Bernh. Orchidaceae

Origins:
From the Greek *zoon* "a living being, animal" and *phoros* "bearing, carrying."

Zootrophion Luer Orchidaceae

Origins:
Greek *zootropheo* "breed animals, keep animals," *zootropheion* "place for keeping animals," the flowers resembling animal heads.

Zornia J. Gmelin Fabaceae

Origins:
After the German pharmacist Johannes Zorn, 1739-1799, botanist, author of *Icones plantarum medicinalium.* Nürnberg 1779 [-1790]; see John H. Barnhart, *Biographical Notes upon Botanists.* 3: 543. 1965; Jonas C. Dryander, *Catalogus bibliothecae historico-naturalis Josephi Banks.* London 1796-1800; Ethelyn Maria Tucker, *Catalogue of the Library of the Arnold Arboretum of Harvard University.* Cambridge, Massachusetts 1917-1933.

Species/Vernacular Names:
Z. glochidiata Rchb. ex DC.
Yoruba: reku reku igbo
Z. latifolia Sm.
Yoruba: emu

Zostera L. Zosteraceae (Potamogetonaceae)

Origins:
Greek *zoster* "a girdle," referring to the leaves, Latin *zoster, eris* for Plinius is a kind of sea shrub, called also *prason*; see Carl Linnaeus, *Species Plantarum.* 968. 1753 and *Genera Plantarum.* Ed. 5. 415. 1754.

Species/Vernacular Names:
Z. capricorni Asch.
English: eel-grass
Z. japonica Asch. & Graebn. (*Zostera nana* Martens ex Roth)
Japan: ko-ama-mo (= small *Zostera*)
Z. marina L.
English: eel-grass, grasswrack, barnacle grass
China: hai dai
Z. mucronata Hartog
English: garweed
Z. muelleri Irmisch ex Asch. (*Zostera novazelandica* Setchell)
English: garweed, dwarf grasswrack

Zosterella Small Pontederiaceae

Origins:
Diminutive of *Zostera.*

Zosterostylis Blume Orchidaceae

Origins:
Greek *zoster* "a girdle" and *stylos* "style, column," referring to the margin of the clinandrium; see also *Cryptostylis* R.Br.

Zoydia Pers. Gramineae

Origins:

See *Zoysia* Willdenow.

Zoysia Willd. Gramineae

Origins:

After the Austrian botanist Karl von Zoys, 1756-1800, plant collector.

Species/Vernacular Names:

Z. japonica Steud.

English: Korean grass, Korean lawn grass, Japanese lawn grass

Z. matrella (L.) Merr. (*Agrostis matrella* L.; *Zoysia pungens* Willd.)

English: Manila grass, Japanese carpet grass, flawn

Japan: Kôshun-shiba, hari-shiba, Yonaguni-shiba

The Philippines: barit-baritan

Z. tenuifolia Willd. ex Trin.

English: Mascarene grass, Korean velvet grass, Korean grass

Zucca Comm. ex Juss. Cucurbitaceae

Origins:

Zucca, Italian for gourd; see M. Cortelazzo & P. Zolli, *Dizionario etimologico della lingua italiana.* 5: 1469-1470. Bologna 1988.

Zuccagnia Cavanilles Caesalpiniaceae

Origins:

For the Italian botanist Attilio Zuccagni, 1754-1807, physician, plant collector and traveler; see John H. Barnhart (1871-1949), *Biographical Notes upon Botanists.* 3: 544. Boston 1965; A. Lasègue, *Musée botanique de Benjamin Delessert.* Paris 1845; Jonas C. Dryander, *Catalogus bibliothecae historico-naturalis Josephi Banks.* London 1796-1800; R. Zander, F. Encke, G. Buchheim and S. Seybold, *Handwörterbuch der Pflanzennamen.* 14. Aufl. Stuttgart 1993; O. Mattirolo, *Cenni cronologici sugli Orti Botanici di Firenze.* in Pubblicazioni R. Ist. Studi Superiori Pratici e di Perfezionamento, Sez. Scienze Fis. Natur. Firenze 1899; P. Luzzi, "L'Orto Botanico *Giardino dei Semplici.*" in *Museo Nazionale di Storia Naturale* a Firenze: ipotesi di insediamento. Alinea Editrice, Firenze 1987; Guido Moggi, "Il Museo Botanico ed il Giardino dei Semplici dell'Università di Firenze." *Boll. Soc. It. Iris.* 26-39. 1975.

Zuccarinia Blume Rubiaceae

Origins:

For the German botanist Joseph Gerhard Zuccarini, 1797-1848, physician, professor of botany and specialist in Cactaceae; see John H. Barnhart (1871-1949), *Biographical Notes upon Botanists.* 3: 544. Boston 1965; Ida Kaplan Langman, *A Selected Guide to the Literature on the Flowering Plants of Mexico.* 1964; A. Lasègue, *Musée botanique de Benjamin Delessert.* Paris 1845; S. Lenley et al., *Catalog of the Manuscript and Archival Collections and Index to the Correspondence of John Torrey.* Library of the New York Botanical Garden. 473. 1973; Ethelyn (Daliaette) Maria Tucker, *Catalogue of the Library of the Arnold Arboretum of Harvard University.* Cambridge, Massachusetts 1917-1933; Stafleu and Cowan, *Taxonomic Literature.* 7: 560-562. Utrecht 1988; Gordon Douglas Rowley, *A History of Succulent Plants.* Strawberry Press, Mill Valley, California 1997; Mariella Azzarello Di Misa, a cura di, *Il Fondo Antico della Biblioteca dell'Orto Botanico di Palermo.* 295. Palermo 1988; R. Zander, F. Encke, G. Buchheim and S. Seybold, *Handwörterbuch der Pflanzennamen.* 14. Aufl. Stuttgart 1993.

Zwackhia Sendtner ex Reichb. Boraginaceae

Origins:

After the German botanist Philipp Franz Wilhelm von Zwackh-Holzhausen, 1825-1903, lichenologist; see John H. Barnhart (1871-1949), *Biographical Notes upon Botanists.* 3: 544. Boston 1965; T. W. Bossert, *Biographical Dictionary of Botanists Represented in the Hunt Institute Portrait Collection.* 451. Boston, Massachusetts 1972.

Zygadenia L.E. Bishop Grammitidaceae

Origins:

From the Greek *zygon, zygos* "yoke" and *aden* "a gland."

Zygalchemilla Rydb. Rosaceae

Origins:

From the Greek *zygon, zygos* "yoke" plus the genus *Alchemilla* L.

Zygella S. Moore Iridaceae

Origins:

From the Greek *zygon, zygos* "yoke," Latin *jugum, iugum,* Akkadian *zunqu, sunqu, suqu* "yoke," *zanaqu, sanaqu* "to yoke, to bind, to fasten," Sumerian *ugug* "yoke"; see also *Cypella* Herbert.

Zygia P. Browne Mimosaceae

Origins:

Greek *zygon, zygos* "yoke," referring to the stamens and leaves; see Patrick Browne (1720-1790), *The Civil and Natural History of Jamaica in Three Parts.* 279. (Mar.) 1756.

Zygocactus Schumann Cactaceae

Origins:

From the Greek *zygon, zygos* "yoke" plus *Cactus*.

Zygocereus Fric & Kreuz. Cactaceae

Origins:

From the Greek *zygon, zygos* "yoke" and *Cereus*.

Zygochloa S.T. Blake Gramineae

Origins:

Greek *zygon, zygos* "yoke" and *chloe, chloa* "grass," referring to the dioecious spikelets.

Species/Vernacular Names:

Z. paradoxa (R. Br.) S.T. Blake (*Neurachne paradoxa* R. Br.; *Spinifex paradoxus* (R. Br.) Benth.)

English: sandhill canegrass

Zygoglossum Reinw. Orchidaceae

Origins:

Greek *zygon, zygos* "yoke, balance" and *glossa* "a tongue," referring to the divisions of the perianth or to the versatile lip; see also *Bulbophyllum*.

Zygogynum Baillon Winteraceae

Origins:

From the Greek *zygon, zygos* "yoke" and *gyne* "a woman, female."

Zygonerion Baillon Apocynaceae

Origins:

From the Greek *zygon, zygos* "yoke" and the genus *Nerium* L., *neros* "wet," *nero* "water."

Zygoon Hiern Rubiaceae

Origins:

Greek *zygon, zygos* "yoke" and *oon* "egg."

Zygopetalum Hook. Orchidaceae

Origins:

From the Greek *zygon, zygos* "yoke" and *petalon* "petal, leaf, sepal," referring to the swelling at the base of the lip.

Zygophlebia L.E. Bishop Grammitidaceae

Origins:

From the Greek *zygon, zygos* "yoke" and *phleps, phlebos* "vein."

Zygophyllidium (Boiss.) Small Euphorbiaceae

Origins:

From the Greek *zygon, zygos* "yoke" and *phyllon* "leaf."

Zygophyllum L. Zygophyllaceae (Rosidae, Sapindales)

Origins:

Greek *zygon* "a yoke" and *phyllon* "leaf," usually the leaves are bifoliate; see Carl Linnaeus, *Species Plantarum.* 385. 1753 and *Genera Plantarum.* Ed. 5. 182. 1754.

Species/Vernacular Names:

Z. fabago L.

English: Syrian bean caper

Z. flexuosum Ecklon & Zeyher

South Africa: spekbossie

Z. fulvum L.

South Africa: bokkos, jakkals pisbos, trekbos

Z. morgsana L.

South Africa: leeubossie, skilpadbossie, skuimbossie, slymbos, spekbossie, vetbossie, vleisbossie

Z. simplex L.

South Africa: brakkies, brakspekbos, rankspek, rankspek-bos, panspekbos, volstruisdruiwe, volstruis-slaai

German: polsterpflanze

Z. stapffii Schinz (after the German (b. Saxony) geologist and mining engineer Friedrich Moritz Stapff (1836-1895, d. Tanzania) who prospected for copper in the Kuiseb Valley in 1885-1886, he wrote *Geologisches Profil des St Gotthard in der Axe des grossen Tunnels während des Banes.* 1873-1880 ... Bern 1880; see Mary Gunn and Leslie Edward W. Codd, *Botanical Exploration of Southern Africa.* 332. A.A. Balkema, Cape Town 1981; Gilbert Westacott Reynolds (1895-1967), *The Aloes of South Africa.* Balkema, Rotterdam 1982)

Namibia: daalderplant (Afrikaans)

German: talerpflanze

Zygoruellia Baillon Acanthaceae

Origins:
From the Greek *zygon* "a yoke" plus the genus *Ruellia* L.

Zygosepalum Reichb.f. Orchidaceae

Origins:
From the Greek *zygon* "a yoke" and Latin *sepalum* "sepal," indicating the base of the lateral sepals, the sepals are basally connate.

Zygosicyos Humbert Cucurbitaceae

Origins:
From the Greek *zygon* "a yoke" plus *sikyos* "wild cucumber, gourd."

Zygospermum Thwaites ex Baill. Euphorbiaceae

Origins:
From the Greek *zygon* "a yoke" and *sperma* "seed."

Zygostates Lindley Orchidaceae

Origins:
Greek *zygostates* "one who weighs," *zygostathmos* "the balance," *zygostasia* "a weighing," *zygostateo* "to weigh," *istemi* "to stand," an allusion to the appearance of the clavate staminodes.

Zygostelma Benth. Asclepiadaceae (Periplocaceae)

Origins:
From the Greek *zygon* "a yoke" and *stelma, stelmatos* "a girdle, belt, crown, garland, wreath."

Zygostigma Griseb. Gentianaceae

Origins:
From the Greek *zygon* "a yoke" and *stigma* "a stigma."

Zygotritonia Mildbr. Iridaceae

Origins:
From the Greek *zygon* "a yoke" plus the genus *Tritonia.*

Zymum Thouars Malpighiaceae

Origins:
From the Greek *zyme* "leaven."

BIBLIOGRAPHY

Eleanor B. Adams, *A Bio-Bibliography of Franciscan Authors in Colonial Central America*. Washington, D.C. 1953.

H.M. Adams, *Catalogue of Books Printed on the Continent of Europe, 1501-1600 in Cambridge Libraries*. Compiled by H.M. Adams. Cambridge University Press, Cambridge 1967.

R.F.G. Adams, "Efik vocabulary of living things." *Nigerian Field*. 11, 1943, II(12): 23-24. 1947 and III(13): 61-67. 1948.

Edouard J. Adjanohoun et al., *Contribution aux études ethnobotaniques et floristiques au Togo*. Editions agence de coopération culturelle et technique (A.C.C.T.), Paris 1986.

Z.M. Agha, *Bibliography of Islamic Medicine and Pharmacy*. London 1983.

W. Aiton, *Hortus kewensis*. London 1789.

J.A. Akinniyi and M.U.S. Sultanbawa, "A glossary of Kanuri names of plants, with botanical names, distribution and uses." *Ann. Borno*. 1. 1983.

J. Albany, *P'tit glossaire. Le piment des mots créoles*. Paris 1974 and 1983.

Albinal and Maljac, *Dictionnaire malgache-français*. Fianarantsoa 1987.

Janis B. Alcorn, *Huastec Mayan Ethnobotany*. University of Texas Press, Austin 1984.

Janis B. Alcorn and Cándido Hernández V., "Plants of the Huastecan region with an analysis of their Huastec names." *Journal of Mayan Linguistics*. 4: 11-18. 1983.

John Alden and Dennis Channing Landis, Eds., *European Americana: A Chronological Guide to Works Printed in Europe Relating to the Americas, 1493-1776*. Six volumes. New York 1980-1997.

Francisco Javier Alegre, S.J., *Historia de la provincia de la Compañia de Jesús de Nueva España*. New edition edited by Ernest J. Burrus and Felix Zubillaga. Rome 1956-1960.

G. Alessio, *Lexicon etymologicum*. Napoli 1976.

J.J.G. Alexander and A.C. de la Mare, *The Italian Manuscripts in the Library of Major J.R. Abbey*. London 1969.

Mea Allan, *The Hookers of Kew*. London 1967.

A. Amiaud and L. Méchineau, *Tableau comparé des écritures babylonienne et assyrienne, archaïques et modernes*. Paris 1902.

Fr. Francisco Antolín, O.P., *Notices of the Pagan Igorots in the Interior of the Island of Manila*. Manila 1988.

Archives. *Archives d'Études Orientales*, publiés par J.A. Lundell. Roma, Lund etc. 1911-1934.

Fernando de Armas Medina, *Cristianización del Perú (1532-1600)*. Escuela de Estudios Hispano-Americanos de Sevilla. Sevilla 1953.

T.H. Arnold and B.C. de Wet, Eds., *Plants of Southern Africa: Names and Distribution*. Pretoria 1993.

M. Arunachalam, *Festivals of Tamil Nadu*. Tiruchitrambalam 1980.

F. Ascarelli, *La tipografia cinquecentina italiana*. Sansoni, Firenze 1953.

F. Ascarelli, *Le cinquecentine romane*. Milano 1972.

G. Aschieri, *Dizionario compendiato di geologia e mineralogia*. Milano 1855.

P.M. Ashburn, *The Ranks of Death: A Medical History of the Conquest of America*. New York 1947.

William Ashworth and Bruce Bradley, *Jesuit Science in the Age of Galileo*. Linda Hall Library, Kansas City 1986.

The Assyrian Dictionary of the Oriental Institute of the University of Chicago. Chicago 1963.

Jean Baptiste Christophe Fusée Aublet, *Histoire des Plantes de la Guiane Françoise*. Paris 1775.

Gabriel Austin, *The Library of Jean Grolier: A Preliminary Catalogue*. New York 1971.

João Lucio de Azevedo, *Os Jesuítas no Grão Pará, Suas missões e colonização*. Coimbra 1930.

Oreste Badellino and Ferruccio Calonghi, *Dizionario Italiano-Latino e Latino-Italiano*. Third edition. Rosenberg & Sellier, Torino 1990.

R.P. Ch. Bailleul, Ed., *Petit dictionnaire français-bambara, bambara-français*. Avebury Publishing Company 1981.

M.J. Balick and P.A. Cox, *Plants, People, and Culture: The Science of Ethnobotany*. Scientific American Press 1996.

S.C. Banerjee, *Flora and Fauna in Sanskrit Literature*. Calcutta 1980.

Antoine Alexandre Barbier, *Dictionnaire des ouvrages anonymes et pseudonymes*. Paris 1872-1879.

C. Bardesono di Rigras, *Vocabolario marinaresco*. Roma 1932.

John H. Barnhart, *Biographical Notes upon Botanists*. Boston 1965.

J. Barrau, "Notes on the significance of some vernacular names of food plants in the South Pacific Islands." *Proc. 9th Pacific Sci. Cong*. 4: 296. 1957 [1962].

A. Barrera Marín, Alfredo Barrera Vázquez and Rosa María López Franco, *Nomenclatura etnobotánica maya: una interpretación taxonómica*. México, D.F. 1976.

Katharine Bartlett, *Prehistoric Pueblo Foods*. Flagstaff, Arizona, Museum Notes, vol. 4, no. 4, October, 1931.

Fray D. Basalenque, *Historia de la Provincia de San Nicolás Tolentino de Michoacán* del Orden de Nuestro Padre Santísimo San Augustín. México 1963.

Henri Louis Baudrier and Julien Baudrier, *Bibliographie lyonnaise*: Recherches sur les imprimeurs, libraires, relieurs et fondeurs de lettres de Lyon au XVIe siècle. VII. Lyons, Paris 1908.

Cecil Beaton, *Indian Diary & Album*. [Originally published by B.T. Batsford Ltd. London as *Far East*, 1945 and *Chinese Album*, Winter 1945-1946] Oxford University Press, Oxford, New York, and Delhi 1991.

Marcos E. Becerra, *Nombres geográficos indígenas del Estado de Chiapas*. Chiapas, México: Tuxtla Gutiérrez 1932.

James Ford Bell, *Jesuit Relations and Other Americana in the Library of James F. Bell.* A catalogue compiled by Frank K. Walter and Virginia Doneghy. Minneapolis 1950.

Eric Temple Bell, *Men of Mathematics.* Simon & Schuster, New York 1965.

G. Bellini, *La tipografia del Seminario di Padova.* Libreria Gregoriana editrice, Padova 1927.

E.C. Bénézit, *Dictionnaire critique et documentaire des peintres, sculpteurs, dessinateurs et graveurs de tous les pays.* Paris 1976.

E. Benveniste, *Origine de la formation des noms en indo-européen.* Paris 1962.

Brent Berlin, "Speculations on the growth of ethnobotanical nomenclature." *Journal of Language in Society.* 1: 63-98. 1972.

Brent Berlin, Dennis E. Breedlove and Peter H. Raven, *Principles of Tzeltal Plant Classification.* Academic Press, New York and London 1974.

G. Bertho, "Quatre dialects Mandé du Nord-Dahomey et de la Nigeria anglaise." *Bull. Inst. Franç. Afr. Noire.* 13: 126-171. 1951.

Elsdon Best, *Maori Agriculture.* Wellington 1976.

N.N. Bhattacharya, *Glossary of Hindu Religious Terms and Concepts.* South Asia Publications, Columbia 1990.

B. Bhattacharyya, *An Introduction to Buddhist Esoterism.* Oxford 1932.

Bibliothèque Nationale, Paris, *Catalogue général des livres imprimés de la Bibliothèque Nationale. Auteurs.* Paris 1897-1981.

Blodwen Binns, *A First Check List of the Herbaceous Flora of Malawi.* Zomba, Malawi 1968.

Blodwen Binns, *Dictionary of Plant Names in Malawi.* Zomba, Malawi 1972.

H. Birnbaum and J. Puhvel, *Ancient Indo-European Dialects.* 1965.

Dor Bahadur Bista, *The People of Nepal.* Kathmandu 1972.

Paul Black, *Aboriginal Languages of the Northern Territory.* Darwin 1983.

E.H. Blair and J.A. Robertson, *The Philippine Islands 1493-1803 [-1898].* Cleveland 1903-1909.

Joseph Blumenthal, *Art of the Printed Book, 1455-1955*: Masterpieces of Typography through Five Centuries from the Collections of The Pierpont Morgan Library, New York. New York and Boston 1973.

F. Boerner and G. Kunkel, *Taschenwörterbuch der botanischen Pflanzennamen.* 4. Aufl. Berlin & Hamburg 1989.

J. Bondet de la Bernadie, "Le dialecte des Kha Boloven." *Bulletin de la Societé des Études Indochinoises.* tome XXIV, no. 3, 1949.

Sarita Boodhoo, *Kanya Dan — The Why's of Hindu Marriage Rituals.* Mauritius Bhojpuri Institute, Port Louis 1993.

Franz Bopp, *Glossarium Sanscritum.* Second edition. Berlin 1847.

Woodrow Borah, *Early Trade and Navigation between Mexico and Peru.* Berkeley 1954.

Rubens Borba de Moraes, *Bibliographia Brasiliana.* Los Angeles and Rio de Janeiro 1983.

G. Borsa, *Clavis typographorum librariorumque Italiae 1465-1600.* Aureliae Aquensis [Baden-Baden] 1980.

L. Bossi, *Spiegazione di alcuni vocaboli geologici.* Milano 1817.

L. Bouquiaux, "Les noms des plantes chez les Birom." *Afrika und Ubersee.* 55. 1971-1972.

C.R. Boxer, *The Dutch Seaborne Empire 1600-1800.* Harmondsworth 1973.

W.J. Boyd, *Satana and Mara. Christian and Buddhist Symbols of Evil.* Leiden 1975.

Pascal Boyeldieu, *La langue lua, groupe Boua, Moyen-Chari, Tchad.* CUP 1985.

Leonard E. Boyle, O.P., *Medieval Latin Palaeography: A Bibliography.* Toronto Medieval Bibliographies 8. University of Toronto Press, Toronto 1984.

A. Brebion, *Bibliographie des voyages dans l'Indochine française du IXe au XIXe siècle.* Paris 1910.

J.B. Brebner, *The Explorers of North America 1492-1806.* Cleveland 1964.

Dennis E. Breedlove, *History of Botanical Exploration in Chiapas, Mexico.* 1971.

Dennis E. Breedlove and Robert M. Laughlin, *The Flowering of Man — A Tzotzil Botany of Zinacantán.* Smithsonian Institution Press, Washington, D.C. 1993.

J.A. Brendon, *Great Navigators and Discoverers.* Books for Libraries Press, Freeport 1967.

Emil Bretschneider, *Mediaeval Researches from Eastern Asiatic Sources.* Fragments towards the knowledge of the geography and history of Central and Western Asia from the 13th to the 17th century. [Reprint of the 1888 edition.] London [1950?].

Emil Bretschneider, *History of European Botanical Discoveries in China.* [Reprint of the original edition, St. Petersburg 1898.] Leipzig 1981.

British Library, *The British Library General Catalogue of Printed Books to 1975.* London 1979.

British Library, *Catalogue of Books from the Low Countries 1601-1621 in the British Library.* Compiled by Anna E.C. Simoni. London 1990.

British Museum, *Short-title Catalogue of Books Printed in France and of French Books Printed in Other Countries from 1470 to 1600* now in the British Museum. London 1966.

British Museum, *Short-title Catalogue of Books Printed in German-speaking Countries and German Books Printed in Other Countries from 1455 to 1600* now in the British Museum. London 1958.

British Museum, *Short-title Catalogue of Books Printed in Italy and of Italian Books Printed in Other Countries from 1465 to 1600* now in the British Museum. London 1958.

British Museum, *Short-title Catalogue of Books Printed in the Netherlands and Belgium and of Dutch and Flemish Books Printed in Other Countries from 1470 to 1600* now in the British Museum. London 1965.

British Museum (Natural History), *Catalogue of the Books, Manuscripts, Maps and Drawings in the British Museum.* London 1903-1933.

C. Brockelmann, *Lexicon Syriacum.* Halis Saxon 1928.

J.S. Bromley and E.H. Kossmann, Eds., *Britain and the Netherlands in Europe and Asia.* London 1968.

H.C. Brooks, *Compendiosa bibliografia di edizioni bodoniane*. Firenze 1927.

R. Brough Smith, *Aborigines of Victoria*. Ferres, Melbourne 1878.

Patrick Browne, *The Civil and Natural History of Jamaica*. London 1756.

R.K. Brummitt, comp., *Vascular Plant Families and Genera*. Royal Botanic Gardens, Kew 1992.

R.K. Brummitt and C.E. Powell, *Authors of Plant Names*. Royal Botanic Gardens, Kew 1992.

J.C. Brunet, *Manuel du libraire et de l'amateur de livres*. 6 vols. Paris 1860-1865 (two vols. of supplements Paris 1878-1880).

Otto Brunner, *Land und Herrschaft*. Wien 1965.

A.T. Bryant, "Zulu medicine and medicine-men." *Annals of the Natal Museum*. 2(1): 1-104. 1909.

H.M. Burkill, *The Useful Plants of West Tropical Africa*. Edition 2. Royal Botanic Gardens, Kew 1985-1997.

I.H. Burkill, *A Dictionary of the Economic Products of the Malay Peninsula*. Kuala Lumpur 1966.

J. Burney, *A Chronological History of Discoveries in the South Sea or Pacific Ocean*. Amsterdam 1967.

Ernest J. Burrus, S.J., Ed., *Misiones Norteñas Mexicanas de la Compañia de Jesús, 1751-1757*. México 1963.

Robert Byron, *The Road to Oxiana*. Oxford University Press, Oxford 1982.

Antoine Cabaton, "Dix dialectes indochinois recueillis par Prosper Oden'hal, étude linguistique." *Journal Asiatique*, 10e série, 5: 265-344. 1905.

Lydia Cabrera, *Vocabulario Congo* (el Bantu que se habla en Cuba). Downtown Book Center, Miami 1984.

Armando Cáceres, *Plantas de uso medicinal en Guatemala*. Universidad de San Carlos de Guatemala 1996.

L. Cadière, *Croyances et pratiques religieuses des Annamites dans les environs de Hué*. Hanoi 1944.

María de los Angeles Calatayud Arinero, *Catalogo de las Expediciones y Viajes Cientificas a America y Filipinas* (siglos XVIII y XIX). Fondos del Archivo del Museo Nacional de Ciencias Naturales. Madrid 1984.

J. Callander, *Terra Australis Cognita or Voyages to the Terra Australis*. London 1768 [Amsterdam 1967].

J.C. Campbell, *An Illustrated Encyclopaedia of Traditional Symbols*. Thames and Hudson Ltd., London 1982.

John F. Campbell, *History and Bibliography of the New American Practical Navigator and the American Coast Pilot*. Peabody Museum, Salem 1964.

Joseph Campbell, *The Mythic Image*. Princeton University Press, Princeton, New Jersey 1974.

Marinus F.A. Campbell, *Annales de la Typographie Néerlandaise au XVe siècle*. The Hague 1874.

Elias Canetti, *Masse und Macht*. Hamburg 1960.

M.A. Canini, *Dizionario etimologico dei vocaboli italiani d'origine ellenica*. UTET, Torino 1925.

Adriano Capelli, *Cronologia, cronografia e calendario perpetuo, dal principio dell'èra cristiana ai nostri giorni*. Edd. Daniela Stroppa Salina, Diego Squarcialupi, Enrico Quigini Puliga. Milano 1978.

Auguste Carayon, *Bibliographie historique de la Compagnie de Jésus*, ou catalogue des ouvrages relatifs à l'histoire des jésuites depuis leur origine jusqu'à nos jours. Paris 1864.

G.G. Carlson and V.H. Jones, "Some notes on uses of plants by the Comanche Indians." *Pap. Michigan Acad. Sci. Arts, and Letters*, vol. 25: 517-542. 1940.

A. Carnoy, *Dictionnaire étymologyque des noms grecs des plantes*. Louvain 1959.

Jean-Marie Carré, *Voyageurs et écrivains français en Égypte*. Le Caire 1956.

John Carter and Percy Muir, *Printing and the Mind of Man*. Compiled and edited by ..., assisted by Nicholas Barker, H.A. Feisenberger, Howard Nixon and S.H. Feinberg. With an introductory essay by D. Hay. Munich 1983.

Carlo Castellani, *La stampa in Venezia dalla sua origine alla morte di Aldo Manuzio seniore*. Venezia 1889.

A.C. Cavicchioni, *Vocabolario Italiano-Swahili*. Zanichelli Editore, Bologna 1923.

Adriano Ceresoli, *Bibliografia delle opere italiane latine e greche su la caccia, la pesca e la cinologia*. Bologna 1969.

L. Chabuis, *Petite histoire naturelle de la Polynésie français*. Papeete n.d.

John Chadwick, *The Decipherment of Linear B*. Cambridge University Press, New York 1958.

Lokesh Chandra, *Buddhist Iconography*. New Delhi 1987.

Lokesh Chandra, *Tibetan–Sanskrit Dictionary*. New Delhi 1958-1961.

Chandra Moti, "Cosmetics and coiffure in ancient India." *Journal of the Indian Society of Oriental Art*, VIII. 1940.

P. Chantraine, *Dictionnaire étymologique de la langue grecque*. 1968-1980.

P. José Chantre y Herrera, *Historia de las Misiones de la Compañia de Jesús en el Marañon Español*. Madrid 1901.

Arthur D. Chapman, ed., *Australian Plant Name Index*. Canberra 1991.

Louis Charbonneau-Lassay, *Il bestiario di Cristo*. Edizioni Arkeios, Roma 1994.

Louis Charbonneau-Lassay, *Il giardino del Cristo ferito*. Il Vulnerario e il Florario di Cristo. Edizioni Arkeios, Roma 1995.

Carlos E. Chardón, *Los naturalistas en la América Latina*. Ciudad Trujillo 1949.

K.N. Chaudhuri, *The English East India Company ... 1600-1640*. London 1965.

K.N. Chaudhuri, *Trade and Civilisations in the Indian Ocean*. Cambridge University Press, Cambridge 1985.

Nanimadhab Chaudhuri, "A pre-historic tree cult." *India Historical Quarterly*, XIX, 4. 1943.

Huguette and Pierre Chaunu, *Séville et l'Atlantique (1504-1650)*. Paris 1955-1960.

U. Chevalier, *Répertoire des sources historiques du Moyen Age. Bio-bibliographie*. Paris 1905-1907.

U. Chevalier, *Répertoire des sources historiques du Moyen Age. Topo-bibliographie*. Montbéliard 1903.

C. Christensen et al., *Index filicum*, with four supplements. Hagerup, Copenhagen, etc. 1906-1965.

C.M. Churchward, *Tongan Dictionary*. Government Printing Press, Nuku'alofa, Tonga 1959.

Paolo Bartolomeo Clarici, *Istoria e coltura delle piante*. Venezia 1726.

Inga Clendinnen, *Ambivalent Conquests — Maya and Spaniard in Yucatan, 1517-1570*. Cambridge University Press, Cambridge 1987.

Terry Clifford, *Tibetan Buddhist Medicine and Psychiatry: The Diamond Healing*. York Beach, Maine 1986.

Michael D. Coe, *The Maya Scribe and His World*. The Grolier Club, New York 1973.

M.R. Cohen and I.E. Drabkin, *A Source Book in Greek Science*. New York 1948.

George S. Cole, *A Complete Dictionary of Dry Goods and History of Silk, Cotton, Linen, Wool and Other Fibrous Substances*. Chicago 1892.

Don Miguel Colmeiro, *Diccionario de los diversos nombres vulgares de muchas plantas usuales o notables del antiguo y nuevo mundo*. Madrid 1871.

Georges Condominas, "Enquête linguistique parmi les populations montagnardes du Sud-Indochinois." *Bulletin de l'École Française d'Extrême-Orient*, tome XLVI, 2, 1954.

Giovanni Consolino, *Vocabolario del dialetto di Vittoria*. Pisa 1986.

W.A. Copinger, *Supplement to Hain's Repertorium bibliographicum*. London 1895-1902.

Henri Cordier, *Bibliotheca Indosinica*. Dictionnaire bibliographique des ouvrages relatifs à la péninsule indochinoise. Paris 1912-1932.

Henri Cordier, *Bibliotheca Japonica*. Dictionnaire bibliographique des ouvrages relatifs à l'Empire Japonais. Paris 1912.

Henri Cordier, *Bibliotheca Sinica*. Dictionnaire bibliographique des ouvrages relatifs à l'Empire chinois. Paris 1904-1924.

Diego de Córdoba Salinas, O.F.M., *Crónica franciscana de las provincias del Perú*. Ed. by Lino G. Canedo, O.F.M. Washington 1957.

Nuñez Corona, *Mitología tarasca*. México 1957.

Corpus of Maya Hieroglyphic Inscriptions. Vols. 1-6. Cambridge, Massachusetts: Peabody Museum of Archaeology and Ethnology. Harvard University 1977-.

Paul Alan Cox and Sandra Anne Banack, Eds., *Islands, Plants, and Polynesians*. An introduction to Polynesian ethnobotany. Dioscorides Press, Portland, Oregon 1991.

G. Crabb, *Universal Historical Dictionary*, or explanation of the names of persons and places in the departments of biblical, political, and ecclesiastical history, mythology, heraldry, biography, bibliography, geography, and numismatics. Enlarged edition. Baldwin & Cradock, London 1833.

J. Cretinau-Joly, *Histoire religieuse, politique et littéraire de la Compagnie de Jésus*. Paris 1844-1846.

A.J. Cronquist, *An Integrated System of Classification of Flowering Plants*. Columbia University Press, New York 1981.

A.J. Cronquist, *The Evolution and Classification of Flowering Plants*. Second edition. The New York Botanical Garden, New York 1988.

Laureano de la Cruz, *Nuevo descubrimiento del Marañon, 1651: Varones ilustres de la Orden Seráfica en el Ecuador*. Quito 1885.

J.S. Cumpston, *Shipping Arrivals and Departures, Sydney*. Canberra 1977.

Eustachio D'Afflitto, *Memorie degli scrittori del Regno di Napoli*. Napoli 1782.

G. D'Africa, *Il R. Istituto-Orto Botanico ed il R. Giardino Coloniale di Palermo*. Palermo 1945.

E.W. Dahlgren, *Les Relations commerciales et maritimes entre la France et les Côtes de l'Océan Pacifique* ... Paris 1909.

R.M. Dahlgren, H.T. Clifford and P.F. Yeo, *The Families of the Monocotyledons: Structure, Evolution and Taxonomy*. Berlin 1985.

E.G. Dale, "Literature on the history of botany and botanic gardens 1730-1840." *Huntia*. 6: 1-121. 1985.

Dam Bo, *Les populations montagnardes du Sud-Indochinois*. Lyon 1950.

William Darlington, *Memorials of John Bartram and Humphry Marshall*. Philadelphia 1849.

Vaidya Bhagvan Dash, *Alchemy and Metallic Medicines in Ayurveda*. Concept Publishing Company, New Delhi 1986.

Hugh William Davies, *Devices of the Early Printers, 1457-1560: Their History and Development*. London 1935.

Warren R. Dawson, *The Banks Letters*. London 1958.

J. Day, "Agriculture in the life of Pompeii." *Yale Classical Studies*. 3: 166-208. 1932.

A. De Baker and A. Carayon, *Bibliothèque de la Compagnie de Jésus*. Paris 1890-1900.

William T. De Bary, *Sources of the Indian Tradition*. New York 1964.

Gerard Decorme, *La Obra de Los Jesuítas Mexicanos Durante la Epoca Colonial, 1572-1767*. México 1941.

V.M. De Grandis, *Dizionario etimologico-scientifico*. Napoli 1824.

F.C. Deighton, *Vernacular Botanical Vocabulary for Sierra Leone*. London 1957.

F. Delitzsch, *Assyrisches Handwörterbuch*. Reprint of the 1896 edition. Leipzig 1968.

William Denevan, Ed., *The Native Population of the Americas in 1492*. University of Wisconsin Press, Madison 1976.

G.R. Dent and C.L.S. Nyembezi, *Scholar's Zulu Dictionary*. Shuter and Shooter, Pietermaritzburg 1995.

Desmond C. Derbyshire and Geoffrey K. Pullum, Eds., *Handbook of Amazonian Languages*. Berlin 1986.

L. Dermigny, *La Chine et l'Occident: le commerce à Canton au XVIIIe siècle*. Paris 1964.

Giovanni Bernardo De Rossi, *Annales hebraeo-typographici sec XV*. Parmae 1795.

Ray Desmond, *Dictionary of British & Irish Botanists and Horticulturists*. London 1994.

Dictionary of Scientific Biography. Editor in Chief Charles Coulston Gillispie. New York 1981.

G.W. Dimbleby, *Plants and Archaeology*. London 1978.

Wilfred Douglas, *The Aboriginal Languages of the South-West of Australia*. Canberra 1976.

P. Dourisboure, *Les sauvages Ba-Hnars* (Cochinchine orientale). Paris 1929.

John Dowson, *A Classical Dictionary of Hindu Mythology and Religion, Geography, History, and Literature*. London 1968.

John Dreyfus et al., Eds., *Printing and the Mind of Man: An Exhibition of Fine Printing in the King's Library of the British Museum*. The British Museum 1963.

A.H. Driver, *Catalogue of Engraved Portraits in the Royal College of Physicians of London*. London 1952.

Jonas C. Dryander, *Catalogus bibliothecae historico-naturalis Josephi Banks*. London 1796-1800.

Abbé Dubois, *Hindu Manners, Customs and Ceremonies*. Oxford University Press, Oxford 1906.

James Alan Duke and Rodolfo Vasquez, *Amazonian Ethnobotanical Dictionary*. CRC Press, Boca Raton, Florida 1994.

James Alan Duke, *CRC Handbook of Medicinal Herbs*. CRC Press, Boca Raton, Florida 1985.

Peter M. Dunne, *Pioneer Jesuits in Northern Mexico*. Berkeley 1944.

Peter M. Dunne, *Early Jesuit Missions in Tarahumara*. Berkeley 1948.

Denis I. Duveen, *Bibliotheca Alchemica et Chemica*. London 1949.

Ursula Dyckerhoff, "Mexican toponyms as a source in regional ethnohistory," in H.R. Harvey and Hanns J. Prem, Eds., *Explorations in Ethnohistory. Indians of Central Mexico in the Sixteenth Century*. 229-252. University of New Mexico Press, Albuquerque 1984.

R.A. Dyer, *The Genera of Southern Africa Flowering Plants*. Pretoria 1975-1976.

Frederic Adolphus Ebert, *A General Bibliographical Dictionary*. Oxford University Press, Oxford 1837.

James Edge-Partington, *Ethnographical Album of the Pacific Islands*. [Originally published as *Album of the Weapons, Tools, Ornaments, Articles of Dress of Natives of the Pacific Islands*] Second edition expanded and edited by Bruce L. Miller. SDI Publications, Bangkok 1996.

M.B. Emeneau, *The Strangling Figs in Sanskrit Literature*. California 1949.

Iris H.W. Engstrand, *Spanish Scientists in the New World*. University of Washington Press, Seattle 1981.

A. Ernout, *Morphologie historique du latin*. Paris 1974.

A. Ernout and A. Meillet, *Dictionnaire étymologique de la langue latine*. Paris 1985.

P. Lucas Espinosa, *Los tupí del oriente peruano*: estudio lingüístico y etnográfico. Madrid 1935.

E.E. Evans-Pritchard, *Witchcraft, Oracles and Magic among the Azande*. Oxford 1937.

Filippo Evola, *Storia tipografico-letteraria del secolo XVI in Sicilia*. Palermo 1878.

Alfonso Fabila, *Las tribus yaquis de Sonora*, su cultura y anhelada autodeterminación. Instituto Nacional Indigenista, México 1978.

William Falconer, *An Universal Dictionary of the Marine*. London 1781.

D.B. Fanshawe, revised by, *Check List of Vernacular Names of the Woody Plants of Zambia*. Lusaka 1965.

D.B. Fanshawe and C.D. Hough, *Poisonous Plants of Zambia*. Forest Research Bulletin, no. 1, revised. Ministry of Rural Development, Government Printer, Lusaka 1967.

D.B. Fanshawe and J.M. Mutimushi, *A Check List of Plant Names in the Nyanja Languages*. Forest Research Bulletin, no. 21, Ministry of Rural Development, Government Printer, Lusaka 1969.

E. Farr, J.A. Leussink and F.A. Stafleu, Eds., *Index nominum genericorum (plantarum)*. 3 vols. Utrecht, etc. 1979.

E. Farr, J.A. Leussink and G. Zijlstra, Eds., *Index nominum genericorum (plantarum)*. Supplementum I. Utrecht, etc. 1986.

J.A. Farrell, "A Hlengwe botanical dictionary of some trees and shrubs in Southern Rhodesia." *Kirkia*. 4: 165-172. 1964.

Mariano Fava and Giovanni Bresciano, *La stampa a Napoli nel XV secolo*. Leipzig 1912.

John Ferguson, *Bibliotheca Chemica*. Glasgow 1906.

L. Ferrari, *Onomasticon. Repertorio bibliografico degli scrittori italiani dal 1501 al 1850*. Milano 1947.

Francisco de Figueroa, *Relación de las Misiones de la Compañia de Jesús en el país de los Maynas*. Madrid 1904.

H.R. Fletcher, *Story of the Royal Horticultural Society, 1804-1968*. Oxford University Press, Oxford 1969.

H.R. Fletcher and W.H. Brown, *Royal Botanic Garden Edinburgh, 1670-1970*. Edinburgh 1970.

Lázaro Flury, *Tradiciones, leyendas y vida de los Indios del norte*. Editorial Ciordia y Rodríguez, Buenos Aires 1951.

Egidio Forcellini, *Lexicon totius Latinitatis*. Patavii 1940.

Karen Cowan Ford, *Las yerbas de la gente: A Study of Hispano-American Medicinal Plants*. Ann Arbor, Michigan 1975.

Richard I. Ford, Ed., *The Nature and Status of Ethnobotany*. Ann Arbor, Michigan 1978.

C. Fossey, *Manuel d'assyriologie. Fouilles, écriture, langues, littérature, géographie, histoire, religion, institutions, art*. Paris 1904-26.

F.W. Foxworthy, *I. Timbers of British North Borneo. II. Minor Forest Products and Jungle Produce*. Sandakan 1916.

Dora Franceschi Spinazzola, *Catalogo della biblioteca di Luigi Einaudi*. Opere economiche e politiche dei secoli XVI-XIX. Fondazione Luigi Einaudi, Torino 1981.

Fathers Franciscan, *An Ethnologic Dictionary of the Navaho Language*. St. Michaels, Arizona 1910.

J.D. Freeman, *Iban Agriculture*. London 1955.

H. Friis, Ed., *The Pacific Basin: A History of Its Geographical Exploration*. New York 1967.

G. Fumagalli, *Lexicon typographicum Italiae*. Dictionnaire géographique d'Italie pour servir à l'histoire de l'imprimerie dans ce pays. Leo S. Olschki, Firenze 1966.

V.A. Funk and S.A. Mori, "A bibliography of plant collectors in Bolivia." *Smithsonian Contr. Bot.* 70: 1-20. 1989.

M.M. Gabriel, *Livia's Garden Room at Prima Porta*. New York 1955.

F.N. Gachathi, *Kikuyu Botanical Dictionary of Plant Names and Uses*. Nairobi 1993.

Andrew Thomas Gage, *A History of the Linnean Society of London*. London 1938 (revised by W.T. Stearn 1985).

A. Galletti et al., *The Dutch in Malabar*. Madras 1911.

G. Garbini, *Le lingue semitiche*. Napoli 1984.

H. García-Barriga, *Flora medicinal de Colombia — botánica médica*. Bogotá 1974-1975.

Z.O. Gbile, *Vernacular Names of Nigerian Plants (Hausa)*. For. Res. Inst. Nigeria, Ibadan 1980.

Z.O. Gbile and M.O. Soladoye, "Plants in traditional medicine in West Africa." *Monographs in Systematic Botany from Missouri Botanical Garden.* 25: 343-349. 1985.

Deno John Geanakoplos, *Greek Scholars in Venice: Studies in Dissemination of Greek Learning from Byzantium to Western Europe.* Harvard University Press, Cambridge 1962.

Deno John Geanakoplos, *Interaction of the "Sibling" Byzantine and Western Cultures in the Middle Ages and Italian Renaissance (330-1600).* Yale University Press, New Haven 1976.

I.J. Gelb, *Glossary of Old Akkadian.* The University of Chicago Press, Chicago 1957.

I.J. Gelb, *Materials for the Assyrian Dictionary.* Chicago 1952-70.

Helmut Genaust, *Etymologisches Wörterbuch der botanischen Pflanzennamen.* Birkhäuser Verlag, Basel 1996.

J/F.M. Genibrel, *Dictionnaire Annamite-Français.* Imprimerie de la Mission à Tan Dinh, Saigon 1898.

Alwyn H. Gentry, *A Field Guide to the Families and Genera of Woody Plants of Northwest South America (Colombia, Ecuador, Peru).* Washington, D.C. 1993.

P. Gerhard, *A Guide to the Historical Geography of New Spain.* Berkeley 1972.

J. Gildemeister, *Bibliothecae sanskritae.* Bonn 1847.

W. Gilges, *Some African Poison Plants and Medicines of Northern Rhodesia.* The occasional papers of the Rhodes–Livingstone Museum, no. 11. Livingstone, Northern Rhodesia 1955.

M.R. Gilmore, *Uses of Plants by the Indians of the Missouri River Region.* 33rd Ann. Rep. Bur. Amer. Ethnol. 1919.

Monique Girardin, *Bibliographie de l'Ile de la Réunion 1973-1992.* 1994.

Lorenzo Giustiniani, *Saggio storico critico sulla tipografia del Regno di Napoli.* Napoli 1793.

Heidi K. Gloria, *The Bagobos: Their Ethnohistory and Acculturation.* New Day Publishers, Quezon City, Philippines 1987.

Harry E. Godwin, *The History of the British Flora.* Cambridge University Press, Cambridge 1956.

Frederick R. Goff, *Incunabula in American Libraries.* A third census. Bibliographical Society of America, New York 1964.

Peter H. Goldsmith, *A Brief Bibliography of Books in English, Spanish and Portuguese, Relating to the Republics Commonly Called Latin American,* with comments. New York 1915.

V.F. Goldsmith, *A Short Title Catalogue of Spanish and Portuguese Books 1601-1700 in the Library of the British Museum* (The British Library–Reference Division). Folkestone and London 1974.

Mario Gongora, *Los grupos de conquistadores en Tierra Firme.* Santiago, Chile 1962.

M. Gongora, *Studies in the Colonial History of Spanish America.* Cambridge 1975.

E.R. Goodenough, *Jewish Symbols in the Greco-Roman Period.* Bollingen Series XXXVII. Pantheon Books, New York and Princeton University Press, Princeton 1953-1965.

E.J. Goodman, *The Explorers of South America.* London 1972.

Jean S. Gottlieb, *A Checklist of the Newberry Library's Printed Books in Science, Medicine, Technology, and the Pseudosciences circa 1460-1750.* New York and London 1992.

S.W. Gould and D.C. Noyce, *International Plant Index.* New York 1962, 1965.

Johann Georg Theodor Graesse, *Trésor de livres rares et précieux.* Dresde 1859-1869.

G. Grandidier, *Bibliographie de Madagascar.* Paris 1906-1957.

Edward Lee Greene, *Landmarks of Botanical History.* Edited by Frank N. Egerton. Stanford, California 1983.

P.J. Greenway, *A Swahili–Botanical–English Dictionary of Plant Names.* Dar es Salaam 1940.

R. Gregorio, *L'Orto botanico di Palermo.* Palermo 1821.

Pierre Grimal, *Les Jardins Romains.* Fayard 1984.

Barbara F. Grimes, Ed., *Ethnologue. Languages of the World.* Twelfth Edition. Dallas, Texas 1992.

J.W. Grimes and B.S. Parris, *Index Thelypteridaceae.* Royal Botanic Gardens, Kew 1986.

Arthur Eric Gropp, *A Bibliography of Latin American Bibliographies.* Scarecrow Press, Metuchen, New Jersey 1968-1971.

J.A. Gruys and C. De Wolf, *Thesaurus 1473-1800.* Dutch Printers and Booksellers with places and years of activity. Nieuwkoop 1989.

Antonio Guasch and Diego Ortiz, *Diccionario Castellano-Guarani, Guarani-Castellano.* Centro de Estudios Paraguayos Antonio Guasch, Asunción 1996.

T. Guignard, *Dictionnaire laotien-français.* Hong Kong 1912.

Mary Gunn and Leslie E. Codd, *Botanical Exploration of Southern Africa.* Cape Town 1981.

N. Gunson, *Messengers of Grace: Evangelical Missionaries in the South Seas.* Melbourne 1972.

H.B. Guppy, "The Polynesians and their plant-names." *Victoria Inst., Journ. of Trans.,* vol. 29: 135-170. 1895-1896.

Shaki M. Gupta, *Plant, Myths and Traditions in India.* Leiden 1971.

R. Gusmani, "Di alcuni prestiti greci in latino." *Bollettino di studi latini.* III: 76-88. 1973.

Malcolm Guthrie, *The Classification of the Bantu Languages.* Oxford University Press, Oxford 1948.

Malcolm Guthrie, *Comparative Bantu.* Farnborough 1971.

Miles Hadfield et al., *British Gardeners: A Biographical Dictionary.* London 1980.

K. Haebler, *Bibliografía iberica del siglo XV.* La Haya, Leipzig 1903-1917.

Ludwig Hain, *Repertorium bibliographicum* in quo libri omnes ab arte typographica inventa usque ad annum MD typis expressi ordine alphabetico vel simpliciter enumerantur vel adcuratius recensentur. Stuttgardiae 1826-1838.

S. Halkett and J. Laing, *A Dictionary of the Anonymous and Pseudonymous Literature of Great Britain.* Edinburgh 1882-1888.

J.B. Hall, P. Pierce and G. Lawson, *Common Plants of the Volta Lake.* Dept. of Botany, University of Ghana, Legon 1971.

E. Charles B. Hallam, *Oriya Grammar for English Students*. Baptist Mission Press, Calcutta 1874.

A. von Haller, *Bibliotheca Botanica*. Zürich 1771-1772.

Carl A. Hanson, *Dissertations on Iberian and Latin American History*. New York 1975.

C.H. Haring, *Trade and Navigation between Spain and the Indies in the Time of the Habsburgs*. Cambridge, Massachusetts 1918.

G.W. Harley, *Native African Medicine with Special Reference to Its Practice in the Mano Tribe of Liberia*. Cambridge, Massachusetts 1941.

Michael J. Harner, Ed., *Hallucinogens and Shamanism*. Oxford University Press, New York 1973.

Regina Harrison, *Signs, Songs, and Memory in the Andes — Translating Quechua Language and Culture*. University of Texas Press, Austin 1989.

John Harvey, *Early Gardening Catalogues*. Phillimore, London 1972.

J.G. Hawkes, "On the origin and meaning of South American Indian potato names." *Journal of the Linnean Society of London*, Botany. 53: 205-250. 1947.

Nicola Francesco Haym, *Biblioteca Italiana o sia Notizia dei Libri Rari nella Lingua Italiana*. Venezia 1728.

Charles B. Heiser, "Cultivated plants and cultural diffusion in nuclear America." *American Anthropologist*. 67: 930-949. 1965.

Blanche Elizabeth Edith Henrey, *British Botanical and Horticultural Literature before 1800*. Oxford University Press, Oxford 1975.

G. Heyd, *Storia del commercio del Levante nel Medio Evo*. Torino 1913.

L.R. Hiatt, Ed., *Australian Aboriginal Mythology*. Essays in honour of W.E.H. Stanner. Australian Institute of Aboriginal Studies, Canberra 1975.

M.E. Hoare, *The Tactless Philosopher*. Melbourne 1976.

George M. Hocking, "The doctrine of signatures." *Quarterly Journal of Crude Drug Research*. 15: 198-200. 1977.

Harry Hoijer, *Navaho Phonology*. University of New Mexico Publications in Anthropology. 1945.

William R. Holland, *Medicina Maya en los Altos de Chiapas*. Instituto Nacional Indigenista, México n.d.

J. Hollyman and A. Pawley, Eds., *Studies in Pacific Languages and Cultures*. Auckland 1981.

M. Holmes, *Captain James Cook: A Bibliographical Excursion*. London 1952.

J.H. Hospers, *A Basic Bibliography for the Study of the Semitic Languages*. Leiden 1973-74.

A.W. Howitt, *The Native Tribes of South-East Australia*. Macmillan, London 1904.

C. Huelsen, "Piante iconografiche encise in marmo." *Mitteilungen des deutschen archäologischen Instituts, Römische Abteilung*. 5: 46-63. 1890.

The Hunt Botanical Library, *Catalogue of Botanical Books in the Collection of Rachel McMasters Miller Hunt*. Pittsburgh 1958-1961.

Francis Huxley, *Affable Savages*. The Viking Press, New York 1957.

Index Aureliensis, *Index aureliensis*: catalogus librorum sedecimo saeculo impressorum [volumes I-IX, A-Coq]. Baden-Baden 1962-1991.

Benjamin Daydon Jackson, *A glossary of botanic terms* with their derivation and accent. Fourth edition. London 1928.

Benjamin Daydon Jackson, *Guide to the Literature of Botany*. Otto Koeltz Science Publishers, Königstein 1974.

Benjamin Daydon Jackson et al., *Index Kewensis plantarum phanerogamarum*. Oxford University Press, Oxford [1893-] 1895, 15 supplements 1901-1974.

T. Jacobsen, *The Sumerian King Lists*. Chicago University Press, Chicago 1939.

S.K. Jain, "Studies in Indian ethnobotany — origin and utility of some vernacular plant names." *Proc. Natl. Acad. Sci.* 33B: 525-530. 1963.

S.K. Jain, Ed., *Glimpses of Indian Ethnobotany*. Oxford and IBH Publishing Co., New Delhi 1981.

F.M. Jarrett et al., *Index Filicum. Supplementum Quintum pro annis 1961-1975*. Oxford University Press, Oxford 1985.

Wilhelmina F. Jashemski, *The Gardens of Pompeii*. Herculaneum and the Villas Destroyed by Vesuvius. New Rochelle, New York 1979.

Ch. F. Jean, *Sumer et Akkad*. Paris 1923.

Susan Jellicoe, "The development of the Mughal Garden." in *The Islamic Garden*. Dumbarton Oaks Colloquium on the History of Landscape Architecture. IV: 107-129, 134-135. Edited by Elisabeth B. MacDougall and Richard Ettinghausen. Washington 1976.

Hans Jensen, *Sign, Symbol, and Script*. New York 1969.

Volney H. Jones, "The nature and status of ethnobotany." *Chronica Botanica*. 6(10): 219-221. 1941.

Elviro Jorde Pérez, *Catálogo bio-bibliográfico de los Religiosos Agustinos de la Provincia del Santísimo Nombre de Jesús de las Islas Filipinas*. Manila 1901.

José Juanen, *Historia de la Compañia de Jesús en la Antigua Provincia de Quito, 1570-1774*. Quito 1941-1943.

M.W. Jurriaanse, *Catalogue of the Archives of the Dutch Central Government of Coastal Ceylon 1640-1796*. Colombo 1943.

H. and R. Kahane and L. Bremner, *Glossario degli antichi portolani italiani*. Traduzione e note di M. Cortelazzo. Firenze 1968.

Lama Dawasamdup Kazi, *English–Tibetan Dictionary*. Baptist Mission Press, Calcutta 1919.

Howard Kelly, *Some American Medical Botanists Commemorated in our Botanical Nomenclature*. Southworth, Troy 1914.

Hayward Keniston, *List of Works for the Study of Hispanic–American History*. The Hispanic Society of America, New York 1920 [Reprint New York 1967 Kraus Reprint Society].

K. Kerényi, *Asklepios: Archetypal Image of the Phisician's Existence*. Bollingen Series LXV. 3. Pantheon Books, New York 1959.

J. Kerharo and J.G. Adam, *La pharmacopée sénégalaise traditionnelle. Plantes médicinales et toxiques*. Paris 1974.

J. Kerharo and A. Bouquet, *Plantes médicinales et toxiques de la Côte d'Ivoire — Haute Volta*. Paris 1950.

M.C. Kiemen, *The Indian Policy of Portugal in the Amazon Region, 1614-1693*. New York 1973.

Konrad Kingshill, *Ku Daeng — The Red Tomb*. A village study in northern Thailand. Suriyaban Publishers, Bangkok, Thailand 1976.

E.F.K. Koerner and R.E. Asher, Eds., *Concise History of the Language Sciences from the Sumerians to the Cognitivists*. Pergamon, Cambridge 1995.

J.O. Kokwaro, *Luo-English Botanical Dictionary*. E. African Publishing House, Nairobi 1972.

Mamadou Koumaré et al., *Contribution aux études ethno-botaniques et floristiques au Mali*. Agence de coopération culturelle et technique (A.C.C.T.), Paris 1984.

Samuel N. Kramer, *Sumerian Mythology*. The American Philosophical Society, Philadelphia 1944.

A.L. Kroeber, *Cultural and Natural Areas of Native America*. University of California Press, Berkeley 1939.

A.L. Kroeber, *Peoples of the Philippines*. Handbook, American Museum of Natural History, no. 8. New York 1919.

G. Kunkel, *Geography through Botany. A Dictionary of Plant Names with Geographical Meanings*. The Hague 1990.

Peter Kunstadter, E.C. Chapman and Sanga Sabhasri, Eds., *Farmers in the Forest*. The University Press of Hawaii, Honolulu 1978.

W. LaBarre, "Potato taxonomy among the Aymara Indians of Bolivia." *Acta Americana*. 5: 83-102. 1947.

R. Labat, *Manuel d'épigraphie akkadienne (signes, syllabaire, idéogrammes)*. Troisième édition. [Reprint of the 1948 lithographed edition]. Paris 1963.

R. Labat, *Traité akkadien de diagnotics et pronostics médicaux*. Leiden 1951.

Y. Laissus, "Catalogue des manuscrits de Philibert Commerson (1727-1773) conservés à la Bibliothèque centrale du Muséum nationale d'Histoire Naturelle (Paris)." *Rev. Hist. Sci., Paris*. 31: 131-162. 1978.

Imre Lakatos, *Proofs and Refutations*. Cambridge University Press, New York 1976.

John Landwehr, *Studies in Dutch Books with Coloured Plates Published 1662-1875*. The Hague 1976.

J. Lang, *Conquest and Commerce: Spain and England in the Americas*. New York 1975.

H. Lanjouw and H. Uittien, "Un nouvel herbier de Fusée Aublet découvert en France." *Recueil des Travaux botaniques néerlandais*. 37: 133-170. 1940.

Daniel N. Lapedes, Editor in Chief, *McGraw-Hill Dictionary of Scientific and Technical Terms*. McGraw-Hill, New York 1978.

Antoine Lasègue, *Musée botanique de M. Benjamin Delessert*. Paris, Leipzig 1845.

G.H.M. Lawrence et al., *B-P-H Botanico-Periodicum-Huntianum*. Hunt Botanical Library, Pittsburgh 1968.

W.F. Leemans, *Foreign Trade in the Old Babylonian Period as Revealed by Texts from Southern Meṣopotamia*. Leiden 1960.

A.J.M. Leeuwenberg, compiler, *Medicinal and Poisonous Plants of the Tropics*. Proceedings of Symposium 5-35 of the 14th International Botanical Congress, Berlin, 24 July – 1 August 1987.

Emile Legrand, *Bibliographie hellénique, ou description raisonnée des ouvrages publiés en grec par des grecs au XVe et XVIe siècles*. Paris 1885-1906.

J. Lemoine, *Un village Hmong vert du Haut-Laos*. Paris 1972.

Rodolfo Lenz, *Diccionario etimolójico de las voces chilenas derivadas de lenguas indijenas americanas*. Santiago de Chile 1904-1910.

Lewis and Short, *A Latin Dictionary*. Oxford University Press, Oxford 1995.

Henry George Liddell and Robert Scott, *A Greek-English Lexicon*. With a Revised Supplement. New edition, revised and augmented by H.S. Jones and R. McKenzie. Oxford University Press, Oxford 1996.

G. Liebert, *Iconographic Dictionary of the Indian Religions — Hinduism, Buddhism, Jainism*. Leiden 1976.

Carl Linnaeus (Carl von Linnaeus, Carl von Linné, Karl af Linné), *Species Plantarum*. Stockholm 1753 and *Genera Plantarum*. Ed. 5. Stockholm 1754.

Linnean Society of London, *Catalogue of the Printed Books and Pamphlets in the Library*. London 1925.

Jorge A. Lira, *Medicina Andina*. Peru, Cusco 1985.

Emilio Lisson Chávez, ed., *La Iglesia de España en el Perú*. Sevilla 1943-1947.

Alfredo López Austin, *Textos de Medicina Náhuatl*. México 1971.

Elias Avery Lowe, *Codices Latini Antiquiores*. III. Clarendon Press, Oxford 1938 and XI. Clarendon Press, Oxford 1966.

Katharine Luomala, *Ethnobotany of the Gilbert Islands*. Bernice P. Bishop Museum, Bulletin 213, Honolulu, Hawaii 1953.

Madeleine Ly-Tio-Fane, *Mauritius and the Spice Trade: The Odyssey of Pierre Poivre*. Port Louis 1958.

Ma Touan-Lin, *Ethnographie des peuples étrangers à la Chine*. Geneva 1876.

D.J. Mabberley, *The Plant-Book*. Second edition. Cambridge 1997.

D. Macdonald, *Moeurs et Coutumes des Tibétains*. Paris 1930.

Arthur Anthony Macdonell, *A Practical Sanskrit Dictionary*. Oxford University Press, Oxford 1976.

Arthur Anthony Macdonell and Arthur Berriedale Keith, *Vedic Index of Names and Subjects*. Motilal Banarsidass, Delhi, Varanasi (U.P.) and Patna (Bihar) 1967.

Elisabeth B. MacDougall, "Ars hortulorum: Sixteenth century garden iconography and literary theory in Italy." in *The Italian Garden*. Dumbarton Oaks Colloquium on the History of Landscape Architecture. I. Edited by David R. Coffin. Washington 1972.

Peter MacPherson, "The doctrine of signatures." *Glasgow Naturalist*. 20(3): 191-210. 1982.

José María Magalli de Pred, *Colección de Cartas sobre las Misiones Dominicanas del Oriente*. Quito 1890.

R.E. Magill, Ed., *Glossarium Polyglottum Bryologiae. A Multilingual Glossary for Bryology*. St. Louis, Missouri 1990.

R. Maire, "Études sur la flore et la végétation du Sahara central." *Mém. Soc. Hist. Nat. Afr. du Nord*. 65. Algiers 1933.

R. Maire and Th. Monod, "Études sur la flore et la végétation du Tibesti." *Mém. Inst. Franç. Afr. Noire*. 8. 1950.

G.P. Majumdar, *Vanaspati. Plants and Plant-life as in Indian Treatises and Traditions*. Calcutta 1927.

Trilok Chandra Majupuria and Rohit Kumar (Majupuria), *Gods and Goddesses — An Illustrated Account of Hindu, Buddhist, Tantric, Hybrid and Tibetan Deities*. Lalitpur Colony, Lashkar (Gwalior) 1994.

M. Maldonado-Koerdell, *Bibliografía Geológica y Paleontológica de América Central*. Mexico 1958.

M.E.L. Mallowan, *Early Mesopotamia and Iran*. McGraw-Hill, New York 1965.

M.A. Mandango and M.B. Bandole, "Contribution à la connaissance des plantes médicinales des Turumbu de la zone de Basoko (Zaire)." *Monographs in Systematic Botany from Missouri Botanical Garden.* 25: 373-384. 1988.

Clements Robert Markham, "A list of the tribes of the valley of the Amazons, ..." *Journal of the Anthropological Institute.* 40: 73-140. 1910.

Sidney David Markman, compiled and annotated by, *Colonial Central America. A Bibliography.* Arizona State University, Center for Latin American Studies, Tempe, Arizona 1977.

A.W. Martin and P. Wardle, *Members of the Legislative Assembly of New South Wales, 1856-1901.* Australian National University, Canberra 1959.

J.E. Martin-Allanic, *Bougainville navigateur et les Découvertes de son Temps.* Paris 1964.

John Alden Mason and George Agogino, "The ceremonialism of the Tepecan." Eastern New Mexico University, Paleo-Indian Institute, *Contributions in Anthropology.* vol. 4(1), October 1972.

C. Massin, *La médecine Tibétaine.* Paris 1982.

R.H. Mathews, *Ethnological Notes on the Aboriginal Tribes of New South Wales and Victoria.* F.W. White, Sydney 1905.

N. Mathur, *Red Fort and Mughal Life.* New Delhi 1964.

S. Matsushita, *A Comparative Vocabulary of Gwandara Dialects.* Tokyo 1974.

Washington Matthews, "Navaho names and uses for plants." *American Naturalist.* 20(9): 767-777. 1886.

M.G. de la Maza, *Diccionario botánico de los nombres vulgares cubanos y puertorriqueños.* 1889.

E.H. McCormick, *Tasman and New Zealand: A Bibliographical Study.* Wellington 1959.

Isaac McCoy, *History of Baptist Indian Missions.* New York 1840.

E. McKay, Ed., *Studies in Indonesian History.* Carlton (Vic.) 1976.

Norman A. McQuown, "The classification of the Mayan languages." *International Journal of American Linguistics.* 22: 191-195. 1956.

Norman A. McQuown, "The indigenous languages of Latin America." *American Anthropologist.* 57: 501-570. 1955.

José Toribio Medina, *Biblioteca hispano-americana (1493-1810).* Santiago de Chile 1897-1907.

José Toribio Medina, *Diccionario biográfico colonial de Chile.* Santiago de Chile 1906.

José Toribio Medina, *Los aboríjenes de Chile.* Santiago 1882.

C.K. Meek, *Tribal Studies in Northern Nigeria.* Oxford University Press, Oxford 1925.

Betty J. Meggers and Clifford Evans, edited by, *Aboriginal Cultural Development in Latin America: An Interpretative Review.* The Smithsonian Institution, Washington, D.C. 1963.

M.A.P. Meilink-Roelofsz, *Asian Trade and European Influence in the Indonesian Archipelago between 1500 and about 1630.* The Hague 1962.

Gaetano Melzi, *Dizionario di opere anonime e pseudonime di scrittori italiani o come che sia aventi relazione all'Italia.* Milano 1848-1859.

Sidney Mendelssohn, *Mendelssohn's South African Bibliography.* London 1910.

Manuel de Mendiburu, *Diccionario histórico-biográfico del Perú.* Lima 1931-1935.

Elmer D. Merrill, *A Dictionary of the Plant Names of the Philippine Islands.* Manila 1903.

Ellen Messer, *Zapotec Plant Knowledge. Memoirs of the Museum of Anthropology,* University of Michigan Press, Ann Arbor 1978.

W. Meyer-Lübke, *Romanisches etymologisches Wörterbuch.* Heidelberg 1935.

S.P. Michel and P.H. Michel, *Répertoire des ouvrages imprimés en langue italienne au XVIIe siècle conservés dans les bibliothèques de France.* Paris 1967-1984.

Frederick William Hugh Migeod, *Mende Natural History Vocabulary.* London 1913.

G.B. Milner, *Samoan Dictionary.* Oxford University Press, London 1966.

J.D. Milner, *Catalogue of Portraits of Botanists Exhibited in the Museums of the Royal Botanic Gardens.* Royal Botanic Gardens, Kew, London 1906.

Giuseppe Maria Mira, *Bibliografia siciliana* ovvero Gran dizionario bibliografico delle opere edite e inedite, antiche e moderne di autori siciliani e di argomento siciliano stampate in Sicilia e fuori. Palermo 1875-1881.

J. Mirsky, *The Westwards Crossings: Balboa, Mackenzie, Lewis and Clark.* Chicago 1970.

Sarat Chandra Mitra, "Worship of the Pipal Tree in North Bihar." *Journal of the Bihar and Orissa Society.* VI. 1920.

A. Mongitore, *Bibliotheca Sicula* sive De scriptoribus Siculis qui tum vetera, tum recentiora saecula illustrarunt, ... notitiae locupletissimae. Panormi 1708-1714.

M. Monier-Williams, *A Sanskrit–English Dictionary.* Motilal Banarsidass, Delhi 1990.

M. Monier-Williams, *English–Sanskrit Dictionary.* Munshiram Manoharlal, New Delhi 1976.

Franco Montanari, *Vocabolario della lingua Greca.* Loescher Editore, Torino 1995.

François-Auguste de Montequin, Ed., *Aspects of Ancient Maya Civilization.* Hamline University, St. Paul, Minn. 1976.

Vayaskara N.S. Mooss, *Ayurvedic Flora Medica.* Kottayam, S. India 1978.

L. Moranti, *Le cinquecentine della biblioteca universitaria di Urbino.* Firenze 1977.

Magnus Mörner, Ed., *The Expulsion of the Jesuits from Latin America.* New York 1967.

Edward E. Morris, *Austral English.* London 1898.

H.B. Morse, *The Chronicles of the East India Company Trading to China 1635-1834.* Oxford University Press, Oxford 1926-1929.

S. Moscati et al., *An Introduction to Comparative Grammar of the Semitic Language: Phonology and Morphology.* Edited by S. Moscati. Wisbaden 1969.

G. Murray, *History of the Collections Contained in the Natural History Departments of the British Museum.* London 1904.

Gary Paul Nabhan, *Gathering the Desert.* The University of Arizona Press, Tucson, Arizona 1985.

A. Narbone, *Bibliografia sicola sistematica.* Palermo 1850-1855.

National Library of Scotland, *A Short-Title Catalogue of Foreign Books, Printed up to 1600.* Edinburgh 1970.

The National Union Catalogue, pre 1965 imprints. London 1968-1981.

D.H. Nicolson, "Orthography of names and epithets." *Taxon.* 23: 549-561, 843-851. 1974.

D.H. Nicolson and R.A. Brooks, "Orthography of names and epithets." *Taxon.* 23: 163-177. 1974.

[Nigeria], *Vocabulary of Nigerian Names of Trees, Shrubs and Herbs.* Government Printer, Lagos 1936.

C. Nigra, *Saggio lessicale di basso latino curiale compilato su Estratti di Statuti medievali piemontesi.* Torino 1920.

Claus Nissen, *Die Botanische Buchillustration. Ihre Geschichte und Bibliographie.* Stuttgart 1966.

Erland Nordenskiöld, *Analyse ethno-géographique de la culture matérielle de deux tribus Indiennes du Gran Chaco.* Paris 1929.

Haskell F. Norman, *The Haskell F. Norman Library of Science and Medicine.* Compiled by Diana H. Hook and Jeremy M. Norman. San Francisco 1991.

Kalervo Oberg, *Indian Tribes of Northern Mato Grosso, Brazil.* Smithsonian Institution, Institute of Social Anthropology, Publication no. 15. Washington 1953.

[Catalogo del Fondo Fiammetta Olschki], *Viaggi in Europa. Secoli XVI-XIX.* Gabinetto scientifico e letterario G.P. Viesseux. Firenze 1990.

Martha Ornstein, *The Role of Scientific Societies in the Seventeenth Century.* University of Chicago Press, Chicago [1938].

F.L. Orpen, "Botanical–Vernacular and Vernacular–Botanical Names of Some Trees and Shrubs in Matabeleland." *Rhod. Agr. Journ.* XLVIII, 2. 1951.

Bernard R. Ortiz de Montellano, "Empirical Aztec Medicine." *Science.* 188: 215-220. 1975.

G. Ortolani, *Biografia degli uomini illustri della Sicilia.* Napoli 1827-1831.

William Osler, *Bibliotheca Osleriana.* Clarendon Press, Oxford 1929.

José I. Otero, Rafael A. Toro and Lydia Pagán de Otero, *Catálogo de los Nombres Vulgares y Científicos de Algunas Plantas Puertorriqueñas.* Universidad de Puerto Rico, Rio Piedras, Puerto Rico 1945.

Giuseppe Ottino and Giuseppe Fumagalli, eds., *Bibliotheca bibliographica italica.* Roma and Torino 1889-1902.

Keith Coates Palgrave, *Trees of Southern Africa.* Struik Publishers, Cape Town 1990.

Eve Palmer and Norah Pitman, *Trees of Southern Africa.* Cape Town 1972.

M. Parenti, *Dizionario dei luoghi di stampa falsi, inventati o supposti.* Sansoni, Firenze 1951.

M. Parenti, *Esercitazioni filologiche.* Modena 1844-1857.

A. Parrot, *Sumer, the Dawn of Art.* Golden Press, New York 1961.

J.H. Parry, *Trade and Dominion: The European Overseas Empires in the Eighteenth Century.* London 1971.

Claire D.F. Parsons, ed., *Healing Practices in the South Pacific.* The Institute for Polynesian Studies 1985.

Michele Pasqualino, *Vocabolario etimologico siciliano italiano e latino.* Palermo 1783-1795.

Giambattista Passano, *Dizionario di opere anonime e pseudonime in supplemento a quello di Gaetano Melzi.* Ancona 1887.

Edward R. Pease, *The History of the Fabian Society.* London 1925.

G.B. Pellegrini, *Gli arabismi nelle lingue neolatine con speciale riguardo all'Italia.* Brescia 1972.

Antonio Penafiel, *Nombres geographicos de Mexico.* Mexico 1885.

Campbell W. Pennington, *The Tepehuan of Chihuahua — Their Material Culture.* University of Utah Press, Salt Lake City 1969.

Campbell W. Pennington, Ed., *The Pima Bajo of Central Sonora, Mexico.* University of Utah Press, Salt Lake City 1979-1980.

Rev. Father Fray Angel Perez of the Order of San Agustin, *Igorots, Geographic and Ethnographic Study of Some Districts of Northern Luzon.* Cordillera Studies Center, University of Philippines, Baguio City 1988.

David Perini, *Bibliographia augustiniana. Scriptores itali.* Firenze 1929-1935.

Luciano Petech, *I missionari italiani nel Tibet e nel Nepal.* Roma 1952-1956.

A. Pételot, "Bibliographie botanique de l'Indochine." *Arch. Regn. Agron. Pastor. Vietnam* no. 24, Saigon 1955.

W.J. Peters, *Landscape in Romano–Campanian Painting.* Assen 1963.

J. Petzholdt, *Bibliotheca bibliographica.* Leipzig 1866.

J.L. Phelan, *El Reino Milenario de los Franciscanos en el Nuevo Mundo.* México 1972.

Sandro Piantanida, Lamberto Diotallevi and Giancarlo Livraghi, eds., *Autori Italiani del '600.* Libreria Vinciana, Milano 1948-1951.

Rodolfo E.G. Pichi Sermolli et collab., "Index Filicum, Supplementum quartum pro annis 1934-1960." *Regnum Veget.* 37: I-XIV + 1-370. Utrecht 1965.

Rodolfo E.G. Pichi Sermolli, comp., *Authors of Scientific Names in Pteridophyta.* Royal Botanic Gardens, Kew 1996.

Arpad Plesch, *The Magnificent Botanical Library of the Stiftung für Botanik, Vaduz Liechtenstein,* collected by the late Arpad Plesch. Part I [-III]. 3 vols. London, Sotheby & Co., 16 June 1975 – 15 March 1976.

L. Polgar, *Bibliography of the History of the Society of Jesus.* Rome 1957.

Alfred W. Pollard and G.R. Redgrave, Eds., *A Short-Title Catalogue of Books Printed in England, Scotland and Ireland, and of English Books Printed Abroad 1475-1640.* Second edition, revised and enlarged, begun by W.A. Jackson and F.S. Ferguson, completed by Katharine F. Pantzer. Bibliographical Society, London 1976-1991.

R. Portères, "La sombre aroidée cultivée: *Colocasia antiquorum* Schott ou taro de Polynésie: essai d'etymologie sémantique." *J. Agric. Trop. Botan. Appl.* 7: 169-192. 1960.

R. Portères, "Les appelations des céréales en Afrique." *J. Agric. Trop. Botan. Appl.* 5 and 6. 1958-1959.

F.N.C. Poynter, Ed., *Medicine and Culture*. London 1969.

Efren C. del Pozo, "Estudios farmacológicos de algunas plantas usadas en medicina azteca." *Bol. Indigenista*. 6: 350-365. 1946.

A.H.J. Prins, *The Swahili-speaking Peoples of Zanzibar and the East African Coast (Arabs, Shirazi and Swahili)*. London 1961.

E.H. Pritchard, *Anglo-Chinese Relations during the Seventeenth and Eighteenth Centuries*. Urbana (Ill.) 1929.

E.H. Pritchard, *The Crucial Years of Early Anglo–Chinese Relations 1750-1800*. New York 1936.

G.A. Pritzel, *Thesaurus literaturae botanicae omnium gentium*. Lipsiae 1871-1877.

Robert Proctor, *The Printing of Greek in the Fifteenth Century*. Bibliographical Society Illustrated Monograph 8. Oxford University Press for The Bibliographical Society, London 1900.

Brian Pullan, *Rich and Poor in Renaissance Venice: The Social Institutions of a Catholic State, to 1620*. Oxford, Cambridge 1971.

Quaritch, *Catalogue of a Most Important Collection of Publications of the Aldine Press, 1494-1595*. Quaritch, London 1929.

Joseph-Marie Quérard, *La France littéraire*. Paris 1827-1864.

Jacobus Quétif and Jacobus Echard, *Scriptores Ordinis Praedicatorum*. Lutetiae Parisiorum 1719-1721.

A. Rainaud, *Le Continent Austral*: Hypothèses et Découvertes. Paris 1894.

Sri Ramakrishna Centenary Committee, Ed., *Cultural Heritage of India*. Calcutta 1937.

José Ramírez, *Sinonimia vulgar y científica de las plantas mexicanas*. México 1902.

W.D. Raymond, "Native poisons and native medicines of Tanganyika." *J. Trop. Med. & Hyg.* 42: 295-303. 1939.

C.F. Reed, "Index Thelypteridis." *Phytologia*. 17(4): 249-328. 1968.

Gerardo Reichel-Dolmatoff, *Amazonian Cosmos*. The University of Chicago Press, Chicago 1971.

Gerardo and Alicia Reichel-Dolmatoff, *The People of Aritama*. The University of Chicago Press, Chicago 1970.

Dietrich Reichling, *Appendices ad Hainii–Copingeri Repertorium bibliographicum*. Additiones et emendationes. Monachii 1905-1911.

Blas Pablo Reko, *Mitobotánica Zapoteca*. Tacubaya, D.F. 1945.

Jane Renfrew, *Palaeoethnobotany*. Columbia University Press, New York 1973.

Antoine-Augustin Renouard, *Annales de l'imprimerie des Aldes*. Paris 1834.

Antoine-Augustin Renouard, *Annales de l'imprimerie des Estienne*. Paris 1843.

Philippe Renouard, *Imprimeurs parisiens*: Libraires, fondeurs de caractères et correcteurs d'imprimerie; depuis l'introduction de l'imprimerie à Paris (1470) jusqu'à la fin du XVIe siècle. Paris 1898.

T.W. Rhys-Davids and William Stede, *The Pali Text Society's Pali–English Dictionary*. London 1979.

Darcy Ribeiro and Mary Ruth Wise, *Los grupos étnicos de la Amazonía Peruana*. Yarinacocha, Pucallpa 1978.

Massimo Ricciardi and G.G. Aprile, "Preliminary data on the floristic components of some carbonized plant remains found in the archaeological area of Oplontis near Naples." *Annali della Facoltà di Scienze Agrarie dell'Università di Napoli in Portici*, ser. 4, 12: 204-212. 1978.

Tania Maura Nora Riccieri, *Bibliografia de Plantas Medicinais*. Jardim Botânico do Rio de Janeiro. Rio de Janeiro 1989.

Roberto Ridolfi, *La stampa in Firenze nel secolo XV*. Firenze 1958.

C.L. Riley, J.C. Kelley, C.W. Pennington and R.L. Rands, Eds., *Man Across the Sea: Problems of Pre-Columbian Contacts*. University of Texas Press, Austin 1971.

Rechung Rinpoche, *Tibetan Medicine*. Berkeley 1973.

C.H. Robinson, *Dictionary of the Hausa Language*. 2 vols. Cambridge University Press, Cambridge 1913.

R.J. Rodin, *The Ethnobotany of the Kwanyama Ovambos*. Missouri Botanical Garden 1985.

F.L.O. Roehrig and J.G. Cogswell, compiled by, *Catalogue of Books in the Astor Library Relating to the Languages and Literature of Asia, Africa and the Oceanic Islands*. New York 1854.

G.N. Roerich, *Le parler de l'Amdo*. Roma 1958.

J. Roscoe, *The Northern Bantu: An Account of Some Central Africa Tribes of the Uganda Protectorate*. Cambridge 1915.

Walter E. Roth, *Ethnological Studies Among the North-West-Central Queensland Aborigines*. Brisbane 1897.

Leslie B. Rout, Jr., *The African Experience in Spanish America*. 1502 to the present day. Cambridge University Press, Cambridge 1976.

The Royal College of Physicians of London. *Portraits*. Edited by G.W. Wolstenholme. Described by David Piper. London 1964.

Ralph L. Roys, *The Ethnobotany of the Mayas*. Tulane University, Middle American Research Institute, New Orleans 1931.

J. Rzedowski, "Nombres regionales de algunas plantas de la Huasteca Potosina." *Actas Científicas Potosinas*. 6: 7-58. 1966.

P.A. Saccardo, *Cronologia della flora italiana*. Padova 1909.

Ch. Sacleux, *Dictionnaire Français-Swahili*. Zanzibar, Paris 1891.

R.N. Salaman, *The History and Social Influence of the Potato*. Cambridge 1949.

Contessa Anna di San Giorgio, *née* Harley d'Oxford, *Catalogo poliglotto delle piante*. Stabilimento tipografico di Giuseppe Pellas, Firenze 1870.

G. Sansom, *The Western World and Japan, 1334-1615*. London 1950.

G. Sansom, *A History of Japan*. London 1961.

Francisco J. Santamaría, *Diccionario de mejicanismos*. México 1978.

G. Saporetti, *Onomastica medio-assira*. Roma 1970.

G. Sapori, *Il fondo di medicina antica della Biblioteca Ginecologica di Milano*. Milano 1975.

G. Sapori, *Le cinquecentine dell'Università di Milano*. Milano 1969.

B.K. Sarkar, *The Folk Elements in Hindu Culture*. New Delhi 1981.

Swami Satchidananda, *Dictionary of Sanskrit Names*. Integral Yoha Publications, Buckingham, Virginia 1989.

Linda Schele, *Maya Glyphs: The Verbs*. University of Texas Press, Austin 1982.

G. Schembari, *La scienza orientale*. Torino 1924.

Amelie Schenk and Holger Kalweit, Eds., *Heilung des Wissens*. Munich 1987.

Stephan Schuhmacher and Gert Woerner, Eds., *The Encyclopedia of Eastern Philosophy and Religion. Buddhism, Hinduism, Taoism, Zen*. Shambhala, Boston 1994.

Hans Wolfgang Schumann, *Immagini buddhiste*. Edizioni Mediterranee. Roma 1989.

D. Scinà, *Prospetto della storia letteraria di Sicilia nel secolo XVIII*. Palermo 1969.

William Henry Scott, *The Discovery of the Igorots — Spanish Contacts with the Pagans of Northern Luzon*. New Day Publishers, Quezon City 1987.

Giovanni Semerano, Le origini della cultura europea. Dizionari Etimologici. Basi semitiche delle lingue indeuropee. *Dizionario della lingua Greca*. Leo S. Olschki Editore, Firenze 1994.

Giovanni Semerano, Le origini della cultura europea. *Dizionario della lingua Latina e di voci moderne*. Leo S. Olschki Editore, Firenze 1994.

María Teresa Sepúlveda y H., *La medicina entre los purépecha prehispánicos*. Universidad Nacional Autónoma de México, México 1988.

R. Shafer, *Ethnography of Ancient India*. Wiesbaden 1954.

J. Shapera, ed., *The Bantu-speaking Tribes of South Africa*. London 1937.

P.V. Sharma, *Indian Medicine in the Classical Age*. India 1972.

H. Sharp et al., *Siamese Rice Village: A Preliminary Study of Bang Chan, 1948-1949*. Bangkok 1953.

Moodeen Sheriff, *A Catalogue of Indian Synonymes of the Medicinal Plants, Products, Inorganic Substances, etc., Proposed to be Included in the Pharmacopoeia of India*. [Published in 1869, 1st reprint edition 1978] Printed at Jayyed Press, Billimaran, Delhi 1978.

Dorothy Shineberg, *They Came for Sandalwood*. Melbourne 1967.

Davi Shorto, Judith Jacob and E.H.S. Simmons, *Bibliographies of Mon-Khmer and Thai Linguistics*. Oxford University Press, Oxford 1963.

M.F. da Silva, P.L. Braga Lisbôa and R.C. Lobato Lisbôa, *Nomes vulgares de plantas amazonicas*. Manaus, Brasil 1977.

D.R. Simmons, *The Great New Zealand Myth*. Wellington 1967.

C. Earle Smith, Jr., Ed., *Man and His Foods: Studies in the Ethnobotany of Nutrition — Contemporary, Primitive and Prehistoric Non-European Diets*. University of Alabama Press, Alabama 1973.

Christo Albertyn Smith, *Common Names of South African Plants*. Edited by E. Percy Phillips and Estelle van Hoepen. Botanical Survey Memoir, no. 35. 1966.

William Henry Smyth and Sir E. Belcher, *Dictionary of Nautical Terms*. Glasgow 1867.

W. von Soden and W. Rollig, *Das akkadische Syllabar*. Roma 1967.

C. Sommervogel, S.J., *Bibliothèque de la Compagnie de Jésus*. Bruxelles and Paris 1890-1916.

C. Sommervogel, S.J., *Dictionnaire des ouvrages anonymes et pseudonymes publiés par des religieux de la Compagnie de Jésus*. Paris 1884.

Jacques Soustelle, *The Four Suns*. Grossman Publishers, New York 1971.

O.H.K. Spate, *The Spanish Lake*. University of Minnesota Press, Minnesota 1979 [Italian translation: *Storia del Pacifico — Il lago spagnolo*. A cura di Gianluigi Mainardi. Giulio Einaudi editore. Torino 1987].

O.H.K. Spate, *Monopolists and Freebooters*. 1983 [Italian translation: *Storia del Pacifico — Mercanti e bucanieri*. A cura di Gianluigi Mainardi. Giulio Einaudi editore. Torino 1988].

O.H.K. Spate, *Paradise Found and Lost*. 1988 [Italian translation: *Storia del Pacifico — Un paradiso trovato e perduto*. A cura di Gianluigi Mainardi. Giulio Einaudi editore. Torino 1993].

Jonathan Spence, *The Memory Palace of Matteo Ricci*. New York 1983.

Baldwin Spencer, *Native Tribes of the Northern Territory of Australia*. Macmillan, London 1914.

Baldwin Spencer and F.J. Gillen, *The Native Tribes of Central Australia*. London 1899.

Frans A. Stafleu and Richard S. Cowan, *Taxonomic Literature*. 7 vols. 1976-1988.

Frans A. Stafleu and E.A. Mennega, *Taxonomic Literature*. 5 suppls. 1992-1998.

Paul C. Standley, *Trees and Shrubs of Mexico*. Washington, D.C. 1920-1926.

Frederick Starr, *Notes upon the Ethnography of Southern Mexico*. [Davenport] 1900-1902.

W.T. Stearn, *Botanical Latin*. London 1966.

W.T. Stearn, *Stearn's Dictionary of Plant Names for Gardeners*. London 1993.

M.J. van Steenis-Kruseman, "Malaysian plant collectors and collections." Supplement 2. Page I-CXV. in *Flora Malesiana* 8(1). Nordhoof International Publishing, Leyden, The Netherlands 1974.

George Steiner, *After Babel: Aspects of Language and Translation*. Oxford University Press, New York 1975.

E. Steinkellner and H. Tauscher, Eds., *Contributions on Tibetan Language History and Culture*. Wien 1983 (Reprint Delhi 1995).

A.F. Stenzler, *Primer of the Sanskrit Language*. Translated into English with some revision by Renate Söhnen. School of Oriental and African Studies. University of London 1992.

Julian Haynes Steward, Ed., *Handbook of South American Indians*. U.S. Government Printing Office, Washington, D.C. 1946-.

George Stewart, *Names on the Globe*. Oxford University Press, Oxford 1975.

Margaret Bingham Stillwell, *Incunabula and Americana 1450-1800. A Key to Bibliographical Study*. Columbia University Press, New York 1931.

Margaret Bingham Stillwell, *The Awakening Interest in Science during the First Century of Printing 1450-1550: An Annotated Checklist of First Editions*. New York 1970.

R. Stillwell, Ed., *The Princeton Encyclopedia of Classical Sites*. Princeton University Press, Princeton 1976.

Margaret Stutley and James Stutley, *A Dictionary of Hinduism*. London 1977.

M. Swadesh, *Indian Linguistic Groups of Mexico*. México, D.F. 1961.

M. Swadesh, "Interrelaciones de las lenguas Mayas." *Anales del Instituto Nacional de Antropología e Historia*. Mexico 1961.

E. Taillemite, *Bougainville et ses Compagnons autour du monde*. Paris 1977.

Avenir Tchemerzine, *Bibliographie d'éditions originales et rares d'auteurs français des XV, XVI, XVII et XVIII siècles*. Paris 1927-1933.

Saladin S. Teo, *The Life-style of the Badjaos — A Study of Education and Culture*. Manila 1989.

H. Ternaux-Compans, *Bibliothèque asiatique et africaine*. Paris 1841.

Cyrus Thomas, assisted by John R. Swanton, *Indian Languages of Mexico and Central America and Their Geographical Distribution*. Smithsonian Institution, Bureau of American Ethnology, Bulletin 44. Washington 1911.

P.L. Thomas, "A Chitonga-Botanical dictionary of some species occurring in the vicinity of the Mwenda Estuary, Lake Kariba, Rhodesia." *Kirkia*. 7, 2: 269-284. 1970.

Dorothy Burr Thompson, "Ancient gardens in Greece and Italy." *Archaeology*. 4. Spring 1951.

Dorothy Burr Thompson, "The Garden of Hephaistos." *Hesperia*. 6: 396-425. 1937.

J. Eric S. Thompson, *Ethnology of the Mayas of Southern and Central British Honduras*. Chicago 1930.

R.C. Thompson, *A Dictionary of Assyrian Botany*. First edition. London 1949.

R.C. Thompson, *A Dictionary of Assyrian Chemistry and Geology*. Oxford University Press, Oxford 1936.

K. Thomson and G. Serle, *A Biographical Register of the Victorian Parliament, 1859-1900*. Australian National University Press, Canberra 1972.

F. Thureau-Dangin, *Le Syllabaire Accadien*. Librairie Orientaliste Paul Geuthner, Paris 1926.

N.B. Tindale, *Aboriginal Tribes of Australia*. University of California Press, Berkeley 1974.

Julio Tobar Donoso, *Historiadores y Cronistas de las Misiones*. Puebla, Mexico 1960.

S. Tolkowsky, *Hesperides, A History of the Culture and Use of Citrus Fruits*. London 1938.

G. Toscano, *Alla scoperta del Tibet. Relazioni dei missionari del sec. XVII*. Bologna 1977.

Margaret A. Towle, *The Ethnobotany of Pre-Columbian Peru*. Aldine Publishing Company, Chicago 1961.

L. Trabut, *Répertoire des noms indigènes des plantes spontanées, cultivées et utilisées dans le Nord de l'Afrique*. Alger 1935.

Hugh Trevor-Roper, *Hermit of Peking. The Hidden Life of Sir Edmund Backhouse*. 1978.

Rolla Milton Tryon and Alice Faber Tryon, *Ferns and Allied Plants*. New York, Heidelberg, Berlin 1982.

Ethelyn (Daliaette) Maria Tucker, *Catalogue of the Library of the Arnold Arboretum of Harvard University*. Cambridge, Massachusetts 1917-1933.

Peter J. Ucko and G.W. Dimbleby, Eds., *The Domestication and Exploitation of Plants and Animals*. Aldine Publishing Co., Chicago 1969.

Ruth M. Underhill, "The Papago Indians of Arizona and their relatives the Pima." *Sherman Pamphlet* no. 3 (Indian Life and Customs), U.S. Office of Indian Affairs, Lawrence, Kansas 1940.

Emerenziana Vaccaro, *Le marche dei tipografi ed editori italiani del sec. XVI nella Biblioteca Angelica di Roma*. Leo S. Olschki, Firenze 1983.

Luis E. Valcárcel, *Mirador Indio*. Apuntes para una filosofía de la cultura incaica. Lima 1937-1941.

Paul Valkema Blouw, *Typographia Batava 1541-1600*. A repertorium of books printed in the Northern Netherlands between 1541 and 1600. Nieuwkoop 1998.

Alvaro Valladares, *Cartas sobre las Misiones Dominicanas del Oriente del Ecuador*. Quito 1912.

David Vancil, *Catalog of Dictionaries, Word Books, and Philological Texts, 1440-1900: Index of the Cordell Collection Indiana State University*. Westport [1993].

John Venn and J.A. Venn, *Alumni Cantabrigienses. A Biographical List*. Cambridge 1922-1954.

F. Verdoorn, Ed., *Plants and Plant Science in Latin America*. Waltham, Massachusetts 1945.

Dominique D. Vérut, *Precolombian Dermatology and Cosmetology in Mexico*. Schering Corporation USA 1973.

J.E. Vidal, "Bibliographie botanique indochinoise de 1955 à 1969." *Bull. Soc. Étud. Indoch.* n.s. 47: 657-748. Saigon 1972.

J.E. Vidal and B. Wall, "Contribution à l'ethnobotanique des Nya Hön." *Journal d'agriculture tropicale et de botanique appliquée*, tome XV(7-8): 243-264. 1968.

Jules Vidal, *La végétation du Laos*. Toulouse 1960.

J.E. Vidal, Y. Vidal and Pham Hoang Hô, *Bibliographie botanique indochinoise de 1970 à 1985*. Paris 1988.

Odette Viennot, *Le culte de l'arbre dans l'Inde ancienne*. Annales du Musée Guimet, LIX. Paris 1954.

Evon Z. Vogt, *Zinacantán: A Maya Community in the Highlands of Chiapas*. Cambridge, Massachusetts 1969.

H.R. Wagner, *The Cartography of the Northwest Coast of America to the Year 1800*. Berkeley 1937.

H.R. Wagner, *Spanish Voyages to the Northwest Coast of America in the Sixteenth Century*. Amsterdam 1966.

A. Walker and R. Sillans, *Les plantes utiles de Gabon*. Paris 1961.

Barbara Wall, *Les Nya Hön*. Étude ethnographique d'une population du Plateau des Bolovens (Sud Laos). Vithagna, Vientiane, Laos 1975.

Alan C. Wares, *Bibliography of the Summer Institute of Linguistics 1935-1985*. Summer Institute of Linguistics, Dallas 1979, 1985 and 1986.

David Watkins Waters, *The Art of Navigation in England in Elizabethan and Early Stuart Times*. Yale University Press, New Haven 1958.

D.B. Waterson, *A Biographical Register of the Queensland Parliament, 1860-1929*. Australian National University Press, Canberra 1972.

J. Watt et al., *Starving Sailors*. London 1981.

S.H. Weber, *Voyages and Travels in the Near East during the XIX Century*. Princeton 1952-1953.

Waldo R. Wedel, "Environment and Native Subsistence Economies in the Central Great Plains." *Smithsonian Miscellaneous Collections*, volume 101, number 3, August 20, 1941.

Ernest Weekley, *An Etymological Dictionary of Modern English*. In two volumes. Dover Publications, New York 1967.

John M. Weeks, comp., *Mesoamerican Ethnohistory in United States Libraries*. Reconstruction of the William E. Gates Collection of Historical and Linguistic Manuscripts. Culver City, California, Labyrinthos 1990.

D. Westermann and M.A. Bryan, *Handbook of African Languages*. London 1970.

Keith Whinnom, *Spanish Contact Vernaculars in the Philippine Islands*. Hong Kong 1956.

K.D. White, *Agricultural Implements of the Roman World*. Cambridge 1967.

R.A. White et al., *Bibliography of American Pteridology*. 1976-1987.

J.H. Wiersema, J.H. Kirkbride and C.R. Gunn, *Legume (Fabaceae) Nomenclature in the USDA Germplasm System*. USDA Tech. Bull. 1757. 1990.

H. Wild, *A Rhodesian Botanical Dictionary of African and English Plant Names*. Revised and enlarged by H.M. Biegel and S. Mavi. Salisbury, Rhodesia 1972.

H. Wild, *A Southern Rhodesian Botanical Dictionary of Native and English Plant Names*. Salisbury, Southern Rhodesia 1953.

W. Willcocks, *Plans of Irrigation of Mesopotamia*. Survey Department, Cairo 1911.

Herbert W. Williams, *A Dictionary of the Maori Language*. Wellington 1975.

Jessie Williamson, *Useful Plants of Malawi*. Zomba, Malawi 1972.

Jessie Williamson, *Useful Plants of Nyasaland*. Zomba, Nyasaland 1955.

K. Williamson, "Some food-plant names in the Niger Delta." *Int. J. Amer. Linguistics*. 36: 156-167. 1970.

J.E. Wills, *Pepper, Guns and Parleys: The Dutch East India Company and China 1622-1681*. Harvard University Press, Cambridge, Massachusetts 1974.

Simon Winchester, *The Professor and the Madman*. 1998.

G.P. Winship, "The Coronado expedition, 1540-1542." *Bureau of American Ethnology, Annual Report 14*, 1: 329-613. 1892-1893 [1896].

L. Wittmack, "Die in Pompeji gefundenen pflanzlichen Reste." *Beiblatt zu den Botanischen Jahrbuchern*. 73. 1903.

Johannes Christophorus Wolff, *Bibliotheca hebraea*. Hamburgi, Lipsiae 1715-1733.

G. Marshall Woodroow, *Gardening in the Tropics*. London n.d.

Yale University, School of Medicine, *Collected Papers on Bibliography and the History of Medicine*. New Haven 1940-1945.

S. Yerasimos, *Les voyageurs dans l'Empire Ottoman (XIVe — XVIe siècles). Bibliographie, itinéraires et inventaire des lieux habités*. Ankara 1991.

Francisco Zambrano, *Diccionario bio-bibliográfico de la Compañia de Jesús en México*. Mexico 1961-1969.

R. Zander, F. Encke, G. Buchheim and S. Seybold, *Handwörterbuch der Pflanzennamen*. 14. Aufl. Stuttgart 1993.

G. Zappella, *Le marche dei tipografi e degli editori italiani del Cinquecento*. Repertorio di figure, simboli e soggetti e dei relativi motti. Milano 1986.

H. Zimmer, *Myths and Symbols in Indian Art and Civilization*. Bollingen Series VI. Pantheon Books, New York and Princeton University Press, Princeton, New Jersey 1946.

Printed and bound by CPI Group (UK) Ltd, Croydon, CR0 4YY

29/10/2024

01780831-0001